THE FAMILIES
AND GENERA
OF VASCULAR PLANTS

Edited by K. Kubitzki

For further volumes see list at the end of the book and:
http://www.springer.com/series/1306

The Families and Genera of Vascular Plants

Edited by K. Kubitzki

XV

Flowering Plants · Eudicots

Apiales, Gentianales (except Rubiaceae)

Volume Editors:
Joachim W. Kadereit • Volker Bittrich

With 85 Figures

Springer

Editors
Joachim W. Kadereit
Johannes Gutenberg
Universität Mainz
Mainz
Germany

Volker Bittrich
Campinas
Brazil

Series Editor

Prof. Dr. Klaus Kubitzki
Universität Hamburg
Biozentrum Klein-Flottbek und Botanischer Garten
22609 Hamburg
Germany

The Families and Genera of Vascular Plants
ISBN 978-3-319-93604-8 ISBN 978-3-319-93605-5 (eBook)
https://doi.org/10.1007/978-3-319-93605-5

Library of Congress Control Number: 2018961008

This Springer imprint is published by the registered company Springer Nature Switzerland AG
The registered company address is: Gewerbestrasse 11, 6330 Cham, Switzerland

Preface

It is with great pleasure that I present a volume with family treatments of the orders Apiales and Gentianales (except Rubiaceae). An immense amount of evidence recently accrued had to be taken into account to present up-to-date treatments of these families, including evidence from taxonomic revisions of large groups, new comparative morphological studies, but also data from karyological and phytochemical findings and, above all, results from an increasing number of molecular systematic studies that have changed our concepts of relationships considerably. In the two large families Apiaceae and Apocynaceae, which comprise 466 and 378 genera respectively, the establishment of a phylogenetic framework obviously has been a comprehensive effort for many years by several authors. Although much smaller, the same is true for Loganiaceae where establishment of family limits proved to be particularly difficult. The authors of the Apiaceae treatment found it difficult to devise a single usable key to the genera of this family. Therefore, separate keys were constructed for the native and naturalized genera in seven major geographic regions of the world. The treatments of Apiaceae and Apocynaceae also include many references to various keys and taxonomic treatments of parts of the families that are difficult to find, which makes these treatments even more useful.

Altogether, the volume contains a wealth of interesting information, and I am greatly indebted to all authors for their ardent desire to complete their contributions, and to the volume editors who took care that the project was brought to an end. I am also grateful to all copyright holders who so kindly gave permission to reproduce illustrations published under their responsibility, and to all artists who contributed original illustrations. The artist Bobbi Angell, New York, again deserves our special thanks for the generosity with which she authorized the use of fine illustrations published under her authorship. Doris Franke, Mainz, is gratefully acknowledged for preparing all figures for print.

Finally, I have great pleasure in thanking the copy editor of the present volume, Dr. Monique Delafontaine, for her dedicated editorial work. My warm thanks also go to Dr. Sabine von Mering, who kindly took care of preparing the index of scientific names. As always, it is a pleasure to acknowledge the cordial collaboration with Dr. Andrea Schlitzberger from Springer Verlag.

Hamburg
2 May 2018

Klaus Kubitzki

Contents

List of Contributors

Volker Bittrich	Rua Dr. Mario de Nucci 500, 13083 290 Campinas, SP, Brazil. folcar2007@gmail.com
Roger C. Carolin	Oak Park, P.O. Box 25, Berry 2535, NSW, Australia. billarr@shoalhaven.net.au
Barry J. Conn	John Ray Herbarium, School of Life and Environmental Sciences, University of Sydney, Sydney, NSW 2006, Australia. tugiri1975@yahoo.com.au
Michael O. Dillon	Botany Department, The Field Museum, 1400 South Lake Shore Drive, Chicago, IL 60605, USA. mdillon@fieldmuseum.org
Mary E. Endress	Institute of Systematic and Evolutionary Botany, University of Zurich, Zollikerstrasse 107, 8008 Zurich, Switzerland. mendress@systbot.uzh.ch
Pedro Fiaschi	Departamento de Botânica, CCB, Universidade Federal de Santa Catarina, Campus Universitário–Trindade, 88040-900 Florianópolis, Santa Catarina, Brazil. pedrofiaschi@gmail.com
Kerry L. Gibbons	Royal Botanic Gardens and Domain Trust, Mrs Macquaries Rd, Sydney, NSW 2000, Australia. kerry.gibbons@rbgsyd.nsw.gov.au
Jennifer M. Hart	School of Life and Environmental Sciences, University of Sydney, Camperdown, NSW 2006, Australia. hart.j.m@gmail.com
Murray J. Henwood	School of Life and Environmental Sciences, University of Sydney, Camperdown, NSW 2006, Australia. murray.henwood@sydney.edu.au
Joachim W. Kadereit	Institut für Organismische und Molekulare Evolutionsbiologie, Johannes Gutenberg-Universität Mainz, 55099 Mainz, Germany. kadereit@uni-mainz.de
Eugene V. Kljuykov	Botanical Garden, Lomonosov Moscow State University, 1-12 Leninskie Gory, 119991 Moscow, Russia. kljuykov@gmail.com
Byoung-Yoon Lee	Plant Resources Division, National Institute of Biological Resources, Incheon 404-708, Republic of Korea. bylee80@korea.kr

Sigrid Liede-Schumann	University of Bayreuth, Universitätstraße 30, 95440 Bayreuth, Germany. sigrid.liede@uni-bayreuth.de
Porter P. Lowry II	Missouri Botanical Garden, P.O. Box 299, St. Louis, MO 63166-0299, USA Institut de Systématique, Évolution et Biodiversité (ISYEB), Unité Mixte de Recherche 7205, Muséum national d'Histoire naturelle, Sorbonne Universités, CP 39, 57 rue Cuvier, 75231 Paris CEDEX 05, France. pete.lowry@mobot.org
Ulrich Meve	Department of Plant Systematics, University of Bayreuth, Universitätstraße 30, 95440 Bayreuth, Germany. ulrich.meve@uni-bayreuth.de
David J. Middleton	Singapore Botanic Gardens, National Parks Board, 1 Cluny Road, Singapore 259569, Singapore. david_middleton@nparks.gov.sg
Anthony D. Mitchell	University of Otago, 2 Riccarton Avenue, P.O. Box 4345, Christchurch 8140, New Zealand. anthony.mitchell@otago.ac.nz
Timothy J. Motley	Department of Biological Sciences, Old Dominion University, 45th Street, Norfolk, VA 23529, USA
Bernard Muckensturm	14 rue Sainte-Anne, 98200 Mulhouse, France
Tatiana A. Ostroumova	Botanical Garden, Lomonosov Moscow State University, 1-12 Leninskie Gory, 119991 Moscow, Russia. ostroumovata@gmail.com
Michael G. Pimenov	Botanical Garden, Lomonosov Moscow State University, 1-12 Leninskie Gory, 119991 Moscow, Russia. mgpimenov@mail.ru
Gregory M. Plunkett	Cullman Program for Molecular Systematics, New York Botanical Garden, 2900 Southern Blvd., Bronx, NY 10458-5126, USA. gplunkett@nybg.org
Martin J. Potgieter	Department of Biodiversity, University of Limpopo, Private Bag X1106, Sovenga 0727, South Africa. martin.potgieter@ul.ac.za
James S. Pringle	Royal Botanical Gardens, 680 Plains Road West, Burlington, ON L7T 4H4, Canada. jpringle@rbg.ca
Fa-Ding Pu	Chengdu Institute of Biology, Chinese Academy of Sciences, P.O. Box 416, Chengdu, Sichuan 610041, People's Republic of China
Jean-Pierre Reduron	Via Apia 10 rue de l'Arsenal, 68100 Mulhouse, France. jp.reduron@hrnet.fr
George E. Schatz	Missouri Botanical Garden, P.O. Box 299, St. Louis, MO 63166-0299, USA. george.schatz@mobot.org

Krzysztof Spalik	Department of Molecular Phylogenetics & Evolution, Faculty of Biology, University of Warsaw Biological & Chemical Research Centre, 101 Żwirki i Wigury St., Warszawa PL-02-096, Poland. spalik@biol.uw.edu.pl
Lena Struwe	Department of Ecology, Evolution, & Natural Resources, Rutgers University, 237 Foran Hall, 59 Dudley Road, New Brunswick, NJ 08901, USA. lena.struwe@rutgers.edu
Patricia M. Tilney	Department of Botany and Plant Biotechnology, University of Johannesburg, P.O. Box 524, Auckland Park 2006 Johannesburg, South Africa. pmtilney@uj.ac.za
Ben-Erik van Wyk	Department of Botany and Plant Biotechnology, University of Johannesburg, P.O. Box 524, Auckland Park 2006 Johannesburg, South Africa. bevanwyk@uj.ac.za
Mark F. Watson	Royal Botanic Garden Edinburgh, 20a Inverleith Row, Edinburgh EH3 5LR, UK. m.watson@rbge.org.uk
Colin J. Webb	Tertiary Education Commission, Wellington, New Zealand
Jun Wen	Department of Botany, National Museum of Natural History, MRC 166, Smithsonian Institution, Washington, DC 20013-7012, USA. wenj@si.edu
Qiu-Yun (Jenny) Xiang	Department of Plant and Microbial Biology, North Carolina State University, Raleigh, NC 27695-7612, USA. jenny_xiang@ncsu.edu

Introduction to the Orders of this Volume

V. Bittrich and J.W. Kadereit

The present volume of this book series completes the treatment of the Asterids. Asterids are now contained in Vols. VI (Cornales, Ericales, 2004), VII (Lamiales, 2004), VIII (Asterales, 2007), XIV (Aquifoliales, Boraginales, Bruniales, Dipsacales, Escalloniales, Garryales, Paracryphiales, Solanales, Icacinaceae, Metteniusaceae, Vahliaceae, 2016) and the present volume, which contains the orders Apiales and Gentianales (except Rubiaceae). The only families of Asterids not treated in the series are Acanthaceae (Lamiales), Convolvulaceae (Solanales) and Rubiaceae (Gentianales).

Conspectus of the Families of Apiales

1. Plants without secretory canals; leaves simple; nectary absent or present; fruit fleshy, 1-seeded 2
– Plants with secretory canals throughout; leaves simple or composed; gynoecial nectary mostly present; fruit fleshy or dry, sometimes schizocarpic, 1- to many-seeded 4
2. Petiole not sheathing; petals valvate; nectary absent; ovary superior, unilocular **Pennantiaceae**
– Petiole ± sheathing; petals (sub-)imbricate, rarely induplicate-valvate; nectary present or absent; ovary inferior, 3-locular at least in upper part 3
3. Leaves distichous or subopposite, blade minutely pellucid-punctate; ovary 3-locular above, 1-locular below **Griseliniaceae**
– Leaves spiral, blade not punctate; ovary 3-locular throughout **Torricelliaceae**
4. Leaves simple, petiole not sheathing at base; inflorescences not umbellate; petals 4–40 mm long; stamen filaments straight in bud; style simple, ovary superior; nectary at base of ovary (rarely absent); fruit a capsule or berry, rarely woody indehiscent **Pittosporaceae**
– Leaves simple or compound, petiole usually sheathing at base; inflorescences or their ultimate units often umbellate; petals usually shorter; stamen filaments inflexed in bud; stylodia 2–many, ovary inferior; nectary on ovary roof; fruit a schizocarp or drupaceous 5
5. Plants herbaceous, rarely woody; leaves mostly compound, rarely entire, usually membraneous; petals usually clawed and with an inflexed tip; ovules 2 per locule, one abortive; fruit usually dry and schizocarpic **Apiaceae**
– Plants mostly shrubs or small trees; leaves simple to compound, usually subcoriaceous to coriaceous; petals usually broadly inserted, rarely clawed, and without inflexed tip; ovule 1 per locule; fruit usually drupaceous, rarely dry schizocarpic 6
6. Calyx with evident lobes; sepals and petals 5, petal apex without pointed adaxial thickening, stamens 5, carpels 2; fruit with secretory vesicles located immediately adjacent to the endocarp **Myodocarpaceae**
– Calyx with minute to obscure lobes, often forming a truncate rim; sepals and petals (3–)5(–12), petal apex with pointed adaxial thickening, stamens 5 (3–numerous), carpels 2–5(–100+); fruit without secretory vesicles **Araliaceae**

Seven families (about 522 genera, 6000 spp.) are currently included: Apiaceae (466 / 3820), Araliaceae (40 / 1900), Griseliniaceae (1 / 7), Myodocarpaceae (2 / 17), Pennantiaceae (1 / 4), Pittosporaceae (9 / 250) and Torricelliaceae (s.l. = incl. Aralidiaceae and Melanophyllaceae; 3 / 10). The phylogeny currently accepted is (Pennantiaceae + (Torricelliaceae + (Griseliniaceae + (Pittosporaceae + (Araliaceae + (Myodocarpaceae + Apiaceae)))))).

Few of the families now included in the Apiales were placed in the order before data from DNA sequencing became available. Undisputed were only Apiaceae + Araliaceae (see

J.W. Kadereit, V. Bittrich (eds.), *Flowering Plants. Eudicots*, The Families and Genera of Vascular Plants 15,
https://doi.org/10.1007/978-3-319-93605-5_1

below). Diversification and evolution of the order is discussed in detail in Stevens (2001 onwards), Kårehed (2003), Chandler and Plunkett (2004) and Nicolas and Plunkett (2014). The current phylogenies (e.g. Soltis et al. 2011) show the monogeneric Pennantiaceae as sister to the remainder of Apiales. Kårehed (2001), based on DNA sequence data, was the first to point out that *Pennantia* does not belong in Icacinaceae, as believed since the mid-19th century (cf. Miers 1852), but in Apiales. Because of the position of *Pennantia* in the phylogeny, recognition of the genus as a separate family had to be accepted. The next branch in the Apiales phylogeny is formed by the small but heterogeneous Torricelliaceae. The three genera of this family were often included in separate monogeneric families and only DNA sequence data suggested that they form a clade in the Apiales (Plunkett et al. 1996; Savolainen et al. 2000; Kårehed 2003). Although *Aralidium* was often included in Araliaceae in the 19th century, the lack of schizogenous secretory canals suggested different positions—for example, inclusion in Cornales (e.g. Thorne 1983). The other two genera, *Melanophylla* and *Torricellia*, were usually included in Cornales or Rosales. The monogeneric Griseliniaceae, forming the next branch in the Apiales phylogeny, was usually included in Cornales or even Cornaceae (Hooker 1867; Cronquist 1981), although already Endlicher (1850) had placed *Griselinia* in Araliaceae. Chase et al. (1993), in the first large angiosperm phylogeny based on DNA (*rbc*L) sequence data, recovered a clade composed of Araliaceae, Apiaceae, Griseliniaceae [as Cornaceae p.p., that family was shown to be polyphyletic] and Pittosporaceae (see also Olmstead et al. 1993). Pittosporaceae here is sister to Araliaceae + (Myodocarpaceae + Apiaceae), a large clade of Apiales, which is also morphologically well characterized. van Tieghem (e.g. 1884) had already early emphasized that schizogenic resin canals in many parts of the plants and a root anatomy with a characteristic arrangement of the canals and lateral roots suggested a close relationship among Pittosporaceae, Apiaceae and Araliaceae. The root anatomy of Myodocarpaceae apparently has not been investigated. Hegnauer (1969) was one of few who endorsed the proposal of van Tieghem and added some supporting evidence, especially chemical data (see also Judd et al. 1994). Ignorance of this evidence by most specialists in angiosperm classification suggests that these characters were given low weight. Cronquist (1981) insisted on the traditional Rosales affinity of Pittosporaceae and did not even mention root anatomical data in the family descriptions of Pittosporaceae, Apiaceae and Araliaceae, and he cited neither van Tieghem (1884) nor Hegnauer (1969). Takhtajan (1980) preferred Saxifragales for placement of Pittosporaceae, and Thorne (1983) his superorder Rosiflorae. Only Dahlgren (1983) relatively early classified Pittosporales (assessing the position of Tremandracae and Byblidaceae in this order as highly uncertain) close to Araliales (Apiaceae and Araliaceae). The apparently considerable difference in the ovary position, inferior in Apiaceae and Araliaceae, but superior in Pittosporaceae, possibly was another reason for the resistance of many authors to accept a close relationship among these families. As Erbar and Leins (2010) pointed out, however, in this case the difference is of lesser importance than thought by most systematists, as the intercalary growth in the floral axis resulting in the inferior ovaries is only a gradual difference. Furthermore, one surely can make an argument that the superior ovary of Pittosporaceae is derived.

Apiaceae and Araliaceae were early considered closely related (e.g. Jussieu 1789; Lindley 1830). While this close relationship has long been accepted due to the obvious similarities between the two, their circumscription and delimitation has proved much less simple. Without DNA sequence data, phylogenetic relationships could not be recovered correctly (e.g. Judd et al. 1994). The two genera of Myodocarpaceae as circumscribed here were originally placed in Apiaceae (*Myodocarpus*) and Araliaceae (*Delarbrea*) respectively, and sometimes considered to link these families, and later were mostly placed in tribe Myodocarpeae of Araliaceae. The need to place the two genera in their own family has only recently been realized based on DNA sequence data (Plunkett and Lowry 2001). Nevertheless, the first attempts to clarify the delimitation of Apiaceae and Araliaceae by using DNA sequence data still gave ambiguous results (e.g. Plunkett and Lowry 2001). Especially problematic was the circumscription of subfamily Hydrocotyloideae, which apparently was polyphyletic (Plunkett et al. 1997). Based on extensive taxon sampling and two DNA plastid markers, Nicolas and Plunkett (2009) finally could convincingly show that the subfamily as traditionally circumscribed (e.g. Drude 1898) is indeed polyphyletic and composed of several lineages dispersed among the Apiaceae and Araliaceae. Hydrocotyloideae s.str. (only four genera) appeared as sister group to the Araliaceae s.str. and today is included in that family. Some systematists, like Thorne (e.g. 1983), had complained that the taxonomic world kept Apiaceae separate from Araliaceae in spite of "numerous non-missing links". Only the new molecular data could clarify details of the relationships between the two families.

In the phylogenies of Winkworth et al. (2008) and Soltis et al. (2011), the Apiales appear as sister to the Paracryphiales (only family Paracryphiaceae) + Dipsacales clade. These orders are part of the subclass Campanulidae.

Woodiness occurs in all families of Apiales and is certainly plesiomorphic; large trees are very rare, however, and the predominantly herbaceous Apiaceae have by far the largest number of species. Plants are evergreen or deciduous. Schizogenous canals containing ethereal oils, resin or gums occur in the large majority of species, but are restricted to the core Apiales (= Pittosporaceae, Araliaceae, Myodocarpaceae, Apiaceae) and are absent in the three basal families of the order. Vessel elements have scalariform or simple perforation plates, helical thickenings in vessels are sometimes present, and fibres are septate and non-septate. Nodes are normally 3-lacunar to multilacunar. Phyllotaxy is usually spiral, the leaves are simple to compound and their margins entire or not. Stomata are paracytic in Pennantiaceae as well as in the core Apiales, but anisocytic, anomocytic, diacytic and encyclocytic types are also reported for the order. The petiole is nearly always exstipulate and very often sheathing at the base (but not in the basal Pennantiaceae and some Pittosporaceae).

The inflorescences are mostly described as paniculate, more rarely cymose. In the Araliaceae / Myodocarpaceae / Apiaceae clade at least the ultimate units are mostly umbellate. The pedicel is often articulated, the flowers are unisexual or bisexual, typically pentamerous with the exception of the ovary. Dioecy occurs at least in some species of most families of the Apiales. The calyx often has more or less strongly reduced lobes; the petals are mostly free at anthesis, but Erbar and Leins (2004, 2011) found early sympetaly in Araliaceae (incl. *Hydrocotyle*) and Pittosporaceae, although not (yet) in Apiaceae (only members of Saniculoideae and Apioideae studied). Stamens have free filaments and anthers open by longitudinal slits (except in a few Pittosporaceae). Pollen is nearly exclusively tricolporate and usually more or less spherical. Small flowers, nectaries or nectar-less but fragrant flowers (in *Pennantia*) suggest insect pollination for Apiales.

Superior ovaries exist in Pennantiaceae and, apparently derived from the inferior condition, in Pittosporaceae. All other members of the order have inferior ovaries. Free or only shortly connate stylodia are characteristic for the order, with the exception of Pittosporaceae, part of the Araliaceae and a few Apiaceae. Trans-septal bundles in the ovary occur in Torricelliaceae and Griseliniaceae. A nectary is absent in Pennantiaceae, but it is usually present in the other families and of gynoecial origin, as also in most Asterales. At anthesis, it is situated on the lower carpel flanks in the superior ovaries of Pittosporaceae and on the ovary roof, typically in an enlarged stylopodium, in part of the Torricelliaceae (*Aralidium* and *Torricellia*, absent in *Melanophylla*), Apiacae, Araliaceae and Myodocarpaceae. According to Erbar and Leins (2010), the different position of the nectary in most Apiales families compared with Pittosporaceae is related to the intercalary growth in the floral axis which results in the inferior ovaries in these families. The nectary is always formed, however, at the dorsal base of the carpels. Families of the basal grade of Apiales (Pennantiaceae, Torricelliaceae, Griseliniaceae) apparently have a 3-carpellate gynoecium, but only one carpel is fertile. Pennantiaceae have pseudomonomerous 1-locular ovaries with a 3-lobed stigma. The bicarpellate condition is derived and is found in most core Apiales, including nearly all Apiaceae, and all carpels are fertile. In Pittosporaceae the ovary is 2(3–5)-carpellate, the placentation axile or parietal, and there are several ovules. Other core Apiales have one fertile ovule per carpel, and the ovule is unitegmic and tenuinucellate or more rarely crassinucellate.

One-seeded drupes, as in the Pennantiaceae, are common in the order and are probably the plesiomorphic condition, but schizocarpic fruits with dry mericarps on a carpophore are typical for many Apiaceae, *Myodocarpus* and a few Araliaceae. The drupes suggest endozoochory but reports of this exist only for Pennantiaceae and some Araliaceae. The dry mericarps are dispersed either by autochory or, when winged, by wind and/or water, when with barbs or spines by epizoochory. Most Pittosporaceae have many-seeded capsules, and the sticky fluid around the seeds of *Pittosporum* (produced by multicellular epidermal hairs in the septal region (Erbar and Leins 1995)) suggests bird dispersal which may explain the wide distribution of the genus in the Indo-Pacific region. Seeds contain copious endosperm, apparently mostly oily, and a small straight embryo.

There are little phytochemical data for several families, making meaningful comparisons difficult. Fruit or seed oils containing petroselinic acid were found in *Griselinia*, Araliaceae and Apiaceae, and may occur also in Myodocarpaceae (no data). They are apparently lacking in Pittosporaceae. As Pennantiaceae and Torricelliaceae have not yet been investigated, the distribution and phylogenetic significance of this fatty acid in the Apiales is unclear. Polyacetylenes are known from *Torricellia*, Pittosporaceae, Araliaceae, Myodocarpaceae and Apiaceae. Cyanogenic compounds are very rare in the order. The lack of proanthocyanidins and ellagic acid (or their restriction to the seed coat) is also typical for Apiales. Many Pittosporaceae, Apiaceae and Araliaceae contain triterpenoid saponins and flavonoids (especially the flavonols kaempferol and quercetin). The occurrence of the iridoid glucoside griselinoside in *Aralidium*, *Torricellia* and *Griselinia* supports their close relationship, but data for Pennantiaceae are needed to conclude whether iridoids are plesiomorphic in the order. A basic chromosome number of $x = 6$ was suggested for the order by Yi et al. (2004). The Apiaceae are extremely variable in their chromosome numbers (see Plunkett et al., this volume).

The historical biogeography of the Apiales clade has recently been analyzed in detail by Nicolas and Plunkett (2014), who also estimated the divergence times for all major clades of the order. They concluded that the order originated in Australasia in the Early Cretaceous, with Australia as the likely centre of origin. Many palaeoendemics of the core Apiales today occur in Australia and especially New Caledonia. This island apparently offered refugia to "relictual lineages" because, unlike Australia, it suffered much less aridification and was much later colonized by humans. However, it was submerged at least partly during the Palaeogene. The early diversification of Apiales was probably due to vicariance events during the break-up of Gondwana. Younger lineages more probably developed after long-distance dispersal to Madagascar, Asia and the Americas (*Pittosporum* and Araliaceae). Apiaceae probably also originated in Australia and migrated to South America and South Africa before the complete Gondwana break-up, and later to north temperate regions.

Conspectus of the Families of Gentianales[1]

1. Ovary inferior (rarely semi-inferior and very rarely superior); plants usually with interpetiolar stipules (sometimes deciduous); internal phloem absent **Rubiaceae**
– Ovary superior, rarely semi-inferior; interpetiolar stipules usually absent (often with interpetiolar ridges); internal phloem present 2
2. Latex in non-articulated laticifers present in vegetative parts, usually white, less often translucent or yellowish or reddish; ovary mostly apocarpous, less often (congenitally or postgenitally) syncarpous **Apocynaceae**
– Latex absent in vegetative parts; ovary syncarpous, less often partially apocarpous (then Australian herbaceous species) 3
3. Style apically divided into 4 long stigmatic lobes, rarely with two free styles (then a tree with alternate leaves and fruit a 1-seeded samara) **Gelsemiaceae**
– Style single, rarely two free styles (then leaves not alternate and fruit not a 1-seeded samara) 4
4. Corolla lobes mostly contorted in bud; rarely valvate or imbricate (then plants either herbaceous and from N temperate or palaeotropical regions, or shrubby with pitcher-shaped leaves, or tropical trees with very long and narrow corolla tubes); placentation usually parietal, rarely axile **Gentianaceae**

[1] Key to families by J.W. Kadereit and L. Struwe.

– Corolla lobes imbricate, valvate or exduplicate-valvate in bud; placentation usually axile **Loganiaceae**

As understood here, the well-supported **Gentianales**, a name first used by Lindley (1833) to rename Bartling's (1830) 'Contortae', comprise Rubiaceae (incl. Dialypetalanthaceae and Theligonaceae; 611 genera / 13150 species fide Stevens 2001 onwards; not treated in this volume), Apocynaceae (incl. Asclepiadaceae; 378 / ca. 5350), Gelsemiaceae (3 / 13), Gentianaceae (102 / ca. 1750) and Loganiaceae (16 / ca. 460).

Following the account by Backlund et al. (2000), this circumscription of the order, apart from details, was essentially established by Downie and Palmer (1992) and Olmstead et al. (1993). Whereas close relationships among Apocynaceae (incl. Asclepiadaceae), Gentianaceae and Loganiaceae (at that time incl. part of Gelsemiaceae as well as various elements since segregated; see below) had been accepted for a long time (Bentham and Hooker 1862–1883), inclusion of Rubiaceae was first suggested by Wagenitz (1959). Two families placed near (or sometimes in) Gentianales even in the more recent past (Dahlgren 1980; Takhtajan 1987), Menyanthaceae and Oleaceae, rather belong to Asterales and Lamiales respectively (Downie and Palmer 1992; Olmstead et al. 1992).

Of all families in the order (apart from inclusion of Asclepiadaceae in Apocynaceae; Judd et al. 1994; Endress et al. 1996; for historical account and further references, see family account in this volume), most changes took place in Loganiaceae in the highly para- and polyphyletic circumscription of Leeuwenberg and Leenhouts (1980). These resulted not only in recognition (Struwe et al. 1994) and later recircumscription (Struwe et al. 2014) of Gelsemiaceae (for details, see account of Gelsemiaceae in this volume) as part of the order, but also in the exclusion of several genera (*Buddleja* – Scrophulariaceae / Lamiales; *Nuxia*, *Stilbe* (*Retzia*) – Stilbaceae / Lamiales; *Polypremum* – Tetrachondraceae / Lamiales; *Plocosperma* – Plocospermataceae / Lamiales; *Peltanthera*, *Sanango* – Gesneriaceae / Lamiales; *Desfontainia* – Columelliaceae / Bruniales; *Anthocleista*, *Fagraea*, *Potalia* – Gentianaceae / Gentianales; generic, familial and ordinal assignment according to Stevens 2001 onwards; for details and further references, see account of Loganiaceae in this volume).

Vegetatively, the order almost always has opposite leaves with interpetiolar lines or interpetiolar stipules (interpetiolar stipules are almost always present in Rubiaceae but far less common in the other four families), colleters, defined as multicellular secretory structures, are common and most often found in the leaf axils and on the sepals, and all families except Rubiaceae have bicollateral vascular bundles (internal phloem). Habit is very diverse in the order, ranging from annual or ephemeral herbs to large trees and lianas, and including several mycoheterotrophic and sometimes achlorophyllous genera in Gentianaceae. With very few exceptions flowers are 4- or 5-merous, tetracyclic and sympetalous with epipetalous stamens on the petal tube (Endress 2011) and a bicarpellate apo- (and then often postgenitally fused (Endress et al. 1983)) or syncarpous ovary which is superior or inferior (Rubiaceae). The corolla is often but by no means always contorted in bud. Ovules are tenuinucellar or sometimes reduced tenuinucellar (Endress 2011).

Route I iridoids, seco-iridoids and indole alkaloids, all part of one biosynthetic pathway (Jensen et al. 2002), are common in the order. The intensely bitter gentiopicrin of, amongst others, *Gentiana lutea* (Gentianaceae) represents an example for seco-iridoids, and the well-known strychnine (*Strychnos nux-vomica* / Loganiaceae) and chinine (*Cinchona* spp. / Rubiaceae) are examples for indole alkaloids. Cardenolides and steroid alkaloids are common in Apocynaceae, and xanthones are common in Gentianaceae.

Whereas Gelsemiaceae and Loganiaceae are exclusively pantropical in distribution, Apocynaceae and Rubiaceae are predominantly pantropical but have reached temperate regions with some genera or suprageneric groups. In Rubiaceae, this is mainly tribe Rubieae (Bremer and Eriksson 2009). In Gentianaceae, the majority of tribes are tropical in distribution but its largest tribe Gentianeae, with more than half of the species diversity of the family and including *Gentiana* as the largest genus of the family, is primarily temperate-alpine (Struwe et al. 2002). Considering this distribution of the order, an origin in tropical areas appears very likely. Several estimates of the stem age of the order converge on

±100 million years ago (Bremer et al. 2004; Janssens et al. 2009; Wikström et al. 2015), but younger estimates of 83–89 (Wikström et al. 2001), 76 (56–97; Lemaire et al. 2011), 85.9 (Magallón et al. 2015) and <76.3 (Tank et al. 2015) have also been published.

Apart from the position of Rubiaceae as sister to a clade of the remaining families, already found in early molecular analyses (Downie and Palmer 1992; Olmstead et al. 1993; Bremer et al. 1999; Backlund et al. 2000), relationships among families, considering sampling of both taxa and data as well as support in various studies (e.g. Refulio-Rodriguez and Olmstead 2014; Nazaire et al. 2014; Struwe et al. 2014; Yang et al. 2016), are best considered unresolved. This, considering the taxonomic history of Loganiaceae, Gentianaceae and Gelsemiaceae (see above), clearly reflecting difficulties in their morphological circumscription, is not entirely surprising. Only Apocynaceae are well characterized by rather clear-cut autapomorphies, mainly the presence of latex and the usually apocarpous gynoecium.

As regards relationships of Gentianales within Asterids I (Lamiidae), the order is part of a well-supported clade also containing Boraginales, Lamiales and Solanales. However, relationships among these four orders, considering sampling of both taxa and data as well as support in various studies (Ku et al. 2013; Martínez-Alberola et al. 2013; Refulio-Rodriguez and Olmstead 2014; Weigend et al. 2014; Magallón et al. 2015; Stull et al. 2015; Chen et al. 2016), again are best considered unresolved.

For Rubiaceae, which are not included in this volume, Robbrecht and Manen (2006) and Bremer (2009) provided extensive accounts of phylogeny, evolution and classification. Whereas Robbrecht and Manen (2006) suggested subdivision into two subfamilies, Cinchonoideae and Rubioideae, of which the former is paraphyletic by inclusion of Coptosapelteae (containing only *Luculia*, *Acranthera* and *Coptosapelta*: Rydin et al. 2009 - these authors treated *Luculia* as the monogeneric tribe Luculieae), Bremer (2009) recognized three subfamilies, Cinchonoideae, Ixoroideae and Rubioideae with Coptosapelteae (+ *Luculia*) unclassified. In a more recent analysis of mitochondrial data (Rydin et al. 2017), Coptosapelteae (excl. *Luculia*) were found to be sister to Cinchonoideae plus Ixoroideae.

Bibliography

Backlund, A., Bremer, B. 1997. Phylogeny of the Asteridae s. str based on *rbc*L sequences, with particular reference to the Dipsacales. Plant Syst. Evol. 207: 225–254.

Backlund, M., Oxelman, B., Bremer, B. 2000. Phylogenetic relationships within Gentianales based on *ndh*F and *rbc*L sequences, with particular reference to the Loganiaceae. Amer. J. Bot. 87: 1029–1043.

Bartling, F.G. 1830. Ordines naturales plantarum eorumque characteres et affinitates. Göttingen: Dieterichianus.

Bentham, G., Hooker, J.D. 1862–1883. Genera Plantarum, vols. 1–3. London: Lovell Reeve.

Bremer, B. 2009. A review of molecular phylogenetic studies of Rubiaceae. Ann. Missouri Bot. Gard. 96: 4–26.

Bremer, B., Eriksson, T. 2009. Time tree of Rubiaceae: phylogeny and dating of the family, subfamilies, and tribes. Int. J. Plant Sci. 170: 766–793.

Bremer, B., Jansen, R.K., Oxelman, B., Backlund, M., Lantz, H., Kim, K.J. 1999. More characters or more taxa for a robust phylogeny - case study from the coffee family (Rubiaceae). Syst. Biol. 48: 413–435.

Bremer, K., Friis, E.M., Bremer, B. 2004. Molecular phylogenetic dating of asterid flowering plants shows Early Cretaceous diversification. Syst. Biol. 53: 496–505.

Chandler, G.T., Plunkett, G.M. 2004. Evolution in Apiales: nuclear and chloroplast markers together in (almost) perfect harmony. Bot. J. Linn. Soc. 144: 123–147.

Chase, M.W., Soltis, D.E., Olmstead, R.G., Morgan, D., Les, D.H., Mishler, B.D., Duvall, M.R., Price, R.A., Hills, H.G., Qiu, Y.-L., Kron, K.A., Rettig, J.H., Conti, E., Palmer, J.D., Manhart, J.R., Sytsma, K.J., Michaels, H.J., Kress, W.J., Karol, K.G., Clark, W.D., Hedrén, M., Gaut, B.S., Jansen, R.K., Kim, K.-J., Wimpee, C.F., Smith, J.F., Furnier, G.R., Strauss, S. H., Xiang, Q.-Y., Plunkett, G.M., Soltis, P.S., Swensen, S.M., Williams, S.E., Gadek, P.A., Quinn, C.J., Eguiarte, L.E., Golenberg, E., Learn, G.H., Jr., Graham, S.W., Barrett, S.C.H., Dayanandan, S., Albert, V.A. 1993. Phylogenetics of seed plants: an analysis of nucleotide sequences from the plastid gene *rbc*L. Ann. Missouri Bot. Gard. 80: 528–580.

Chen, Z.-D., Yan, T., Lin, L., Lu, L.-M., Li, H.-L., Sun, M., Liu, B., Chen, M., Niu, Y.-T., Ye, J.-F., Cao, Z.-Y., Liu, H.-M., Wang, X.-M., Wang, W., Zhang, J.-B., Meng, Z., Cao, W., Li, J.-H., Wu, S.-D., Zhao, H.-L., Liu, Z.-J., Du, Z.-Y., Wan, Q.-F., Guo, J., Xin-Xin Tan, X.-X., Su, J.-X., Zhang, L.-J., Yang, L.-L., Liao, Y.-Y., Li, M.-H., Zhang, G.-Q., Chung, S.-W., Zhang, J., Xiang, K.-L., Li, R.-Q., Soltis, D.E., Soltis, P.S., Zhou, S.-L., Ran, J.-H., Wang, X.-Q., Jin X.-H., Chen, Y.-S., Gao, T.-G., Li, J.-H., Zhang, S.-Z., Lu, A.M., China Phylogeny Consortium. 2016. Tree of life for the genera of Chinese vascular plants. J. Syst. Evol. 54: 277–306.

Cronquist, A. 1981. An Integrated System of Classification of Flowering Plants. New York: Columbia University Press.

Dahlgren, R. 1980. A revised system of classification of the angiosperms. Bot. J. Linn. Soc. 80: 91–124.

Dahlgren, R. 1983. General aspects of angiosperm evolution and macrosystematics. Nord. J. Bot. 3: 119–149.

Downie, S.R., Palmer, J.D. 1992. Restriction site mapping of the chloroplast DNA inverted repeat: a molecular phylogeny of the Asteridae. Ann. Missouri Bot. Gard. 79: 266–283.

Drude, O. 1898. Umbelliferae. In: Engler, A., Prantl, K. (eds.) Die Natürlichen Pflanzenfamilien, vol. III. Leipzig: Verlag von Wilhelm Engelmann, pp. 63–250.

Endlicher, S.L. 1850. Genera Plantarum, suppl. 4(3). (*Griselinia*, p. 16, n. 4562). Wien: Fr. Beck.

Endress, P.K. 2011. Evolutionary diversification of the flowers in angiosperms. Amer. J. Bot. 98: 370–396.

Endress, P.K., Jenny, M., Fallen, M.E. 1983. Convergent elaboration of apocarpous gynoecia in higher advanced dicotyledons (Sapindales, Malvales, Gentianales). Nord. J. Bot. 3: 293–300.

Endress, M.E., Sennblad, B., Nilsson, S, Civeyrel, L., Chase, M.W., Huysmans, S., Grafström, E., Bremer, B. 1996. A phylogenetic analysis of Apocynaceae s. str. and some related taxa in Gentianales: a multidisciplinary approach. Op. Bot. Belgica 7: 59–102.

Erbar, C., Leins, P. 1995. An analysis of the early floral development of *Pittosporum tobira* (Thunb.) Aiton and some remarks on the systematic position of the family Pittosporaceae. Feddes Repert. 106: 463–473.

Erbar, C., Leins, P. 2004. Sympetaly in Apiales (Apiaceae, Araliaceae, Pittosporaceae). S. Afr. J. Bot. 70: 458–467.

Erbar, C., Leins, P. 2010. Nectaries in Apiales and related groups. Plant Divers. Evol. 128: 269–295.

Erbar, C., Leins, P. 2011. Synopsis of some important, non-DNA character states in asterids with special reference to sympetaly. Plant Divers. Evol. 129: 93–123.

Hegnauer, R. 1969. Chemical evidence for the classification of some plant taxa. In: Harborne, J.B., Swain, T. (eds.) Perspectives in Phytochemistry. London: Academic Press, pp. 121–138.

Hooker, J.D. 1867. Cornaceae. In: Bentham, G., Hooker, J. D. (eds.) Genera Plantarum 1(3). London: Reeve & Co, pp. 947–952.

Janssens, S.B., Knox, E.B., Huysmans, S., Smets, E.F., Merckx, V.S.F.T. 2009. Rapid radiation of *Impatiens* (Balsaminaceae) during Pliocene and Pleistocene: result of a global climate change. Molec. Phyl. Evol. 52: 806–824.

Jensen, S.R. Franzyk, H., Wallander, E. 2002. Chemotaxonomy of the Oleaceae: iridoids as taxonomic markers. Phytochemistry 60: 213–231.

Judd, W.S., Sanders, R.W., Donoghue, M.J. 1994. Angiosperm family pairs: preliminary phylogenetic analyses. Harvard Pap. Bot. 5: 1–51.

Jussieu, A.-L. de. 1789. Genera Plantarum. Paris: Herissant & Barrois.

Kårehed, J. 2001. Multiple origin of the tropical forest tree family Icacinaceae. Amer. J. Bot. 88: 2259–2274.

Kårehed, J. 2003. The family Pennantiaceae and its relationship to Apiales. Bot. J. Linn. Soc. 141: 1–24.

Ku, C., Chung, W.-C., Chen, L.-L., Kuo, C.-H. 2013. The complete plastid genome sequence of Madagascar periwinkle *Catharanthus roseus* (L.) G. Don: plastid genome evolution, molecular marker identification, and phylogenetic implications in Asterids. PLoS ONE 8: e68518.

Leeuwenberg, A.J.M., Leenhouts, P.W. 1980. Loganiaceae. In: Natürl. Pflanzenfam., ed. 2, 28b (1). Berlin: Duncker & Humblot, pp. 211–237.

Lemaire, B. Vandamme, P., Merckx, V., Smets, E., Dessein, S. 2011. Bacterial leaf symbiosis in angiosperms: host specificity without co-speciation. PLoS ONE 6: e24430.

Lindley, J. 1830. An introduction to the natural system of botany: or, A systematic view of the organisation, natural affinities, and geographical distribution, of the whole vegetable kingdom: together with the uses of the most important species in medicine, the arts, and rural or domestic economy (1st ed.). London: Longman.

Lindley, J. 1833. Nixus Plantarum. London: Ridgeway & Sons.

Magallón, S., Gómez-Acevedo, S., Sánchez-Reyes, L.L., Hernández-Hernández, T. 2015. A metacalibrated time-tree documents the early rise of flowering plant phylogenetic diversity. New Phytol. 207: 437–453.

Martínez-Alberola, F., del Campo, E.M., Lázaro-Gimeno, D., Mezquita-Claramonte, S., Molins, A., Mateu-Andrés, I., Pedrola-Monfort, J., Casano, L.M., Barreno, E. 2013. Balanced gene losses, duplications and intensive rearrangements led to an unusual regularly sized genome in *Arbutus unedo* chloroplasts. PLoS ONE 8: e79685

Miers, J. 1852. On some genera of the Icacinaceae. Ann. Mag. Nat. Hist. II, 9: 485–492.

Nazaire, M., Wang, X.-Q., Hufford, L. 2014. Geographic origins and patterns of radiation of Mertensia (Boraginaceae). Amer. J. Bot. 101: 104–118.

Nicolas, A.N., Plunkett, G.M. 2009. The demise of subfamily Hydrocotyloideae (Apiaceae) and the re-alignment of its genera across the whole order Apiales. Molec. Phyl. Evol. 53: 134–151.

Nicolas, A.N., Plunkett, G.M. 2014. Diversification times and biogeographic patterns in Apiales. Bot. Review 80: 30–58.

Olmstead, R.G., Michaels, H., Scott, K.M., Palmer, J.D. 1992. Monophyly of the Asteridae and identification of their major lineages inferred from DNA sequences of *rbc*L. Ann. Missouri Bot. Gard. 79: 249–265.

Olmstead, R.G., Bremer, B., Scott, K.M., Palmer, J.D. 1993. A parsimony analysis of the Asteridae *sensu lato* based on *rbc*L sequences. Ann. Missouri Bot. Gard. 80: 700–722.

Plunkett, G.M., Lowry, P.P. Jr. 2001. Relationships among "ancient araliads" and their significance for the systematics of Apiales. Molec. Phyl. Evol. 19: 259–276.

Plunkett, G.M., Soltis, D.E., Soltis, P.S. 1996. Higher level relationships of Apiales (Apiaceae and Araliaceae) based on phylogenetic analysis of *rbc*L sequences. Amer. J. Bot. 83: 499–515.

Plunkett, G.M., Soltis, D.E., Soltis, P.S. 1997. Clarification of the relationship between Apiaceae and Araliaceae based on *mat*K and *rbc*L sequence data. Amer. J. Bot. 84: 565–580.

Refulio-Rodriguez, N.F., Olmstead, R.G. 2014. Phylogeny of Lamiidae. Amer. J. Bot. 101: 287–299

Robbrecht, E., Manen, J.-F. 2006. The major evolutionary lineages of the coffee family (Rubiaceae, Angiosperms). Combined analysis (nDNA and cpDNA) to infer the position of *Coptosapelta* and *Luculia*, and supertree construction based on *rbc*L, *rps*16, *trn*L-*trn*F and *atp*B-*rbc*L data. A new classification in two subfamilies, Cinchonoideae and Rubioideae. Syst. Geogr. Pl. 76: 85–146.

Rydin, C., Kainulainen, K., Razafimandimbison, S.G., Smedmark, J.E.E., Bremer, B. 2009. Deep divergences in the coffee family and the systematic position of *Acranthera*. Plant Syst. Evol. 278:101–123.

Rydin, C., Wikström, N., Bremer, B. 2017. Conflicting results from mitochondrial genomic data challenge current views of Rubiaceae phylogeny. Amer. J. Bot. 104:1522–1532.

Savolainen, V., Fay, M.F., Albach, D.C., Backlund, A., van der Bank M., Cameron, K.M., Johnson, S.A., Lledó, M.D., Pintaud, J.-C., Powell, M., Sheahan, M.C., Soltis, D.E., Soltis, P.S., Weston, P., Whitten, W.M., Wurdack, K.J., Chase, M.W. 2000. Phylogeny of the eudicots: a nearly complete familial analysis based on *rbc*L gene sequences. Kew Bull. 55: 257–309.

Soltis, D.E., Smith, S.A., Cellinese, N., Wurdack, K.J., Tank, D.C. , Brockington, S.F., Refulio-Rodriguez, N.F., Walker, J.B., Moore, M.J., Carlsward, B.S., Bell, C.D., Latvis, M., Crawley, S., Black, C., Diouf, D., Xi, Z., Rushworth, C.A., Gitzendanner, M.A., Sytsma, K. J., Qiu, Y.-L., Hilu, K.W., Davis, C.C., Sanderson, M. J., Beaman, R.S., Olmstead, R.G., Judd, W.S., Donoghue, M.J., Soltis, P.S. 2011. Angiosperm phylogeny: 17 genes, 640 taxa. Amer. J. Bot. 98: 704–730.

Stevens, P.F. (2001 onwards). Angiosperm Phylogeny Website. Version 14, July 2017.

Struwe, L., Albert, V.A., Bremer, B. 1994. Cladistics and family level classification of the Gentianales. Cladistics 10: 175–205.

Struwe, L., Kadereit, J.W., Klackenberg, J., Nilsson, S., Thiv, M., Hagen, K.B. von, Albert, V.A. 2002. Systematics, character evolution, and biogeography of Gentianaceae, including a new tribal and subtribal classification. In: Struwe, L., Albert, V.A. (eds.) Gentianaceae - Systematics and Natural History. Cambridge: Cambridge University Press, pp. 21–309.

Struwe, L. Soza, V.L., Sugumaran, M., Olmstead, R.G. 2014. Gelsemiaceae (Gentianales) expanded to include the enigmatic Asian genus *Pteleocarpa*. Bot. J. Linn. Soc. 175: 482–496.

Stull, G.W., Duno de Stefano, R., Soltis, D.E., Soltis, P.S. 2015. Resolving basal lamiid phylogeny and the circumscription of Icacinaceae with a plastome-scale data set. Amer. J. Bot. 102: 1794–1813.

Takhtajan, A. 1980. Outline of the classification of flowering plants (Magnoliophya). Bot. Review 46(3): 225–359.

Takhtajan, A. 1987. Systema Magnoliophytorum. Leningrad: Nauka.

Tank, D.C., Eastman, J.M., Pennell, M.W., Soltis, P.S., Soltis, D.E, Hinchliff, C.E., Brown, J.W., Sessa, E.B., Harmon, L.J. 2015. Nested radiations and the pulse of angiosperm diversification: increased diversification rates often follow whole genome duplications. New Phytol. 207: 454–467.

Thorne, R.F. 1983. Proposed new alignments in the angiosperms. Nord. J. Bot. 3: 85–117.

van Tieghem P. 1884. Sur la structure et les affinités des Pittosporées. Bull. Soc. Bot. France 31: 383–385.

Wagenitz, G. 1959. Die systematische Stellung der Rubiaceae—ein Beitrag zum System der Sympetalen. Bot. Jahrb. Syst. 79: 17–35.

Weigend, M., Luebert, F., Gottschling, M., Couvreur, T.L. P., Hilger, H.H., Miller, J.S. 2014. From capsules to nutlets—phylogenetic relationships in the Boraginales. Cladistics 30: 508–518

Wikström, N., Savolainen, V., Chase, M.W. 2001. Evolution of the angiosperms: calibrating the family tree. Proc. R. Soc. Lond. B 268: 2211–2220.

Wikström, N., Kainulainen, K., Razafimandimbison, S.G., Smedmark, J.E.E., Bremer, B. 2015. A revised time tree of the Asterids: establishing a temporal framework for evolutionary studies of the coffee family (Rubiaceae). PLoS ONE 10: e0126690.

Winkworth, R.C., Lundberg, J., Donoghue, M.J. 2008. Toward a resolution of campanulid phylogeny, with a special reference to the placement of Dipsacales. Taxon 57: 53–65.

Yang, L.-L., Li, H.-L., Wei, L., Kuang, D.-Y., Li, M.-H., Liao, Y.-Y., Chen, Z.-D., Wu, H., Zhang, S.-Z. 2016. A supermatrix approach provides a comprehensive genus-level phylogeny for Gentianales. J. Syst. Evol. 54: 400–415.

Yi, T., Lowry, P.P. II, Plunkett, G.M., Wen, J. 2004. Chromosomal evolution in Araliaceae and close relatives. Taxon 53: 987–1005.

Apiaceae

Apiaceae Lindl., Intr. Nat. Syst. Bot., ed. 2: 21 (1836), nom. cons. et nom. alt.
Umbelliferae Juss., Gen. Pl.: 218 (1789), nom. cons. et nom. alt.

G.M. Plunkett, M.G. Pimenov, J.-P. Reduron, E.V. Kljuykov, B.-E. van Wyk, T.A. Ostroumova, M.J. Henwood, P.M. Tilney, K. Spalik, M.F. Watson, B.-Y. Lee, F.-D. Pu, C.J. Webb, J.M. Hart, A.D. Mitchell, and B. Muckensturm

Annual, biennial or perennial, monocarpic or polycarpic, caulescent or acaulescent, glabrous or pubescent or glandular-pubescent herbs, rarely suffrutescent or woody subshrubs, shrubs or trees, evergreen or deciduous; with taproots, rhizomes, or unbranched to branched rootstocks, sometimes swollen and tuberiform; with schizogenous secretory canals throughout the plant. Stems erect, ascending, decumbent, prostrate or rarely creeping; branched or unbranched; with multilacunar (or rarely trilacunar) nodes, internodes sometimes fistulose. Leaves alternate, rarely opposite or verticillate; petioles usually present and typically sheathing at base, often inflated, without (or rarely with) stipules; blades ternately or pinnately compound, or pinnately, ternately, or palmately lobed, or simple and entire to toothed, to deeply divided, rarely sessile or perfoliate; venation pinnate or less commonly palmate or parallel. Plants andromonoecious or hermaphroditic, rarely gynodioecious to dioecious. Inflorescences compound-umbellate (less commonly simple-umbellate, capitulate or cymose), arranged in cymose or racemose synflorescences; umbels typically subtended by an involucre of one to many bracts at the base of the rays (secondary axes), bracts entire or dissected, or lacking; umbellules typically subtended by involucel of one to many bracteoles at the base of the pedicels, bracteoles entire or dissected, or lacking. Flowers perfect or staminate (rarely functionally pistillate), epigynous, actinomorphic (or outer flowers of the umbel or umbellule zygomorphic). Perianth and androecium 5-merous (rarely 5–6-merous). Calyx lobes linear to lanceolate, triangular, or ovate, sometimes spiny or pinnatisect, or small to obscure. Petals valvate, usually basally clawed and having a narrowed, inflexed apex (less commonly with broad insertion and/or lacking the inflexed apex), sometimes lobed (or shortly notched to deeply bifid), dorsally glabrous or puberulent; sometimes with one to several secretory ducts. Stamens alternate with the petals; anthers dorsifixed or basifixed, tetrasporangiate, dehiscing by longitudinal slits; filaments filiform, inflexed in bud. Ovary syncarpous of 2 carpels (rarely 1 or 2–4), each carpel unilocular with apical to axile placentation; stigmas on stylodia; stylodia distinct, rarely connate, frequently reflexed in fruit, usually swollen at the base to form a nectariferous disc or stylopodium. Ovules anatropous, pendulous, two per locule (but one always abortive), unitegmic, tenuinucellate or pseudocrassinucellate. Fruits dry (rarely fleshy), usually schizocarpic with two mericarps attached by a commissural face, usually with a bifurcated or entire carpophore (becoming free from or remaining fused to the mericarps at maturity); fruits with filiform, keeled, corky, winged, dentate, spiny or obscure ribs formed over the vascular bundles (primary ribs), and sometimes formed in the furrows or valleculae (secondary ribs); one or more secretory canals found in association with the vascular strands (rib oil ducts, sometimes obscure) and/or in the valleculae and commissural face (vittae, sometimes arranged cyclically); endocarp parenchymatous (soft to lignified) or sclerified. Seeds straight, endosperm at commissural face plane to deeply concave or sulcate, copious and oily; embryo minute but often well differentiated.

A cosmopolitan family comprising 466 genera and about 3,820 species, most diverse in temperate Eurasia and North America.

J.W. Kadereit, V. Bittrich (eds.), *Flowering Plants. Eudicots*, The Families and Genera of Vascular Plants 15,
https://doi.org/10.1007/978-3-319-93605-5_2

Vegetative Morphology. The typical umbellifers are herbaceous annuals, biennials or perennials, but some species may become woody at the base, and a few species are truly woody trees, shrubs, or subshrubs. Most species are characterized by leafy or scapose stems, often branched and frequently hollow at the internodes. In some species, the branches may be highly reduced, or largely prostrate and rooting at the nodes, or forming a branched or unbranched caudex at or just beneath the surface of the ground; cushion plants are also known among some of the montane and alpine genera. Unbranched or sparsely branching taproots predominate in Apiaceae, ranging from slender to fleshy or tuberous; tubers may be hypocotylar or radical. The leaves of most umbellifers (especially in subfamily Apioideae) are alternate (rarely opposite, as in *Bowlesia*, *Drusa* and *Spananthe*), pinnate, ternate, decompound, or more rarely undivided. The ultimate leaf segments (leaflets or lobes) range from broad and entire, to toothed or lobed, to filiform. A "rachis-leaf" morphology (Kaplan 1970) is found in several lineages of apioids (*Lilaeopsis*, *Harperella*, *Ottoa*, *Tiedemannia*, and some species of *Oenanthe*, *Oreomyrrhis*, and *Perissocoeleum*) and saniculoids (e.g., some species of *Eryngium*), in which the leaflets are lost and the rachis is septate and either inflated or compressed. In some genera, the leaves may be spinescent (e.g., *Aciphylla*, *Echinophora*, *Eryngium*, and *Pycnocycla*), or reduced to simple or branched phyllodes (e.g., *Anginon*). Most genera of subfamilies Saniculoideae, Azorelloideae, and Mackinlayoideae have simple leaves, variously crenate, toothed or lobed (either palmately, ternately, or pinnately). Most umbellifer leaves have pinnate venation, but palmate or parallel venation is known in some genera. They are usually petiolate, but some or all leaves may be sessile or perfoliate (and in some species of *Eryngium*, the leaves may approach a monocot morphology, with a basal intercalary meristem and parallel venation in long blades that lack petioles). The petiole bases are usually sheathing and often inflated; these alate bases are generally not treated as stipules, which are therefore considered lacking in all but a few genera of Azorelloideae and Mackinlayoideae.

Vegetative Anatomy. The stems of umbellifers are often ribbed and fistulose, with the pith lost at the internodes, or the pith may be present and spongy. Strands of collenchyma are often found in the cortex (under or within the ribs), but these may be lacking, or sometimes replaced by a ring of sclerenchyma (Hoar 1915). Vascular strands are usually arranged in a ring of isolated bundles (or less commonly fused and continuous); cortical and medullary bundles are also common (Hoar 1915). Most species are herbaceous, but the woody habit may result from conventional or anomalous secondary growth in some taxa. The xylem is characterized by libriform fibers and diffuse pores, scanty vasicentric parenchyma, heterogeneous IIB rays (uniseriate or biseriate), vessel elements with almost exclusively simple, elliptical or nearly round perforations (rarely also scalariform) that are nearly at right angles to the lateral walls, simple and bordered pits, and sometimes other modifications, such as helical thickenings (Hoar 1915; Rodríguez 1971; Watson and Dallwitz 1992). Schizogenous secretory canals are found associated with vascular strands, originating from the intercellular space formed by 3 or 4 adjacent cells, and expanding to form elongated, circular canals surrounded by an endothelium of 10–20 cells in transverse section (Mittal 1961), which secrete essential oils, mucilage, balsam, gums and resins. In the roots, concentric rings of amphicribral bundles may be present (as in species of *Angelica*, *Anthriscus*, *Apium*, *Carum*, *Chaerophyllum*, *Cicuta*, *Daucus*, *Oenanthe*, and *Sium*; Metcalfe and Chalk 1983). The sieve-tube plastids of Apiaceae belong to the S-type (see Wagenitz 1992).

The nodes are usually multilacunar, with 7 to 15–17 traces per leaf (Mittal 1961), but trilacunar nodes have also been reported for *Bowlesia* and *Drusa* (Pimenov and Sdobnina 1984).

The petioles may be solid or fistulose, glabrous or pubescent, grooved or ungrooved on the adaxial side; with or without medullary vascular bundles; with collenchyma strands that may or may not be lignified. Petiolar anatomy may vary considerably in some polymorphic genera (e.g., *Angelica*, *Seseli*), where it may be useful for infrageneric taxonomy, whereas in some large genera (e.g., *Ferula*) the petioles are much less variable, and in some unrelated genera, the petioles may be quite similar (Pimenov et al. 1982, 1986).

The primary roots of most Apiaceae exhibit a typical dicot development, in which the central xylem core is diarch to polyarch (with 2–many protoxylem ridges) developing centripetally and alternating with areas of primary phloem, surrounded by 1–2 layers of parenchyma (from which the pericycle develops) and an endodermis (Courchet 1884; Solereder 1908; Havis 1939; Esau 1940; Metcalfe and Chalk 1950). Schizogenous secretory canals are located just interior of the endodermis, opposite both the xylem archs and phloem groups, leading to a peculiar development of lateral roots from the pericycle regions but between the resin canals (and thus between the xylem archs and phloem groups), a feature shared with both Araliaceae and Pittosporaceae (van Tieghem 1872; Solereder 1908; Esau 1940). Secondary growth is initiated from the vascular cambium, after which the cortex and endodermis are shed and the periderm (which develops from the pericycle) becomes the protective layer (Warning 1934; Esau 1940). Schizogenous secretory canals are present in the secondary phloem of older roots (and rarely also in secondary xylem in some species of *Myrrhis* and *Opopanax*) (Metcalfe and Chalk 1950). In the tuberous roots of some species (esp. *Oenanthe*, but also in *Angelica*, *Anthriscus*, *Apium*, *Carum*, *Chaerophyllum*, *Cicuta*, *Daucus*, *Ferula*, *Magydaris*, and *Sium*), numerous centric bundles are found in concentric rings, each with central xylem (Metcalfe and Chalk 1950).

The indumentum is rarely taxonomically important, and many species of Apiaceae are entirely glabrous. In those taxa having an indumentum, it ranges from scarcely puberulent to densely tomentose, and the trichomes range from short papillae to thin and slender hairs, and may be unicellular or multicellular, multiseriate, glandular, and/or stellate (including hairs, scales or glochids); many taxa are glaucescent, which may be associated with different types of wax crystalloids on adjacent epidermal cells, as in *Bupleurum* (Barthlott et al. 1998). The stomata are anomocytic, hemiparacytic, paracytic, and/or diacytic (Guyot 1966, 1971, 1978; Ostroumova 1987, 1990; Ostroumova and Kljuykov 1991, 2007), and in some cases there is a mixture of stomatal types on a single leaf blade. For those groups in which it has been studied, the dominant stomatal type can be useful as a taxonomic character at the generic or sectional level. Bicytic stomata are relatively rare, but they are known from some species of *Carum*, *Chaerophyllum*, *Ostericum*, and *Vanasushava* (in proportions of 10–50%) and in *Daucus*, *Pimpinella*, *Pternopetalum* and infrageneric groups of *Angelica*, *Laserpitium*, and *Scandix* (in proportions > 50%).

Inflorescence Structure and Floral Morphology. Umbellate inflorescences predominate in Apiaceae, although capitulate and cymose inflorescences are also found, particularly in Saniculoideae, Azorelloideae, and Mackinlayoideae. In Apioideae, the inflorescence is most commonly a compound umbel. The peduncles may be terminal, axillary, or both, terminating in one to many secondary axes known as rays. Each ray terminates in one or more pedicels, forming umbellules. The umbels are often subtended by one or more variously shaped bracts, forming an involucre, and the umbellules by one or more bracteoles, forming an involucel. Bracts and/or bracteoles may be entire or dissected, herbaceous, membranaceous, or spiny. Pseudanthia (anthoids) may be formed by the enlargement of the marginal petals of the outer umbellules or flowers (especially in compound-umbellate species of genera such as *Orlaya*, *Lisaea*, *Coriandrum*, *Tordylium*, *Artedia*, and *Scandix*), or by the enlargement of the involucral bracts (especially in simple-umbellate and capitulate species, as in *Astrantia*, *Actinolema*, *Hacquetia*, *Alepidea*, *Hymenolaena*, *Bupleurum*, *Xanthosia*). Rarely, umbellules may be uniflorous (as in some species of *Centella*, *Xanthosia*, *Bupleurum* and *Lagoecia*). The plants are frequently andromonoecious (with perfect and functionally staminate flowers variously arranged), or completely hermaphroditic, or less commonly gynodioecious or dioecious.

The flowers are epigynous, and development proceeds acropetally through a variety of initiation patterns, but generally ending in the same number and positions of the floral organs (Erbar and Leins 1997; Leins and Erbar 2004). The calyx may be represented by a rim of 5 (4–6) variously sized lobes or teeth, often rather small or sometimes obscure, or linear to lanceolate, triangular or ovate, and sometimes spiny or pinnatisect (*Lagoecia*) or rarely petaloid (*Hermas*). The corolla appears choripetalous at maturity, but the petals are initiated from a continuous ring primordium ("early sympetaly"; Erbar and Leins 1996, 2004), and comprises 5 (4–6) white, yellow, greenish to rose or purple, valvate petals. The petal bases are usually clawed and the apices are typically narrowed and strongly inflexed; this inflection may also result in the appearance of a shallow notch or deep cleft, and the apex may be further connate to the body of the petal through a midrib "bridge". Less commonly, the petals have a broader base and/or a plane to merely recurved apex, or may be deeply dissected (*Sinocarum*), have attenuate hornlets (*Lagoecia*), or a long attenuate apex (*Acronema*); they may also be filiform (resembling staminodes, as in *Hermas*), or sepaloid (as in the female flowers of *Arctopus*). They are typically equal to subequal, but the marginal petals may become enlarged and radiate.

The androecium is formed by 5 (4–6) stamens, alternate with the petals and opposite to the calyx lobes. The filaments are typically long and slender, strongly inflexed in bud but ascending at anthesis. The anthers are basifixed to dorsifixed and longitudinally dehiscent, with two thecae (tetrasporangiate).

The syncarpous gynoecium is almost always bicarpellate and bilocular (rarely unilocular through carpel abortion, or up to 4-carpellate and 4-locular in *Apiopetalum*). The apex of the ovary and the bases of the styles are frequently enlarged into a nectariferous disc or stylopodium (Erbar and Leins 2010) of various shapes and colors, terminating in two (rarely up to 4) styles with wet, non-papillate stigmas. In many New World Umbellifers, the stylopodia (or their upper apocarpic part) may be reduced. The placentation is axile (but appears apical). Each locule encloses a single functional pendulous ovule and a second entirely abortive ovule.

Embryology. The endothecium of the anther walls develops fibrous thickenings, but the middle layer is ephemeral (Davis 1966). The cells of the tapetum are glandular, becoming binucleate. Cell wall formation after meiosis is simultaneous. The microspore tetrads are tetrahedral, isobilateral, or decussate, and the pollen, when shed, is 3-celled (Davis 1966).

The ovules are non-arillate, epitropous (with a ventral raphe), anatropous, unitegmic, and tenuinucellate or pseudocrassinucellate, with a differentiated endothelium. The micropyle is very long due to a massive integument and small nucellus. Several megaspore mother cells may form, but generally only one develops to maturity. Meiosis and cytokinesis of the megaspore mother cell usually occur together. The resulting tetrads are linear, with the chalazal megaspore usually developing into a Polygonum-type embryo sac, although the tetrasporic 16-nuclear Drusa-type (*Drusa* and related genera in Azorelloideae) and Penaea-type (*Azorella trifurcata*) are also represented; in *Sanicula* and *Bupleurum aureum*, the Allium-type and Adoxa-type have been observed alongside the Polygonum-type. The synergids are broad and short, the polar nuclei fuse during or before fertilization, and the 3–11 antipodal cells persist into the early stages after fertilization. Embryogeny is of the solanad type and long suspensors are common (Davis 1966).

Pollen Morphology. The pollen grains of Apiaceae are isopolar, tricolporate, small to large (polar axis 15–70 μm), prolate or perprolate (P/E 1.3 to 2.5), rarely subspheroidal, and triangular in polar view. Cerceau-Larrival (1971) distinguished five pollen types: subrhomboidal, subcircular, ovoid, sub-rectangular, and equatorially constricted. The exine comprises a tectum that appears smooth using light microscopy but is typically striate (or rugulate or psilate) using SEM, and smooth in cross-section. The underlying columellae are arranged in a reticulate pattern and are usually straight (rarely branched). Umbelliferous pollen is triaperturate, with three longitudinal colpi (ectoapertures) and three endoapertures; the furrows are angular or interangular. The pores have distinct costae and may be circular, or elliptical to rectangular, and are arranged perpendicularly to the longitudinal axis of the grain (Cerceau-Larrival 1971; Punt 1984).

Karyology. Chromosome counts have been completed for nearly 75% of umbellifer genera and 45% of species, and these data have been treated at length by Pimenov et al. (2003). With the transfer of *Hydrocotyle*, *Neosciadium*, and *Trachymene* to Araliaceae, haploid chromosome numbers in the remaining Apiaceae range from $n = 3$ (in *Sium suave* and *Zosima korovinii*) to

$n = 77$ (in *Lomatium columbianum*). Among members of subfamily Apioideae, $n = 11$ is most common (53%); haploid numbers based on $x = 11$ ($n = 22, 33, 44, 55, 66$, and 77) together account for an additional 8%. A dysploid or aneuploid series extends without interruption down to $n = 3$; each of these numbers also represents the base of an additional polyploid series (Pimenov et al. 2003; see also Moore 1971). In Saniculoideae, the haploid number ranges from $n = 4$ to $n = 48$. The most common number is $n = 8$ (44%), followed by $n = 7$ (17%). Among the genera assigned to Azorelloideae and Mackinlayoideae, haploid numbers range from $n = 5$ to $n = 48$; although $n = 10$ and $n = 8$ are most common (16% each). Together with their polyploid series (i.e., $x = 8$, $x = 10$), these numbers account for 65% of chromosome counts in these subfamilies. Some large genera are uniform in chromosome numbers (e.g., *Ferula* with $2n = 22$), but other (even some small) genera exhibit dysploidy (e.g., *Bupleurum* and *Bunium*) or polyploidy. *C*-values in Apiaceae have been determined for 46 species in 24 genera, ranging from $C = 0.63$ pg in *Oenanthe fistulosa* ($n = 11$) to $C = 5.48$ pg in *Daucus montanus* ($n = 3x = 33$); a similar value of $C = 5.18$ pg was found in *Astrodaucus littoralis* ($n = 10$; Pimenov et al. 2003; Bennett and Leitch 2005). B-chromosomes have been documented from 40 species in 24 genera (Pimenov et al. 2003).

Pollination and Reproductive Systems. Apiaceae are largely insect pollinated (Koul et al. 1993). The apparent "monotony" of floral morphology among umbellifers, together with the wide range of insect visitors (Diptera, Coleoptera, Hymenoptera, Hemiptera, Lepidoptera) suggested promiscuous pollination, but for those taxa studied more intensively, relatively few of the visitors, mostly Hymenopterans, are effective pollinators (Bell 1971; Bell and Lindsey 1982). Pollinators are attracted by nectar (produced by the stylopodium) and pollen, and presumably by the coloration of the petals and stylopodia; in at least one species (*Angelica triquinata*), the nectar also has a narcotic effect (Bell 1971). Where studied, floral fragrance is dominated by volatile monoterpene hydrocarbons (especially α-pinene, β-pinene, *cis*-*β*-ocimene, *trans*-*β*-ocimene, limonene, sabine, and myrcene), which is consistent with hymenopteran pollinators (Borg-Karlson et al. 1994).

Most apioids are andromonoecious; the distribution of staminate flowers varies widely, but they are most commonly distributed in the center of the outer umbellules and become progressively predominant in the inner umbellules, often with entirely staminate central umbellule(s) (Bell 1971; Webb 1981). The flowering sequence is usually centripetal, with the outer flowers (and outer umbellules) opening before the inner flowers (and inner umbellules; Webb 1981). Protandry also dominates in Apiaceae. If this is strongly developed, it may provide complete outcrossing or at least geitonogamy; when weakly developed, some self-pollination (within flowers) is possible, especially from late-dehiscing anthers (Bell 1971; Webb 1981; Bell and Lindsey 1982; Koul et al. 1993), and self-compatibility seems to be widespread (Lindsey 1982). Protogyny, while less common, has been described for many North American taxa (e.g., *Thaspium* and *Zizia*), but even among protogynous taxa, the distribution and sequence of staminate flowers (predominating in earlier-developing umbels) and hermaphrodite flowers (concentrated in late-developing umbels) provides an overall protandrous development to the entire plant (Webb 1981; Schlessman and Graceffa 2002). Less common still are gynodioecy (as in *Scandia*, *Gingidia*, and *Lignocarpa*) and dioecy (e.g., *Aciphylla*, *Arctopus*, *Anisotome*, and *Trinia*), in which the pistillate flowers often have only rudimentary or non-functional staminodes (Dawson 1971). Even more complex patterns can be found (as in *Centella*), with monoecious, androdioecious and andromonoecious species having umbellules of exclusively staminate, pistillate, or hermaphrodite umbellules.

Fruits and Seeds. The fruit is a cremocarp, developing from an inferior, syncarpous ovary usually formed by two carpels (Figs. 1–5). At maturity, the fruits are dry (rarely fleshy, as in *Apiopetalum* and *Mackinlaya*) and indehiscent, usually schizocarpic, splitting along a central commissure and yielding two mericarps. The two mericarps are usually equal, but in some taxa, they may be heteromericarpic (e.g., *Heteromorpha*, *Komarovia*, and some species of *Tordylium*), and in others pseudo-monocarpellate (e.g., *Lagoecia*, *Petagnaea*, *Echinophora*, *Symphyoloma* and

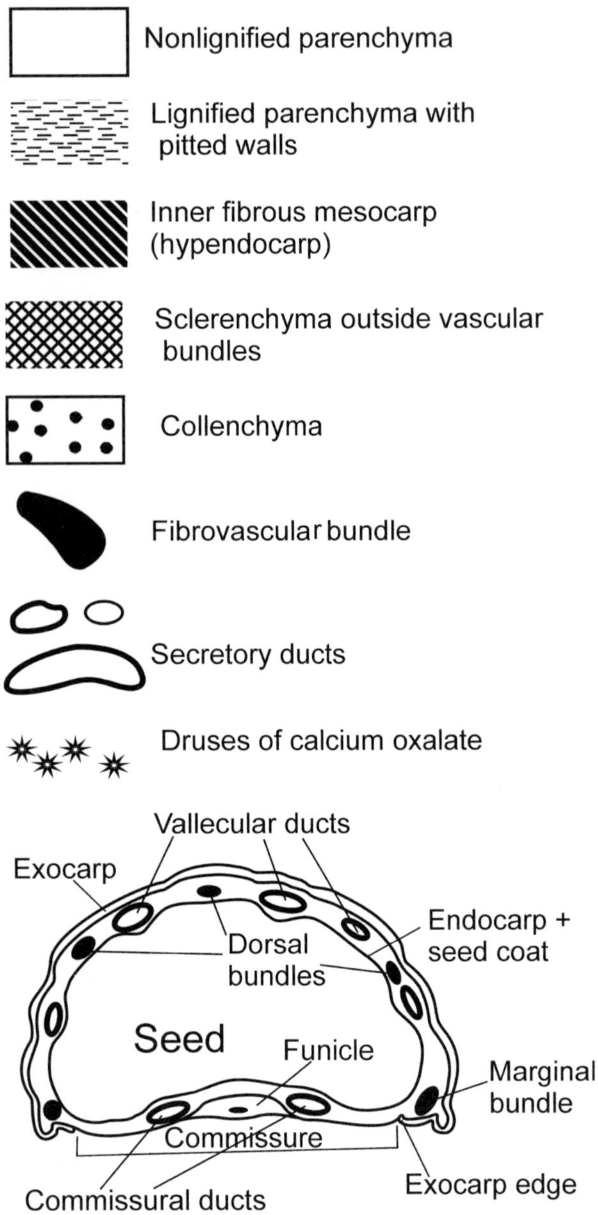

Fig. 1. Apiaceae. Line drawings of a generalized mericarp transverse section, highlighting major features illustrated in Figs. 2 and 3. (Orig., illustrations by Tatiana Ostroumova)

Schtschurowskia) (Baumann-Bodenheim 1955). The commissure ranges from very broad (as wide as the mericarps themselves) to very narrow (less than 5% of the mericarp width). Fruits in which the commissure is narrow and the two mericarps are almost spherical are said to be didymous. One or two vascular bundles in the center of the pistil lead to the formation of the

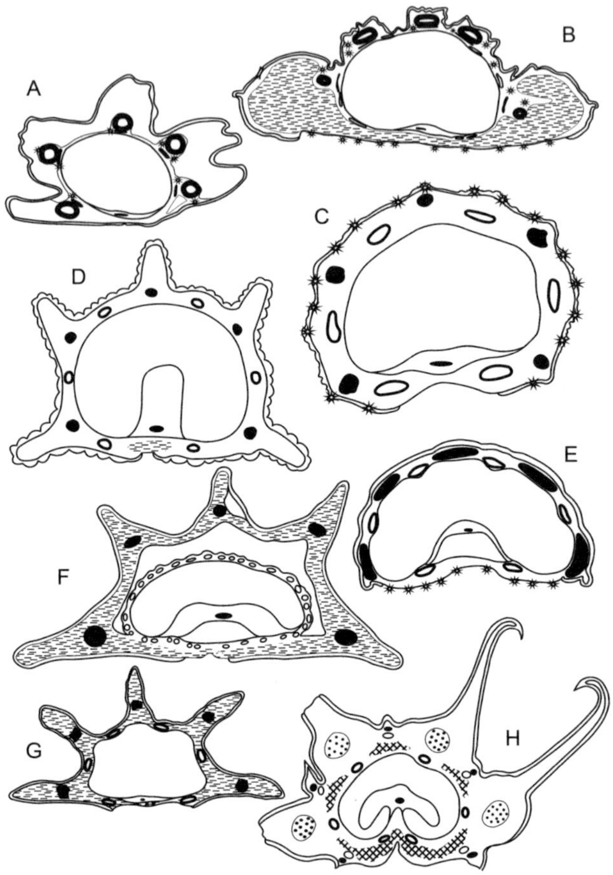

Fig. 2. Apiaceae. Line drawings highlighting major features of the transverse sections of mericarps of the following species: **A** *Astrantia major*. **B** *Eryngium maritimum*. **C** *Ammi majus*. **D** *Aulacospermum anomalum*. **E** *Chaerophyllum prescottii*. **F** *Angelica decurrens*. **G** *Cnidium monnieri*. **H** *Caucalis platycarpos*. Scale bars = 1 mm. (Orig., illustrations by Tatiana Ostroumova)

carpophore. In many species, the mericarps are borne at the apex of free carpophores (which may be entire, partially bifid, or entirely bifurcating), but in other species, the carpophores remain fused to the mericarps (or the carpophores may be absent and represented by small vascular bundles in the commissure). The fruits range from transversely terete to strongly compressed, either dorsally or laterally. In some taxa, the upper part of the mericarp is extended into a narrow beak (e.g., *Scandix* and *Anthriscus*). Each mericarp typically has five (or rarely more) primary ribs, of which three are dorsal (one median rib and two laterals) and two are closest to the commissure (marginal ribs). The primary ribs are separated by furrows called valleculae. Secondary ribs

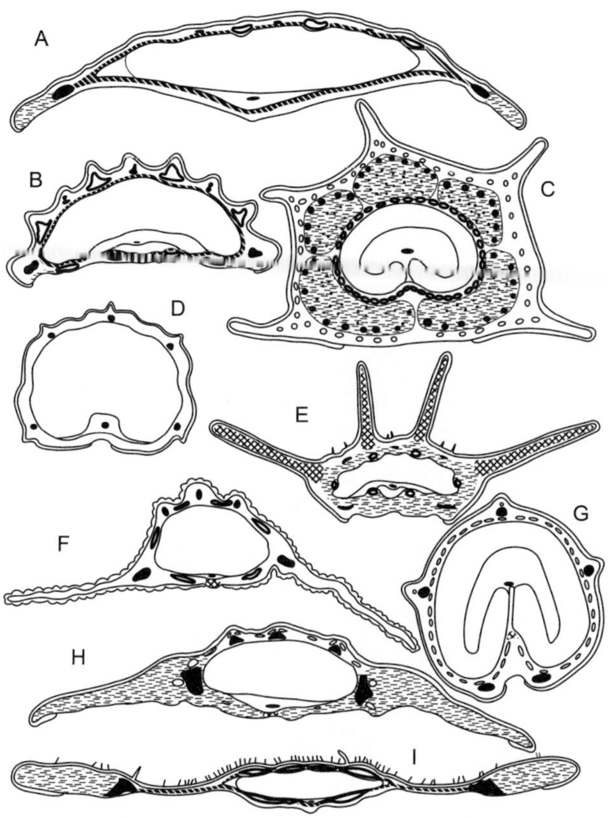

Fig. 3. Apiaceae. Line drawings highlighting major features of the transverse sections of mericarps of the following species: **A** *Heracleum chorodanum*. **B** *Laser trilobum*. **C** *Prangos ferulacea*. **D** *Bupleurum rotundifolium*. **E** *Laserpitium latifolium*. **F** *Ostericum tenuifolium*. **G** *Smyrnium olusatrum*. **H** *Rhizomatophora aegopodioides*. **I** *Zosima absinthiifolia*. Scale bars = 1 mm. (Orig., illustrations by Tatiana Ostroumova)

develop from the valleculae in some groups. Either set of ribs may be filiform to very broad, or develop narrow to broad wings, or be variously ornamented (e.g., with hairs, barbs or scales), or obscure. Each primary rib encloses one to several vascular bundles and often one to several associated schizogenous secretory canals called rib oil ducts (also known as intrajugal oil ducts or companion canals). Rib oil ducts are usually well developed in Saniculoideae (where they are typically rather large), Azorelloideae, and Mackinlayoideae; in most Apioideae, they are small to absent, but may be quite large in some apioids (e.g., *Johrenia*, *Oreoselinum*, and some species of *Peucedanum* s. lat.). Exceptions are found in *Steganotaenia* and *Polemanniopsis* (where they are replaced by large cavities in the wings), in *Phlyctidocarpa* (where they co-occur with vittae), in *Lichtensteinia* and *Marlothiella* (where they are large and characteristically surrounded by concentric rings of cells), and in *Choritaenia* (where the segments of the four marginal rib oil ducts are uniquely separated to form rows of globose vesicles) (Van Wyk et al. 2013). A second set of schizogenous secretory canals, not associated with the vasculature, are called vittae and are found in the valleculae and/or the commissural face of the fruits (less commonly cyclical, forming a near-continuous ring, as in *Berula*) in Apioideae, and they are often septate, and rarely anastomosing. More rarely, the vittae of Apioideae may be represented by very short, vesicular oil cavities dispersed in the mesocarp (known only in *Bilacunaria* and *Smyrniopsis*), or lacking altogether (as in *Polemanniopsis*, *Steganotaenia*, *Lichtensteinia*, *Marlothiella*, *Choritaenia*, and the mature fruits of *Aegopodium* and *Conium*). Vittae are also lacking in the fruits of Saniculoideae (but rarely found in *Alepidea*), Azorelloideae, and Mackinlayoideae (although small, vittae-like oil ducts may be scattered in the mesocarp in these subfamilies).

Three layers of pericarp are usually distinguished, but their delimitation can be rather complicated in Apiaceae. The exocarp is derived from the epidermis; it is usually a single layer of small cells with thickened outer walls, but in some genera, the cells may be rather large (30–40 μm; e.g., *Physotrichia*, *Taeniopetalum*). The mesocarp has several layers and is parenchymatous (although the cells of the inner layers may be thickened with lignified cell walls). Transformed parenchyma may be variously localized in the fruits of many apioids, especially aquatic and xerophytic species. The mesocarp may also be crystalliferous. When present, the crystals of Azorelloideae and Mackinlayoideae are rhomboidal (and found only in the inner layer of the mesocarp, often forming a continuous ring around the endocarp), but druses in Saniculoideae and Apioideae (either scattered throughout the mesocarp, as in Saniculoideae and some early diverging Apioideae, or on the commissural face only, sometimes localized around the carpophore, as in some Apioideae). The endocarp/inner pericarp of Apioideae and Saniculoideae may be parenchymatous, but in some apioids (e.g., in *Heracleum*, *Pastinaca*, *Leiotulus*, *Ferula*

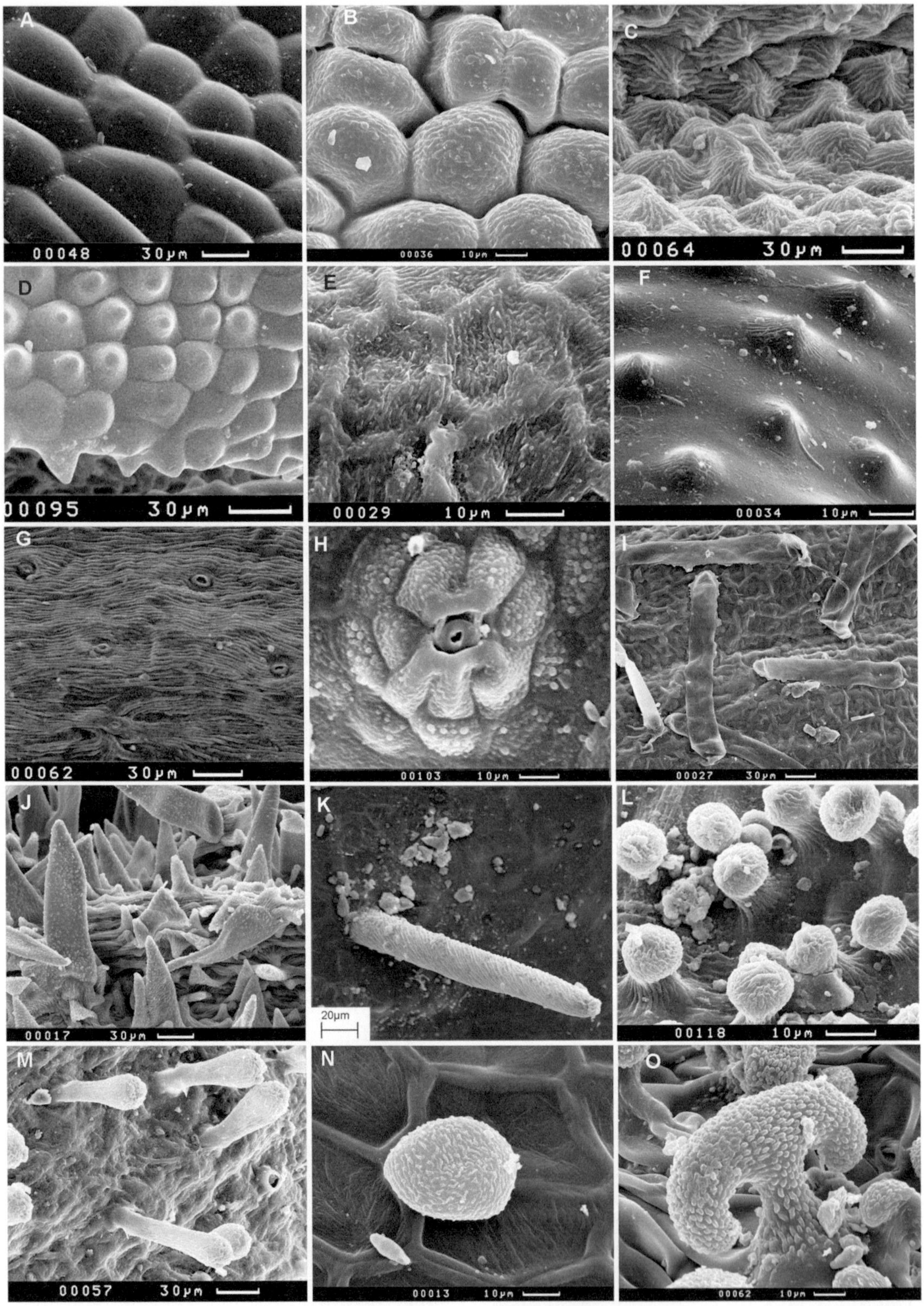

Fig. 4. Apiaceae. Variation in fruit micromorphology. **A–D** Cells with distinctly sunken borders, outer periclinal walls convex. **A** *Ostericum palustre*. **B** *Taeniopetalum arenarium*. **C** *Katapsuxis silaifolia*. **D** *Eryngium planum*. **E** Cells with raised borders (*Pastinaca pimpinellifolia*). **F, G** Cells with indistinct borders. **F** *Geocaryum macrocarpum*.

and *Oenanthe*) it consists of lignified fibers or prosenchyma cells (often called the inner mesocarp or hypendocarpium; Koso-Poljansky 1916) or 2–3 layers of sclerified longitudinal fibers (as in *Hacquetia* of Saniculoideae). In Azorelloideae and Mackinlayoideae, however, the inner pericarp is usually referred to as endocarp (e.g., Tseng 1967) and is sclerified. The testa is thinly membranous, but tends to be thicker when the endocarp is parenchymatous. The seed may be adnate to the wall of the locule to entirely free, sometimes occupying only the upper part of the locule (as in *Ligusticum* s.str. and *Haloselinum*). In transverse section, seeds vary from terete to strongly compressed either dorsally or laterally, and are sometimes sulcate under the vittae; the commissural face may be convex to plane or strongly concave to deeply sulcate (rarely with revolute margins, in *Caucalis*, *Turgenia*). The endosperm is copious, oily, rich in starch, and smooth or ruminate, starting as nuclear but maturing into cellular endosperm with thick cell walls. The embryo is small and germination is usually dicotyledonous, with equal to unequal, ovate ("round") or linear cotyledons (known as R- and L-types, respectively) (Cerceau-Larrival 1962) (Fig. 6). In some taxa (e.g., *Acronema*, *Astomaea*, *Bunium*, *Conopodium*, *Erigenia*, *Geocaryum*, *Hellenocarum*, *Kozlovia*, *Orogenia*, *Stefanoffia* and some species of *Elaeosticta*, *Neomuretia*, *Scaligeria* and *Sinocarum*), the embryo and seedlings are pseudo-monocotyledonous, in which the single cotyledon is usually entire and narrowly linear to ovate (but in some species of *Acronema*, it is more leaf-like, with a deeply dissected lamina) (Kljuykov et al. 2014). The radicle is usually short and always superior (Lubbock 1892).

Dispersal. Modifications to the schizocarp in Apiaceae are generally interpreted as adaptations for dispersal. Many taxa have winged ribs or exhibit compressed and/or winged mericarps, which apparently serve as adaptations for wind dispersal (but see Jongejans and Telenius 2001); another adaptation for wind dispersal includes the thick but very light and corky pericarps of *Prangos*. In some taxa, the primary or secondary ribs may be spiny or barbed, indicating ectozoochory (e.g., *Daucus* and relatives), or the fruits may be fleshy (*Apiopetalum* and *Mackinlaya*), suggesting endozoochory. In other cases, the dispersal unit may be the umbellule or entire inflorescence (umbel or capitulum), as in *Petagnaea* and *Thecocarpus* (see Rodríguez 1971), or even entire lax synflorescences (e.g., *Eriosynaphe* and some species of *Seseli*). Dispersal through vegetative means is also known in Apiaceae, especially by creeping rhizomes, stoloniferous roots, and bulbules (e.g., *Sium ninsi* and *Cicuta bulbifera*).

Phytochemistry. The distinctive phytochemistry of Apiaceae makes the family easily recognizable (Crowden et al. 1969; Hegnauer 1971, 1973, 1989; Stuhlfauth et al. 1985; Holub et al. 1987; Reduron and Muckensturm 2007–2008). Mono- and sesquiterpenoids are well represented and are at the origin of essential oils (some important for medicinal or perfumery uses). Higher molecular-weight lactones that are not steam volatiles (e.g., archangelolide, gradolide, isomontanolide, laserolide, polhovolide, thapsigargin, and trilobolide) are specific markers of Apiaceae. Diterpenoids are generally absent, but kaurene-type diterpenoids are found rarely (as in *Alepidea* and *Arctopus*). Non-volatile sesquiterpenoids (e. g., sesquiterpene lactones) are found in some apioid genera (e.g., *Ferula* and *Laserpitium*). Phtalides are restricted to only a few genera (e. g., *Apium*, *Ligusticum* s.l., *Angelica* and *Levisticum*). Phenylpropanoids are often encountered, most commonly apiole, asarone, dillapiole, elemicin, estragole, eugenol, and myristicin (plus pseudoeugenol derivatives in *Pimpinella* only). Many coumarins occur in the family, but especially characteristic are the furanocoumarins and pyranocoumarins, whose presence is limited mainly to Apiaceae and one other family

Fig. 4. (Continued) G *Chaerophyllum aureum*. **H** A stoma on a multicellular bulge, *Astrantia pontica*; such multicellular bulges can be 50–100 microns high in *Astrantia* and some other genera in the family. **I–O** Unicellular trichomes. **I** Thin-walled and strap-like, *Pastinaca pimpinellifolia*. **J** Thick-walled, solitary and tufted, *Seseli petraeum*. **K** Thick-walled with helical striations, *Ferula rutbaensis*. **L** Capitate, *Astrodaucus littoralis*. **M** Clavate, *Lagoecia cuminoides*. **N** Vesicular, *Tordylium lanatum*. **O** T-shaped, *Trachyspermum paktianum*. (Orig., SEM photographs by Tatiana Ostroumova)

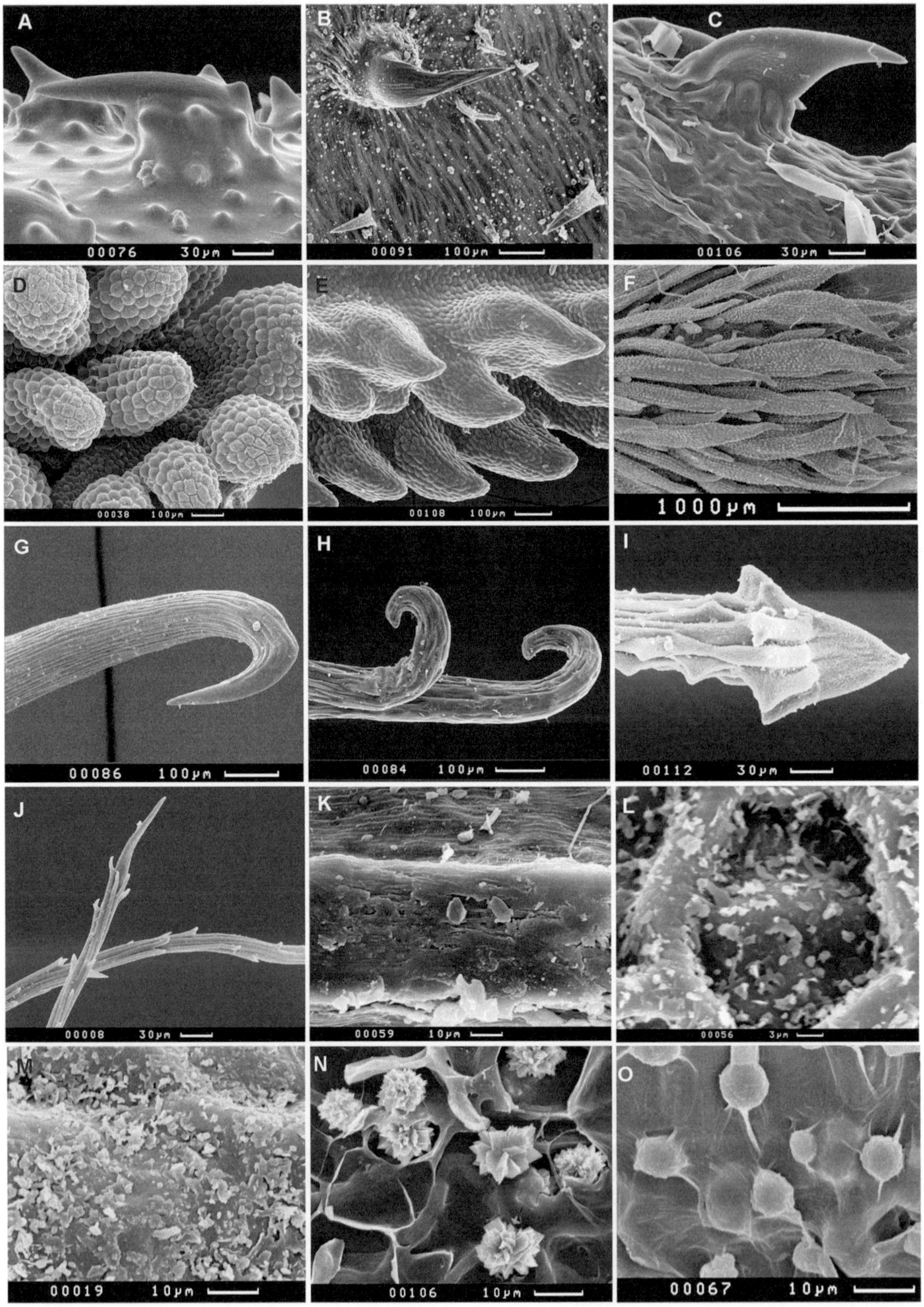

Fig. 5. (in the Apiaceae treatment). Variation in fruit micromorphology. **A–C** Trichomes with multiseriate bases. **A** *Anthriscus sylvestris*. **B** *Heracleum grandiflorum*. **C** *Heracleum sosnowskyi*. **D–F** Multicellular

Fig. 6. Apiaceae. Seedlings, highlighting cotyledon variation. **A–D** Typical dicotyledonous seedlings. **A** *Ferula tingitana*, with narrowly oblong cotyledons. **B** *Oenanthe pimpinelloides*, with narrowly lanceolate cotyledons. **C** *Scandix grandiflora*, with linear cotyledons. **D** *Mediasia macrophylla*, with ovate-lanceolate cotyledons. **E–G** Pseudo-monocotyledonous seedlings. **E** *Scaligeria napiformis*, with a single ovate cotyledon, bilobed at the tip. **F** *Acronema commutatum*, with a single leaf-like, ternately dissected cotyledon. **G** *Bunium microcarpum*, with a single narrowly oblong cotyledon. (Orig., illustrations by Svetlana Petrova)

(Rutaceae). These substances are known for their phototoxicity (inducing dermatitis) but also for their antifungal, bacteriostatic, insect antifeedant and generally repulsive properties. The most representative compounds of these classes are angelicin, byakangelicol, bergapten, imperatorin, peucedanin, pimpinellin, psoralen and xanthotoxin. Chromones are also represented, for example khellin, an important medicinal source from *Visnaga daucoides*. Polyacetylenes are present in more than 100 species, including useful compounds such as falcarinol and falcarindiol (often present in the roots, and apparently beneficial to humans in small amounts), as well as toxic compounds such as aethusin (*Aethusa*), cicutoxin (*Cicuta virosa*), oenanthotoxin (*Oenanthe crocata*), responsible for the poisonous properties of these taxa. Alkaloids are rarely found in the genera of Apiaceae, but the toxic coniine (and related compounds) from poison hemlock (*Conium maculatum*) is a famous exception. Other constituents include saponins (noticeably present in certain genera, including *Bupleurum*, *Centella*, *Eryngium*, and *Sanicula*), sugars such as the pentose sugar apiose (first isolated from parsley) and the trisaccharide umbelliferose (specific to umbellifers), and fatty acids such as petroselinic acid (also found in Araliaceae). Flavonoids are present in most Apiaceae (especially kaempferol, luteolin and quercetin), and sometimes have taxonomic value. Several species are sources of oleo-gum-resins, particularly in the genus *Ferula*. Umbellifers lack iridoids, cyanogenic compounds, proanthocyanidins and ellagic acid.

Subdivision and Relationships Within the Family. The systems of the 19th and early 20th centuries (e.g., Koch 1824; Bentham 1867; Drude 1898; Koso-Poljansky 1916) differed dramatically in their intergeneric treatments of Apiaceae, but only that of Drude (1898) has had a lasting influence on umbellifer taxonomy. Drude recognized three subfamilies: Apioideae, Saniculoideae, and Hydrocotyloideae. These were further divided into 12 tribes (eight in Apioideae and two each

Fig. 5. (Continued) tubercles and scales. **D** *Bupleurum papillosum*. **E** *Astrantia maxima*. **F** *Eryngium planum*. **G–J** Spines and bristles. **G** *Orlaya daucoides*. **H** *Sanicula rubriflora*. **I** *Astrodaucus littoralis*. **J** *Chaetosciadium trichospermum*. **K–M** Epicuticular waxes. **K** *Angelica czernaevia*. **L** *Bupleurum lophocarpum*. **M** *Bupleurum sibiricum*. **N, O** Calcium oxalate crystals on the commissural face. **N** *Eryngium planum*. **O** *Chaerophyllum aureum*. (Orig., SEM photographs by Tatiana Ostroumova)

in Saniculoideae and Hydrocotyloideae). This system has been updated and enhanced by Pimenov and Leonov (1993), who expanded the number of tribes to 16 with the addition of four new apioid tribes. Subfamilies Apioideae and Saniculoideae have largely held up as natural groups, although some taxonomic transfers (e.g., the transfer of *Lagoecia* and *Oligocladus* to Apioideae) were necessary to restore monophyly. Magee et al. (2010), however, have shown that several, mostly African genera form a grade at the base of Apioideae. Almost all of these "protoapioids" have druse crystals scattered around the mesocarp. Given the placement of the protoapioids relative to Saniculoideae, these authors suggested the inclusion of Saniculoideae within an expanded Apioideae, providing a more accurate reflection of the major morphological discontinuities, but the recognition of the two subfamilies is maintained in the current treatment until greater consensus emerges. Even more difficult have been the Hydrocotyloideae, which are grossly polyphyletic. The type genus *Hydrocotyle*, along with *Trachymene* (including *Uldinia*) and *Neosciadium* have been transferred to Araliaceae, while other members have been transferred to Apioideae (e.g., *Naufraga* and *Notiosciadium*) or Saniculoideae (viz., *Arctopus*). The remaining genera are currently treated in two new subfamilies, Azorelloideae and Mackinlayoideae (Plunkett et al. 2004; Nicolas and Plunkett 2009). Subfamily Mackinlayoideae unites *Centella*, *Micropleura*, *Xanthosia*, *Actinotus*, and several other "former hydrocotyloids" with two genera transferred to Apiaceae from Araliaceae (*Mackinlaya* and *Apiopetalum*). The majority of genera, however, are now placed in subfamily Azorelloideae, which most closely resembles Drude's (1898) circumscription of Hydrocotyloideae, and includes (among others) the large genera *Azorella*, *Bowlesia*, and *Eremocharis*. *Hermas* and *Klotzschia* remain *incertae sedis*; they are tentatively assigned here to Azorelloideae, but new subfamilies may be needed to accommodate one or both genera.

The four subfamilies recognized here are distinguished primarily on the basis of fruit characters. Apioideae have vittae in the valleculae and on the commissure, but rib oil ducts are usually very small or lacking (rarely large). The inner pericarp of Apioideae and Saniculoideae is typically parenchymatous, but when "sclerified", the inner fiber-like cells are always arranged transversely and the outer fiber-like cells (if present) longitudinally. Saniculoideae typically have distinct (and often very large) rib oil ducts, and usually lack vallecular vittae (although small, branched, vittae-like rib oil ducts may be found scattered in the mesocarp). In addition, they often have outgrowths on their exocarps (e.g., scales, bristles, or prickles) and druse crystals scattered throughout the mesocarps. Druses surrounding the seeds are also found in the protoapioids; otherwise, druses are rare in Apioideae, and are then found only on the commissural side of the mericarp. Subfamilies Azorelloideae and Mackinlayoideae share many features, notably an endocarp of fiber-like sclereids (with the inner fibers longitudinal and the outer fibers transverse), rhomboidal crystals limited to the inner layer of the mesocarp (and often forming a ring around the endocarp). Like Saniculoideae, vallecular vittae are lacking (and likewise, if vittae are present, they are scattered in the mesocarp). Rib oil ducts are usually present and distinct (but not generally large). The mericarps of most Azorelloideae are compressed dorsally, typically by a broadening of the two lateral ribs (or less commonly of the marginal ribs, as in *Azorella* sect. *Laretia*), or rarely terete (*Oschatzia*). In Mackinlayoideae, the fruits are laterally compressed.

Few of the tribes and subtribes recognized by Drude (1898) have proved monophyletic, although Tordyliinae and the small Echinophoreae are notable exceptions. Downie et al. (2001, 2010) provided synopses of currently recognized "tribes and clades" in Apioideae, but the assignment of genera to molecular clades remains incomplete, with several groups recognized only informally, and many genera left unsampled or non-monophyletic. In Saniculoideae, after the transfer of *Lagoecia* to Apioideae, the remaining genera of tribe Lagoecieae have been united with those of tribe Saniculeae. Calviño and Downie (2007) transferred two protoapioids (*Steganotaenia* and *Polemanniopsis*) to Saniculoideae, which they recognized as a new tribe, Steganotaenieae, but Magee et al. (2010) have more recently proposed several additional tribes for other protoapioids (see also Van Wyk et al. 2013). The tribes (and subtribes) of Drude's Hydrocotyloideae, like the subfamily itself, are largely polyphyletic, and no new tribes have yet been proposed for subfamilies Azorelloideae or Mackinlayoideae (see Nicolas and Plunkett 2009). The difficulty in circumscribing tribes in Apioideae is connected with the lack of morphological discontinuities in many characters across a great many taxa, coupled with the frequent occurrence of convergences and parallelisms for many other morphological characters. Given that the recognition and circumscription of tribes remains largely unsettled, the genera of Apiaceae are organized herein by subfamily, and thence alphabetically. Unless otherwise noted, phylogenetic relationships cited in the generic treatments below follow Downie et al. (2010) for Apioideae, Calviño and Downie (2007) for Saniculoideae, and Nicolas and Plunkett (2009) for Azorelloideae and Mackinlayoideae.

Affinities. In groups known variously as Umbelliflorae, Umbellales or Apiales, the Apiaceae have traditionally been allied with Araliaceae, with which they share many floral features, such as frequently pentamerous perianths, reduced calyces, free petals and stamens, inferior ovaries, nectariferous discs or stylopodia, a single pendant, anatropous, functional ovule, seeds with copious endosperms and small, straight embryos, and frequently umbellate inflorescences (Rodríguez 1971). Additionally, the families share schizogenous secretory canals with essential oils and resins, and a host of other phytochemical similarities (e.g., the shared presence of polyacetylenes, seed fats with petroselinic acid, oleanene- and ursene-type triterpenic sapogenins, and the lack of iridoids and true tannins) (Hegnauer 1971). The current circumscription of the order Apiales (Plunkett et al. 2004) also includes Pittosporaceae, which shares with Apiaceae and Araliaceae schizogenous secretory canals (van Tieghem 1884), a similar ovule structure (Jurica 1922), phytochemistry (Jay 1969; Hegnauer 1971), and "early sympetalous" corolla development (Erbar and Leins 2004), although Pittosporaceae differ in several characters, notably large and showy petals, superior ovaries, and the complete lack of umbels. Several other small families have been added to Apiales, including the araliad segregate family Myodocarpaceae, plus Torricelliaceae (defined broadly to include *Torricellia*, *Melanophylla*, and *Aralidium*), Griseliniaceae, and Pennantiaceae. The order is placed in subclass Asteridae near Dipsacales and Asterales, with which it shares early sympetaly, unitegmic and mostly tenuinucellate ovules, S-type sieve-tube plastids, and a series of phytochemical characters (e.g., similar alkaloids, sesquiterpene lactones, falcarinone-type polyacetylenes, acetate-derived anthraquinones, triterpenic sapogenins, flavonols, isopentenyl-substituted coumarins, and ethereal oils, plus the absence of iridoids and tannins; reviewed in Plunkett 2001; Erbar and Leins 2004).

Distribution and Habitats. Umbellifers are nearly cosmopolitan, but biogeographic studies (based on time-dated molecular phylogenies; Nicolas and Plunkett 2014) suggest that Apiaceae had an Australasian origin dating to the late Cretaceous/early Eocene (median minimum crown age of ~87 Ma). The major clades corresponding to subfamilies Mackinlayoideae, Azorelloideae, and Saniculoideae+Apioideae all originated at roughly the same time (median minimum crown ages of ~66 Ma), but in different regions (Australasia, southern South America, and southern Africa, respectively). Today, they are most diverse in the temperate (and especially arid) regions of the Northern Hemisphere. By comparison, they are relatively rare in the humid tropics, where they are typically restricted to higher elevations. Habitats vary widely from dry, rocky fields to aquatic environments, from open to woodland settings, and from low elevations (sea level) to high-alpine sites.

Asia has the greatest number of genera (289) and the largest number of endemics (177 genera). Geographic connections of the Asian genera are greatest with Europe (with 101 shared genera) and Africa (55 genera), and much less so with the New World (22 genera shared with North and South America combined). Europe and Africa have a comparable number of genera. Europe has a total of 126 genera, but only 17 are endemic. Europe shares the greatest number of genera with Asia (101) and Africa (58), and relatively few with the New World (16 genera). Africa has a total of 121 genera, of which 24 are shared between northern and sub-Saharan Africa. Northern Africa (with a total of 82 genera and 13 endemics) shares the greatest number of genera with Europe (56) and Asia (52), but only

10 with the New World. Sub-Saharan Africa (with 63 total genera and 36 endemics) shares the greatest number of genera with northern Africa (24), Asia (18) and Europe (15), and only nine with the New World. North America (including Central America) shows the greatest generic diversity in the New World, with 80 genera, of which 44 are endemic (including nine genera found only in Mexico and/or Central America). Geographic connections of North America are greatest with Asia (sharing 21 genera), and slightly less so with South America (19 genera), Europe (15 genera), and Africa (11 genera). The South American genera number 35; of the 15 endemics, most (9 genera) are from subfamily Azorelloideae. Nineteen of the South American genera are shared with North America, and only five with the north temperate regions of the Old World. Finally, Oceania has a total of 27 genera, of which 18 are endemic, almost entirely from temperate Australia and/or New Zealand, and the majority (14 genera) from subfamilies Azorelloideae and Mackinlayoideae.

In general, members of subfamilies Azorelloideae and Mackinlayoideae are best represented in the Southern Hemisphere, particularly temperate South America and Australia, followed by southern Africa, but some species (e.g., the widespread *Centella asiatica*) and even entire genera (*Dickinsia* from China, and *Drusa* from the Canary Islands) are from the Northern Hemisphere. The Saniculoideae are largely temperate, in both the Northern and Southern Hemispheres. The greatest generic diversity of saniculoids is in Eurasia and southern Africa, whereas those of the New World are represented mostly by species of the widespread genera *Eryngium* and *Sanicula*. Subfamily Apioideae is represented in almost all geographic regions, but proportionately much more frequent in the north temperate zones of Eurasia (and to a lesser degree in North America), although many of the earliest lineages of apioids appear in southern Africa.

The largest genus of Apiaceae is *Eryngium*, with approximately 250 species distributed throughout the Old and New Worlds, in both the Northern and Southern Hemispheres. *Bupleurum*, with roughly 195 species, is also widespread, but best represented in Eurasia. The next three largest genera, *Ferula* (185 spp.), *Pimpinella* (180 spp.) and *Seseli* (140 spp.), are all from the Old World. *Heracleum* (130 spp.) and *Angelica* (120 spp.) are broadly north-temperate, while *Lomatium* (86 spp.) is restricted to North America. *Azorella* (58 spp., mostly from South America and New Zealand), *Arracacia* (55 spp., from Mexico to South America) and *Ferulago* (50 spp., from Europe, SW Asia, and northern Africa) round out the genera with 50 or more species. Notwithstanding these large genera, Apiaceae are dominated by an abundance of very small genera. Forty percent (187 of the 466 genera) are monospecific, over three-quarters (351 genera) have five or fewer species, and over 85% of genera have 10 or fewer species. These patterns are almost certainly skewed by an abundance of problems in the taxonomic circumscriptions of umbellifer genera. Phylogenetic studies have already demonstrated that many of the largest genera (e.g., *Angelica*, *Heracleum*, *Pimpinella*, and *Lomatium*) are polyphyletic or paraphyletic (often in relation to the small genera), but in most groups, our knowledge has not progressed sufficiently to provide revised circumscriptions.

Parasites and Herbivores. Fungal parasites of Apiaceae have been documented from two genera of Basidiomycota, *Nyssopsora*, and *Puccinia* (both Uredinales), and three genera of Ascomycota, *Erysiphe* (Erysiphales), *Fabraea* (Helotiales), and *Cercosporidium* (Capnodiales) (Durrieu 1982). Other species of *Nyssopsora* have been documented from Araliaceae and Pittosporaceae (also Apiales), suggesting a phylogenetic connection of susceptibility (although related species in *Nyssopsora* also attack Sapindaceae and Meliaceae, both Sapindales; Henderson 1973).

Several insect groups have developed specialized herbivore relationships with umbellifer species, particularly in subfamily Apioideae (Berenbaum 1990). In Hemiptera, two families are represented, the aphids (Aphididae) and mirids (Miridae), all phloem suckers. Dipterans are represented by leaf miners (in two families, Agromyzidae, Tephritidae) and root miners (in Psilidae). Coleopterans are restricted to one family (Curculionidae), and may be either seed miners or root borers. The insect group best represented among the herbivores is Lepidoptera, including seed and leaf miners (from Incurvariidae and

Epermeniidae), stem borers (from Noctuidae), and leaf and flower rollers and chewers (from Oecophoridae and Papilionidae). In most cases, the herbivore species are specialists on umbellifers, either generally, or on particular apioid genera or species.

Palaeobotany. Macrofossils of Apiaceae are rare, but a fossil of *Steganotaenia araliacea* has been dated to the early Pleistocene (or earlier) from Ethiopia (Deschamps 1982). Microfossils are more abundant, dating back at least to the lower Eocene, including pollen attributable to *Bupleurum* (lower Eocene, France), *Pleurospermum* (upper Eocene, France), the fossil genus *Umbelliferoipollenites* (≈ *Echinophora*, upper Eocene, India), *Sanicula* (undifferentiated Eocene, England), *Eryngium* (upper Miocene, Spain, and Pliocene, Portugal), *Ferulago* (or *Opopanax* or *Petroselinum*), *Selinum*, *Helosciadium*, *Heteromorpha*, *Sium*, and *Conium* (Pliocene, Portugal; Diniz 1969; Muller 1981; Gruas-Cavagnetto and Cerceau-Larrival 1982). Microfossils from undescribed species of Apiaceae have been tentatively dated to the Oligocene in Australia, and the early Miocene in New Zealand (MacPhail et al. 1994). In the New World, umbellifer fossils have been dated to the Oligocene (*Oxypolis*, from Colorado), the Miocene (the fossil genus *Umbelliferospermum* from Washington), and the Pleistocene (*Zizia* from Louisiana, and *Cymopterus*, *Daucus*, and *Oenanthe* from California; Mathias 1965).

Economic Importance. Many species of Apiaceae are used as foods, flavorings, and folk remedies, primarily due to the phytochemical diversity of the family. Many umbellifers are edible, but few approach the status of staple crop, and in many parts of the world, the plants are not cultivated but rather collected from the wild (French 1971). Root vegetables (from swollen or tuberous roots) are known from many species, most notably carrots (*Daucus carota*), parsnips (*Pastinaca sativa*), and celeriac (*Apium graveolens*), but also many others, including many wild-collected species from Eurasia (e.g., *Bunium bulbocastanum*, *Chaerophyllum bulbosum*, *Conopodium majus*, and *Sium sisarum*), North America (*Cryptotaenia canadensis*, *Lomatium* spp., and *Perideridia gairdneri*), South America (notably the arracacha, *Arracacia xanthorrhiza*), and Africa (*Annesorhiza* and *Chamarea* species). Leafy parts (especially swollen petioles) may be used as vegetables, as in celery (*Apium graveolens*), lovage (*Levisticum officinale*), and fennel (*Foeniculum vulgare*), but more typically as aromatic herbs, such as dill (*Anethum graveolens*), parsley (*Petroselinum crispum*), anise (*Pimpinella anisum*), chervil (*Anthriscus cerefolium*), angelica (*Angelica* spp.), cilantro (*Coriandrum sativum*), alexanders (*Smyrnium olusatrum*), and sweet cicely (*Myrrhis odorata*). Many species also yield edible fruits (commonly marketed as "seeds"), including anise, celery, dill, and fennel, as well as ajowan (*Trachyspermum ammi*), caraway (*Carum carvi*), coriander (*Coriandrum sativum*), cumin (*Cuminum cyminum*), and zira (*Elwendia persica*). The leaves and/or fruits of some species may be used to flavor beverages, alcoholic or otherwise (e.g., *Pimpinella anisum* in anisette, ouzo, and the Turkish "raki", *Carum carvi* in akvavit and Kümmel, *Angelica archangelica* in vermouth and Chartreuse). Many of these same species (plus others collected from the wild) are used as folk medicines, including stomachics, carminatives, laxatives, diuretics, stimulants or sedatives, antispasmodics, and topical applications; *Ferula* (esp. *F. assa-foetida*, *F. galbaniflua*, and *F. tingitana*), and *Dorema ammoniacum* are important sources of gums and resins (French 1971). Pharmacological herbs used in various countries include *Angelica archangelica*, *Carum carvi*, *Visnaga daucoides*, *Centella asiatica* ("gotu kola"), *Phlojodicarpus sibiricus*, *Ammi majus*, *Foeniculum vulgare*, *Pimpinella anisum* and *Bupleurum* spp., among others. Despite the large number of edible umbellifers, many species are mildly to strongly toxic, including the infamous poison hemlock (*Conium maculatum*), the water hemlocks (*Cicuta* spp.), and fool's-parsley (*Aethusa cynapium*). Some taxa have also become popular as ornamental garden plants, such as species of *Astrantia*, *Angelica*, *Bupleurum*, *Eryngium*, and *Heracleum*.

Authors. Due to the size and great complexity of Apiaceae, a collaboration among many experts and students of the family was required to complete the following generic treatments. Two authors provided their expertise on taxonomic groups traditionally assigned to tribe Scandiceae of subfamily Apioideae, namely B.-Y. Lee (22 genera) and K. Spalik (12 genera). Most authors, however, provided their expertise on the umbellifers of various geographic areas. Most notably, M.G. Pimenov wrote treatments for 131 genera, primarily from Eurasia, and these were rounded out by contributions by E.V. Kljuykov (32 genera) and T.A. Ostroumova (19 genera); these three authors also provide the three keys for Asia. Many genera endemic to or centered in Europe were treated by J.-P. Reduron (47 genera), who provided the key for Europe and North Africa. M. F. Watson provided additional genera from Europe as well as some from eastern Asia (for a total of 23 genera); F.-D. Pu provided another 15 treatments for East Asian genera. Most genera from sub-Saharan Africa were provided by B.-E. van Wyk and P.M. Tilney (48 genera together), and the former contributed the key for this region. Genera from Oceania (especially Australia and New Zealand) were written by M.J. Henwood and J.M. Hart (11 genera), C.J. Webb (8 genera), and A.D. Mitchell (1 genus); the key for this region was written by M.J. Henwood. Most genera centered in North and South America (plus the key for these regions) and taxa not otherwise covered were contributed by G.M. Plunkett (98 genera), with a great intellectual debt to the late Professors M.E. Mathias and L. Constance, who did so much to advance our knowledge of New World umbellifers. Finally, B. Muckensturm kindly provided the section on Phytochemistry for the family treatment.

Classification of Apiaceae

I. Subfam. Mackinlayoideae G.M. Plunkett & Lowry (2004).
II. Subfam. Azorelloideae G.M. Plunkett & Lowry (2004).
III. Subfam. Saniculoideae Burnett (1835).
IV. Subfam. Apioideae Seem. (1866).

Keys to the Genera of Apiaceae

A single key to all of the world's umbellifer genera proved difficult to devise, and even more difficult to use. Therefore, seven separate keys were constructed for the native and naturalized genera of Apiaceae in each of the following major regions of the world (Fig. 7).

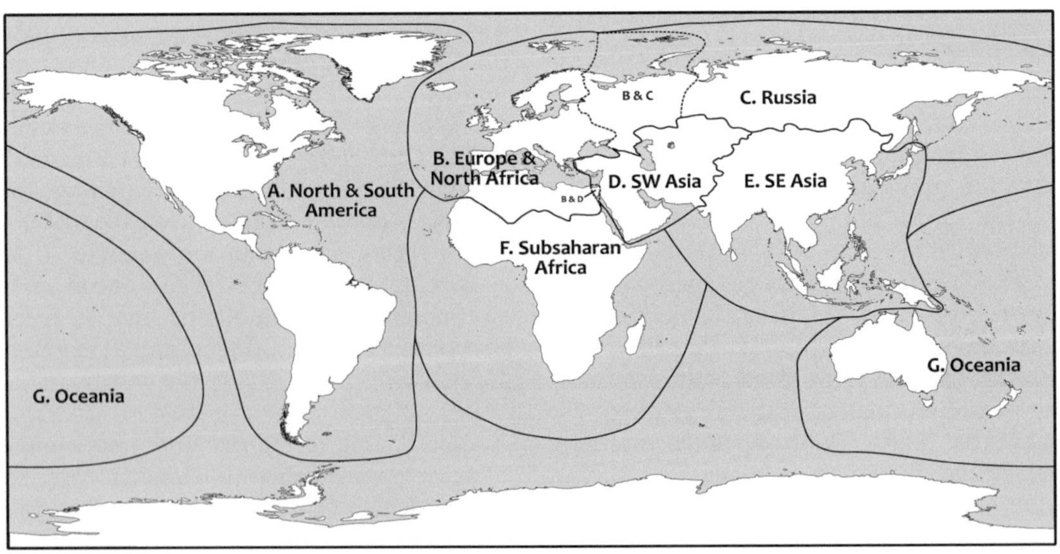

Fig. 7. Apiaceae. Map showing regions of the world used for the subdivision of the keys

A. *Key to the Native and Naturalized Genera of Apiaceae in North and South America*

1. Inflorescence capitulate or reduced to 1 or 2 flowers 2
– Inflorescence umbellate 3
2. Inflorescence a capitulum; widespread **35. *Eryngium***
– Inflorescence reduced to 1 or 2 flowers; restricted to Cuba **73. *Asciadium***
3. Inflorescence a simple umbel (or irregularly compound) 4
– Inflorescence a regular, compound umbel 24
4. Carpophore lacking 5
– Carpophore present 13
5. Leaves septate phyllodes, vallecular vittae present **246. *Lilaeopsis***
– Leaves laminar, vallecular vittae lacking 6
6. Fruits covered in prickles, spicules, squamae, or tubercles; rib oil ducts large **38. *Sanicula***
– Fruits glabrous or pubescent; rib oil ducts small or lacking 7
7. Fruits compressed laterally 8
– Fruits terete, subterete, or compressed dorsally 11
8. Endosperm strongly compressed laterally in transverse section, the face plane to convex; petals lacking a median dorsal gland 9
– Endosperm cuneate-rectangular in transverse section, intrusively sulcate at the commissure; petals with a median dorsal gland 10
9. Plants low, stoloniferous, rooting at the nodes **4. *Centella***
– Plants erect and caulescent **8. *Micropleura***
10. Plants annual or herbaceous perennial; stylodia equal to or shorter than the stylopodium; petal apex plane to shortly inflexed **21. *Domeykoa***
– Plants shrubby or woody-based perennial; stylodia longer than the stylopodium; petal apex strongly inflexed and long-acuminate **23. *Eremocharis***
11. Plants stellate-pubescent **16. *Bowlesia***
– Plants glabrous or pubescent (but never with stellate trichomes) 12
12. Leaves succulent; petals with a median dorsal gland; plants rosulate **29. *Pozoa***
– Leaves membranaceous to subcoriaceous or indurate (sometimes spiny), but not succulent; petals lacking a median dorsal gland; plants rosulate or forming cushion or mat **14. *Azorella***
13. Vallecular vittae present; rib oil ducts lacking 14
– Vallecular vittae lacking; rib oil ducts usually present 16
14. Plants annual; involucral bracts lacking **132. *Cyclospermum***
– Plants perennial or biennial; involucral bracts conspicuous 15
15. Carpophore entire (not splitting), calyx teeth prominent **279. *Neogoezia***
– Carpophore bifid or bifurcating (splitting at least apically); calyx teeth obsolete **296. *Oreomyrrhis***
16. Fruits winged 17
– Fruits unwinged 21
17. Petals with a dorsal median gland; styles long and slender 18
– Petals lacking a dorsal median gland; styles short and stout 19
18. Stems leafless at anthesis, fruit wings very wide and thin, formed from the extension of the lateral rib **24. *Gymnophyton***
– Stems leafy at anthesis, fruits wings thick, lateral ribs reflexed **13. *Asteriscium***
19. Plants delicate herbs growing from a globose tuberous root **20. *Diposis***
– Plants coarse, forming shrubs, subshrubs, cushions or mats, growing from a stout taproot 20
20. Plants stellate pubescent **15. *Bolax***
– Plants glabrous to pubescent (but never with stellate trichomes) **14. *Azorella***
21. Plants annual 22
– Plants perennial 23
22. Plants stellate-pubescent; calyx teeth of plumose bristles **26. *Homalocarpus***
– Plants glabrous (rarely pubescent, but without stellate trichomes); calyx teeth prominent, ovate **30. *Spananthe***
23. Styles usually short; commissure usually narrow, plants cushion- or mat-forming, or rosulate and having a short scape or 3–7-parted leaves **14. *Azorella***
– Styles long, commissure wide, plants rosulate with an elongated scape, leaves unlobed or shallowly 3–5-lobed **27. *Klotzschia***

24. Mericarps with 2 or 4 secondary ribs (located on the valleculae between the vascular bundles) 25
– Mericarps lacking secondary ribs 32
25. Secondary ribs with spines or bristles 26
– Secondary ribs lacking spines or bristles 31
26. Fruits dorsally compressed; seed face plane to slightly convex or concave 27
– Fruits laterally compressed to subterete; seed face deeply sulcate 28
27. Fruits with barbed or glochidiate spines **141. *Daucus***
– Fruits with setulose bristles (neither barbed nor glochidiate) **129. *Cuminum***
28. Stems and rays with retrorsely appressed hairs; spines of primary ribs recurved toward the stylar end **443. *Torilis***
– Stems and rays hispid, with spreading hairs, or glabrous; spines of primary ribs ± erect 29
29. Spines of primary ribs smooth, arranged in a single row **99. *Caucalis***
– Spines of primary ribs tuberculate, arranged in 2–3 rows 30
30. Spines of primary ribs arranged in 3 rows; spines of secondary ribs free at the base (not confluent) **463. *Yabea***
– Spines of primary ribs arranged in 2 rows; spines of secondary ribs confluent at the base **454. *Turgenia***
31. Plants perennial, acaulescent (but pseudoscapose), leaves with ovate divisions; primary ribs 5, conspicuous; secondary ribs 2; southeastern Oregon **370. *Rhysopterus***
– Plants annual, caulescent, leaves with short, linear divisions; primary ribs obsolete; secondary ribs 4; southeastern United States **447. *Trepocarpus***
32. Fruits dorsally compressed 33
– Fruits terete, subterete, or laterally compressed 78
33. Fruits unwinged 34
– Fruits winged 36
34. Vittae 3 in the valleculae and 6 on the commissure; petal apex recurved but not narrowed; Andes of Argentina **290. *Oligocladus***
– Vittae 1 in the valleculae and 2 on the commissure; petal apex narrowed and strongly inflexed; Peru or introduced from Eurasia 35
35. Plants perennial, growing from tuberous roots; stylopodium depressed; Andes of Peru **316. *Paraselinum***
– Plants annual or biennial, growing from taproots; stylopodium subconical to conical; plants introduced from Eurasia **51. *Ammi***
36. Vallecular vittae lacking; endocarp sclerified; Chile and Argentina **13. *Asteriscium***
– Vallecular vittae present; endocarp typically parenchymatous; North America and northern South America 37
37. Plants annual, growing from taproots or fascicled fibrous roots 38
– Plants biennial or perennial, roots various 41
38. Petals yellow; bracts and bracteoles both lacking **56. *Anethum***
– Petals white; bracts or bracteoles (or both) present 39
39. Calyx teeth lacking; leaves at least twice compound; northwestern North America **115. *Cnidium***
– Calyx teeth prominent; leaves simple to once compound; south-central United States 40
40. Plants growing from taproots; fruits scabrous; petal apex narrowed and inflexed, styles long and slender **166. *Eurytaenia***
– Plants growing from fascicles of fibrous roots; fruits glabrous; petal apex plane, styles short and stout **247. *Limnosciadium***
41. Three to five ribs of the mericarp winged 42
– Only the two marginal ribs winged (the dorsal and lateral ribs obsolete or filiform to keeled) 56
42. Plants low, acaulescent to short-caulescent, stems scapose to subscapose 43
– Plants caulescent, stems leafy and usually tall 48
43. Leaves densely white-tomentose beneath **183. *Glehnia***
– Leaves glabrous or hirtellous beneath 44
44. Calyx lobes prominent, linear-lanceolate and acuminate **360. *Pteryxia***
– Calyx lobes obsolete, or evident but ovate to deltate 45
45. Commissural face of the seed concave to deeply sulcate 46
– Commissural face of the seed plane 47
46. Bracteoles basally connate and white or purple scarious, with prominent nerves **457. *Vesper***
– Bracteoles not connate and white or purple, without prominent nerves **134. *Cymopterus***
47. Involucral bracts usually lacking; peduncles conspicuously hirtellous-pubescent distally; southwestern United States **350. *Pseudocymopterus***
– Involucral bracts present; peduncles glabrous; Alaska **306. *Orumbella***
48. Stylopodium lacking 49
– Stylopodium subconical to conical 52
49. Carpophore lacking; plants growing from a fascicle of fibrous roots; eastern United States **433. *Thaspium***

- Carpophore present and bifurcating; plants growing from slender taproot or a tuberous root; western United States or Mexico 50
50. Calyx lobes prominent, linear-lanceolate and acuminate **360. *Pteryxia***
- Calyx lobes obsolete, or evident but ovate 51
51. Peduncles glabrous; bracteoles of the involucel entire (not dimidiate), filiform to linear-lanceolate; Mexico **346. *Prionosciadium***
- Peduncles conspicuously hirtellous-pubescent distally; bracteoles of the involucel dimidiate, oval to filiform and connate at base; western United States **350. *Pseudocymopterus***
52. Umbellules compact, sub-capitate, pedicels reduced to a tomentose disc; fruits cuneate-obovoid **57. *Angelica***
- Umbellules with distinct pedicels; fruits oblong, ellipsoid, ovoid or obovoid 53
53. Flowers few per umbellule (1–6); petals purple or greenish-yellow **368. *Rhodosciadium***
- Flowers many per umbellule; petals white or sometimes pink or purple 54
54. Endosperm not sulcate under the vittae **117. *Conioselinum***
- Endosperm sulcate under the vittae 55
55. Bracteoles of involucel linear and entire, or lacking; North America south to Guatemala **57. *Angelica***
- Bracteoles of involucel usually toothed to cleft or divided (sometimes entire); Costa Rica to northern South America **273. *Myrrhidendron***
56. Plants woody and arborescent **138. *Dahliaphyllum***
- Plants herbaceous above ground 57
57. Cauline leaves opposite **160. *Enantiophylla***
- Cauline leaves alternate, or all leaves basal 58
58. Stylopodium subconical to conical 59
- Stylopodium lacking 71
59. Petals acute and scarcely incurved 60
- Petal apex narrowed and strongly inflexed 61
60. Carpophore bifid (splitting only apically); Andes of Argentina **290. *Oligocladus***
- Carpophore bifurcating (splitting along the entire length); Andes of Colombia **323. *Perissocoeleum***
61. Outer petals radiate **201. *Heracleum***
- All petals of equal size 62
62. Plants scabrous to scaberulous 63
- Plants glabrous to pubescent (but not scabrous) 65
63. Petals white; plants biennial; northwestern Argentina **81. *Austropeucedanum***
- Petals yellow or purple; plants perennial 64
64. Flowers few (1–6) per umbellule; Mexico **368. *Rhodosciadium***
- Flowers many per umbellule; not native (introduced from Europe) **241. *Levisticum***
65. Flowers few (1–6) per umbellule; Mexico **368. *Rhodosciadium***
- Flowers several to many (> 10) per umbellule; native or introduced plants of temperate North America 66
66. Petals yellow 67
- Petals white, pink or purple 68
67. Bracts and bracteoles conspicuous; seed face concave **241. *Levisticum***
- Bracts and bracteoles usually lacking, seed face plane **318. *Pastinaca***
68. Plants slender, growing from fascicled tubers; marginal wings nerved dorsally at the inner margin, giving the appearance of mericarps with 5 filiform ribs 69
- Plants stout, growing from stout taproots or tuberous to branched rhizomes, marginal mericarp wings lacking dorsal nerves 70
69. Leaves once-pinnate or once-ternate **310. *Oxypolis***
- Leaves phyllodes, hollow and septate **436. *Tiedemannia***
70. Leaves biternate or pinnate; leaflets usually three-parted, sharply serrate and incised **209. *Imperatoria***
- Leaves ternately pinnate, 1–4-pinnate, or 2–3-ternate; leaflets variously dentate, serrate, or incised **57. *Angelica***
71. Petals acute and scarcely incurved **323. *Perissocoeleum***
- Petal apex narrowed and strongly inflexed 72
72. Involucre and involucel large and showy, of 3–5 broadly obovate and overlapping bracts and bracteoles **259. *Mathiasella***
- Involucre lacking or bracts minute; bracteoles various, but not large, broadly ovate, and overlapping 73
73. Styles short **354. *Pseudotaenidia***
- Styles long and slender 74
74. Seed face concave; Mexico **346. *Prionosciadium***
- Seed face plane; temperate North America 75
75. Leaves (at least cauline leaves) decompound, ultimate segments filiform to narrowly lanceolate 76
- Leaves pinnately, ternately, or ternate-pinnately compound, ultimate segments broad 77
76. Peduncles conspicuously hirtellous-pubescent at the base **350. *Pseudocymopterus***

- Peduncles glabrous, or pubescent along the entire length **251. *Lomatium***

77. Calyx teeth obsolete or minute; cauline leaves few, western North America **251. *Lomatium***
- Calyx teeth evident (ovate to acuminate); cauline leaves several; eastern and central North America **341. *Polytaenia***

78. Fruits winged 79
- Fruits unwinged 93

79. Stylopodium lacking 80
- Stylopodium present, subconical to conical 84

80. Plants low, acaulescent to subcaulescent, above-ground stems scapose 81
- Plants caulescent and erect, stems leafy 83

81. Calyx teeth obsolete; styles short **305. *Orogenia***
- Calyx teeth conspicuous; styles long and slender 82

82. Carpophore lacking **300. *Oreoxis***
- Carpophore present and bifurcating **49. *Aletes***

83. Petals white; leaflets linear; carpophore present and bifurcating; Ecuador and Colombia **124. *Cotopaxia***
- Petals yellow or purple; leaflets broad (or leaves simple below); carpophore lacking; eastern North America **433. *Thaspium***

84. Outer petals radiate; bracteoles of involucel well developed, asymmetrical, linear and long recurved **44. *Aethusa***
- All petals of equal size; bracteoles lacking to conspicuous, but not long recurved 85

85. Plants annual 86
- Plants perennial 89

86. Plants growing from taproots; petals elongated with a broad, 2-lobed and inflexed apex; carpophore bifurcating (splitting to the base) **140. *Daucosma***
- Plants growing from fascicles of fibrous roots; petal apex plane or inflexed, short and entire; carpophore subentire or merely bifid (splitting only apically) 87

87. Fruits ovoid, tapering to a prominent beak at the apex; cauline leaves palmately compound with 3–5 leaflets **137. *Cynosciadium***
- Fruits ovoid to orbicular but lacking a prominent beak at the apex; cauline leaves pinnately compound with 2–9 leaflets, or hollow, septate phyllodes 88

88. Leaves pinnately compound with 2–9 leaflets **361. *Ptilimnium***
- Leaves phyllodes, hollow and septate **195. *Harperella***

89. Fruits heteromorphic, one mericarp having two winged lateral ribs but an inconspicuous dorsal rib, the second mericarp having a winged dorsal rib but inconspicuous lateral ribs **116. *Coaxana***
- Fruits homomorphic, having an equal number of wings on both mericarps 90

90. Fruits clavate with a stipe-like base; endosperm transversely nearly terete **125. *Coulterophytum***
- Fruits linear to oblong-ovoid, lacking a stipe-like base; endosperm transversely compressed dorsally 91

91. Calyx lobes obsolete; styles long and slender; Costa Rica to northern South America **273. *Myrrhidendron***
- Calyx lobes minute but evident; styles short; North America as far south as northern Mexico 92

92. Rays scabrous **437. *Tilingia***
- Rays glabrous **245. *Ligusticum***

93. Carpophore lacking 94
- Carpophore present 99

94. Plants caulescent; stylopodium conical; endosperm transversely subterete 95
- Plants acaulescent; stylopodium lacking; endosperm transversely compressed dorsally 97

95. Plants biennial or perennial; commissural vittae 2, vallecular vittae 1; calyx lobes prominent, fruits oblong-ovoid to subglobose with a broad commissure **289. *Oenanthe***
- Plants annual; vittae indistinct; calyx lobes reduced; fruits didymous (the 2 mericarps almost spherical with a narrow commissure) 96

96. Involucre present, ribs filiform (native) **79. *Atrema***
- Involucre lacking, ribs obscure (adventive) **85. *Bifora***

97. Corky rib-like projection running the length of the commissural face of each mericarp; seed face deeply sulcate; calyx teeth obsolete; petal apex plane; woodlands of eastern North America **164. *Erigenia***
- Commissural face of each mericarp lacking a corky rib-like projection; seed face plane; calyx teeth prominent; petal apex strongly inflexed; Rocky Mountains of western North America 98

98. Umbellules with sessile perfect flowers and sharply distinct pedicellate staminate flowers; fruits scaberulous **397. *Shoshonea***
- Umbellules of perfect flowers only (sometimes also having some poorly distinguished staminate flowers); fruits glabrous to slightly pubescent **300. *Oreoxis***

99. Stylopodium lacking 100
- Stylopodium present 109

100. Bracteoles of involucel lacking 101
- Bracteoles of involucel present 103

101. Carpophore bifid only to half its length; petals golden yellow; fertile flowers several to many per umbellule **465. *Zizia***
- Carpophore fully bifurcating; petals pale yellow or purple; fertile flowers few per umbellule 102

102. Styles short; rays 1–8; Mexico and Central America **150. *Donnellsmithia***
- Styles long; rays 15–20; eastern North America **425. *Taenidia***

103. Plants distinctly caulescent 104
- Plants acaulescent or low and short-caulescent 105

104. Plants growing from long, slender taproot; styles short; leaves pinnately decompound, ultimate segments linear; central Rocky Mountains **194. *Harbouria***
- Plants growing from fascicle of somewhat fleshy roots; styles slender; leaves simple or ternately compound, leaflets lanceolate to ovate or orbicular; eastern North America **465. *Zizia***

105. Vittae many, of various sizes, and scattered throughout the pericarp; endosperm slightly compressed dorsally **281. *Neoparrya***
- Vittae discrete, 1–many, but not scattered; endosperm subterete 106

106. Plants growing from a stout, branched caudex covered with marcescent leaf bases **49. *Aletes***
- Plants growing from taproots or tuberous roots 107

107. Umbellules with subsessile perfect flowers and staminate flowers having long, persistent pedicels **297. *Oreonana***
- Umbellules lacking persistent long-pedicellate staminate flowers 108

108. Plants scabrous (at least in the inflorescence); styles slender; seed face usually plane to concave **271. *Musineon***
- Plants glabrous or pubescent, but not scabrous; styles short; seed face usually deeply concave to sulcate **429. *Tauschia***

109. Plants annual 110
- Plants biennial or perennial 122

110. All leaves simple and entire **91. *Bupleurum***
- At least some leaves deeply lobed, dissected or compound 111

111. Marginal ribs from each mericarp closely appressed and appearing to form a single broad rib (often covering the commissure with corky tissue) **54. *Ammoselinum***
- Marginal ribs filiform or broad, but not closely appressed and not appearing to form a single broad rib 112

112. Vallecular vittae obscure; mericarps not separating readily at maturity **120. *Coriandrum***
- Vallecular vittae distinct; mericarps separating at maturity 113

113. Fruits conspicuously beaked, the beak differentiated from the fruit body 114
- Fruits not beaked, or the beak not differentiated from the fruit body 115

114. Involucel of entire bracteoles; beak shorter than the fruit body; commissure narrow; carpophore bifid (splitting apically) **64. *Anthriscus***
- Involucel of lobed or dissected bracteoles; beak several times longer than the fruit body; commissure broad; carpophore bifurcating (splitting entirely) **382. *Scandix***

115. Involucel lacking 116
- Involucel present 120

116. Leaves mostly opposite; fruits papillate-scabrate to glabrate **66. *Apiastrum***
- Leaves alternate; fruits glabrous, puberulent, scaberulous, or setulose 117

117. Petals yellow or purple **150. *Donnellsmithia***
- Petals white 118

118. Petal apex inflexed and narrowed **334. *Pimpinella***
- Petal apex plane or scarcely incurved, not narrowed 119

119. Fruits ovoid to oblong with a rounded to acute apex; ribs glabrous to setulose; widespread **132. *Cyclospermum***
- Fruits oblong to subclavate with a truncate apex; ribs papillose-roughened; pampas of Argentina **286. *Notiosciadium***

120. Seed face plane; commissure narrow **334. *Pimpinella***
- Seed face concave to deeply sulcate; commissure broad or only slightly constricted 121

121. Fruits linear-oblong to ovoid with a short (or obsolete) beak; carpophore bifurcating (splitting along its length) **104. *Chaerophyllum***
- Fruits ovoid, but not beaked; carpophore bifid (splitting only at the apex) **410. *Spermolepis***

122. All leaves simple and entire 123
- At least some leaves toothed, lobed, dissected, or divided 124

123. Basal leaves often petiolate, cauline leaves sessile, auriculate to perfoliate with parallel venation **91. *Bupleurum***
- Leaves chiefly basal and reduced to fistulose, transversely septate phyllodes **309. *Ottoa***

124. Vittae obscure in fruit 125
- Vittae distinct and evident in fruit 126

125. Fruits delicately bristled with an aculeate cuticle, narrow-elliptic with a short beak; ribs angular (thus the mericarps transversely stellate); seed face deeply sulcate **274. *Myrrhis***
- Fruits glabrous, oblong-ovoid and not beaked; ribs filiform (thus the mericarps transversely subterete); seed face plane **43. *Aegopodium***

126. Petal apex obtuse to acute, not narrowed and inflexed 127
- Petals apex conspicuously narrowed and inflexed 128

127. Involucral bracts present; vallecular vittae 1; páramos of South America and Costa Rica **282. *Niphogeton***
- Involucral bracts lacking; vallecular vittae 2–3; Dominican Republic **321. *Pedinopetalum***

128. Plants acaulescent **338. *Podistera***
- Plants caulescent 129

129. Vittae numerous and cyclic (closely surrounding the seed cavity in a continuous ring) **84. *Berula***
- Vittae few to many, but not cyclic (1 or in groups under each vallecula and the commissure) 130

130. Leaves compound, the leaflets mostly distinct, large and broad 131
- Leaves decompound, the leaflets somewhat confluent, small and filiform to ovate 142

131. Fruits clavate to obovoid, contracted below the seed to form a short, winged, stipe-like base **125. *Coulterophytum***
- Fruits lacking a winged, stipe-like base 132

132. Ribs of fruits prominent and corky; plants mostly aquatic or semi-aquatic 133
- Ribs of fruits filiform or obscure (not prominent) and not corky; plants of drier habitats 135

133. Plants growing from a chambered rhizome **113. *Cicuta***
- Plants growing from taproots or solid rhizomes 134

134. Leaflets linear-lanceolate to ovate; petals inflexed and emarginate; terminal and lateral peduncles mostly long **406. *Sium***
- Leaflets lanceolate to suborbicular; petals inflexed but not emarginate; lateral peduncles mostly subsessile **67. *Apium***

135. Fruits ellipsoid-cordate; involucel of several filiform bracteoles extending beyond the fruits **280. *Neonelsonia***
- Fruits orbicular to linear-oblong; bracteoles of involucel not filiform, or filiform and shorter than the fruits 136

136. Leaflets often falcate, finely serrate with callous margins **169. *Falcaria***
- Leaflets not falcate and lacking callous-margined serrations 137

137. Calyx teeth conspicuous; bracteoles present, usually scarious or colored **322. *Perideridia***
- Calyx teeth obsolete or minute; bracteoles lacking or foliaceous, but neither scarious nor colored 138

138. Fruits linear-oblong to narrowly obovate, usually several times longer than broad; plants growing from fascicles of roots 139
- Fruits orbicular to oblong, usually only 2–3 times longer than broad; plants growing from taproots, tubers, or creeping rootstocks 140

139. Fruit ribs angular, bristled or glabrous; seed face sulcate **307. *Osmorhiza***
- Fruit ribs filiform, glabrous; seed face plane **128. *Cryptotaenia***

140. Endosperm transversely terete or pentagonal; bracteoles lacking; outer petals usually radiate; seed face plane to concave; plants introduced from Eurasia **334. *Pimpinella***
- Endosperm transversely subterete to compressed dorsally; bracteoles usually present; all petals alike; seed face concave to sulcate; plants native to the New World 141

141. Endosperm transversely subterete; bracteoles few (or lacking); Mexico to tropical South America **71. *Arracacia***
- Endosperm transversely compressed dorsally; bracteoles several; western North America **135. *Cynapium***

142. Ribs prominent and corky; plants mostly aquatic or semi-aquatic **406. *Sium***
- Ribs obscure to prominent but never corky; plants mostly of dry land 143

143. Plants glaucous, having strong anise scent when crushed; involucel lacking **174. *Foeniculum***
- Plants green (sometimes glaucous in *Perideridia*), lacking a strong anise scent; involucel often present 144

144. Bracts conspicuous, usually numerous 145
- Bracts lacking or inconspicuous 146

145. Stems spotted; fruit ribs crenate and undulate; bracts short, ovate-acuminate **118. *Conium***
- Stems not spotted; fruit ribs neither crenate nor undulate; bracts narrow **322. *Perideridia***

146. Stylopodium depressed; involucel usually lacking **150. *Donnellsmithia***
- Stylopodium conical or subconical; involucel usually present 147

147. Seed face plane 148
- Seed face concave to sulcate 150

148. Petals yellowish **325. *Petroselinum***
- Petals white or pinkish 149

149. Fruits densely pubescent; plants puberulent at the nodes and puberulent or pubescent in the inflorescence **395. *Seseli***
- Fruits glabrous; plants glabrous throughout **98. *Carum***

150. Outer petals often radiate; introduced biennials from Europe to eastern North America **104. *Chaerophyllum***
- Petals all alike; native plants of Mexico to tropical South America **71. *Arracacia***

B. *Key to the Native and Naturalized Genera of Apiaceae in Europe and North Africa*

Note: Imperfect knowledge of the characters of the European genus *Cyathoselinum* precludes its inclusion in this key.

1. Inflorescence capitulate; plants generally spiny with a blue iridescence **35. *Eryngium***
- Inflorescence not capitulate, generally umbellate, sometimes flowers and fruits grouped in heads; plants rarely spiny and lacking blue iridescence 2

2. Inflorescence (all or partly) a simple umbel or appearing so, or inflorescence irregularly compound [see also 13 (compound umbel but umbellules 1-flowered)] 3
- Inflorescence a regular compound umbel 15

3. Leaves (or phyllodes) simple and narrowly linear to linear-lanceolate 4
- Leaves simple and pentagonal, ovate or suborbicular, or ± deeply lobed, or compound 5

4. Leaves septate phyllodes, inserted on a creeping rhizome; umbel loose, pedunculate **246. *Lilaeopsis***
- Leaves not septate, in rosettes; umbel globose, sessile in leaf axils **204. *Hohenackeria***

5. Inflorescence ± irregularly dichotomously cymose 6
- Inflorescence a regular simple umbel, sometimes much reduced and lateral, or ± compact and terminal 7

6. Fruits glabrous; each unit of inflorescence having a central, sessile carpellate or hermaphroditic flower and 2–4 pedicellate staminate flowers with pedicels adnate to the ovary of the central flower **37. *Petagnaea***
- Fruits covered with hooked bristles; each unit of inflorescence subglobose with staminate flowers free, not adnate to the ovary of the central flower **38. *Sanicula***

7. Plants (incl. fruit) with stellate or forked hairs 8
- Fruits glabrous or covered with scales or prickles 9

8. Fruits dorsally concave, unwinged **16. *Bowlesia***
- Fruits not dorsally concave, winged, the lateral ribs extended into wings with glochidiate bristles **22. *Drusa***

9. Fruits covered with vesicular scales; umbel stelliform **34. *Astrantia***
- Fruits glabrous, or somewhat hairy or prickled; umbel not stelliform 10

10. Flowers yellow, clustered and surrounded by broad leaf-like bracts; leaves digitately tripartite **36. *Hacquetia***
- Flowers whitish, pinkish or greenish, without broad leaf-like bracts; leaves pinnate or ternate 11

11. Fruits clearly narrow-elongated, usually 20–85 mm long, with a beak 40–60 mm **382. *Scandix***
- Fruits ovoid, not beaked 12

12. Fruits covered with prickles **443. *Torilis***
- Fruits glabrous 13

13. Leaves divided into narrowly linear segments; plants annual, usually 20–60 cm **132. *Cyclospermum***
- Leaves divided into ovate entire or lobed segments; plants very low, 1–4 cm 14

14. Leaves stipulate; bracts absent; fruit primary ribs inconspicuous **277. *Naufraga***
- Leaves exstipulate; bracts present; fruit primary ribs well marked and visible **199. *Helosciadium***

15. Leaves simple with entire margin, always elongated, often graminiform, sometimes perfoliate **91. *Bupleurum***
- Leaves simple with a dentate or lobed margin, never graminiform, or divided 16

16. Fruiting umbellule shell- or cage-like (bracteoles fused with pedicel bases) enclosing the fruit; only central fruit developing, usually monocarpellate **155. *Echinophora***
- Fruiting umbellule not shell- or cage-like; fruit not enclosed, developing anywhere on the umbellule, usually bicarpellate 17

17. Bracts, bracteoles and calyx teeth ± pectinate; petals with a long aristate bristle on either side of inflexed tip; umbel globose; umbellule 1-flowered; fruit of only one mericarp **232. *Lagoecia***
- Calyx teeth, if present, not pectinate; petals without bristles; umbellule generally many-flowered; fruit of two mericarps 18

18. Fruits didymous, hard; each mericarp ± globose, ± rugose, without conspicuous ribs and commissural face perforated **85. *Bifora***

- Fruits not as above; commissural face not perforated 19
19. Fruits globose, hard; mericarps not separating at maturity; ribs not prominent, primary ribs wavy, secondary ribs ± straight **120. *Coriandrum***
- Fruits generally not hard; mericarps separating at maturity; ribs different 20
20. Fruits strongly flattened; mericarps with 5 ribs, the 3 dorsal filiform, the laterals bearing large, flattened, spathulate, scarious projections; central flowers of the umbel forming a tuft of purplish or blackish bristles **72. *Artedia***
- Fruits without lateral ribs bearing spathulate scarious projections; central flowers of the umbel different 21
21. Fruits not flattened; ribs with sinuate lobes, giving the fruit an appearance of a human brain **375. *Rumia***
- Fruits different, not having the appearance of a human brain 22
22. Plants 3–15 cm; leaves glaucous with lamina orbicular, crenate or lobed, or 1-pinnate with orbicular palmately lobed segments; rays (2)4–7; bracts and bracteoles absent; fruits flattened **304. *Ormosolenia***
- Plants different 23
23. Fruits covered with prickles 24
- Fruits without prickles, but glabrous, scabrous, tuberculate or hairy (sometimes densely) 38
24. Fruits with primary ribs only 25
- Fruits with primary and secondary ribs (often hidden by the bristles) 27
25. Fruits ovoid-pyriform, 2–4 mm, shortly beaked; primary ribs visible only on the beak; annuals **64. *Anthriscus***
- Fruits ovoid, 7–20 mm; primary ribs visible on the whole fruit; perennials 26
26. Flowers white; ultimate leaf segments elliptic, 12–18 mm wide; mericarps incurved longitudinally **236. *Lecokia***
- Flowers yellow; ultimate leaf segments linear, ca. 1 mm wide; mericarps not incurved longitudinally **92. *Cachrys***
27. Endosperm involute 28
- Endosperm not involute (seed commissural face plane, slightly concave to sulcate) 29
28. Leaves 1-pinnate; bracts several, ovate to oblong with membranous margins **454. *Turgenia***
- Leaves 2-3-pinnate; bracts lacking or very few **99. *Caucalis***
29. Bracts divided, often pinnatisect, sometimes leaf-like or only trifid 30
- Bracts entire or lacking 31
30. Spines arranged in a single row on the secondary ribs (one species, *D. setifolius*, with a single row of branched spines) **141. *Daucus***
- Spines arranged in (1) 2–3 rows on the secondary ribs (if only one row, then fruit clearly compressed dorsally and commissure wide, plants 3–20 cm) **352. *Pseudorlaya***
31. Involucre lacking or much reduced 32
- Involucre present 34
32. Primary ribs with branched hairs; petiole base conspicuously and widely sheathing **76. *Astrodaucus***
- Primary ribs with simple, unbranched hairs; petiole base narrowly sheathing 33
33. Fruits ca. 8 mm; secondary ribs strongly developed, broad, prominent, nearly contiguous **455. *Turgeniopsis***
- Fruits usually 2–6 mm, rarely up to 8 mm; secondary ribs not so developed, less prominent **443. *Torilis***
34. Prickles not arranged in distinct rows **443. *Torilis***
- Prickles arranged in 1–3 distinct row(s) on the secondary ribs 35
35. Fruits compressed dorsally, commissure distinctly wider than the thickness of the mericarp 36
- Fruits not or laterally compressed, sometimes slightly compressed dorsally, commissure subequal to or slightly wider than the thickness of the mericarp 37
36. Outer petals strongly enlarged, up to 8 times as long as the central ones; spines not glochidiate **301. *Orlaya***
- Outer petals only weakly enlarged; spines glochidiate **352. *Pseudorlaya***
37. Spines arranged in a single row on the secondary ribs (one species, *D. setifolius*, with one row of branched spines) **141. *Daucus***
- Spines arranged in (1)2–3 rows on the secondary ribs **352. *Pseudorlaya***
38. Fruits winged, incl. fruits with a thin marginal wing-like rib 39
- Fruits unwinged, but sometimes with narrow ribs (but which cannot be interpreted as a wing) or marginal enlarged ribs too thickened to be form a wing 143
39. Fruits with primary and secondary ribs 40
- Fruits with primary ribs only 57
40. Flowers yellow or greenish-yellow 41
- Flowers white, whitish, pink or pinkish 43

41. Commissural seed face involute **432. *Thapsia***
- Commissural seed face plane to concave 42

42. Flowers yellow, often bright yellow; bracteoles lacking or very few (very rarely present); dorsal wings weakly developed **432. *Thapsia***
- Flowers greenish-yellow, sometimes reddish; bracteoles several; dorsal wings well developed **235. *Laserpitium***

43. Fruits hairy 44
- Fruits glabrous or nearly so 48

44. Dorsal primary ribs of the fruits low or filiform 45
- Dorsal primary ribs of the fruits prominent 47

45. Monocarpic hispid or pubescent plants; rootstock without fibers; umbels somewhat contracted at maturity **401. *Silphiodaucus***
- Perennial glabrous or pubescent plants; rootstock with fibers; umbels spreading at maturity 46

46. Bracts absent or few; fruit lateral wings 0.4–0.9 mm; umbellules distinctly globose **432. *Thapsia***
- Bracts numerous; fruit lateral wings (1)1.5–3 mm; umbellules convex or hemispherical **235. *Laserpitium***

47. Fruits 2.5–3 mm, wings dentate; plants prostrate or ascending, 0.1–0.2 m **311. *Pachyctenium***
- Fruits 6–12 mm, wings entire at margin; plants erect, 0.5–1.3 m **432. *Thapsia***

48. Involucre lacking or much reduced **432. *Thapsia***
- Involucre present 49

49. Fruit wing margins clearly dentate, sometimes spinulate 50
- Fruit wing margins entire, sometimes shallowly crenulate 52

50. Fruits 12–14 mm, blackish; stem up to 3 m **263. *Melanoselinum***
- Fruits 2–7 mm, generally brownish; stem 0.3–1 m 51

51. Fruits 6–7 mm; upper leaves strongly reduced, sometimes only consisting of the sheath **141. *Daucus***
- Fruits 2–5 mm; upper leaves not reduced or only slightly so, similar to the lower ones **444. *Tornabenea***

52. Stem woody; fruits up to 2 mm, beaked, with lateral wings weakly developed **444. *Tornabenea***
- Stem herbaceous; fruits 3.5–16 mm, not beaked, with lateral wings well developed 53

53. Root scaly at top; plants ascending; fruits strongly compressed dorsally **374. *Rouya***
- Root crowned with fibrous remains; plants erect; fruits not or slightly compressed dorsally 54

54. Endosperm plane or slightly concave; fruit wings silvery white, greenish or brownish 55
- Endosperm involute; fruits with golden wings **432. *Thapsia***

55. Bracts numerous 56
- Bracts absent or few, deciduous **432. *Thapsia***

56. Fruit lateral wings 0.5–1 mm; ultimate segments with entire margin **400. *Siler***
- Fruit lateral wings (1)1.5–3 mm; ultimate segments with lobed, dentate or entire margin **235. *Laserpitium***

57. Fruits hairy 58
- Fruits glabrous or nearly so 68

58. Dorsal ribs of the fruits clearly prominent 59
- Dorsal ribs of the fruits low, filiform, weakly prominent or indistinct 62

59. Fruits narrowly elliptic, with an attenuate apex; petals narrowly lanceolate; leaves minutely divided into narrow, lanceolate-cuneiform, lobed segments **438. *Todaroa***
- Fruits elliptic or ovoid, without an attenuate apex; petals orbicular, ovate or obovate; leaves divided into ovate lobed segments 60

60. Fruits very strongly flattened; lower leaves generally 3–4-pinnate **460. *Xanthoselinum***
- Fruits only slightly flattened; lower leaves 2–3-pinnate 61

61. Fruits 3–4 mm long with 3 vallecular vittae; rays short-haired **395. *Seseli***
- Fruits (4)5–9 mm long, with 1 vallecular vitta; rays with long thin curly hairs **328. *Phlojodicarpus***

62. Basal leaves simple, lobed **201. *Heracleum***
- Basal leaves divided 63

63. Fruits with the inner part of margin pellucid and translucent, the outer part inflated **466. *Zosima***
- Fruits different, without a partly pellucid and translucent margin 64

64. Flowers yellow, green or greenish 65
- Flowers white or pinkish 67

65. Involucre lacking or much reduced 66
- Involucre present **318. *Pastinaca***

66. Fruits with a downy-tomentose commissural face; leaves divided into long, linear segments; vallecular vittae 3–4; umbels with 5–9 rays **165. *Eriosynaphe***
- Fruits with a glabrous commissural face; leaves divided into generally broadly ovate or lanceolate, lobed segments; vallecular vittae 1; umbels with 10–30 rays **201. *Heracleum***

67. Fruits with a thick margin; lower leaves 2–3-pinnate with narrowly elliptic or linear-lanceolate segments; involucre usually present **153. *Ducrosia***
- Fruits with a thin margin; lower leaves 1-pinnate or 1-ternate (rarely 2-) with usually wide, ± lobed,

suborbicular, ovate or ovate-oblong segments; involucre often lacking (sometimes present) **201. *Heracleum***

68. Lower leaves simple, sometimes pinnatifid 69
- Lower leaves divided 71

69. Involucre present; fruits slightly compressed dorsally; endosperm involute **200. *Heptaptera***
- Involucre lacking or much reduced; fruits strongly compressed dorsally; endosperm face plane 70

70. Plants hairy; outer petals radiate; flowers bright white or pinkish, rarely whitish **201. *Heracleum***
- Plants glabrous; petals equal; flowers yellow or yellowish-green **318. *Pastinaca***

71. Fruits not strongly dorsally flattened 72
- Fruits strongly dorsally flattened 85

72. Flowers white, whitish or pinkish 73
- Flowers yellow, yellow-green, green or greenish 79

73. Fruits narrowly elliptic, with an attenuate apex; sepals linear-subulate **438. *Todaroa***
- Fruits subglobular, ovoid, ovoid-oblong or ellipsoid-oblong, without an attenuate apex; sepals not developed, or ± broadly ovate, rarely shortly subulate 74

74. Involucre present 75
- Involucre lacking or much reduced 76

75. Plants 1–2 m; ultimate leaf segments 20–30 mm long; fruits ovoid-oblong or ellipsoid-oblong; vallecular vittae 1 **268. *Molopospermum***
- Plants 0.5–1 m; ultimate leaf segments 2–6 mm long; fruits broadly ovoid, subglobular; vallecular vittae 2–3 **80. *Aulacospermum***

76. Lower leaves with broadly ovate and merely dentate segments 77
- Lower leaves with segments wide in outline but deeply lobed 78

77. Calyx teeth not developed or much reduced; leaf rachis straight **57. *Angelica***
- Calyx teeth developed, ovate, 0.3–0.5 mm long; leaf rachis geniculate **308. *Ostericum***

78. Upper cauline leaves with broad inflated sheaths diverging from stem **117. *Conioselinum***
- Upper cauline leaves with narrow sheaths lying close to the stem **393. *Selinum***

79. Fruits narrowly elliptic, sometimes with an attenuate apex 80
- Fruits globose, ovoid or ellipsoid, sometimes subquadrangular, without an attenuate apex 81

80. Plants leafy; leaves 2–4-pinnately dissected into lanceolate-ovate lobes; bracts present; fruits with an attenuate apex **438. *Todaroa***
- Plants with rod-like leafless flowering stems; leaves early deciduous, ultimate divisions filiform; bracts absent; fruits not attenuate at top **384. *Schoenoselinum***

81. Involucre lacking or much reduced **57. *Angelica***
- Involucre present 82

82. Leaves much divided into linear or filiform segments; endosperm involute 83
- Leaves divided into ovate-rhomboidal or lanceolate-acuminate lobed or dentate segments; endosperm face plane to slightly concave 84

83. Leaf lobes rigid and almost spiny; fruits 7–10 mm, with dentate-cristate (sometimes prickly) wing-like ribs **92. *Cachrys***
- Leaf lobes not rigid; fruits 12–30 mm, with straight to slightly undulate wings, sometimes with erose margins **345. *Prangos***

84. Leaf segments ovate-rhomboidal, lobed at apex; rays 8–20; fruits 5–7 mm **241. *Levisticum***
- Leaf segments lanceolate-acuminate, deeply lobed all along; rays 18–45; fruits 6.5–12 mm **268. *Molopospermum***

85. Fruit dorsal ribs clearly prominent 86
- Fruit dorsal ribs low, filiform, weakly prominent or indistinct 99

86. Flowers white or pinkish 87
- Flowers yellow, yellow-green, green or greenish-cream 93

87. Involucre lacking or much reduced 88
- Involucre present 91

88. Annual, 0.2–0.4 m; rays 2–5; mericarps transversely rugose **228. *Krubera***
- Perennial or monocarpic, 0.5–3 m; rays 7–90; mericarps not transversely rugose 89

89. Ultimate leaf segments linear with an entire margin; rays usually 7–14 **326. *Peucedanum***
- Ultimate leaf segments ovate-oblong pinnately incised, or ovate or lanceolate and only dentate; rays 15–90 90

90. Ultimate leaf segments ovate or lanceolate, only dentate; commissural vittae 2 **57. *Angelica***
- Ultimate leaf segments ovate-oblong pinnately incised; commissural vittae 4–8 **117. *Conioselinum***

91. Leaf segments ovate or lanceolate, margin only serrate; wings of each mericarp standing apart; monocarpic **57. *Angelica***
- Leaf segments ovate, deeply lobed; wings of each mericarp closely appressed to one another; not monocarpic 92

92. Fruits 7–12 mm with lateral wings 1–2.5 mm; petiole triangular **325. *Pteroselinum***
- Fruits 3–7 mm with lateral wings 0.4–1 mm; petiole terete **460. *Xanthoselinum***

93. Involucre lacking or much reduced 94
- Involucre present 97

94. Plants very stout but small (0.1–0.3 m); ultimate leaf segments and lobes lanceolate **461. *Xatartia***
- Plants taller (0.5–3 m); ultimate leaf segments linear or ± broadly ovate 95

95. Leaf segments and lobes linear; plants slender; rays not > 12 **56. *Anethum***
- Leaf segments ovate, serrate, rarely with very few irregular lobes; plants robust; rays 12–90 96

96. Umbels verticillately arranged; rays 12–20, rarely 30 **439. *Tommasinia***
- Umbels not verticillately arranged; rays 20–90 **57. *Angelica***

97. Ultimate leaf lobes linear to lanceolate-linear, sometimes short; vallecular vittae 4–12, commissural 20–32 (giving a striate aspect of the fruit, especially on the commissural face) **172. *Ferulago***
- Ultimate leaf segments ± broadly ovate, lobed or not; vallecular vittae 1, commissural 2 98

98. Leaf segments lobed; fruit wings 0.4–1 mm wide; rays usually 6–20 **460. *Xanthoselinum***
- Leaf segments entire, serrate; fruit wings 1.2–3.5 mm wide; rays 15–60 **57. *Angelica***

99. Fruits with the inner part of margin pellucid and translucent, the outer part inflated 100
- Fruits different, without a partly pellucid and translucent margin 101

100. Lower leaves 2–3-pinnate with lobed segments 3–5 mm long; fruit wing ca. 2 mm wide **466. *Zosima***
- Lower leaves 1-pinnate with simple or 3-partite segments 20–30 mm long; fruit wing ca. 1 mm wide **239. *Leiotulus***

101. Flowers white, whitish or pinkish 102
- Flowers yellow, yellow-green, green, greenish-white or rarely brownish 119

102. Involucre lacking or much reduced 103
- Involucre present 110

103. Lower leaves with segments and lobes linear with an entire margin, often 15–90 mm long, sometimes shorter 4–15 mm 104
- Lower leaves with segments ± broadly ovate or lanceolate in outline, dentate or (sometimes deeply) lobed 106

104. Commissural vittae invisible, hidden by the pericarp; petals whitish **435. *Thysselinum***
- Commissural vittae visible; petal white, pink or pinkish 105

105. Lower leaves oblong in outline (sometimes the very next triangular), 1–2-pinnate; calyx teeth inconspicuous **145. *Dichoropetalum***
- Lower leaves triangular in outline, usually 2–4-ternate or 2–4-pinnate; calyx teeth short, triangular-acuminate **326. *Peucedanum***

106. Ultimate leaf lobes up to 5 mm wide, generally numerous 107
- Ultimate leaf segments or lobes 15–70 mm wide; lobes generally few or almost lacking (segments entire or serrate instead), but sometimes segments pinnately lobed 108

107. Lower leaves triangular in outline, with basal primary segments stalked, not sessile; fruits 8–10 mm **201. *Heracleum***
- Lower leaves oblong in outline, with basal primary segments sessile; fruits generally 3–7 mm, rarely more **145. *Dichoropetalum***

108. Leaf segments unlobed (or sometimes with very few lobes), entire and serrate; petals not radiate 109
- Leaf segments lobed, sometimes only the lower or the terminal; petals radiate, or rarely ± regular (in this case, rays 2–8) **201. *Heracleum***

109. Lateral ultimate leaf segments mostly stalked; inflexed lobule of petal with an obtuse or emarginate apex **326. *Peucedanum***
- Lateral ultimate leaf segments mostly sessile; inflexed lobule of petal with an acute apex **209. *Imperatoria***

110. Lower leaves with segments and lobes linear with an entire margin 111
- Lower leaves with segments ± broadly ovate or lanceolate in outline, dentate or (sometimes deeply) lobed 112

111. Commissural vittae invisible, hidden by the pericarp; ultimate leaf lobes 15–100 mm long; petals whitish **435. *Thysselinum***
- Commissural vittae visible; ultimate leaf lobes 4–15 mm long; petals pure white **358. *Pteroselinum***

112. Lower leaves with main or all segments entire (sometimes with very few ± irregular lobes) 113
- Lower leaves with lobed segments, generally lobes numerous within the leaf 116

113. Lower leaves 2-ternate; fruit wings 1.5–2 mm wide **326. *Peucedanum***
- Lower leaves 1–3-pinnate; fruit wings 0.4–1 mm wide 114

114. Leaves 2–3-pinnate; margin of the leaf segments irregularly dentate with teeth sharp and spinescent (with awn-like point) **102. *Cervaria***
– Leaves 1-pinnate; margin of the leaf segments ± regularly dentate, teeth without awn-like point 115
115. Petals subequal, not radiating; fruits 5–7 mm; plants glabrous, 40–100 cm, stem branched **253. *Macroselinum***
– Petals clearly radiate; fruits 7–11 mm; usually plants pubescent, usually 10–40 cm; stem simple or with few branches **201. *Heracleum***
116. Plants huge and very robust, up to 2–5 m; leaf lobes several cm wide; rays 40–170 **201. *Heracleum***
– Plants more slender, 0.25–1.5 m, rarely more; leaf lobes 1–15 mm wide; rays 5–30, rarely 40 117
117. Commissural vittae invisible, hidden by the pericarp; stem hollow, with milky exudate **435. *Thysselinum***
– Commissural vittae visible; stem solid, without milky exudate 118
118. Leaf rachis geniculate; fruit wings 0.6–1 mm wide; petiole terete **299. *Oreoselinum***
– Leaf rachis straight; fruit wings 1–2.5 mm wide; petiole triangular in transverse section **358. *Pteroselinum***
119. Involucre present 120
– Involucre lacking or much reduced 127
120. Ultimate leaf segments narrow, linear or lanceolate, entire, 121
– Ultimate leaf segments wide in outline, dentate or ± deeply lobed, rarely entire 124
121. Vittae very numerous (vallecular 6–12, commissural 14–36) giving a striate aspect to the fruit, especially on the commissural face; leaf segments usually 2–30 mm long **172. *Ferulago***
– Vallecular vittae 1, commissural 2(4); leaf segments 15–100 mm long 122
122. Commissural vittae invisible, hidden by the pericarp; stem hollow, with milky exudate; petals greenish or reddish **435. *Thysselinum***
– Commissural vittae visible; stem solid, without milky exudate; petals yellow 123
123. Lower leaves divided into segments ovate to oblong in outline, ± deeply lobed; lobes ± broadly linear, 4–14(20) mm long; petals dark-nerved **426. *Taeniopetalum***
– Lower leaves divided into linear, entire, unlobed, 20–100 mm long segments; petals not dark-nerved **326. *Peucedanum***
124. Vallecular vittae 2–12, rarely 1, commissural 6–36, giving a striate aspect to the fruit, especially on the commissural face 125
– Vallecular vittae 1, commissural 2–4 126
125. Leaves without stellate hairs, glabrous or scabrous; ultimate leaf lobes 2–10 mm, lanceolate; vittae very numerous (vallecular 6–12, commissural 14–36) **172. *Ferulago***
– Leaves with stellate hairs; ultimate leaf segments 20–130 mm, ovate-lanceolate; vittae less numerous (vallecular (1)2–3, commissural 6–8) **292. *Opopanax***
126. Leaves with broad, obovate-cuneate, incised segments; rays 6–15 **77. *Astydamia***
– Leaves with ovate-oblong, deeply dentate or pinnately lobed segments; rays 5–6 **426. *Taeniopetalum***
127. Ultimate leaf segments of lower leaves broad in outline 128
– Ultimate leaf segments or lobes of lower leaves narrow in outline, lanceolate, linear or filiform 137
128. Involucel present 129
– Involucel lacking or much reduced 133
129. Leaves with stellate hairs; commissural vittae 6–14 **292. *Opopanax***
– Leaves without stellate hairs; commissural vittae 2–4, rarely 6 130
130. Plants hairy 131
– Plants glabrous 132
131. Fruit wings proximally thin and distally thickened; plants 0.1–0.6 m; flowers bright yellow **239. *Leiotulus***
– Fruit wings thin, not distally thickened; plants 0.6–1.2 m; flowers green, sometimes pale green-yellow **201. *Heracleum***
132. Stem solid; ultimate leaf segments up to 2 cm; fruits 6–12 mm long; vallecular vittae 1, commissural 4 **426. *Taeniopetalum***
– Stem hollow; ultimate leaf segments 6–12 cm or more; fruits 5–6 mm long, small vittae arranged into a ring **57. *Angelica***
133. Flowers polygamous, central umbel with hermaphrodite flowers, surrounded by lateral umbels usually with male flowers only **171. *Ferula***
– Umbel arrangement different: almost all umbels with hermaphrodite flowers (except sometimes male flowers on late, ± reduced umbels) 134
134. Dorsal fruit ribs distant from marginal ones, included in the vascularized wings; leaves generally 1-pinnate (rarely 2-pinnate) **318. *Pastinaca***

- All fruit ribs equidistant, wings not vascularized; leaves 2–4-pinnate 135

135. Rays 1–8, frequently proliferating **171. *Ferula***
- Rays 10–30; rays usually not proliferating 136

136. Ultimate leaf segments 7–15 mm, pinnately divided; stem solid; fruits 5–6 mm long **145. *Dichoropetalum***
- Ultimate leaf segments 25–80 mm, entire, dentate; stem hollow; fruits 7–9 mm long **439. *Tommasinia***

137. Flowers polygamous, central umbel with hermaphrodite flowers, surrounded by lateral umbels usually with male flowers only **171. *Ferula***
- Umbels arrangement different: almost all umbels with hermaphrodite flowers (except sometimes male flowers on late, ± reduced umbels) 138

138. Involucel lacking or much reduced 139
- Involucel present 140

139. Plants annual; ultimate leaf lobes up to 15–20 mm; rays usually 15–35 **56. *Anethum***
- Plants perennial; ultimate leaf lobes usually 1–7 mm, up to 9 mm; rays 1–15 **171. *Ferula***

140. Stems hollow; commissural vittae invisible, hidden by the pericarp **435. *Thysselinum***
- Stems solid; commissural vittae visible 141

141. Lower leaves oblong in outline **145. *Dichoropetalum***
- Lower leaves triangular in outline 142

142. Lower leaves divided into ovate to oblong segments, ± deeply lobed; lobes linear to narrowly oblanceolate, 4–14(20) mm long; petals dark-nerved **426. *Taeniopetalum***
- Lower leaves divided into linear, entire, unlobed, (10)20–70(100) mm long segments; petals not dark-nerved **326. *Peucedanum***

143. Fruits clearly narrow-elongated, ca. 3.5–4 times as long as broad 144
- Fruits not elongated, but ellipsoid, ovoid-oblong, ovoid or globular, sometimes flattened 164

144. Fruit ribs clearly prominent 145
- Fruit ribs low, filiform, weakly prominent or indistinct 149

145. Flowers bright yellow **229. *Kundmannia***
- Flowers white or pinkish 146

146. Geophyte with a ± globose tuber; stem with a flexuous underground part; fruits glabrous and 3–7 mm long **180. *Geocaryum***
- Plants not tuberous; stem without a flexuous underground part; fruits hairy or glabrous, longer than 7 mm 147

147. Fruits densely pubescent **78. *Athamanta***
- Fruits glabrous or only locally scabrous 148

148. Fruits 15–30 mm; ultimate leaf divisions ovate or lanceolate incised **274. *Myrrhis***
- Fruits 4.5–6 mm; ultimate leaf divisions filiform, entire **372. *Rivasmartinezia***

149. Fruits with an obvious beak (sometimes long but not abruptly separate), 1.8–24 mm, sometimes up to 65 mm 150
- Fruits without an obvious beak (beak absent or present, but almost indistinct or very short) 151

150. Beak up to 4 mm; fruits not curved when ripe, smooth and shiny on the seed-bearing part (rarely hairy), the ribs apparent only on the beak **64. *Anthriscus***
- Beak usually 7–65 mm, very rarely less; fruits curved when ripe, usually not smooth and shiny on the seed-bearing part, the ribs discernible (especially the marginal ones) on the seed-bearing part **382. *Scandix***

151. Fruits obviously or densely hairy or scabrous 152
- Fruits glabrous or nearly so (very sparsely or only locally hairy, scabrous or papillose) 155

152. Fruits covered with antrorse tuberculate generally acute bristles; endosperm deeply sulcate 153
- Fruits covered often densely with straight to flexuous-arachnoid hairs; endosperm plane or concave 154

153. Stem nodes not or weakly inflated; ribs not discernible on the fruit body, only apparent on the short beak **64. *Anthriscus***
- Stem nodes strongly inflated; ribs discernible on the fruit body **275. *Myrrhoides***

154. Fruits with inconspicuous ribs, pubescent throughout; stylopodium widely conical (as high as wide) **78. *Athamanta***
- Fruits with visible pubescence on filiform ribs, but nearly or quite glabrous between them; stylopodium narrowly conical (twice as high as wide) **89. *Bubon***

155. Tuberous geophyte 156
- Hemicryptophyte perennial with a rootstock, biennial or annual plant 160

156. Involucre present; endosperm plane, convex or slightly concave 157
- Involucre lacking (very rarely present) or much reduced; endosperm concave 158

157. Flowers white, pink or reddish **90. *Bunium***
- Flowers yellow **157. *Elaeosticta***

158. Fruit ribs wide and low; stem without a subterranean, flexuous part; plants up to 2 m, usually not less than 0.5 m **104. *Chaerophyllum***
- Fruit ribs narrow and slightly prominent; stem with a subterranean, flexuous part (sometimes short); plants up to 1 m but generally less (0.15–0.5 m) 159

159. Fruit surface covered with small cones, ribs ± sharp **180. *Geocaryum***
- Fruit surface without small cones, ribs ± rounded **119. *Conopodium***

160. Ribs not discernible on the often shiny (rarely dull) fruit body 161
- Ribs discernible on the often dull (rarely shiny) fruit body, even if they are filiform or very obtuse 162

161. Lower leaves 2–4-pinnate or -ternate with pinnately lobed segments; involucel developed with lanceolate, ciliate bracteoles; endosperm deeply sulcate **64. *Anthriscus***
- Lower leaves 1-ternate with broadly ovate, entire, serrate segments; bracteoles lacking or few and minute; endosperm plane **128. *Cryptotaenia***

162. Leaves subcoriaceous, 1-2-ternate with very long (up to 30 cm) linear-lanceolate segments with a cartilaginous serrate margin **169. *Falcaria***
- Leaves very different, soft and with ovate segments without a cartilaginous margin 163

163. Plants glabrous; rays 2–7, very unequal; petals entire, obovate **128. *Cryptotaenia***
- Plants usually hairy, at least at base; rays usually 6–32, subequal or slightly unequal; petals notched **104. *Chaerophyllum***

164. Fruits dorsally flattened 165
- Fruits not dorsally flattened 177

165. Fruits hairy, often with vesicular hairs 166
- Fruits glabrous or nearly so (sometimes very shortly pubescent or papillose) 170

166. Fruit margin ± strongly thickened, smooth or corrugated; secondary ribs absent 167
- Fruit margin thin, subalate, densely long setose-hairy; secondary ribs present **52. *Ammodaucus***

167. Annual or biennial plant; outer petals radiate **442. *Tordylium***
- Perennial; outer petals not radiate 168

168. Bracts and bracteoles absent; commissural face of mericarp downy-tomentose **165. *Eriosynaphe***
- Bracts and bracteoles present; commissural face of mericarp glabrous 169

169. Plants glabrous; rays 4–9; fruits ca. 3.5 mm **153. *Ducrosia***
- Plants cinereous-pubescent; rays 10–25; fruits 8–12 mm **466. *Zosima***

170. Fruits with the inner part of margin pellucid and translucent, the outer part inflated 171
- Fruits different, without a partly pellucid and translucent margin 172

171. Perennial; petals slightly unequal; all fruits similar, flat **466. *Zosima***
- Annual; petals clearly radiating; fruits dimorphic, flat and orbicular or semi-spherical **442. *Tordylium***

172. Flowers yellow or green 173
- Flowers white, whitish or pinkish 175

173. Plants stout but small (0.1–0.3 m); flowers green; dorsal fruit ribs thick and clearly prominent **461. *Xatartia***
- Plants more slender, 0.3–1 m; flowers yellow or greenish; dorsal fruit ribs inconspicuous or slightly prominent 174

174. Fruit ribs inconspicuous, immersed in the mesocarp; involucre lacking; fruits ca. 3 mm; rays 3–5 **211. *Johrenia***
- Fruit ribs visible; involucre present; fruits 4–6 mm; rays 6–20 **88. *Bonannia***

175. Fruits with secondary ribs; stem woody, up to 1.5 m; rays 20–25 **269. *Monizia***
- Fruits with primary ribs only; stem herbaceous, up to 0.5 m; rays 2–16 176

176. Perennial with a junciform aspect; umbels not cymosely arranged; fruits 8–10 mm, elliptic-oblong, not transversely rugose **431. *Thamnosciadium***
- Annual, not junciform; umbels cymosely arranged; fruits 4–7 mm, ovate, transversely rugose, with rugose marginal ribs **228. *Krubera***

177. Fruit dorsal ribs clearly prominent, sharp or rounded 178
- Fruit dorsal ribs low, filiform, weakly prominent or indistinct 289

178. Fruits with primary and secondary ribs 179
- Fruits with primary ribs only 182

179. Plants dioecious (or rarely monoecious), up to 0.25 m; stem branched from the base; rays 3–11 **450. *Trinia***
- Plants hermaphrodite, often taller; stem not branched from the base; rays 11–50 180

180. Leaves divided into wide, ovate, obovate or suborbicular, entire or lobed segments; involucre mostly lacking; fruits glabrous 181
- Leaves divided into narrow, linear or lanceolate segments; involucre present, with usually divided bracts; fruits finely and densely papillate-granulate **141. *Daucus***

181. Ultimate leaf segments obtusely dentate, up to 8 cm long, 2.5–6.5 cm wide; flowers often pale yellow, sometimes white; bracteoles usually absent; commissural vittae 2–6 **234. *Laser***
- Ultimate leaf segments acutely dentate, up to 4 cm long, 2–3 cm wide; bracteoles usually present; commissural vittae 2 **432. *Thapsia***

182. Flowers yellow, yellowish, green or greenish 183
- Flowers white, whitish, pink or pinkish, sometimes reddish 202
183. Plants dioecious (or rarely monoecious); leaves divided into linear segments **450. *Trinia***
- Plants hermaphrodite; leaves divided into segments of various shape 184
184. Involucre lacking or much reduced 185
- Involucre present 191
185. Leaves divided into ovate or ovate-lanceolate entire segments (rarely with very few lobes) 186
- Leaves divided into narrow, lanceolate, linear-lanceolate to setaceous segments or lobes, or into ovate and pinnately lobed segments 187
186. Umbels cymosely arranged; leaf segments broadly ovate; calyx teeth absent or very short; endosperm involute **409. *Smyrnium***
- Umbels arranged in a lax panicle; leaf segments ovate-lanceolate, acuminate; calyx teeth triangular, acute; endosperm plane **452. *Trochiscanthes***
187. Involucel lacking **174. *Foeniculum***
- Involucel present 188
188. Plants stout but small (0.1–0.3 m); fruits dorsally compressed with marginal ribs distinctly larger than the dorsal **461. *Xatartia***
- Plants slender, usually 0.2–1 m; fruits not dorsally compressed with marginal ribs equal to or slightly larger than the dorsal 189
189. Leaves narrowly lanceolate in outline; ultimate leaf segments 2–5(–8) mm **98. *Carum***
- Leaves triangular in outline; ultimate leaf segments (5–)8–100 mm 190
190. Vallecular vittae 1, visible; ultimate leaf segments linear, 0.5–3 mm wide **395. *Seseli***
- Vallecular vittae 3–10, inconspicuous; ultimate leaf segments entire, lanceolate or linear-lanceolate, usually 3–6 mm wide, rarely less **399. *Silaum***
191. Ultimate leaf segments or lobes linear, often narrowly so 192
- Ultimate leaf segments lanceolate, sometimes narrowly so or ovate 197
192. Fruits broadly ovoid, sometimes subdidymous, usually 7–17 mm; endosperm involute **92. *Cachrys***
- Fruits ovoid to ovoid-oblong, 2–5(–7) mm; endosperm plane 193
193. Fruits obviously hairy 194
- Fruits glabrous 195
194. Rays 30–50; fruits hispid and with stellate hairs, filiform secondary ribs present; leaves 4–5-pinnate **343. *Portenschlagiella***
- Rays 6–18; fruits without stellate hairs, secondary ribs absent; leaves 3-pinnate **395. *Seseli***
195. Lower leaves clearly oblong in outline; fruit ribs medium-developed, obtuse 196
- Lower leaves triangular in outline; fruit ribs well developed, sharp **377. *Rutheopsis***
196. Rays subequal, usually 3–8; stylopodium oblong-conical **98. *Carum***
- Rays very unequal, usually 10–15; stylopodium low-conical to conical **179. *Gasparrinia***
197. Fruits hairy, ovoid-oblong with an attenuate apex; plants hairy or glabrescent 198
- Fruits glabrous, ovoid or ellipsoid without a marked attenuate apex; plants glabrous, sometimes locally puberulent 199
198. Plants hairy; ultimate leaf segments ovate-lanceolate to lanceolate; calyx lobes triangular, 0.25–0.5 mm, Europe, N Africa and the Canary Islands **78. *Athamanta***
- Plants glabrescent; ultimate leaf segments usually narrowly lanceolate; calyx lobes narrowly triangular, 0.5–1 mm, endemic to the Canary Islands **438. *Todaroa***
199. Lower leaves clearly oblong in outline; fruits with medium-developed ribs 200
- Lower leaves triangular in outline; fruits with well-developed ribs 201
200. Lower leaves 2-pinnate; fruits compressed dorsally; vallecular vittae 3–4 **88. *Bonannia***
- Lower leaves 1-pinnate; fruits not compressed dorsally; vallecular vittae 1 **377. *Rutheopsis***
201. Robust plants up to 2 m; fruits 6.5–12 mm **268. *Molopospermum***
- Slender plants up to 0.5 m; fruits ca. 3 mm **377. *Rutheopsis***
202. Lower leaves simple 203
- Lower leaves divided 204
203. Plants glabrous, up to 0.5 m; fruits glabrous, 3–4 mm **314. *Pancicia***
- Plants hairy, 0.6–2.5 m; fruits hairy, 5.5–9.5 mm **255. *Magydaris***
204. Involucre present 205
- Involucre lacking or much reduced 236
205. Fruits hairy 206
- Fruits glabrous or nearly so 209
206. Ultimate leaf segments broadly ovate, entire, crenate-denticulate, widely obtuse **255. *Magydaris***
- Ultimate leaf segments ± acuminate, mucronate, strongly dentate or lobed, or if entire, oblong, linear-lanceolate or linear 207

207. Fruits 5–13 mm, with an attenuate apex 208
- Fruits 2–4.5 mm, not attenuate at apex **395. *Seseli***
208. Hairy plant; ultimate leaf segments ovate-lanceolate to lanceolate; calyx lobes triangular, 0.25–0.5 mm **78. *Athamanta***
- Glabrescent plant; ultimate leaf segments usually narrowly lanceolate; calyx lobes narrowly triangular, 0.5–1 mm **438. *Todaroa***
209. Leaves fleshy; fruits straw-colored, spongy, nearly octagonal in transverse section; vittae numerous, most invisible, forming a ring inside the fruits around endosperm **127. *Crithmum***
- Leaves not fleshy and fruits different 210
210. Geophyte with a single tuber; subterranean part of stem flexuous **90. *Bunium***
- Plants with a rootstock or a fusiform root or with many tubers or with a creeping stem rooting at nodes; subterranean part of stem not flexuous 211
211. Lower leaves divided into narrow linear to filiform ultimate segments or lobes, 0.5–2 mm wide 212
- Lower leaves divided into ovate or lanceolate, dentate or (sometimes narrowly) lobed ultimate segments 3–40 mm wide 220
212. Vallecular vittae 1, commissural vittae 2 213
- Vallecular vittae 2–7, commissural vittae 4–14 215
213. Styles long, clearly longer than stylopodium; fruit dorsal ribs acute (or rarely ± obtuse) 214
- Styles short, up to slightly longer than stylopodium; fruit dorsal ribs obtuse or rounded **395. *Seseli***
214. Rays 16–35; fruits 2–2.5 mm long; stem hollow; calyx teeth not developed **212. *Kadenia***
- Rays 4–12; fruits 3–6 mm long; stem solid; calyx teeth usually developed, lanceolate-acuminate **143. *Dethawia***
215. Bracts frequently divided or dentate at apex 216
- Bracts entire 217
216. Low plant, usually 5–15 cm; stylopodium dark; flowers generally pink, red or whitish-greenish **312. *Pachypleurum***
- Plants 20–60 cm; stylopodium whitish; flowers white **121. *Coristospermum***
217. Lower leaves narrowly lanceolate in outline; stylopodium oblong-conical **98. *Carum***
- Lower leaves triangular or ovate in outline; stylopodium low-conical or conical 218
218. Stem procumbent or ascending; rays 2–3(4) **245. *Ligusticum***
- Stem usually erect; rays 4–10(–15) 219
219. Rootstock without fibers; bracteoles often longer than umbellules; stem hollow **272. *Mutellina***
- Rootstock with fibers; bracteoles often shorter than umbellules; stem solid **143. *Dethawia***
220. Plants without main stem, either the umbels born close to the ground, in the center of a leaf rosette, or with a creeping stem rooting at nodes 221
- Plants with a main stem, the umbels ending it or the branches, or sometimes umbels lateral, shortly pedunculate on a ± erect stem 222
221. Annual plants with a taproot; main umbel acaulous, sessile in the center of a leaf rosette; leaves 2–3-pinnate **389. *Sclerosciadium***
- Perennial plant, with a creeping stem rooting at nodes; umbels pedunculate, leaf-opposed; leaves 1-pinnate **199. *Helosciadium***
222. Fruits 6–13 mm long 223
- Fruits 1–5.5 mm long 226
223. Lower leaves 1–2-ternate with wide, ovate or rhombic segments, shallowly lobed in the upper half; plants rarely exceeding 0.5 m; terminal umbel 3–7 cm in diameter **245. *Ligusticum***
- Lower leaves 2–4-divided (pinnate or sometimes ternate), rarely less, with ovate to lanceolate, lobed segments; plants 0.5–2 m; terminal umbel 7–20 cm in diameter 224
224. Fruit ribs unequal, the dorsal developed, subalate, the commissural reduced; endosperm ± concave petals lanceolate, acuminate; commissural vittae invisible **268. *Molopospermum***
- Fruit ribs subequal; endosperm ± deeply sulcate; petals ovate-suborbicular, rounded or obtuse at top; commissural vittae visible, 2–4 225
225. Vallecular vittae 1; rays usually 20–40 **337. *Pleurospermum***
- Vallecular vittae 3–4; rays 12–22 **187. *Grafia***
226. Fruits broadly ovate with usually undulate-crispate ribs; vittae inconspicuous in the ripe fruit **118. *Conium***
- Fruits different, with smooth ribs; vittae visible 227
227. Lower leaves with ultimate leaf segments entire, dentate (sometimes deeply), rarely with very few lobes 228
- Lower leaves with ultimate leaf segments incised, with many lobes 233
228. Fruits subterete; styles very long, erect, as long as or ⅔ as long as the fruit; geophyte, with several tubers distant from the base of stem **289. *Oenanthe***
- Fruits ovoid or ovoid-oblong; styles shorter than ⅔ of the fruit, divergent or reflexed at maturity; roots different 229
229. Rays 2–6 230
- Rays 8–35 231

230. Fruits ovoid-oblong; plants 10–25 cm **362. *Ptychotis***
- Fruits ovoid-globular or ovoid; plants 30–200 cm **405. *Sison***

231. Lower leaves 1–2-ternate or 1-pinnate with 1–2 pairs of lateral segments; fruits ovoid oblong, 4–8 mm 232
- Lower leaves (excluding submerged leaves) 1-pinnate with 3–9(–16) pairs of lateral segments; fruits ovoid, 2.5–4 mm **406. *Sium***

232. Leaves all 1–2-ternate; fruit ribs sharp; vallecular and commissural vittae numerous **245. *Ligusticum***
- Leaves 1-ternate or 1-pinnate; fruit ribs obtuse; vallecular vittae 1, commissural 2 **203. *Hladnikia***

233. Vallecular vittae 3–7, commissural 6–14; plants 5–50 cm 234
- Vallecular vittae 1, commissural 2–4; plants 50–200 cm, rarely less 235

234. Stem simple; bracts often 2–3-fid; fruit ribs very prominent, subalate **312. *Pachypleurum***
- Stem much branched; bracts entire, narrowly linear; fruit ribs weakly or moderately prominent, not subalate **198. *Hellenocarum***

235. Root crowned with fibers; umbels terminal; fruits ovoid; plants often hairy or sometimes glabrous **395. *Seseli***
- Root not crowned with fibers (aquatic plant); umbels terminal or leaf-opposed; fruits ovoid-oblong to subterete; plants always glabrous **289. *Oenanthe***

236. Plants dioecious (or rarely monoecious) **450. *Trinia***
- Plants hermaphrodite 237

237. Geophyte with a single ± globose tuber; stem with a flexuous underground part **90. *Bunium***
- Plants not tuberous, or tuberous but without a stem with a flexuous underground part, or with several fasciculate tubers 238

238. Plants annual or biennial 239
- Plants perennial, often polycarpic, sometimes monocarpic 247

239. Bracteoles present, deflexed, inserted on the outer side of the umbellule or with long soft-ciliate margins, never connate 240
- Bracteoles lacking or different, sometimes connate 241

240. Bracteoles glabrous, not scarious, deflexed, inserted on the outer side of the umbellule; fruits 2–4 mm long **44. *Aethusa***
- Bracteoles with long soft-ciliate margins, broadly scarious, not inserted as above; fruits 5–6 mm long **413. *Sphallerocarpus***

241. Plants stout, low, 6–16 cm; central umbel surrounded by lateral umbels 242
- Plants taller, slender (20–)30–120(–200) cm, rarely less; umbels differently arranged 243

242. Central umbel sessile, with 25–30 rays, surrounded by prostrate stems; fruits glabrous **98. *Carum***
- Central umbel pedunculate, with 6–12 rays; fruits sparsely puberulent **395. *Seseli***

243. Lower leaves 1-pinnate with ovate, dentate or lobed segments 244
- Lower leaves 2–4-pinnate or -ternate with linear or linear-lanceolate segments 245

244. Umbels quite conspicuous, with equal or slightly unequal rays; margin of leaves not cartilaginous **362. *Ptychotis***
- Umbels weakly visible, with very unequal rays; margin of leaves cartilaginous **405. *Sison***

245. Bracteoles present; umbels pedunculate 246
- Bracteoles lacking; umbels sessile or shortly pedunculate **132. *Cyclospermum***

246. Root crowned with fibrous remains; umbels erect before flowering **395. *Seseli***
- Root not crowned with fibrous remains; umbels bent downward before flowering **362. *Ptychotis***

247. Lower (aerial, for aquatic plants) leaves with segments ± broadly ovate or suborbicular, sometimes cuneate, dentate, entire or with very few lobes 248
- Lower leaves with segments and lobes linear to lanceolate with an entire margin, or ovate or lanceolate in outline but trifid or pinnately incised 254

248. Lower leaves 1–4-ternate, triangular in outline 249
- Lower leaves 1–3-pinnate, often oblong in outline but sometimes triangular 250

249. Umbels arranged in a lax panicle; lower leaves 3–4-ternate; plants 1–2 m **452. *Trochiscanthes***
- Umbels alternate, not forming a lax panicle; lower leaves 1–2-ternate; plants up to 0.5–0.9 m **245. *Ligusticum***

250. Vallecular vittae 1, commissural 2, rarely 4 251
- Vallecular vittae 2–6, commissural 4–10 253

251. Rays 20–30 **395. *Seseli***
- Rays 2–15 252

252. Fruits subterete, crowned by obvious, acute calyx teeth; tubers distant from the base of the stem; lower leaves triangular in outline **289. *Oenanthe***
- Fruits ovoid, ± globular or oblong; calyx teeth absent or much reduced; tubers lacking; lower leaves oblong in outline **199. *Helosciadium***

253. Fruit ribs sharp, subalate; stoloniferous plant, glabrous **334. *Pimpinella***
- Fruit ribs obtuse; tufted plant, hairy (at least at the base) **161. *Endressia***

254. Bracteoles lacking or 1–2 255
- Bracteoles several (at least 3) to many, sometimes connate 257

255. Fruits 1.5–4 mm; stem not rigid nor junciform, simple or with few branches 256
- Fruits 8–10 mm; stem rigid, junciform, much divaricately branched **431. *Thamnosciadium***

256. Stem branched; styles short, a little longer than stylopodium; commissural vittae 2 **392. *Selinopsis***
- Stem simple; styles long, notably longer than stylopodium; commissural vittae 4–6 **161. *Endressia***

257. Bracteoles clearly connate, sometimes forming a cup **395. *Seseli***
- Bracteoles free or connate only at the very base 258

258. Fruits hairy 259
- Fruits glabrous or nearly so 260

259. Fruits ovoid-oblong, attenuate at apex, 5–8 mm; ultimate leaf segments ovate, pinnately divided into ovate lobes **78. *Athamanta***
- Fruits ovoid or ellipsoid (sometimes ± narrowly), not attenuate at apex (rarely weakly attenuate), 2–6 mm; ultimate leaf segments often filiform, or linear to lanceolate, more rarely ovate, sometimes with ovate, lanceolate or linear-lanceolate lobes **395. *Seseli***

260. Fruits 7–10 mm; stem junciform, divaricate or leaves ultra-divided into very numerous short capillary, apparently verticillate segments 261
- Fruits 2–7 mm; if 7 mm, stem not junciform and leaves divided into lanceolate or linear, not apparently verticillate segments 262

261. Stem not junciform or divaricate; leaves ultra-divided into very numerous short capillary, apparently verticillate segments; rays usually 5–15; vallecular vittae (2)3–6 **264. *Meum***
- Stem junciform, divaricate; leaves different, less divided into fewer segments; rays 2–5; vallecular vittae 1 **431. *Thamnosciadium***

262. Commissural vittae 2 263
- Commissural vittae 3–8 278

263. Rachis of the cauline leaves not plane, geniculately curved below **100. *Cenolophium***
- Rachis of the cauline leaves plane, not geniculately curved 264

264. Lower leaves with ultimate leaf segments ovate, sometimes obovate-cuneate pinnately incised into ovate, lanceolate or linear lobes 265
- Lower leaves with ultimate leaf segments linear to lanceolate with an entire margin 271

265. Rays numerous, 20–45 266
- Rays less numerous, 2–16 267

266. Lower leaves 2–4-pinnate; calyx teeth not developed **219. *Katapsuxis***
- Lower leaves 1(2)-pinnate; calyx teeth triangular-lanceolate **395. *Seseli***

267. Roots with ovoid, obovoid or cylindrical tubers **289. *Oenanthe***
- Roots without tubers (but sometimes with a strong taproot) 268

268. Lower leaves oblong in outline **98. *Carum***
- Lower leaves triangular in outline 269

269. Lower (submerged) leaves divided into oblong-cuneate segments; fruits ovoid-oblong, 5–6.5 mm; calyx teeth conspicuous, acute **289. *Oenanthe***
- Lower leaves 3-pinnate into lanceolate or linear segments or lobes; fruits ovoid, usually 2.5–4 mm; calyx teeth not developed 270

270. Plants 50–110 cm; fruit ribs clearly prominent, almost alate; stylodia exceeding the stylopodium **393. *Selinum***
- Plants 12–35 cm; fruit ribs not so prominent; stylodia equaling the stylopodium **98. *Carum***

271. Rays numerous, usually 20–30 272
- Rays less numerous, 3–20 273

272. Rays subequal; fruits ovoid; stem often hollow in central part **212. *Kadenia***
- Rays clearly unequal; fruits ovoid- or ellipsoid-oblong; stem solid **395. *Seseli***

273. Roots with ovoid or cylindrical tubers **289. *Oenanthe***
- Roots without tubers 274

274. Lower leaves oblong in outline 275
- Lower leaves triangular or ovate in outline 276

275. Root without fibrous remains of leaves; rays usually 5–6, glabrous; ultimate leaf segments ± divaricate **98. *Carum***
- Root crowned with fibrous remains of leaves; rays often 6–15, up to 20, rarely only 4–6, often scabrous or puberulent, more rarely glabrous; ultimate segments usually not divaricate **395. *Seseli***

276. Petals spreading; weak secondary ribs sometimes present on fruits 277
- Petals rolled inward, ovate or suborbicular, apex inflexed; secondary ribs absent on fruits **395. *Seseli***

277. Fruits 2–3.2 mm; leaves ternately divided; secondary ribs absent on fruits **372. *Rivasmartinezia***
- Fruits 3–6 mm; leaves pinnately divided; weak secondary ribs present on fruits **143. *Dethawia***

278. Rays numerous, 20–50 279
– Rays less numerous, usually 2–20 280
279. Lower leaves 1(2)-pinnate; vallecular vittae 1 **395. *Seseli***
– Lower leaves 3–5-pinnate; vallecular vittae 3–5 **121. *Coristospermum***
280. Lower leaves oblong in outline 281
– Lower leaves triangular or ovate in outline 283
281. Umbels lax; rays 3–8; stylodia equaling the stylopodium **98. *Carum***
– Umbels quite dense; rays usually 6–20; stylodia exceeding the stylopodium (sometimes only a little) 282
282. Sheaths of the lower leaves sulphur-yellow; stem single, always simple, not branched; calyx teeth linear-lanceolate (but sometimes weakly visible) **161. *Endressia***
– Sheaths of the lower leaves green; stem usually with at least some branches; calyx teeth not developed **272. *Mutellina***
283. Ultimate leaf segments ovate, pinnately incised into ovate, lanceolate or linear lobes 284
– Ultimate leaf segments linear with an entire margin 287
284. Vallecular vittae 2; cauline leaves clearly different (linear lobes) from the lower ones **393. *Selinum***
– Vallecular vittae 3–5; cauline leaves similar to lower ones (but reduced) 285
285. Rays 2–3(4), 3–9 mm long; stems several, procumbent to ascending **245. *Ligusticum***
– Rays (5)6–15, 10–40 mm long; stem usually single, erect 286
286. Plants rhizomatous, 0.8–1.2 m; leaf lobes (not segments) 2–4 mm wide, ovate **427. *Tamamschjanella***
– Plants with a rootstock, 0.1–0.6 m; leaf lobes (not segments) 0.5–1.5 mm wide, lanceolate to linear **272. *Mutellina***
287. Leaf segments positioned at the upper part of the lamina (basal part of rachis free); calyx teeth usually lanceolate-acuminate 288
– All parts of the lamina with segments; calyx teeth absent or shortly triangular **272. *Mutellina***
288. Fruits 2–3.2 mm, secondary ribs absent; leaves ternately divided **372. *Rivasmartinezia***
– Fruits 3–6 mm, weak secondary ribs present; leaves pinnately divided **143. *Dethawia***
289. Flowers yellow, yellowish, green or greenish 290
– Flowers white, whitish, pink or pinkish 306
290. Fruits hairy 291
– Fruits glabrous or nearly so 292
291. Plants woody, glabrous; leaf segments narrow, oblong-linear to filiform; vallecular vittae 1 **144. *Deverra***
– Plants herbaceous, pubescent; leaf segments ovate, ± cuneate; vallecular vittae (2)3(4) **334. *Pimpinella***
292. Fruits 10–30 mm **345. *Prangos***
– Fruits 1.2–7 mm 293
293. Plants dioecious (or rarely monoecious) **450. *Trinia***
– Plants hermaphrodite 294
294. Involucre and involucel lacking 295
– Involucel present, involucre lacking or present 299
295. Ultimate segments of the lower leaves wide, oblong-cuneate, ovate or orbicular 296
– Ultimate segments or lobes of the lower leaves narrow, lanceolate to filiform 297
296. Plants robust; upper cauline leaves undivided, ovate-cordate or -truncate, amplexicaulous; endosperm involute **409. *Smyrnium***
– Plants usually slender; upper cauline leaves generally reduced to a linear sheath; endosperm plane or slightly concave **334. *Pimpinella***
297. Ultimate segments of the lower leaves filiform, 0.2–0.5 mm wide; plants annual; vallecular vittae 1 298
– Ultimate segments or lobes of the lower leaves wider; plants biennial or perennial; vallecular vittae (2)3(4) **334. *Pimpinella***
298. Fruit ribs not apparent; rays usually 25–50; fruits ovoid-oblong **371. *Ridolfia***
– Fruit ribs apparent; rays usually 5–10; fruits ovoid **351. *Pseudoridolfia***
299. Lower leaves with segments suborbicular, ovate or ovate-lanceolate, entire (rarely with very few ± irregular lobes), dentate or crenate 300
– Lower leaf dissected into linear segments or lobes 302
300. Lower leaves 3–4-ternate into ovate-lanceolate, markedly acuminate segments; fruits 4–6 mm **452. *Trochiscanthes***
– Lower leaves 1-pinnate into suborbicular or ovate, not markedly acuminate segments; fruits 2.5–4 mm 301
301. Rays 3–6; lower leaves absent at flowering time; vallecular vittae 2–5 **334. *Pimpinella***
– Rays 10–20; lower leaves present at flowering time; vallecular vittae 1 **377. *Rutheopsis***
302. Lower leaves clearly oblong in outline 303
– Lower leaves triangular or ovate in outline 304

303. Plants with a ± globose tuber; fruits subcylindrical, ca. 1 mm wide **157. *Elaeosticta***
- Plants with a taproot and sometimes a rhizome; fruits ellipsoid, sometimes narrowly, usually wider **98. *Carum***

304. Biennial plant; ultimate segments of the lower leaves ovate, often cuneate, lobed **325. *Petroselinum***
- Perennial plant; ultimate segments of the lower leaves linear 305

305. Lower leaves 2–3-ternate; ultimate segments few, 20–100 mm long; vallecular vittae 1, commissural 2 **395. *Seseli***
- Lower leaves 3-pinnate; ultimate lobes very numerous, 2–5 mm long; vallecular vittae 3–4, commissural 4–6 **313. *Palimbia***

306. Fruit secondary ribs present **129. *Cuminum***
- Fruit secondary ribs absent 307

307. Bracteoles present, one spathulate-inflated, the others filiform **53. *Ammoides***
- No spathulate bracteoles present 308

308. Fruiting umbellule cage-like; umbels globose, long-pedunculate; umbellules sessile, with a central bisexual flower surrounded by male flowers **363. *Pycnocycla***
- Flowering and fruiting umbel organization different 309

309. Lower leaves reduced to the fistulate, septate rachis, segments absent or highly reduced; cauline leaves very elongate in outline, with apparently verticillate segments **97. *Caropsis***
- Leaf morphology different; if divided, segments developed 310

310. Plants dioecious (or rarely monoecious) **450. *Trinia***
- Plants hermaphrodite 311

311. Fruits hairy 312
- Fruits glabrous, but sometimes papillate or scabrous 320

312. Involucel lacking or much reduced 313
- Involucel present 316

313. Vallecular vittae 2–8; lower leaves sometimes simple **334. *Pimpinella***
- Vallecular vittae 1; lower leaves never simple 314

314. Perennial shrubby plant, broom-like **144. *Deverra***
- Annual or biennial plant 315

315. Biennial plants 0.5–1.3 m; lower leaves 1-pinnate, oblong in outline; fruits 1.5–2 mm, hairy, not papillate **334. *Pimpinella***
- Annual plants usually 0.1–0.4 m; lower leaves 2–3-pinnate, ovate in outline; fruits 0.6–1.4 mm, hairy-papillate **421. *Stoibrax***

316. Plants shrubby, tortuose divaricately branched or broom-like **144. *Deverra***
- Plants herbaceous 317

317. Lower leaves with ultimate segments ovate to rhombic, dentate or lobed 318
- Lower leaves with ultimate segments linear to lanceolate, entire 319

318. Fruits pubescent throughout; stylopodium low-conical; ultimate segments ± deeply lobed **78. *Athamanta***
- Fruits only pubescent on the ribs, almost glabrous between them; stylopodium narrowly conical; ultimate segments dentate **89. *Bubon***

319. Styles deflexed; fruits not narrowed at apex; rays 2–15, rarely up to 25 **395. *Seseli***
- Styles erect to patent; fruits narrowed at apex; rays 15–50 **78. *Athamanta***

320. Annual or biennial plant 321
- Perennial plant 341

321. Endosperm deeply sulcate, ± strongly arcuate in transverse section 322
- Endosperm convex, plane to weakly concave 324

322. Fruits 1.5–2 mm, ribs thickened above **380. *Scaligeria cretica***
- Fruits 3–6 mm, ribs not thickened above 323

323. Bracts present; fruits 3–3.5 mm long; vittae lacking in ripe fruit; sepals obsolete **118. *Conium***
- Bracts usually absent; fruits 5–6 mm; vittae present in ripe fruit; sepals conspicuous, subulate **413. *Sphallerocarpus***

324. Fruits covered with papillae 325
- Fruits without papillae 327

325. Fruits ovoid, 1.5–2.5 mm **446. *Trachyspermum***
- Fruits broadly ovoid-subdidymous, 1–1.5 mm 326

326. Bracteoles absent **421. *Stoibrax***
- Bracteoles present, 6–7 **266. *Modesciadium***

327. Bracts divided 328
- Bracts entire or lacking 329

328. Ultimate leaf segments narrowly linear, 1–1.5 mm wide, margin entire; one species with nest-like mature umbel **459. *Visnaga***
- Ultimate leaf segments usually wider (up to 40 mm wide), ovate to ± narrowly lanceolate, margin serrate, rarely linear; umbel never nest-like **51. *Ammi***

329. Central umbel sessile, surrounded by erect stems bearing lateral umbels **98. *Carum***
- General disposition of umbels different; main umbel (if present) pedunculate, or umbels lateral and shortly pedunculate or sessile but without a central sessile umbel 330

330. Lower leaves with ultimate segments or lobes linear to linear-lanceolate, entire 331
- Lower leaves with ultimate leaf segments wider, ovate to widely lanceolate, dentate or lobed 334
331. Lower leaves oblong in outline 332
- Lower leaves triangular in outline 333
332. Calyx teeth not developed; young umbels erect before flowering; stem hollow **98. *Carum***
- Calyx teeth short, acute; young umbels reclinate before flowering; stem solid **362. *Ptychotis***
333. Involucre of 5–7 bracts; fruits narrowly elliptic **198. *Hellenocarum***
- Involucre lacking or much reduced; fruits ovoid **395. *Seseli***
334. Ultimate segments of the lower leaves lobed (several lobes, sometimes ± narrow) 335
- Ultimate segments of the lower leaves entire-dentate (rarely with a few lobes) 338
335. Umbels all or partly short-pedunculate or sessile 336
- All umbels pedunculate 337
336. Fruits ovoid; calyx lobes short, narrowly triangular; vittae wide, visible; plants aquatic **289. *Oenanthe***
- Fruits ovoid-pyriform, beaked; calyx lobes absent; vittae invisible; plants terrestrial **64. *Anthriscus***
337. Lower leaves oblong in outline, with segments widely ovate or suborbicular, primary segments sessile; fruits brown **362. *Ptychotis***
- Lower leaves widely triangular in outline, with segments ovate to ovate-triangular acuminate, primary segments long-stalked; fruits blackish or dark-brown **128. *Cryptotaenia***
338. Fruits 4–6 mm; rays usually 10–20, rarely fewer **51. *Ammi***
- Fruits 1.2–3 mm; rays usually 3–12 339
339. Umbels sessile or shortly pedunculate; stem hollow **67. *Apium***
- Umbels pedunculate; stem solid 340
340. Plants glabrous; bracteoles 1–5; petals glabrous **405. *Sison***
- Plants hairy; bracteoles absent; petals hairy **334. *Pimpinella***
341. Plants with a globose or globose-ovoid single tuber; stem often with a flexuous subterranean part 342
- Plants not a geophyte, or if a geophyte, rhizomatous, or with several fasciculate tubers, or tuber single but fusiform, not globose 346
342. Fruits 1–1.5 mm, globose or broader than long 343
- Fruits (2–)2.5–6(–6.5) mm, ovoid- or ellipsoid-oblong 344
343. Fruits broader than long, each mericarp globose; bracts linear-setaceous; stylopodium depressed, almost disc-like **74. *Astomaea***
- Fruits globose or ovoid-globose, each mericarp oblong; bracts usually 2–3-fid; stylopodium conical **415. *Stefanoffia***
344. Fruits always brown-black to black, surface covered with small cones; ribs sharp, quite prominent **180. *Geocaryum***
- Fruits red-brown to brown, rarely brown-black, surface smooth; ribs rounded or filiform, or ± not prominent 345
345. Endosperm plane or slightly sinuous (orthospermous); ribs ± prominent; bracts present, rarely few **90. *Bunium***
- Endosperm sulcate, arcuate in transverse section (campylospermous); ribs not prominent, often filiform; bracts usually lacking (rarely present) **119. *Conopodium***
346. Lower leaves 1–2-ternate into very long (up to 30 cm) linear-lanceolate segments strongly, sharply and regularly cartilaginous-serrate **169. *Falcaria***
- Lower leaves different 347
347. Leaves fleshy; fruits straw-colored, spongy, nearly octagonal in cross section; vittae numerous, most of them invisible, forming a ring inside the fruit around endosperm **127. *Crithmum***
- Leaves not fleshy and fruits different 348
348. Lower leaves entire, lobed or 1-ternate into wide segments 349
- Lower leaves often more divided, 2–4-ternate or 1–4-pinnate into wide to linear segments 350
349. Plants hairy; umbels regular, alternate; lower leaves entire or ternate; rays 8–16 **334. *Pimpinella***
- Plants glabrous; umbels ± irregular, arranged in a leafless panicle or an irregular cymose inflorescence; lower leaves always ternate; rays 2–7(–10) **128. *Cryptotaenia***
350. Endosperm sulcate, ± strongly arcuate (horseshoe-shaped) in transverse section 351
- Endosperm convex, plane or somewhat concave (sometime sinuous), never arcuate in transverse section 353
351. Bracts present; vallecular vittae 1 **331. *Physospermum***
- Bracts lacking; vallecular vittae 2–4 352
352. Fruits 4–6 mm; root tubers several, fasciculate **334. *Pimpinella***
- Fruits 1.5–2 mm; root tuber single, fusiform **380. *Scaligeria***

353. Bracts and bracteoles lacking 354
- Bracteoles present 355
354. Vittae not visible in the ripe fruit; lower leaves 2-ternate **43. *Aegopodium***
- Vittae always visible in the ripe fruit, vallecular vittae 2–6; lower leaves 1–3-pinnate **334. *Pimpinella***
355. Styles erect or slightly spreading (never reflexed), becoming hard at maturity, long, equaling usually ½–1 the length of the fruit; calyx teeth always developed and persisting **289. *Oenanthe***
- Styles usually horizontal or reflexed at maturity, much shorter, equaling usually not more than ¼ the length of the fruit; calyx teeth poorly developed, minute, shortly triangular (rarely developed, ovate or lanceolate, acute) 356
356. Ultimate segments of the lower (aerial) leaves ovate or ovate-lanceolate, sometimes cuneate, entire, dentate, sometimes with few lobes; plants not growing underground 357
- Ultimate segments of the lower (aerial) leaves divided, or ± wide in outline but ± deeply lobed, or entire and linear or linear-lanceolate, sometimes serrate; plants growing partially underground or submerged 360
357. Lower leaves 3–4-ternate into ovate-lanceolate, markedly acuminate segments; umbels arranged in a lax panicle **452. *Trochiscanthes***
- Lower (aerial) leaves 1-pinnate into ovate segments; umbels not arranged in a lax panicle, ± alternate, often leaf-opposed or terminal 358
358. Plants creeping or creeping and then erect, 5–40 cm; rays 2–7; leaf segments up to 1.2 cm long; vallecular vittae 1, commissural 2 **199. *Helosciadium***
- Plants erect, (30)40–150 cm; rays 7–30; leaf segments up to 5–10 cm long; vallecular vittae 3, commissural 4–6, or numerous forming a ring inside the fruit, around the endosperm 359
359. Rhizomatous stoloniferous plant; bracts often dentate or incised; vittae sunken in the pericarp **84. *Berula***
- Root with fasciculate tubers; bracts entire; vittae superficial, not sunken in the pericarp **406. *Sium***
360. Involucre lacking or much reduced 361
- Involucre present 366
361. Lower leaves triangular in outline 362
- Lower leaves oblong in outline 365
362. Root hollow, septate and tuberous, without fibrous remains of leaves; lower leaves with ultimate segments 3–9 cm long, narrowly lanceolate, 0.5–1 cm wide, strongly serrate **113. *Cicuta***
- Root different, crowned with fibrous remains of leaves or solid; lower leaves with ultimate segments or lobes, either much shorter than 3 cm, or linear, up to 3–12 cm long, 1–4.5 mm wide, their margin entire 363
363. Ultimate lobes of the lower leaves 3–10 mm long 364
- Ultimate lobes of the lower leaves 10–120 mm long **395. *Seseli***
364. Plants 12–35 cm; umbels terminal, long-pedunculate; fruits ca. 2.5 mm **98. *Carum***
- Plants 50–150 cm, sometimes more; umbels terminal and lateral, the lateral shortly pedunculate, sometimes sessile; fruits 3.5–6.5 mm **289. *Oenanthe***
365. Lower leaves forming a rosette; umbels long pedunculate, terminal; fruits 2–3 mm **98. *Carum***
- Lower leaves not forming a rosette; umbels shortly pedunculate, leaf-opposed; fruits 1–1.5 mm **199. *Helosciadium***
366. Plants subterranean for the most part (including stem and petioles); lower leaves 3-ternate **205. *Horstrissea***
- Plants not subterranean except the root (no part of stem or petioles underground); lower leaves 1–4-pinnate 367
367. Vallecular vittae 1, commissural 2 368
- Vallecular vittae 3–4, commissural 4–12 369
368. Lower leaves with very numerous (> 20) pairs of primary segments, the lobes appearing as if whorled; herbaceous plant; rays 6–15 **451. *Trocdaris***
- Lower leaves with 2–3 pairs of primary segments; plants woody at base; rays 3–5 **395. *Seseli***
369. Cauline leaves reduced to scale-like petioles; fruits 5–6.5 mm, lateral ribs thick, prominent, the dorsal rib not prominent **313. *Palimbia***
- Cauline leaves like the basal but smaller; fruits (2.5–) 3–5 mm; ribs subequal, filiform, slightly prominent 370
370. Lower leaves triangular in outline, primary segments usually stalked; rays (12–)15–35 **198. *Hellenocarum***
- Lower leaves oblong in outline, primary segments sessile; rays 3–8(–10) **98. *Carum***

C. *Key to the Native and Naturalized Genera of Apiaceae in Russia*

1. Flowers in simple umbels or in dense multi-flowered capitula 2

- Flowers in compound umbels; umbellules with 2--several flowers 4
2. Bracts spiny **35. *Eryngium***
- Bracts soft, not spiny 3
3. Fruits globose, mericarps covered with hooked prickles; ribs obsolete; bracts narrow, herbaceous **38. *Sanicula***
- Fruits elongate, covered with inflated vesicular scales; bracts broad, scarious, eventually translucent **34. *Astrantia***
4. Leaves simple, margin entire **91. *Bupleurum***
- Leaves pinnate or ternate, or pinnatisect or ternatisect; if simple, then toothed or lobed at the margin 5
5. Plants annual with thin roots 6
- Plants perennial or biennial with thickened sometimes tuberous roots, or rhizomes 22
6. Fruits glabrous, smooth or tuberculate 7
- Fruits hairy 14
7. Fruits covered with tubercles 8
- Fruits smooth 9
8. Fruits didymous, mericarps spherical **85. *Bifora***
- Fruits elongate, not didymous, length exceeding width 4–8 times **64. *Anthriscus cerefolium***
9. Bracts prominent, 3 or more 10
- Bracts lacking or 1–2, often caducous 11
10. Bracts pinnatisect, their terminal lobes filiform, long **459. *Visnaga***
- Bracts entire, short **115. *Cnidium monnieri***
11. Bracteoles 6–12; fruits elongate, 3–6 times longer than broad **104. *Chaerophyllum temulum***
- Bracteoles lacking or 1–3; fruits globose to elliptic, as long as broad, or 1.5 times longer than broad 12
12. Terminal lobes of radical leaves filiform **132. *Cyclospermum***
- Terminal lobes of radical leaves broader, linear to ovate 13
13. Mericarp ribs inconspicuous; terminal lobes of upper leaves filiform **120. *Coriandrum***
- Mericarp ribs keeled; terminal segments of stem leaves ovate, divided into linear-oblong lobes **44. *Aethusa***
14. Fruits covered with spines (spines longer than half of fruit width) 15
- Fruits covered with short hairs 18
15. Outer petals in umbellules radiate, 4–7 mm long; petals always white **301. *Orlaya***
- Outer petals not radiate; petals white or pink 16
16. Leaves pinnate **454. *Turgenia***
- Leaves 2–3-pinnate 17
17. Secondary mericarp ribs filiform, covered with prickles arranged in 1–3 rows, with free bases; endosperm curved at commissural side, with deep groove **443. *Torilis***
- Secondary ribs with prickles, fused at their bases; endosperm edges incurved, with deep mushroom-like groove at commissural side **99. *Caucalis***
18. Fruits strongly compressed dorsally **442. *Tordylium***
- Fruits not compressed 19
19. Mericarps beaked 20
- Mericarps not beaked 21
20. Mericarp beak much longer than fertile part; fruits 5–20 times longer than broad; terminal leaf lobes elongate, linear or filiform, $\leq$ 1 mm broad **382. *Scandix***
- Mericarp beak considerably shorter than fertile part; fruits 1.5–8 times longer than broad; terminal leaf lobes broader than 1 mm **64. *Anthriscus***
21. Blades of lower leaves entire, round to reniform, or ternate; endosperm flat at commissural side; stems not swollen below nodes **334. *Pimpinella anisum***
- All leaves pinnatisect; their terminal segments lobed; endosperm edges incurved, with deep mushroom-like groove at commissural side; stems swollen below nodes **275. *Myrrhoides***
22. Mericarps with primary and secondary ribs; secondary ribs winged, keeled to filiform, sometimes covered with series of prickles or bristles 23
- Mericarps with primary ribs only; ribs sometimes inconspicuous 29
23. Secondary mericarp ribs winged, keeled or filiform; without spines, prickles or bristles 24
- Secondary mericarp ribs covered with spines, prickles or bristles 27
24. Leaves 2–4-pinnatisect or 2–4-ternatisect; all flowers with long pedicels; vittae under secondary ribs and at commissure prominent 25
- Leaves pinnate; central flowers in umbellules almost sessile; vallecular and commissural vittae lacking **176. *Froriepia***
25. All mericarp ribs keeled; bracts and bracteoles absent **234. *Laser***
- Primary mericarp ribs sub-inconspicuous; secondary ribs winged; bracts and bracteoles numerous 26
26. Monocarpic hispid or pubescent plants; umbels somewhat contracted at maturity **401. *Silphiodaucus***
- Polycarpic glabrous or pubescent plants; umbels spreading at maturity **235. *Laserpitium***

27. Bracts and bracteoles usually pinnatisect; secondary mericarp ribs with spines in a single series **141. *Daucus***
- Bracts and bracteoles, if present, simple; spines in 1–3 series 28
28. Secondary mericarp ribs filiform, covered with spines or bristles in 1–3 series, not fused at their bases **443. *Torilis***
- Spines in secondary mericarp ribs in a single series, fused at their bases **76. *Astrodaucus***
29. Plants with tubers (entire or lobed) of hypocotyloid or main root origin 30
- Plants with vertical or adventitious roots, vertical or horizontal rhizomes or branched rootstocks, never with tubers of hypocotyloid or main root origin (sometimes tuber-like storage thickenings are situated at lateral roots) 33
30. Leaves pinnate or 2-3-pinnatisect; their segments sessile; petals yellow **157. *Elaeosticta***
- Leaves 2-4-pinnatisect; their segments petiolulate; petals white or brownish-green 31
31. Roots napiform, slightly branched, occurring near soil surface; terminal leaf lobes narrow, filiform; synflorescence paniculate, considerably branched, of numerous umbels; umbel rays strongly unequal **450. *Trinia***
- Tubers usually globose or fusiform, sometimes deeply situated in soil; umbel rays ± equal 32
32. Stems usually hispid, rarely nearly glabrous, stout in lower part (near tubers); bracteoles ciliate **104. *Chaerophyllum***
- Stems glabrous, thin near the tubers and easily breaking; tubers situated deeply in soil; bracteoles glabrous **90. *Bunium***
33. Terminal leaf lobes filiform to linear (5 or more times longer than broad) 34
- Terminal leaf lobes broader, lanceolate to almost round 79
34. Petals yellow or greenish 35
- Petals white, pink or purplish 51
35. Fruits not compressed dorsally 36
- Fruits strongly compressed dorsally 43
36. Bracts absent or 1–2 37
- Bracts 3 or more 39
37. Terminal leaf lobes filiform, $\leq$ 0.5 mm broad **174. *Foeniculum***
- Terminal leaf lobes linear, 1–3 mm broad 38
38. Stems 50–120 cm high, 3–8 mm in diameter; petioles of radical leaves circular in cross section **399. *Silaum***
- Stems up to 40 cm tall, 2–5 mm in diameter, petioles with prominent adaxial groove **458. *Vicatia***
39. Mericarp ribs inconspicuous 40
- Mericarp ribs prominent to filiform, keeled or winged 41
40. Fruits 6–20 mm long, smooth **345. *Prangos***
- Fruits 1.5–3 mm long, covered with tubercles **86. *Bilacunaria***
41. Fruits 10–25 mm long, with winged ribs **345. *Prangos ferulacea***
- Fruits 3–6 mm long, with filiform to keeled ribs 42
42. Plants with branched rootstocks, several stems and fleshy leaves; terminal leaf lobes terete; 25–50 mm long, 5–8 mm broad; vittae numerous; plants of seaside rocks and steeps **127. *Crithmum***
- Plants with short rhizomes and solitary stems; leaves not fleshy, their terminal lobes flat, 3–10 mm long, 0.5–1 mm broad; vallecular vittae 1–2; commissural 2–4; forest and grassland plants **179. *Gasparrinia***
43. Bracts lacking or 1–2 44
- Bracts 3 or more 48
44. Upper stem leaves reduced to sheaths 45
- Upper stem leaves with developed blades 46
45. Central umbel large, 15–20 cm in diameter; mericarp commissure densely covered with fibers; vallecular vittae hardly distinguishable in mature fruits **165. *Eriosynaphe***
- Central umbel smaller, 3–7 cm in diameter; mericarp commissure glabrous; vallecular vittae prominent in mature fruits **171. *Ferula nuda, F. tatarica***
46. Umbel rays strongly unequal; the longest terminal lobes of radical leaves no longer than 35 mm **145. *Dichoropetalum***
- Umbel rays ± equal; the longest terminal lobes of radical leaves exceed 30 mm 47
47. Stylopodia flat or short-conic; vascular bundles 2 or more in marginal mericarp ribs **171. *Ferula***
- Stylopodia conic; vascular bundles solitary in marginal mericarp ribs, situated near rib base **326. *Peucedanum***
48. Bracts and bracteoles leafy; calyx teeth prominent, persistent; vittae numerous, almost cyclic; upper branches verticillate **172. *Ferulago***
- Bract and bracteoles short, linear, sometimes caducous; calyx teeth short, never persistent; vallecular vittae 1–5, commissural 1–4; upper branches alternate, sometimes opposite 49
49. Terminal leaf lobes 1–2 mm long; leaves early withering (at end of flowering time) **313. *Palimbia***

- Terminal leaf lobes longer than 4 mm; leaves persistent 50
50. Umbellules compact, dense, ± globose, with very short rays; exocarp cells large, with thickened convex outer walls; mericarp surface granular-polished **426. *Taeniopetalum***
- Umbellules lax, usually corymbose, their rays ± long, distinct; exocarp cells small, similar to mesocarp cells; mericarp surface dull, smooth **326. *Peucedanum***
51. Ovaries and fruits pubescent 52
- Ovaries and fruits glabrous 53
52. Bracts 3 or more; stems unbranched or with few branches; umbels 1–3 **328. *Phlojodicarpus***
- Bracts 1–2, or lacking; stems branched with several or numerous umbels **395. *Seseli***
53. Bracts 3 or more 54
- Bracts 1–2 or absent 68
54. Fruits strongly compressed dorsally 55
- Fruits terete, not compressed dorsally or laterally 58
55. Stem leaves reduced to sheaths **313. *Palimbia***
- Lower and middle stem leaves with blades 56
56. Exocarp cells large, with thickened convex outer walls; mericarp surface granular polished; terminal leaf lobes 2–8 mm broad; plants with long thin horizontal rhizomes **308. *Ostericum maximowiczii***
- Exocarp cells small, similar to mesocarp cells; fruit surface smooth; terminal leaf lobes 0.5–3 mm broad; thin horizontal rhizomes absent 57
57. Stems and petioles solid; stems unwinged; terminal leaf lobes puberulent along veins, narrow, 0.5 mm broad; bracts absent; bracteoles membranous at margin; calyx teeth obsolete; fruits 2–3 mm long, 1.5–2.5 mm broad **326. *Peucedanum puberulum***
- Stems and petioles hollow; stems ribbed, ribs narrowly winged; terminal leaf lobes glabrous, 1–3 mm broad; bracts entire, reflexed; bracteoles linear, herbaceous; calyx teeth short, triangular; fruits 4–6 mm long, 2.5–3.5 mm broad **435. *Thysselinum***
58. Bracts pinnatisect 59
- Bracts entire or toothed 60
59. Plants 10–45 cm tall; umbels 1 or rarely 2–3; terminal leaf lobes 1.5–2 mm long; vallecular vittae 2–6 **387. *Schulzia***
- Plants 15–100 cm tall; umbels numerous; terminal leaf lobes 5–30 mm long; vallecular vittae solitary **459. *Visnaga***
60. Terminal leaf lobes serrate, large, > 30 mm long 61
- Terminal leaf lobes entire, 5–50 mm long 62
61. Leaves pinnate, with toothed segments; petioles round in cross section; synflorescences corymbose with central umbel larger than lateral ones **406. *Sium suave***
- Leaves 2–3-pinnatisect; their lobes serrate at the margin; petioles grooved at adaxial side; synflorescences semi-spherical in outline, repeatedly branching divaricately **169. *Falcaria***
62. Leaf segments 20–50 mm long 63
- Leaf segments less than 20 mm long 64
63. Leaf segments 1–3 mm broad, thin, mericarp ribs winged; vallecular vittae 1–2 **117. *Conioselinum longifolium***
- Leaf segments 5–8 mm broad; fleshy; mericarp ribs keeled, thick; vallecular vittae numerous **127. *Crithmum***
64. Terminal leaf lobes narrow, 0.2–0.3 mm broad 65
- Terminal leaf lobes broader, 0.5 mm broad 66
65. Mericarp ribs not prominent; radical leaves 2–3-pinnatisect, early withering; petioles of radical leaves prostrate, with broad adaxial groove; umbels solitary; the Caucasus **177. *Fuernrohria***
- Mericarp ribs keeled; radical leaves 3–4-pinnatisect, long persistent; petioles of radical leaves upright, circular in cross section, with small adaxial groove; umbels 1–3; Far East **376. *Rupiphila***
66. Calyx teeth in fruit large, with acute apex, > 0.7 mm long; carpophore lacking **289. *Oenanthe***
- Calyx teeth in fruit inconspicuous or very small, < 0.5 mm long, obtuse; carpophore present 67
67. Fruits large, 5–8 mm long, 2–4 mm broad; dorsal mericarp ribs thick, keeled; polycarpic plants with branching rootstocks **328. *Phlojodicarpus sibiricus***
- Fruits small, 2–4 mm long, 1–2 mm broad; dorsal mericarp ribs thin, narrowly winged; monocarpic plants with taproots **98. *Carum meifolium***
68. Fruits strongly compressed dorsally in transverse section 69
- Fruits not compressed dorsally in transverse section 70
69. Terminal leaf lobes 2–10 mm long; plants with vertical taproot **326. *Peucedanum***
- Terminal leaf lobes > 10 mm long; plants with creeping rhizomes **308. *Ostericum***
70. Mericarp ribs inconspicuous or filiform 71
- Mericarp ribs keeled to winged 73
71. Plants with vertical, hollow rhizomes, divided by transverse partitions; terminal leaf lobes lanceolate to lanceolate-linear, 30–80 mm long, incised at the margins; leaf-blade rachis with ring-shaped

bolsters; umbels globose at flowering and fruiting time **113. *Cicuta***
- Plants with solid, vertical rhizomes or rootstocks; leaf-blade rachis without ring-shaped bolsters; terminal leaf lobes entire at the margin; 2–60 mm long; umbels usually corymbose 72

72. Terminal leaf lobes 2–5 mm long; fruits linear, 8–11 mm long **104. *Chaerophyllum borodinii***
- Terminal leaf lobes 15–60 mm long; fruits ovoid, 2–5 mm long **395. *Seseli glabratum***

73. Leaf rachis geniculate; primary leaf segments reflexed 74
- Leaf rachis not geniculate; leaf segments in one plane 76

74. Umbels with short peduncles, opposite to leaves; calyx teeth in fruit large, > 0.5 mm long; aquatic plant **289. *Oenanthe aquatica***
- Umbels with long peduncles; calyx teeth indistinct or small, < 0.2 mm long 75

75. Stems unbranched or with 1–2 branches, 1–2.5 mm in diameter; petioles with peripheral vascular bundles only **395. *Seseli alpinum***
- Stems branched, 3–5 mm in diameter, with several or numerous umbels; petioles with central (medullar) and peripheral vascular bundles **100. *Cenolophium***

76. Petals dark red or violet **272. *Mutellina***
- Petals white or pink 77

77. Mericarp ribs keeled **395. *Seseli***
- Mericarp ribs winged 78

78. Terminal leaf lobes 3–7 mm long **458. *Vicatia***
- Terminal leaf lobes 20–50 mm long **117. *Conioselinum longifolium***

79. Bracteoles 1–3 or lacking 80
- Bracteoles prominent 96

80. Fruits strongly compressed dorsally 81
- Fruits not compressed, round or elliptic in cross section 88

81. Petals white, greenish or pink 82
- Petals yellow 86

82. Dorsal mericarp ribs keeled to winged **57. *Angelica***
- Dorsal mericarp ribs filiform 83

83. Plants dwarf, 5–14 cm tall, of alpine belt of Caucasus; stems reduced to 2 cm high; fruits not separating into mericarps; vallecular vittae lacking **422. *Symphyoloma***
- Plants > 30 cm tall, growing in lower and middle mountain belts; stems leafy, branching; fruits separating into mericarps; vallecular vittae prominent 84

84. Outer petals in umbels radiate; marginal mericarp ribs slightly thickened in their distal part, with vascular bundles situated in middle rib part; plants 30–70 cm high; stem diameter 3–6 mm 85
- Outer petals not radiate; marginal mericarp ribs not thickened in their distal part, with vascular bundles situated near the rib base; plants 40–120 cm high; stem diameter 6–15 mm **145. *Dichoropetalum caucasicum***

85. Leaves pinnate; their primary segments with large teeth at the margin; vallecular vittae ca. ¾ as long as mericarp; commissural ones < ½ of the mericarp length **201. *Heracleum apiifolium***
- Leaves bipinnate; their segments with small teeth or crenate at the margin; vallecular and commissural vittae as long as the mericarp **256. *Mandenovia***

86. Leaves 3–5-pinnatisect or 3–5-ternatisect **171. *Ferula***
- Leaves pinnate or bipinnatisect 87

87. Plants glabrous (sometimes only stem pubescent under umbel); stem base with collar of remains of petioles of previous years; petioles with peripheral vascular bundles only; vascular bundle of marginal mericarp rib situated near the rib base **145. *Dichoropetalum carvifolia***
- Plants pubescent; stem base without collar of remains of petioles of previous years; petioles with central (medullar) and peripheral vascular bundles; vascular bundle of marginal mericarp rib situated in the middle part of rib **318. *Pastinaca***

88. Petals yellow; upper stem leaves orbicular to ovoid, sessile, amplexicaulous, 3–10 cm long, 2–8 cm broad, endosperm with deep, mushroom-like groove in cross section at commissural side **409. *Smyrnium***
- Petals white, pink or red; upper stem leaves small, not amplexicaulous; endosperm flat or not deeply grooved at commissural side 89

89. Fruits linear, 10–25 mm long; endosperm slightly grooved at commissural side **307. *Osmorhiza***
- Fruits not > 7 mm long; endosperm flat at commissural side 90

90. Umbel rays 2–7, stiffly ascending; strongly unequal, 1–6 cm long; leaves ternate **128. *Cryptotaenia***
- Umbel rays ± equal, not ascending; umbels hemispheric or corymbose; leaves simple, pinnate, bipinnatisect or biternatisect 91

91. Vallecular vittae solitary, persistent in mature fruits 92
- Vallecular vittae 2–5, sometimes becoming obsolete in mature fruits 93

92. Umbels 1.5–2.5 cm in diameter, with short peduncles; primary leaf segments petiolulate **67. *Apium***

- Umbels 3–5 cm in diameter, with long peduncles; primary leaf segments sessile **98. *Carum***

93. Plants with long creeping rhizomes 94
- Plants with vertical roots, branching rootstocks or short rhizomes 95

94. Petals white; vittae in young fruits small cyclic, in mature fruits obsolete; leaves 2–3-ternate; their segments with long petiolules **43. *Aegopodium***
- Petals pink-violet; vittae 3–6 at valleculas, persistent in mature fruits; leaves bipinnatisect with sessile segments **427. *Tamamschjanella***

95. Calyx teeth obsolete **334. *Pimpinella***
- Calyx teeth in mature fruits prominent **414. *Spuriopimpinella***

96. All mericarp ribs equal, filiform or inconspicuous 97
- All or only marginal ribs of mericarps keeled to winged 113

97. Plants stemless (central umbel sessile) or with leafless stem up to 10 cm tall **106. *Chamaesciadium***
- Stems leafy, > 10 cm long 98

98. Terminal umbels compact, almost capitate, surrounded by scarious enlarged sheaths of upper stem leaves and bracts, with violet veins; umbel rays covered with long hairs; petals pale violet **378. *Sajanella***
- Sheaths of upper stem leaves narrow and non-scarious; umbel rays much longer than bracts, glabrous or shortly puberulent; petals white or yellow, rarely red 99

99. Fruits elongate, 2 or more times longer than broad 100
- Fruits globose, ovoid or elliptic; not > 2 times longer than broad 102

100. Petals pink-violet; mericarp ribs filiform **427. *Tamamschjanella***
- Petals white or pink; mericarp ribs broad and low, sometimes hardly distinguishable 101

101. Commissure narrow; mature fruits black, with short beak **64. *Anthriscus***
- Commissure broad; fruits brown or brownish, without beak **104. *Chaerophyllum***

102. Fruits pubescent **395. *Seseli seseloides***
- Fruits glabrous 103

103. Endosperm with deep groove at commissural side 104
- Endosperm flat at commissural side 106

104. Terminal leaf lobes 2–3 mm long; fruits light yellow, 6–10 mm long **345. *Prangos odontalgica***
- Terminal leaf lobes > 10 mm long and broad; fruits 3–6 mm long, brownish or black 105

105. Petals yellow; fruits 5–6 mm long; petioles with small cavity, with peripheral vascular bundles only **158. *Eleutherospermum***
- Petals white; fruits 3–4 mm long; petioles solid, with central (medullar) and peripheral vascular bundles **331. *Physospermum***

106. Leaves pinnate; their primary segments toothed; some species with submerged multisected leaves 107
- Leaves bi- to tripinnate; if pinnate, their segments deeply, ⅓ or more, dissected 108

107. Umbels terminal, situated on long peduncles; umbel rays nearly equal; vallecular vittae 1 or 2–3 **406. *Sium***
- Umbels with short peduncles, opposite to leaves; umbel rays strongly unequal; vittae numerous, cyclic **84. *Berula***

108. Primary leaf segments with long petiolules 109
- Primary leaf segments sessile or with very short (up to 3 mm) petiolules 110

109. Plants with long horizontal stolons; umbels with short peduncles (up to 3 cm), considerably shorter than lateral branches; fruits ovoid **289. *Oenanthe***
- Plants with thick vertical or oblique hollow rhizomes with transepts; all umbels situated on long peduncles > 5 cm; fruits globose **113. *Cicuta***

110. Umbels 0.5–1 cm in diameter, adpressed to main stem; vallecular vittae obsolete **176. *Froriepia***
- Umbels > 1 cm in diameter; synflorescence branches not adpressed to main stem; vallecular vittae prominent 111

111. Umbels 5–8 cm in diameter, with 6–15 rays; calyx teeth prominent; vallecular vittae 3–6 **414. *Spuriopimpinella***
- Umbels ≤ 4 cm in diameter; with 4–7 rays, calyx teeth inconspicuous; vallecular vittae solitary 112

112. Plants 30–110 cm tall; synflorescence of several dozen umbels; leaf primary segments 10–55 mm long, 5–25 mm broad; vittae broad, clavate, half as long as mericarps **405. *Sison***
- Plants 4–35 cm tall; branches and umbels not numerous (1–7); leaf segments 3–8 mm long, 1–3 mm broad, vittae narrow, band-like, almost as long as mericarps **98. *Carum caucasicum***

113. Dorsal mericarp ribs filiform or inconspicuous 114
- Dorsal mericarp ribs keeled to winged 131

114. Vascular bundles of marginal mericarp ribs situated in middle part of rib; petioles with central and peripheral vascular bundles 115
- Vascular bundles of marginal mericarp ribs situated near the base of rib; petioles with or without central vascular bundles 118

115. Vallecular vittae broad and long, filling whole vallecula; marginal ribs with markedly thickened margin **466. *Zosima***
- Vallecular vittae narrower than vallecula; marginal ribs thin or slightly thickened 116

116. Vallecular vittae never clavate, band-like 117
- Vallecular vittae always clavate in lower part **201. *Heracleum***

117. Petals yellow; calyx teeth obsolete; fruits glabrous or sparsely hairy; margin of mericarp marginal ribs thin **318. *Pastinaca***
- Petals white; calyx teeth prominent; fruits covered with hispid appressed hairs; margin of marginal mericarp ribs thin or moniliform **442. *Tordylium***

118. Petals yellow 119
- Petals white, greenish or reddish 120

119. Petioles round in cross section **171. *Ferula***
- Petioles with broad groove on adaxial side **460. *Xanthoselinum***

120. 120 Petioles with central and peripheral vascular bundles 121
- Petioles with peripheral vascular bundles only 125

121. Exocarp cells large with thickened convex outer walls, mericarp surface granular polished **308. *Ostericum***
- Exocarp cells small, similar to mesocarp cells; fruit surface smooth, dull or rugulose 122

122. Rachis of leaves geniculate; primary segments reflexed **299. *Oreoselinum***
- Rachis straight or arched, never geniculate; leaf segments not reflexed 123

123. Petioles with narrow groove on adaxial side; dorsal mericarp ribs inconspicuous **102. *Cervaria***
- Petioles round in cross section; dorsal mericarp ribs prominent, filiform 124

124. Leaves coriaceous **253. *Macroselinum***
- Leaves membranaceous **57. *Angelica***

125. Stems with winged ribs in upper part **435. *Thysselinum***
- Stems not winged 126

126. Exocarp cells large with thickened convex outer walls; mericarp surface granular polished; calyx teeth prominent **308. *Ostericum***
- Exocarp cells small, similar to mesocarp cells; fruit surface smooth or somewhat wrinkled; calyx teeth present or absent 127

127. Terminal leaf segments large, 30–70 mm long, 20–40 mm broad; toothed, entire or with 1–2 lobes **367. *Rhizomatophora***
- Terminal leaf segments < 30 mm long, pinnatilobed or pinnatipartite 128

128. Vallecular vittae solitary **223. *Kitagawia***
- Vallecular vittae 2–4 129

129. Plants polycarpic; bracteoles longer than pedicels **145. *Dichoropetalum affine***
- Plants monocarpic; bracteoles shorter than pedicels 130

130. Petioles circular in transection; leaf blades lanceolate in outline; ultimate segments 3–10 mm long, 0.5 mm broad, acute; calyx teeth inconspicuous **326. *Peucedanum vaginatum***
- Petioles with adaxial groove; leaf blades ovate in outline; ultimate segments 10–30 mm long, 1–6 mm broad, acute; calyx teeth triangular **190. *Haloselinum***

131. Endosperm with deep groove on commissural side 132
- Endosperm flat on commissural side 138

132. Involucel of 1–6 one-sided entire bracteoles, connate at the base; stems round in cross section, red-spotted in lower part **118. *Conium***
- Bracteoles obsolete or arranged regularly, not connate at the base; stems without red spots 133

133. Bracts obsolete; fruits linear, 2.5–10 times longer than broad 134
- Bracts prominent; fruits globose to elliptic; not > 2 times longer than broad 136

134. Plants perennial, polycarpic; fruits 10–25 mm long 135
- Biennials; fruits 5–7 mm long **413. *Sphallerocarpus***

135. Umbels loose, with 2–9 rays; umbellules 3–12-flowered **307. *Osmorhiza***
- Umbels compact, corymbose; central umbels with 10–20 rays; umbellules 20–32-flowered **274. *Myrrhis***

136. Petals yellow; bracts and bracteoles entire **158. *Eleutherospermum***
- Petals white or greenish; bracteoles often toothed or dissected 137

137. Terminal leaf lobes broad, toothed; stems thick, 1–2 cm in diameter, with winged ribs; upper branches opposite or verticillate; stylopodia light-green **337. *Pleurospermum***
- Leaf segments divided into narrow lobes; stems up to 1 cm in diameter, finely striate; branches alternate; stylopodia emerald-green at anthesis **80. *Aulacospermum***

138. Stems with winged ribs 139
- Stems without winged ribs 140

139. Central umbels 8–20 cm in diameter; petioles round in cross section, hollow; terminal leaf segments pinnately incised and toothed **114. *Cnidiocarpa***

- Central umbels 5–7 cm in diameter; petioles with groove on adaxial side, solid or with small cavity; terminal leaf segments pinnatifid **393. *Selinum***
140. Petioles of radical leaves solid 141
- Petioles of radical leaves hollow 156
141. Terminal leaf lobes 5 mm or more broad, entire or with 1–2 incisions 142
- Terminal leaf lobes pinnatisect 148
142. Petioles round in cross section 143
- Petioles with groove on adaxial side 145
143. Terminal leaf lobes entire, fleshy **127. *Crithmum***
- Terminal leaf lobes toothed, membranous 144
144. Terminal leaf segments sessile or decurrent **57. *Angelica***
- Terminal leaf segments with petiolules **241. *Levisticum***
145. Exocarp cells large with thickened convex outer walls; mericarp surface granular polished; calyx teeth prominent **308. *Ostericum***
- Exocarp cells small, similar to mesocarp cells, fruit surface smooth, dull or somewhat wrinkled; calyx teeth obsolete or minute (0.1 mm) 146
146. Terminal leaf lobes 5–20 mm long; petioles with peripheral vascular bundles only **395. *Seseli buchtormense***
- Terminal leaf lobes 25–50 mm long; petioles with central and peripheral vascular bundles 147
147. Plants glabrous **47. *Agasyllis***
- Stems, leaves and fruits pubescent **183. *Glehnia***
148. Petioles round in cross section; ovaries covered with numerous transverse whitish outgrowths (disappearing in mature fruits) **379. *Saposhnikovia***
- Petioles grooved on adaxial side; ovaries glabrous or puberulent 149
149. Plants monocarpic with vertical taproots and solitary stems, or polycarpic with unbranched rootstocks 150
- Plants polycarpic with branched rootstocks and several stems 154
150. All mericarp ribs keeled 151
- Marginal mericarp ribs winged; dorsal ribs keeled to winged 152
151. Central umbel rays strongly unequal; stems leafless, 2–15 cm tall **417. *Stenocoelium***
- Umbel rays ± equal; stems short, leafless, or tall, leafy **395. *Seseli***
152. Marginal mericarp ribs considerably broader than dorsal ribs 153
- Dorsal and marginal mericarp ribs equal, winged **115. *Cnidium***
153. Stem leaves similar to radical ones, bipinnatisect; umbel rays 20–35; commissure narrow **212. *Kadenia***
- Stem leaves considerably smaller than radical leaves; umbel rays 9–15; commissure broad **223. *Kitagawia komarovii***
154. Stem bases covered with rigid, straight petiole remains of preceding years **173. *Ferulopsis***
- Petiole remains at stem bases fibrous, soft 155
155. Fruits 2.5–4.5 mm long, 1–2 mm broad, dorsal and marginal ribs nearly equal **395. *Seseli***
- Fruits 4–9 mm long, 2–5 mm broad, mericarp ribs thick and obtuse: marginal mericarp ribs markedly broader than dorsal ones **328. *Phlojodicarpus***
156. Leaf terminal segments broad, entire or with 1–2 deep incisions 157
- Leaf segments divided into several small lobes 166
157. Leaves ternate or pinnate 158
- Leaves 2–4-pinnatisect or 2–3-ternatisect 161
158. Fruits pubescent **254. *Magadania***
- Fruits glabrous 159
159. Petals dark-red **189. *Halosciastrum***
- Petals white 160
160. Radical leaves ternate **69. *Arafoe***
- Radical leaves pinnate, with 5 or more pairs of lateral segments **406. *Sium***
161. Marginal mericarp ribs winged, considerably broader than keeled or narrow-winged dorsal ones 162
- Marginal and dorsal mericarp ribs equal, either all keeled or all winged 163
162. Umbels numerous; calyx teeth obsolete **57. *Angelica***
- Umbels 1–3; calyx teeth present, short **437. *Tilingia***
163. All mericarp ribs keeled 164
- All mericarp ribs winged 165
164. Plants monocarpic with vertical roots; stems branched from the base **250. *Lithosciadium***
- Plants polycarpic with horizontal or oblique rhizomes; stems branched only in upper part **114. *Cnidiocarpa***
165. Leaf segments broadly ovoid, with obtuse tips, coriaceous **245. *Ligusticum***
- Leaf segments broadly lanceolate, acute, herbaceous **191. *Hansenia***
166. Petioles round in cross section 167
- Petioles with groove on adaxial side 170
167. Plants monocarpic with vertical roots; stems branched from the base **250. *Lithosciadium***
- Plants polycarpic with short rhizomes; stems branched in middle and upper parts 168

168. All mericarp ribs keeled **114. *Cnidiocarpa***
- All mericarp ribs winged 169

169. Calyx teeth obsolete **117. *Conioselinum***
- Calyx teeth prominent **191. *Hansenia***

170. All mericarp ribs keeled 171
- All or only marginal ribs winged 172

171. Umbel rays strongly unequal; carpophore present; calyx teeth absent or < 0.1 mm; mesocarp fragile, without sclerenchyma layer **179. *Gasparrinia***
- Umbels hemispheric or corymbose, with ± equal rays; carpophore absent; calyx teeth > 0.2 mm, persistent; mesocarp solid, with sclerenchyma layer **289. *Oenanthe***

172. Vallecular vittae solitary **212. *Kadenia***
- Vallecular vittae 2–5 or all vittae small, cyclic 173

173. Petals reddish 174
- Petals white 176

174. Styles reflexed on dorsal mericarp side; calyx teeth absent 175
- Styles divergent; calyx teeth present, minute **306. *Orumbella***

175. Mericarp narrow, < 2 mm broad, 2 times longer than broad; ribs filiform; vallecular vittae 2–4, commissural vittae 3–6; commissural endosperm face with broad furrow **458. *Vicatia***
- Mericarp elliptic in outline, > 2 mm broad, 1.5 times longer than broad; ribs keeled or narrowly winged; vittae small cyclic; commissural endosperm face plane **272. *Mutellina***

176. Stems leafless or with one leaf, unbranched; some bracteoles toothed **312. *Pachypleurum***
- Stems leafy, simple or branched; all bracteoles entire **395. *Seseli***

D. *Key to the Native and Naturalized Genera of Apiaceae in Southwestern and Middle Asia*

Note: Imperfect knowledge of the characters of the European genus *Cyathoselinum* precludes its inclusion in this key.

1. Small trees or true shrubs; stems highly woody. Yemen **202. *Heteromorpha***
- Perennial or annual herbs or undershrubs; stems ligneous only in their basal parts 2

2. Umbels always simple, or both simple and compound umbels on the same plant 3
- Umbels always compound 11

3. Simple umbels with peduncles terminal or lateral, petals white 4
- Simple umbels arranged alternately, or verticillately along branches of synflorescence, or simple and compound umbels on the same plant; petals yellow 9

4. Indumentum of stellate hairs. Pakistan, adventive **16. *Bowlesia***
- Indumentum of simple hairs, bristles or outgrowths 5

5. Stems creeping, usually rooting at the nodes, or procumbent; leaves stipulate **4. *Centella***
- Stems erect, stout; never rooting at the nodes, leaves without stipules 6

6. Inflorescence capitulate; fruit dorsal surface covered with whitish, lanceolate or triangular scale-like outgrowths; bracts and often leaves spiny at margins; calyx with prominent spiny teeth **35. *Eryngium***
- Inflorescence umbellate; fruits without scale-like outgrowths, glabrous or covered with hairs, bristles, or vesicles; bracts, leaves and calyx not spiny 7

7. Fruits globose; mericarps covered with hooked bristles; ribs inconspicuous **38. *Sanicula***
- Fruits elongate to ovoid, slightly compressed laterally, covered with scarious or vesiculose outgrowths 8

8. Plants perennial; umbels terminal, many-flowered; all flowers of umbel hermaphrodite, pedicellate; calyx teeth subulate **34. *Astrantia***
- Plants annual; umbels lateral, axillary; central flower of umbel hermaphrodite, sessile, outer flowers male, pedicellate; calyx teeth broadened in upper part, tridentate, persistent **31. *Actinolema***

9. Simple umbels with short peduncles arranged alternately along branches of synflorescence **151. *Dorema***
- Synflorescence loose; all simple umbels arranged verticillately along branches of synflorescence or simple and compound umbels on the same plant 10

10. Fruits scarcely compressed dorsally, with heteromorphic mericarps (one with 4 prominent ribs, the other with 3 prominent ribs); inner mesocarp layer not lignified; plants polycarpic. Uzbekistan **224. *Komarovia***
- Fruits strongly compressed dorsally; mericarps equal, with filiform dorsal and two winged marginal ribs; inner mesocarp layer of prosenchymatous lignified cells; plants monocarpic **171. *Ferula***

11. All leaves entire, often indistinctly differentiated into petiole and lamina; lamina with entire margin and ± parallel curved veins 12
- Leaves pinnately, ternately or digitately dissected, if blades entire, then with distinct petioles, margins toothed or lobed, with pinnate or reticulate veins 13

12. Dwarf, stemless plants; all leaves rosulate; umbels axillary; flowers sessile; calyx teeth prominent, hardened; vittae indistinct in mature fruits **204. *Hohenackeria***
- Plants with prominent stems (rarely almost stemless); flowers in terminal and lateral compound umbels, pedicellate; calyx teeth absent; vittae usually prominent **91. *Bupleurum***

13. Usually only the central flower of the umbellule hermaphrodite, bearing a fruit with a single developed mericarp; outer flowers male, rarely hermaphrodite, their pedicels often fusing with each other and with fruit, rarely free; calyx teeth in mature fruits becoming hard 14
- Flowers in umbellules hermaphrodite and male (rarely only one flower fertile) and fruit with two developed mericarps; bracteoles and pedicels of male flowers not connate with each other and with fruit; calyx teeth not hardening 19

14. Perennials 15
- Annuals 16

15. Flowers in capitate inflorescences; a single fertile flower in each umbellule surrounded by the free pedicels of male flowers; bracteoles free at fruiting; rays very short; dorsal vittae 9 or more **363. *Pycnocycla***
- Flowers in umbellate inflorescences, a single fertile flower in each umbellule connate up to half its length with prominent pedicels of male flowers and bracteole bases; rays not shortened; dorsal vittae 4 **155. *Echinophora***

16. Bracteoles large, triangular-lanceolate, with prominent white margins, completely hiding the umbellules **163. *Ergocarpon***
- Bracteoles never hiding the umbellules, rarely white-margined 17

17. Central umbellules in umbel sterile, dark purple; fruits densely pubescent **146. *Dicyclophora***
- Central sterile umbellule in umbel lacking; fruits glabrous 18

18. Leaves pinnate; bracteoles free; male flowers hardening and completely fused with central fruit at fruiting time; bracteoles and calyx teeth without leafy appendices; endosperm at the commissural side flat to slightly concave **434. *Thecocarpus***
- Leaves bipinnate; central fruit of one fertile mericarp, free, surrounded by male and hermaphrodite flowers and bracteoles adnate to them; some bracteoles and calyx teeth of male flowers with leafy appendices; endosperm deeply sulcate at the commissural side **61. *Anisosciadium***

19. Leaf-blade margins cartilaginous, regularly serrate, with subulate-pointed teeth 20
- Leaf-blade margin without subulate-pointed teeth, not regularly serrate 21

20. Radical and lower stem leaves ternatisect, biternatisect, pinnatisect or bipinnatisect; bracts and bracteoles present; vallecular vittae solitary, commissural 2 **169. *Falcaria***
- Radical and lower stem leaves entire or ternate; bracts and bracteoles absent; vallecular vittae 3–4, commissural 4–8 **185. *Gongylosciadium***

21. Fruits with primary and secondary ribs 22
- Fruits with primary ribs only 45

22. Bracteoles absent 23
- Bracteoles present 24

23. Roots thickened, tuber-like; terminal leaf segments entire, linear; secondary mericarp ribs narrowly winged, marginal ribs broadly winged **432. *Thapsia***
- Roots vertical, cylindrical, not or slightly thickened; terminal leaf segments broad, ovoid to almost round, deeply trilobed, with crenate margins; secondary ribs slightly prominent, ± equal, keeled; marginal ribs keeled, slightly larger then secondary ribs **234. *Laser***

24. All primary and secondary mericarp ribs approximately equal, strongly undulate or folded; bracteoles broadly elliptic, membranous with colored veins **156. *Ekimia***
- Primary mericarp ribs straight; secondary ribs sometimes undulate; bracteoles neither broad, linear to lanceolate, not membranous 25

25. Plants perennial, rarely biennial; primary mericarp ribs slightly prominent, secondary ribs winged, marginal secondary ribs without spathulate projections 26
- Plants annual, rarely biennial; primary and secondary mericarp ribs ± equal, low, or marginal secondary ribs with spathulate projections 28

26. Ultimate leaf segments linear to filiform; bracts inconspicuous; commissural seed face involute **432. *Thapsia***
- Ultimate leaf segments ovate to rhomboid; bracts and bracteoles often conspicuous; commissural seed face plane to slightly concave 27

27. Monocarpic hispid or pubescent plants, rarely perennial; umbels somewhat contracted at maturity **401. *Silphiodaucus***
- Perennial glabrous or pubescent plants; umbels spreading at maturity **235. *Laserpitium***

28. Fruits glabrous 29
- Fruits covered with hairs, bristles, or scales 31

29. Bracts and bracteoles spiny, long, triangular-lanceolate; pedicels short, thickened, flowers almost sessile; primary and secondary mericarp ribs filiform, without spathulate projections **167. *Exoacantha***
- Bracts and bracteoles never spiny; flowers with prominent thin pedicels; primary and secondary mericarp ribs filiform, or marginal secondary ribs with spathulate projections 30

30. Bracts and bracteoles pinnatisect, their terminal segments setaceous or filiform; central flowers of the umbels sterile, forming a turf of blackish or purplish bristles; fruits strongly compressed dorsally; dorsal primary and secondary ribs filiform, marginal secondary ribs with spathulate projections **72. *Artedia***
- Bracts and bracteoles entire, linear to lanceolate, all flowers in the umbel hermaphrodite, fruits ovoid, compressed laterally; primary and secondary mericarp ribs keeled **176. *Froriepia***

31. Secondary mericarp ribs covered with folded scales, arranged in a single series **424. *Szovitsia***
- Fruit surface without folded scales 32

32. Leaves pinnate 33
- Leaves 2–3-pinnatisect 34

33. Dorsal primary mericarp ribs strongly prominent, bearing triangular spines or teeth, arranged in a single series; secondary ribs sub-inconspicuous, covered with a few much smaller bristles or tubercles **249. *Lisaea***
- Dorsal primary and secondary ribs ± similar, distinctly prominent, secondary ribs bearing equally long rigid spines in one or two series **454. *Turgenia***

34. Secondary mericarp ribs with one (rarely 3) series of rigid, glochidiate spines 35
- Secondary mericarp ribs with 3–5 series of soft bristles or T-like hairs, never glochidiate 42

35. Mericarps not compressed dorsally; secondary mericarp ribs with 1–3 series of bristles, not confluent at the base or with two series of bristles, bracts usually absent 36
- Mericarps clearly compressed dorsally; secondary mericarp ribs setose; bristles arranged in one (rarely 2) series; bracts present 39

36. Secondary mericarp ribs with 1–3 series of bristles, not confluent at the base 37
- Secondary mericarp ribs with bristles, confluent at the base and arranged in two series 38

37. Plants glabrous; terminal leaf segments filiform; calyx teeth inconspicuous **455. *Turgeniopsis***
- Plants pubescent; terminal leaf segments linear to ovate, lobed; calyx teeth prominent, narrowly triangular, sometimes unequal **443. *Torilis***

38. Umbel rays 2–3, short; leaf blades sessile on short sheaths; bracteoles herbaceous, glabrous **99. *Caucalis***
- Umbel rays 8–15, long; sheaths of upper stem leaves elongate, broadened, membranous; bracteoles membranous, with ciliate margins **76. *Astrodaucus***

39. Secondary mericarp ribs bristly; bristles rigid, their bases forming a thick wing; bracts entire or rarely trifid 40
- Bristles on secondary mericarp ribs thin, forming a thin wing at their bases; bracts dissected 41

40. Outer petals in umbellules prominently radiate; stems almost glabrous, branching from middle part, bracts entire **301. *Orlaya***
- All petals ± equal (outer not radiate); plants low, densely pubescent, stems branching almost from the base; bracts partly trifid **352. *Pseudorlaya***

41. Umbel rays contracted in fruit; calyx teeth obtuse, bristles on the secondary mericarp ribs straight or glochidiate **141. *Daucus***
- Umbel rays not contracted in fruit; calyx teeth with pointed apex; bristles on the secondary mericarp glochidiate. Yemen **48. *Agrocharis***

42. Bracts absent; bristles pointed, long, exceeding seed diameter 5–7 times **105. *Chaetosciadium***
- Bracts present; bristles shorter, smooth or clavate 43

43. Stems and leaves densely puberulent, terminal leaf segments elongate, cuneate at base; fruits densely covered with long, thin, hammer-like bristles **347. *Psammogeton***
- Stems and leaves glabrous; terminal leaf segments setaceous to oblong, not cuneate at base; fruits covered with thin, long, pointed hairs 44

44. Terminal leaf segments oblong, toothed, cuneate at base; petals dorsally pubescent; stylopodia hairy; secondary mericarp ribs hardly visible. Afghanistan **365. *Registaniella***
- Terminal leaf segments setaceous to filiform; petals and stylopodia glabrous; secondary mericarp ribs more or less prominent **129. *Cuminum***

45. Plants annual or biennial 46
- Plants perennial, monocarpic or polycarpic 79

46. Leaves narrowly oblong in outline; umbels globose, dense; umbellules uniflowered; calyx teeth, bracts and bracteoles pectinate; fruits of one mericarp;

vallecular vittae very short, commissural vittae reaching half the fruit length **232. *Lagoecia***
- Leaves broad, ovate to lanceolate in outline, umbels not globose; umbellules of several–many flowers; calyx teeth entire, sometimes obsolete; bracts and bracteoles obsolete or of various shape, but never pectinate 47

47. Umbels of 2–5 rays, umbellules of 2–3 flowers 48
- Umbel rays and umbellule flowers more numerous 49

48. Plants glabrous or sometimes minutely scabrous; leaves 3–4-pinnatisect; pedicels short, thickened; fruits covered with warts **228. *Krubera***
- Plants pubescent; leaves bipinnatisect; pedicels 2–4 mm long, thin; fruits pubescent **369. *Rhopalosciadium***

49. Bracteoles unilaterally arranged 50
- Bracteoles arranged regularly around umbellules, or obsolete 52

50. Bracteoles longer than pedicels; fruits almost globose with keeled ribs **44. *Aethusa***
- Bracteoles shorter or almost equal to pedicels; fruits ovoid with short-winged or filiform ribs 51

51. Bracteoles not connate at the base, one spathulate-inflated, the others filiform; terminal leaf segments linear; fruits with filiform ribs **53. *Ammoides***
- Bracteoles connate at the base, elliptic to lanceolate, equal; terminal leaf segments ovate; fruits with short-winged ribs **118. *Conium***

52. Leaves ternate; umbel rays 2–7, very unequal, stiffly ascending; pedicels ascending; bracteoles obsolete **128. *Cryptotaenia***
- Leaves pinnate, umbels with larger number of equal or slightly unequal rays, umbel rays and pedicels not ascending; bracteoles usually present 53

53. Fruits strongly compressed dorsally 54
- Fruits not or slightly compressed dorsally, or compressed laterally, or terete 59

54. Dorsal mericarp ribs inconspicuous; vallecular vittae obsolete; endosperm with deep broad groove at commissural side 55
- Dorsal ribs filiform, prominent, vallecular vittae usually present; endosperm almost flat on commissural side 56

55. Marginal mericarp ribs strongly thickened **215. *Kalakia***
- Marginal mericarp ribs filiform **133. *Cymbocarpum***

56. Margins of marginal mericarp ribs corky, thickened, strongly moniliform to smooth 57
- Marginal mericarp ribs thin, or, if slightly thickened, then not moniliform 58

57. Terminal leaf segments filiform to setaceous; always entire, inner lignified mesocarp layer obsolete; vittae in mature fruits absent **303. *Ormosciadium***
- Terminal leaf segments ovate to cordate, sometimes dissected; inner mesocarp layer lignified, of prosenchyma cells; vallecular vittae 1–3, commissural 2–10 **442. *Tordylium***

58. Leaves pinnate, their segments sessile, ovate to elliptic, dentate; inner mesocarp layer lignified, of prosenchyma cells **318. *Pastinaca***
- Leaves 3–4-ternatisect; their terminal segments linear to filiform, petiolulate; inner lignified mesocarp layer obsolete **56. *Anethum***

59. Fruits glabrous 60
- Fruits covered with hairs or vesiculose outgrowths 70

60. Bracts and bracteoles obsolete 61
- Bracts and bracteoles, or only bracteoles, prominent 66

61. Plants biennial with vertical, slightly thickened roots; leaves ternate or pinnate, segments broadly ovate to elliptic, cuneate at the base; upper leaves often perfoliate 62
- Plants annual with thin roots, rarely biennial; leaves pinnatisect, with linear to filiform terminal segments, upper leaves never perfoliate 63

62. Umbels opposite leaves or axillary; leaf margins dentate; fruits not didymous; ovoid in outline; mericarp ribs keeled **67. *Apium***
- Umbels terminal and lateral; leaf margins entire or crenate; fruits almost didymous; consist of two almost globose mericarps with narrow commissure, dorsal mericarp ribs filiform; marginal ribs inconspicuous **409. *Smyrnium***

63. Petals yellow; fruits elongate-cylindric **371. *Ridolfia***
- Petals white; fruits ovoid, globose or almost didymous 64

64. Fruits didymous; mericarp ribs and vittae inconspicuous **85. *Bifora***
- Fruits ovoid to globose, never didymous; mericarp ribs prominent; vallecular vittae solitary, commissural 2 65

65. Leaves almost triternatisect; their terminal segments with prominent petiolules; fruits globose **132. *Cyclospermum***
- Leaves 2–3-pinnatisect; their terminal segments sessile; fruits ovoid to elongate **98. *Carum***

66. Bracts present 67
- Bracts absent 69

67. Bracts entire; fruits slightly compressed laterally; dorsal mericarp ribs keeled to winged **405. *Sison***
- Bracts pinnatisect or ternate; fruits slightly compressed dorsally; mericarp ribs filiform 68
68. Terminal leaf segments linear to filiform, entire at margin, up to 1 mm wide; umbel rays thickened; rigid at fruiting **459. *Visnaga***
- Terminal segments of radical and lower stem leaves obovate to elliptic, 4–20(–30) mm wide, serrate at margins; umbel rays not thickened, thin at fruiting **51. *Ammi***
69. Fruits elongate to almost linear; lower leaves with filiform, terminal segments entire at margin, upper leaves with lanceolate terminal segments; mericarp ribs extending at apex into sepal-like protrusions; vallecular vittae solitary, commissural vittae 2 **265. *Microsciadium***
- Fruits almost globose; terminal segments of lower stem leaves broad, toothed; ribs at apex without sepal-like protrusions; vittae in mature fruits absent **120. *Coriandrum***
70. Bracteoles absent **334. *Pimpinella***
- Bracteoles prominent 71
71. Bracts present 72
- Bracts absent 77
72. Bracts and bracteoles entire, extremely rarely dentate 73
- Bracts and bracteoles dissected or dentate 76
73. Bracts and bracteoles reflexed in fruit; umbel rays thickened, curved toward the outside; fruits strongly compressed laterally, covered with big vesiculose, often lobed outgrowths **162. *Eremodaucus***
- Bracts and bracteoles not reflexed in fruit; umbel rays thin, straight in fruit; fruits terete, convex dorsally, scarcely compressed laterally, glabrous, or covered with unicellular hairs or bladder-like whitish papillae 74
74. Petals glabrous **65. *Aphanopleura***
- Petals dorsally pubescent 75
75. Fruits covered with bladder-like whitish papillae; mericarp ribs filiform, almost obscure. S Asia **446. *Trachyspermum***
- Fruits covered with unicellular hairs; mericarp dorsal ribs filiform, marginal ribs broader. Yemen, Oman **41. *Adenosciadium***
76. Plants procumbent, pilose; roots slender; bracts ternate, petals pubescent; fruits densely hairy. Iran, Turkey, Iraq, Syria, Jordan **291. *Oliveria***
- Plants with straight stems, glabrous; roots fleshy; bracts pinnatisect; petals glabrous; fruits vesiculose. Saudi Arabia, Yemen **298. *Oreoschimperella***
77. Stems swollen below nodes; fruits without beak, slightly constricted below stylopodia **275. *Myrrhoides***
- Stems not swollen below nodes; fruits beaked, beak prominent, ± long 78
78. Fruits compressed laterally, their beaks considerably longer than or at least as long as fertile part; fruit ribs prominent **382. *Scandix***
- Fruits not compressed laterally; their beaks ⅓–½ as long as fertile part; fruit ribs obscure **64. *Anthriscus***
79. Subterranean part of plant ending in hypocotyloid tuber or solitary tuberous root, entire or lobed 80
- Plants with vertical roots, vertical or horizontal rhizomes, or branched rootstocks; sometimes there are tuberiform thickenings only at lateral roots 105
80. Primary leaf segments sessile 81
- Primary leaf segments petiolulate 86
81. Bracts and bracteoles ciliate at margins; fruit ribs transversely folded **302. *Ormopterum***
- Bracts and bracteoles glabrous; fruit ribs without transverse folds 82
82. Petals and fruits pubescent, hairs short; stems covered with thin hairs; roots branching; lateral roots swollen, bearing tuber-like fusiform thickenings. Afghanistan **186. *Gongylotaxis***
- Petals and fruits glabrous; very rarely fruits covered with papillae; stems glabrous, rarely puberulent; root tubers of principal root globose, ovoid, fusiform, or cylindrical, usually not branching 83
83. Mericarp ribs obscure; exocarp cells large, considerably larger than mesocarp cells; commissure narrow **157. *Elaeosticta***
- Mericarp ribs prominent; exocarp cells small, similar to those of mesocarp; commissure broad 84
84. Dorsal and marginal mericarp ribs approximately equal, short-keeled, not thickened, both sometimes undulate; exocarp slightly extending at the commissural mericarp side; petals with single secretory ducts and free tip **206. *Hyalolaena***
- Dorsal mericarp ribs filiform, marginal ribs broadly winged, never undulate with thickened distal part; exocarp interrupted near the ends of marginal ribs; petals with several secretory ducts; their tip attached to petal blade 85
85. Fruits elliptic in outline; terminal leaf segments white-mucronulate **267. *Mogoltavia***
- Fruits oblong to oblanceolate-linear; teeth of terminal leaf segments without white prickles **288. *Oedibasis***
86. Fruits elliptic in outline, strongly compressed dorsally; dorsal mericarp ribs filiform, marginal ribs winged 87

- Fruits globose, ovoid, elongate or lanceolate, not compressed dorsally, all ribs equal, rarely unequal 88
87. Leaves pinnate or bipinnatisect; terminal umbels considerably broader than lateral ones; vallecular vittae solitary, commissural 2; mericarp marginal ribs slightly inflated, with solitary vascular bundles **239. *Leiotulus***
- Leaves repeatedly ternatisect; all umbels approximately equal; vittae 3 per vallecula, 6 at the commissure; mericarp marginal ribs not inflated, mericarp ribs with several vascular bundles **171. *Ferula sibirica***
88. Roots napiform, slightly branched, occurring near soil surface; terminal leaf segments narrow, filiform; synflorescence paniculate, considerably branched, of numerous umbels; umbel rays strongly unequal; mericarp ribs thickened at the base **450. *Trinia***
- Underground tubers usually globose, sometimes situated deeply in soil; terminal leaf segments usually broader, rarely filiform; synflorescence corymbose, umbel rays ± equal; mericarp ribs not thickened at the base 89
89. Upper part of fruit narrowed in short beak; vittae in mature fruits indistinct 90
- Upper part of fruit not narrowed in beak; vallecular vittae always prominent 92
90. Fruits covered with entire or branched rigid bristles, sitting on scabrous cylindric tubercles; mericarps slightly compressed dorsally; outer petals radiate **226. *Kozlovia***
- Fruits glabrous or covered with entire hairs or tubercles; mericarps compressed laterally, convex dorsally; petals non radiate or outer ones slightly radiate 91
91. Fruits pyriform, shining, covered with tubercles; exocarp of 2–3 layers of cells with lignified walls. Middle Asia **227. *Krasnovia***
- Fruits almost cylindrical, brown, glabrous or sparsely covered with entire hairs; exocarp unilayered. Himalaya **278. *Neoconopodium***
92. Stems and/or sheaths of stem leaves pubescent 93
- Plants glabrous 94
93. Stems usually ± hispid, rarely almost glabrous; bracteoles ciliate; mericarp ribs obtuse; vallecular vittae solitary, commissural 2; crown of bristles below fruit base absent **104. *Chaerophyllum***
- Stems glabrous; bracteoles glabrous; mericarp ribs acute; vittae obsolete; crown of bristles below fruit base present **180. *Geocaryum***
94. Tuberous root globose, entire, rarely fusiform 95
- Tuberous root elongate to cylindric, often lobed 102
95. Fruits globose, or almost didymous 96
- Fruits ovoid to elongate, never didymous 99
96. Calyx teeth present; stylopodia long-conical; terminal leaf segments ovate to rhombic **380. *Scaligeria***
- Calyx teeth absent; stylopodia short-conical; terminal leaf segments filiform to linear, rarely ovate 97
97. Endosperm flat at commissural side; terminal leaf segments filiform, long; upper part of fruit constricted under stylopodia; synflorescence paniculate; central umbel ± approximately equal to lateral ones. W Turkey **344. *Postiella***
- Endosperm deeply grooved on commissural side; terminal leaf segments linear, short; upper part of fruit not constricted under stylopodia; synflorescence corymbose; central umbel markedly broader than lateral ones 98
98. Bracts and bracteoles obsolete or 1–2. Styles up to 1 mm long. Mericarps in maturity fruits straight; ovary and seedlings dicotyledonous **75. *Astomatopsis***
- Bracts and bracteoles 3 or more. Styles more 1.5 mm long. Mericarps in maturity across curved; ovary and seedlings monocotyledonous **74. *Astomaea***
99. Tuberous roots globose or fusiform; mericarps clearly ribbed; their ribs shortly winged to terete; vittae numerous, nearly cyclic **198. *Hellenocarum***
- Tuberous roots always globose; mericarp ribs inconspicuous or filiform; vallecular vittae solitary to several, commissural two to several 100
100. Bracts and bracteoles dissected; petals yellow (rarely white), their tips attached to petal blades **415. *Stefanoffia***
- Bracts and bracteoles entire, or bracts absent; petals white, their tips free 101
101. Commissure narrow, seedlings monocotyledonous **90. *Bunium***
- Commissure broad, seedlings dicotyledonous **159. *Elwendia***
102. Fruit bases cordate, mericarp ribs sub-inconspicuous; commissural side of endosperm sulcate; petals emarginate; their tips attached to petal blades **225. *Korshinskia***
- Fruit bases not cordate; their ribs prominent to winged; commissural side of endosperm almost flat; petals not emarginate, with free tips 103
103. Petals yellow, very rarely white (*G. gracilis*); dorsal mericarp ribs thin, marginal ribs winged, or all mericarp ribs inconspicuous to filiform; commissure broad **178. *Galagania***

- Petals white or rose; mericarp ribs keeled to winged, commissure narrow 104

104. Bracts obsolete or 1–2 **396. *Seselopsis***
- Bracts well developed, several to numerous **320. *Paulita***

105. Upper leaves frequently simple, sometimes perfoliate **409. *Smyrnium***
- Upper leaves dissected, not perfoliate 106

106. Leaf rachis geniculate; terminal leaf segments 15–60 mm long 107
- Leaf rachis not geniculate; when geniculate (*Seseli alpinum*), terminal leaf segments 2–8 mm long 108

107. Stems solid; mericarp ribs keeled, marginal ribs somewhat broader than dorsal; exocarp cells small, similar to mesocarp cells **100. *Cenolophium***
- Stems hollow; marginal mericarp ribs winged, dorsal ribs filiform; exocarp cells large, with considerably thickened outer walls **308. *Ostericum***

108. Leaf sheaths remarkably inflated, herbaceous, amplexicaulous; terminal leaf segments broad, ovate to rhombic, few-toothed or crenate at margin; bracts absent; petals white or rarely greenish-white 109
- Leaf sheaths usually not inflated, with terminal segments of various shape; bracts and bracteoles present or absent; if sheaths inflated, then petals yellow; petals white, yellow or pink 111

109. Commissure broad, exocarp interrupted near bases of marginal ribs. Georgia **112. *Chymsydia***
- Commissure narrow, exocarp interrupted near carpophore 110

110. Marginal mericarp ribs broadly winged; dorsal keeled or terete, acute; vallecular vittae 1, commissural 2, or vittae cyclic **57. *Angelica***
- Dorsal mericarp ribs triangular in transverse section or shortly keeled, obtuse, marginal slightly broader; vittae numerous, almost cyclic **47. *Agasyllis***

111. Stems solitary, with considerably shortened lateral branches; leaves numerous (up to 25); radical and lower stem leaves 5–6-ternatisect, other stem leaves considerably reduced; bracts and bracteoles usually rigid, membranous; late flowering plants (August) E Iran, W Afghanistan **258. *Mastigosciadium***
- Stem branches well developed; leaves less numerous, usually with developed blades, rarely with reduced blades; bracts and bracteoles not rigid, glumaceous; plant usually flowering in spring or early summer 112

112. Stem leaves reduced to lanceolate sheaths 113
- All or only the lower stem leaves with developed blades; sometimes only uppermost leaves reduced to lanceolate sheaths 114

113. All umbels nearly equal, with bracts and bracteoles, or with bracteoles only; leaves pinnatisect; their terminal segments short, lanceolate to linear; roots vertical, slightly and evenly thickened; vallecular vittae 2–5, commissural vittae 6 **313. *Palimbia***
- Central umbel significantly larger than lateral ones; all umbels without bracts and bracteoles; leaves ternatisect with long (ca. 2 cm) filiform terminal segments; principal and lateral roots with tuberiform thickenings; vallecular vittae solitary, small, commissural 2, in mature fruits scarcely visible **165. *Eriosynaphe***

114. Terminal umbels compact, almost capitate, surrounded by enlarged sheaths of upper stem leaves and bracts, white-membranous with violet veins; petals pale violet. Mongolia **378. *Sajanella***
- All umbels usually loose, not capitate; sheaths of upper stem leaves narrow and non-membranous; bracts linear to lanceolate, herbaceous, rarely ovate and white-membranous; petals white or yellow, rarely red 115

115. Fruits didymous with very narrow commissure, with filiform or inconspicuous ribs 116
- Fruits not didymous 123

116. Bracts and bracteoles absent; umbels globose **412. *Sphaerosciadium***
- Bract and bracteoles present; umbels corymbose 117

117. Terminal leaf segments ovate to broadly lanceolate, usually lobed 118
- Terminal leaf segments filiform or linear to narrow lanceolate 119

118. Plants with swollen, thickened roots; petals with tips curved inward and adnate (attached) to petal lamina; vittae in immature fruits very large; seeds on commissural side with narrow, cleft-like groove **225. *Korshinskia***
- Plants with cylindrical taproots; petals with tips not adnate to lamina; vittae in immature fruits small; seeds on commissural side deeply sulcate **331. *Physospermum***

119. Stylopodia in mature fruits covered by enlarged pericarp tissue; mericarp ribs inconspicuous, mesocarp bistratous, inner mesocarp layer consisting of five groups of large parenchyma cells with porous walls **345. *Prangos***

- Stylopodia clearly visible, mericarp ribs filiform to shortly keeled; pericarp with layer of strongly lignified sclerenchymous cells 120
120. Umbellules with a single central sessile hermaphroditic flower, surrounded by pedicellate lateral males flowers 121
- Umbellules with several hermaphroditic flowers, and all flowers pedicellate **385. *Schrenkia***
121. Fruits glabrous **386. *Schtschurowskia***
- Fruits covered with rigid bristles or short prickles 122
122. Fruits easily separating into mericarps, dorsally covered with rigid bristles; stems prominent; stylopodia short conical. Pamiro-Alai **248. *Lipskya***
- Fruits not separating into mericarps at fruit maturity, dorsally covered with short prickles; plants almost stemless; stylopodia conical. N Tianschan **390. *Sclerotiaria***
123. Fruits terete, or compressed laterally 124
- Fruits strongly compressed dorsally 203
124. Endosperm with broad or ± deep groove on the commissural side 125
- Endosperm almost flat on the commissural side 147
125. Bracteoles absent **317. *Parasilaus***
- Bracteoles present 126
126. Bracteoles broad, ovoid, exceeding umbellules, usually scarious 127
- Bracteoles herbaceous, linear to lanceolate, usually shorter than pedicels 129
127. Bracts present **207. *Hymenidium***
- Bracts absent 128
128. Leaves pinnate with almost sessile segments **208. *Hymenolaena***
- Leaves ternate; their basal segments long-petiolulate **207. *Hymenidium corydalifolium***
129. Bracts present 130
- Bracts absent 134
130. Fruits ovoid, densely covered with hooked prickles; vittae numerous, almost cyclic; sheaths of stem leaves inflated, membranous **236. *Lecokia***
- Fruits elongate to elongate-linear, glabrous or covered with tubercles and bristles; sheaths of stem leaves not inflated; vallecular vittae solitary or indistinct 131
131. Plants glabrous; basal leaf segments with very short petiolules, sometimes subsessile; vallecular vittae always present, solitary **458. *Vicatia***
- Plants often pubescent, rarely almost glabrous; basal leaf segments with prominent petiolules; vallecular vittae indistinct in mature fruits 132
132. Mericarps at base with long, narrowed, sterile caudate part, almost half as long as fruit; umbel rays long, thin, divaricate, 12–25 cm long **307. *Osmorhiza***
- Mericarps at base without long, narrowed, sterile caudate part; umbels usually corymbose, rays shorter, not divaricate 133
133. Fruits glabrous, elongate, without beak; mericarp ribs obtuse and flat, but visible; vallecular vittae 1; commissure broad **104. *Chaerophyllum***
- Fruits with short beak, glabrous or covered with hairs or bristles, sitting on tubercles, oblong-linear to almost linear; mericarp ribs sub-inconspicuous (visible only in beak); vittae in mature fruits obsolete; commissure narrow **64. *Anthriscus***
134. Petals white 135
- Petals yellow 138
135. Leaves triternatisect; their terminal segments narrowly linear to cylindric, rigid, divaricate **388. *Sclerochorton***
- Leaves pinnate, terminal leaf segments flat, soft, not divaricate 136
136. Mericarps with ribs inflated at the base and narrow furrows; rib secretory ducts always present, more than one per each rib; leaves usually puberulent along veins on the lower surface and at the margin **337. *Pleurospermum***
- Mericarps with ribs not inflated at the base and broad furrows; rib secretory ducts solitary, not visible in each rib; leaves usually glabrous 137
137. Bracteoles usually entire, lanceolate to lanceolate-linear, or, if toothed, mericarp with solitary vittae per furrow and seed with deep groove on the commissural side **80. *Aulacospermum***
- Bracteoles toothed or pinnatifid; if entire, mericarp with 2–3 vittae per furrow and seed flat or almost flat on the commissural side **207. *Hymenidium***
138. Terminal leaf segments ovate to lanceolate, crenate or few-toothed at the margin; fruits glabrous, sometimes with cuneate bases, heteromericarpous; marginal mericarp ribs winged, dorsal ribs either filiform (in one mericarp) or winged (in the other mericarp); seed occupies only ⅓–⅔ of upper part of the mericarp, lower part sterile **200. *Heptaptera***
- Terminal leaf segments narrowly linear, entire or ovate to lanceolate, incised; fruits glabrous or pubescent; both mericarps of a fruit similar, not cuneate at the base; all mericarp ribs equal, filiform to winged; seed occupies the whole mericarp cavity 139
139. Terminal leaf segments narrow, often long, linear, rarely lanceolate, toothed 140

– Terminal leaf segments ovate to lanceolate, sometimes decurrent, lobed or toothed at the margin 144

140. Plants glabrous; synflorescence paniculate, composed of numerous small umbels; mesocarp homogeneous, of parenchymatic cells; vallecular vittae solitary to several. E Iran, Afghanistan **93. *Calyptrosciadium***

– Plants pubescent; synflorescences corymbose, with larger central umbel; mesocarp often of two distinct layers, with inner layer of prosenchymatous lignified cells; vittae numerous, cyclic 141

141. Fruits tuberculate or vesiculose; mericarp ribs obtuse, keeled, with large vascular bundles; mesocarp of similar parenchymatous cells, having no lignified walls; seed with deep groove with 2(3) large secretory cavities in the funicle **86. *Bilacunaria***

– Fruits smooth, glabrous or tomentose or with vesiculose outgrowths; fruit ribs winged, straight or undulate; mesocarp two-layered; vascular elements dispersed in inner mesocarp layer; large secretory cavities on the commissural side in the funicle obsolete 142

142. Fruit ribs straight, dentate and crested; mericarps smooth, without furrows; large sclerenchyma tissue situated in distal parts of mericarp of ribs; inner mesocarp layer consists of large parenchymatous cells **92. *Cachrys***

– Fruit ribs keeled, undulate or straight, or obsolete; mericarp furrows sometimes with vesiculose outgrowths; inner mesocarp layer consisting of five groups of large parenchyma cells with porous walls; sclerenchyma tissue in mericarp ribs obsolete 143

143. Terminal leaf segments rigid; bracts and bracteoles tomentose; fruits almost sessile, tomentose, covered with dense multicellular hairs, sitting on tubercles **50. *Alococarpum***

– Terminal leaf segments soft; bracts and bracteoles not tomentose; fruits with well-developed pedicels, glabrous or covered with papillae, sometimes didymous **345. *Prangos***

144. Tall (> 1 m high), robust plants with thick stems and thick vertical taproots 145

– Low (< 1 m high) plants with short, thin, branched rhizomes 146

145. Pedicels long, considerably exceeding fruits; terminal leaf segments crenate at the margins; mericarp ribs straight, corky; mesocarp cells large with lignified porous walls; vittae cyclic; stylopodia flat, dish-shaped, undulate at the margin **324. *Petroedmondia***

– Pedicels short, ca. as long as fruit; terminal leaf segments toothed; mericarp ribs thin, undulate; mesocarp parenchymatous; vallecular mericarp ribs short, thick; stylopodia short-conical **408. *Smyrniopsis***

146. Stems solid; leaves pinnate, their segments sessile; fruits covered with verrucose outgrowths; vallecular vittae solitary, commissural vittae 2 **355. *Pseudotrachydium***

– Stems hollow; primary leaf segments petiolulate; fruits without verrucose outgrowths; vallecular vittae 2–3, commissural vittae 4 **158. *Eleutherospermum***

147. Mericarp ribs undulate or folded, covered with vesiculose tubercles or rigid hairs 148

– Mericarp ribs straight, glabrous or covered with short prickles 149

148. Stems short, leafless, puberulent; petals pubescent; bracts several, membranous, puberulent; mericarps covered with rigid hairs and tubercles **417. *Stenocoelium***

– Stems leafy, glabrous; petals glabrous; bracts obsolete; mericarps covered with vesiculose tubercles **237. *Ledebouriella***

149. Plants with hollow vertical rhizomes with transversal partitions; rhachis of leaf blades with ring-shaped bolsters; umbels at flowering and fruiting time globose **113. *Cicuta***

– Plants with solid rhizomes; leaf-blade rhachis without ring-shaped bolsters; umbels corymbose 150

150. Main vertical root reduced; roots thin, sometimes swollen or thickened like tubers; usually aquatic or hydrophilic plants 151

– Plants with vertical roots, short or long horizontal rhizomes; usually terrestrial plants 154

151. Leaves 2–3-pinnatisect; lateral roots often swollen; calyx teeth lanceolate, persistent, broadening at fruit maturity **289. *Oenanthe***

– Leaves pinnate; lateral roots not swollen, calyx teeth inconspicuous or very short 152

152. Bracts absent **199. *Helosciadium***

– Bracts present 153

153. Umbels terminal, at stem and branch tops, their rays ± equal; leaf segments lanceolate, with serrate margin **406. *Sium***

– Umbels on short peduncles, alternate to leaves; their rays strongly unequal; leaf segments ovate to ovate-lanceolate, at base with unequal sides, with incised margin and acute teeth **84. *Berula***

154. Stems twiggy, junciform, branching almost from the base; all stem leaves scale-like 155

– Stems and branches not twiggy or junciform; stem leaves, especially lower ones, with developed blades 156

155. Stems lignified in lower part; fruits elliptic, pubescent; vallecular vittae solitary, commissural vittae 2. Kuwait, Saudi Arabia **144. *Deverra***

– Stems completely herbaceous; fruits linear to cylindric, glabrous; vittae numerous, small, scarcely visible **366. *Rhabdosciadium***

156. At least some bracteoles dissected 157

– All bracteoles entire or obsolete 161

157. Terminal leaf segments ovate to elliptical, toothed, petiolulate; bracteoles scarious at the margins, ciliate, dissected **294. *Oreocome vaginata***

– Terminal leaf segments setaceous, filiform to narrowly linear, sessile; bracteoles herbaceous, glabrous, pinnatisect 158

158. Plants dwarf, almost stemless; mericarp ribs nearly equal, keeled or winged 159

– Stems well developed; mericarp ribs inconspicuous or filiform, or dorsal filiform and marginal winged 160

159. Petals yellow **106. *Chamaesciadium***

– Petals white **387. *Schulzia***

160. Calyx teeth absent; fruits ovoid, their ribs filiform to inconspicuous **177. *Fuernrohria***

– Calyx teeth present; fruits linear or oblong, often compressed dorsally; dorsal mericarp ribs filiform, marginal ribs filiform or winged **188. *Grammosciadium***

161. Main stem almost obsolete; central umbel situated almost at soil surface, its rays considerably longer than leafless peduncles of lateral umbels, situated in axils of radical leaves **147. *Dimorphosciadium***

– Main stem well developed; all umbels approximately equal or central umbel slightly larger than lateral ones 162

162. Bracteoles absent, or sometimes 1–2 163

– Bracteoles present, usually more than 3 165

163. Terminal leaf segments narrowly linear to filiform, entire; fruits terete, dorsal mericarp ribs filiform, marginal ribs slightly broader; vallecular vittae solitary, large **174. *Foeniculum***

– Terminal leaf segments ovate to rhombic, incised; fruits ovoid to elongate, with ± equal ribs; vallecular vittae several or inconspicuous 164

164. Leaves pinnate or bipinnatisect, their basal segments sessile; rhizomes short, vertical, sometimes branching; fruits pubescent or velvety, rarely glabrous; commissure broad; vittae numerous, cyclic or several in the valleculas **334. *Pimpinella***

– Leaves bi- to tripinnatisect with petiolulate basal segments; rhizomes horizontal, short or long; fruits glabrous; commissure narrow; vittae indistinct in mature fruits **43. *Aegopodium***

165. Plants of seaside rocks and cliffs with fleshy leaves; terminal leaf segments terete; styles very short; vittae numerous, cyclic **127. *Crithmum***

– Leaves not fleshy, their terminal segments not terete; styles often well developed, vittae solitary, or several 166

166. Fruits densely (or rarely sparsely) pubescent 167

– Fruits glabrous 172

167. Plants tall (up to 1.5–2 m high); terminal leaf segments large, broadly ovate, 3–8 cm long, with unequal sides at base, acute-toothed or crenulate at the margins **260. *Mediasia***

– Plants more low-growing; terminal leaf-segments various; if ovate, then smaller (up to 1.5 cm long), with unequal teeth or lobed 168

168. Fruits covered with vesiculose papillae or hammer-like hairs **446. *Trachyspermum***

– Fruits covered with smooth pointed hairs 169

169. Bracts absent **395. *Seseli***

– Bracts present 170

170. Stems leafy; if leafless, fruits with short, almost filiform ribs **395. *Seseli***

– Stems almost leafless or with 1–2 leaves; mericarp ribs winged or thickened, ± inflated 171

171. Stems solid; bracts and bracteoles glabrous; rootstocks almost entire or few-branched; mericarp ribs winged; fruits with scattered short prickles; vittae numerous, nearly cyclic **312. *Pachypleurum***

– Stems hollow, bracts and bracteoles puberulent; rootstocks strongly branched; mericarp ribs short, inflated, densely covered with thin hairs; vallecular vittae solitary, commissural 2 **328. *Phlojodicarpus***

172. Umbels globose, with short rays and short peduncles, alternately arranged along branches of synflorescence. Iran **197. *Haussknechtia***

– Umbels not globose, with peduncles more than 2 cm long; umbel rays distinctly visible 173

173. Petals yellow 174

– Petals white, rarely brown, red or pink 183

174. Marginal mericarp ribs broadly winged, dorsal ribs narrow winged; terminal leaf segments ovate to rhombic **241. *Levisticum***

- Dorsal mericarp ribs filiform, marginal slightly broader; terminal leaf segments linear, rarely ovate 175
175. Main root reduced, short, with adventitious cord-like thin roots; terminal leaf segments sessile or short-petiolulate; umbel rays strongly unequal **179. *Gasparrinia***
- Rhizomes and roots vertical, not shortened; terminal leaf segments with long petiolules; umbel rays equal or almost so 176
176. Terminal leaf segments rigid **464. *Zeravschania membranacea***
- Terminal leaf segments soft 177
177. Calyx teeth present; terminal leaf segments of radical and lower stem leaves broad, ovate, fruits cylindrical, ribs obtuse **229. *Kundmannia***
- Calyx teeth absent or rarely present; terminal leaf segments linear to lanceolate, rarely ovate; fruits ovoid or oblong, ribs keeled to winged 178
178. Bracts obsolete; vittae absent in mature fruits **399. *Silaum***
- Bracts prominent; vittae present in mature fruits 179
179. Petioles of basal stem and rosulate leaves enlarged, rigid, very upright, long-persistent, forming cone-like stem base **101. *Cephalopodum***
- Petioles of basal stem and rosulate leaves fibrillose or fibrose and soft after leaves senesce 180
180. Mericarp ribs filiform **395. *Seseli foliosum***
- Mericarp ribs keeled to winged, marginal ribs broader than dorsal 181
181. Leaves bipinnate with almost sessile segments, petals brownish or greenish yellow, glabrous; mericarp commissure narrow, mericarp ribs equal 182
- Leaves bi- to triternatisect, their segments with long petiolules, petals light yellow, short-pubescent; mericarp commissure broad, mericarp ribs unequal, dorsal keeled, marginal slightly broader **411. *Sphaenolobium***
182. Monocarpic perennials, umbels rays distinctly unequal, petals brownish; mericarp dorsal ribs winged **230. *Kuramosciadium***
- Polycarpic perennials, umbel rays nearly equal, greenish yellow; mericarp ribs keeled **453. *Tschulaktavia***
183. Stylopodia long-conic 184
- Stylopodia flat to short-conic 186
184. Plants pubescent; radical leaves entire; stem leaves ternate; their segments cordate to oval, serrate at the margins. Georgia **69. *Arafoe***
- Plants glabrous or puberulent only under umbels; leaves pinnately or ternately dissected, their segments ovate to lanceolate, incised 185
185. Stems solid; rhizomes vertical, thickened; terminal leaf segments cuneate at the bases, deeply incised, sheaths of stem leaves elongate, not inflated **219. *Katapsuxis***
- Stems hollow; rhizomes short, horizontal, slightly creeping; terminal leaf segments not cuneate, with large teeth near the tip, sheaths of stem leaves often inflated, ovate in outline **114. *Cnidiocarpa***
186. Bracteoles absent **334. *Pimpinella***
- Bracteoles present 187
187. Petals dark purple; plants with long horizontal rhizomes **427. *Tamamschjanella***
- Petals white; plants with vertical roots or with short vertical rhizomes 188
188. Bracts absent 189
- Bracts present 191
189. Mericarp ribs thin, filiform or almost inconspicuous **395. *Seseli***
- Mericarp ribs prominent, marginal ribs winged; dorsal ribs filiform, or ribs keeled or winged 190
190. Dorsal mericarp ribs keeled, marginal narrowly winged, calyx teeth short **212. *Kadenia***
- All ribs narrowly winged, equal or marginal ribs slightly enlarged, calyx teeth obsolete **117. *Conioselinum***
191. Bracts and bracteoles lanceolate, margins scarious 192
- Bracts and bracteoles almost completely herbaceous, usually linear or filiform 193
192. Leaves bi- to triternately or -pinnately dissected; their basal segments petiolulate **464. *Zeravschania***
- Leaves pinnate; their segments sessile. Iran **142. *Demavendia***
193. Plants growing from much-branching rootstocks; fruits with broadening base, narrowed cuneately toward stylopodia; mericarp ribs entirely inconspicuous; inner mesocarp layer with many additional small vittae; calyx teeth prominent **42. *Aegokeras***
- Plants growing from unbranched or sparsely branched rootstocks; fruits not broadened basally and cuneate toward stylopodia; mericarp ribs ± prominent, filiform to winged; additional small vittae in inner mesocarp layer lacking; calyx teeth obsolete, rarely small 194
194. Terminal leaf segments linear, rigid, entire; mericarp ribs broad at the base; mericarp valleculas narrow; vittae small, cyclic. Afghanistan, Pakistan **420. *Stewartiella***

- Terminal leaf segments soft, rarely rigid and spiny, lanceolate, oval to ovate, toothed, rarely linear; mericarp ribs filiform, keeled to narrowly winged; mericarp valleculas broad, vallecular vittae solitary or several 195

195. Stems leafless, branching only at the base **252. *Lomatocarpa***
- Stems usually leafy, branching in middle and upper parts 196

196. Terminal leaf segments spiny; fruits pyramid-shaped or truncate, cuneate at the base, with heteromorphic mericarps: one mericarp with 2 ribs, the other with 3 winged ribs, mericarps remain fused. Afghanistan **364. *Pyramidoptera***
- Terminal leaf segments soft; fruits ovate to elliptic, divided into two ± equal mericarps 197

197. Mericarp ribs filiform to short-keeled 198
- Mericarp ribs winged 199

198. Plants with vertical few-branched roots; umbel rays 4–7, strongly unequal; vallecular vittae solitary **98. *Carum***
- Plants with short, branched rhizomes; umbel rays 15–30, ± equal; vallecular vittae several **395. *Seseli***

199. Bracteoles ovate to lanceolate, often broadly white-membranous, exceeding umbellules **207. *Hymenidium***
- Bracteoles small, shorter than umbellules, margin narrowly white-membranous or completely herbaceous 200

200. Terminal leaf segments narrowly lanceolate to linear, entire **272. *Mutellina***
- Terminal leaf segments oblong to obovate, toothed at the margin 201

201. Plants low, < 30 cm tall; rhizomes creeping; leaves pinnate. Turkey **126. *Crenosciadium***
- Plants > 1 m tall, with taproots; leaves 3–4-pinnatisect 202

202. Calyx teeth obsolete; umbel rays and pedicels glabrous; rib secretory ducts large; commissure broad **315. *Paraligusticum***
- Calyx teeth prominent; umbel rays and pedicels densely pubescent; rib secretory ducts small to inconspicuous; commissure narrow **294. *Oreocome***

203. Plants with long horizontal rhizomes **367. *Rhizomatophora***
- Plants without long horizontal rhizomes, with taproots or short rhizomes 204

204. Dwarf glabrous alpine plants up to 20 cm high; stems creeping, almost leafless; leaves mainly radical, simple or ternate. W Turkey **304. *Ormosolenia***
- Taller plants, usually puberulent or pubescent; stems not creeping, leafy; stem leaves dissected 205

205. One to several lateral umbels below the terminal, larger umbel; marginal mericarp ribs with several small vascular bundles; petals yellow; sheaths of stem leaves usually inflated **171. *Ferula***
- All umbels terminal; vascular bundles in marginal mericarp ribs solitary or rarely paired, petals white, rarely yellow; sheaths of stem leaves usually not inflated 206

206. Leaf segments large, cordate, round or broadly elliptic, coriaceous, entire; plants glabrous 207
- Leaf segments usually small, soft; if larger, then deeply lobed; plants usually ± pubescent 212

207. Exocarp of broad cells with thickened outer walls (mericarp surface appears grainy) **308. *Ostericum***
- Exocarp of small cells similar to mesocarp cells; mericarp surface not grainy 208

208. Leaves pinnate, their segments spiny at the margins **83. *Azilia***
- Leaves 2–3-pinnatisect, margins not spiny 209

209. Stems hollow; petals greenish-yellow or yellow; mericarp marginal ribs inflated and spongy. Iran **270. *Mozaffariania***
- Stems solid; petals white; mericarp marginal ribs thin, not inflated 210

210. Leaves blue-glaucous, subcoriaceous; terminal leaf segments broadly ovate to cordate; vallecular vittae 1–4; commissural absent; rib ducts clearly visible; calyx teeth obsolete. W Turkey **182. *Glaucosciadium***
- Leaves green, soft; terminal leaf segments linear to lanceolate; calyx teeth triangular, vallecular vittae solitary, commissural 2, rib ducts small or absent. Eastern and Central Asia 211

211. Leaf segments lanceolate; dorsal mericarp ribs keeled, marginal ribs winged **217. *Karatavia***
- Leaf segments pinnatisect to linear or with oblong lobes; dorsal mericarp ribs filiform, marginal ribs winged **223. *Kitagawia***

212. Roots tuberously thickened, swollen **449. *Trigonosciadium***
- Roots vertical, slightly thickened 213

213. Bracteoles absent 214
- Bracteoles present 225

214. Bracts present **82. *Autumnalia***
- Bracts absent 215

215. Distal parts of marginal mericarp ribs thickened 216

- Marginal mericarp ribs not thickened or slightly thickened 218
216. Leaves pinnate; segments sessile, vallecular vittae 3–4, commissural 8, narrow. Afghanistan **216. *Kandaharia***
- Leaves pinnate or bi- to tripinnatisect, basal segments petiolulate; vallecular vittae solitary, commissural 2 217
217. Calyx teeth large, stylopodium conical; vittae narrow **256. *Mandenovia***
- Calyx teeth absent, stylopodium short conical; vittae very broad **466. *Zosima***
218. Stems strongly ribbed, grooved; sheaths of stem leaves amplexicaulous, almost cylindrical **326. *Peucedanum zedelmeieranum***
- Stems finely furrowed or rounded; sheaths of stem leaves ovate, triangular or lanceolate, diverging from stem 219
219. Stemless alpine plants; petals white; fruits pseudomonocarpellate, not separating into mericarps; all ribs short, triangular; vittae indistinct. Georgia **422. *Symphyoloma***
- Stems prominent; petals yellow, pink or white; fruits separating into mericarps, dorsal mericarp ribs filiform or keeled, marginal ones broader; vittae present or absent 220
220. Vallecular vittae absent 221
- Vallecular vittae present, ± numerous 222
221. Stems thin, strongly flexuose; leaves with sub-inconspicuous sheaths, long petioles, small blades considerably shorter than petioles, their segments small, umbel rays 3–6; mericarp commissural surface smooth. Afghanistan, Pakistan **391. *Scrithacola***
- Stems stout, firm; leaf sheaths triangular to lanceolate; petioles short, shorter than leaf blades; segments ovate to obovate, larger, umbel rays 11–25; mericarp commissural surface scabrous **231. *Ladyginia***
222. Mericarp vascular bundles situated in middle part of marginal ribs; vallecular vittae solitary to several 223
- Mericarp vascular bundles situated in basal part of marginal ribs; vittae cyclic, numerous. Iran 224
223. Vallecular vittae several, narrow, commissural 4–6, calyx teeth short **419. *Stenotaenia***
- Vallecular vittae solitary, broad, commissural vittae 2, calyx teeth obsolete **239. *Leiotulus***
224. Plants glabrous; leaves bipinnatisect, their segments with short petiolules, oblong-elliptic, 7–10 cm long **221. *Kelussia***
- Plants pubescent; leaves pinnate, their segments sessile, round, 3–6 cm long and broad **293. *Opsicarpium***
225. Marginal mericarp ribs inflated, dorsal filiform or indistinct 226
- Marginal mericarp ribs winged, thin;, dorsal filiform to keeled 227
226. Petals white; commissural vittae situated near carpophore. Pakistan **448. *Tricholaser***
- Petals yellow; commissural vittae situated near marginal rib bases **153. *Ducrosia***
227. Terminal leaf segments large, ovate to elongate, often coriaceous, with finely serrate margin 228
- Terminal leaf segments often filiform to lanceolate, membranous, or if ovate, then roughly dentate 229
228. Leaves almost glabrous, pinnate or bipinnate with petiolulate segments; petals white; calyx teeth prominent **253. *Macroselinum***
- Leaves bipinnate, covered with stellate or glochidiate hairs, sometimes glabrous beneath, with almost sessile segments; petals yellow; calyx teeth lacking **292. *Opopanax***
229. Robust plants, up to 1.5–3 m high, rarely lower; stems covered with rigid, spreading hairs; leaves entire, ternate or pinnate with large, usually lobed segments, 10–40 cm long; vittae not reaching fruit base, usually with clavate ends **201. *Heracleum***
- Lower plants, up to 50 (rarely 100) cm high; stems glabrous or shortly pubescent; leaves pinnate or bipinnate; their segments small or of medium size, 0.5–5 cm long; vittae reaching fruit base or absent 230
230. Bracts present 231
- Bracts absent 239
231. Ovaries and frequently fruits pubescent 232
- Ovaries and fruits glabrous 234
232. Vallecular vittae absent; stems not branching; petals purple. N Tianschan **319. *Pastinacopsis***
- Vallecular vittae solitary or numerous; stems branching in the upper part; petals white or yellowish 233
233. Mericarp vascular bundles situated in middle part of marginal ribs; vallecular vittae solitary, rarely two, rib secretory ducts small, solitary or absent **394. *Semenovia***
- Mericarp vascular bundles situated in basal part of marginal ribs; vittae cyclic, numerous, rib secretory ducts large, 1–4 per rib **333. *Pilopleura***
234. Terminal leaf segments long, cylindric or filiform, similar to petioles; all mericarp ribs filiform, or

dorsal ribs filiform and marginal slightly enlarged **240. *Leutea***
- Terminal leaf segments ovate, elliptic or linear, not cylindric or filiform; fruits usually with winged marginal ribs 235

235. Stems hollow; leaf segments rhombic to ovate, coarsely toothed; exocarp not penetrating at the commissural side of mericarps **102. *Cervaria***
- Stems solid; terminal leaf segments narrowly filiform to linear, entire, rarely toothed; exocarp penetrating slightly or up to the middle at the commissural side of mericarps 236

236. Bracts narrow, linear-lanceolate, not reflexed 237
- Bracts leafy, rather broad, green, reflexed 238

237. Plants polycarpic with branched rootstocks; terminal leaf segments linear, 3–9 cm long; central umbel larger than lateral ones **326. *Peucedanum***
- Plants monocarpic with solitary stems thickened toward the base; terminal leaf segments short, up to 10 mm long; stems many-branched with numerous approximately equal umbels **460. *Xanthoselinum***

238. Petals white; fruits oblong, broad at the base and cuneately narrowing to stylopodia; mericarp ribs nearly equal, keeled to winged; marginal ribs slightly broader than dorsal; vallecular vittae 1–3, commissural 3–4 **149. *Diplotaenia***
- Petals yellow; fruits ovoid to elongate, not narrowing to stylopodia, marginal mericarp ribs winged, dorsal winged, keeled or rarely filiform; vittae numerous, cyclic **172. *Ferulago***

239. Umbellules compact, dense, ± globose, with very short rays; exocarp cells large, with thickened outer and inner walls. W Turkey **426. *Taeniopetalum***
- Umbellules loose, usually corymbose, their rays ± long, distinct; exocarp cells small, with thin walls 240

240. Leaf segments long-petiolulate, narrowly linear, entire, 1.5–9 cm long 241
- Leaf segments sessile or shortly petiolulate, toothed or lobed; ≤ 1 cm long 242

241. Remains of radical leaf petioles few, soft, fibrous; leaf segments reflexed; calyx teeth prominent; inner mesocarp layers consisting partly of prosenchyma cells with lignified walls **326. *Peucedanum***
- Remains of radical leaf petioles numerous, rigid, persistent; leaf segments not reflexed; calyx teeth sub-inconspicuous; inner mesocarp layers parenchymatous. Pamiro-Alai **213. *Kafirnigania***

242. Terminal leaf segments obovate, lobed, rigid, decurrent at the base; bracteoles fused at the base. N Pamiro-Alai **170. *Fergania***
- Terminal leaf segments lanceolate to linear, soft, not decurrent; bracteoles free 243

243. Leaf sheaths long, linear; dorsal mericarp ribs indistinct; mesocarp at the commissural side thick, commissural face white with two grooves of crushed cells near carpophore; vallecular and commissural vittae obsolete; rib secretory ducts solitary, large **211. *Johrenia***
- Leaf sheaths short, linear or triangular; dorsal mericarp ribs filiform to keeled; mesocarp at the commissural side thin, semitranslucent, flat; vallecular vittae 1–3 or absent, commissural usually 2; dorsal ribs without large secretory ducts **145. *Dichoropetalum***

E. *Key to the Native and Naturalized Genera of Apiaceae in Eastern, Southeastern and South Asia*

1. True shrubs with highly lignified stems. Vietnam, Laos **462. *Xyloselinum***
- Herbaceous plants or half-shrubs with stems lignified only near the bases 2

2. Umbels simple 3
- Umbels compound 6

3. Stems creeping, rooting at the nodes, rarely erect; leaves simple, leaf blade entire, reniform or rounded-cordate; fruits glabrous 4
- Stems usually erect, stout, or if decumbent, not rooting at the nodes; leaves usually compound, pinnately or palmately dissected and often lobed; fruits glabrous or covered with hooked prickles or tubercles 5

4. Stems creeping, rooting at the nodes, fruits compressed laterally; carpophore lacking **4. *Centella***
- Stems erect or decumbent, not rooting at the nodes; fruits compressed dorsally; carpophore shortly bifid at the apex **18. *Dickinsia***

5. Fruits covered with hooked prickles or tubercles; vallecular vittae obsolete; rib ducts prominent; stylopodia very small; stems stout or slender **38. *Sanicula***
- Fruits glabrous; vallecular vittae solitary, prominent; rib ducts small to inconspicuous; stylopodia prominent, low conical to conical; plants stemless **296. *Oreomyrrhis***

6. Umbellules uniflowered, forming general synflorescences of cylindrical, ovoid or almost globose capitula; fruit dorsal surface covered with whitish, lanceolate or triangular scale-like outgrowths; bracts and often leaves spiny at margins; calyx with prominent spiny teeth **35. *Eryngium***
– Umbellules of several to many flowers, rarely uniflowered, not forming capitula; fruits without whitish, lanceolate or triangular scales, glabrous or covered with hairs, bristles or bladders; bracts, leaves and calyx usually without spines, if leaves, bracts and bracteoles are spiny, only one central flower in umbellule fertile 7
7. All leaves undivided and entire, often indistinctly divided between petiole and lamina; lamina with ± parallel, curved veins and entire margin, without petioles **91. *Bupleurum***
– Leaves pinnate, ternate, or palmately dissected; if lamina undivided, then toothed or lobed at the margin; petioles prominent 8
8. Leaves spiny; only one central flower in umbellule fertile, sessile, free, surrounded by elongate, pubescent pedicels of male flowers; bracts and bracteoles spiny. India **363. *Pycnocycla***
– Leaves herbaceous, not spiny; flowers usually with prominent pedicels, rarely sessile, in each umbellule hermaphroditic or unisexual, usually more than one flower in umbellule fertile; bracts and bracteoles not spiny 9
9. Terminal leaf lobes cartilaginous and serrate at the margin, the teeth subulate acute, pointed **169. *Falcaria***
– Margins of terminal leaf lobes without serrate acute, pointed teeth at margin 10
10. Fruits with primary and secondary ribs 11
– Fruits with primary ribs only (secondary ribs obsolete) 16
11. Plants perennial, glabrous; fruits glabrous **107. *Chamaesium***
– Plants annuals, pubescent; fruits usually densely covered with spines or glochidiate bristles 12
12. Leaves pinnate; commissural face of endosperm deeply emarginated, with involute margins **454. *Turgenia***
– Leaves bi- to tripinnatisect; commissural face of endosperm flat or with broad groove 13
13. Bracts trifid to pinnatisect; secondary mericarp ribs with spines, arranged in one series, basally connate; mericarps compressed dorsally **141. *Daucus***
– Bracts simple or obsolete; secondary mericarp ribs with spines, arranged in several series, basally not connate; mericarps not compressed dorsally 14
14. Spines of secondary ribs broadened at the base, glochidiate or sharpened at tip, covered with rigid short-toothed prickles **443. *Torilis***
– Secondary mericarp ribs with thin, long, smooth hairs, not sharpened at the tip 15
15. Stems and leaves densely covered with short hairs; terminal leaf lobes oblong, cuneate at the base; fruits densely covered with thin clavate hairs **347. *Psammogeton***
– Stems and leaves glabrous; terminal leaf lobes subulate to filiform; fruits densely covered with thin, long, not clavate hairs **129. *Cuminum***
16. Annuals or biennials 17
– Monocarpic or polycarpic perennials 35
17. Bracteoles unilaterally arranged **118. *Conium***
– Bracteoles arranged regularly around umbellules, or obsolete 18
18. Leaf blades ternate; umbel rays 2–7, strongly unequal, stiffly ascending; umbellules similar **128. *Cryptotaenia***
– Leaf blades pinnate, umbels with more than 8 rays, umbel rays and pedicels not stiffly ascending 19
19. Fruits strongly compressed dorsally 20
– Fruits not or slightly compressed dorsally, or compressed laterally, or terete 21
20. Leaves pinnate, their primary segments sessile, ovate to elliptic, toothed; inner mesocarp layer of lignified prosenchyma cells **318. *Pastinaca***
– Leaves 3–4-ternatisect, their primary segment petiolulate, terminal segments filiform, entire; mesocarp entirely parenchymatous **56. *Anethum***
21. Umbellules with 2–4(5) flowers; one or several flowers sessile; umbel rays capillary **356. *Pternopetalum***
– Umbellules with 6 and more flowers, all having prominent pedicels; umbel rays stout 22
22. Fruits glabrous 23
– Fruits covered with hairs or wart-like outgrowths 31
23. Bracteoles lacking 24
– Bracteoles prominent 26
24. Leaves ternate; terminal leaf lobes broad, ovate to oblong, cuneate at the base **67. *Apium***
– Leaves bi- to triternate or pinnatisect; terminal leaf lobes linear to filiform 25
25. Leaves almost triternatisect, their terminal lobes with prominent petiolules; fruits globose **132. *Cyclospermum***

- Leaves bi- to tripinnatisect, their terminal lobes sessile; fruits ovoid **98. *Carum***
26. Bracts present 27
- Bracts absent 29
27. Bracts and bracteoles entire; fruits slightly compressed laterally, with keeled-winged dorsal ribs **115. *Cnidium***
- Bracts pinnate or trifid; fruits slightly compressed dorsally, with filiform dorsal ribs 28
28. Terminal leaf lobes linear to filiform, up to 1 mm broad, entire; umbel rays thickened, rigid at fruit maturity **459. *Visnaga***
- Terminal lobes of radical and lower stem leaves obovate to elliptic, 4–20(30) mm broad, serrate at the margin; umbel rays not thickened, remaining thin even at fruit maturity **51. *Ammi***
29. Plants glabrous; fruits almost globose; vittae in mature fruits inconspicuous; lower stem leaves with broad, terminal lobes **120. *Coriandrum***
- Plants pubescent; fruits ellipsoid to ovoid, with 1 to many small vittae, persistent at fruit maturity; terminal lobes of lower stem leaves lanceolate 30
30. Vittae numerous; endosperm with groove on commissural side **413. *Sphallerocarpus***
- Vallecular vittae solitary, commissural 2; endosperm on commissural side flat, without groove **103. *Chaerophyllopsis***
31. Fruits long-beaked **382. *Scandix***
- Fruits without beak, or with a very short beak 32
32. Bracts and bracteoles absent **334. *Pimpinella***
- Bracteoles present; bracts absent or present 33
33. Fruits covered with scattered warts or papillae **196. *Harrysmithia***
- Fruits covered with vesiculose outgrowths or hairs 34
34. Fruits densely covered with thin clavate hairs; mericarps without beak **65. *Aphanopleura***
- Fruits covered with vesiculose outgrowths or scabrous hairs; mericarps with very short beak **446. *Trachyspermum***
35. Subterranean part of plant ending in hypocotyloid tuber or solitary tuberous root, entire or lobed 36
- Plants with vertical roots, vertical or horizontal rhizomes, or branched rootstocks; sometimes with tuberiform thickenings only at lateral roots 55
36. Primary segments of leaf blades sessile; bracts always present 37
- Primary segments of leaf blades with prominent petiolules; bracts often obsolete 38
37. Mericarp ribs inconspicuous; exocarp of tangentially elongated large cells; commissure narrow **157. *Elaeosticta***
- Mericarp ribs prominent; exocarp of small cells, not differing from mesocarp cells; commissure broad **206. *Hyalolaena***
38. Fruits strongly compressed dorsally; dorsal ribs filiform, marginal ribs winged 39
- Fruits ovoid, oblong to lanceolate, not compressed dorsally; dorsal and marginal ribs ± equal 42
39. Tuberous root cylindrical, bracteoles absent **111. *Chuanminshen***
- Tuberous root ovoid or globose; bracteoles present 40
40. Terminal leaf lobes linear; fruits densely pubescent; marginal mericarp ribs broadly winged **171. *Ferula karelinii***
- Terminal leaf lobes lanceolate to ovate, toothed; fruits glabrous, marginal ribs narrowly winged. S India 41
41. Outer petals of marginal flowers in umbellules only slightly, if at all, broader than inner ones; marginal mericarp ribs thin, winged; vallecular vittae reaching fruit base **342. *Polyzygus***
- Outer petals of marginal flowers in umbellule clearly radiate; marginal mericarp ribs prominent but never winged; vallecular vittae not reaching fruit base **336. *Pinda***
42. Fruits narrowed into a short beak below stylopodia; vittae inconspicuous in mature fruits 43
- Fruits without beak; vallecular vittae always present 44
43. Fruits pyriform, shining, covered with tubercles; exocarp of 2–3 layers of cells with lignified walls. Middle Asia **227. *Krasnovia***
- Fruits almost cylindrical, brown, glabrous or covered with simple hairs; exocarp unilayered. Himalaya **278. *Neoconopodium***
44. Stems densely pubescent; tuberous roots globose **104. *Chaerophyllum prescottii***
- Plants glabrous, rarely minutely scabrous; if pubescent, then tubers fusiform, never globose 45
45. Petal tips linear to long-caudate **40. *Acronema***
- Petal tips plane, acuminate or shortly involute 46
46. Tuberous roots palmately lobed, rarely fusiform; petals with narrowly inflexed apex 47
- Tuberous roots globose; if palmately lobed, then petals without narrowly inflexed apex 48

47. Mericarp ribs keeled or thinly winged; calyx teeth absent; vallecular vittae solitary, commissural 2. W China **396. *Seselopsis***
- Mericarp ribs filiform; calyx teeth present; vallecular vittae 2–3, commissural 4. Himalaya **261. *Meeboldia***
48. Petals yellow, calyx teeth well developed; mericarp ribs keeled. S India **407. *Sivadasania***
- Petals white, calyx teeth usually absent, rarely prominent; mericarp ribs filiform 49
49. Petals plane, entire or dissected 50
- Petals with attenuate and inflexed tip, entire 52
50. Tuberous roots globose; umbels with 2–4 rays 51
- Tuberous roots fusiform; umbels with more than 5 rays **402. *Sinocarum***
51. Leaves pinnate, leaflets deeply 3-lobed into lanceolate lobes; fruits minutely scabrous when young, subdidymous **373. *Rohmooa***
- Leaves 2–3-ternate into linear segments; fruits glabrous, ovoid-oblong **423. *Synclinostyles***
52. Calyx teeth absent 53
- Calyx teeth present 54
53. Tuberous roots fusiform **402. *Sinocarum***
- Tuberous roots globose **159. *Elwendia***
54. Commissure narrow; bracteoles entire; petal apex shorter than blade **90. *Bunium***
- Commissure broad; bracteoles entire and toothed; petal apex narrow, long, approximately equal to petal blade in length **218. *Karnataka***
55. Stems profusely branched, forming corymbose or paniculate synflorescence; leaves often rosulate, stem leaves reduced to lanceolate sheaths **313. *Palimbia***
- Stems variously branched or unbranched; all or lower stem leaves with developed blades 56
56. Stems slender, procumbent, creeping, rooting at the nodes; leaves digitate; carpophore absent. S India **456. *Vanasushava***
- Stems usually stout or ascending, not rooting at the nodes; leaves ternately or pinnately dissected; carpophore present, rarely lacking 57
57. Fruits compressed laterally or slightly dorsally; dorsal and marginal mericarp ribs approximately equal, sometimes marginal ribs narrower; commissure narrow 58
- Fruits strongly compressed dorsally, flat or slightly concave; dorsal mericarp ribs sub-inconspicuous or filiform to narrowly keeled, marginal ribs winged to broadly winged; commissure broad 137
58. Bracts dissected or toothed 59
- Bracts entire, or absent 69
59. Bracts 1–2, bi- to tripinnatisect; bracteoles filiform, shorter than pedicels **222. *Keraymonia***
- Bracts more than 3; bracteoles linear or larger, equal or longer than pedicels 60
60. Mericarp ribs inflated at the base; valleculae narrow; rib secretory ducts large, more than one in each rib; inner pericarp layer crushed in mature fruits **337. *Pleurospermum***
- Mericarp ribs not inflated at the base; valleculae broad; rib secretory ducts solitary, small, visible only in some ribs, or absent; pericarp persistently adhering to the seed 61
61. All mericarp ribs ± equal and broadly winged, or marginal ribs broadly winged and dorsal ribs keeled to narrowly winged 62
- All mericarp ribs equal and filiform to keeled, or rarely all ribs narrowly winged 66
62. Bracts leafy, with small pinnatisect blade and long petiole; bracteoles entire, linear to subulate, 2–3 times exceeding umbellules, herbaceous **295. *Oreocomopsis***
- Bracts dissected, petiole not prominent; bracteoles pinnatisect, equal or slightly exceeding the flowers; margins white-membranous 63
63. Vallecular vittae not distinct **294. *Oreocome vaginata***
- Vallecular vittae always present 64
64. Mesocarp cells parenchymatous, their walls without pores; commissure narrow; basal primary leaf segments with prominent (rarely short) petiolules **207. *Hymenidium***
- Mesocarp cells with lignified porous walls; commissure broad; basal primary leaf segments sessile 65
65. Carpophore lacking; umbels compact with short rays; pedicels thickened; mericarps with unequal, broad and narrow ribs **123. *Cortiella***
- Carpophore prominent, bifid; umbels loose, with long, curved rays; pedicels thin; mericarp ribs ± equal, winged **122. *Cortia***
66. Commissure broad; exocarp interrupted near distal ends of marginal ribs **220. *Kedarnatha***
- Commissure narrow; exocarp interrupted near middle of commissural mericarp side 67
67. Carpophores reduced; bracts with narrow, distinct petiole; mericarp ribs undulate **445. *Trachydium***
- Carpophores present; bracts without distinct petiole, sometimes cuneate at the base; mericarp ribs straight 68
68. Petals with several secretory ducts, apex emarginate and incurved; bracteoles not cuneate at the base, with narrow filiform lobes **387. *Schulzia***

– Petals with solitary secretory ducts, almost flat; bracts toothed at the margin, cuneate at the base **330. *Physospermopsis***

69. Terminal umbels compact, almost capitate, surrounded by white-membranous enlarged and ovate sheaths of upper stem leaves and bracts with violet veins; petals pale violet **378. *Sajanella***
– Sheaths of upper stem leaves narrow and non-membranous; bracts linear to lanceolate, rarely ovate; petals white or yellow, rarely red 70

70. Vertical rhizomes chambered; terminal leaf lobes lanceolate to lanceolate-linear, incised at the margins; leaf-blade rachis with ring-shaped bolsters; umbels globose in flower and fruit **113. *Cicuta***
– Vertical rhizomes solid; leaf-blade rachis without ring-shaped bolsters; terminal leaf lobes diverse; umbels usually corymbose 71

71. Main vertical root reduced; roots thin, mainly adventitious, sometimes swollen or thickened like tubers; usually aquatic or hydrophilic plants 72
– Plants with vertical roots, short or long horizontal rhizomes; lateral roots without tuber-like thickenings 75

72. Leaves 2–3-pinnatisect; roots mainly adventitious, frequently thickened, forming fusiform or globose tuber-like thickenings; calyx teeth lanceolate, persistent, enlarging in mature fruits **289. *Oenanthe***
– Leaves pinnate; calyx teeth obsolete or rarely shortly prominent; roots without tuber-like thickenings 73

73. Umbels with short peduncles, leaf-opposed; rays strongly unequal; terminal leaf lobes ovate to ovate-lanceolate, with unequal sides at the base, incised, with pointed teeth at the margin, sometimes lobed **84. *Berula***
– Umbels terminal on stems and branches, rays roughly equal; terminal leaf lobes lanceolate, with equal sides at the base, serrate 74

74. Plants slender, decumbent; calyx teeth minute to obsolete; petals without prominent narrowly inflexed apex; fruit ribs corky-thickened, triangular in cross section; valleculae narrow. Japan **68. *Apodicarpum***
– Plants with stout stems, erect; calyx teeth evident to prominent; petals with conspicuously narrowed and inflexed apex; mericarp ribs thinly keeled or narrowly winged; valleculae broad **406. *Sium***

75. Bracteoles absent 76
– Bracteoles present 80

76. Petals yellow 77
– Petals white 78

77. Plants glaucous; terminal leaf lobes narrowly linear to filiform **174. *Foeniculum***
– Plants dark green; terminal leaf lobes lanceolate to linear-lanceolate **131. *Cyclorhiza***

78. Leaves pinnate or bipinnate, their basal segments sessile; fruits often pubescent, rarely glabrous **334. *Pimpinella***
– Leaves 2–3-pinnatisect; their basal segments with prominent petiolules; fruits glabrous 79

79. Plants with short or long horizontal rhizomes; vittae indistinct in mature fruits **43. *Aegopodium***
– Roots fusiform, sometimes with adventitious shoots; vittae prominent in mature fruits **440. *Tongoloa***

80. Leaflets large, rhomboid to ovate-rotund, more than 3 cm long, with regularly toothed margins (serrate or crenate). Mericarps long, up to 20 mm long, lanceolate or lanceolate-linear **357. *Pterocyclus***
– Leaflets usually smaller, linear to lanceolate-ovate, or if large, with incised or irregularly toothed margins. Mericarps usually smaller, ovate, elliptic to oblong 81

81. Bracts and bracteoles prominent, or if bracts absent, then bracteoles entire and linear or dissected with linear to filiform terminal lobes 82
– Bracts absent or soon falling; bracteoles present, usually entire or dentate, lanceolate to ovate 111

82. Umbels capitate, with very short rays; petals purple-brown **193. *Haplosphaera***
– Umbels not capitate, with prominent rays; petals white, or rarely yellow, greenish or red 83

83. Petals yellow 84
– Petals white, rarely greenish or red 85

84. Terminal leaf lobes linear to filiform; fruits with thick corky pericarp, convex dorsally; endosperm deeply grooved on commissural side, with involute ends; vittae numerous, of two kinds: small, scattered in mesocarp, and cyclic in inner mesocarp layer **345. *Prangos***
– Terminal leaf lobes ovate, toothed; fruits compressed dorsally, with thin pericarp; endosperm plane on commissural side; vallecular vittae 1, commissural 2 **241. *Levisticum***

85. Endosperm on commissural side with broad or narrow and deep groove; if plane, then bracteoles large, exceeding or nearly equal to umbellules 86
– Endosperm at the commissural side nearly plane; bracts not exceeding umbellules 88

86. Petals almost plane, with short apex; bracteoles narrow, entire linear or dissected with linear to filiform terminal lobes **404. *Sinolimprichtia***
- Petals inflexed with narrowed apex; bracteoles broad, lanceolate to ovate, entire, toothed or pinnatifid 87

87. Bracteoles usually entire, lanceolate to lanceolate-linear; if toothed, then endosperm with a deep groove on the commissural side **80. *Aulacospermum***
- Bracteoles usually toothed or pinnatifid, lanceolate to ovate; if entire (rare) then endosperm flat on the commissural side **207. *Hymenidium***

88. Stylopodia long conical; terminal leaf lobes filiform to narrowly linear, entire at the margin **376. *Rupiphila***
- Stylopodia flat to low conical; terminal leaf lobes lanceolate to ovate, often toothed 89

89. Fruits almost didymous 90
- Fruits not didymous 92

90. Bracts subulate; commissure broad; fruits glabrous or covered with smooth hairs **334. *Pimpinella***
- Bracts linear or lanceolate-linear; commissure narrow, fruits pubescent 91

91. Fruits narrowly short-beaked below stylopodia, covered with vesiculose, clavate or scabrous hairs; vallecular vittae solitary, commissural 2 **446. *Trachyspermum***
- Fruits not narrowed below stylopodia, glabrous or covered with smooth hairs; vallecular vittae several **285. *Nothosmyrnium***

92. Fruits pubescent 93
- Fruits glabrous 98

93. Mericarp ribs $\pm$ equal, winged 94
- Mericarp ribs corky-thickened, keeled or filiform, equal, or dorsal ribs filiform to keeled, and marginal ribs broader and narrowly winged 95

94. Fruits covered with small prickles; vallecular vittae solitary, commissural 2; primary leaf segments petiolulate; sheaths of stem leaves broadened. Himalaya, Tibet **214. *Kailashia***
- Fruits sparsely covered with soft hairs; vittae numerous, cyclic; leaf segments sessile; sheaths of stem leaves narrow. N Eurasia **312. *Pachypleurum***

95. Stems solitary, short, leafless, fruits covered with rigid scales, dorsal ribs undulate, thickened, obtuse **417. *Stenocoelium***
- Stems leafy; mericarps hairy; dorsal ribs not undulate, thin 96

96. Mericarp ribs slightly inflated, corky-thickened, keeled, marginal ribs somewhat broader **328. *Phlojodicarpus***
- Mericarp ribs thin, filiform to narrowly winged 97

97. Calyx teeth prominent, triangular to lanceolate, connate at the base; styles long, as long as the fruit; mericarp ribs nearly indistinct **96. *Carlesia***
- Calyx teeth lacking or short, not connate at the base; styles considerably shorter than fruits, equal or slightly longer than stylopodia; mericarp ribs filiform to keeled, or dorsal ribs filiform to keeled, and marginal ribs broader **395. *Seseli***

98. Fruits strongly compressed dorsally; dorsal mericarp ribs filiform, keeled or short-winged, marginal ribs broadly winged 99
- Fruits slightly compressed or not compressed dorsally; all ribs equal, filiform to winged 100

99. Calyx teeth present; commissure narrow; exocarp of large cells with thick outer walls; mesocarp cells without lignified pitted walls **308. *Ostericum***
- Calyx teeth absent; if prominent, commissure broad; exocarp cells small; mesocarp cells with lignified pitted walls **57. *Angelica***

100. Petals red or reddish-brown 101
- Petals white 102

101. Leaves ternate, lobes cuneate at the base, with a rounded tooth near the tip; vittae cyclic **189. *Halosciastrum***
- Leaves 2–3-pinnatisect, terminal lobes laciniate; vallecular vittae solitary **250. *Lithosciadium***

102. Main stem nearly reduced; central umbel situated nearly on soil surface, considerably exceeding lateral umbels, which have long leafless peduncles situated in axils of basal leaves **147. *Dimorphosciadium***
- Main stem well developed; all umbels on prominent peduncles, approximately equal or central umbel slightly larger than lateral ones 103

103. Stems leafless or with one leaf; terminal leaf lobes linear to lanceolate; dorsal mericarp ribs keeled, marginal ribs narrowly winged 104
- Stems with several leaves; terminal lobes ovate to lanceolate, toothed; mericarp ribs subequal, filiform or winged, or dorsal ribs keeled to narrowly winged, and marginal ribs winged 106

104. Plants monocarpic; rootstocks unbranched **252. *Lomatocarpa***
- Plants polycarpic; rootstocks well branched 105

105. Stem bases with collar of soft petiolar remains; stem leaves 1–2; leaf blades triangular to ovate in outline;

umbels terminal and lateral; umbellule rays 8–16; dorsal mericarp ribs slightly inflated; vallecular vittae solitary **328. *Phlojodicarpus sibiricus***
- Stem bases covered with rigid, straight petiolar remains of preceding years; leaves only basal, blades elongate-lanceolate in outline; umbels only terminal; umbellule rays 15–20; dorsal mericarp ribs filiform, marginal ribs winged; vallecular vittae 2–4 **173. *Ferulopsis***

106. Mericarp ribs filiform **98. *Carum***
- All, or at least some mericarp ribs winged 107

107. Terminal leaf lobes narrowly linear, 2–10 cm long; mericarp ribs inflated at base **359. *Pterygopleurum***
- Terminal leaf lobes ovate, often toothed; mericarp ribs not inflated at base 108

108. Rootstocks strongly branched; terminal leaf lobes obovate, serrate or minutely toothed at margin; seed occupies only upper part of mericarps **245. *Ligusticum***
- Plants with fusiform roots or slightly branched rhizomes; terminal leaf lobes usually deeply dissected, toothed or lobed; seed occupies whole fruit cavity 109

109. Calyx teeth obsolete; vallecular vittae solitary **115. *Cnidium***
- Calyx teeth prominent; vallecular vittae often several, rarely solitary 110

110. Calyx teeth long, lanceolate, triangular or filiform, exceeding stylopodia; petioles almost round in cross section; stems well developed, with several (> 2) leaves **294. *Oreocome***
- Calyx teeth triangular, shorter than stylopodia; petioles grooved on adaxial side; stem leaves few, (0)1–2, with reduced blades **437. *Tilingia***

111. Mericarp ribs obtuse, thick, covered with vesiculose tubercles **237. *Ledebouriella***
- Mericarp ribs glabrous, smooth or pubescent, never with vesiculose tubercles 112

112. Stems procumbent or ascending; leaves bipinnatisect; their terminal lobes round to ovate, 2.5–3 cm long and broad, coriaceous, tomentose beneath, shiny above; mericarp ribs winged, with several vascular bundles; stylopodia lacking **183. *Glehnia***
- Stems erect and stout; leaves membranous, with narrower and smaller terminal lobes, never tomentose; mericarp ribs filiform to keeled, rarely narrowly winged, with solitary vascular bundles; stylopodia present, flat to conical 113

113. Fruits compressed dorsally; all mericarp ribs winged, roughly equal, or dorsal ribs narrower than marginal, narrowly winged, keeled or filiform 114
- Fruits concave or slightly compressed dorsally and laterally; all mericarp ribs filiform or keeled, or marginal ribs slightly broader than dorsal 121

114. Mericarps strongly compressed dorsally; marginal mericarp ribs broadly winged, dorsal filiform to keeled 115
- Mericarps slightly compressed dorsally; all ribs equally winged, or marginal somewhat broader 116

115. Calyx teeth inconspicuous; exocarp consisting of small cells, similar to those of mesocarp **57. *Angelica***
- Calyx teeth prominent; exocarp consisting of large cells with thickened outer walls **308. *Ostericum***

116. Bracteoles large, ovate, exceeding umbellules, almost completely scarious **208. *Hymenolaena***
- Bracteoles linear to lanceolate, herbaceous, not exceeding umbellules 117

117. Calyx teeth absent 118
- Calyx teeth present 119

118. Commissure broad; exocarp interrupted near distal ends of marginal ribs; rib secretory ducts large; vascular bundles situated in basal parts of mericarp ribs **315. *Paraligusticum***
- Commissure narrow (exocarp interrupted near carpophore); rib secretory ducts small or lacking; vascular bundles situated in distal parts of mericarp ribs **117. *Conioselinum***

119. Plants completely glabrous; exocarp of large cells with concave outer walls; petals usually yellow, rarely white or brown **191. *Hansenia***
- Plants densely pubescent in inflorescence; exocarp of small cells; petals always white 120

120. All mericarp ribs ± equal, winged, commissure narrow; valleculae narrow; all stem leaves similar **154. *Dystaenia***
- Marginal mericarp ribs winged, dorsal ribs shorter, or all ribs keeled; commissure usually broad, rarely narrow; vallecullae broad; upper stem leaves differ markedly from lower leaves, with reduced blades **294. *Oreocome***

121. Umbellules with 2–4(5) flowers; one or several flowers almost sessile, others pedicellate **356. *Pternopetalum***
- Umbellules with greater number of flowers; all flowers pedicellate 122

122. Leaf rachis geniculate 123
- Leaf rachis not geniculate 124

123. Terminal leaf lobes linear to linear-lanceolate; vallecular vittae solitary; endosperm flat at commissural side **100. *Cenolophium***
- Terminal leaf lobes ovate to rhombic; vallecular vittae 3–4; endosperm grooved at commissural face **262. *Melanosciadium genuflexum***

124. Plants with short horizontal rhizomes; calyx teeth present **414. *Spuriopimpinella***
- Plant with vertical roots; if rhizomes present, then calyx teeth absent 125

125. Terminal leaf lobes ovate or rhombic, with small teeth at margin; mericarp ribs keeled to narrowly winged **262. *Melanosciadium***
- Terminal leaf lobes linear to ovate, entire, or lobed, or with big teeth at margin; mericarp ribs inconspicuous or filiform, rarely keeled 126

126. Endosperm broadly or rather narrowly grooved on the commissural face 127
- Endosperm flat on the commissural face 133

127. Fruits didymous; pericarp with a layer of strongly lignified cells **385. *Schrenkia***
- Fruits ovoid, elongate, lanceolate, but never didymous; pericarp parenchymatous 128

128. Base of mericarps with long, narrowed, sterile caudate part, almost as long as fertile part; umbel rays long, thin, divaricate **307. *Osmorhiza***
- Base of mericarps without long, narrowed, sterile caudate part; umbels usually corymbose, rays not divaricate 129

129. Fruits narrowed into a short beak below stylopodium; mericarps oblong, linear to lanceolate; ribs inconspicuous; vittae absent in mature fruits **64. *Anthriscus***
- Fruits not narrowed into a beak below stylopodium; mericarps ovoid to oblong, without ribs or with filiform to broader ribs; vittae always present 130

130. Mericarp ribs indistinct; vittae numerous, cyclic **109. *Changium***
- Mericarp ribs evident, filiform to broad; vallecular vittae 1–4 131

131. Plants often pubescent; mericarp ribs broad; commissure broad, exocarp interrupted near distal ends of marginal ribs; vallecular vittae solitary, commissural 2 **104. *Chaerophyllum***
- Plants completely glabrous; mericarp ribs filiform, terete or narrowly winged; commissure narrow, exocarp reaching carpophore; vallecular vittae 2–4, commissural 3–6 132

132. Calyx teeth absent; plants with short or creeping rhizomes, or with adventitious fibrillose roots; terminal leaf lobes with short petiolules **458. *Vicatia***
- Calyx teeth present; roots thickened, palmately branched; petiolules of leaf lobes long **403. *Sinodielsia***

133. Fruits almost didymous **334. *Pimpinella***
- Fruits ovoid or ellipsoid, never didymous 134

134. Petals yellow; all mericarp ribs keeled; vittae lacking in mature fruits **399. *Silaum***
- Petals white, rarely yellowish; mericarp ribs filiform to keeled, equal, or dorsal keeled, and marginal winged; vittae prominent in mature fruits 135

135. Umbels numerous, mainly lateral, forming multi-branching hemispheric synflorescence; ovaries covered with whitish transversal outgrowths; rib secretory ducts large, solitary, exceeding vallecular vittae **379. *Saposhnikovia***
- Umbels several, terminal and lateral, ovaries without whitish transversal outgrowths, some of them glabrous; rib secretory ducts indistinct or very small 136

136. Dorsal mericarp ribs keeled, marginal ribs winged; fruits glabrous; vallecular vittae solitary **212. *Kadenia***
- All mericarp ribs keeled or filiform, or dorsal ribs filiform, and marginal ribs keeled; fruits often pubescent; if glabrous, then vallecular vittae 2–3 **395. *Seseli mucronatum***

137. Plants with one to several lateral umbels below larger, terminal umbel; marginal mericarp ribs with several small vascular bundles; petals yellow; sheaths of stem leaves usually inflated **171. *Ferula***
- Plants with terminal umbels only on stem; rib vascular bundles 1–2; petals usually white; sheaths of stem leaves usually not inflated 138

138. Fruits with narrow commissure 139
- Fruits with broad commisure 140

139. Calyx teeth inconspicuous; exocarp consisting of small cells, similar to those of mesocarp **57. *Angelica***
- Calyx teeth prominent; exocarp consisting of large cells with thickened outer walls **308. *Ostericum***

140. Bracteoles absent 141
- Bracteoles present 142

141. Leaves pinnate or ternate, their primary lobes sessile or with very short petiolules; umbel rays stout, thicker than filiform; marginal mericarp ribs narrowly winged; inner part of mesocarp with a layer of lignified cells **233. *Lalldhwojia***
- Leaves bi- to triternate with long-petiolulate segments; umbel rays filiform; marginal mericarp ribs broadly winged; mesocarp parenchymatous **70. *Arcuatopterus***

142. Plants almost completely glabrous; marginal mericarp ribs with vascular bundles at the base of the wing 143
- Plants pubescent, or partly (stems) glabrous; marginal mericarp ribs with vascular bundles in middle part of the wing 148
143. Terminal leaf lobes filiform to lanceolate, entire, 2–9 cm long; petiolules long 144
- Terminal leaf lobes ovate to lanceolate, often toothed, rarely linear, short; petiolules short 145
144. Leaf blades lanceolate, terminal segments 3–10 × 0.5 mm, acuminate, calyx teeth obscure, mericarps 2–5 × 1.5–3 mm **326. *Peucedanum vaginatum***
- Leaf blades ovate, terminal segments 10–30 × 1–6 mm, blunt, calyx teeth broadly triangular, mericarps 5–7 × 4–5 mm **190. *Haloselinum***
145. Marginal mericarp ribs inflated; dorsal ribs filiform; commissural vittae situated very close to the carpophore. N India **448. *Tricholaser***
- Marginal mericarp ribs thin, not inflated; dorsal ribs usually short-keeled; commissural vittae situated in the central part of the commissure 146
146. Bracteoles often dissected, calyx teeth filiform to linear, or lacking; dorsal mericarp ribs filiform to keeled **244. *Ligusticopsis***
- Bracteoles entire or absent, calyx teeth triangular to lanceolate, sometimes persistent; dorsal mericarp ribs sub-inconspicuous to filiform 147
147. Terminal leaf lobes long-petiolulate, rib secretory ducts large, exceeding vallecular vittae; vallecular vittae 1–3, commissural 2–4 **398. *Sillaphyton***
- Terminal leaf lobes sessile, sometimes decurrent; rib secretory ducts inconspicuous or small; vallecular vittae always solitary, commissural 2 **223. *Kitagawia***
148. Distal parts of marginal mericarp ribs thickened and separated from proximal parts by a thin half-translucent constriction; terminal leaf lobes narrow, linear to lanceolate **466. *Zosima***
- Distal parts of marginal mericarp ribs thin, winged and without a thin half-translucent constriction; terminal leaf lobes filiform to ovate, often lobed or toothed 149
149. Stems simple, not branching; vittae lacking. E Tianschan **319. *Pastinacopsis***
- Stems branching; vittae present 150
150. Bracts leafy, reflexed, about half as long as rays or longer. Himalaya **441. *Tordyliopsis***
- Bracts small, considerably shorter than umbel rays, or lacking 151
151. Bracts present; plants low, to 50(–100) cm tall; stems glabrous or puberulent; leaves pinnate or bipinnate; their segments usually small (rarely of medium size), filiform to ovate, dissected; vittae reaching fruit base or nearly so **394. *Semenovia***
- Bracts often falling; plants larger, to 1.5–3 m tall (rarely smaller); stems covered with rigid spreading hairs; leaves simple, ternate or pinnate with large-lobed segments; vittae usually not reaching fruit base, often clavate 152
152. Petals white, outer petals often radiate; vallecular vittae 1 (rarely more); dorsal mericarp ribs filiform **201. *Heracleum***
- Petals yellow or yellowish-green, not radiate; vallecular vittae several, commissural usually 4; dorsal mericarp ribs keeled **430. *Tetrataenium***

F. *Key to the Native and Naturalized Genera of Apiaceae in Sub-Saharan Africa and Madagascar*

1. Stems with 4-rayed glochids and sessile, 4–8-rayed stellate hairs **22. *Drusa***
- Stems without glochids and stellate hairs 2
2. Umbels 1–3-fruited; fruits laterally compressed (mericarp wider in lateral than in dorsal view); leaves simple **4. *Centella***
- Umbels (4–)∞-fruited; fruits not laterally compressed (mericarp isodiametric, as wide in lateral as in dorsal view or often much narrower in lateral view than in dorsal view); leaves simple or compound 3
3. Sepals petaloid; petals terete and filiform (resembling staminodes); fruits with pseudo-wings (the lateral ribs of both mericarps extended into wing-like margins; true wings absent); leaves often densely hairy below (the surface completely obscured); peduncles densely hairy, at least toward the base **25. *Hermas***
- Sepals, if present, small and triangular, not petaloid; petals laminate; fruits without pseudo-wings (but true wings often present); leaves usually glabrous (sometimes hairy but the surface visible between the hairs); peduncle glabrous, rarely sparsely hairy 4
4. Pseudanthia present (female-fertile flower of each umbellule sessile and surrounded by large bracts or bracteoles so that the whole resembles a flower); leaves simple, usually markedly dentate, often with aristate or spiny teeth 5

- Pseudanthia absent (flowers pedicellate or, if rarely subsessile, then with inconspicuous bracts or bracteoles); leaves mostly compound, occasionally simple and/or markedly dentate, rarely with aristate or spiny teeth 9
5. Leaves all radical; plants dioecious; roots single, thick, tuberous; fruits usually with one fertile and one aborted mericarp; perianth of 10 equal segments **33. *Arctopus***
- Leaves both radical and cauline; plants andromonoecious or hermaphroditic; roots several, thin; fruits with two fertile mericarps; perianth differentiated into calyx and corolla 6
6. Leaves oblong, dentate-serrate but not lobed or cleft 7
- Leaves rounded, palmately lobed or cleft 8
7. Bracts large, leafy and spiny; flower heads oblong; foetid smell **35. *Eryngium***
- Bracts small, not leafy or spiny; flower heads rounded; resinous smell **32. *Alepidea***
8. Leaves mostly cauline, the lobes broad; pseudanthia small, few-flowered, the bracts inconspicuous; fruits with spiny bristles; widespread **38. *Sanicula***
- Leaves mostly radical, the lobes linear; pseudanthia relatively large, several-flowered, the bracts conspicuous; fruits without spiny bristles; highly localized, known from a single locality in South Africa **32. *Alepidea***
9. Fruits with large blisters in several longitudinal rows; annual; Namibia **329. *Phlyctidocarpa***
- Fruits glabrous or with papillae, warts or hairs (but not blisters); annual or perennial; Namibia and/or widespread 10
10. Fruits orbicular and discus-shaped; rib oil ducts absent (represented by globose oil vesicles in the wings only); vittae absent; prostrate annual; confined to central parts of South Africa **110. *Choritaenia***
- Fruits not as above; rib oil ducts and/or vittae present (absent in *Conium*); annual or perennial, South Africa and/or elsewhere 11
11. Regular vittae absent and rib oil ducts present and large (or represented by large cavities within the wings) 12
- Regular usually vittae present and rib oil ducts absent (or if vittae absent or very small, then rib oil ducts also inconspicuous or absent) 15
12. Perennial herbs with annual, non-woody flowering stems; leaves all radical; stylopodium prominent, conical; fruits somewhat flattened but not conspicuously winged **242. *Lichtensteinia***
- Trees, shrubs or shrublets with persistent, woody flowering stems; leaves all cauline; stylopodium not prominent; fruits usually flattened and conspicuously winged (isodiametric and not winged in *Marlothiella*) 13
13. Leaves small, succulent, persistent; fruits small (< 5 mm long), without wings **257. *Marlothiella***
- Leaves large, non-succulent, deciduous; fruits large (> 10 mm long), with broad wings 14
14. Mericarps heteromorphic (one with three wings, the other with two); leaves mostly digitately compound; localized in southern Africa **340. *Polemanniopsis***
- Mericarps homomorphic (both with two marginal wings); leaves pinnately compound, rarely digitate or simple; widespread **416. *Steganotaenia***
15. Leaves simple or palmately lobed or digitate 16
- Leaves pinnate, pinnately lobed or decompound, or basal leaves simple and cauline leaves pinnately dissected 29
16. Plants woody, not dying down to ground level after flowering 17
- Plants herbaceous, with stems dying down to ground level after flowering (the radical leaves may sometimes be persistent and the stems may be creeping, with only the inflorescences dying) 23
17. Fruits hairy or densely and evenly papillate **144. *Deverra***
- Fruits glabrous (sometimes wrinkled or irregularly warty) 18
18. Mature leaves acicular (pine needle-like) **58. *Anginon***
- Mature leaves broad, laminate 19
19. Mericarps heteromorphic (one with three wings, the other with two) **202. *Heteromorpha***
- Mericarps homomorphic (both with narrow marginal wings or no wings) 20
20. Bracts and bracteoles persistent along the woody stems, conspicuously veined **283. *Nirarathamnos***
- Bracts and bracteoles confined to the inflorescence, not conspicuously veined 21
21. Mericarps dorsally compressed (wider in dorsal than in lateral view), with narrow marginal wings; leaflets irregular, cuneate; southern Africa **339. *Polemannia***
- Mericarps isodiametric, without wings; leaflets simple, broad, elliptic to lanceolate; Madagascar 22
22. Rays and pedicels long and slender; climber; leaves with marginal teeth aristate **349. *Pseudocarum***
- Rays and pedicels relatively short and thick; shrub; leaves with marginal teeth not aristate **55. *Andriana***

23. Leaf or leaflet margins entire 24
- Leaf or leaflet margins not entire (dentate, serrate or lobed) 27
24. Leaves (or phyllodes) narrow, linear 25
- Leaves broad, lanceolate to ovate 27
25. Plant aquatic, rooting at the nodes; Madagascar **246. *Lilaeopsis***
- Plant terrestrial, not rooting at the nodes; southern Africa 26
26. Leaves grass-like, tapering to an acute tip; flowers yellow **91. *Bupleurum***
- Leaves linear or spathulate, blunt; flowers white **210. *Itasina***
27. Leaves with distinct petioles; flowers white **334. *Pimpinella***
- Leaves sessile, ± amplexicaulous; flowers yellow **91. *Bupleurum***
28. Leaves palmately lobed, cordate or peltate; margins irregularly and deeply dentate or more often lobed or cleft **152. *Dracosciadium***
- Leaves and leaflets lanceolate to oblong; margins evenly dentate or serrate 29
29. Fruits densely covered with minute papillae **332. *Physotrichia***
- Fruits glabrous or hairy but not papillate **334. *Pimpinella***
30. Plants woody, not dying down to ground level after flowering 31
- Plants herbaceous, with stems dying down to ground level after flowering (the radical leaves may sometimes be persistent and the stems may be creeping, with only the inflorescences dying) 49
31. Plants only slightly woody; umbels simple, short, leaf-opposed; petals papillose; vittae accompanied by a row of square or upright cells; confined to the South African coast, on dunes **139. *Dasispermum***
- Plants distinctly woody, at least at the base; umbels usually compound, long, terminal or axillary; petals usually glabrous; vittae not as above; widespread 32
32. Mericarps heteromorphic **202. *Heteromorpha***
- Mericarps homomorphic 33
33. Fruits hairy or densely and evenly papillate 34
- Fruits glabrous (sometimes wrinkled or irregularly warty) 36
34. Fruits densely and evenly papillate **332. *Physotrichia***
- Fruits hairy 35
35. Leaflets evenly serrate along the margins (resembling those of *Cannabis sativa*); calyx lobes triangular, well developed; mericarps perfectly round in transverse section **94. *Cannaboides***
- Leaflets entire or irregularly dentate; calyx lobes usually obsolete; mericarps angular and/or dorsally slightly compressed **148. *Diplolophium***
36. Upper cauline leaves and leaflets narrowly linear or acicular, the lower ones may have well-developed laminas 37
- Upper cauline leaves with broad, flat lamina (rarely modified to spines in *Pycnocycla*) 43
37. Leaves minutely hairy (scabrid or pilulose, visible at least along the margins and midrib below); root tuberous **184. *Glia***
- Leaves glabrous; root not tuberous 38
38. Rays and pedicels exceptionally long **284. *Normantha***
- Rays and pedicels not particularly long 39
39. Fruits with distinct ribs or wings 40
- Fruits without distinct ribs or wings 42
40. Mericarps dorsally compressed, with vascular bundles near the edge of the ribs; Madagascar **87. *Billburttia***
- Mericarps ± isodiametric, with vascular bundles within or below the ribs; southern Africa 41
41. Fruit ribs or wings very broad at the base (as seen in transverse section); stems with white ribs; suffrutex **418. *Stenosemis***
- Fruit ribs inconspicuous or absent, narrow if present; stems not ribbed; woody shrubs or shrublets **58. *Anginon***
42. Leaflets acicular, not linear, terete (often pine needle-like); fruits with rib oil ducts small or absent; southern Africa **58. *Anginon***
- Leaflets linear, flat; fruits with rib oil ducts relatively large; Madagascar **428. *Tana***
43. Mericarps ± isodiametric in transverse section; wings absent 44
- Mericarps dorsally compressed, with narrow to broad marginal wings 47
44. Fruit surface densely and evenly verrucose; leaves pinnately decompound, with two opposite pairs of leaflets and three terminal leaflets; flowers and fruits subsessile **60. *Anisopoda***
- Fruit surface glabrous or irregularly and minutely warty; leaves not as above; flowers and fruits on long or short pedicels 45
45. Rays and pedicels exceptionally long; woody climbers **349. *Pseudocarum***
- Rays and pedicels short; shrubs 46

46. Leaflets oblong with evenly serrate margins; flowers in long, compound umbels; large, erect shrubs; Madagascar **94. *Cannaboides***
- Leaflets cuneate, with dentate apices; flowers in short simple umbels; dwarf, decumbent shrublets; Somalia **395. *Seseli***
47. Leaf segments with striate venation; fruits without wings; fruit ribs with vascular bundle in the tips **87. *Billburttia***
- Leaf segments with pinnate venation (or venation invisible); fruits with marginal wings; fruit ribs with vascular bundle within or near the base 48
48. Dwarf shrublets; inflorescences much taller than the plants; fruits large, > 10 mm long **276. *Nanobubon***
- Shrubs or shrublets; inflorescences usually much shorter that the plants; fruits small, < 9 mm long **287. *Notobubon***
49. Perennials (sometimes biennial) 50
- Annuals (sometimes flowering in the second year) 78
50. Leaves all radical 51
- Leaves both radical and cauline 59
51. Leaves present at time of flowering 52
- Leaves usually withered or not yet produced at time of flowering 57
52. Fruits not dorsally flattened; marginal wings absent 53
- Fruits dorsally flattened; marginal wings usually present 55
53. Flowering stems and petioles densely squamose (covered with scales) **46. *Afrosciadium***
- Flowering stems and petioles not squamose 54
54. Leaves narrowly oblong, pinnate; umbels simple; eastern Africa **192. *Haplosciadium***
- Leaves triangular to rounded, bi- to tripinnate; umbels compound; southern Africa **63. *Annesorhiza***
55. Fruits with three broadly winged dorsal ribs (in addition to the two marginal wings) **353. *Pseudoselinum***
- Fruits with marginal wings only 56
56. Leaflet segments less than 8 mm wide (or if rarely > 8 mm wide, then plant less than 1 m tall, with few umbels on long thin peduncles, in *A. caffrum*, *A. eylesii*, or *A. platycarpum*), lamina coriaceous, veins ± obscured below; mericarps invariably winged **46. *Afrosciadium***
- Leaflet segments of basal leaves usually wider than 8 mm (or, if less than 8 mm wide, then leaf segment margin revolute, in *A. runssoricum*), lamina thin, veins conspicuous below; mericarps winged or wingless (*A. elliotii*, *A. thodei*) **45. *Afroligusticum***
57. Fruits large (10–25 mm long); marginal wings broad (4 mm or more) **136. *Cynorhiza***
- Fruits small (< 10 mm long); marginal wings, if present, narrow (< 2 mm) 58
58. Fruits oblong, usually conspicuously ribbed, 4–10 mm long; mericarps homomorphic or heteromorphic **63. *Annesorhiza***
- Fruits ovoid or flask-shaped (< 4 mm long), usually not ribbed; mericarps homomorphic **108. *Chamarea***
59. Fruits hairy, bristly, spiny or papillate 60
- Fruits glabrous 63
60. Fruits papillate **332. *Physotrichia***
- Fruits hairy, bristly or spiny 61
61. Fruits with long spines **48. *Agrocharis***
- Fruits hairy 62
62. Flowers in condensed umbels ("*Scabiosa*-like"), the umbellules with a single, sessile, central hermaphroditic flower surrounded by pedicellate male flowers; pedicels of outer flowers often indurated after anthesis, forming an hirsute cage around the fruit **363. *Pycnocycla***
- Flowers not as above **334. *Pimpinella***
63. Leaves hairy (sometimes minutely so) 64
- Leaves glabrous 66
64. Leaves minutely hairy (scabrid or pilulose, visible at least along the margins and midrib); root tuberous **184. *Glia***
- Leaves densely hairy on the lower surface; root not tuberous 65
65. Bracts reduced or absent **334. *Pimpinella***
- Bracts present **327. *Phellolophium***
66. Mericarps dorsally compressed (less so in *Bunium* and *Foeniculum*) 67
- Mericarps ± isodiametric 70
67. Small geophyte with a tuberous (fusiform) taproot; rays few (4–8) **90. *Bunium***
- Large herbs; roots not tuberous or fusiform; rays numerous 68
68. Mericarps ribbed; marginal wings absent; plants with anise smell; leaf segments filiform; fruit small (± 5 mm long) **174. *Foeniculum***
- Mericarps not ribbed; marginal wings prominent; plants without anise smell; leaf segments filiform or broad; fruit large (15–20 mm long) 69
69. Plant very robust (2–4 m high); sheathing leaf bases very large; leaflet segments filiform; leaves mostly 3- or 4-pinnate **171. *Ferula***

- – Plant not very robust (usually < 2 m high); sheathing leaf bases not very large; leaflet segments broad; leaves mostly 1-pinnate **201. *Heracleum***
- 70. Fruits spongy and corky (fruit wall thick or with blunt, corky ribs); calyx lobes small but present (minute or absent in *Helosciadium*) 71
- – Fruits not spongy (fruit wall thin); calyx lobes minute or absent 75
- 71. Leaves small, fleshy; confined to sand dunes along the coast **139. *Dasispermum***
- – Leaves relatively large, not fleshy; not on sand dunes along the coast 72
- 72. Rhizomes not rooting at the nodes; terrestrial plants **327. *Phellolophium***
- – Rhizomes often rooting at the nodes; usually aquatic herbs, often along streams and in marshy places 73
- 73. Calyx lobes minute or absent; fruit mesocarp spongy but relatively thin **199. *Helosciadium***
- – Calyx lobes distinct; fruit mesocarp usually thick 74
- 74. Leaves pinnate; bracteoles not exceeding the flowers in length **84. *Berula***
- – Leaves at least partly bipinnate; bracteoles often exceeding the flowers in length **289. *Oenanthe***
- 75. Bracts and bracteoles conspicuous, lanceolate-acuminate; leaflet segments serrate (resembling those of *Cannabis sativa*) **348. *Pseudocannaboides***
- – Bracts and bracteoles absent or inconspicuous; leaflets not as above 76
- 76. Basal leaves finely divided (resembling those of *Ceratophyllum*); pinnae sessile, much divided into slender, filiform ultimate segments **175. *Frommia***
- – Basal leaves simple or with broad segments 77
- 77. Stem long, creeping below the ground; basal leaves pinnatisect; flowers up to 7 per umbellule **128. *Cryptotaenia***
- – Stem short, erect, not creeping below the ground; basal leaves mostly simple; flowers > 7 per umbellule **334. *Pimpinella***
- 78. Fruits spiny, hairy or warty 79
- – Fruits glabrous 89
- 79. Bracts and bracteoles pinnate **141. *Daucus***
- – Bracts and bracteoles simple or absent 80
- 80. Fruits with spines (with or without hairs) 81
- – Fruits without spines (hairy or warty only) 83
- 81. Fruits with spines and hairs; petals hairy **443. *Torilis***
- – Fruits with spines only; petals not hairy 82
- 82. Bracts and bracteoles present; spines barbed **48. *Agrocharis***
- – Bracts and bracteoles inconspicuous or absent; spines not barbed **99. *Caucalis***
- 83. Mericarps dorsally compressed; hairs long, thick (bristly), dense **52. *Ammodaucus***
- – Mericarps ± isodiametric; hairs short, thin, dense or sparse 84
- 84. Ultimate leaf segments broad, ovate to lanceolate 85
- – Ultimate leaf segments narrow, oblong, linear to acicular 86
- 85. Fruits oblong and tapering, several times longer than broad; surface covered with sparse, minute bristles **64. *Anthriscus***
- – Fruits rounded or transversely oblong, shorter than wide in lateral view; surface covered with minute warts **298. *Oreoschimperella***
- 86. Petals papillose; umbels sessile or subsessile **139. *Dasispermum***
- – Petals glabrous; umbels usually pedunculate, rarely sessile (*Ezosciadium*) 87
- 87. Fruits oblong in lateral view; surface covered with short thin hairs; umbels sessile (peduncle absent) **168. *Ezosciadium***
- – Fruits rounded or broadly ovate; surface covered with short hairs or warts; umbels pedunculate 88
- 88. Fruits globose; surface densely covered with short hairs; Angola **59. *Angoseseli***
- – Fruits ovoid in lateral view; surface covered with sparse warts or hairs; horn of Africa **446. *Trachyspermum***
- 89. Fruits slender, acicular, beak-like, ± 50 mm long **382. *Scandix***
- – Fruits not as above 90
- 90. Bracts and bracteoles pinnate **51. *Ammi***
- – Bracts and bracteoles simple or absent 91
- 91. Fruits globose; outer petals of outer flowers radiate **120. *Coriandrum***
- – Fruits not globose; outer petals not enlarged 92
- 92. Mericarps ± isodiametric 93
- – Mericarps dorsally compressed 97
- 93. Calyx lobes small, triangular or irregular; escapes from cultivation 94
- – Calyx lobes minute or absent 95
- 94. Fruits ovate in lateral view; umbels compound, pedunculate; petals yellow; leaf segments laminate **325. *Petroselinum***
- – Fruits narrowly elliptical in lateral view; umbels simple, sessile or subsessile; petals white or pink; leaf segments acicular **129. *Cuminum***
- 95. Leaf segments acicular; umbels sessile or subsessile **132. *Cyclospermum***

- Leaf segments flat, laminate; umbels pedunculate 96
96. Bracts absent; fruits with corky ribs; vittae present; seed flat on commissural face; erect or procumbent herb with broad leaf segments **67. *Apium***
- Bracts conspicuous; fruits smooth or ribbed but the ribs not corky; vittae inconspicuous or absent; seed sulcate on commissural face; erect herb with relatively narrow leaf segments **118. *Conium***
97. Fruits with narrow marginal wings; dorsal surface strongly convex; vittae usually not clearly visible on the dorsal surface 98
- Fruits with broad marginal wings; dorsal surface ± flat; vittae usually clearly visible on the dorsal surface 100
98. Petals yellow; robust erect herb; strongly aromatic **56. *Anethum***
- Petals white; procumbent herbs; not strongly aromatic 99
99. Bracts and bracteoles absent; rays and raylets scabrous; fruits with indistinct median and lateral ribs; additional vittae in the marginal wings; commissural vittae close together; ultimate leaflet segments > 1.5 mm broad (never subterete), green **383. *Scaraboides***
- Bracts and bracteoles present; rays and raylets glabrous; fruits with prominent median and lateral ribs; additional vittae in the marginal wings absent; commissural vittae widely separate; ultimate leaflet segments less than 1 mm broad (often subterete), glaucous **95. *Capnophyllum***
100. Stems usually purple-spotted near base; rib oil ducts present; root not fleshy **238. *Lefebvrea***
- Stems not purple-spotted; rib oil ducts inconspicuous; root fleshy, white, carrot-like **318. *Pastinaca***

G. *Key to the Native and Naturalized Genera of Apiaceae in Oceania*

1. Endocarp soft parenchymatous 2
- Endocarp woody 41
2. Free carpophores present 3
- Free carpophores absent 38
3. Petals keeled abaxially 4
- Petals not keeled abaxially 5
4. Carpophores bifid (splitting about ⅔ of their length); petals varying in size and shape **120. *Coriandrum***
- Carpophores bifurcated (splitting entirely); petals uniform in size and shape **174. *Foeniculum***
5. Fruits with a beak 6
- Fruits without a beak 7
6. Involucel of entire bracteoles; beak shorter than the fruit body; commissure narrow; carpophore bifid (splitting apically) **64. *Anthriscus***
- Involucel of lobed or dissected bracteoles; beak several times longer than the fruit body; commissure broad; carpophore bifurcating (splitting entirely) **382. *Scandix***
7. Primary ribs of mericarps without bristles or spines 8
- Primary ribs of mericarps with bristles or spines 36
8. Bracts present 9
- Bracts absent 33
9. Plants of predominantly dry places, never emergent 10
- Plants of wet places, sometimes emergent 32
10. Stems with purple blotches 11
- Stems without purple blotches 13
11. Stems hollow; petal apices acute 12
- Stems solid; petal apices with two acute tips **262. *Melanoselinum***
12. Fruits compressed laterally (mericarps pentagonal); commissural surface sulcate **118. *Conium***
- Fruits and mericarps compressed dorsally; commissural surface plane **326. *Peucedanum***
13. Vittae conspicuous 14
- Vittae inconspicuous 30
14. Stems hollow 15
- Stems solid 16
15. Stems exuding a milky sap when cut **39. *Aciphylla***
- Stems not exuding sap when cut **181. *Gingidia***
16. Commissure narrower or as wide as the median lateral plane of the mericarp 17
- Commissure wider than the median lateral plane of the mericarp 18
17. Commissural surface width about as wide as the median lateral plane width 25
- Commissural surface much narrower than the median lateral plane width 27
18. Marginal ribs winged 19
- Marginal ribs not winged 21
19. Inflorescences terminal; commissural surface flat; mericarps with 5 ribs, secondary ribs absent 20
- Inflorescences leaf-opposed; commissural surface sulcate; mericarps with 5 ribs separated by secondary ribs **95. *Capnophyllum***
20. All ribs conspicuous **201. *Heracleum***
- All ribs inconspicuous **442. *Tordylium***
21. Pedicels all ± the same length 22
- Pedicels of unequal lengths **128. *Cryptotaenia***

22. Staminodes and fertile stamens present **62. *Anisotome***
- Stamens all fertile 23
23. Lateral ribs winged **171. *Ferula***
- Lateral ribs not winged 24
24. Rays equal **98. *Carum***
- Rays unequal in length **405. *Sison***
25. Leaves simple and entire **91. *Bupleurum***
- Leaves compound 26
26. Leaves glaucous; carpophores bifurcating (splitting entirely) **127. *Crithmum***
- Leaves not glaucous; carpophores bifid (splitting less than ½ their length) **410. *Spermolepis***
27. Petals uniform in size and shape 28
- Outer petals radiate **51. *Ammi***
28. Inflorescences simple **296. *Oreomyrrhis***
- Inflorescences compound, sometimes much reduced 29
29. Bracts and/or bracteoles present; mericarps with air chambers at maturity **199. *Helosciadium***
- Bracts lacking or inconspicuous; mericarps without air chambers at maturity **325. *Petroselinum***
30. Plants herbaceous; marginal ribs not winged 31
- Plants subshrubs or shrubs; marginal ribs winged **381. *Scandia***
31. Plants perennial; mericarps smooth **43. *Aegopodium***
- Plants annual; mericarps sculptured **85. *Bifora***
32. Vittae cyclic **84. *Berula***
- Vittae not cyclic **406. *Sium***
33. Commissural surface wider than the median lateral plane width; fruits dorsally flattened 34
- Commissural surface much narrower than the median lateral plane width; fruits laterally flattened 35
34. Dorsal ribs filiform, not winged; lateral ribs filiform; marginal ribs broad (± winged) **318. *Pastinaca***
- Dorsal ribs filiform to keeled or narrowly winged; lateral ribs filiform to narrowly winged; marginal ribs not winged **56. *Anethum***
35. Inflorescences terminal **67. *Apium***
- Inflorescences leaf-opposed **132. *Cyclospermum***
36. Outermost petals often larger than innermost petals (± radiate) 37
- All petals uniform in size and shape, the outermost petals not radiate **443. *Torilis***
37. Outer (radiate) petals in umbellules deeply bilobed **301. *Orlaya***
- Outer petals (radiate) in umbellules not deeply bilobed **141. *Daucus***
38. All petals uniform in size and shape, the outermost petals not radiate 39
- Outermost petals often larger than innermost petals (± radiate) **289. *Oenanthe***
39. Plants of wet places, sometimes emergent **246. *Lilaeopsis***
- Plants of predominantly dry places, never emergent 40
40. Leaves spinulose-dentate; flowers arranged in a capitulum **35. *Eryngium***
- Leaves not spinulose-dentate; flowers arranged in umbels **38. *Sanicula***
41. Free carpophores present; ovary bilocular 42
- Free carpophores absent; ovary bilocular or pseudo-unilocular 47
42. Plants herbaceous 43
- Plants subshrubs or shrubs **10. *Platysace***
43. Leaves in basal rosettes 44
- Leaves basal or cauline, but not in rosettes 46
44. Inflorescences terminal; petals not inflexed 45
- Inflorescences leaf-opposed; petals inflexed **243. *Lignocarpa***
45. Flowers in umbels; fruits dorsally flattened **19. *Diplaspis***
- Flowers solitary; fruits laterally flattened **28. *Oschatzia***
46. All ribs conspicuous; plants perennial **14. *Azorella***
- All ribs inconspicuous; plants annual **6. *Homalosciadium***
47. Fruits consisting of a single "mericarp" **1. *Actinotus***
- Fruits always with 2 mericarps 48
48. Fruits dry 49
- Fruits fleshy 56
49. Sepals absent; fruits obscurely ribbed; plants stellate-pubescent; umbels simple **16. *Bowlesia***
- Sepals present, or if absent then fruits lacking ribs; plants rarely stellate-pubescent, but then with compound umbels and well-developed sepals 50
50. Sepals petaloid with a broad base, white, identical to petals; rosettes atop a well-developed fleshy taproot; restricted to alpine habitats of Australia **17. *Dichosciadium***
- Sepals large and peltate, reduced or absent, rarely similar to petals; herbs or subshrubs, not forming rosettes; occupying a range of habitats but never alpine 51
51. Fruits ribbed; sepals present 52
- Fruits not ribbed; sepals absent 55
52. Fruits with obvious reticulation between the primary ribs **4. *Centella***

- Fruits without obvious reticulation between the primary ribs 53
53. Leaves, linear, terete (probably phyllodenous), bracts and bracteoles cartilaginous **11. *Schoenolaena***
- Leaves flat, bracts and bracteoles foliaceous, occasionally petaloid 54
54. Petals not inflexed; styles thick, constricted at the base **9. *Pentapeltis***
- Petals with an inflexed appendage; styles not thickened or constricted at the base **12. *Xanthosia***
55. Bracts and bracteoles cartilaginous, ovate; nectaries free from the styles **5. *Chlaenosciadium***
- Bracts and bracteoles foliaceous, linear; nectaries adnate to styles **3. *Brachyscias***
56. Shrubs or small trees; leaves simple with pinnate venation or compound; locules 2–4 57
- Robust herbs; leaves simple with palmate venation; locules 3–4; restricted to subantarctic islands **14. *Azorella***
57. Densely hirsute shrubs; leaves simple, penniveined, strongly dentate; styles united; locules 2–4; endemic to New Caledonia **2. *Apiopetalum***
- Glabrous shrubs; leaves unifoliolate or palmately compound; styles free; locules 2; Malesia and western Pacific **7. *Mackinlaya***

I. **Subfam. Mackinlayoideae** G.M. Plunkett & Lowry (2004).

Caulescent and often erect, herbaceous annuals or perennials, or woody subshrubs to shrubs. Leaves basal and/or cauline, usually petiolate, the lamina entire to divided. Inflorescence simple to compound umbellate, or pseudanthial and capitulate; peduncles terminal and/or lateral. Petals glabrous, the base usually clawed and the apex narrowed and inflexed; ovaries usually bicarpellate. Fruits typically dry and schizocarpic (rarely fleshy and indehiscent); vallecular vittae lacking (but vittae-like oil ducts may be scattered in the mesocarp); rib oil ducts usually present and distinct (but generally not large). Rhomboidal crystals limited to the inner layer of the mesocarp (often forming a ring around the endocarp). Endocarp of fiber-like sclereids (the inner fibers arranged longitudinally, the outer fibers transversely). Fruits often strongly compressed laterally.

1. ***Actinotus*** Labill. [Hart & Henwood]

Actinotus Labill., Nov. Holl. Pl. 1: 67, t. 92 ('1804', 1805).

Pilose or pubescent, herbaceous annuals or perennials, or suffruticose. Leaves simple or ternately dissected, petioles scarcely sheathing; ultimate divisions narrowly oblong. Inflorescence often a many-flowered pseudanthial capitulum, or a simple umbel; bracts in one or two whorls, usually lanate or sericeous. Calyx lobes prominent, free or basally fused; petals white, pink or purple, the apex plane, or petals lacking; stamens 3 or 5; ovary unilocular; styles elongate; stylopodium compressed. Fruits pubescent or glabrous, oblong to ellipsoid or ovoid, comprising a single mericarp; ribs 4 or 5, filiform; carpophore lacking; rib oil ducts 1 under each rib or lacking. Seeds transversely compressed laterally. $n = 10$.

21 species, Australia (all states except South Australia) and New Zealand, ranging from alpine to semi-arid habitats. *Actinotus helianthi* Labill. is floriculturally significant.

2. ***Apiopetalum*** Baill. Fig. 8 [Plunkett]

Apiopetalum Baill., Adansonia 12: 133 (1878).

Densely hirsute, woody, tall shrubs. Leaves simple, strongly dentate, oblong-lanceolate to spathulate-oblanceolate, thick and subcoriaceous; petiole bases slightly sheathing. Umbels 2–3-compound (to irregularly umbellate or cymose); bracts and bracteoles several, lanceolate to spathulate; rays ascending to spreading. Calyx lobes narrow to subulate; petals greenish-white; styles united; stylopodium low-conical to conical; ovaries 2–4-carpellate. Fruits puberulent, narrow, transversely subterete (not compressed laterally); ribs and carpophore lacking; exocarp leathery, mesocarp fleshy; rib oil ducts associated with vascular bundles. $n = 12$.

Two species, New Caledonia.

3. ***Brachyscias*** J.M. Hart & Henwood [Hart & Henwood]

Brachyscias J.M. Hart & Henwood, Austral. Syst. Bot. 12: 176 (1999).

Glabrous annual or ephemeral herbs with short, unbranched stem and slender taproot. Leaves basal, ternately dissected, petiole sheathing only shortly; the ultimate divisions linear. Umbels compound, composed of rayed umbellules and single flowers; bracts 3; rays 3, long, spreading;

Fig. 8. Apiaceae, Mackinlayoideae. *Apiopetalum.* **A–C** *A. glabratum.* **A** Branch with leaves. **B** Juvenile leaf. **C** Branch of infructescence with mature fruit. **D, E** *Apiopetalum velutinum.* **D** Fruiting branch with leaves. **E** Flowers at anthesis. (Orig., illustrations by J. K. Myers; © P. P. Lowry II)

bracteoles 3, the lateral asymmetrical. Calyx lobes absent; petals white; styles short; stylopodium flat. Fruits glabrous, broadly ovate to spherical, base cordate, black, only weakly laterally compressed; ribs obscure, mericarp surface undulate; carpophore absent. Seeds transversely compressed laterally.

One species, *B. verecundus* J.M. Hart & Henwood, Australia (temperate SW Western Australia), known from only two localities.

4. ***Centella*** L. [Van Wyk & Tilney]

Centella L., Sp. Pl. ed. 2: 1393 (1763), Gen. Pl. ed. 6, 458 (1764); Adamson, J. S. Afr. Bot. 17: 1–48 (1951), rev.; Cannon, Fl. Zambes. 4: 561–565 (1978), reg. rev.; Eichler, Feddes Repert. 98, 1–2: 1–51 (1987), rev.; Schubert & Van Wyk, Nord. J. Bot. 17: 301–314 (1997), sec. rev.; Schubert, unpubl. Ph.D. thesis, Rand Afrikaans University (2000), rev.; Van Wyk et al., Afr. Apiaceae: 126–153 (2013).

Glabrous or pubescent annuals or mostly perennials; stems herbaceous, prostrate and creeping, or woody and stout or ascending. Leaves petiolate or sessile, simple, lamina broad to acicular; petiole sheathing or not. Umbels usually simple; bracts and bracteoles few, lanceolate to ovate, sometimes foliaceous; rays few. Flowers solitary or in small imperfect or perfect capitate umbellules, usually with a 1-flowered perfect umbellule and several 1–7-flowered staminate umbellules (usually andromonoecious or androdioecious) or with a single perfect (functionally pistillate) flower surrounded by 4 staminate flowers, rarely all flowers perfect; calyx annular; petals white, yellow or purplish; styles relatively long and divergent, rarely short; stylopodium variable. Fruits glabrous or pubescent, variable in shape; primary and secondary ribs often forming narrow ridges; carpophore lacking; rib oil ducts small or obsolete. Seeds strongly compressed laterally, the face plane to convex. $n = 9, 11$; $2n = 14$, 18, 20, 22, 36, 38.

45 species, mostly S Africa. *Centella asiatica* (L.) Urban pantropical and subtropical, and an important medicinal plant.

5. ***Chlaenosciadium*** C. Norman [Hart & Henwood]

Chlaenosciadium C. Norman, J. Bot. 76: 198 (1938).

Sericeous pubescent, herbaceous perennials with a taproot. Leaves mostly basal, simple, elliptic to oblanceolate or linear, petioles scarcely sheathing. Umbels compound, pseudanthial; bracts and bracteoles few to several, ovate with acuminate tips, cartilaginous (lateral bracteoles asymmetrical); rays 3, compressed, spreading, plus one central, rayless umbellule. Calyx lobes obscure; petals white; styles long; stylopodium flat. Fruits glabrous, rugose, broadly ovate, base cordate; ribs obscure; carpophores lacking; rib oil ducts lacking. Seeds transversely slightly compressed laterally. $n = 10$.

One species, *C. gardneri* C. Norman, SW Australia, semi-arid regions.

6. ***Homalosciadium*** Domin [Hart & Henwood]

Homalosciadium Domin, Beih. Bot. Centralbl. 23(2): 294 (1908).

Glabrous, herbaceous annuals. Leaves cauline, ternately dissected, deltoid; petioles sheathing. Umbels simple, axillary and sessile; bracts linear, inconspicuous. Flowers few per umbel, pedicles unequal; calyx lobes obscure; petals white or reddish; stylopodium flat. Fruits glabrous, oblong; ribs present, median and lateral ribs prominent; carpophore bifid less than half the length; rib oil ducts present in some ribs. Seeds transversely cruciform. $n = 12$.

One species, *H. verticillatum* Domin, temperate SW Australia.

7. ***Mackinlaya*** F. Muell. Fig. 9 [Plunkett]

Mackinlaya F. Muell., Fragm. 4: 119 (1864).
Anomopanax Harms (1904).

Glabrous woody shrubs (or epiphytes or treelets). Leaves unifoliolate or palmately compound, the central leaflet (or three leaflets) sometimes palmately lobed; petioles dilated and sheathing. Umbels 2–3-compound or cymose; bracts several, lanceolate; rays ascending to spreading; bracteoles several, small, linear to lanceolate. Flowers 5–6-merous (except for bilocular ovary); calyx lobes triangular to lanceolate; petals white to greenish-white; styles free, subulate, recurved in fruit; stylopodium prominent, flat, with crenate margins. Fruits glabrous, often maturing to pale blue, transversely narrow-elliptical; ribs and carpophore lacking; exocarp leathery, mesocarp fleshy; vittae scattered, rib oil ducts usually associated with vascular bundles. Seeds transversely strongly compressed laterally.

Five species, Malesia (New Guinea, Celebes, Philippines), Solomon Islands, Australia (Queensland), understories of rain forests, montane forests, or secondary growth.

8. ***Micropleura*** Lag. [Plunkett]

Micropleura Lag., Ocios Esp. Emigr. 4: 347 (1825); Mathias & Constance, N. Amer. Fl. 28B(1): 59–60 (1944), reg. rev.

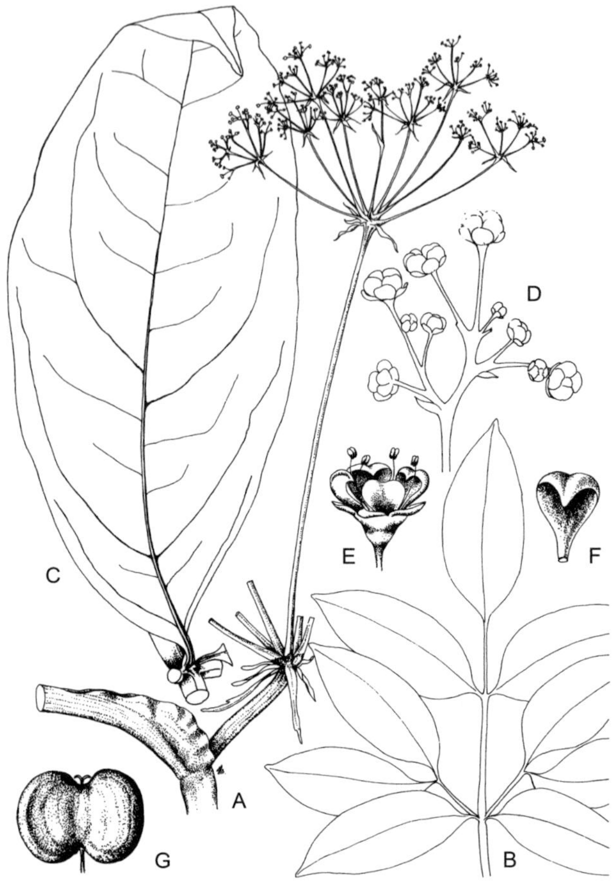

Fig. 9. Apiaceae, Mackinlayoideae. *Mackinlaya celebica*. **A** Part of inflorescence. **B** Part of leaf. **C** Leaflet. **D** Ultimate inflorescence unit. **E** Flower. **F** Petal. **G** Fruit. (Philipson 1979; illustrations by H. Mulder, reproduced with permission from Flora Malesiana)

Villous to glabrous, herbaceous perennials with tuberous roots. Leaves simple, repand to crenate, reniform to broadly ovate with cordate bases; petioles scarious-sheathing. Umbels simple or irregularly compound; bracts one or more, small; rays few, very unequal, slender, ascending or weakly spreading; bracteoles several, linear or lanceolate. Calyx lobes obsolete; petals white; styles short; stylopodium depressed. Fruits glabrous, ellipsoid, cordate, membranaceous; commissure constricted; ribs filiform, curved; carpophore lacking; rib oil ducts solitary in each rib. Seeds transversely strongly compressed laterally, the face plane. $n = 10$.

Two species, N America (Mexico), C America (Costa Rica, Guatemala, Honduras), S America (Colombia).

9. *Pentapeltis* (Endl.) Bunge [Hart & Henwood]

Pentapeltis (Endl.) Bunge in Lehm., Pl. Preiss. 1: 292 (1845).

Procumbent, glabrescent, herbaceous perennials with a fleshy taproot. Leaves simple, depressed ovate or trullate, dentate; petioles scarcely sheathing. Umbels compound, peduncles terminal or lateral; bracts few, linear, foliaceous, equal to or longer than the rays; rays few, spreading, terete, plus one central, rayless umbellule; bracteoles few, elliptic to linear, foliaceous. Flowers few to several per umbellule; calyx lobes prominent, petaloid and peltate; petals white or pink, apex acute but not inflexed; styles stout, basally constricted; stylopodium flat. Fruits glabrous or minutely papillate, ovate, base cordate; primary and secondary ribs keeled; carpophores lacking; rib oil ducts indistinct. Seeds transversely compressed laterally to subterete. $n = 5$.

Two species, temperate SW Australia.

10. *Platysace* Bunge [Hart & Henwood]

Platysace Bunge in Lehm., Pl. Preiss. 1: 285 (1845); Norman, J. Bot. 77: 207–211 (1939), rev.

Glabrous to pubescent perennial herbs, subshrubs or shrubs (stems sometimes woody and apparently leafless, then terete or compressed and photosynthetic). Leaves reduced to scales, or fully expanded and basal or cauline, petiolate or sessile, simple or ternately dissected, narrowly linear to lanceolate or orbicular, narrowly sheathing. Umbels compound (occasionally simple); bracts simple, small, subulate, sometimes caducous; rays few, thyrsoid; bracteoles usually 3, prominent. Calyx lobes minute or obscure; petals white, pink or yellow, apex early inflexed, later plane; styles short to long and slender, ascending to reflexed; stylopodium conical. Fruits glabrous or hirsute, smooth or tuberculate, elliptic, oblong, ovoid or orbicular; ribs prominent (sometimes scarcely winged) to obscure; carpophores entire or apically bifid; rib oil ducts obscure. Seeds transversely compressed laterally. $n = 8$. 27 species, Australia (all states except Tasmania).

Two sections recognized (Norman 1939).

11. *Schoenolaena* Bunge [Hart & Henwood]

Schoenolaena Bunge in Lehm., Pl. Preiss. 1: 289 (1845).

Glabrous herbaceous perennials with rhizomes. Leaves mostly basal, linear, terete and hollow (probably phyllodinous), with a conspicuous apical hydathode, base sheathing. Umbels compound, pseudanthial. Bracts few to several, white and cartilaginous, elliptic to rhombic with an acute to acuminate apex; rays few to several, very short, slightly compressed; bracteoles 4, white and cartilaginous, the lateral bracteoles asymmetrical. Calyx lobes petaloid, ovate or circular, $\pm$ peltate; petals white; styles long; stylopodium flat. Fruits glabrous, narrowly oblong; ribs filiform; carpophores lacking; rib oil ducts indistinct. Seeds transversely strongly compressed laterally. $n = 5$.

One species, *S. juncea* Bunge, SW Australia, in winter-wet swamps, the phyllodes absent at summer flowering.

12. *Xanthosia* Rudge [Hart & Henwood]

Xanthosia Rudge, Trans. Linn. Soc. London, Bot. 10: 301 (1811).

Hirsute to pilose, herbaceous, suffrutescent or frutescent perennials (rarely ephemeral plants). Leaves petiolate to nearly sessile, simple, ovate to triangular, ternately dissected or compound, petioles narrowly sheathing. Umbels compound (occasionally appearing simple), often pseudanthial; bracts $\pm$ petaloid, one per ray; rays solitary to several, plus often one central, rayless umbellule, or all umbellules sessile (and thus the umbel appearing simple); bracteoles few, $\pm$ petaloid, green, white, pink or yellow, lateral bracteoles often asymmetrical. Calyx lobes large, petaloid, rarely peltate; petals white or reddish; styles long and slender to short and stout; stylopodium flat or raised, glabrous or hirsute. Fruits $\pm$ hirsute above, ovate or narrowly oblong, base cordate; primary ribs prominent, often keeled; secondary ribs frequently evident and keeled; carpophore lacking; rib oil ducts usually solitary under each rib. Seeds transversely compressed laterally. $n = 10, 20$.

20 species, S temperate Australia.

II. **Subfam. Azorelloideae** G.M. Plunkett & Lowry (2004).

Typically caulescent (rarely acaulescent) with a slender to stout or fleshy taproot (or a stout caudex, thick rootstock or rhizome). Stems alternately or dichotomously branched (or scapose and/or unbranched). Leaves usually petiolate, with sheathing bases. Umbels simple (sometimes arranged in higher orders of umbels, panicles, or corymbs); flowers few to many per umbel; petals usually glabrous, base usually clawed and apex narrowed and inflexed. Fruits bicarpellate and schizocarpic (fleshy with 2–4 carpels in *Azorella* sect. *Stilbocarpa*); compressed dorsally (laterally in *Domeykoa* and *Eremocharis*), typically widest at the two lateral ribs (less commonly at the marginal ribs, as in *Laretia*, or rarely terete, as in *Oschatzia*); endocarp of fiber-like sclereids (the inner fibers arranged longitudinally, the outer fibers transversely); vallecular vittae lacking (vittae-like oil ducts may be scattered in the mesocarp); rib oil ducts usually solitary under each rib (not generally large); rhomboidal crystals limited to the inner layer of the mesocarp (often forming a ring around the endocarp).

13. ***Asteriscium*** Cham. & Schltdl. [Plunkett]

Asteriscium Cham. & Schltdl., Linnaea 1: 354, pl. 5 (1826); Mathias & Constance, Univ. Calif. Publ. Bot. 33(2): 107–128 (1962), rev.

Glabrous, glaucous, herbaceous to subshrubby annuals or perennials. Leaves basal and sometimes cauline, simple and undulate to spinulose and dentate or lobed, to ternately or pinnately dissected, ovate to orbicular-reniform, sometimes subcoriaceous or succulent. Umbels simple, solitary or in corymbs, panicles, or compound umbels (one umbel of each group often sessile); bracts several, linear to obovate, entire or lobed, free to slightly connate. Calyx lobes triangular to ovate-lanceolate; petals greenish-white or pinkish to yellow, with a median dorsal gland; styles slender to short and stout; stylopodium conical to depressed-conical. Fruits glabrous, oblong to oblong-elliptical, transversely obtuse-tetragonal to narrowly winged, with a convex commissural face and concave to plane dorsal face; dorsal ribs prominent to obscure, lateral ribs winged or forming a spreading or reflexed flange, marginal ribs filiform or obscure; carpophore filiform, entire. Seeds transversely lunate (to nearly terete), the face convex. $n = 10$.

Nine species, temperate S America (Chile, Argentina, esp. Andes). Closely related to *Gymnophyton* and *Pozoa*.

14. ***Azorella*** Lam. [Plunkett, Webb, Mitchell]

Azorella Lam., Encycl. Meth. Bot. 1: 344 (1783); Nicolas & Plunkett, Taxon 61: 826–840 (2012), phylog.; Plunkett & Nicolas, Brittonia 69: 31–61 (2016), rev.
Huanaca Cav. (1800).
Mulinum Pers. (1805).
Laretia Gillies & Hook. (1830).
Stilbocarpa (Hook. f.) Decne. & Planch. (1854).
Schizeilema (Hook. f.) Domin (1908).
Kirkophytum (Harms) Allan (1961).

Glabrous to strongly pubescent, herbaceous to very hard and woody perennials, often caespitose or forming dense cushions, mats, subshrubs, shrubs, or rarely mega-herbs. Leaves basal or cauline (uppermost sometimes opposite), sessile or petiolate (rarely peltate), small to large, simple and entire to deeply 3–7-lobed to compound (the divisions sometimes further trilobed), membranaceous to coriaceous or fleshy, often thickened or spinose. Umbels simple, sometimes accompanied by 1 to several lateral umbellules or arranged in dichasia (rarely compound with 3 orders of branching); bracts few to several, linear, ovate, triangular or lanceolate, to foliaceous and sometimes basally connate, entire or with a few teeth or lobes (sometimes fused); bracteoles (if present) foliaceous. Calyx lobes evident to obsolete; petals white, greenish, yellowish, reddish-yellow, or purplish (or with a purple center), ovate to oblong, apex plane or somewhat incurved; styles usually short to longer and slender; stylopodium depressed to conical; carpels usually 2 (rarely 3–4; sect. *Stilbocarpa* only). Fruits dry and glabrous (rarely scaly or minutely rugose), or rarely succulent (sect. *Stilbocarpa* only), ellipsoid to ovoid or obovoid to linear-oblong or subglobose, transversely compressed dorsally or terete, the commissure often narrow; ribs inconspicuous, or filiform and unwinged, or lateral sometimes strongly angled to winged, or marginal broadly winged; carpophore absent or present, sometimes free and persistent, apically bifid to bifurcating;

rib oil ducts sometimes inconspicuous. Seeds transversely dorsally compressed to dorsally elliptic or subterete to terete (rarely 3–4 angular), the face convex to plane (more rarely concave). $n = 8, 9, 16, 24$; $2n = 16, 18, 32, 48, 64, 80$.

58 species, S America (esp. Andes and Patagonia), C America (Costa Rica), New Zealand, Australia (SE), and several subantarctic islands in the Atlantic (Falkland Islands, Is.), Indian (Heard Is., Kerguelen Is., Crozet Is., Marion and Prince Edward Is.), and Pacific (Macquarie Is., Auckland Is., Campbell Is., Antipodes Is.) oceans. Ten sections recognized (Plunkett and Nicolas 2016). Closely related to *Diplaspis* and *Dickinsia* (Nicolas and Plunkett 2012).

15. ***Bolax*** Comm. ex Juss. [Plunkett]

Bolax Comm. ex Juss., Gen. Pl. 226 (1789); Constance in Correa, Fl. Patagonica 8(5): 332–334 (1988), reg. rev.

Cushion-forming, stellate-pubescent, herbaceous perennials. Leaves simple, entire to trilobed. Umbels simple, bracts sessile or subsessile. Calyx lobes conspicuous; petals greenish-yellow, apex plane or slightly incurved and bifid; styles short; stylopodium conical. Fruits stellate-pubescent, orbicular to obovoid, transversely tetragonal; dorsal and marginal ribs obscure, lateral ribs narrowly winged; mericarps dorsally compressed, narrowed at the commissure, easily separated; carpophore filiform, entire and persistent. Seeds transversely cuneate-tetragonal to triangular, the face convex. $n = 16$.

Two species, temperate S America (Subantarctic Argentina and Chile) and Falkland Islands. Closely related to *Drusa* and *Dichosciadium*.

16. ***Bowlesia*** Ruiz & Pav. Fig. 10 [Plunkett]

Bowlesia Ruiz & Pav., Fl. Peruv. Prodr. 44, t. 34 (1794); Mathias & Constance, Univ. Calif. Publ. Bot. 38: 13–55 (1965), rev.

Caulescent, stellate-pubescent herbaceous perennials or annuals. Leaves opposite or alternate, simple and orbicular-reniform to ovate-acuminate, palmately lobed or parted. Umbels simple, subcompact; bracts few, subulate. Calyx lobes hyaline or lacking; petals whitish (to variously tinged), dorsally stellate-pubescent or glabrous, apex not inflexed; styles rather short; stylopodium conical. Fruits stellate-pubescent or prominently glochidiate, broadly ovoid to oblong-ovoid, quadrangular, subwinged to unwinged, strongly compressed dorsally to terete; ribs obsolete; carpophore lacking; rib oil ducts sometimes obsolete. Seeds transversely compressed dorsally, the face slightly to strongly convex. $n = 8, 16$, $2n = 16$.

15 species, S America (esp. high Andes), C America (C and N Mexico), N America (S United States); naturalized in Australasia, Asia, and Europe; various habitats from low to (especially) high elevations. Closely related to *Drusa* and *Homalocarpus*.

17. ***Dichosciadium*** Domin [Hart & Henwood]

Dichosciadium Domin, Repert. Spec. Nov. Regni Veg. 5: 104 (1908).

Low-caulescent, hirsute, herbaceous perennials. Leaves rosulate, ternately to palmately lobed, orbicular to broadly ovate, slightly fleshy. Umbels simple; bracts foliose, unequal, basally united. Calyx lobes petaloid; petals equal to sepals, white, base clawed or truncate, apex rounded and plane; styles short, reflexed; stylopodium low-conical. Fruits ellipsoid; median and marginal ribs obscure to filiform, lateral ribs prominent; carpophore lacking. Seeds transversely compressed dorsally, the face convex.

One species, *D. ranunculaceum* (F. Muell.) Domin, SE Australia, alpine regions.

18. ***Dickinsia*** Franch. [Watson]

Dickinsia Franch., Nouv. Arch. Mus. Hist. Nat. II, 8: 244 (1885); She et al., Fl. China 14: 18 (2005), rev.

Scapose, glabrous, herbaceous annuals or biennials. Stems unbranched, leafless. Leaves petiolate, mostly basal, simple, reniform or rounded and deeply cordate, margin irregularly crenate. Umbels simple; bracts 2, like the leaves (but $\pm$ sessile). Flowers several to many per umbel; calyx lobes minute or obsolete; petals white or greenish, apex obtuse, plane; styles very short, recurved; stylopodium conical. Fruits glabrous, rectangular-cubic; median rib filiform and prominent, lateral ribs obscure, marginal ribs winged; carpophore shortly bifid at the apex, persistent.

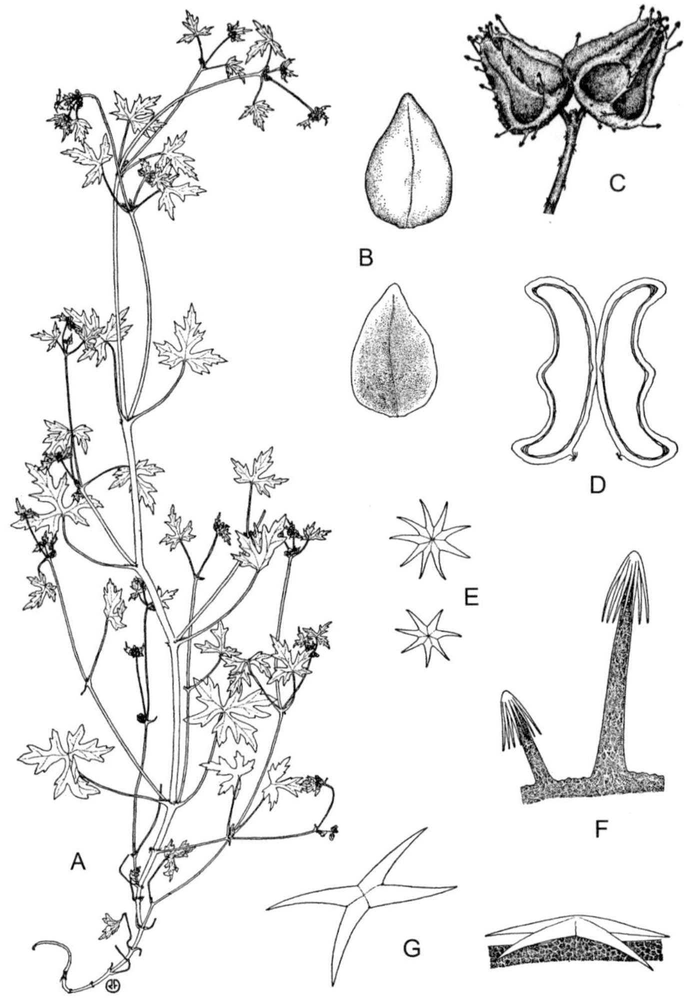

Fig. 10. Apiaceae, Azorelloideae. *Bowlesia palmata* (*Beetle 9088*, UC). **A** Habit. **B** Ventral and dorsal petal surfaces. **C** Fruits. **D** Fruit transverse section. **E** Stellate trichomes from fruit. **F** Glochids from fruit. **G** Stellate processes from petiole. (Mathias and Constance 1965; illustrations by Reino Alava)

Seed transversely compressed dorsally, the face plane. $2n = 22$.

One species, *D. hydrocotyloides* Franch., Asia (S and SW China). Damp shady forests and sides of streams.

19. ***Diplaspis*** Hook.f. [Hart & Henwood]

Diplaspis Hook.f., London J. Bot. 6: 468 bis (1847); Van den Borre & Henwood, Austral. Syst. Bot. 11: 1–12 (1998), rev.

Low-caulescent, glabrous to pubescent herbaceous perennials. Leaves rosulate, simple, ovate to cordiform; petiole sheaths with filiform extensions at the apex. Umbels simple; bracts linear, simple, apex obtuse. Calyx lobes lacking; petals green, white or yellow; styles stout, incurved to spreading; stylopodium subconical. Fruits glabrous and smooth, lateral ribs prominent, median and marginal ribs obscure; carpophore shortly bifid. Seeds transversely dorsally compressed.

Three species, SE Australia, alpine regions.

20. ***Diposis*** DC. [Plunkett]

Diposis DC., Coll. Mém. 5: 33, t. 3 (1829); Constance in Correa, Fl. Patagonica 8(5): 343–344 (1988), reg. rev.

Glabrous herbaceous perennials with a globose tuber. Leaves weakly rosulate and cauline; basal leaves 3–5-parted, segments 3–5-parted or lobed, or ternate-pinnately compound; the cauline leaves small, the uppermost simple and bracteate. Umbels simple or appearing compound; bracts conspicuous, each containing three flowers (1 perfect, 2 staminate). Calyx lobes minute or obsolete; petals white, apex plane or slightly incurved; styles short; stylopodium low conical. Fruits glabrous, ellipsoid to orbicular; lateral ribs winged, dorsal and marginal wings obscure; carpophore present, free, bifurcated; commissure very narrow, carpels easily separated; rib oil ducts displaced toward the margin of the wing in the lateral rib.

Three species, temperate S America (C Chile, Uruguay, and Argentina).

21. *Domeykoa* Phil. [Plunkett]

Domeykoa Phil., Fl. Atacam. 25, pl. 2c (1860); Mathias & Constance, Univ. Calif. Publ. Bot. 33(2): 173–182 (1962), rev.

Glabrous and glaucous annuals or woody-based perennials. Leaves rosulate and cauline (sometimes opposite), petiolate or sessile, lobed or 1–3-ternately or pinnately divided. Umbels simple, compact-globose, solitary or in terminal umbellate clusters; bracts several, linear-lanceolate to orbicular, free to partially connate. Calyx lobes ovate to obovate, entire or trifid; petals white, yellowish, or purplish, with a median dorsal gland, apex plane or slightly incurved or very shortly inflexed; styles short to long; stylopodium depressed to low-conical. Fruits glabrous, orbicular to elliptic, truncate at apex, transversely cuneate-subtetragonal with prominent, unwinged margins, compressed laterally; commissure constricted; lateral ribs prominent, sharply angled, dorsal and marginal ribs filiform; carpophore lacking. Seeds transversely cuneate-rectangular, intrusively sulcate at the commissure.

Five species, W tropical and temperate S America (Peru and Chile). Mostly restricted to lomas of NW Chile and SW Peru. Closely related to *Eremocharis*.

22. *Drusa* DC. [Reduron]

Drusa DC. Ann. Mus. Natl. Hist. Nat. 10: 466, 470, pl. 38 (1807); Mathias & Constance, Univ. Calif. Publ. Bot. 38: 55–58 (1965), rev.

Stellate-pubescent and scabrous-glochidiate, herbaceous annuals. Leaves opposite, simple, ovate to orbicular, palmately lobed; petiole not sheathing. Umbels simple, bracts absent. Calyx lobes shortly obtuse, scarcely visible; petals white, apex acute, plane to scarcely inflexed; styles short; stylopodium conical. Fruits stellate-pubescent, orbicular, strongly compressed dorsally; median and marginal ribs obscure, lateral ribs prominent, expanded into glochidiate-dentate wings; carpophore free, entire; vittae obsolete; rib oil ducts solitary in some ribs. Seeds transversely strongly compressed dorsally, the face plane. $2n = 16$.

One species, *D. glandulosa* (Poir.) H. Wolff ex Engl., Africa (Canary Is., Morocco and Somalia). Forest margins.

23. *Eremocharis* Phil. [Plunkett]

Eremocharis Phil., Fl. Atacam. 25, pl. 2b (1860); Mathias & Constance, Univ. Calif. Publ. Bot. 33(2): 153–173 (1962), rev.

Glabrous and glaucous, sometimes spinescent, suffrutescent or woody-based herbaceous perennials. Leaves 1–3-pinnately lobed or divided (rarely entire), cuspidate, often succulent. Umbels simple, often arranged in umbellate clusters, bracts several, small, lanceolate to ovate, entire. Calyx lobes ovate-lanceolate to triangular-ovate; petals yellow or greenish-white to deep maroon or reddish violet, with median dorsal gland; styles short to slender; stylopodium low-conical. Fruits glabrous, oblong-elliptic to orbicular, truncate at apex, transversely subtetragonal, compressed laterally; commissure constricted; ribs filiform; carpophore lacking. Seeds transversely cuneate-rectangular, intrusively sulcate at the commissure. $n = 5, 8, 10$.

Nine species, W tropical and temperate S America (Peru and Chile); mostly restricted to lomas along the Peruvian coast. Closely related to *Domeykoa*.

24. *Gymnophyton* Clos [Plunkett]

Gymnophyton Clos, Fl. Chil. 3: 102 (1848); Mathias & Constance, Univ. Calif. Publ. Bot. 33(2): 137–153 (1962), rev.

Glabrous, usually glaucous, sparsely leafy (or appearing leafless), often spinescent shrubs.

Leaves entire and linear-filiform to linear-lanceolate, or trifid or ternate (rarely biternate) and spathulate, greatly reduced upward. Umbels simple, globose, solitary or in terminal racemes, panicles, or corymbose to umbellate clusters, bracts small, linear; bracteoles several, small, subulate to lanceolate, often scarious. Calyx lobes ovate-lanceolate to triangular-ovate; petals yellow or greenish-yellow to purplish-red, with a prominent to inconspicuous median dorsal gland; styles short to slender; stylopodium conical to subconical. Fruits glabrous, oblong-elliptic to orbicular-obovate, 4-winged (sometimes appearing 2-winged); commissural face plane or convex, dorsal face plane or concave or convex; dorsal ribs evident or obscure, lateral ribs with conspicuously spreading wings (narrower or broader than the fruit body), marginal ribs obscure; carpophore filiform, entire or apically bifid. Seeds transversely convex to elliptical, the face convex. $n = 10$.

Six species, temperate S America (Chile, Argentina, esp. Andes). Closely related to *Asteriscium*.

25. ***Hermas*** L. [Van Wyk & Tilney]

Hermas L., Mant. Pl. 2: 163 (1771); Van Wyk et al., Afr. Apiaceae: 208–213 (2013); Magee et al., Syst. Bot. 40, 352–365 (2015), rev.

Densely pubescent, woody perennials. Leaves subsessile to petiolate, simple, basal (and sometimes also cauline), obovate to elliptic, often large and coriaceous. Umbels compound, capitate, peduncles basally pubescent; bracts and bracteoles large, lanceolate. Umbellules with 2–4 perfect flowers surrounded by many staminate flowers; calyx lobes petaloid; petals white to purple, filiform; styles short; stylopodium depressed. Fruits glabrous, ovate or oblong; mericarps flat or convex on dorsal and commissural faces, usually deeply channeled on each margin; ribs 5 or 6, the dorsal often keeled, marginal ribs winged; carpophore free, entire; rib oil ducts several, irregular. Seeds transversely elliptic, face convex. $2n = 14$.

Nine species, South Africa (W and E Cape provinces). Fynbos, often in rocky places at high elevations. Fruit anatomy suggests a placement in Azorelloideae, but results from phylogenetic studies are ambiguous.

26. ***Homalocarpus*** Hook. & Arn. [Plunkett]

Homalocarpus Hook. & Arn., Bot. Miscel. 3: 348 (1833); Mathias & Constance, Univ. Calif. Publ. Bot. 38: 58–71 (1965), rev.

Stellate-pubescent, herbaceous annuals. Leaves opposite, simple, narrowly oblong to orbicular-reniform, entire or distally toothed, or lobed to parted, sometimes subcoriaceous. Umbels simple, subcompact, densely stellate-tomentose; bracts a few plumose bristles, or lacking. Calyx lobes forming plumose bristles; petals white to red, purple, or yellowish, oval to ovate, obtuse, dorsally stellate-pubescent, apex not inflexed; styles short; stylopodium conical. Fruits stellate-tomentose or scurfy, oblong-oval to broadly oval, unwinged, sometimes subterete, commissure constricted; dorsal and marginal ribs obsolete, lateral ribs strongly angled; carpophore filiform, entire, persistent. Seeds transversely compressed dorsally, lunate to half-terete, triangular or polygonal, the face slightly to strongly convex. $n = 8$.

Six species, temperate S America; deserts and mountains of N and C Chile, from low to high elevations. Closely related to *Bowlesia*, *Bolax*, *Dichosciadium* and *Drusa*.

27. ***Klotzschia*** Cham. [Plunkett]

Klotzschia Cham., Linnaea 8: 327 (1833).

Caulescent or acaulescent, glabrous to sparsely pubescent, herbaceous perennials. Leaves alternate or rosulate, simple, peltate, triangular to ovate, unlobed to 3–5-lobed, margin callose-dentate to callose-denticulate. Umbels simple, in panicles of capituliform glomerules; bracts 2, entire and opposite. Calyx lobes evident, triangular ovate to ovate-lanceolate; petals canaliculate, apex inflexed (but not narrowed); styles large, recurved and diverging; stylopodium plane to depressed-conical. Fruits rugose, ovoid; commissure half the width of the mericarps; ribs prominent, rounded; carpophore entire; vittae sometimes scattered, rib oil ducts sometimes indistinct. Seed transversely dorsally compressed to subtriangular, the face concave to plane to convex.

Three species, S America (E tropical Brazil).

28. ***Oschatzia*** Walp. [Hart & Henwood]

Oschatzia Walp., Ann. Bot. Syst. 1: 340 (1848).

Glabrous, herbaceous perennials. Leaves mostly basal, spathulate, simple, ternately dissected or toothed, the petioles sheathing. Inflorescence irregular, thyrsoid, with few to several single flowers, the terminal usually the most mature; bracts lacking, bracteoles cuneate, lobed, one subtending each flower. Calyx lobes free, small; petals white to pink, apex rounded to acute, plane; styles short, reflexed; stylopodium flat to subconical. Fruits glabrous, compressed laterally; mericarps subterete to pentagonal; ribs filiform to inconspicuous; carpophore free, entire or bifurcating; rib oil ducts 1–2 under each rib. Seeds transversely terete or subterete to pentagonal, the face plane.

Two species, SE Australia, alpine regions.

29. ***Pozoa*** Lag. Fig. 11 [Plunkett]

Pozoa Lag., Gen. Sp. Pl. 13 (1816); Mathias & Constance, Univ. Calif. Publ. Bot. 33(2): 128–137 (1962), rev.

Caespitose and cushion-forming, subcaulescent, glabrous and glaucous herbaceous perennials. Leaves strongly rosulate and overlapping on the short caudex, simple, undulate to shallowly toothed or lobed, orbicular to cordate or obovate, coriaceous to succulent. Umbels simple, compact-globose; bracts connate, forming a concave, dentate to lobed disc. Calyx lobes triangular-ovate, obtuse; petals greenish-yellow to purple, with a linear (but inconspicuous) median dorsal gland, apex plane or slightly incurved; styles short, recurved; stylopodium low-conical or depressed. Fruits glabrous, the surface wrinkled, oblong-elliptic to cuneate-oblong, transversely tetragonal, commissural face convex, dorsal face concave; dorsal ribs prominent, lateral ribs forming a reflexed flange, marginal ribs obscure; carpophore lacking (occasionally weakly developed but rarely separating from the carpel). Seeds transversely lunate, not sulcate under the vittae, the face convex. $n = 10$.

Two species, temperate S America (Andean Chile and Argentina), rocky or arid habitats. Closely related to *Asteriscium* and *Gymnophyton*.

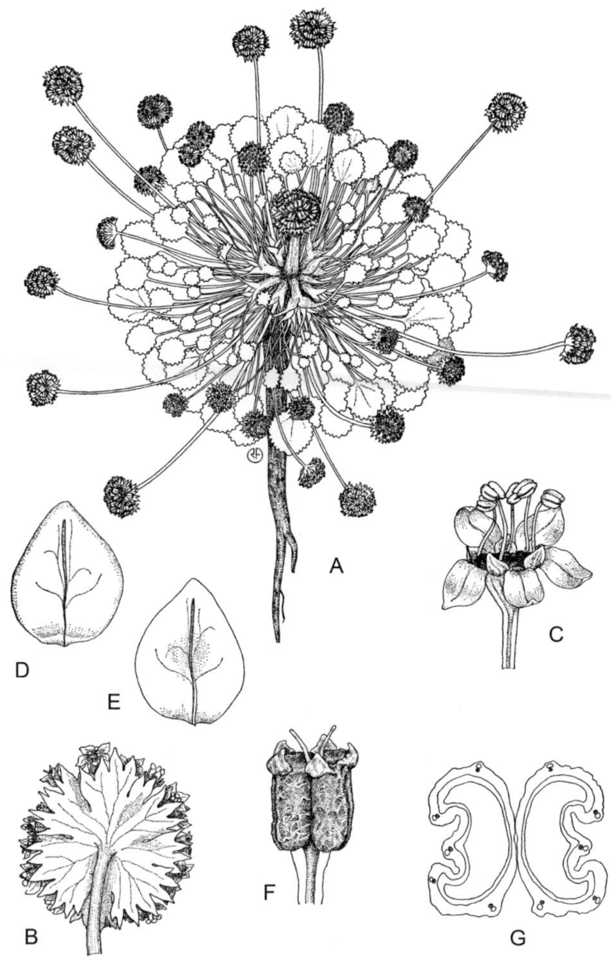

Fig. 11. Apiaceae, Azorelloideae. *Pozoa volcanica* (*Constance & Sparre 3559*, UC). **A** Habit. **B** Umbel from below (showing involucre). **C** Flower. **D, E** Petals. **F** Fruit. **G** Fruit, transverse section. (Mathias and Constance 1962; illustrations by Reino Alava)

30. ***Spananthe*** Jacq. [Plunkett]

Spananthe Jacq., Collectanea 3: 247 (1789 [1791]); Mathias & Constance, Fl. Peru XIII, V-A, 1: 94–96 (1962), reg. rev.

Glabrous (to pubescent), herbaceous annuals. Leaves petiolate, opposite, simple, deltoid-ovate, crenate-dentate; petiole bases sheathing. Umbels simple, arranged in panicles, bracts several, small and narrow, largely entire. Calyx lobes prominent, petals white or greenish-white, apex plane or scarcely incurved; styles short, recurved; stylopodium depressed-conical. Fruits glabrous, ovoid, constricted at the commissure, ribs filiform, carpophore entire; rib oil ducts minute and solitary in the ribs or lacking. Seed transversely compressed dorsally, the face plane. $n = 8$, $2n = 16$.

One species, *S. paniculata* Jacq., N America (C and S Mexico), C America, to S America (Brazil, Peru, Bolivia) and the W Indies, weedy habitats. Related to *Azorella*, *Schizeilema*, *Diplaspis*, and their allies.

III. **Subfam. Saniculoideae** Burnett (1835).

Herbaceous. Umbels typically simple, peduncles terminal and/or lateral. Calyx lobes often large; petals usually glabrous. Fruits schizocarpic; vallecular vittae usually lacking (small, branched, vittae-like oil ducts may be found scattered in the mesocarp); rib oil ducts distinct and typically very large. Exocarp often with outgrowths (e.g., scales, bristles, or prickles). Druse crystals scattered throughout the mesocarp. Inner pericarp typically parenchymatous (sometimes sclerified with inner, fiber-like cells arranged transversely and, if present, the outer fiber-like cells longitudinally).

31. ***Actinolema*** Fenzl Fig. 12 [Pimenov]

Actinolema Fenzl, Pug. Pl. Nov. Syr. 16 (1842).

Glabrous, herbaceous annuals. Stems dichotomously branched, ribbed. Leaves simple, obcordate to broadly lanceolate, spiny at margin, petioles alate. Bracts broad, entire, membranous, reticulate, persistent. Simple umbels few-flowered, central flower hermaphroditic and sessile, others staminate and pedicellate; calyx lobes leafy, pentagonal, persistent, enlarged at tip, spinulose; petals whitish, cuneate, apex inflexed; styles short, stylopodium flat. Fruits ovoid-oblong, sessile, slightly compressed dorsally; ribs dentate or tuberculate, carpophore not free; small, vittae-like ducts scattered in the mesocarp; rib ducts large, solitary. Seed face plane. $n = 5$.

Two species, SW Asia. Ornamental. Morphological and molecular data show a close affinity to *Astrantia*.

Fig. 12. Apiaceae, Saniculoideae. *Actinolema macrolema*. (Illustration by E.S. Gaskevich in A.A. Grossheim (ed.) 1967, Flora Kavkaza (Flora of Caucasus), ed. 2, vol. VII, p. 17, fig. II.1; reproduced with permission from Nauka Publ., St. Petersburg)

32. ***Alepidea*** F. Delaroche [Van Wyk & Tilney]

Alepidea F. Delaroche, Eryng. Hist. 19, t. 1 (1808); Dümmer, Trans. Roy. Soc. S. Afr. 3: 1–21 (1913), rev.; Weimarck, Bot. Notiser 1949: 217–268 (1949), rev.; Townsend, Fl. Trop. E. Afr. 19 (1989), reg. rev.; Van Wyk et al., Afr. Apiaceae: 58–69 (2013).

Acaulescent, often coarsely ciliate, setose or pilose perennials with resinous rhizomes. Leaves petiolate (basal) or sessile (upper), usually simple, orbicular, ovate to oblong or elliptic, usually coarsely ciliate or setose on the margins. Simple umbels small, capitulate; bracts usually colored; rays few, short or long. Flowers few, sessile, in andromonoecious pseudanthia; calyx lobes conspicuous, variable; petals white, apex long and inflexed; styles short; stylopodium usually flat, with a lobed margin. Fruits glabrous or pustulate, tuberculate, rugose or squamose, globose, ovoid or obovoid, terete or slightly compressed laterally; ribs prominent, obtuse; carpophore reduced, entire; vittae rarely present, small; rib oil ducts large. Seeds slightly to deeply sulcate, commissural face ± plane. $2n = 16$.

Ca. 27 species, Africa (mainly S and SE, extending N to Ethiopia). Grasslands and open woodlands. Important in traditional medicine. Closely related to *Arctopus*.

33. ***Arctopus*** L. [Van Wyk & Tilney]

Arctopus L., Sp. Pl.: 1058 (1753); Gen. Pl. ed. 5: 480 (1754); Magee et al., Ann. Missouri Bot. Gard. 95: 471–486 (2008), rev.; Van Wyk et al., Afr. Apiaceae: 106–109 (2013).

Dwarf, acaulescent, setaceous to spiny, dioecious, summer-deciduous perennials growing from thick, tuberous roots. Leaves petiolate, rosulate, simple, ovate, lobed, with setose and mostly spiny, toothed margins; coriaceous; petiole not prominently sheathing. Umbels compound; bracts spine-tipped, linear to ovate; rays few, short, ascending; bracteoles ovate or obovate and naviculate, forming a pseudanthium (female plants), similar to bracts (male plants). Flowers few, sessile (female) or many, pedicellate (male); calyx lobes linear-oblong (male) or triangular (female); petals white (male) or greenish (female), oblong-oblanceolate (male) or triangular and sepaloid (female) with apex inflexed (male) or not (female); stylopodium compressed. Fruits spiny to pustulate, oblong, rostrate, pseudo-monocarpellate, very rarely bicarpellate, fused to the bracteoles, slightly compressed dorsally; ribs inconspicuous; rib oil ducts small, up to 12. Seeds transversely lunate, the face concave. $2n = 18$.

Three species, South Africa (N, W and E Cape provinces). Fynbos and renosterveld. The resinous roots are used in traditional medicine. Related to *Alepidea*.

34. ***Astrantia*** L. Figs. 2A, 13 [Reduron]

Astrantia L. Sp. Pl. 235 (1753); Gen. Pl. ed. 5: 110 (1754); Grintzesco, Ann. Conserv. Jard. Bot. Genève 13–14: 66–194 (1910), monogr.; H. Wolff in Engl., Pflanzenr. 4 (228) Heft 61: 80–92 (1913), rev.; Wörz, Bot. Jahrb. Syst. 121 (4): 507–536 (1999), syst. evol.; Wörz, Acta Bot. Fenn. 162: 141–143 (1999), distrib.

Caulescent, glabrous perennials. Leaves trisect, palmatipartite or palmately lobed, polygonal or reniform. Umbels stellate, appearing simple (but a much reduced compound umbel); bracts large, white or pink. Flowers many per umbel; calyx lobes well developed, lanceolate; petals white or pink, narrow, apex inflexed; styles long; stylopodium annular. Fruits covered with vesicular scales, ellipsoid-oblong, ovoid-oblong, sometimes nearly cylindrical, slightly compressed; ribs inflated; carpophore free or absent; commissure narrow; rib oil ducts solitary in the ribs. Seeds transversely reniform, the face plane. $2n = 14, 16, 28$.

Fig. 13. Apiaceae, Saniculoideae. *Astrantia trifida*. (Illustration by E.S. Gaskevich in A.A. Grossheim (ed.) 1967, Flora Kavkaza (Flora of Caucasus), ed. 2, vol. VII, p. 15, fig. I.2; reproduced with permission from Nauka Publ., St. Petersburg)

Nine species, Eurasia (from the Pyrenees to Turkey, the Caucasus, and NW Iran) mountains. Ornamental. Closely related to *Actinolema.*

35. ***Eryngium*** L. Figs. 2B, 14 [Plunkett]

Eryngium L., Sp. Pl. 1: 232 (1753); Gen. Pl., ed. 5: 108 (1754). Wörz, Biblioth. Bot. 159: 1–498, rev.

Creeping to low, caulescent or acaulescent, usually glabrous, often spinescent and frequently blue-iridescent biennials or perennials with stout taproots or rootstocks bearing fibrous roots. Leaves basal and/or cauline, entire or pinnately or palmately lobed to divided (or rarely forming rachis-leaves), coriaceous to membranaceous, often ciliate to spinose, with parallel or reticulate venation, often lanceolate to oblanceolate; petioles sheathing and sometimes septate. Capitula solitary or in dichasia (less commonly in monochasia or racemes); bracts entire or lobed, often spinescent; bracteoles entire or lobed. Flowers sessile; calyx lobes prominent and persistent, rigid, ovate to lanceolate, entire or rarely spinescent; petals greenish-white, white, blue to purple, apex variously inflexed and lobed to fimbriate; styles shorter or longer than the sepals; stylopodium lacking. Fruits covered with hyaline scales, tubercles, or papillae, globose to obovoid, scarcely compressed laterally, the commissure broad; ribs absent; carpophore absent; vittae scattered; rib oil ducts 5. Seeds transversely subterete, the face plane or slightly concave. $n =$ 5, 6, 7, 8, 9, 10, 11, 14, 16, 24, 32, 40, 48, $2n =$ 14, 15, 16, 18, 28, 32, 48, 64, 96.

About 250 species, Europe, Asia, Africa (N and tropical), N, C, and S America, Australia, Australasia.

Fig. 14. Apiaceae, Saniculoideae. *Eryngium bungei.* **A** Habit. **B** Fruit. (Orig., illustration by Elena Mzhelskaya)

36. ***Hacquetia*** Neck. ex DC. [Reduron]

Hacquetia Neck. ex DC., Prodr. 4: 85 (1830); H. Wolff in Engl., Pflanzenr. 4 (228) Heft 61: 47–48, (1913), rev.; Froebe, Beitr. Biol. Pfl. 40: 325–388 (1964), morphol.; Calviño & Downie, Molec. Phylogen. Evol. 44: 175–191 (2007), phylog.

Low, caulescent, glabrous perennials with short creeping rhizomes. Leaves digitately 3(–5)-dissected, suborbicular; petiole broadly sheathing; leaflets sessile, ovate-cuneate. Simple umbels compact, pseudanthial; bracts large, foliaceous. Flowers many per umbel; pedicels unequal; calyx lobes narrowly triangular to ovate; petals yellow, apex narrowed and inflexed; styles very long, exserted; stylopodium annular. Fruits glabrous, ovoid-cordate, subdidymous; slightly compressed laterally; ribs low, obtuse, rather wide; carpophore lacking; large rib oil ducts solitary in each rib, and smaller, vittae-like oil ducts scattered in the mesocarp. Seeds transversely subterete, the face convex. $2n = 16$.

One species, *H. epipactis* (Scop.) DC., SE and C Europe. Forests. Ornamental. Molecular data suggest a close affinity and even inclusion in *Sanicula* (Calviño and Downie 2007).

37. *Petagnaea* Caruel [Reduron]

Petagnaea Caruel in Parl., Fl. Ital. 8: 199 (1889); H. Wolff in Engl., Pflanzenr. 4 (228) Heft 61: 273–274 (1913), rev.; Magin, Pl. Syst. Evol. 133 (3–4): 239–259 (1980), morphol.; Rauschert, Taxon 31(3): 554–563 (1982), nomencl.
Petagnia Guss. (1827), non Raf. (1814).

Caulescent, glabrous, aromatic perennials with a rhizome. Leaves simple, pentagonal, ± deeply 3–4-lobed (like *Ranunculus* or *Geranium*); petiole not sheathing. Inflorescence dichasial, each unit having a central, sessile carpellate or hermaphroditic flower and 2–4 pedicellate staminate flowers with pedicels adnate to the ovary of the central flower; bracts and bracteoles small; "rays" few, unequal. Flowers few; calyx lobes lanceolate; petals whitish, oblong-cuneate; apex long, inflexed; ovary unilocular; styles long, exserted; stylopodium compressed. Fruits glabrous, hard and nut-like, ovoid-subglobose; ribs prominent above, obsolete below; carpophore absent; rib oil ducts sometimes present. Seed solitary, broadly elliptic. $2n = 42$.

One species, *P. gussonei* (Spreng.) Rauschert, Europe (Sicily). Forest brooks.

38. *Sanicula* L. [Pimenov]

Sanicula L., Sp. Pl. 235 (1753); Gen. P. ed. 5: 109 (1754); Wolff in Engl., Pflanzenr. IV, 228 (Hf. 61): 1–305 (1913), rev.; Shan Ren Hwa, Sinensia 7: 477–489 (1936), reg. rev.; Shan & Constance, Univ. Calif. Publ. Bot. 25: 1–78 (1951), rev.

Glabrous or puberulent perennials or biennials, with taproots, rootstocks, rhizomes, tubers or fascicled fibrous roots. Leaves petiolate or subsessile, pinnately or palmately lobed, rarely entire; lobes obcordate to lanceolate, toothed, serrate or biserrate; petioles sheathing; cauline leaves often forming whorl. Umbels capitate; bracts lanceolate to linear, sometimes lobed; calyx lobes setiform to ovate, distinct or connate; petals white, greenish-white, yellow, pinkish, red or purple, apex narrowed, inflexed; styles erect, divergent or recurved; stylopodium compressed, or lacking. Fruits with prickles, spicules, squamae, or tubercles, ribs sub-inconspicuous; carpophore absent; commissure narrow; outer mesocarp layer parenchymous, inner layer of cells with lignified walls. Commissural vittae 2 to several; rib oil ducts large. Seed face plane. $2n = 8, 16, 32, 48, 64$.

41 species, Europe, Asia (SW, Caucasus, N, E, S, SE), N and C America, S America (trop. and temp.), Africa, Oceania. Deciduous, mixed and coniferous forests, river and stream valleys. Five sections (Shan and Constance 1951). Used in Chinese and folk medicines.

IV. Subfam. Apioideae Seem. (1866).

Typically erect (sometimes decumbent, creeping, or scandent), caulescent (to acaulescent), herbaceous (rarely woody), annuals, biennials or perennials, usually growing from taproots or branching rootstocks. Stems typically branched. Leaves basal and/or cauline, often reduced at distal portion of the stems, petioles typically sheathing, lamina divided (to lobed). Umbels compound (rarely simple), peduncles terminal and/or lateral. Calyx lobes obscure or minute (to conspicuous); petals usually glabrous (sometimes pubescent), base usually clawed and the apex narrowed and inflexed. Fruits schizocarpic, each mericarp having 5 primary ribs (and sometimes ca. 4 secondary ribs); vittae arranged under each vallecula (or secondary rib) and on the commissure; rib oil ducts usually very small or lacking (rarely large). Druse crystals, if present, restricted to the commissural side of the mericarp or in the endosperm. Inner pericarp typically parenchymatous (sometimes sclerified with inner, fiber-like cells arranged transversely and, if present, the outer fiber-like cells longitudinally).

39. *Aciphylla* J.R. Forst. & G. Forst. [Webb]

Aciphylla J.R. Forst. & G. Forst., Char. Gen. Pl. 135 (1776); Allan, Fl. New Zealand 1: 465–487 (1961), reg. rev.; Dawson, New Zealand J. Bot. 17: 339–351 (1979), group rev.; Dawson & Le Comte, Tuatara 23: 49–67 (1978), inform. groups.
Coxella Cheeseman & Hemsl. (1911).

Glabrous, dioecious, perennials. Leaves rosulate, usually 1–4-pinnately (or sometimes palmately) compound, or simple, often rigid and pungent-pointed and often with a stipule-like pair at sheath apex; segments unifacial, adaxial surface reduced to a groove and abaxial face secondarily compressed. Bracts and bracteoles spinose; rays few to many. Petals white to yellowish, with a median oil tube (larger in staminate flowers); pistillate flowers with staminodes; stylopodium depressed (staminate) or conical (pistillate). Fruits glabrous, narrowly oblong to elliptic, subterete to strongly compressed dorsally; ribs 2–5 (mericarps rarely heteromorphic), usually distinctly winged, sometimes merely filiform; carpophore bifid almost to base; vallecular vittae (0) 1 to several, commissural 2–5; rib oil ducts solitary. Seeds transversely subterete, 3–5-angular, or compressed dorsally, the face plane, convex or concave. $2n = 22$.

About 42 species, New Zealand and Australia. Mainly montane and alpine. Some species cultivated as rock-garden plants.

40. ***Acronema*** Falc. ex Edgew. [Watson]

Acronema Falc. ex Edgew., Proc. Linn. Soc. London 1: 252 (1845).

Essentially glabrous biennials or perennials, rhizome tuberous, often globose. Leaves 1–3-ternate-pinnate, blade broadly triangular-ovate; ultimate divisions often narrow. Bracts and bracteoles often lacking, sometimes few, linear; rays few to several, slender, spreading. Petals white or purple-red with a characteristic long-linear or aristate apex (rarely acute or obtuse); styles short, reflexed; stylopodium depressed or low-conic. Fruits ovoid or broadly so, subterete, slightly compressed laterally, smooth; ribs filiform; carpophore bifid to bifurcated; vallecular vittae 1–3, commissural 2–4. Seeds transversely subterete, the face plane. $2n = 18, 20$.

About 25 species, Asia (Nepal to SW China and Thailand); high woodlands, often in moss carpets on rocks, trees and soil banks. A problematic genus, boundaries with *Tongoloa* and *Sinocarum* blurred.

41. ***Adenosciadium*** H. Wolff [Pimenov]

Adenosciadium H. Wolff in Engl., Pflanzenr. IV, 228 (Heft 90): 364 (1927).

Puberulent annuals. Leaves ternate, biternate, or 3–5-lobed, densely hirsute to glabrous; ultimate segments lanceolate to linear, obtuse or acute; upper leaves pinnatisect. Bracts narrowly linear, hirsute; rays 5–12, unequal; bracteoles filiform or narrowly linear, tomentose. Petals white, dorsally sparsely puberulent, marginal radiating; styles short, reflexed; stylopodium conical. Fruits densely covered with unicellular hairs, with attenuate apex; mericarps slightly compressed laterally; ribs filiform, sub-inconspicuous; carpophore bifid; vallecular vittae 1, commissural 2. Seed face plane.

One species, *A. arabicum* (T. Anderson) H. Wolff., SW Asia (Yemen, Oman). Related to *Trachyspermum*.

42. ***Aegokeras*** Raf. [Pimenov]

Aegokeras Raf., Good book: 51 (1840).
Olymposciadium H. Wolff (1922).

Glabrous or minutely puberulent perennials. Leaves rosulate, pinnatisect, leaflets ovate to oblong, toothed. Bracts and bracteoles linear, acuminate, unequal; rays 2–10. Petals white, broadly elliptic to obcordate; styles short, reflexed; stylopodium low-conical. Fruits glabrous, oblong, slightly compressed dorsally; commissure narrow; ribs filiform; carpophore bifid; mesocarp of two layers: outer parenchymatous, inner of cells with lignified walls; vallecular vittae 1, commissural 2, many additional small vittae in inner mesocarp layer. Seed face plane. $2n = 22$.

One species, *A. caespitosa* (Sm.) Raf., SW Asia (Turkey), high-montane, on limestone.

43. ***Aegopodium*** L. [Pimenov]

Aegopodium L., Sp. Pl. 265 (1753); Gen. Pl., ed. 5: 128 (1754); Zakharova, Degtjareva et al., Willdenowia 42: 159 (2012), rev.
Chamaele Miq. (1867).
Pseudopimpinella F. Ghahrem., Khajepiri & Mozaff. (2010).

Stout perennials, rhizomes usually long. Leaves ternate to biternate, glabrous to puberulent; leaflets broad, ovate to ovate-lanceolate, toothed. Bracts and bracteoles absent. Petals white, obcordate, with one to several secretory ducts; styles reflexed; stylopodium conical. Fruits glabrous, oblong to ovoid, compressed laterally; commissure narrow; carpophore bifid; pericarp not lignified; commissure narrow; vittae reduced at maturity. Seeds transversely subterete, the face plane. $2n = 18, 22, 38, 40, 42, 44, 50, 54, 56, 66, 88$.

Ten species, Europe, Asia (N, Middle, C, Caucasus, SW, E, S), N America (naturalized). Folk-medicinal plants, weeds, ornamentals. Molecular data confirm its placement near *Carum*.

44. ***Aethusa*** L. [Reduron]

Aethusa L., Sp. Pl. 256 (1753); Gen. Pl. ed. 5: 123 (1754).

Glabrous annuals or biennials. Leaves 2–3-pinnately divided, triangular; leaflets ovate or triangular, deeply lobed. Bracts absent; rays several to many, rarely few, unequal, spreading; bracteoles well developed, asymmetrical, long recurved. Petals white or slightly reddish, marginals weakly radiate; styles short, equaling or slightly exceeding the conical stylopodium. Fruits glabrous, broadly ovoid to ovoid-globular, not compressed; ribs keeled, rounded, the marginal narrowly winged; carpophore free, bifurcated; commissure broad; vittae small, vallecular 1, commissural 2. Seeds transversely elliptic, the face plane. $2n = 20, 22$.

One species, *A. cynapium* L., Eurasia (western Europe to Caucasus), forests, cultivated ground, waste places. Toxic or at least suspicious.

45. ***Afroligusticum*** C. Norman [Van Wyk & Tilney]

Afroligusticum C. Norman in De Wild., Pl. Bequaert. 4: 301 (1927); Townsend, Kew Bull. 38(2): 311–315 (1983); Townsend, Fl. Trop. E. Afr.: 78–80 (1989), reg. rev.; Winter et al., Taxon 57: 347–364 (2008), taxon.; Van Wyk et al., Afr. Apiaceae: 42–47 (2013).

Acaulescent, occasionally ramentaceous (covered in thin brown scales) perennials with woody rhizomes. Leaves pinnate to sub-bipinnate; leaflets broad, toothed (rarely entire and revolute), with prominent abaxial veins. Bracts and bracteoles small; rays few to several. Flowers sometimes polygamous (inner staminate, outer perfect); petals greenish or yellow; styles short; stylopodium conical. Fruits glabrous, ellipsoid-fusiform to elliptic or ovate, compressed dorsally; ribs narrow to broad, obtuse, marginal ribs broadly winged (rarely unwinged); carpophore bifid; vittae large, vallecular 1, commissural 2. Seeds transversely compressed dorsally, deeply sulcate beneath the dorsal vittae, the face slightly sulcate.

13 species, tropical Africa. Afromontane grassland or forest. Related to *Afrosciadium*.

46. ***Afrosciadium*** P.J.D. Winter [Van Wyk & Tilney]

Afrosciadium P.J.D. Winter in Winter et al., Taxon 57: 347–364 (2008); Van Wyk et al., Afr. Apiaceae: 48–55 (2013).

Acaulescent, glabrous perennials with woody rhizomes. Leaves pinnate to sub-bipinnate; leaflets narrow, toothed or entire, with veins obscure below. Bracts and bracteoles small; rays few to several. Flowers sometimes polygamous (inner staminate, outer perfect); petals greenish or yellow; styles short; stylopodium conical. Fruits glabrous, oblong-elliptic or ovate, slightly to markedly compressed dorsally; ribs narrow to broad, obtuse, marginal ribs broadly winged; carpophore bifid; vittae large, vallecular 1, commissural 2. Seeds transversely compressed dorsally, deeply sulcate beneath the dorsal vittae, the face slightly sulcate.

18 species, tropical E Africa. Afromontane grasslands and forest margins. Related to *Afroligusticum.*

47. ***Agasyllis*** Spreng. [Pimenov]

Agasyllis Spreng., Neue Schrift. Naturf. Ges. Halle (Pl. Umbell. Prodr.) 2(1): 22 (1813).

Stout perennials. Leaves bipinnate or biternate; leaflets ovate, decurrent. Bracts absent; terminal umbel many-rayed; bracteoles linear, unequal. Petals white; styles reflexed; stylopodium low conical. Fruits glabrous, ovoid, commissure narrow; ribs terete; carpophore bifid; mesocarp of lignified parenchyma partly destroyed at maturity; vittae cyclical, in inner mesocarp layer. Seed face plane. $2n = 22$.

One species, *A. latifolia* (M. Bieb.) Boiss., Asia (Caucasus). Subalpine high-herbaceous meadows. Closely related to *Angelica* and *Chymsydia*.

48. ***Agrocharis*** Hochst. [Lee]

Agrocharis Hochst., Flora 27(1): 19 (1844).
Gynophyge Gilli (1973).

Pubescent to subglabrous annuals or perennials. Leaves 2–4-pinnatisect, broadly ovate, ultimate divisions linear, lanceolate, ovate. Bracts several, linear to lanceolate with membranous margins, simple to pinnatisect (like the leaves); rays many, sessile to long, not contracted in fruit, glabrous or hispid; bracteoles many, simple, linear. Calyx lobes conspicuously evident with needle-like tips; petals greenish-white to pinkish, the outer sometimes slightly radiate; styles subsessile, shorter than the conical stylopodium. Fruits spinescent, oblong, ellipsoid, cylindrical; mericarps slightly compressed dorsally, with both primary and secondary ribs prominent; primary ribs with tuberculated, ± erect hairs arranged in 2–3 rows (the 2 commissural ribs rarely inconspicuous); secondary ribs 4, with glabrous spines arranged in a single row, swollen or simple at the base and conspicuously glochidiate at the apex; vittae large, vallecular 1 (under each secondary rib), commissural 2. Seeds transversely depressed-pentagonal, the face narrowly sulcate. $2n = 44$.

Four species, tropical Africa. Closely related to *Daucus*, under which it has recently been placed into synonymy by Banasiak et al. (2016).

49. ***Aletes*** J.M. Coult. & Rose [Plunkett]

Aletes J.M. Coult. & Rose, Rev. N. Amer. Umbell. 125 (1888); Theobald et al., Brittonia 16: 296–315 (1964), rev.

Densely caespitose, glabrous to pubescent perennials; branched caudex stout. Leaves pinnatifid to pinnately or bipinnately compound; ultimate divisions linear to ovate, distinct to confluent, often lobed and spinulose-dentate or entire. Bracts lacking (rarely solitary, linear, entire); rays few to many, widely spreading (sometimes reflexed in fruit); bracteoles subdimidiate, linear or lanceolate, free or basally connate. Calyx lobes evident to very conspicuous, deltoid-ovate to linear; petals yellow, whitish (rarely purple); styles slender, spreading; stylopodium lacking (spongy disc sometimes present). Fruits oblong to ovoid-oblong, terete to dorsally compressed; ribs subequal, prominent and corky-winged or inconspicuous; carpophore bifurcated, sometimes readily deciduous; vallecular vittae 1, commissural 2, rib oil ducts sometimes present in the apex of each rib. Seeds transversely subterete, sometimes slightly sulcate under the vittae, the face ± plane. $2n = 22$.

21 species, N America (C and W United States and adjacent Mexico), mid to high elevations of the Rocky Mountains.

50. ***Alococarpum*** Riedl & Kuber [Pimenov]

Alococarpum Riedl & Kuber, Anz. Österr. Akad. Wiss., Math.-Naturwiss. Kl. 101: 363 (1964); Leute in Rechinger, Fl. Iran. 162: 210–211 (1987), descr. emend.

Tomentose perennials. Leaves ternately compound, glabrous or scabrid; leaflets linear, rigid; upper cauline leaves reduced to sheaths. Bracts and bracteoles linear-triangular, tomentose; rays few, thickened, tomentose. Umbellules compact; calyx lobes lanceolate, tomentose; petals yellowish, dorsally tomentose; styles long, divergent; stylopodium flat, wavy at margins. Fruits tomentose, oblong to ovate; commissure narrow, ribs subequal, obtuse, corky; vascular elements many, scattered; vittae many, cyclical. Seed face sulcate, with involute margins.

One species, *A. erianthum* (DC.) Riedl & Kuber, Asia (W & C Iran). Stony slopes. Close to *Prangos*, to which recent phylogenetic results suggest it should be transferred (Lyskov et al. 2017).

51. ***Ammi*** L. Fig. 2C [Pimenov]

Ammi L., Sp. Pl. 243 (1753); Gen. Pl., ed. 5: 113 (1754).

Glabrous annuals or biennials. Lower cauline leaves pinnate to bipinnate with elliptic to obovate, toothed or serrate leaflets; upper cauline leaves with narrower lobes. Bracts many, trifid to pinnatisect or entire; rays many; bracteoles subulate. Petals white or yellowish; styles slender, long; stylopodium low conical. Fruits slightly compressed dorsally, commissure narrow; ribs filiform; carpophore entire to bifurcated; pericarp not lignified; vittae minute, vallecular 1, commissural 2. Seed face plane. $2n = 22$.

Three to four species, Europe (SW, S, SE), Africa (N, NW), Asia (SW), naturalized in S Africa, N America, S America (temperate). Weeds. *A. majus* L. used for treatment of vitiligo. Closely related to *Petroselinum* and *Sclerosciadium* (Jiménez-Mejías and Vargas 2015). See also *Visnaga*.

52. *Ammodaucus* Coss. & Durieu [Lee]

Ammodaucus Coss. & Durieu in Coss., Bull. Soc. Bot. France 6: 393 (1859).

Pubescent annuals. Leaves 2–3-pinnatisect, fleshy, ovate, ultimate divisions linear-oblong. Bracts and bracteoles few to several, trifid to pinnatifid with membranous margins; rays few to several, densely papillate. Petals white to purplish-white, outer one radiate; styles subsessile, shorter than the semiconical stylopodium. Fruits spinescent, oblong; mericarps smooth, compressed dorsally, both primary and secondary ribs prominent; primary ribs papillate, with ± erect hairs arranged in 1–2 rows; secondary ribs 4, more prominent, with long bristles in several rows; carpophore entire, free; vittae large, transversely elliptical, vallecular 1 (under secondary ribs), commissural 2. Seed face furrowed. $2n = 16$.

Two species, Canary Islands and NW Africa. Probably related to *Cuminum* (Banasiak et al. 2016).

53. *Ammoides* Adans. [Reduron]

Ammoides Adans., Fam. Pl. 2: 96, 516 (1763); H. Wolff in Engl., Pflanzenr. 4 (228) Heft 90: 124–127 (1927), rev.

Glabrous annuals or perennials. Lower leaves 2–3-pinnately divided, oblong; ultimate divisions linear. Bracts absent or rarely few and linear; rays several to many, unequal, very thin, spreading; involucel dimorphic, composed of linear and spathulate bracteoles. Petals white; styles slightly exceeding the weakly mammillate stylopodium. Fruits glabrous, smooth or scabridulous, ovoid, not compressed; ribs filiform; carpophore free, apically bifid or bifurcated; vittae large, vallecular 1, commissural 2. Seeds transversely broadly ovate, the face plane. $2n = 12$.

Two species, W Mediterranean area, SW Asia (Saudi Arabia), dry habitats (slopes, open places). Related to *Ptychotis*.

54. *Ammoselinum* Torr. & A. Gray [Plunkett]

Ammoselinum Torr. & A. Gray, Pacif. Railr. Rep. 2(4): 165 (1857) ("1855"); Mathias & Constance, N. Amer. Fl. 28B (1): 103–104 (1944), reg. rev.

Scabrous annuals. Leaves ternately or ternately-pinnately decompound, ovate; ultimate leaflet segments linear to spathulate. Bracts usually lacking; rays few, spreading-ascending or spreading, unequal; bracteoles several, narrow, entire or toothed. Petals white, lacking a narrowed inflexed tip; styles short; stylopodium low-conical. Fruits oblong-ovoid to ovoid to ellipsoid, compressed laterally; ribs prominent, acute or rounded, coarsely scabrous to glabrous; marginal ribs of each mericarp closely appressed and appearing to form a single broad rib (often covering the commissure with corky tissue); carpophore apically bifid to bifurcated; pericarp composed almost exclusively of strengthening cells; vittae small, vallecular 1–3, commissural 2–4. Seeds transversely slightly compressed dorsally, the face plane to concave. $n = 12, 19, 22$.

Four species, N America (SW United States and adjacent Mexico) and S America (Uruguay).

55. *Andriana* B.-E. van Wyk [Van Wyk & Tilney]

Andriana B.-E. van Wyk in Van Wyk et al., Taxon 48: 739 (1999); Sales & Hedge, Flore de Madagascar et Comores. Fam. 157 (2009), rev.; Van Wyk et al., Afr. Apiaceae: 74–75 (2013).

Glabrous, woody shrubs. Leaves variable, simple or palmately1–5-foliate, margins serrate; leaflets oblong-lanceolate to elliptic. Bracts lanceolate, 5–6-veined; rays relatively few; bracteoles large, often fused. Petals white or greenish-white; styles short, reflexed; stylopodium conical. Fruits glabrous, ovoid to ellipsoid; slightly compressed laterally; mericarps isodiametric; ribs usually conspicuous, sometimes narrowly winged; carpophore poorly developed or absent; vittae large, vallecular usually 1, commissural 2; rib oil ducts relatively large, invariably present. Seeds often transversely pentagonal, slightly to deeply sulcate, the face usually concave.

Three species, Madagascar. Localized on isolated mountains, in ericoid vegetation. Related to *Heteromorpha*.

56. *Anethum* L. [Pu]

Anethum L. Sp. Pl. 263 (1753); Gen. Pl. ed. 5:127 (1754).

Glabrous and glaucous annuals. Leaves 3–4-pinnately dissected; ultimate segments filiform. Bracts and bracteoles absent; rays many, unequal, spreading-ascending. Petals yellow; styles short, reflexed; stylopodium conical. Fruits glabrous, ovoid to ellipsoid, ± compressed dorsally, the mericarps easily separating at maturity; dorsal and lateral ribs filiform to narrowly winged, the marginal ribs slightly broader; carpophore bifurcated; vallecular vittae 1, commissural 2. Seed transversely compressed dorsally, the face plane to slightly concave. $2n = 22$.

One species, *A. graveolens* L., Mediterranean, and widely adventive in Europe, Asia and Africa. The source of dill, whose fruits and leaves are used as an aromatic spice and herb, and as a stomachic. Molecular data show an affinity to *Foeniculum* (Jiménez-Mejías and Vargas 2015).

57. ***Angelica*** L. Fig. 2F [Pimenov]

Angelica L., Sp. Pl. 250 (1753); Gen. Pl., ed. 5: 119 (1754); Kitag., J. Jap. Bot. 12(4–5): 1–24 (1935), rev.; Pimenov, Bjull. Moskovsk. Obšč. Isp. Prip. Otd. Biol. 73, 1: 124–139 (1968), sect. rev.; Yuan Chang Qi & Shan Ren Hwa, Bull. Nanjing Bot. Gard. Mem. Sun Yat Sen, 1983: 1–17 (1983), rev. Chinese spp.; Vasilieva & Pimenov, Pl. Syst. Evol. 177: 117–138 (1991), sect. rev.; Spalik, Reduron & Downie, Pl. Syst. Evol. 243: 189–210 (2004), molec. phylog.
Epikeros Raf. (1840).
Sphenosciadium A. Gray (1865).

Glabrous to tomentose or scabrous perennials with taproots, rootstocks, or rhizomes. Leaves 1–4-pinnate or 2–3-ternate; leaflets narrowly lanceolate to ovate. Bracts absent or few, linear; rays usually many; bracteoles linear. Petals white, or greenish, rarely red or purplish; styles reflexed; stylopodium flat to conical. Fruits glabrous (rarely tomentose), lanceolate to ovoid, compressed dorsally, commissure usually narrow; dorsal and lateral ribs terete or keeled, rarely short-winged, marginal ribs thin- or corky-winged; mesocarp cells with lignified, pitted walls, exocarp cells small; carpophore bifid (or bifurcated); vallecular vittae 1, commissural 2, or vittae cyclic. Seeds transversely compressed dorsally, the face plane. $2n = 22, 28, 44, 66$.

Ca. 120 species, Europe, Asia, N Africa, N America. Forests, meadows, some species on rocks. Medicinal (widely used especially in Eastern medicine), aromatic, culinary. Classification of Eurasian species (Pimenov 1968; Vasilieva and Pimenov 1991) into 3 subgenera, 6 sections. In its current circumscription, the genus is not monophyletic; closely related to *Agasyllis* and *Chymsydia*, see also *Ostericum*.

58. ***Anginon*** Raf. [Van Wyk & Tilney]

Anginon Raf., Good book 56 (1840); Burtt, Notes Roy. Bot. Gard. Edinb. 45: 90 (1988); Allison & B.-E. van Wyk, Nord. J. Bot. 17: 561–577 (1997), rev.; Van Wyk et al., Afr. Apiaceae: 78–87 (2013).

Glabrous (sometimes glaucous), woody shrubs. Leaves very variable, tripinnatisect; leaflets simple or divided and laminate (juvenile and basal), often reduced to junciform phyllodes. Umbels compound, sometimes verticillate; bracts and bracteoles small; rays many, long. Petals yellow; styles very short; stylopodium conical. Fruits smooth, rugose or tuberculate, ovoid to ellipsoid, laterally somewhat compressed; mericarps isodiametric to slightly compressed dorsally; ribs narrow or obtuse, rarely only the marginal ribs expanded; carpophore bifid; vittae large, vallecular 1 (2–3), commissural 2 (1–3). Seeds transversely pentagonal, the face concave. $2n = 22$.

13 species, S Africa. Fynbos, renosterveld or arid succulent karoo. Related to *Glia*.

59. ***Angoseseli*** Chiov. [Plunkett]

Angoseseli Chiov., Boll. Soc. Bot. Ital., 1924: 38 (1924); Van Wyk et al., Afr. Apiaceae: 88 (2013).

Hispidulous biennials. Leaves 2–3-ternately parted, obovate; ultimate segments irregularly dissected, linear with acute or bifid tips. Bracts and bracteoles few to several, subulate; rays several, subequal, slender and spreading-ascending. Petals white, abaxially pubescent; styles slender and strongly angular-reflexed in fruit; stylopodium conical. Fruits densely pubescent along the entire surface with long, flat, spreading hairs, ovoid-oblong, slightly compressed laterally; mericarps transversely pentagonal; ribs filiform, low and rounded; carpophore apparently lacking; vallecular vittae 1, commissural 2. Seeds transversely subterete to pentagonal (?), the face nearly plane or slightly sulcate.

One species, *A. mazzocchii-alemannii* Chiov., Africa (W tropical, Angola).

60. ***Anisopoda*** Baker [Van Wyk & Tilney]

Anisopoda Baker, J. Linn. Soc., Bot. 25: 318, t. 52 (1890); H. Wolff, Pflanzenr. IV, 228 (Heft 90): 1–398 (1927); Sales & Hedge, Flore de Madagascar et Comores. Fam. 157 (2009), rev.; Van Wyk et al., Afr. Apiaceae: 89 (2013).

Acaulescent, glabrous perennial, rhizome woody. Leaves radical, pinnately decompound, with two opposite pairs of leaflets and three terminal leaflets; petiole not sheathing; leaflets obovate, entire or crenate. Bracts small, lanceolate; rays few, rather short; bracteoles lanceolate-linear. Petals purple (?); styles short; stylopodium short conical. Fruits verrucose, ovoid, somewhat compressed laterally; ribs inconspicuous; carpophore unknown; vallecular vittae 1, commissural 2. Seeds transversely semi-terete, the face plane.

One species, *A. bupleuroides* Baker, Madagascar. Endangered. Very poorly known; similar to *Tana* (both possibly related to *Heteromorpha*).

61. ***Anisosciadium*** DC. [Plunkett]

Anisosciadium DC., Coll. Mém.: 63 t. 15 (1829); DC., Prodr. 4: 234 (1830); Boiss., Fl. Or. 2: 950–951 (1892), rev.; Townsend, Kew Bull. 17: 427 (1964), rev.
Echinosciadium Zohary (1948).

Glabrous to pubescent annuals. Leaves bipinnate, ovate to lanceolate; ultimate segments oblong-ovate, bi- or trifid. Bracts few to several, narrow, oblong, acuminate (reflexing in fruit); rays few to several, becoming ± compressed in fruit, subequal, spreading to divaricating; bracteoles several, foliaceous or spinescent, recurved. Umbellules with a central, sessile, bicarpellate, perfect flower, surrounded by two rows of outer flowers, which are either staminate, or monocarpellate, perfect and fused to the pedicel of a staminate flower; calyx lobes present or absent on the central flower, present on outer flowers (sometimes with two foliaceous outer lobes); petals white, some radiate or all alike; styles long, ascending, distally reflexed; stylopodium elongate-conical. Fruits enclosed in a "cage" formed from the hardened outer pedicels, fruit of central flower bicarpellate, slightly sunken into the receptacle, fruits of the outer flowers monocarpellate, ovoid-oblong, subterete, the mericarps semi-circular; ribs filiform; carpophore lacking; dorsal vittae 4 or 9, commissural vittae 2. Seeds transversely semi-circular to compressed dorsally, the face deeply sulcate. $n = 11$, $2n = 18$.

Three species, Asia (Syria, Jordan, Arabia, Iraq, Iran, Afghanistan), deserts, steppes, riverbanks.

62. ***Anisotome*** Hook. f. [Webb]

Anisotome Hook. f., Fl. Antarctica 1: 16 (1844); Dawson, Univ. Calif. Publ. Bot. 33: 1–98 (1961), rev.

Glabrous, dioecious, herbaceous perennials, caudex multicipital (with many heads from the main crown) and thickened taproots, sometimes rhizomatous. Leaves rosulate, 1–4-pinnately or ternately compound; leaflets or ultimate segments ovate or deltate to lanceolate or filiform, sometimes hair-tipped; cauline leaves few, reduced. Bracts and bracteoles few to several, membranaceous to foliaceous, ovate to linear, or rarely lacking; rays few to many, spreading to ascending, sometimes obsolete. Calyx lobes usually conspicuous, unequal; petals usually white to cream, rarely red or purplish, with a median oil tube and an acute but not inflexed apex, larger in staminate flowers; pistillate flowers with conspicuous staminodes; styles mostly slender; stylopodium low-conical. Fruits glabrous, oblong-ellipsoid to ovoid, subterete or slightly compressed dorsally; ribs (3–)5, subequal, filiform to narrowly winged; carpophore bifid almost to base, sometimes lacking; vallecular vittae 1(2–3), commissural 2–4(5–6); rib oil ducts solitary. Seeds transversely subterete to slightly compressed dorsally, sometimes sulcate under the vittae, the face plane to slightly concave. $2n = 22$.

About 16 species, New Zealand and Australia (Tasmania). Coastal to alpine.

63. ***Annesorhiza*** Cham. & Schltdl. [Van Wyk & Tilney]

Annesorhiza Cham. & Schltdl., Linnaea 1: 398 (1826); Tilney & Van Wyk, Nord. J. Bot. 21: 615–649 (2001), rev.; Magee et al., Syst. Bot. 36, 508–519 (2011), rev.; Van Wyk et al., Afr. Apiaceae: 90–101 (2013).

Acaulescent, rarely pubescent perennials with 1–3(–5) or many fleshy, pencil-like roots. Leaves basal, hysteranthous, pinnatipartite or multisect, or pinnate with the pinnae lobed; leaflets variable, irregularly ovate to oblong or narrowly linear. Bracts small or rarely large, persistent; rays rather long; bracteoles small. Calyx lobes rarely large, ovate; petals yellow; styles short or long; stylopodium conical. Fruits glabrous, oblong, not compressed; mericarps homomorphic or heteromorphic (one 3-ribbed, the other 4-ribbed); ribs prominent, sclerified, often winged; carpophore bifid; vittae large, vallecular 1, commissural 2. Seeds transversely dorsally compressed, terete or concave-convex, the face flat or concave. $2n = 22, 24$.

22 species, S Africa. Fynbos, renosterveld, grasslands or arid semi-deserts. Roots of some species are edible. Closely related to *Chamarea* and *Itasina.*

64. ***Anthriscus*** Pers. Fig. 15 [Spalik]

Anthriscus Pers., Syn. Pl. 1: 320 (1805), nom. cons.; Spalik, Polish Bot. Stud. 13: 1–69, (1997), rev.; Spalik et al., Edinb. J. Bot., 58: 331–346 (2001), phylog., taxon.
Chaerefolium Haller (1768).

Glabrous to pubescent annuals, biennials, or perennials. Leaves petiolate, 2–4-pinnately dissected, triangular; leaflets linear-oblong to ovate. Bracts usually absent; rays few to many, usually unequal, spreading or slightly ascending; bracteoles few, usually ciliate. Petals white, rarely pinkish, outer ones often radiate; styles short to relatively long; stylopodium flat-conical in flower, conical in fruit. Fruits glabrous or bristled, with an aculeate cuticle; linear-oblong to ovoid with a short beak, laterally compressed; mericarps terete; ribs only on beak, angular; commissure narrow; carpophore apically bifid; vittae compressed at maturity, vallecular usually 1, commissural 2. Seed transversely terete, the face deeply sulcate. $2n = 14, 16, 18$.

Nine species, Eurasia, N Africa, and mountains of E Africa (adventive in North America and Oceania). From montane screes and primary forests to lowland meadows, waste places, and arable land. *Anthriscus cerefolium* (L.) Hoffm. cultivated for fresh leaves and used as condiment. Four sections recognized. Closely related to *Kozlovia*, *Geocaryum*, *Myrrhis*, and *Osmorhiza* (Spalik et al. 2001).

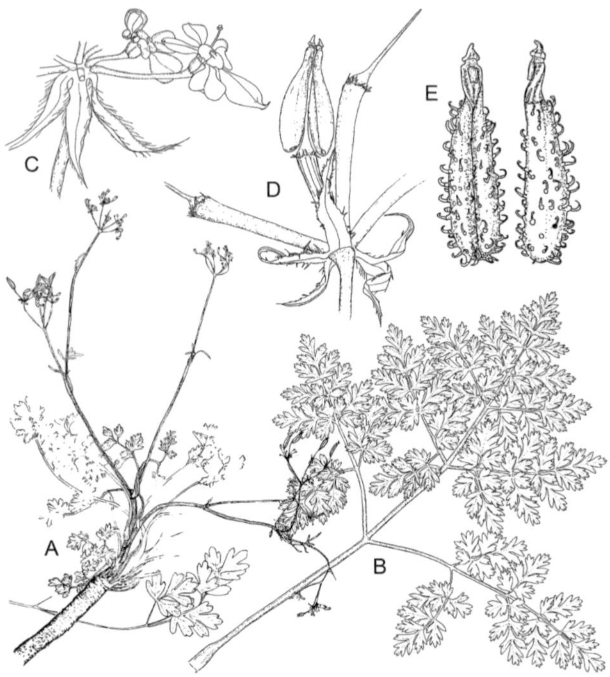

Fig. 15. Apiaceae, Apioideae. *Anthriscus.* **A** *A. kotschyi*, habit. **B, C** *A. sylvestris.* **B** Basal leaf. **C** Perfect flowers at the end of male stage. **D, E** *A. caucalis.* **D** Fruiting umbellet of var. *gymnocarpa.* **E** Mericarps of var. *caucalis.* (Spalik 1997; illustration by Krzysztof Spalik, reproduced with permission)

65. ***Aphanopleura*** Boiss. [Lee]

Aphanopleura Boiss. Fl. Orient. 2: 855 (1872).

Pilose to glabrous annuals. Leaves petiolate, ternate or sometimes biternate, broadly ovate, ultimate divisions linear to lanceolate, petiole bases not sheathing. Bracts and bracteoles several to many, linear to lanceolate with membranous margins, simple; rays several, ± equal, not contracted in fruit, glabrous. Petals white to pinkish, the outer sometimes slightly radiate; styles longer than the conical stylopodium. Fruits pubescent with clavate-tipped bristles, ovoid to globose; mericarps compressed laterally, ribs rounded to obscure; carpophore free, apically bifid; commissure broad; vittae large, transversely elliptical or triangular, vallecular 1 (under each secondary rib), commissural 2. Seeds transversely pentagonal, the face plane. $2n = 22$.

Four species, SW and C Asia. Molecular data suggest a placement close to *Pimpinella.*

66. ***Apiastrum*** Nutt. ex Torr. & A. Gray [Plunkett]

Apiastrum Nutt. ex Torr. & A. Gray, Fl. North Amer. 1: 643 (1840); Mathias & Constance, N. Amer. Fl. 28B(1): 71 (1944), reg. rev.

Glabrous annuals. Leaves mostly opposite, membranaceous, ternately decompound; ultimate segments narrow and entire. Bracts and bracteoles lacking (or involucre formed by subtending foliage leaves); rays few, unequal, ascending, or some umbellules sessile. Petals white, apex acute but scarcely inflexed; styles very short; stylopodium depressed. Fruits papillate-scabrate to glabrate, ellipsoid-cordate, compressed laterally; commissure strongly constricted; ribs filiform, undulate, and inconspicuous; carpophore bifurcated; vallecular vittae 1, commissural 2; rib oil ducts solitary. Seeds transversely subterete, the face concave or shallowly sulcate. $n = 11$.

One species, *A. angustifolium* Nutt. ex Torr. & A. Gray, N America (California to Arizona and Baja California). A second species, *Apiastrum patens* Coult. & Rose, has been referred to *Spermolepis: S. inermis* (Nutt.) Math. & Const.

67. ***Apium*** L. [Pimenov]

Apium L., Sp. Pl. 264 (1753); Gen. Pl., ed. 5: 128 (1754); Short, J. Adelaide Bot. Gard. 1, 4: 205–235 (1979), rev. Australasian spp.; Ronse et al., Pl. Syst. Evol. 287: 1–17 (2010), rev.

Glabrous, aromatic perennials or biennials. Leaves 1–2-ternate, or 1–2-pinnatisect, segments broad, lanceolate to ovate, serrate, or lobed. Peduncles often leaf-opposed; bracts and bracteoles absent; rays few. Petals white or greenish; styles short or reflexed; stylopodium low-conical. Fruits glabrous, oblong to ovoid, compressed laterally; commissure narrow; ribs subequal, keeled; carpophore entire or only apically bifid; pericarp not lignified; vallecular vittae 1–3, commissural 2–6. Seed face plane. $2n = 18, 22$.

Ca. 10 species, Europe, Asia (except N), Africa, Australia, Australasia, S America, Oceania. *Apium graveolens* L. (celery) is widely cultivated as an edible plant. Medicinal, aromatic. See also *Cyclospermum* and *Helosciadium*.

68. ***Apodicarpum*** Makino [Plunkett]

Apodicarpum Makino, Illustr. Fl. Jap. 1: 1, pl. 58 (1891); Hiroe & Constance, Univ. Calif. Publ. Bot. 30: 1–144 (1958), reg. rev.

Glabrous perennials, roots fascicled and fleshy. Leaves once pinnately compound and oblong; leaflets ovate to lanceolate, margins serrate or dentate to slightly incised or lobed toward the base. Bracts 1 to few, narrow; rays few, spreading to ascending and unequal; bracteoles several, narrow. Petals white, orbicular, with a short, acute, slightly incurving apex; styles minute; stylopodium depressed. Fruits glabrous; ovoid, slightly compressed laterally, the commissure slightly constricted; ribs triangular, prominently corky-thickened; carpophore lacking; vallecular vittae 1, commissural 2. Seeds transversely subterete, the face plane. $2n = 22$.

One species, *A. ikenoi* Makino, E Asia (Japan).

69. ***Arafoe*** Pimenov & Lavrova [Pimenov]

Arafoe Pimenov & Lavrova in Bot. Zhurn. 74(1): 102 (1989).

Stout, pubescent perennials, rhizomes vertically thickened. Basal leaves entire, cauline leaves ternate; leaflets cordate to ovate, serrate, abaxially pubescent, almost glabrous adaxially. Bracts several, linear, hairy; rays many, unequal, densely pubescent; bracteoles linear-subulate. Petals white, dorsally pubescent; styles reflexed; stylopodium conical. Fruits almost glabrous, ovoid, commissure rather broad; mesocarp of slightly lignified cells; ribs keeled; carpophore bifurcated; vallecular vittae 3, commissural 4–6. Seed face plane. $2n = 22$.

One species, *A. aromatica* Pimenov & Lavrova, Asia (W Caucasus). Subalpine high-herbaceous meadows. Aromatic. Related to *Pimpinella*.

70. ***Arcuatopterus*** M.L. Sheh & R.H. Shan [Watson]

Arcuatopterus M.L. Sheh & R.H. Shan, Bull. Bot. Res., Harbin 6(4): 11 (1986).

Glabrous perennials. Leaves 2–3-pinnatisect; ultimate segments ovate or obovate to linear. Bracts and bracteoles lacking, or occasionally bracts solitary and deciduous; rays few to several, slender, unequal. Petals white, purplish or dull cream, with red-brown costa; styles short; stylopodium conic or low-conic, margin slightly undulate. Fruits glabrous, oblong or ellipsoid, strongly compressed dorsally; dorsal ribs obscure or very slightly raised, marginal ribs broadly winged, wings thin to corky, margins often incurved; carpophore bifid; vallecular vittae 1, commissural 2. Seeds transversely compressed dorsally, the face plane. $2n = 22$.

Six species, Asia (E Himalaya to SW China). Mid-elevation forests and grasslands.

71. ***Arracacia*** Bancroft [Plunkett]

Arracacia Bancroft, Trans. Agric. Soc. Jamaica 1825: 3 (1825); Mathias & Constance, N. Amer. Fl. 28B(1): 90–102 (1944), reg. rev.; Mathias & Constance, Fl. Panama VII, 4: 243–245 (1959), reg. rev.; Mathias & Constance, Fl. Peru XIII, V-A, 1: 13–19 (1962), reg. rev.

Glabrous to pubescent, sometimes somewhat woody perennials (roots sometimes tuberous). Leaves 1 to several times ternately, pinnately, or ternate-pinnately dissected or compound; leaflets various. Bracts lacking, or solitary and foliose, or several and inconspicuous; fertile rays few to many, spreading-ascending to divaricate and reflexed; bracteoles few, short to long, narrow, or lacking. Petals white to purple, greenish-yellow, or greenish, oblanceolate to obovate; styles short to long, erect to spreading or reflexed; stylopodium conical and conspicuous to depressed and indistinct. Fruits glabrous or pubescent, lanceoloid to ovoid, usually narrowed at apex, laterally compressed; ribs prominent, acute to obtuse, or filiform and indistinct; carpophore bifurcated or apically bifid; vallecular vittae 1 to several, commissural 2 to several. Seeds transversely subterete, often sulcate under the vittae, the face sulcate or concave. $n = 22$.

55 species, N America (California, Mexico), C America, S America (W tropical). Roots of some species (esp. *A. xanthorrhiza* Bancroft) cultivated as food source in Latin America. Monophyly questioned; close to *Rhodosciadium*, *Coulterophytum*, and their allies.

72. ***Artedia*** L. [Lee]

Artedia L., Sp. Pl. 242 (1753); Gen. Pl. ed. 5: 112 (1754).

Glabrous annuals. Leaves 2–3-pinnatisect, broadly ovate, ultimate divisions filiform. Bracts and bracteoles several to many, filiform with membranous margins, pinnatisect (like the leaves), deflexed in fruits; rays many, long. Central umbellules of sterile flowers appearing as a blackish to purplish tuft of bristles; petals white, the outer conspicuously radiate and plane, the inner much smaller and inflexed; styles subsessile, but longer than the depressed stylopodium. Fruits glabrous, orbicular, strongly compressed dorsally, with both primary and secondary ribs; primary ribs filiform; secondary ribs 4, the two dorsal ones filiform, the lateral ones broadly winged with discontinuous, spathulate projections; vittae minute or obsolete, transversely elliptical, vallecular 1 (under each secondary rib), commissural 2. Seeds transversely compressed, the face plane. $2n = 16$.

One species, *A. squamata* L., S (adv.) and SE (Turkey) Europe, W Asia.

73. ***Asciadium*** Griseb. [Plunkett]

Asciadium Griseb., Cat. Pl. Cub. 118 (1866); Mathias & Constance, N. Amer. Fl. 28B(1): 51 (1944), reg. rev.

Prostrate, glabrous annuals with creeping, filiform stems, rooting at the nodes. Leaves usually rosulate at the nodes, obovate, pinnately dissected; leaf segments five (lower ones reduced), obovate, crenate to lobed. Umbels simple, reduced (1–2 flowers); bracts and bracteoles obsolete. Petals white or greenish, apex not inflexed; styles short; stylopodium depressed-conic. Mature fruits unknown; immature fruits glabrous, ovoid, laterally compressed.

One species, *A. coronopifolium* Griseb., Cuba; known only from the type.

74. ***Astomaea*** Rchb. [Kljuykov]

Astomaea Rchb., Handb. Nat. Pflanzensyst. 218 (1837); Pimenov & Kljuykov, Bot. Zhurn. (Moscov & Leningrad) 66(4): 465–482 (1981), rev. gen.

Glabrous, ephemeral perennials with deep, globose tubers. Leaves 2–3-pinnately divided; leaflets lanceolate to linear, lobed. Bracts and bracteoles entire, herbaceous. Petals white; styles reflexed; stylopodium low-conical. Fruits glabrous, didymous, the commissure narrow; mericarps globose, ribs thin; carpophore bifid; vittae many, slender. Seed face with a broad, deep groove; seedlings monocotyledonous. $2n = 18$.

One species, *A. seselifolia* (DC.) Rauschert. Near East. Segetal plants. Closely related to *Bunium*.

75. ***Astomatopsis*** Korovin [Kljuykov]

Astomatopsis Korovin, Bot. Mater. Gerb. Inst. Bot. Zool. Akad. Nauk Uzbeksk. S.S.R. 12: 29 (1948); Pimenov & Kljuykov, Bot. Zhurn. (Moscov & Leningrad) 66(4): 465–482 (1981), rev. gen.

Glabrous ephemeral perennials with deep, globose tubers. Leaves 2–3-pinnately divided; leaflets lanceolate to ovate, lobed. Bracts absent, bracteoles few or absent. Calyx lobes obsolete; petals white; styles reflexed; stylopodium low-conical. Fruits glabrous, didymous, the commissure narrow; mericarps globose, ribs thin; carpophore bifid; vittae many, slender, anastomosing. Seed face deeply and broadly grooved. $2n = 12$, 22.

One species, *A. galiocarpa* Korovin, Middle Asia (Pamiro-Alai). Scrub, subalpine meadows, under rocks, gravel slopes and screes. Closely related to *Elwendia*.

76. ***Astrodaucus*** Drude [Lee]

Astrodaucus Drude in Engl. & Prantl, Nat. Pflanzenfam. 3 (8): 156 (1898).

Glabrous to pubescent biennials (or annuals). Leaves petiolate, 3–4-pinnatisect, triangular, ultimate divisions linear to oblong. Bracts absent or rarely several, linear to lanceolate, caducous; rays several to many, nearly equal, slightly contracted in fruit, glabrous; bracteoles several, simple or bifid, oblong to lanceolate, ciliate with membranous margins. Petals yellowish-white to white, bilobed, the outer radiate; styles very long; stylopodium conical. Fruits spinescent, ellipsoid to oblong; mericarps terete to slightly compressed laterally; primary ribs filiform, with much-branched, papillate, tuberculate, or striate hairs or bristles, arranged in 1–2 rows; secondary ribs 4, between the primaries, very prominent, spinate with triangular or pyramidal, densely papillated spines, confluent at the base and conspicuously glochidiate at the tip, arranged in 1–2 rows; commissure of intermediate width; vittae minute or obsolete, transversely elliptical, vallecular 1 (under each secondary rib), commissural 2. Seeds transversely semi-orbicular to slightly compressed laterally, the face plane to somewhat concave. $2n = 20$.

Two species, SW Asia and SE Europe.

77. ***Astydamia*** DC. [Reduron]

Astydamia DC., Coll. Mém. 5: 53, pl. 1, fig. D (1829), Prodr. 4: 190 (1830).

Glabrous perennials growing from a woody, swollen rootstock. Leaves petiolate, deeply lobed or pinnately divided, ovate, fleshy; leaflets broad, cuneate. Bracts and bracteoles many; rays many, subequal, spreading. Petals bright yellow; styles short; stylopodium conical. Fruits glabrous, somewhat corky, sub-rectangular, strongly compressed dorsally; dorsal ribs slightly prominent, the lateral expanded into a thick margin; carpophore bifurcated; vittae large, vallecular 1, commissural not visible. Seeds transversely compressed dorsally, the face plane or slightly concave. $2n = 20$, 22.

One species, *A. latifolia* (L. f.) Kuntze, Canary Islands, Mauritania and Morocco, seashore.

78. ***Athamanta*** L. [Spalik]

Athamanta L., Sp. Pl. 244 (1753); Tutin, Feddes Repert. 79: 18–20 (1968), reg. rev.; Spalik et al., Edinb. J. Bot. 58: 331–346 (2001), phylog., sect. rev.
Tinguarra Parl. (1843).

Glabrous to pubescent perennials. Leaves petiolate, 3–4-pinnately dissected, triangular to ovate, lobes linear to broadly ovate. Bracts few; rays few to many, equal to subequal, spreading or ascending; bracteoles many, pubescent, ciliate. Petals white or yellow, pubescent; styles long; stylopodium conical in flower, rounded in fruit. Fruits pubescent, oblong ovoid to ovoid with a short beak, laterally compressed; mericarps terete or somewhat dorsally compressed; commissure constricted; carpophore bifurcated; ribs filiform; vascular bundles small, dorsal and marginal usually similar in size; vittae prominent, vallecular 1–3, commissural 2–6. Seeds terete, the face sulcate. $2n = 22$.

Seven species, Europe, N Africa and the Canary Islands. Open montane habitats. Two sections recognized (Spalik et al. 2001). Closely related to *Conopodium* and *Todaroa* (Spalik et al. 2001); Banasiak et al. (2016) proposed to transfer the Lybian endemic *A. dellacellae* to *Daucus*.

79. ***Atrema*** DC. [Plunkett]

Atrema DC., Coll. Mém. 5: 71 (1829); Mathias & Constance, N. Amer. Fl. 28B(1): 150 (1944), reg. rev.

Glabrous to scaberulous annuals. Leaves ternate-pinnately decompound, ovate-oblong; ultimate divisions filiform. Bracts and bracteoles few, small, entire to pinnatifid, linear; rays few to several. Petals white, subequal or radiate; styles elongate; stylopodium conical. Fruits glabrous, didymous, the 2 mericarps subglobose; ribs filiform, prominent; carpophore bifurcated; vittae indistinct or reduced; commissural face generally perforated with 2 pores. Seeds transversely subterete, the face deeply and broadly concave. $n = 10$.

One species, *A. americanum* DC., N America (south-central United States), re-segregated from *Bifora* on the basis of molecular data.

80. ***Aulacospermum*** Ledeb. Fig. 2D [Kljuykov]

Aulacospermum Ledeb., Fl. Altaic. 4: 334 (1833); Kljuykov et al., Bull. Soc. Nat. Mosc. Div. Biol. 81(4): 75–89 (1976), limits and syst. gen., 81(5): 61–68 (1976), rev. spp.; Pimenov & Kljuykov, Feddes Repert. 111: 517–534 (2000), rev. gen.

Glabrous perennials. Leaves 1–4-pinnate; leaflets linear to ovate, lobed, usually sessile. Bracts and bracteoles linear to linear-lanceolate, herbaceous, entire or toothed. Petals white or rarely yellow with claw; styles reflexed; stylopodium low-conical. Fruits glabrous or with wart-like outgrowths, slightly compressed laterally, the commissure narrow; ribs equal, winged or rarely keeled; exocarp cells large with convex outer walls; carpophore bifid; vallecular vittae 1–3, commissural several (to many). Seed face narrowly to broadly and deeply grooved. $2n = 16, 18$.

15 species, Asia (Middle, NW Himalaya, NW China, Mongolia, N Siberia), E Europe. On mountains up to the snow line. Three sections have been recognized (Kljuykov et al. 1976). Closely related to *Pleurospermum* s.l.

81. ***Austropeucedanum*** Math. & Const. [Plunkett]

Austropeucedanum Math. & Const., Bull. Torrey Bot. Club 79: 365 (1952).

Scabrous biennials. Leaves pinnately decompound, leaflets entire, serrate or lobed. Bracts lacking or few, filiform and inconspicuous; fertile rays several (7–10), spreading-ascending, unequal; bracteoles several, linear. Petals white; styles short, spreading; stylopodium low-conical. Fruits glabrous, oblong, strongly compressed dorsally; dorsal and lateral ribs filiform, marginal ribs with membranous wings (about as wide as the fruit body); carpophore bifurcated; vittae large, vallecular 1, commissural 2. Seed transversely compressed dorsally, sulcate under the vittae, the face plane.

One species, *A. oreopansil* (Griseb.) Math. & Const., S America, restricted to the mountains of NW Argentina. Mathias and Constance (1952) suggest a close relationship to *Rhodosciadium* and *Prionosciadium*.

82. ***Autumnalia*** Pimenov Fig. 16 [Pimenov]

Autumnalia Pimenov, Bot. Zhurn. 74(10): 1492 (1989).

Stout perennials with thin, moniliformly thickened taproots. Leaves biternate; leaflets lanceolate to ovate, petiolulate. Bracts lanceolate to ovate, membranous; bracteoles absent; pedicels slender. Marginal flowers perfect with subequal, comparatively long pedicels, central flowers staminate on short pedicels; petals yellowish, ovate, entire; styles reflexed; stylopodium low-conical. Fruits glabrous, ovoid, strongly compressed dorsally; commissure broad; mesocarp non-lignified; dorsal ribs almost inconspicuous, marginal ribs narrowly winged, slightly thickened; carpophore

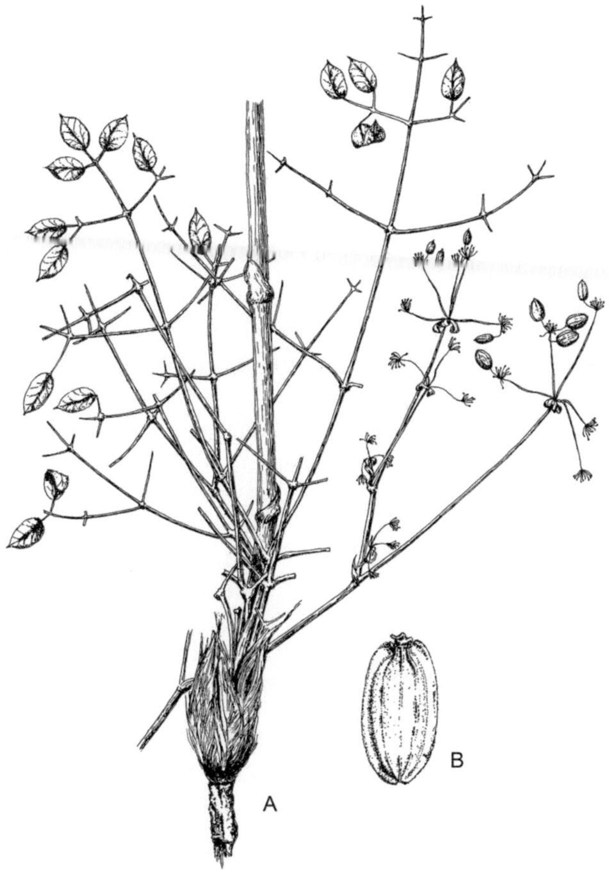

Fig. 16. Apiaceae, Apioideae. *Autumnalia innopinata.* **A** Habit. **B** Fruit. (Orig., illustration by Elena Mzhelskaya)

bifurcated, vallecular vittae 1, commissural 2. Seed face plane. $2n = 22$.

Two species, Middle Asia. In xerophilic scrub-ephemeroid vegetation. Related to *Ferula.*

83. ***Azilia*** Hedge & Lamond [Pimenov]

Azilia Hedge & Lamond in Rech.f., Fl. Iran. 162: 386 (1987).

Stout, glabrous perennials with branched lignified rootstocks. Basal leaves pinnate or sometimes simple, segments thick, broad, almost orbicular to reniform, with spiny margins; upper leaves much reduced. Bracts and bracteoles triangular to lanceolate, reflexed; rays thickened. Petals white or greenish; styles short, divergent; stylopodium compressed, wavy at margin. Fruits compressed dorsally, commissure broad; mesocarp parenchymatous; dorsal and lateral ribs inconspicuous, marginal ribs narrowly winged, thickened; carpophore bifurcated, persistent; vallecular vittae 1–2, commissural 2–3. Seed face plane. $2n = 22$.

One species, *A. eryngioides* (Pau) Hedge & Lamond), Asia (W Iran). Screes, among xerophilic scrub and small trees.

84. ***Berula*** W.D.J. Koch Fig. 17 [Pimenov]

Berula W.D.J. Koch, Deutschl. Fl., ed. 3, 2: 25, 433 (1826); Spalik, Downie & Watson, Taxon 58: 735–748 (2009), phylog., taxon.
Siella Pimenov (1978).
Afrocarum Rauschert (1982).

Glabrous perennials with stolons and fascicled fibrous roots; stems rooting at nodes. Leaves submerged and aerial; submerged leaves 3-4-pinnate with narrow segments; aerial leaves pinnate with lanceolate to ovate, serrate, or irregularly lobed leaflets. Bracts 3–5, lanceolate, entire to dissected, reflexed; rays few; bracteoles linear-lanceolate. Calyx lobes triangular-lanceolate; petals white, styles short; stylopodium low-conical. Fruits subdidymous, compressed laterally, commissure narrow; mesocarp bistratous, the outer cell layers thin-walled, the inner of lignified cells; carpophore bifurcated; vittae large, many, cyclic. Seed face plane. $2n = 12, 18, 20$.

Seven species, Europe, Asia, Africa (incl. St. Helena), N America, C America. Along streams, marshes. Used in folk medicine and for salads, but potentially poisonous. Molecular data suggest affinity to *Sium, Helosciadium, Cryptotaenia* and several other hydrophilous genera (Spalik et al. 2009). An intergeneric hybridization with *Helosciadium nodiflorum* was established by karyological and molecular studies, resulting in the name × *Beruladium* A.C. Leslie.

85. ***Bifora*** Hoffm. [Reduron]

Bifora Hoffm., Gen. Pl. Umbell. ed. 2: XXXIV, 191 (1816), nom. cons.; Guérin, Bull. Soc. Bot. France 70: 481–487 (1923), anat.

Glabrous annuals. Leaves 1–3-pinnately divided, triangular or ovate; ultimate divisions linear. Bracts and bracteoles absent or very few; rays few to several. Petals white, obovate to deeply bifid, subequal or radiate; styles short to very

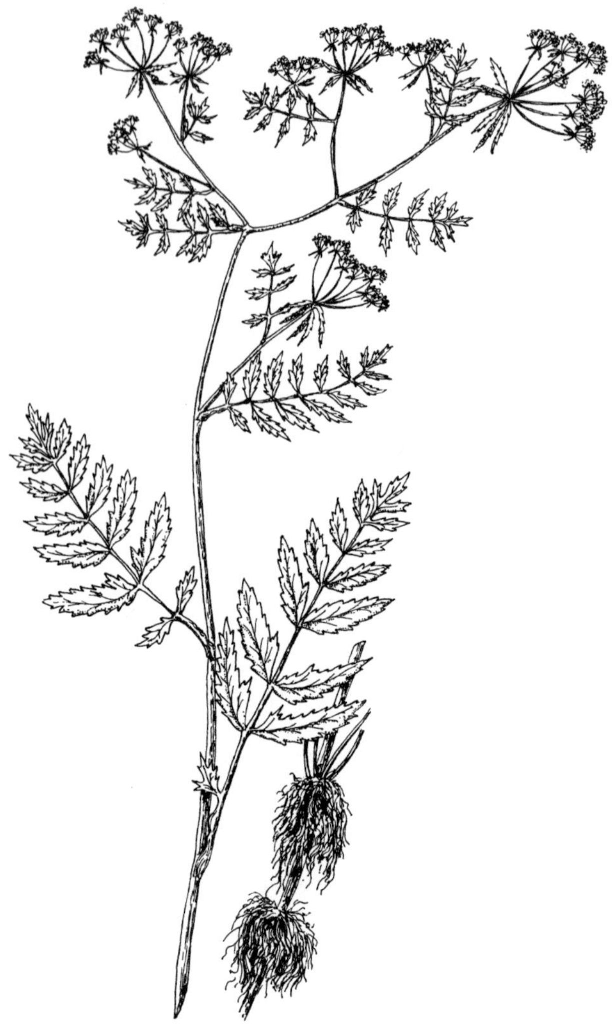

Fig. 17. Apiaceae, Apioideae. *Berula erecta*. (Orig., illustration by Elena Mzhelskaya)

elongate; stylopodium conical or reduced. Fruits glabrous, didymous, the 2 mericarps almost spherical; ribs scarcely visible; carpophore absent; vittae indistinct or reduced; commissural face generally perforated with 2 pores. Seeds transversely subterete, the face deeply and broadly concave. $2n = 20, 22$.

Two species, mostly Mediterranean region and the Caucasus to SW Asia, sometimes adventive elsewhere. See also *Atrema*.

86. ***Bilacunaria*** Pimenov & V.N. Tikhom. [Pimenov]

Bilacunaria Pimenov & V.N. Tikhom., Feddes Repert. 94: 151 (1983).

Glabrous or hispidulous perennials. Leaves 4–5-ternate, leaflets filiform to linear. Bracts and bracteoles linear to lanceolate, entire. Calyx lobes obtuse; petals yellow, glabrous or dorsally puberulent; styles short; stylopodium low-conical, often wavy. Fruits tuberculose or verrucose, elliptic to globose, commissure narrow; pericarp parenchymatous, its inner layer broken down at maturity; ribs keeled, with large vascular bundles; carpophore bifurcated; vittae cyclic, many, rib oil ducts short, a few in each rib. Seed face deeply grooved, with two to three large, central funicular secretory canals. $2n = 22$.

Five species, SW Asia, Caucasus. Dry stony slopes. Culinary (as condiment). Closely related to *Cachrys*, *Prangos*, and *Alococarpum*.

87. ***Billburttia*** Magee & B.-E. van Wyk [Van Wyk & Tilney]

Billburttia Magee & B.-E. van Wyk in Magee et al., Pl. Syst. Evol. 283: 241 (2009); Van Wyk et al., Afr. Apiaceae: 114–115 (2013).

Glabrous, woody shrubs or subshrubs with woody rootstock. Leaves 1–3-pinnate; leaflets filiform-linear or linear-elliptic to elliptic, margins entire. Bracts and bracteoles numerous, linear-oblong, unequal; rays several, relatively long. Petals greenish yellow, yellow or white; styles short; stylopodium broadly conical. Fruits glabrous, oblong-ellipsoid, compressed dorsally; mericarps dorsally compressed; ribs conspicuous, marginal ones narrowly winged (with vascular bundles along the edge); carpophore bifurcated; vittae large, vallecular 1–2, commissural 6. Seeds transversely pentagonal, slightly to deeply sulcate, flat or concave on commissural face.

Two species, Madagascar. Related to *Ammi*, *Sclerosciadium* and *Petroselinum* (Jiménez-Mejías and Vargas 2015).

88. ***Bonannia*** Guss. [Watson]

Bonannia Guss., Fl. Sicul. Syn. 1: 355 (1843), nom. cons.

Puberulent perennials, rhizomes thick. Leaves nearly all basal, bipinnate; ultimate segments oblong-lanceolate, acute, irregularly dentate or lobed; cauline leaves usually reduced to sheathing petioles. Bracts and bracteoles several, short and linear; rays several, spreading. Petals yellow; styles short; stylopodium low-conic. Fruits ovoid, dorsally compressed, pruinose; ribs thickened; carpophore bifid; vallecular vittae ca. 5, commissural 8–9. Seed compressed dorsally, the face plane. $2n = 22$.

One species, *B. resinifera* Guss., S Europe (Italy, Sicily, S Greece).

89. ***Bubon*** L. [Spalik]

Bubon L., Sp. Pl. 253 (1753); Gen. Pl. ed. 5: 121 (1754); Spalik & Downie, Ann. Missouri Bot. Gard. 88: 270–301 (2001), phylog.

Stout, pubescent perennials. Leaves 2–3-pinnately dissected, triangular-ovate; lobes lanceolate to ovate. Bracts few; rays few to many, unequal, ascending; bracteoles few, pubescent. Calyx lobes prominent; petals white; stylopodium conical in flower, rounded in fruit. Fruits pubescent, particularly along ribs, ovate with a short beak, laterally compressed; mericarps terete; ribs filiform; commissure broad; vittae prominent, vallecular 2–3, commissural 2. Seed face plane.

One species, *B. macedonicum* L., Europe. Open montane habitats. Once included in *Athamanta*, it is instead more closely related to *Arafoe* and *Pimpinella* (Downie et al. 2000).

90. ***Bunium*** L. [Kljuykov]

Bunium L., Sp. Pl. 243 (1753); Gen. Pl., ed. 5: 128 (1754); Wolff, Pflanzenr. (Engler) IV, 228 (Heft 90): 186–214 (1927), sect. & subsect. rev.; Korovin, Bull. Middle As. State Univ. 15: 117–129 (1927), reg. surv.; Vasileva et al., Pl. Syst. Evol. 149: 71–88 (1985), karyol.; Kljuykov, Bjull. Moskovsk. Obšč. Isp. Prip. Otd. Biol. 93: 76–88 (1988), rev.
Wallrothia Spreng. (1815).

Glabrous perennials with globose, usually deep tubers. Leaves 3–4-pinnate; leaflets lobed. Bracts and bracteoles filiform to linear, entire or absent. Petals white; styles straight to reflexed; stylopodium flat to conic. Fruits glabrous, ovate or oblong, slightly compressed laterally or dorsally, commissure narrow to broad; ribs equal, filiform or terete, or marginal ribs rarely narrowly winged; carpophore bifid or rarely reduced; vallecular vittae 1–3, commissural several. Seed face plane or sulcate. Seedlings mono- or dicotyledonous. $2n = 12, 24$.

27 species, Europe (W, S), Africa (NW), Asia (SW). From semi-deserts to alpine meadows.

91. ***Bupleurum*** L. Fig. 3D [Pimenov]

Bupleurum L., Sp. Pl. 236 (1753); Gen. Pl., ed. 5, 110 (1754); Wolff, Pflanzenr. IV, 228 (Heft 43): 36–173 (1910), rev. gen.; Koso-Poljansky, Acta Horti Bot. Petersb. 30: 135–333 (1913), rev. reg.; Shan Ren Hwa & Li Yin, Acta Phytotax. Sin. 12(3): 261–294 (1974), rev. reg.; Cauwet-Marc, Biosystématique des espèces vivaces de *Bupleurum* L. (Umbelliferae) du bassin Méditerranéen occidental, Facs. 1–3, Perpignan: 1–426, 1–172, 1–230 (1976), rev. reg.; Pimenov & Sdobnina, Bjull. Moskovsk. Obšč. Isp. Prip. Otd. Biol. 88, 1: 105–117, 5: 82–94 (1983), rev. reg.; Snogerup & Snogerup, Willdenowia 31: 205–308 (2001), rev. reg. annual spp.; Neves & Watson, Ann. Bot. 36: 379–398 (2004), phylog.

Glabrous, herbaceous annuals to perennial shrubs with taproots or branched rootstocks. Leaves entire, usually deciduous (or evergreen), linear to broadly elliptic, or perfoliate, margins rarely serrulate. Bracts linear-lanceolate to ovate, sometimes yellowish, or rarely absent; bracteoles linear to roundish, sometimes yellowish. Petals usually glabrous, yellow, or dorsally purplish; styles short; stylopodium flat to low-conical. Fruits smooth, rarely tuberculate, slightly compressed laterally; commissure narrow; pericarp not lignified; ribs inconspicuous to narrowly winged; carpophore bifurcated; vallecular vittae 1–4, commissural several, rarely inconspicuous. Seed face plane or slightly emarginate. $2n = 8, 12, 14, 16, 20, 22, 24, 26, 28, 29, 30, 32, 36, 42, 60, 64$.

185–195 species, Eurasia, N Africa, and one species each in N America and S Africa; adventive in Australasia. Medicinal, especially in E Asian countries for treatment of hepato-biliary diseases. Closely related to *Hohenackeria*.

92. ***Cachrys*** L. [Pimenov]

Cachrys L., Sp. Pl. 246 (1753); Gen. Pl., ed. 5: 117 (1754).

Stout, glabrous or slightly scabrid perennials. Leaves 2–6-pinnatisect; leaflets linear, entire, or dentate, or trifid, mucronate, scabrid on the margin. Bracts linear-triangular or pinnatifid, rays scabrous, thickened near the base; bracteoles entire, bifid or pinnatisect. Calyx lobes lanceolate; petals yellowish; styles long; stylopodium flat. Fruits covered by irregular tubercles or papillae; commissure narrow, ribs prominent, with big sclerenchyma groups, more or less dentate-cristate, subequal; mesocarp bistratous; fibro-vascular bundles small, many, situated in outer mesocarp layer, near border with inner layer; vittae many, cyclical in inner mesocarp layer; scattered secretory ducts many in outer mesocarp. Seed face deeply emarginate, with involute margins. $2n = 22$.

Four species, Mediterranean (S Europe, N Africa, SW Asia). Dry stony slopes, among scrub, sand dunes. Closely related to *Prangos* and *Bilacunaria*.

93. ***Calyptrosciadium*** Rech.f. & Kuber [Pimenov]

Calyptrosciadium Rech.f. & Kuber, Anz. Math-Nat. Kl. Oesterr. Akad. Wiss. 101: 362 (1964); Pimenov & Kljuykov, Candollea 59(1): 95–101 (2004), rev.

Slender glabrous perennials with branching lignified rootstocks. Leaves 4–5-ternatisect, glaucous; leaflets linear-filiform, divaricate. Bracts and bracteoles few, herbaceous; rays 1–5, unequal. Calyx lobes ovate to triangular, acuminate; petals yellow, obovate, entire; styles deflexed; stylopodium low-conical or depressed. Fruits slightly compressed dorsally, mesocarp of cells with non-lignified walls; ribs approximately equally winged; carpophore bifurcated; vallecular vittae 1–2, commissural 4. Seeds transversely lunate; the face deeply and broadly grooved. $2n = 22$.

Two species, Asia (E Iran, Afghanistan). Rocky crevices. Molecular data suggest an affinity to *Komarovia*, *Parasilaus* and *Cyclorhiza*.

94. ***Cannaboides*** B.-E. van Wyk [Van Wyk & Tilney]

Cannaboides B.-E. van Wyk in Van Wyk et al., Taxon 48: 737–745 (1999); Van Wyk et al., Afr. Apiaceae: 118–119 (2013).

Pubescent, woody perennials. Leaves pinnate to bipinnate; leaflets oblong-elliptic or linear, with serrate margins (similar to *Cannabis sativa* L.). Bracts obovate to oblong-elliptic; rays relatively few, short or long; bracteoles small. Umbels usually densely hairy, perfect or staminate; petals yellow-green or white, broadly obovate to semiquadrangular; styles relatively long; stylopodium disc-shaped. Fruits usually densely hairy, ovoid to ellipsoid, compressed laterally; mericarps terete; ribs inconspicuous; carpophore bifid; vittae very large, vallecular 1, commissural 2; rib oil ducts prominent. Seeds transversely stellate and deeply sulcate beneath the vittae, the face plane or convex.

Two species, S Madagascar. Ericoid vegetation. Similar to *Diplolophium* but related to other woody Madagascan genera.

95. ***Capnophyllum*** Gaertn. [Van Wyk & Tilney]

Capnophyllum Gaertn., Fruct. Sem. Pl. 2: 32, t. 85 (1790); Magee et al., Syst. Bot. 34: 580–594 (2009), phylog.; Magee et al., S. Afr. J. Bot. 75: 283–291 (2009), rev.; Van Wyk et al., Afr. Apiaceae: 120–123 (2013).

Glabrous, glaucous annuals. Leaves pinnate to bipinnate; leaflets linear-oblong. Bracts few, ovate to lanceolate, often connate (rarely absent); rays few, relatively long; bracteoles small (rarely absent). Petals white, oblong to obovate; styles very short to very long; stylopodium flat or conical. Fruits tuberculate or glabrous, elliptic to oblong, dorsally compressed, marginal ribs winged, dorsal ribs prominent, rarely indistinct, secondary ribs present; mericarps strongly compressed dorsally with median keel on the commissural face; carpophore bifid; vittae large, vallecular 1, commissural 2. Seeds transversely strongly compressed dorsally, sulcate beneath the vittae, the face plane.

Four species, South Africa. Mostly deep coastal sands. Related to *Scaraboides* and *Dasispermum*.

96. ***Carlesia*** Dunn [Pu]

Carlesia Dunn, Hook. Icon. Pl. 28(2): t. 2739 (1902); She et al., Fl. China 14: 114 (2005), rev.

Glabrous perennials. Basal leaves clustered, 3-pinnately compound, oblong to ovate; ultimate segments linear; cauline leaves few, similar to the basal, gradually reduced; uppermost leaves smaller, trilobed. Bracts and bracteoles linear and entire; rays many, spreading. Calyx lobes conspicuous, triangular; petals white; styles as long as the fruit; stylopodium conical. Fruits scabrous, ovoid-oblong, slightly compressed dorsally, ribs obtuse; carpophore apically bifid;

vallecular vittae 3, commissural 4. Seed face plane. $2n = 22$.

One species, *C. sinensis* Dunn, Asia (NE China).

97. ***Caropsis*** (Rouy & E.G. Camus) Rauschert [Reduron]

Caropsis (Rouy & E.G. Camus) Rauschert, Taxon 31: 555 (1982).
Thorella Briq., Ann. Conserv. Jard. Bot. Genève 17: 274 (1914), non Gaillon (1833); Briquet, *Thorella* Etude monogr. 1–43 (1914); H. Wolff in Engl., Pflanzenr. 4 (228) Heft 90: 136–137 (1927), rev.

Dwarf, glabrous perennials with a rhizome. First leaves reduced to a subulate petiole and rachis, other leaves pinnately divided, very elongate; leaflets short, spathulate, sometimes pinnatisect. Bracts several; bracteoles present; rays few, slightly unequal, spreading. Petals white; styles subequal to the conical stylopodium. Fruits glabrous, ovoid; ribs stout, prominent; carpophore free, bifurcated; vittae large, vallecular 1, commissural 2. Seeds sub-rectangular, the face plane. $2n = 20$.

One species, *C. verticillatoinundata* (Thore) Rauschert, Europe (W France, W Portugal). Flooded habitats. Related to hydrophytic umbellifers including *Helosciadium*, *Cryptotaenia*, *Sium* and *Berula* (Spalik et al. 2014).

98. ***Carum*** L. [Kljuykov]

Carum L., Sp. Pl. 263 (1753); Gen. Pl., ed. 5: 127 (1754); Wolff, Pflanzenr. (Engler) IV, 228 (Heft 90): 143–167 (1927), rev.

Glabrous biennials to perennials. Leaves petiolate, 2–4-pinnate; leaflets filiform to lanceolate, basal segments sessile or rarely petiolulate. Bracts and bracteoles entire, filiform to lanceolate or absent. Petals white, rarely pinkish; styles reflexed; stylopodium low-conical. Fruits glabrous, ovoid or oblong, slightly compressed laterally; commissure narrow; ribs equal, filiform to terete; carpophore bifurcated; vallecular vittae 1 (2–3), commissural 2–4. Seed face plane. $2n = 18$, 20, 22, 23, 24, 25.

30 species (about half of which should be moved to other genera), Europe, temperate regions of Asia, Africa. Meadows, forest margins, rocks, road edges, pastures, and weedy areas. *Carum carvi* L. widely cultivated for its aromatic fruits; also used in folk medicine.

99. ***Caucalis*** L. Fig. 2H [Lee]

Caucalis L., Sp. Pl. 241 (1753); Gen. Pl., ed. 5: 112 (1754).

Sparsely pubescent annuals. Leaves 2–3-pinnate, broadly ovate, ultimate divisions lanceolate to oblong. Bracts absent (rarely 1–2, linear to lanceolate); rays several, long, not contracted in fruit; bracteoles several, simple, linear to lanceolate, glabrous, with ciliate margins. Petals white to pinkish, the outer slightly radiate; styles subsessile, shorter than the conical stylopodium. Fruits spinescent, oblong-ellipsoid; mericarps compressed laterally, with both primary and secondary ribs prominent; primary ribs with smooth, ± erect hairs, swollen at the base, arranged in a single row; secondary ribs 4, with glabrous spines, not confluent at the base and simple at the tip, arranged in a single row; carpophore free; commissure broad; vittae small, transversely elliptical, vallecular 1 (under each secondary rib), commissural 2. Seeds transversely elliptical, the face deeply furrowed and curved inward. $2n = 20$.

One species, *C. platycarpos* L., C Europe and SW Asia. Closely related to *Turgenia* and *Lisaea*, sharing common characters of fruit anatomy.

100. ***Cenolophium*** W.D.J. Koch [Watson]

Cenolophium W.D.J. Koch, Nova Acta Phys.-Med. Acad. Caes. Leop.-Carol. Nat. Cur. 12(1): 103 (addit.) (1824).

Glabrous perennials. Leaves petiolate, diffusely 3–4-pinnate; ultimate segments linear or linear-lanceolate, entire. Bracts lacking (occasionally few and linear); rays several; bracteoles several, linear, equaling pedicels. Petals white; styles reflexed, twice the length of the conical stylopodium. Fruits glabrous, oblong-ellipsoid, slightly compressed dorsally; ribs prominent to narrowly winged and hollow; carpophore bifurcated; commissure narrow; vallecular vittae 1, commissural 2. Seeds transversely compressed dorsally, the face plane. $2n = 22$.

One species, *C. denudatum* (Hornem.) Tutin, E Europe, C Asia, N Asia (Siberia), E Asia (China). Forests, grasslands, often in damp or wet areas.

101. ***Cephalopodum*** Korovin [Pimenov]

Cephalopodum Korovin, Izv. Akad. Nauk Tadzhiksk. SSR, Otd. Estestv. Nauk 1 (50): 17 (1973).

Glabrous perennials with many marcescent petioles at the base of the stems. Leaves tripinnate; leaflets filiform to linear. Bracts linear-lanceolate, entire, flaccid; rays many; bracteoles lanceolate, mucronulate. Petals yellow or yellowish; styles reflexed; stylopodium conical. Fruits slightly compressed dorsally, commissure broad, pericarp not lignified; dorsal and lateral ribs filiform, marginal ribs narrow-winged; carpophore bifurcated; vallecular vittae 1, commissural 2. Seed face plane. $2n = 22$.

Two species, Asia (Tadzhikistan, Afghanistan, Uzbekistan). Rocky crevices. Presumably related to *Seseli*.

102. ***Cervaria*** Wolf [Reduron]

Cervaria Wolf, Gen. Pl. 28 (1776); Pimenov in Rechinger, Fl. Iran. 162: 451–454 (1987), taxon., nomencl.; Reduron et al., J. Bot. Soc. Bot. France 1: 93 (1997), taxon.; Spalik et al., Pl. Syst. Evol. 243: 189–210 (2004), phylog.

Glabrous perennials with a rootstock or rhizome. Leaves 1-3-ternately or 1-3-pinnately divided, ovate or triangular; leaflets rhomboidal, ovate or lanceolate with a crenate, roughly serrate or lobed margin. Bracts and bracteoles several to many; rays several to many, unequal, spreading. Petals white, greenish or purplish; styles equaling or exceeding the short-conical stylopodium. Fruits glabrous or scabridulous, ovate, ellipsoid or suborbicular, strongly compressed dorsally; dorsal ribs filiform, ± prominent, the marginal ribs narrowly to broadly winged; carpophore free, bifurcated; commissure broad; vittae small to large, vallecular 2, commissural 2–4. Seeds transversely compressed dorsally, the face plane. $2n = 22$.

Two species, Eurasia, N Africa (Algeria), open slopes to shadowy forests. Segregated from *Peucedanum* s.lat.

103. ***Chaerophyllopsis*** H. Boissieu [Pu]

Chaerophyllopsis H. Boissieu, Bull. Soc. Bot. France 56: 353 (1909); Fl. China 14: 29 (2005), rev.

Squamose-pubescent annuals. Leaves 2–3-pinnately dissected to compound, ovate-oblong, papyraceous; primary leaflets 5–6 pairs; ultimate segments nearly triangular or rhombic, base cuneate or truncate, apex acute, margins coarsely serrate or pinnatifid. Bracts absent or occasionally solitary; rays many; bracteoles few, linear-lanceolate, persistent. Calyx lobes conspicuous, lanceolate, longer than the stylopodium; petals white or pinkish; styles short; stylopodium conic, slightly undulate on the margin. Fruits glabrous, oblong-ovoid or cylindroid, small, compressed laterally; ribs filiform; carpophore apically bifid; vallecular vittae 1–2, commissural 2. Seeds transversely pentagonal, the face plane.

One species, *C. huai* H. Boissieu, Asia (SW China).

104. ***Chaerophyllum*** L. Fig. 2E [Spalik]

Chaerophyllum L., Sp. Pl. 258 (1753); Gen. Pl., ed. 5: 125 (1754); Spalik & Downie, Ann. Missouri Bot. Gard. 88: 270–301 (2001), sect. rev.; Chung et al., Amer. J. Bot. 92: 2054–2071 (2005), phylog.; Chung et al., Syst. Bot. 32: 671–681 (2007), taxon.; Piwczyński et al., Bot. J. Linn. Soc. 178: 298–313 (2015), phylog.

Glabrous to pubescent annuals, biennials, or perennials. Leaves 2–4-pinnately dissected, triangular to triangular-ovate; lobes linear-oblong to ovate. Bracts usually absent, rarely many; rays few to many, ascending or rarely spreading; bracteoles few, usually ciliate. Petals white, rarely pinkish or yellow, outer often radiate, sometimes pubescent or ciliate; styles short to relatively long; stylopodium conical in flower, raised conical or rounded in fruit. Fruits glabrous, rarely setose or hispid, linear-oblong to ovoid with a short or obsolete beak, often scarcely compressed laterally; mericarps terete to semi-circular, sometimes scarcely compressed dorsally; ribs broad, round; carpophore bifurcate, commissure broad; vascular bundles prominent; vittae prominent, vallecular 1, commissural 2. Seed transversely terete to lunate, the face concave to deeply sulcate. $2n =$ 12, 14, 22.

About 35 species, Eurasia, N Africa, Azores, N America. Primary and secondary open and shaded lowland and montane habitats. One species, *C. bulbosum* L., is rarely cultivated for its tuberous roots, used as vegetable. Four sections are recognized by Spalik and Downie (2001); molecular data (Chung et al. 2005, 2007) suggest that the genus is paraphyletic with respect to S Hemisphere *Oreomyrrhis*, and the latter has been placed into synonymy of *Chaerophyllum* in some treatments.

105. ***Chaetosciadium*** Boiss. [Plunkett]

Chaetosciadium Boiss., Fl. Orient. 2: 1078 (1872); Post, Fl. Syria, ed. 2, 1: 567 (1932), reg. rev.; Zohary, Fl. Palaest. 2 (1): 393–394 (1972), reg. rev.

Glabrous or setulose herbaceous annuals. Leaves 2–3-pinnately dissected, oblong to ovate; lobes oblong, acute, pinnately incised to lobed. Bracts absent; rays few, ascending to spreading; bracteoles many, subulate, longer than pedicels. Petals white or pinkish, outer slightly radiate; calyx teeth short, subulate; styles very long; stylopodium long-conical. Fruits irregularly covered by many long, purple or white scabrous bristles, oblong, slightly compressed dorsally; mericarp ribs obsolete; carpophore bifurcate; vallecular vittae 1, commissural 2. Seed face concave. $n = 6$, $2n = 12$.

One species, *C. trichospermum* (L.) Boiss., SW Asia (Levant). Mediterranean woodlands and scrub, shrubby steppe, and desert, in shaded rocky areas. Sometimes treated in synonymy under *Torilis* based on molecular and morphological data (Lee et al. 2001).

106. ***Chamaesciadium*** C.A. Mey. [Pimenov]

Chamaesciadium C.A. Mey., Verz. Pfl. Cauc.: 122 (1831); Papini, Fl. Medit. 16: 45–55 (2006), phylog.

Acaulescent or shortly caulescent, glabrous perennials. Leaves rosulate, bipinnate, oblong; leaflets linear, acute. Peduncles aerial and short or underground; bracts linear to lanceolate, membranaceous at the margin, or pinnatisect; rays few or several, unequal, sometimes long; bracteoles entire, lanceolate, or pinnatisect. Petals yellowish; styles long, reflexed; stylopodium flat or low conical, undulate at the margin. Fruits slightly compressed laterally; sometimes only one mericarp developed; ribs filiform, equal; carpophore entire; commissure narrow; vallecular vittae 3–4, commissural 6–8. Seed face plane. $2n = 22$.

One species, *C. acaule* (M. Bieb.) Boiss., SW Asia (Caucasus). High-mountain vegetation, screes, alpine meadows, stony slopes. Molecular data (Papini 2006) suggest a relationship to *Carum*, *Fuernrohria*, and *Grammosciadium*.

107. ***Chamaesium*** H. Wolff [Kljuykov]

Chamaesium H. Wolff, Notizbl. Bot. Gard. Berlin 9: 275 (1925).
Dolpojestella Farille & Lachard (2002).

Glabrous perennials. Leaves pinnate; leaflets sessile, ovate to linear, entire, toothed or serrate. Bracts and bracteoles entire to divided, or absent. Calyx lobes connate or obsolete; petals usually greenish-white, almost plane; styles short, reflexed; stylopodium massive, depressed, dark-green. Fruits glabrous; ovoid to oblong, commissure narrow; ribs doubled (10), short-winged to obscure; carpophore stout, bifurcated; vallecular vittae 1(2–3), commissural several. Seed face plane to concave. $2n = 12$.

Ten species, E Himalaya, China (SW). Scrub, forest margins, alpine and subalpine meadows, moist turf, among dwarf shrubs, pastures.

108. ***Chamarea*** Eckl. & Zeyh. [Van Wyk & Tilney]

Chamarea Eckl. & Zeyh., Enum. Pl. Afr. austr. extratrop. 3: 346 (1837); Burtt, Edinb. J. Bot. 48: 200, 261 (1991); Vessio, unpubl. M.Sc. thesis, Rand Afrikaans University (2001), phylog.; Van Wyk et al., Afr. Apiaceae: 154–159 (2013).

Acaulescent, glabrous perennials with tuberous roots. Leaves radical, hysteranthous, first-formed leaves finely dissected or pinnate with broad, dentate lobes, later ones variously dissected, usually finely 2–3-ternate; leaflets variable, the segments often dentate. Bracts lanceolate to linear or absent; rays several, relatively short; bracteoles small or absent. Petals yellowish; styles relatively long, divergent; stylopodium large, conical, often bulging. Fruits glabrous, oblong to ovoid, usually small, often narrowing toward apex, terete or somewhat laterally compressed; mericarps terete; ribs usually prominent, obtuse; carpophore bifid; vittae large, vallecular 1, commissural 2. Seeds transversely pentagonal, the face plane to concave.

About 11 species (5 not yet described), S Africa (South Africa and Lesotho). Edible roots. Very closely related to *Annesorhiza*.

109. ***Changium*** H. Wolff [Watson]

Changium H. Wolff, Repert. Spec. Nov. Regni Veg. 19: 314 (1924).

Glabrous perennials. Leaves ternate to 2–3-pinnate, reduced upward to linear or bladeless sheaths; leaflets pinnatifid, ultimate segments oblong-lanceolate. Bracts lacking or few; rays several, spreading; bracteoles few, small and linear. Petals white (pale purple when young); styles reflexed; stylopodium low-conical. Fruits glabrous, ovoid-globose to ovoid-oblong, slightly compressed laterally; commissure constricted; ribs inconspicuous; carpophore bifid; vittae many (ca. 20) and scattered through the mesocarp. Seed transversely subterete, the face deeply sulcate. $2n = 20$.

One species, *C. smyrnioides* Fedde ex H. Wolff (or perhaps two), Asia (E China, possibly also in W China). Low-elevation hills and rocky areas. Root is used in traditional Chinese medicine.

110. ***Choritaenia*** Benth. [Van Wyk & Tilney]

Choritaenia Benth. in Benth. & Hook.f., Gen. Pl. 1: 907 (1867); Liu et al., S. Afr. J Bot. 73: 184–189 (2007), rev.; Van Wyk et al., Afr. Apiaceae: 160–161 (2013).

Prostrate, glabrous annuals. Leaves small, upper subsessile, irregularly subternate; leaflets small, linear, succulent. Umbels simple or compound, peduncles absent; bracts foliaceous, deeply dissected (resembling the upper leaves); rays few, short; bracteoles small. Petals white or purplish; styles short; stylopodium small, conical. Fruits with long, stiff, clavate hairs mixed with thin hairs, orbicular, strongly compressed dorsally; mericarps somewhat biconvex; endocarp woody; ribs absent but with two interjugal marginal wings; carpophore very short, bifid, upper part reduced to soft tissue, lower part short, thick, hygroscopic and myxogenic; vittae and rib oil ducts lacking; large globose oil vesicles present at regular intervals in the wings. Seeds transversely strongly compressed dorsally, the face slightly convex to concave.

One species, *C. capensis* Benth., S Africa (C and NE parts). Dry grassland. Phylogenetically isolated but related to *Marlothiella* and *Lichtensteinia*.

111. ***Chuanminshen*** M.L. Sheh & R.H. Shan [Watson]

Chuanminshen M.L. Sheh & R.H. Shan, Acta Phytotax. Sin. 18: 47 (1980).

Glaucous perennials. Leaves many, mainly basal, 2–3-ternate-pinnate; ultimate segments ovate to long-ovate. Synflorescence many-branched, spreading; bracts and bracteoles usually lacking, rarely 1–3, linear and deciduous; rays few to several, very unequal. Petals white or purplish, dark midrib conspicuous; styles much longer than stylopodium, often reflexed; stylopodium conic. Fruits glabrous, ellipsoid, strongly compressed dorsally; ribs prominent and filiform, marginal ribs narrowly thick-winged; carpophore bifid; vallecular vittae 2–3, commissural 4–6. Seed somewhat compressed dorsally, the face plane. $2n = 22$.

One species, *C. violaceum* M.L. Shen & R.H. Shan, Asia (endemic to Hubei and Sichuan, China). Low-elevation open grassy areas, often by streams. Roots used medicinally.

112. ***Chymsydia*** Albov [Pimenov]

Chymsydia Albov, Prodr. Fl. Colch.: 110 (1895); ejusd. in Bull. Herb. Boiss. 3: 233 (1895).

Scabrous or almost glabrous perennials. Leaves biternate; leaflets coriaceous, lanceolate-ovate, decurrent. Bracts and bracteoles absent; rays several. Petals white; styles reflexed; stylopodium conical. Fruits glabrous, ovoid, slightly compressed dorsally, commissure narrow, mesocarp cells with thin walls or with lignified pitted walls; ribs winged; carpophore bifurcated; vallecular vittae 1, commissural 2–4. Seed face plane. $2n = 22$.

One species, *C. agasylloides* (Albov) Albov, Asia (W Caucasus). Limestone rocks and subalpine high-herbaceous meadows. Related to *Agasyllis.* Molecular data suggest affinity to two Caucasian species of *Angelica* sect. *Xanthogalum* (Spalik et al. 2004).

113. ***Cicuta*** L. [Pu]

Cicuta L., Sp. Pl. 255(1753); Gen. Pl., ed. 5: 123 (1754); Mathias & Constance, N. Amer. Fl. 28B(1): 154–157 (1944), reg. rev.; T.G. Tutin et al., Fl. Europ. 2: 352

(1968), reg. rev.; She et al., Fl. China 14: 77 (2005), reg. rev.; Lee & Downie, Can. J. Bot. 84: 453–468 (2006), phylog.

Glabrous perennials with a branched, swollen caudex bearing fascicled fibrous or tuberous roots. Leaves 1–3-pinnately compound, oblong or ovate-lanceolate or deltoid; leaflets narrow, serrate to incised at margin. Bracts absent (or 1-several and inconspicuous); rays many; bracteoles several, rarely absent. Calyx lobes conspicuous; petals white; styles short; stylopodium low conical to depressed. Fruits glabrous; ovoid to globose, compressed laterally; ribs obtuse, wide, corky, subequal; carpophore bifurcated; commissure broad; vallecular vittae 1, commissural 2. Seed transversely subterete, the face plane to slightly concave. $2n = 22$.

Eight species, circumboreal: N America, Europe, and Asia; adventive in Australia. Most species are highly toxic. Related to *Oenanthe* and some other hydrophytic umbellifers (Spalik et al. 2014).

114. ***Cnidiocarpa*** Pimenov [Pimenov]

Cnidiocarpa Pimenov, Bot. Zhurn. 68(1): 103 (1983).
Macrosciadium V.N. Tikhom. & Lavrova (1988).

Glabrous perennials, often with horizontal rhizomes. Leaves 2–3-pinnatisect, broad-triangular to triangular-ovate; leaflets ovate, deeply serrate or lobate. Bracts and bracteoles absent or several, caducous; rays many. Petals white (sometimes pink); styles reflexed; stylopodium conical. Fruits glabrous, ellipsoid to ovoid or oblong-ovoid, slightly compressed dorsally, pericarp not lignified, commissure narrow; ribs subequal, keeled or obtuse; carpophore bifid or bifurcated; vallecular vittae 1–3, commissural 2–5. Seed face plane. $2n = 22$.

Five species, Asia (Middle, SW, Caucasus). Subalpine meadows, forest openings, wet places. Molecular data suggest a close relationship to *Selinum* and *Katapsuxis*.

115. ***Cnidium*** Cusson ex Juss. Fig. 2G [Pimenov]

Cnidium Cusson ex Juss., Hist. Mém. Soc. Roy. Med. Paris: 280 (1787).
Pinasgelon Raf. (1840).

Glabrous or scabrous annuals, biennials or perennials. Leaves 2–3-pinnatisect or ternatisect; leaflets linear-lanceolate to ovate, deeply lobed. Bracts and bracteoles filiform to lanceolate, subulate, glabrous or ciliate, white-membranous at the margins; rays few to many. Petals white; styles reflexed; stylopodium low-conical or conical. Fruits glabrous or covered with small wart-like outgrowths, slightly compressed dorsally, commissure narrow, pericarp non-lignified; ribs subequal, keeled; carpophore bifid; vallecular vittae 1, commissural 2. Seed face plane. $2n = 12, 14, 20, 22, 24, 33, 36, 44, 48, 72$.

Five species, Asia, Europe (S & SE, adventive). Used in folk and Chinese medicines. Molecular data suggest that this genus is polyphyletic.

116. ***Coaxana*** J.M. Coult. & Rose [Plunkett]

Coaxana J.M. Coult. & Rose, Contr. U. S. Natl. Herb. 3: 297 (1895); Mathias & Constance, N. Amer. Fl. 28B(2): 165–166 (1945), reg. rev.; Mathias & Constance, Contr. Univ. Michigan Herb. 11(1): 1–24 (1973), taxon.

Glabrous to pubescent, glaucous perennials. Basal leaves large, thrice ternately compound, with sharply biserrate margins; cauline leaves smaller, alternate, reduced upward, ternately to ternate-bipinnately compound, triangular-ovate; leaflets ovate-lanceolate, margins serrate to spinose-serrate. Bracts solitary or lacking; rays many, subequal to unequal, ascending to spreading-ascending; bracteoles few, conspicuous, foliar and toothed, or lacking. Petals purple; styles slender, spreading to recurved; stylopodium conical to low-conical. Fruits glabrous, glaucous, ovoid to oblong, subterete to slightly compressed laterally, heteromorphic, each mericarp having 2 winged marginal ribs and either 1 winged dorsal rib or 2 winged lateral ribs (remaining ribs obscure), the commissure narrow; carpophore bifurcated to the base, slender; vallecular vittae 2–4, commissural 2–6. Seeds subterete, sulcate under vittae, face plane to concave or sulcate. $2n = 22$.

Two species, N America (Oaxaca and Guerrero, Mexico) and C America (Guatemala). Restricted to high-elevation cloud forests near mountain summits. Molecular data place it with *Arracacia*, *Rhodosciadium*, and relatives.

117. ***Conioselinum*** Hoffm. [Pimenov]

Conioselinum Hoffm., Gen. Pl. Umbell., ed. 1: 180 (1814), ed. 2: 185 (1816), descr. emend.; Pimenov et al., Willdenowia 33: 353–377 (2003), rev.
Vvedenskya Korovin (1947).

Almost glabrous perennials with short horizontal rhizomes or rootstocks. Leaves 2–3-pinnate; leaflets linear-lanceolate to ovate. Bracts and bracteoles linear to lanceolate, subulate, glabrous, or bracts caducous; rays many. Petals white or greenish, sometimes with secretory ducts; styles reflexed; stylopodium conical. Fruits slightly compressed dorsally; carpophore bifurcated; commissure narrow, dorsal ribs narrowly winged, marginal ribs slightly enlarged; mesocarp cell walls pitted, slightly lignified; vascular bundles situated in the distal part of the ribs; vallecular vittae 1–3, commissural 2–4 (or vittae absent). Seed face plane or slightly concave. $2n = 14, 22, 44$.

23 species, Europe (N, C, E), Asia, N America. Tall-herbaceous meadows, forests, along streams. Used in Oriental and folk medicines. Molecular data support a relationship to *Lithosciadium*, *Paulita*, and their allies.

118. ***Conium*** L. [Reduron]

Conium L., Sp. Pl. 243 (1753); Gen. Pl., ed. 5: 114 (1754); Leute, Ann. Naturhist. Mus. Wien 75: 91–98 rev. (1971); Hilliard & Burtt, S. Afr. J. Bot. 51(6): 465–474 (1985), rev. South African spp.; Van Wyk et al., Afr. Apiaceae: 162–165 (2013).

Glabrous biennials, with green or purple-spotted stems. Leaves 2–4-pinnately divided, triangular; leaflets ovate, serrate or pinnatifid. Bracts absent to several; bracteoles present; rays several, spreading. Petals white or pale yellow; styles as long or longer than the broadly conical stylopodium. Fruits glabrous, broadly ovate to ovate, slightly compressed laterally; ribs flattened or keeled, undulate-crenulate; carpophore bifurcated; commissure narrow; vittae many but indistinct at maturity. Seeds transversely suborbicular to reniform, the face sulcate. $2n = 22$.

Seven species, Eurasia, N & S Africa. Open lands, fields, often ruderal. *C. maculatum* L., the deadly hemlock of Socrates' poison.

119. ***Conopodium*** W.D.J. Koch [Spalik]

Conopodium W.D.J. Koch, Nov. Acta Acad. Leop.-Carol. Nat. Cur. 12: 118 (1824); López Udias & Mateo Sanz, An. Jard. Bot. Madrid 57: 466–474 (2000), reg. rev.; Silvestre, Lagascalia 3: 3–48 (1973), reg. rev.

Glabrous to slightly pubescent perennials with globose tubers. Leaves 2–3-pinnately dissected, triangular-ovate; lobes linear-oblong to ovate. Bracts usually absent; rays few to many, subequal or unequal, slightly ascending; bracteoles few, glabrous. Petals white, rarely pinkish; styles short to relatively long; stylopodium conical in flower, rounded in fruit. Fruits glabrous, oblong to ovoid with a short beak, laterally compressed; mericarps terete to pentagonal, scarcely compressed dorsally; ribs filiform; commissure constricted; carpophore bifurcated; vascular bundles small; vittae extant at maturity, vallecular 1–3, commissural 2–4. Seed transversely terete, the face sulcate. $2n = 22, 44$.

Eight species, Europe and Africa (W Mediterranean, with *C. majus* (Gouan) Loret extending N to Scandinavia). Open places. Tubers of *C. majus* were once eaten and used as fodder. Closely related to *Athamanta* and *Todaroa* (Spalik et al. 2001).

120. ***Coriandrum*** L. [Reduron]

Coriandrum L., Sp. Pl. 256 (1753); Gen. Pl., ed. 5: 124 (1754); Purseglove et al., Spices (London) 2: 736–813 (1981).
Coriandropsis H. Wolff (1921).

Glabrous, aromatic annuals. Leaves 1–3-pinnately or ternately divided, the lowest sometimes simply lobed, triangular; ultimate divisions ovate to linear. Bracts usually absent; rays few to several, unequal; bracteoles present. Calyx lobes acute; petals white to purple, marginals generally radiate and deeply 2-lobed; styles slender, spreading; stylopodium conical. Fruits hard, glabrous, ovoid or globose, the mericarps often not separating; ribs very low; carpophore free, bifurcated; commissure broad; vallecular vittae obscure, commissural 0–2. Seeds transversely orbicular or dorsally compressed, the face slightly concave or furrowed. $2n = 22$.

Two species, Mediterranean and SW Asia. Fields, waste grounds. *Coriandrum sativum* L. is widely used as a condiment.

121. ***Coristospermum*** Bertol. [Reduron]

Coristospermum Bertol. Fl. Ital. 3: 466 (1838); Leute, Ann. Naturhist. Mus. Wien 74: 474 (1970), rev. sub *Ligusticum* subg. *Coristospermum*; Tammaro, Ann. Bot. 47: 215–225 (1989), taxon.; Reduron et al., J. Bot. Soc. Bot. France 1: 97–98 (1997), taxon.

Glabrous perennials. Leaves 3–5-pinnately divided, triangular; ultimate divisions linear, lanceolate or obovate. Bracts absent or present; rays many, subequal or unequal, spreading; bracteoles several. Petals white; styles exceeding the hemispheric or conical stylopodium. Fruits glabrous, ovoid; ribs very prominent (narrowly winged), subequal; carpophore free, bifurcated; vittae small, many and cyclic. Seeds transversely polygonal, the face plane. $2n = 22$.

Three species, S Europe. Screes, pastures, forest clearings. Segregate from *Ligusticum* s.l.

122. ***Cortia*** DC. [Watson]

Cortia DC., Prodr. 4: 186 (1830); Pimenov & Kljuykov, Feddes Repert. 116: 80–91 (2005).

Acaulescent to caulescent, glabrous to puberulent (foliage) and scaberulous (inflorescence) dwarf perennials. Leaves rosulate, 2–3-pinnatisect, oblong; ultimate segments linear. Bracts and bracteoles few to several, foliaceous and 1–2-pinnate; rays many, very unequal. Calyx lobes long, linear or narrowly lanceolate, unequal; petals white, purplish to deep purple with yellow-brown midrib; styles short, slightly longer than stylopodium, elongated in fruit; stylopodium conical. Fruits glabrous, ovoid to orbicular, compressed dorsally; ribs prominent, dorsal ribs filiform or narrowly winged; marginal ribs broadly winged; carpophore bifurcated; vallecular vittae 1–2, commissure 2–4. Seed compressed dorsally, the face slightly concave. $2n = 22$.

Two species, Asia (Pakistan, Himalaya, C Asia, and SC Xizang, China). High-altitude alpine pastures, semi-stable screes and moraines.

123. ***Cortiella*** C. Norman [Pu]

Cortiella C. Norman, J. Bot. 75: 94 (1937); H. Hara & L.H.J. Williams, Fl. Pl. Nepal 2: 135 (1979); She et al., Fl. China 14: 153–154 (2005), reg. rev.

Caespitose, acaulescent, pubescent or puberulent perennials. Leaves rosulate, pinnately decompound, narrowly oblong to oblong-ovate; ultimate divisions linear. Bracts several, foliaceous, 1–2-pinnate; rays many, unequal, spreading-ascending; bracteoles many, linear, entire or trilobate at the apex. Calyx lobes prominent, linear or triangular; petals white; styles slender, diverging; stylopodium low-conical. Fruits glabrous, subquadrate-orbicular, compressed dorsally, pale yellow or slightly purplish when mature; ribs unequal, the lateral and 1 or 2 of the other ribs broadly winged and loosely corky, or dorsal ribs filiform and marginal broadly winged; carpophore obsolete; vallecular vittae 1, commissural (0–)2. Seed transversely compressed dorsally, the face plane.

Four species, Eastern Himalaya (Nepal, Sikkim, Bhutan) and China (Xizang and Qinhai). Resembles *Cortia* DC., but differs in its unequal, winged and corky fruits and vittae number.

124. ***Cotopaxia*** Math. & Const. [Plunkett]

Cotopaxia Math. & Const., Bull. Torrey Bot. Club 79: 362 (1952); Mathias & Constance, Fl. Ecuador 5: 50–51 (1976), reg. rev.

Scaberulous or glabrous (or puberulent at nodes) perennials. Leaves ternate-pinnately to pinnately decompound, ultimate segments linear, entire. Bracts few, filiform to foliaceous, or lacking; rays several; bracteoles several, linear or slightly foliaceous. Petals white, apex recurved but not strongly inflexed; styles short to slender, spreading; stylopodium lacking. Fruits glabrous, oblong-ovoid, subterete, the commissure narrow; all ribs obtusely or narrowly winged; carpophore bifurcated; vallecular vittae 1, commissural 2. Seeds transversely compressed dorsally, sulcate beneath the vittae, the face plane to shallowly concave or convex.

Two species, S America, high elevations, páramos of Ecuador, and Sierra Nevada de Santa Maria of Colombia.

125. ***Coulterophytum*** B.L. Rob. [Plunkett]

Coulterophytum B.L. Rob., Proc. Amer. Acad. Arts 27: 168 (1892); Mathias & Constance, N. Amer. Fl. 28B(2): 166–168 (1945), reg. rev.

Pubescent, herbaceous or woody perennials. Leaves ternately or pinnately compound, the upper opposite; leaflets ovate to lanceolate, crenate to serrate. Bracts lacking or inconspicuous; rays several to many, spreading to reflexed; bracteoles several, small, linear. Petals purple; styles short, erect or spreading; stylopodium conical. Fruits glabrous or puberulent, clavate to obovoid, contracted below the seed to form a short, winged, stipe-like base, slightly compressed laterally or subterete; ribs filiform and prominent or with short wings; carpophore bifurcated or quadrafid; vallecular vittae 1, commissural ca. 4. Seed transversely nearly terete, sometimes sulcate under the vittae, the face concave. $2n = 22$.

Five species, N America (Mexico). Molecular data place the genus with other meso-American genera, including *Arracacia*, *Rhodosciadium*, and relatives.

126. ***Crenosciadium*** Boiss. & Heldr. [Pimenov]

Crenosciadium Boiss. & Heldr. in Boiss., Diagn. Pl. Orient. Nov. I, 2(10): 30 (1849).

Glabrous perennials with creeping rootstocks. Leaves pinnatisect; leaflets oblong to obovate, crenate, the margin cartilaginous. Bracts 1–2, linear-lanceolate, rays few, unequal; bracteoles few, linear. Petals white; styles strongly reflexed; stylopodium conical. Fruits slightly compressed dorsally, commissure narrow; ribs prominent, narrowly winged, with groups of large round cells with thin lignified pitted walls, marginal ribs broader than dorsal ribs; carpophore bifurcated; vallecular vittae 3–4, commissural 7–8. Seed face slightly convex, with a broad, shallow groove.

One species, *C. siifolium* Boiss. & Heldr., SW Asia (Turkey).

127. ***Crithmum*** L. [Reduron]

Crithmum L., Sp. Pl. 246 (1753); Gen. Pl., ed. 5: 116 (1754); Borde, Etude Pharmacogn. 1–100 (1910), pharm.; Briquet, Compte Rendu Séances Soc. Phys. Hist. Nat. Genève 40 (3): 115–121 (1923), carpol.; Vazart, Rev. Cytol. Biol. Vég. 21(4): 339–371 (1960), karyol.; Coiffard, Rev. Hist. Pharm. 38(290): 313–317 (1991), hist. uses; Coiffard et al., Int. J. Cosm. Sci. 15: 15–21 (1993), chem.

Glabrous, succulent perennials with a creeping rootstock. Leaves 2–3-pinnately divided, triangular, fleshy; ultimate divisions subterete. Bracts many, lanceolate; bracteoles several; rays many, subequal, spreading. Petals white or greenish, entire; styles short; stylopodium conical. Fruits glabrous, ovoid, spongy; ribs thick and prominent; carpophore free, bifurcated; commissure rather broad; vittae many, but small and not easily visible (sunk deeply in mesocarp). Seeds transversely elliptic, the face plane. $2n = 20, 22$.

One species, *C. maritimum* L. Eurasia, Africa (Atlantic to Mediterranean and Black Sea coasts), Macaronesia, seashores. Used as a condiment.

128. ***Cryptotaenia*** DC. [Pimenov]

Cryptotaenia DC., Coll. Mém. 5: 42 (1829), nom. cons.; Mathias & Constance, N. Amer. Fl. 28B(1): 104–105 (1944), reg. rev.; Spalik & Downie, J. Biogeogr. 34 (12): 2039–2054 (2007), phylog., geogr.
Lereschia Boiss., Ann. Sci. Nat. III, 1: 127 (1884).

Glabrous or pubescent biennials or short-lived perennials with taproots and adventitious roots or a rhizome. Leaves ternate or biternate, glabrous; leaflets obovate, doubly serrate. Bracts few, linear, or absent; rays very unequal, glabrous or pubescent; bracteoles few or lacking. Petals white or greenish white; styles short to long, erect or reflexed; stylopodium (narrowly) conical or flat. Fruits glabrous, sometimes didymous or subclavate; commissure narrow; pericarp non-lignified; ribs filiform; carpophore bifurcated; vallecular vittae 1–4, commissural 2–4. Seed face plane. $2n = 16, 18, 20, 22$.

Seven species, Asia (Caucasus, E), N America, Africa, Europe (S, adventive W), and Oceania (adventive). Ornamental (*C. canadensis* (L.) DC.). Molecular data indicate that the genus is polyphyletic, with *Cryptotaenia* s. str. (four species) closely related to *Sium*, *Berula* and *Helosciadium* while the remaining species are related to *Pimpinella* (Spalik and Downie 2007). The former Macaronesian congener was recently transferred to *Daucus* (Banasiak et al. 2016).

129. ***Cuminum*** L. [Lee]

Cuminum L., Sp. Pl. 1: 254 (1753); Gen. Pl., ed. 5: 121 (1754); Banasiak et al., Taxon 65: 563–585 (2016), phylog.

Glabrous annuals. Leaves 1–2-ternate, broadly ovate, ultimate divisions filiform. Bracts several, filiform with membranous margins, simple to trifid or ternate; rays few to several, not contracted in fruit; bracteoles few to several, simple, linear to lanceolate with membranous margins. Calyx lobes conspicuous with needle-like tips; petals white to pinkish, the outer radiate; styles short, as long as the conical stylopodium. Fruits hairy (rarely glabrous), oblong, ovoid; mericarps compressed dorsally, with both primary and secondary ribs prominent; primary ribs with striate, unbranched hairs spread laterally and arranged in 1–2 rows (the 2 commissural ribs rarely inconspicuous); secondary ribs 4, with long, tuberculate, multicellular bristles, simple at the tip and arranged in 4 or more rows; carpophore entire, not easily separated; commissure broad; vittae large, transversely triangular, vallecular 1 (under each secondary rib), commissural 2. Seeds transversely compressed dorsally, the face slightly concave. $2n = 14$.

Three species, Europe (cult. & adv.), SW and S Asia, N Africa. *C. cyminum* L. is the spice cumin.

130. ***Cyathoselinum*** Benth. [Watson]

Cyathoselinum Benth. in Benth. & Hook.f., Gen. Pl. 1: 912 (1867).

Essentially glabrous perennials. Leaves 2–4-pinnate; ultimate segments linear-setaceous, rigid. Bracts lacking; rays several, rather short, tomentose; bracteoles several, connate almost to apex. Petals white. Fruits tomentose, ovoid, slightly compressed; vallecular vittae 2–3, commissural 6–8.

One species, *C. tomentosum* Benth. ex B.D. Jacks., Europe (endemic to Croatia). Limestone cliffs. Sometimes included in *Seseli*.

131. ***Cyclorhiza*** M.L. Sheh & R.H. Shan [Pu]

Cyclorhiza M.L. Sheh & R.H. Shan, Acta phytotax. Sin. 18 (1): 45 (1980); Shu & Sheh, Pollen Photog. Fl. Umbell. China 1: 25 (2001), palyn.; She et al., Fl. China 14: 53 (2005), rev.

Glabrous perennials. Basal leaves ternate to 3–4-pinnately dissected, ovate-triangular; ultimate segments lanceolate or linear-lanceolate. Bracts and bracteoles usually lacking (rarely present and few); rays few to several, unequal. Petals yellow to greenish-yellow; styles short, reflexed; stylopodium low-conical. Fruits glabrous, ovoid to oblong-ovoid, slightly compressed laterally; mericarp pentagonal in cross section; ribs somewhat protruding, undulate to subalate; carpophore bifurcated; vallecular vittae 1–3, commissural 2–6. Seed face concave or deeply sulcate. $2n = 22$.

Two species, Asia (SW China). Molecular data show an affinity to *Komarovia*, *Parasilaus*, and *Calyptrosciadium*.

132. ***Cyclospermum*** Lag. [Webb]

Cyclospermum Lag., Amen. Nat. Españ. 2: 101 (1821), nom. cons.; Constance, Brittonia 42: 276–278 (1990), taxon.

Glabrous annuals. Leaves oblong-ovate, 1–4-pinnately or ternately compound; ultimate segments lanceolate to linear or filiform. Umbels irregular (simple, compound, or mixed); bracts and bracteoles lacking; rays lacking to several. Petals whitish with an acute apex (neither narrowed nor inflexed); styles very short; stylopodium low-conical. Fruits glabrous to setulose along the ribs, ovoid to oblong, slightly compressed laterally; ribs filiform to broad and thickened (but not winged); carpophore apically bifid; commissure narrow; vittae large, vallecular 1, commissural 2. Seeds transversely terete to pentagonal, the face plane. $n = 7$, $2n = 14$.

Three species, N, C, and S America; *C. leptophyllum* (Pers.) Sprague ex Britton & P. Wilson widely adventive.

133. ***Cymbocarpum*** DC. ex C.A. Mey. [Pimenov]

Cymbocarpum DC. ex C.A. Mey., Verz. Pfl. Cauc. 132 (1831).

Glabrous or sparsely papillose annuals or biennials. Leaves 2–4-pinnate or ternate; petiole sheaths white-membranous; leaflets linear, divaricate. Bracts linear to linear-lanceolate, simple or bifid, caducous; bracteoles linear or absent. Petals white or pink; styles reflexed; stylopodium low-conical. Fruits glabrous or covered with glandular outgrowths, ovoid to ellipsoid, transversely lunate; mericarps strongly compressed dorsally, commissure broad, mesocarp cells non-lignified; dorsal ribs nearly inconspicuous, marginal ribs filiform, with large cells with lignified walls; carpophore bifurcated; vittae absent in mature fruits. Seed face deeply and broadly emarginate. $2n = 22$.

Four species, Asia (Caucasus, SW). Arid gravel slopes.

134. *Cymopterus* Raf. [Plunkett]

Cymopterus Raf., J. Phys. Chim. Hist. Nat. Arts 89: 100 (1819); Mathias, Ann. Missouri Bot. Gard. 17: 213–476 (1930), rev.; Mathias & Constance, N. Amer. Fl. 28B(2): 170–183 (1945), reg. rev.

Acaulescent or subcaulescent, glabrous to pubescent or scabrous perennials. Leaves all basal, or basal and 1 to few low-cauline, ternately, palmately, pinnately or bipinnately divided to ternate-pinnately decompound; ultimate divisions variously toothed or lobed to entire, usually narrow. Bracts lacking, or scarious to foliaceous; rays few and spreading, or abortive; bracteoles usually conspicuous, dimidiate, foliaceous or scarious. Petals white, yellow, or purple; styles slender, spreading; stylopodium lacking. Fruits glabrous to pubescent, ovoid to oblong, compressed dorsally; marginal and usually one or more lateral/dorsal ribs broadly winged, the wings membranaceous to spongy-corky, broader or narrower than the fruit body; carpophore bifurcated or obsolete; vallecular vittae 1 to several, commissural 2 to several, rib oil ducts sometimes found at wing bases. Seeds transversely compressed dorsally, the face concave to deeply sulcate. $n = 11$, $2n = 22$.

About 42 species, N America (W United States and adjacent Canada and Mexico); sandy or rocky habitats, prairies, plains, mountains, deserts and sand dunes, gravelly hills.

135. *Cynapium* Nutt. ex Torr. & A. Gray [Plunkett]

Cynapium Nutt. ex Torr. & A. Gray, Fl. North Amer. 1: 640 (1840); Mathias & Constance, N. Amer. Fl. 28B(1): 148 (1944).

Glabrous to pubescent perennials. Leaves mostly basal, ternate-pinnately compound, ovate to orbicular, leaflets ovate to oblong, coarsely toothed to deeply pinnatifid, the divisions obtuse or acute, mucronate, scabrous or glabrate on the margins and veins. Bracts 1 to several, narrow, or lacking, rays 12–20, spreading-ascending, unequal, bracteoles several, linear to lanceolate (shorter than flowers and fruits). Petals white or pinkish; styles short, spreading; stylopodium low-conical. Fruits ovoid to orbicular, slightly compressed laterally, ribs filiform to prominent but unwinged; carpophore bifurcated; vallecular vittae 3–6, commissural 6–8. Seeds compressed dorsally, the face concave, with prominent central longitudinal ridge. $n = 11$.

One species, *C. apiifolium* Nutt. ex Torr. & A. Gray, North America (Washington to N California). Meadows and shaded banks, mid to low elevations. North American manuals treat this genus in *Ligusticum* (= *L. apiifolium* (Nutt. ex Torr. & A. Gray) A. Gray), but Pimenov and Leonov (1993) recognize it as distinct; molecular data suggest that *Ligusticum* is polyphyletic, but *L. apiifolium* has not been sampled.

136. *Cynorhiza* Eckl. & Zeyh. [Van Wyk & Tilney]

Cynorhiza Eckl. & Zeyh., Enum. Pl. Afr. Austr. Extratrop. 3: 351 (1837); Winter et al., Taxon 57: 347–364 (2008), phylog.; Magee et al., S. Afr. J. Bot. 74: 726–734 (2008), rev.; Van Wyk et al., Afr. Apiaceae: 172–173 (2013).

Acaulescent, glabrous perennials with leafless, annual scapes and tuberous roots. Leaves radical, large, hysteranthous (appearing after flowers have senesced), pinnatipartite or multisect; petiole not sheathing; leaflets cuneate, obovate or linear. Bracts many, linear-acuminate; rays many, often exceptionally long; bracteoles small. Petals yellow or greenish yellow; styles short or long; stylopodium conical. Fruits glabrous, very large, broadly elliptic to broadly obovate, strongly compressed dorsally, apex truncate to deeply notched and auriculate; ribs indistinct on dorsal side, marginal ribs broadly winged; carpophore bifid; vallecular vittae 1, commissural 2; rib oil ducts small, solitary in each wing. Seeds transversely elliptic, the face plane to convex. $2n = 22$.

Three species, South Africa (coastal regions). Sand dunes, limestone or clay. The root of *C. typica* Eckl. & Zeyh. is traditionally used to make honey beer (mead). Related to *Lefebvrea*.

137. ***Cynosciadium*** DC. [Plunkett]

Cynosciadium DC., Coll. Mém. 5: 44 (1829); Mathias & Constance, Amer. J. Bot. 28: 162 (1941), emend. descr.; Mathias & Constance, N. Amer. Fl. 28B(2): 164 (1945).

Glabrous annuals with fascicled fibrous roots. Basal leaves entire and septate; cauline leaves palmately compound, broadly ovate; leaflets linear-lanceolate to linear. Bracts and bracteoles few and linear or lacking; rays few, unequal, spreading-ascending. Petals white; styles very short with divergent tips; stylopodium conical. Fruits glabrous, ovoid, tapering to a prominent beak at the apex, rounded at the base, slightly compressed dorsally or subterete; dorsal and lateral ribs narrow but prominent, marginal ribs winged, broad and corky; carpophore apically bifid; vallecular vittae 1, commissural 2. Seeds transversely slightly compressed dorsally, the face plane. $n = 6$.

One species, *C. digitatum* DC., N America (Missouri and Oklahoma to Mississippi and Texas). Swamps and low wet woodlands. Close to *Limnosciadium*.

138. ***Dahliaphyllum*** Constance & Breedlove [Plunkett]

Dahliaphyllum Constance & Breedlove, Acta Bot. Mex. 26: 84 (1994).

Scaberulous, arborescent and woody perennials. Leaves once pinnately compound (lower leaflets lobed to pinnatifid); leaflets ovate to lanceolate, finely doubly spinose-serrate, minutely scaberulous. Bracts few, prominent, scarious-sheathing, foliar; rays many, spreading to reflexed; bracteoles several, linear, inconspicuous. Petals maroon; styles long, reflexed; stylopodium conical, minute. Fruits glabrous, orbicular, compressed dorsally; dorsal ribs filiform; marginal ribs broadly thin-winged (wings twice as broad as the fruit body); carpophore bifurcated; vallecular vittae 1–2, commissural 2. Seeds transversely lunate, the face deeply sulcate. $2n = 22$.

One species, *D. almedae* Constance & Breedlove, N America (Mexico). Wet montane forests. Constance and Breedlove (1994) suggest a relationship to *Prionosciadium*, *Mathiasella*, and *Rhodosciadium*, and molecular data support this placement.

139. ***Dasispermum*** Neck. ex Raf. [Van Wyk & Tilney]

Dasispermum Neck. ex Raf., Good Book 56 (1840); Magee et al., S. Afr. J. Bot. 76: 308–323 (2010), rev.; Van Wyk et al., Afr. Apiaceae: 174–179 (2013).
Sonderina H. Wolff (1927).

Glabrous or scabrid annuals or short-lived perennials. Leaves mostly pinnate to bipinnate, sometimes fleshy; leaflets linear to oblong, sometimes conduplicate and fleshy. Bracts few, small, or absent; rays few to many, relatively short; bracteoles small. Petals white; styles long, deflexed; stylopodium shortly conical. Fruits glabrous, pustulate, scabrid or pilose; variable in shape; mericarps rarely heteromorphic (then often with 3–5 prominent wings); ribs prominent; carpophore bifid; vittae small or large, vallecular usually 1, commissural 2 (rarely up to 12 per mericarp and irregular in size); rib oil ducts absent (rarely very small, 1–2). Seeds transversely terete to slightly compressed dorsally, the face convex, plane or concave. $2n = 16, 18$.

Seven species, South Africa. Widespread along the coastline or in fynbos, renosterveld, succulent karoo or grasslands. Related to *Capnophyllum*.

140. ***Daucosma*** Engelm. & A. Gray [Plunkett]

Daucosma Engelm. & A. Gray in A. Gray, Boston J. Nat. Hist. 6: 210 (1850); Mathias & Constance, N. Amer. Fl. 28B (2): 161 (1945), reg. rev.

Glabrous annuals. Leaves ternate-pinnately dissected, deltoid; ultimate segments lanceolate and laciniate. Bracts and bracteoles conspicuous, pinnate-parted, equaling or longer than the rays or pedicels; rays few, spreading-ascending. Calyx lobes conspicuous and subulate; petals white; styles slender, reflexed, and longer than the conical stylopodium. Fruits glabrous, ovoid-oblong, compressed laterally; dorsal and lateral ribs broad and low, marginal ribs prominent, forming acute ridge around the fruit; carpophore bifurcated; vallecular vittae 1, commissural 2. Seeds transversely subterete or slightly compressed dorsally, the face plane. $n = 8$.

One species, *D. laciniata* Engelm. & A. Gray, N America (W Texas).

141. ***Daucus*** L. [Lee, Reduron]

Daucus L., Sp. Pl. 1: 242 (1753); Gen. Pl. ed. 5: 113 (1754); Banasiak et al., Taxon 65: 563–585 (2016), phylog., taxon.

Glabrous to sparsely or densely pubescent biennials (rarely annuals). Leaves 2–3-pinnatisect to 2–3-pinnately or verticillately dissected, triangular or ovate to oblong, ultimate divisions filiform to oblong. Bracts several, filiform, linear to lanceolate, simple, trifid to pinnatisect, deflexed or spreading in fruit; rays few to many, sessile to long, slightly or rarely strongly contracted in fruits, glabrous or hispid, rarely muricate; bracteoles many, simple, trifid, pinnatisect, filiform to lanceolate. Petals white to pinkish (rarely yellowish), the outer sometimes radiate; styles subsessile to elongated, rarely divergent; stylopodium conical, subconical, or depressed. Fruits spinescent, ovoid, oblong, ellipsoid, or cylindrical; mericarps tuberculate or papillate, ± compressed dorsally, with prominent primary and secondary ribs; primary ribs with papillate, unbranched, semi-erect or laterally spreading hairs, arranged in 2–4 rows; secondary ribs 4, with glabrous, tuberculate or papillate spines, confluent or simple at the base, conspicuously glochidiate or simple at the apex, arranged in a single row; carpophore entire or apically bifid; commissure narrow; vittae large or ruminate, transversely triangular or elliptical, vallecular 1 (under each secondary rib), commissural 2. Seeds transversely ± compressed dorsally, kidney-shaped or trapeziform, the face plane to concave. $2n = 12, 18, 20, 22, 44, 66$.

22 species (in the traditional circumscription adopted here), widespread in the N Hemisphere, with one species each in S America, and Australia and New Zealand. Molecular evidence indicates that the genus is polyphyletic; to restore monophyly, the following genera were recently included in *Daucus*: *Agrocharis*, *Melanoselinum*, *Monizia*, *Pseudorlaya*, *Pachyctenium*, *Rouya* and *Tornabenea* (Banasiak et al. 2016). *Daucus carota* L. subsp. *sativus* (Hoffm.) Schübl. & G. Martens is the economically important carrot.

142. ***Demavendia*** Pimenov [Pimenov]

Demavendia Pimenov in Rech.f., Fl. Iran. 162: 450 (1987).

Glabrous perennials. Leaves mainly basal, pinnatisect; leaflets broad, oblong, with irregularly dentate margins; upper cauline leaves few, very reduced, with enlarged sheaths. Bracts and bracteoles white-membranous, irregularly toothed at margin; rays unequal. Pedicels slightly thickened. Petals small, white or dingy-white, with attenuate tip; styles short, reflexed; stylopodium low-conical or conical. Fruits scarcely compressed dorsally, dorsal ribs filiform, marginal narrow-winged; commissure broad; carpophore bifid; vittae large, vallecular 1, commissural 2. Seed face slightly hollowed out. $2n = 22$.

One species, *D. pastinacifolia* (Boiss. & Hausskn.) Pimenov, Asia (Iran). Stony mountain slopes, in rock crevices. Closely related to *Zeravschania* and *Haussknechtia*.

143. ***Dethawia*** Endl. [Reduron]

Dethawia Endl., Gen. Pl. 10: 775 (1839); Villar, Fl. Iber. 10: 200–202 (2003), rev.

Glabrous perennials growing from a rootstock. Leaves 3(4)-pinnately divided, triangular or ovate; ultimate divisions linear. Bracts few; rays several, subequal, spreading; bracteoles usually several. Flowers many per umbellule, pedicels scabrous; petals white; styles clearly exceeding the low-conical stylopodium. Fruits glabrous, glossy, ovoid; ribs prominent, obtuse or acute; carpophore free, bifurcated; vittae large, vallecular 1–4, commissural 2–6. Seed transversely subterete to slightly compressed dorsally, the face plane or with two shallow grooves. $2n = 22 + 4B$.

One species, *D. splendens* (Lapeyr.) Kerguélen, Europe (France and Spain; Pyrenees, Cordillera Cantabrica), calcareous rocks.

144. ***Deverra*** DC. [Van Wyk & Tilney]

Deverra DC., Coll. Mém. 5: 45 (1829); Pfisterer & Podlech, Mitt. Bot. Staatssamml. München 22: 571–604 (1986), rev.; Van Wyk et al., Afr. Apiaceae: 182–183 (2013).

Glabrous or pubescent, broom-like, woody perennials. Leaves aborted or basal and tripinnately decompound, cauline leaves few or scale-like; leaflets obsolete. Bracts few, small; rays few; bracteoles few, small, often absent. Petals yellow; styles long, reflexed; stylopodium conical. Fruits pilose, scabrid, papillose or scaly, elliptic, slightly compressed laterally or dorsally; mericarps subterete; ribs inconspicuous; carpophore bifid; vittae large, vallecular 1, commissural 2. Seeds usually transversely slightly compressed dorsally, sulcate beneath the vittae, the face plane or somewhat sulcate. $2n = 20, 22, 40, 44, 66$.

Seven species, N Africa to the Arabian Peninsula and S Africa. Related to *Billburtia*, *Ammi* and *Petroselinum* (Jiménez-Mejías and Vargas 2015).

145. ***Dichoropetalum*** Fenzl [Pimenov]

Dichoropetalum Fenzl, Pug. Pl. Nov. Syr. 17 (1842); Pimenov et al., Willdenowia 37: 465–502 (2007), rev.
Chabraea Raf. (1840), non *Chabraea* Adans. (1763), nec DC. (1812).
Caroselinum Griseb. (1843).
Schlosseria Vucotinovic (1857).
Johreniopsis Pimenov (1987).
Holandrea Reduron et al. (1997).
Siculosciadium C. Brullo, Brullo, S.R. Downie & Giusso (2013).

Caulescent (rarely almost acaulescent), glabrous (rarely puberulent) perennials. Leaves mainly radical, bipinnatisect; sheaths short, linear or triangular; upper leaves reduced to sheaths only; leaflets linear, lanceolate or ellipsoid. Bracts usually reduced; rays unequal; bracteoles linear-lanceolate. Petals yellow, greenish-yellow or white; styles reflexed; stylopodium flat or conical. Fruits glabrous, compressed dorsally; dorsal ribs filiform to keeled, marginal narrowly winged, thin; commissure broad; vallecular vittae 1–3 or absent, commissural 0 or 2–4. Seed face plane. $2n = 22$.

32 species, Asia (SW, Middle, the Caucasus), Europe (SE), Africa (N, NE). Stony slopes, limestone, xerophilic vegetation, forest margins and openings, scrub. Molecular data suggest that *Dichoropetalum* forms a separate clade from *Peucedanum* s. str. (Spalik et al. 2004).

146. ***Dicyclophora*** Boiss. [Plunkett]

Dicyclophora Boiss., Ann. Sci. Nat. Bot. III, 2: 89 (1844); Hedge & Lamond in Rechinger, Fl. Iran. 162: 67–68 (1987), reg. rev.

Asperulous annuals. Leaves 2–3-pinnatisect, ovate-lanceolate; ultimate segments ovate to oblong, dentate to deeply dissected. Bracts and bracteoles few to several, large, narrowly triangular, white-margined, not hardening in fruit; rays several, unequal (outer exceeding the inner), spreading to divaricating; central umbellule sterile, purplish-black and clavate; outer umbellules with central, perfect, sessile flowers surrounded by several staminate flowers. Calyx absent; petals white, outer radiate; styles long and slender, ascending, tips recurved; stylopodium slender, long-conical. Fruits densely pubescent, slightly compressed laterally, often bicarpellate (or monocarpellate through abortion), pyriform, terete, enclosed in a "cage" formed by the persistent, hardened pedicels of the outer staminate flowers (the pedicels of the inner staminate flowers persistent and capillary); mericarps semi-circular, ribs inconspicuous; carpophore present; vittae 9 on the dorsal face, 2 on the commissure. Seeds transversely subterete, the face deeply sulcate.

One species, *D. persica* Boiss., Asia (SW Iran), in sandy deserts and hillsides.

147. ***Dimorphosciadium*** Pimenov [Pimenov]

Dimorphosciadium Pimenov, Bjull. Moskovsk. Obšč. Isp. Prip. Otd. Biol. 80(3): 83 (1975).

Acaulescent or subacaulescent, glabrous perennials (main stem solitary, short, partly submerged in soil, with large central umbel; lateral stems thin, ascending, with smaller umbels). Leaves rosulate, bipinnatisect; leaflets lanceolate, sessile, acute. Bracts of central umbel linear; rays 12–15, unequal; bracteoles linear-lanceolate, unequal. Petals white; styles long, reflexed; stylopodium low-conical. Fruits compressed dorsally; ribs filiform, carpophore reduced; pericarp not lignified; commissure broad; vittae small, many, cyclic in inner mesocarp layer. Seed face nearly plane. $2n = 22$.

Two species, Asia (Middle, Central). High-montane cryophyton vegetation. Related to *Cortia*, *Cortiella*, *Ligusticopsis*, and *Oreocome* (Downie et al. 2010).

148. ***Diplolophium*** Turcz. [Van Wyk & Tilney]

Diplolophium Turcz., Bull. Soc. Nat. Mosc. 20(1): 173 (1847); Cannon, Fl. Zambes. 4: 600–603 (1978), reg. rev.; Townsend, Fl. Trop. E. Afr. 80–83 (1989), reg. rev.; Van Wyk et al., Afr. Apiaceae: 184–187 (2013).

Partly pubescent, woody perennials. Leaves usually tripinnate or multifid; leaflets filiform, elliptic or sub-rotund. Bracts large and foliaceous, ovate, oblong-acuminate to linear; rays many, long; bracteoles conspicuous, similar to bracts but smaller. Petals white or cream-colored; styles long, clavate; stylopodium conical. Fruits densely bristly, oblong, subterete, compressed dorsally; ribs obtuse, prominent, equal; carpophore entire or bifid; vittae large, vallecular usually 1, commissural 2 or 4. Seeds transversely compressed dorsally, sulcate beneath vittae, the face plane or somewhat sulcate. $2n = 22$.

Seven species, tropical Africa. Open woodlands and grasslands. Phylogenetically isolated but similar to *Physotrichia*.

149. ***Diplotaenia*** Boiss. [Pimenov]

Diplotaenia Boiss., Ann. Sci. Nat. III, 1: 308 (1844); Pimenov et al., Willdenowia 41: 67–74 (2011), rev.

Glabrous perennials with branched rootstocks. Basal leaves 3–4-ternate or 3–4-pinnate; leaflets linear to oblong-ovate, or trifid. Bracts and bracteoles several or many, lanceolate; rays long. Petals white; styles rigid, reflexed; stylopodium conical. Fruits strongly compressed dorsally; ribs prominent but unwinged; carpophore bifurcated; commissure broad; vittae large, vallecular 1–3, commissural 2–4. Seed face concave.

Four species, SW Asia (E & S Anatolia, N Iran). Limestone ravines, stony mountain meadows. Molecular data suggest *Prangos*, *Bilacunaria*, and *Azilia* are related genera.

150. ***Donnellsmithia*** J.M. Coult. & Rose [Plunkett]

Donnellsmithia J.M. Coult. & Rose, Bot. Gaz. 15: 15 (1890); Mathias & Constance, N. Amer. Fl. 28B(1): 75–81 (1944), reg. rev.; Mathias & Constance, Brittonia 33: 342–346 (1981), taxon.

Glabrous to scaberulous or puberulent annuals, biennials or perennials with taproots or globose tubers. Leaves 1–4-ternate or ternate-pinnately decompound; ultimate segments ovate to filiform. Bracts and bracteoles usually lacking; rays few, spreading. Petals yellow or purple; styles spreading or loosely reflexed; stylopodium lacking or depressed. Fruits glabrous to scaberulous or puberulent; oval to ellipsoid to obcordate, compressed laterally; mericarps subterete; ribs narrow to filiform, prominent to obscure; carpophore bifurcated to the base; vittae small, vallecular 1–several, commissural 2–several. Seed nearly terete, the face plane or sulcate. $n = 11, 14, 20, 21, 22, 40, 42, 44$.

15–20 species, C and N America (Mexico). Molecular data suggest a close relationship to several other meso-American genera, including *Arracacia*, *Enantiophylla*, and close relatives.

151. ***Dorema*** D. Don [Pimenov]

Dorema D. Don, London & Edinb. Phil. Mag. 9: 47 (1831); Pimenov, Bjull. Moskovsk. Obšč. Isp. Prir. Otd. Biol. 93, 2: 76–90 (1988), rev.

Glabrous or puberulent perennials. Leaves mainly basal, pinnatisect; leaflets lanceolate to rhombic. Umbels simple, in branched panicles, sessile or with short peduncles; bracts reduced, or several, caducous. Petals yellow or greenish-yellow; styles short; stylopodium flat. Fruits glabrous or pubescent, dorsally strongly compressed; dorsal ribs filiform, marginal narrow-winged; carpophore bifid; commissure broad; vallecular vittae solitary or several, commissural 2–8, or vittae reduced. Seed face plane. $2n = 22$.

12 species, Asia (SW to Middle and Pakistan). Sandy deserts, salt slopes, gypsum limestone, in xerophilic vegetation. Two sections (Pimenov 1988). The source of the medicinal aromatic gum ammoniac; leaves used as fodder in desert pastures. Phytochemical and some molecular data support maintaining the genus as distinct, but Panahi et al. (2015) suggested that the genus is nested within *Ferula*.

152. ***Dracosciadium*** Hilliard & B.L. Burtt [Van Wyk & Tilney]

Dracosciadium Hilliard & B.L. Burtt, Notes Roy. Bot. Gard. Edinb. 43(2): 220 (1986); Burtt, Edinb. J. Bot. 48: 206 (1991); Van Wyk et al., Afr. Apiaceae: 188–189 (2013).

Glabrous perennials with a resinous caudex. Leaves peltate-palmately divided, margins dentate-serrate. Bracts linear-lanceolate, entire; rays few, very long; bracteoles small. Petals cream-colored or flushed purple; styles relatively long, divergent; stylopodium hemispherical or conical. Fruits glabrous, ovoid to shortly oblong, somewhat compressed laterally; mericarps terete; ribs inconspicuous or prominent, dorsal ribs warty or smooth, marginal ribs slightly more prominent; carpophore slender, bifid; vittae large, vallecular usually 1, commissural 2 (sometimes irregular and up to 18); rib oil ducts small, always present, usually 1 in each rib. Seeds sulcate beneath the vittae, the face slightly concave.

Two species, South Africa. Related to *Anginon* and *Heteromorpha.*

153. ***Ducrosia*** Boiss. [Ostroumova]

Ducrosia Boiss., Ann. Sci. Nat. Bot. III, 1: 341 (1844); Alava, Notes Bot. Gard. Edinb. 34: 183–193 (1975), rev.

Glaucous biennials or perennials. Leaves simple or 2–3-ternate; ultimate divisions broadly reniform to linear. Bracts and bracteoles small. Petals white or yellow, pubescent; styles short, reflexed; stylopodium conical. Fruits glabrous or pubescent, elliptic or orbicular, strongly compressed dorsally; dorsal ribs filiform, marginal ribs thickened; fibrous inner mesocarp present; commissure broad; vittae narrow, vallecular 1, commissural 2, arcuate, near the outer edge of the mericarp cavity. Seed face plane. $2n = 22$.

Six species. Asia (SW), Africa (Egypt). Molecular data suggest a relationship to *Cymbocarpum.*

154. ***Dystaenia*** Kitag. [Plunkett]

Dystaenia Kitag., Bot. Mag. Tokyo 51: 805 (1937); Pfosser et al., Pl. Syst. Evol. 256(1): 159–170 (2005), evol.

Glabrous (except hirsutulous in the inflorescence) perennials. Leaves biternately compound, ovate to orbicular; leaflets few to many, linear to obovate or ovate or cuneate, sessile, entire to coarsely serrate, incised. Bracts and bracteoles several and linear, or wanting; rays few to many, subequal, stout, spreading-ascending. Calyx lobes linear-lanceolate; petals white or pinkish; styles slender, reflexed; stylopodium conical. Fruits glabrous (or the ribs slightly pilose), oblong, slightly compressed laterally; mericarps slightly compressed dorsally, the apex and base somewhat emarginate, the ribs narrowly winged (marginal wings slightly wider than dorsal and lateral wings); carpophore bifurcated; vittae small, vallecular 2–4, commissural 4–6, rib oil ducts thin, solitary in each rib. Seeds transversely sulcate under the valleculae, transversely somewhat dorsally compressed, the face plane. $2n = 22, 44$.

Two species, Asia (Japan, Korea). A segregate of *Ligusticum*, in which it is often placed.

155. ***Echinophora*** L. [Pimenov]

Echinophora L., Sp. Pl. 239 (1753), Gen. Pl., ed. 5: 111 (1754); Hedge & Lamond, Notes Roy. Bot. Gard. Edinb. 32 (2): 167–188 (1973), rev.; ejusd., ibid. 37(1): 61–66 (1978), additions.

Glabrous, often spiny perennials. Leaves 1–4-pinnate or triternate, usually coriaceous; leaflets filiform or linear-lanceolate to rhombic and ovate, often spine-tipped. Bracts often flaccid; rays few or several, bracteoles hardened, sometimes spinose. Central flower in umbellule sessile, hermaphroditic, outer flowers staminate, with hardened pedicels; pedicels usually connate with bracteoles; calyx lobes broadly ovate to triangular, sometimes unequal; petals white, yellow, or reddish, glabrous or dorsally pubescent; styles long, persistent, woody; stylopodium inconspicuous or saucer-shaped. Fruits formed by glabrous or pubescent, pyriform carpel submerged in receptacle; carpophore absent; ribs indistinct; some mesocarp cells with lignified walls; vittae narrow, dorsal various, commissural 2. Seed face grooved. $2n = 22, 60, 64$.

11 species, Europe (W, SW, S, SE, E), Asia (SW, Middle), Africa (N). Dry stony slopes, xerophilic vegetation, maritime sands. Three sections (Hedge and Lamond 1973). Aromatic. Related to *Dicyclophora* and *Pycnocycla* (Downie et al. 2010).

156. ***Ekimia*** H. Duman & M.F. Watson [Watson]

Ekimia H. Duman & M.F. Watson, Edinb. J. Bot. 56(2): 200 (1999); Banasiak et al., Taxon 65: 563–585 (2016), phylog., taxon.

Glabrous, glaucous perennials. Leaves petiolate, mostly basal, finely triternately divided; ultimate divisions filiform, acute. Bracts lacking or few and small; rays few, long, spreading; bracteoles several, conspicuous, broadly elliptic and membranaceous. Petals whitish-yellow; styles spreading, longer than depressed-conical stylopodium. Fruits ovoid-oblong, symmetrical, slightly dorsally compressed; primary and secondary ribs all narrowly thick-winged, equally undulate-pleated; carpophore bifid; vallecular vittae 1, commissural 2. Seeds slightly compressed dorsally, the face plane.

One species, *E. bornmuelleri* (Hub.-Mor. & Reese) H. Duman & M.F. Watson, Asia (SW Turkey). Serpentine *Quercus* woodlands, stony slopes. On the basis of molecular data only, two former species of *Laserpitium* were recently transferred to this genus (Banasiak et al. 2016).

157. ***Elaeosticta*** Fenzl Fig. 18 [Kljuykov]

Elaeosticta Fenzl, Flora 26(2): 458 (1843); Kljuykov et al., Bull. Soc. Nat. Mosc. Div. Biol. 81(4): 75–89 (1976), rev.; Pimenov et al., Bot. Zhurn. 66, 3: 328–340 (1981), rev.; Kljuykov, Novit. Syst. Pl. Vasc. (Leningrad) 20: 140–154 (1983), rev.

Glabrous, ephemeral perennials usually with globose tubers. Leaves 2–3-pinnate with sessile segments; leaflets ovate to filiform, toothed. Bracts and bracteoles entire, herbaceous, lanceolate to linear, rarely elliptic, with scarious margins. Petals white or rarely yellow; styles reflexed; stylopodium low-conical. Fruits usually glabrous, slightly compressed laterally, globose to oblong-linear; commissure narrow; ribs obscure or lacking; carpophore bifurcated; exocarp cells large; vittae many, often anastomosing. Seed face plane to concave. The seedlings mono- or dicotyledonous. $2n = 20, 22, 40$.

26 species, Asia (SW, Middle, S) and SE Europe. Semi-deserts, dry steppes, loess and gravel slopes, scrub, forests, subalpine and rarely alpine meadows. *Elaeosticta lutea* (Hoffm.) Kljuykov, Pimenov & V.N. Tikhom. is poisonous. Three sections have been recognized (Pimenov et al. 1981). Closely related to *Hyalolaena*, *Galagania* and *Oedibasis*.

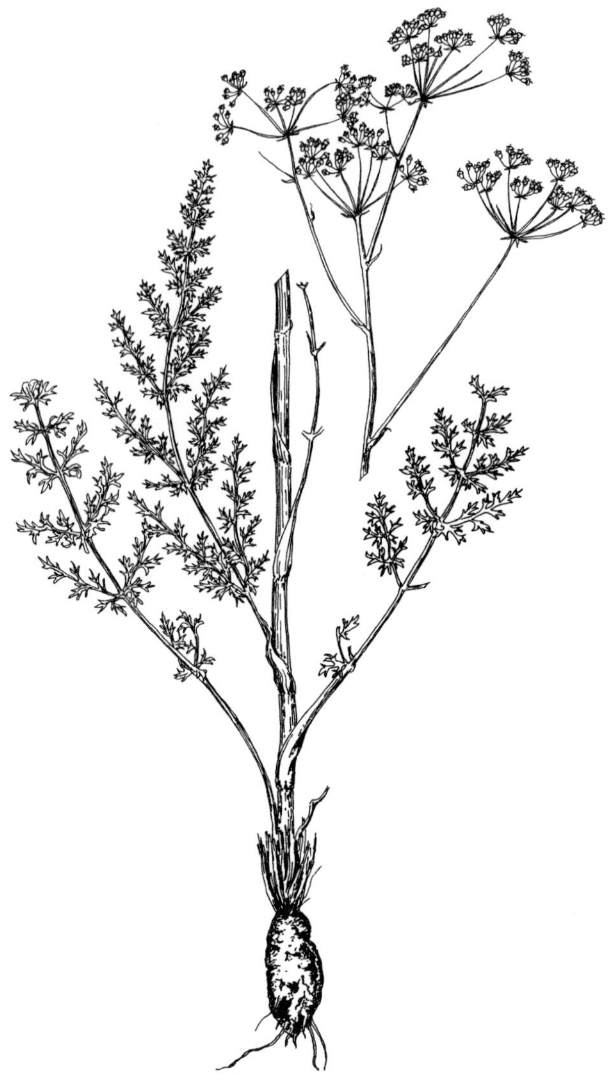

Fig. 18. Apiaceae, Apioideae. *Elaeosticta knorringiana*. (Orig., illustration by Elena Mzhelskaya)

158. ***Eleutherospermum*** K. Koch [Kljuykov]

Eleutherospermum K. Koch, Linnaea 16: 365 (1842).

Glabrous perennials. Leaves 2–3-pinnate with petiolulate primary basal segments; leaflets lanceolate to ovate, toothed or lobed. Bracts and bracteoles linear and herbaceous, deflexed. Petals greenish; styles reflexed; stylopodium conical. Fruits glabrous, ovoid, slightly compressed laterally; commissure narrow; ribs short-winged; carpophore bifurcated; vallecular vittae slender, 3–5, commissural 5–6. Seed face deeply concave. $2n = 22$.

Two species, Caucasus, NE Turkey, W Iran. Mountain forests, subalpine meadows. Closely related to *Pleurospermum*, *Hymenidium*, *Aulacospermum*.

159. ***Elwendia*** Boiss. [Pimenov]

Elwendia Boiss., Ann Sci. Nat. ser. 3, 1: 140 (1844).
Buniella Schischk. (1960).

Glabrous, geophilous perennials with deeply burried, swollen taproots. Leaves 3–4-pinnatisect, basal and cauline leaves sometimes differing; leaflets filiform and linear to lanceolate or cordate. Bracts and bracteoles absent or several, linear. Petals white; stylopodium flat to conical. Fruits compressed laterally or rarely slightly compressed dorsally, ovate to oblong; commissure narrow to rarely intermediate in width; ribs equal or subequal, filiform or keeled; vallecular vittae usually 1 (rarely several); commissural vittae 2. Seed face plane or sulcate. $2n = 12, 14, 16, 18, 20, 22$.

24 species. Asia (Iraq to W China, Mongolia, and W India). Arid mountains from foothills to alpine, xerophilous scrub and grasslands, rocky slopes, gypseous hills. Fruits of *E. persica* (Boiss.) Pimenov & Kljuykov (zira) used as flavoring and in folk medicine. The species of *Elwendia* are often included in *Bunium*, but recent studies restored *Elwendia* as a separate genus, differing in morphology, geography, and chromosome numbers; molecular data (Degtjareva et al. 2013) also suggest *Elwendia* and *Bunium* are distinct.

160. ***Enantiophylla*** J.M. Coult. & Rose [Plunkett]

Enantiophylla J.M. Coult. & Rose, Bot. Gaz. 18: 55 (1893); Mathias & Constance, N. Amer. Fl. 28B(2): 218 (1945), reg. rev.

Pubescent, perennial, large herb. Leaves opposite, pinnately or ternately compound, ovate; leaflets lanceolate to ovate-lanceolate. Bracts and bracteoles several, linear, scarious-margined, shorter than the rays; rays many (15–20), spreading to reflexed. Petals cream to yellow to yellowish-green; styles furrowed on ventral face, more or less thickened above; stylopodium conical. Fruits glabrate, oblanceolate and contracted below to form a stipe-like base, strongly compressed dorsally; dorsal and lateral ribs filiform, acute; marginal ribs narrowly winged; carpophore bifurcated to the base; vallecular vittae 1, commissural 2. Seed transversely compressed dorsally, sulcate under the vittae, the face shallowly concave. $n = 22$.

One species, *E. heydeana* J.M. Coult. & Rose, C and N America (S Mexico, Guatemala, El Salvador). Closely related to *Arracacia*, *Rhodosciadium*, and close relatives.

161. ***Endressia*** J. Gay [Reduron]

Endressia J. Gay, Ann. Sci. Nat. (Paris) I, 26: 223 (1832); Rey, Rév. *Endressia* pyr.-cantabr. Doc. Cart. Prod. Vég. sér. Pyr. tome Gén. 3(3): 1–30 (1945), rev.; Leute, Ann. Naturhist. Mus. Wien 73: 85–88 (1969), rev.; Montserrat & Villar, Fl. Iber. 10: 304–308 (2003), reg. rev.

Glabrous or ± pubescent perennials. Leaves pinnately divided, elongate; leaflets ovate, pinnatifid or palmatipartite. Bracts and bracteoles absent or present; rays several to many, subequal, spreading. Petals white, sometimes pink; styles exceeding the conical stylopodium. Fruits smooth, glabrous, ovoid or ellipsoid; ribs prominent, obtuse; carpophore free, bifurcated; vittae small, vallecular 2–5, commissural 4–8. Seeds transversely elliptic or pentagonal, the face plane. $2n = 22$.

Two species, Europe (Pyrenees and North Spain). Pastures.

162. ***Eremodaucus*** Bunge Fig. 19 [Kljuykov]

Eremodaucus Bunge, Delect. Sem. Hort. Dorpat. 6 (1843).

Glabrous annuals. Leaves 2–3-pinnate with petiolulate lower leaflets, lanceolate, toothed or lobed. Umbels with reddish outgrowth in the center; bracts and bracteoles entire, oval or oblong, deflexed, with scarious margins; marginal rays hardened, deflexed in fruit. Pedicels spinescent at maturity; petals white; styles reflexed;

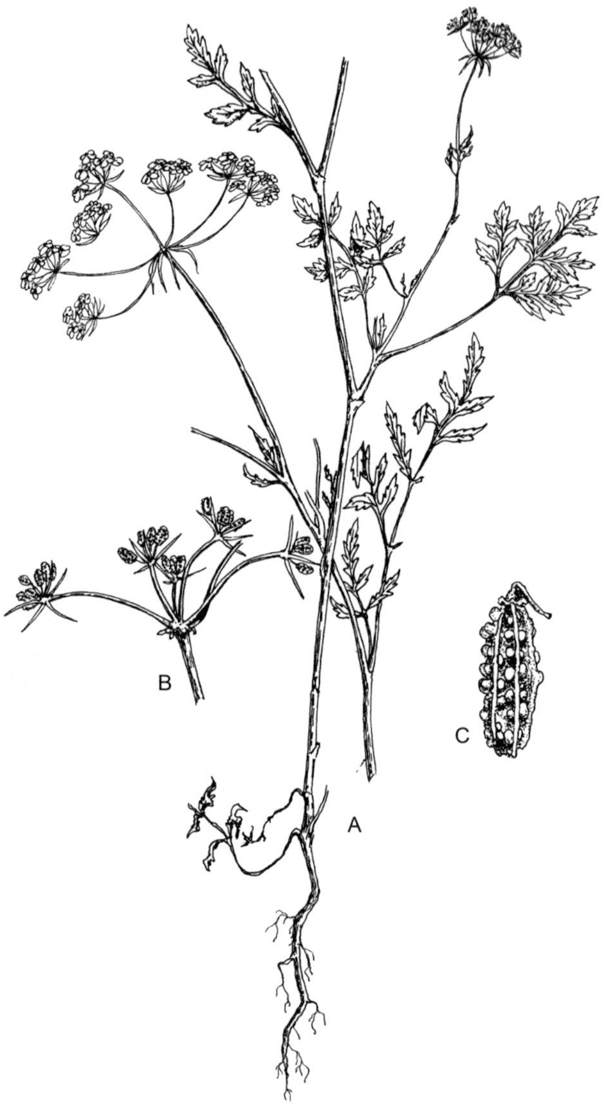

Fig. 19. Apiaceae, Apioideae. *Eremodaucus lehmannii*. **A** Habit. **B** Fruiting umbel. **C** Mericarp. (Orig., illustration by Elena Mzhelskaya)

stylopodium low-conical. Fruits with white outgrowths, suborbicular, compressed laterally; commissure narrow; ribs acute; carpophore entire, spinescent; vallecular vittae 1 (but obsolete at maturity), commissural 2. Seed face narrowly and deeply grooved. $2n = 18$.

One species, *E. lehmannii* Bunge, Middle Asia, Afghanistan, N, E Iran and Azerbaidzhan (adventive). Semi-deserts and dry steppes, and weeds of fields, road edges, and waste areas.

163. ***Ergocarpon*** C.C. Towns. [Plunkett]

Ergocarpon C.C. Towns., Kew Bull. 17: 437 (1964); Hedge & Lamond in Rechinger, Fl. Iran. 162: 80–81 (1987), reg. rev.

Glabrous annuals. Leaves pinnately dissected, soon withering, pinnately compound, ovate to lanceolate; leaflets ovate, pinnatisect, the ultimate segments linear. Bracts few, large, triangular-lanceolate, with broad white margins; rays few, unequal, stout, inflated distally, spreading-ascending; bracteoles few, stiff (larger than the bracts, and exceeding the length of the umbellules, hiding the flowers), with broad white margins; umbellules of a central, perfect and sessile flower surrounded by a ring of staminate flowers, surrounded in turn by a ring of reduced, clavate flowers. Petals white; styles short; stylopodium conical. Fruits deeply sunk into the receptacle, often bicarpellate (or monocarpellate through abortion), obovoid-oblong, subterete; ribs filiform, not obvious; carpophore lacking; vittae 5, slender. Seed face deeply sulcate.

One species, *E. cryptanthum* (Rech. f.) C.C. Towns., Asia (E Iraq, SW Iran), hillsides, rocks, and cliffs.

164. ***Erigenia*** Nutt. [Plunkett]

Erigenia Nutt., Gen. North Amer. Pl. 1: 187 (1818); Harper, Torreya 32: 141–146 (1932), taxon.; Mathias & Constance, N. Amer. Fl. 28B(1): 74 (1944), reg. rev.

Nearly acaulescent, glabrous perennials with deep-seated globose tubers. Leaves ternately decompound, broadly ovate; ultimate segments linear to spathulate. Bract a reduced foliage leaf or lacking; rays few, spreading-ascending; bracteoles oblong or spathulate, foliaceous, entire or toothed. Petals white, spathulate to obovate, without an inflexed apex; styles long and slender, recurved; stylopodium lacking. Fruits glabrous, orbicular to ellipsoid, incurved at apex and cordate at base (rendering each mericarp nearly reniform), strongly compressed laterally; ribs filiform, evident, a corky rib-like projection running the length of the commissural surface in the middle of each mericarp; carpophore lacking; vittae small, vallecular 1–3, commissural 9–11. Seeds transversely laterally compressed, the face deeply sulcate. $n = 10$.

One species, *E. bulbosa* (Michx.) Nutt., N America (E United States); rich woodlands.

165. ***Eriosynaphe*** DC. [Pimenov]

Eriosynaphe DC., Coll. Mém. 5: 50 (1829); Prodr. 4: 175 (1830).

Glabrous perennials; taproots forming tuber-like thickenings. Leaves tripinnatisect; leaflets linear, long. Bracts and bracteoles absent; rays many, divaricate. Petals greenish-yellow; styles reflexed; stylopodium low-conical, undulate at margin. Fruits compressed dorsally, dorsally glabrous, commissure tomentose, with an additional rib; dorsal ribs filiform, marginal thickened; commissure broad; cells of mesocarp in marginal ribs and on the commissure with pitted, slightly lignified walls; vallecular vittae 1, commissural 2, small, sometimes invisible. Seed face plane or slightly concave. $2n = 22$.

One species, *E. longifolia* (Fisch. ex Spreng.) DC., Europe (E), Asia (Kazakhstan). Steppes, sometimes saline. Distantly related to *Ferula*.

166. ***Eurytaenia*** Torr. & A. Gray [Plunkett]

Eurytaenia Torr. & A. Gray, Fl. North Amer. 1: 633 (1840); Mathias & Constance, N. Amer. Fl. 28B(2): 203–204 (1945), reg. rev.; Mathias & Constance, Contr. Texas Res. Found. 1: 2–3 (1950), taxon.

Somewhat scaberulous annuals. Basal leaves lobed or pinnatifid with obtuse, crenate-serrate lobes; cauline leaves pinnately or ternate-pinnately dissected, the ultimate segments narrow (often elongate), entire or serrate. Bracts and bracteoles several, tricleft, reflexed; rays several, spreading-ascending, unequal, scaberulous. Calyx lobes prominent; petals white; styles slender; stylopodium depressed. Fruits scaberulous, oblong-oval to oval to orbicular, strongly compressed dorsally; dorsal and lateral ribs filiform to narrowly winged, marginal ribs broadly thick-winged (wings narrower than the mericarp body), prominently nerved near the middle, contiguous between adjacent mericarps, or the nerve carinate on the dorsal surface, giving the appearance of a 5-winged mericarp; carpophores bifurcated; vittae large and compressed, vallecular 1, commissural 2. Seeds transversely compressed dorsally, the face plane. $n = 7$.

Two species, N America (Texas).

167. ***Exoacantha*** Labill. [Plunkett]

Exoacantha Labill., Icon. Pl. Syr. 1: 10 (1791); Townsend, Kew Bull. 17: 3 (1964), emend descript.; Hedge & Lamond in Davis, Fl. Turkey 4: 537–538 (1972), reg. rev.

Glabrous annuals. Leaves once pinnate (lower leaves soon deciduous), membranaceous to subcoriaceous; ultimate segments ovate below, to linear-lanceolate, entire or incised upward. Bracts several, lanceolate, spinose, reflexed, equal to subequal, thickened in fruit; rays several, unequal, thickened, spreading to divaricating in flower, becoming strongly incurved and whitish in fruit; bracteoles several, lanceolate to subulate, spinose, spreading, unequal. Petals white, with an oil duct; styles slender, sharply reflexed in fruit; stylopodium low-conical. Fruits glabrous, oblong-ovate, slightly compressed laterally; primary ribs conspicuous, about as broad as the valleculae, secondary ribs filiform, in the depression of each vallecula; carpophore lacking; vittae several, indistinct. Seeds transversely subterete, sulcate under the secondary ribs, the face convex. $n = 11$.

One species, *E. heterophylla* Labill., Asia (Turkey, W Syria, Lebanon, Israel, Jordan), roadsides and pastures.

168. ***Ezosciadium*** B.L. Burtt [Van Wyk & Tilney]

Ezosciadium B.L. Burtt, Edinb. J. Bot. 48: 207, 268 (1991); Magee et al., Pl. Syst. Evol. 276: 167–175 (2008), phylog., rev.; Van Wyk et al., Afr. Apiaceae: 193 (2013).

Sparsely to densely pubescent annuals. Leaves slender, ternately dissected; lobes 2–3-fid, linear-oblong, acute, margins entire. Umbels simple and sessile, 1–3(–5) at each node; bracts and bracteoles few, linear-lanceolate; rays few, very unequal. Petals white; styles short; stylopodium depressed. Fruits sparsely to densely pilose, narrowly oblong, mericarps $\pm$ isodiametric; ribs prominent, blunt; carpophore bifid; vittae rather small, vallecular 1, commissural 2. Seeds transversely $\pm$ orbicular, the face concave.

One species, *E. capense* (Eckl. & Zeyh.) B.L. Burtt, South Africa (E Cape province). Highly localized endemic, rocky places. Phylogenetically isolated but closest to *Annesorhiza*, *Astydamia* and *Molopospermum*.

169. **Falcaria** Fabr. [Reduron]

Falcaria Fabr., Enum. 34 (1759), nom. cons.; H. Wolff in Engl., Pflanzenr. 4 (228) Heft 90: 129–133 (1927), rev.

Glabrous perennials. Leaves 1–2-ternately divided, triangular or rhombic, coriaceous; ultimate divisions very long, linear-lanceolate or linear with a serrate and cartilaginous margin. Bracts and bracteoles several; rays many, subequal; spreading; flowers many per umbellule. Petals white; styles slightly exceeding the conical stylopodium. Fruits glabrous, oblong, subterete; ribs filiform; carpophore free, bifurcated; commissure narrow; vittae small, vallecular 1, commissural 2(4). Seeds pentagonal, the face plane. $2n = 22$.

One species, *F. vulgaris* Bernh., Eurasia (France to C Asia). Field margins, dry slopes.

170. **Fergania** Pimenov [Pimenov]

Fergania Pimenov, Novosti Sist. Vyssh. Rast. 19: 120 (1982).

Glabrous perennials. Leaves bipinnate, coriaceous; petioles with slightly enlarged sheaths, terete; leaflets obovate, obtuse, lobate. Bracts absent; rays few, unequal; bracteoles linear-lanceolate, connate at base. Petals yellowish; styles short, reflexed; stylopodium low-conical. Fruits strongly compressed dorsally; dorsal ribs filiform, marginal ribs narrow-winged, thickened at distal ends; commissure broad; mesocarp mainly of non-lignified cells, in marginal ribs of large cells with lignified walls; carpophore bifid; vittae large, vallecular 1, commissural 2; secretory ducts in marginal ribs larger than vittae. Seed face plane.

One species, *F. polyantha* (Korovin) Pimenov, Asia (Middle). Xerophilic scrub-ephemeral vegetation, stony and conglomerate slopes, sometimes slightly saline, screes. Related to *Ferula*.

171. **Ferula** L. [Pimenov]

Ferula L., Sp. Pl. 246 (1753), Gen. Pl. ed. 5: 117 (1754); Korovin, Gen. Ferula Monogr. Illustr. (1947), rev.; Pimenov & Baranova, Bjull. Moskovsk. Obšč. Isp. Prip. Otd. Biol. 84(3): 82–92 (1979), rev. sect.; Pimenov, Bjull. Moskovsk. Obšč. Isp. Prip. Otd. Biol. 84(5): 106–111 (1979), rev. sect.

Usually robust, sometimes megaherbs, glabrous or puberulent perennials; roots sometimes with tuber-like thickenings. Leaves 2–6-dissected; leaflets filiform, linear, lanceolate, elliptic to ovate, glabrous, or puberulent beneath or on both sides. Umbels partly of staminate flowers, or umbellules clustered verticillately in proliferating umbels; bracts absent; bracteoles absent, or few, caducous. Petals yellow or yellowish, glabrous or hairy dorsally; styles short or elongate, reflexed; stylopodium flat or low-conical. Fruits glabrous, rarely puberulent, strongly compressed dorsally; dorsal ribs filiform, the marginal winged; carpophore bifid; commissure broad; mesocarp bistratous (inner layer of lignified prosenchyma cells); vallecular vittae 1–3, commissural 2–6, or vittae small and arranged almost cyclically. Seed face plane. $2n = 22$.

180–185 species, Europe (mainly S), Asia, Africa (N, NE, NW, trop.), Australia (adv.). Arid and subarid vegetation, sandy, salt and stony deserts, xerophilic scrub, gypsum slopes. Six subgenera (Korovin 1947), some polyphyletic. Used in traditional oriental and folk medicine, many species are essential fodders in arid pastures. Some species are foliar ornamentals. Closely related to *Dorema*, *Leutea*, *Kafirnigania* and *Fergania* (Panahi et al. 2015).

172. **Ferulago** W.D.J. Koch [Pimenov]

Ferulago W.D.J. Koch, Nov. Acta Acad. Leop.-Carol. Nat. Cur. 12(1): 97 (1824); Bernardi, Boissiera 30: 1–182 (1979), rev.; Tomkovich & Pimenov, Bot. Zhurn. 72(7): 964–971 (1987), subg. classif.; ejusd. in Feddes Repert. 100 (3–4): 119–129 (1989), distr.

Glabrous or hispidulous perennials. Leaves petiolate, 3–4-pinnatisect; leaflets filiform to broadly lanceolate. Bracts and bracteoles filiform to triangular-orbicular, entire; rays usually many. Calyx lobes often triangular, usually glabrous (rarely puberulent); petals yellow, glabrous or dorsally puberulent; styles reflexed; stylopodium low-conical. Fruits glabrous, compressed dorsally; ribs sometimes undulate, dorsal ribs keeled to filiform, marginal winged; carpophore bifid; commissure broad; pericarp parenchymatous, in ribs partly of lignified cells with pitted walls; vittae small, many, often cyclical. Seed face plane. $2n = 22, 44, 66$.

50 species, Europe, Asia (SW, Middle, the Caucasus), Africa (N, NW). Two subgenera, 8 sections (Tomkovich and Pimenov 1987). Some species used in folk medicine. Molecular data suggest a close relationship to *Prangos*.

173. ***Ferulopsis*** Kitag. [Pimenov]

Ferulopsis Kitag., J. Jap. Bot. 46(9): 283 (1971); Pimenov, Bot. Zhurn. 76(10): 1387–1391 (1991), rev.

Subglabrous perennials with branching rootstocks. Leaves mainly basal, bipinnate; leaflets lanceolate. Bracts and bracteoles glabrous, linear to lanceolate, acuminate, white-membranous at the margin, rays scabrous. Petals white; styles short, reflexed; stylopodium low-conical. Fruits compressed dorsally; dorsal ribs keeled, marginal winged; carpophore bifid; commissure broad; mesocarp of cells with pitted walls; vallecular vittae 2–4, commissural 2–6. Seed face plane. $2n = 20$.

Two species, Asia (S Siberia, Mongolia, NE China). Steppe slopes, sometimes subalpine stony pastures. A segregate of *Peucedanum*.

174. ***Foeniculum*** Mill. [Pu]

Foeniculum Mill., Gard. Dict. Abr., ed. 4 (1754); She et al., Fl. China 14: 134 (2005), reg. rev.

Glabrous and glaucous perennials or biennials. Leaves pinnately decompound; ultimate segments filiform. Bracts and bracteoles lacking; rays many, ascending. Petals yellow; styles short, recurved; stylopodium conical. Fruits glabrous, ovoid-oblong, slightly compressed laterally to subterete; mericarps pentagonal and slightly compressed dorsally; ribs prominent, subequal or the lateral slightly broader than the dorsal; carpophore bifurcated; commissure broad; vallecular vittae 1, commissural 2. Seed transversely compressed dorsally, the face plane or slightly concave. $2n = 22$.

4–5 species, Mediterranean area of Europe and N Africa, Asia, and widely cultivated and adventive. The leaves of *F. vulgare* Mill. are used as an herb or vegetable, and the fruits as a spice; the plant is used medicinally as a stomachic. Related to *Anethum*, *Pseudoridolfia* and *Ridolfia* (Jiménez-Mejías and Vargas 2015).

175. ***Frommia*** H. Wolff [Van Wyk & Tilney]

Frommia H. Wolff, Bot. Jahrb. 48: 266 (1912); Townsend, Fl. Trop. E. Afr. 48–51 (1989), reg. rev.

Glabrous to puberulous perennials with woody rootstock. Leaves tripinnate, mostly basal, feathery with 2-3-forked linear segments, upper cauline leaves much reduced to a small sheath with ciliate margins; ultimate segments linear. Bracts and bracteoles absent; rays rather few, short. Petals yellow; styles short, divergent; stylopodium shortly conical. Fruits glabrous, ovoid with a short apical neck, ± terete; mericarps compressed dorsally; ribs indistinct; carpophore bifid at apex; vittae fairly large, vallecular usually 3, commissural (2–)4–6; rib oil ducts solitary in each rib, variably present. Seeds transversely compressed dorsally to reniform, the face concave. $2n = 22$.

One species, *F. ceratophylloides* H. Wolff, tropical Africa (S Tanzania, Malawi and Zambia). Grasslands or rarely open woodlands. Doubtfully distinct from *Pimpinella*.

176. ***Froriepia*** K. Koch [Pimenov]

Froriepia K. Koch, Linnaea 16: 362 (1842); Leute, Notes Roy. Bot. Gard. Edinb. 31(2): 317–320 (1972), rev.

Glabrous annuals or biennials. Basal leaves shortly petiolate, pinnatisect, sheathing, with sessile, ovate, serrate leaflets; cauline with few reduced leaflets, upper without petiole and lamina. Bracts and bracteoles linear-lanceolate, at the margin white-membranaceous, or absent; rays few, markedly unequal. Petals white, greenish or reddish; styles short, reflexed; stylopodium low conical or flat. Fruits laterally compressed; primary and secondary ribs filiform; carpophore bifid; commissure narrow; vallecular vittae 1 or reduced, commissural 2 or reduced. Seed face plane.

Two species (but inclusion of *F. gracillima* Leute is dubious), Asia (SW, the Caucasus). In scrub habitats, meadows and at forest margins.

177. ***Fuernrohria*** K. Koch [Pimenov]

Fuernrohria K. Koch, Linnaea 16: 356 (1842); Koso-Pol., Monit. Tifl. Bot. Gard. 3–4: 136–170 (1915), rev.

Glabrous perennials. Leaves 2–3-pinnatisect, lanceolate; leaflets filiform, verticillate. Bracts linear or pinnate with filiform segments; bracteoles linear. Petals white; styles short, divergent; stylopodium flat to low-conical. Fruits slightly compressed laterally; ribs inconspicuous; carpophore bifid; commissure broad, mesocarp thin, lignified; vallecular vittae 1 (reduced in mature fruits), commissural 2 (large and persistent). Seed face plane. $2n = 22$.

One species, *F. setifolia* K. Koch, Asia (SW, the Caucasus). Subalpine meadows, forest openings. Molecular studies suggest a close relationship to *Aegokeras*, *Aegopodium*, *Carum*, *Cyclospermum*, *Falcaria*, *Grammosciadium*, and *Rhabdosciadium*.

178. ***Galagania*** Lipsky [Kljuykov]

Galagania Lipsky, Acta Horti Petrop. 18: 62 (1900); Kljuykov et al., Bjull. Moskovsk. Obšč. Isp. Prip. Otd. Biol. 83 (6): 100–107 (1978), rev.
Korovinia Nevski & Vved. (1937).

Erect, glabrous, short-lived, monocarpic, herbaceous perennials with globose to cylindrical tubers. Stems solitary, corymbosely or paniculately branched. Leaves 3–4-pinnate with narrow segments; leaflets filiform to lanceolate; upper stem leaves entire. Bracts and bracteoles lanceolate, herbaceous, entire. Petals yellow or rarely white; styles reflexed; stylopodium low-conical. Fruits glabrous, obovoid to oblong, compressed dorsally or laterally; commissure broad; ribs equal, filiform to obscure, or marginal ribs narrowly winged; carpophore bifurcated; vallecular vittae 1–many, often anastomosing. Seed face plane. $2n = 22$.

Seven species, Middle Asia, Afghanistan and NE Iran. Semi-deserts, dry steppes, scrub, mountain forests, weeds in fields. Closely related to *Elaeosticta*, *Hyalolaena*, and *Bunium*.

179. ***Gasparrinia*** Bertol. [Pimenov]

Gasparrinia Bertol., Fl. Ital. 3: 614 (1839).

Glabrous perennials with short rhizomes. Leaves bi- to tripinnatisect; leaflets linear to oblong, clustered, scabrous at margin. Bracts several, linear; rays markedly unequal, straight in mature fruits; bracteoles linear to filiform, acuminate, free. Petals greenish-yellowish; styles reflexed, short; stylopodium low-conical or conical. Mericarp ribs equal, keeled; carpophore bifid; commissure narrow; exocarp cells with thickened outer walls; mesocarp parenchymatous; vallecular vittae 1–2, commissural 2–4. Seed face plane. $2n = 22$.

One species, *G. peucedanoides* (M. Bieb.) Thell., Europe, Asia (SW, the Caucasus). Meadows, pine and oak forests, scrub, calcareous slopes. Molecular data suggest a relationship to *Katapsuxis* and *Cnidiocarpa*.

180. ***Geocaryum*** Coss. [Spalik]

Geocaryum Coss., Not. Pl. Crit. 112 (1851); Engstrand, Biosys. Tax. *Geocaryum*, 1–126 (1977), rev.
Huetia Boiss., Diagn. Pl. Orient. ser. 2, 2: 103 (1856); Ball, Feddes Repert. 79: 3–18 (1968), reg. rev.

Glabrous to slightly pubescent perennials with globose tubers. Leaves 2–4-pinnately dissected, triangular; lobes filiform to broadly elliptical. Bracts usually absent; rays few to many, subequal or unequal, spreading or slightly ascending; bracteoles few, often ciliate. Petals white, sometimes pinkish, outer somewhat radiate; styles short to relatively long; stylopodium conical. Fruits glabrous with an aculeate cuticle, linear-oblong with a short beak, laterally compressed; mericarps ± pentagonal, usually as broad as wide; ribs angular, sometimes denticulate; commissure constricted; carpophore bifurcate; vittae small, extant or compressed at maturity, vallecular 1, commissural 2. Seed transversely terete, the face sulcate. $2n = 10, 18, 20$.

13 species, Eurasia (NE Mediterranean). Thickets and open woods, open montane habitats. Closely related to *Anthriscus*, *Kozlovia*, *Myrrhis*, and *Osmorrhiza* (Spalik et al. 2001).

181. ***Gingidia*** J.W. Dawson [Webb]

Gingidia J.W. Dawson, Kew Bull. 29: 476 (1974); Dawson, New Zealand J. Bot. 5: 84–116 (1967), reg. rev.; Mitchell et al., New Zealand J. Bot. 36: 417–424 (1998), phylog.

Glabrous, gynodioecious or andromonoecious perennials with caudex usually multicipital (with many heads from the main crown) with thickened taproots (sometimes rhizomatous). Leaves rosulate, 1–2-pinnately or pinnate-ternately compound, often sub-fleshy, sometimes glaucous; leaflets petiolulate to sessile, mostly ovate, rhomboid or flabellate, simple to pinnatifid or pinnate with serrate, dentate or crenate margins. Bracts and bracteoles few to several, membranous or foliaceous, narrowly linear, free to partly fused; rays few to many. Calyx lobes distinct, unequal; petals white, with a median oil tube, larger in staminate flowers; pistillate flowers with rudimentary staminodes; styles long and slender, divergent; stylopodium low-conical. Fruits glabrous, ovoid to oblong, subterete to slightly compressed dorsally; dorsal and lateral ribs acute, rounded or narrowly winged; marginal ribs similar to dorsal or broadly winged; carpophore bifid almost to base; vittae large, vallecular 1, commissural 2; rib oil ducts 1 in each rib. $n = 11$, $2n = 22$.

12 species, New Zealand and Australia. Molecular evidence indicates that *Gingidia* (as currently circumscribed) may be polyphyletic.

182. ***Glaucosciadium*** B.L. Burtt & P.H Davis [Reduron]

Glaucosciadium B.L. Burtt & P.H. Davis, Kew Bull. 1949: 225 (1949); Meikle, Fl. Cyprus 766–767 (1977), rev.

Glabrous perennials growing from a rhizomatous rootstock. Leaves blue-glaucous, ternate or biternate, triangular, subcoriaceous; leaflets large, broadly ovate-orbicular or cordate, almost entire. Bracts scale-like, caducous; bracteoles ± obsolete; rays few to several, unequal, spreading. Petals white; styles equaling the short-conical stylopodium. Fruits glabrous, oblong-elliptic, dorsally compressed; dorsal ribs slender, not very prominent, the lateral winged; carpophore free, bifurcated; vittae small, vallecular 1–4, commissural lacking; rib oil ducts solitary, broad. Seeds transversely compressed dorsally, the face plane. $2n = 18$.

One species, *G. cordifolium* (Boiss.) B.L. Burtt & P.H. Davis, Eurasia (S and C Turkey, Cyprus). Stony river banks, limestone screes. Closely related to *Mozaffariania* (Kurzyna-Młynik et al. 2008).

183. ***Glehnia*** F. Schmidt ex Miq. [Watson]

Glehnia F. Schmidt ex Miq., Ann. Mus. Bot. Lugduno-Batavi 3: 61 (1867); Mathias & Constance, N. Amer. Fl. 28B(2): 169 (1945), reg. rev.; She et al., Fl. China 14: 173 (2005), reg. rev.

Subcaulescent perennials, white-pubescent throughout. Leaves 1–2-ternate; ultimate segments oblong to broadly obovate, incised-serrate with white cartilaginous margins. Bracts lacking; rays few to many, unequal; bracteoles several, lanceolate. Flowers crowded in (sub-)capitate umbellules; petals white or purple red, pubescent on the dorsal face; styles short; stylopodium lacking. Fruits densely hirsute and velutinous, obovoid to subglobose, slightly compressed dorsally; ribs equally corky-winged or marginal ribs slightly broader than the dorsal; carpophore bifurcated; commissure narrow; vittae almost cyclic, vallecular 1–3, commissural 6–8. Seeds compressed dorsally, the face slightly concave. $2n = 22$.

Two species, E Asia and W N America. Sandy beaches in coastal areas. Roots of *G. littoralis* F. Schmidt ex Miq. are used in traditional Chinese medicine.

184. ***Glia*** Sond. [Van Wyk & Tilney]

Glia Sond., Fl. Cap. 2: 547 (1862); Wijnands, Bot. Commelins: 199 (1983); Van Wyk et al., S. Afr. J. Bot. 76: 259–271 (2010), rev.; Van Wyk et al., Afr. Apiaceae: 198–201 (2013).

Minutely pubescent, woody perennials, often summer-deciduous geophytes with a tuberous root. Leaves pinnate, variable, dimorphic; basal with broad segments, upper sparse, smaller, usually with linear segments; leaflets variable, broad and laminar to short and almost acicular, minutely scabrous or pilulose, at least along the margins and/or veins. Bracts several, linear to lanceolate; rays several, rather long; bracteoles small. Petals yellow; styles short, divergent; stylopodium conical. Fruits glabrous, ovate-oblong; mericarps subterete; ribs distinct, with narrow wing-like ridges; carpophore bifid; vittae large, vallecular 1, commissural 2. Seeds transversely semi-terete or dorsally compressed, slightly to deeply sulcate beneath vittae, the face plane. $2n = 22$.

Three species, South Africa. Fynbos or renosterveld. Closely related to *Anginon*.

185. ***Gongylosciadium*** Rech.f. [Pimenov]

Gongylosciadium Rech.f., Fl. Iran. 162: 308 (1988).

Glabrous perennials or biennials. Basal and lower cauline leaves simple; lamina narrow-ovate to oblong-linear, at base cordate, coriaceous and serrate at margin; upper leaves with long narrow sheaths and trisect lamina, uppermost without petiole and lamina. Bracts and bracteoles absent; rays several, unequal. Calyx lobes triangular, persistent; petals white, papillose; styles straight; stylopodium low-conical. Fruits not compressed, without prominent ribs; carpophore bifid; commissure narrow; vittae small, many, 12–16 on the dorsal side, 4–8 on the commissure. Seed face plane. $2n = 22$.

One species, *G. falcarioides* (Bornm. & H. Wolff) Rech.f., Asia (SW). Related to *Falcaria.*

186. ***Gongylotaxis*** Pimenov & Kljuykov [Pimenov]

Gongylotaxis Pimenov & Kljuykov, Edinb. J. Bot. 53(2): 188 (1996).

Pubescent, short-lived perennials; branched taproots with chain-like tubers on lateral branches. Leaves 3-pinnatisect with sessile or subsessile basal segments; leaflets linear. Bracts and bracteoles entire, herbaceous or membranous, lanceolate to subulate, shortly pubescent. Petals white, shortly pubescent; styles reflexed; stylopodium low-conical. Ovaries and young fruits pubescent, mature fruits subglabrous, slightly compressed laterally, constricted below stylopodium; commissure narrow; ribs inconspicuous; carpophore bifurcated; exocarp cells minute; mesocarp parenchymatous; vallecular vittae 1 broad and 2 minute, commissural 2 broad and 2–4 minute. Seed face slightly concave.

One species, *G. rechingeri* Pimenov & Kljuykov, Asia (Central Afghanistan). Dry slopes, granite rocks. Closely related to *Elaeosticta, Hymenolaena, Galagania.*

187. ***Grafia*** Rchb. [Watson]

Grafia Rchb., Handb. Nat. Pfl.-Syst. 219 (1837).

Glabrous and glaucous perennials. Leaves 3–4-pinnate or ternate; ultimate segments lanceolate to ovate, nearly rhombic, pinnately lobed, dentate or coarsely serrate; upper cauline leaves reduced. Bracts many, ovate-oblong, acute, sometimes bilobed or pinnatifid at apex, reflexed, often deciduous with age; rays many, long, spreading-ascending; bracteoles several, linear to lanceolate, entire or dissected, usually on the outer side of the umbellules. Petals white; styles rather short, almost as long as the low-conical stylopodium. Fruits ovate-oblong, compressed laterally; ribs prominent, almost narrowly winged; vallecular vittae 1, commissural 2. $2n = 22$.

One species, *G. golaka* Rchb., Europe (Italy, Bosnia-Herzegovina, Slovenia). Calcium-rich areas. Sometimes included in *Pleurospermum.*

188. ***Grammosciadium*** DC. [Pimenov]

Grammosciadium DC., Coll. Mém. 5: 62 (1829); Koso-Poljansky, J. Bot. Russe 1–2: 1–22 (1915), rev.; Tamamschian & Vinogradova, Taxon, 18(5): 546–548 (1969), rev.; ejusd. in Bot. Zhurn. 54, 8: 1197–1212 (1969), rev.

Glabrous or scabrous perennials or biennials. Leaves 2–3-pinnatisect; leaflets filiform, bristle-like. Bracts pinnatisect; rays subequal, thickened in fruit; bracteoles setaceous or trisected. Calyx lobes elongate or triangular, often rigid, persistent; petals white, outer usually radiate; styles divergent or straight and rigid, stylopodium conical. Fruits glabrous, linear to oblong, hardly separating into mericarps; dorsal ribs filiform, marginal filiform or winged; carpophore reduced; vallecular vittae 1–3, commissural 2. Seed face plane. $2n = 20$.

Nine species, Asia (SW, the Caucasus). Subalpine meadows, forest margins, dry slopes. Some species can be a source of essential oil. Closely related to *Rhabdosciadium.*

189. ***Halosciastrum*** Koidz. [Pimenov]

Halosciastrum Koidz., Acta Phytotax. Geobot. 10: 54 (1941); Pimenov & Tichomirov, Biol. Nauki (Sci. Rep. High. School, Moscow) 5: 97–104 (1968), rev.

Glabrous or scabrous short-lived perennials. Leaves ternate; leaflets broadly ovate, cuneate, toothed. Bracts and bracteoles linear, subulate; rays few, unequal. Calyx lobes ovate to triangular-lanceolate, acute; petals purple-violet; styles reflexed; stylopodium low-conical. Fruits glabrous, slightly compressed dorsally, usually heteromericarpic, one mericarp with three prominent ribs, the other with two ribs; exocarp of large cells with thickened outer walls; carpophore bifid; commissure narrow; mesocarp often crushed at maturity; vittae small, many and cyclical. Seed face plane. $2n = 22$.

One species, *H. crassum* (Nakai) Koidz., E Asia (Russian Far East, Korea, NE China). Rich mixed oak and deciduous forests.

190. ***Haloselinum*** Pimenov [Pimenov]

Haloselinum Pimenov in Pimenov & Ostroumova, Umbelliferae Russia: 300 (2012).

Glabrous perennials. Leaves bipinnatisect, leaflets linear-lanceolate to lanceolate, obtuse. Bracts lacking or 1–3, deciduous; rays 7–20, unequal; bracteoles linear to lanceolate, membranous at margin. Calyx lobes triangular; petals white; stylopodium flat, dark violet. Fruits compressed dorsally; dorsal ribs filiform, marginal ribs broadly winged; commissure broad; vallecular vittae 2–4, commissural 4–6, some not reaching the mericarp base. Seeds occupying only the upper part of the mericarp, the face plane. $2n = 22$.

One species, *H. falcaria* (Turcz.) Pimenov. Asia (S Siberia, Kazakhstan, Mongolia, W China). Salt meadows, salt marshes. Segregate of *Peucedanum* s.lat. Molecular data suggest an affinity to *Seseli*, *Stenocoelium*, and *Ledebouriella*.

191. ***Hansenia*** Turcz. [Pimenov]

Hansenia Turcz., Bull. Soc. Imp. Naturalistes Moscou 17 (4): 754 (1844); Pimenov & Lavrova, Bot. Zhurn. 74(1): 96–103 (1989), taxon.; Pimenov et al., Willdenowia 37(2): 155–172 (2008), rev.
Notopterygium H. Boissieu (1903).

Glabrous perennials. Leaves biternate, rarely triternate; leaflets lanceolate to suborbicular or ovate. Bracts absent or few, linear to lanceolate, white-membranaceous at the margin; rays several, unequal; bracteoles entire, linear to lanceolate. Petals white or pink; styles reflexed, short; stylopodium low-conical. Fruits slightly compressed dorsally; ribs winged, almost equal; carpophore bifid, commissure narrow; mesocarp parenchymatous of cells with pitted walls; vallecular vittae 2–5, commissural 4–6. Seed face moderately concave. $2n = 22$.

Six species, Asia (S Siberia, Mongolia, China). Subalpine stony meadows, forest openings, shrubby areas.

192. ***Haplosciadium*** Hochst. [Van Wyk & Tilney]

Haplosciadium Hochst., Flora 27: 20 (1844); Townsend, Fl. Trop. E. Afr. 35–37 (1989), reg. rev.; Van Wyk et al., Afr. Apiaceae: 202–203 (2013).

Dwarf, acaulescent, glabrous or hirtellous perennials with a woody rootstock. Leaves all basal, pinnate with pinnatisect to pinnatipartite segments, feathery; ultimate segments linear. Umbels compound (often sessile, without a scape and appearing simple); bracts and bracteoles large, linear; rays few, long. Petals white, cream-colored or greenish; styles long, divergent; stylopodium flat, disc-shaped. Fruits glabrous, oblong-ellipsoid, subterete; mericarps angular and slightly compressed dorsally; ribs distinct or indistinct; carpophore entire; vittae large, vallecular 1, commissural 2. Seeds transversely subreniform, sulcate beneath vittae, the face plane to somewhat concave. $2n = 22$.

One species, *H. abyssinicum* Hochst., E Africa. Moist medium- to high-elevation Afroalpine habitats.

193. ***Haplosphaera*** Hand.-Mazz. [Watson]

Haplosphaera Hand.-Mazz., Anz. Akad. Wiss. Wien, Math.-Naturwiss. Kl. 57: 143 (1920); P.K. Mukherjee & Constance, Umbelliferae India 1: 96 (1993), reg. rev.

Glabrous perennials with a stout rootstock. Leaves petiolate, ternate-pinnate; ultimate segments broad, ovate, irregularly dentate, serrate or pinnatisect. Umbels compound but appearing simple, very compact, globose and capitate at anthesis; bracts several, linear-lanceolate, rays very short, stout (sometimes elongating in fruit). Petals dark brown or purplish-brown; styles short; stylopodium depressed. Fruits glabrous, obovoid-oblong, slightly compressed dorsally; ribs equally narrow-winged; carpophore bifid; vallecular vittae (1–)3, commissural 3–6. Seeds slightly compressed, the face plane. $2n = 22$.

Two species, Asia (E Himalaya to SW China). High-elevation forests, shrublands, and open areas.

194. ***Harbouria*** J.M. Coult. & Rose [Plunkett]

Harbouria J.M. Coult. & Rose, Rev. North Amer. Umbellif. 125 (1888); Mathias & Constance, N. Amer. Fl. 28B(1): 124 (1944), reg. rev.

Subcaespitose, nearly glabrous perennials. Leaves mostly basal, pinnately decompound, ovate-oblong; ultimate segments linear, mucronulate. Bracts usually lacking, or few, lanceolate and inconspicuous; rays few to many, spreading, subequal; bracteoles few, linear and inconspicuous (usually shorter than the pedicels). Petals yellow; styles short; stylopodium lacking. Fruits granular-roughened, ovoid, slightly compressed laterally, the commissure constricted (and having 1–2 accessory ribs); dorsal, lateral and marginal ribs obtuse, corky, subequal; carpophore entire; vittae large, vallecular usually 1, commissural 1–3. Seed transversely subterete, the face plane. $n = 10$, $2n = 22$.

One species, *H. trachypleura* (A. Gray) J.M. Coult. & Rose, N America (C Rocky Mountains), mid to lower elevations.

195. ***Harperella*** Rose [Plunkett]

Harperella Rose, Proc. Biol. Soc. Washington 19: 96 (1906); Feist et al., Taxon 61: 402–418 (2012), rev.

Glabrous annuals with fascicles of fibrous roots. Leaves reduced to terete, hollow, septate rachis-leaves. Bracts and bracteoles few, small, entire, or lacking; rays few to many, spreading-ascending to spreading. Calyx lobes conspicuous; petals white; styles spreading to reflexed; stylopodium conical. Fruits glabrous, ovoid, compressed laterally; dorsal and lateral ribs filiform, rounded or acute, marginal ribs inconspicuous but distinct from the corky tissue that forms an unwinged, conspicuous longitudinal band around the commissural edge of each mericarp; carpophore bifid; vallecular vittae 1, commissural 2. Seeds transversely subterete, the face plane. $n = 6$.

One species, *H. nodosa* Rose, N America (SE United States); wet places.

196. ***Harrysmithia*** H. Wolff [Watson]

Harrysmithia H. Wolff, Acta Horti Gothob. 2: 310 (1926).

Essentially glabrous annuals. Leaves 2–3-pinnately dissected; ultimate segments ovate or linear; upper leaves sometimes heteromorphic with long linear segments. Bracts lacking (occasionally solitary); rays short, spreading; bracteoles few and small. Petals white; styles spreading or reflexed, about twice the length of the low-conical stylopodium. Fruits sparsely verrucose or papillose, ovoid-globose, slightly compressed dorsally; mericarps transversely sub-pentagonal; ribs prominent or equally narrowly erose-winged, margins irregularly denticulate or entire and valleculae rather broad; carpophore apically bifid; vallecular vittae 1, commissural 2. Seeds transversely terete, the face subplane.

Two species, Asia (Sichuan, SE Xizang and N Yunnan in SW China). High-elevation forests and open areas.

197. ***Haussknechtia*** Boiss. [Plunkett, Pimenov]

Haussknechtia Boiss., Fl. Orient. 2: 960 (1872); Pimenov et al., Nord. J. Bot. 24: 555–565 (2007), rev.

Glabrous perennials. Basal leaves large, bipinnate; petioles long; leaflets elliptical to ovate or obovate; cauline leaves reduced to scarious scales. Bracts obsolete; rays few and very short; bracteoles several, whitish and membranaceous. Petals white; styles divergent, long; stylopodium conical. Fruits compressed dorsally; carpophore bifurcated; dorsal mericarp ribs filiform, marginal ribs very narrowly winged; commissure broad; multilayer hypodermal parenchyma present; vallecular vittae 1, commissural 2; rib oil ducts 1, small; seed face plane or slightly concave.

One species, *H. elymaitica* Boiss. Asia (S. Iran), in *Quercus brantii* Lindl. forests, openings, margins. Carpological and molecular data suggest a relationship to *Zeravschania* and *Demavendia*.

198. ***Hellenocarum*** H. Wolff [Kljuykov]

Hellenocarum H. Wolff in Engl., Pflanzenr. IV, 228 (Heft 90): 167 (1927); Kljuykov, Biol. Nauki (Sci. Rep. High. School, Moscow) 8: 60–63 (1985), rev.
Neomuretia Kljuykov, Degtjareva & Zakharova (2016).

Glabrous perennials with tuberous roots. Leaves 2–3-pinnate; leaflets lanceolate or linear. Bracts and bracteoles linear-lanceolate, entire. Petals white or yellow, with one or several secretory ducts; styles reflexed; stylopodium low-conical. Fruits glabrous, lanceolate, slightly compressed laterally; commissure narrow; ribs narrowly winged or terete; vallecular and commissural vittae many. Seed face plane. The seedlings mono- or dicotyledonous. $2n = 20$.

Five species, SE Europe, SW Asia (Turkey, W Iran and Iraq). Rock crevices, limestone. Related to *Bunium*.

199. ***Helosciadium*** W.D.J. Koch [Pimenov]

Helosciadium W.D.J. Koch, Nova Acta Phys.-Med. Acad. Caes. Leop.-Carol. Nat. Cur. 12, 1: 125 (1824); Ronse et al., Pl. Syst. Evol. 287: 1–17 (2010), rev.

Glabrous perennials with rhizomes, rooting at the nodes. Leaves homomorphic or heteromorphic, 1–2-ternate, or 1–2-pinnatisect; segments broad, lanceolate to ovate, serrate, or lobed. Bracts few or several; bracteoles lanceolate, often white-margined, rarely absent. Petals white; styles short, recurved; stylopodium low-conical. Fruits separating into mericarps or not; commissure narrow; ribs subequal; vascular bundles situated distally; carpophore entire; pericarp with sclerenchyma cell groups and air cavities; vallecular vittae 1, commissural 2, rib secretory ducts small. Seed face plane. $2n = 22$.

Five species, Macaronesia, Europe, Asia (SW, Middle), Africa (N, E tropical). Wet places. Molecular data show affinity with *Sium*, *Berula* and *Cryptotaenia* (Spalik et al. 2009). An intergeneric hybridization with *Berula erecta* was established by karyological and molecular studies, resulting in the name × *Beruladium* A.C. Leslie.

200. ***Heptaptera*** Margot & Reut. [Pimenov]

Heptaptera Margot & Reut., Mém. Soc. Phys. Genève 8: 302 (1839); Bornmüller, Bot. Jahrb. 67: 290–295 (1936), reg. rev.; Herrnstadt & Heyn, Notes Roy. Bot. Gard. Edinb. 31(1): 91–107 (1971), rev.

Glabrous or papillose perennials. Leaves 1–4-pinnatisect, leaflets ovate to lanceolate, margin crenate to serrate. Bracts and bracteoles ovoid, falling; rays glabrous, angled. Petals yellow; styles short, reflexed; stylopodium flat. Fruits glabrous, sometimes cuneate at base, slightly compressed dorsally, mericarps usually heteromorphic, or rarely homomorphic; marginal ribs often larger than dorsal; carpophore bifid; commissure narrow; exocarp of small cells, mesocarp cells with thin pitted walls; vittae small, many, almost cyclical in the inner mesocarp layer. Seed face with a deep groove and involute margins. $2n = 22$.

Six species, S Europe, SW Asia. Deciduous forests, openings and margins, sclerophyllous scrub, stony slopes. Two sections (Herrnstadt and Heyn 1971).

201. ***Heracleum*** L. Fig. 3A [Ostroumova]

Heracleum L., Sp. Pl. 249 (1753); Mandenova, Notes Plant Syst. Geogr. Tbil. Bot. Inst. 41: 43–49 (1981), gener. rev.; Sazyperova, *Heracleum* species of the USSR – new fodder plants (1984), reg. gener. rev.

Pubescent, sometimes megaherbs, biennials or perennials. Leaves 1–2-pinnate or simple, usually pubescent or tomentose abaxially. Bracts and bracteoles small, linear. Petals white, yellow, greenish or pink, marginal often radiate; styles recurved; stylopodium conical. Fruits densely pubescent to glabrous, orbicular, ellipsoid or ovoid, strongly compressed dorsally; dorsal ribs filiform, marginal ribs broad; fibrous inner mesocarp present; commissure broad; vittae short or long, rather narrow, sometimes thick, often clavate, vallecular 1, commissural 2 (sometimes absent), straight and parallel. Seed face plane. $2n = 22, 24, 40, 44$.

120–130 species, Europe, Asia, N America, Africa. The genus includes fodder, edible, melliferous, medicinal, ornamental plants, and some species can cause contact dermatitis. Eight sections (Mandenova 1981) based mainly on secretory-duct morphology. *Heracleum* is paraphyletic and close to *Pastinaca* and *Tetrataenium*, among other genera.

202. ***Heteromorpha*** Cham. & Schltdl. [Van Wyk & Tilney]

Heteromorpha Cham. & Schltdl., Linnaea 1: 385, t. 5 (1826), nom. cons.; Winter & Van Wyk, Kew Bull. 51: 225–265 (1996), rev.; Van Wyk et al., Afr. Apiaceae: 214–219 (2013).
Aframmi C. Norman (1929).

Glabrous or pubescent, woody shrubs (rarely subshrubs) or sometimes small to medium-sized trees or woody climbers. Leaves simple or trisect to pinnately and/or palmately compound; leaflets variable, ovate to obovate to linear or irregularly cleft. Bracts conspicuous, leaf-like; rays many, often long; bracteoles small. Petals yellow; styles short, erect; stylopodium conical. Fruits glabrous or pubescent, elliptical or somewhat pyriform; mericarps heteromorphic, one 2-winged, the other 3-winged, usually slightly compressed dorsally; primary ribs prominent, winged, secondary ribs sometimes present; carpophore bifid; vittae fairly large, vallecular 1(–3), commissural 2; rib oil ducts often present, sometimes large, 1–2 in some ribs. Seeds transversely subterete, usually sulcate beneath vittae, slightly compressed, the face convex or concave. $2n = 22$.

Seven species, sub-Saharan Africa and Yemen. Grasslands, open woodlands or forest margins. Belongs to an early branching clade of Apioideae.

203. *Hladnikia* Rchb. [Watson]

Hladnikia Rchb., Icon. Bot. Pl. Crit. 9: 9, t. 825 (1831).

Glabrous biennials or perennials. Leaves petiolate, mostly basal, ternate or pinnate; ultimate segments ovate, dentate to pinnatifid; cauline leaves few. Bracts many, entire or trilobed, appressed becoming deflexed; rays several, spreading, scabrid on inner angles; bracteoles several, entire, appressed and becoming deflexed. Calyx lobes conspicuous, broadly triangular; petals creamy white with a conspicuous brown midrib. Styles ca. twice as long as stylopodium, reflexed. Fruits glabrous, oblong, angled, slightly compressed laterally; ribs broad and rounded; carpophore bifid almost to the base; vittae large, vallecular 1, commissural 2. $2n = 22$.

One species, *H. pastinacifolia* Rchb., Europe (W Slovenia). Calcareous rocks. Sometimes included in *Pleurospermum* or *Falcaria.*

204. *Hohenackeria* Fisch. & C.A. Mey. [Pimenov]

Hohenackeria Fisch. & C.A. Mey., Ind. Sem. Horti Petrop. 2: 38 (1836); Tamamschjan in Sov. Bot. 14, 4: 219–238 (1946), rev.

Dwarf, glabrous annuals or perennials. All leaves rosulate, entire, with linear-lanceolate blades and dentate scabrid margins, narrowed in lower part. Umbels simple, globose, sessile in leaf axils; bracts leaf-like. Flowers sessile; calyx lobes rigid, spinose, longer than petals; petals whitish, minute; styles long, erect; stylopodium inconspicuous. Fruits slightly compressed laterally, undivided; ribs inconspicuous or terete; carpophore entire or absent; mericarps thickened below, with distinct beak; ribs corky, sub-inconspicuous; commissure narrow; mesocarp of two layers: outer of large cells, inner of cells with lignified pitted walls; vallecular vittae 1, commissural 2–4, inconspicuous at maturity. Seed face plane. $2n = 32$.

Two species, Asia (the Caucasus, SW), Africa (NW, N), Europe (SW). Arid slopes. Closely related to *Bupleurum.*

205. *Horstrissea* Greuter, P. Gerstberger & B. Egli [Watson]

Horstrissea Greuter, P. Gerstberger & B. Egli, Willdenowia 19(2): 391 (1990).

Diminutive, extreme geophytic, glabrous perennials with a deeply buried, swollen taproot. Leaves hysteranthous (appearing after flowers senesce), arising from the crown of the taproot, petioles and rachis mostly below soil surface, the blades thus appearing as scattered above the soil, small, and 2–3-pinnate. Compound umbels densely compact, small, globular, peduncles held just above the soil surface; bracts several, leaf-like; rays several, short and stout; bracteoles several, linear-spathulate with hyaline margins. Petals pink with white margin; styles reflexed, 2–3 times longer than the broadly conical stylopodium. Fruits glabrous, broadly ellipsoid, slightly compressed dorsally; ribs filiform; carpophore bifurcated; vittae large, vallecular 1, commissural 2. Seeds transversely compressed dorsally, the face bisulcate. $2n = 14$.

One species, *H. dolinicola* Greuter, P. Gerstberger & B. Egli, Europe (Crete). Limestone sinks of Mt. Ida (Psiloritis). Critically endangered.

206. *Hyalolaena* Bunge [Kljuykov]

Hyalolaena Bunge, Mem. Acad. Imp. Sci. St. Petersb. Div. Sav. 7: 304 (1854); Pimenov & Kljuykov, Bot. Zhurn. 67 (7): 873–889 (1982), rev.
Hymenolyma Korovin (1948).

Glabrous, ephemeroid perennials with globose to cylindrical tubers. Leaves 2–4-pinnate; leaflets filiform to lanceolate. Bracts and bracteoles entire, lanceolate to ovate, herbaceous to white-membranaceous. Petals white or yellow; styles reflexed; stylopodium low-conical. Fruits glabrous, oblong, compressed dorsally; commissure broad; ribs equal, slightly short-keeled, or the marginal ribs narrowly winged; carpophore bifurcated; vallecular and commissural vittae several, wide or narrow, often short, anastomosing. Seed face plane. $2n = 20, 22$.

11 species, Middle Asia, NW China, NW Afghanistan and NE Iran. Semi-deserts, salt plains, dry steppes, gravel and stony slopes, scrub and forests. Closely related to *Elaeosticta*, *Galagania*, and *Bunium*.

207. ***Hymenidium*** Lindl. [Kljuykov]

Hymenidium Lindl. in Royle, Ill. Bot. Himal. Mts. 233 (1835); Pimenov et al., Feddes Repert. 111: 499–515 (2000), gen. limits; Pimenov & Kljuykov, Feddes Repert. 111: 535–552 (2000), rev.
Pleurospermopsis C. Norman (1938).

Glabrous perennials. Leaves 1–3-pinnate or rarely ternate with basal primary segments; leaflets filiform to ovate, lobed. Bracts many (rarely 1–2), entire to pinnate; bracteoles entire to pinnatifid, linear to ovate, usually equal or longer than umbellules. Petals white; stylopodium flat to conical. Fruits glabrous or verrucose, slightly compressed dorsally; commissure narrow; ribs winged equally, or marginal broadly winged; carpophore bifurcated or rarely reduced; vallecular vittae 1–3, commissural 2–6. Seed face plane or broadly sulcate (rarely narrowly and deeply sulcate). $2n = 22, 33, 44$.

42 species, Asia (from Afghanistan and Middle Asia to China and Burma). Scrub, forests, alpine and subalpine meadows, stony slopes. Closely related to *Pleurospermum* s.l.

208. ***Hymenolaena*** DC. Fig. 20 [Kljuykov]

Hymenolaena DC., Prodr. 4: 244 (1830); Pimenov & Kljuykov, Feddes Repert. 111: 517–534 (2000), rev.

Glabrous perennials. Leaves pinnate; segments ovate, dentate or incised. Bracts few, entire; bracteoles many, almost completely membranaceous, usually entire. Calyx lobes triangular or subulate; petals white, nearly flat; styles reflexed;

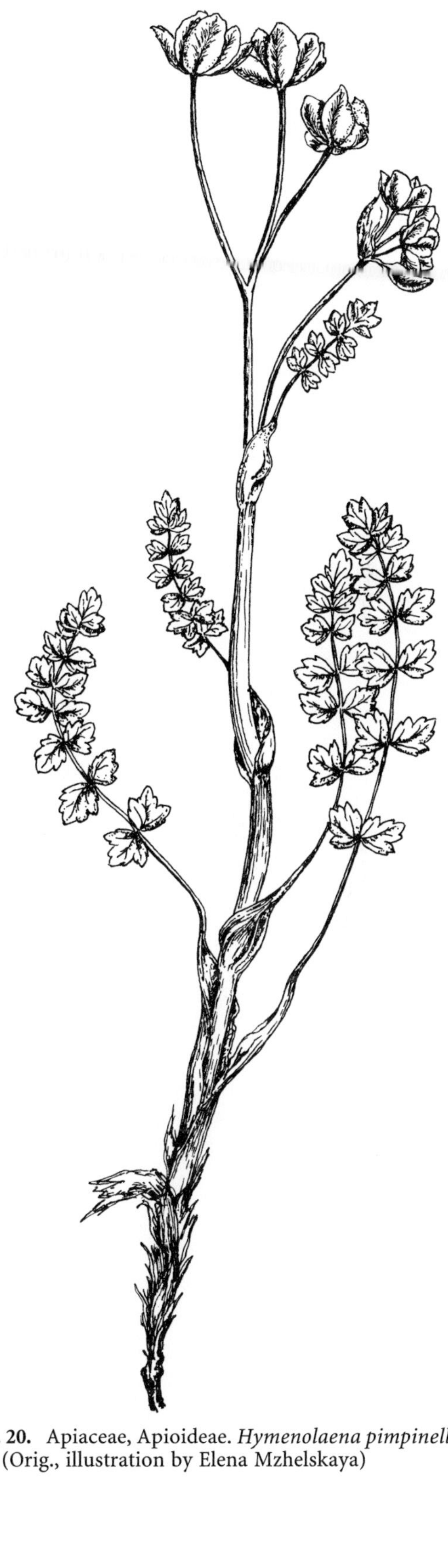

Fig. 20. Apiaceae, Apioideae. *Hymenolaena pimpinellifolia*. (Orig., illustration by Elena Mzhelskaya)

stylopodium short-conical. Fruits glabrous, oblong, slightly compressed laterally; commissure narrow; ribs equal, narrowly winged; carpophore bifurcated; vallecular vittae 2–3, commissural 5–7. Seed face with broad deep groove. $2n = 22$.

Three species, Middle Asia, E Afghanistan, N Pakistan and NW Himalaya. Alpine meadows and stony slopes. Closely related to *Hymenidium*, *Pleurospermum*, *Aulacospermum*, and *Eleutherospermum*.

209. ***Imperatoria*** L. [Pimenov]

Imperatoria L., Sp. Pl. 259 (1753); Gen. Pl., ed. 5, 125 (1754).

Glabrous perennials with branched rootstocks and rhizomes. Leaves biternate or pinnate; leaflets broad, ovate to lanceolate, irregularly toothed. Bracts absent or few, linear; rays many, puberulent; bracteoles several, setaceous. Petals white or pinkish, minute, with several secretory ducts; styles reflexed; stylopodium low-conical, undulate at margin. Fruits compressed dorsally; dorsal ribs keeled, marginal winged; carpophore bifurcated; commissure broad; mesocarp cells with pitted walls; vittae large, vallecular 1, commissural 2. Seed face plane. $2n = 22$.

Three species, Europe, Macaronesia (Madeira). Mountain forests, subalpine meadows. Used in mediaeval and folk medicines. Closely related and sometimes included in *Peucedanum* (Spalik et al. 2004).

210. ***Itasina*** Raf. [Van Wyk & Tilney]

Itasina Raf., Good Book 51 (1840); Burtt, Notes Roy. Bot. Gard. Edinb. 45: 92 (1988); Burtt, Edinb. J. Bot. 48: 243 (1991); Van Wyk et al., Afr. Apiaceae: 220–221 (2013).

Acaulescent, glabrous perennials (sometimes appearing annual) growing from a fleshy, cylindric root. Leaves simple, filiform or grass-like, rarely a few cauline, linear. Bracts few, linear; rays few, short; bracteoles small. Calyx lobes ovate-acuminate, sometimes linear, recurved or subspinescent; petals white; styles long, erect to slightly divergent; stylopodium conical to narrowly conical. Fruits glabrous, oblong; mericarps oblong, semi-terete; ribs prominent; carpophore bifid; vittae large, vallecular 1, commissural 2. Seeds transversely sulcate beneath the vittae, the face somewhat plane, concave or convex. $2n = 24$.

One species, *I. filifolia* (Thunb.) Raf., South Africa. Sandy places in fynbos. Closely related to *Annesorhiza* and *Chamarea*.

211. ***Johrenia*** DC. [Pimenov]

Johrenia DC., Coll. Mém. 5 54 (1829); Bornmüller in Feddes Repert. 28: 33–53 (1930), reg. rev.; Pimenov et al., Willdenowia 37, 2: 465–502 (2007), rev.

Glabrous perennials. Leaves petiolate, 1–2(–3)-pinnate; leaflets oblong-ovate to linear. Bracts absent; rays few to many, often very unequal; bracteoles linear or setaceous. Petals yellow; styles reflexed; stylopodium low-conical to almost plane. Fruits dorsally compressed; dorsal ribs inconspicuous, submerged into the spongy pericarp; marginal ribs spongy, thickened; mesocarp cells parenchymatous with lignified pitted walls, thickened on the commissural side; carpophore bifid; vallecular and commissural vittae absent, or vallecular vittae rarely very small; rib oil ducts solitary, large. Seed face plane. $2n = 22$.

Five species, Asia (SW, the Caucasus), Europe (SE). On dry stony slopes, forest margins, xerophilic scrub, limestone rocks. Related to *Dichoropetalum* and *Zeravschania*.

212. ***Kadenia*** Lavrova & V.N. Tikhom. [Pimenov]

Kadenia Lavrova & V.N. Tikhom., Bjull. Moskovsk. Obšč. Isp. Prip. Otd. Biol. 91(2): 93 (1986).

Almost glabrous perennials with taproots or short rhizomes and adventive roots. Leaves petiolate, 2–3-pinnate; leaflets linear to lanceolate, acuminate. Bracts 0–4, subulate; rays glabrous or scabrous, subequal; bracteoles subulate. Calyx lobes short, triangular, acuminate, or nearly inconspicuous; petals white; styles long, reflexed; stylopodium low-conical. Fruits glabrous, broadly lanceolate to ovate, slightly compressed dorsally; ribs prominent, dorsal keeled, marginal narrowly winged; carpophore bifid; commissure of intermediate width; mesocarp cells with pitted walls; vallecular vittae 1, commissural 2. Seed face plane. $2n = 20, 22$.

Two species, Europe, Asia (N, Central). Forest margins and openings, meadows, salt lake shores. Segregate of *Cnidium.*

213. ***Kafirnigania*** Kamelin & Kinzik. [Pimenov]

Kafirnigania Kamelin & Kinzik., Fl. Tadzh. SSR 7: 523 (1984).

Glabrous perennials with vertical roots and lignified rootstocks. Leaves 4-pinnate; leaflets narrow-linear, flat, acute. Bracts absent; rays many; bracteoles linear, acute, herbaceous. Petals light yellow; styles short, reflexed; stylopodium low-conical. Fruits compressed dorsally; dorsal mericarp ribs filiform, marginal narrow winged; carpophore bifid; commissure broad; mesocarp of parenchymous cells, not lignified; vallecular vittae 1, commissural 2. Seed face plane. $2n = 22$.

One species, *K. hissarica* (Korovin) Kamelin & Kinzik., Middle Asia (Tadzhikistan). In rock crevices. Molecular data suggest an affinity to *Ferula.*

214. ***Kailashia*** Pimenov & Kljuykov [Pimenov]

Kailashia Pimenov & Kljuykov, Feddes Repert. 116(1–2): 82 (2005).

Glabrous or scabridulous perennials. Leaves 3-pinnate; leaflets rhombic to lanceolate, acute. Bracts entire, glabrous, narrowly lanceolate; rays scabrous; bracteoles entire, filiform or dissected, scabridulous. Calyx lobes long, filiform to linear; petals white; styles reflexed; stylopodium flat to low conical. Fruits glabrous to sparsely puberulent, compressed dorsally; ribs subequal, narrowly winged; carpophore bifid; commissure broad; mesocarp of cells with pitted walls; vallecular vittae 1, commissural 2. Seed face plane.

Two species, W Himalaya (India and China). High-montane vegetation, alpine meadows, stony slopes. Related to *Oreocome* and *Ligusticopsis.*

215. ***Kalakia*** Alava [Ostroumova]

Kalakia Alava, Notes Roy. Bot. Gard. Edinb. 34(2): 190 (1975).

Fetid, glabrous annuals. Leaves triternate, ovate; ultimate divisions linear. Bracts and bracteoles simple, sometimes bifid or pinnate. Petals white, marginal radiate; styles short, arcuate; stylopodium flat with a crenate margin. Fruits glabrous or pubescent, oblong, compressed dorsally; dorsal ribs filiform, marginal ribs winged and swollen; fibrous inner mesocarp present; commissure broad; vittae absent. Seed face broadly sulcate.

One species, *K. marginata* (Boiss.) Alava, Asia (Iran).

216. ***Kandaharia*** Alava [Ostroumova]

Kandaharia Alava, Candollea 31(1): 92 (1976).

Pubescent perennials. Leaves bipinnate; ultimate divisions cuneate. Bracts and bracteoles small. Petals white, dorsally pubescent; styles recurved; stylopodium broadly depressed. Fruits hirsute, orbicular or elliptic, notched at the apex, strongly compressed dorsally; dorsal ribs filiform, marginal ribs broadly winged (proximally thin and distally thickened); fibrous inner mesocarp present; commissure broad; vallecular vittae 3, commissural 8. Seed face plane.

One species, *K. rechingerorum* Alava, SW Asia(Afghanistan).

217. ***Karatavia*** Pimenov & Lavrova [Pimenov]

Karatavia Pimenov & Lavrova, Bot. Zhurn. 72(1): 36 (1987).

Glabrous perennials. Leaves pinnate or almost bipinnate; leaflets lanceolate to ovate, toothed. Bracts absent; rays several, unequal; bracteoles linear. Calyx lobes triangular, acute; petals whitish; styles long, reflexed, stylopodium low-conical, undulate at margin. Fruits compressed dorsally; dorsal ribs filiform, marginal winged; carpophore bifid; commissure broad; mesocarp cells with pitted walls; vallecular vittae 1, commissural 2; secretory ducts in ribs large, equal or exceeding vallecular vittae. Seed face plane. $2n = 22$.

One species, *K. kultiassovii* (Korovin) Pimenov & Lavrova, Asia (S Kazakhstan). In mesophilic scrub, stony slopes. Molecular data suggest a relationship to *Endressia*, *Dystaenia*, *Paraligusticum*, and *Tommasinia.*

218. ***Karnataka*** P.K. Mukh. & Constance [Pimenov]

Karnataka P.K. Mukh. & Constance, Brittonia 38(2): 145 (1986).

Glabrous or scaberulous perennials with swollen and tuberous taproots. Leaves 2–3-pinnate; leaflets ovate, crenate-dentate, sparsely scaberulous. Bracts 2–5, linear-lanceolate, entire, white-membranaceous at margins; rays few, subequal; bracteoles like bracts but sometimes irregularly toothed. Umbellules with 1–3 central hermaphroditic and several marginal staminate flowers; calyx lobes linear, persistent; petals white; styles recurved; stylopodium low conical. Fruits glabrous, narrowly ovoid, compressed dorsally; ribs filiform; carpophore bifid; vallecular vittae 1, commissural usually 6. Seed face concave or almost plane.

One species, *K. benthamii* (C.B. Clarke) P.K. Mukh. & Constance, S Asia (India).

219. ***Katapsuxis*** Raf. [Reduron]

Katapsuxis Raf., Good Book 58 (1840); Leute, Ann. Naturhist. Mus. Wien 74: 498 (1970), carpol., distrib.; Reduron et al., J. Bot. Soc. Bot. France 1: 99 (1997), taxon.

Glabrous perennials. Leaves 2–4-pinnately divided, triangular; ultimate divisions ovate, deeply lobed. Bracts absent or very few; bracteoles present, linear; rays many, subequal. Petals white; styles equaling to clearly exceeding the conical stylopodium. Fruits glabrous but shortly papillate, ovoid-globular or ellipsoid; ribs subequal, prominent, narrow and thin; carpophore free, bifurcated; vittae small, vallecular 1, commissural 2. Seeds transversely pentagonal, the face plane. $2n = 22$.

One species, *K. silaifolia* (Jacq.) Reduron, Charpin & Pimenov., S Europe to SW Asia (S France to Turkey). Woody slopes. Segregate of *Ligusticum* s. lat.

220. ***Kedarnatha*** P.K. Mukh. & Constance [Kljuykov]

Kedarnatha P.K. Mukh. & Constance, Brittonia 38(2): 147 (1986); Pimenov & Kljuykov, Feddes Repert. 115(3–4): 230–238 (2004), rev.
Indoschulzia Pimenov & Kljuykov (1995).

Caulescent to acaulescent, mainly glabrous perennials with creeping and slender rhizomes. Leaves linear to narrowly oblong, 2–3-pinnate; basal primary segments sessile or shortly petiolulate; leaflets filiform to linear, small. Bracts and bracteoles mostly pinnatisect with linear lobes. Petals white; styles reflexed; stylopodium short conical or conical. Fruits glabrous, ovoid or ovoid-lanceolate; commissure broad; ribs keeled, equal to subequal; carpophore bifurcated; vallecular vittae 2–4, commissural 3–6. Seed face plane or slightly concave. $2n = 22$.

Five species, Himalaya (from Kashmir to Burma). Alpine and subalpine meadows, forests.

221. ***Kelussia*** Mozaff. [Pimenov]

Kelussia Mozaff., Bot. Zhurn. 88(2): 88 (2003).

Glabrous perennials; stems covered by leaf sheaths when young (naked when older). Basal leaves large, petiolate, bipinnate, broadly elliptic; leaflets imbricate, oblong-elliptic, undivided or rarely trilobate, serrate at margin, reticulate. Bracts lanceolate, caducous; rays several, equal, spreading; bracteoles usually absent (or like bracts). Petals yellow; styles short, reflexed; stylopodium flat to low conical. Fruits glabrous, elliptic to orbicular, strongly compressed dorsally; dorsal ribs keeled, lateral narrow-winged; commissure broad; vittae many, septate, vallecular 3, commissural 12–14. Seed face plane.

One species, *K. odoratissima* Mozaff., SW Asia (Iran).

222. ***Keraymonia*** Farille [Kljuykov]

Keraymonia Farille in Farille, Cauwet-Marc & Malla, Candollea 40(2): 528 (1985); Pimenov & Kljuykov, Feddes Repert. 111: 517–534 (2000), rev.

Acaulescent to caulescent, nearly glabrous perennials. Leaves 1–3-pinnate with shortly petiolulate primary segments; leaflets linear to oblong. Bracts 1–3, 2–3-pinnate; bracteoles entire or toothed, filiform to oblanceolate. Petals white or purple, with a plane or inflexed apex; stylopodium low-conical, massive. Fruits glabrous, oblong, slightly compressed laterally; commissure narrow; ribs equal, narrowly winged; carpophore bifurcated; mesocarp parenchymatous;

vallecular vittae many, rib oil ducts large, solitary. Seed face broadly and deeply sulcate.

Three species, Asia (S Tibet and Nepal). Among boulders and dwarf *Rhododendron*, screes and low grassy alpine meadows. Related to *Pleurospermum*, *Pterocyclus*, *Trachydium*, *Aulacospermum*, and *Hymenolaena*.

223. ***Kitagawia*** Pimenov [Pimenov]

Kitagawia Pimenov, Bot. Zhurn. 71(7): 943 (1986); Pimevov, Turczaninowia 20(2): 165–168 (2017), reg. rev.

Mostly glabrous (but puberulent below the umbels) perennials. Leaves bipinnate or biternate; leaflets linear, lanceolate to obovate or almost orbicular. Bracts absent or filiform; rays several, almost equal; bracteoles narrowly lanceolate with white margins. Calyx lobes triangular or lanceolate; petals white or pinkish; styles reflexed; stylopodium conical. Fruits glabrous or minutely puberulent, compressed dorsally; dorsal ribs filiform, marginal narrowly winged; carpophore bifid; commissure broad; mesocarp cells with pitted walls; vallecular vittae solitary, commissural 2. Seed face plane. $2n = 22$.

Ten species (but some Chinese species of *Peucedanum* presumably belong to this genus), E, N and C Asia. Forests, scrub, stony slopes, meadows. Used in Tibetan and Chinese medicine. Segregate of *Peucedanum*.

224. ***Komarovia*** Korovin Fig. 21 [Pimenov]

Komarovia Korovin, Commem. Vol. 70th Ann. V. L. Komarov 427 (1939); Pimenov et al., Komarovia (St. Petersburg) 1: 61–73 (1999), taxon., molec. phylog.

Glabrous perennials with massive thickened rootstocks. Leaves 2–3-ternate; leaflets lanceolate. Bracts and bracteoles absent; rays few or several, long, thin. Petals yellow; styles short, divergent; stylopodium conical, undulate at margin. Fruits slightly compressed dorsally, of two unequal mericarps: one having 3, the other 4 prominent ribs; carpophore bifid; commissure rather narrow; mesocarp parenchymatous; vallecular vittae 1, commissural 2; rib oil ducts in prominent ribs. Seed face semi-lunar, with a large deep groove. $2n = 20$.

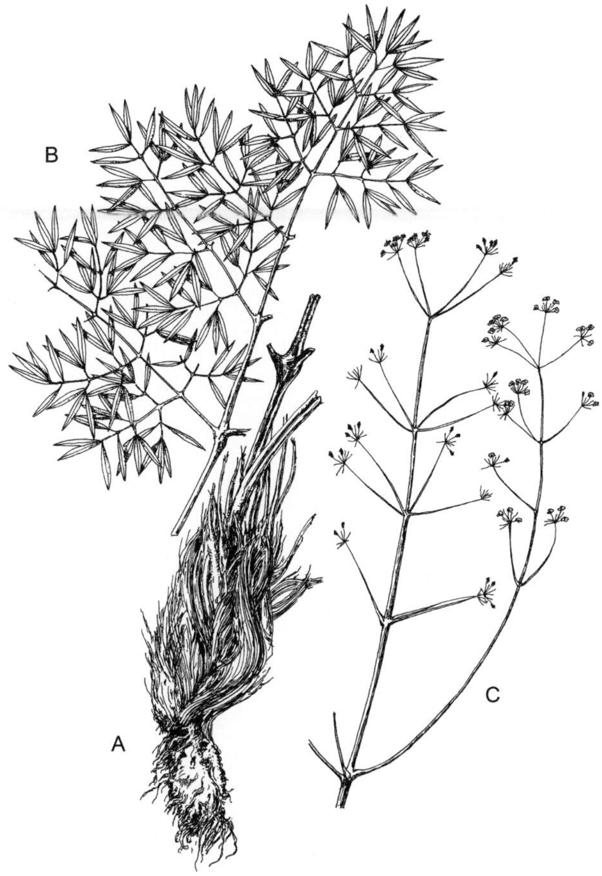

Fig. 21. Apiaceae, Apioideae. *Komarovia anisosperma*. **A** Plant base. **B** Leaf fragment. **C** Fragment of inflorescence. (Orig., illustration by Elena Mzhelskaya)

One species, *K. anisosperma* Korovin, Asia (Uzbekistan, Tadzhikistan). Rocks, stony slopes. Molecular studies suggest a relationship to *Parasilaus*, *Cyclorhiza*, and *Calyptrosciadium*.

225. ***Korshinskia*** Lipsky [Pimenov]

Korshinskia Lipsky, Acta Horti Petrop. 18: 60 (1900); Pimenov & Kljuykov, Bot. Zhurn. 66, 4: 465–482 (1981), rev.; ejusd. in Edinb. J. Bot. 52(3): 337–342 (1995), rev.
Kamelinia F.O. Khass. & I.I. Maltzev (1992).

Glabrous perennials with swollen, thickened, finger-like roots. Leaves 2–3-pinnate; petioles long, sheathing; leaflets lanceolate, undivided or trilobate; upper leaves entire, lanceolate. Bracts 3–6, lanceolate; rays unequal; bracteoles few, short. Petals greenish-yellow or white; styles short, divergent; stylopodium flat or low-conical. Fruits shortly beaked, slightly compressed laterally; ribs sub-inconspicuous; carpophore bifid; commissure narrow; vittae in immature fruits very large, vallecular solitary, commissural 2, in mature fruits rather reduced, dorsally compressed. Seed face with deep narrow groove. $2n = 14, 18$.

Six species, Asia (Middle, SW). Open stony and steppe slopes, scrub, forest margins and openings. Molecular data suggest an affinity to *Aulacospermum*, *Pleurospermum*, and *Physospermum*.

226. ***Kozlovia*** Lipsky [Spalik]

Kozlovia Lipsky, Trudy Imp. St.-Peterburgsk. Bot. Sada 23: 146 (1904); Schischkin, Flora SSSR 16: 117–118, 151–152, 591 (1950), reg. rev.; Pimenov & Kljuykov, Feddes Repert. 98: 377 (1987), taxon.; Spalik & Downie, Ann. Missouri Bot. Gard. 88: 270–301 (2001), phylog., taxon.

Pubescent perennials with globose tubers. Leaves 2–3-pinnately dissected, triangular; lobes oblong to broadly ovate. Bracts usually absent; rays few to many, subequal or unequal, spreading or slightly ascending; bracteoles few, ciliate or scabrous at margin. Petals white, outer often radiate; styles relatively long, deflexed in fruit; stylopodium conical. Fruits tuberculate and covered by simple or branched bristles, oblong to ovoid, laterally compressed, mericarps terete or pentangular in transverse section, as broad as wide; beak short; ribs angular-filiform; commissure narrow; carpophore bifurcate; vascular bundles small, dorsal and commissural similar in size; vallecular vittae 1, commissural 2, reduced at maturity. Seed round in transverse section, sulcate. $2n = 22$.

One species, *K. paleacea* (Regel & Schmalh.) Lipsky, Middle Asia and Afghanistan, thickets and open woods, open montane habitats. Closely related to *Neoconopodium* and *Krasnovia*, which are sometimes reduced in synonymy under *Kozlovia* (Spalik and Downie 2001).

227. ***Krasnovia*** Popov ex Schischk. [Pimenov]

Krasnovia Popov ex Schischk., Fl. URSS 16: 591 (1950).

Softy puberulent perennials with globose tubers. Basal leaves 2–3-pinnate; leaflets oblong to linear. Bracts 1–2, caducous, or absent; rays few, unequal; bracteoles lanceolate to ovate-lanceolate, later reflexed. Petals white, papillose, marginal slightly enlarged; styles long, reflexed; stylopodium low-conical, undulate at the margin. Fruits glabrous, ovoid, slightly compressed laterally, with attenuate apex; ribs obtuse; carpophore bifid; exocarp with inner 2–3 layers of lignified cells; commissure narrow; mesocarp parenchymatous with several lignified cells in outer layer; vallecular vittae 1, sub-inconspicuous at maturity. Seed face not deeply grooved. $2n = 22$.

One species, *K. longiloba* (Kar. & Kir.) Popov ex Schischk., Asia (Middle, Kazakhstan, W China). Meadows and stony slopes. Molecular data indicate an affinity to *Kozlovia*, into which this genus is sometimes placed in synonymy (Spalik and Downie 2001).

228. ***Krubera*** Hoffm. [Reduron]

Krubera Hoffm., Gen. Pl. Umbell. XXIV: 103 (1814); Meikle, Fl. Cyprus 751–752 (1977), rev.

Glabrous or sparsely scabridulous annuals. Leaves 3–4-pinnately compound, triangular; ultimate divisions lanceolate or linear. Bracts absent or few, deltoid to subulate; bracteoles short, ovate; rays few to several, subequal to unequal, spreading. Petals white; styles short; stylopodium conical. Fruits glabrous, ovoid, strongly compressed dorsally; ribs very prominent, the lateral thick, sometimes narrowly winged, all transversely rugose-scabrid; carpophore free, bifurcated; vittae obscure. Seeds compressed dorsally, the face plane. $2n = 20$.

One species, *K. peregrina* (L.) Hoffm., Europe and Africa (Mediterranean and Macaronesia). Arable fields.

229. ***Kundmannia*** Scop. [Reduron]

Kundmannia Scop., Intr. Hist. Nat. 116 (1777), nom. cons.; Knees, Fl. Iber. 10: 234 (2003), rev.

Glabrous or basally pubescent perennials growing from a stout rootstock. Leaves 1–2-pinnately divided, ovate-triangular; leaflets ovate or rounded, a pair of supplementary leaflets on rachis. Bracts and bracteoles many, linear; rays several to many, ± unequal, spreading. Petals bright yellow; styles short; stylopodium conical. Fruits glabrous, nearly cylindrical, weakly compressed; ribs slender, obtuse, prominent;

carpophore free, bifurcated; vittae small, vallecular 1–3, commissural 4–12. Seeds transversely compressed dorsally, the face plane. $2n = 22$.

Three species, S Europe and N Africa, including many Mediterranean islands, SW Asia (Turkey).

230. ***Kuramosciadium*** Pimenov, Kljuykov & Tojibaev [Pimenov]

Kuramosciadium Pimenov, Kljuykov & Tojibaev, Syst. Bot. 36 (2): 491 (2011).

Glabrous perennials. Leaves petiolate, bipinnatisect, lanceolate; leaflets ovoid or ovoid-lanceolate, dentate. Bracts and bracteoles triangular, acute, white membranous; rays distinctly unequal. Petals dark-brownish; styles reflexed; stylopodium low-conical. Fruits slightly compressed dorsally; dorsal ribs narrowly winged, marginal slightly broader; carpophore bifid; commissure narrow; mesocarp parenchymatous, in ribs with pitted walls; vallecular vittae 1, thin, commissural 2; rib secretory ducts sometimes broad, or small to inconspicuous. Seed face almost plane to slightly emarginate.

One species, *K. corydalifolium* Pimenov, Kljuykov & Tojibaev, Middle Asia (W Thian-Shan). Stony mountain slopes, scree. Related to *Sphaenolobium*.

231. ***Ladyginia*** Lipsky [Pimenov]

Ladyginia Lipsky, Acta Horti Petrop. 23: 150 (1904); Pimenov & Kljuykov, Edinb. J. Bot. 49(2): 213–218 (1992), rev.
Spongiosyndesmus Gilli (1959).

Glabrous or scabrous perennials. Leaves ternate or biternate, coriaceous; leaflets broadly ovate to obovate, at base cordate or truncate. Bracts and bracteoles absent; rays many, glabrous. Petals yellow or yellow-brown, glabrous or dorsally puberulent; styles divergent, short; stylopodium cushion-shaped, undulate at the margin. Fruits glabrous, strongly compressed dorsally; dorsal ribs filiform, marginal narrowly winged; carpophore bifid; commissure broad; mesocarp bistratous, inner layer of prosenchymatous cells with lignified pitted walls; vallecular vittae 1 in immature fruits, later obliterated. Seed face plane. $2n = 22$.

Three species, Asia (Middle, Afghanistan). Low-mountain scrub and open (sometimes salty) slopes. Molecular data suggest an affinity to *Mozaffariania* and *Glaucosciadium* (Degtjareva et al. 2018).

232. ***Lagoecia*** L. [Pimenov]

Lagoecia L., Sp. Pl. 203 (1753); Gen. Pl., ed. 5: 95 (1754); Wolff in Engl., Pflanzenr. IV, 228 (Heft 61): 1–305 (1913), rev.

Glabrous annuals. Leaves pinnate, narrow-oblong; cauline blades on lanceolate sheaths with white membranaceous margins; leaflets ovate, toothed to deeply incised, cartilaginous at the margin. Umbels compound but appearing simple, globose and axillary; bracts pectinate; rays many, short; umbellules uniflorous; bracteoles pectinate with setaceous lobes. Calyx lobes pectinate; petals white, oblong, with a long aristate bristle on either side of inflexed apex; styles solitary, short, erect; stylopodium flat. Fruits of a single mericarp, terete, slightly compressed dorsally, covered with minute white, clavate, brittle, unicellular hairs; ribs inconspicuous; vittae apical, 4 dorsal (short), and 2 ventral (up to the middle part of mericarp). Seed face plane. $2n = 16$.

One species, *L. cuminoides* L., S Europe, N Africa (Libya), SW Asia. Dry hillsides, sometimes weedy. Aromatic, used as condiment. Originally placed in Saniculoideae, the genus was transferred to Apioideae on the basis of molecular data.

233. ***Lalldhwojia*** Farille [Kljuykov]

Lalldhwojia Farille, Rev. Gén. Bot. 91: 27 (1984); Pimenov & Kljuykov, Willdenowia 32: 93–97 (2002), rev.

Appressed-pubescent perennials with vertical rootstocks or slightly thickened taproots; caudex entire or divided. Leaves mostly basal, trifoliolate to pinnate; leaflets ovate to orbicular, dentate or lobed. Bracts and bracteoles lacking. Petals dark red; styles short, reflexed; stylopodium conical. Fruits sparsely pubescent to glabrous, compressed dorsally, obovoid to oblanceolate or ovoid; commissure broad; dorsal ribs short-keeled to obscure, marginal ribs narrowly winged; carpophore bifurcated; mesocarp with lignified inner layer; vallecular vittae slender, unequal in length, 1–3, commissural 1–7. Seed face plane. $2n = 22$.

Two species, Asia (S Tibet, C Himalaya). Among dwarf *Rhododendron*, stony slopes, coniferous forests and scrub.

234. ***Laser*** Borkh. ex G. Gaertn., B. Mey. & Scherb. Fig. 3B [Lee]

Laser Borkh. ex G. Gaertn., B. Mey. & Scherb., Oekon. Fl. Wetterau 1: 244, 384 (1799); Banasiak et al., Taxon 65: 563–585 (2016), phylog. taxon.

Glabrous perennials. Leaves 2–3-ternately dissected, triangular or broadly ovate, ultimate divisions ovate or cordate, crenate-dentate and often broadly lobed, glaucous. Bracts 0–2, ovate to lanceolate, simple, caducous; rays several to many, long, divaricate, slightly contracted in fruit; bracteoles few, simple, lanceolate, caducous. Petals white; styles elongated, rarely divergent; stylopodium conical to subconical. Fruits glabrous, ovoid to oblong, compressed dorsally; primary ribs prominent and thickened, lateral ribs somewhat winged; secondary ribs 4, conspicuous but less prominent than the primary ribs; carpophore free; commissure broad; vittae conspicuous, transversely elliptical, vallecular 1 (under each secondary rib), commissural 2–4(–6). Seeds transversely compressed dorsally, the face plane. $2n = 22$.

One species, *L. trilobum* Borkh. ex G. Gaertn., B. Mey. & Scherb., C and E Europe and SW Asia (Turkey, Caucasus, and N Iran). Closely related to *Laserpitium*. Based on molecular data (Banasiak et al. 2016), the genus was recently enlarged to include five other species formerly placed in *Laserpitium*.

235. ***Laserpitium*** L. Fig. 3E [Lee, Reduron]

Laserpitium L. Sp. Pl. 248 (1753); Gen. Pl., ed. 5: 118 (1754); Banasiak et al., Taxon 65: 563–585 (2016), phylog. taxon.

Pubescent to glabrous perennials. Rootstock often with fibers. Leaves 2–5-pinnately to ternately dissected, broadly triangular or ovate, rarely oblong-lanceolate; ultimate divisions suborbicular, ovate, rhombic to linear or setaceous. Bracts and bracteoles several to many, oblong, filiform to ovate or lanceolate, simple; rays several to many (rarely few), long, glabrous to hispid. Petals white, yellowish, greenish or pinkish; styles short to elongate, often recurved; stylopodium conical, subconical, or depressed. Fruits glabrous, ovoid to oblong or ellipsoid; primary ribs filiform, usually with papillate unbranched hairs (or prominent and narrowly winged), arranged in 2 rows ; secondary ribs 4, winged, unequal, the marginal often wider than dorsal, shiny, often undulate; carpophore free; commissure broad; vittae large or rarely minute, transversely rounded or elliptic, vallecular 1 (under each secondary rib), commissural 2–4. Seeds trapeziform, kidney-shaped to compressed dorsally, the face plane to slightly concave. $2n = 22$.

About 20 species in its traditional circumscription, Europe, NW Africa, and SW Asia. Recent revision based on molecular data (Banasiak et al. 2016) retained only six species, transferring the remaining species to *Thapsia*, *Ekimia*, *Laser*, a restored *Siler*, and the newly described *Silphiodaucus*.

236. ***Lecokia*** DC. [Pimenov]

Lecokia DC., Coll. Mém. 5: 67 (1829).

Glabrous perennials with thick horizontal rhizomes. Leaves 2–4-pinnate or 2–4-ternate; upper leaves opposite; leaflets ovate to ellipsoid. Bracts few, small, caducous, or absent; rays thickened, divergent; bracteoles few, subulate. Petals white; styles erect, hardened; stylopodium conical. Fruits slightly compressed laterally, constricted at the commissure, shortly beaked, covered with hooked bristles; ribs subequal, obtuse, with broad ridges of lignified parenchyma; carpophore bifid; commissure narrow; vittae many in inner mesocarp, cyclical. Seed face deeply grooved, with involute margins. $2n = 22$.

One species, *L. cretica* (Lam.) DC., Asia (SW, the Caucasus), Europe (Crete). Shady deciduous forests, limestone rocks. Molecular data suggest an affinity to *Smyrnium*.

237. ***Ledebouriella*** H. Wolff [Pimenov]

Ledebouriella H. Wolff in Engl., Pflanzenr. IV, 228 (Heft 43): 191 (1910); Fedoronczuk, Systematics, geography and phylogeny of the genera *Trinia*, *Rumia* and *Ledebouriella* [in Russ.] (1983), rev.; Vinogradova, Novosti Sist. Vyssh. Rast. 24: 148–155 (1987), reg. rev.

Glabrous perennials. Leaves 2–3-pinnate, ovate; leaflets linear. Bracts absent, or single, linear, soon caducous; rays unequal, thickened; bracteoles few, minute, lanceolate to ovate. Central flowers in umbellule often sessile; calyx lobes prominent or absent; petals whitish; styles reflexed or straight; stylopodium low-conical or conical. Fruits slightly compressed dorsally, undivided; ribs covered with inflated vesicular outgrowths, obtuse, prominent, marginal slightly enlarged; carpophore reduced; commissure narrow; mesocarp parenchymatous, with large cells with lignified walls; vittae small, vallecular 1, commissural 2; rib oil ducts large. Seed face plane.

Two species, Asia (Kazakhstan). Dry mountain steppes, limestone slopes and cliffs. Closely related to *Trinia* and *Rumia*.

238. ***Lefebvrea*** A. Rich. [Van Wyk & Tilney]

Lefebvrea A. Rich., Ann. Sc. Nat. II, 14: 260, t. 15, fig. 1 (1840); Cannon, Fl. Zambes. 4: 614–616 (1978), reg. rev.; Townsend, Fl. Trop. E. Afr.: 109–115 (1989), reg. rev.; Winter in Winter et al., Taxon 57: 347–364 (2008), rev.; Van Wyk et al., Afr. Apiaceae: 222–225 (2013).
Erythroselinum Chiov. (1911).

Glabrous biennials with a fusiform to subdigitately branched or tuberous rootstock. Leaves biternate to irregularly bipinnate; leaflets ovate to linear, acute, toothed, with a characteristic vein midway between the midvein and margin. Bracts small, linear, inconspicuous or absent; rays several, usually slender; bracteoles small or absent. Petals yellow, greenish, reddish or dark purple; styles long, reflexed; stylopodium conical or hemispherical. Fruits glabrous, broadly elliptic to obovate, strongly compressed dorsally; ribs indistinct on dorsal side, marginal ribs broadly winged; carpophore bifid; vittae large, vallecular 1, commissural 2; rib oil ducts small, 1 in each dorsal rib and/or 1–3 in each wing, variably present. Seeds transversely compressed dorsally, slightly to deeply sulcate beneath the vittae, the face ± plane or bisulcate. $2n = 22$.

Ten species, tropical Africa (South Africa to Tanzania). Open woodlands or grasslands.

239. ***Leiotulus*** Ehrenb. [Ostroumova]

Leiotulus Ehrenb., Linnaea 4: 400 (1829); Pimenov & Ostroumova, Feddes Repert. 105: 141–155 (1994), gen. rev., nomencl.

Pubescent perennials with taproots, tubers or rootstocks. Leaves 1–3-pinnate. Bracts and bracteoles caducous or absent. Petals yellow; styles recurved or erect; stylopodium conical. Fruits pubescent, ellipsoid of ovoid, strongly compressed dorsally; dorsal ribs filiform, marginal ribs broad (proximally thin and distally thickened); fibrous inner mesocarp present; commissure broad; vittae narrow, vallecular 1, commissural 2, straight and parallel. Seed face plane. $2n = 20, 22, 44$.

Ten species, Asia, Europe, N Africa. Many species were described or treated as *Malabaila*, the type species of which (*M. graveolens* M. Bieb.) was recently transferred to *Pastinaca*, leaving *Leiotulus* Ehrenb. as the valid generic name of the remaining species of "*Malabaila*".

240. ***Leutea*** Pimenov [Pimenov]

Leutea Pimenov in Rech. f., Fl. Iran. 162: 445 (1987); Mozaffarian, Bot. Zhurn. 88 (4): 104–123 (2003), taxon.; Panahi et al., Taxon 64: 770–783 (2015), phylog.

Glabrous perennials with branching, often lignified rootstocks. Leaves 2–3-ternate; leaflets linear, cylindrical, rarely compressed. Bracts absent, or few, ovate to elliptic, soon caducous; rays subequal, slightly thickened; bracteoles few, oblong, or absent. Petals yellowish, greenish-grey, or rarely whitish; styles reflexed; stylopodium conical or low-conical. Fruits strongly compressed dorsally; dorsal ribs almost inconspicuous, marginal narrowly winged; carpophore bifid; commissure broad; mesocarp bistratous; inner mesocarp fibrous, lignified, of cells with pitted walls; vallecular vittae 1, commissural 2. Seed face plane. $2n = 22$.

11 species, Asia (Iran, Iraq, Turkmenistan). On arid stony mountain slopes, in scree and rock crevices. Closely related to *Ferula*.

241. ***Levisticum*** Hill [Reduron]

Levisticum Hill, Brit. Herb. 42: 423 (1756), nom. cons.; Thellung, Verh. Naturf. Ges. Basel 35(1): 27–33 (1923), taxon.; Lavrova et al., Bull. Soc. Nat. Moscow Div. Biol. 88 (2): 107–122 (1983), carpol.

Glabrous perennials. Leaves 2–3-pinnately divided, triangular-rhombic; leaflets ovate-rhombic, cuneate, lobed in the upper part. Bracts many; bracteoles several; rays several to many, subequal or unequal, spreading. Petals greenish-yellow; styles equaling or clearly exceeding the stylopodium; stylopodium conical. Fruits glabrous, ovoid-oblong; somewhat compressed dorsally; dorsal and lateral ribs prominent, obtuse or acute, the marginal thickly winged; carpophore free, bifurcated; commissure narrow; vittae medium-sized, vallecular 1, commissural 2(–4). Seeds transversely compressed dorsally, the face slightly concave. $2n = 22$.

One species, *L. officinale* W.D.J. Koch, Asia (Afghanistan and Iran). Wet mountain habitats. Cultivated (esp. C Europe) for its aromatic and medicinal properties.

242. ***Lichtensteinia*** Cham. & Schltdl. [Van Wyk & Tilney]

Lichtensteinia Cham. & Schltdl., Linnaea 1: 394, t. 5 (1826), nom. cons.; Tilney et al., S. Afr. J. Bot. 75: 64–82 (2009), phylog.; Van Wyk et al., Afr. Apiaceae: 226–231 (2013).

Scapose, glabrous (rarely setaceous or minutely pubescent) perennials with resinous rhizomes. Leaves persistent or mostly summer-deciduous, in basal rosettes, simple and entire, pinnatifid or bipinnatisect, toothed along the margins; leaflets dentate or serrate. Bracts large and foliaceous; rays few to many, relatively long; bracteoles similar to bracts or usually much smaller. Calyx lobes thick; petals yellow or white; styles long, prominent, erect; stylopodium long-conical. Fruits glabrous, ovoid or oblong, nearly terete, crowned with persistent calyx and long stylopodium; mericarps semi-terete or compressed dorsally; ribs prominent or inconspicuous; carpophore bifid; vittae absent; rib oil ducts large, solitary in each rib. Seeds usually transversely compressed dorsally, deeply sulcate beneath rib oil ducts, the face convex, plane or grooved. $2n = 22$.

Seven species, South Africa. Fynbos, renosterveld or grassland. Rhizomes have been used in traditional medicine. Phylogenetically isolated but closest to *Marlothiella* and *Choritaenia*.

243. ***Lignocarpa*** J.W. Dawson [Webb]

Lignocarpa J.W. Dawson, New Zealand J. Bot. 5: 400 (1967); Mitchell et al., New Zealand J. Bot. 36: 417–424 (1998), phylog.

Thick and fleshy, glabrous and early glaucous, dioecious or gynodioecious perennials. Leaves mostly rosulate, 3–4-ternately or ternate-pinnately compound; ultimate segments linear. Bracts foliaceous, simple or 2–3-ternately compound; rays many, stout; bracteoles simple or compound. Calyx lobes distinct, unequal; petals white to cream or tinged pink, with 1 central and several lateral oil tubes; female flowers with rudimentary petals and staminodes; styles stout; stylopodium depressed to low-conical. Fruits narrowly ovoid to oblong, slightly to strongly compressed dorsally; endocarp heavily lignified; ribs 5, ± equal or marginal broader, obtuse to subacute; carpophore bifid almost to base; vallecular vittae 1, commissural 2; rib oil ducts solitary in each rib. Seeds transversely subterete to compressed dorsally, the face concave. $n = 11$, $2n = 22$.

Two species, New Zealand. In mobile alpine rock scree. Molecular evidence indicates that *Lignocarpa* may belong to one of two clades of the polyphyletic *Gingidia* (Mitchell et al. 1998).

244. ***Ligusticopsis*** Leute [Pimenov]

Ligusticopsis Leute, Ann. Hist. Mus. Wien 73: 66 (1969).

Glabrous perennials. Leaves 2–4-pinnate; leaflets linear, lanceolate to oblong, glabrous or scabrous at the margin. Bracts often absent, more rarely pinnatisect or entire; rays subequal to evidently unequal; bracteoles pinnatisect, or rarely undivided, white-membranaceous, glabrous or scabrid. Calyx lobes filiform, linear to lanceolate, often longer than stylopodia, or lacking; petals white; styles reflexed; stylopodium low-conical. Fruits compressed dorsally; dorsal ribs subinconspicuous to keeled, marginal winged; carpophore bifid; commissure broad; mesocarp parenchyma cells in ribs with pitted walls; vallecular vittae (1–)2–3, commissural (2–)4–10. Seed face plane or slightly concave. $2n = 22$.

19 species, E Asia (China, Himalaya). In subalpine meadows, forest openings, rocks and stony

hillsides. Closely related to *Oreocome*, *Cortia*, and *Cortiella*.

245. ***Ligusticum*** L. [Pimenov]

Ligusticum L., Sp. Pl. 250 (1753); Gen. Pl., ed. 5, 119 (1754); Leute, Ann. Naturhist. Mus. Wien 74: 457–519 (1970), rev.; Pimenov & Lavrova, Bot. Zhurn. 74(1): 96–103 (1989), rev.

Glabrous perennials. Leaves 2–3-ternate; petioles hollow, sheathing; leaflets broadly ovoid or rhombic. Bracts and bracteoles entire, linear to lanceolate. Petals white or reddish; styles short, divergent; stylopodium low-conical. Fruits slightly compressed dorsally, with narrowly winged ribs, marginal slightly larger than dorsal; carpophore bifid; commissure narrow; mesocarp parenchymatous, partly crushed, with cells having pitted walls; vallecular vittae 2–5, commissural 4–10. Seed occupies only the upper part of the mericarp. Seed face plane. $2n = 22$.

One species, *L. scoticum* L. (see Pimenov and Lavrova 1989), Europe, Asia, N America. Sea shores. In broader circumscriptions (e.g., Leute 1970) the genus comprises about 40 species, but molecular data suggest *Ligusticum* s. lat. is polyphyletic. *Ligusticum scoticum* is used in folk medicine and is an edible plant.

246. ***Lilaeopsis*** Greene [Plunkett]

Lilaeopsis Greene, Pittonia 2: 192 (1891); Affolter, Syst. Bot. Monogr. 6: 1–140 (1985), rev.; Bone et al., Syst. Bot. 36: 789–805 (2011), phylog.

Glabrous perennials with creeping rhizomes (sometimes also with vertical branches). Leaves produced singly or in clusters at the rhizome nodes or apex; reduced to linear, spathulate, or subulate, terete or compressed, hollow or solid, septate rachis-leaves. Umbels simple (and rarely sessile), arising from rhizome nodes singly or in groups of 2–4; bracts ovate to lanceolate. Petals translucent, white, greenish-white or maroon, broad or narrowed at the base, apex shortly acuminate (not inflexed); styles short; stylopodium depressed to conical. Fruits glabrous, ovoid, obovoid, ellipsoid, or globose, slightly compressed laterally (to subterete); dorsal and lateral ribs filiform, marginal ribs filiform to slightly winged, often with corky thickenings adjacent to the commissural margin; pericarp also with a distinctive peripheral layer of cells that expand upon hydration; carpophore obsolete; vallecular vittae usually 1(–2), commissural 2(–several). Seeds transversely terete to subterete, the face plane to convex. $n = 11, 22, 2n = 22, 44$.

15 species, S America (Andes, Patagonia), N America (interior Mexico, E and W coastal United States), Africa (Mascarene Is., possibly Madagascar), Australia, New Zealand, Europe (Portugal and Spain); in soggy soils of swales, bogs, marshes, at pond or stream margins, to partially or entirely submerged, from coasts to high montane habitats.

247. ***Limnosciadium*** Mathias & Constance [Plunkett]

Limnosciadium Mathias & Constance, Amer. J. Bot. 28: 162 (1941); Mathias & Constance, N. Amer. Fl. 28B(2): 164–165 (1945), reg. rev.

Glabrous annuals with fascicled fibrous roots. Basal leaves entire and septate or pinnate; cauline leaves either entire or pinnate with a few filiform to linear-lanceolate leaflets. Bracts and bracteoles several, linear to linear-lanceolate, or lacking; rays few, unequal, spreading-ascending. Calyx lobes prominent; petals white, apices not inflexed; styles very short with divergent tips; stylopodium conical. Fruits glabrous, oblong-elliptical to orbicular, rounded at the apex and base, slightly compressed dorsally; dorsal and lateral ribs filiform, marginal ribs winged, broad and corky; carpophore bifid at the apex; vallecular vittae 1, commissural 2. Seeds transversely compressed dorsally, the face plane. $n = 6$.

Two species, N America (S Missouri and Kansas to Louisiana and Texas). In low, moist areas. A segregate of *Cynosciadium*.

248. ***Lipskya*** (Koso-Pol.) Nevski [Pimenov]

Lipskya (Koso-Pol.) Nevski, Acta Inst. Bot. Acad. Sci. URSS I, 4: 272 (1937).
Anidrum Neck. ex Raf. sect. *Lipskya* Koso-Pol. (1920).

Glabrous or sparsely scabrous perennials. Leaves bipinnate; leaflets elliptic. Bracts unequal, lanceolate, sometimes pinnatisect; rays very unequal; bracteoles linear-lanceolate. Central flowers in umbellules hermaphroditic, sessile, other flowers staminate, pedicellate; calyx lobes lanceolate, subulate, persistent, later hardening; petals white, outer slightly radiate; styles reflexed; stylopodium flat. Mericarps barrel-shaped; ribs with hard teeth; carpophore adnate to mericarps; outer mesocarp layer of large lignified cells, inner sclerified; vittae many, thin, cyclical, situated outside of sclerified layer. Seed face plane.

One species, *L. insignis* (Lipsky) Nevski, Middle Asia (W Pamiro-Alai). Gypsum clay and sandstone slopes. Molecular data indicate a close affinity to *Schrenkia*, *Sclerotiaria*, and *Schtschurowskia* (Terentieva et al. 2015).

249. ***Lisaea*** Boiss. [Lee]

Lisaea Boiss., Ann. Sci. Nat. Bot. III, 2: 54 (1844).

Pubescent, glochidiate annuals. Leaves petiolate, simple to bipinnate, pubescent on both sides, broadly ovate, ultimate divisions lanceolate to oblong. Bracts several, linear to oblong or lanceolate, with membranous margins, simple; rays few to many, unequal, scabrous and glochidiate; bracteoles many, simple, trifid, pinnatisect, lanceolate to obovate. Petals pinkish-white to white, the outer strongly radiate; styles conspicuously elongated; stylopodium conical. Fruits spinescent or tuberculate or papillate, globose or subglobose, slightly compressed laterally; primary ribs prominent, with broadly based, erect spines arranged in a single row; secondary ribs 4, $\pm$ inconspicuous with tubercles, or with much smaller spines; carpophore free or entire; vittae obsolete. The seed face shallowly to deeply sulcate, often strongly incurved. $2n = 12$.

Three species, Asia (SW, Caucasus). Closely related to *Turgenia*.

250. ***Lithosciadium*** Turcz. [Pimenov]

Lithosciadium Turcz., Bull. Soc. Imp. Naturalistes Moscou 17: 730 (1844).

Glabrous perennials. Leaves 2–3-ternate; leaflets suborbicular to ovoid, incised-dentate, obtuse. Bracts entire, glabrous, acute; rays several, subequal; bracteoles linear, white-membranaceous at the margin. Petals greenish or reddish-brown; styles reflexed; stylopodium flat. Fruits slightly compressed dorsally; ribs equal, keeled; carpophore bifid; commissure narrow; mesocarp cells with pitted walls; vallecular vittae 1, commissural 2. Seed face plane. $2n = 22$.

Two species, Asia (S Siberia, Mongolia, NW China). Stony slopes, screes, sometimes in subalpine meadows near timberline. Segregate of *Cnidium.*

251. ***Lomatium*** Raf. [Plunkett]

Lomatium Raf., J. Phys. Chim. Hist. Nat. Arts 89: 101 (1819); Mathias & Constance, N. Amer. Fl. 28B(2): 222–258 (1945), rev.; Schlessman, Syst. Bot. Monogr. 4: 1–55 (1984), part. reg.; Sun & Downie, Pl. Divers. Evol. 128: 151–172 (2010), phylog.

Acaulescent to caulescent, glabrous to pubescent perennials with slender taproots (sometimes tuber-like) or a branched, woody caudex, and occasionally a short pseudoscape. Leaves mostly basal, ternately, pinnately, quinately, or ternate-pinnately divided or decompound; ultimate divisions obovate to filiform, entire or toothed, lobed or divided. Bracts lacking or inconspicuous; rays few to many, spreading to ascending or strict, rarely reflexed; bracteoles filiform to obovate, foliaceous to subscarious, distinct or connate (rarely lacking). Petals yellow, yellowish-pink, greenish-white, white, or purplish; styles slender, often curved or coiled; stylopodium lacking. Fruits glabrous, roughened, or pubescent, linear to orbicular or obovate, compressed dorsally; dorsal and lateral ribs filiform or obsolete (or occasionally with rudimentary wings at the base), marginal ribs with membranaceous or corky wings (broader or narrower than the fruit body), rarely obsolete; carpophore bifurcated; vittae small or large, vallecular 1 to many, commissural 2 to several, sometimes obscure. Seeds transversely compressed dorsally, the face plane to slightly concave. $n = 11, 22, 66, 2n = 22$.

86 species, N America (W United States and adjacent Canada). Molecular data suggest the genus is polyphyletic with regard to *Cymopterus* and other perennial genera from western North America (Sun and Downie 2010).

252. ***Lomatocarpa*** Pimenov [Pimenov]

Lomatocarpa Pimenov, Bjull. Moskovsk. Obšč. Isp. Prip. Otd. Biol. 87(1): 115 (1982).
Alposelinum Pimenov (1982).

Usually acaulescent, glabrous perennials. Leaves mainly rosulate, pinnate or bipinnate; leaflets elliptic to linear. Bracts and bracteoles linear to lanceolate; rays several, subequal. Petals white; styles reflexed; stylopodium low-conical to flat. Fruits slightly compressed dorsally; dorsal ribs keeled to filiform, marginal narrowly winged; carpophore bifid; commissure broad; mesocarp in ribs of cells with pitted walls; vallecular vittae 1–4, sometimes inconspicuous, commissural 2–10. Seed face plane. $2n = 22$.

Three species, Asia (Kazakhstan, Afghanistan, Pakistan, W China). Alpine plants, growing in subalpine meadows, screes, stony slopes.

253. ***Macroselinum*** Schur [Pimenov]

Macroselinum Schur, Verh. Siebenbürg. Vereins Naturwiss. 4: 30 (1853).

Glabrous perennials. Leaves pinnate or bipinnate; leaflets ovoid, irregularly toothed or crenate, coriaceous. Bracts many, lanceolate or linear-lanceolate, often reflexed; rays many, unequal, scabridulous; bracteoles many, linear. Petals white; styles long, reflexed; stylopodium low-conical. Fruits compressed dorsally; dorsal ribs filiform or inconspicuous, marginal broadly winged; carpophore bifid; commissure broad; vallecular vittae 1, commissural 2(–4). Seed face plane. $2n = 22$.

One species, *M. latifolium* (M. Bieb.) Schur, Europe (S, SE, E), Asia (N Caucasus). In steppe depressions, on saline soils. Segregate of *Peucedanum.*

254. ***Magadania*** Pimenov & Lavrova [Pimenov]

Magadania Pimenov & Lavrova, Bot. Zhurn. 70(4): 531 (1985).
Ochotia A.P. Khokhr. (1985).

Glabrous or puberulent perennials. Leaves mainly basal, pinnate to 2-3-pinnatisect; leaflets ovate to broadly lanceolate. Bracts few, scabrous, or absent; rays scabrous; bracteoles lanceolate, ciliate, often reflexed. Petals white or pinkish; styles reflexed; stylopodium conical. Fruits glabrous or papillose, slightly compressed dorsally; ribs equal, winged, covered with sparse papillae; carpophore bifid; commissure narrow; mericarp cells with pitted walls; vallecular vittae 1, commissural 2. Seed face plane. $2n = 14$.

Two species, NE Asia. Stony tundras, *Pinus pumila* scrub, along streams. Morphological and molecular data suggest a relationship to *Cnidium* and *Pachypleurum.*

255. ***Magydaris*** W.D.J. Koch ex DC. [Reduron]

Magydaris W.D.J. Koch ex DC., Coll. Mém. 5: 68 (1829); Knees, Fl. Iber. 10: 155–158 (2003), reg. rev.

Pubescent perennials. Leaves simple, lobed or pinnately divided, ovate; lobes or leaflets wide, obtuse. Bracts and bracteoles several, lanceolate to linear-lanceolate; rays many, subequal or unequal, spreading. Petals white, villous beneath; styles very long; stylopodium conical. Fruits hairy, ovoid, slightly compressed dorsally, spongy; ribs wide, rounded; carpophore bifurcated; vittae small and many, cyclic. Seeds transversely reniform, the face deeply sulcate. $2n = 22$.

Two species, Europe and N Africa (W and C Mediterranean). Dry rocky places.

256. ***Mandenovia*** Alava [Ostroumova]

Mandenovia Alava, Notes Roy. Bot. Gard. Edinb. 32(2): 191 (1975).

Pubescent perennials or biennials. Leaves biternate or bipinnate; ultimate divisions round to elliptic; petioles subterranean. Bracts and bracteoles absent. Calyx lobes large; petals white; styles recurved; stylopodium conical with undulating margin. Fruits pubescent, elliptic, strongly compressed dorsally; dorsal ribs filiform, marginal ribs broad and thick; fibrous inner mesocarp present; commissure broad; vittae narrow, vallecular 1, commissural 2, straight and parallel. The seed face plane. $2n = 22$.

One species, *M. komarovii* (Manden.) Alava, Asia (the Caucasus). A segregate of *Tordylium*, but molecular data place it closer to *Heracleum* (Logacheva et al. 2008).

257. ***Marlothiella*** H. Wolff [Van Wyk & Tilney]

Marlothiella H. Wolff, Bot. Jahrb. 48: 263 (1912); Liu et al., S. Afr. J. Bot. 73: 208–213 (2006), rev.; Van Wyk et al., Afr. Apiaceae: 234–235 (2013).

Glabrous shrublet. Leaves sparsely pinnate to bipinnate, subsucculent; leaflets subterete, acute. Bracts and bracteoles ovate and navicular; rays few, short. Flowers sessile; calyx lobes broadly ovate-triangular; petals yellowish brown; styles short; stylopodium conical. Fruits minutely pubescent with unicellular, stellate trichomes, broadly ovoid, not compressed; mericarps somewhat heteromorphic; ribs blunt; carpophore present or absent; vittae absent; rib oil ducts very large, usually solitary in each rib, sometimes 2 in a marginal rib. Seeds transversely orbicular, not sulcate, the face somewhat plane.

One species, *M. gummifera* H. Wolff, Africa (S Namibia). Coastal deserts. Related to *Lichtensteinia* and *Choritaenia.*

258. *Mastigosciadium* Rech.f. & Kuber [Plunkett]

Mastigosciadium Rech.f. & Kuber, Anz. Oesterr. Akad. Wiss., Math.-Naturwiss. Kl. 101: 364 (1964); Leute in Rechinger, Fl. Iran. 162: 214–215 (1987), reg. rev.

Glabrous perennials, stems glaucescent to subpruinose. Basal leaves 5–6-ternatisect, flabelliform, subcoriaceous; ultimate segments subulate-filiform. Bracts and bracteoles few to several and linear; rays several, unequal. Calyx lobes well developed, triangular; petals yellow; styles deflected; stylopodium low-conical (campanulate-verruciform). Mature fruits not known; immature fruits glabrous; the ribs prominent, transversely triangular; carpophore unknown; vallecular vittae 3 (the central larger than the two laterals), commissural 6, rib oil ducts solitary in each rib. Seeds transversely hexagonal, somewhat attenuate at the commissure.

One species, *M. hysteranthum* Rech. f. & Kuber, Asia (SW Afghanistan, E Iran).

259. *Mathiasella* Constance & C.L. Hitchc. [Plunkett]

Mathiasella Constance & C.L. Hitchc., Amer. J. Bot. 41: 56 (1954).

Glabrous (to minutely scaberulous) perennials with a woody caudex. Leaves deltoid, ternate-pinnate; leaflets large, serrate and often lobed. Bracts and bracteoles few to several, large and conspicuous, foliose, ovate or obovate; rays few to several. Petals of carpellate flowers lacking, of staminate flowers greenish with purple mottling; styles slender, spreading; stylopodium lacking. Fruits glabrous, oblong-ovate with blunt base and apex, strongly compressed dorsally; dorsal and lateral ribs filiform, marginal ribs broadly thin-winged; vallecular vittae 1, commissural ca. 4. Seeds transversely strongly compressed dorsally, slightly sulcate under the vittae, the face plane.

One species, *M. bupleuroides* Constance & C. L. Hitchc., N America (Mexico, Tamaulipas). Molecular data suggest a placement with *Arracacia*, *Rhodosciadium*, and other Mesoamerican genera.

260. *Mediasia* Pimenov [Pimenov]

Mediasia Pimenov, Novosti Sist. Vyssh. Rast. 11: 254 (1974).

Puberulent perennials. Leaves 2–3-pinnate; leaflets broadly ovate, large, puberulent at the veins. Bracts and bracteoles linear-subulate to narrowly lanceolate, reflexed; rays many, puberulent. Calyx lobes prominent, triangular; petals white or greenish-yellow, dorsally puberulent; styles short, reflexed; stylopodium almost flat, undulate at the margin. Fruits densely puberulent, ovoid to pyriform, ribs short-keeled; carpophore bifid; commissure broad; mesocarp cells with lignified pitted walls; vittae large, vallecular 1, commissural 2. Seed face plane. $2n = 22$.

One species, *M. macrophylla* (Regel & Schmalh.) Pimenov, Middle Asia, N Afghanistan. Scrub, sparse forests, near rocks, on screes. Used by local population to preserve milk products at pastures.

261. *Meeboldia* H. Wolff [Pu]

Meeboldia H. Wolff, Feddes Repert. 19: 313 (1924); Mukh. & Constance, Umbelliferae India 1: 116–117 (1993), reg. rev.; Liou in Wu (ed.), Fl. Yunnan. 7: 405 (1997), reg. rev.

Mostly glabrous (to scabrescent) perennials with digitate, tuber-like roots. Basal leaves pinnately or ternately decompound, ovate; leaflets petiolulate and pinnatifid with linear-subulate ultimate segments. Bracts one to several, linear-lanceolate; rays subequal, spreading-ascending to divaricate; bracteoles several, linear to linear-lanceolate. Petals white; styles recurved; stylopodium conical. Fruits glabrous, oblong-ovoid, compressed laterally, rounded at base; mericarps transversely pentagonal, slightly constricted at the commissure; ribs prominent, filiform; carpophore bifid to half or more of its length; vallecular vittae 2–3, commissural 4. Seed transversely subterete to slightly compressed dorsally, the face concave to sulcate. $2n = 22$.

One species, *M. selinoides* H. Wolff, Asia (Himalayas, from Bhutan and Nepal to India). *Sinodielsia* is sometimes treated in synonymy under *Meeboldia*.

262. ***Melanosciadium*** H. Boissieu [Pu]

Melanosciadium H. Boissieu, Bull. Herb. Boiss. II, 2: 803 (1902); Pimenov & Kljuykov, Feddes Repert. 117: 466–475 (2006), rev.

Pubescent perennials. Leaves biternate, ovate-triangular; ultimate segments ovate, margin serrate, subglabrous except for pilose veins. Bracts caducous; rays several, short or long, scarcely pubescent; bracteoles persistent, linear, pilose. Petals deeply purple, white, or yellowish; styles purple, recurved; stylopodium purple, depressed-conical, the margin undulate. Fruits glabrous, subglobose to ovate, compressed laterally, ribs keeled or filiform; carpophore bifurcated; vallecular vittae 3–5, commissural 6–8. Seed face nearly plane to sulcate. $2n = 22$.

Three species, Asia (SW and C China), in grasslands at forest edges.

263. ***Melanoselinum*** Hoffm. [Lee]

Melanoselinum Hoffm., Gen. Pl. Umbell., ed. 1: 156. (1814).

Pubescent perennials or biennials with woody rootstocks and stems. Leaves petiolate, 2–3-pinnately dissected, broadly ovate, the ultimate divisions ovate to lanceolate, sharply dentate. Bracts several, lanceolate to ovate, trifid to pinnatisect (like the leaves), deflexed in fruit; rays many, long, ± pubescent; bracteoles many, simple, lanceolate. Petals whitish to purplish; styles short to elongated, rarely divergent; stylopodium subconical. Fruits pubescent, blackish, oblong, mericarps compressed dorsally; primary ribs inconspicuous; secondary ribs 4, the lateral ribs winged and strongly denticulate, much larger than the dorsal ribs; carpophore free; vittae transversely elliptical, vallecular 1 (under each secondary rib), 2 or indistinct on the commissure. The seed face plane.

One species, *M. decipiens* (Schrad. & J.C. Wendl.) Hoffm., Macaronesia (endemic to Madeira Is.). Closely related to *Monizia*, and molecular data suggest that both genera are nested within *Daucus*, under which they were recently synonymized by Banasiak et al. (2016).

264. ***Meum*** Mill. [Reduron]

Meum Mill., Gard. Dict. Abr., ed. 4 (1754); Leute, Ann. Naturhist. Mus. Wien 73: 91–97 (1969), rev.; Lavrova et al., Bjull. Moskovsk. Obšč. Isp. Prip. Otd. Biol. 88(2): 107–122 (1983), carpol.

Glabrous perennials. Leaves 3–4-pinnately divided, triangular to oblong-elliptic; ultimate divisions very many, crowded, short and filiform. Bracts absent or few, setaceous; bracteoles present; rays several to many, unequal, spreading. Petals white, sometimes purplish, entire, lanceolate; styles equaling or exceeding the stylopodium; stylopodium short conical. Fruits glabrous, ovoid-oblong; slightly compressed dorsally; ribs very prominent; carpophore free, bifurcated; commissure narrow; vittae small, vallecular 2–6, commissural 4–8. Seeds transversely lunate, the face concave. $2n = 22$.

1–3 species, Europe (W and C, extending locally to Calabria and C Bulgaria), NE Africa (Morocco). Mountain pastures. Medicinal and aromatic plant.

265. ***Microsciadium*** Boiss. [Pimenov]

Microsciadium Boiss., Ann. Sci. Nat. Bot. III, 1: 141 (1844).

Glabrous annuals. Leaves 2–3-pinnate; segments deeply divided into filiform to lanceolate lobes. Bracts absent; rays few, unequal, capillary in flower, thickening later; bracteoles linear-lanceolate, minute, acute. Pedicels thickening in fruit; calyx lobes inconspicuous but ten sepal-like segments without vasculature found in their place (interpreted as protrusions of the ovary ribs; Hedge and Lamond 1972); petals white; styles short, straight or divergent; stylopodium substipitate, wavy at the margin. Mericarp ribs filiform; carpophore undivided or bifid; commissure narrow; mesocarp parenchymatous, with a layer of large cells with lignified walls over vittae; vallecular vittae 1, commissural 2. Seed face plane. $2n = 12$.

One species, *M. minutum* Briq., E Mediterranean (Aegean Isls., W Turkey). Stony slopes among shrubs. Related to *Carum* and *Bunium*.

266. ***Modesciadium*** P. Vargas & Jim.-Mejías [Reduron]

Modesciadium P. Vargas & Jim.-Mejías, Phytotaxa 212(1): 75 (2015).

Glabrous annual. Leaves 2–3-ternately divided, triangular-ovate, ultimate divisions filiform; sheaths white-margined. Bracts few; rays several, ± unequal; bracteoles several. Flowers several to many per umbellule, petals white, broadly obcordate; styles reflexed, longer than stylopodium. Fruits papillate, broadly rounded-ovoid, didymous; ribs filiform; carpophore free, bifurcated at apex; vittae large, vallecular 1, commissural 2. Seed face sinuous, almost plane.

One species, *M. involucratum* (Maire) P. Vargas & Jim.-Mejias, endemic to Morocco. Similar to *Stoibrax*.

267. ***Mogoltavia*** Korovin [Kljuykov]

Mogoltavia Korovin, Bot. Mat. Herb. Inst. Bot. Zool. Acad. Sci. Uzbekistan 8: 11 (1947).

Glabrous, ephemeroid perennials with cylindrical, branched tubers. Leaves rosulate, tripinnate with numerous sessile segments; terminal lobes filiform; upper leaves entire. Bracts and bracteoles lanceolate, entire. Petals white or pink; styles reflexed; stylopodium short-conical. Fruits glabrous, ovoid, strongly compressed dorsally; commissure broad; dorsal ribs filiform, marginal ribs winged; carpophore bifurcated; mesocarp cells in marginal ribs with lignified pitted walls; vittae in two layers of the mesocarp: outer vittae solitary and broad in the valleculae, inner vittae many, slender, and anastomosing. Seed face plane. $2n = 20$.

Two species, Middle Asia. Semi-desert mountain slopes. Closely related to *Oedibasis*.

268. ***Molopospermum*** W.D.J. Koch [Reduron]

Molopospermum W.D.J. Koch, Nova Acta Phys.-Med. Acad. Caes. Leop.-Carol. Nat. Cur. 12: 108 (1824); Ullmann, Flora 179: 253–270 (1987), rev.; Krähenbühl & Küpfer, Bauhinia 10: 75–84 (1992), caryol.; Villar, Fl. Iber. 10: 141–143 (2003), reg. rev.

Glabrous perennials. Leaves 3–4-pinnately divided, large, triangular; ultimate divisions long, lanceolate, deeply incised-dentate. Bracts and bracteoles many; rays many, unequal, spreading. Petals yellowish or white, lanceolate; styles shorter or longer than the conical stylopodium. Fruits glabrous, ovoid or ellipsoid, ± oblong; slightly compressed laterally; ribs winged, unequal (the marginal shorter); carpophore free, bifurcated; vittae medium-sized, vallecular 1, commissural absent. Seeds transversely pentagonal, the face concave. $2n = 44, 46$.

One species, *M. peloponnesiacum* (L.) W.D.J. Koch, Europe (southern W and C Europe). Mountains. Medicinal; young shoots edible.

269. ***Monizia*** Lowe [Ostroumova, Lee]

Monizia Lowe in Hookers J. Bot. 8: 57 (1856).

Pubescent woody perennials with woody rootstocks. Leaves fern-like, 3–4-pinnately dissected, shiny. Bracts linear to lanceolate; bracteoles many, lanceolate. Calyx lobes small; petals dull white; styles short, divergent; stylopodium subconical. Fruits pubescent, oblong to ellipsoid; mericarps compressed dorsally; primary dorsal ribs thickened and corky, marginal filiform; secondary ribs 4, median thickened, lateral ribs winged, thickened, tumid and obtuse; vallecular vittae 1, commissural 2. Seed face plane. $2n = 22$.

One species, *M. edulis* Love, Africa (Madeira Is.). High sea cliffs. Roots edible. Rare. Distinguished from *Melanoselinum* by the presence of corky primary and secondary ribs. Based on molecular data, both genera are nested within *Daucus*, under which they were recently synonymized by Banasiak et al. (2016).

270. ***Mozaffariania*** Pimenov & Maassoumi [Pimenov]

Mozaffariania Pimenov & Maassoumi, Bot. Zhurn. 87 (11): 96 (2002).

Glabrous perennials. Basal leaves bipinnate, subcoriaceous; leaflets petiolulate, broadly ovate, orbicular or ovate-cordate, coarsely crenate-dentate or shallowly lobed. Bracts many, linear to subulate; rays many, subequal; bracteoles like bracts, smaller. Petals greenish-yellowish or yellow; styles short, straight, soon caducous; stylopodium conical. Fruits dorsally compressed, dorsal ribs filiform, marginal narrow-winged, thickened, spongy; carpophore bifid; mesocarp bistratous, inner layer of prosenchymatous cells with lignified walls; vallecular vittae 1, commissural 2. Seed face plane.

One species, *M. insignis* Pimenov & Maassoumi, SW Asia (Iran), arid slopes. Molecular data suggest an affinity to the *Ferula-Leutea* clade (Valiejo-Roman et al. 2006), *Glaucosciadium* (Kurzyna-Młynik et al. 2008), and *Ladygynia* (Degtjareva et al. 2018).

271. ***Musineon*** Raf. [Plunkett]

Musineon Raf., J. Phys. Chim. Hist. Nat. Arts 91: 71 (1820); Mathias & Constance, N. Amer. Fl. 28B(1): 124–126 (1944), rev.

Short-caulescent or acaulescent, glabrous to scabrous perennials. Leaves 1–3-pinnately or ternately divided; the ultimate divisions linear, distinct to confluent. Bracts usually lacking; rays few, spreading; bracteoles dimidiate or subdimidiate, usually distinct, linear, acute, occasionally scarious-margined. Calyx lobes conspicuous; petals white or yellow; styles slender, spreading; stylopodium lacking. Fruits glabrous to scabrous, ovoid to linear-oblong, compressed laterally; commissure somewhat constricted; ribs acute, prominent; carpophore entire, or bifid at the apex or bifurcated; vallecular vittae 1–4, commissural 2–6, rib oil ducts sometimes 1 in each rib. Seeds transversely subterete, the face plane or concave. $2n = 22$.

Four species, N America (W, especially Rocky Mountains); rocky fields and plains.

272. ***Mutellina*** Wolf [Reduron]

Mutellina Wolf, Gen. Pl. 31 (1776); Leute, Ann. Naturhist. Mus. Wien 74: 457–519 (1970), rev. *Ligusticum* s. lat.; Pimenov & Lavrova, Bot. Zhurn. 74(1): 96–103 (1989), taxon.; Lavrova, Bjull. Moskovsk. Obšč. Isp. Prip. Otd. Biol. 98(6): 93–98 (1993), taxon.

Glabrous perennials. Leaves 2–4-pinnately divided, oblong or triangular; ultimate divisions lanceolate or linear. Bracts absent to several; bracteoles several; rays several to many, spreading. Petals white, pink or purple; styles exceeding the stylopodium, rarely shorter; stylopodium conical. Fruits glabrous, ovoid or ovoid-oblong; slightly compressed dorsally; ribs equal, prominent, nearly winged; carpophore free, bifurcated; commissure narrow; vittae small, many and cyclic. Seeds transversely pentagonal, the face plane. $2n = 14, 22, 44$.

Three species, C and E Europe (France to Bulgaria), Asia (Caucasus). Mountain grasslands, rocky places. Segregate of *Ligusticum* s. lat.

273. ***Myrrhidendron*** J.M. Coult. & Rose [Webb]

Myrrhidendron J.M. Coult. & Rose, Bot. Gaz. 19: 465 (1894); Coulter & Rose, J. Wash. Acad. 17: 213–215 (1927), rev.; Mathias & Constance, N. Amer. Fl. 28B(2): 204–205 (1945), reg. rev.; Mathias & Constance, Fl. Panama VII, 4: 245–246 (1959), reg. rev.

Glabrous or pubescent, andromonoecious perennials or distinctly woody shrubs to small trees. Leaves ternate-pinnately to pinnately decompound; leaflets large, often incised or lobed, acute to acuminate, serrate. Bracts few, membranous to foliaceous or lacking; rays many, spreading-ascending; bracteoles few to many, usually cleft or divided. Petals white to cream; styles slender; stylopodium low-conical. Fruits glabrous, linear to oblong, subterete or compressed dorsally; dorsal and lateral ribs narrowly winged; marginal ribs similar or more broadly winged; carpophore bifurcated; vallecular vittae 1; commissural usually 2(–3). Seeds transversely compressed dorsally, sulcate beneath the vittae, the face plane or concave. $n = 22$.

Four species, C and S America (Costa Rica to Ecuador).

274. ***Myrrhis*** Mill. [Spalik]

Myrrhis Mill., Gard. Dict. Abr. ed. 4 (1754).

Pubescent perennials. Leaves 2–4-pinnately dissected, triangular; lobes lanceolate to oblong-ovate. Bracts usually absent; rays many, subequal, slightly ascending; bracteoles few, pubescent, ciliate. Petals white, outer somewhat radiate; styles relatively long; stylopodium conical in flower, high-conical in fruit. Fruits glabrous or delicately bristled with an aculeate cuticle, narrowly ellipsoid with a short beak, laterally compressed, mericarps transversely stellate, as broad as wide; ribs angular; commissure narrow; carpophore bifurcate; vittae many in young fruits, obsolete at maturity. Seed transversely terete, the face sulcate. $2n = 22$.

One species, *M. odorata* (L.) Scop., C Europe. Open montane habitats, elsewhere naturalized and cultivated as a condiment. Closely related to *Anthriscus*, *Geocaryum*, *Kozlovia*, and *Osmorhiza* (Spalik et al. 2001).

275. ***Myrrhoides*** Heist. ex Fabr. [Pimenov]

Myrrhoides Heist. ex Fabr., Enum. 37 (1759); ibid. ed. 2: 66 (1763); Piwczyński et al., Bot. J. Linn. Soc. 178: 298–313 (2015), phylog.
Physocaulis (DC.) Tausch (1834).

Pubescent annuals. Stems swollen under nodes, covered with rigid, tuberculate, retrorse bristles. Leaves puberulent, 2–3-ternate or 2–3-pinnate; leaflets oblong to linear-lanceolate, lobate. Bracts solitary or absent; rays few, thick; bracteoles 5, linear to lanceolate, pubescent, reflexed. Petals white, covered dorsally with sparse bristles; styles ± inconspicuous; stylopodium conical. Fruits beaked, covered with bristles and hairs; ribs equal, obtuse; carpophore bifid; commissure narrow; vallecular vittae 1, commissural 2. Seed face with deep groove. $2n = 14, 22$.

One species, *M. nodosa* (L.) Cannon, Mediterranean Europe and NW Africa, SW and Middle Asia, Caucasus. In forests, sometimes weedy. Closely related to and sometimes included in *Chaerophyllum* (Spalik et al. 2001).

276. ***Nanobubon*** Magee [Van Wyk & Tilney]

Nanobubon Magee in Winter et al., Taxon 57: 356 (2008); Magee et al., S. Afr. J. Bot. 74: 713–719 (2008), rev.; Van Wyk et al., Afr. Apiaceae: 236–238 (2013).

Glabrous, evergreen shrublets. Leaves 3–5-pinnate, shape variable, sclerophyllous; leaflets erect, rigid, linear, subterete. Bracts and bracteoles many, lanceolate to narrowly ovate; rays many. Calyx lobes relatively large; petals yellow or cream-colored; styles short, becoming elongated and often reflexed; stylopodium broadly conical. Fruits glabrous, elliptic to broadly obovate, strongly compressed dorsally; dorsal ribs prominent, marginal ribs very broadly winged, thick; carpophore bifid; vittae relatively small, vallecular 1, commissural 2. Seeds transversely compressed dorsally, somewhat sulcate under the vittae, the face plane.

Three species, South Africa (W Cape province). Fynbos or renosterveld vegetation. Related to *Notobubon*.

277. ***Naufraga*** Constance & Cannon [Reduron]

Naufraga Constance & Cannon, Feddes Repert. 74(1–2): 1 (1967); Duvigneaud, Bull. Soc. Roy. Bot. Belgique 103: 31–38 (1970), ecol.; Froebe, Trop. subtrop. Pflanzenwelt 29: 1–181 (1979), morphol.; Friedlender, J. Bot. Soc. Bot. France 13: 7–24 (2001), rev.

Dwarf, glabrous perennials with a short rhizome. Leaves ternately or pinnately compound, ovate or ovate-oblong; the apex of the sheath stipulate or auriculate; leaflets ovate. Umbels simple; bracts lacking. Petals whitish, with a plane or slightly inflexed apex; styles equaling the flat stylopodium. Fruits glabrous, truncate-ovoid, didymous, laterally compressed; ribs filiform, inconspicuous; carpophore absent; vittae small, vallecular 1, commissural absent. Seeds sub-rectangular, the face plane to convex. $2n = 20$.

One species, *N. balearica* Constance & Cannon, Europe (Balearic Islands, Corsica). Shadowy and wet coastal rock crevices, near the sea. Closely related to *Apium* (Jiménez-Mejías and Vargas 2015).

278. ***Neoconopodium*** (Koso-Pol.) Pimenov & Kljuykov [Kljuykov]

Neoconopodium (Koso-Pol.) Pimenov & Kljuykov, Feddes Repert. 98: 377 (1987).

Glabrous to pubescent, short-lived perennials with globose tubers. Leaves 2–3-ternate; triangular-ovate, leaflets oblong to ovate, entire or lobed. Bracts absent, bracteoles several, lanceolate to ovate and semiscarious. Petals white; styles reflexed; stylopodium low conical. Fruits glabrous or pilose, linear-oblong, apically attenuated into a well-developed beak; mericarps transversely subterete; commissure narrow; ribs filiform; carpophore apically bifid; vittae obscure. Seed face narrowly and deeply sulcate.

Two species, Asia (E Afghanistan, N Pakistan, NW India). Low grassy meadows in middle mountain belts. Closely related to *Kozlovia* and *Krasnovia,* and sometimes treated under the former (Spalik and Downie 2001), but differs in fruit form and indumentum, as well as the seed face.

279. ***Neogoezia*** Hemsl. [Plunkett]

Neogoezia Hemsl., Bull. Misc. Inform. Kew 1894: 354 (1894); Mathias & Constance, N. Amer. Fl. 28B(1): 62 (1944), reg. rev.

Acaulescent, biennial or perennial herbs growing from clustered tuberous roots. Stems erect, unbranched. Leaves pinnately compound (or lobed), narrowly elliptic to lanceolate, membranaceous; petioles sheathing; leaflets often 1–2-pinnate or pinnatisect. Umbels simple, loose, peduncles scapose, pedicels many (25–70), filiform, ascending to reflexed; bracts many, entire or lobed, reflexed and shorter than the pedicels. Calyx lobes prominent, purplish; petals yellow or yellow-white, obovate, with narrowed, inflexed apex; styles short; stylopodium depressed-conical. Fruits glabrous, ovoid to ellipsoid, cordate, constricted at commissure, compressed laterally; mericarps subterete, pentagonal, ribs filiform or obsolete; carpophore entire; vittae small, several in each valleculae, several on the commissure. Seeds transversely subterete, the face deeply sulcate. $n = 22, 28, 33$.

Five species, Mexico. Closely related to *Oenanthe*, *Sium*, and *Perideridia*.

280. ***Neonelsonia*** J.M. Coult. & Rose [Plunkett]

Neonelsonia J.M. Coult. & Rose, Contr. U.S. Natl. Herb. 3: 306, t. 10 (1895); Mathias & Constance, N. Amer. Fl. 28B (1): 73–74 (1944), reg. rev.; Mathias & Constance, Fl. Peru XIII, V-A, 1: 77–78 (1962), reg. rev.

Essentially glabrous perennials growing from woody roots. Leaves ternately or ternate-pinnately compound; leaflets ovate or ovate-lanceolate, crenate, serrate, or lobed. Bracts lacking; rays few to several, slender and spreading; bracteoles several, filiform, longer than fruit. Petals greenish-yellow; styles short; stylopodium depressed-conical. Fruits glabrous, wrinkled, ellipsoid-cordate, strongly compressed laterally; ribs filiform; carpophore bifurcated to the base; vallecular vittae 2–3, unequal, commissural 4–6. Seeds subterete, the face involute. $2n = 44$.

Two species, N America (Mexico), C America, S America (W tropical).

281. ***Neoparrya*** Mathias [Plunkett]

Neoparrya Mathias, Ann. Missouri Bot. Gard. 16: 393 (1929); Theobald et al., Brittonia 16: 296–315 (1964), rev.

Glabrous, acaulescent, caespitose perennials. Leaves oblong, pinnately compound, somewhat rigid; leaflets linear, entire or toothed to lobed. Bracts lacking; rays few to several (5–10), subequal or equal, spreading, reflexed in flower and fruit; bracteoles inconspicuous, linear-lanceolate or short-lanceolate (sometimes minutely scaberulous-margined). Petals yellow; styles long, reflexed in fruit; stylopodium lacking. Fruits glabrous, oblong, terete; ribs all prominent and unwinged; carpophore bifurcated; vittae many and of various sizes, scattered throughout the pericarp. Seeds transversely slightly compressed dorsally, the face plane or slightly convex. $2n = 22$.

One species, *N. lithophila* Mathias, N America (Colorado); forming dense clumps among rocks at high elevations in the Rocky Mountains (known only from type locality).

282. ***Niphogeton*** Schltdl. [Plunkett]

Niphogeton Schltdl., Linnaea 28: 481 (1856); Mathias & Constance, Fl. Peru XIII, V-A, 1: 79–84 (1962), reg. rev.

Caulescent to acaulescent, caespitose, glabrous to puberulent, sometimes glaucous or scaberulous perennials with a woody base. Leaves simple to ternate, pinnate, or pinnately decompound; subcoriaceous; basal leaves densely rosulate to loosely clustered; ultimate segments linear-oblong, lanceolate, orbicular, or ovate. Bracts few to several, linear-filiform to obovate, entire, lobed or pinnatifid; rays few to many, angled, spreading to ascending, subequal or unequal; bracteoles 1 to several, linear to obovate, entire to pinnatifid. Petals white with colored midvein, the apex plane; styles short, spreading or erect; stylopodium depressed- or low-conical. Fruits glabrous, oblong to ovoid, compressed laterally, ribs prominent, acute to obtuse or corky-winged; carpophore stout, bifurcated, the halves divergent at apex; vallecular vittae 1, commissural 2; rib oil ducts sometimes present. Seed half-terete to terete, often sulcate under the vittae; the face plane or concave. $n = 11$.

18 species, S America (W tropical) and C America, páramos, especially Andean. Aromatic herbage used locally as a medicine, especially for fevers and stomach ailments.

283. ***Nirarathamnos*** Balf.f. [Van Wyk & Tilney]

Nirarathamnos Balf.f., Proc. Roy. Soc. Edinb. 11: 513 (1882); Wolff, Pflanzenr. IV, 228 (Heft 43): 173 (1910), rev.; Miller & Morris, Ethnofl. Soqotra Archipelago: 702–703 (2004), taxon.

Glabrous, woody shrublets. Leaves crowded at the branch ends, simple, orbicular, conspicuously veined, shallowly and irregularly dentate; petiolar sheaths persistent on the stems. Bracts and bracteoles large, persistent, oblong-elliptic, conspicuously veined; rays few to several, short. Petals white or greenish-yellow; styles short; stylopodium conical. Fruits glabrous, verrucose, ellipsoid, slightly compressed dorsally; mericarps semi-lunate; ribs conspicuous; carpophore free, entire; vittae large, vallecular 1, commissural 2; rib oil ducts conspicuous, solitary in each rib. Seeds transversely slightly compressed dorsally, markedly sulcate beneath the vittae, the face plane to somewhat concave. $2n = 22$.

One species, *N. asarifolius* Balf.f., Africa (Socotra). Between rocks on steep cliffs. Molecular data suggest a relationship to *Echinophora* and its relatives (Downie et al. 2010).

284. ***Normantha*** P.J.D. Winter & B.-E. van Wyk [Van Wyk & Tilney]

Normantha P.J.D. Winter & B.-E. van Wyk, Phytotaxa 298 (1): 74 (2017); Van Wyk et al., Afr. Apiaceae: 239 (2013). *Aframmi* sensu C. Norman, pro parte minore.

Perennial herb with a woody rootstock; stems woody, erect, glabrous. Leaves pinnate or bipinnate to biternate, finely divided, with linear segments. Bracts prominent, entire or slightly pinnatifid; rays several, long, wiry; bracteoles entire; Petals yellowish green; styles short, divergent. Stylopodia low-conical, disc with an undulate margin. Fruits ovoid to narrowly obovoid, slightly compressed laterally, glabrous; commissure narrow; carpophore bifurcate; ribs prominent, obtuse; rib oil ducts 5, large; vittae usually 6. Seeds pentagonal to polygonal, sulcate beneath vittae, very slightly concave on the commissural face.

One species, *N. filiformis* P.J.D. Winter, Africa (Angola, and the border of the Democratic Republic of Congo, Zambia and Tanzania). The woody habit, slender elongated rays and prominent rib oil ducts suggest a relationship with *Dracosciadium*, *Pseudocarum* and other woody Apiaceae of Madagascar.

285. ***Nothosmyrnium*** Miq. [Watson]

Nothosmyrnium Miq., Ann. Mus. Bot. Lugduno-Batavi 3: 58 (1867).

Sparsely pubescent perennials with a stout taproot bearing fascicled fibrous roots. Leaves 1–3-pinnate, ternate-pinnate, or ternate-bipinnate; ultimate segments broadly ovate to ovate-lanceolate, margin serrate or incised. Bracts several, lanceolate; rays slender, spreading; bracteoles several, ovate. Petals white, the marginal slightly radiate; styles reflexed; stylopodium conic. Fruits glabrous or pilose, ovoid, slightly compressed dorsally, constricted at the commissure; dorsal ribs filiform, marginal ribs obscure; carpophore bifurcated; vallecular vittae 1–2 or 3–6, commissure 2 or 4–8. Seed transversely terete, the face plane or slightly concave. $2n = 18, 20$.

Two species, Asia (widespread in China; one species, *N. japonicum* Miq., cultivated and adventive in Japan). Low- to mid-elevation forests and grasslands.

286. ***Notiosciadium*** Speg. [Plunkett]

Notiosciadium Speg., Comun. Mus. Nac. Hist. Nat. Buenos Aires 2: 79 (1924).

Prostrate to low, glabrous annuals. Leaves ovate, ternately to pinnately decompound; divisions lanceolate. Bracts and bracteoles lacking; rays few, very short and stout. Petals white, apex merely incurved slightly; styles short and spreading; stylopodium low-conical. Fruits glabrous, oblong to subclavate, apex truncate, compressed laterally; mericarps transversely subterete and pentagonal; commissure not constricted; ribs equal, slightly raised, papillose-roughened; carpophore entire but longitudinally grooved, with 2 apical teeth, vallecular vittae 1, commissural 2. Seeds subterete, the face plane to slightly concave.

One species, *N. pampicola* Speg., S America (Argentina, Buenos Aires state), in marshy grasslands.

287. ***Notobubon*** B.-E. van Wyk [Van Wyk & Tilney]

Notobubon B.-E. van Wyk in Winter et al., Taxon 57: 355 (2008); Magee et al., Syst. Bot. 34: 220–242 (2009), rev.; Van Wyk et al., Afr. Apiaceae: 240–247 (2013).

Glabrous, evergreen, woody perennial shrubs or small trees. Leaves 2–4-pinnate, variable, sclerophyllous, persistent; leaflets oblong to linear, entire, lobed or pinnatisect, sometimes modified to a rigid spine. Bracts and bracteoles many, linear to lanceolate; rays often many (up to 180), short to long. Calyx lobes minute to large; petals yellow; styles short, elongating and often becoming reflexed; stylopodium broadly conical. Fruits glabrous, broadly elliptic to rotund, strongly compressed dorsally; mericarps homomorphic (rarely heteromorphic); dorsal ribs indistinct, marginal ribs usually winged; carpophore bifid; vittae small to large, vallecular 1, often with an additional vitta below each vascular bundle, commissural 2–3 (often much larger). Seeds transversely compressed dorsally, sulcate beneath the vittae, the face plane to slightly sulcate. $2n = 22$.

12 species, South Africa. Mostly fynbos or renosterveld vegetation. Related to *Nanobubon*.

288. ***Oedibasis*** Koso-Pol. [Pimenov]

Oedibasis Koso-Pol., Bull. Soc. Imp. Naturalistes Moscou, n.s., 29: 175 (1916).

Glabrous or puberulent, short-lived perennials with globose or fusiform tubers. Leaves pinnate with ellipsoid sessile segments, or 2–3-pinnate with filiform leaflets. Bracts and bracteoles linear to lanceolate, acuminate, with white-membranaceous margins; rays unequal. Petals white or pinkish; styles short, reflexed; stylopodium conical. Fruits compressed dorsally; dorsal ribs filiform, marginal narrowly winged, whitish; carpophore bifid; commissure broad; mesocarp parenchymatous, in marginal ribs cells with pitted walls; vittae of two types: large (vallecular 1 and commissural 2) and narrower, sometimes anastomosing. Seed face plane. $2n = 20$.

Four species, Asia (Middle, Afghanistan). Lowland deserts, stony slopes, scrub, dry meadows, steppes. Closely related to *Bunium*, *Mogoltavia*, *Hyalolaena*.

289. ***Oenanthe*** L. [Pimenov]

Oenanthe L., Sp. Pl. 254 (1753); Gen. Pl., ed. 5, 122 (1754); Bertova, Biol. Prac. 19, 4: 5–73 (1973), reg. rev.; Froebe, Bot. Jahrb. 96(1–4): 84–89 (1975), morphol., carpol.; Townsend, Kew Bull. 38(2): 311–315 (1983), reg. rev.

Glabrous biennials or perennials with rhizomes, stolons and adventive roots, often thickened, tuberous. Leaves 2–3-pinnate; leaflets linear to ovoid, or filiform (in submerged leaves), or reduced to fistulose rachis-leaves. Bracts absent, or rarely few, linear; rays sometimes thickening in fruit; bracteoles entire, linear to lanceolate. Calyx lobes prominent, lanceolate; petals white or pinkish, the outer sometimes radiate; styles erect or spreading; stylopodium conical. Fruits slightly compressed laterally, not separating into mericarps; dorsal ribs filiform or sub-inconspicuous, marginal often thickened; carpophore lacking; commissure broad; mesocarp outer layer of large lignified parenchyma cells; inner layer of sclerenchyma; vallecular vittae 1, commissural 2. Seed face plane. $2n = 16, 18, 20, 22, 42, 44, 66$.

40–45 species, Europe, Asia, Australia, Australasia, Africa, N. America (adv.). Moist places, swamps, herbaceous slopes, rarer in wet forests. Some species are poisonous, others were used in folk medicine and as salad plants. Molecular data suggest a close relationship to *Cicuta* (Spalik et al. 2014).

290. ***Oligocladus*** Chodat & Wilczek [Plunkett]

Oligocladus Chodat & Wilczek, Bull. Herb. Boiss. II, 2: 527 (1902); Pérez-Moreau, Physis 12: 94–95 (1936), taxon.

Glabrous perennials. Leaves cauline, ternately dissected into ternate, simple or bifid leaflets, the ultimate segments narrowly spathulate or sublinear. Bracts few, linear and elongate; rays few, subequal; bracteoles few, linear-spathulate.

Petals white; styles stout; stylopodium long conical. Fruits papillose and verruculose, elliptic to ovate, compressed dorsally; ribs subequal, or the marginal ribs narrowly winged; carpophore apically bifid; vallecular vittae 3, commissural 6. Seeds compressed dorsally, the face plane. $2n = 44$.

Two species, S America (Andes of Argentina).

291. ***Oliveria*** Vent. [Plunkett]

Oliveria Vent., Descript. Pl. Nouv. Jard. Cels.: t. 21 (1801); Rechinger, Fl. Iran. 162: 309–310 (1987), reg. rev.

Pilose annuals with whitish stems. Basal and lower cauline leaves small, generally withered at anthesis, pinnately dissected, ovate to lanceolate; leaflets ternate-pinnately dissected, the ultimate segments lanceolate-linear. Bracts (bi-)ternately dissected, pilose; rays few, equal, short, stout, and spreading; bracteoles few, trifid, densely pilose. Calyx lobes elongate-subulate; petals white or pink, dorsally pubescent; styles stout and spreading; stylopodium conical. Fruits densely pubescent, oblong-ovoid, laterally compressed; mericarps terete, ribs filiform, low and rounded; vittae large, vallecular 1, commissural 2. Seeds transversely subterete, sulcate under the vittae, the face deeply sulcate.

One species, *O. decumbens* Vent., SW Asia (Iran, Iraq, Turkey).

292. ***Opopanax*** W.D.J. Koch [Pimenov]

Opopanax W.D.J. Koch, Nov. Acta Acad. Leop.-Carol. Nat. Cur. 12(1): 96 (1824).

Puberulent or scabrous perennials. Leaves bipinnate; petioles covered with white, chaffy hairs; leaflets oblong or lanceolate, coriaceous, with stellate or glochidiate hairs or glabrous beneath. Bracts linear to lanceolate, sometimes few; rays glabrous or hairy; bracteoles like bracts or setaceous. Petals yellow; styles ± short, reflexed; stylopodium low-conical, with undulate margin. Fruits glabrous, compressed dorsally; dorsal ribs filiform, marginal narrowly winged, slightly thickened; carpophore bifid; commissure broad; mesocarp cells with lignified pitted walls; vallecular vittae 1 or 2–4(–5), commissural (4–)6–8(–14). Seed face almost flat. $2n = 20, 22$.

Three species, Europe, Asia (SW, the Caucasus). Stony slopes, meadows, roadsides. Molecular data suggest an affinity to *Smyrniopsis*.

293. ***Opsicarpium*** Mozaff. [Pimenov]

Opsicarpium Mozaff., Bot. Zhurn. 88(2): 89 (2003).

Shortly pubescent perennials. Stems solid. Leaves shortly pubescent or glabrous, pinnate, with 3–5 segment pairs; leaflets entire, serrate-dentate. Bracts and bracteoles absent; rays few, glabrous. Petals golden yellow, pubescent; styles short, straight, soon caducous; stylopodium conical. Fruits glabrous, dorsally compressed, dorsal ribs sub-inconspicuous, marginal shortly keeled; carpophore bifid; vittae many, septate, vallecular 3–6, 1–2 under rib vascular bundles, commissural 2–8, vittae partly reticulated. Seed face plane.

One species, *O. insignis* Mozaff., SW Asia (W Iran). Molecular data suggest an affinity to *Pimpinella*.

294. ***Oreocome*** Edgew. [Pimenov]

Oreocome Edgew., Proc. Linn. Soc. London 1: 252 (1845); Pimenov et al., Willdenowia 31(1): 101–124 (2001), rev.; Pimenov & Kljuykov, Willdenowia 38: 93–99 (2009), rev.

Glabrous or scabrous perennials. Leaves 3–4-pinnate or 3–4-ternate; leaflets ovate to lanceolate-linear. Bracts lacking, or several, entire, linear, or pinnatifid, puberulent, white-membranaceous at margin; rays many; bracteoles entire, linear to lanceolate, or pinnatisect. Calyx lobes lanceolate-triangular to filiform, exceeding stylopodia; petals white; styles reflexed; stylopodium low-conical to conical. Fruits glabrous; dorsal ribs keeled or narrowly winged, marginal narrowly winged, rarely keeled; carpophore bifid; commissure narrow or intermediate; mesocarp cells with pitted walls; vallecular vittae 1–2(–3), commissural 2–4(–6). Seed face plane. $2n = 22$.

Ten species, Asia (China, Himalaya, Thailand, Vietnam, Afghanistan). Subalpine meadows, forest margins, stony slopes. Related to *Ligusticopsis*, *Cortia*, and *Cortiella*.

295. ***Oreocomopsis*** Pimenov & Kljuykov [Kljuykov]

Oreocomopsis Pimenov & Kljuykov, Acta Phytotax. Sin. 34 (1): 2 (1996).

Nearly glabrous perennials. Leaves 3–4-pinnate; leaflets rhombic to ovate, toothed or lobed. Bracts foliaceous, with long petioles and small pinnatifid blades; bracteoles linear to filiform, usually entire, deflexed, 2–3 times longer than umbellules. Petals white or dark brown; styles reflexed; stylopodium conical. Fruits glabrous, elliptic to oblong; commissure narrow, mericarps slightly compressed dorsally; ribs equally winged or the marginal ribs broader; carpophore bifurcated; mesocarp in ribs with lignified and pitted walls; vallecular vittae 1–3, commissural 4–6. Seed face plane to slightly concave. $2n = 22$.

Three species, Asia (SW China, Nepal). Stony and grassy slopes, scrub and among rocks. Related to *Ligusticopsis*.

296. ***Oreomyrrhis*** Endl. [Webb]

Oreomyrrhis Endl., Gen. Pl. 787 (1839); Mathias & Constance, Univ. Calif. Publ. Bot. 27: 347–416 (1955), rev.; Chung et al., Amer. J. Bot. 92: 2054–2071 (2005), phylog.; Piwczyński et al., Bot. J. Linn. Soc. 178: 298–313 (2015), phylog.

Acaulescent to caulescent, often caespitose or pulvinate, glabrous or pubescent perennials, sometimes rhizomatous. Leaves usually densely rosulate, narrowly oblong to elliptic, 1–2-pinnately dissected or compound, or sometimes reduced to rachis-leaves; ultimate segments linear to flabellate, entire to pinnatisect. Umbels simple; bracts few to several, linear-lanceolate to obovate, entire to pinnatifid, ± connate, erect to reflexed in fruit. Petals white (often with a colored midvein), yellow, pink or purplish, often dorsally pubescent; styles short to slender, erect to divaricating; stylopodium depressed-conical to conical. Fruits glabrous or pubescent, oblong to ovoid, slightly compressed laterally or dorsally; ribs filiform, sometimes slightly thickened or corky but unwinged; carpophore bifurcated to apically bifid; vallecular vittae 1 to several, commissural 2 to several. Seeds transversely subterete, the face nearly plane to concave or deeply sulcate. $n = 6$, $2n = 12$.

About 25 species, N, C and S America (Mexico, Guatemala, Andes, Falkland Is.), Australia (SE), New Zealand, New Guinea, Borneo, Taiwan. On the basis of molecular data, Chung et al. (2005) treat this genus under *Chaerophyllum*.

297. ***Oreonana*** Jeps. [Plunkett]

Oreonana Jeps., Madroño 1: 140 (1923); Shevock & Constance, Madroño 26: 128–134 (1979), rev.

Caespitose, acaulescent, densely greyish-hirsute or white tomentose (to glabrate) perennials. Leaves of two types, bladeless sheath leaves and laminate leaves; laminate leaves pinnately or ternately decompound, ultimate divisions small, lanceolate to oblong, crowded, cuspidate to mucronate. Bracts lacking; rays few to many, short, stout, spreading, winged or wingless; bracteoles 1, dimidiate with 3–5 ovate to lanceolate, acuminate to attenuate lobes. Petals white or purplish; styles short or slender, compressed; stylopodium lacking. Fruits tomentose to glabrate, ovoid to orbicular, slightly compressed laterally; ribs filiform, rounded, evident; carpophore bifid to half or entire length; vallecular vittae 3–5, commissural 3–5. Seeds transversely subterete, the face sulcate. $n = 11$, $2n = 22$.

Three species, N America, restricted to montane/alpine habitats of S California.

298. ***Oreoschimperella*** Rauschert [Van Wyk & Tilney]

Oreoschimperella Rauschert, Taxon 31(3): 556 (1982); Townsend, Fl. Trop. E. Afr.: 46–48 (1989), reg. rev.; Hedberg & Hedberg, Fl. Ethiopia & Eritrea: 23–25 (2003), reg. rev.; Van Wyk et al., Afr. Apiaceae: 250 (2013).

Glabrous annuals (or biennials). Leaves pinnate to bipinnate, oblong-lanceolate; leaflets lanceolate, broad, irregularly serrate. Bracts often foliaceous with a broad sheathing base and a trifid to pinnate apex; rays several to many, long, wiry; bracteoles small. Petals white; styles rather short, slender; stylopodium depressed-conical, with an expanded, saucer-shaped crenulate disc. Fruits glabrous, densely verrucose, ovoid to sub-rotund; laterally compressed; mericarp semi-lunate, commissure very narrow; ribs inconspicuous, obtuse; carpophore bifid; vittae relatively small and irregular, variable in number in both the valleculae and the commissure; rib oil ducts small. Seeds transversely subterete, sulcate beneath vittae, the face ± plane.

Two species, Africa (Kenya, Ethiopia) and Yemen. Poorly known; related to *Trachyspermum*.

299. ***Oreoselinum*** Mill. [Reduron]

Oreoselinum Mill., Gard. Dict. ed. 4 (1754); Reduron et al., J. Bot. Soc. Bot. France 1: 94 (1997), taxon.

Glabrous perennials. Leaves 2–4-pinnately divided, triangular; rachis flexuous; leaflets ovate-cuneiform, pinnately lobed. Bracts and bracteoles usually many, linear to long-triangular; rays many, subequal, spreading. Petals white, rarely pinkish; styles slightly to clearly exceeding the short-conical stylopodium. Fruits glabrous, suborbicular or broadly elliptic, strongly compressed dorsally; dorsal ribs prominent but weak, the marginal winged; carpophore free, bifurcated; commissure broad; vittae medium-sized, vallecular 1, arcuate, commissural 2. Seeds transversely compressed dorsally, the face plane. $2n = 22$.

One species, *O. nigrum* Delarbre, Europe (Spain to Russia). Sunny slopes, forest margins. Segregate of *Peucedanum* s. lat.

300. ***Oreoxis*** Raf. [Plunkett]

Oreoxis Raf., Bull. Bot., Geneva 1: 217 (1830); Mathias & Constance, N. Amer. Fl. 28B(2): 168 (1945), reg. rev.; Welsh & Goodrich, Great Basin Natur. 45: 34–36 (1985), taxon.

Caespitose, acaulescent, glabrous to pubescent, scabrous, and/or glandular perennials. Leaves pinnately or bipinnately dissected; ultimate segments linear and acute. Bracts usually lacking; rays few, short, spreading; bracteoles usually dimidiate, linear to obovate, entire or toothed. Calyx lobes prominent; petals yellow or whitish; styles slender; stylopodium lacking. Fruits slightly pubescent to glabrous, oblong to ovoid-oblong, slightly compressed laterally; ribs subequal, or the marginal ribs broader and corky-winged (wings broadly linear to subovate transversely, narrower than the fruit body); carpophore lacking; vittae small, vallecular 1 to several, commissural 2 to several. Seeds transversely slightly compressed dorsally, the face plane or slightly concave. $2n = 60$, ca. 88.

Four species, N America (C Rocky Mountains).

301. ***Orlaya*** Hoffm. [Lee]

Orlaya Hoffm., Gen. Pl. Umbell. XXVI, 58 (1814).

Glabrous or rarely pubescent annuals (rarely procumbent). Leaves 2–3-pinnatisect, broadly ovate; the ultimate divisions lanceolate to oblong. Bracts and bracteoles few, lanceolate to oblong or obovate, simple, white-membranous and narrowly scabrous margined; rays few to many, slightly contracted or not in fruit. Petals white, the outer strongly radiate; styles elongated; stylopodium subconical to depressed. Fruits spinescent, ovoid to ellipsoid, compressed dorsally; primary ribs prominent with striate, unbranched, more or less erect hairs, arranged in 2–4 rows; secondary ribs 4, with large spines, rarely confluent at the base, simple and sometimes glochidiate at the apex, arranged in 1–3 rows; vittae large, transversely elliptical, vallecular solitary (under each secondary rib), commissural 2. Seeds strongly compressed dorsally, the face ± deeply sulcate to incurved. $2n = 14, 16, 20$.

Three species, Europe, N Africa, SW Asia. Closely related to *Daucus* (Banasiak et al. 2016).

302. ***Ormopterum*** Schischk. [Pimenov]

Ormopterum Schischk., Fl. URSS 16: 597 (1950).

Glabrous biennial or short-lived perennials with tuberous roots. Leaves tripinnate; leaflets sessile, linear, mucronulate. Bracts ovoid to lanceolate, white-membranaceous and ciliate at the margin; rays few; bracteoles short, linear-lanceolate to ellipsoid, sometimes ciliate. Petals white or rose; styles short, thickened, divergent; stylopodium low-conical with undulate margin. Fruits glabrous, oblong; ribs equal, filiform or corky, undulate, folded; carpophore reduced, sometimes bifid; commissure narrow; mesocarp parenchymous, partly crushed at fruit maturity; vittae almost cyclical, few on the commissure. Seed face shallowly sulcate. $2n = 22$.

Two species, Asia (Turkmenistan, Afghanistan, Pakistan). Sandy desert hills, in pine forests. Related to *Elaeosticta*, *Bunium*, *Galagania* and their relatives.

303. ***Ormosciadium*** Boiss. [Ostroumova]

Ormosciadium Boiss., Ann. Sci. Nat. Bot. III, 2: 95 (1844).

Nearly glabrous annuals. Leaves sessile or short-petiolate, bipalmate or bipinnate; ultimate divisions filiform. Bracts and bracteoles long, numerous, filiform, simple or trifid. Petals white or pink, marginal strongly radiate; styles reflexed; stylopodium short-conical. Fruits sparsely pubescent, elliptic, compressed dorsally; dorsal ribs filiform, marginal ribs moniliform, expanded on the lateral valleculae; commissure very broad; vittae obliterated; fibrous inner mesocarp absent. Seed face concave.

One species, *O. aucheri* Boiss., SW Asia. Relationships uncertain.

304. ***Ormosolenia*** Tausch [Pimenov]

Ormosolenia Tausch, Flora 17: 348 (1834); Pimenov, Edinb. J. Bot. 49(2): 219–223 (1992), rev.

Dwarf, glaucous perennials with branched rootstocks and creeping rhizomes and stolons. Leaves mainly radical, simple, orbicular-cuneate to reniform, or ternate, or partly pinnate with large upper leaflet. Bracts absent; rays few, unequal; bracteoles absent, or 1–2, setaceous. Petals greenish-yellow or brown; styles short, reflexed; stylopodium low-conical. Fruits compressed dorsally; dorsal ribs inconspicuous, marginal narrowly winged; carpophore bifid; commissure broad; mesocarp of two distinct layers: outer of parenchymous cells, partly with pitted walls, inner of prosenchymatous lignified cells; vallecular vittae 3–6, commissural 6–8. Seed face plane. $2n = 22$.

One species, *O. alpina* (Sieber ex Schultes) Pimenov, E Mediterranean (Crete, W Turkey). Stony slopes and screes in high mountains.

305. ***Orogenia*** S. Watson [Plunkett]

Orogenia S. Watson, U.S. Geol. Explor. 40th Parallel, Bot. 120 (1871); Mathias & Constance, N. Amer. Fl. 28B(1): 74–75 (1944), reg. rev.

Acaulescent or very shortly caulescent, glabrous perennials with tuberous roots. Leaves 1–3-ternately divided (rarely simple); leaflets narrow, elongate and usually entire. Bracts lacking; rays few, spreading, unequal; bracteoles few, minute, narrow and entire, or lacking. Petals white; styles short, spreading; stylopodium lacking. Fruits glabrous, oblong to oval, slightly compressed laterally, the commissural face of each mericarp with a corky, rib-like projection running down its length; dorsal and lateral ribs prominent to obscure, marginal ribs broadly and involutely corky winged; carpophore lacking; vallecular vittae several, commissural several. Seeds transversely compressed dorsally, the face slightly concave.

Two species, N America (NW United States), mountains at low to middle elevations.

306. ***Orumbella*** J.M. Coult. & Rose [Pimenov]

Orumbella J.M. Coult. & Rose, Contr. U.S. Natl. Herb. 12: 445 (1909); Mathias & Constance, N. Amer. Fl. 28B(1): 149–150 (1944), reg. rev.

Dwarf, caespitose, glabrous perennials. Leaves basal, bipinnate; leaflets oval, coarsely dentate. Bracts linear to linear-oval, entire, dentate; rays few; bracteoles entire, linear to narrowly lanceolate. Petals red or purplish; styles reflexed; stylopodium low-conical, undulate. Fruits slightly compressed dorsally; ribs prominent, broadly winged, inflated, marginal broader than dorsal; carpophore bifurcated; mesocarp parenchymatous, in ribs consisting of large cells with lignified walls; commissure narrow; vallecular vittae 2–4, commissural 6. Seed face plane. $2n = 22$.

One species, *O. macounii* (J.M. Coult. & Rose) J.M. Coult. & Rose, N America (Alaska and islands), NE Asia (Chukotka). Arctic tundras, stony mountain slopes, scree.

307. ***Osmorhiza*** Raf. [Spalik]

Osmorhiza Raf., Amer. Monthly Mag. & Crit. Rev. 4: 192 (1819); Lowry & Jones, Ann. Missouri Bot. Gard. 71: 1128–1171 (1984), rev.; Wen et al., Ann. Missouri Bot. Gard. 89: 414–428 (2002), phylog.

Pubescent to glabrate perennials. Leaves 2–3-pinnately or ternately dissected, triangular; lobes lanceolate to ovate. Bracts absent or few; rays few to many, subequal or unequal, ascending or spreading; bracteoles absent or few, ciliate. Petals white, pinkish or yellow, denticulate at margin or ciliate, outer sometimes radiate; styles short to relatively long; stylopodium conical in flower, conical or rounded in fruit. Fruits glabrous or bristled, with an aculeate cuticle; linear-oblong to narrow-obovate, usually with caudate appendage, laterally compressed; mericarps subterete to pentagonal, as broad as wide or scarcely dorsally compressed, the beak short or obsolete; ribs angular, glabrous or bristled; commissure narrow; carpophore bifurcate; vittae compressed at maturity, vallecular 1, commissural 2. Seed transversely terete, the face sulcate. $2n = 22, 44$.

Ten species, N and S America (nine species,) and Asia (one species). Moist habitats, forests to open montane places. Closely related to *Myrrhis*, *Anthriscus*, *Geocaryum*, and *Kozlovia* (Spalik et al. 2001).

308. ***Ostericum*** Hoffm. Fig. 3F [Pimenov]

Ostericum Hoffm., Pl. Umbell. Gen., ed. 2: 162 (1816).
Gomphopetalum Turcz. (1841).

Usually glabrous perennials (rarely with horizontal rhizomes). Leaves 2–3-pinnate or 2–3-ternate; leaflets ovate or broadly lanceolate to almost filiform. Bracts absent or few, linear; bracteoles linear. Calyx lobes ovate to triangular; petals white or greenish; styles reflexed; stylopodium low-conical or conical. Fruits compressed dorsally; dorsal ribs filiform to keeled, marginal broadly winged; carpophore bifid; exocarp of large cells with thickened outer walls; commissure narrow; mesocarp parenchymatous, often partly crushed at fruit maturity; vallecular vittae 1–3, commissural 2–6, or almost cyclical. Seed face plane. $2n = 22, 44$ (and dubious reports of $2n = 18$ and 110).

12 species, Europe, Asia. Meadows, scrub, coniferous and beach forests, marshes.

309. ***Ottoa*** Kunth [Plunkett]

Ottoa Kunth in Humb., Bonpl. & Kunth, Nov. Gen. Sp. (quarto ed.) 5: 20 (1821); Mathias & Constance, N. Amer. Fl. 28B(1): 90 (1944), reg. rev.

Subscapose, glabrous to scaberulous perennials. Leaves chiefly basal, entire, reduced to fistulose, transversely septate rachis-leaves. Bracts usually lacking; rays few to several, spreading-ascending; bracteoles inconspicuous or lacking. Petals white; styles slender, recurved; stylopodium depressed-conic. Fruits glabrous, oblong, compressed laterally; ribs filiform, obtuse; carpophore bifurcated; vallecular vittae 1; commissural 2. Seeds transversely subterete, the face deeply sulcate. $2n = 22$.

One species, *O. oenanthoides* Kunth, N American (Mexico), C America, and S America (W tropical).

310. ***Oxypolis*** Raf. [Plunkett]

Oxypolis Raf., Neogenyton 2 (1825); Feist et al., Taxon 61: 402–418 (2012), phylog., rev.

Glabrous perennials growing from fascicled tubers. Leaves once pinnately or ternately compound; leaflets mostly distinct, broad or narrow, serrate to incised. Bracts few and slender, or lacking; rays few to many, usually spreading-ascending; bracteoles like the bracts, or lacking. Calyx lobes conspicuous to minute; petals white; styles slender, spreading; stylopodium conical. Fruits glabrous, oblong to ovoid, strongly compressed dorsally; dorsal and lateral ribs filiform; marginal ribs broadly thin-winged (nerves on inner margin of lateral wings give the appearance of 5 dorsal ribs); carpophore bifurcated; vallecular vittae 1, commissural 4–8. Seeds transversely compressed dorsally, the face plane. $n = 16, 18$, $2n = 36$.

Four species. N America; wet habitats.

311. ***Pachyctenium*** Maire & Pamp. [Lee]

Pachyctenium Maire & Pamp., Arch. Bot. (Forli) 12: 176 (1936).

Prostrate, pubescent perennials. Leaves 2–3-pinnatisect, oblong to lanceolate, the ultimate divisions linear to lanceolate. Bracts and bracteoles several, linear to lanceolate, simple; rays several, unequal, hispid. Calyx lobes conspicuous; petals white to pinkish, the outer often radiate; styles elongated; stylopodium depressed, rounded. Fruits partly pubescent, ovoid; mericarps $\pm$ compressed laterally; primary ribs variously lobed, with papillate, semi-erect hairs arranged in 2 rows; secondary ribs 4, dentate-winged, arranged in a single row at the lower half but naked above; carpophore free; vittae ruminate, vallecular 1 (under each secondary rib), commissural 2. The seed face plane.

One species, *P. mirabile* Maire & Pamp., N Africa (Libya). Nested within (and recently transferred to) *Daucus* (Banasiak et al. 2016).

312. ***Pachypleurum*** Ledeb. [Pimenov]

Pachypleurum Ledeb., Fl. Alt. 1: 296 (1829); Leute, Ann. Naturhist. Mus. Wien 74: 457–519 (1970), rev.; Pimenov, Bjull. Moskovsk. Obšč. Isp. Prip. Otd. Biol. 80(3): 76–84 (1975), rev.

Glabrous or sparsely puberulent perennials. Leaves mostly basal, tripinnate; leaflets linear to lanceolate, acute. Bracts entire, linear to lanceolate, sometimes forked at the tip; rays puberulent; bracteoles entire, lanceolate. Petals white; styles long, reflexed; stylopodium flat. Fruits sparsely puberulent, slightly compressed dorsally; ribs prominent, almost equal, narrowly winged; carpophore bifid; commissure narrow; mesocarp of cells with pitted walls; vittae cyclical. Seed face plane. $2n = 22, 44$.

Three species, C Europe, Asia. High mountain vegetation, tundras, alpine meadows, stony slopes. Molecular data suggest a relationship to *Meum*, *Seseli* and *Phlojodicarpus*.

313. ***Palimbia*** Besser ex DC. [Pimenov]

Palimbia Besser ex DC., Prodr. 4: 175 (1830).

Glabrous perennials. Leaves tripinnate, glabrous or scabrous; leaflets linear, acuminate. Bracts and bracteoles few, linear or setaceous; rays few or several, unequal. Petals white or whitish-pale; styles reflexed; stylopodium conical. Fruits compressed dorsally; ribs straight, dorsal keeled, marginal narrowly winged, or all ribs nearly equal and keeled; carpophore bifid; commissure broad; mesocarp cells parenchymatous, not lignified; vallecular vittae 3–5, commissural 6. Seed face plane. $2n = 22$.

Three species, E Europe, Asia (Kazakhstan, S Siberia, W China). Dry steppes, stony hillsides, limestone, saline depressions.

314. ***Pancicia*** Vis. [Watson]

Pancicia Vis., Sem. Hort. Bot. Patav. 1857: 9 (1858); Linnaea 29: 732 (1857–58).

Glabrous perennials with a corky, branched rhizome. Basal leaves simple, cordate and serrate; middle cauline leaves deeply lobed and sessile; upper leaves palmately divided to the base; ultimate segments setaceous. Bracts several, scabrid and linear, often as long as peduncle and sometimes caducous; rays several, spreading; bracteoles several and linear. Petals creamy white; styles reflexed, about as long as stylopodium; stylopodium low-conical. Fruits ovoid, slightly compressed laterally; ribs very narrowly winged; vallecular and commissural vittae several.

One species, *P. serbica* Vis., Europe (former Yugoslavia and Albania). Mountain meadows. Sometimes included in *Pimpinella*.

315. ***Paraligusticum*** V.N. Tikhom. [Pimenov]

Paraligusticum V.N. Tikhom., Bjull. Moskovsk. Obšč. Isp. Prip. Otd. Biol. 78(1): 107 (1973).

Almost glabrous perennials. Leaves tripinnatisect; leaflets narrowly ovate, pinnatifid. Bracts linear, scabrous, soon caducous; rays many, unequal, scabrous; bracteoles filiform. Petals white; styles elongate, reflexed; stylopodium low-conical. Fruits slightly compressed laterally; dorsal ribs narrowly winged, marginal slightly broader winged; carpophore bifid; commissure broad; mesocarp cells with pitted walls, partly crushed at maturity; vallecular vittae 3, commissure 4–8; rib oil ducts large. Seed face almost flat, with shallow groove. $2n = 22$.

One species, *P. discolor* (Ledeb.) V.N. Tikhom., Asia (Altai Mts., Mongolia to Tadzhikistan, NW China). High-herbaceous vegetation, near rocks. Molecular data show a relationship to *Endressia*, *Dystaenia*, *Karatavia*, and *Tommasinia*.

316. ***Paraselinum*** H. Wolff [Plunkett]

Paraselinum H. Wolff, Feddes Repert. 17: 174 (1921); Mathias & Constance, Bull. Torrey Bot. Club 84: 197 (1957), rev.; Mathias & Constance, Fl. Peru XIII, V-A, 1: 89–90 (1962), reg. rev.

Somewhat scaberulous perennials. Leaves ternate-pinnately or pinnately dissected, ultimate segments linear. Bracts conspicuous; fertile rays several, unequal, spreading-ascending and thickened in fruit; bracteoles several, narrowly linear. Petals unknown (thought to be white); styles abruptly shortened, reflexed in fruit; stylopodium depressed. Fruits minutely setose-scabrous, ovoid with a cordate base, slightly compressed dorsally, commissure with a deep, central, longitudinal furrow; ribs thick, prominent, equal in width to the intervals; carpophore bifurcated to the base (but appearing only apically cleft); vittae large, vallecular 1, commissural 2. Seed transversely slightly compressed dorsally, deeply sulcate under the vittae, the face plane.

One species, *P. weberbaueri* H. Wolff, S America (Peru). Restricted to high elevations in the Peruvian Andes. Mathias and Constance (1957) consider it closely related to *Cotopaxia* and *Niphogeton.*

317. ***Parasilaus*** Leute [Pimenov]

Parasilaus Leute, Oesterr. Bot. Zeitschr. 120: 290 (1972); Pimenov, Bull. Main Bot. Gard. Acad. Sci URSS 109: 30–33 (1978), rev.
Scaphospermum Korovin (1951), non *Scaphespermum* Edgew. (1846).

Glabrous perennials. Leaves 3–4-pinnatisect; leaflets ovate, inciso-lobate. Bracts absent or few, linear; rays few; bracteoles absent. Petals yellow; styles long, reflexed; stylopodium low-conical. Fruits slightly compressed laterally; ribs narrowly winged, almost equal; carpophore bifid; commissure narrow; mesocarp cells parenchymatous, partly crushed at fruit maturity; vallecular vittae 3(–4), commissural 4–6. Seed face with broad deep groove. $2n = 22$.

One species, *P. asiaticus* (Korovin) Pimenov, Asia (Tadzhikistan, Afghanistan). Dry gravel slopes, scree, among xerophilic shrub vegetation. Molecular data suggest a close relationship to *Komarovia*, *Cyclorhiza*, and *Calyptrosciadium.*

318. ***Pastinaca*** L. [Ostroumova]

Pastinaca L., Sp. Pl. 262 (1753); Gen. Pl., ed. 5: 126 (1754).

Pubescent (rarely glabrous) perennials or biennials. Leaves 1–2-pinnate; the ultimate divisions round or elliptic; basal primary segments sessile. Bracts and bracteoles absent or sometimes small and persistent. Petals yellow, orange or red; styles recurved; stylopodium depressed-conical. Fruits pubescent or glabrous, broadly ovoid or ellipsoid, strongly compressed dorsally; dorsal ribs filiform, marginal ribs rather broad, slightly thickened; fibrous inner mesocarp present; commissure broad; vittae long and narrow, vallecular 1, commissural 2, straight and parallel. Seed face plane. $2n = 16, 22, 44$.

16 species, Europe, Asia, adventive in Australia, America, and Africa. *Pastinaca sativa* L. is cultivated as a root vegetable. Molecular data confirm an affinity with *Leiotulus* and *Heracleum.*

319. ***Pastinacopsis*** Golosk. [Ostroumova]

Pastinacopsis Golosk., Bot. Mat. Herb. Bot. Inst. Acad. Sci. URSS 12: 198 (1950).

Pubescent perennials. Leaves pinnate; segments round, pinnatifid. Bracts triangular or linear, or lacking; bracteoles linear or lacking. Petals purple, pubescent; styles recurved; stylopodium conical. Fruits circular, pubescent, strongly compressed dorsally; dorsal ribs filiform, marginal ribs broad (proximally thin and distally slightly thickened); fibrous inner mesocarp present; commissure broad; vittae absent; rib oil ducts comparatively large; seed face plane. $2n = 22$.

One species, *P. glacialis* Golosk., Middle Asia (SE Kazakhstan, Kirghyzia), W China (Xinjiang), scree and stony slopes in high mountains and glacial cirques. Related to *Semenovia.*

320. ***Paulita*** Sojak [Kljuykov]

Paulita Sojak, Čas. Nár. Mus. (Prague) 150: 216 (1981); Pimenov & Kljuykov, Bull. Mosc. Univ., Ser. Biol. 1: 13–19 (1979), rev.
Paulia Korovin (1973), non Fée (1836), nec C.G. Lloyd (1916).
Neopaulia Pimenov & Kljuykov (1983).

Glabrous or puberulent perennials with lobed tubers. Leaves 2–3-pinnate, leaflets rhombic, toothed or lanceolate to linear. Bracts and bracteoles lanceolate to linear, entire, herbaceous to broadly membranaceous at margin. Petals white; styles reflexed; stylopodium low conical. Fruits glabrous, ovoid, slightly compressed dorsally; commissure narrow; ribs keeled to winged; carpophore bifurcated; mesocarp parenchymatous; vallecular vittae 1–3, commissural 2–4. Seed face plane. $2n = 22$.

Three species, Middle Asia, scrub, grassy and stony slopes, among rocks, subalpine and alpine meadows. Related to *Seselopsis.*

321. ***Pedinopetalum*** Urban & H. Wolff [Plunkett]

Pedinopetalum Urban & H. Wolff, Ark. Bot. Stockh. 22A (10): 43 (1929); Mathias & Constance, N. Amer. Fl. 28B(1): 73 (1944), reg. rev.

Glabrous perennials. Leaves ternate-pinnately or ternately decompound; ultimate segments linear to linear-oblong, or obtuse. Bracts usually lacking; rays few, very short; bracteoles few, linear with an obtuse apex. Petals white, apex obtuse, not inflexed; styles very short; stylopodium subconical. Fruits roughened when young, glabrate at maturity, ovoid, narrowed at apex, compressed laterally; commissure not constricted; ribs filiform, conspicuous, obtuse; carpophore bifid at apex; vallecular vittae 2–3, commissural 2. Seeds subterete, the face plane.

One species, *P. domingense* Urban & H. Wolff, Caribbean (Dominican Republic), known only from type locality.

322. ***Perideridia*** Rchb. [Plunkett]

Perideridia Rchb., Handb. Nat. Pflanzensyst. 219 (1837); Chuang & Constance, Univ. Calif. Publ. Bot. 55: 1–74 (1969), rev.; Chuang, Amer. J. Bot., 57: 495–503 (1970), anat.

Glabrous, sometimes glaucous perennials with tuberous or fibrous roots. Basal leaves ternately to biternately or pinnately to tetrapinnately divided; cauline leaves ternate or pinnate; ultimate divisions linear, linear-lanceolate, or ovate. Bracts lacking, or few to many and entire, setaceous or linear to ovate-lanceolate; rays few to many, spreading-ascending or ascending, unequal to subequal; bracteoles several to many, linear, linear-lanceolate, or spathulate. Calyx lobes evident, green or scarious; petals white; styles slender, recurved at maturity; stylopodium conical or low-conical. Fruits glabrous, linear-oblong to orbicular, compressed laterally; ribs filiform (or corky), prominent or inconspicuous; carpophore bifurcated; vallecular vittae 1–5, commissure 2–8. Seeds transversely subterete, the face plane or broadly concave (sometimes with a central longitudinal ridge). $n = 8, 9, 10, 13, 17, 18, 19, 20, 22, 40, 60$, $2n = 40$.

14 species, N America (W and C United States); typically in open fields or meadows. Tuberous roots used as food source by Amerindians.

323. ***Perissocoeleum*** Mathias & Constance [Plunkett]

Perissocoeleum Mathias & Constance, Bull. Torrey Bot. Club 79: 360 (1952).

Glabrous to sparsely puberulent or scaberulous perennials with a tuberous base. Leaves compound, pinnate, ternate, or quinate-ternate with linear and entire leaflets, or the leaf reduced to a cylindrical, septate rachis-leaf. Bracts several, pinnatifid, lobed or entire, somewhat connate at base; fertile rays several to many, spreading-ascending, unequal; bracteoles several, similar to bracts. Petals white or greenish-white, usually with colored midvein, apex reflexed (but not inflexed); styles slender to short and stout, ascending-spreading; stylopodium low-conical to lacking. Fruits glabrous, ovoid or ellipsoid to oblong-ellipsoid, strongly compressed dorsally; dorsal and lateral ribs prominent, thickened or thin, obtuse or acute, marginal ribs with thick or thin wings (narrower than the fruit body); carpophore bifurcated; vallecular vittae solitary, commissural 2. Seed transversely compressed dorsally, sulcate under the vittae, the face slightly concave to slightly convex. $2n = 22$.

Four species, S America (Colombia), restricted to high elevations of the Andean páramos.

324. ***Petroedmondia*** Tamamsch. [Pimenov]

Petroedmondia Tamamsch. in Rech.f., Fl. Iran. 162: 167 (1987).

Glabrous or papillate herbaceous perennials. Leaves pinnatisect, papillate, coriaceous; leaflets broad, lanceolate, decurrent at base, crenate at margins. Bracts and bracteoles triangular, coriaceous; rays many, long, angled. Petals yellow, with broad incurved tip; styles short, slightly exceeding the depressed, disc-shaped stylopodium. Fruits glabrous, prismatic, cuneate at base, compressed laterally; ribs prominent, almost equal, corky; mesocarp of large cells with lignified walls; vittae cyclical, many, small. Seed face with deep groove and involute margins.

One species, *P. syriaca* (Boiss.) Tamamsch., SW Asia (from Syria to Iran). Clay slopes, fields, dry hills.

325. ***Petroselinum*** Hill [Reduron]

Petroselinum Hill, Brit. Herb. 424 (1756); H. Wolff in Engl., Pflanzenr. 4 (228) Heft 90: 63–68 (1927), rev.; Danert, Kulturpfl. 7: 73–81 (1959), taxon. infrasp.; Jork, Univ. Saarlandes, Math.-Wissensch. Fak., Dokt. Dissert., Saarbrücken, 1–172 (1963), chemotaxon.

Glabrous biennials or annuals. Leaves 2–3-pinnately divided, triangular; leaflets broadly ovate, frequently cuneate, lobed. Bracts few, sometimes absent; bracteoles many, short; rays several to many, subequal, spreading. Petals yellowish; styles equaling or rarely exceeding the conical stylopodium. Fruits glabrous, ovoid, oblong to suborbicular, slightly compressed laterally; ribs filiform; carpophore free, bifurcated; vittae small, vallecular 1, commissural 2. Seeds transversely suborbicular or broadly subelliptic, the face plane. $2n = 22$.

1(–3) species, (*P. segetum* (L.) W.D.J. Koch better treated as *Sison segetum* L.), geographic origin unknown, probably W Mediterranean and Macaronesia. Widely cultivated and used as an aromatic herb (parsley). Closely related to *Ammi* and *Sclerosciadium* (Jiménez-Mejías and Vargas 2015).

326. ***Peucedanum*** L. [Pimenov]

Peucedanum L., Sp. Pl. 245 (1753), Gen. Pl., ed. 5: 116 (1754), s. str.

Glabrous perennials. Leaves 2–4-ternate, rhachis geniculate; leaflets filiform to linear, rarely lanceolate. Bracts few, linear-subulate, caducous, or lacking; rays many; bracteoles linear-subulate or filiform, sometimes reflexed. Petals yellow or yellowish, rarely white; styles reflexed; stylopodium low-conical. Fruits strongly compressed dorsally; dorsal ribs filiform, marginal narrowly winged; carpophore bifid; commissure broad; mesocarp cells parenchymatous, partly with pitted walls; vallecular vittae 1, commissural 2. Seed face plane. $2n = 22, 44, 66$.

Seven species, Mediterranean region, SW Asia, Siberia, Kazakhstan, NW China. Steppes, limestone rocks and slopes. This narrow concept excludes many species from Eurasia, Africa, and Hawaii. In part, they have been transferred to *Imperatoria*, *Oreoselinum*, *Cervaria*, *Taeniopetalum*, *Kitagawia* and other satellite genera. The species contain peucedanin, used in skin therapy; some species are used as folk medicines.

327. ***Phellolophium*** Baker [Van Wyk & Tilney]

Phellolophium Baker, J. Linn. Soc. Bot. 21: 349 (1894); Sales & Hedge, Flore de Madagascar et Comores. Fam. 157 (2009), rev.

Acaulescent, sparsely pubescent perennials with a short rhizome. Leaves bipinnate; leaflets large, oblong, margins serrate. Bracts linear, entire; rays many; bracteoles small, caducous (usually absent at anthesis). Petals greenish-yellow; styles short, falcate, attenuate; stylopodium conical. Fruits glabrous, oblong-ellipsoid; mericarps transversely terete; ribs thick, corky or inconspicuous; carpophore bifid; vittae very large, vallecular 1–3, commissural 2–4. Seeds transversely sulcate beneath vittae, the face slightly sulcate.

Two species, Madagascar. Poorly known; doubtfully distinct from *Pimpinella*.

328. ***Phlojodicarpus*** Turcz. ex Ledeb. [Pimenov]

Phlojodicarpus Turcz. ex Ledeb., Fl. Ross. 2, 2: 331 (1844); Tamamschjan, Bot. Mat. Herb. Bot. Inst. Acad. Sci. URSS 14: 265–277 (1951), crit. acc.; Siplivinsky, Novosti Sist. Vyssh. Rast. 7: 257–268 (1970), rev.

Glabrous or puberulent perennials. Leaves 3-pinnate; leaflets linear to broadly lanceolate. Bracts and bracteoles glabrous or puberulent, long and sharply pointed, white-membranaceous at margin; rays scabrous or soft hairy. Petals white; styles reflexed; stylopodium low-conical. Fruits almost glabrous to densely puberulent, slightly compressed dorsally; dorsal ribs keeled, thickened, marginal narrowly winged; carpophore bifid; commissure broad; mesocarp cells with pitted walls; vallecular vittae 1, commissural 2. Seed face plane. $2n = 22, 44, 48, 60, 80, 84, 90$.

Three species, Asia (N, Central). Steppes, stony tundra, sand and rocks. *Phlojodicarpus sibiricus* (Fischer ex Spreng.) Koso-Pol. is used as a spasmolytic. Molecular data suggest an affinity to *Seseli* and some *Peucedanum* species.

329. ***Phlyctidocarpa*** Cannon & Theobald [Van Wyk & Tilney]

Phlyctidocarpa Cannon & Theobald, Mitt. Bot. Staatssamml. Münch. 6: 479 (1967); Burtt, Edinb. J. Bot. 48: 240 (1991), taxon.; Magee et al., Taxon 59: 567–580 (2010), taxon.; Van Wyk et al., Afr. Apiaceae 254–255 (2013).

Glabrous annuals. Leaves petiolate, tripinnate, broadly triangular; leaflets linear-lanceolate, sparsely toothed. Bracts conspicuous, linear-lanceolate, unequal, resembling the leaflets; rays few, long and slender; bracteoles small. Petals yellow; styles short; stylopodium obsolete. Fruits densely covered with rows of cone-shaped papillae, oblong, slightly compressed dorsally; mericarps transversely semi-circular; ribs inconspicuous; carpophore absent; vittae large, vallecular 1, commissure 2 and larger; rib oil ducts large, solitary in each rib. Seeds transversely slightly compressed dorsally, slightly sulcate beneath vittae, the face ± plane.

One species, *P. flava* Cannon & Theobald, Africa (Namibia). A phylogenetically isolated protoapioid.

330. ***Physospermopsis*** H. Wolff [Kljuykov]

Physospermopsis H. Wolff, Notizbl. Bot. Gart. Berlin 9: 276 (1925); Pimenov & Kljuykov, Feddes Repert. 111: 535–552 (2000), rev.

Acaulescent or caulescent, glabrous or puberulent perennials. Leaves 1–3-pinnate, with usually sessile segments; leaflets ovate to lanceolate, toothed or rarely entire. Bracts and bracteoles many, herbaceous, pinnatifid or rarely toothed at tip, cuneate at base. Petals white or rarely purple, or emerald green, nearly plane; styles reflexed; stylopodium flat to conical. Fruits glabrous or verrucose, ovoid, slightly compressed laterally; commissure narrow; ribs filiform to narrowly winged; mesocarp parenchymatous; vallecular vittae 2–3 (rarely 1 or many), commissural 2–4. Seed face plane or sulcate. $2n = 18, 22$.

15 species, Asia (SW China, Himalaya). Scrub, forests, among rocks, scree, among boulders, low grassy subalpine and alpine meadows. Related to *Hymenidium*.

331. ***Physospermum*** Cusson ex Juss. [Kljuykov]

Physospermum Cusson ex Juss., Hist. Soc. Roy. Med. Paris 279 (1787); Pimenov & Kljuykov, Bot. Zhurn. 66: 465–482 (1981), taxon.

Glabrous perennials with cylindrical taproots. Basal leaves tripinnate; leaflets ovate, lobed; cauline leaves reduced. Bracts and bracteoles entire, lanceolate, herbaceous. Calyx lobes lanceolate; petals white; styles reflexed; stylopodium conical. Fruits glabrous, ovoid, didymous; commissure very narrow; ribs filiform; carpophore bifurcated; mesocarp parenchymatous; vallecular vittae 1, commissural 2. Seed face deeply sulcate. $2n = 14, 16, 22$.

Two species, Europe, N Africa, SW Asia and Caucasus. Scrub and forests, meadows.

332. ***Physotrichia*** Hiern [Van Wyk & Tilney]

Physotrichia Hiern, J. Bot. 11: 161 (1873); Cannon, Fl. Zambes. 4: 603–605 (1978), reg. rev.; Townsend, Fl. Trop. E. Afr.: 83–85 (1989), reg. rev.; Van Wyk et al., Afr. Apiaceae: 256–259 (2013).
Spuriodaucus C. Norman (1930).

Glabrous or bristly biennials or woody perennials. Leaves mostly basal, variable, simple to ternatisect or pinnatisect, the upper ones much reduced; leaflets orbicular to oblong, usually large. Umbels compound; bracts leaf-like, linear, entire or minutely toothed; rays several, relatively long, spreading or ascending; bracteoles similar to bracts. Flowers several, with staminate flowers in lateral umbels; calyx lobes conspicuous; petals white to cream; styles short and stout; stylopodium shortly conical. Fruits densely verrucose, elliptic to ovoid; mericarps dorsally compressed; commissure broad; ribs usually prominent, the lateral ones sometimes winged; carpophore bifid; vittae large, usually solitary in the valleculae, 2 on the commissure; rib oil ducts large, solitary in each rib (lacking in some species?). Seeds transversely compressed dorsally, sulcate beneath vittae, the face ± plane. $2n = 22$.

Six species, C Africa. Grassy places in open woodlands. Poorly known; relationships unclear, but similar to *Diplolophium*.

333. ***Pilopleura*** Schischk. [Pimenov]

Pilopleura Schischk., Fl. USSR 17: 358 (1951); Pimenov, Bull. Main Bot. Gard. Acad. Sci. URSS 101: 44–51 (1976), rev.
Merwiopsis Saphina (1975).

Puberulent perennials. Basal leaves pinnate or bipinnate; leaflets ovate, toothed, sometimes lobed. Bracts and bracteoles many, linear to lanceolate, reflexed; rays many, hairy. Pedicels hairy; calyx lobes conspicuous, lanceolate, acute; petals white, dorsally puberulent; styles reflexed; stylopodium low-conical. Fruits compressed dorsally, densely covered with soft white hairs; dorsal ribs filiform, marginal winged; carpophore bifid; commissure broad; mesocarp cell walls with pitted walls; vittae cyclical; rib oil ducts large, 1–4 per rib. Seed face plane. $2n = 22$.

Two species, Asia (Middle, Kazakhstan). Stony hillsides, among rocks. Molecular data show a close affinity to *Seseli.*

334. ***Pimpinella*** L. [Pimenov]

Pimpinella L., Sp. Pl. 263 (1753); Gen. Pl., ed. 5, 128 (1754); Wolff, Pflanzenr. (Engler) IV, 228 (Heft 90): 1–398 (1927), rev.; Abebe Dawit, Bot. J. Linn. Soc. 110 (4): 327–372 (1992), reg. rev.
Afrosison H. Wolff (1912).
Albovia Schischk. (1950).
Similisinocarum Cauwet & Farille (1984).

Puberulent (rarely glabrous) perennials, biennials, or annuals, sometimes with lignified roots or rhizomes. Leaves simple to 3-pinnate; leaflets linear to ovoid, often dissected or toothed. Bracts and bracteoles lacking, or few, linear to lanceolate. Petals white, cream, yellow, red, purple, or pink, sometimes dorsally pubescent; styles short; stylopodium flat to long-conical. Fruits glabrous, pubescent, squamose, or tuberculate, usually laterally compressed and shortly beaked at apex; ribs filiform or inconspicuous; carpophore bifid or entire; commissure rather broad; mesocarp parenchymatous, thin; vittae many, cyclical, or vallecular (1)2–5, commissural (2)3–6. Seed face plane to concave. $2n = 16, 18, 20, 22, 24, 36, 40, 44$.

170–180 species, Eurasia, Africa, N and S America (adv.). Three sections and many subsections (Wolff 1927). Monophyly of the genus is doubtful. *Pimpinella anisum* is the aromatic anise, and some other species are used as folk medicines. Closely related to *Aphanopleura*, *Arafoe*, *Bubon*, *Psammogeton*, and *Registaniella.*

335. ***Pinacantha*** Gilli [Plunkett]

Pinacantha Gilli, Feddes Repert. 61: 207 (1959); Hedge & Lamond in Rechinger, Fl. Iran. 162: 527–528 (1987), reg. rev.

Glabrous and glaucous, probably perennials. Lower leaves ± fleshy, 3–4-pinnately dissected, ultimate segments linear with acute apices; upper leaves progressively reduced to broad, membranously margined sheaths, blades lobed to pinnate, reduced to linear segments.

One species, *P. porandica* Gilli, Asia (eastern Afghanistan); being known only from insufficient material of the type collection, Hedge and Lamond (1987) indicated that it was not possible to assign this taxon to an existing genus, but molecular data suggest its placement within *Ferula* (Degtjareva et al. 2018).

336. ***Pinda*** P.K. Mukh. & Constance [Plunkett]

Pinda P.K. Mukh. & Constance, Kew Bull. 41: 224 (1986); Mukherjee & Constance, Umbelliferae India 1: 248–249 (1993), rev.

Hirsute to hispidulous perennials with a swollen tuberous base. Leaves basal and cauline, ternate-pinnately or bipinnately compound, ovate to triangular-ovate; leaflets ovate-lanceolate to ovate, with large serrations and often lobed. Bracts few, broad and sheathing, or lacking; rays few, slender, unequal, spreading-ascending; bracteoles dimidiate, several, oblong or ovate. Calyx of outer flowers with two marginal, persistent outer lobes (inner ones obsolete); petals white with colored veins, the outer of marginal flowers greatly enlarged; styles very short; stylopodium massive, low-conical. Fruits glabrous, ellipsoid, strongly compressed dorsally; ribs low and rounded, the marginal slightly broader than the dorsal and lateral (but none winged); carpophore bifurcated; vittae prominent, vallecular 1–3, commissural 2 or more; rib oil ducts present. Seeds transversely dorsally compressed, the face slightly concave.

One species, *P. concanensis* (Dalzell) P.K. Mukh. & Constance, Asia (India, W Ghats). A segregate of *Heracleum.*

337. ***Pleurospermum*** Hoffm. [Kljuykov]

Pleurospermum Hoffm., Gen. Pl. Umbell. ed. 1: VIII (1814); Pimenov et al., Feddes Repert. 111: 499–515 (2000), taxon.; Pimenov & Kljuykov, Feddes Repert. 111: 517–534 (2000), rev.
Physospermum Lag. (1821), non Cusson ex Juss. (1782).

Subglabrous perennials. Leaves 1–3-ternate or pinnate; leaflets ovate or lanceolate, dentate. Bracts many, apically toothed, bracteoles entire, linear or lanceolate. Petals white, apex nearly plane; styles reflexed; stylopodium conical. Fruits glabrous, ovoid, slightly compressed laterally; commissure narrow; ribs keeled, inflated at the base; carpophore bifurcated; exocarp cells with strongly convex outer walls, separating from mesocarp at maturity; vallecular vittae 1, commissural 2; rib oil ducts 2 in each rib, large. Seed face broadly and deeply sulcate. $2n = 18, 22, 44, 50$.

Two species, Europe, N Asia. Scrub, forests, subalpine meadows. Many Chinese and Himalayan species traditionally ascribed to this genus are herein treated in *Aulacospermum, Hymenidium, Hymenolaena, Physospermopsis* and *Pterocyclus.*

338. ***Podistera*** S. Watson [Pimenov, Plunkett]

Podistera S. Watson, Proc. Amer. Acad. Arts 22: 475 (1887); Mathias & Constance, N. Amer. Fl. 28B(1): 148–150 (1944), reg. rev.

Dwarf, caespitose, subacaulescent or acaulescent, glabrous, scabrous or shortly puberulent perennials. Leaves pinnate (rarely bipinnate); leaflets lanceolate to orbicular, entire or deeply incised or lobed. Bracts 1 to many, linear, entire or trifid, glabrous, or absent; rays few, very short; bracteoles entire, foliaceous or subscarious, narrowly lanceolate, or 2–3-toothed, or 3–5-cleft, puberulent or glabrous. Calyx lobes conspicuous; petals yellow or greenish-yellow; styles short or longer but slender, compressed, divergent; stylopodium low-conical. Fruits glabrous, ovoid-oblong to ovoid, slightly compressed dorsally; ribs subinconspicuous or terete, almost equal; carpophore bifurcated; commissure narrow; mesocarp of cells with pitted walls; vallecular vittae 2–4, commissural 5–8; some species with broad rib oil ducts. Seed face plane. $2n = 22$.

Three species, N America (Pacific, Yukon). Alpine vegetation, stony slopes and screes. Molecular data confirm a close relationship to *Cymopterus, Lomatium* and their allies. See also *Orumbella.*

339. ***Polemannia*** Eckl. & Zeyh. [Van Wyk & Tilney]

Polemannia Eckl. & Zeyh., Enum. Pl. Afr. Extratrop. 347 (1837), nom. cons.; Hilliard & Burtt, Notes Roy. Bot. Gard. Edinb. 43: 225 (1986), reg. rev.; Burtt, Edinb. J. Bot. 48: 242 (1991), taxon.; Van Wyk et al., Afr. Apiaceae: 276–279 (2013).

Glabrous woody shrubs or small trees. Leaves palmately 3–5-foliolate, deltoid to orbicular, somewhat coriaceous; petiole not sheathing; leaflets obovate to elliptic, cuneate, sometimes 3-lobed and serrate, with distinct intramarginal nerve. Bracts few, large, usually elliptic; rays several to many, relatively long, spreading; bracteoles few, small, linear. Petals yellow, elliptic; styles very short; stylopodium conical. Fruits glabrous, oblong or ovoid, compressed dorsally; commissure fairly broad; ribs obtuse, prominent, dorsal ribs small but distinct, marginal ribs slightly winged; carpophore bifid; vittae large, vallecular 1, commissural 2, larger; rib oil ducts variable in size, solitary in each rib, variably present. Seeds transversely compressed dorsally, sulcate beneath vittae, the face convex, plane or slightly concave.

Three species, S Africa. Grasslands and woodlands. Closely related to *Heteromorpha.*

340. ***Polemanniopsis*** B.L. Burtt [Van Wyk & Tilney]

Polemanniopsis B.L. Burtt, Notes Roy. Bot. Gard. Edinb. 45: 498 (1988); Burtt, Edinb. J. Bot. 48: 243 (1991); Van Wyk et al., S. Afr. J. Bot. 76: 153–157 (2010); Van Wyk et al., Afr. Apiaceae: 280–282 (2013).

Glabrous, summer-deciduous, woody shrubs with long fleshy roots. Leaves hysteranthous (appearing after flowers senesce), crowded on branch ends, palmately (1–)3–5-foliolate, obovate, somewhat coriaceous; leaflets obovate, serrate, minutely aristate. Bracts few, caducous, linear-lanceolate; rays several, relatively long; bracteoles similar to bracts but smaller. Petals yellow; styles very short; stylopodium conical, small. Fruits glabrous, oblong, not compressed; mericarps heteromorphic, one 2-winged, the other 3-winged; commissure very narrow; primary ribs broadly winged, the wings hollow; secondary ribs sometimes present; carpophore bifid; vittae lacking; rib ducts large, without oil, solitary within each wing, forming cavities in the mature fruit. Seeds transversely elliptic, the face strongly sulcate in the three-winged mericarp.

Two species, S Africa (S Namibia and South Africa). Dry fynbos or desert. Related to *Steganotaenia.*

341. ***Polytaenia*** DC. [Plunkett]

Polytaenia DC., Coll. Mém. 5: 53 (1829); Mathias & Constance, N. Amer. Fl. 28B(2): 258–259 (1945), reg. rev.

Puberulent perennials. Leaves bipinnate or ternate-pinnate; leaflets large, crenate to incised or lobed. Bracts lacking; rays few, spreading-ascending, puberulent; bracteoles several, linear to filiform, entire, puberulent. Calyx lobes ovate, acute or acuminate; petals yellow; styles slender, spreading; stylopodium lacking. Fruits glabrous, broadly ellipsoid to globose or obovoid, strongly compressed dorsally; dorsal and lateral ribs filiform, marginal ribs corky-winged; carpophore bifurcated; vittae distinct or indistinct, vallecular 1 to several, commissural 2 to several, and scattered throughout the pericarp. Seeds transversely strongly compressed dorsally, the face plane. $2n = 22$.

Two species, N America (central); open plains, prairies, rocky glades. Molecular data suggest a close relationship to *Cymopterus*, *Lomatium* and their allies.

342. ***Polyzygus*** Dalzell [Plunkett]

Polyzygus Dalzell, J. Bot. Kew Bot. Misc. 2: 260 (1850); Mukherjee & Constance, Umbelliferae India 1: 247 (1993), reg. rev.

Hirtellous to hispidulous perennials with a short pseudoscape emerging from a tuberous root. Leaves basal and cauline, once or twice ternately or ternate-pinnately compound, triangular-ovate; leaflets lanceolate to ovate, mucronate-serrate and often lobed. Bracts usually lacking; rays few, unequal, slender, spreading-ascending; bracteoles dimidiate, several, narrow. Petals whitish, the outer often slightly enlarged; styles short to slender; stylopodium low-conical. Fruits glabrous, ellipsoid, strongly compressed dorsally; dorsal and lateral ribs low and filiform, marginal ribs narrowly thin-winged; carpophore bifurcated; vittae prominent, vallecular 1–2, commissural 2–4. Seeds transversely dorsally compressed, the face plane to concave. $n = 11$, $2n = 22$.

One species, *P. tuberosus* Walp., Asia (India, W Ghats). A segregate of *Heracleum.*

343. ***Portenschlagiella*** Tutin [Watson]

Portenschlagiella Tutin, Feddes Repert. 74(1–2): 32 (1967).
Portenschlagia Vis. (1850), non Trattinick (1818).

Pubescent perennials with a subglobose rootstock. Leaves 4–5-pinnate; ultimate segments linear-filiform, acuminate. Bracts many and lanceolate; rays many, puberulent; bracteoles many. Petals yellow, ciliate beneath. Fruits ovoid-cylindrical, slightly compressed dorsally, hispid with stellate hairs; primary ribs stout and prominent; secondary ribs filiform; carpophore bifid; vallecular vittae 1 (under each secondary rib), commissural 2; rib oil ducts 1 under each primary rib. Seed slightly compressed dorsally, the face plane. $2n = 22$.

One species, *P. ramosissima* (Port. ex Spreng.) Tutin, Europe (S Italy, former Yugoslavia, NW Albania). Rocky places.

344. ***Postiella*** Kljuykov [Kljuykov]

Postiella Kljuykov, Bull. Soc. Nat. Mosc., Div. Biol. 90: 103 (1985).

Glabrous perennials with small, globose tubers. Leaves 2–3-pinnate, segments filiform and drooping; leaflets filiform. Bracts and bracteoles herbaceous, subulate; rays few. Petals white; styles reflexed; stylopodium flat. Fruits glabrous, globose, slightly compressed laterally, constricted at the apex; commissure narrow; ribs obscure; mesocarp parenchymatous; vittae cyclic and anastomosing. Seed face plane.

One species, *P. capillifolia* (Post ex Boiss.) Kljuykov, SW Asia (S Turkey). Woods and thickets. Closely related to *Bunium*.

345. ***Prangos*** Lindl. Figs. 3C, 22 [Pimenov]

Prangos Lindl., Quart. J. Sci. Lit. Art 19: 7 (1825); Herrnstadt & Heyn, Boissiera 26: 1–91 (1977), rev.; Pimenov & Tikhom., Feddes Repert. 94(3–4): 145–164 (1983), taxon.; Lyskov et al., Pl. Syst. Evol. 303(7): 815–826 (2017), rev.
Koelzella M. Hiroe (1958).
Neocryptodiscus Hedge & Lamond (1987).

Slightly scabrid to softly pubescent perennials. Leaves 3-many-pinnatisect, glabrous or scabrid; leaflets linear to lanceolate, entire, mucronate, scabrid. Bracts linear; rays scabrous, thickened in fruits; bracteoles entire, often puberulent. Calyx lobes lanceolate, or absent; petals usually yellow, puberulent dorsally; styles divergent or reflexed; stylopodium flat. Fruits glabrous, puberulent, or covered by papillae, sometimes didymous; commissure broad or narrow, ribs subequal, winged, straight or slightly wavy, or inconspicuous; mesocarp bistratous; five columns of large parenchyma cells with lignified walls in inner layer, with isolated vascular elements; vittae cyclical in inner mesocarp layer; many scattered secretory ducts in outer layer. Seed face deeply furrowed with involute margins. $2n = 22, 34, 36, 44, 66$.

47 species, Mediterranean Europe, and SW, Middle and C Asia. From sandy and clay deserts on the plains to subalpine meadows, stony slopes, among shrubs. Fodder plants in arid pastures and as winter cattle forage. Molecular data suggest the inclusion of *Alococarpum* and a close relationship to *Ferulago*, *Azilia* and *Bilacunaria* (Lyskov et al. 2017).

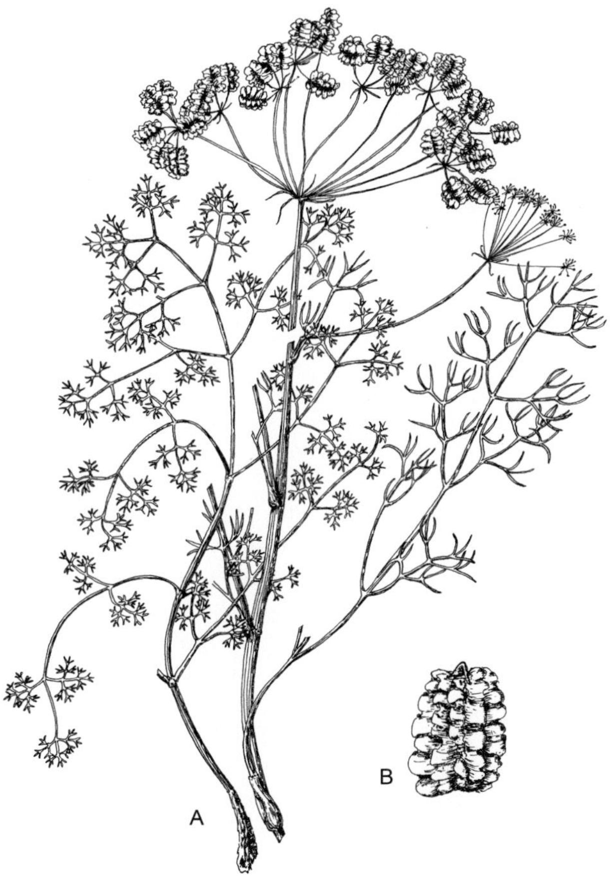

Fig. 22. Apiaceae, Apioideae. *Prangos latiloba*. **A** Leaf and inflorescence fragments. **B** Fruit. (Orig., illustration by Elena Mzhelskaya)

346. ***Prionosciadium*** S. Watson [Plunkett]

Prionosciadium S. Watson, Proc. Amer. Acad. Arts 23: 275 (1888); Mathias & Constance, N. Amer. Fl. 28B(2): 205–212 (1945), reg. rev.; Mathias & Constance, Contr. Univ. Michigan Herb. 11: 15–19 (1973), taxon.

Glabrous or pubescent perennials growing from stout taproots or tubers. Leaves ternate-pinnate to ternate-pinnately decompound; rachis frequently winged; leaflets filiform to ovate, serrate to pinnately divided (rarely entire). Bracts lacking or few and filiform; fertile rays few to many, spreading-ascending, spreading, or divaricate; bracteoles several, filiform to linear-lanceolate, entire. Petals white, greenish-yellow, or purple; styles slender and recurved; stylopodium lacking. Fruits glabrous or rarely hispidulous, oblong to globose or ellipsoid, frequently retuse at apex and base, strongly compressed dorsally; dorsal and lateral ribs prominent, occasionally narrowly winged; marginal ribs broadly thin-winged; carpophore bifurcated, each half usually apically bifid; vallecular vittae 1 to several, commissural 4–12. Seed transversely dorsally compressed, frequently sulcate under the vittae, the face slightly concave to involute. $n = 21, 22$.

Ten species, N America (Mexico). Molecular data place the genus with *Arracacia*, *Rhodosciadium*, and their allies (Downie et al. 2010).

347. ***Psammogeton*** Edgew. [Lee]

Psammogeton Edgew., Proc. Linn. Soc. 1: 253 (1845).

Pubescent or glabrous annuals. Leaves 1–2-ternate, broadly triangular, the ultimate divisions linear to lanceolate. Bracts and bracteoles several, linear to lanceolate with scabrous margins, simple, glabrous or pubescent; rays several, unequal, hispid. Petals pinkish-white to pinkish, the outer rarely radiate; styles elongated; stylopodium conical. Fruits bristly, ovoid to oblong, terete to slightly compressed laterally; primary ribs filiform, with hammer-like, erect hairs arranged randomly; secondary ribs 4, filiform, hairs like those of the primary ribs; carpophore free; vittae large, transversely triangular, vallecular 1 (under each secondary rib), commissural 2. Seeds transversely subterete, the face plane to slightly concave. $2n = 18$.

Seven species, SW, S, and Middle Asia. Molecular evidence suggests a close relationship to *Pimpinella*. Molecular data suggest a close affinity to *Aphanopleura*, as well as *Pimpinella* and *Arafoe* (Katz-Downie et al. 1999).

348. ***Pseudocannaboides*** B.-E. van Wyk [Van Wyk & Tilney]

Pseudocannaboides B.-E. van Wyk in Van Wyk et al., Taxon 48: 742 (1999); Sales & Hedge, Flore de Madagascar et Comores. Fam. 157 (2009), rev.; Van Wyk et al., Afr. Apiaceae: 283 (2013).

Acaulescent, glabrous perennials with a fleshy rhizome. Leaves crowded in the basal part of the stem, tripinnate, triangular leaflets linear, strongly acuminate, serrate (similar to those of *Cannabis sativa* L.). Bracts conspicuous, linear to lanceolate-acuminate; rays many, long, slender; bracteoles conspicuous, lanceolate-acuminate. Petals yellow; styles short; stylopodium broadly conical. Fruits glabrous, ellipsoid, compressed laterally; mericarps transversely terete; commissure very narrow; ribs conspicuous, narrowly winged; carpophore bifid; vittae very large, vallecular 1, commissural 2; rib oil ducts small, solitary. Seeds transversely terete but sulcate beneath the vittae, the face bisulcate.

One species, *P. andringitrensis* (Humbert) B.-E. van Wyk, Madagascar (Andringitra Mountains). Poorly known.

349. ***Pseudocarum*** C. Norman [Van Wyk & Tilney]

Pseudocarum C. Norman, J. Bot. 62: 333 (1924); Townsend, Fl. Trop. E. Afr.: 50–53 (1989); Van Wyk et al., Taxon 48: 737–745 (1999), rev.; Van Wyk et al., Afr. Apiaceae: 284–285 (2013).

Glabrous, woody perennial climbers. Leaves ternate to bi- or subtriternate, triangular; petioles not sheathing; petioles and petiolules sometimes prehensile; leaflets ovate to lanceolate-ovate, margins serrate, with setaceous-mucronate teeth. Bracts and bracteoles conspicuous, ovate to lanceolate, sometimes with 3–5 longitudinal veins, sometimes finely ciliate along the margins; rays up to 20, slender, wiry. Petals greenish-white or yellowish; styles short; stylopodium broadly conical. Fruits glabrous, ovoid to ellipsoid, scarcely compressed laterally; mericarps ± terete; commissure narrow; ribs absent or narrowly winged; carpophore bifid or apparently entire; vittae large, anastomosing, vallecular 1(2), commissural 2(3); rib oil ducts small, 1 in each rib or absent. Seeds transversely ± pentagonal, the face plane to slightly sulcate.

Two species, NE Africa and Madagascar. Related to *Heteromorpha*.

350. ***Pseudocymopterus*** J.M. Coult. & Rose [Plunkett]

Pseudocymopterus J.M. Coult. & Rose, Rev. North Amer. Umbellif. 74 (1888); Mathias & Constance, N. Amer. Fl. 28B(2): 189–190 (1945), reg. rev.; Sun et al., J. Torrey Bot. Soc. 133: 499–512 (2006), taxon.

Subacaulescent to caulescent, glabrous to somewhat scaberulous, ciliate or hirsute-pubescent, perennials. Leaves 1–3-pinnately dissected or compound; leaflets or ultimate segments lanceolate to filiform. Bracts usually lacking; rays few to many, spreading-ascending or spreading, subequal to equal; bracteoles dimidiate, oval to filiform, connate at base. Petals yellow or purple; styles slender; stylopodium lacking. Fruits glabrous, ovoid to ovoid-oblong, compressed dorsally; marginal ribs broadly winged (wings spongy), one or more of the dorsal and lateral ribs similar to the marginal, or reduced to obsolete; carpophore bifurcated; vallecular vittae 1–5, commissural 2–6, rib oil ducts rare, at the base of the wings. Seeds transversely compressed dorsally, the face plane. $n = 11$, $2n = 22$.

Two species (all other species transferred to *Cymopterus* or *Pteryxia*), N America (C and S Rocky Mountains and adjacent Mexico to Durango); mountains. Molecular data suggest a close relationship to *Cymopterus*, *Lomatium* and their allies.

351. ***Pseudoridolfia*** Reduron, Mathez & S.R. Downie [Reduron]

Pseudoridolfia Reduron, Mathez & S.R. Downie, Acta Bot. Gall. 156(3): 496 (2009).

Glabrous annuals. Leaves 2–3-pinnately divided, triangular; ultimate divisions filiform. Bracts and bracteoles lacking; rays several, subequal. Petals bright yellow, ovate, involute with a truncate apex; stylopodium conical. Fruits glabrous, smooth, ovoid; ribs filiform; carpophore free, bifurcated; vittae quite large, vallecular 1, commissural 2. Seeds transversely almost reniform, the face slightly concave. $2n = 22$.

One species, *P. fennanei* Reduron, Mathez & S.R. Downie, N Africa (Morocco). Rocky open forests. Closely related to *Ammi* and *Sclerosciadium* (Jiménez-Mejías and Vargas 2015).

352. ***Pseudorlaya*** (Murb.) Murb. [Lee]

Pseudorlaya (Murb.) Murb., Acta Univ. Lund. 33: 86 (1897); Saenz de Rivas, Anal. Inst. Bot. Cavanilles 31(2): 191–204 (1974), fruit anat., taxon.

Densely pubescent annuals, rarely biennials. Leaves 2–3-pinnately divided, ovate, ultimate divisions ovate to oblong. Bracts and bracteoles several, linear to lanceolate, sometimes trifid; rays few to several, subequal to unequal. Calyx lobes conspicuous; petals purple, pinkish to white, the outer slightly radiate; styles short; stylopodium conical. Fruits spinescent, ellipsoid, not compressed to strongly compressed dorsally; primary ribs filiform, prominent, hispid; secondary ribs 4, with 1–3 rows of glochidiate spines; carpophore free, entire; vittae large, vallecular 1 (under each secondary rib), commissural 2. Seeds strongly compressed dorsally or transversely trapeziform, the face plane to slightly concave. $2n = 14, 16$.

Three species, Mediterranean region, S Europe, N Africa, and SW Asia, sandy places. Nested within *Daucus*, and recently placed into its synonymy (Banasiak et al. 2016).

353. ***Pseudoselinum*** C. Norman [Van Wyk & Tilney]

Pseudoselinum C. Norman, J. Bot. 66, Suppl. 1: 201 (1929); Van Wyk et al., Afr. Apiaceae: 286 (2013).

Glabrous perennials with a rhizome. Leaves bipinnatisect, radical, triangular; leaflets irregularly obovate to rounded, margins serrate. Bracts many, linear-acuminate; rays many, long and slender; bracteoles small. Petals purplish, lanceolate-acuminate; styles short; stylopodium conical. Fruits glabrous, ovoid, strongly compressed dorsally; ribs broadly winged, dorsal ribs slightly narrower than the marginal; carpophore bifid; vallecular vittae 1, commissural 2. Seeds transversely compressed dorsally, the face ± plane.

One species, *P. angolense* (C. Norman) C. Norman, S tropical Africa (Angola). Poorly known, with no obvious relatives.

354. ***Pseudotaenidia*** Mack. [Plunkett]

Pseudotaenidia Mack., Torreya 3: 158 (1903); Cronquist, Brittonia 34: 365–367 (1982), rev.

Glabrous perennials with subfusiform tubers. Leaves 2–3-ternately divided to compound; leaflets oblong to oval, entire or rarely with a basal lobe, acute and mucronate. Bracts and bracteoles lacking (or bracts solitary and linear); rays few, slender, spreading-ascending, unequal. Petals yellow; styles short, spreading; stylopodium lacking. Fruits glabrous, oblong-ovoid to suborbicular, compressed dorsally; dorsal and lateral ribs filiform, marginal ribs of adjacent mericarps contiguous, corky-winged (wings narrower than the fruit body); carpophore bifurcated; vittae usually 1 in each vallecula, 2–4 on the commissure. Seeds transversely compressed dorsally, the face plane. $2n = 22$.

One species, *P. montana* Mack., N America (E); restricted to shale barrens of the Appalachian Mountains. Often considered congeneric with *Taenidia*, despite differences in the direction of fruit compression (Cronquist 1982).

355. ***Pseudotrachydium*** (Kljuykov, Pimenov & V.N. Tikhom.) Pimenov & Kljuykov [Kljuykov]

Pseudotrachydium (Kljuykov, Pimenov & V.N. Tikhom.) Pimenov & Kljuykov, Feddes Repert. 111: 526 (2000).
Aulacospermum Ledeb. sect. *Pseudotrachydium* Kljuykov, Pimenov & V.N. Tikhom. (1976).

Glabrous perennials. Leaves 2–3-pinnate with sessile segments; leaflets linear to lanceolate. Bracts and bracteoles entire, linear, herbaceous. Petals yellow, base short-clawed or broad; styles reflexed; stylopodium low conical. Fruits verrucose, ovoid, compressed laterally; commissure narrow; ribs narrowly winged or keeled; carpophore bifurcated; vallecular vittae 1–3 or many. Seed face deeply sulcate. $2n = 18$.

Five species, Asia (from Turkey and Iraq to Middle Asia, Afghanistan and Pakistan). Mountain meadows, stony slopes, rocks and scree. Closely related to *Aulacospermum* and its allies.

356. ***Pternopetalum*** Franch. [Pu]

Pternopetalum Franch., Nouv. Arch. Mus. Hist. Nat. Paris II, 8: 246 (1885); M.A. Farille, Candollea 40: 557 (1985), reg. rev.; Mukherjee & Constance, Umbelliferae India 1: 91–93 (1993), reg. rev.; She et al., Fl. China 14: 85–92 (2005), reg. rev.; Wang, J. Syst. Evol. 50: 550–572 (2012), rev.

Glabrous to hirsute annuals or perennials. Basal leaves ternately or ternate-pinnately compound or dissected; cauline leaves similar, reduced, or lacking; leaflets or ultimate divisions broadly or narrowly dissected. Bracts usually absent; rays few to many, unequal, spreading-ascending; bracteoles few, linear-lanceolate. Calyx lobes prominent, unequal to subequal (rarely minute to obscure); petals white or purplish, base attenuate, thickened; styles elongate or short, erect to spreading; stylopodium conical or low-conical. Fruits glabrous, oblong-ovoid to broadly ovoid, apex narrowed, base often cordate, compressed laterally; ribs denticulate, scabridulous or filiform; carpophore bifid to entirely bifurcated; vallecular vittae 1–3, commissural 2–6. Seed transversely subterete, the face plane to slightly concave. $2n = 36$.

20 species, Asia (China, Nepal, Bhutan, N India, Korea and Japan). Mountains.

357. ***Pterocyclus*** Klotzsch [Kljuykov]

Pterocyclus Klotzsch in Klotzsch & Garcke, Bot. Ergebn. Reise Prinz Waldemar v. Preussen 150 (1862); Pimenov & Kljuykov, Feddes Repert. 111: 517–534 (2000), rev.

Glabrous perennials. Leaves 1–3-ternate or pinnate; basal primary segments long-petiolulate; leaflets rhomboid to almost round, entire to pinnatifid, serrate, or crenate. Bracts one to several, usually entire; bracteoles many, entire, linear or lanceolate. Petals white to purple; styles reflexed; stylopodium conical. Fruits glabrous, oblong; commissure narrow; mericarps slightly compressed dorsally; ribs winged, equal; carpophore bifurcated; vallecular vittae 1, commissural 2. Seed face concave. $2n = 18, 22$.

Four species, Asia (SW China, Himalaya). Scrub, forests and subalpine meadows. Related to *Hymenidium*, *Pleurospermum*, and *Hymenolaena*.

358. ***Pteroselinum*** (Rchb.) Rchb. [Reduron]

Pteroselinum (Rchb.) Rchb., Fl. Germ. Excurs. 453 (1832); Spalik et al., Pl. Syst. Evol. 243: 189–210 (2004), phylog.; Reduron, Bull. Soc. Bot. Centre-Ouest n° spécial 29: 2149–2158, monogr.

Glabrous perennials. Leaves 3–4-pinnately divided, triangular; leaflets ovate, deeply cut into lanceolate to linear lobes. Bracts and bracteoles many; rays many, subequal to unequal. Petals white, the outer cordate, notched; stylopodium short-conical. Fruits glabrous, ovate or ellipsoid, strongly compressed dorsally; dorsal ribs filiform but prominent, the marginal ribs broadly winged; carpophore free, bifurcated; commissure broad; vittae medium-sized, vallecular 1, commissural 2–4. Seeds transversely compressed dorsally, the face plane. $2n = 22$.

One species, *P. austriacum* (Jacq.) Rchb., C & E Europe. Mountain and hill slopes, forest margins. Segregated from *Peucedanum* s. lat.

359. ***Pterygopleurum*** Kitag. [Watson]

Pterygopleurum Kitag., Bot. Mag. (Tokyo) 51: 654 (1937).

Essentially glabrous perennials, stems sometimes rooting at basal nodes. Leaves 1–2-pinnate or ternate to 2–3-pinnate; ultimate segments linear or linear-lanceolate, entire. Bracts and bracteoles several and linear; rays rather few, spreading. Calyx lobes conspicuous, lanceolate; petals white; styles slightly longer than conical stylopodium. Fruits glabrous, ovoid, slightly compressed laterally; ribs prominent, corky, dilated at base, narrowly winged; carpophore bifurcated; vallecular vittae 1, commissural 2. Seed transversely subterete, the face plane. $n = 11$.

One species, *P. neurophyllum* (Maxim.) Kitag., Asia (China, Korea and Japan). Stream sides and damp areas.

360. *Pteryxia* (Nutt. ex Torr. & A. Gray) J.M. Coult. & Rose [Plunkett]

Pteryxia (Nutt. ex Torr. & A. Gray) J.M. Coult. & Rose, Contr. U. S. Natl. Herb. 7: 170 (1900); Mathias & Constance, N. Amer. Fl. 28B(2): 183–187 (1945), reg. rev.

Caespitose, acaulescent or caulescent, essentially glabrous perennials. Leaves bipinnate to pinnately or ternate-pinnately decompound; ultimate segments rigid or herbaceous, linear, oblong, or subcuneate. Bracts usually lacking; rays few to many, spreading-ascending, usually unequal; bracteoles usually dimidiate, linear to lanceolate, entire or occasionally toothed, sometimes scarious. Calyx lobes prominent, often unequal; petals yellow, whitish, or purple; styles slender; stylopodium lacking. Fruits glabrous or pubescent, narrowly oblong to ovoid, compressed dorsally; marginal ribs membranously winged (wings linear in cross section), some or all dorsal and lateral ribs like the marginal or shorter; carpophore bifurcated; vittae small, vallecular 1--several, commissural several. Seeds transversely compressed dorsally, the face somewhat concave to plane. $n = 11$, $2n = 44$.

Five species, N America (W United States).

361. *Ptilimnium* Raf. [Plunkett]

Ptilimnium Raf., Neogenyton 2 (1825); Easterly, Brittonia 9: 136–145 (1957), rev.; Kral, Sida 9: 124–134 (1981), taxon.; Feist et al., Taxon 61: 402–418 (2012), phylog., rev.

Glabrous annuals or perennials with fascicles of fibrous roots, corms or rhizomes. Leaves pinnately decompound; ultimate divisions filiform. Bracts and bracteoles entire or pinnatifid, conspicuous or inconspicuous; rays few to many, spreading-ascending to spreading. Petals white (rarely pink); styles spreading to reflexed; stylopodium conical. Fruits glabrous, ovoid, compressed laterally; dorsal and lateral ribs filiform, rounded or acute, marginal ribs inconspicuous but distinct from the corky tissue that forms a winged, conspicuous or inconspicuous, longitudinal band around the commissural edge of each mericarp; carpophore bifid at the apex; vallecular vittae 1, commissural 2. Seeds transversely subterete, the face plane. $n = 7, 14, 16$.

Four species, N America (E United States); wet places.

362. *Ptychotis* W.D.J. Koch [Reduron]

Ptychotis W.D.J. Koch, Nova Acta Phys.-Med. Acad. Caes. Leop.-Carol. Nat. Cur. 12(1): 124 (1824); Wolff in Engl., Pflanzenr. 4 (228) Heft 90: 127–129 (1927), rev.

Glabrous biennials. Rosette leaves pinnately divided, oblong; leaflets ovate or suborbicular, ± deeply lobed; upper leaves divided into linear segments. Bracts absent or few, linear (rarely bifid); bracteoles several, setaceous; rays several to many, subequal to unequal, spreading. Petals white; styles equaling the short-conical stylopodium. Fruits glabrous, ovoid-oblong; slightly compressed laterally; ribs narrow but prominent, acute; carpophore free, entire or bifid; vittae small to medium-sized, vallecular 1, commissural 2. Seeds transversely sub-pentagonal, the face plane. $2n = 22$.

One or two species, Europe (SW to Switzerland, Italy, Corsica, Sardinia). Calcareous screes.

363. *Pycnocycla* Lindl. [Van Wyk & Tilney]

Pycnocycla Lindl. in Royle, Ill. Bot. Himal. Mount. 232, t. 51 (1834); Boiss., Fl. Or. 2: 951–954 (1872); Hedge & Lamond, Notes Roy. Bot. Gard. Edinb. 37(1): 61–66 (1978), rev.; Hedberg & Hedberg, Fl. Ethiopia & Eritrea: 7–8 (2003), reg. rev.; Van Wyk et al., Afr. Apiaceae: 287 (2013).

Glabrous or partially pubescent perennials with thick rhizomes, or shrubs. Leaves 1–4-pinnate, subpalmatisect or sometimes simple, coriaceous, often spiny; petiole sometimes sheathing; leaflets narrowly linear. Bracts and bracteoles narrowly lanceolate to filiform, often spiny; rays many, very short, forming condensed capitula; umbellules with a single, sessile, central perfect flower surrounded by pedicellate staminate flowers, pedicels of which often indurate after anthesis, forming a hirsute cage around the fruit. Petals white or flushed with pink, outer ones sometimes radiate, apex pubescent; styles longer than shortly conical stylopodium. Fruits pubescent, oblong-pyramidal, terete, sometimes with a single carpel; commissure broad; ribs inconspicuous, rounded; carpophore bifid or absent; vallecular vittae several, commissural several; rib oil ducts present, relatively large. Seeds transversely subterete, the face plane to deeply sulcate. $2n = 22$.

14 species, N Africa and SW and S Asia. Arid scrub and open woodlands.

364. ***Pyramidoptera*** Boiss. [Pimenov]

Pyramidoptera Boiss., Diagn. Pl. Orient. Nov. II, 3(2): 106 (1856); Koso-Poljansky, Monit. Tifl. Bot. Gard. 11(2): 3–13 (1915), carpol.

Glabrous perennials. Leaves mainly basal, bipinnatisect, coriaceous; petiole sheaths white-membranous; leaflets linear or deeply slashed into narrow divisions, spiny. Bracts and bracteoles linear-lanceolate, white-membranous at the margin, deflected; rays many, thickened. Calyx lobes subulate, elongate, spreading; petals white; styles divergent; stylopodium conical. Fruits heteromericarpic (one mericarp sterile, one fertile), pyramid-shaped or truncate with winged, papery ribs, indehiscent; carpophore absent; mesocarp parenchyma in ribs of lignified cells; vittae developed in the fertile mericarp, short, dorsal 1, ventral 2. Seed face almost plane or slightly concave.

One species, *P. cabulica* Boiss., E Afghanistan. Rocks, limestone in subalpine belt, scree. Molecular studies suggest a relationship to *Bunium*, *Crithmum*, *Elaeosticta*, *Lagoecia*, *Oedibasis*, *Scaligeria*, and (in part) *Trachyspermum*.

365. ***Registaniella*** Rech.f. [Plunkett]

Registaniella Rech.f., Fl. Iran. 162: 152 (1987).

Glabrescent annuals. Leaves apparently pinnate or ternate (known only from fragments); segments ternately dissected, sharply serrate with setose-apiculate teeth. Bracts lanceolate, long-pilose, margins long-ciliate; rays several, subequal, slender, erect in flower, spreading in fruit; bracteoles few, oblong-ovate, long-pilose. Petals white, the marginal broadly obcordate with longitudinal rows of reddish hairs; styles slender, erect to spreading; stylopodium conical. Fruits pubescent, ellipsoid, compressed laterally, ribs low and rounded, valleculae covered with longitudinal rows of simple hairs; carpophore unknown; vittae very large, vallecular irregularly shaped and distributed, commissural 2. Seeds transversely compressed laterally, the face plane.

One species, *R. hapaxlegomena* Rech.f., Asia (SE Afghanistan), in the shifting sandy deserts of Registan.

366. ***Rhabdosciadium*** Boiss. [Pimenov]

Rhabdosciadium Boiss., Ann. Sci. Nat. II, 2: 68 (1844); Lyskov et al., Phytotaxa 331(2): 253–262 (2017), phylog.

Glabrous perennials; stems almost leafless, junciform. Basal leaves shortly petiolate, cauline with reduced petioles and oblong sheaths, pinnatisect, obovate; segments few, divaricate, filiform to oblong, entire to deeply lobed. Bracts ovate to lanceolate; rays few, unequal; bracteoles lanceolate, triangular to linear. Central flower in umbellule hermaphroditic, sessile, others staminate; petals white, greenish or rose; styles long, divergent; stylopodium short, conical. Fruits linear-cylindrical, slightly compressed laterally; ribs filiform, marginal slightly winged; carpophore bifid; commissure narrow; vittae many, hardly distinguished. Seed face plane. $2n = 22$.

Six species, SW Asia (Turkey, Iran). Dry rocky and herbaceous slopes, scree. Closely related to *Grammosciadium*.

367. ***Rhizomatophora*** Pimenov Fig. 3H [Pimenov]

Rhizomatophora Pimenov in Pimenov & Ostroumova, Umbelliferae Russia: 284 (2012).

Perennials with long horizontal rhizomes. Leaves 2–3-ternatisect; leaflets petiolulate, broad, ovate to oblong; margins dentate or incised. Bracts absent or 3–6, lanceolate to subulate; rays many, velvety; bracteoles many, subulate; pedicels many, minutely pubescent. Calyx lobes triangulate, membranous at margin; petals white or pinkish; stylopodium plane to conical, wavy at margin. Fruits compressed dorsally, glabrous;

dorsal ribs filiform, marginal ribs broadly winged; commissure broad; vittae minute, vallecular 1, commissural 2; rib secretory ducts large, 2 in each rib. Seed face plane. $2n = 22$.

One species, *R. aegopodioides* (Boiss.) Pimenov, Europe (Balkans), W Asia (Turkey, the Caucasus), along streams, rocky rivers, shady moist forests. Segregate of *Peucedanum*.

368. ***Rhodosciadium*** S. Watson [Plunkett]

Rhodosciadium S. Watson, Proc. Amer. Acad. Arts 25: 151 (1890); Mathias & Constance, N. Amer. Fl. 28B(2): 213–218 (1945), reg. rev.; Mathias & Constance, Contr. Univ. Michigan Herb. 11: 19–23 (1973), taxon.

Glabrous (often glaucous) to puberulent, scaberulous or hispidulous perennials, sometimes with tuberous roots. Leaves 1–2-pinnate, ternate, ternate-pinnate to ternate, or ternate-pinnately decompound. Bracts lacking or 1 to several and filiform, foliaceous or minute; fertile rays few, spreading-ascending or divaricate; bracteoles several, lanceolate to filiform. Petals purple or greenish-yellow; styles long or short, spreading or recurved; stylopodium low-conical. Fruits glabrous, ellipsoid to obovoid, strongly compressed dorsally; dorsal and lateral ribs prominent to filiform, marginal ribs broadly thin-winged; carpophore bifurcated; vittae small, vallecular 1–3, commissural 6–9. Seeds transversely dorsally compressed, the face involute, sulcate, or plane. $n = 21, 22, 42$.

12 species, N America (Mexico). Molecular data suggest placement with other *Arracacia*, *Coulterophytum* and other meso-American genera (Downie et al. 2010).

369. ***Rhopalosciadium*** Rech.f. [Plunkett]

Rhopalosciadium Rech.f., Anz. Oesterr. Akad. Wiss., Math.-Naturwiss. Kl. 89: 240 (1952); Rechinger, Fl. Iran. 162: 130–131 (1987), rev.; Spalik & Downie, Ann. Missouri Bot. Gard. 88: 270–301 (2001), phylog.

Pubescent annuals. Basal leaves withered at time of fruiting; cauline leaves subappressed-setulose, bipinnately dissected, oblong-triangular or oblong-ovate; ultimate segments linear-lanceolate, flat, apex acute, the bases often connected. Bracts lacking; rays few (2–3), unequal, the marginal rays longer; bracteoles several, very narrowly linear, acute, spreading or reflexed, ± appressed-setulose. Calyx lobes unequal, laciniate-subulate; petals white with a central purplish stripe, often asymmetric; styles filiform, erect-divergent. Mature fruits unknown; immature fruits appressed-setose, linear-clavate, thicker above with an attenuate base.

One species, *R. stereocalyx* Rech.f., Asia (W Iran). Poorly known and possibly based on an aberrant individual of *Torilis* (Spalik and Downie 2001).

370. ***Rhysopterus*** J.M. Coult. & Rose [Plunkett]

Rhysopterus J.M. Coult. & Rose, Contr. U.S. Natl. Herb. 7 (1): 185 (1900); Mathias & Constance, N. Amer. Fl. 28B(1): 128–129 (1944), rev.

Acaulescent (but appearing caulescent, with a pseudoscape having reduced leaves), glabrous perennials. Leaves rosulate, ternate-subpinnately divided, subcoriaceous; divisions ovate, obtuse, confluent, toothed or lobed. Bracts lacking; rays few, stout, spreading and reflexed, subequal; bracteoles conspicuous, dimidiate, foliaceous. Calyx lobes conspicuous, scarious-margined; petals white; styles very short, spreading; stylopodium lacking. Fruits glabrous, ovoid to orbicular, compressed laterally; ribs 7 (1 dorsal, 2 lateral, 2 marginal, plus 2 accessory ribs adjacent to the commissural margin), equal, conspicuous, obtuse, corky, appearing crenulate-winged on immature fruits; carpophore obsolete; vallecular vittae solitary, commissural 2, rib oil ducts solitary in the apex of each rib. Seeds transversely compressed dorsally, the face concave.

One species, *R. plurijugus* J.M. Coult. & Rose, N America (SE Oregon); loose dry ground. Sometimes treated in *Cymopterus*.

371. ***Ridolfia*** Moris [Reduron]

Ridolfia Moris, Enum. Sem. Hort. Taurin. 43 (1841); Briquet, Rev. Gén. Bot. 25(2): 61–82 (1914), carpol.

Glabrous, fetid annuals. Leaves 3–4-pinnately divided, triangular; ultimate divisions linear. Bracts and bracteoles lacking; rays many, subequal, spreading. Petals bright yellow, involute with a truncate apex; styles equaling the short-conical stylopodium. Fruits glabrous, smooth, ovoid-cylindrical; slightly compressed laterally; ribs very slender, scarcely visible; carpophore bifurcated; vittae small, vallecular 1, commissural 2. Seeds transversely broadly subelliptic, the face plane. $2n = 22$.

One species, *R. segetum* (Guss.) Moris, Mediterranean region (Europe, N Africa, and SW Asia) and Macaronesia. Open places, weedy in arable fields. Closely related to *Pseudoridolfia* and *Foeniculum* (Jiménez-Mejías and Vargas 2015).

372. ***Rivasmartinezia*** Fern. Prieto & Cires [Reduron]

Rivasmartinezia Fern. Prieto & Cires, Pl. Biosyst. 148(5): 982 (2014).

Glabrous perennials. Leaves 3–5(6)-ternately divided, broadly triangular, ultimate divisions filiform. Bracts few to several; rays several, unequal or subequal; bracteoles several to many. Flowers several to many per umbellule, pedicels glabrous and smooth; petals white; styles subequal or exceeding the stylopodium. Fruits smooth, ovoid to narrowly ellipsoidal; ribs prominent; vallecular vittae 1–5, commissural 2–10. Seed face plane or slightly concave.

Two species, Europe (Spain), in rocky areas. Related to *Dethawia*.

373. ***Rohmooa*** Farille & Lachard [Reduron]

Rohmooa Farille & Lachard, Acta Bot. Gallica 149(4): 377 (2002).

Dwarf, glabrous perennials with tuberous roots. Leaves pinnate, triangular; leaflets deeply 3-lobed into lanceolate lobes. Bracts lacking or few; rays few, unequal; bracteoles few. Petals dark purple, the apex plane; styles divergent in mature fruit; stylopodium low-conical. Fruits minutely scabrous on the ribs when young, subdidymous, compressed laterally; ribs prominent; carpophore unknown; vittae large, vallecular 1–2, commissural 2–4. Seeds transversely elliptic, commissural face plane, very narrow.

One species, *R. kirmzii* Farille & Lachard, Himalaya (Nepal), high mountains in moss ground cover. Imperfectly known genus probably related to *Acronema* and *Synclinostyles*.

374. ***Rouya*** Coincy [Ostroumova]

Rouya Coincy, Naturaliste (Paris) II, 349: 213 (1901).

Nearly glabrous perennials. Leaves 2–3-pinnate, ultimate divisions ovate to linear, petiole sheathing. Bracts many, oblong-lanceolate, often bi- to trifid; bracteoles many, oblong-lanceolate. Calyx lobes acute, persistent; petals white; styles long, divergent; stylopodium long-conical. Fruits glabrous, oblong-ellipsoid, compressed dorsally; primary ribs filiform or narrowly winged, secondary ribs prominently winged, undulate, the lateral wings broader than median; commissure rather broad; vallecular vittae 1, commissural 2. Seed face plane or slightly concave. $2n = 20$.

One species *R. polygama* (Desf.) Coincy, Europe, N Africa. Maritime sands. Nested within *Daucus* on the basis of molecular data, and recently treated in synonymy under the latter by Banasiak et al. (2016).

375. ***Rumia*** Hoffm. [Pimenov]

Rumia Hoffm., Pl. Umbell. Gen., ed. 2: 171 (1816); Fedoronczuk, Systematics and phylogeny of the genera *Trinia*, *Rumia* and *Ledebouriella* (in Russ.). Kiev (1983), rev.

Glabrous, dioecious perennials or biennials with tuberous taproots. Leaves bipinnate; leaflets filiform. Bracts absent; bracteoles lanceolate to linear, membranaceous at the margins. Petals white, lanceolate in staminate flowers, broadly ovoid in female ones; styles divergent, stylopodium low-conical. Fruits with prominent, obtuse, rugose, undulate ribs, covered in spongy, folded outgrowths; carpophore bifid; commissure narrow; mesocarp parenchymatous of cells with pitted walls; vallecular vittae 2–3, commissural 4–6; rib oil ducts large. Seed face plane. $2n = 18, 20$.

One species, *R. crithmifolia* (Willd.) Koso-Pol., E Europe (Crimea). Open stony slopes, often limestone. Related to *Trinia* s. lat.

376. ***Rupiphila*** Pimenov & Lavrova [Pimenov]

Rupiphila Pimenov & Lavrova, Bjull. Moskovsk. Obšč. Isp. Prip. Otd. Biol. 91(2): 97 (1986); Pimenov & Kljuykov, Bot. Zhurn. 87(1): 122–127 (2002), reg. rev.

Glabrous perennials. Leaves 3–4-pinnatisect with geniculate rhachis; leaflets filiform or linear-subulate. Bracts linear, scabrous; bracteoles filiform, often unequal, exceeding pedicels. Calyx lobes narrowly triangular to lanceolate, persistent; petals white or pinkish; styles thin, divergent; stylopodium conical. Mericarp ribs filiform to keeled; carpophore bifid; commissure narrow; mesocarp parenchymatous; vallecular vittae 1, commissural 2. Seed face plane. $2n = 22$.

One species, *R. tachiroei* (Franch. & Sav.) Pimenov & Lavrova, E Asia (Japan, Korea, Russian Far East, NE and E China). Limestone cliffs. Molecular data suggest a relationship to *Sinocarum*, then to *Laser*, *Orlaya*, *Turgenia*, *Halosciastrum*, *Ligusticum*, *Pternopetalum*, *Meeboldia*, and *Tilingia*.

377. *Rutheopsis* A. Hansen & Kunkel [Watson, Reduron]

Rutheopsis A. Hansen & Kunkel, Cuad. Bot. Canar. 26–27: 61 (1976).
Gliopsis Rauschert (1982).
Canaria Jim.-Mejías & P. Vargas, Phytotaxa 212(1): 73 (2015).

Glabrous perennials. Leaves 1–3-pinnate; ultimate segments rhombic-ovate or ovate-lobed or dentate, to linear. Bracts and bracteoles several; rays several to many, spreading. Petals yellow; styles short, adpressed reflexed; stylopodium long-conical. Fruits glabrous, ovoid to oblong; ribs thickened, prominent; carpophore bifid; vallecular vittae 1, commissural 2.

Two species (the type species *R. herbanica* (Bolle) A. Hansen & G. Kunkel, and a segregate of Seseli, *S. webbeli* Coss.); NW Africa (Canary Islands).

378. *Sajanella* Soják [Pimenov]

Sajanella Soják, Čas. Nár. Mus. Praze Rada Přír. 148(3–4): 209 (1980).
Sajania Pimenov (1974), non A.G. Vologdin (1962).

Pubescent perennials. Leaves pinnate or ternate, amplexicaulous on stems, sheaths dark-veined and spreadingly hairy; leaflets sessile, ovate to suborbicular, toothed; upper sheaths surrounding umbels. Bracts entire, similar to upper sheaths; rays hollow, puberulent; bracteoles pubescent, elliptic to ovoid. Calyx lobes short, triangular, ciliate; petals pale violet, not inflexed; styles straight; stylopodium flat. Fruits soft-hairy, oblong, slightly compressed dorsally, with 7 filiform ribs; carpophore absent; exocarp detached of mesocarp at fruit maturity; commissure rather broad; vallecular vittae 1, commissural 2. Seed face with broad, shallow groove. $2n = 22$.

One species, *S. monstrosa* (Willd. ex Spreng.) Soják, Asia (S Siberia, NE Kazakhstan, N Mongolia). Subalpine meadows, mountain tundra. Molecular data show an affinity to some *Seseli* species.

379. *Saposhnikovia* Schischk. [Pimenov]

Saposhnikovia Schischk., Fl. USSR 17: 359 (1951); Tamamschjan, Taxon 10(7): 221–225 (1961).

Glabrous perennials. Leaves bipinnate; leaflets elliptic, broadly lanceolate or obovate, cuneate. Bracts absent; rays unequal; bracteoles linear to lanceolate. Petals white; styles straight, later reflexed; stylopodium conical. Ovaries covered with transverse whitish outgrowths; fruits glabrous, slightly compressed dorsally; ribs keeled; carpophore bifid; commissure broad; mesocarp with cells with pitted walls; vallecular vittae 1, commissural 2; rib oil ducts large. Seed face plane. $2n = 16$.

One species, *S. divaricata* (Turcz.) Schischk., Asia (E Siberia, Russian Far East, Mongolia, Korea, China). Stony steppes, scrub, forest margins. Used in Chinese medicine.

380. *Scaligeria* DC. [Kljuykov]

Scaligeria DC., Coll. Mém. 5: 70 (1829), nom. cons., non *Scaligera* Adans. (1763); Korovin, Acta Univ. As. Med. VII-b, Bot. 2: 1–96 (1928), rev.; Engstrand, Bot. Notiser 123: 505–511 (1973), reg. rev.; Kljuykov et al., Bjull. Moskovsk. Obšč. Isp. Prip. Otd. Biol. 81: 75–89 (1976), taxon.

Glabrous perennials with globose to oblong tubers. Leaves 2–3-pinnate; the segments petiolulate or sessile, and ovate, lobed or toothed; uppermost leaves entire. Bracts absent, bracteoles few, entire. Petals white; styles slender, long, and reflexed; stylopodium conical-cylindrical. Fruits glabrous, almost globose or ellipsoid; commissure narrow; ribs terete; carpophore bifurcated; vallecular vittae 2–4, commissural 4. Seed face concave to plane. $2n = 20, 22$.

Three species, S Europe (Greece, Cyprus, Albania, former Yugoslavia), N Africa (Libya) and SW Asia. Open scrub, forests, shady canyons, and limestone cliffs. Molecular data suggest a relationship to *Bunium*, *Elaeosticta*, *Galagania* and related genera.

381. **Scandia** J.W. Dawson [Webb]

Scandia J.W. Dawson, New Zealand J. Bot. 5: 407 (1967); Mitchell et al., New Zealand J. Bot. 36: 417–424 (1998), phylog.

Glabrous, gynodioecious, perennial subshrubs or lianas. Leaves all cauline, simple or once pinnately compound, sub-fleshy or subcoriaceous; leaflets ovate to obovate, serrate or crenate. Bracts sometimes foliaceous; rays few to many, slender; bracteoles simple, narrowly linear. Petals white, with a median oil tube; female flowers with conspicuous petals and rudimentary staminodes; styles slender; stylopodium conical. Fruits glabrous, ovate-ellipsoid, slightly compressed dorsally or subterete; dorsal and lateral ribs acute or narrowly winged; marginal ribs broadly winged; carpophore bifid almost to base; vittae large, vallecular 1, commissural 2; rib oil ducts 1 in each rib. Seeds transversely subterete to slightly compressed dorsally, sulcate under the vittae, the face convex. $2n = 22$.

Two species, New Zealand. Molecular evidence indicates *Scandia* is polyphyletic and not distinct from *Gingidia* (Mitchell et al. 1998).

382. **Scandix** L. [Spalik]

Scandix L., Sp. Pl. 256 (1753); Gen. Pl., ed. 5: 124 (1754); Schischkin, Flora SSSR 16: 139–148, (1950), reg. rev.; Cannon in Tutin et al., Fl. Eur. 2: 326–327 (1968), reg. rev.; Hedge & Lamond in Davis, Fl. Turkey 4: 325–330 (1972), reg. rev.

Pubescent annuals. Leaves 2–3-pinnately dissected, triangular to ovate-oblong; leaflets linear-lanceolate to lanceolate. Bracts usually absent; rays few, usually unequal, spreading or ascending; bracteoles few, entire, dentate or dissected, glabrous or hairy, ciliate. Petals white, the outer often radiate; styles short to relatively long; stylopodium flat-conical in flower, high-conical in fruit. Fruits glabrous or delicately bristled, the cuticle often aculeate, linear-oblong to oblong-ovoid with a very long beak, compressed laterally; mericarps transversely terete to semi-circular, scarcely compressed dorsally; ribs broad or filiform; commissure broad; carpophore bifurcate; vittae small, extant at fruit maturity, vallecular 1, commissural 2. Seed transversely terete, the face sulcate. $2n = 14, 16, 18, 20, 26, 34$.

About ten species (but *S. pecten-veneris* L. and *S. australis* L. with several subspecies, often treated as distinct species), Eurasia and N Africa, widespread as weeds, open, often disturbed habitats, arable land, waste places, thickets, open woods. Closely related to *Anthriscus*, *Geocaryum*, *Kozlovia*, *Myrrhis*, and *Osmorhiza* (Spalik and Downie 2001).

383. **Scaraboides** Magee & B.-E. van Wyk [Van Wyk & Tilney]

Scaraboides Magee & B.-E. van Wyk in Magee et al., Syst. Bot. 34: 591 (2009); Van Wyk et al., Afr. Apiaceae: 291 (2013).

Glabrous annuals. Leaves pinnate; leaflets broadly ovate, segments narrowly oblong. Bracts and bracteoles absent; rays few, slightly scabrous. Petals white, orbicular, papillose; stylopodium flat. Fruits broadly elliptic, dorsally compressed; mericarps strongly compressed dorsally with a median keel on the deeply concave commissural face; median and lateral ribs indistinct; marginal ribs involute, winged; carpophore bifid; vittae small, vallecular 1, commissural 2, one additional vitta in each wing. Seeds transversely strongly compressed dorsally, sulcate beneath the vittae, the face plane.

One species, *S. manningii* Magee & B.-E. van Wyk, South Africa. Known only from two localities in succulent karoo. Related to *Capnophyllum* and *Dasispermum*.

384. **Schoenoselinum** Jim.-Mejías & P. Vargas [Reduron]

Schoenoselinum Jim.-Mejías & P. Vargas, Phytotaxa 212 (1): 75 (2015).

Glabrous perennials. Shrubby habit with rod-like leafless flowering stems. Leaves early deciduous, 2–3-pinnatifid, ultimate divisions filiform, much reduced on the upper part of the stem, often reduced to sheaths. Bracts absent; rays several to many, slightly unequal; bracteoles absent. Flowers several per umbellule, petals gold yellow, rounded; styles reflexed, equal to a little longer than the subconical stylopodium. Fruits glabrous, ellipsoid-oblong, compressed dorsally; dorsal ribs distinctly prominent, marginal ribs developed, wing-like; vittae large, vallecular 1, commissural 2. Seed face plane.

One species, *S. foeniculoides* (Maire & Wilczek) Jim.-Mejías & P. Vargas, endemic to Morocco and Mauritania. Close to *Foeniculum*, *Ridolfia* and *Pseudoridolfia*.

385. ***Schrenkia*** Fisch. & C.A. Mey. Fig. 23 [Pimenov]

Schrenkia Fisch. & C.A. Mey. in Schrenk, Enum. Pl. Nov. 1: 65 (1842); Terentieva et al., Phytotaxa 195 (4): 251–271 (2015), phylog., rev.
Kosopoljanskia Korovin (1923).

Glabrous or puberulent perennials or low subshrubs with lignified rootstocks or taproots. Leaves bipinnate; leaflets ovate, lanceolate, or linear, deeply toothed. Bracts entire or pinnatifid, sometimes caducous; rays frequently rigid; bracteoles short, linear. Petals white; styles short, reflexed; stylopodium conical. Fruits glabrous, mericarps globose to narrowly ovoid; ribs inconspicuous or filiform, covered with prickles or short hairs, or glabrous; commissure rather broad; inner mesocarp layer hard, lignified; vittae many, obscure in mature fruits. Seed face plane to concave. $2n = 22$.

14 species, Middle Asia, NW China. Stony hillsides, among xerophilic shrubs, limestone and sandstone cliffs, scree. Closely related to *Lipskya* and *Schtschurowskia*.

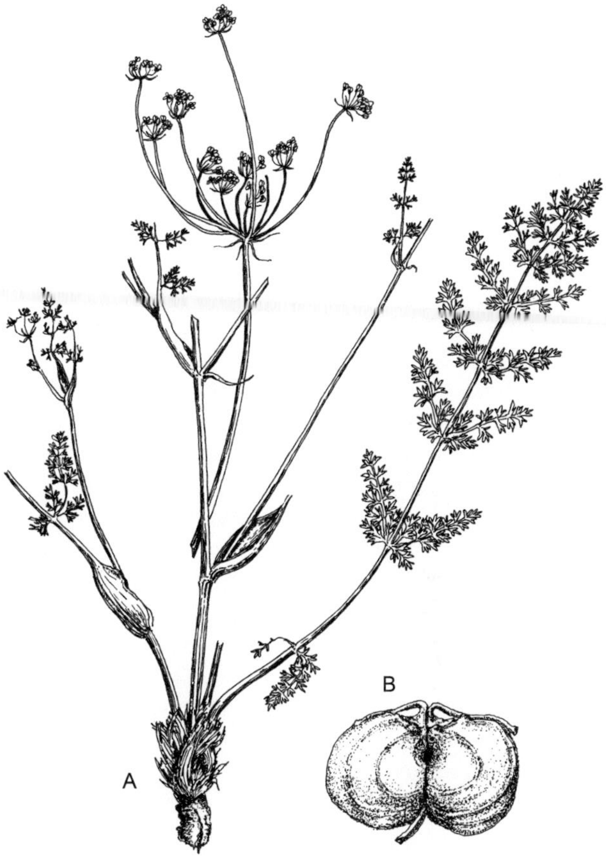

Fig. 23. Apiaceae, Apioideae. *Schrenkia vaginata*. **A** Habit. **B** Fruit. (Orig., illustration by Elena Mzhelskaya)

386. ***Schtschurowskia*** Regel & Schmalh. [Pimenov]

Schtschurowskia Regel & Schmalh., Izv. Imp. Obsc. Ljubit. Estestv. Moskovsk. Univ. 34(2): 40 (1881).

Glabrous perennials. Leaves bipinnatisect; leaflets linear, divaricate, mucronulate. Bracts linear; rays unequal; bracteoles linear. Central hermaphroditic flower of each umbellule sessile, outer flowers staminate, with longer pedicels; calyx lobes lanceolate-subulate; petals white; styles erect, elongate; stylopodium conical. Fruits with mericarps not separating at maturity, without ribs; inner mesocarp layer lignified, absent on commissural side but accreting into the two mericarps; vittae many, situated outside lignified mesocarp, obscure in mature fruits, commissural absent. Seed face slightly concave. $2n = 22$.

Two species, Asia (Middle). Alpine slopes, scree, dry slopes, among xerophilic shrubs. Closely related to *Lipskya* and *Schrenkia*.

387. ***Schulzia*** Spreng. [Pimenov]

Schulzia Spreng., Neue Schriften Naturf. Ges. Halle 2(1): 30 (1813), nom. cons., non *Schultzia* Rafin. (1808).

Low caulescent or dwarf acaulescent, glabrous perennials. Leaves 2–3-pinnate; leaflets linear to lanceolate, acuminate. Bracts and bracteoles bipinnatisect with acuminate lobes; rays subequal. Petals white; styles straight; stylopodium low-conical. Fruits slightly compressed laterally; ribs filiform, carpophore entire or bifid only in the uppermost part; commissure narrow; mesocarp parenchymatous; vallecular vittae 3–6, commissural 7–10. Seed face plane, or slightly emarginate. $2n = 20, 22$.

Three species, Asia (Siberia, Mongolia, Kazakhstan, Tian-Schan, W China up to N Pakistan). Subalpine meadows, tundra, among rocks, scree, stony hillsides.

388. ***Sclerochorton*** Boiss. [Plunkett]

Sclerochorton Boiss., Fl. Orient. 2: 968 (1872); Rechinger, Fl. Iran. 162: 366 (1987), reg. rev.

Glabrous perennials with a branched caudex. Leaves crowded toward the base and well spaced upward, ternate-pinnately dissected, coriaceous; ultimate segments linear, thick, grooved, rigid, compressed to cylindrical. Bracts and bracteoles few, linear; rays few, slender and spreading. Petals white; styles slender and reflexed; stylopodium depressed to conical. Fruits glabrous, oblong-ovoid, subterete; ribs rounded, prominent but narrower than the wide valleculae; carpophore bifurcated; vittae slender, vallecular 5–7. Seeds transversely slightly compressed dorsally, the face concave.

One species, *S. haussknechtii* Boiss., Asia (Iran and Iraq), on high mountains.

389. ***Sclerosciadium*** W.D.J. Koch ex DC. [Pimenov]

Sclerosciadium W.D.J. Koch ex DC., Coll. Mém. 5: 43 (1829).

Decumbent, scabrous annuals. Leaves 2–3-pinnatisect, leaflets lanceolate to linear, lobed or entire. Umbels axillary, almost sessile, with globose umbellules; bracts absent or reduced; bracteoles linear. Calyx teeth lanceolate to triangular, hardening; petals white, keeled; styles medium-sized, spreading; stylopodium conical. Fruits almost globose to elliptic, not separating into mericarps, glabrous to finely scabrous, subsessile, with entire carpophore; mericarps terete, slightly compressed laterally; commissure intermediate; dorsal ribs keeled, inflated, marginal ribs slightly thickened; vallecular vittae 1, commissural 2. Seed face plane or slightly convex.

One species, *S. nodiflorum* (Schousb.) Ball, NW Africa (Morocco). Sandy dunes and pastures.

Molecular data suggest a close relationship to *Ammi* and *Petroselinum* (Jiménez-Mejías and Vargas 2015).

390. ***Sclerotiaria*** Korovin [Pimenov]

Sclerotiaria Korovin, Trudy Inst. Bot. Akad. Nauk Kazakh. SSR 13: 243 (1962).

Acaulescent, glabrous perennials. Leaves rosulate, 3-pinnatisect; leaflets linear, divaricate, mucronulate, crowded. Central umbel sessile, lateral ones shortly pedunculate; bracts and bracteoles entire, subulate; rays unequal. Calyx lobes large, triangular, becoming hard, hook-like, and divaricate in fruits; petals white; styles erect, hardened in fruit; stylopodium conical. Fruits sessile or subsessile, covered by short prickles, undivided, ribs obtuse; mericarps not separating at maturity; commissure broad; inner layer of mesocarp lignified, vittae many, narrow, situated outside lignified layer of mesocarp, obscure in mature fruits, commissural absent. Seed face slightly concave.

One species, *S. pentaceros* (Korovin) Korovin, Asia (W Tian-Schan). Stony slopes, scree, alpine vegetation. Closely related to *Schrenkia*.

391. ***Scrithacola*** Alava [Plunkett]

Scrithacola Alava, Notes Roy. Bot. Gard. Edinb. 38: 260 (1980); Alava in Rechinger, Fl. Iran. 162: 518–519 (1987), reg. rev.

Glabrous and glaucous perennials with extensive growth of underground stems. Leaves trifoliolate, long petiolate, broadly ovate, succulent; leaflets small, broadly ovate, often deeply tripartite. Bracts and bracteoles obsolete or rudimentary; rays few, subequal, slender and spreading-ascending. Petals greenish-yellow; styles short, divergent; stylopodium low-conical. Fruits glabrous, oblong-elliptic, compressed dorsally, dorsal ribs inconspicuous, marginal ribs narrowly and nearly transparently winged; carpophore unknown; vittae lacking, rib oil ducts solitary in each rib. Seeds transversely compressed dorsally, the face plane.

One species, *S. kuramensis* (Kitam.) Alava, Asia (Afghanistan, W Pakistan), among scree of steep, unstable slopes, mid to high elevations.

392. ***Selinopsis*** Coss. & Durieu ex Batt. [Kljuykov]

Selinopsis Coss. & Durieu ex Batt. in Batt. & Trab., Fl. Alger. Tunisie: 139, 141 (1905); Kljuykov, Bull. Mosc. Univ. XVI, Biol. 1: 21–25 (1986), rev.

Glabrous perennials. Leaves 2–3-pinnate; leaflets lanceolate to oblong or linear. Bracts and bracteoles lacking. Petals white; styles reflexed; stylopodium low-conical. Fruits glabrous; ovoid, slightly compressed laterally; commissure broad; ribs equal, prominent, terete; exocarp 2-layered; mesocarp parenchymatous; vallecular vittae 1, commissural 2. Seed transversely pentagonal, the face plane to slightly concave.

Two species, NW Africa (Algeria), SW Europe (Spain). Limestone, clay, herbaceous slopes.

393. ***Selinum*** L. [Pimenov]

Selinum L., Sp. Pl., ed. 2: 350 (1762), nom. cons.; Leute, Ann. Naturhist. Mus. Wien 74: 457–519 (1970), rev.; Lavrova et al., Bot. Zhurn. 72(1): 25–38 (1987), rev.; Pimenov et al., Willdenowia 31(1): 101–124 (2001), rev.

Glabrous perennials. Leaves 3–4-pinnate; leaflets lanceolate to ovoid, deeply dissected, acuminate. Bracts absent or 1–2, caducous; rays many, scabrous; bracteoles many, linear. Petals white or reddish; styles reflexed; stylopodium conical or low-conical. Fruits slightly compressed dorsally; dorsal ribs winged, marginal broadly winged; carpophore bifid; commissure very narrow; mesocarp parenchyma of cells with pitted walls; vallecular vittae 1–4, commissural 3–5; rib vascular bundles stretched radially. Seed face plane or slightly convex. $2n = 22$.

Two or three species, Europe, W Siberia. Deciduous, mixed and coniferous forests, meadows, among shrubs, in moist places. Treated here in a narrow circumscription, following Lavrova et al. (1987) and Pimenov et al. (2001); see also *Oreocome*, *Sphaenolobium*, *Karatavia*.

394. ***Semenovia*** Regel & Herder [Ostroumova]

Semenovia Regel & Herder, Bull. Soc. Nat. Mosc. 39(3): 78 (1866); Mandenova, Acta Tbil. Bot. Inst. 20: 3–57 (1959), rev.
Platytaenia Nevski & Vved. (1937) non Kuhn (1882).
Neoplatytaenia Geldykh. (1990).

Pubescent perennials. Leaves 1–2-pinnate, ultimate divisions round to linear. Bracts and bracteoles linear-lanceolate. Petals yellow, white or purple, often radiate; styles recurved; stylopodium conical with an undulate margin. Fruits pubescent, ellipsoid, strongly compressed dorsally; dorsal ribs keeled, marginal ribs winged; fibrous inner mesocarp present; commissure broad; vallecular vittae 1, commissural 2. Seed face plane. $2n = 22, 24$.

29 species, Asia. Two sections recognized (Mandenova 1959). Molecular data suggest a close relationship to *Zosima* and *Tetrataenium*.

395. ***Seseli*** L. [Pimenov]

Seseli L., Sp. Pl. 259 (1753); Gen. Pl., ed. 5: 126 (1754); Calest., Bull. Soc. Bot. Ital. 6: 193–201 (1905). reg. rev.; Pimenov & Sdobnina, Bot. Zhurn. 60(8): 1108–1122 (1975), part. rev.; Pimenov, Novosti Sist. Vyssh. Rast. 15: 188–200 (1978), reg. rev.; Pardo, Lazaroa 3: 163–188 (1981), reg. rev.; V.M. Vinogr., Nov. Syst. Pl. Vasc. 26: 121–128 (1989), reg. rev.; Pimenov & Kljuykov, Bot. Zhurn. 85(10): 96–109 (2000), part. rev.

Puberulent (rarely glabrous) perennials, rarely subshrubs. Leaves pinnate to 2–4-pinnatisect; leaflets filiform, linear or lanceolate to suborbicular, glabrous or puberulent. Bracts linear to lanceolate, or absent; rays glabrous or hairy; bracteoles free or connate at the base, linear to lanceolate. Petals white, pink, or yellowish; styles straight to divergent or reflexed; stylopodium conical to flat. Fruits glabrous or pubescent to tomentose, ribs filiform to keeled, or unequal, dorsal keeled, marginal narrowly winged; carpophore bifid; commissure different; mesocarp parenchymatous, sometimes with lignified cells; vallecular vittae 1 or 2–5, commissural 2–6. Seed face plane. $2n = 16$, 18, 20, 22, 26, 32, 33, 40, 42, 44, 132.

125–140 species, Eurasia, Africa (N, NW, trop.), N America (adv.), Australia (adv.). Steppes, mountain deserts, cliffs, limestone, dry stony slopes, scree, etc.; the species of sect. *Libanotis* grow in meadows, forests, among shrubs, etc.; those of sect. *Condensata* in wet meadows, along mountain streams. Sixteen sections (Pimenov and Sdobnina 1975; Pimenov 1978; Pimenov and Kljuykov 1999). Some species are used as folk medicines. Aromatic. Molecular data suggest the genus is polyphyletic.

396. ***Seselopsis*** Schischk. [Pimenov]

Seselopsis Schischk., Bot. Mat. Herb. Bot. Inst. Acad. Sci. URSS 13: 159 (1950).

Glabrous perennials or biennials with swollen, finger-like roots. Leaves 2–3-ternate or pinnate; leaflets lanceolate to linear. Bracts absent or 1–2, linear; rays few or several; bracteoles linear, sometimes clearly unequal. Petals white; styles long, reflexed; stylopodium low-conical. Mericarp ribs keeled, equal; carpophore bifid; commissure narrow; mesocarp parenchymatous, with several cells having lignified pitted walls; vittae large, vallecular 1, commissural 2. Seed face plane. $2n = 22$.

Two species, Asia (Kazakhstan, Kirghizia, NW China). Stony mountain slopes, among shrubs.

397. ***Shoshonea*** Evert & Constance [Plunkett]

Shoshonea Evert & Constance, Syst. Bot. 7(4): 471 (1982).

Caespitose-pulvinate, acaulescent, (sub-)scaberulous perennials with a woody taproot. Leaves ovate or oblong, pinnately divided, subcoriaceous; leaflets linear or oblanceolate, cuspidate. Bracts lacking; rays few, angled, spreading; bracteoles dimidiate, several, basally connate, linear or lanceolate. Calyx lobes prominent; petals yellow; styles slender, spreading; stylopodium lacking. Fruits scaberulous, sessile, oblong or ovoid-ellipsoid, subterete or slightly compressed laterally; commissure not constricted; ribs subequal, prominent to subprominent, obtuse, unwinged, ovate in transection; carpophore obsolete; vittae small, vallecular 2–6, commissural 2–6, rib oil ducts often 1 in each rib. Seeds transversely compressed dorsally, the face plane to concave. $2n = 22$.

One species, *S. pulvinata* Evert & Constance, N America (NW Wyoming), forms dense, extensive cushions in crevices and talus of calcareous outcroppings. Molecular data suggest a close relationship to *Cymopterus*, *Lomatium*, and their allies.

398. ***Sillaphyton*** Pimenov [Pimenov]

Sillaphyton Pimenov, Botanica Pacifica 5(2): 37 (2016).
Wangsania B.Y. Lee & Hyun (2017).

Glabrous perennials. Leaves long-petiolate, 2–3-ternate; terminal segments long-petiolulate, elliptic. Terminal umbels loose, with 6–14 thin rays; bracts absent or 1–4, subulate to filiform. Petals white; stylopodium conical. Fruits compressed dorsally, dorsal ribs sub-inconspicuous, marginal ribs narrowly winged; commissure broad; mesocarp in mature fruits partly crushed; vascular bundles in marginal ribs cleft into several groups; vittae small, vallecular 1–3, commissural 2–4; rib secretory ducts broad, sometimes exceeding vallecular vittae. Seed face plane.

One species, *S. podagraria* (H. Boissieu) Pimenov. E Asia (Korea), mountain forests. Segregate of *Peucedanum* s.lat. Molecular data suggest an affinity to *Arcuatopterus*, a Sino-Himalayan endemic genus.

399. ***Silaum*** Mill. [Reduron]

Silaum Mill., Gard. Dict. ed. 4 (1754); Mandenova, Not. Pl. Syst. Geogr. Georg. Fil. Acad. Sci. USSR 10: 73–78 (1941), taxon.; Lavrova et al., Bot. Zhurn. 72(1): 25–38 (1987), taxon.

Glabrous perennials with a woody rootstock. Leaves 2–4-pinnately divided, triangular; ultimate divisions linear or linear-lanceolate. Bracts absent or very few; bracteoles several; rays several to many, spreading. Petals yellow; styles equaling or slightly exceeding the conical stylopodium. Fruits glabrous, ovoid-oblong; not compressed; ribs subequal, prominent, acute, the lateral ± narrowly winged; carpophore free, bifurcated; commissure narrow; vittae small, many and cyclic, obliterated in mature fruit. Seeds transversely pentagonal to ± concave, the face plane. $2n = 22$.

Two species, Eurasia (from Western Europe to Central Asia), damp meadows.

400. ***Siler*** Mill. [Reduron]

Siler Mill., Gard. Dict. Abr. Ed.4 (1754).

Glabrous and glaucous perennials. Leaves petiolate, 2–4-pinnately dissected, triangular; ultimate divisions narrowly to broadly elliptical to lanceolate. Bracts and bracteoles many; rays many, spreading. Petals white; styles elongated; stylopodium hemispherical to conical. Fruits glabrous, shiny, ellipsoid-oblong; primary ribs weakly prominent; secondary ribs 4, with narrow, slightly unequal wings; carpophore free, divided to the base; vallecular vittae 1 (under each secondary rib), triangular in transection, commissural 2. Seed face slightly concave. $2n = 22$.

One species, *S. montanum* Crantz, S Europe. Segregated from *Laserpitium* (Banasiak et al. 2016).

401. ***Silphiodaucus*** (Koso-Pol.) Spalik, Wojew., Banasiak, Piwczyński & Reduron [Reduron]

Silphiodaucus (Koso-Pol.) Spalik et al., Taxon 65(3): 563–585 (2016).

Hispid monocarpic biennials. Leaves 2–3-pinnately dissected, triangular or broadly ovate; ultimate divisions broadly ovate to lanceolate, deeply lobed. Bracts and bracteoles many; rays several to many, pubescent to hairy, somewhat contracted at maturity. Petals white, sometimes becoming yellowish later; styles exceeding the stylopodium. Fruits ovoid to ellipsoid-oblong; primary ribs filiform, hairy; secondary ribs 4, the marginal with broader wings than the dorsal; carpophore free; vallecular vittae 1 (under each secondary rib), commissural 2. Seed face plane. $2n = 22$.

Two species, Europe from Iberian Peninsula to Russia, Caucasus, Turkey, Georgia, Azerbaijan. Segregated from *Laserpitium* (Banasiak et al. 2016).

402. ***Sinocarum*** H. Wolff ex R.H. Shan & F.T. Pu [Watson]

Sinocarum H. Wolff ex R.H. Shan & F.T. Pu in R.H. Shan et al., Acta Phytotax. Sin. 18: 374 (1980).

Glabrous (rarely sparsely pubescent) perennials with an elongate, fusiform rhizome. Leaves petiolate, ternate to 1–3-pinnate or simply 1–3-pinnate; ultimate segments linear to ovate. Umbels compound, subcompact, peduncles terminal; bracts lacking (occasionally few, small and linear); rays slender, spreading; bracteoles several, linear and entire (occasionally lobed at apex). Calyx lobes obsolete or conspicuous; petals white or purple (rarely 2–3-lobed or palmately 3–5-lobed); styles short; stylopodium compressed. Fruits glabrous, oblong-ovoid, slightly compressed laterally; ribs filiform; carpophore bifid or bifurcated; vallecular vittae 1–3, commissural 2–6. Seeds transversely subterete, the face plane. $2n = 18, 20$.

21 species, Asia (Sino-Himalaya, Nepal to SW China), high-elevation forests and open alpine turf. A taxonomically complex genus with unclear generic boundaries with *Acronema* and *Tongoloa*.

403. ***Sinodielsia*** H. Wolff [Kljuykov]

Sinodielsia H. Wolff, Notizbl. Bot. Gart. Berlin 9: 278 (1925); Kljuykov, Feddes Repert 97: 753–757 (1986), rev.; Pimenov et al., Feddes Repert. 102: 375–384 (1991), taxon.; Pimenov & Kljuykov, Feddes Repert. 110: 481–491 (1999), part. rev.

Perennials with thickened, branched roots. Leaves 2–4-pinnate; leaflets linear to ovate or rhombic. Bracts 1–2 or lacking; bracteoles few, entire, linear. Calyx lobes triangular or inconspicuous; petals white; styles reflexed; stylopodium low-conical. Fruits glabrous, ovoid; commissure narrow; ribs terete or narrowly winged; carpophore bifurcated; mesocarp parenchymatous; vallecular vittae 2–4, commissural 4–5. Seed face concave to plane. $2n = 22, 24$.

Four species, Asia (SW China, Nepal), open mountain slopes, scrub, forest margins. Carpological and molecular data do not support the inclusion of *Sinodielsia* in *Meeboldia*.

404. ***Sinolimprichtia*** H. Wolff [Pu]

Sinolimprichtia H. Wolff, Feddes Repert. 12: 448 (1922); S. L. Liou, Fl. Reipub. Popul. Sin. 55(1): 192 (1979); P. Shu & M.L. Sheh, Pollen Photog. Fl. Umbell. China 1: 24 (2001), palyn.; Fl. China 14: 55–56 (2005), rev.

Glabrous perennials. Basal and lower leaves compact, bipinnately compound, lanceolate; primary leaflets 6–7 pairs, ultimate segments oblong-ovate, the margins much incised or irregularly serrulate. Bracts absent; rays stout, several to many, subequal; bracteoles several, linear, entire, with narrowly membranous margin, or 2–3-pinnate, ovate-triangular. Petals white; styles slightly recurved; stylopodium low-conic. Fruits glabrous, oblong-ovoid, compressed laterally; mericarps transversely pentagonal to semi-circular; dorsal and intermediate ribs filiform, lateral ribs narrowly winged; carpophore very slender, bifurcated; vittae large, vallecular 2–3, commissural 2. Seed face concave to sulcate.

One species, *S. alpina* H. Wolff, Asia (SW China). Closely related to *Hymenidium* (*Pleurospermum* s. ampl.).

405. ***Sison*** L. [Reduron]

Sison L., Sp. Pl. 252 (1753); Gen. Pl., ed. 5: 120 (1754); H. Wolff in Engl., Pflanzenr. 4 (228) Heft 90: 71–74 (1927), rev.

Glabrous annuals or biennials. Leaves pinnately divided, oblong to elongate; leaflets many, ovate or ovate-oblong, ± deeply incised. Bracts and bracteoles present, but sometimes few; rays few to several, unequal, spreading. Petals whitish or sometimes pinkish; styles short; stylopodium conical. Fruits glabrous, ovoid or ovoid-subglobose, somewhat compressed laterally; ribs ± prominent and broad, obtuse; carpophore free, bifurcated; commissure broad; vittae medium-sized to large, vallecular 1, commissural 2. Seeds transversely sub-rectangular or sub-pentagonal, the face plane, sometimes slightly concave or sinuous. $2n = 14, 16, 18$.

Three species [incl. *S. segetum* L. = *Petroselinum segetum* (L.) W.D.J. Koch]. Europe (W, S, Macaronesia, E Mediterranean) and Asia (SW, Caucasus), forest margins, scrub, river banks, arable fields.

406. ***Sium*** L. [Pimenov]

Sium L., Sp. Pl. 251 (1753); Gen. Pl., ed. 5: 120 (1754); Wolff in Engl., Pflanzenr. IV, 228 (Hf. 90): 341–358 (1927), rev.; Spalik, Downie & Watson, Taxon 58: 735–748 (2009), phylog., taxon.

Glabrous perennials with rhizomes and fascicles of fleshy-tuberous or fibrous roots; stems often rooting at the lower nodes, sometimes with bulblets in the leaf axils. Leaves pinnate; leaflets sessile, ovate to lanceolate, filiform in submerged leaves. Bracts linear to lanceolate, entire or incised, often reflexed; bracteoles linear to lanceolate. Petals white, with several secretory ducts; styles short, reflexed; stylopodium low-conical. Mericarp ribs filiform to narrowly corky winged; carpophore bifurcated or adnate to the mericarps; commissure narrow; mesocarp parenchymatous; vallecular vittae 1 or 2–4, commissural 2–6. Seed face plane. $2n = 6, 12, 18, 20, 22, 40$.

Nine species, Eurasia, N America, Africa. Swamps to ± dry meadows. Molecular data show affinity to *Berula*, *Cryptotaenia* and *Helosciadium* (Spalik et al. 2009).

407. ***Sivadasania*** Mohanan & Pimenov [Pimenov]

Sivadasania Mohanan & Pimenov, Bot. Zhurn. 92(6): 901 (2007).

Glabrous perennials. Leaves mainly rosulate, 2–3-pinnatisect; leaflets ovate, usually trilobed, crenate-dentate. Bracts and bracteoles 2–5, linear to lanceolate; rays unequal. Peripheral flowers bisexual, inner ones staminate and short-pedicellate; petals yellow; styles reflexed; stylopodium conical, slightly constricted near the base. Fruits slightly compressed laterally; ribs subequal, keeled; carpophore bifid; commissure narrow. Mesocarp parenchymatous; vallecular vittae 1, commissural 2; rib oil ducts solitary, visible. Seed face deeply grooved.

One species, *S. josephiana* (Wadhwa & H.J. Chowdhery) Mohanan & Pimenov, Asia (Kerala in Peninsular India). Grasslands, rocky habitat on mountains at mid elevations.

408. ***Smyrniopsis*** Boiss. [Pimenov]

Smyrniopsis Boiss., Ann. Sci. Nat. Bot. III, 2: 72 (1844); Tamamschjan, Bull. Acad. Sci. Arm. SSR 5-6: 47–62 (1945), rev.; Shneyer et al., Pl. Syst. Evol. 182: 135–148 (1992), serotaxon.

Glabrous or papillose perennials. Leaves 2–3-ternate; leaflets ellipsoid to ovoid, coriaceous; upper leaves simplified. Bracts lanceolate to lanceolate-linear, unequal, reflexed; bracteoles small, few, linear. Petals yellow or yellowish; styles long, reflexed; stylopodium low-conical to flat, with wavy margin. Fruits glabrous at maturity and sparsely papillate when young; ribs narrowly keeled, wavy; valleculae vesiculose; carpophore bifid; commissure rather narrow; mesocarp parenchymatous; vittae large and short, vallecular 2–3, commissural 3–4, 2–3 in the funicle. Seed face deeply sinuate, with involute margins. $2n = 22$.

Two species, Asia (SW, Caucasus), subalpine stony meadows. Immunochemical (Shneyer et al. 1992) and molecular data suggest affinity to

Prangos, *Bilacunaria*, *Heptaptera*, *Opopanax*, *Ferulago*, and *Azilia*. See also *Petroedmondia*.

409. ***Smyrnium*** L. Fig. 3G [Reduron]

Smyrnium L., Sp. Pl. 262 (1753); Gen. Pl., ed. 5: 127 (1754); Stevens in Davis, Fl. Turkey 4: 337–340 (1972), taxon.; Shneyer et al., Pl. Syst. Evol. 182: 135–148 (1992), serotaxon.

Glabrous biennials, sometimes with tuberous taproots. Basal leaves 2–4-ternately or 2–4-pinnately divided, triangular; leaflets ovate or oblong-cuneate; upper leaves frequently simple, sometimes perfoliate. Bracts and bracteoles usually absent or few; rays several to many, subequal to unequal, spreading. Petals yellow or yellowish; styles equaling to clearly exceeding the conical stylopodium. Fruits glabrous, ovoid or subglobose, subdidymous; terete or slightly compressed laterally; dorsal ribs $\pm$ prominent, the lateral usually inconspicuous; carpophore free, bifurcated; commissure narrow; vittae small, many and cyclic. Seeds transversely compressed laterally and involute, the face deeply sulcate. $2n = 22$.

Seven species, Europe, Asia (SW, Middle, Caucasus), N Africa. Rocky slopes, forest margins, waste places. Several species used as vegetables in the past. The genus is well defined phytochemically (furanodiene).

410. ***Spermolepis*** Raf. [Plunkett]

Spermolepis Raf., Neogenyton 2 (1825); Mathias & Constance, N. Amer. Fl. 28B(1): 71–73 (1944), reg. rev.

Glabrous annuals. Leaves ternately to ternate-pinnately decompound; ultimate divisions linear to filiform. Bracts lacking (rarely few, linear); rays few, erect to spreading; bracteoles few, linear. Petals white, lacking an inflexed apex; styles very short; stylopodium low-conical. Fruits smooth, tuberculate or echinate, ovoid, compressed laterally, slightly constricted at the commissure; ribs filiform, rounded; carpophore bifid only at the apex; vallecular vittae 1–3, commissural 2. Seeds transversely subterete, the face sulcate. $n = 8, 10, 11, 32$.

Five species, N America (E United States to N Mexico), S America (Argentina: Rio Negro, Neuquén, Mendoza), and Oceania (Hawaii); sandy fields, rocky prairies, glades, and sandy or gravelly ground along streams.

411. ***Sphaenolobium*** Pimenov [Pimenov]

Sphaenolobium Pimenov, Novosti Sist. Vyssh. Rast. 12: 243 (1975).

Glabrous perennials. Leaves 2–3-ternate; leaflets coriaceous, rhombic to narrowly lanceolate. Bracts linear-subulate, minutely scabrous; rays subequal; bracteoles lanceolate to linear, short, acute, scabrous. Petals white, light yellow or greenish, dorsally minutely puberulent; styles reflexed; stylopodium low-conical, with wavy margin. Fruits slightly compressed dorsally, ribs subequal, or dorsal keeled and marginal slightly enlarged; carpophore bifid; commissure broad; mesocarp parenchymatous; vittae large, vallecular 1, commissural 2; rib oil ducts small, solitary. Seed face plane. $2n = 22$.

Three species, Asia (Middle: W Tianshan Mts.). Stony slopes, limestone cliffs, among xerophilic shrubs.

412. ***Sphaerosciadium*** Pimenov & Kljuykov [Pimenov]

Sphaerosciadium Pimenov & Kljuykov, Bot. Zhurn. 66(4): 480 (1981).

Glabrous perennials. Leaves tripinnate; petioles of cauline leaves reduced, sheaths amplexicaulous; leaflets ovate, lobate. Bracts and bracteoles absent; rays long. Petals white, flat, apex mucronate, not inflexed; styles reflexed; stylopodium flat. Fruits didymous, with suborbicular mericarps; ribs inconspicuous; commissure narrow; carpophore bifid; exocarp multilayered; mesocarp parenchymatous; endocarp of many layers; vittae small, vallecular 1–3 (only one of them long), commissural up to 4. Seed face deeply grooved, with involute margins. $2n = 20$.

One species, *S. denaense* (Schischk.) Pimenov & Kljuykov, Middle Asia (W Pamiroalai), among rocks, in scree, sclerophyllous scrub, at lower mountain elevations. Molecular data suggest affinity to the *Komarovia* clade.

413. ***Sphallerocarpus*** Besser ex DC. [Spalik]

Sphallerocarpus Besser ex DC., Coll. Mém. 5: 60 (1829); Spalik & Downie, Ann. Missouri Bot. Gard. 88: 270–301 (2001), phylog.; She et al., Fl. China 14: 25–26 (2005), reg. rev.

Pubescent biennials or perennials. Leaves 3–4-pinnately dissected, triangular-ovate; lobes lanceolate. Bracts usually absent; rays many, unequal, spreading and slightly ascending; bracteoles few, ciliate. Petals white, the outer somewhat radiate; styles short; stylopodium short-conical, valvate at margin. Fruits glabrous, ellipsoid to ovoid with a short to obsolete beak, compressed laterally; ribs filiform; commissure narrow; carpophore bifurcate; vittae small, vallecular 3–5, commissural 4–6. Seed transversely round, the face sulcate. $2n = 22$.

One species, *S. gracilis* (Besser ex Trevir.) Koso-Pol., Asia (Siberia, Mongolia, Russian Far East, Korea, and China). Open montane habitats, waste places, arable land.

414. ***Spuriopimpinella*** Kitag. [Pimenov]

Spuriopimpinella Kitag., J. Jap. Bot. 17(10): 558 (1941).

Glabrous to puberulent perennials with short horizontal rhizomes and adventive roots. Leaves biternate or ternate; leaflets ovate to lanceolate, irregularly serrate, with attenuate apex, hirsute on the margin and veins beneath, rarely deeply dissected. Bracts absent or small, solitary; rays thin, scabrous; bracteoles linear. Calyx lobes lanceolate, persistent; petals white, with several secretory ducts; styles slender, reflexed; stylopodium conical. Mericarp ribs filiform; carpophore bifid; commissure narrow; mesocarp parenchymatous; vallecular vittae 2–6, commissural 6–10. Seed face plane. $2n = 22, 24$.

Five species, E Asia (Russian Far East, Japan, Korea, China). Shady deciduous and mixed forests.

415. ***Stefanoffia*** H. Wolff [Kljuykov]

Stefanoffia H. Wolff, Notizbl. Bot. Gart. Berlin 9: 278 (1925); Pimenov & Kljuykov, Notes Roy. Bot. Gard. Edinb. 38: 267–272 (1980), rev.; Kljuykov, Novit. Syst. Pl. Vasc. (Leningrad) 23: 89–92 (1986), part. rev.

Glabrous perennials with globose or ovate tubers. Leaves 2–3-pinnate; leaflets linear to oblong. Bracts and bracteoles entire, linear or dissected. Petals white or greenish yellow; styles reflexed; stylopodium low-conical. Fruits glabrous, ovoid to oblong; commissure narrow; ribs filiform; mesocarp parenchymatous; vallecular vittae 1–3, commissural 2–4. Seed face plane. The seedlings monocotyledonous. $2n = 20$.

Three species, SW Asia (Turkey), S Europe (Bulgaria, Greece). Dry grasslands, thickets, limestone gorges.

416. ***Steganotaenia*** Hochst. [Van Wyk & Tilney]

Steganotaenia Hochst., Flora 27(1): 4 (1844); C. Norman, J. Linn. Soc., Bot. 49: 514 (1934); Cannon, Fl. Zambes. 4: 616–619 (1978), reg. rev.; Townsend, Fl. Trop. E. Afr. 115–118 (1989), reg. rev.; Thulin, J. Linn. Soc., Bot. 107 (2): 163–169 (1991); Thulin in Fl. Somalia 2: 81–286 (1999), reg. rev.; Van Wyk et al., Afr. Apiaceae: 293–297 (2013).

Glabrous woody shrubs, trees or pyrophytic subshrubs with yellowish papery bark. Leaves crowded at the branch ends, hysteranthous, pinnate, trifoliolate or simple; leaflets ovate and acuminate to rounded and emarginate, dentate-aristate. Bracts and bracteoles few, small, filiform; rays short, spreading. Petals yellowish or flushed with green or purple; styles short; stylopodium short and broad, disc-shaped. Fruits glabrous, obovate, strongly compressed dorsally; commissure narrow; ribs distinct, dorsal ribs rarely winged, marginal ones broadly winged; carpophore bifid; vittae absent; rib ducts large, without oil, solitary within each wing, forming cavities in the mature fruit. Seeds transversely strongly compressed dorsally, the face plane or sulcate. $2n = 24$.

Three species, tropical Africa. Dry bushveld. Related to *Polemanniopsis*.

417. ***Stenocoelium*** Ledeb. [Pu]

Stenocoelium Ledeb., Fl. Alt. 1: 297 (1829); Pimenov, Fl. Sib. 10: 161 (1996), reg. rev.; She et al., Fl. China 14: 139 (2005), reg. rev.

Acaulescent or short-caulescent hirsute perennials. Basal leaves many, bipinnate, lanceolate; primary leaflets of several pairs, ultimate segments ovate to ovate-lanceolate. Bracts and bracteoles linear, or lanceolate-linear; rays stout, unequal, violet. Calyx lobes conspicuous, acutely triangular; petals white or pale violet, dorsally pubescent or glabrous; styles recurved; stylopodium low-conical. Fruits densely pubescent or glabrous, ovoid-oblong, compressed dorsally; ribs thick, obtuse, markedly protruding or densely covered with stiff membranes; carpophore bifurcated; commissure rather broad; vallecular vittae 1, commissural 2. Seed face plane or slightly concave. $2n = 20$.

Three species, C Asia (Siberia, Kazakhstan, Mongolia and NW China), on stony slopes of mountains and subalpine habitats.

418. ***Stenosemis*** E. Mey. ex Harv. & Sond. [Van Wyk & Tilney]

Stenosemis E. Mey. ex Harv. & Sond., Fl. Cap. 2: 551 (1862); Burtt, Edinb. J. Bot. 48: 250 (1991), taxon.; Van Wyk et al., Afr. Apiaceae: 298–300 (2013).

Glabrous (except for the minutely scabrid pedicels and bracts) perennial suffrutices with a woody base. Leaves triternate, tripinnate to multifid, slender, coriaceous; petiole not or slightly sheathing; leaflets nearly terete, oblong to filiform. Bracts and bracteoles narrowly triangular to lanceolate; rays several, unequal, spreading. Calyx lobes large, triangular; petals white, tinged with purple; styles long, spreading; stylopodium very broadly cuneate to disc-shaped. Fruits glabrous, roundish to ovoid; mericarps subterete; commissure broad; ribs with thick, corky, undulate wings; carpophore bifid; vittae very large, vallecular 1, commissural 2; rib oil ducts small, 1 or 2 sporadically present. Seeds transversely pentagonal, the face plane.

Two species, South Africa (Eastern Cape province to KwaZulu-Natal). Grasslands.

419. ***Stenotaenia*** Boiss. [Ostroumova]

Stenotaenia Boiss., Ann. Sci. Nat. Bot. I, 1: 339 (1844).

Pubescent perennials. Leaves pinnate; segments ovate, sessile. Bracts lacking; rays unequal, several; bracteoles lacking or few. Petals yellow or purple, pubescent; styles recurved; stylopodium flat with an undulating margin. Fruits pubescent, obovoid, strongly compressed dorsally; dorsal ribs filiform, marginal ribs winged; fibrous inner mesocarp present; commissure broad; vittae unequal, vallecular 3–5, commissural 4–6. Seed face plane. $2n = 22$.

Six species, Asia (Turkey, Iran, Transcaucasia), mountain meadows and forests, stony slopes, rocky hillsides, scree, scrub. Closely related to *Pastinaca*, *Leiotulus* and *Trigonosciadium*.

420. ***Stewartiella*** Nasir [Plunkett]

Stewartiella Nasir, Fl. West Pakist. 20: 152 (1972); Hedge & Lamond in Rechinger, Fl. Iran. 162: 213–214 (1987), reg. rev.

Glabrous and glaucescent perennials with a hardened, fibrous base. Leaves mostly basal, broadly triangular, 4–5-ternately to pinnately divided, coriaceous and stiff; ultimate divisions linear, acute. Bracts and bracteoles few, linear, caducous; rays several, slender, unequal, spreading-ascending. Petals white; styles long, slender, divaricate; stylopodium compressed. Fruits glabrous, oblong, compressed dorsally; ribs winged equally, broader at the base; carpophore not known; vittae minute, many around the seed and a few in the ribs. Seeds transversely slightly compressed dorsally, the face plane.

One species, *S. crucifolia* (Gilli) Hedge & Lamond, Asia (Afghanistan and Pakistan). Dry, rocky areas of mountains.

421. ***Stoibrax*** Raf. [Van Wyk & Tilney]

Stoibrax Raf., Good Book 52 (1840); Sond., Fl. Cap. 2: 538 (1862), under *Ptychotis*; Burtt, Edinb. J. Bot. 48: 250 (1991), taxon.; Magee et al., Syst. Bot. 34: 580–594 (2009), phylog.
Tragiopsis Pomel (1874), nom. illeg., non H. Karst. (1859).
Brachyapium (Baill.) Maire (1932).

Glabrous or minutely tuberculate annuals. Leaves tripinnate with many filiform, glabrous segments. Bracts absent; rays spreading; bracteoles usually absent. Petals white; styles relatively short; stylopodium conical. Fruits glabrous or minutely tuberculate, broader than long, distinctly didymous; mericarps ovate, subterete; commissure narrow; ribs indistinct; carpophore bifid; vittae large, vallecular 1, commissural 2. Seeds transversely subterete, the face convex. $2n = 18, 20, 22$.

Three species, S Europe (Iberian Peninsula) and N Africa (Morocco, Algeria). Sandy, disturbed places. Despite superficial similarities with *Dasispermum*, *Stoibrax* does not appear to be part of the African peucedanoid clade.

422. ***Symphyoloma*** C.A. Mey. [Plunkett]

Symphyoloma C.A. Mey., Verz. Pfl. Cauc. 128 (1831); Tamamschjan, Bot. Zhurn. 35(4): 335–342 (1950), rev.

Acaulescent to barely caulescent, pilose perennials with a thickened, chambered taproot. Leaves rosulate, once pinnate, ovate, subcoriaceous; lateral leaflets pilose, in 2–3 pairs, irregularly serrate to incised, ovate to oblong, the terminal leaflet much larger than the laterals. Bracts lacking; rays several, unequal, pubescent, ascending to spreading or lax; bracteoles few and subulate, or obsolete. Petals whitish or pinkish-purple with white margins, dorsally pubescent or almost glabrous, the outer slightly radiate; styles short and stout in flower, long, slender and sharply reflexed-angled in fruit; stylopodium conical. Fruits densely glabrous, oblong-ovoid, compressed dorsally, single-seeded (due to abortion of ovule in one mericarp), mericarps fused at inner sclerenchyma layers; carpophore lacking; ribs low and indistinct; vittae lacking. Seeds transversely compressed dorsally, the face convex. $2n = 22$.

One species, *S. graveolens* C.A. Mey., Asia (Caucasus). Screes in high-mountain belt. Molecular and phytochemical data show close affinity to *Heracleum*.

423. ***Synclinostyles*** Farille & Lachard [Reduron]

Synclinostyles Farille & Lachard, Acta Bot Gallica 149(4): 376 (2002).

Dwarf, glabrous perennials, with partly subterranean tuberous roots. Leaves triangular, 2–3-ternate into linear segments. Bracts lacking; bracteoles few; rays few, unequal. Petals white or purple, plane; styles curved with opposite stigmas at anthesis; stylopodium low-conical, blackish. Fruits glabrous, ovoid-oblong, laterally compressed; (immature) ribs obtuse; commissure narrow; vallecular vittae 2–3, commissural 4–6. Seeds transversely elliptic, the face plane.

Two or three species, Himalayas (Nepal, Burma-Tibet frontier), alpine *Rhododendron* shrubberies and moss ground cover. Imperfectly known genus probably related to *Sinocarum* and *Acronema*.

424. ***Szovitsia*** Fisch. & C.A. Mey. [Lee]

Szovitsia Fisch. & C.A. Mey., Index Sem. Hortus Bot. Petrop. 1: 39 (1835).

Glabrous annuals. Leaves 2–3-pinnatisect, broadly triangular or ovate, ultimate divisions filiform to linear. Bracts absent; rays few to several, ± equal; bracteoles several, simple, linear to lanceolate. Calyx lobes conspicuous, triangular; petals white, the outer often radiate; styles elongated; stylopodium subconical. Fruits spinescent, ovoid to oblong, slightly compressed laterally; primary ribs filiform, with papillate, unbranched hairs arranged in a single row; secondary ribs 4, broadly elevated, with spatulate, plicate appendages arranged in a single row; carpophore free; vittae large, transversely elliptical, vallecular 1 (under each secondary rib), commissural 2. Seeds semi-lunate, the face plane or slightly sulcate. $2n = 20$.

One species, *S. callicarpa* Fisch. & C.A. Mey., SW Asia (Caucasus, Turkey, Iran). Close to *Astrodaucus* and related genera.

425. ***Taenidia*** (Torr. & A. Gray) Drude [Plunkett]

Taenidia (Torr. & A. Gray) Drude in Engl. & Prantl, Nat. Pflanzenfam. 1, 3(8): 195 (1898); Cronquist, Brittonia 34: 365–367 (1982), rev.

Glabrous and glaucous perennials. Leaves 2–3-ternately or 2–3-pinnately compound; petioles sheathing; leaflets lanceolate to ovate, entire, acute and shortly mucronulate. Bracts and bracteoles lacking; rays several to many, spreading-ascending, unequal. Petals yellow; styles slender; stylopodium lacking. Fruits glabrous, oblong-ellipsoid, slightly compressed laterally; ribs filiform; carpophore bifurcated; vallecular vittae 3, commissural 4. Seeds transversely subterete, the face plane or slightly concave. $n = 11$.

One species, *S. integerrima* (L.) Drude, N America (E United States and Canada), dry woods, rocky hillsides. Molecular data suggest placement in the *Angelica* clade, which might confirm the inclusion of *Pseudotaenidia*.

426. ***Taeniopetalum*** Vis. [Pimenov]

Taeniopetalum Vis., Fl. Dalmat. 3: 49 (1849); Ostroumova et al., Skvortsovia 3(3): 32 (2016).

Glabrous perennials. Leaves 3–4-pinnatisect; leaflets linear, lanceolate or narrowly oblanceolate. Bracts linear to lanceolate, white-membranaceous at margin, or lacking; bracteoles linear-lanceolate, short. Petals whitish, greenish or yellowish; styles divergent; stylopodium low-conical. Fruits compressed dorsally; dorsal ribs filiform, marginal narrowly winged; carpophore bifid; exocarp of large cells with thickened outer walls; mesocarp parenchymatous; commissure broad; vallecular vittae 1–2, commissural 4. Seed face plane. $2n = 22$.

Three species and two subspecies, C, SE, and E Europe, SW Asia, sandy coasts and steppes, dry pine forests. Segregate of *Peucedanum*.

427. ***Tamamschjanella*** Pimenov & Kljuykov [Pimenov]

Tamamschjanella Pimenov & Kljuykov, Bot. Zhurn. 81 (8): 75 (1996); Zakharova, Degtjareva & Pimenov, Willdenowia 42: 159 (2012), rev., nomencl.

Glabrous perennials with horizontal rhizomes. Leaves 2–3-pinnate; leaflets ovoid to oblong, decurrent. Bracts and bracteoles few, lanceolate or subulate, unequal, or absent; rays scabrous. Petals rose-violet or dark-violet; styles reflexed; stylopodium low-conical to flat. Fruits slightly compressed laterally; ribs keeled to narrow-winged, slightly wavy; carpophore bifid; commissure narrow; mesocarp parenchymatous; vallecular vittae 3–5, commissural 6–8. Seed face plane. $2n = 20$.

Two species. Asia (Caucasus, NE Turkey), SE Europe (Greece). Subalpine meadows, forest openings. Molecular data suggest an affinity to *Bunium* and related geophytic taxa, despite considerable morphological dissimilarity.

428. ***Tana*** B.-E. van Wyk [Van Wyk & Tilney]

Tana B.-E. van Wyk in Van Wyk et al., Taxon 48: 743 (1999); Sales & Hedge, Flore de Madagascar et Comores. Fam. 157 (2009), rev.; Van Wyk et al., Afr. Apiaceae: 301 (2013).

Mostly glabrous, woody perennials. Leaves subpalmately compound; leaflets filiform, margins entire. Bracts and bracteoles ovate to lanceolate; rays few, short, spreading, hairy. Calyx lobes triangular; petals yellow, broadly ovate; styles relatively short, spreading; stylopodium very broadly conical. Fruits glabrous, tuberculate, ovoid to ellipsoid; mericarps subterete in transverse section; commissure narrow; ribs inconspicuous; carpophore bifid; vittae very large, vallecular 1, commissural 2; rib oil ducts large, solitary in each rib. Seeds transversely subterete, sulcate beneath vittae, the face $\pm$ plane.

One species, *T. bojeriana* (Baker) B.-E. van Wyk, Madagascar. Poorly known; similar to *Anisopoda* and both related to *Heteromorpha*.

429. ***Tauschia*** Schltdl. [Plunkett]

Tauschia Schltdl., Linnaea 9: 607 (1835), nom. cons., non Preissler (1828); Mathias & Constance, N. Amer. Fl. 28B (1): 81–89 (1944), reg. rev.

Acaulescent to short-caulescent, glabrous to pubescent perennials, sometimes with tuberous roots. Leaves mostly basal (sometimes cauline), entire, pinnate, or ternate to ternately decompound; leaflets various. Bracts usually lacking (occasionally scarious to foliaceous); rays few to many, spreading-ascending or spreading; bracteoles dimidiate, several and prominent, linear to ovate, free to connate, entire or toothed to divided. Calyx lobes sometimes prominent; petals yellow, white, or purplish; styles short, terete or compressed; stylopodium lacking. Fruits glabrous, oblong to orbicular or ellipsoid, slightly compressed laterally; ribs prominent to filiform, unwinged, obtuse or acute; carpophore bifid to the middle or base; vittae large to small, vallecular 1 to several, commissural 2 to several, or numerous and cyclic. Seeds transversely subterete, the face deeply concave or sulcate. $n = 11, 22, 44$.

31 species, N, C, and S America, dry to moist mountainous habitats.

430. ***Tetrataenium*** (DC.) Manden. [Plunkett]

Tetrataenium (DC.) Manden., Trudy Tbilissk. Bot. Inst. 20: 16 (1959); Mandenova, Bot. Zhurn. 80(4): 90–96 (1995), rev.

Pubescent to subglabrous perennials. Basal and lower cauline leaves, 1–2-pinnatisect or ternatisect, or rarely undivided or sublobed, orbicular to obovate; leaflets ovate to broadly ovate or ovate-oblong, the margins serrate or dentate to nearly entire. Bracts few and linear, persistent or caducous, or absent; rays many, unequal; bracteoles few to several, linear. Petals yellow or greenish; styles slender; stylopodium conical. Fruits densely pubescent to glabrate, obovoid (to suborbicular), strongly compressed dorsally, mericarps dorsally compressed; the dorsal and lateral ribs narrowly carinate, filiform, approximately equidistant, the marginal ribs extended into wings; carpophore bifurcated; commissure broad; vittae extending from apex two-thirds to three quarters the length of the fruit, the bases acute (not clavate, as in many *Heracleum*), vallecular 1–3, commissural 2–6, some short. Seeds transversely compressed dorsally, the face plane but sulcate under the vittae. $2n = 22, 38, 40, 42, 44, 46, 66$.

Seven species, Asia (India, Peninsular and Himalayas). A segregate of *Heracleum* accepted here in the narrow sense (Mandenova 1995).

431. ***Thamnosciadium*** Hartvig [Watson]

Thamnosciadium Hartvig, Willdenowia 14(2): 321 (1985) ('1984').

Glabrous and glaucous perennials with a long taproot and short woody rhizome. Stems several, branched, rigid. Leaves rigid, lower 2–3-pinnate; ultimate segments short, linear and fleshy. Bracts and bracteoles few, short and linear; rays few, long. Petals white, densely papillose; styles spreading, longer than the conical stylopodium. Fruits glabrous, ovoid-oblong, compressed dorsally; carpophore bifid; ribs prominent, marginal ribs thickened; vallecular vittae 1, commissural 2–4(5). Seed transversely compressed dorsally, the face plane. $2n = 22$.

One species, *T. junceum* (Sm.) Hartvig, Europe (S & C Greece), high-elevation rocky areas. Previously included in *Sclerochorton* or *Seseli*.

432. ***Thapsia*** L. [Lee, Reduron]

Thapsia L., Sp. Pl. 261 (1753); Gen. Pl., ed. 5: 126 (1754); Brullo et al., Inform. Bot. Ital. 40 Suppl. 3 (2008), rev.; Weitzel et al., Bot. J. Linn. Soc. 174: 620–636 (2014), phylog. taxon.; Banasiak et al., Taxon 65: 563–585 (2016), phylog. taxon.
Elaeoselinum Koch ex DC. (1830).
Margotia Boiss. (1838).
Guillonea Coss. (1851).
Distichoselinum F. García Martín & Silvestre (1983).

Pubescent to glabrous perennials. Leaves mostly basal, sometimes cauline, 1–5-pinnately dissected, broadly triangular, broadly ovate or oblong, the ultimate divisions suborbicular, ovate, ovate-oblong, elliptical to linear. Bracts and bracteoles lacking or several to many; rays several to many, ± equal, spreading or rarely contracted in fruit. Calyx lobes very reduced to conspicuous with subulate tips; petals yellow or white (rarely pinkish); styles short to long; stylopodium conical to depressed. Fruits glabrous or villous, orbicular, ovoid, ellipsoid to oblong, ± compressed dorsally; primary ribs filiform or slender with inconspicuous appendages or rarely winged; secondary ribs 4, the dorsal ribs filiform, sometimes narrowly to broadly winged (rarely unwinged), the lateral ribs with broad (rarely short), often undulate and shiny wings; carpophore free, bifurcated (rarely undivided); vittae small to large, transversely elliptical, vallecular 1 to several, commissural 2(4). Seed face convex, plane to slightly concave, or deeply involute. $2n = 22, 44, 66$.

21–24 species, Europe and N Africa (especially Mediterranean). On the basis of molecular data, broadened to include *Elaeoselinum*, *Distichoselinum*, *Guillonea* and *Margotia* (Weitzel et al. 2014) and two species of *Laserpitium* (Banasiak et al. 2016).

433. ***Thaspium*** Nutt. [Plunkett]

Thaspium Nutt., Gen. North Amer. Pl. 1: 196 (1818); Mathias & Constance, N. Amer. Fl. 28B(2): 187–189 (1945), reg. rev.; Cooperrider, Castanea 50: 116–119 (1985), reg. rev.

Glabrous or pubescent perennials with a fascicle of fibrous roots. Leaves ternately or ternate-pinnately compound or dissected (or simple below); ultimate divisions serrate, dentate, or lobed and incised. Bracts lacking; rays several (6–16), spreading-ascending; bracteoles small, narrow, entire (shorter than the flowers). Petals yellow or purple; styles slender; stylopodium lacking. Fruits glabrous or pubescent, ovoid to oblong, subterete or slightly compressed dorsally; several or all ribs prominently winged; carpophore obsolete; vallecular vittae 1, commissural 2. Seeds transversely subterete or somewhat compressed dorsally, sulcate beneath the valleculae, the face plane. $n = 11$.

Three species, N America (E United States and Canada); dry or moist woodlands, prairies.

434. ***Thecocarpus*** Boiss. [Plunkett]

Thecocarpus Boiss., Ann. Sci. Nat. Bot. III, 2: 93 (1844); Hedge & Lamond, Notes Roy. Bot. Gard. Edinb. 32(2): 167–188 (1973); Hedge & Lamond in Rechinger, Fl. Iran. 162: 81–82 (1987), reg. rev.

Glabrous biennials or perennials. Leaves 1(2)-pinnate, ovate to elliptical; leaflets deeply divided, ultimate segments short and linear. Bracts and bracteoles few, narrowly triangular; rays several, unequal, stout, spreading to divaricate; umbellules with 1 (to several) central, perfect, sessile flower(s) surrounded by several pedicellate staminate flowers, the hardened pedicels of which are fused to the central flower(s). Calyx evident, triangular and hardened in central perfect flowers (sometimes lacking in staminate flowers); petals white; styles short and stout, ascending to spreading; stylopodium conical. Fruits hardened "false nuts", glabrous, bicarpellate or monocarpellate through abortion, appearing angular with spreading protrusions due to adnation of hardened pedicels of outer staminate flowers; ovoid to orbicular, terete; ribs inconspicuous; carpophore present; vittae several, slender. Seeds transversely compressed dorsally, the face plane (but minutely sulcate under the carpophore).

Two species, Asia (S Turkey, W Iran), woodlands, open stony slopes, low to middle elevations.

435. ***Thysselinum*** Adans. [Pimenov]

Thysselinum Adans., Fam. Pl. 2: 100, 615 (1763); Hoffmann, Gen. Pl. Umbell.: 154, 179 (1814), taxon.; Reduron & Muckensturm, Omb. France, Bull. Soc. Bot. Centre-Ouest no. sp. 30: 2435 (2007–2008), nomen.
Thyselium Raf. (1840).

Glabrous perennials. Leaves 2–4-pinnate; leaflets linear to linear-lanceolate, mucronulate, entire or lobed. Bracts lanceolate, apices attenuate, white-membranous at margin, reflexed; rays many, scabrous; bracteoles linear. Petals white to greenish; styles divergent; stylopodium almost flat or low conical. Fruits compressed dorsally, with obtuse dorsal and winged, slightly thickened marginal ribs; carpophore bifid; commissure broad; mesocarp parenchyma of large cells, partly with lignified pitted walls; vallecular vittae 1, commissural inconspicuous. Seed face plane. $2n = 22$.

Two species, Europe, Siberia. Swamps, marshy meadows, lake and river shores and wet forests. *Thysselinum palustre* (L.) Hoffm. is used in folk medicine. Segregate of *Peucedanum*.

436. ***Tiedemannia*** DC. [Plunkett]

Tiedemannia DC., Coll. Mém. 5: 51 (1829); Feist et al., Taxon 61: 402–418 (2012), phylog., rev.

Glabrous perennials with fascicled roots, or rhizomes. Leaves reduced to hollow, acute, septate rachis-leaves. Bracts several, filiform to lanceolate; rays few to several, usually spreading-ascending; bracteoles linear to lanceolate. Calyx lobes evident to prominent (sometimes caducous); petals white or maroon; styles slender, spreading; stylopodium conical. Fruits glabrous, orbicular, ovoid, or obovoid, strongly compressed dorsally; dorsal and lateral ribs filiform; marginal ribs broadly thin-winged (nerves on inner margin of lateral wings give the appearance of 5 dorsal ribs); carpophore bifurcated; vittae large, vallecular 1, commissural 2. Seeds transversely compressed dorsally, the face plane. $n = 14$.

Two species, N America (SE United States), Cuba and Bahamas; wet habitats and pine flatwoods.

437. ***Tilingia*** Regel [Pimenov]

Tilingia Regel, Nouv. Mém. Soc. Imp. Nat. Mosc. 11 (Fl. Ajan.): 97 (1859); Gorovoy, Actes 2e Symp. Int. Ombellifères (Perpignan): 653–661 (1978), taxon.

Glabrous perennials. Leaves ternate or ternate-pinnate to 2–3-pinnate; leaflets oblong to orbicular, deeply pinnately incised. Bracts linear to lanceolate, unequal, scarious, or lacking; rays unequal; bracteoles linear-lanceolate to filiform. Petals white or pink; styles short, divergent or reflexed; stylopodium low-conical. Fruits slightly compressed laterally; ribs subequal, thick-winged, or dorsal keeled, marginal thick-winged; carpophore bifurcated; commissure narrow; mesocarp cells slightly lignified; vallecular vittae 1–4, commissural 3–6. Seed face plane or slightly concave. $2n = 22, 24$.

Three species, E Asia (from NE China and central Japan to Chukotka), N America (Alaska). Stony, mountainous tundras, rocks. Used as folk medicine. Molecular data suggest a relationship to *Sinocarum*, *Rupiphila*, *Halosciastrum*, etc.

438. ***Todaroa*** Parl. [Spalik]

Todaroa Parl. in Webb & Berthel., Hist. Nat. Iles Canar. 3, 2: 155 (1843).

Glabrous or pubescent perennials. Leaves 2–4-pinnately dissected, triangular, lobes lanceolate-ovate. Bracts few to many; rays many, equal to subequal, spreading or ascending; bracteoles few, pubescent, ciliate. Petals white or yellow, pubescent; styles short; stylopodium conical in flower, ovoid in fruit. Fruits glabrous to pubescent, narrowly ellipsoid with a short to obsolete beak, compressed dorsally; commissure broad; ribs winged, the marginal much wider than dorsal; vittae prominent, vallecular 1, commissural 2. Seeds transversely semi-circular, the face concave to almost plane. $2n = 22$.

One species, *T. aurea* (Sol.) Parl., Canary Islands. Open montane habitats. Closely related to *Conopodium* and *Athamanta* (Spalik et al. 2001).

439. ***Tommasinia*** Bertol. [Reduron]

Tommasinia Bertol., Fl. Ital. 3: 414 (1838); Leute, Ann. Naturhist. Mus. Wien 69: 69–79 (1966), taxon.; Vasil'eva & Pimenov, Pl. Syst. Evol. 177: 117–138 (1991), karyo-taxon.

Nearly glabrous perennials. Leaves large, 2–3-pinnately divided, triangular; leaflets ovate or ovate-oblong. Umbels verticillate; bracts absent or very few; rays many, unequal, spreading; bracteoles few. Petals greenish-yellow; styles clearly exceeding the short-conical stylopodium. Fruits glabrous, suborbicular to broadly ellipsoid, strongly compressed dorsally; dorsal ribs weakly prominent, the marginal broadly winged; carpophore free, bifurcated; vittae small, vallecular 1, commissural 2. Seeds transversely subelliptic-compressed, the face plane. $2n = 22$.

One species, *T. altissima* (Mill.) Reduron (= *T. verticillaris* (L.) Bertol.), C Europe. Fresh rocky habitats, forest margins. Related to *Imperatoria* and *Angelica*.

440. ***Tongoloa*** H. Wolff [Pu]

Tongoloa H. Wolff, Notizbl. Bot. Gart. Berlin Dahlem 9: 279 (1925); P.K. Mukherjee & Constance, Umbelliferae India 1: 134–136 (1993), reg. rev.; She et al., Fl. China 14: 34–37 (2005), reg. rev.

Glabrous perennials, sometimes with creeping rhizomes. Leaves ternately to pinnately decompound, narrowly to broadly triangular or ovate; ultimate segments linear, rarely ovate-lanceolate; cauline leaves reduced upward. Bracts and bracteoles absent, or few and small; rays few to many, usually unequal. Petals white or reddish-purple; styles short, spreading or recurved; stylopodium depressed. Fruits glabrous, ovoid, compressed laterally, cordate at the base; mericarps subterete; ribs filiform; carpophore bifid to bifurcated; vallecular vittae 2–3, commissural 2–4. Seeds transversely subterete to slightly compressed dorsally, the face plane or concave. $2n = 16, 22$.

15 species, Asia (N and E China to E Himalaya of Nepal, India, and Bhutan), mostly high elevations, moist forests, alpine meadows, or marshlands. Poorly defined genus, closely related to *Pimpinella*, *Sinocarum*, and *Vicatia*.

441. ***Tordyliopsis*** DC. [Ostroumova]

Tordyliopsis DC., Prodr. 4: 199 (1830); Pimenov et al., Willdenowia 30 (2): 361–367 (2000), taxon.

Pubescent perennials. Leaves pinnate, oblong; ultimate divisions elliptic, subsessile. Bracts and bracteoles many, large, ovate-lanceolate; rays short. Calyx lobes linear, unequal; petals purple, marginal radiate; styles long, recurved; stylopodium narrowly conical. Fruits sparsely pubescent, ellipsoid, strongly compressed dorsally; dorsal ribs keeled, marginal ribs broadly winged; fibrous inner mesocarp present; commissure broad; vallecular vittae 1, commissural 2, rather short, or absent. Seed face plane.

One species, *T. brunonis* DC., S Asia (the Himalayas). High mountain meadows and among *Rhododendron* shrubs. Sometimes treated in *Heracleum*, but differs in large bracts and bracteoles.

442. ***Tordylium*** L. [Ostroumova]

Tordylium L., Sp. Pl. 239 (1753); Gen. Pl., ed. 5: 111 (1754); Al-Eisawi & Jury, Bot. J. Linn. Soc. 97: 357–403 (1988), rev.

Setose annuals or biennials. Leaves simple to tripinnate; segments ovate to cordate; primary segments sessile. Bracts and bracteoles lacking or few. Calyx lobes present and often unequal, or lacking; petals white, pinkish or yellow, pubescent or glabrous, marginal radiate; styles erect or recurved; stylopodium broadly conical. Fruits pubescent, orbicular, ovoid or ellipsoid, strongly compressed dorsally; dorsal ribs filiform, marginal ribs broad, thick, smooth or moniliform; fruits uniform or dimorphic (peripheral fruits bicarpellate and compressed, central fruits unicarpellate and hemispherical); fibrous inner mesocarp present; commissure broad; vallecular vittae 1–3, commissural 2–10, sometimes very short. Seed face plane. $2n = 8, 14, 16, 18, 20, 22$.

20 species, Europe, Asia, Africa (N), S America (adv.), Oceania (adv.). Two subgenera and three sections (Al-Eisawi and Jury 1988).

443. ***Torilis*** Adans. [Lee]

Torilis Adans., Fam. Pl. 2: 99 (1763).

Pubescent annuals (rarely biennials). Leaves 2–3-pinnate, broadly triangular or oblong to ovate, the ultimate divisions linear, lanceolate, or ovate. Bracts absent or several, linear to lanceolate; rays few, with appressed bristles; bracteoles many, linear. Calyx lobes evident to conspicuous; petals white to pinkish, often pubescent, the outer sometimes slightly radiate; styles short to long; stylopodium conical to subconical. Fruits spinescent and tuberculate or papillate, ovoid to oblong, ellipsoid, or linear-cylindrical, subterete to slightly compressed laterally; mericarps homomorphic or heteromorphic; primary ribs prominent with tuberculate, unbranched trichomes, appressed toward stylar ends, arranged in 1–3 rows; secondary ribs 4, prominent, with tuberculate spines, not confluent at the base, conspicuously glochidiate or simple at the tip, arranged in 1–3 rows; carpophore bifid to bifurcated; commissure narrow; vittae large or minute, transversely elliptical, vallecular 1(3) under each secondary rib, commissural 2. Seeds semi-lunate, the face sulcate. $2n = 12, 16, 18, 22, 24, 36$.

15 species, widespread in the N Hemisphere. Closely related to *Astrodaucus*, *Caucalis*, and their allies.

444. ***Tornabenea*** Parl. ex Webb [Ostroumova]

Tornabenea Parl. ex Webb, Hooker's J. Bot. Kew Bot. Misc. 2: 370 (1850); Brochmann et al., Sommerfeltia 24: 77–91 (1997), rev.

Pubescent or glabrous annuals, biennials or perennials. Leaves 1–4-pinnate, segments ovate to linear. Bracts and bracteoles many, linear, entire or 2–3 divided. Petals white, slightly radiating; styles long, erect; stylopodium conical. Fruits glabrous or sparsely pubescent, ellipsoid; mericarps slightly compressed dorsally; primary ribs filiform; secondary ribs equal and winged, or the median filiform and the lateral winged, the wings sometimes dentate; commissure rather narrow; carpophore entire or bifid; vittae narrow, vallecular 1, commissural 2. Seed face plane. $2n = 16$.

Six species, Africa (Cape Verde Is.), rare. Molecular data indicate that the genus is nested within *Daucus*, to which it has been transferred by Banasiak et al. (2016).

445. ***Trachydium*** Lindl. [Kljuykov]

Trachydium Lindl. in Royle, Ill. Bot. Himal. Mount. 232 (1835); Norman, J. Bot. (London) 72: 205–206 (1938), rev.; Pimenov et al., Feddes Repert. 111: 499–515 (2000), taxon.; Pimenov & Kljuykov, Feddes Repert. 111: 517–534 (2000), rev.

Acaulescent, glabrous perennials. Leaves 2–3-pinnate, oblong to oblong-lanceolate, basal primary segments sessile; leaflets small, lanceolate. Bracts few to several, foliaceous, oblong to ovate (similar to leaves); bracteoles several, oblanceolate to obovate, pinnatisect. Calyx lobes evident; petals white, apex inflexed to nearly plane; styles reflexed; stylopodium low-conical. Fruits verrucose, suborbicular; mericarps transversely terete; commissure narrow; ribs equal, narrowly winged, hollow and inflated, rugulose; carpophore lacking; vallecular vittae 1–2, commissural 2–4. Seed face broadly sulcate. $2n = 22$.

One species, *T. roylei* Lindl., Asia (Tibet, NW Himalaya). Alpine zone. Many Asian high-montane species with shortened stems and prostrate habit have been attributed to this genus, but must be transferred to other genera.

446. ***Trachyspermum*** Link [Pu]

Trachyspermum Link, Enum. Pl. Hort. Berol. Alt. 1: 267 (1821), nom. cons.; H. Wolff in Engl., Pflanzenr. 90 (IV. 228): 87 (1927); Hedge & Lamond in Rechinger, Fl. Iran. 162: 336–343 (1987), reg. rev.; Mukherjee & Constance, Umbelliferae India 1: 160–164 (1993), reg. rev.; She et al., Fl. China 14: 77–79 (2005), reg. rev.

Glabrous or hirtellous, sometimes glaucous annuals or perennials, sometimes caespitose. Leaves ternate or pinnate to pinnately decompound, ovate to ovate-triangular or oblong; ultimate segments linear or linear-lanceolate to ovate, entire or serrate at margins. Bracts and bracteoles similar, several (rarely absent), linear-lanceolate or linear, occasionally trifid at apex, pubescent, scarious or ciliate; rays few to many, filiform, unequal, glabrous or hirsutulous. Petals white or pinkish, dorsally pubescent; styles recurved; stylopodium conical. Fruits hispid or scaberulous, papillose, or almost glabrous, ovoid to broadly ovoid, strongly compressed laterally, constricted into a short neck at the apex; mericarps transversely pentagonal-subterete; ribs filiform; carpophore bifid to bifurcated or entire; vallecular vittae 1–3, commissural 2–4. Seed transversely subterete, the face plane. $2n = 16$, 18, 20, 22, 40, 42, 44.

16 species, N and NE Africa to S and SE Asia. Two species widely cultivated as condiments and naturalized.

447. ***Trepocarpus*** Nutt. ex DC. [Plunkett]

Trepocarpus Nutt. ex DC., Coll. Mém. 5: 56 (1829); Mathias & Constance, N. Amer. Fl. 28B(1): 115–116 (1944), reg. rev.

Glabrous annuals. Leaves pinnately decompound; the divisions short and linear. Bracts and bracteoles 1 to several, foliaceous to linear, entire or divided; rays few to several, unequal, ascending to spreading. Calyx lobes prominent, linear and unequal; petals white; styles very short; stylopodium conical. Fruits glabrous, oblong-linear, slightly compressed laterally; mericarps slightly compressed dorsally, commissural face with a prominent corky layer; primary ribs obsolete, secondary ribs prominent and corky; carpophore bifurcated; vittae small and embedded in (and adhering to) the seed, vallecular 1 (under the secondary ribs), commissural 2. Seeds transversely slightly compressed dorsally, the face plane or slightly concave. $n = 9$.

One species, *T. aethusae* Nutt. ex DC., N America (SE United States); margins of swampy forests and sandy river bottoms.

448. ***Tricholaser*** Gilli [Ostroumova]

Tricholaser Gilli, Feddes Repert. 61: 205 (1959).

Pubescent perennials. Leaves biternate, ultimate divisions orbicular to lanceolate. Bracts and bracteoles absent or present. Petals white, dorsally pubescent; styles long, recurved; stylopodium flat with a crenate margin. Fruits pubescent, ellipsoid, notched at the apex, strongly compressed dorsally; dorsal ribs filiform, marginal ribs broad and thick; fibrous inner mesocarp present; vallecular vittae 1, commissural 2 (toward the center), straight and parallel. Seed face plane. $2n = 22$.

Two species, Asia (Afghanistan, Pakistan, NW India). Himalayan mountains.

449. ***Trigonosciadium*** Boiss. [Ostroumova]

Trigonosciadium Boiss., Ann. Sci. Nat. Bot. III, 1: 344 (1844).

Pubescent perennials with inflated, fusiform tuber. Leaves 1–2-pinnate; ultimate divisions ovate or oblong. Bracts filiform to lanceolate-subulate, or absent; bracteoles linear. Petals white, marginal radiate; styles long, erect; stylopodium narrow-conical and very long. Fruits orbicular, obcordate or ellipsoid with a notched apex, strongly compressed dorsally; dorsal ribs filiform, marginal ribs very broad (proximally thin and distally thickened); fibrous inner mesocarp present; vittae narrow, short or long, vallecular 1, commissural 2 (toward the center). Seed face plane. $2n = 22$.

Five species, SW Asia (Turkey, Iraq, Iran). Closely related to *Pastinaca* and *Leiotulus*.

450. ***Trinia*** Hoffm. [Pimenov]

Trinia Hoffm., Gen. Pl. Umbell., ed. 1: 92 (1814), nom. cons.; Fedoronczuk, Systematics and phylogeny of the genera *Trinia*, *Rumia* and *Ledebouriella* (in Russ.). Kiev (1983), rev.

Glabrous or puberulent, monoecious or dioecious perennials or biennials. Leaves 2–3-pinnate; leaflets filiform to linear. Bracts absent to several, linear; rays glabrous, rather unequal; bracteoles entire, lanceolate, almost completely membranaceous, or lacking. Petals white, sometimes hairy; styles reflexed; stylopodium low-conical. Fruits glabrous or scabrid; carpophore bifid; ribs subequal, straight, thickened, obtuse; commissure narrow; mesocarp parenchyma of cells with pitted walls; vallecular vittae 1, small (sometimes lacking), commissural 2; rib oil ducts large. Seed face plane. $2n = 18, 20$.

Eight to ten species, Europe, Asia (SW, W Siberia, Kazakhstan). Steppes, stony hillsides. Closely related to *Rumia* and *Ledebouriella*.

451. ***Trocdaris*** Raf. [Reduron]

Trocdaris Raf., Good Book 50 (1840); Zakharova et al., Willdenowia 42: 149–168 (2012), taxon.

Glabrous, tuberous perennials. Leaves narrowly oblong in outline, mostly basal, divided into deeply palmatisect segments and filiform lobes, appearing whorled. Bracts and bracteoles many; rays several to many, subequal. Petals white; styles long, exceeding the conical stylopodium. Fruits glabrous, ellipsoid, a little oblong; ribs low and narrow; vittae wide, vallecular 1, commissural 2. Seeds sub-pentagonal, the face plane. $2n = 20, 22$.

One species, *T. verticillatum* (L.) Raf., W Europe, Morocco. Segregated from *Carum* on the basis of morphological and molecular data (Zakharova et al. 2012).

452. ***Trochiscanthes*** W.D.J. Koch [Reduron]

Trochiscanthes W.D.J. Koch, Nova Acta Acad. Leop.-Carol. Nat. Cur. 12 (1): 103 (1824); Ferrarini, Webbia 41 (1): 45–60 (1987), chorol.

Glabrous perennials. Leaves 3–4-ternately divided, triangular; leaflets large, ovate-lanceolate. Umbels arranged in panicles with opposite and verticillate branches; bracts absent; rays several, subequal to unequal, spreading; bracteoles present. Petals greenish-white; styles short; stylopodium short-conical. Fruits glabrous, ovoid; slightly compressed laterally; ribs equal, slender, prominent, obtuse to nearly winged; carpophore free, bifurcated; vittae small, many. Seeds transversely pentagonal, the face plane. $2n = 22$.

One species, *T. nodiflora* (All.) W.D.J. Koch, C Europe (S alpine areas of SE France, SW Switzerland, N Italy; records from the Pyrenees erroneous). Forest margins.

453. ***Tschulaktavia*** Bajtenov ex Pimenov & Kljuykov [Pimenov]

Tschulaktavia Bajtenov ex Pimenov & Kljuykov, Bot. Zhurn. 97(5): 651 (2012).

Glabrous perennials. Leaves bipinnatisect, basal segments petiolulate, leaflets oblanceolate. Bracts and bracteoles several, lanceolate, short, slightly brown; rays 4–6, slightly unequal. Petals yellowish-green, entire, with secretory duct; stylopodium conical. Fruits ovate-lanceolate, slightly compressed laterally; all ribs keeled; commissure narrow; vascular bundles large, in distal parts of ribs; vallecular vittae 1 or sometimes 2 long and 2 short in each furrow, commissural 2 long and 2 short. Seed face plane. $2n = 22$.

One species, *T. saxatilis* (Bajtenov) Bajtenov ex Pimenov & Kljuykov, Asia (SE Kazakhstan), cliff crevices in arid mountains. Molecular data suggest distant affinity to *Ostericum*, *Pterygopleurum*, and *Halosciastrum*.

454. ***Turgenia*** Hoffm. [Lee]

Turgenia Hoffm., Gen. Pl. Umbell. 59 (1814).

Pubescent annuals. Leaves once-pinnate, broadly ovate; ultimate divisions linear to oblong. Bracts and bracteoles several, ovate to oblong, simple, pubescent (bracts), margins scabrous; rays few to several, glabrous. Calyx lobes conspicuous; petals white to purple, the outer conspicuously radiate; styles short; stylopodium conical. Fruits tuberculate, ovoid to ellipsoid, subterete to slightly compressed laterally; primary ribs prominent, with tuberculate, unbranched, ± erect spines arranged in 2 rows; secondary ribs 4, prominent, with tuberculate spines, confluent at the base, simple (or rarely glochidiate) at the tip, arranged in a single row; carpophore free, bifid above; commissure narrow; vallecular 1 (under each secondary rib), commissural 2. Seeds transversely subterete, the face deeply sulcate and incurved. $2n = 18, 20, 24, 32$.

Two species, widespread in Europe, W Asia, and N Africa. Closely related to *Lisaea*.

455. ***Turgeniopsis*** Boiss. [Lee]

Turgeniopsis Boiss., Ann. Sci. Nat. Bot. III, 2: 53 (1844).

Glabrous annuals. Leaves 3–4-pinnatisect, 2–3-pinnate, broadly ovate, ultimate divisions filiform. Bracts absent or rarely 1, linear, glabrous; rays few, long, not contracted in fruits; bracteoles few, simple. Petals white, the outer conspicuously radiate; styles short, as long as the conical stylopodium. Fruits spinescent, ellipsoid, slightly compressed laterally, covered with peg-like hairs; primary ribs obscure, bearing 2–3 rows of short bristles; secondary ribs 4, well developed, with papillate, unbranched, unequal spines with a conspicuously glochidiate apex, arranged in 2–3 rows; carpophore free; vittae small, transversely elliptical, vallecular 1 (under each secondary rib), commissural 2. Seeds transversely subterete to pentagonal, the face sulcate.

One species, *T. foeniculacea* (Fenzl) Boiss., SE Europe and SW Asia. Molecular data show relationship to *Szovitsia* and *Astrodaucus*.

456. ***Vanasushava*** P.K. Mukh. & Constance [Plunkett]

Vanasushava P.K. Mukh. & Constance, Kew Bull. 29: 595 (1974); Blasco et al., Actes 2e Symp. Int. Ombellifères (Perpignan): 663–674 (1978), taxon.

Hispidulous perennials with trailing rootstock and stems rooting at the nodes. Leaves trifoliolate (but appearing palmately 5-foliate), orbicular to obovate; leaflets large, ovate to lanceolate, serrate with acuminate apices and cuneate bases. Bracts usually lacking; rays few, unequal, slender and spreading-ascending; bracteoles few to several, short, narrow, reflexed (occasionally lacking). Petals white, the marginal sometimes slightly radiate; styles slender and reflexed; stylopodium conical to low-conical. Fruits glabrous, ovoid-oblong, the apex narrowed and the base cordate, transversely subterete; mericarps dorsally compressed; ribs subequal, filiform, low and rounded; carpophore lacking; vittae arranged cyclically, vallecular 2–3, commissural 2–6. Seeds transversely compressed dorsally, the face sulcate. $2n = 46$.

One species, *V. pedata* (Wight) P.K. Mukh. & Constance, Asia (S India), moist hill forests. A segregate of *Heracleum*.

457. ***Vesper*** R.L. Hartm. & G.L. Nesom [Plunkett]

Vesper R.L. Hartm. & G.L. Nesom, Phytoneuron 2012–94: 2 (2012).

Acaulescent (but appearing caulescent with a pseudoscape), glabrous to scaberulous, often glaucous perennials. Leaves mostly basal (a few reduced leaves on the pseudoscapes), somewhat fleshy, lanceolate to ovate or triangular, 1–3-pinnate-pinnatifid. Bracts few to several, minute or larger and oblong to ovate, scarious and white or white-margined or purple, sometimes fused; rays few to many; bracteoles few to several, spatulate to ovate or orbicular, scarious and usually white or purple, fused below, prominently nerved. Petals white to cream, pink, or purple; styles slender, spreading; stylopodium lacking. Fruits glabrous, oblong to broadly elliptical or suborbicular, compressed dorsally; (4–)5 of the ribs usually broadly winged, the wings membranaceous, broader than the fruit body, straight (or wavy); carpophore bifurcated or obsolete; vallecular vittae 2–4, commissural 4–7. Seeds transversely compressed dorsally, the face concave. $2n = 22$.

Six species, N America (SW United States and adjacent Mexico); sandy, loamy or chalky habitats, prairies, plains, mountains, and hillsides.

458. ***Vicatia*** DC. [Pimenov]

Vicatia DC., Prodr. 4: 243 (1830).

Glabrous perennials, sometimes with short rhizomes and adventive roots. Leaves 1–3-pinnate or ternate; petioles long, hollow; leaflets linear to lanceolate. Bracts lacking; rays few, unequal; bracteoles linear-lanceolate to linear. Petals white, greenish or reddish-brown; styles reflexed;

stylopodium low-conical. Fruits oblong to ovoid, narrowed toward apex, compressed laterally; ribs filiform; carpophore bifid; commissure rather narrow; mesocarp parenchymatous; vallecular vittae 2–4, commissural 3–6. Seed face broadly furrowed. $2n = 22$.

Three species, Middle Asia, Kazakhstan, China, Himalaya. Mountain meadows, forests, stony slopes. Closely related to *Carum*, *Tongoloa*, and possibly *Sinodielsia*.

459. ***Visnaga*** Mill. [Pimenov]

Visnaga Mill., Gard. Dict. Abr. ed. 4 (1754); Pimenov & Pogorelova, Plant Resources 10(2): 216–219 (1974), rev.

Glabrous annuals or biennials. Leaves 2–3-pinnate; leaflets linear to linear-filiform, entire, divaricate. Bracts many, pinnatisect or bipinnatisect, with subulate lobes; rays many, spreading, becoming contracted and thickened at fruiting time; bracteoles entire, subulate, contracted in fruit; pedicels thickened. Petals white, unequal, the outer radiate; styles reflexed; stylopodium low-conical. Mericarp ribs filiform, approximately equal; carpophore entire; mesocarp parenchymatous; vallecular vittae 1, commissural 2; rib oil ducts 1, large. Seed face plane. $2n = 20, 22$.

One or two species, Mediterranean region, S Europe, SW Asia, N and NW Africa, fallow fields, frequently adventive, plains and foothills. *Visnaga daucoides* Gaertn. used as a spasmolytic. Often treated under *Ammi*, but fruit and phytochemical features (Pimenov and Pogorelova 1974) and molecular data (Jiménez-Mejías and Vargas 2015) support its segregation.

460. ***Xanthoselinum*** Schur [Pimenov]

Xanthoselinum Schur, Enum. Pl. Transsilv. 264 ("*Xantholoselinum*"), 981 (1866); Reduron et al., J. Bot. Soc. Bot. France 1: 91–104 (1997), taxon.

Glabrous or minutely pubescent perennials. Leaves 3–4-pinnate or 3–4-ternate; leaflets linear to lanceolate. Bracts linear to lanceolate, white-membranaceous at margin; rays unequal; bracteoles unequal, linear. Calyx lobes triangular; petals light-yellow or white; styles reflexed; stylopodium low-conical. Fruits dorsally compressed; dorsal ribs filiform, marginal winged; carpophore bifid; commissure broad; mesocarp parenchymatous; vallecular vittae 1, commissural 2; rib oil ducts large. Seed face plane. $2n = 22$.

Two species, Europe, Asia (Siberia, N Caucasus, Kazakhstan). Oak and pine forests, forest margins and openings, steppes, dry meadows, sometimes weedy. Molecular data suggest a relationship to *Endressia*, *Karatavia*, *Dystaenia*, and *Paraligusticum*.

461. ***Xatartia*** Meisn. & Zeyh. ex Meisn. [Reduron]

"*Xatartia*" Meisn. & Zeyh. ex Meisn., Pl. Vasc. Gen. 1: 145, 2: 105 (1838); Baudière & Serve, An. Inst. Bot. Cavanilles 32(2): 537–556 (1975), ecol.; Saenz de Rivas, An. Inst. Bot. Cavanilles 34(1): 133–138 (1977), carpol.; Baudière & Serve, Bull. Soc. Bot. France (Lettres Bot.) 127(1): 71–79 (1980), chorol. ecol. Dedicated to Barthélemy Xatart (1774–1846); the wrong original spelling here corrected into Xatartia (Code art. 60.1 ex. 7).

Scabrous perennials. Leaves rosulate, 2–3-pinnately divided, triangular; ultimate divisions lanceolate. Bracts absent or very few; bracteoles several; rays many, very unequal, spreading. Petals green or yellowish; styles equaling to clearly exceeding the subconical stylopodium. Fruits glabrous, ovoid, strongly compressed dorsally; ribs stout, prominent, the lateral nearly winged; carpophore free, bifurcated; vittae medium-sized, vallecular 1, commissural 2. Seeds transversely compressed, the face plane. $2n = 22$.

One species, *X. scabra* (Lapeyr.) Meisn., Europe (E Pyrenees). Screes at high elevations.

462. ***Xyloselinum*** Pimenov & Kljuykov [Pimenov]

Xyloselinum Pimenov & Kljuykov, Komarovia (St. Petersburg) 4: 129 (2006); Pimenov et al., Phytotaxa 244(3): 252 (2016), rev., phylog.

Almost glabrous, shrubby polycarpic perennials with massive taproots, stems lignescent. Leaves 2–3-pinnatisect; leaflets rhombic or ovate, minutely toothed or deeply dissected. Bracts solitary or absent; rays several, densely covered with scattered short prickles, these remote in fruit; bracteoles several, linear, entire, toothed, or dissected. Petals whitish or yellowish; styles thin, reflexed; stylopodium low-conical or conical. Fruits compressed dorsally; dorsal ribs filiform, marginal ribs broadly winged, thin; carpophore bifurcated; commissure broad; mesocarp parenchymatous; vallecular vittae 1–2, commissural 3–4; rib oil ducts small, solitary. Seed face nearly plane.

Three species, SE Asia (N Vietnam, Laos). Limestone ridges, in disturbed and mixed forests. Molecular data suggest an affinity to *Kitagawia*, some E Asian species of *Peucedanum* s. lat., *Oreocome*, and *Ligusticopsis* (Pimenov et al. 2016).

463. ***Yabea*** Koso-Pol. [Lee]

Yabea Koso-Pol., Bull. Soc. Nat. Mosc., n.s., 28: 202 (1915).

Pubescent annuals. Leaves 2–3-pinnatisect, broadly ovate, the ultimate divisions filiform; petiole bases not sheathing. Bracts few to several, trifid to pinnatisect (like the leaves); rays few to several, unequal; bracteoles many, simple and linear or rarely trifid. Petals white; styles short; stylopodium conical to subconical. Fruits spinescent and tuberculate, ovoid to oblong, slightly compressed laterally, primary ribs prominent, with tuberculate, unbranched, ± erect hairs arranged in 3 rows; secondary ribs 4, prominent, with tuberculate or papillate spines, arranged in a single row, simple at the tip; carpophore free; vittae large, transversely elliptical, vallecular 1 (under each secondary rib), commissural 2. Seeds transversely subterete, the face sulcate. $2n = 10, 12$.

One species, *Y. microcarpa* (Hook. & Arn.) Koso-Pol., N America (W). Cytological and molecular evidence suggests a closely relationship to *Torilis*.

464. ***Zeravschania*** Korovin [Pimenov]

Zeravschania Korovin, Bot. Mat. Syst. Herb. Inst. Bot. Zool. Acad. Sci. Uzbek. 12: 28 (1948); Pimenov, Bjull. Moskovsk. Obšč. Isp. Prip. Otd. Biol. 93(4): 75–80 (1988), rev.

Glabrous or scabrous perennials. Leaves 2–3-pinnate or 2–3-ternate; leaflets elliptic, dissected, glabrous or scabrous. Bracts and bracteoles lanceolate, broadly white-membranaceous. Petals white or yellowish, apex short-acuminate; styles long, reflexed; stylopodium conical, undulate at the margin. Fruits glabrous, compressed dorsally; dorsal ribs narrow, terete, marginal slightly enlarged; carpophore bifid; commissure broad; vallecular vittae 1, commissural 2; rib oil ducts developed or absent. Seed face plane. $2n = 22$.

11 species, Asia (Middle, Iran, Afghanistan, S Transcaucasus), dry mountain slopes, rocky crevices. Morphological and molecular data suggest an affinity to *Demavendia* and *Haussknechtia*.

465. ***Zizia*** W.D.J. Koch [Plunkett]

Zizia W.D.J. Koch, Nov. Acta Acad. Leop.-Carol. Nat. Cur. 12(1): 128 (1824); Mathias & Constance, N. Amer. Fl. 28B (1): 122 (1944), reg. rev.; Cooperrider, Castanea 50: 116–119 (1985), reg. rev.

Glabrous (or nearly so) perennials with a fascicle of somewhat fleshy roots. Leaves simple or ternately compound; leaves or leaflets serrate or dentate with hyaline margins. Bracts lacking; rays few to several, spreading-ascending or spreading; bracteoles few, small and linear, or lacking. Calyx lobes prominent; petals golden yellow; styles slender, erect or spreading; stylopodium lacking. Fruits glabrous, oblong to ellipsoid, slightly compressed laterally; mericarps transversely pentagonal; ribs filiform; carpophore bifid to ca. half its length; vallecular vittae 1, commissural 2. Seeds transversely subterete, sulcate under the vittae, the face plane or slightly concave. $n = 11$, $2n = 22$.

Four species, N America; moist fields, meadows, and open woods.

466. ***Zosima*** Hoffm. Fig. 3I [Ostroumova]

Zosima Hoffm., Gen. Pl. Umbell., ed. 1: 145 (1814).

Pubescent perennials. Leaves 2–3-pinnately dissected; ultimate divisions small, linear, dentate, lobed or pinnatifid. Bracts and bracteoles linear or lanceolate. Petals white, yellow or pink, pilose, marginal often radiate; styles recurved; stylopodium short-conical. Fruits pubescent, orbicular to oblong with a notched apex, strongly compressed dorsally; dorsal ribs filiform, marginal ribs very broad (proximally thin and distally thickened); fibrous inner mesocarp present; commissure broad; vittae very broad, vallecular 1, commissural 2 (toward the middle). Seed face plane. $2n = 6, 10, 12$.

Four species, Asia. Closely related to *Heracleum*, *Leiotulus*, *Pastinaca*, *Tetrataenium*, *Semenovia*, and *Tordylium*.

Selected Bibliography

Banasiak, Ł., Wojewódzka, A., Baczyński, J., Reduron, J.-P., Piwczyński, M., Kurzyna-Młynik, R., Gutaker, R., Czarnocka-Cieciura, A., Kosmala-Grzechnik, S., Spalik, K. 2016. Phylogeny of Apiaceae subtribe Daucinae and the taxonomic delineation of its genera. Taxon 65: 563–585.

Barthlott, W., Neinhuis, C., Cutler, D., Ditsch, F., Meusel, I., Theisen, I., Wilhelmi, H. 1998. Classification and terminology of plant epicuticular waxes. Bot. J. Linn. Soc. 126: 237–260.

Baumann-Bodenheim, M.G. 1955. Ableitung und Bau bicarpellat-monospermer und pseudomonocarpellater Araliaceen- und Umbelliferen-Früchte. Ber. Schweiz. Bot. Ges. 65: 481–510.

Behnke, H.-D. 1971. Sieve-tube plastids of Magnoliidae and Ranunculidae in relation to systematics. Taxon 20: 723–730.

Bell, C.R. 1971. Breeding systems and floral biology of the Umbelliferae, or evidence for specialization in unspecialized flowers. In: Heywood, V.H. (ed.) The Biology and Chemistry of the Umbelliferae, Bot. J. Linn. Soc. 64, Suppl. 1. London & New York: Academic Press, pp. 93–108.

Bell, C.R., Lindsey, A.H. 1982. The umbel as a reproductive unit in the Apiaceae. In: Cauwet, A.M., Carbonnier, J. (eds.) Contributions Pluridisciplinaires à la Systématique; actes du 2ème Symposium International sur les Ombellifères, Centre Universitaire de Perpignan. St. Louis: Missouri Botanical Garden, pp. 739–748.

Bennett, M.D., Leitch, I.J. 2005. Plant DNA C-values Database (release 4.0). http://www.rbgkew.org.uk/cval/database1.html

Bentham, G. 1867. Umbelliferae. In: Bentham, G., Hooker, J.D. (eds.) Genera Plantarum 1. London: Reeve & Co., pp. 859–931.

Berenbaum, M.R. 1990. Evolution of specialization in insect-umbellifer associations. Annu. Rev. Entomol. 35: 319–343.

Borg-Karlson, A.-K., Valterova, I., Nilsson, L.A. 1994. Volatile compounds from flowers of six species in the family Apiaceae: bouquets for different pollinators? Phytochemistry 35: 111–119.

Calviño, C.I., Downie, S.R. 2007. Circumscription and phylogeny of Apiaceae subfamily Saniculoideae based on chloroplast DNA sequences. Molec. Phylogen. Evol. 44: 175–191.

Cerceau-Larrival, M.-T. 1962. Plantules et pollens d'ombellifères. Leur intérêt systématique et phylogénique. Mém. Mus. Natl Hist. Nat., sér. B, Bot. 14: 1–166.

Cerceau-Larrival, M.-T. 1971. Morphologie pollinique et corrélations phylogénétiques chez les Ombellifères. In: Heywood, V.H. (ed.) The Biology and Chemistry of the Umbelliferae, Bot. J. Linn. Soc. 64, Suppl. 1. London & New York: Academic Press, pp. 109–135.

Corner, E.J.H. 1976. The seeds of dicotyledons, Vol. 1. Cambridge: Cambridge University Press.

Courchet, L. 1884. Étude anatomique sur les Ombellifères et sur les principales anomalies de structure que présentent leurs organes végétatifs. Ann. Sci. Nat., Bot., sér. 6, 17: 107–129.

Crowden, R.K., Harbourne, J.B., Heywood, V.H. 1969. Chemosystematics of the Umbelliferae - a general survey. Phytochemistry 8: 1963–1984.

Davis, G.L. 1966. Systematic embryology of the angiosperms. New York: Wiley.

Dawson, J.W. 1971. Relationships of the New Zealand Umbelliferae. In: Heywood, V.H. (ed.) The Biology and Chemistry of the Umbelliferae, Bot. J. Linn. Soc. 64, Suppl. 1. London & New York: Academic Press, pp. 43–62.

Deschamps, R. 1982. Un cas intéressant d'évolution entre une espèce fossile arborescente d'Afrique et son correspondant actuel *Steganotaenia araliacea* (Ombellifères). In: Cauwet-Marc, A.-M., Carbonnier, J. (eds.) Les Ombellifères: Contributions Pluridisciplinaires à la Systématique: actes du 2ème Symposium International sur les Ombellifères, Centre Universitaire de Perpignan. St. Louis: Missouri Botanical Garden, pp. 207–212.

Degtjareva, G.V., Kljuykov, E.V., Samigullin, T.H., Valiejo-Roman, C.M., Pimenov, M.G. 2013. ITS phylogeny of Middle Asian geophilic Umbelliferae-Apioideae genera with comments on their morphology and utility of *psbA-trnH* sequences. Pl. Syst. Evol. 299: 985–1010.

Degtjareva, G.V., Pimenov, M.G., Samigullin, T.H. 2018. Molecular data allow to elucidate the taxonomic placement of some Umbelliferae from Middle Asia and Afghanistan (*Pinacantha, Ladyginia, Peucedanum mogoltavicum*). Phytotaxa, 350: 42–50.

Desjardins, S.D., Leslie, A.C., Stace, C.A., Schwarzacher, T., Bailey, J.P. 2015. Intergeneric hybridization between *Berula erecta* and *Helosciadium nodiflorum* (Apiaceae). Taxon 64(4): 784–794.

Diniz, F. 1969. Ombellifères pliocènes de Rio Major (Portugal). Naturalia Monspeliensia sér. Bot. Fasc. 20: 77–88.

Downie, S.R., Katz-Downie, D.S., Spalik, K. 2000. A phylogeny of Apiaceae tribe Scandiceae: evidence from nuclear ribosomal DNA internal transcribed spacer sequences. Amer. J. Bot. 87(1): 76–95.

Downie, S.R., Plunkett, G.M., Watson, M.F., Spalik, K., Katz-Downie, D.S., Valiejo-Roman, C.M., Terentieva, E.I., Troitsky, A.V., Lee, B.-Y., Lahham, J., El-Oqlah, A. 2001. Tribes and clades within Apiaceae subfamily Apioideae: the contribution of molecular data. Edinb. J. Bot. 58: 301–330.

Downie, S.R., Spalik, K., Katz-Downie, D.S., Reduron, J.-P. 2010. Major clades within Apiaceae subfamily Apioideae as inferred by phylogenetic analysis of nrDNA ITS sequences. Pl. Divers. Evol. 128: 111–136.

Drude, C.G.O. 1898. Umbelliferae. In: Engler, A., Prantl, K. (eds.) Die natürlichen Pflanzenfamilien ed. 1, 3(8). Leipzig: Wilhelm Engelmann, pp. 63–250.

Durrieu, G. 1982. Les Champignons parasites et leur apport à la systématique des Ombellifères. In: Cauwet, A.-M., Carbonnier, J. (eds.) Contributions Pluridisciplinaires à la Systématique; actes du 2ème Symposium International sur les Ombellifères, Centre Universitaire de Perpignan. St. Louis: Missouri Botanical Garden, pp. 549–561.

Erbar, C., Leins, P. 1996. Distribution of the character states "early" and "late sympetaly" within the "Sympetalae Tetracyclicae" and presumably related groups. Bot. Acta 109: 427–440.
Erbar, C., Leins, P. 1997. Different patterns of floral development in whorled flowers, exemplified by Apiaceae and Brassicaceae. Int. J. Pl. Sci. 158: S49–S64.
Erbar, C., Leins, P. 2004. Sympetaly in Apiales (Apiaceae, Araliaceae, Pittosporaceae). S. Afr. J. Bot. 70: 458–467.
Erbar, C., Leins, P. 2010. Nectaries in Apiales and related groups. Pl. Divers. Evol. 128: 269–295.
Esau, K. 1940. Developmental anatomy of the fleshy storage organ of *Daucus carota*. Hilgardia 13: 175–209.
Fernandez-Prieto, J.A., Cires, E. 2014. Phylogenetic placement of *Dethawia*, *Meum*, and *Rivasmartinezia* (Apioideae, Apiaceae): evidence from nuclear and plastid DNA sequences. Pl. Biosyst. 148(5): 975–987.
French, D.H. 1971. Ethnobotany of the Umbelliferae. In: Heywood, V.H. (ed.) The Biology and Chemistry of the Umbelliferae, Bot. J. Linn. Soc. 64, Suppl. 1. London, New York: Academic Press, pp. 385–412.
Grossheim, A.A. 1967. Flora Kavkaza (Flora of Caucasus), ed. 2, vol 7. Leningrad: Nauk.
Gruas-Cavagnetto, C., Cerceau-Larrival, M.T. 1982. Présence de pollens d'Ombellifères fossiles dans le Paléogène du bassin Anglo-Parisien: premiers résultats. In: Cauwet, A.-M., Carbonnier, J. (eds.) Contributions Pluridisciplinaires à la Systématique; actes du 2ème Symposium International sur les Ombellifères, Centre Universitaire de Perpignan. St. Louis: Missouri Botanical Garden, pp. 255–267.
Guyot, M. 1966. Les stomates des Ombellifères. Bull. Soc. Bot. France 113: 244–273.
Guyot, M. 1971. Phylogenetic and systematic value of stomata in the Umbelliferae - Caucalideae. Bot. J. Linn. Soc. 64: 199–214.
Guyot, M. 1978. Intérêt des études de phytodermologie dans la famille des Ombellifères. In: Cauwet, A.M., Carbonnier, J. (eds.) Contributions Pluridisciplinaires à la Systématique; actes du 2ème Symposium International sur les Ombellifères, Centre Universitaire de Perpignan. St. Louis: Missouri Botanical Garden, Louis., pp.133–148.
Havis, L. 1939. Anatomy of the hypocotyl and roots of *Daucus carota*. J. Agric. Res. 58: 557–564.
Hedge, I.C., Lamond, J.M. 1972. *Microsciadium*. In: Davis, P.H. (ed.) Flora of Turkey and the East Aegean Islands, vol. 4, pp. 420–421. Edinburgh: Edinburgh University Press.
Hegnauer, R. 1971. Chemical patterns and relationships of Umbelliferae. In: Heywood, V.H. (ed.) The Biology and Chemistry of the Umbelliferae, Bot. J. Linn. Soc. 64, Suppl. 1. London, New York: Academic Press, pp. 267–277.
Hegnauer, R. 1973. Umbelliferae. In: Chemotaxonomie der Pflanzen, Vol. VI. Basel: Birkhäuser, pp. 554–629.
Hegnauer, R. 1989. Umbelliferae. In: Chemotaxonomie der Pflanzen, Vol. IX. Basel: Birkhäuser, pp. 663–714.
Henderson, D.M. 1973. The rust fungus genus *Nyssopsora* and its host relations. Notes Roy. Bot. Gard. Edinb. 32: 217–221.
Hoar, C.S. 1915. A comparison of the stem anatomy of the cohort Umbelliflorae. Ann. Bot. 29: 55–63.
Holub, M., Toma, J., Herout, V. 1987. The phylogenetic relationships of the Asteraceae and Apiaceae based on phytochemical characters. Biochem. Syst. Ecol. 15: 321–326.
Jarvis, C.E., Reduron, J.-P., Spencer, M.A., Cafferty, S. 2006. Typification of Linnaean plant names in Apiaceae. Taxon 55(1): 207–216.
Jay, M. 1969. Chemotaxonomic researches on vascular plants. XIX. Flavonoid distribution in the Pittosporaceae. Bot. J. Linn. Soc. 62: 423–429.
Jiménez-Mejías, P., Vargas, P. 2015. Taxonomy of the tribe Apieae (Apiaceae) revisited as revealed by molecular phylogenies and morphological characters. Phytotaxa 212: 57–79.
Jongejans, E., Telenius, A. 2001. Field experiments on seed dispersal by wind in ten umbelliferous species (Apiaceae). Pl. Ecol. 152: 67–78.
Jurica, H.S. 1922. A morphological study of the Umbelliferae. Bot. Gaz. 74: 292–307.
Kaplan, D.R. 1970. Comparative development and morphological interpretation of "rachis-leaves" in Umbelliferae. In: Robson, N.K.B., Cutler, D.F., Gregory, M. (eds.) New Research in Plant Anatomy, Suppl. 1, Bot. J. Linn. Soc. 63. London & New York: Academic Press, pp. 101–125.
Katz-Downie, D.S., Valiejo-Roman, C.M., Terentieva, E.I., Troitsky, A.V., Pimenov, M.G., Lee, B.-Y., Downie, S. R. 1999. Towards a molecular phylogeny of Apiaceae subfamily Apioideae: additional information from nuclear ribosomal DNA ITS sequences. Pl. Syst. Evol. 216: 167–195.
Kljuykov, E.V., Zakharova, E.A., Petrova, S.E., Tilney, P. M. 2014. On the unusual structure of the monocotyledonous embryo and seedling of *Acronema commutatum* H. Wolff (Apiaceae) and related species. Pl. Divers. Evol. 131: 1–10.
Koch, W.D.J. 1824. Generum tribuumque plantarum Umbelliferarum nova dispositio. Nova Acta Acad. Caes. Leop.-Carol. German. Nat. Cur. 12: 55–156.
Koso-Poljansky, B. 1916. Sciadophytorum systematis lineamenta. Bull. Soc. Imp. Naturalistes Moscou 29: 93–221.
Koul, P., Sharma, N., Koul, A.K. 1993. Pollination biology of Apiaceae. Curr. Sci. 65: 219–222.
Kurzyna-Młynik, R., Oskolski, A.A., Downie, S.R., Kopacz, R., Wojewódzka, A., Spalik, K. 2008. Phylogenetic position of the genus *Ferula* (Apiaceae) and its placement in tribe Scandiceae as inferred from nrDNA ITS sequence variation. Pl. Syst. Evol. 274: 47–66.
Lee, B.-Y., Levin, G.A., Downie, S.R. 2001. Relationships within the spiny-fruited umbellifers (Scandiceae subtribes Daucinae and Torilidinae) as assessed by phylogenetic analysis of morphological characters. Syst. Bot. 26: 622–642.
Leins, P., Erbar, C. 2004. Floral organ sequences in Apiales (Apiaceae, Araliaceae, Pittosporaceae). S. Afr. J. Bot. 70: 468–474.
Lindsey, A.H. 1982. Floral phenology patterns and breeding systems in *Thaspium* and *Zizia* (Apiaceae). Syst. Bot. 7: 1–12.
Logacheva, M.D., Valiejo-Roman, C.M., Pimenov, M.G. 2008. ITS phylogeny of West Asian *Heracleum* species and related taxa of Umbelliferae–Tordylieae W.

D.J. Koch, with notes on evolution of their *psbA-trnH* sequences. Pl. Syst. Evol. 270: 139–157.
Lubbock, J. 1892. A contribution to our knowledge of seedlings, Vol. II. New York: D. Appleton.
Lyskov, D.F., Samigullin, T.H., Pimenov, M.G. 2017. Molecular and morphological data support the transfer of the monotypic Iranian genus *Alococarpum* to *Prangos* (Apiaceae). Phytotaxa 299: 223–233.
MacPhail, M.K., Alley, N.F., Truswell, E.M., Sluiter, I.R. K. 1994. Early Tertiary vegetation: evidence from spores and pollen. In: Hill, R.S. (ed.) History of the Australian vegetation: Cretaceous to Recent. New York: Cambridge University Press, pp. 189–261.
Magee, A.R., Calviño, C.I., Liu, M., Downie, S.R., Tilney, P. M., Van Wyk, B.-E. 2010. New tribal delimitations for the early diverging lineages of Apiaceae subfamily Apioideae. Taxon 59: 567–580.
Mathias, M.E. 1965. Distribution patterns of certain Umbelliferae. Ann. Missouri Bot. Gard. 52: 387–398.
Mathias, M.E., Constance, L. 1962. A revision of *Asteriscium* and some related Hydrocotyloid Umbelliferae. Univ. Calif. Publ. Bot. 33: 99–184.
Mathias, M.E., Constance, L. 1965. A revision of the genus *Bowlesia* Ruiz & Pav. (Umbelliferae-Hydrocotyloideae) and its relatives. Univ. Calif. Publ. Bot. 38: 1–73.
Metcalfe, C.R., Chalk, L. 1950. Anatomy of the Dicotyledons, vol. 1. Oxford: Clarendon Press.
Metcalfe, C.R., Chalk, L. 1983. Anatomy of the Dicotyledons, 2nd ed. Oxford: Clarendon Press.
Mittal, S.P. 1961. Studies in the Umbellales. II. The vegetative anatomy. J. Indian Bot. Soc. 40: 424–443.
Moore, D.M. 1971. Chromosome studies in the Umbelliferae. In: Heywood, V.H. (ed.) The Biology and Chemistry of the Umbelliferae, Bot. J. Linn. Soc. 64, Suppl. 1. London & New York: Academic Press, pp. 233–254.
Muller, J. 1981. Fossil pollen record of extant angiosperms. Bot. Rev. 47: 1–142.
Nicolas, A.N., Plunkett, G.M. 2009. The demise of subfamily Hydrocotyloideae (Apiaceae) and the re-alignment of its genera across the entire order Apiales. Molec. Phylogen. Evol. 53: 134–151.
Nicolas, A.N., Plunkett, G.M. 2012. Untangling *Azorella*, *Laretia*, and *Mulinum* (Apiaceae, Azorelloideae): insights from phylogenetics and biogeography. Taxon 61: 826–840.
Nicolas, A.N., Plunkett, G.M. 2014. Diversification times and biogeographic patterns in Apiales. Bot. Rev. 80: 30–58.
Ostroumova, T.A. 1987. The types of stomata in representatives of the Apiaceae family. Bot. Zhurn. 72: 1479–1488. (in Russ.)
Ostroumova, T.A. 1990. Stomatal types in the Umbelliferae in relation to taxonomy: tribes Corinadreae and Scandiceae. Feddes Repert. 101: 409–417.
Ostroumova, T.A., Kljuykov, E.V. 1991. Stomatal types as a taxonomic character in the Umbelliferae: tribe Apieae, subtribe Apiinae. Feddes Repert. 102: 105–114.
Ostroumova, T.A., Kljuykov, E.V. 2007. Stomatal types in Chinese and Himalayan Umbelliferae. Feddes Repert. 118: 84–102.
Panahi, M., Banasiak, Ł., Piwczyński, M., Puchałka, R., Oskolski, A.A., Spalik, K. 2015. Phylogenetic relationships among *Dorema*, *Ferula* and *Leutea* (Apiaceae, Scandiceae, Ferulinae) inferred from nrDNA ITS and cpDNA noncoding sequences. Taxon 64: 770–783.
Philipson, W.R. 1979. Araliaceae. Flora Malesiana, ser. 1, 9 (1): 1–105.
Pimenov, M.G., Leonov, M.V. 1993. The Genera of Umbelliferae: a Nomenclator. Kew: Royal Botanic Gardens.
Pimenov, M.G., Pogorelova, O.V. 1974. Taxonomic position of the genera *Ammi* L. and *Visnaga* Mill. Rastit. Resur 10: 216–19.
Pimenov, M.G., Sdobnina, L.I. 1984. Nodal anatomy as a taxonomic character in the family Umbelliferae. Bot. Zhurn. 69: 283–294.
Pimenov, M.G., Ostroumova, T.A., Tomkovich, L.P. 1982. Petiole structure of the Caucasian Umbelliferae. Byull. Moskovsk. Obshch. Isp. Prir., Otd. Biol. 87: 57–75.
Pimenov, M.G., Ostroumova, T.A., Tomkovich, L.P., Kljuykov, E.V. 1986. La structure anatomique du petiole d'Ombellifères d'Asie Moyenne. Bull. Mus. Natl. Hist. Nat. IV (8) B, Adansonia 1: 77–99.
Pimenov, M.G., Vasil'eva, M.G., Leonov, M.V., Daushkevich, J.V. 2003. Karyotaxonomical analysis in the Umbelliferae. Enfield, New Hampshire: Science Publishers Inc.
Pimenov, M.G., Degtjareva, G.V., Ostroumova, T.A., Samigullin, T.H., Averyanov, L.V. 2016. *Xyloselinum laoticum* (Umbelliferae), a new species from Laos, and taxonomic placement of the genus in the light of nrDNA ITS sequence analysis. Phytotaxa 244: 248–262.
Plunkett, G.M. 2001. Relationship of the order Apiales to subclass Asteridae: a re-evaluation of morphological characters based on insights from molecular data. Edinb. J. Bot. 58: 183–200.
Plunkett, G.M., Nicolas, A.N. 2016. Assessing *Azorella* (Apiaceae) and its allies: phylogenetics and a new classification. Brittonia 69: 31–61.
Plunkett, G.M., Chandler, G.T., Lowry II, P.P., Pinney, S. M., Sprenkle, T.S. 2004. Recent advances in understanding Apiales with a revised classification. S. Afr. J. Bot. 70: 371–381.
Punt, W. 1984. Umbelliferae. In: Punt, W., Clarke, G.C.S. The Northwest European Pollen Flora 37. Amsterdam: Elsevier, pp. 155–363.
Reduron, J.-P., Muckensturm, B. 2007–2008. Ombellifères de France. Monographie des Ombellifères (*Apiaceae*) et plantes alliées, indigènes, naturalisées, subspontanées, adventices ou cultivées de la flore française, vol. 1–5. Jarnac: Soc. bot. Centre-Ouest.
Rodríguez, R.L. 1971. The relationships of the Umbellales. In: Heywood, V.H. (ed.) The Biology and Chemistry of the Umbelliferae, Bot. J. Linn. Soc. 64, Suppl. 1. London & New York: Academic Press, pp. 63–91.
Roth, I. 1977. Fruits of angiosperms. Berlin: Gebrüder Borntraeger.
Schlessman, M.A., Graceffa, L.M. 2002. Protogyny, pollination, and sex expression of andromonoecious *Pseudocymopterus montanus* (Apiaceae, Apioideae). Int. J. Pl. Sci. 163: 409–417.

Solereder, H. 1908. Systematic anatomy of the dicotyledons, vol. 1. Oxford: Clarendon Press.

Spalik, K. 1997. Revision of *Anthriscus* (Apiaceae). Polish Bot. Stud. 13(1): 1–69.

Spalik, K., Downie, S.R. 2001. The utility of morphological characters for inferring phylogeny in Scandiceae subtribe Scandicinae (Apiaceae). Ann. Missouri Bot. Gard. 88: 270–301.

Spalik, K., Downie, S.R. 2007. Intercontinental disjunctions in *Cryptotaenia* (Apiaceae, Oenantheae): an appraisal using molecular data. J. Biogeogr. 34: 2039–2054.

Spalik, K., Wojewódzka, A., Downie, S.R. 2001. Delimitation of genera in Apiaceae with examples from Scandiceae subtribe Scandicinae. 58: 331–346.

Spalik, K., Reduron, J.-P., Downie, S.R. 2004. The phylogenetic position of *Peucedanum* sensu lato and allied genera and their placement in tribe Selineae (Apiaceae, subfamily Apioideae). Pl. Syst. Evol. 243: 189–210.

Spalik, K., Downie, S.R., Watson, M.F. 2009. Generic delimitations within the *Sium* alliance (Apiaceae tribe Oenantheae) inferred from cpDNA *rps16-5'trnK*$^{(UUU)}$ and nrDNA ITS sequences. Taxon 58: 735–748.

Spalik, K., Banasiak, Ł., Feist, M.A.E., Downie, S.R. 2014. Recurrent short-distance dispersal explains wide distributions of hydrophytic umbellifers (Apiaceae tribe Oenantheae). J. Biogeogr. 41: 1559–1571.

Stuhlfauth, T., Fock, H., Huber, H., Klug, K. 1985. The distribution of fatty acids including petroselinic and tariric acids in the fruit and seed oils of the Pittosporaceae, Araliaceae, Umbelliferae, Simarubaceae, and Rutaceae. Biochem. Syst. Ecol. 13: 447–453.

Sun, F.-J., Downie, S.R. 2010. Phylogenetic relationships among the perennial, endemic Apiaceae subfamily Apioideae of western North America: additional data from the cpDNA *trnF-trnL-trnT* region continue to support a highly polyphyletic *Cymopterus*. Pl. Divers. Evol. 128: 151–172.

Terentieva, E.I., Valiejo-Roman, C.M., Samigullin, T.H., Pimenov, M.G., Tilney, P.M. 2015. Molecular phylogenetic and morphological analyses of the traditional tribe Coriandreae (Umbelliferae-Apioideae). Phytotaxa 195(4): 251–271.

Tseng, C.C. 1967. Anatomical studies of flowers and fruits in the Hydrocotyloideae (Umbelliferae). Univ. Calif. Publ. Bot. 42: 1–58.

Valiejo-Roman, C.M., Terentieva, E.I., Samigullin, T.H., Pimenov, M.G., Ghahremani-Nejad, F., Mozaffarian, V. 2006. Molecular data (nrITS-sequencing) reveal relationships among Iranian endemic taxa of the Umbelliferae. Feddes Repert. 117: 367–388.

van Tieghem, P. 1872. Sur les canaux oléo-résineaux des Ombellifères et des Araliacées. Bull. Soc. Bot. France 19: 113–129.

van Tieghem, P. 1884. Sur la structure et les affinités des Pittosporées. Bull. Soc. Bot. France 31: 384–385.

Van Wyk, B.-E., Tilney, P.M., Magee, A.R. 2013. African Apiaceae: a synopsis of the Apiaceae/Umbelliferae of sub-Saharan Africa and Madagascar. Pretoria: Briza Academic Books.

Wagenitz, G. 1992. The Asteridae: evolution of a concept and its present status. Ann. Missouri Bot. Gard. 79: 209–217.

Warning, W.C. 1934. Anatomy of the vegetative organs of the parsnip. Bot. Gaz. 96: 44–72.

Watson, L., Dallwitz, M.J. 1992 onwards. The families of flowering plants: descriptions, illustrations, identification, and information retrieval. Version: 9th September 2008. http://delta-intkey.com/angio/index.htm

Webb, C.J. 1981. Andromonoecism, protandry, and sexual selection in Umbelliferae. New Zealand J. Bot. 19: 335–338.

Weitzel, C., Rønsted, N., Spalik, K., Simonsen, H.T. 2014. Resurrecting deadly carrots: towards a revision of *Thapsia* (Apiaceae) based on phylogenetic analysis of nrITS sequences and chemical profiles. Bot. J. Linn. Soc. 174: 620–636.

Apocynaceae

Apocynaceae Jussieu, Gen. Pl. 143 (1789) (Apocinae), nom. cons.
Asclepiadaceae Borkh. (1797) (Asclepiadeae), nom. cons.
Periplocaceae Schltr. (1905).

M.E. Endress, U. Meve, D.J. Middleton, and S. Liede-Schumann

Woody climbers, vines, perennial herbs, trees or shrubs, more rarely annuals, sometimes with large water-storing tubers or a xylopod, sometimes succulent, with large grappling hooks and/or tendrils in several lianoid genera of Willughbeieae; latex in non-articulated laticifers present, most commonly white, but in some genera usually translucent and in others yellowish or reddish. Leaves simple and usually entire, very rarely dentate or repand, usually isophyllous, but often anisophyllous in Tabernaemontaneae-Tabernaemontaninae, sometimes with distinctly different juvenile and adult foliage, normally petiolate, sometimes sessile, usually opposite, less frequently alternate or whorled (whorled phyllotaxis characteristic for a number of Rauvolfioid genera); stipules usually absent or small and caducous, sometimes enlarged and fused into dentate interpetiolar collars (a few Periplocoid genera), commonly with interpetiolar lines or ridges, sometimes the petioles of a leaf pair connate at the node, forming a short ocrea, which may be expanded into small intrapetiolar flaps clasping the stem (Tabernaemontaneae), almost always with colleters in the axil of the leaf, sometimes on the petiole, in a cluster adaxially at the juncture of petiole and lamina or along the midrib above, occasionally with abaxial domatia in the axils of the secondary veins (mainly in Apocynoids). Flowers perfect, rarely functionally dioecious, often scented, sessile or more commonly pedicellate, in solitary or more commonly in axillary, extra-axillary or terminal multi-flowered cymes, panicles or thyrses, sometimes appearing as an axillary fascicle. Perianth almost always actinomorphic, very rarely slightly zygomorphic; calyx almost always 5- (rarely 4- or 6–7)-merous, lobes normally quincuncially arranged, synsepalous or aposepalous, commonly with colleters, in Periplocoideae, Secamonoideae and Asclepiadoideae these are usually in the sinuses, but in some Rauvolfioids and several Apocynoids colleters in a continuous ring, in multiple rows in some Tabernaemontaneae and Hunterieae, or a single antesepalous colleter (especially in Echiteae), and in several genera of Rauvolfioids and Apocynoids colleters are absent; corolla sympetalous, rarely apopetalous (a few Ceropegieae), salverform, infundibuliform, tubular, urceolate or rotate, lobes almost always 5 (very rarely 4), usually contorted in bud, either dextrorse or sinistrorse, more rarely valvate; corolline or gynostegial coronas often present; stamens 5 (rarely 4), filaments mostly straight, sometimes geniculate, sometimes connate around the style (some species of *Forsteronia*, *Thoreauea*), sometimes coiled around the style (*Dewevrella*, some species of *Parsonsia* and *Thenardia*), inserted on the corolla tube, on prominent staminal feet (broadened filament base fused with corolla tube) or forming a staminal tube, included to exserted; anthers introrse, rarely latrorse, in almost all Apocynoids, Secamonoideae and Asclepiadoideae with highly elaborated and lignified guide rails (lignified guide rails absent in most Rauvolfioids and in Periplocoideae) and often with an apical connective

Rauvolfioids and Apocynoids contributed by David J. Middleton and Mary E. Endress. Periplocoideae, Secamonoideae and Asclepiadoideae contributed by Sigrid Liede-Schumann and Ulrich Meve.
Ulrich Meve and Sigrid Liede-Schumann wish to thank Prof. Dr. Focke Albers, Münster, for his input in the initial planning of the then Asclepiadaceae treatment in this series; he and his students obtained several preliminary results we built on.

J.W. Kadereit, V. Bittrich (eds.), *Flowering Plants. Eudicots*, The Families and Genera of Vascular Plants 15,
https://doi.org/10.1007/978-3-319-93605-5_3

appendage, thecae 4, unequal in most Apocynoids, with dorsal ones smaller through presence of guide rails, reduced to 2 in Asclepiadoideae, dehiscence longitudinal, attached to the style-head forming a gynostegium (gynostegium absent in Rauvolfioids); nectaries in alternistaminal pockets on the staminal tube, on sides of staminal feet or 5 (rarely 2) lobes encircling the base of the ovary, these often fused to varying degrees into an (often deeply lobed) ring (in some Rauvolfioids and early-branching Apocynoids nectaries are adnate to the outer wall of the ovary at the base or are sometimes nonfunctional or absent); gynoecium normally of two carpels (very rarely up to five); ovary mostly apocarpous, sometimes congenitally (Rauvolfioids only) or postgenitally syncarpous (several Apocynoids), in some genera only one carpel developing, superior to subinferior; placentation marginal when the ovary is apocarpous, parietal or axile when syncarpous, when apocarpous upper part of the carpels fusing postgenitally to form a complex style-head that produces adhesive for pollen transport, with a pollen-trapping basal collar and/or pollen-presenting upper crest present in many Rauvolfioids and Apocynoids; stigma mostly on the underside of the style-head, often restricted to five chambers behind the guide rails, but style-head scarcely morphologically differentiated and nearly uniformly receptive in some Rauvolfioids; adhesive a sticky foam or mucilage, or differentiated into five translators with a scoop-like pollen receptacle and sticky base, or as five hard clips (corpuscles) usually accompanied by five pairs of flexible arms (caudicles) forming a pollinarium. Fruit in Rauvolfioids diverse: drupes, berries, follicles or capsules; seeds usually without a coma, naked, arillate, or winged or fimbriate at the margin very rarely with a coma (*Haplophyton*); in the remainder of the family, fruit almost always a pair of ventrally dehiscent follicles (often only one due to abortion or due to postgenital fusion; rarely a septicidally dehiscent capsule) with small seeds with a micropylar coma, rarely with a chalazal coma, coma at both ends (only in early-branching Apocynoids), or fringed with long trichomes circumferentially (a few Periplocoid and *Hoya* species), or without a coma.

The classification used here employs both formal and informal ranks for the main divisions of the family. The two subfamilies comprising the traditional Apocynaceae—Rauvolfioideae and Apocynoideae—are paraphyletic and thus used here informally, as Rauvolfioids and Apocynoids, respectively, whereas formal ranks are maintained for Periplocoideae, Secamonoideae and Asclepiadoideae, which are all nested within Apocynoids. The clade consisting of the Apocynoids along with Periplocoideae, Secamonoideae and Asclepiadoideae is here referred to as the APSA clade. The taxa in the APSA clade have many features in common that set them apart from the less derived Rauvolfioids. The family comprises 378 genera and about 5,350 species and is predominantly pantropical, with some genera reaching temperate regions.

Vegetative Morphology. Apocynaceae are primarily woody plants. A large percentage of the Rauvolfioids and several genera in the early-branching Apocynoid tribes Wrightieae, Nerieae and Malouetieae are trees or shrubs, whereas the great majority of the remainder of the family is lianas or vines. Herbs are rare in Rauvolfioids and Apocynoids, except in some genera that are adapted to drier or colder habitats, where they are usually perennial (e.g., *Apocynum*); some have a water-storing xylopod, which produces herbaceous shoots (e.g., several species of *Mandevilla*). The majority of Periplocoideae, Secamonoideae and Asclepiadoideae are herbaceous or only basally woody; thus twiners, perennial herbs and subshrubs are the dominant growth forms. In the Periplocoideae, small trees occur only in *Decalepis*. In Secamonoideae, Lahaye et al. (2005) report the evolution of a shrub-like growth form from lianoid ancestors. No trees or large shrubs are known in Asclepiadoideae (except *Calotropis*), but secondary woodiness has been proposed for *Marsdenia erecta* (Yaman and Tumen 2012). Geophytes with root and hypocotyl tubers are frequent in African Periplocoideae (*Raphionacme*, *Schlechterella*, *Pentopetia*) and Asclepiadoideae (diverse Asclepiadeae, *Ceropegia*). Succulents are rare in Rauvolfioids and Apocynoids; in Rauvolfioids, *Plumeria* and *Himatanthus* are subsucculent. In Apocynoids two genera from Africa/Madagascar are stem succulent: *Adenium* and *Pachypodium*. Ceropegieae include the often bizarre-looking, stem succulent Stapeliads (e.g., *Stapelia*, *Caralluma*, *Duvalia*, etc., and some *Ceropegia* species). *Cynanchum* (Asclepiadeae) includes ca. 50 stem succulent species from Madagascar, and the "*Sarcostemma* group", widespread in the Paleotropics and tropical Australia. True epiphytes are unknown in Rauvolfioids and Apocynoids. There are reports of *Mandevilla boliviensis* growing in trees in Costa Rica and it is therefore sometimes said to be epiphytic (e.g., Gentry and Dodson 1987; Morales 1998). However, this species is normally terrestrial, and is probably best interpreted as merely an accidental epiphyte (Kress 1986). In Periplocoideae epiphytes are restricted to *Epistemma* and *Sarcorrhiza*, but are more frequent in Marsdenieae (Asclepiadoideae), including numerous species of *Hoya* and *Dischidia*. In addition, some species of *Dischidia* and *Hoya* have a close association with ants; the leaves of many species are modified into inflated structures (imbricate or pitcher leaves) that house ants. In turn, the waste products of the ants provide additional nutrients, which are absorbed by the plant (Peeters and Wiwatwitaya 2014).

Leaves are simple and entire (except in *Emicocarpus*), almost always with an entire smooth margin (very rarely dentate in *Alyxia* or repand in *Parsonsia*), and predominantly opposite. Alternate or whorled leaves are characteristic for certain genera in Rauvolfioids, but these become infrequent in the more derived APSA clade. The leaves of immature plants can sometimes be of a remarkably different shape and/or size from those of mature plants (e.g., in *Parsonsia* and *Micrechites*), although this has been recorded in very few genera and its prevelance is as yet unknown. Venation is typically pinnate, and mostly brochidodromous. Stipules in the usual sense are normally lacking (although stipules have been reported in some species of *Rauvolfia* and petioles of a leaf pair may be connate at the base forming a short ocrea, which may be expanded in the intrapetiolar region, forming flaps, which are characteristic for some Tabernaemontaneae and found occasionally in other Rauvolfioid genera); other outgrowths, especially stipular or interpetiolar ridges, are common throughout the family. Whereas stipular glands are the common state in Ceropegieae, spiny stipules characterize the Stapeliad *Tavaresia*. Paired, often branched, spines in the leaf axils are characteristic for *Carissa* (Carisseae), deciduous, blunt, conical spines occur on the trunk in some species of *Lacmellea* and ternate spines (the middle one often very reduced or absent) are found in *Pachypodium*. Except for *Boucerosia frerei*, leaves are rudimentary in all Stapeliads, and sometimes form spines (*Hoodia*, *Edithcolea*). Tendrils are common in Willughbeieae, where they are often modified inflorescences. In most genera, young leaves show flat to curved ptyxis (Cullen 1978); only in Rauvolfioids is conduplicate ptyxis found (*Alyxia*, *Carissa*, *Cerbera*, *Plumeria*), and involute ptyxis is known from *Vinca* (Vinceae) and *Landolphia* (Willughbeieae).

Tooth-like colleters are usually present in the leaf axil throughout the family (decreasing in number and size in derived Asclepiadoideae), sometimes spreading along the adaxial side of the petiole (e.g., some species of *Rauvolfia*, and Periplocoideae); in many species of *Mandevilla* small, round, black colleters are scattered along the adaxial side of the petiole. A cluster of flat, deltoid-shaped colleters at the base of the leaf blade on the adaxial surface is found in most Mesechiteae and is widespread in Asclepiadoideae. In Baisseeae, colleters are absent in the leaf axils, but tooth-like colleters are present along the adaxial side of the petiole, and/or as a cluster at the base of the adaxial side of the leaf blade. Abaxial domatia are found in the axils of the secondary veins with the midrib of the leaf in some genera of Malouetieae (e.g., *Funtumia*, *Malouetia*), Apocyneae (e.g., *Urceola*), Odontadenieae (e.g., *Pinochia*), many Mesechiteae (e.g., *Tintinnabularia*, *Forsteronia*) and are characteristic for Baisseeae.

Vegetative Anatomy. The presence of non-articulated, branched or unbranched laticifers containing white (or sometimes differently colored) latex is one of the distinguishing characteristics of Apocynaceae (Mahlberg 1980). White latex is the most common state, but many Wrightieae and Echiteae (Apocynoids), all Ceropegieae, and some genera of Asclepiadeae (*Astephanus*, *Microloma*, *Oncinema*, *Pentatropis*, and many species of *Vincetoxicum*) and some species of *Hoya* have translucent latex. Many species of *Aspidosperma* (Rauvolfioids) as well as *Cryptolepis sanguinolenta* (Periplocoideae) have red or orange latex, and some species of *Cynanchum* (Asclepiadoideae) have yellow latex. Leaves are normally dorsiventral, but isobilateral leaves have been reported in *Nerium*. In stem succulent Ceropegieae and *Cynanchum* species, leaves are almost always reduced to scales. Stomata are ranunculaceous or rubiaceous, and sometimes both types occur in a single species (Metcalfe and Chalk 1972). In Asclepiadoideae, they can be tetracytic, and in Stapeliads even hexacytic (Metcalfe and Chalk 1979). Trichomes of various types occur; they may be simple, uniseriate, and several-celled (e.g., *Chonemorpha*), unicellular (e.g., *Vinca*, *Allamanda*), sometimes uniseriate, but with a series of compressed basal cells (e.g., *Echites*); in *Anechites* hooked trichomes with enlarged multicellular bases occur (Fallen 1983). The typical trichome in Asclepiadoideae is multicellular, eglandular, transparent to whitish or occasionally colored, antrorse or retrorse. Glandular trichomes are known in *Dischidia* (Solereder 1899), *Vincetoxicum hirsuta* (Trivedi and Upadhyay 1984, as *Tylophora*), *Araujia* (Metcalfe and Chalk 1972), and in a wide range of Gonolobinae (e.g., *Gonolobus*, *Macroscepis*, *Matelea*, Metcalfe and Chalk 1972).

Internal phloem is nearly always present, either as a continuous ring or as separate bundles at the margin of the pith; the pericycle is generally in the form of a continuous ring or as separate strands of white cellulose fibers. Cork is present in Rauvolfioids and Apocynoids, but its origin has not been investigated. Cork, as far as is known, arises from a phellogen in the endodermis in Periplocoideae (except for *Cryptolepis*) and Marsdenieae (Asclepiadoideae), but from the epidermis in Ceropegieae and Asclepiadeae (except for succulent *Cynanchum* (former *Sarcostemma*); Treiber 1891). Wood features that are uniform throughout Apocynaceae include simple vessel perforations, sometimes with short, scalariform perforation plates, alternate vestured intervessel pits, and vessel-ray pits that are similar in shape and size to intervessel pits (Lens et al. 2009), and vestured perforate tracheary elements either septate with simple pits, or non-septate and with bordered pits. Wood parenchyma is often apotracheal, sometimes also with some paratracheal, in some genera predominantly paratracheal, and absent in genera with septate fibers. There are a number of general evolutionary trends evident from Rauvolfioids to the more derived Apocynaceae of the APSA clade including: shorter vessel elements (on average 700–1000 μm in Rauvolfioids versus 200–500 μm in Periplocoideae, Secamonoideae and Asclepiadoideae); a higher proportion of vessels in multiples and more vessels per multiple (solitary vessel elements or radial vessel multiples in Rauvolfioids and Wrightieae, Nerieae and Malouetieae of the Apocynoids versus large vessel clusters in more derived members of the ASPA clade). There is a close correlation between vessel grouping pattern and the type of imperforate tracheary cells in the ground tissue: tracheids coevolve with solitary vessels, whereas fibers are linked with vessel multiples and clusters (Carlquist 1984). Reduction of vessel element length and the trend toward large vessel clusters are correlated with vasicentric tracheid abundance, increase of paratracheal parenchyma, and decrease in number of cells per axial parenchyma strand (6–12 in Rauvolfioids versus 2–5 in Periplocoideae, Secamonoideae and Asclepiadoideae; Lens et al. 2009). Most of these evolutionary trends are typically correlated with habitat shift to more arid regions and/or a shift to climbing habit (Baas et al. 1983; Lens et al. 2008). Other features, such as vessel grouping, vessel element length, fiber type, frequency of uniseriate rays and fused multiseriate rays, vary and, in combination, are indicative of specific tribes in Rauvolfioids and Apocynoids and in Periplocoideae (Lens et al. 2009). Sieve-element plastids are of the derived S-type (Behnke 1981).

Inflorescence Structure. Terminal thyrsoidal systems are regarded as the plesiomorphic condition throughout the family; the diversity of inflorescences found in the family results from different degrees of reduction (Liede and Weberling 1995). Inflorescences in Rauvolfioids and Apocynoids are mostly cymose, simple or compound in dichasia, thyrses or panicles, sometimes fasciculate (e.g., *Kibatalia*, *Malouetia*, *Pleiocarpa*) or umbelloid cymes (sciadioids, e.g., *Thenardia*). In some lianoid genera of Apocynoids, inflorescences are racemose (e.g., *Mandevilla*, *Asketanthera*). In some genera, an inflorescence may be reduced to 1–2 flowers (e.g., *Haplophyton*, *Vinca*, *Catharanthus*, *Rhabdadenia*, *Salpinctes*). Inflorescences may be axillary and/or terminal.

Synflorescences in Periplocoideae and Secamonoideae show some deviations from terminal and/or axillary branching thyrses (e.g., *Tacazzea*) to extra-axillary branching ones with racemose, corymbose or sciadioidal inflorescences (e.g., *Raphionacme*), and/or cyme formation and reduction of the number of flowers to two or one (e.g., *Secamone*). Inflorescence types are most diverse in Asclepiadoideae, where extra-axillary inflorescences predominate, whereas the axillary inflorescences found in some American Asclepiadeae are best regarded as secondary (e.g., *Orthosia*, *Oxypetalum*, *Peplonia*; cf. Liede and Weberling 1995). In consequence, inflorescences are usually solitary (one inflorescence per node) and only rarely paired (two inflorescences per node). Compound inflorescences are rare in the subfamily, while simple, many-flowered sciadioids are typical for the horticulturally attractive groups of Marsdenieae (*Dregea*, *Hoya*) and Ceropegieae (e.g., *Boucerosia*, *Caralluma*). By reduction of paracladia in number and size, originally dichasially branching cymes become few-flowered bostrices and sciadioids, or geminiflorous helices (e.g., *Cynanchum*, *Stapelia*). In many species of *Tassadia*, the α-bract supports the continuation of the inflorescence, producing sympodial shoots of very regular length, while the β-bract produces a few-flowered, very dense bostryx (Liede and Weberling 1995).

Floral Structure and Anatomy. Flowers are bisexual, very rarely functionally unisexual in a few Rauvolfioids (e.g., *Carissa*, Schroeder 1951; *Rauvolfia*, Koch et al. 2002), and andromonoecious in *Cynanchum* (*Glossonema*) *varians* (Ali and Ali 1996) and *Cynanchum hemsleyanum* (= *Metaplexis japonica*, Tanaka et al. 2006). They are actinomorphic or rarely slightly zygomorphic (monosymmetric) due to a gibbous, arcuate (some species of *Rauvolfia* and *Mandevilla*) or strongly bent corolla tube (some species of *Lacmellea*) and/or 3 + 2 organization of the corolla lobes (*Allamanda schottii*, *Adenium*). The number and arrangement of floral organs is quite uniform: flowers are almost always pentamerous (rarely 4-merous), with each type of organ arranged in a whorl, except for the gynoecium, which, with few exceptions, is of two carpels. A calyx is always present, and is almost always 5-merous, but is consistently 4-merous in *Leuconotis* and *Parahancornia*. Three species of *Aspidosperma* (*A. illustre*, *A. megalocarpon* and *A. myristicifolium*) have a 4-merous calyx; in the first two the four lobes are decussate, with the outer two significantly larger and fused at the base, and the inner two much reduced, and the calyx is reported to be 6–7-merous in *Aspidosperma darienense*, though more thorough study may show some of these to be modified bracts. The calyx lobes are fused into a 2-lipped tube or partial tube in *Alyxia kabaenae*. Usually the calyx is green, though large, showy, unequal and overlapping white or pink calyx lobes are frequent in certain *Tabernaemontana* species from South America, and in *Rhodocalyx* the large petaloid calyx lobes are the same wine-red color as the corolla; yet, even in genera where it may be foliose or petaloid, the calyx is normally easily differentiated from the corolla. The calyx lobes are typically split nearly to the base; sometimes they may be fused in the lower part or for most of the length into a cup (e.g., *Chonemorpha*, *Voacanga*). In the majority of the genera, colleters are found at the base of the calyx on the adaxial side. Most frequently calycine colleters are few and in alternisepalous position, quincuncially arranged, but in several Apocyneae colleters are more numerous and in a single ring, most Echiteae are characterized by solitary antesepalous colleters, many Tabernaemontaneae have multiple rows of colleters, and in many Rauvolfioids calycine colleters are absent.

A corolla is always present, though it may be minute (e.g., many *Urceola* species, *Cynanchum roulinioides*, *Orthosia meridensis*), and is almost always 5-merous (4-merous in *Leuconotis* and two species of *Parsonsia*: *P. kimberleyensis* and *P. lanceolata*). It comprises a congenitally fused lower corolla-staminal tube (stapet), and a postgenitally fused upper corolla tube (Nishino 1982, 1983). Fusion of the upper tube occurs relatively late in development, and occurs from the apex to the base; in some Rauvolfioids the flower reaches maturity before the two epidermal edges are completely fused, leaving gaps just above the insertion of the stamens, which are characteristic for certain genera (e.g., *Aspidosperma*, *Stephanostegia*; Endress and Bruyns 2000). Corolla shape is extremely variable. Corollas are most commonly salverform in Rauvolfioids and Apocynoids, although lignification of the anthers in the latter protects the pollen from desiccation, making possible more open corolla types, including rotate corollas with the reproductive organs completely exserted (e.g., *Forsteronia, Thenardia, Pottsia*, many species of *Parsonsia* and *Wrightia*). Infundibuliform corollas are also common, especially in the Apocynoids. Urceolate corollas are found in *Urceola* and *Ecua*.

Corolla lobes are normally contorted and the direction of contortion is a systematically important character (Endress et al. 1990). In Rauvolfioids, corolla lobe aestivation is almost always sinistrorse (exceptions include *Geissospermum, Kopsia, Ochrosia* and some species of *Tabernaemontana, Haplophyton, Alstonia* and *Carissa*, which all have dextrorse corolla lobe aestivation). In Apocynoids, Periplocoideae, Secamonoideae and Asclepiadoideae, in contrast, corolla lobe aestivation is typically dextrorse, with a few exceptions (e.g., all Wrightieae as well as some species of *Urceola* (Apocynoids), and all *Genianthus* and *Pervillaea* (Secamonoideae), all of which have sinistrorse corolla lobe aestivation, and some *Urceola* and *Parsonsia* species, which have valvate corolla lobe aestivation). Valvate corolla lobe aestivation is characteristic for some derived Asclepiadoideae, especially the complex flowers of some Ceropegieae (e.g., *Ceropegia*), in which complicated structures arise through postgenital fusions of petals during ontogeny. Corolla lobes are typically not inflexed in bud, but strongly inflexed corolla lobes are characteristic for some genera of Alyxieae and rarely in Apocynoids, and weakly inflexed corolla lobes are found in *Tabernaemontana* (Leeuwenberg 1991, 1994b).

Coronas are an important feature of the flower and play a key role in the characterization of genera in the more derived APSA clade (Fishbein 2001; Wanntorp and Forster 2007). They are simple to very elaborate outgrowths, which may be grouped into two main types depending on their location: corolline and gynostegial. It should be emphasized, however, that the same location does not necessarily imply homology. In the Rauvolfioids and Apocynoids coronas are mostly corolline (Kunze 1990, 2005), and are most commonly found in the staminal (alternipetalous) sectors, either in the corolla throat or somewhat lower down, near the anthers and the region of confluence between the congenitally (stapet) and postgenitally fused parts of the corolla. There is a great variety of different forms of corolline coronas (see e.g., Endress and Bruyns 2000). They may be in the form of free lobes, or fused to varying degrees into annular structures. In *Wrightia* the corona is alternipetalous or antepetalous or both, depending on the species. In some species of *Wrightia* and in *Pleioceras* the corona consists of many fimbriate lobes. They may be petaloid (as in *Melodinus, Nerium* or *Strophanthus*); they may be in the form of a knob-like protuberance behind the anther (as in *Artia* or *Cycladenia*); often they are rather inconspicuous pocket-like flaps that join two petals across the sinus (e.g., *Vinca*). In *Prestonia* (Apocynoids) a thick annular corona is usually present in the corolla mouth, and in addition, in many species, knob-like free lobes are present just above the anthers. More rarely corolline coronas are in the alternistaminal (antepetalous) sectors (e.g., *Apocynum*). In addition, ledge-like outgrowths of the corolla are found in some genera (e.g., *Baissea, Ixodonerium*). The homology of these various types of coronas has not been determined. In Periplocoideae the corona lobes are typically initiated near the petal sinuses, in the region of confluence between the congenitally (stapet) and postgenitally fused regions of the corolla (Fig. 24).

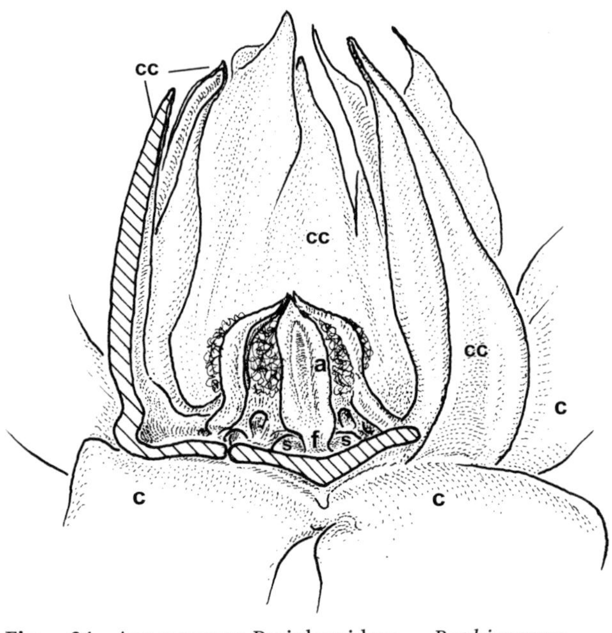

Fig. 24. Apocynaceae-Periplocoideae. *Raphionacme*. Gynostegium with corona (front corona lobe removed). (a = anther, c = corolla lobe (mostly cut away), cc = corolline corona lobe, f = filament, s = staminal foot). (Drawn by U. Meve)

Gynostegial coronas, which have their origin as outgrowths of the back of the stamens, are the rule in Asclepiadoideae (Fig. 25), but absent in Rauvolfioids and rare in Apocynoids (e.g., *Kibatalia*, *Vallaris*) and Periplocoideae (e.g., *Finlaysonia obovata*, where they occur as simple staminal corona lobes). Interstaminal elements, which are usually combined with the staminal elements, are apomorphic in Asclepiadoideae (Kunze 1990). A system of classification of the multitude of sometimes extravagant corona forms has been developed by Liede and Kunze (1993). Basic corona composition (e.g., whether just in the staminal sectors, or in both staminal and interstaminal sectors) has been used widely for recognizing larger taxonomic groups, but has been found to be misleading in several examples (e.g., *Cynanchum* (former *Sarcostemma*) and *Funastrum*, Liede and Täuber 2000, circumscription of *Cynanchum*, Liede and Täuber 2002). (Fig. 25).

Stamens are morphologically clearly differentiated into an anther and filament in the Rauvolfioids, Apocynoids and Periplocoideae, although the anthers are often subsessile, and thus a distinct filament is not present in many taxa (e.g., *Thevetia*, *Vallaris*, *Voacanga*). In Secamonoideae and Asclepiadoideae, the anthers are ± sessile on a tube. The tube has often been called a staminal tube or filament tube, the latter implying that it is formed of the fused filaments (Kunze 1990). Endress and Bruyns (2000), conversely, hypothesized that the filaments have been lost, and that the tube is formed by the gynostegial corona. This was based for the most part on the presence of a staminal tube in *Hemidesmus* and *Phyllanthera grayi* (Periplocoideae), both of which have distinct staminal filaments arising from atop the tube. Conclusive evidence, however, is lacking for both interpretations. The anthers are introrse (to latrorse in some Plumerieae-Thevetiinae). Each anther contains two thecae, and in all of the family except Asclepiadoideae, each theca contains two pollen sacs. In Asclepiadoideae, the dorsal pollen sac in each theca has been lost, so that in each anther only the two ventral pollen sacs develop. In most Tabernaemontaneae (Rauvolfioids), Apocynoids, Secamonoideae and Asclepiadoideae the anthers are morphologically and histologically highly specialized. Pollen production is restricted to the upper part of the anther. The lower part is enlarged and sterile; thick, lignified guide rails develop laterally. Lignified guide rails are absent in Rauvolfioids (except Tabernaemontaneae), and in Periplocoideae. Based on character optimization, this is interpreted as plesiomorphic in Rauvolfioids, and as a secondary loss in Periplocoideae. Older suggestions that the absence of lignified guide rails in Periplocoideae implies a closer relationship to Rauvolfioids (Wanntorp 1988) have been shown to be incorrect (Nilsson et al. 1993; Sennblad and Bremer 2000; Potgieter and Albert 2001; Livshultz et al. 2007). Conversely, the homology of lignified guide rails in Tabernaemontaneae and Apocynoids with those in Secamonoideae and Asclepiadoideae has not been proven. Although character optimization implies that at least those within Apocynoids are homologous, there are some differences in the detailed morphology and histology of the anthers (Kunze 1996) that suggest this interpretation may be overly simplistic. In higher Asclepiadoideae, the guide rails usually consist of two ridges with upwardly directed bristles between them (Liede 1996a).

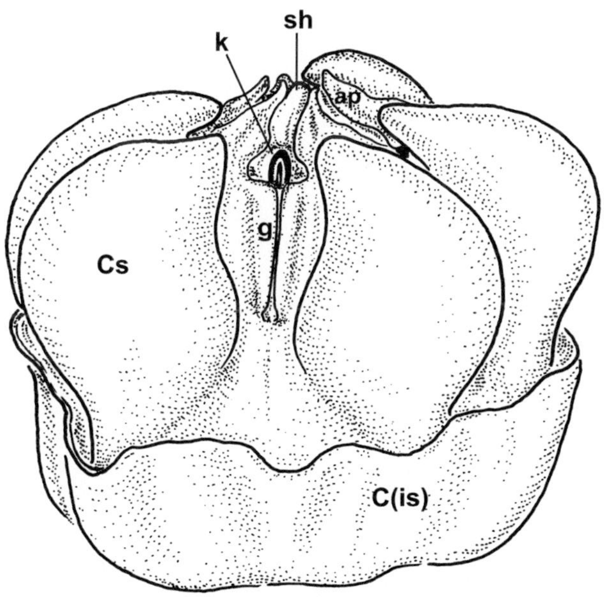

Fig. 25. Apocynaceae-Asclepiadeae. *Cynanchum*. Gynostegium with corona. (ap = anther appendage, Cs = staminal corona lobe, C(is) = ring-shaped gynostegial corona of connate staminal and interstaminal parts, g = guide rail, k = corpusculum, sh = style-head). (Drawn by U. Meve)

There is a high degree of synorganization both within the androecium as well as between the androecium and the gynoecium, intensified by postgenital fusions. In Apocynoids, Secamonoideae and Asclepiadoideae, the guide rails of adjacent anthers are separated only by a narrow slit, and are synorganized to function in guiding the pollinators' proboscis or leg into the pollination unit. In addition, adjacent anthers may be laterally postgenitally united, forming a ring (e.g., *Hemidesmus*, some species of *Mandevilla*; Nilsson et al. 1993; Moré et al. 2007). The anthers are postgenitally united with the style-head, forming a gynostegium except in Rauvolfioids. This results in a pentamerous revolver flower with five separate nectar/pollination chambers. Rarely, in some basal Apocynoids and Periplocoideae, the union occurs farther down on the filament, and may then be very weak (e.g., *Holarrhena*). In most Apocynoids a special pad of hairs is present on the ventral face of the connective, by which the anthers are attached to the style-head. In some genera of Mesechiteae (e.g., *Mandevilla*, *Mesechites*; Fallen 1986; Nilsson et al. 1993) the ventral face of the anther connective is smooth, and the anthers and style-head become postgenitally united by cellular fusion, as is the case in Periplocoideae, Secamonoideae and Asclepiadoideae. The connective apex is sometimes drawn out into a long filiform appendage (e.g., *Skytanthus*, *Nerium*, *Strophanthus*, *Pentalinon*), or a narrowly triangular to deltoid appendage (most Tabernaemontaneae, many Apocynoids), or a more or less membranous flap (most Asclepiadoideae). In Rauvolfioids, although a gynostegium is absent, the anthers may become stuck to the copious style-head secretions; the anthers sometimes closely surround the style-head (in e.g., *Allamanda*, *Cerbera*, *Vinca* and many other genera) forming a cone over its top, with their thecae resting against the slopes of the sterile upper part.

The majority of genera have distinct nectaries or nectar-producing tissue. In a number of genera of Rauvolfioids a distinct nectary disc or lobes cannot be distinguished, yet studies have found nectar to be present. Microtome sectioning and appropriate staining has revealed that in some of these taxa the lower part of the outer carpel wall is thickened and comprises cytoplasm-rich cells with stomata in the epidermis that secrete nectar (e.g., several genera of Tabernaemontaneae, *Hancornia* (Willughbeieae), *Chilocarpus*, *Condylocarpon* and *Lepinia* (Alyxieae), *Picralima* (Hunterieae) and *Acokanthera* (Carisseae); Haber et al. 1981; Zarucchi 1987; Endress et al. 1996, 1997; Darrault and Schlindwein 2005; Simões et al. 2007, 2016; Moura et al. 2011). In detailed studies of *Condylocarpon isthmicum* by Morokawa et al. (2015), nectar was found to be secreted by modified stomata with guard cells that have lost the ability to close their pores (Fahn 1979; Davis and Gunning 1992). Other genera without a discernible nectary disc, or with a much reduced one, are nectarless. This has been demonstrated in *Aspidosperma* (Aspidospermateae) and in *Plumeria* (Plumeriinae) in the Rauvolfioids as well as in some genera of the two early-branching tribes of Apocynoids (Wrightieae and Nerieae) and in *Holarrhena* and *Malouetiella* (Malouetieae). In all other tribes of the Apocynoids as well as in the Periplocoideae, Secamonoideae, and Asclepiadoideae, nectaries are characteristic. In Rauvolfioids and Apocynoids nectaries, when present, surround the base of the ovary (Endress and Bruyns 2000). Typically there are five, alternate to the staminal filaments. These may be distinct lobes (e.g., *Cleghornia*, *Trachelospermum*, *Apocynum* (Apocyneae), some species of *Forsteronia* (Mesechiteae); Nilsson et al. 1993; Middleton 2007), but often are fused to varying degrees into a disc, which may be of nearly free lobes, which are fused only near the base (e.g., many species of *Prestonia* (Echiteae) or *Mandevilla* (Mesechiteae); Simões et al. 2006; Morales et al. 2017b), or the segments may be ± completely fused into a ring, often with a lobed or crenulate top (e.g., all Baisseeae and many Apocyneae such as *Aganosma*, *Amphineurion*, *Anodendron*, *Epigynum*; De Kruif 1983; Middleton 2007, 2014). The nectary is of two lobes that are alternate to the carpels in several genera of Vinceae (e.g., *Catharanthus*, *Kopsia*, *Vinca* (Rauvolfioids) and in a few genera of Apocynoids, such as *Spirolobium*, *Carruthersia* (Malouetieae), *Salpinctes* (Odontadenieae) and several species of *Mandevilla*; Woodson 1936; Simões et al. 2006; Middleton 2007; Endress et al. 2007a.

In Periplocoideae, Secamonoideae and Asclepiadoideae distinct nectary lobes or a disc, such as found in Rauvolfioids and Apocynoids, are absent. Instead, nectar is secreted by specialized patches of secretory epithelium found in five pockets (chambers) on the staminal tube beneath the guide rail base (Kunze 1997). Another region of epithelium is often found along the anther flanks within the guide rails, or on the sides of the thickened corona lobes (or region between them) below the filaments in Periplocoideae. In Asclepiadoideae, nectar is frequently collected and presented in cups formed by the gynostegial corona; in *Asclepias* and its relatives, elaborate nectar conducting systems have evolved (Galil and Zernoni 1965; Kunze 1997).

The gynoecium is of two carpels, with a few rare exceptions in Rauvolfioids (up to five carpels in *Lepinia*, *Lepiniopsis* and *Pleiocarpa*), and in the great majority of taxa it is apocarpous. In Rauvolfioids, congenital syncarpy is characteristic for all genera of the Willughbeieae and Carisseae and the non-arillate members of the Tabernaemontaneae. Sometimes the gynoecium is syncarpous in the lower part, but apocarpous above (e.g., many species of *Rauvolfia*). In the APSA clade congenital syncarpy is absent; rather, in genera with syncarpous gynoecia, these come about by postgenital fusion (e.g., *Amalocalyx*, *Artia*, *Beaumontia*, *Nerium*, *Parepigynum*, *Parsonsia*, *Temnadenia*, *Thenardia* and *Vallaris*). In the great majority of genera the gynoecium is differentiated into ovaries, style and style-head. However, genera of some of the derived tribes of Asclepiadoideae typically lack a style. In genera with apocarpous gynoecia, the carpel apices undergo a temporary postgenital fusion, which lasts at least through anthesis. Where the flanks of the two carpels come together, the epidermis is obliterated and a uniform tissue is formed (Walker 1975, 1978; Sage et al. 1990). The uppermost part of this postgenitally fused region develops into an enlarged style-head, the enlarged apices of which sometimes contribute to flower attraction and pollinator control (*Araujia*, *Oxypetalum*). The style-head is a complex structure, which functions not only as a stigma, but also has a secretory epithelium, which produces adhesive components that play an important role in pollination biology (Schick 1980, 1982a; Fallen 1986). The adhesive is a heterogeneous mixture of polysaccharides and free sugars, terpenes and cutin derivatives, with the different components expressed to various degrees in the different parts of the family (Schick 1982b). Except in a few genera of Rauvolfioids, in which the style-head is uniformly receptive and secretory all over (e.g., *Alyxia*), the style-head is differentiated into zones: the five regions between the adnate anthers secrete the adhesive used in pollen transport. In the Periplocoideae, Secamonoideae and Asclepiadoideae, the secretions harden into acellular bodies that aid in the transport of pollen, and are called translators. Generally the stigmatic region is located beneath or on the lower half of the style-head. In Asclepiadoideae the receptive region is restricted to five stigmatic chambers on the underside of the style-head alternating with the stamens and behind the guide rails.

In Rauvolfioids the adhesive secreted forms a continuous ring of foamy or viscous substances around the style-head. In Apocynoids, the adhesive is interrupted in five regions due to the adnate anthers, so that it is restricted to five alternistaminal sectors. In both Rauvolfioids and Apocynoids two different components of the adhesive can often be distinguished in serial sections based on differences in staining and consistency. Sometimes both are frothy with considerable air spaces; sometimes one is frothy and the other is firm and compact. Secretion of these different components is sequential, and may be either temporally (epithelial cells producing first one, then switching to a second component) or spatially differentiated (i.e., specialized regions on the style-head secrete only one or the other component; Endress, unpubl. data). Rarely in Apocynoids the secreted substances become firm through desiccation, forming five simple band-like translators (e.g., *Apocynum*, some species of *Forsteronia*; Nilsson et al. 1993). In Periplocoideae, differentiation of the two adhesive components is both temporal and spatial: the denser adhesive component forms a firm spoon-like translator that is morphologically differentiated into three regions: (1) a broad, upper, flattened pollen-presenting scoop; (2) an inwardly rolled stalk; and (3) a broadened, flattened base (Fig. 26A). The foamy component, which is secreted over the scoop, helps pollen tetrads to adhere. At the base of the translator, the foamy component is secreted copiously, and is pushed outward forming a sticky glob (sometimes called a sticky disc or viscidium) that projects from the periphery of the style-head, and by which the translator adheres to the pollinator. In Secamonoideae and Asclepiadoideae, the translators are formed from the firmer secretory components; the foamy, sticky component has been

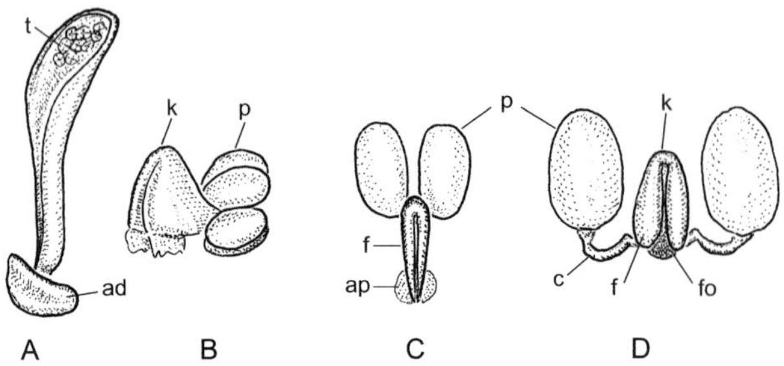

Fig. 26. Translators and pollen in Periplocoideae, Secamonoideae and Asclepiadoideae. **A** Periplocoideae. Spoon-shaped translator with pollen tetrads. **B** Secamonoideae. Polliniarium with four pollinia. **C** Asclepiadoideae-Fockeeae. Pollinarium with two pollinia and without caudicles, corpusculum with flanks and adhesive pads. **D** Asclepiadoideae-Marsdenieae. Pollinarium with two pollinia, connected via caudicles to corpusculum, corpusculum with flanks flanks and floor. (ad = adhesive disc, ap = adhesive pad, c = caudicle, f = flank of corpuscle, fo = floor of corpuscle, k = corpusculum, p = pollinium, t = pollen tetrad). (Figures not to scale; drawn by U. Meve)

interpreted to have been greatly decreased (Demeter 1922; Schick 1982b; Kunze et al. 1994). The secretions are differentiated into a hard clip-like corpuscle to which the pollinia are attached (Fig. 26B, D). In Asclepiadoideae, in addition to the corpuscle, two slender, more flexible arms (caudicles) are secreted, by which the pollinia are attached to the corpuscle (Fig. 26D; Kunze 1994). In some Secamonoideae slender thread-like caudicles are present (Civeyrel 1995; Civeyrel and Rowe 2001), but their homology to those in Asclepiadoideae is uncertain, particularly since the first branch of Asclepiadoideae, Fockeeae, with the two genera *Fockea* and *Cibirhiza*, lacks the caudicles characteristic for all other Asclepiadoideae (Fig. 26C). In these two genera the pollinia are ± sessile on a flat plate of adhesive on the dorsal side of the corpuscle, similar to the situation in most Secamonoideae (Fig. 26B; Kunze 1993; Civeyrel 1994, 1995; Kunze et al. 1994).

Embryology. The embryology of the Apocynaceae has been summarized by Johri et al. (1992). Maheswari Devi (1964) gives an overview for Periplocoideae and Asclepiadoideae. The anther tapetum is secretory. The pollen mother cells undergo simultaneous as well as successive cytokinesis in Rauvolfioids, Apocynoids and Periplocoideae (sometimes within the same species), and successive in Asclepiadoideae. In Rauvolfioids and Apocynoids the tetrads are tetrahedral and isobilateral. Decussate, T-shaped and linear tetrads are known from Periplocoideae (e.g., *Hemidesmus indicus*). The formation of linear tetrads is the rule in Asclepiadoideae, but T-shaped tetrads occur as well (e.g., *Pergularia*). In Rauvolfioids and Apocynoids pollen grains are triporate and three-celled at maturity. Two-celled pollen has been reported for *Catharanthus*, *Plumeria* and *Holarrhena*. In Asclepiadoideae pollen grains are three-celled at maturity and are aggregated into pollinia. The pollinium is covered by a persistent tapetal membrane of sporopollenin, which can be discontinuous in one region, which is the germination pore. The ovules are hemianatropous or anatropous, unitegmic, tenuinucellate, usually with an integumentary tapetum. Of the original one-layered nucellus, the lateral cells become crushed while the apical cells persist. Ategmic ovules have been reported in the following genera of Asclepiadoideae: *Araujia*, *Asclepias*, *Gomphocarpus*, *Cynanchum* and *Marsdenia*. But these are older studies (e.g., Maheswari Devi 1964, and references therein) that need to be reconfirmed. The archesporium is one-celled, rarely 2–3-celled. The chalazal megaspore develops into an embryo sac, and development is of the Polygonum type. A report of bisporic development in *Cynanchum* (Asclepiadoideae; Maheswari Devi 1964) should be reconfirmed. The polar nuclei fuse before fertilization; the antipodal cells are ephemeral and secondary multiplication is found in Asclepiadoideae. Starch grains accumulate in the embryo sac.

Endosperm development is of the nuclear type. Embryogenesis of the Asterad, Onagrad, Caryophyllad and Chenopodiad type has been reported in Rauvolfioids and Apocynoids, and of the Solanad type in Asclepiadoideae. Apomixis, as far as is known, occurs only in *Vincetoxicum* as integumentary polyembryony (Naumova 1992). Endosperm is ruminate in some Rauvolfioids (e.g., Tabernaemontaneae-Tabernaemontaninae and some Alyxieae such as *Alyxia*, *Lepinia*, *Lepiniopsis*, *Condylocarpon*, some species of *Chilocarpus*; Periasamy 1963; Corner 1976). The seed is typically albuminous and exotestal.

Pollen Morphology. Palynological characters have long been the most important morphological features in distinguishing larger taxonomic units in the family (Brown 1810; Civeyrel et al. 1998; Endress 2001). Pollen is shed as monads in Rauvolfioids and Apocynoids, with a few sporadic exceptions where it remains in tetrads. Tetrads are known from five genera of Rauvolfioids, where they occur sporadically in Tabernaemontaneae, Melodineae and Alyxieae, and in Apocynoids only in *Apocynum* (Erdtman 1952; Lodder et al. 2007). Pollen is typically shed as tetrads in Periplocoideae (Nilsson et al. 1993). In seven (mostly Asian) genera of Periplocoideae (*Decalepis*, *Epistemma*, *Finlaysonia*, *Gymnanthera*, *Hemidesmus*, *Schlechterella* and *Streptocaulon*) each anther produces four pollinia (Verhoeven and Venter 1998, 2001; Ionta and Judd 2007). Likewise, in Secamonoideae each anther produces four pollinia, whereas in Asclepiadoideae each anther produces two pollinia (Civeyrel et al. 1998).

In general, Rauvolfioids have 3–5-colporate pollen grains (Nilsson 1986), but pollen of *Craspidospermum* is in 5–10-periporate tetrads, and most Alyxieae have distinctive, barrel-shaped (2–)3–5-porate, triangular pollen grains (sometimes 1-porate in *Chilocarpus* and *Plectaneia*, and inaperturate tetrads in *Condylocarpon*; Nilsson 1990; Van der Ham et al. 2001; Endress et al. 2007a). Pollen grains in Apocynoids are generally (2–)3(–6)-porate or 2–6-stephanoporate, but (1–)2(–4)-porate grains have been reported for *Mascarenhasia* and *Pachypodium* (Lienau et al. 1986), and 4–8(–22)-periporate grains have been reported in *Isonema*, *Carruthersia*, *Apocynum*, *Trachelospermum* and *Micrechites* (Endress et al. 1990; Lodder et al. 2007). In both Rauvolfioids and Apocynoids, the outer exine surface is usually smooth and perforate, whereas the inner surface is variously patterned. Rauvolfioids show much more variation as to the inner exine patterning than do Apocynoids, with endocracks, colpal and mesocolpial plates, or granules (Nilsson 1986). In most New World species of *Tabernaemontana* (and some paleotropical species and a few species of *Voacanga*) endoapertures are fused into an endocingulum (Van der Weide and Van der Ham 2012). In Apocynoids the inner exine is always granular. Tetrads in both Rauvolfioids and Apocynoids are acalymmate. In Rauvolfioids tetrahedral, colporate tetrads are found in *Callichilia*, at least one species of *Tabernaemontana* and some species of *Melodinus*. *Condylocarpon* has non-tetrahedral, inaperturate tetrads with a reduced exine, a suite of features considered to be neotonic, and *Craspidospermum* has non-tetrahedral, irregularly pantoporate tetrads, which are quite similar to those found in *Apocynum* (Van de Ven and Van der Ham 2006).

In Periplocoideae, in contrast, the tetrads are calymmate, and are generally rhomboidal or decussate, although other arrangements are present in some genera. The grains are 4–6-porate in most genera, although *Raphionacme*, one of the largest genera in the subfamily, has 8–16-porate grains. The pores are typically lined up pair-wise where two pollen grains abut. The inner walls of the tetrads have the same structure as the outer wall, but are interrupted by intine wall bridges. Periplocoideae pollinia (known from seven genera) are always four per anther and comprise loosely coherent tetrads that show a reduction of the number of pores in the distal wall. A pollinium wall is absent (Verhoeven and Venter 2001). Pollinia in Periplocoideae have evolved at least three times independently, and also separately from those in the other subfamilies (Ionta and Judd 2007).

Secamonoideae pollinia (four per anther) are small, globose, and comprise loosely coherent calymmate, inaperturate tetrads. The inner walls of the tetrads are reduced compared to the outer walls, and are interrupted by intine wall bridges. A pollinium wall is absent (Verhoeven and Venter 2001).

Asclepiadoideae pollinia (two per anther—the two dorsal pollen sacks are reduced; Kunze 1982) comprise single, inaperturate pollen grains, and are surrounded by a pollinium wall (Frye 1901; Verhoeven and Venter 2001). The only exception known is *Fockea*, in which the pollinia consist of coherent inaperturate calymmate tetrads not covered by a pollinium wall (Verhoeven et al. 2003). There is great variation in the size and shape of the pollinia (El-Gazzar et al. 1974; Schill and Jäkel 1978). In Secamonoideae and Asclepiadoideae, pollinia of neighboring anthers are joined in a pollinarium by an acellular translator apparatus secreted by the style-head (Fig. 26; Schnepf et al. 1979). Within the Secamonoideae, this structure normally lacks translator arms but displays considerable variation (Civeyrel 1994; Civeyrel et al. 1998). Orientation details of pollinaria morphology are the basis of tribal delimitation in Asclepiadoideae. In Fockeeae, the two erect pollinia are attached dorsally directly to the corpusculum, which has broad outer flanks at least basally isolated and not united by a floor (Fig. 26C). These flanks continue basally into adhesive pads glued to the anther wings (Kunze et al. 1994). Marsdenieae and Ceropegieae have erect pollinia (Fig. 26D), and sterile insertion crests on the outer side in Marsdenieae, if present (Swarupanandan et al. 1996), but always present on the inner side in Ceropegieae. Asclepiadeae have pendent pollinia, except for most Gonolobinae and some *Vincetoxicum* species with horizontally oriented pollinia. Some species of *Vincetoxicum* (*Sphaerocodon* group) and *Cynanchum verrucosum* even have erect pollinia, caused by upward-curved caudicles. Sterile insertion crests on the top or the outer margin are rare, but characteristic for most Gonolobinae.

Karyology. Karyological information is available for about 190 genera. The most comprehensive reviews are those of Van der Laan and Arends (1985) and Albers and Meve (2001). The base chromosome number of $x = 11$ is predominant, and is found in 60% of the genera of Rauvolfioids and Apocynoids, without exceptions in Periplocoideae and Secamonoideae, and in 96% of the genera of Asclepiadoideae. In Rauvolfioids, no deviations from the base number were found in Willughbeieae, Tabernaemontaneae, Hunterieae, and Carisseae. Most genera have only one base number, but both $x = 11$ and $x = 8$ have been reported for *Apocynum*. Significant deviations include $x = 9$, 10 for all of the genera of Plumerieae counted (six out of ten), and $x = 6$, 9 for all of the Echiteae counted (five genera out of 20). In addition, sporadic variation in base number has been found in some genera of all other tribes of Apocynoids. In Asclepiadoideae, reduction to $x = 9$, 10 is restricted to Asclepiadeae; $x = 10$ is characteristic for *Funastrum*, *Microloma* and *Orthosia* (and is also found in around ten more genera), whereas an increased base number occurs sporadically in all tribes.

In Rauvolfioids and Apocynoids polyploidy has been reported in nine genera (ca. 13%), and in Asclepiadoideae only about 6% of the taxa are polyploid, with tetraploidy being by far most common (Albers and Meve 2001).

Chromosome morphology is similar throughout the family, with (sub)metacentric chromosomes being predominant. Usually, each genome possesses one pair of chromosomes with secondary constrictions and satellites (Van der Laan and Arends 1985; Albers and Meve 2001).

Average chromosome length is ca. 1.5 μm in Rauvolfioids and Apocynoids, and gradually decreases to an average of 1.3 μm in Periplocoideae, 1.2 μm in Secamonoideae, and somewhat less than 1 μm in Asclepiadoideae. Within Asclepiadoideae, chromosome length on the average is greatest in Fockeeae (1.2 μm), and gradually diminishes from the more basal to the most derived tribes: 1.1 μm in Marsdenieae, 1.0 μm in Ceropegieae, and 0.9 μm in Asclepiadeae (Albers and Meve 2001).

Reproductive Biology and Pollination. Flowers are bisexual. In a few genera of Rauvolfioids, gender dimorphism and associated functional dioecy have been reported (e.g., *Carissa*, Schroeder 1951, and *Rauvolfia*, Koch et al. 2002), and in *Cynanchum* (*Glossonema*) *varians* (Ali and Ali 1996) and *Cynanchum hemsleyanum* (= *Metaplexis japonica*, Tanaka et al. 2006), both Asclepiadoideae, andromonoecy has been demonstrated. The flowers are protandrous. In Rauvolfioids and Apocynoids pollen is shed as single grains, or rarely as tetrads, shortly before anthesis onto the sides of the sterile upper style-head or into a pollen chamber formed by the cone of anthers above it, where it is secondarily presented (Church 1908; Fallen 1986). In Periplocoideae the tetrads or pollinia are shed onto morphologically and histologically differentiated spoon-like translators, which are arranged radially around the style-head between adjacent anthers. The translators include an adhesive-covered scoop onto which the pollen is shed, a stalk and a sticky disc (viscidium) by which they become attached to pollinators (Nilsson et al. 1993). In Secamonoideae and Asclepiadoideae the pollinia more or less remain in the anther at anthesis, although they have already become attached to the translator. In Rauvolfioids, in which the anthers are free from the style-head, the corolla is sometimes shed after the male phase (Boiteau and Allorge 1978). In the APSA clade the antepetalous anthers are adnate to the style-head, thus the corolla remains attached longer, until the style breaks off. Longevity of flowers has rarely been investigated in Rauvolfioids and Apocynoids. In *Aspidosperma quebracho-blanco*, the male phase lasts 1.5–2 days, followed by the female phase, which lasts 1 day (Lin and Bernardello 1999). *Nerium* flowers are said to last up to 7 days if unpollinated (Herrera 1991). In the night-flowering *Kamettia chandeei*, no open flowers can be observed on plants during the day but early each morning fallen corollas carpet the ground, suggesting the corolla lasts less than 1 day and is dropped immediately after pollination, presumably by moths (Middleton et al. 2006). In Periplocoideae and Asclepiadoideae flowers last from less than 1 day (e.g., *Schlechterella abyssinica*, *Piaranthus* p.p.) to 8 or even 14 days (*Hoya*, *Marsdenia* s.l.; Meve 1994, unpubl. data).

The complex flower construction suggests that the family is predominantly out-crossing (Albers and van der Maesen 1994). However, relatively few detailed studies have been conducted on breeding systems in the family except in Asclepiadoideae. In *Aspidosperma quebracho-blanco* (Rauvolfioids), the plants were found to be self-compatible when artificially selfed; due to dichogamy, however, autogamy is unlikely (Lin and Bernardello 1999). Similarly, *Mandevilla pentlandiana* and *Wrightia arborea* (Apocynoids) were found to be self-compatible, but not autogamous (Torres and Galetto 1999; Barman et al. 2018). *Apocynum* apparently has a late-acting self-incompatibility system (Lipow and Wyatt 1999); the same is true for *Gonolobus suberosus* and *Periploca aphylla* (Lipow and Wyatt 1998). Stapeliads are usually self-incompatible when diploid, but tetraploidy is correlated with self-compatibility (Meve 1997; Meve et al. 2004). The bulk of the breeding system literature is focused on *Asclepias*. Wyatt and Broyles (1994) recognized two main types of breeding systems in *Asclepias*: species that are (almost) completely self-incompatible and species that are (almost) completely self-compatible. Most species of *Asclepias* have been shown to possess a late-acting self-incompatibility system similar to that in *Apocynum* (Wyatt 1976, 1981; Kahn and Morse 1991; Sage and Williams 1993, 1995; Wyatt et al. 1998). However, *Asclepias incarnata* (Kephart 1981; Ivey et al. 1999), *A. curassavica* and *A. fruticosa* have been shown to be self-compatible (Wyatt and Broyles 1997), and self-compatibility is assumed to have arisen multiple times within the genus (Wyatt and Broyles 1994). Even spontaneous self-pollination was observed in *A. curassavica* (Wyatt and Broyles 1997). For *Vincetoxicum* autogamy and apomixis have been demonstrated (Chaturvedi 1988; Lumer and Yost 1995; Leimu 2004; Yamashiro et al. 2008).

Apocynaceae are entomophilous, with two possible exceptions both involving pollination by birds. The first is in *Mandevilla* (Apocynoids), in which two species, *M. hirsuta* and *M. veraguasensis*, were found to be pollinated by trap-lining hummingbirds (Feinsinger 1978; Linhart and Feinsinger 1980). Hummingbirds have been observed visiting other species of *Mandevilla* for nectar as well—e.g., *M. guanabarica*, *M. pentlandiana* and *M. scabra* (Torres and Galetto 1998; Machado 2009; Abrahamczyk and Kessler 2010; Fonseca et al. 2015)—but in these species it was not shown that they effected pollination; in the case of the small-flowered *M. pentlandiana* they are considered to be nectar thieves, not pollinators (Galetto, pers. comm.), which is supported by the small diameter of the openings between adjacent anthers (see Moré et al. 2007) and as no pollen was found in the stigmatic region after a visit (Endress, pers. obs.). The second, well-documented case is *Microloma* (Asclepiadoideae), where sunbirds (Nectarinidae) have been shown to take up pollinia on their tongue (Pauw 1998). Visits of the sunbird *Nectarinia sovimanga* to *Pleurostelma cernuum* on Aldabra have been observed as well (Woodell 1979). Two other species of *Nectarina*, *N. cuprea* and *N. chloropygia*, frequently visit flowers of *Voacanga africana*, but they puncture the corolla tube and steal nectar, without effecting pollination; the real pollinators for this species is more likely Lepidoptera (Albers and van der Maesen 1994).

The floral reward is consistently nectar. The chemical composition of nectar in some species has been reported (Galetto 1997). In some genera flowers produce no floral reward and are considered to be pollinated by deceit (Herrera 1991); these have visual and olfactory cues that promise a reward, but offer none. They are often floral mimics, and depending on the model, they may be a specialized or generalized mimic (sometimes called non-model mimicry); in the latter plants use cues that potential pollinators innately recognize, and which are not tied to a specific model (Haber 1984). The plants are often mass-flowering and/or may have an extended flowering period, producing a surplus of flowers in a (sometimes showy) display to compensate for the low visitation/pollination rates once pollinators realize there is no reward. Deceit pollination of this type requires naïve or inexperienced pollinators, and thus pollination events are uncommon. In Rauvolfioids deceit flowers have been reported in Plumerieae-Plumeriinae. *Plumeria* has flowers with general adaptations typical for sphingid pollination and it has been suggested it is a generalized mimic (Haber 1984). In addition to being mass-flowering, the large showy flowers are long-lived and have a strong sweet scent (Haber 1984). In the closely related genus *Himatanthus*, although few flowers open each day, flowering can continue for more than a year (Schlindwein et al. 2004). Both *Plumeria* and *Himatanthus* are host plants to caterpillars of the same sphingids that pollinate their flowers (Plumel 1991). *Aspidosperma pyrifolium* and *A. quebracho-blanco* (Aspidospermateae) are also mass-flowering and pollinated by deceit, but in this genus the individual flowers are small and inconspicuous, and thus they are grouped into many-flowered inflorescences in order to make a greater impact; they open toward evening and have a sweet scent. The pollinators in these two species of *Aspidosperma* were shown to be moths (Lin and Bernardello 1999; Queiroz 2009). In Apocynoids deceit flowers are suspected in *Nerium* and *Strophanthus* (Nerieae). Both genera have mass flower displays and additional optical attractants, such as a showy, dissected petaloid corona and long, hairy, filiform anther appendages (*Nerium*) and/or thin, extremely elongate, dangling petal appendages (*Strophanthus*), which has made them popular in cultivation. The movements of the long, thin appendages in the breeze presumably attract pollinators. As with the deceit flowers in Plumeriinae, pollination events are very rare, and to date the pollinators in these two genera are unknown. In the Asclepiadoideae-Ceropegieae, deceit flowers are common (Meve and Liede 1994).

In Rauvolfioids and Apocynoids night-flowering occurs sporadically in many tribes, whereas in Asclepiadoideae night-flowering seems to be restricted to Marsdenieae (*Stephanotis*, *Telosma*) and Asclepiadeae (*Vincetoxicum*, *Pergularia*), though it may occur more frequently. Bhatnagar (1986) reported Lepidoptera (Noctuidae) as pollinators. In Rauvolfioids and Apocynoids the pollen is transferred on the proboscis, and many genera have flowers with a slender, salverform corolla, in which nectar accumulates at the base. To reach the nectar, the pollinator must have a proboscis long enough to reach nearly to the bottom of the tube. The most likely pollinators are, therefore, butterflies, hawkmoths, moths or large bees. Relatively few actual observations of pollination have been made, however. *Rauvolfia grandiflora* was shown to be pollinated by one species of long-tongued bee in northeastern Brazil (Lopes and Machado 1999). Some genera of Apocynoids with more open flowers are pollinated by flies (e.g., *Apocynum androsaemifolium*, Ludwig 1880). Wasps have been identified as the pollinators of *Parsonsia alboflavescens* (Livshultz, unpubl. obs.). Night-flowering species of *Mandevilla* with white corollas with a slender tube up to 15 cm long are adapted for sphingid pollination, in which not only length, but also width of the proboscis restrict the assemblage of pollinators that can reach the nectar (Moré et al. 2007). Sphingids have been observed to pollinate *Cryptostegia grandiflora* (Periplocoideae) in Madagascar (Walther 1994) and *Oxypetalum* species (Asclepiadoideae), which are also characterized by long corolla tubes in Argentina (Liede-Schumann, unpubl. obs.). Often more than one type of insect may visit flowers of the same plant. Darrault and Schlindwein (2005) observed 33 different insect species visiting flowers of *Hancornia speciosa*, but that fruit set was correlated with pollen load, so that some pollinators, in this case hawkmoths, are more efficient than others. Thus, even though a taxon may exhibit a particular floral syndrome associated with a specific type of pollinator, this often does not exclude visitation (and in some cases pollination) by other types as well. For example, Torres and Galetto (1998) observed flowers of *Mandevilla pentlandiana* to be visited by bumblebees and honeybees, as well as the hummingbird *Chlorostilbon aureoventris*. Waddington (1976) concluded that, although the flowers of *Apocynum sibiricum* are visited by a variety of insects, only Lepidoptera (including species from seven different families) actually pollinate it. In Periplocoideae, the majority of the genera have a nearly rotate corolla, which suggests fly pollination, and fly pollination has been reported for some species of *Periploca* (Schick 1982b). The more open flower construction makes possible the attachment of the translator in regions other than the proboscis (e.g., on the leg, back or head of the insect). Some genera of Periplocoideae have a salverform corolla and these are presumably not fly-pollinated. In Australia, Forster (1991a) reported a possible transfer of *Gymnanthera* pollinia by mosquitos. *Cryptostegia*, with a corolla ca. 5–6 cm long, was observed to be pollinated not only by sphingids (Walther 1994), but also by large *Xylocopa* bees (S. Vogel, pers. comm.).

In Asclepiadoideae, some species tend to be specialists, being pollinated exclusively or nearly so by one pollinator, whereas others are generalists, with several types of insects visiting the same flower, which may or may not be pollinators (Fishbein and Venable 1996). The main morphological changes associated with pollination mode shifts are found in the corolla and/or corona (Yamashiro et al. 2008). In Marsdenieae (Asclepiadoideae), Hesperidae butterflies have been reported to transfer *Hoya australis* pollinia (Forster 1992a), but several visitors and possible pollinators have been observed for *Marsdenia cymulosa* (Forster 1992b). *Cynanchum hemsleyanum* (= *Metaplexis japonica*) was found to be moth-pollinated (Sugiura and Yamazaki 2005). Examples of the specialist flowers from South Africa include *Asclepias woodii* and *Sisyranthus trichostomus*, which are pollinated almost exclusively by fruit beetles (Ollerton et al. 2003), or species of *Pachycarpus* and *Miraglossum* that are pollinated exclusively or nearly so by large spider-hunting wasps (Ollerton et al. 2003; Shuttleworth and Johnson 2006). Wasps play an important but hitherto understudied role in Asclepiadoideae pollination. A species of *Pepsis* was observed on several New World Asclepiadoideae *Morrenia* and *Philibertia* (Baranzelli et al. 2014; Cocucci et al. 2014). The *Hemipepsis*-wasp-pollination-guild was found to include 17 South African species of *Asclepias*, *Aspidoglossum*, *Miraglossum*, *Pachycarpus*, *Periglossum*, *Woodia* and *Xysmalobium* (Shuttleworth and Johnson 2012). Bitter nectar only palatable to wasps can play an important role as a filter in the pollination system of some of these Asclepiadinae species (e.g., Shuttleworth and Johnson 2009). *Polybia ignobilis* pollinates four *Oxypetalum* species in Brazil (Vieira and Shepherd 1999), and various wasps were found to carry pollinaria of Old World *Cynanchum* species (Kugler 1973). Reports of more generalist flowers have been published by Pant et al. (1982), who studied the flowers of *Calotropis procera*, *Leptadenia reticulata*, *Oxystelma secamone* and *Wattakaka volubilis* and recorded 17 species of insect visitors, several of which take nectar but do not carry pollinia. In Brazilian *Ditassa capillaris* and *D. hastata* not only Hymenoptera, Diptera and Lepidoptera were reported as pollinators but also ants of the genera *Cephalotes* and *Dorymyrmex* (Domingos-Melo et al. 2017). Flowers of *Xysmalobium gerrardii* were found to be pollinated by large pompilid wasps, fruit beetles and lygaeid bugs (Ollerton et al. 2003). Pollinaria of *Calotropis procera* were found to be transported not only by large carpenter bees (*Xylocopa*), which carry them on the forelegs (Wanntorp 1974), but also by small species of *Micrapis* and *Apis*, which are small enough to creep through the stigmatic slits to reach the nectar; upon leaving the flower, pollinia become attached to the back legs. Both Ramakrishna and Arekal (1979) and Eisikowitch (1986), however, found that only species of *Xylocopa* are pollinators of *Calotropis*. Pollination requires not only removal, but also correct insertion of the pollinia into the germination slit of another flower, and it is doubtful that this can be achieved by all the insects that carry pollinia of any given generalist species. According to a study by Ollerton et al. (2003) in South Africa, the majority of flower visitors did not pick up pollinia and must therefore be considered nectar thieves, rather than pollinators. A myophilous pollination syndrome has evolved independently in the Stapeliads (Ceropegieae) and in the Gonolobinae (Asclepiadeae) (Meve and Liede 1994; Kunze 1995). Flowers in these two groups show both visual and olfactory specializations mimicking typical substrates of calyptrate flies (e.g., dung, carrion, decaying organic matter) by imitating odors and surfaces that are glossy or warty, often reddish brown and occasionally blotched (Jürgens et al. 2006). Small, kleptoparasitic flies (1–3 mm) are attracted by scent mimicking hymenopteran alarm pheromones emitted by the pitfall flowers of *Ceropegia sandersonii*—a pollination system described as kleptomyiophily

(Heiduk et al. 2016). Similar, but not yet understood selective pressures have led to striking similarities, especially in floral structure, between unrelated genera of both hemispheres (e.g., between *Funastrum* and some *Cynanchum* species (formerly *Sarcostemma*), and between *Philibertia*, *Oxystelma* and another *Cynanchum* species (formerly *Platykeleba*); Liede and Täuber 2000). A survey of pollination patterns in Periplocoideae, Secamonoideae and Asclepiadoideae is given in Ollerton and Liede (1997).

The pollination mechanism of the Apocynaceae is complex and centers around the specialized secretions produced by the style-head. In Rauvolfioids and Apocynoids, in which the secretions usually remain undifferentiated, the most common flower shape is a salverform corolla with a constricted orifice, although large infundibuliform corollas can be found in genera that have anthers with hardened guide rails (Tabernaemontaneae, several tribes of Apocynoids). The inner surface of the corolla tube often has various sorts of emergences (e.g., corona lobes, protruding ridges below the stamens, rings or ridges of downward-directed hairs, etc.) that guide the proboscis of the pollinating insect into one of the five nectar channels of the revolver flower construction. The proboscis extends, searching into the nectar channel, with relatively little resistance. When it is withdrawn, however, it is directed inward, where it first encounters the base of the enlarged style-head. In many Rauvolfioids, the style-head is broadest at the base, and is equipped with a collar (like an inverted cup) or a wreath of longer hairs, which act as a scraper to scrape off the pollen that was introduced on the proboscis, where it accumulates in the stigmatic region at the base of the style-head. As it moves upward on its way out of the flower, the proboscis next slides through the region where the pollen transport adhesive is secreted, and becomes coated with sticky adhesive. Finally, it slides up through an upper chamber into which the anthers have shed pollen; the pollen grains adhere to the sticky adhesive, coating the proboscis. When the pollinator visits the next flower, the process is repeated. In Apocynoids, the mechanism is more precise due to the adnate anthers. Upon being withdrawn, the pollinator's proboscis is threaded between the lignified guide rails of adjacent anthers and forced through one of the five pollination chambers. Once threaded inside the guide rails, the insect must pull its proboscis upward in order to escape (Fallen 1986).

In Periplocoideae, the floral structure is usually more open, and the secretions harden into distinctly shaped translators. Here the nectar is generally openly presented in shallow pockets, into which the insect can crawl or easily reach with its proboscis. Dangling directly above the nectar pocket is the sticky base of the translator, which becomes stuck to the pollinator. When the pollinator flies away, it pulls the translator (and its load of pollen tetrads or pollinia) free from the style-head. When it visits the next flower, some pollen will hopefully be rubbed off in the vicinity of the nectar pocket, which is also the stigmatic region (Nilsson et al. 1993). In Secamonoideae and Asclepiadoideae, the construction is reminiscent of that in Apocynoids (Safwat 1962), with the proboscis or leg of the pollinator being caught between the guide rails of adjacent anthers (Kunze 1991). In order to free itself it must proceed upward. In these two subfamilies, one part of the style-head secretions develops into a hard narrow clamp (corpuscle), into which the proboscis or leg is threaded and becomes wedged. In order to escape, the insect must pull the corpuscle (with its attached pollinia) free from the style-head. When it visits the next flower, pollination is achieved when a pollinium becomes wedged in one of the five pollination chambers, which lie just behind the narrow slit between adjacent anther wings. Usually a basal or apical part of the pollinium, rarely the entire pollinium, is inserted into the pollination chamber. A specialized protruding insertion crest has evolved independently in the Paleotropics and Australia in the Ceropegieae and some Marsdenieae (e.g., *Hoya*), and in the New World in more derived Gonolobinae (Asclepiadeae), and only this crest is inserted into the pollination chamber. In Ceropegieae, this crest is also the germination pore, through which the pollen tubes germinate (Kunze 1995). In Stapeliads, the size and shape of structures involved in pollination were shown to operate in a key (pollinium, germination crest) and lock (guide rail) manner causing mechanical prezygotic isolation in most cases (Meve et al. 2004).

Fruit and Seed. In the great majority of taxa of the APSA clade the 2-carpellate, apocarpous gynoecium results in a fruit that is apocarpous and consists of a pair of follicles, each one of which at maturity dehisces along the ventral suture, releasing the seeds. In some genera of Apocynoids, instead of the ventral margins of each individual carpel fusing, the two carpels fuse postgenitally along their ventral margins. This postgenital fusion occurs early in floral ontogeny, and thus can be seen only with a microscope and in sections of very early stages of carpel development. In these cases, instead of a pair of follicles, each flower produces a single postgenitally syncarpous "double-follicle", which at maturity splits apart along the lines of fusion. The fusion between the united follicles in this type of fruit varies from relatively weak to a union indistinguishable from that of congenital fusion. Genera with flowers that produce a single postgenitally syncarpous fruit include *Amalocalyx*, *Beaumontia*, *Parepigynum* and *Vallaris* (Apocyneae), *Artia*, *Ecua*, *Parsonsia*, *Temnadenia*, *Thenardia* and *Thoreauea* (Echiteae), *Nerium* (Nerieae) and several species of *Wrightia* (Wrightieae; Candolle 1844; Thomas and Dave 1994; Li et al. 1995b; Middleton 2007, 2010, 2014; Simões et al. 2007), and in a few genera of Asclepiadoideae (e.g., *Heterostemma* p.p.). In certain alliances there is a tendency for only one of the carpels to develop, and thus the fruit consists of a single follicle. In the majority of genera the follicles are slenderly fusiform or linear, non-fleshy, less frequently fleshy, sometimes woody at maturity—e.g., many genera of Apocynoids, *Fischeria* (Asclepiadoideae). In some genera of Asclepiadoideae (e.g., *Gonolobus* and *Marsdenia*) the follicles are rather stout and winged. The fruit wall is most commonly smooth, but may be ornamented with irregular corky protuberances, prickles, ridges or wings. Single rheophytic species of various genera, e.g., *Cynanchum*, *Vincetoxicum* and *Oxystelma*, form an inflated pericarp capable of flotation; in these cases, dehiscence is delayed or incomplete. *Emicocarpus* is the only genus in the family with one-seeded, spiny nutlets. In all taxa with follicular fruits, these often remain fused at the apices for some time after anthesis, until the bond is ruptured by the enlarging follicles. Only rarely are the apices still fused by the time the fruit reaches maturity (e.g., some species of *Echites*, *Prestonia*, *Jobinia*).

In contrast to this fixed uniformity in the more derived APSA clade, fruit and seed morphology are complex and homoplasious in the Rauvolfioids. Though usually 2-carpellate, fruits are up to 5-carpellate in some genera of Hunterieae and Alyxieae. Fruits are apocarpous, congenitally syncarpous or more rarely postgenitally syncarpous (*Allamanda*, *Alstonia rostrata*; Fallen 1985; Sidiyasa 1998) and include follicles, capsules, drupes and berries. There is a great evolutionary potential, and selective pressures acting on this lability have resulted in a complex array of fruit types with very similar morphology evolving independently in different lineages. For example, drupes have evolved in Vinceae, Plumerieae and Alyxieae, and berries have evolved in Willughbeieae, Tabernaemontaneae, Melodineae, Hunterieae and Carisseae (Endress et al. 2007a). In both *Petchia* (Vinceae) and *Alyxia* (Alyxieae) a distinctive type of drupaceous fruit has evolved in parallel. The fruiting carpels consist of a number of articles, each formed by an individual seed surrounded by a stony endocarp; the regions between the articles are very narrow, so that the fruiting carpels are strongly torulose, resembling beads on a string (Simões et al. 2016). In addition, in both tribes the fleshy outer fruit wall is most often bright orange or red, sometimes black, making these some of the showiest fruits in the family. In other cases, genera that are very closely related may have very different fruit types. For example, in the small tribe Aspidospermateae *Aspidosperma* with a dry follicle and *Geissospermum* with a fleshy berry are otherwise very similar morphologically and are sisters in phylogenetic analyses (Simões et al. 2007). Similarly, within Vinceae, about half of the genera have follicular fruits, but drupes have arisen independently at least three times (Simões et al. 2016). In Plumerieae, fruit types include dry, spiny capsules or follicles, drupes, nutlets or samaras (Simões et al. 2007). Also in Alyxieae genera with woody drupes occur next to ones with follicles. Some of the most unusual fruits in the family are found in this tribe. The fruit of *Chilocarpus* is syncarpous with a somewhat fleshy mesocarp and hard, dry endocarp (like a drupe), but splits apart into two halves at maturity. *Chilocarpus* also has perhaps the most variable fruits in the family: from more or less globose to up to 40+ cm long and slenderly moniliform and everything in-between (illustrated in Markgraf 1971: 157; Middleton 2007; Endress et al. 2007a). Probably the strangest fruits of all are those of *Lepinia*, which are of 4–5 carpels that are congenitally fused into a stipe at the base, and remain firmly postgenitally fused at the apex, with the region in-between apocarpous, and bowed outward, so that the fruit resembles a Chinese lantern (Endress et al. 1997; illustrated in Lorence and Wagner 1997, and Middleton 2007).

The number of seeds per fruiting carpel is highly variable, ranging from one to a few hundred, with Rauvolfioids being the most variable. The tendency is for an increase in seed number in the genera with small seeds with a coma (i.e., the APSA clade), although in a few cases there has been an extreme reduction in the number of seeds per fruiting carpel in the more highly derived subfamilies (e.g., 2-seeded in *Stigmatorhynchus*, and a 1-seeded nutlet in *Emicocarpus*, both Asclepiadoideae).

Except in Rauvolfioids, seeds in Apocynaceae are generally small, lightweight, one end of which almost always has a coma—a (mostly easily detached) tuft of hairs. In the early-branching Apocynoids (Wrightieae, Nerieae and Malouetieae) the position of the coma is not fixed; it may be on the micropylar end (e.g., *Carruthersia*, *Holarrhena*, *Spirolobium*) or the chalazal end (Wrightieae, *Kibatalia*), on both the micropylar and chalazal ends, where the chalazal one is often smaller and quickly deciduous (e.g., *Adenium*, *Farquharia*, *Funtumia*, *Isonema*, *Strophanthus*), or it may be lacking completely (*Eucorymbia*, some *Malouetia* species). In higher Apocynoids, Periplocoideae, Secamonoideae and Asclepiadoideae the coma is always micropylar. In Apocynoids, the color of the coma varies from brown to yellow, cream or white, whereas it is often yellowish in Periplocoideae. In almost all Asclepiadoideae, the coma is silky and white. Sporadically, in various subfamilies, the coma has been lost in at least some species of a genus (e.g., *Marsdenia* p.p., *Sarcolobus* species). In some *Malouetia* species, a normal coma is lacking; rather, the whole seed testa is covered by long lanate hairs. Similarly, in *Batesanthus*, *Finlaysonia*, and some *Raphionacme* species, a coma in the normal sense is absent. Instead, hairs occur around the whole seed margin. *Cynanchum adalinae* has hairs firmly attached along the margin of the chalazal half of the seed. *Diplolepis* (*Grisebachiella*) *hieronymi* is the only Asclepiadoideae species with heavy discoid seeds with a much reduced coma on the hilar side. Fruits and seeds of Asclepiadoideae are often parasitized by Tephritidae (Landolt 1994; Meve 1995; Solbreck 2000). The seeds of Apocynoids are variable in shape, from flat and ovoid to fusiform, often inrolled or flattened longitudinally, sometimes with a rostrum or beak, sometimes slightly curved, usually smooth and glabrous, and rarely with hairs (e.g., *Nerium*, *Urceola*). When a rostrum is present the coma may arise only at the end of the rostrum (e.g., *Chonemorpha*), along the rostrum with the hairs pointing away from the seed (e.g., *Anodendron*), or along the rostrum with the hairs pointing back toward the seed (e.g., *Kibatalia*). The testa of the cylindrical to flat seeds of Periplocoideae, Secamonoideae and Asclepiadoideae can be smooth, papillate, distinctly tuberculate or hairy, usually identical on both sides, but occasionally more pronounced on the side without the raphe (sometimes termed aseta side). In many species of Periplocoideae, Secamonoideae and Asclepiadoideae the seeds are distinctly winged (all around except the micropylar region). Anatomy of these wings is three-layered with (1) a one cell layer thick dorsal testa, (2) a thin central layer of simple parenchymatic tissue, and (3) a ventral testa layer. In Ceropegieae this ventral testa consists of voluminous, elongated and intensively pitted cells allowing quick and efficient water uptake, whereas in *Asclepias* this layer, like the dorsal layer, is simple. In *Vincetoxicum*, in contrast, the wing is reduced to a simple margin derived from collapsed testa cells (Sylla and Albers 1989). The endosperm is almost completely reduced; the chlorophyllous embryo has fairly large cotyledons. Polyembryony occurs in *Vincetoxicum*, e. g., *V. nigrum* and *V. rossicum* (Denis and Capuccino 2004). In the stem succulent Stapeliads the cotyledons are more or less reduced, while the hypocotyl is succulent and oil-rich.

Seeds in Rauvolfioids are diverse, and the seed type is usually correlated with fruit type. If the fruit is a dry, dehiscent follicle or capsule, the seed is most commonly compressed and winged or ciliate at the margin (e.g., *Alstonia*, *Aspidosperma*, *Cameraria*, *Craspidospermum*, *Plumeria*, *Tonduzia*), although in *Catharanthus* and *Vinca* (Vinceae) and *Amsonia* (Amsonieae) with thin-walled, dry follicles, seeds are not compressed and are naked. There is a great diversity of winged seeds, and at least one genus with winged seeds is found in the majority of Rauvolfioid tribes (though none occur in Carisseae, Hunterieae, Tabernaemontaneae and Willughbeieae). The wings may be concentric (e.g., *Aspidosperma*, *Dyera*), they may be only on one end (e.g., *Diplorhynchus*, *Plumeria*) or at both ends of the seed (e.g., *Gonioma*, *Kamettia*, *Strempeliopsis*). Seeds with long fimbria around the margin have evolved in parallel in *Alstonia* (Alstonieae) and *Tonduzia* (Vinceae). Seeds with a coma are rare in Rauvolfioids, but seeds of *Haplophyton* have both a micropylar and a chalazal coma. In Tabernaemontaneae-Tabernaemontaninae (with fleshy follicular fruits) seeds are globular with a deep furrow on the hilar side, and lines of shallower grooves on the other side; in addition, they have an (often brightly colored) aril and strongly ruminate endosperm. In berry-fruited taxa of the same tribe (Tabernaemontaneae-Ambelaniinae), the seeds are often globular or somewhat flattened, are unadorned and have smooth endosperm. In *Chilocarpus* (Alyxieae), seeds have a corky aril (the only known occurrence of arils outside of Tabernaemontaneae), and some species have smooth endosperm whereas in others it is strongly ruminate. In the genera with drupaceous fruits, the seeds are enclosed in a lignified or stony endocarp; in most Alyxieae, seeds are cylindrical with a long hilar furrow and thick, ruminate endosperm. In Vinceae, the seeds of drupaceous taxa are somewhat flattened and the endosperm is not ruminate. In Plumerieae the seeds of drupaceous taxa may be ± globular (e.g., *Thevetia*) or thin and flattened (*Anechites*; Alvarado-Cárdenas and Ochoterena 2007), and endosperm is not ruminate. In general, germination follows the epigeous type, but in some species of derived Asclepiadeae (*Vincetoxicum*) hypogeal germination has been observed.

Dispersal. Wind dispersal is the rule for the majority of genera of the APSA clade, with the dispersal unit being the seed, which in almost all cases has a coma. Exceptionally, the coma has been lost (often in conjunction with a shift to water-dispersal), and this has occurred in parallel at least once in Apocynoids (*Malouetia*), Periplocoideae (*Finlaysonia*) and Asclepiadoideae (*Marsdenia rubrifusca*, *M. oblanceolata*, several *Sarcolobus* species, *Asclepias perennis*, *Matelea*). In Rauvolfioids, many genera are wind- or water-dispersed. In the former, the dispersal unit is normally the seed, which is usually winged or has long hairs around the margin; the samaras in *Cameraria* and *Cerberiopsis* (Plumerieae) are also presumably wind-dispersed. Most water-dispersed genera of Rauvolfioids have drupaceous fruits and the dispersal unit is often the whole fruit, one fruiting carpel, or, in moniliform fruits, one or more one-seeded segments. Specializations for sea water dispersal are found mainly in genera with drupaceous fruits in Plumerieae, Vinceae and Alyxieae, and include hollow flotation chambers in the endocarp or a mesocarp with a thick layer of fibers, which serve to protect the seeds, so that they remain viable over long distances in sea water (e.g., *Cerbera*, *Lepiniopsis* and *Ochrosia*; Simões et al. 2016). In Alyxieae, the hard endocarp of *Pteralyxia* forms a complex framework shaped like a boat around the seed (Endress et al. 2007a), and in *Lepinia*, each long-stipitate fruiting carpel of the 4–5-carpellate fruits has an air-filled cavity; the carpels are connate at the base and apex, so that expansion can only occur in the free middle part, causing the long slender fruiting carpels to curve strongly outward, forming a light-weight flotation dispersal unit (Endress et al. 1997). The genera of Rauvolfioids with fleshy fruits are primarily mammal- and/or bird-dispersed. In the New World, monkey and/or bird dispersal has been reported for *Couma*, *Lacmellea*, *Macoubea*, *Pacouria* and *Parahancornia*. *Geissospermum* (Aspidospermateae) is dispersed by spider monkeys and woolly monkeys (Defler and Defler 1996; Nevo et al. 2015), and *Lacmellea* by monkeys and kinkajous (Van Roosmalen 1985). Reports of animal dispersal in the Paleotropics have also mostly been for primates. In Africa *Cylindropsis*, *Dictyophleba*, *Landolphia*, *Hunteria* and *Tabernaemontana* are reported to be consumed and dispersed by cercopithecine monkeys (Gautier-Hion and Michaloud 1989; Astaras and Waltert 2010), in Africa and Madagascar *Carissa*, *Landolphia* and *Saba* are reported to be consumed and dispersed by a number of primate species including lemurs and chimpanzees (Wrangham and Waterman 1983; Birkinshaw 2001; Britt and Iambana 2003) and at least one species of *Petchia* is dispersed by lemurs (Razafindratsima et al. 2014); *Picralima* is dispersed by various sorts of rodents as well as elephants (Gautier-Hion et al. 1985; Beaune et al. 2013). In Asia, orangutans eat and disperse *Willughbeia* (Leighton 1993) and long-tailed macaques are reported to disperse *Leuconotis* and *Willughbeia* (Corlett and Lucas 1990). In the dehiscent-fruited Tabernaemontaneae with arillate seeds, aril color ranges from red to orange, through white to translucent-white (Leeuwenberg 1991, 1994b). Birds are probably the primary dispersers of the seeds with red or orange arils. This has been demonstrated in *Tabernaemontana* in Central America (McDiarmid 1977). At least in the New World both birds and monkeys often disperse the same species of *Tabernaemontana* (Cant 1979; Van Roosmalen 1985). It is also likely that birds are the primary seed dispersers of genera with brightly colored fruits on low shrubs or small twiners such as *Carissa*, *Alyxia* and *Petchia*.

Phytochemistry. Apocynaceae are well known as a family that contains poisonous and bioactive compounds; these evolved as a response to stress encountered by the plants (Sabir et al. 2016)—for example, as defense against herbivores—and diversified into a broad spectrum of different secondary metabolites (Hegnauer 1970, 1989). In Rauvolfioids, taxa in the tribes Aspidospermateae, Alstonieae, Vinceae, Tabernaemontaneae, Amsonieae, Melodineae and Hunterieae produce a vast array of monoterpenoid indole alkaloids (Bisset 1958, 1961; Hegnauer 1964, 1989; Taylor and Farnsworth 1975; Kisakürek et al. 1983; Van Beck and Van Gessel 1988; Zhu et al. 1990; Van der Heijden et al. 2004), a plesiomorphic feature they share with some other families of Gentianales, such as Rubiaceae, Gelsemiaceae and Loganiaceae (Kisakürek et al. 1983; Sabir et al. 2016). A single species often contains a great number of indole alkaloids; for example, more than 100 indole alkaloids have been isolated from *Amsonia stricta* (Gilani et al. 2007), and some 200 from *Rauvolfia serpentina* (Pathania et al. 2015). In Willughbeieae, indole alkaloids are apparently restricted to the two genera of subtribe Leuconotidinae, *Bousigonia* and *Leuconotis* (Goh et al. 1989; Abe and Yamauchi 1994; Fu et al. 2012, 2014). In Rauvolfioid tribes Carisseae, Plumerieae and Alyxieae indole alkaloids have been lost and the role of protection against predators has been taken over by other secondary metabolites (Endress et al. 1990, 2007a). Plumerieae contain iridoid glycosides (e.g., *Allamanda*, *Plumeria*) or cardenolides (e.g., *Cerbera*, *Cerberiopsis*, *Thevetia*; Hegnauer 1964, 1970, 1989; Coppen and Cobb 1983; Jensen 1992). *Acokanthera* (Carisseae) is rich in cardenolide glycosides (Hegnauer 1970; Kingston and Reichstein 1974). In Alyxieae no cardenolides or iridoid glycosides have been found; however, coumarins have been isolated from *Alyxia*, *Lepinia* and *Lepiniopsis* (Johns et al. 1968; Hegnauer 1970, 1989). In Apocynoids, the early-branching tribes Wrightieae and Malouetieae are characterized by steroidal alkaloids (e.g., *Funtumia*, *Holarrhena*, *Kibatalia* and *Malouetia*; Endress et al. 1990; Bisset 1992), and Nerieae are characterized by cardenolides and/or steroidal alkaloids (e.g., *Adenium*, *Nerium*, *Strophanthus*; Bisset 1987; Yamauchi and Abe 1990). In the higher Apocynoids (comprising Odontadenieae, Mesechiteae, Apocyneae, Rhabdadenieae, Echiteae and Baisseeae), characteristic secondary compounds in the first three tribes seem to be mainly cardenolides (Hegnauer 1964; Abe and Yamauchi 1985, 1989; Yamauchi et al. 1990; Burrows and Tyrl 2013), though steroidal alkaloids have been reported in *Amalocalyx* and *Chonemorpha* (Apocyneae; Hegnauer 1964; Endress et al. 1990; Hu et al. 1992), but detailed analyses are still lacking for a number of genera. In Echiteae, parsonsine type pyrrolizidine alkaloids (Colegate et al. 2016) are found in *Echites*, *Parsonsia*, *Peltastes*, *Prestonia* and *Temnadenia*, suggesting that parsonsine type pyrrolizidine alkaloids may have replaced cardenolides as the characteristic secondary metabolites in the tribe (Morales et al. 2017a). A number of genera in Echiteae are an important source of pyrrolizidine alkaloids for ithomiine and danainae butterflies and arctiid moths, which have evolved a tolerance to these toxins and use them as a defense against predators (Boppré 1990). This has been best studied in *Parsonsia* and *Prestonia* (Edgar 1984; Trigo and Brown 1990). Certain lineages in both groups of these insects have evolved a dependency on pyrrolizidine alkaloids as the basis for the production of mating pheromones (Boppré and Schneider 1985; Boppré 1995; Burzynski et al. 2015). In Baisseeae spermidine alkaloids have been isolated from *Oncinotis* (Hegnauer 1989).

In Periplocoideae, Secamonoideae and Asclepiadoideae cardenolides are widespread. In Asclepiadoideae, they are found predominantly in *Asclepias* and its relatives (Abisch and Reichstein 1962). Coevolutionary relationships between the monarch butterfly *Danaus* and *Asclepias* (e.g., Morse 1985) and between the milkweed beetle *Tetraopes* (Coleoptera: Cerambycidae) and *Asclepias* (Farrell and Mitter 1998) have been studied. In both Periplocoideae and Asclepiadoideae, steroidal glycosides are the most important components. Glycosides with pregnane skeletons, pregnane-12-esters and pregnane-12,20-esters are widely distributed in Marsdenieae, Ceropegieae (*Caralluma* s.l.) and Asclepiadeae. The occurrence of 14,15-seco-pregnane skeletons and 13,14:14,15-diseco-pregnane skeletons is rare and was the first hint to demonstrate the close relationship between *Tylophora* and *Vincetoxicum* (Liede 1996b). Steroidal alkaloids were found in *Cryptolepis* (Periplocoideae; Paulo et al. 2000), *Marsdenia* (Summons et al. 1972) and *Cynanchum* sect. *Rhodostegiella* (Lee et al. 2000). From *Cryptolepis*, indoloquinoline alkaloids are also reported (Sharaf et al. 1996). A furopyridine, Ceropegin, was isolated from *Ceropegia juncea* (Sukumar et al. 1995) and confirmed for *Ceropegia pusilla* (Kalimuthu and Prabakaran 2013). Hordenin, an alkaloid of the phenethylamine class, was found in *Stapelia gigantea* (Meyer et al. 1981). Phenanthroindolizidine alkaloids are found in *Telosma* and *Vincetoxicum* (Govindchari 1967). Salicylic acid is found in *Dischidia* (Chen et al. 1993) and 4-methoxy-salicylaldehyde in several Periplocoideae genera (Hegnauer 1964).

Subdivision and Relationships Within the Family. Robert Brown (1810) split the Asclepiadaceae from Jussieu's (1789) "Apocinae" based on what he called their "essential character": pollen shed in tetrads or fused into pollinia attached to a translator (Endress 2001). For most of the period since then until the mid-1990s the Apocynaceae and Asclepiadaceae have been recognized as two separate families. It has long been known, however, that in a number of morphological characters many genera of Apocynaceae sensu stricto are more similar to Asclepiadaceae than to other Apocynaceae. With a few exceptions, those Apocynaceae with dextrorse corolla lobe aestivation, presence of a gynostegium, apocarpous gynoecium, seeds with a coma, and secondary compounds steroidal alkaloids or cardenolides (never indole alkaloids) are more similar to Asclepiadaceae than to those Apocynaceae with sinistrorse corolla lobe aestivation, lacking a gynostegium, seeds without a coma, and secondary compounds indole alkaloids, sometimes iridoids or cardenolides (Endress and Bruyns 2000). Nevertheless, most traditional treatments recognized Apocynaceae and Asclepiadaceae as two very closely related but separate families with Apocynaceae comprising the two subfamilies Rauvolfioideae and Apocynoideae, and Asclepiadaceae comprising the three subfamilies Periplocoideae, Secamonoideae and Asclepiadoideae. Because their floral structure and pollination mechanism differ significantly from those of Secamonoideae and Asclepiadoideae, Schlechter (1905) raised Periplocoideae to family rank, and up until at least the mid-1990s Periplocaceae were still sometimes recognized as a distinct family (e.g., Nilsson et al. 1993; Swarupanandan et al. 1996).

One of earliest proposals for a unified Apocynaceae was by Judd et al. (1994), who discussed the "family pair" Apocynaceae/Asclepiadaceae in a phylogenetic context using non-molecular characters and clearly showed that Apocynaceae s. str. are paraphyletic unless Asclepiadaceae are included. The advent of molecular phylogenetics, with its emphasis on recognizing only monophyletic groups, dramatically changed the way we do systematics. These corroborated that Asclepiadaceae are firmly embedded with strong support in Apocynoideae, rendering Apocynaceae paraphyletic; furthermore, the three major lineages recognized in the traditional Asclepiadaceae since Brown (1810) do not come out together in Apocynoideae. Rather than Periplocoideae, *Baissea* (Baisseeae, Apocynoideae) is supported as sister to Secamonoideae + Asclepiadoideae. In larger molecular analyses Baisseeae is strongly supported as sister to Secamonoideae + Asclepiadoideae, whereas Periplocoideae, which are undisputedly monophyletic (Ionta and Judd 2007), are one of five main clades in an unresolved polytomy comprising the crown clade of Apocynaceae (together with Rhabdadenieae, Apocyneae, Mesechiteae + Odontadenieae + Echiteae, and Baisseeae + Secamonoideae + Asclepiadoideae) (Lahaye et al. 2007; Livshultz et al. 2007; Livshultz 2010; Straub et al. 2014). The relationship of Periplocoideae to the other crown clade Apocynaceae remains one of the most enigmatic systematic puzzles in the family. There are various solutions in order to make Apocynaceae monophyletic (Civeyrel et al. 1998), but the simplest is to unite Apocynaceae and Asclepiadaceae into a single family, as was done by Endress and Bruyns (2000).

The two main groups of the traditional Apocynaceae were described as the subfamilies Plumerioideae and Echitoideae for the first time by Schumann (1895), who recognized three tribes and six subtribes in Plumerioideae, based mainly on fruit and seed characters: Arduineae (including Melodininae and Landolphiinae), Pleiocarpeae, Plumerieae (including Alstoniinae, Tabernaemontaninae, Rauwolfiinae and Cerberinae), whereas Echitoideae were divided into two tribes (Echiteae and Parsonsieae) based on whether the anthers were included or exserted.

The next person to make significant contributions to our understanding of Apocynaceae s. str. as a whole was Pichon who, between 1948–1950, published a number of seminal papers (for a more detailed discussion, see Endress and Bruyns 2000, pp. 5–7). In addition to Plumerioideae and Echitoideae, Pichon recognized two further subfamilies: Tabernaemontanoideae (Pichon 1949), which corresponded to Schumann's subtribe Tabernaemontaninae, and Cerberoideae (Pichon 1948b), which contained half of the genera in Schumann's Cerberinae, together with *Skytanthus* and *Cameraria* (from Schumann's Alstoniinae). In addition to splitting and redefining Schumann's Cerberinae, Pichon also, for example, recognized that Schumann's Pleiocarpeae contained taxa of two unrelated groups, which had been brought together only because they had more than two carpels. He recognized that those with drupaceous fruits were related to *Alyxia* rather than *Pleiocarpa*, and split and redefined the group accordingly. Similarly, he recognized that *Ambelania*, *Allamanda*, and *Craspidospermum* were out of place in Schumann's Arduineae, and moved them to more plausible positions in other tribes. Pichon's Plumerioideae contained six tribes: Allamandeae, Alstonieae, Ambelanieae, Carisseae, Chilocarpeae and Rauvolfieae (Pichon 1948a, 1949, 1950b). Overall, Pichon's classification was an improvement over that of Schumann's due to more careful observation of characters. But, like Schumann, Pichon also used mainly fruit and seed characters to delimit tribes in his Plumerioideae and thus overlooked more subtle convergences, with the result that most of his tribes also contained at least one or two genera placed elsewhere in recent phylogenetic studies.

Pichon's classification of the Echitoideae (Pichon 1950a) completely abandoned Schumann's two tribes and instead divided the subfamily into four tribes based on the manner in which the anthers are attached to the style-head, a feature he called the "retinacle" (Pichon 1948c).

The last classification of Apocynaceae s. str. published before the molecular revolution was that by Leeuwenberg (1994a). His Plumerioideae differed relatively little from that of Pichon; the main difference being that Pichon's Tabernaemontanoideae and Cerberoideae were again reduced to tribes of Plumerioideae and *Macoubea* was placed in its own tribe, Macoubeeae. Further, application of the principle of priority (McNeill et al. 2012) resulted in name changes for one subfamily and two tribes: Echitoideae became Apocynoideae, Alstonieae became Plumerieae, and Rauvolfieae became Alyxieae. Leeuwenberg's Plumerioideae thus comprised the nine tribes Allamandeae, Alyxieae, Ambelanieae, Carisseae, Cerbereae, Chilocarpeae, Macoubeeae, Plumerieae and Tabernaemontaneae. Leeuwenberg divided the genera of Apocynoideae into three tribes: Echiteae and Wrightieae (which corresponded for the most part to Pichon's Parsonsieae and Nerieae, respectively), and Apocyneae (which contained elements of Pichon's Ecdysanthereae and Ichnocarpeae).

The first system to classify all genera of the Apocynaceae into subfamilies and tribes after inclusion of Asclepiadaceae, and which incorporated molecular data, was that of Endress and Bruyns (2000). In that treatment, further name changes following the Priority Principle were made for one subfamily and one tribe: Plumerioideae became Rauvolfioideae and Rauvolfieae became Vinceae. They also recognized nine tribes in Rauvolfioideae (Alstonieae, Alyxieae,

Carisseae, Hunterieae, Melodineae, Plumerieae, Tabernaemontaneae, Vinceae and Willughbeieae), but their compositions differed significantly from those of Leeuwenberg (1994a). They recognized five tribes in Apocynoideae: Apocyneae, Echiteae, Malouetieae, Mesechiteae and Wrightieae, and again the composition differed from that of Leeuwenberg (1994a).

Since 2000, two more classifications have been published. The first was by Endress et al. (2007b) and incorporated results from publications published between 2001 and 2007, in particular larger phylogenetic analyses (e.g., Simões et al. 2004, 2006, 2007; Livshultz et al. 2007). The last update to the classification was that of Endress et al. (2014), which took into consideration larger phylogenetic studies published after 2007 (e.g., Simões et al. 2010; Middleton and Livshultz 2012). Since 2014 two further tribal level phylogenetic studies providing insights into relationships in Apocynaceae s. str. have been published: Simões et al. (2016) and Morales et al. (2017a).

The first molecular-based study of Periplocoideae was conducted by Meve and Liede (2004a), who investigated relationships among tuberous genera from Africa and Madagascar. This was followed by a large-scale molecular phylogenetic study by Ionta and Judd (2007), which supported Periplocoideae as monophyletic, but did not attempt to determine its position within the family.

The question whether or not Secamonoideae should be included in Asclepiadoideae had been raised by Robert Brown (1810), who distinguished *Secamone* as an unnamed group from the "Asclepiadeae verae". Molecular results (Civeyrel et al. 1998; Sennblad and Bremer 2000; Livshultz et al. 2007) firmly place Secamonoideae as sister to Asclepiadoideae, but they do not answer the question about the appropriate rank for the group. Based on the current phylogeny, it could be treated as either a subfamily or as a tribe within Asclepiadoideae. Here we recognize Secamonoideae as a distinct subfamily.

Asclepiadoideae were divided by Endlicher (1838) into three tribes, depending on the orientation of the pollinia, "Cynancheae" (pendulous), "Gonolobeae" (horizontal) and "Pergularieae" (erect). "Pergularieae" were divided in two subtribes, the "Hoyeae" and the "Stapelieae", with and without membranous connective appendages, respectively. Decaisne (1844) raised Gonolobeae and Stapelieae to tribal status, and Bentham (1876) divided Asclepiadoideae and Secamonoideae into a total of six tribes. Schumann (1895) and later Wagenitz (1964) essentially followed the system of Bentham (1876). Rosatti (1989) gave a detailed overview of the classification history of Periplocoideae, Secamonoideae and Asclepiadoideae. Molecular studies have largely confirmed tribal circumscriptions based on pollinarium characters, with Marsdenieae and Ceropegieae having erect and Asclepiadeae pendulous (to horizontal) pollinia. The two genera of Fockeeae with pollinia directly attached to the corpusculum were removed from Marsdenieae with well-developed caudicles and are sister to the remainder of Asclepiadoideae (Kunze et al. 1994; Verhoeven et al. 2003). Marsdenieae and Ceropegieae are sister groups, distinguished by the presence of a pellucid germination zone on the proximal margin or the apex of the pollinia and the absence of connective appendages (except for *Caudanthera*) in the latter. Thus, Marsdenieae could also be included within an expanded Ceropegieae, as was proposed by Swarupanandan et al. (1996). More recently, a fifth tribe, Eustegieae, was proposed for two small genera with a 3-seriate corona, pendent pollinia and somewhat ambiguous molecular signal, placing *Eustegia* in a sister group relationship either to Asclepiadeae (Surveswaran et al. 2014) or to Marsdenieae + Ceropegieae (Rapini et al. 2003).

Of the three larger Asclepiadoideae tribes, the further subdivision of Ceropegieae and Asclepiadeae has been confirmed by molecular methods. Four subtribes (Anisotominae, Heterostemminae, Leptadeniinae and Stapeliinae) have been found in Ceropegieae (Meve and Liede 2004b), and twelve in Asclepiadeae. Among the latter, Liede (1997), following Swarupanandan et al. (1996), included the Gonolobeae of Endlicher in Asclepiadeae as subtribe Gonolobinae, because they share true styles and essentially pendulous pollinia.

The classification here follows that of Endress et al. (2014), which was an update of Endress et al. (2007b) and included 364 genera divided among five subfamilies, 25 tribes and 49 subtribes, with the following differences: (1) both Rauvolfioideae and Apocynoideae have repeatedly been shown to be paraphyletic; the first is a grade, and the second has the three formal subfamilies of the traditional Asclepiadaceae nested within it (Sennblad and Bremer 1996, 2000; Sennblad et al. 1998; Potgieter and Albert 2001; Livshultz et al. 2007; Simões et al. 2007); therefore, here the two subfamilies of the traditional Apocynaceae are treated as informal ranks as proposed in Simões et al. (2016); (2) Echiteae are recircumscribed to comprise 14 genera following Morales et al. (2017a). The five major divisions of the family are thus Rauvolfioids, Apocynoids, Periplocoideae, Secamonoideae and Asclepiadoideae.

Biogeography. The age of the family is still a matter of debate. Rapini et al. (2007) calibrated the age of crown-group Apocynaceae at ca. 54 Ma, but Bell et al. (2010) suggested an age of only ca. 21 Ma. The most recent dating (Tank et al. 2015) gave a stem age for the family of 47 (43–62) Ma.

The fossil record of Apocynaceae is poor (Martínez-Millán 2010), but supports the higher estimates for family age. The oldest pollen of the *Alyxia* type is known from the Paleocene of Borneo (Muller 1968). Further records of *Alyxia* type pollen are known from the Eocene of Europe (Muller 1981). Pollen of the *Rauvolfia* type is known from the Upper Eocene and the Lower Oligocene (Muller 1981). A seed, *Cypselites*, resembling the seeds of some modern genera of Apocynoids, was retrieved from the Middle Eocene (ca. 47 Ma) Messel biota in Germany (Collinson et al. 2010, 2012). Several apocynaceous seeds have been retrieved from the Bembridge flora of England (Reid and Chandler 1926). In contrast, the Lower Eocene fossils from the London Clay cannot be assigned to Apocynaceae with certainty (Kirchheimer 1957).

For Periplocoideae, the oldest described fossil is *Polyporotetradites laevigatus* from the Oligocene and the Lower Miocene of Cameroon.

Recently, a distinctive pollinia-bearing fossil, *Discoflorus neotropicus*, has been retrieved from Dominican amber (dated 20–15 Ma or 45–30 Ma, depending on the dating method used; Poinar 2017). A pollinarium with two or even four pollinia is attached to the head of a termite (Termitidae: Isoptera) adjacent to the flower. The flower is somewhat similar to that of *Metastelma parviflorum*, but smaller and of a different fine structure.

The origin of the family is certainly tropical, with the first-branching tribe Aspidospermateae restricted to the New World, whereas the second-branching Alstonieae is restricted to the Asia-Pacific region (Simões et al. 2007). Studies focusing on broadscale biogeographic patterns in Rauvolfioids and Apocynoids are lacking, including attempts to assess the age of various clades. Biogeographic patterns can, however, be seen in well-sampled phylogenies of particular groups. In *Tabernaemontana*, for example, Simões et al. (2010) found that the first divergence in the genus is between neotropical and paleotropical clades. Within the paleotropical clade, the Madagascan species are sister to the African + Asian species. The only endemic species from New Caledonia sampled, *Tabernaemontana cerifera*, is nested in the Asian clade. Simões et al. (2016) found a similar divergence into neotropical and paleotropical clades in *Rauvolfia* but, unlike in *Tabernaemontana*, within the paleotropical clade the first divergence was between an Asian clade and an African + Madagascan clade. The Hawaiian *Rauvolfia sandwicensis* is nested in the neotropical clade and must be assumed to be the result of long-distance dispersal. Phylogenetic studies are lacking for other large Rauvolfioid genera such as *Alyxia*, which occurs in Asia, Australasia and the Pacific Islands, although Middleton (2002) speculated that patterns of variation in *Alyxia stellata*, a species widespread across the Pacific Islands, may in part be due to human influence.

Livshultz et al. (2007: their Fig. 1A) suggest an African origin for the entire APSA clade and for the "Crown clade" and "Malouetia clade", which comprise all taxa in the APSA clade except for Wrightieae and Nerieae. An African origin is also suggested for Wrightieae even though the highest diversity is now in Asia. Within the Crown clade, distribution patterns of Apocynoids are strongly aligned with phylogeny. Apocyneae is a clade of Asian origin, although *Apocynum* itself is distributed from temperate Asia to southeastern Europe and North America. A large American clade comprises the taxa now placed in Mesechiteae, Odontadenieae and Echiteae, the last of which also includes *Parsonsia* and its relatives, which are most diverse in Australasia but reach into continental Southeast Asia, suggesting Gondwanan vicariance and long-distance dispersal.

Goyder (2006) showed that southern Africa and Madagascar are key areas of diversity for the subfamilies Asclepiadoideae, Secamonoideae and Periplocoideae. This conclusion was also reached by Livshultz (2010), who showed that pollinia have evolved several times independently in Africa, in *Epistemma* and *Schlechterella* of Periplocoideae and in Secamonoideae + Asclepiadoideae. This African origin is documented by the restriction of the small relictual Fockeeae, Eustegieae, and Astephaninae to the continent. Both Ionta and Judd (2007) and Joubert et al. (2016) retrieved the Australian *Phyllanthera grayi* and the southern African *Pentopetia natalensis* as the first two branches of Periplocoid taxa, making it difficult to narrow down an area of origin for the group. For Secamonoideae, no comprehensive phylogeny that would allow a biogeographic interpretation is available. Asclepiadoideae are firmly rooted in Africa, with the first-branching Fockeeae ranging from the southern tip of the continent to the Arabian peninsula (Bruyns and Klak 2006). For the sister tribes Marsdenieae and Ceropegieae, the ancestral area lies in the Old World, but cannot be identified more precisely at present. For Ceropegieae, Meve et al. (2017) reconstructed Southeast Asia as the ancestral area, with subsequent dispersal and vicariance events via India and the Arabian peninsula to Africa, from where the most derived subtribe, Stapeliinae, radiated to Madagascar and back to India and Southeast Asia several times (Bruyns et al. 2015). The stem succulent Stapeliads expanded their area of occurrence twice across Africa, once from north to south and later again from south to north (Meve and Liede 2002a; Bruyns et al. 2014).

Asclepiadeae, again, are firmly rooted in Africa, both with their sister tribe Eustegieae and the first-branching Astephaninae, both relictual groups of southern Africa (Surveswaran et al. 2014). For Asclepiadinae, Cynanchinae, and Tylophorinae, the first-branching taxa are all African. However, beyond this fact, the biogeographic histories of the three subtribes are widely different. In Asclepiadinae, the New World taxa branch off early, as sister to a species-rich and yet poorly resolved eastern and southern African group. The role of *Asclepias* (*Trachycalymma*) *pseudofimbriata* from Ethiopia either as sister of both American *Asclepias* and the African group or unresolved with both (Chuba et al. 2017) needs further study, but points in the direction of a northern East African ancestral area for the group. In Cynanchinae, an early split separates the species-rich Madagascan lineage, out of which some succulent taxa have spread from Madagascar into Africa and Asia (*Cynanchum gerrardii* and former "*Sarcostemma*" species, see Meve and Liede 2002b) from the remainder of the genus. Two clades have radiated in Asia, one of them extending into Europe, the other one into Southeast Asia. The New World species of *Cynanchum* are monophyletic and sister to the northeastern African and Arabian "thick-fruited" clade (Khanum et al. 2016). In Tylophorinae, the most recent radiations have taken place in Asia (Liede-Schumann et al. 2012) and the subtribe has reached the New World only by human introduction (e.g., Sheeley and Raynal 1996). Shifts between tropical and temperate habitats, accompanied by a change in habit from twining evergreen lianas to rhizomatous perennials, have occurred at least twice (Liede-Schumann et al. 2016). While one of these temperate radiations is restricted to the Far East, a continuous westward expansion along the Asian mountain ranges into Europe can be reconstructed for the crown clade.

Biogeography in the second-branching Asclepiadeae clade, comprising all exclusively New World genera, is not yet completely understood. The first-branching New World genus, *Pentacyphus*, shows a probably relictual distribution along the Andes from southern Peru to Venezuela (Meve and Liede-Schumann 2015), while the second-branching genus, *Diplolepis*, is restricted to southwestern South America, reaching from northern Patagonia to northern Chile, with radiations in forest, near-desert and desert, and high elevations (Hechem et al. 2011a, b). *Monsanima*, sister to the remaining, widespread South American Orthosiinae, is relictual with two species restricted to a single Brazilian mountain top each. Ribeiro et al. (2014) date the origin of Metastelmatinae to the Late Miocene and suppose an origin most likely in open savanna-like vegetation. The biogeography of *Minaria*, the second-branching lineage in Metastelmatinae, provides an insight into the complex migrations and adaptations of an Asclepiadoid genus during climatic oscillations of the Pleistocene. Spatial, ecological and dating analyses suggest that the distribution of *Minaria* seems to be mainly governed by an overall retraction (larger ancestral distributions resulting in smaller, disjunct distributions), in parallel with a general trend of decreasing temperatures and increasing aridity through the Quaternary. Biogeographical analysis of *Metastelma*, one of the few Asclepiadoid genera with a center of distribution in Central America and the Caribbean, revealed that the colonization of Central America most likely did not take the Panamanian isthmus route, but went from northern South America via the Caribbean to the Yucatán isthmus of Tehuantepec area. From here, radiations northward along the Sierra Madre mountain massifs and southward to southern Central America occurred (Liede-Schumann et al. 2014).

In *Tweedia*, the first-branching genus of the Oxypetalinae, two clades were retrieved, one from Argentina, the other one from Chile, west of the Andes, comprising the species from a Mediterranean climate. Because the Mediterranean climate was only established in the late Neogene, dispersal across the Andean barrier and the invasion of new habitats that provided novel opportunities for speciation is likely for *Tweedia* (Calviño et al. 2014).

Affinities. Apocynaceae (often as two families, Apocynaceae and Asclepiadaceae) has typically been placed within the Gentianales, together with Gentianaceae, Loganiaceae and Rubiaceae. In the 1980s to 1990s Saccifoliaceae was also included in the order (Thorne 1992; Nicholas and Baijnath 1994), but is now included in Gentianaceae (Thiv et al. 2000). Sometimes Retziaceae (Cronquist 1981) was also included, a family that is currently placed in Stilbaceae (Lamiales). Based on more recent molecular evidence Gentianales comprises five families: Apocynaceae, Gelsemiaceae, Gentianaceae, Loganiaceae and Rubiaceae (Backlund et al. 2000; APG IV 2016). Within Gentianales, Rubiaceae is usually considered the most different, and is often the first-branching family, followed by Gentianaceae. The closest relatives of Apocynaceae seem to be either Loganiaceae or Gelsemiaceae, depending on which species are sampled and what type of characters are analyzed (Struwe et al. 1994; Endress et al. 1996; Bremer et al. 1999; Backlund et al. 2000; Yang et al. 2016).

Distribution and Habitat. Apocynaceae are a predominantly pantropical family, with relatively few genera ranging north or south into temperate regions. For Rauvolfioids and Apocynoids the main centers of diversity are tropical South America, Asia and, to a lesser extent, tropical Africa, although notable radiations of particular genera have also taken place on islands (e.g., *Parsonsia* on Australia and on New Guinea, *Alyxia* on New Caledonia and on New Guinea and *Petchia* on Madagascar). Two subfamilies, Periplocoideae and Secamonoideae, are restricted to the Paleotropics (extending to Australia).

Middleton (2007) mapped the numbers of genera, species and endemic species in each of the regions/islands of Malesia for Rauvolfioids and Apocynoids. Species diversity was, unsurprisingly, shown to be highest in the everwet regions of Peninsular Malaysia, Borneo and New Guinea and lower in the more seasonal areas of the Lesser Sunda Islands and Philippines. The relatively lower than expected diversity on Sumatra could be an artefact of lack of exploration. Generic diversity declined from West to East. Endemism was by far the highest in New Guinea but was also high in the Philippines.

Good (1947 and later editions, 1952) mapped the distributions of many taxa in the subfamilies Asclepiadoideae, Secamonoideae and Periplocoideae, and concluded that southern Africa and Madagascar are key areas of diversity for these groups. While diversity in southern Africa is made up of a large number of genera, diversity in Madagascar is largely due to a few genera with species-rich radiations, such as *Pentopetia* in Periplocoideae (Klackenberg 1999, ca. 20 species), *Secamone* in Secamonoideae (ca. 60 species, Klackenberg 1992; Lahaye et al. 2005) and *Cynanchum* in Asclepiadoideae (ca. 100 species, Liede-Schumann unpubl. data).

Asclepiadoideae, Rauvolfioids and Apocynoids are found in both the Paleotropics and Neotropics though many tribes tend to be predominantly in either one or the other. For example, in Rauvolfioids all genera of Aspidospermateae are New World, whereas Alstonieae, Melodineae, Hunterieae and Carisseae are restricted to Africa and the Asia-Pacific region, as is Alyxieae except for *Condylocarpon*. *Rauvolfia* and *Tabernaemontana* are pantropical although clades within these two genera tend to be exclusively paleotropical or neotropical. In Apocynoids the two earliest branching tribes, Wrightieae and Nerieae, are found in Eurasia and the Paleotropics, and the next, Malouetieae, is both paleo- and neotropical. Apocyneae are also restricted to the Paleotropics (most of them in Asia) except for *Apocynum*, which has a disjunct distribution between temperate Eurasia and North America (Rosatti 1989); Baisseeae are restricted to tropical Africa and Madagascar (De Kruif 1983). Conversely, Rhabdadenieae, Mesechiteae and Odontadenieae are restricted to the New World, as are Echiteae with the exception of *Artia*, *Ecua* and *Parsonsia*.

In Asclepiadoideae, only Marsdenieae and Asclepiadeae have representatives in the New World, and only three genera are presently considered to have representatives in both hemispheres, *Cynanchum* (Liede and Täuber 2002; Khanum et al. 2016), *Asclepias* and *Marsdenia*. The species-rich (> 1050 species) clade of Pentacyphinae, Diplolepinae, Orthosiinae, Metastelmatinae, Tassadiinae, Oxypetalinae, Gonolobinae and Topeinae is restricted to the New World (Rapini et al. 2003; Liede-Schumann et al. 2005).

In Rauvolfioids and Apocynoids, the largest genera are *Mandevilla* (ca. 170 species), *Tabernaemontana* (ca. 122 species), *Alyxia* (ca. 106 species), *Parsonsia* (ca. 82 species), *Rauvolfia* (ca. 60 species), *Prestonia* (ca. 58 species), *Aspidosperma* (ca. 50 species) and *Alstonia* (ca. 45 species) (Middleton 2007, 2014; Pereira et al. 2016, 2017; Morales and Zamora 2017; Morales et al. 2017b). In Periplocoideae, *Raphionacme* is the largest genus (36 species, Venter 2009), in Secamonoideae, *Secamone* s.str. comprises ca. 90 species (Klackenberg 2001). In Asclepiadoideae, *Hoya* (ca. 350–450 species) is by far the largest genus in Marsdenieae (Rodda 2015), and *Ceropegia* (ca. 330 species) in Ceropegieae. In Asclepiadeae, *Cynanchum* (Cynanchinae, ca. 250 species, Meve and Liede-Schumann 2017), *Vincetoxicum* (Tylophorinae, ca. 140 species, Liede-Schumann et al. 2016) and *Asclepias* (ca. 100 species) are the most species-rich genera. In the New World Clade, *Gonolobus* (Gonolobinae, 120–140 species) and *Oxypetalum* (Oxypetalinae, ca. 130 species, Farinaccio and Mello-Silva 2006) are the most species-rich genera since the former largest genus *Matelea* (Gonolobinae) has now been split into various smaller genera (Morillo 2012, 2013, 2015).

In general, Rauvolfioids and Apocynoids are most diverse in humid lowland tropical habitats, although some species are found at high elevations. In lowland evergreen forest in Asia some of the largest trees belong to *Dyera* and *Alstonia* (Middleton 2007). Other tree genera, such as *Kibatalia*, *Kopsia* and *Wrightia* are part of the forest understory (Middleton 2007). *Alstonia* is particularly diverse in habit and habitat with trees over 60 m tall in lowland forest, small understory trees in lowland and montane forest, small and large trees in swamp forest, and shrubs on karst limestone (Sidiyasa 1998). Many of the lianas, particularly in genera such as *Chilocarpus*, *Leuconotis*, *Melodinus*, *Micrechites* and *Willughbeia*, can grow to enormous heights and reach the canopy in tropical lowland rainforest. Some genera (e.g., *Adenium*, *Pachypodium* and some species of *Alyxia*, *Aspidosperma*, *Himatanthus* and *Mandevilla*) have radiated into drier habitats, others into swamp forest and then often developing pneumatophores to survive the waterlogged soils (some species of *Alstonia* and *Dyera*), and others are littoral or on the edge of mangrove (e.g., *Cerbera*, *Lepinia* and *Ochrosia*). In the case of *Mandevilla*, most of the species at the extremes of its distribution, in the deserts of Mexico and the southwestern USA to the North, and in the open cerrado vegetation of Brazil, Argentina and Paraguay to the South, have a water-storing xylopod. Of the handful of genera that are restricted to temperate climates, both *Amsonia* and *Apocynum* are perennial herbs with a disjunct distribution in temperate Eurasia and North America, *Haplophyton* is restricted to dry desert habitats in Mexico and the extreme southwestern USA, and *Cycladenia* is a dwarf, subsucculent perennial herb, restricted to a few mountain peaks in California, Arizona and Utah. In addition, in the Old World, *Vinca* and *Nerium* occur in temperate to subtropical climates, respectively. They remain evergreen in temperate climates. *Malouetia*, which has an African-Neotropical disjunct distribution, has several species in Amazonia that are adapted to the seasonally flooded forest (mostly blackwater rivers) and/or nutrient-poor white sand soils, and are able to tolerate long periods of inundation. Another white sand specialist is the beautiful *Galactophora*, which is found in the white sand savannas of the Guiana Shield. Periplocoideae, Secamonoideae and Asclepiadoideae have also been successful in expanding and diversifying into drier or cooler habitats, and to an extent that surpasses the outer limits of the adaptability of Rauvolfioids and Apocynoids. Periplocoideae and Asclepiadoideae both have members with special adaptations to arid habitats such as tubers (e.g., *Raphionacme* (Periplocoideae), Ceropegieae) or succulent stems as in many Ceropegieae and about half of the Malagasy *Cynanchum* species. Ceropegieae have their highest diversity in Africa, the species-rich stem succulent Stapeliads and *Ceropegia* have centers of distribution in East Africa, South Africa and India. Marsdenieae are most diverse in Asia and include many species that are truly epiphytic, a habit unknown in the Rauvolfioids, Apocynoids, and Secamonoideae. In Asclepiadoideae, *Vincetoxicum* has radiated in Asia and Europe, reaching as far north as southern Sweden. In the New World, members of *Asclepias* reach southern Canada (Woodson 1954), and *Diplolepis* extends south to Patagonia (Hechem et al. 2011a, b).

Asclepiadeae in Africa often have storage roots, and are typically confined to grasslands. The majority of Asclepiadoideae species, however, are leafy twiners and lianas in the tropics and subtropics, where they inhabit different kinds of forests and scrub, often in slightly disturbed habitats. In lowland rainforests of Africa and the Americas, however, they are rare, whereas they are most diverse in this habitat in Southeast Asia. Abundances are usually low with single scattered plants and without forming thickets or smothering other plants.

Uses and Economic Importance. Due to the vast array of bioactive compounds in the family, numerous genera of Apocynaceae figure prominently in folk medicines. One genus is often used to treat a broad spectrum of diverse ailments such as fever, malaria, jaundice, cancer, hypertension, inflammation, pain, diarrhea, skin diseases, gastrointestinal problems, as a purgative, to promote healing of wounds, etc. (e.g., *Allamanda, Alstonia, Alyxia, Amsonia, Aspidosperma, Catharanthus, Cerbera, Kibatalia, Melodinus, Ochrosia, Picralima, Rauvolfia, Strophanthus, Tabernaemontana, Thevetia, Vallaris, Vinca, Voacanga*; Bisset 1958, 1961, 1989, 1991, 1992; Schultes 1979; Van Beck et al. 1984; Hutchings 1989; Neuwinger 1994a; Metzner 1998; Hendrian 2001a, b; Rahayu 2001; Rudjiman 2001; Teo 2001; Tran 2001; Chua and Horsten 2001; Sangat-Roemantyo and Middleton 2001; Van Valkenburg and Hendrian 2001; Van Valkenburg and Horsten 2001; Gilani et al. 2007; Wong et al. 2013). Indeed, the names of several species emphasize their medicinal use, such as *Wrightia antidysenterica, Rauvolfia vomitoria* and *Allamanda cathartica*. The toxic latex of some of the genera that contain cardenolides or steroidal alkaloids has long been used to make arrow poisons by indigenous peoples in both Africa and South America; the three most widely used are *Acokanthera* (Watt and Breyer-Brandwijk 1962), *Strophanthus* (Beentje 1982) and *Malouetia* (Schultes 1979; Bisset 1992). *Anisopus* (Neuwinger 1994b), *Marsdenia* and *Vincetoxicum* (*Tylophora*) species are used as canine and rat poisons. In southern Africa and Australia, several Asclepiadoideae species are serious toxic weeds for cattle and sheep (Watt and Breyer-Brandwijk 1962; Hall 1964; Everist 1981). The cardenolide-containing and commonly planted ornamentals *Cerbera, Nerium* and *Thevetia* have been involved in human poisoning (Gaillard et al. 2004; Bandara et al. 2010). Hallucinogenic drinks are made from *Tabernanthe* in Africa (Bisset 1958) and *Malouetia* in Amazonia (Schultes 1979). An important role in traditional medicine has been reported for *Mondia whitei* (Periplocoideae), including use as an antidepressant and aphrodisiac (Aremu et al. 2011). Many Periplocoideae and Asclepiadoideae species are used in folk medicine in China (e.g., Hong et al. 2015; Brand et al. 2017; Chang et al. 2017), tropical Asia (Perry 1980; Kiew 1994, 2001), Africa (Burkill 1985; Hutchings 1989), and Central America (Williams 1981). *Calotropis* is still the most important medicinal plant in Asclepiadoideae, with a variety of uses across its distribution area (Neuwinger 1994a). Several species of *Asclepias* are used medicinally because of their cardenolide glycosides (Usher 1974). Many species of *Vincetoxicum* contain phenanthroindolizidine alkaloids and are used in folk medicine (Schultes and Raffauf 1990). Members of *Cynanchum* sect. *Rhodostegiella* have anti-epileptic properties (Mu et al. 1986). *Hoodia* contains appetite-suppressing pregnane glycosides, which were expected to become a drug of high economic value (Van Heerden 2008), but market introduction failed. Recently, also *Caralluma adscendens* s.l. has been identified as an appetite suppressant and is offered as a dietary supplement in India and online (Dutt et al. 2012). *Gymnema sylvestre* (Asclepiadoideae–Marsdenieae) contains gymnemosides (triterpene glycosides, Yoshikawa et al. 1997), acting as powerful sweetness inhibitors (Suttisri et al. 1995).

Few genera of Apocynaceae are of broadscale economic importance in Western medicine. In the last century, pharmacognosists or ethnopharmacologists (often working for pharmaceutical companies) made excursions to the tropics to observe which plants were being used by indigenous peoples (often performing simple tests in the field to determine presence or absence of alkaloids or other secondary metabolites), and in this way brought the first apocynaceous drugs to the Western world. One of the first was extracted from *Strophanthus* seeds, which are used to make arrow or dart poison in various parts of Africa; strophanthin was administered as a cardiac stimulant to treat heart failure. By 1970, several tons of *Strophanthus* seeds were being imported to Europe per year (Beentje 1982). Today strophanthin is still sometimes prescribed as an alternative treatment for heart disease (Bisset 1991). Ibogaine, extracted from *Tabernanthe iboga*, is used in private clinics in Europe and North America to treat opioid withdrawal symptoms (Alper et al. 2008). Today, compounds of plant origin still provide the template for many synthetically produced drugs (De Luca et al. 2012). A number of genera show promise as anticancer or antimalarial drugs. Two genera of Vinceae (Rauvolfioids) have been the source of drugs based on indole alkaloids that have become commercially available: *Catharanthus roseus* is the source of vinblastine and vincristine, which are used to treat leukemia and other types of cancer (Taylor and Farnsworth 1975; Wong et al. 2013), and *Rauvolfia serpentina* is the source of reserpine, which has long been used in traditional and Ayurveda medicines and was earlier used in Western medicine for treating hypertension and mental illness. *Rauvolfia serpentina* has also recently shown great promise as a drug against diabetes (Pathania et al. 2013). *Hemidesmus indicus* (Indian Sarsaparilla; Periplocoideae) is widely used in India in Ayurveda medicine and is used to treat a variety of diseases and in cosmetics. The alkaloids of *Cryptolepis sanguinolenta* (Periplocoideae) are known for their hypoglycemic properties (Bierer et al. 1998) and are being tested for tumoristatic effects (Bonjean et al. 1998); Buhner (2012) reported on their considerable activity against multi-resistant bacteria. Both *Calotropis procera* in Africa and Arabia (Neuwinger 1994a) and *Calotropis gigantea* on the Indian subcontinent (Kumar et al. 2013) are widely used as poison and in folk medicine.

Apocynaceae contain too many toxic secondary compounds to be an economically important food source, although local uses are reflected in several plant names such as *Carissa edulis* and *Willughbeia edulis*. Most genera of Willughbeieae have edible fruits, which are consumed by people in the regions where they grow and are sometimes sold in local markets (e.g., several species of *Landolphia*, *Hancornia*). Fruits of *Carissa* can be used to make jam. The latex of *Couma* makes a refreshing drink and that of *Dyera* was once an important source of latex for chewing gum. The flowers of *Echites pandurata* (Apocynoids) are popular as a condiment in cooking in parts of Central America (Morton et al. 1990), and flower buds of *Telosma cordata* are used similarly in parts of Asia. Green fruits of *Gonolobus edulis* and related species are consumed in a sugared preparation called "guayote" in Costa Rica. The young fruits, seeds and sometimes leaves of *Telectadium edule* (Periplocoideae), *Morrenia odorata* (Arenas 1999) and *Cynanchum* (*Glossonema*) *boveanum* (Asclepiadoideae-Asclepiadeae, Burkill 1985) are cooked and eaten in a variety of ways. The raw or cooked stems of *Caudanthera edulis* (Stapeliinae) are popular as a vegetable in southeastern Africa and parts of the Indian subcontinent.

A number of genera are cultivated as ornamentals, particularly in tropical and subtropical climates (e.g., *Adenium*, *Allamanda*, *Amsonia*, *Araujia*, *Asclepias*, *Beaumontia*, *Catharanthus*, *Cerbera*, *Chonemorpha*, *Cryptostegia*, *Dischidia*, *Hoya*, *Mandevilla*, *Nerium*, *Pachypodium*, *Pergularia*, *Periploca*, *Plumeria*, *Tabernaemontana*, *Telosma*, *Thevetia*, *Wrightia* and *Vinca*). Many genera of Ceropegieae are cultivated worldwide by enthusiasts of succulent plants for their beautiful and often bizarre flowers.

Some species are used in religious and other traditional ceremonies. *Wrightia religiosa* and some species of *Plumeria* are commonly planted in temple and burial grounds in Southeast Asia. *Alyxia* species are used in the making of leis in Polynesian cultures. The hard endocarps of *Thevetia peruviana* and *T. thevetioides* are strung together to make folk instruments such as rattles, which can be hand-held, collected into bracelets or worn as "ayoyotes", the traditional leg shakers used by Aztec dancers (Alvarado-Cárdenas et al. 2017).

The wood of *Alstonia* and *Aspidosperma* is used for construction, especially for cabinetmaking, furniture, floors, tool handles and carving (Ezcurra et al. 1992; Middleton 2007). Before *Hevea* dominated the rubber industry, the latex of Apocynaceae was used, particularly various genera of Willughbeieae (certain species of *Landolphia* and *Willughbeia*). During World War II *Cryptostegia grandiflora* (Periplocoideae), a native of Madagascar, was introduced into several other countries as a source of rubber. It has since become an aggressive, invasive weed, and has greatly impacted the ecology where it has been introduced (McFadyen and Harvey 1990; Rodríguez-Estrella et al. 2010). *Araujia sericifera* and *Morrenia odorata* (Asclepiadoideae-Asclepiadeae-Oxypetalinae) are known as noxious weeds in *Citrus* groves in California, Florida, Southern Africa, Australia, New Zealand, Israel, Italy, and Spain (Spellman and Gunn 1976). For *Morrenia odorata* a mycoherbicide based on an isolate of *Phytophthora palmivora* is available (Winks and Fowler 2000), for *Araujia sericifera*, the *Araujia mosaic virus* is discussed as a potential biological control agent (Winks and Fowler 2000). Three species of *Vincetoxicum* (Asclepiadoideae-Asclepiadeae-Tylophorinae), *V. hirundinaria*, *V. nigrum* and *V. rossicum*, have been introduced into North America, and two of them are invasive in natural areas. *Hypena opulenta* (Lepidoptera: Erebidae) has been tested as an efficient biological control (Young and Weed 2014).

Conservation. Apocynaceae, being predominantly tropical, are threatened mainly by the destruction of their natural habitats. Some of the more showy species, such as in *Pachypodium*, and those used in traditional medicines, such as *Rauvolfia serpentina*, are listed in CITES, but this is only the tip of the iceberg. The majority of Apocynaceae species have not yet been assessed for the IUCN Red List of Threatened Species and large numbers of those that have been need to be updated in light of taxonomic changes and better distribution data. In both the Neotropics and the Paleotropics, many taxa are narrow ecological endemics, sometimes highly specialized and known from a single locality (e.g., the blackwater–white sand endemics of the Guianas and NW Amazonia (e.g., many species of *Malouetia*) and karst limestone endemics in Southeast Asia), where destruction of habitat equates to species extinction. It has recently been suggested that all three species of *Thevetia* endemic to Mexico be considered for protection as critically endangered, vulnerable or near threatened (IUCN categories CR, VU and NT, respectively: Alvarado-Cárdenas et al. 2017). *Dyera polyphylla*, a species restricted to peat swamp forest, is considered vulnerable due to habitat loss and over-exploitation (IUCN category VU A1cd; Middleton 2007). *Kibatalia* is particularly diverse in the Philippines, where there has been major forest loss, and many species have not been collected for decades. As forests are cleared the large timber species, such as in *Dyera* and *Alstonia*, are lost but so are the understory tree species in genera such as *Kopsia*, *Wrightia* and *Kibatalia*. Along with the loss of these large and small trees, the lianas, twiners and predominantly epiphytic species, such as in *Hoya* and *Dischidia*, are also lost to logging and clear-cutting; many species are already extinct and many more threatened with extinction (Kleijn and van Donkelaar 2001).

Islands present especially fragile ecosystems. In addition to habitat destruction due to human activity such as deforestation and fire, island taxa are often threatened by introduced species such as feral animals (e.g., goats, rats) and invasive plants. This has been well documented for Apocynaceae on various islands in French Polynesia. On Reunion Island, for example, the endangered endemic *Carissa xylopicron* has been almost completely decimated by introduced giant African land snails (*Achatina* species.; Meyer and Picot 2001), while on Mauritius the last remaining individuals of the endemic *Tabernaemontana persicariifolia* are threatened by road construction and listed as endangered (IUCN category EN; Pynee et al. 2013). *Lepinia taitensis*, endemic on

Tahiti and Moorea, is losing the battle against the introduced garden plant *Miconia calvescens* (Meyer 1996) and is considered critically endangered (IUCN category CR). In the Marquesas *Ochrosia* and *Rauvolfia* are represented by two endemic species each, *O. brownii* and *O. nukuhivensis* and *R. nukuhivensis* and *R. sachetiae*, respectively. All four species are considered critically endangered (IUCN category CR). In many cases the remaining individuals of a species can be counted on one hand. *Rauvolfia sachetiae* is known from a single tree on Hiva Oa, but has not been seen at the type locality since 1977; the area is overgrazed by goats and has been invaded by *Syzygium cumini* (Meyer and Butaud 2009; Lorence and Butaud 2011). On Madagascar, several of the endemic *Pachypodium* species have a restricted distribution and are endangered, mainly due to over-exploitation by hobby gardeners and plant poachers as well as to habitat destruction. Stronger conservation measures have been urged for at least three species: *P. baronii*, *P. windsorii*, and *P. decaryi* (Burge et al. 2013).

A number of genera have diversified on karst limestone in Southeast Asia and these areas are particularly vulnerable to habitat destruction, particularly through over-exploitation for the manufacture of concrete. Other limestone sites, even in protected areas, receive unsustainable numbers of tourists. Consequently, some species in genera such as *Wrightia*, *Alstonia* and *Dischidia* are in rapid decline. As narrowly endemic species in these genera continue to be discovered as karst limestone areas are explored, there is considerable potential for species loss even before those species have ever been collected and catalogued. This problem, of course, is not confined to species of Apocynaceae. Also in temperate regions, some taxa are rare or endangered. In southwestern USA, for example, a number of species of *Amsonia* found in arid habitats, and also *Cycladenia*, which is restricted to three or four mountain peaks in California, Arizona and Utah are rare or threatened. While Africa is home to a number of extremely threatened—or maybe extinct—paleoendemics such as *Emicocarpus fissifolius*, the Americas, in particular the Andes, are home to swarms of very narrow neoendemics, e.g., species of *Orthosia* and *Scyphostelma* known from a single valley each.

Many species of all larger genera of Asclepiadoideae are known only from the type collection. In the horticulturally popular and mostly succulent Ceropegieae nearly every second species deserves conservation concern (Albers and Meve 1997). Commercial collecting activities like those in *Hoodia* (see "Uses and Economic Importance") severely increase the risks for some of the species.

Classification of Apocynaceae

I. **Rauvolfioids**

1. Tribe Aspidospermateae Miers (1878).
   Gen. 1–6
2. Tribe Alstonieae G. Don (1837).
   Gen. 7–8
3. Tribe Vinceae Duby (1828).
   a. Subtribe Kopsiinae Leeuwenb. (1994).
      (Gen. 9).
   b. Subtribe Ochrosiinae Pichon ex Boiteau (1981).
      (Gen. 10).
   c. Subtribe Tonduziinae M.E. Endress (2014).
      (Gen. 11–12).
   d. Subtribe Vincinae M.E. Endress (2014).
      (Gen. 13).
   e. Subtribe Catharanthinae Pichon ex Boiteau (1981).
      (Gen. 14–16).
   f. Subtribe Rauvolfiinae Benth & Hook.f. (1876).
      (Gen. 17).
4. Tribe Willughbeieae A. DC. (1844).
   a. Subtribe Leuconotidinae Pichon ex Leeuwenb. (1994).
      (Gen. 18–20).
   b. Subtribe Willughbeiinae A. DC. (1844).
      (Gen. 21).
   c. Subtribe Landolphiinae K. Schum. (1895).
      (Gen. 22–31).
   d. Subtribe Lacmelleinae Pichon ex Leeuwenb. (1994).
      (Gen. 32–35).
5. Tribe Tabernaemontaneae G. Don (1837).
   a. Subtribe Ambelaniinae A.O. Simões & M.E. Endress (2010).
      (Gen. 36–42).
   b. Subtribe Tabernaemontaninae A. DC. (1844).
      (Gen. 43–50).
6. Tribe Amsonieae M.E. Endress (2014).
   (Gen. 51).
7. Tribe Melodineae G. Don (1837).
   (Gen. 52–56).

8. Tribe Hunterieae Miers (1878).
   (Gen. 57–60).
9. Tribe Alyxieae G. Don (1837).
   a. Subtribe Condylocarpinae Pichon ex Leeuwenb. (1994).
      (Gen. 61–63).
   b. Subtribe Alyxiinae A. DC. (1844).
      (Gen. 64–67).
10. Tribe Plumerieae E. Mey. (1838).
    a. Subtribe Allamandinae A. DC. (1844).
       (Gen. 68).
    b. Subtribe Plumeriinae Pichon ex Leeuwenb. (1994).
       (Gen. 69–71).
    c. Subtribe Thevetiinae A. DC. (1844).
       (Gen. 72–77).
11. Tribe Carisseae Dumort (1829).
    (Gen. 78–79).

II. **Apocynoids**
1. Tribe Wrightieae G. Don (1837).
   (Gen. 80–82).
2. Tribe Nerieae Baill. (1889).
   a. Subtribe Neriinae Benth. & Hook.f. (1876).
      (Gen. 83–84).
   b. Subtribe Alafiinae Pichon ex Leeuwenb. (1994).
      (Gen. 85–88).
3. Tribe Malouetieae Müll. Arg. (1860).
   a. Subtribe Galactophorinae Pichon ex M.E. Endress (2014).
      (Gen. 89).
   b. Subtribe Pachypodiinae Pichon ex Leeuwenb. (1994).
      (Gen. 90–91).
   c. Subtribe Malouetiinae Pichon (1950).
      (Gen. 92–101).
4. Tribe Rhabdadenieae Pichon ex M.E. Endress (2014).
   (Gen. 102).
5. Tribe Odontadenieae Miers (1878).
   (Gen. 103–111).
6. Tribe Mesechiteae Miers (1878).
   (Gen. 112–117).
7. Tribe Echiteae Bartl. (1830).
   a. Subtribe Laubertiinae J.F. Morales, M.E. Endress & Liede (2017).
      (Gen. 118–119).
   b. Subtribe Peltastinae Pichon ex M.E. Endress (2014).
      (Gen. 120–122).
   c. Subtribe Echitinae Kitt. (1843).
      (Gen. 123–127).
   d. Subtribe Parsonsiinae Benth. & Hook.f. (1876).
      (Gen. 128–130).
   e. Subtribe Prestoniinae Pichon ex M.E. Endress (2014).
      (Gen. 131).
8. Tribe Apocyneae Rchb. (1831).
   a. Subtribe Papuechitinae Pichon ex M.E. Endress (2014).
      (Gen. 132–134).
   b. Subtribe Amphineuriinae Pichon ex M.E. Endress (2014).
      (Gen. 135–137).
   c. Subtribe Beaumontiinae Pichon ex M.E. Endress (2014).
      (Gen. 138–140).
   d. Subtribe Apocyninae Pichon ex Leeuwenb. (1994).
      (Gen. 141–142).
   e. Subtribe Urceolinae Pichon ex M.E. Endress (2014).
      (Gen. 143).
   f. Subtribe Chonemorphinae Pichon ex M.E. Endress (2014).
      (Gen. 144–147).
   g. Subtribe Ichnocarpinae Benth. & Hook.f. (1876).
      (Gen. 148–152).
9. Tribe Baisseeae (Pichon ex De Kruif) M.E. Endress (2007).
   (Gen. 153–156).

III. **Subfam. Periplocoideae** Endl. (1838).
     (Gen. 157–189).
IV. **Subfam. Secamonoideae** Endl. (1838).
    (Gen. 190–197).
V. **Subfam. Asclepiadoideae** Burnett (1835).
1. Tribe Fockeeae H. Kunze, Meve & Liede (1994).
   (Gen. 198–199).
2. Tribe Marsdenieae Benth. (1868).
   (Gen. 200–225).
3. Tribe Ceropegieae Orb. (1843).
   a. Subtribe Heterostemminae Meve & Liede (2004).
      (Gen. 226).

b. Subtribe Leptadeniinae Meve & Liede (2004). (Gen. 227–230).
c. Subtribe Anisotominae Meve & Liede (2004). (Gen. 231–235).
d. Subtribe Stapeliinae G. Don (1837). (Gen. 236–271).

4. Tribe Eustegieae Rchb. ex Meve & Liede (2014). (Gen. 272–273).
5. Tribe Asclepiadeae Duby (1828).
   a. Subtribe Astephaninae Endl. ex Meisn. (1840). (Gen. 274–276).
   b. Subtribe Asclepiadinae Decne. ex Miq. (1857). (Gen. 277–302).
   c. Subtribe Cynanchinae K. Schum. (1895). (Gen. 303–304).
   d. Subtribe Tylophorinae K. Schum. (1895). (Gen. 305–306).
   e. Subtribe Pentacyphinae Liede & Meve (2014). (Gen. 307).
   f. Subtribe Diplolepinae Liede & Meve (2014). (Gen. 308).
   g. Subtribe Orthosiinae Liede & Rapini (2005). (Gen. 309–312).
   h. Subtribe Metastelmatinae Endl. ex Meisn. (1840). (Gen. 313–324).
   i. Subtribe Tassadiinae Liede & Meve (2014). (Gen. 325).
   j. Subtribe Oxypetalinae E. Fourn. (1885). (Gen. 326–331).
   k. Subtribe Gonolobinae Liede (1997). (Gen. 332–376).
   l. Subtribe Topeinae H.A. Keller & Liede (2017). (Gen. 377).

   Genus of uncertain subtribal placement in Asclepiadeae (Gen. 378).

Key to the Major Divisions of Apocynaceae

1. Anthers adnate to the style-head; corolla lobe aestivation in bud typically dextrorse (overlapping to the right) or valvate, rarely sinistrorse; fruit dehiscent, almost always apocarpous, a pair of follicles, sometimes reduced to one by abortion or postgenitally fused; seeds small, almost always with a coma at one end 2
– Anthers free from style-head; corolla lobe aestivation in bud typically sinistrorse (overlapping to the left), rarely dextrorse; fruit dehiscent or indehiscent, syncarpous or apocarpous, a berry, drupe, follicle, or capsule; seeds naked, with wings, or arils, but almost never with a coma at one end
I. **Rauvolfioids** (p. 253)
2. Nectaries, if present, encircling the base of the ovary, as five distinct lobes, or these fused to varying degrees into a lobed or crenulate ring, rarely 2 distinct lobes alternating with the carpels; anthers 4-locular; pollen almost always shed as monads; style-head secretions normally a foamy adhesive or gummy, undifferentiated translators II. **Apocynoids** (p. 281)
– Nectaries located in alternistaminal pockets on staminal feet or staminal tube; anthers 2–4-locular; pollen in tetrads or united in pollinia; style-head secretions forming differentiated translators with a sticky pad (viscidium) or consisting of a corpuscle and two caudicles 3
3. Anthers 4-locular, pollen in tetrads or, when in pollinia, then these without a waxy outer wall 4
– Anthers 2-locular, pollen in pollinia, these almost always covered by a waxy outer wall (ectexine)
V. **Subfam. Asclepiadoideae** (p. 324)
4. Translators with a sticky end (viscidium) that adheres to pollinator for removal; pollen usually shed in tetrads, sometimes in pollinia, from anthers onto a spoon- or cornet-shaped receptacle of the translator
III. **Subfam. Periplocoideae** (p. 309)
– Translators with a hardened clip-like corpusculum in which some part of pollinator body becomes caught for removal; pollen in 4 minute pollinia attached directly or indirectly to the corpusculum
IV. **Subfam. Secamonoideae** (p. 321)

Key to the Genera of Rauvolfioids and Apocynoids (Apocynaceae s. str.)

1. Plants erect: trees, shrubs, subshrubs or herbs 2
– Plants twining: vines, lianas, scrambling shrubs or rarely stoloniferous perennials 113
2. Plants with paired, often branched, sharp spines in the axils; leaves opposite; fruit a berry **79. *Carissa***

- Plants without paired sharp spines in the axils or, if spines present, then leaves not opposite and fruit not a berry 3
3. Herbs (often with a woody base) or subshrubs usually < 2 m tall 4
- Shrubs or trees, generally > 2 m tall but sometimes shorter 23
4. Leaves alternate; calycine colleters absent 5
- Leaves opposite or verticillate; calycine colleters absent or present 7
5. Corolla broadly bowl-shaped; anthers exposed, firmly attached to the style-head; nectaries present **142. *Apocynum***
- Corolla salverform; anthers hidden, free from the style-head; nectaries absent 6
6. Inflorescences thyrsiform, several-flowered; flowers blue; seeds naked **51. *Amsonia***
- Inflorescences consisting of only 1–4 flowers in the upper leaf axils; flowers yellow; seeds with a coma at each end **3. *Haplophyton***
7. Ovary completely syncarpous; fruit a berry (Willughbeieae-Landolphiinae) 8
- Ovary apocarpous (sometimes partially syncarpous at the base); fruit a pair of follicles 10
8. Inflorescences elongated, branched terminal panicles; fruit densely velutinous on the outer surface **26. *Ancylobothrys***
- Inflorescences short and usually congested, terminal or axillary cymes; fruit glabrous 9
9. Corolla tube 9–13 mm long; anthers inserted in the lower quarter of the corolla tube **27. *Chamaeclitandra***
- Corolla tube 4–11 mm long; anthers inserted in the upper half of the corolla tube **31. *Landolphia***
10. Leaf blades adaxially with colleters clustered at the base of the midrib, or scattered along its length; calycine colleters present **117. *Mandevilla***
- Leaf blades without colleters; calycine colleters absent or present 11
11. Nectaries or nectary disc absent 12
- Nectaries or nectary disc present 13
12. Petaloid corona present in the corolla mouth; corolla lobes often caudate; seeds with a rostrate coma at one end and a deciduous coma at the other **85. *Strophanthus***
- Petaloid corona absent; corolla lobes not caudate; seeds with a sessile coma at each end **3. *Haplophyton***
13. Nectaries 2, alternating with the carpels 14
- Nectaries in a ring surrounding the base of the ovary 16
14. Corolla lobe aestivation sinistrorse in bud; seeds naked **16. *Catharanthus***
- Corolla lobe aestivation dextrorse in bud; seeds with a micropylar coma 15
15. Corolla infundibuliform; anthers cordate at the base, only very weakly attached to the style-head; plants of SE Asia and Borneo **92. *Spirolobium***
- Corolla salverform; anthers sagittate at the base and firmly attached to the style-head; plants of Amazonia **109. *Salpinctes***
16. Leaves in whorls of 3–5 (rarely opposite at some nodes); fruit indehiscent, drupaceous, often only one carpel developing; seeds one per carpel, naked **17. *Rauvolfia***
- Leaves opposite; fruit dehiscent, follicular, usually both carpels developing; seeds several per carpel, with a micropylar coma 17
17. Bracts very conspicuous, large and foliaceous to subpetaloid, often purplish red tinged; annular corona present in the corolla mouth **121. *Rhodocalyx***
- Bracts small, inconspicuous, green or scarious; annular corona absent 18
18. Corolla usually showy, > 2 cm long; plants of the Neotropics 19
- Corolla smaller, < 2 cm long; plants of temperate regions 22
19. Sepals foliaceous to subfoliaceous, without colleters on the inner surface; corolla usually pink, purple, red or magenta 20
- Sepals smaller, not foliaceous, with colleters on the inner surface; corolla yellow or cream (Odontadenieae) 21
20. Calyx and corolla tube often with glandular hairs on the outer surface; corolla tube with five longitudinal, alternipetalous ridges on the outer surface ; seed coma sessile; follicles fused from the base to about 2 cm (stipitate), free above **89. *Galactophora***
- Calyx and corolla tube without glandular hairs; corolla tube without five longitudinal, alternipetalous ridges on the outer surface; seed coma rostrate; follicles free or fused only at the very base (not stipitate) **102. *Rhabdadenia***
21. Anthers with a long, slender spiraled apical appendage **111. *Pentalinon***
- Anthers without a long apical appendage **110. *Angadenia***
22. Corolla white, greenish white or cream, sometimes with pink stripes, tubular, urceolate or broadly campanulate, with small v-shaped corona lobes near the base of the tube in the alternistaminal sectors **142. *Apocynum***

- Corolla reddish purple to violet, infundibuliform, with small corona lobes just above the anthers in the staminal sectors **105. *Cycladenia***
23. Leaves alternate 24
- Leaves opposite or verticillate 38
24. Corolla lobe aestivation dextrorse in bud; plants succulent or not; seeds either naked or with a well-developed coma at one end 25
- Corolla lobe aestivation sinistrorse in bud; plants subsucculent or not; seeds almost always without a coma; if hairs present, then these distributed over the seeds and not restricted to one end 27
25. Plants succulent; flowers large and showy; fruit a pair of dehiscent follicles; seeds with a coma 26
- Plants not succulent; flowers small and inconspicuous; fruit a pair of leathery or fleshy berries; seeds without a coma **5. *Geissospermum***
26. Plants with spines; anther connective without a long terminal appendage **91. *Pachypodium***
- Plants without spines; anther connective with a long, hairy apical appendage **84. *Adenium***
27. Ovary of 3–5 carpels (Alyxieae-Alyxiinae) 28
- Ovary of 2 carpels 29
28. Carpels long-stipitate, fused at the base and the apex, but free in the middle **65. *Lepinia***
- Carpels sessile, completely syncarpous **66. *Lepiniopsis***
29. Corolla with a corona lobe directly above each anther; sepals mostly large, subfoliaceous, usually spreading or reflexed, sometimes caducous; flowers relatively showy, the corolla tube (7–1)13–33(–55) mm long, white, often with red or yellow markings in the throat or rose-tinged, to cream, bright yellow or salmon-orange; fruit indehiscent (Plumerieae-Thevetiinae) 30
- Corona absent; sepals usually smaller; flowers showy or not, usually white, rarely with yellow or red markings; fruit dehiscent or indehiscent 32
30. Calycine colleters present; nectary disc fleshy, distinct, surrounding the ovary; corolla mostly yellow or orangish, tubular-campanulate, rarely cream-colored and ± salverform; fruit drupaceous, red or black **75. *Thevetia***
- Calycine colleters absent; nectary disc absent, adnate or indistinct; corolla mostly salverform, sometimes slightly subcampanulate above, white to cream (rarely red), sometimes with yellow, red or rose markings in the throat; fruit various 31
31. Corolla tube 7–14 mm long; fruit a pair of samaroid drupes with two lateral wings **77. *Cerberiopsis***
- Corolla tube (15–)20–45 mm long; fruit a fleshy, globose or ovoid drupe **76. *Cerbera***
32. Ovary sunken into the receptacle; stamens inserted at the base of the corolla tube; flowers mostly large and showy, often waxy in appearance; calyx lobes small, often rudimentary or caducous; fruit a pair of follicles (Plumerieae-Plumeriinae) 33
- Ovary not sunken into the receptacle; flowers usually small; stamens inserted in the upper half of the corolla tube; calyx lobes normally developed; fruit various 35
33. Floral bracts large, foliaceous, with numerous colleters at the base; calyx rudimentary, persistent, with 1–5 lobes of varying sizes, these eglandular at the apex; seeds normally with a large concentric papery wing **71. *Himatanthus***
- Floral bracts inconspicuous, scarious, caducous; calyx of 5 ± equal lobes, these glandular or eglandular at the apex; seeds with a narrow flattened rim around the margin, or if wing present, then not concentric 34
34. Petiole with a cluster of 2–3 colleters about midway on the adaxial side; calyx lobes caducous, eglandular at the apex; seeds unwinged, with merely a narrow, flattened rim around the margin **69. *Mortoniella***
- Petiole without colleters on the adaxial side; calyx lobes persistent, glandular at the apex; seeds with a wing at one end **70. *Plumeria***
35. Nectary lobes 2, alternating with the carpels; both carpels developing normally; fruit a pair of long and slender follicles; seeds flattened and with long brown hairs at least at the margins **12. *Laxoplumeria***
- Nectary absent, adnate or indistinct; normally only one carpel developing; fruit an indehiscent drupe, if dehiscent, then not a pair of long, slender follicles; seeds naked or winged 36
36. Fruit dehiscent, a usually compressed, stout follicle with thick, woody pericarp; seeds with a papery concentric or excentric wing; corolla not fused (with short slits) just above stamen insertion; shrubs to big trees up to 40 m tall **6. *Aspidosperma***
- Fruit indehiscent, drupaceous; seeds naked; corolla tube fused (without slits) above stamen insertion; shrubs or small trees 37
37. Leaves coriaceous, oblong to broadly ovate, ≥ 4 cm wide; fruit bright red, 5–7 cm long **64. *Pteralyxia***
- Leaves membranous to subcoriaceous; fruit opalescent, subreniform, < 2 cm long **2. *Vallesia***

38. Leaves whorled, occasionally opposite at some nodes 39
– Leaves strictly opposite 57
39. Corolla infundibuliform, large and showy, colorful; fruit syncarpous 40
– Corolla salverform, smaller, usually white, cream or yellow, sometimes pinkish; fruit apocarpous (rarely syncarpous) 41
40. Anther connective with a long, filiform, hairy terminal appendage; mouth of the corolla with a dissected petaloid corona; nectary disc absent; leaves coriaceous, ternate; fruit smooth; seeds pubescent with a coma at one end, not flattened **83. *Nerium***
– Anther without a long hairy appendage; petaloid corona absent; nectary disc present around ovary; leaves thin, usually 4 per whorl; fruit usually with spines; seeds flattened, winged **68. *Allamanda***
41. Corolla lobe aestivation dextrorse in bud 42
– Corolla lobe aestivation sinistrorse in bud 43
42. Ovules few (up to 6 per carpel); fruit indehiscent, a pair of ovoid to globose, red, yellow or green drupes; pericarp fleshy; seeds without hairs or cilia **10. *Ochrosia***
– Ovules numerous (up to 100 or more); fruit dehiscent, a pair of long, slender, dry, green to brown follicles; seeds with long hairs or cilia around the margins **7. *Alstonia***
43. Ovary syncarpous; fruit many-seeded 44
– Ovary apocarpous (rarely hemi-syncarpous in some *Tabernaemontana* species); fruit few- to many-seeded 46
44. Petioles with interpetiolar flap of tissue in the axils; fruit indehiscent, a berry **34. *Couma***
– Petioles without a flap of tissue in the axils; fruit dehiscent, a capsule or two follicles fused along the ventral side 45
45. Leaves with 20–60 pairs of secondary veins; branches terete; stamens inserted in the upper half of the corolla tube; fruit long and cylindrical, of fused follicles, 15–30 cm long **7. *Alstonia***
– Leaves with 10–20 pairs of secondary veins; branches quadrangular or triangular; stamens inserted near the base of the tube; fruit small and squat, a capsule up to 6 cm long **55. *Craspidospermum***
46. Ovary 2–5-carpellate; inflorescences ramiflorous, fasciculate, the flowers sessile in the leaf axils; fruit of 2–5 one-seeded berries **60. *Pleiocarpa***
– Ovary 2-carpellate; inflorescences normally not ramiflorous, the flowers usually pedicillate; fruit of 2 (or sometimes 1 by abortion) follicles or drupes 47
47. Calycine colleters present, in a row at the base of each sepal 48
– Calycine colleters absent 49
48. Giant trees; leaves in whorls of 4–8; corolla rotate; seeds surrounded by a membranous wing **8. *Dyera***
– Shrubs ca. 2 m. tall; leaves in whorls of three with occasional opposite pairs at some nodes; corolla salverform; seeds surrounded by an aril **50. *Tabernaemontana***
49. Style-head with a membranous collar at the base 50
– Style-head without a membranous collar at the base 54
50. Nectaries well-developed, distinct from the ovary 51
– Nectaries indistinct, adnate or absent 52
51. Nectaries forming a fleshy ring surrounding the ovary; fruit indehiscent, one-seeded drupes, often only one developing **17. *Rauvolfia***
– Nectary with 2 conspicuous deltoid to narrowly triangular lobes alternating with the carpels; fruit dehiscent, many-seeded follicles, usually both developing **7. *Alstonia***
52. Colleter(s) in the leaf axils deltoid or triangular; leaves in whorls of (3–)4–9; fruit a pair of dehiscent follicles; seeds with long cilia, sometimes more than 1 cm long, at the ends; large trees up to 70 m tall; Africa, Asia and Pacific Islands **7. *Alstonia***
– Colleters in the leaf axils small, ligulate to linear; leaves in whorls of 3; fruit a pair of drupes or follicles; seeds with cilia or not; shrubs or small trees up to 20 m tall; Central America, Africa, Madagascar, Indian Ocean Islands and Sri Lanka (Vinceae) 53
53. Anthers inserted at about the middle of the corolla tube; fruit a pair of thinly woody follicles; seeds ciliate at the margin; plants of Central America **11. *Tonduzia***
– Anthers inserted just below the orifice of the corolla tube; fruit indehiscent, a pair of fleshy, torulose, red or orange drupes; plants of the Old World **15. *Petchia***
54. Leaf axils lacking colleters (Aspidospermateae) 55
– Leaf axils with colleters 56
55. Leaves less than 5 cm long, with a spine at the tip; corolla glabrous externally; fruit usually a single (through abortion) thick, woody orbicular to broadly obovoid follicle; seeds compressed with a membranous, concentric wing **6. *Aspidosperma***
– Leaves more than 5 cm long, without a spine at the tip; corolla densely silvery green villous externally;

fruit a pair of slender cylindrical follicles; seeds cylindrical, unwinged **4. *Microplumeria***

56. Ovules attached in 2 rows, 2–6(–7) per carpel; fruit indehiscent, a pair (sometimes only 1 by abortion) of strongly torulose drupes; seeds not compressed; endosperm ruminate **67. *Alyxia***
- Ovules attached in more than 2 rows, ca. 20 per carpel; fruit dehiscent, a pair of non-torulose follicles; seeds compressed with a papery wing at one end; endosperm non-ruminate **57. *Gonioma***

57. Corolla lobe aestivation dextrorse; anthers often adnate to the style-head; seeds naked or with a coma, rarely with a brightly colored aril 58
- Corolla lobe aestivation sinistrorse; anthers mostly free from the style-head; seeds naked, winged, arillate (arils sometimes brightly colored) or, rarely, with a coma 75

58. Anthers free from the style-head; fruit apocarpous, rarely partially syncarpous, dehiscent or indehiscent 59
- Anthers adnate to the style-head; fruit apocarpous and dehiscent 65

59. Colleters absent on the inner surface of the sepals; fruit a pair of drupes (rarely solitary) 60
- Colleters present near the base on the inner surface of the sepals; fruit a pair of follicles 61

60. Sepals with a colleter just below the apex on the outer surface; disc of 2 well-developed lobes alternating with the carpels, ± as high as the carpels or higher; drupes almost always with a hollow spur-like appendage **9. *Kopsia***
- Sepals without a colleter on the outer surface; disc absent or of 2 tiny rudimentary lobes alternating with the carpels and much shorter than them; drupes without a spur-like appendage **10. *Ochrosia***

61. Colleters centered at the base of the sepals on the inner surface; seeds arillate, without a coma 62
- Colleters few, in the sepal sinuses corresponding to the quincuncial pattern; seeds with a coma, aril absent 64

62. Inflorescences pendent; calycine colleters 5–30 per sepal in 2–4 rows; corolla infundibuliform **43. *Callichilia***
- Inflorescences erect; calycine colleters 1–several in a single row 63

63. Petiole bases of a leaf pair fused into a short ocrea around the node; corolla white, the tube 17–36 mm long; plants of SW India **50. *Tabernaemontana***
- Petiole bases of a leaf pair not forming an ocrea; corolla yellow to cream-yellow, the tube 4–5 mm long; plants of tropical C and E Africa and the Comoro Islands **48. *Schizozygia***

64. Corolla infundibuliform, the tube 17–35 cm long; nectary of 2 well-developed lobes alternating with the carpels **92. *Spirolobium***
- Corolla salverform, tube 9–22 mm long; nectary absent **96. *Holarrhena***

65. Nectaries absent or adnate or indistinct from the base of the ovary 66
- Nectaries present, of 2 or 5 lobes at the base of the ovary 68

66. Corolla infundibuliform to subcampanulate; petaloid corona present in the corolla mouth; petals often caudate; seeds with a rostrate coma **85. *Strophanthus***
- Corolla salverform or narrowly tubular-campanulate; petaloid corona absent; petals not caudate; seed coma sessile or absent (Malouetieae-Malouetiinae) 67

67. Anthers with lignified basal appendages; seeds without a coma **100. *Malouetiella***
- Anthers without lignified basal appendages; seeds with a sessile coma **96. *Holarrhena***

68. Nectaries 2, distinct, alternating with the carpels; small shrubs up to 2 m **92. *Spirolobium***
- Nectaries 5, separate or fused at the base into a ring surrounding the carpels; mostly trees, more rarely shrubs 69

69. Calycine colleters solitary, centered at the base of the sepals; anthers completely exserted from the corolla tube; Amazonia **119. *Hylaea***
- Calycine colleters mostly few in the sinuses or few to many in a ring at the base of the sepals or rarely absent; anthers included to exserted; tropics of the Old and New World and South Pacific (Malouetieae) 70

70. Inflorescences bostrychoid cymes or reduced to a solitary flower; leaves without domatia; colleters absent or much reduced, rarely few and alternisepalous; nectaries separate or connate just at the base; seeds with sessile coma at the micropylar end; plants of the West Indies **90. *Neobracea***
- Inflorescences usually in alternate leaf axils and ± fasciculate, subsessile or shortly pedunculate; leaves often with abaxial domatia in the axils of the secondary veins; calycine colleters few and alternisepalous or several to many in a ring; nectaries fused into a 5-lobed ring; seeds with coma at one end, at both ends or without a coma and seed body naked or with a sparse scattering of hairs or a thick indument of long hairs; plants of the tropics in Africa, Asia,

America and the Solomon Islands (Malouetieae-Malouetiinae) 71

71. Corolla lobes induplicate in bud, the margins often long-ciliate or crispate; seeds with a well-developed sessile micropylar coma and a small deciduous chalazal coma; domatia absent **95. *Mascarenhasia***

– Corolla lobes not induplicate in bud, the margins not long-ciliate or crispate; seeds either without a coma or with a chalazal coma only; abaxial domatia often present in axils of secondary veins 72

72. Seeds with a rostrate chalazal coma with retrorse hairs; testa almost always glabrous; calycine colleters few and alternisepalous or up to 100 in a ring; plants of tropical Africa and Asia 73

– Seeds without a coma; testa glabrous to densely velutinous or villous; calycine colleters few and alternisepalous; plants of tropical Africa, America and the South Pacific 74

73. Corolla thick and fleshy, salverform, the tube constricted at the orifice; anthers completely included; plants of tropical Africa **101. *Funtumia***

– Corolla mostly thinner, membranous, the tube usually with a narrow cylindrical lower part and an expanded cupular upper part (tubular-campanulate); anthers included to nearly completely exserted; plants of SE Asia and Malesia **98. *Kibatalia***

74. Leaves mostly with abaxial domatia in the axils of the secondary veins; small bifid corona lobe present behind each anther; anthers included to nearly completely exserted; seed body naked, very sparsely pilose or densely velutinous to villous; plants of the American and African tropics, often growing in white sand and/or blackwater habitats and seasonally inundated forest **97. *Malouetia***

– Leaves without abaxial domatia; corona lobes absent; anthers included; testa glabrous; plants endemic to the Solomon Islands, growing in well-drained forest **99. *Allowoodsonia***

75. Anthers adnate to the style-head; petaloid or much dissected corona normally present in the corolla mouth; seeds with a sessile chalazal coma; plants of the Old World (Wrightieae) 76

– Anthers free from the style-head; petaloid or dissected corona usually absent; seeds without a coma; plants of the Old and New Worlds 78

76. Corona bright yellow, of two sorts of lobes, the alternipetalous ones of 3–4 filiform segments on a long, slender stalk; inflorescences multi-flowered panicles with many simultaneously open flowers, often thyrsoid **81. *Pleioceras***

– Corona variously colored, entire, crenulate, or lobed in various ways, rarely on long, slender stalks; inflorescences usually few-flowered, or if multi-flowered usually with only few open flowers or, if multi-flowered and many simultaneously open flowers, then never thyrsoid and corona either absent or united into a cup 77

77. Flower resembling a miniature daffodil; corolla yellow; corona white, the lobes united for nearly their entire length into a crenulate cup much surpassing the anthers **80. *Stephanostema***

– Flower not daffodil-like; corona and corolla variously colored, sometimes concolorous; corona lobes either free or united to varying degrees, if forming a cup around the anthers this only short **82. *Wrightia***

78. Anthers with lignified, sagittate basal appendages; fruit an indehiscent berry with naked seeds, or a pair of follicles (these sometimes partially to nearly completely fused) and seeds mostly arillate (Tabernaemontaneae) 79

– Anthers simple, ovate or, if sagittate at base, then not lignified; fruit various; seeds not arillate (except in some species of *Tabernaemontana*) 94

79. Corolla infundibuliform to tubular-campanulate; calyx lobes often foliaceous 80

– Corolla salverform; calyx lobes foliaceous or not 84

80. Corolla small, white with red stripes, the tube up to 10 mm long **47. *Carvalhoa***

– Corolla large, yellow, cream, white, or rarely orange or mauve, not striped, the tube 11–135 mm long 81

81. Bracts very large, 4–11 cm long, completely covering the calyx; inflorescences long-pedunculate, of 1–2 large, white pendulous flowers **46. *Crioceras***

– Bracts smaller, never completely covering the calyx; inflorescences various 82

82. Corolla yellow, cream, white or orange, the tube almost always twisted at the level of the anthers, the lobes inflexed in bud; style-head with a basal flange **50. *Tabernaemontana***

– Corolla white, the tube not twisted at the level of the anthers, lobes not (or only scarcely) inflexed in bud; style-head almost always without a basal flange 83

83. Peduncle very short, up to 2 mm long; calycine colleters few or absent; fruit 2/3 or more syncarpous **45. *Calocrater***

- Peduncle longer, 5–230 mm long; calycine colleters several, multiseriate; fruit usually apocarpous (except in *C. orientalis*) **43. *Callichilia***

84. Style-head shed with the corolla; sepals often united for up to 2/3 their length into a tubular, circumscissile calyx; petals not inflexed in bud **44. *Voacanga***
- Style-head not shed with the corolla; sepals usually free or united for less than 2/3 their length and not forming a circumscissile calyx; petals inflexed in bud or not 85

85. Ovary completely syncarpous; fruit a berry; seeds arillate or not 86
- Ovary apocarpous, sometimes partially syncarpous; fruit usually a pair of follicles (rarely these partially to nearly totally fused); seeds arillate or covered in a translucent membrane 92

86. Nectary disc present; seeds with a white aril; plants of the tropical W and C Africa **49. *Tabernanthe***
- Nectary disc absent; plants of Amazonia (Tabernaemontaneae-Ambelaniinae) 87

87. Calycine colleters present 88
- Calycine colleters absent 91

88. Corolla tube completely glabrous within, short (3–9 mm); fruit few-seeded, with thin pericarp; seeds embedded in fluffy white pulp **37. *Molongum***
- Corolla tube pilose within in some regions, short to long (3.5–50 mm); fruit few- to many-seeded, with a thick or thin and leathery or crustaceous pericarp; seeds embedded in pulp or not 89

89. Calycine colleters in a single series; corolla tube 10–30 mm long; fruit unilocular; seeds not embedded in pulp, attached to the fruit wall by fleshy funicles **38. *Spongiosperma***
- Calycine colleters multiseriate; corolla tube 3.5–15 mm long; fruit bilocular; seeds embedded in stringy pulp 90

90. Inflorescences axillary; young stems compressed; peduncles mostly reduced; fruit many-seeded, with a thick, granular pericarp **40. *Mucoa***
- Inflorescences terminal; young stems terete; peduncles elongate; fruit few- to many-seeded, with thin crustaceous pericarp **41. *Neocouma***

91. Inflorescences axillary; young stems ± terete; seeds not embedded in pulp, attached to fleshy funicles **36. *Ambelania***
- Inflorescences terminal; young stems sharply quadrangular; seeds embedded in reddish pulp **39. *Rhigospira***

92. Ovary pubescent; fruit indehiscent with a thick, somewhat indurated pericarp; usually only one carpel developing, with the rudimentary second carpel present at the base, sometimes both carpels developing, these often partially syncarpous; seeds covered in a translucent membrane that liquefies when the fruit is mature; plants of Amazonia and southern Central America **42. *Macoubea***
- Ovary glabrous; fruit almost always dehiscent (sometimes only tardily so, very rarely indehiscent), with a thick or thin, non-indurated pericarp; both carpels usually developing, or, if only one, then not with the second rudimentary carpel attached at the base 93

93. Petals not inflexed in bud; calycine colleters single in the sepal sinuses; nectary distinct; fruit completely apocarpous to partly syncarpous; aril white **49. *Tabernanthe***
- Petals inflexed in bud; calycine colleters mostly in groups (occasionally multiseriate) in the center of the sepal; nectary indistinct, adnate or absent; fruit apocarpous; aril red, orange, white or gray **50. *Tabernaemontana***

94. Ovary syncarpous; fruit a solitary berry 95
- Ovary apocarpous (sometimes partially syncarpous at the base); fruit dehiscent or indehiscent, various 100

95. Calyx usually 4-merous **32. *Parahancornia***
- Calyx 5-merous 96

96. Coralline corona present in throat **56. *Melodinus***
- Corona absent in throat 97

97. Inflorescences terminal, few-flowered; ovules numerous; fruit up to 5 cm in diameter, pubescent or not 98
- Inflorescences axillary, few- to many-flowered; ovules few to numerous; fruit smaller, glabrous 99

98. Corolla tube > 2 cm long; anthers inserted near the orifice of the corolla tube; fruit glabrous to pubescent, 1–6-seeded **33. *Hancornia***
- Corolla tube to 1 cm long; anthers inserted in the lower 1/3 of the corolla tube; fruit densely velutinous, many-seeded **26. *Ancylobothrys***

99. Ovary with several ovules per placenta; fruit yellow to orange; plants of the Neotropics **35. *Lacmellea***
- Ovary with 1 ovule per locule; fruit black to purplish black; plants of Africa **78. *Acokanthera***

100. Calycine colleters present, mostly in a group in the center of the sepal, sometimes in multiple series 101
- Calycine colleters absent 103

101. Calycine colleters covering 1/3 or more of the adaxial surface of the sepals; ovules 1–6(–30) per carpel; fruit indehiscent, 1–9(–20)-seeded berries **58. *Hunteria***

- Calycine colleters usually covering less than 1/3 of the adaxial surface of the sepals, typically in a single row at the base; ovules 70–150 per carpel; fruit dehiscent or indehiscent, 10–80 seeded 102

102. Fruit dehiscent, 2 follicles (these rarely partially syncarpous or indehiscent) with a fleshy, thin or thick, but not hard and fibrous, wall; seeds arillate; plants of tropical Africa, S and SE Asia and S and South West Pacific **50. *Tabernaemontana***
- Fruit indehiscent, 2 berries with a very hard and fibrous wall, sometimes only one developing; seeds without arils; plants of tropical W and C Africa **59. *Picralima***

103. Anthers with sublateral dehiscence and filiform apical appendages; corolla tube with small appendages below the anthers (Plumerieae-Thevetiinae) 104
- Anthers with introrse dehiscence, without filiform apical appendages; corolla tube without small appendages below the anthers 105

104. Flowers yellow; ovules 40–70 in several series in each carpel; fruit a pair of long, slender corkscrew-like, several-seeded follicles; plants of Brazil and Chile **74. *Skytanthus***
- Flowers white; ovules 2–6 in 2 series in each carpel; fruit a pair of strongly compressed, single-seeded samaroid drupes; plants of the Greater Antilles and Central America **72. *Cameraria***

105. Nodes without intrapetiolar or interpetiolar colleters; fruit normally consisting of a single, thick, woody pyriform to nearly circular follicle; seeds with a membranous, concentric wing; plants of the Neotropics **6. *Aspidosperma***
- Nodes with intrapetiolar and/or interpetiolar colleters; fruit and seeds various; plants of the Paleotropics and Oceania (except *Strempeliopsis*) 106

106. Colleters in the leaf axils deltoid or triangular; seeds with long cilia at least at each end, usually also around the margin **7. *Alstonia***
- Colleters in the leaf axils small, ligulate to linear; seeds without cilia 107

107. Inflorescences ramiflorous, the flowers in sessile fascicles in the leaf axils; ovary (2–)3–5-carpellate; fruit of (2–)3–5 separate, few-seeded, yellow to bright orange berries **60. *Pleiocarpa***
- Inflorescences normally with a distinct peduncle; ovary always 2-carpellate; fruit follicles or drupes, never berries 108

108. Ovules 2–4(–7) per carpel; style-head without a basal collar 109
- Ovules up to 20 per carpel; style-head with or without a basal collar 110

109. Corolla tube 1.5–3 mm long, shorter than the lobes; petioles usually 1–2 cm long; ovules 4 per carpel; fruit of 2 follicles less than 5 cm long; seeds compressed with a wing at one end **52. *Diplorhynchus***
- Corolla tube 1.5 mm or longer, almost always noticeably longer than the lobes; petioles often less than 1 cm long; ovules up to 7 per carpel; fruit indehiscent, of a pair of strongly torulose drupes with 1–7 articles; seeds ellipsoid (not compressed) and unwinged **67. *Alyxia***

110. Style-head with free apical appendages longer than the body; corolla tube with visible slits just above stamen insertion and a callous ring in the mouth; fruit dehiscent, a pair of non-compressed follicles; seeds with a thin wing around the margin **54. *Stephanostegia***
- Style-head with free apical appendages shorter than the body; corolla tube without visible slits, with or without a callous ring in the mouth; fruit and seeds various 111

111. Style-head with a short collar at the base; corolla tube with a callous ring in the mouth; fruit indehiscent, a pair of torulose, red or orange drupes; seeds without a wing **15. *Petchia***
- Style-head without a collar at the base; corolla tube without a callous ring in the mouth; fruit dehiscent, a pair of non-torulose follicles; seeds winged 112

112. Seeds with a wing at each end; leaf blades 6–16 × 2–7 cm, obovate to oblong-obovate; peduncles up to 11 cm long; plants of Jamaica and Cuba **1. *Strempeliopsis***
- Seeds with a wing at one end; leaf blade 2.5–12 × 0.5–3 cm, narrowly elliptic to oblong-elliptic; peduncles up to 1 cm long; plants of southern Africa or Madagascar **57. *Gonioma***

113. Plants with paired, often branched spines in the axils **79. *Carissa***
- Plants without paired spines in the axils 114

114. Leaf blades adaxially with colleters clustered at the base of the midrib (these sometimes caducous) or scattered along its length, or sometimes colleters scattered along the length of the petiole (colleters in the axil of the petiole absent or present), abaxial domatia often present in axils of secondary veins; anthers adnate to the style-head; seeds with a coma 115
- Leaf blades and petioles adaxially without colleters (but colleters in the axil of the petiole often present), usually without abaxial domatia, more rarely present; anthers free from the style-head or adnate to it; seeds with or without a coma 122

115. Ovary half-inferior; leaf blades adaxially with colleters scattered along the petiole, with hair-filled or cilia-rimmed abaxial domatia in the axils of the secondary veins; corolla tube 1–6 mm long, with small knob-like corona lobes in the mouth or above the anthers; plants of Africa and Madagascar (Baisseeae) 116
- Ovary superior; leaf blades adaxially with colleters clustered at the base of the midrib or with colleters scattered along the length of the midrib, but not along the length of the petiole; abaxial domatia mostly absent, if present then usually glabrous; corolla tube often longer than 6 mm and without knob-like corona lobes; plants of the Neotropics (Mesechiteae) 118

116. Anthers with an enlarged apical connective appendage covered with coarse hairs; inflorescences terminal **153. *Motandra***
- Anthers without an enlarged hairy apical connective appendage; inflorescences axillary and terminal 117

117. Corolline corona present in the mouth of the corolla tube consisting of 5 alternipetalous lobes; corolla mouth usually constricted **155. *Oncinotis***
- Corolline corona present below the mouth of the corolla tube consisting of 5 alternipetalous lobes, one just behind each anther; corolla tube mouth not constricted **154. *Baissea***

118. Inflorescences simple, unbranched (racemose); colleters scattered along the length of the midrib adaxially or clustered at its base **117. *Mandevilla***
- Inflorescences branched 2–3 times (corymbose or dichasial cymes or thyrses); colleters clustered at the base of the midrib only 119

119. Leaves mostly with abaxial domatia in at least some axils of the secondary veins; stamens almost always inserted at the base of the corolla tube, with well-developed filaments 120
- Leaves without abaxial domatia; stamens inserted higher in the corolla tube, often just at the base of the expanded upper part; stamens without a well-developed filament, anthers ± sessile 121

120. Corolla infundibuliform to tubular-campanulate, the tube > 1 cm long; anthers with pubescent, filiform apical appendages; staminal filaments 10–30 mm long (or if without filiform apical appendages and long staminal filaments, then corolla tube > 2 cm long); anthers included or only apical appendages exserted **115. *Tintinnabularia***
- Corolla tubular to subcampanulate or rotate, the tube < 3 mm long; anthers without filiform apical appendages; staminal filaments ≤ 2 mm long; anthers mostly ± completely exserted from the corolla tube **116. *Forsteronia***

121. Corolla salverform **114. *Mesechites***
- Corolla infundibuliform **113. *Allomarkgrafia***

122. Leaves verticillate, sometimes opposite at some nodes 123
- Leaves strictly opposite 133

123. Corolla lobe aestivation dextrorse in bud; anthers united with the style-head; fruit dry; seeds with a coma 124
- Corolla lobe aestivation sinistrorse in bud; anthers free from the style-head; fruit dry or fleshy; seeds mostly without a coma 127

124. Petaloid corolline corona present in the throat; petals often with a caudate appendage at the apex; seeds with a rostrate coma **85. *Strophanthus***
- Petaloid corona absent; petals never with caudate appendages; seed coma sessile 125

125. Flowers small, the corolla tube tubular to urceolate or ± rotate and almost always < 10 mm long; knob-like corona lobes present or absent; anthers usually at least partially exserted; ovary and fruit syncarpous; filiform, stipule-like outgrowths absent 126
- Flowers larger, the corolla tube salverform, 10–15 mm long; knob-like corona lobes absent; anthers included; ovary and fruit apocarpous; long, filiform stipules present **108. *Odontadenia***

126. Knob-like corolline corona lobe present just above each anther **130. *Artia***
- Knob-like corolline corona lobes absent **129. *Parsonsia***

127. Petaloid corona present in the corolla mouth; ovary syncarpous; fruit a hard-walled solitary berry **56. *Melodinus***
- Petaloid corona absent; ovary apocarpous; fruit berries, drupes or follicles 128

128. Sepals with multiple rows of colleters at the base within; fruit a pair of berries **58. *Hunteria***
- Sepals without colleters within; fruit various 129

129. Corolla lobes with a long ligulate appendage; fruit a pair of indehiscent, torulose drupes **62. *Condylocarpon***
- Corolla lobes without a long ligulate appendage; fruit dehiscent or indehiscent, berries, drupes or follicles 130

130. Corolla campanulate, the tube up to 2 mm long, ca. 1/3 the length of the lobes; anthers with a short sterile apex and sagittate (but non-lignified) base; seeds with a concentric papery wing **53. *Pycnobotrya***

- Corolla salverform to tubular, the tube up to 22 mm long, usually longer than the lobes; anthers without sterile apex and sagittate base; seeds various 131
131. Carpels usually 3–5, rarely only two; fruit 1-seeded, subglobose to ovoid berries **60. *Pleiocarpa***
- Carpels 2; fruit follicles or drupes 132
132. Corolla tube curved; fruit dehiscent, a pair of slender cylindrical follicles (usually both carpels developing); mesocarp dry; seeds compressed, with a wing at each end; carpels conspicuously notched at the apex **14. *Kamettia***
- Corolla tube straight; fruit indehiscent, carpels forming torulose chains of up to 7 articles, sometimes reduced to a single article (often only 1 carpel developing); mesocarp fleshy; seeds naked, globose to ovoid, not compressed; carpels not conspicuously notched at the apex **67. *Alyxia***
133. Stoloniferous herbs; flowers 1 (rarely 2) in the axils; corolla usually bluish violet, occasionally white; nectary lobes 2, alternating with the carpels **13. *Vinca***
- Woody plants without stolons; inflorescence type, number of flowers and corolla color various; nectaries absent or present 134
134. Leaves peltate; seeds with the coma at the end of a glabrous rostrum 135
- Leaves not peltate; seed coma rostrate or not 136
135. Corolla infundibuliform, usually yellow to green; plants mostly with ferruginous pubescence; calyx lobes foliaceous to subfoliaceous, up to 3 cm long **122. *Macropharynx***
- Corolla salverform, pinkish; plants glabrous; calyx lobes smaller, not foliaceous, $\leq$ 2 mm long **106. *Stipecoma***
136. Corolla lobe aestivation sinistrorse; anthers mostly free from the style-head, more rarely adnate to the style-head; fruit various, often syncarpous and indehiscent 137
- Corolla lobe aestivation dextrorse or valvate; anthers adnate to the style-head (sometimes only weakly so); fruit usually apocarpous, of 2 follicles (these sometimes postgenitally syncarpous and weakly to firmly fused) (Apocynoids) 164
137. Anthers adnate to the style-head; fruit a pair of follicles; seeds with a coma **143. *Urceola***
- Anthers free from the style-head; fruit usually berries, less frequently drupes or follicles; seeds without a coma (Rauvolfioids) 138
138. Corolla 4-merous; leaves punctate abaxially **19. *Leuconotis***
- Corolla 5-merous; leaves punctate abaxially or not 139
139. Corolline corona present in the corolla mouth; ovary syncarpous **56. *Melodinus***
- Corolline corona absent; ovary apocarpous or syncarpous 140
140. Anthers with sterile, sagittate, lignified bases and guide rails; calycine colleters present or absent (Tabernaemontaneae-Tabernaemontaninae) 141
- Anthers without sterile sagittate bases and guide rails, or, if present, then not lignified; calycine colleters mostly absent 142
141. Corolla infundibuliform; stamens inserted in the upper part of the corolla tube; nectary disc present **43. *Callichilia***
- Corolla salverform; stamens inserted near the base of the corolla tube; nectary disc adnate or indistinct **50. *Tabernaemontana***
142. Ovary syncarpous; fruit mostly berries, rarely capsules 143
- Ovary apocarpous (rarely partially syncarpous at the base); fruit various 158
143. Corolla infundibuliform to tubular-campanulate, bright yellow or purple, the tube $>$ 3.5 cm long; nectary disc well-developed; fruit a dry (usually spiny) capsule; seeds with a concentric wing or flattened margin; pollen colporate **68. *Allamanda***
- Corolla tube salverform, smaller; mostly cream or white, greenish white or pale yellow or orange; nectary disc absent or present; fruit dehiscent or indehiscent, but never a spiny capsule; seeds and pollen various 144
144. Plants with grappling tendrils often formed from modified inflorescences (Willughbeieae) 145
- Plants without tendrils 155
145. Calycine colleters numerous, 3–5-seriate across the base of the sepals **24. *Vahadenia***
- Calycine colleters absent or, more rarely, few in the sepal sinuses 146
146. Inflorescences terminal, laxly branched, with an elongate peduncle and flowers $\pm$ sessile 147
- Inflorescences axillary or terminal, often short-pedunculate and congested and appearing fasciculate, sometimes with longer peduncles 150
147. Ovary pubescent 148
- Ovary glabrous 149
148. Anthers dorsally with 2 vertical ridges (keeled); fruit velutinous on the outer surface; plants of tropical Africa, Madagascar and the Comoro Islands **26. *Ancylobothrys***
- Anthers dorsally without vertical ridges (unkeeled); fruit glabrous on the outer surface; plants of the

Guianas, Amazonian Brazil and Bolivia **30. *Pacouria***

149. Corolla lobes with hairs 1.2–5 mm long on the margin that is covered in bud; plants of Africa **25. *Dictyophleba***
- Corolla lobes either completely glabrous or with only short cilia on the margin; plants of Indomalaysia **21. *Willughbeia***

150. Anthers with a dorsal ridge (keeled); wall of the corolla tube not thickened above the level of stamen insertion; secondary veins 16–32 pairs, closely spaced **29. *Orthopichonia***
- Anthers without a dorsal ridge (not keeled); wall of the corolla tube thickened above the level of stamen insertion or not; secondary veins typically 5–9(–19) pairs, more widely spaced 151

151. Corolla tube thickened above the level of stamen insertion **31. *Landolphia***
- Corolla tube not thickened above the level of stamen insertion 152

152. Inflorescences terminal only **28. *Saba***
- Inflorescences mostly axillary, sometimes also terminal 153

153. Endosperm thin (less than 1 mm); cotyledons thick (2–3 mm), cartilaginous; plants of Asia and Africa 154
- Endosperm thick (ca. 20 mm), cartilaginous; cotyledons thin (ca. 0.2 mm); plants of Africa **23. *Clitandra***

154. Ovary glabrous; plants of Indomalaysia **21. *Willughbeia***
- Ovary pubescent; plants of Nigeria and C Africa **22. *Cylindropsis***

155. Nectary disc present, a fleshy ring surrounding the base of the ovary; calycine colleters usually present; each placenta with 1–6 ovules; pollen colporate 156
- Nectary disc absent, adnate or indistinct; calycine colleters absent; each placenta normally with numerous ovules; pollen porate 157

156. Nectary disc up to half as high as the ovary; inflorescences axillary cymes; seeds not compressed, embedded in pulp; plants of SE Asia and southern China **18. *Bousigonia***
- Nectary disc higher than the ovary; inflorescences terminal thyrses; seeds compressed, not embedded in pulp; plants of W and C Africa **20. *Cyclocotyla***

157. Leaves usually punctate beneath; corolla lobes strongly inflexed in bud; fruit fleshy, nearly berry-like at first, becoming leathery to dry at maturity and splitting open on at least one side; seeds angular to ovoid, not compressed, with a pulpy to corky aril on the funicle **61. *Chilocarpus***
- Leaves not punctate beneath; corolla lobes not inflexed in bud; fruit a dry, usually winged or strongly ribbed septicidal capsule; seeds compressed, with a wing at each end **63. *Plectaneia***

158. Leaves with recurved hairs with multicellular bases on the adaxial surface; nectar disc present; fruit a pair of 1-seeded drupes with short recurved hairs **73. *Anechites***
- Leaves without recurved hairs; nectary disc absent or indistinct; fruit various, but never with recurved hairs 159

159. Calycine colleters present, covering at least 1/3 of the inner surface of the sepal; fruit a pair of berries **58. *Hunteria***
- Calycine colleters absent; fruit various, dehiscent or indehiscent 160

160. Carpels 3–5, with 2 ovules each; inflorescences ± sessile, mostly axillary; corolla tube 10–22 mm long; fruit of 3–5 berries **60. *Pleiocarpa***
- Carpels 2, with 2–16 ovules each; inflorescences various; corolla tube mostly ≤ 10 mm long; fruit dehiscent or indehiscent, a capsule or a pair of follicles or drupes 161

161. Leaves abaxially with numerous black dots; secondary veins numerous, ± straight and inconspicuous; anthers with sterile apical appendage and sagittate base; fruit a pair of obliquely ovoid, compressed follicles; seeds with a concentric wing **53. *Pycnobotrya***
- Leaves abaxially without black dots; secondary venation various; anthers without sterile apical appendage and sagittate base; fruit a pair of drupes or a capsule (Alyxieae) 162

162. Corolla lobes inflexed in bud, sometimes with a ligulate appendage on one margin; fruit a pair of woody indehiscent drupes (sometimes only a single carpel developing); pollen seemingly inaperturate, in tetrads at maturity **62. *Condylocarpon***
- Corolla lobes not inflexed in bud, without a ligulate appendage; fruit drupaceous or capsular; pollen 2–3-porate, in single grains at maturity 163

163. Ovary glabrous; ovules 10–16 per carpel; fruit a single dry, septicidal capsule; seeds compressed, with a wing at each end; endosperm not ruminate **63. *Plectaneia***
- Ovary glabrous or pubescent; ovules 2–7 per carpel; fruit a pair of fleshy torulose drupes with 1–7 articles each (sometimes only one carpel developing) **67. *Alyxia***

164. Nectary absent; latex often translucent (Nerieae-Alafiinae) 165
- Nectary present; latex mostly white, sometimes translucent or other colors 168

165. Petaloid corona present in the corolla mouth; seed coma rostrate with long hairs also along the rostrum **85. *Strophanthus***
- Petaloid corona not present in the corolla mouth; seed coma rostrate or not 166

166. Stamens inserted in the mouth of the corolla, with distinct filaments, the anthers almost completely exserted; style-head with a membranous basal collar **86. *Isonema***
- Stamens inserted at various levels in the corolla tube, the filaments very short, the anthers ± subsessile, completely included or just the apices exserted; style-head without a basal collar 167

167. Style filiform, tuberculate, not enlarged at the apex; corolla tube 10–21 mm long, the lobes without conspicuous cilia at the margin; follicles rust-brown puberulent on the outer surface; chalazal coma present **88. *Farquharia***
- Style rather thick and enlarged at the apex, not tuberculate; corolla tube mostly 4–11 mm (rarely up to 21 mm) long, the lobes often conspicuously ciliate at the margin; follicles glabrous on the outer surface; chalazal coma absent **87. *Alafia***

168. Carpels postgenitally united; fruit syncarpous, of two follicles fused (sometimes only weakly so) along their ventral margins, which split apart at or near maturity 169
- Carpels free; fruit apocarpous, of 2 separate follicles, each of which dehisces along its ventral margins 180

169. Anthers dorsally with a rounded hump-like swelling at the base (corona) **138. *Vallaris***
- Anthers dorsally without a rounded swelling 170

170. Corolla infundibuliform to tubular-campanulate, usually relatively large and showy, mostly > 2.5 cm long 171
- Corolla salverform, tubular, urceolate or rotate (if campanulate, then the corolla tube < 1 cm long), often small and inconspicuous, sometimes larger and showy (Echiteae) 174

171. Anthers exserted or included, filaments 1–6 cm long; corolla tube up to 13 cm long **139. *Beaumontia***
- Anthers included, filaments usually shorter; corolla tube less than 10 cm long 172

172. Calycine colleters solitary and centered at the base of each sepal; latex translucent **120. *Temnadenia***
- Calycine colleters alternisepalous, singly or in groups, or spread ± irregularly in a ring around the base of the calyx; latex white; corolla tube pubescent on the outer surface (Apocyneae) 173

173. Corolla lobes 1/5 to 1/7 the length of the corolla tube; calycine colleters in a ring around the base of the calyx; ovary superior, glabrous; style-head with a basal collar; fruit at maturity with corky warts, velvety pubescent **147. *Amalocalyx***
- Corolla lobes longer than 1/5 the length of the corolla tube; calycine colleters few, alternisepalous; ovary partly inferior to inferior, pubescent; style-head without a basal collar; fruit at maturity smooth, without corky warts, not velvety pubescent **140. *Parepigynum***

174. Corolla salverform, the tube at least 9 mm long, with an uninterrupted thickened annulus in the throat **131. *Prestonia***
- Corolla rotate, tubular, urceolate or campanulate, the tube less than 9 mm long, without a thickened annulus in the throat, or if annulus present, then deeply divided, of a ring of linear-lanceolate lobes 175

175. Upcurved knob-like corona lobes present just above anthers **130. *Artia***
- Upcurved knob-like corona lobes absent 176

176. Orifice of the corolla tube with ring of linear-lanceolate corona lobes **124. *Thoreauea***
- Orifice of the corolla tube without a ring of corona lobes 177

177. Inflorescences umbelloid, the showy flowers clustered at the ends of long peduncles; corolla and ovary completely glabrous; plants of Mexico **123. *Thenardia***
- Inflorescences mostly branched several times, usually not umbelloid, or if the small flowers clustered at the end of long peduncles, then ovary pubescent or corolla pubescent either on inner or outer surface; plants of the Old World 178

178. Anthers partially to completely exserted from the corolla tube **129. *Parsonsia***
- Anthers included within the corolla tube 179

179. Corolla urceolate, glabrous in the throat; stamens inserted in the corolla throat, the filaments curving downward and then back up **128. *Ecua***
- Corolla mostly tubular, cylindrical rotate or salverform, mostly with hairs in a ring or in patches in the throat; stamens inserted at various levels in the corolla tube, the filaments straight or upcurved, sometimes strongly geniculate or twisted **129. *Parsonsia***

180. Pedicels with more than 2 bracts; plants usually with an indument, pubescent to hirsute or strigillose; inflorescences not or scarcely branched 181
- Pedicels with 0–2 bracts; plants glabrous or with an indument; inflorescences various 182

181. Corolla campanulate, the tube 6–9 mm long; inflorescences 1–3-flowered; calycine colleters absent; seed coma sessile **112. *Elytropus***
- Corolla salverform to subinfundibuliform, the tube 10–80 mm long; inflorescences up to 20-flowered; calycine colleters solitary, centered at the base of the each sepal adaxially; seed coma rostrate **122. *Macropharynx***

182. Anthers partially to completely exserted from the corolla tube 183
- Anthers included within the corolla tube 191

183. Anther connective with a long filiform apical appendage; upper expanded part of the corolla much longer than the narrow cylindrical basal part; seed coma rostrate **111. *Pentalinon***
- Anther connective without a long filiform apical appendage; ratio of upper and lower part of the corolla tube various; seed coma sessile or rostrate 184

184. Staminal filaments distinct; calycine colleter arrangement various, sometimes obscure or absent 185
- Staminal filaments scarcely developed, the anthers ± sessile; calycine colleters present, several in a row across the base of the calyx abaxially 190

185. Corolla rotate, tube ≤ 3 mm long; anthers conspicuously exserted on long, slender filaments 186
- Corolla tube salverform or tubular at the base with a slightly expanded upper cupular part, usually > 3 mm long 187

186. Staminal filaments not coiled, fused around the style for at least some of their length; calycine colleters present at the margins of the sepals; abaxial domatia often present in the axils of the secondary veins of the leaves **103. *Pinochia***
- Staminal filaments tightly coiled around the style, but not fused; calycine colleters absent; abaxial domatia absent **156. *Dewevrella***

187. Corolla with a thickened annular corona present in the mouth and/or with a finger-like corona lobe just above each anther; style without swollen section; plants of the Neotropics 188
- Corolla without a thickened annular corona in the mouth or a finger-like corona lobes; section of style distinctly swollen; plants of the Paleotropics **149. *Pottsia***

188. Annular corona absent; finger- or knob-like corona lobes present on the corolla just above stamen insertion; anthers completely to partially exserted **119. *Hylaea***
- Annular corona present in the corolla tube mouth; finger- or knob-like free corona lobes just above anther insertion present or absent; only the apices of the anthers exserted 189

189. Sepals each with a solitary colleter in the center at the base adaxially **131. *Prestonia***
- Sepals without colleters **118. *Laubertia***

190. Style-head with a turbinate, upturned cup-like lower part; anthers barely exserted from corolla tube **146. *Micrechites***
- Style-head without an upturned cup-like lower part; anthers distinctly exserted from corolla tube **145. *Trachelospermum***

191. Nectary of 2 separate lobes alternating with the carpels; stamens inserted in the lower 1/5 of the corolla tube **93. *Carruthersia***
- Nectary of 5 free lobes or these fused to varying degrees into an annular nectary disc, this often crenate or lobed; stamens inserted at various levels 192

192. Corolla infundibuliform to tubular-campanulate, (1.5–)2.5–4.5(–6) cm long, large and showy; plants of the Neotropics 193
- Corolla salverform, subsalverform, tubular, urceolate or rotate (if campanulate or tubular-campanulate, then the corolla tube < 1 cm long), sometimes small and inconspicuous, sometimes larger and showy; plants of the Neo- or Paleotropics 197

193. Calycine colleters absent; inflorescences usually with only 2 flowers; calyx lobes foliaceous to subfoliaceous; seed coma rostrate **102. *Rhabdadenia***
- Calycine colleters present; inflorescences normally with > 2 flowers; calyx lobes foliaceous or not; seed coma rostrate or not 194

194. Calycine colleters solitary (but sometimes deeply dissected), centered at the base of each sepal; seed coma rostrate (Echiteae-Echitinae) 195
- Calycine colleters alternisepalous, often quincuncially arranged, sometimes spread irregularly across the base of each sepal; seed coma rostrate or sessile (Odontadenieae) 196

195. Corolla throat with conspicuous, 2–3 mm long white hairs; anther bases obtuse to broadly acute; plants of

Mexico, Central America and the Caribbean **126. *Echites***

- Corolla throat glabrous to inconspicuously puberulent; anther bases slender and acuminate; plants of the Atlantic coast of Brazil **127. *Bahiella***

196. Style-head without a well-developed membranous collar at the base; seed coma sessile; calyx lobes often closely imbricate at anthesis and conspicuously unequal in size; plants of Central and South America (2 species reaching the West Indies) **108. *Odontadenia***

- Style-head with a well-developed membranous collar at the base; seed coma rostrate; calyx lobes not closely imbricate at anthesis or conspicuously unequal in size; plants of Florida and the West Indies **110. *Angadenia***

197. Anthers with a conspicuous tuft of soft hairs at the apex; plants of C China and SE Asia (Apocyneae-Amphineuriinae) 198

- Anthers without a conspicuous tuft of soft hairs at the apex; plants of the Paleo- or Neotropics or subtropics 199

198. Ovary pubescent; nectary disc a ring, sometimes somewhat 5-lobed; calycine colleters in a continuous ring; pedicel not twisted in fruit **136. *Sindechites***

- Ovary glabrous; nectary disc irregularly 2-lobed; calycine colleters alternisepalous; pedicel twisted in fruit **137. *Streptoechites***

199. Leaves usually with abaxial domatia in at least some axils of the secondary veins; ovary pubescent, at least at the apex; seed coma sessile (Apocyneae) 200

- Leaves without abaxial domatia; ovary pubescent or glabrous; seed coma rostrate or sessile 201

200. Calycine colleters alternisepalous or absent; seed body pubescent **143. *Urceola***

- Calycine colleters spread in a ring around the calyx; seed body glabrous **148. *Baharuia***

201. Flowers medium-sized to large and showy; corolla tube at least 1 cm long 202

- Flowers mostly small, inconspicuous or somewhat showy; corolla tube $<$ 1 cm long 215

202. Calycine colleters solitary (sometimes deeply dissected), centered at the base inside each sepal, rarely absent; plants of the New World (Echiteae) 203

- Calycine colleters alternisepalous or spread $\pm$ in a ring around the base inside the calyx (occasionally absent); plants of the Paleotropics (except *Odontadenia*) 206

203. Orifice of the corolla tube with a thickened annular corona in the throat and/or a knob-like corona lobe just above each anther **131. *Prestonia***

- Corolla tube without an annular corona in the throat or a knob-like corona lobe above each anther 204

204. Bracts and calyx lobes lanceolate to ovate, $>$ 5 mm wide, markedly foliaceous; inflorescences simple bostrychoid racemes **125. *Asketanthera***

- Bracts and calyx lobes mostly trigonal, narrowly lanceolate, not foliaceous; inflorescences compound dichasial or helicoid cymes 205

205. Leaves coriaceous with revolute margins and inconspicuous secondary and tertiary venation; anther bases slender and acuminate; plants of E Brazil **127. *Bahiella***

- Leaves membranous, without revolute margins; secondary and tertiary venation conspicuous; anther bases obtuse to broadly acute; plants of S Florida, the West Indies, Mexico and Central America **126. *Echites***

206. Sepals longer than the corolla tube (very rarely same length), mostly subfoliaceous, puberulent to tomentose or strigillose on the outer surface **150. *Aganosma***

- Sepals shorter than the corolla tube, subfoliaceous or not, glabrous or with an indument 207

207. Calyx lobes deciduous; anthers only weakly attached to the style-head; seeds without a coma **94. *Eucorymbia***

- Calyx lobes persistent; anthers firmly attached to the style-head; seeds with a coma 208

208. Calyx lobes connate into a tube at the base **144. *Chonemorpha***

- Calyx lobes $\pm$ free to the base 209

209. Corolla tube inside glabrous above the level of stamen insertion; seed coma sessile; plants of northern South America **108. *Odontadenia***

- Corolla tube inside pubescent above the level of stamen insertion; seed coma rostrate or sessile; plants of East, South, and SE Asia (Apocyneae) 210

210. Nectaries $\pm$ separate; seed coma sessile 211

- Nectaries fused into an often 5-lobed or crenulate disc; seed coma sessile or rostrate 212

211. Anthers inserted in the lower half of the corolla tube; inflorescences flat-topped and umbelliform; style-head with a short basal collar **152. *Epigynum***
- Anthers inserted above the middle of the corolla tube; inflorescences not flat-topped or umbelliform; style-head without a basal collar **145. *Trachelospermum***

212. Calycine colleters few, alternisepalous; seed coma on a long, hairy rostrum **134. *Anodendron***
- Calycine colleters either distributed in a continuous ring or absent; seed coma sessile or at the end of a glabrous rostrum 213

213. Leaves with a marginal vein that is as well-developed as the secondary veins; corolla lobes not or only weakly falcate and strap-shaped **135. *Amphineurion***
- Leaves without a marginal vein that is as strongly developed as the secondary veins; corolla lobes distinctly falcate and widening from base 214

214. Corolla lobes distinctly shorter than corolla tube; seed coma sessile **152. *Epigynum***
- Corolla lobes as long as or longer than corolla tube; seed coma at the end of a glabrous rostrum **144. *Chonemorpha***

215. Calyx lobes usually narrowly lanceolate, longer than the corolla tube **150. *Aganosma***
- Calyx lobes shorter than the corolla tube, mostly deltoid or ovate, sometimes lanceolate 216

216. Calycine colleters spread in a ring; seed coma sessile or long rostrate with the coma arising only at the end 217
- Calycine colleters restricted to the margins of the sepals (alternisepalous), usually in a quincuncial arrangement, rarely absent; seed coma sessile or shortly rostrate or long rostrate but with the coma arising along the rostrum 219

217. Leaves with a marginal vein that is as well-developed as the secondary veins; corolla throat without vertical ridges; seed coma sessile **135. *Amphineurion***
- Leaves without a marginal vein that is as strong as the secondary veins; corolla throat with vertical ridges in the staminal sectors; seed coma sessile or rostrate (Apocyneae-Chonemorphinae) 218

218. Inflorescences panicles; calyx lobes fused into a short cup at the base; nectary disc an entire 5-crenate ring; seed coma rostrate **144. *Chonemorpha***
- Inflorescences cymes; calyx lobes essentially free to the base; nectary of five distinct lobes; seed coma sessile **145. *Trachelospermum***

219. Ovary puberulent to tomentose or hirsute 220
- Ovary glabrous 223

220. Corolla red or pink; thickened annular corona or corona of 5 lobes present in the throat; seed coma rostrate (Apocyneae-Papuechitinae) 221
- Corolla white, greenish white, yellowish, or rarely orange; corolline corona absent; seed coma sessile 222

221. Corona of 5 distinct lobes in corolla throat; inner surface of the corolla tube glabrous above the insertion of the stamens; plants of New Guinea and the islands of Aru and Ambon **132. *Papuechites***
- Corona a thickened annular ring in corolla throat; inner surface of the corolla tube pubescent above the insertion of the stamens; plants endemic to Vietnam **133. *Ixodonerium***

222. Nectary annular; corolla urceolate or campanulate, not constricted at the throat; corolla lobe aestivation dextrorse or valvate; seed body hirsute **143. *Urceola***
- Nectary of 5 distinct narrow lobes; corolla salverform, constricted at the throat; corolla lobe aestivation dextrorse; seed body glabrous **151. *Ichnocarpus***

223. Nectary disc usually annular; seeds with a rostrate coma, the rostrum also with long hairs along its length **134. *Anodendron***
- Nectary of 5 separate lobes, or these fused only at the base; seed coma sessile or at the most with a very short beak and the hairs only at the end 224

224. Corolla tube ca. 2 mm long, the lobes remaining slightly inrolled at anthesis; plants of Sri Lanka, China and SE Asia **141. *Cleghornia***
- Corolla tube 5–18 mm long, the lobes spreading at anthesis; plants of the New World (Odontadenieae) 225

225. Calyx lobes usually aristate and keeled, with calycine colleters in pairs or groups in the sinuses; follicles slender, ca. 4 mm diam.; plants of SE USA **104. *Thyrsanthella***
- Calyx lobes not aristate or keeled, with calycine colleters single in the sinuses; follicles robust, 1.5–3.5 cm. diam.; widespread in South America **107. *Secondatia***

I. Rauvolfioids

Rauvolfioideae Kostel. (1834) (subfamily rank dubious).
Plumerioideae Luerss. (1882).
Cerberoideae Pichon (1948).
Tabernaemontanoideae Stapf ex Boiteau & Sastre (1975).

Trees, shrubs, woody lianas or vines, rarely herbs; latex typically white (this very rarely bluish tinged), rarely red, orange or yellow. Leaves opposite, whorled, or alternate. Inflorescences cymose or racemose. Flowers almost always 5-merous (4-merous in *Leuconotis*), inconspicuous to large and showy; calyx almost always 5-merous, rarely 4- (*Parahancornia*), 6- or 7-merous; calycine colleters often absent, when present usually at the base of the sepals in a single series (but often multiseriate in Tabernaemontaneae, some Hunterieae and paleotropical Willughbeieae); corolla mostly salverform, more rarely with inflated throat (tubular-campanulate or infundibuliform, mainly in some Tabernaemontaneae and Plumerieae); corolla lobe aestivation almost always sinistrorsely contorted (but dextrorsely contorted in *Kopsia*, *Ochrosia*, some species of *Haplophyton*, *Alstonia* and *Tabernaemontana*); corona when present almost always in the staminal sectors, usually of simple lobes or pouches in the petal sinuses (rarely fused into an annulus), sometimes lower down on the corolla tube as small lobes or thickenings above the stamens; stamens inserted at various levels in the corolla tube, almost always included in the corolla tube (partially exserted in some Tabernaemontaneae and Plumerieae), the filaments mostly short or anthers subsessile; anthers usually fertile to the base (fertile in the upper part only and with sterile lignified basal appendages in most Tabernaemontaneae, and with small sterile, non-lignified basal appendages and apical appendage in some Melodineae and Hunterieae, and with non-lignified basal appendages in *Allamanda*), free from the style-head; pollen mostly 3–4-colporate, but typically porate and often with only 2 apertures in Alyxieae and a few Melodineae; nectaries as free lobes surrounding the base of the ovary (these sometimes fused into a crenulate or lobed ring), or indistinct and adnate to the base of the outer ovary wall or absent; gynoecium almost always bicarpellate (but normally with 3–5 carpels in *Pleiocarpa* (Hunterieae), and *Lepinia* and *Lepiniopsis* (Alyxieae)); ovary normally superior, but half-inferior in some genera (e.g., *Plumeria*, *Himatanthus* and *Mortoniella*); ovary congenitally syncarpous or apocarpous (postgenitally syncarpous in *Allamanda*); ovules few to many; style-head differentiated vertically, usually the tips of the carpels unfused and forming two unreceptive free appendages at the top, sometimes with a crest of longer hairs below this and/or a collar or expanded flange at the base, beneath which the stigmatic region is normally found, more rarely the style-head is not noticeably differentiated and the entire body is more or less uniformly receptive. Nectaries free, surrounding the base of the ovary, indistinct and adnate to the base of the outer ovary wall or absent. Fruit apocarpous or syncarpous, dehiscent or indehiscent; pericarp dry or fleshy; endocarp sometimes stony. Seeds generally naked or winged, testa glabrous or hairy, smooth, or pitted, ridged or rugulose, without coma (except in *Haplophyton*); endosperm smooth, sometimes ridged or strongly ruminate. Indole alkaloids often present, cardenolides less frequently. 79 genera, most either in the Paleo- or Neotropics.

Key to the Tribes of Rauvolfioids

1. Anthers sagittate at the base, lignified (except in some species of *Tabernaemontana*); leaves almost always opposite; calycine colleters usually present; fruit fleshy, either a pair of follicles (these rarely hemi-syncarpous or indehiscent) with arillate seeds or a berry **5. Tabernaemontaneae** (p. 266)
– Anthers rarely sagittate at the base, non-lignified; leaves opposite, whorled or alternate; calycine colleters usually absent, sometimes present; fruit fleshy or dry, berries, drupes, follicles or capsules; arils absent (except in *Chilocarpus*) 2
2. Ovary congenitally syncarpous; fruit a berry; phyllotaxis almost always opposite 3
– Ovary apocarpous (syncarpous in *Chilocarpus*, *Lepiniopsis* and *Allamanda* and hemi-syncarpous in *Diplorhynchus*); fruit various, dehiscent or indehiscent; phyllotaxis various (opposite, whorled or alternate) 5
3. Placentas forming a lignified partition in fruit; spines often present in leaf axils (*Carissa*); indole alkaloids absent **11. Carisseae** (p. 280)
– Placentas becoming pulpy in fruit; without spines in leaf axils; indole alkaloids present or absent 4
4. Corolline corona present; lianas without tendrils **7. Melodineae** (p. 271)
– Corolline corona absent; trees, shrubs or lianas (the latter often with tendrils), rarely rhizomatous subshrubs **4. Willughbeieae** (p. 261)

5. Ovary 2–5-carpellate; fruit berries **8. Hunterieae** (p. 273)
- Ovary 2-carpellate (except in some Alyxieae); fruit dehiscent or indehiscent, but not berries 6
6. Anthers with latrorse to sublatrorse dehiscence and mostly with elongated cylindrical apical connective appendages; epistaminal appendages usually present on the corolla **10. Plumerieae** (p. 277)
- Anthers with introrse dehiscence, normally lacking elongated apical appendages (but large flat apical appendages present in *Vinca*); epistaminal appendages normally absent 7
7. Forest trees of the Paleotropics, Australia and Pacific islands; leaves whorled, with up to 9 per node (opposite in some species of *Alstonia*); both carpels usually maturing; fruit a pair of several-seeded follicles; seed margin with cilia, hairs or wings **2. Alstonieae** (p. 257)
- Shrubs, trees, lianas or herbs of the Paleotropics, Australasia, Oceania and temperate Eurasia and the New World; leaves opposite, alternate or if whorled, then usually 3–5 per node; sometimes only one carpel maturing; fruit mostly indehiscent, few-seeded drupes or several-seeded follicles, rarely berries; seed margin usually naked or winged 8
8. Flowers large, showy; corolla usually > 2 cm long, often thick, waxy in appearance; ovary partly inferior in *Plumeria*, *Himatanthus* and *Mortoniella*; leaves usually alternate, rarely opposite (whorled only in *Allamanda*); corolla tube without slits above stamen insertion; indole alkaloids absent; $x = 9$ **10. Plumerieae** (p. 277)
- Flowers small to medium-sized; corolla usually < 2 cm long, usually thin, delicate in texture (but sometimes somewhat thick in *Ochrosia*); ovary almost always superior; leaves often whorled, sometimes alternate or opposite; corolla tube normally with slits above stamen insertion; indole alkaloids present or absent; $x = 10, 11$ 9
9. Pollen porate; style-head without membranous basal collar; seeds not compressed, usually with a hilar groove and ruminate endosperm; indole alkaloids absent **9. Alyxieae** (p. 274)
- Pollen colporate; style-head with or without membranous basal collar; seeds compressed or not, without hilar groove or ruminate endosperm; indole alkaloids present 10
10. Style-head cylindrical, almost always with a basal collar or wreath of hairs at the base (except in *Kamettia*) 11
- Style-head mostly ovoid, usually without a basal collar or wreath of hairs at the base 12
11. Leaves whorled or opposite (if alternate, then seeds covered with long lanate hairs) **3. Vinceae** (p. 258)
- Leaves alternate; seeds naked **6. Amsonieae** (p. 271)
12. Corolline corona often present as small lobes in throat or ridges behind anthers; anthers sometimes with short, sterile apical appendages (*Diplorhynchus*) or with short sterile appendages at apex and thecae bases (*Pycnobotrya*); leaves opposite; fruit follicular, usually both carpels developing; trees or lianas of the Paleotropics, Australasia and Oceania **7. Melodineae** (p. 271)
- Corolline corona absent; anthers without apical or basal extensions; leaves most frequently alternate, rarely whorled or opposite; fruit and seeds very diverse, often only one carpel maturing; trees, shrubs or subshrubs of the tropics and subtropics of the New World **1. Aspidospermateae** (p. 255)

I.1. Tribe Aspidospermateae Miers (1878).

Trees or shrubs, rarely subshrubs; latex in some species of *Aspidosperma* red, orange or yellow. Leaves alternate, less frequently whorled or opposite. Inflorescences axillary and/or terminal, cymose, dichasial or thyrsiform, few- to many-flowered. Flowers mostly small; calyx rarely 4-, 6- or 7-merous (some species of *Aspidosperma*), calycine colleters absent; corolla salverform to tubular with slits in the tube behind the stamens (i.e., incomplete fusion of the postgenitally fused upper corolla tube); corolla lobe aestivation sinistrorse (dextrorse in *Geissospermum* and one species of *Haplophyton*); corona absent; stamens usually inserted about midway or above in the corolla tube; pollen 3-colporate; nectary disc absent, adnate or inconspicuous; ovary apocarpous; ovules few to several per carpel; style-head mostly without apical crest or basal collar (with a short basal collar in *Haplophyton*), concomitantly receptive and secretory. Fruit apocarpous; normally with dry pericarp and dehiscent, a pair of follicles, but often only one carpel reaching maturity, but fruit fleshy in *Geissospermum* and *Vallesia*; endocarp usually not stony, but somewhat woody in *Vallesia*. Seeds various: thin and compressed with papery wing in *Aspidosperma* and *Strempeliopsis*, or seeds cylindrical without wing or hairs (*Vallesia*, *Geissospermum*, *Microplumeria*). Indole alkaloids present.

x = 10, 11. Six genera in the tropics and subtropics of the New World.

1. ***Strempeliopsis*** Benth.

Strempeliopsis Benth. in Benth. & Hook. f., Gen Pl. 2: 702 (1876).

Small trees or shrubs. Leaves opposite. Inflorescences terminal, cymose, many-flowered; corolla salverform, greenish white; tube inside with moniliform hairs above the stamens; ovules 10–14 per carpel. Follicles slender, ± terete. Seeds several per carpel, compressed, with a papery wing at each end.

Two spp., one on Cuba, the other on Jamaica.

2. ***Vallesia*** Ruiz & Pav.

Vallesia Ruiz & Pav., Fl. Peruv. Prodr. 28, t. 5 (1794); Morales, Novon 8: 263–264 (1998), tax.; Morales, Anales Jardin Bot. Madrid 66: 257–258 (2009), reg. rev.

Trees or shrubs. Leaves alternate. Inflorescences axillary, congested, cymose. Corolla salverform, white or yellowish; ovules 6 per carpel. Fruit an opalescent, one-seeded drupe, usually only one carpel maturing. Seeds not compressed, slightly arcuate. 2n = 22.

Nine spp. from Florida to the West Indies, Mexico, Central America, the Galapagos and Colombia, to N Argentina, usually in wooded areas, often along rivers.

3. ***Haplophyton*** A. DC.

Haplophyton A. DC., Prodr. 8: 412 (1844); Williams, Sida 16: 469–475 (1995), rev.

Subshrubs. Leaves alternate, occasionally opposite. Inflorescences axillary or terminal, 1–4-flowered. Sepals glandular at the apex; corolla salverform, cream flushed with pink; corolla lobe aestivation sinistrorse in one species, dextrorse in the other; ovules several. Follicles slender, thin-walled. Seeds with longitudinal grooves and a deciduous tawny to gray coma at the chalazal and the micropylar end.

Two spp. in SW USA to S Mexico, often in rocky or open areas, hillsides or canyons, at low to mid elevations.

4. ***Microplumeria*** Baill.

Microplumeria Baill., Bull. Mens. Soc. Linn. Paris 1: 749 (1889); Zarucchi, Fl. Venez. Guayana 2: 531 (1995), reg. rev.

Small tree or shrub. Leaves in whorls of three or opposite. Inflorescences terminal, cymose, few- to many-flowered. Corolla tubular, densely silvery green villous outside, the lobes much shorter than the tube; ovules numerous. Follicles slender, thin-walled. Seeds cylindrical, diagonal at the ends (as if broken off).

One sp., *M. anomala* (Müll. Arg.) Markgr., in NW South America, usually along seasonally flooded river banks or along streams, at low elevations.

5. ***Geissospermum*** Allemão

Geissospermum Allemão, Pl. Novas Brasil 707 (1846); Gentry, Ann. Missouri Bot. Gard. 71: 1075–1081 (1984), synop.; Zarucchi, Fl. Venez. Guayana 2: 503 (1995), reg. rev.

Trees. Leaves alternate, usually silvery sericeous. Inflorescences normally arising from between nodes, cymose, few- to many-flowered. Corolla tubular to salverform, white, cream to greenish or brownish; corolla lobe aestivation dextrorse; stamens subsessile; ovules many. Berries stout, leathery, containing much latex, often only one developing. Seeds few, somewhat flattened, elliptic; testa pitted, somewhat rugulose.

Five spp. in NW South America.

6. ***Aspidosperma*** Mart. & Zucc. Fig. 27

Aspidosperma Mart. & Zucc., Flora 7 (Beil. 4): 135 (1824), nom. cons.; Marcondes-Ferreira, *Aspidosperma* Mart. nom. cons. (Apocynaceae): Estudios Taxonômicos, Ph. D. Thesis, University of Campinas, Brazil: 1–431(1988), rev.; Potgieter, Phylogenetic study of Apocynaceae Juss. and *Aspidosperma* Mart. and Zucc., Ph.D. Thesis, University of Illinois-Urbana, USA: 1–530 (1999), rev.; Pereira et al., Biota Neotrop. 16: e20150080 (2016), reg. rev.; Morales & Zamora, Phytoneuron 68: 1–13 (2017), reg. rev.

Shrubs to big trees; latex white, yellow, orange or red. Leaves alternate, rarely opposite or verticillate. Inflorescences axillary and/or terminal, compound-cymose, dichasial or thyrsiform, usually

Fig. 27. Apocynaceae-Aspidospermateae. Branches, flowers and fruits. **A** *Aspidosperma excelsum*. **B** *A. pachypterum*. **C** *A. neblinae*. (From Zarucchi et al. 1995, p. 489, with permission from the Missouri Botanical Garden Press, St. Louis; drawn by B. Manara)

many-flowered. Calyx (4–)5(–7)-merous; corolla tubular to salverform, usually white, yellow or greenish; ovules few to many. Follicles dry, woody, usually compressed, nearly cylindrical to falciform or dolabriform, surface smooth, lenticillate, sometimes warty or spiny, frequently only one developing. Seeds 2–many, peltately attached, strongly compressed, with eccentric or concentric papery wing, this rarely thicker and rudimentary.

About 50 spp. with a center of diversity in Brazil, but ranging from Mexico and the Antilles to Argentina. The hard wood of some species is used to make as simple tools, tool handles, wheels small art objects, others for cabinetry and floors.

I.2. **Tribe Alstonieae** G. Don (1837).

Trees or shrubs. Leaves whorled or opposite. Flowers mostly small; calycine colleters absent except in *Dyera*; corolla salverform to tubular (rotate in *Dyera*); corolla lobe aestivation sinistrorse (dextrorse in many species of *Alstonia*); corona absent; stamens inserted midway or above in the corolla tube; filaments filiform; anthers with sterile apical and basal appendages in *Dyera*; pollen 3-colporate; nectary disc absent, adnate or inconspicuous, 2-lobed in some species of *Alstonia*; ovary apocarpous, hemi-syncarpous or rarely completely syncarpous, half inferior in *Dyera*; ovules numerous per carpel; style-head with or without a membranous basal collar. Fruit apocarpous, a pair of follicles with woody pericarp (these fused into a syncarpous "double-follicle" in *Alstonia rostrata*). Seeds thin, compressed, testa glabrous or pubescent with long hairs around the margin (*Alstonia*) or surrounded by a membranous wing (*Dyera*). Indole alkaloids present. x = 10, 11. Two genera in the Paleotropics.

7. ***Alstonia*** R. Br. Fig. 28

Alstonia R. Br., Asclepiadeae 64 (1810), nom. cons.; Monachino, Pacific Sci. 3: 133–182 (1949), rev.; Sidiyasa, Blumea Suppl. 11: 1–230 (1998), rev.; Sidiyasa, Fl. Males. I, 18: 31–64 (2007), reg. rev.; Middleton, Fl. Cambodia, Laos and Vietnam 33: 18–29 (2014), reg. rev.
Blaberopus A. DC. (1844).
Winchia A. DC. (1844).
Paladelpha Pichon (1947).

Trees or shrubs; latex white. Leaves verticillate or opposite. Inflorescences terminal, cymose. Typical calycine colleters absent, but colleters often on the edges of the sepals; corolla white, yellow or red; corolla lobe aestivation sinistrorse or dextrorse; nectary disc adnate, indistinct, sometimes 2-lobed; ovary apocarpous, hemi-syncarpous or syncarpous; ovules numerous; style-head with a basal collar. Follicles long and slender, free or rarely fused (*Alstonia rostrata* C.E.C. Fisch.). Seeds with hairs around the margins, longer at ends. 2n = 22, 42, 44, 80, 84, 88.

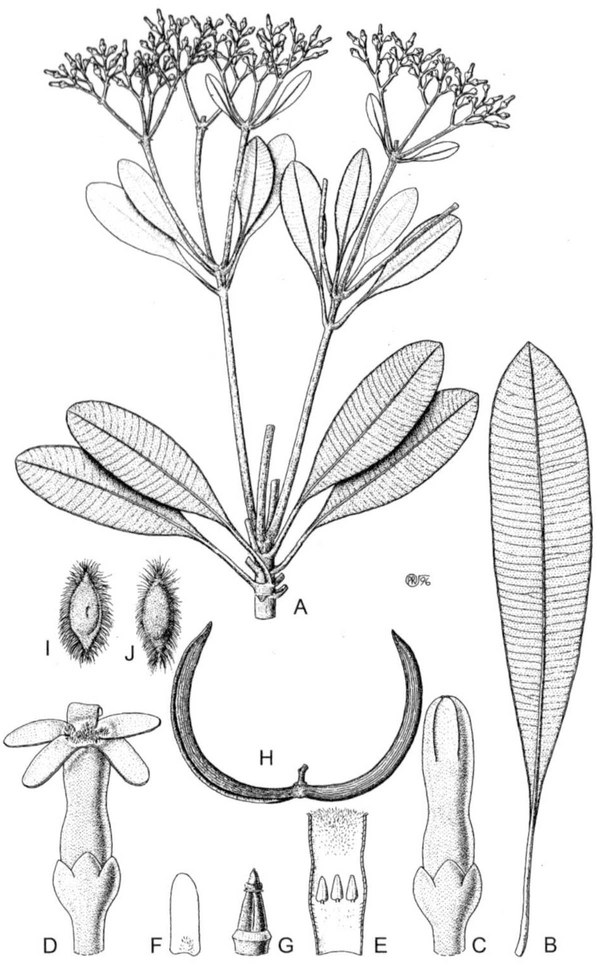

Fig. 28. Apocynaceae-Alstonieae. *Alstonia quaternata*. **A** Flowering branch. **B** Leaf. **C** Flower in bud. **D** Open flower. **E** Opened corolla tube, showing three anthers. **F** Corolla lobe. **G** Gynoecium. **H** Fruit. **I** Seed, raphe side. **J** Seed, antiraphe side. (From Sidiyasa 1998, p. 167, with permission from Naturalis, Leiden; drawn by Priyono)

About 43 spp. from Africa and China and the Himalayas to Australia and Pacific Islands, in primary and secondary forests, swamps or dry habitats, up to 2900 m.

8. ***Dyera*** Hook. f.

Dyera Hook. f., J. Linn. Soc. Bot. 19: 293 (1882); Monachino, Lloydia 9: 174–202 (1946), rev.; Middleton, Gard. Bull. Sing. 55: 209–218 (2003), rev.; Middleton, Fl. Males. I, 18: 186–190 (2007), reg. rev.

Giant trees up to 80 m and 300 cm diameter; latex white. Leaves in whorls of 4–8. Inflorescences axillary, many-flowered cymes. Calycine colleters ca. 4 per sepal; corolla small, white, greenish or pinkish yellow, lobes longer than tube, spreading; nectary disc annular, inconspicuous, adnate; ovary half inferior; ovules many; style very short; style-head scarcely enlarged, undifferentiated, without collar at base. Follicles thick, woody, up to 40 cm long, spreading. Seeds 12–24 per carpel, peltately attached, elliptic, compressed, surrounded by membranous wing.

Two spp. in the far S of Thailand and W Malesia, in lowland evergreen forest. *Dyera costulata* (Miq.) Hook. f. was once cultivated in large plantations as a source of latex ("jelutong") for chewing gum.

I.3. **Tribe Vinceae** Duby (1828).

Trees or shrubs, more rarely lianas, vines or herbs. Leaves whorled or opposite, rarely alternate. Flowers small to medium-sized, often whitish, showy in some genera; typical calycine colleters absent, but colleters at exposed edge of calyx lobes in *Ochrosia*, *Vinca* and several species of *Rauvolfia*; corolla salverform (infundibuliform in *Vinca*), the orifice usually constricted by a thickened ring and hairs; free corona lobes absent; corolla lobe aestivation usually sinistrorse (dextrorse in *Kopsia* and *Ochrosia*); stamens usually inserted at the middle of the corolla tube or above (except in *Kamettia*); filaments normally filiform and anthers without lignified guide rails or sagittate basal appendages (in *Vinca* with spathulate apical appendage); pollen mostly 3-colporate; nectaries when present often consisting of 2 lobes alternating with the carpels (a 5-lobed nectary disc in *Rauvolfia*), sometimes adnate and indistinct or absent; ovary generally apocarpous, (hemi-syncarpous or syncarpous in some species of *Rauvolfia* and *Ochrosia*), usually glabrous, often only one carpel maturing; ovules few to numerous; style usually filiform; style-head cylindrical, almost always with a wreath of hairs around the top and a basal collar or wreath of hairs at the base (except in *Kamettia*). Fruit apocarpous, of 2 separate fruiting carpels (in *Rauvolfia* often partially or wholly syncarpous and sometimes only one carpel developing); either indehiscent and drupaceous with a thick pericarp and stony endocarp or a pair of thin-walled woody to papery follicles. Seeds 1–4(–6) per carpel, rarely more, ovoid or compressed, mostly not winged (the flat margins forming a rudimentary ledge in *Ochrosia*); in *Tonduzia* and *Laxoplumeria* seeds numerous, elliptic-compressed with long hairs around the margin; endosperm usually fleshy and smooth (very thin, seemingly absent in *Kopsia*). Indole alkaloids present. x = 9, 10, 11, 23. Nine genera, pantropical, one in temperate regions of the Old World.

9. ***Kopsia*** Blume

Kopsia Blume, Catalogus 12 (1823), nom. cons.; Pichon, Mém. Mus. Natl. Hist. Nat. 27: 153–252 (1948); Markgraf, Blumea 20: 416–425 (1972), morph.; Forster, Fl. Australia 28: 134–138 (1996), reg. rev.; Middleton, Harvard Pap. Bot. 9: 89–142 (2004), rev.; Middleton, Fl. Males. I, 18: 232–260 (2007), reg. rev.; Middleton, Fl. Cambodia, Laos and Vietnam 33: 116–127 (2014), reg. rev.
Calpicarpum G. Don (1837).
Kentrochrosia K. Schum. & Lauterb. (1900).

Trees or shrubs; latex white. Leaves opposite. Inflorescences terminal, cymose. Sepals with colleter abaxially below apex; corolla often large and showy, white, yellowish or pink, lobes usually shorter than tube; stamens inserted high in corolla tube, very rarely near base; nectaries of two free lobes; ovary apocarpous; ovules 2 per carpel. Drupes one-seeded, usually with cavity forming a hooked or sharp appendage on ventral side, this rarely absent; mesocarp stringy or pulpy. Seeds curved, slightly compressed; endosperm very thin or absent. 2n = 36, 72.

23 spp. from S China and Burma to N Australia and Vanuatu.

10. ***Ochrosia*** Juss.

Ochrosia Juss., Gen. Pl. 144 (1789); Markgraf, Blumea 25: 233–247 (1979); Fosberg & Sachet, Adansonia II, 17: 19–22 (1977), nomen.; Wagner et al., Man. Fl. Pl. Hawai'i 1: 216–218 (1990), rev.; Li et al., Fl. China 16: 22–164 (1995), reg. rev.; Forster, Fl. Australia 28: 134–137 (1996), reg. rev.; Hendrian, Blumea 49: 101–228 (2004), reg. rev.; Hendrian, Fl. Males. I, 18: 289–302 (2007), reg. rev.; Hendrian & Kondo, Chromosome Bot. 2: 127–149 (2007), phyl.; Middleton, Fl. Cambodia, Laos and Vietnam 33: 144–146 (2014), reg. rev.
Lactaria Rumph. (1838).
Neisosperma Raf. (1838).
Bleekeria Hassk. (1855).
Excavatia Markgr. (1927).

Small trees; latex white. Leaves in whorls of 3–6, occasionally opposite. Inflorescences axillary cymes. Colleters at the outside edge of the sepals in most species; corolla white, cream or orange; nectary two small lobes or adnate, indistinct or absent; ovary rarely hemi-syncarpous or syncarpous; ovules 2–6 per carpel. Drupes yellow, red, purple or black; endocarp splitting into fibers or entire around two lateral cavities. Seeds 1–3 per carpel, compressed, with a wing-like structure round the margin. 2n = 22.

About 40 spp. in continental SE Asia, Malesia, N Australia, the West Pacific Islands, Marquesas and Hawaiian Islands.

11. ***Tonduzia*** Pittier

Tonduzia Pittier, Contr. U. S. Natl. Herb. 12: 103 (1908); Morales, Darwiniana 47: 176–177 (2009), reg. rev.
Alstonia section *Tonduzia* (Pittier) Pichon (1947).

Shrubs or trees. Leaves narrowly elliptic, in whorls of 3(–4). Inflorescences terminal, many-flowered, cymose. Corolla small, white to cream; lobes shorter to slightly longer than the tube; nectary disc obscure, annular, adnate to the base of the ovary; ovary apocarpous; ovules numerous. Follicles often rostrate, often stipitate. Seeds numerous, elliptic, with cilia of different shapes and sizes along the margin.

Two spp. in Mexico and Central America, in open woodlands or dense tropical forest, on slopes, hills or on rocky soil, up to 2000 m.

12. ***Laxoplumeria*** Markgr.

Laxoplumeria Markgr., Notizbl. Bot. Gart. Berlin-Dahlem 9: 981 (1926); Monachino, Phytologia 3: 67–70, synop.

Trees. Leaves alternate, unusually long-petiolate. Inflorescences terminal, many-flowered, cymose. Corolla small, white, the lobes linear, longer than the tube; nectary disc a low ring with two lobes alternate to the carpels; ovary apocarpous, glabrous; ovules numerous. Follicles up to 40 cm long, slender. Seeds numerous, compressed, elliptic, acute at the ends, the testa covered by long, brown hairs.

Five spp., in South America, one of them reaching Panama.

13. ***Vinca*** L.

Vinca L., Sp. Pl. 1: 209 (1753); Lawrence, Baileya 7: 113–119 (1959), synop.

Perennial herbs, often procumbent. Leaves opposite. Inflorescences seemingly axillary, 1-flowered. Flowers showy; colleters at outside edge of sepals in most species; corolla infundibuliform, blue, rarely violet or white, with pouch-like corona in petal sinuses, lobes mostly shorter than tube; staminal filaments thick, geniculate; anthers with hairy, spathulate apical appendage; nectary of two lobes; ovary apocarpous; ovules 6–8 per carpel; style gradually thickened toward the apex. Follicles slender, Seeds oblong with a hilar groove, testa warty. 2n = 46, 92, 32.

About seven spp. in SE Europe, SW Asia and NW Africa, some cultivated and naturalized.

14. *Kamettia* Kostel.

Kamettia Kostel., Allg. Med.-Pharm. Fl. 3: 1062 (1834); Middleton et al., Thai Forest Bull. Bot. 33: 75–80, synop.
Ellertonia Wight (1848).

Woody lianas; latex white. Leaves verticillate, sometimes opposite at some nodes. Inflorescences terminal, lax, dichasial, few- to many-flowered. Corolla red outside, white inside, the tube usually curved; stamens inserted in the lower third of the tube; nectary disc absent, adnate or indistinct; ovary apocarpous; ovules numerous; style-head ellipsoid, without upper wreath or basal collar. Follicles fusiform, divergent, often lenticellate. Seeds compressed, with a wing at each end.

Two spp., one in the Western Ghats of India, and one in Thailand.

15. *Petchia* Livera

Petchia Livera, Ann. Roy. Bot. Gard. (Peradeniya) 10: 140 (1926); Huber, Fl. Ceylon 1(1): 1–27 (1973), reg. rev.; Leeuwenberg, Wageningen Agric. Univ. Pap. 97-2: 53–80 (1997), rev.
Cabucala Pichon (1948).

Shrubs or small trees. Leaves in whorls of 3–5, often opposite at branchings. Inflorescences terminal or axillary, cymose, few-flowered. Corolla sometimes slightly zygomorphic, creamy white; nectary disc adnate, indistinct or absent; ovary apocarpous; ovules up to 26 per carpel. Fruits drupaceous, stipitate, torulose, orange or red; mesocarp fleshy. Seeds 1–several per carpel, ellipsoid, not compressed.

Eight spp. in Madagascar, the Comoro Islands, and one each in Cameroon and Sri Lanka, in the forest understory in rainforest, montane forest, or bush, up to 1700 m.

16. *Catharanthus* G. Don

Catharanthus G. Don, Gen. Hist. 4: 95 (1837); Lawrence, Baileya 7: 113–119 (1959), synop.; Plaizier, Meded. Landbouwhogeschool Wageningen 81-9: 1–12 (1981), rev.; Allorge et al., Candollea 70: 61–66 (2015), reg. rev.

Perennial or annual herbs or subshrubs; latex white. Leaves opposite. Inflorescences axillary, sessile, on alternate sides of the nodes, usually 2-flowered. Flowers showy; corolla salverform, pink, red or white, with a ring of strigose hairs in the orifice; nectary of two lobes; ovary apocarpous, usually glabrous, rarely sparsely pilose on top; ovules numerous. Follicles slender. Seeds numerous, black, oblong, with a hilar groove; testa rugose. 2n = 16, 32, 24.

Nine spp., one restricted to India and Sri Lanka, the rest endemic to Madagascar, mostly in exposed, sandy or rocky areas up to 2000 m. One sp., *C. roseus* (L.) G. Don, is a widely cultivated ornamental, and has become naturalized in many regions of the tropics. Because of its many indole alkaloids, this genus is widely used in folk medicine, and compounds from *C. roseus* have been used commercially in treating leukemia.

17. *Rauvolfia* L. Fig. 29

Rauvolfia L., Sp. Pl. 1: 208 (1753); Rao, Ann. Missouri Bot. Gard. 43: 253–355 (1999), reg. rev.; Leeuwenberg & van Dilst, Bull. Jard. Bot. Natl. Belg. 61: 21–69 (1991), reg. rev.; Hendrian & Middleton, Blumea 44: 449–470 (1999) reg. rev.; Koch, Estudos das espécies neotropicais do gênero *Rauvolfia* L. (Apocynaceae), Ph.D. Thesis, University of Campinas, Brazil: 1–292 (2002), reg. rev.; Hendrian & Middleton, Fl. Males. I, 18: 347–359 (2007), reg. rev.; Middleton, Fl. Cambodia, Laos and Vietnam 33: 160–172 (2014), reg. rev.
Podochrosia Baill. (1888).

Subshrubs to trees; latex white. Leaves in whorls of 3–5(–7), rarely opposite, sometimes

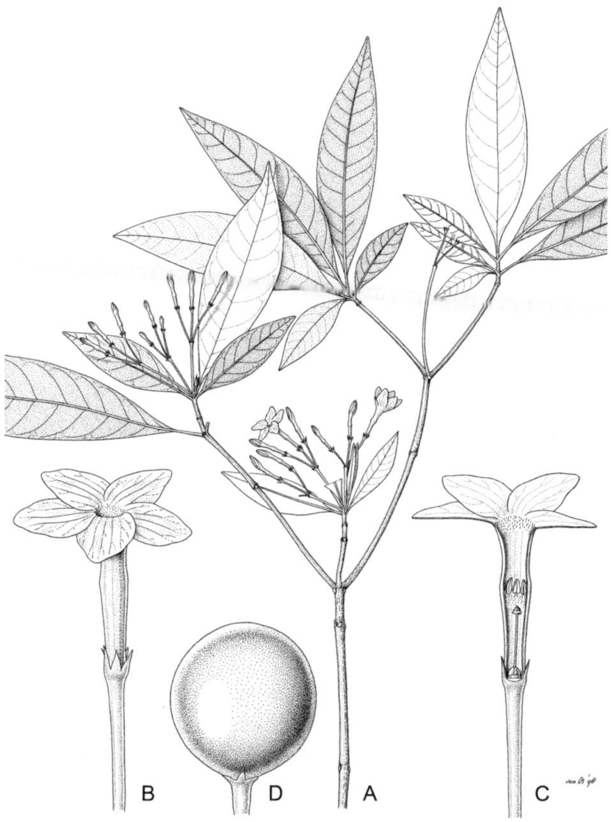

Fig. 29. Apocynaceae-Vinceae. *Rauvolfia kamarora*. **A** Flowering branch. **B** Flower. **C** Opened corolla tube showing spatial relationship between anthers and style-head. **D** Fruit with only one carpel developed. (From Hendrian and Middleton 1999, p. 455, with permission from Naturalis, Leiden; drawn by J. van Os)

anisophyllous; petioles with colleters along adaxial side. Inflorescences usually terminal, cymose. Colleters at outside edge of sepals in several species; corolla sometimes slightly zygomorphic, mostly white, sometimes other colors, the lobes normally (much) shorter than tube; nectary disc annular; ovary apocarpous to syncarpous; ovules 1–2 per carpel. Drupes variously colored, apocarpous, hemi-syncarpous to syncarpous (often only one carpel developing). Seeds 1 per carpel, compressed. 2n = 22, 44, 66, 88.

About 60 spp., pantropical in various habitats. Hundreds of different indole alkaloids have been extracted from *Rauvolfia* species, and the genus is an important plant in folk medicine throughout the regions where it occurs, and is used for treating a wide variety of ailments.

A worldwide revision for this genus is lacking.

I.4. **Tribe Willughbeieae** A. DC. (1844).

Woody lianas, often very large and climbing high into the canopy, sometimes with grappling often with tendrils, less frequently trees, shrubs or pyrophytic subshrubs; latex usually white, sometimes reddish or bluish tinged in *Hancornia*; trunk in some species of *Lacmellea* with deciduous, blunt conical spines. Leaves opposite (verticillate in *Couma*). Flowers small- to medium-sized; 4-merous in *Leuconotis*; calycine colleters mostly absent in neotropical taxa and the great majority of the paleotropical taxa (but numerous and multiseriate in *Bousigonia*, *Leuconotis* and *Cyclocotyla* (subtribe Leuconotidinae), and *Vahadenia* (subtribe Landolphiinae)), and few and alternisepalous in some species of *Ancylobothrys* and *Landolphia*; corolla salverform; corolline corona absent; corolla lobe aestivation sinistrorse, not inflexed in bud except in *Cyclocotyla*; pollen (1–)3(–4)-colporate; nectary disc normally absent, adnate or indistinct from ovary (but well-developed and higher than the ovary in *Cyclocotyla*), ovary congenitally syncarpous; uni- to bilocular; placentation parietal to axile; ovules few to numerous; style usually filiform; style-head often not clearly differentiated into different morphological regions, usually ellipsoid or ovoid to ovoid-conical, without a distinct upper crest or basal collar. Fruit indehiscent, a berry, often edible, with fleshy, non-fibrous pericarp, mostly without a sclerified layer (but sclerified layer characteristic for *Landolphia*, *Clitandra*, *Orthopichonia* and *Saba* (subtribe Landolphiinae)); placentas mostly becoming pulpy in fruit; endocarp not stony. Seeds few to numerous, ovoid or often irregular in shape, not or somewhat compressed, embedded in sweet to acidic, usually edible pulp, with or without a hilar groove; endosperm smooth or superficially ruminate longitudinally, typically either scanty and associated with cartilaginous cotyledons (subtribes Leuconotidinae, Willughbeiinae) and in *Cylindropsis* (subtribe Landolphiinae) or cartilaginous to subcorneous and associated with thin membranous cotyledons (subtribes Lacmelleinae and Landolphiinae); cotyledons usually cordate at the base, sometimes attenuate. Indole alkaloids present in subtribe Leuconotidinae. x = 11. 18 genera, pantropical.

18. ***Bousigonia*** Pierre

Bousigonia Pierre, Bull. Mens. Soc. Linn. Paris II, 1: 35 (1898); Middleton, Fl. Cambodia, Laos and Vietnam 33: 63–66 (2014), reg. rev.

Woody lianas without tendrils. Leaf blades with scattered dots abaxially. Inflorescences axillary or terminal cymes. Calycine colleters numerous, in a row; corolla white, the lobes much shorter than the tube; stamens inserted around middle of corolla tube; ovary bilocular, glabrous; placentas axile; ovules 2 per placenta; style filiform; style-head ellipsoid. Berries few-seeded, ovoid to pyriform. Seeds not or only slightly compressed, without hilar groove, testa thin, smooth; endosperm scanty to absent, cotyledons fleshy, deeply incised at the base.

Two spp. in China, Thailand, Laos and Vietnam, in mixed forest, forest edges and montane forest up to 1600 m.

19. ***Leuconotis*** Jack

Leuconotis Jack, Trans. Linn. Soc. London 14: 121 (1823); Leeuwenberg, Syst. Geogr. Pl. 72: 111–126 (2002), rev.; Middleton, Fl. Males. I, 18: 264–269 (2007), reg. rev.

Woody lianas without tendrils. Leaves usually with scattered black dots abaxially. Inflorescences axillary and terminal cymes. Flowers 4-merous; calycine colleters in single row; corolla cream, yellow or orange; stamens inserted around middle of corolla tube; filaments short; ovary bilocular, glabrous or pubescent; placentation axile; ovules 2–3 per carpel; style-head globose. Berries soft, with pulp. Seeds not or only slightly compressed, without hilar groove, testa thin, smooth; endosperm scanty to absent, cotyledons fleshy, deeply incised at the base.

Four spp. recognized in the most recent revision, in western Malesia, in tropical rainforest up to 1000 m, usually lower.

20. ***Cyclocotyla*** Stapf

Cyclocotyla Stapf, Bull. Misc. Inform. Kew 1908: 259 (1908); Van der Ploeg, Agric. Univ. Wageningen Pap. 85-2: 57–85 (1985), rev.

Woody liana without tendrils. Leaves with closely spaced secondary veins. Inflorescences terminal and axillary, many-flowered. Calycine colleters multiseriate; corolla white, pink or red; corolla lobes inflexed in bud; stamens inserted near base of corolla tube; nectary disc annular, higher than the ovary; ovary bilocular, without pulp, glabrous; placentation axile; ovules 4 per locule; style-head fusiform. Berries pale green, few-seeded. Seeds compressed, with a ridge on the antiraphe side; testa smooth; endosperm thin; cotyledons fleshy, spathulate.

One sp., *C. congolensis* Stapf, found in W and C Africa in primary and secondary forest, near rivers, at low altitudes.

21. ***Willughbeia*** Roxb.

Willughbeia Roxb., Pl. Coromandel 3: 77, t. 280 (1820), nom. cons.; Middleton, Blumea 38: 1–24 (1993), rev.; Middleton, Fl. Males. I, 18: 420–436 (2007), reg. rev.; Middleton, Fl. Cambodia, Laos and Vietnam 33: 230–234 (2014), reg. rev.
Urnularia Stapf (1901).

Woody lianas with branched tendrils formed from modified inflorescences. Inflorescences axillary, congested to lax, thyrsoid or cymose. Corolla white to yellowish; lobes spreading; stamens inserted at various heights; ovary glabrous, superior to partly inferior, unilocular; placentas parietal; ovules numerous; style-head ellipsoid with long apical apex. Berries large, spherical to pear-shaped; pulp edible. Seeds numerous, compressed, without a hilar groove, testa smooth; endosperm scanty to absent; cotyledons fleshy, rounded at base.

16 spp. in Asia from NE India, Sri Lanka, continental SE Asia, the Malay Peninsula, Borneo and Palawan, in primary and secondary forest, and peat swamps, up to 1600 m.

22. ***Cylindropsis*** Pierre

Cylindropsis Pierre, Bull. Mens. Soc. Linn. Paris II, 1: 38 (1898). Pichon, Mém. Inst. Franç. Afrique Noire 35: 329–331 (1953), rev.; Haegens, Bull. Jard. Bot. Nat. Belg. 63: 313–328 (1994), rev.

Woody liana with curled tendrils. Inflorescences axillary, sometimes also terminal, few-flowered, congested cymes. Corolla white, the tube sometimes greenish, tube about twice as long as the lobes; stamens inserted near the orifice of the tube; filaments short, slender; ovary usually pubescent at the apex, unilocular with parietal placentation or incompletely bilocular (placentas meeting below) with seemingly axile placentation; ovules several per carpel; style-head cylindrical-pentagonal. Berries globose to ovoid, orange, edible, few- to several-seeded; pericarp smooth, spongy; pulp slimy. Seeds ellipsoid; endosperm scanty, cotyledons fleshy.

One sp., *C. parvifolia* Pierre, in Nigeria and C Africa, in rainforest and riverine forest up to 700 m.

23. ***Clitandra*** Benth.

Clitandra Benth. in Hook., Niger Fl. 445 (1849); Pichon, Mém. Inst. Franç. Afrique Noire 35: 205–211 (1953); Leeuwenberg & Berndsen, Bull. Jard. Bot. Nat. Belg. 58: 159–168 (1988), rev.

Large liana with curled tendrils. Leaves with secondary veins closely spaced. Inflorescences axillary, often paired in the axils. Corolla white, pinkish or yellowish, pubescent on the outer surface; stamens inserted at about the middle region of corolla tube; filaments short, slender; ovary unilocular, pubescent on top; placentation parietal (sometimes meeting at top and bottom and there placentation seemingly axile); ovules few to several. Berries with a sclerified layer; pulp fibrous, red, edible. Seeds ovate, endosperm thick, cartilaginous, cotyledons membranous.

One sp., *C. cymulosa* Benth. in tropical Africa, from Guinea to Tanzania, in forests up to 1750 m.

24. ***Vahadenia*** Stapf

Vahadenia Stapf, Fl. Trop. Africa 4(1): 26, 29 (1902); Pichon, Mém. Inst. Franç. Afrique Noire 35: 266–272 (1953), rev.; Haegens, Bull. Jard. Bot. Nat. Belg. 63: 313–328 (1994), rev.

Large lianas with tendrils. Inflorescences axillary or terminal, cymose. Calycine colleters numerous, irregularly 3–5-seriate; corolla white, reddish at base; lobes about as long as the tube; stamens inserted ca. midway or somewhat below in the corolla tube; ovary unilocular, pubescent at apex; placentation parietal; ovules many; style-head cylindrical. Berries glabrous, greenish, orange or reddish; pulp edible. Seeds ellipsoid to somewhat irregularly shaped, somewhat compressed; endosperm thick, cartilaginous, cotyledons membranous. 2n = 22.

Two spp. in tropical W Africa, from Guinea to the Ivory Coast, in humid forest, up to 1000 m.

25. ***Dictyophleba*** Pierre

Dictyophleba Pierre, Bull. Mens. Soc. Linn. Paris II, 1: 92 (1898); Pichon, Mém. Inst. Franç. Afrique Noire 35: 250–265 (1953), rev.; Hoogh, Bull. Jard. Bot. Nat. Belg. 59: 207–226 (1989), rev.

Lianas or shrubs with curled tendrils. Inflorescences terminal, cymose, several- to many-flowered, often tendriloid. Corolla white, yellow, pink or red; lobes shorter to longer than the tube, with long cilia at the margins; stamens inserted at various levels in the tube; filaments short; ovary unilocular, glabrous; placentation parietal; ovules several to many; style-head ovoid to truncate-conical. Berries red, orange or yellow; pulp yellow, edible. Seeds somewhat irregularly compressed; endosperm thick, cartilaginous, cotyledons membranous. 2n = 22.

Five spp. in tropical Africa, in forests, often on periodically inundated river banks, in gallery forests and savannas, up to 1300 m.

26. ***Ancylobothrys*** Pierre

Ancylobothrys Pierre, Bull. Mens. Soc. Linn. Paris II, 1: 91 (1898); Vonk et al., Wageningen Agric. Univ. Pap. 94-3: 1–44 (1994), rev.

Large lianas or low shrubs with large, curled, terminal tendrils. Inflorescences terminal panicles, sometimes reduced to a single cyme. Calyx with 1–5 alternisepalous colleters or colleters absent; corolla white to yellowish, sometimes rose-tinted; stamens inserted in lower 1/3 of corolla tube; ovary hairy, unilocular; placentas parietal; ovules several to many; style-head cylindrical to ovoid. Berries velutinous, yellow, orange or reddish, many-seeded; pulp edible; endosperm thick, cartilaginous, cotyledons membranous. 2n = 22, 44, 66.

Seven spp. in Africa, one reaching Madagascar and the Comores.

27. ***Chamaeclitandra*** (Stapf) Pichon

Chamaeclitandra (Stapf) Pichon, Mém. Inst. Franç. Afrique Noire 35: 202–205 (1953), rev.; Leeuwenberg & Berndsen, Bull. Jard. Bot. Nat. Belg. 58: 159–168 (1988), rev.
Clitandra sect. *Chamaeclitandra* Stapf (1902).

Rhizomatous shrub up to 1 m tall with erect shoots, without tendrils. Leaves glaucous, secondary veins relatively closely spaced. Inflorescences axillary and terminal few-flowered cymes. Corolla white to pinkish; ovary pubescent toward the apex, unilocular; placentas parietal; ovules few. Berries globose or pyriform, edible, without a sclerified layer. Seeds 1–33; endosperm thick, cartilaginous, cotyledons membranous.

One sp., *C. henriquesiana* (Hallier f.) Pichon, in tropical Africa in SE DR Congo, Angola and Zambia, at forest edges and woodland, up to 1500 m.

28. ***Saba*** (Pichon) Pichon

Saba (Pichon) Pichon, Mém. Inst. Franç. Afrique Noire 35: 302–324 (1953), rev.; Leeuwenberg & van Dilst, Bull. Jard. Bot. Nat. Belg. 59: 189–206 (1989), rev.
Landolphia sect. *Saba* Pichon (1948).

Lianas (sometimes shrub-like) with curled tendrils. Inflorescences terminal or in forks, rarely axillary. Calycine colleters absent; corolla white or cream with yellow or orange throat; stamens inserted in lower half of corolla tube; ovary mostly pubescent at the top; placentas parietal; ovules numerous; style-head narrowly fusiform. Berries orange to red, with a sclerified layer; placentas becoming pulpy, pulp yellowish or orange, edible. Endosperm thick, cartilaginous, cotyledons membranous. 2n = 22.

Three spp. in Tropical Africa, in riverine vegetation and open woodland, up to 1650 m.

29. ***Orthopichonia*** H. Huber

Orthopichonia H. Huber, Kew Bull. 15: 437 (1962); Pichon, Mém. Inst. Franç. Afrique Noire 35: 202–205 (1953), rev.; Vonk, Wageningen Agric. Univ. Pap. 89-4: 27–50 (1989), rev.

Lianas with tendrils. Leaves usually with numerous, ± parallel secondary veins. Inflorescences usually axillary, if terminal then tendril-like. Corolla usually white; stamens inserted in lower half of corolla tube; ovary glabrous to pubescent, unilocular to imperfectly bilocular; placentation parietal; ovules several to numerous; style-head ovoid. Berries green, yellow or orange, edible, with a sclerified layer. Seeds ellipsoid to reniform, smooth; endosperm thick, cartilaginous, cotyledons membranous. 2n = 22.

Six spp. in Tropical W and C Africa, in tropical rainforest or humid savanna, up to 1500 m.

30. ***Pacouria*** Aubl.

Pacouria Aubl., Hist. Pl. Guiane. 268, pl. 105 (1775); Pichon, Mém. Inst. Franç. Afrique Noire 35: 245–250 (1953), rev.; Monachino, Lloydia 8: 291–317 (1945), rev.

Large woody lianas. Inflorescences terminal and pseudo-axillary, cymose, the axis long and tendriliform; flowers many, subsessile, clustered at the ends of the inflorescence branches. Corolla greenish white to cream; stamens inserted in lower half of corolla tube; ovary unilocular, pubescent; placentas parietal; ovules numerous; style-head spindle-shaped. Berries many-seeded, up to 10 cm diam.; pericarp thick, leathery; placentas not becoming pulpy. Seeds plano-convex, smooth; endosperm thick, cartilaginous, cotyledons membranous.

Two spp. in the Guianas, Amazonian Brazil and Bolivia, in evergreen lowland forests, up to 100 m.

It is disputed whether *Pacouria* should be recognized as distinct from *Landolphia*. The fruits are edible, and the latex is a source of gum.

31. ***Landolphia*** P. Beauv. Fig. 30

Landolphia P. Beauv., Fl. Oware 1: 54 (1806), nom. cons.; Pichon, Mém. Inst. Franç. Afrique Noire 35: 40–202 (1953), reg. rev.; Persoon et al., Wageningen Agric. Univ. Pap. 92-2: 1–232 (1992), reg. rev.; van Dilst, Syst. Geogr. Pl. 69: 91–110 (1999), reg. rev.
Carpodinus R. Br. ex G. Don (1837).
Aphanostylis Pierre (1898).
Anthoclitandra (Pierre) Pichon (1953).

Lianas or shrubs with tendrils, rarely rhizomatous subshrubs. Inflorescences terminal and/or axillary cymes. Corolla white, cream, or yellow, sweet-scented; ovary glabrous or pubescent, unilocular or bilocular, placentation parietal with

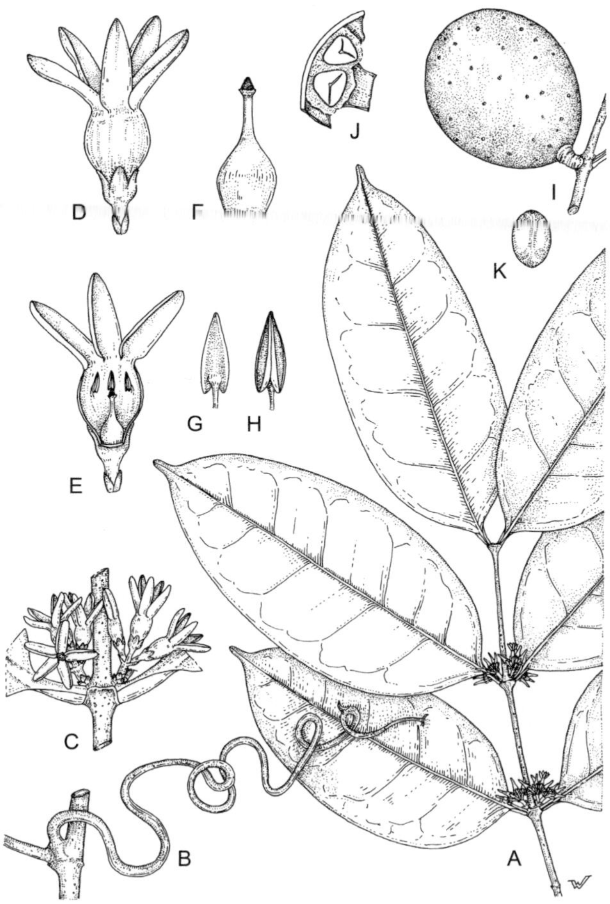

Fig. 30. Apocynaceae-Willughbeieae. *Landolphia letestui.* **A** Branch showing axillary inflorescences. **B** Tendril. **C** Inflorescences. **D** Flower. **E** Opened flower, showing spatial relationship of stamens and style-head. **F** Gynoecium. **G** Anther dorsal side. **H** Anther ventral side. **I** Fruit. **J** Fragment of fruit, showing seeds embedded in pulp. **K** Seed. (From Persoon et al. 1992, p. 124, with permission from the Library of Wageningen University, Wageningen; drawn by W. Wessel-Brand)

prominent placentas; ovules many. Berries mostly yellow, orange or brownish and often with lenticels or rough spots, many-seeded, variously shaped, with a sclerified layer; placentas becoming pulpy, pulp sweet and acidic, edible. Seeds irregularly ellipsoid; endosperm thick, cartilaginous, cotyledons membranous. 2n = 22.

A genus of about 56 spp. in Africa, Madagascar and the Mascarenes.

The delimitation of *Landolphia* has fluctuated greatly; the genus and its many segregate genera are in need of revision based on modern methods.

32. ***Parahancornia*** Ducke

Parahancornia Ducke, Arch. Jard. Bot. Rio de Janeiro 3: 242 (1922); Monachino, Lloydia 6: 240–247 (1944), rev.; Zarucchi, Novon 1: 37–44 (1991), synop.; Zarucchi et al., Fl. Venez. Guayana 2: 543–545 (1995), reg. rev.

Shrubs or trees. Inflorescences terminal, cymose, usually many-flowered. Calyx normally 4-merous; corolla 5-merous, white to cream; stamens inserted in lower half of the tube; ovary unilocular, pubescent; placentation parietal; ovules numerous. Berries many-seeded; pericarp thick, leathery. Seeds not compressed, elliptic to ovate with a hilar groove, embedded in pulpy placentas; endosperm cartilaginous, superficially ruminate, with regular longitudinal furrows, cotyledons thin, tender, cordate at base.

Seven spp. in NW South America, in non-flooded evergreen lowland forests, on seasonally flooded margins of blackwater rivers or on granitic outcrops, sandstone or white sand savannas or gallery forests in these areas, up to 600 m.

33. ***Hancornia*** Gomes

Hancornia Gomes, Mem. Math. Phis. Acad. Real Sci. Lisboa 3: 51 (1812); Monachino, Lilloa 11: 19–48 (1945), rev.

Small tree; latex whitish, sometimes reddish or bluish tinged. Leaves with numerous ± parallel secondary veins. Inflorescences terminal, few-flowered dichasia. Flowers relatively large; corolla white to yellowish; stamens inserted in upper quarter of corolla tube; filaments short; ovary unilocular; glabrous or pubescent at apex, placentation parietal; ovules numerous. Berries few-seeded. Seeds not compressed, elliptic to ovate with a hilar groove, embedded in pulpy placentas; endosperm cartilaginous, superficially ruminate, with regular longitudinal furrows, cotyledons thin, tender, attenuate at the base.

One extremely variable sp., *H. speciosa* Gomes, widespread in Brazil, reaching Paraguay and Argentina, often in sandy or rocky areas, up to 1500 m. The fruits are sold in markets. Extracts of the bark are used in folk medicine to treat internal disorders. The latex is used in a wide variety of folk remedies. The hard wood is used to manufacture small utility items.

34. ***Couma*** Aubl.

Couma Aubl., Hist. Pl. Guiane 2, supp. 39, t. 392 (1775); Monachino, Lloydia 6: 230–247 (1944), rev.

Shrubs or trees. Leaves in whorls of 3–4(5); petiole base with small interpetiolar flaps. Inflorescences axillary, cymose, few- to many-flowered. Corolla white to dark pink; stamens inserted at the middle of the tube or above; ovary unilocular, half-inferior; placentation parietal; ovules many. Glabrous, many-seeded berries with seeds embedded in pulpy placentas; pericarp thick, leathery. Seeds not compressed, elliptic to ovate with a hilar groove; endosperm cartilaginous, superficially ruminate, with regular longitudinal furrows, cotyledons thin, tender, cordate at base.

Six spp., one widespread from Guatemala to Amazonian Bolivia, the rest in Amazonia and Bahia, Brazil, in non-flooded forest, lower montane and montane forest, or on sandy soil in low forest adjacent to savannas, up to 1600 m. The latex is used used as a milk substitute, as a base for chewing gum and for caulking canoes. The fruits are edible, and some species are sold in markets.

35. ***Lacmellea*** H. Karst.

Lacmellea H. Karst., Linnaea 28: 449 (1857); Monachino, Lloydia 275–302 (1944), rev.; Morales, Novon 8: 259–262 (1998), reg. rev.

Shrubs or trees; trunk in some species with deciduous blunt conical spines. Inflorescences axillary, cymose, few- to many-flowered. Corolla sometimes slightly zygomorphic, greenish white, white or cream; stamens inserted about the middle of the tube; ovary glabrous (rarely pubescent), unilocular; placentation parietal; ovules numerous. Berries yellow to orange, 1–3 cm diam., 1- to few-seeded; sweet, edible; pericarp thin. Seeds somewhat compressed, elliptic to ovate with or without a hilar groove, testa thick, tough; endosperm subcorneous, smooth; cotyledons thin, delicate, attenuate at the base.

Around 23 spp. in Central America and NW South America, in white sand savannas, periodically flooded white sand savannas, lower montane forests, and on granitic outcrops, often along rivers, up to 600 m. The latex of some species is used as a milk substitute in tea and coffee; the fruits are eaten fresh and used for making juice or preserves.

The genus is in need of revision.

I.5. **Tribe Tabernaemontaneae** G. Don (1837).

Trees or shrubs (1 species of *Callichilia* and 1–2 species of *Tabernaemontana* lianoid), almost always glabrous, with white latex; young stems often compressed or quadrangular. Leaves opposite, rarely in whorls of 3 (*Tabernaemontana ternifolia*); petioles of leaf pair often connate at the node, forming a short ocrea, this usually expanded into small intrapetiolar flaps, the blades isophyllous in Ambelaniinae, often anisophyllous in Tabernaemontaninae. Inflorescences axillary, terminal, or (usually paired) in the forks of branches. Flowers mostly medium-sized to very large and showy; calyx lobes mostly small, but sometimes larger and foliaceous or thick and fused into a tube in some genera of Tabernaemontaninae; calycine colleters mostly present and centered on the lower part of the sepal, often multiseriate, rarely absent, very rarely (*Carvalhoa*, *Tabernanthe*) alternisepalous; corolla usually salverform, but tubular-campanulate or infundibuliform in some genera of Tabernaemontaninae; corona absent; corolla lobes normally extended in bud, but inflexed in *Tabernaemontana*; corolla lobe aestivation almost always sinistrorse (dextrorse in *Schizozygia*, 1 species of *Callichilia* and 2 species of *Tabernaemontana*); stamens inserted at various levels in the corolla tube; anthers mostly sessile or nearly so, mostly with massive lignified guide rails (these absent in all species of *Tabernaemontana* from Asia and in some species from Australasia and one from Africa), usually included (but the tips exserted in a few species of *Tabernaemontana*); pollen 3–5-colporate, sometimes zonocolporate; in *Callichilia* in tetrads; nectary disc surrounding the base of the ovary adnate, indistinct or more rarely fleshy and free (in some genera of Tabernaemontaninae); ovary syncarpous or apocarpous, glabrous (except in *Macoubea*); ovules numerous in most genera; style slender or thick, sometimes twisted; style-head almost always with a strongly five-ribbed upper crest (star-shaped in cross-section) and the stigmatic region beneath an expanded basal flange (subglobose and topped by two elongate slender appendages, without basal flange or upper crest and body uniformly receptive in all species of *Tabernaemontana* from Asia and in some species from Australasia and one from Africa). Fruit with fleshy, rarely woody, pericarp, endocarp not stony; either a berry with the non-arillate seeds embedded in pulp or projecting into the ovary on enlarged funicles (Ambelaniinae) or a pair of follicles (these sometimes hemi-syncarpous) with arillate seeds (most Tabernaemontaninae). Testa often wrinkled, pitted or with longitudinal ridges, often with a long hilar groove; endocarp not stony; endosperm often ruminate. Highly evolved indole alkaloids of the heynean type present. n = 11. 15 genera, mostly either in Africa or South America, *Tabernaemontana* pantropical.

36. ***Ambelania*** Aubl.

Ambelania Aubl., Hist. Pl. Guiane 1: 265, t. 104 (1775); Monachino, Lloydia 8: 109–130 (1945), rev.; Zarucchi, Agric. Univ. Wageningen Pap. 87-1: 23–40 (1988), rev.

Trees up to 25 m tall; young stems terete to subterete. Petioles of leaf pair often connate at the node, forming a short ocrea, the blades isophyllous. Inflorescences axillary. Calycine colleters absent; corolla white to orange, lobes about as long as tube; stamens inserted about middle of tube or below; ovary bilocular; ovules numerous. Berries bilocular, many-seeded, ovoid to elongate, sometimes ribbed; pericarp thick, leathery. Seeds projecting into ovary cavities on fleshy funicles; testa smooth to pitted.

Three spp., widespread in N South America in humid forest and forest edges up to 1000 m.

37. ***Molongum*** Pichon

Molongum Pichon, Mém. Mus. Natl. Hist. Nat. 24: 167 (1948); Zarucchi, Agric. Univ. Wageningen Pap. 87-1: 66–79 (1988), rev.

Shrubs or small trees up to 12 m tall; young stems terete to subterete. Petioles of leaf pair often connate at the node, forming a short ocrea, the blades isophyllous. Inflorescences terminal. Calycine colleters several, multiseriate; corolla lobes slightly longer than tube; stamens inserted about middle of tube; ovary bilocular; ovules numerous per carpel. Berries bilocular, few-seeded, narrowly ellipsoid to narrowly ovoid; pericarp indurated, thin. Seeds embedded in fluffy, white pulp; testa muriculate and finely reticulate.

Three spp., Amazonia, in sandy soil in riparian forests along blackwater rivers, or in open sandy savannas, up to 200 m.

38. ***Spongiosperma*** Zarucchi

Spongiosperma Zarucchi, Agric. Univ. Wageningen Pap. 87(1): 48–66 (1988), rev.
Molongum sect. *Trichosiphon* Pichon (1948).

Shrubs or small trees up to 6 m tall; young stems terete to subterete. Petioles of leaf pair often connate at the node, forming a short ocrea, the blades isophyllous. Inflorescences terminal. Calycine colleters in a single row; corolla lobes about as long as tube; stamens inserted below middle of tube; ovary bilocular; ovules numerous per carpel. Berries unilocular (due to septum tearing during fruit maturation), many-seeded, globose, obovoid, narrowly ellipsoid to fusiform; pericarp thick, leathery. Seeds attached to fruit wall by fleshy funicles, not embedded in pulp; testa spongy, finely pitted, covered with irregular protuberances.

Six spp. in Amazonia, in sandy soil in seasonally flooded forests along blackwater streams or savannas over sandy soil, up to 600 m.

39. ***Rhigospira*** Miers

Rhigospira Miers, Apocyn. S. Amer. 67, t. 10 (1878); Pichon, Mém. Mus. Natl. Hist. Nat. 24: 169, t. IV, 18, 28, 39 (1948), synop.; Zarucchi, Agric. Univ. Wageningen Pap. 87(1): 79–86 (1988), rev.

Tree up to 30 m tall; young stems strongly quadrangular, becoming terete with age. Petioles of leaf pair often connate at the node, forming a short ocrea, the blades isophyllous. Inflorescences terminal. Calycine colleters absent; corolla lobes less than half the length of the tube; stamens inserted near base of tube; ovary bilocular; ovules numerous per carpel. Berries unilocular (due to septum tearing during fruit maturation), ellipsoid to obovoid, many-seeded; pericarp indurated, grainy. Seeds embedded in red to maroon, acidic pulp; testa irregularly reticulate.

One sp., *R. quadrangularis* (Müll. Arg.) Miers, in C and NW South America, in evergreen lowland and lower montane forests.

40. ***Mucoa*** Zarucchi

Mucoa Zarucchi, Agric. Univ. Wageningen Pap. 87-1: 40–48 (1988), rev.

Trees up to 15 m tall; young stems compressed, becoming terete with age. Petioles of leaf pair often connate at the node, forming a short ocrea, the blades isophyllous. Inflorescences axillary. Calycine colleters many, multiseriate; corolla lobes about as long as tube; stamens inserted just below middle of tube; ovary bilocular; ovules numerous per carpel. Berries bilocular, globose to ellipsoid, many-seeded; pericarp indurated, grainy. Seeds embedded in stringy, aromatic pulp; testa shiny, finely reticulate.

Two spp., Amazonia, in montane forest up to 1400 m.

41. ***Neocouma*** Pierre

Neocouma Pierre, Bull. Soc. Mens. Soc. Linn. Paris II, 1: 33 (1898); Zarucchi, Agric. Univ. Wageningen Pap. 87-1: 86–94 (1988), rev.

Trees up to 25 m tall; young stems terete to subterete, with large petiole scars. Petioles of leaf pair often connate at the node, forming a short ocrea, this usually expanded into small intrapetiolar flaps, the blades isophyllous. Inflorescences terminal. Calycine colleters numerous, multiseriate; corolla lobes shorter to much longer than tube; stamens inserted about middle of tube; ovary bilocular; ovules numerous per carpel. Berries uni- or bilocular, ellipsoid to globose, few- to many-seeded; pericarp crustaceous, thin. Seeds embedded in somewhat stringy pulp; testa finely pitted with irregular protuberances.

Two spp., NW Amazonia, in non-flooded and upland forests up to 900 m.

42. ***Macoubea*** Aubl.

Macoubea Aubl., Hist. Pl. Guiane 2 (Suppl.): 17, t. 378 (1775); Monachino, Lloydia 8: 291–317 (1945), rev.; Morales, Novon 9: 86–88 (1999), synop.

Trees, sometimes exceeding 30 m height; young stems subterete. Petioles of leaf pair often connate at the node forming a short ocrea with interpetiolar flaps, the blades isophyllous. Inflorescences terminal, dichasial, usually many-flowered. Calycine colleters several, 1–3-seriate; corolla rather thin, lobes longer than tube; stamens inserted near base of tube; ovary apocarpous to hemi-syncarpous, puberulent to densely sericeous; placentation marginal; ovules numerous per carpel. Berries many-seeded, up to 2/3 syncarpous or only 1 carpel developing and fruit asymmetric, clog-shaped; pericarp thick, woody. Seeds with thin, translucent membrane that liquefies at maturity; testa pitted, with shallow hilar depression.

Three spp., tropical Central and South America, in humid forest at lower elevations.

43. ***Callichilia*** Stapf

Callichilia Stapf, Fl. Trop. Africa 4(1): 130 (1902); Beentje, Meded. Landbouwhogeschool Wageningen 78-7: 1–32 (1978), rev.
Ephippiocarpa Markgr. (1923).

Mostly erect shrubs (one species said to be climbing); young stems terete. Leaves subsessile or petiolate, blades of a pair isophyllous or anisophyllous. Inflorescences 1–2 in forks, cymose, 1- to several-flowered, pendulous, conspicuously bracteate at base. Calyx lobes foliaceous, often unequal, with 1–4 rows of colleters within; corolla tubular-infundibuliform, white; tube fleshy, longer than lobes; corolla lobe aestivation rarely dextrorse; nectary disc fleshy, shallowly lobed; ovary apocarpous or hemi-syncarpous (*C. orientalis* S. Moore); ovules numerous. Follicles free to nearly completely united, sometimes only tardily dehiscent, pulp juicy. Aril thin, ± translucent. $2n = 22$.

Seven spp. in tropical Africa, in rainforest, secondary or riverine forest, especially in moist localities, near streams up to 1000 m.

44. ***Voacanga*** Thouars — Fig. 31

Voacanga Thouars, Gen. Nov. Madagasc. 10 (1806); Leeuwenberg, Agric. Univ. Wageningen Pap. 85-3: 1–122 (1985), rev.; Middleton, Fl. Males. I, 18: 411–420 (2007), reg. rev.

Trees and shrubs; young stems terete to subterete. Petioles of leaf pair often connate at node, the blades isophyllous or anisophyllous. Inflorescences paired in branch forks. Calyx of free

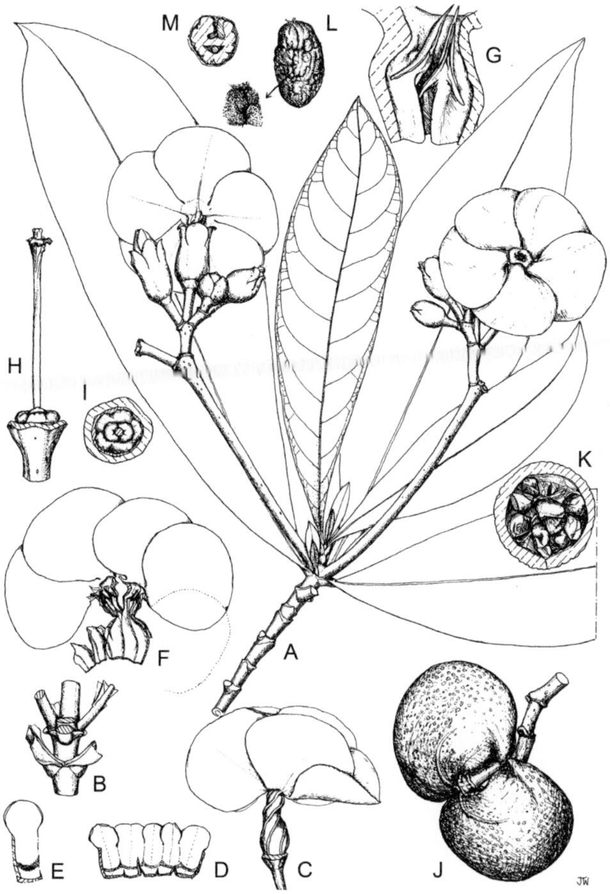

Fig. 31. Apocynaceae-Tabernaemontaneae. *Voacanga grandifolia*. **A** Flowering branch. **B** Part of branch, showing connate leaf bases at nodes. **C** Flower with calyx removed showing twisted corolla tube. **D** Opened calyx with colleters spread across the base. **E** One calyx lobe with colleters near the base. **F** Opened corolla. **G** Section through top of corolla tube showing exserted anther apices. **H** Gynoecium with fleshy lobed nectary disc at base. **I** Ovary surrounded by nectary disc seen from above. **J** Fruit. **K** Section through fruit, showing seeds. **L** Seed (arrow: detail of testa). **M** Transverse section through seed showing deep hilar groove and ruminate endosperm. (From Leeuwenberg 1985, p. 54, with permission from the Library of Wageningen University, Wageningen; drawn by J. Beentje-Williamson)

lobes or fused into tube; calycine colleters usually multiseriate; corolla mostly cream or yellow; tube twisted; stamens inserted in upper half of corolla tube; nectary disc annular, sometimes lobed, adnate to ovary base; ovary apocarpous, hemi-syncarpous or syncarpous; ovules numerous per carpel. Follicles free to completely united, green spotted, white, yellow or orange. Aril yellow to orange-red, pulpy. 2n = 22.

Twelve spp. in tropical Africa, Malesia and Australia (Queensland).

45. *Calocrater* K. Schum.

Calocrater K. Schum. in Engl. & Prantl, Nat. Pflanzenfam. 4(2): 175, t. 58, s–t. (1895); Hallé, Adansonia 5: 507–510 (1965), synop.; Leeuwenberg, Fontqueria 42: 11–16 (1995), rev.

Single-stemmed or once-branched shrub; young stems terete. Leaves subsessile, large and thin; petioles of leaf pair often connate at node, the blades isophyllous. Inflorescences axillary, few-flowered; peduncle with many sepal-like bracts. Flowers large, subsessile; calycine colleters in single row; corolla infundibuliform-campanulate; tube thin; lobes much shorter than tube; stamens inserted in lower 1/3 of corolla tube; ovary glabrous, 2/3 or more syncarpous; ovules several per carpel. Fruit syncarpous, subglobose to ellipsoid, yellow to orange, bicuspidate at the apex. Aril white.

One sp., *C. preussii* K. Schum., found in Cameroon, Gabon and DR Congo, in the shady forest understory, often on creek banks, up to 500 m.

46. *Crioceras* Pierre

Crioceras Pierre, Bull. Mens. Soc. Linn. Paris 1: 1311 (1897); Hallé, Adansonia 11: 301–308 (1971), synop.; Leeuwenberg, Fontqueria 42: 11–16 (1995), rev.

Shrub or small tree up to 8 m tall; young stems terete. Leaves subsessile, large and thin, those of a pair isophyllous or anisophyllous. Inflorescences axillary, long-pedunculate, with 1(–3) large pendulous flowers; pedicels with large foliaceous bracts completely covering sepals. Calycine colleters in single row; corolla tubular-campanulate; tube thin, lobes much shorter than tube; stamens inserted at base of expanded mouth; ovary ± apocarpous or hemi-syncarpous; ovules numerous. Fruit hemi-syncarpous, ellipsoid with recurved apex. Seeds ellipsoid. Aril thin, white.

One sp., *C. dipladeniiflorus* (Stapf) K. Schum., in Cameroon, Gabon, DR Congo and Angola, in the shady forest understory, up to 600 m.

47. *Carvalhoa* K. Schum.

Carvalhoa K. Schum., in Engl. & Prantl, Nat. Pflanzenfam. 4(2): 189 (1895); Leeuwenberg, Agric. Univ. Wageningen Pap. 85-2: 49–55 (1985), rev.

Shrubs or small trees; young stems terete to subterete. Petioles of leaf pair often connate at the node, forming a short ocrea, the blades isophyllous or anisophyllous. Inflorescences axillary, paired in the forks, few-flowered panicles. Calycine colleters alternisepalous, rarely in a single row; corolla campanulate, white to pale yellow; tube thin, with many red stripes distally and at base of lobes; lobes suborbicular, much shorter than tube; stamens inserted in lower part of corolla tube; nectary disc annular, fleshy, adnate to base of ovary; ovary apocarpous; ovules numerous. Follicles yellow or pale orange, soft, thin-walled. Aril dark orange, pulpy.

Two spp. in E Africa, in montane rainforest or secondary forest, up to 1900 m.

48. *Schizozygia* Baill.

Schizozygia Baill., Bull. Mens. Soc. Linn. Paris 1: 752 (1888); Barink, Meded. Landbouwhogeschool Wageningen 83-7: 47–53 (1983), rev.

Shrub or small tree up to 8 m tall; young stems terete to subterete. Leaves subsessile, those of a pair isophyllous. Inflorescences axillary, congested, few-flowered, usually paired in the branch forks. Calyx foliaceous; calycine colleters several, in a single row; corolla yellow; tube thin; the lobes somewhat shorter than tube; corolla lobe aestivation dextrorse; stamens inserted at about middle of corolla tube; nectary disc adnate to lower part of ovary; ovary apocarpous; ovules several per carpel. Follicles dry, thinly coriaceous, laterally compressed, irregularly grooved, yellow to orange. Aril thin, yellow to orange-red, pulpy. 2n = 22.

One sp., *S. coffaeoides* Baill. in C and E Africa, in rainforest understory or bush.

49. *Tabernanthe* Baill.

Tabernanthe Baill., Bull. Mens. Soc. Linn. Paris 1: 783 (1889); Vonk & Leeuwenberg, Wageningen Agric. Univ. Pap. 89-4: 1–26 (1989), rev.

Shrubs or small trees up to 4 m tall; young stems terete to subterete. Leaves of a pair isophyllous or anisophyllous. Inflorescences paired in the branch forks, few- to many-flowered, corymbose. Calycine colleters few, alternisepalous; corolla white to pale yellow or orange, mostly with pink to violet markings in the throat; tube thin, longer than the lobes; stamens inserted about midway in the corolla tube; ovary congenitally syncarpous to nearly completely apocarpous; ovules several to numerous per carpel. Fruiting carpels fused and smooth or separate and with soft prickles, yellow, orange or red. Aril white, pulpy. 2n = 22.

Two spp. in C Africa.

50. *Tabernaemontana* L.

Tabernaemontana L., Sp. Pl. 1: 210 (1753); Leeuwenberg, Rev. Tabernaemontana 1 (1991), 2 (1994), reg. revs.; Morales & Méndez, Candollea 60: 345–371 (2005), reg. rev. (as *Stemmadenia*); Middleton, Fl. Males. I, 18: 371–390 (2007), reg. rev.; Middleton, Fl. Cambodia, Laos and Vietnam 33: 184–202 (2014), reg. rev.
Pandaca Noronha ex Thouars (1806).
Rejoua Gaudich. (1829).
Conopharyngia G. Don (1837).
Bonafousia A. DC. (1844).
Peschiera A. DC. (1844).
Stemmadenia Benth. (1844).
Anacampta Miers (1878).
Anartia Miers (1878).
Gabunia K. Schum. (1896).
Ervatamia (A. DC.) Stapf (1902).
Pagiantha Markgr. (1935).
Stenosolen Markgr. (1937).
Hazunta Pichon (1948).
Muntafara Pichon (1948).
Pandacastrum Pichon (1948).
Woytkowskia Woodson (1960).
Camerunia (Pichon) Boiteau (1976).
Leptopharyngia (Stapf) Boiteau (1976).
Sarcopharyngia (Stapf) Boiteau (1976).

Shrubs or small trees, mostly between 1–15 m tall (but a few species may be much taller); young stems terete to subterete. Leaves opposite, rarely in whorls of 3; petioles of leaf pair often connate at node, forming a short ocrea, this usually expanded into small intrapetiolar flaps, the blades sometimes anisophyllous. Inflorescences usually paired in branch forks. Calycine colleters often numerous, multiseriate; corolla white, yellow or mauve; tube sometimes twisted; lobes inflexed in bud; stamens inserted at various levels in corolla tube; anther tips sometimes exserted; ovary apocarpous, sometimes hemi-syncarpous; ovules several to many. Follicles normally free, sometimes hemi-syncarpous, slender and elongate to globose, sometimes echinate or warty; pericarp thin and delicate to thick and fleshy, spongy or woody, sometimes indehiscent. Seeds few to many; aril white, gray, orange or red, pulpy. 2n = 22, 33.

Pantropical genus of some 122 spp. A few species are planted as ornamentals.

I.6. **Tribe Amsonieae** M.E. Endress (2014).

Subshrubs or perennial herbs with white latex. Leaves alternate. Calycine colleters absent; corolla salverform; corolla lobe aestivation sinistrorse; anthers free from the style-head, wholly fertile, included; nectaries absent, adnate or indistinct; ovary apocarpous; ovules numerous; style-head squat, with a collar at the base and a conspicuous wreath of hairs around the top. Fruit a pair of delicate follicles. Seeds naked, not compressed, oblong to ovoid with slanted ends; testa not smooth. Indole alkaloids present. x = 11. One genus in southern USA, Japan and Europe.

51. ***Amsonia*** Walter

Amsonia Walter, Fl. Carol. 98 (1788); Woodson, Ann. Missouri Bot. Gard. 15: 379–435 (1928), rev.; Woodson, R.E., Jr., North American Fl. 29: 126–131 (1938), reg. rev.; McLaughlin, Ann. Missouri Bot. Gard. 69: 336–350 (1982), reg. rev.
Rhazya Decne (1835).

Subshrubs or perennial herbs. Leaves alternate. Inflorescences terminal, few- to many-flowered thyrses or corymbs. Flowers medium-sized, relatively showy; calycine colleters absent; corolla thin, white, blue or pink, the tube not strongly constricted or with thickened annulus at the orifice; filaments slender; anthers without lignified guide rails or sagittate basal appendages; nectary disc absent, adnate or indistinct; ovary apocarpous; ovules numerous. Follicles terete, thin-walled. Seeds cylindrical, oblong to ovoid (in *A. stricta* flattened at the margins into two irregular ribs); testa rugulose, pitted or corky. 2n = 22, 32 or 16.

Genus with a disjunct distribution: ca. 16 spp. in SE to SW USA, one in Japan and one in Europe, mainly in woods, margins of streams, prairies, fields, sand hills and barrens, rocky ravines and canyons, mountain slopes and desert.

Rhazya is sometimes treated as a genus distinct from *Amsonia*, but molecular evidence to support this is lacking.

I.7. **Tribe Melodineae** G. Don (1837).

Trees, shrubs or woody lianas, tendrils absent; latex white (sometimes yellow in *Diplorhynchus* and *Pycnobotrya*). Leaves opposite or whorled. Inflorescences terminal and in the upper leaf axils. Flowers mostly small- to medium-sized; calycine colleters usually absent (reported in a few species of *Melodinus*); corolla salverform, sometimes unfused (with gaps) just above the insertion of the stamens; corolla lobe aestivation sinistrorse; small corolline corona lobes often present in petal sinuses or behind the anthers (well-developed and usually annular and petaloid in *Melodinus*); stamens included, stamens inserted at various levels in the corolla tube; anther connective with an apical extension in *Diplorhynchus*, and with both apical connective extension and small sterile basal lobes in *Pycnobotrya*; pollen usually 3-colporate (in tetrads in some *Melodinus* species, and porate tetrads in *Craspidospermum*); nectary disc absent or an adnate, indistinct ring around the base of the ovary; ovary apocarpous or congenitally syncarpous (hemi-syncarpous in *Diplorhynchus*); ovules few to numerous per carpel; style slender, often short; style-head mostly without basal collar or upper wreath. Fruit mostly a pair of follicles with woody pericarp (fruit a capsule in *Craspidospermum* and a berry with sclerotic outer pericarp in *Melodinus*); endocarp not stony. Seeds usually numerous, smooth, flat and winged, often peltately attached with a long funicle (somewhat verrucose, only slightly compressed and embedded in pulpy placenta in *Melodinus*); endosperm mostly thin. Indole alkaloids present. x = 10, 11. Five genera, most restricted to Africa or Madagascar, *Melodinus* widespread in the Paleotropics.

This tribe is morphologically heterogeneous, likely non-monophyletic and sorely in need of molecular phylogenetic study.

52. ***Diplorhynchus*** Welw. ex Ficalho & Hiern.

Diplorhynchus Welw. ex Ficalho & Hiern., Trans. Linn. Soc. London, Bot. II, 2: 22, t. 5 (1881); Plaizier, Meded. Landbouwhogeschool Wageningen 80-12: 28–40 (1980).

Shrub or small many-stemmed tree with many raised lenticels (occasionally reported as scrambling). Leaves opposite. Inflorescences thyrses. Corolla white to cream, with a strong jasmine-like scent; stamens inserted about midway in corolla tube; filaments flattened toward the apex; anthers with sterile apical and basal appendages; ovary hemi-syncarpous; placentation parietal; ovules four per carpel. Fruit apocarpous; follicles widely divergent, stout, woody, with much latex. Seeds obliquely oblong, peltately attached, with a large, elongate wing on one end. 2n = 22.

One sp., *D. condylocarpon* (Müll. Arg.) Pichon in tropical and S Africa, usually found in wetter spots in savannas and open forests, up to 1700 m.

53. ***Pycnobotrya*** Benth.

Pycnobotrya Benth. in Benth. & Hook. f., Gen. Pl. 2: 688, 715 (1876); Van der Ploeg, Meded. Landbouwhogeschool Wageningen 83-4: 13–20 (1983), rev.

Large woody liana. Leaves opposite or ternate. Inflorescences many-flowered panicles. Corolla pink, sometimes with pale yellow throat, rotate, fragrant; stamens inserted below the middle of corolla tube; anthers subsessile, with sterile basal appendages and an elongate, apical connective appendage; ovary apocarpous; ovules four per carpel; style very short. Fruit apocarpous; follicles broadly and obliquely elliptic, laterally compressed, warty. Seeds 1–4 per carpel, obliquely oblong, peltately attached, surrounded by a papery wing that is thickened at the margins.

One poorly known sp., *P. nitida* Benth., in W and C Africa, in forest, often on river banks.

54. ***Stephanostegia*** Baill.

Stephanostegia Baill., Bull. Mens. Soc. Linn. Paris 1: 748 (1888); Leeuwenberg, Wageningen Agric. Univ. Pap. 97-2: 95–102 (1997), rev.

Trees. Leaves opposite, coriaceous. Inflorescences lax cymes. Corolla white or lilac, the tube incompletely fused behind the stamens, with a thickened ring constricting the orifice; stamens inserted in upper half of corolla tube, filaments very short; ovary apocarpous; ovules ca. 10 per carpel. Follicles ellipsoid, woody. Seeds few to several per carpel, elliptic to ovate, peltately attached, surrounded by an elliptic papery wing.

Two spp. endemic to Madagascar, in wet or lowland forest, up to 400 m.

55. ***Craspidospermum*** Bojer ex A. DC. Fig. 32

Craspidospermum Bojer ex A. DC., Prodr. 8: 323 (1844); Leeuwenberg, Wageningen Agric. Univ. Pap. 97-2: 11–16 (1997), rev.

Tree; branches angled near apex. Leaves in whorls of 3–4(–6), coriaceous. Inflorescences many-flowered thyrses. Corolla white to rose, often with dark pink or red tube; tube incompletely

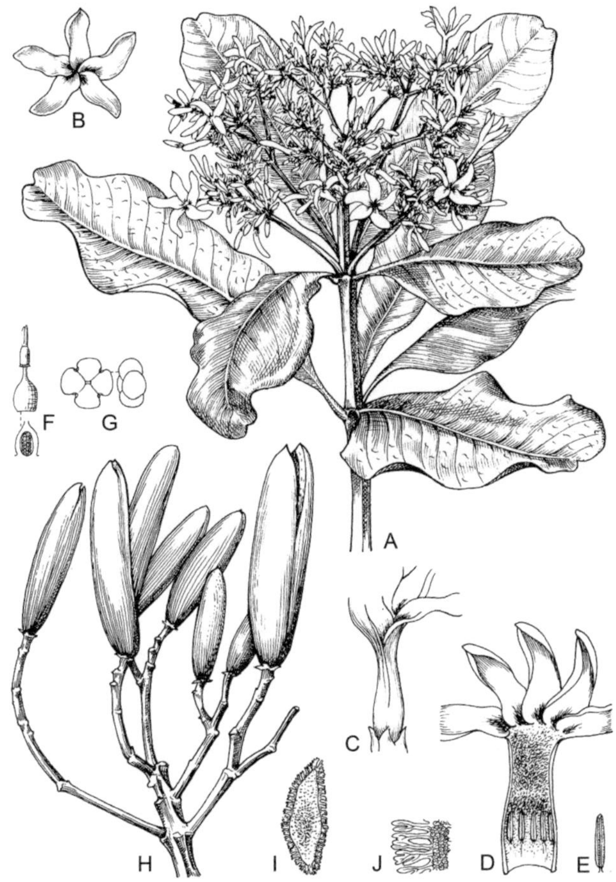

Fig. 32. Apocynaceae-Melodineae. *Craspidospermum verticillatum*. **A** Flowering branch. **B** Flower viewed from above. **C** Corolla, detail of tube and base of lobes. **D** Opened corolla. **E** Stamen ventral side. **F** Gynoecium and longitudinal section through ovary showing median cylindrical placenta with many ovules. **G** Non-tetrahedral, porate pollen tetrad, top and side view. **H** Fruiting branch, showing dehiscence into two halves. **I** Seed. **J** Detail of seed margin showing flattened, fringed emergences. (From Markgraf 1976, p. 99; drawn by M. Rabarijaona)

fused behind the stamens, inside with moniliform hairs above the stamens; stamens inserted near the base of the tube; pollen porate, in tetrads; ovary congenitally syncarpous; placentation parietal; style-head apical appendages long, slender. Fruit syncarpous, a 2-valved woody capsule. Seeds compressed, elliptic, with flattened, fringed emergences around the margin.

One sp., *C. verticillatum* Bojer ex A. DC., endemic to Madagascar, in rainforest up to 1800 m.

56. ***Melodinus*** J.R. Forst. & G. Forst.

Melodinus J.R. Forst. & G. Forst., Char. Gen. Pl. 37, t. 19 (1776); Pichon, Mém. Mus. Natl. Hist. Nat. II, 24: 125–130 (1948), synop.; Markgraf, Blumea 19: 149–166 (1971), reg. rev.; Leeuwenberg, Syst. Geogr. Pl. 73: 3–62 (2003), rev.; Middleton, Fl. Males. I, 18: 270–281 (2007), reg. rev.; Middleton, Fl. Cambodia, Laos and Vietnam 33: 127–139 (2014), reg. rev.
Clitandropsis S. Moore (1923).

Woody lianas. Leaves opposite, rarely ternate. Inflorescences cymose, axillary and/or terminal. Calycine colleters almost always absent; corolla white to orange; petaloid corolline corona in mouth; stamens inserted at about the middle of corolla tube or below; filaments short; ovary congenitally syncarpous, unilocular, placentation parietal; ovules numerous. Fruit a subglobose to pear-shaped berry; pericarp with a sclerotic outer layer. Seeds not compressed, without a hilar groove, embedded in pulpy placentas; endosperm fleshy, superficially irregularly wrinkled, cotyledons thin, cordate at the base. 2n = 22.

About 25 spp. currently recognized, widespread in the Paleotropics, Australasia and Oceania.

The genus is in need of a new revision.

I.8. **Tribe Hunterieae** Miers (1878).

Glabrous trees or shrubs, rarely lianas, with white latex (latex sometimes translucent in *Gonioma*). Leaves opposite, rarely in whorls of 3. Inflorescences cymose. Flowers mostly small to medium-sized, with a strong sweet scent; calycine colleters absent or, when present, several and multiseriate; corolla salverform; corolla lobe aestivation sinistrorse; corona absent; stamens inserted midway or above in the corolla tube; anthers ovate, without lignified guide rails, sometimes shortly sagittate at the base, included; pollen 3-colporate; nectary disc an adnate or indistinct ring around the base of the ovary; ovary apocarpous, 2–5-carpellate, glabrous; ovules few to numerous; style-head ovoid to narrowly clavate, without basal collar or upper crest, but often with slender, elongate non-receptive apices. Fruit apocarpous, indehiscent, of 2–5 broadly divergent berries with fleshy, fibrous pericarp (fruit dehiscent, a pair of woody follicles in *Gonioma*); endocarp not stony. Seeds ovoid, without a longitudinal groove, embedded in pulp (seeds compressed with a papery wing at one end in *Gonioma*); endosperm smooth, mostly thick and rather hard. Indole alkaloids present. n = 11. Four genera in Africa, one in Madagascar and one reaching south and continental South-East Asia.

57. ***Gonioma*** E. Mey.

Gonioma E. Mey., Comm. Pl. Afr. Austr. 188 (1837); Leeuwenberg, Wageningen Agric. Univ. Pap. 97-2: 16–21 (1997), rev.

Shrubs or small slender trees. Leaves in whorls of 3–4, the upper leaves often opposite. Inflorescences terminal, congested. Calycine colleters absent; corolla yellow; staminal filaments short; ovary 2-carpellate; ovules numerous. Fruit a pair of oblong, woody, longitudinally finely ribbed follicles. Seeds numerous, compressed, subrectangular, with a papery wing at one end. 2n = 20.

Two spp., one endemic to the Cape Province of South Africa, the other to Madagascar, in forest understory and dry forest up to 900 m. The hard wood is used to make small objects such as tool handles, shuttles, etc.

58. ***Hunteria*** Roxb. Fig. 33

Hunteria Roxb., Fl. Ind., ed. 1832, 1: 695 (1832); Omino, Wageningen Agric. Univ. Pap. 96-1: 88–128 (1996), rev.; Middleton, Fl. Males. I, 18: 200–203 (2007), reg. rev.; Middleton, Fl. Cambodia, Laos and Vietnam 33: 102–105 (2014), reg. rev.

Trees, shrubs or rarely lianas. Leaves opposite, with numerous straight secondary veins. Inflorescences terminal, rarely axillary, few- to many-flowered dichasial cymes. Calycine colleters several, multiseriate; corolla white, yellow or red; staminal filaments short; ovary glabrous; style-

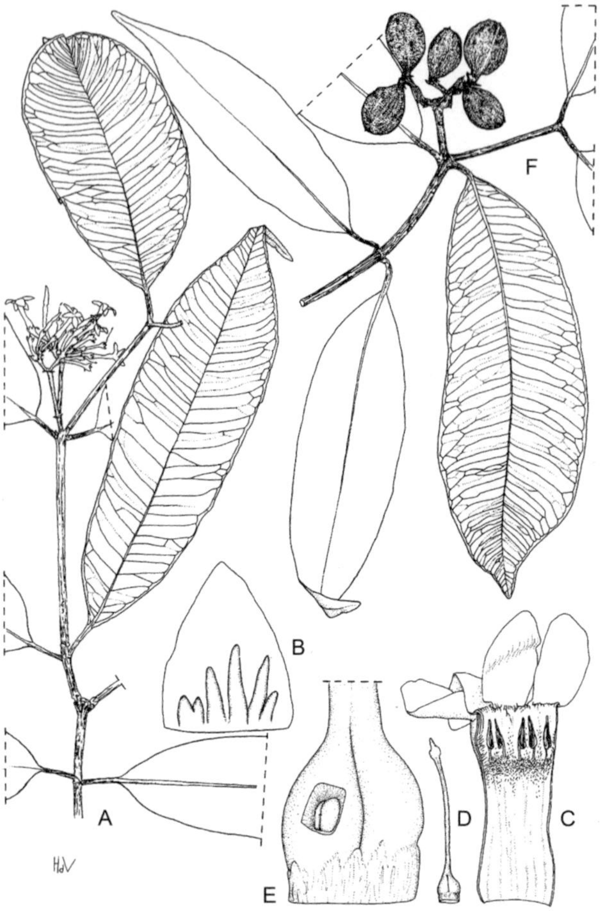

Fig. 33. Apocynaceae-Hunterieae. *Hunteria zeylanica*. **A** Flowering branch. **B** Inner surface of calyx lobe showing colleters at base. **C** Opened corolla showing position of stamens. **D** Gynoecium. **E** Ovary with section cut from wall showing the position of the two ovules. **F** Fruiting branch. (From Omino 1996, p. 122, with permission from the Library of Wageningen University, Wageningen; drawn by H. de Vries)

head ovoid with slender apex; ovary 2-carpellate; ovules 2–30 per carpel. Berries stipitate, subglobose to obovoid, yellow to orange, smooth or warty, often only one developing. Seeds few, smooth or slightly verrucose. 2n = 22.

12 spp. in Africa, one of them, *H. zeylanica* (Retz.) Gardner ex Thwaites, also widespread in S and continental SE Asia; most species are found in primary or secondary forest often along streams or forest edges, usually at low elevations.

59. ***Picralima*** Pierre

Picralima Pierre, Bull. Mens. Soc. Linn. Paris 2: 1278 (1896); Omino, Wageningen Agric. Univ. Pap. 96-1: 81–177 (1996), rev.

Tree or shrub. Leaves opposite. Inflorescences terminal, sometimes axillary, compound umbellate cymes. Calycine colleters numerous, multiseriate; corolla white to yellow, the tube often greenish, more or less the same length as the lobes; anthers with slightly developed basal appendages; ovary 2-carpellate; ovules numerous per carpel. Berries obovoid to ellipsoid or pear-shaped, somewhat compressed, yellow to orange. Seeds somewhat angular and irregularly shaped, embedded in soft pulp. 2n = 22.

One sp., *P. nitida* (Stapf) T. Durand & H. Durand, in tropical Africa.

60. ***Pleiocarpa*** Benth.

Pleiocarpa Benth. in Benth. & Hook. f., Gen. Pl. 2: 699 (1876); Omino, Wageningen Agric. Univ. Pap. 96-1: 134–178 (1996), rev.
Carpodinopsis Pichon (1953).

Trees, shrubs or rarely lianas. Leaves opposite or in whorls of 3–5, glabrous. Inflorescences axillary, rarely terminal, fascicles. Flowers subsessile; calycine colleters absent; corolla white to yellow; ovary 2–5-carpellate; ovules 1–6 per carpel. Berries subglobose to obovoid, or elliptic, yellow to orange, 1–6-seeded. Seeds angular, rather irregularly shaped. 2n = 22.

Five spp. in Africa, in secondary or primary forest, montane forest, low bush, swampy areas, or along river banks up to 2000 m.

I.9. **Tribe Alyxieae** G. Don (1837).

Trees, shrubs or woody lianas; latex white. Leaves often whorled, sometimes opposite or alternate. Inflorescences cymose, terminal and/or axillary, often congested. Flowers mostly small to medium-sized and inconspicuous; calycine colleters absent; corolla salverform (sometimes throat somewhat expanded in *Condylocarpon*); corolla lobes inflexed in bud or not; corolla lobe aestivation sinistrorse;

corona absent (rudimentary corona lobes reported in *Plectaneia*); stamens inserted in the upper half of corolla tube (usually lower in *Chilocarpus*, sometimes lower in *Alyxia*), included; filaments short; anthers ovate, without lignified basal appendages; nectary disc absent, indistinct or adnate; pollen 2–3-porate (sometimes barrel-shaped with two large pores, in tetrads in *Condylocarpon*); ovary apocarpous or syncarpous, normally bicarpellate with axile or marginal placentation (up to 5-carpellate in *Lepinia* and *Lepiniopsis*, and in *Chilocarpus* unilocular with parietal placentation), sometimes stipitate; ovules 2 to several per carpel; style slender; style-head globose, ovoid or fusiform, without basal collar or upper wreath, body uniformly receptive. Fruit usually indehiscent (dehiscent in *Chilocarpus* and *Plectaneia*), with fleshy pericarp and stony endocarp (pericarp dry in *Condylocarpon*, *Chilocarpus* and *Plectaneia*). Seeds usually globular or ovoid (not compressed), with deep hilar groove or depression and thick, strongly ruminate endosperm (in *Chilocarpus* seeds with an aril on the funicle and endosperm reported to be non-ruminate in some spp; in *Plectaneia* seeds compressed with wings and with non-ruminate endosperm). No alkaloids or cardenolides are known to occur in the tribe. x = 9 (known only for *Alyxia*). Seven genera, six in the Paleotropics, one in Central and South America.

61. *Chilocarpus* Blume

Chilocarpus Blume, Catalogus 22 (1823); Markgraf, Blumea 19: 161–166 (1971), reg. rev.; Middleton, Fl. Males. I, 18: 164–178 (2007), reg. rev.; Middleton, Fl. Cambodia, Laos and Vietnam 33: 75–78 (2014), reg. rev.
Neokeithia Steenis (1948).

Woody lianas. Leaves opposite, usually with numerous black dots beneath. Inflorescences terminal and/or axillary; pedicels sometimes with numerous bracteoles. Corolla variously colored, lobes strongly inflexed, forming a globose head in bud; stamens sometimes inserted below the middle of corolla tube; ovary syncarpous, unilocular; placentas parietal; ovules numerous. Fruit syncarpous, variously colored, leathery to woody, ellipsoid to globular or long, narrow and torulose, dehiscent. Seeds globular-ovoid, with a corky aril on the funicle; endosperm ruminate or smooth.

14 spp. in continental SE Asia and Malesia, mainly in evergreen forest or swamp forest up to 650 m.

62. *Condylocarpon* Desf.

Condylocarpon Desf., Mém. Mus. Hist. Nat. 8: 119 (1822); Fallen, Ann. Missouri Bot. Gard. 70: 149–169 (1983), rev.

Woody lianas. Leaves opposite or in whorls of 3 (–4). Inflorescences terminal and axillary. Corolla salverform or narrowly infundibuliform (throat not constricted), white, yellow or orange, lobes often with a ligulate appendage, strongly inflexed and forming conspicuously globose head in bud; ovary apocarpous, ovules 4–6(–16) per carpel. Fruit apocarpous; drupes thinly woody, indehiscent, usually torulose, rarely only one uniovulate carpel developing; endocarp thick, leathery. Seeds fusiform, longitudinally folded with a deep groove from end to end.

Seven spp. in Brazil and the Guianas, one in Central America; found in coastal and lowland forest, usually near water.

63. *Plectaneia* Thouars

Plectaneia Thouars, Gen. Nov. Madagasc. 11 (1806); Leeuwenberg, Wageningen Agric. Univ. Pap. 97-2: 81–95 (1997), rev.

Woody lianas, often shrubby. Leaves opposite. Inflorescences terminal and axillary. Corolla yellowish white, greenish yellow to yellow-brown, with five small corona lobes in the throat in the staminal sectors, the lobes not inflexed in bud; ovary seemingly apocarpous viewed externally, but carpels partially united ventrally; ovules many. Fruit of two fused follicles forming a septicidal capsule with four ribs or wings. Seeds compressed, subrectangular, peltately attached, with a papery wing at each end; endosperm not ruminate.

Three spp. currently recognized, endemic to Madagascar, in coastal forest, dry open forest or brush, up to 1500 m.

64. *Pteralyxia* K. Schum.

Pteralyxia K. Schum. in Engl. & Prantl, Nat. Pflanzenfam. 4(2): 151 (1895); Degener, Fl. Hawaiiensis, fam. 305 (1933), synop.; Wagner et al., Man. Fl. Pl. Hawai'i 1: 219–220 (1990), rev.

Trees; branches thick with prominent lenticels and leaf scars. Leaves alternate. Inflorescences terminal, congested. Corolla waxy, yellow-green, lobes not inflexed in bud; ovary apocarpous; ovules 2 per carpel. Fruit with usually only one carpel maturing, drupaceous, fleshy, bright red; endocarp woody, with a network of crests and wings. Seeds 1 per carpel, not compressed; endosperm ruminate.

Two spp., one each endemic to Kauai and Oahu, Hawaii, on slopes and ridges in mesic to wet forest below 800 m.

65. ***Lepinia*** Decne.

Lepinia Decne., Ann. Sci. Nat. Bot. 3, 12: 194, t. 9 (1849); Lorence & Wagner, Allertonia 7(4): 254–266 (1996), rev.; Endress et al., Allertonia 7(4): 267–272 (1997), fl. morph.; Middleton, Fl. Males.I, 18: 260–262 (2007), reg. rev.

Small trees or shrubs. Leaves alternate. Inflorescences apparently terminal but appearing axillary due to lateral prolongation of the axis, few-flowered. Corolla thick, white to cream, lobes shorter than the tube, strongly inflexed in bud; ovary stipitate, of 3–5 2-ovulate carpels, apocarpous but the apices remaining firmly postgenitally fused. Fruit basket-like, of 3–5 stipitate, indehiscent, woody, uniovulate carpels bowed outward in the middle and fused at the apex. Seeds elongate, with a long hilar groove; endosperm ruminate.

Four spp. disjunct across the Pacific Basin from New Guinea to the Marquesas Islands.

66. ***Lepiniopsis*** Valeton

Lepiniopsis Valeton, Ann. Jard. Bot. Buitenzorg 12: 251, t. 28 (1895); Merrill, Bull. Bur. For. Philip. 1: 48 (1903), reg. rev.; Markgraf, Blumea 30: 169–172 (1984), reg. rev.; Middleton, Fl. Males. I, 18: 262–264 (2007), reg. rev.

Shrubs or small trees. Leaves alternate. Inflorescences axillary, long-pedunculate. Flowers ± sessile. Corolla thick, white, lobes strongly inflexed in bud; ovary syncarpous, 3–5-locular; ovules two per locule. Fruit red to purple-black, plum-like, syncarpous, indehiscent, by abortion 1–3-locular, each locule uniovulate; mesocarp fibrous; endocarp lignified. Seeds elongate, with a long hilar groove; endosperm ruminate.

Two spp., one in the Philippines and E Malesia, the other on the Palau Islands, on sandy beaches or other open areas, lowland rainforest, sago swamps, up to 900 m.

67. ***Alyxia*** Banks ex R. Br. Fig. 34

Alyxia Banks ex R. Br., Prodr. 469 (1810), nom. cons.; Middleton, Blumea 45: 1–146 (2000), Blumea 47: 1–93 (2002), rev.; Middleton, Fl. Males. I, 18: 64–128 (2007), reg. rev.; Middleton, Fl. Cambodia, Laos and Vietnam 33: 30–44 (2014), reg. rev.
Discalyxia Markgr. (1927).

Woody lianas, shrubs, rarely small trees. Leaves in whorls of 3–4 or opposite. Inflorescences axillary, 1- to many-flowered; bracts with colleters immediately below the calyx in some species. Corolla usually white, cream or yellow, the tube sometimes orange or purple, lobes not inflexed in bud; ovary apocarpous; ovules 2–6 per carpel. Fruit apocarpous, a pair of stipitate, torulose drupes with 1–6 articles each, red, orange, black or yellow when ripe; mesocarp fleshy. Endosperm ruminate. 2n = 36.

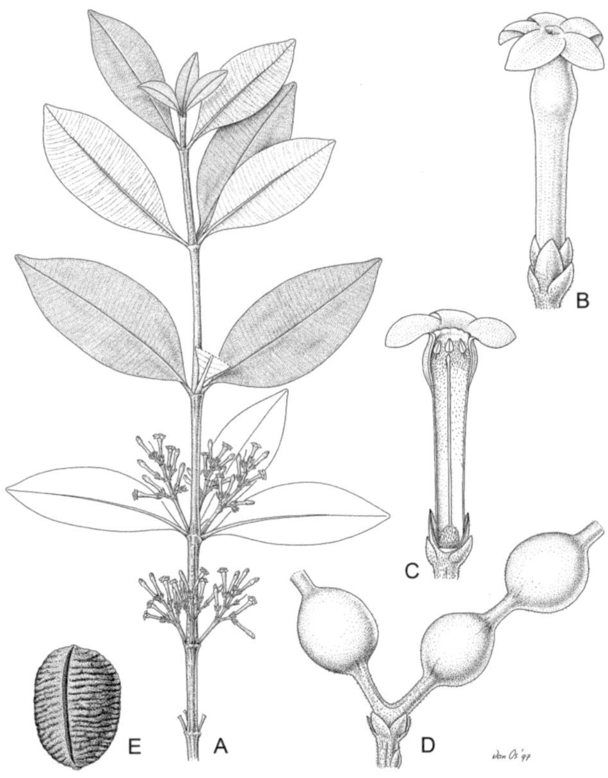

Fig. 34. Apocynaceae-Alyxieae. *Alyxia concatenata*. **A** Flowering branch. **B** Flower. **C** Open flower. **D** Fruit. **E** Seed showing ruminate surface with deep hilar groove. (From Middleton 2000, p. 42, with permission from Naturalis, Leiden; drawn by J. van Os)

Currently 106 spp. are recognized, from NE India, continental SE Asia, Malesia, Australia, and the Pacific Islands in various habitats and elevations.

I.10. **Tribe Plumerieae** E. Mey. (1838).

Trees or shrubs, rarely lianas; latex white. Leaves usually alternate, sometimes opposite (whorled in *Allamanda*). Inflorescences usually terminal in erect species and axillary in those that are lianoid. Flowers mostly large to medium-large and showy (smaller in *Cameraria* and *Anechites*); calycine colleters mostly absent (present in *Anechites*, *Thevetia* and some species of *Allamanda)*; corolla salverform, infundibuliform or tubular-campanulate, the lobes not inflexed in bud; corolline corona often present below the petal sinuses behind stamens (usually much dissected in *Allamanda*) and infrastaminal appendages generally present; corolla lobe aestivation sinistrorse; stamens inserted at various levels in the corolla tube; anther connective often broadened resulting in sublatrorse dehiscence and with an elongate filiform apical appendage, these often exserted, anthers otherwise included, without lignified guide rails; pollen 3-colporate; nectary disc mostly absent (but well-developed in *Allamanda*, *Anechites* and *Thevetia*); ovary normally apocarpous (hemi-syncarpous in some species of *Thevetia* and *Cerbera* and postgenitally syncarpous in *Allamanda*), partly inferior in *Plumeria*, *Himatanthus* and *Mortoniella*; placentation normally marginal (axile at the base in hemi-syncarpous taxa, in *Allamanda* unilocular with parietal placentation); ovules many in the genera with woody dehiscent fruits, and 2–6 in genera with indehiscent fruits; style slender or thick; style-head with free apices conspicuously enlarged, mostly with basal collar or lobes (no distinct basal collar in *Plumeria*, *Himatanthus*, and *Mortoniella*), mostly without upper wreath (well-developed upper wreath present in *Allamanda*). Fruit usually apocarpous, sometimes hemi-syncarpous or syncarpous (in *Allamanda* postgenitally syncarpous), either indehiscent, and a pair of drupes or samaroids with fleshy or dry pericarp and stony endocarp, or dehiscent and a pair of dry, woody follicles (in *Allamanda* fruit a capsule, usually with spines). Seeds typically many in the genera with dehiscent fruits and 1–4 in those with indehiscent fruits, usually compressed with a narrow wing-like margin or distinct papery wing. Cardenolides or iridoid glycosides present. x = 9, 10. Ten genera, eight in the Neotropics, one endemic on New Caledonia and one widespread in the Paleotropics.

68. ***Allamanda*** L.

Allamanda L., Mant. Pl. 2: 146, 214 (1771); Sakane & Shepherd, Revista Bras. Bot. 9: 125–149. (1986), rev.; Morales, Phytotaxa 162: 51–56 (2014), reg. rev.

Shrubs, sometimes scandent. Leaves in whorls of (3–)4–5. Inflorescences terminal, sometimes also axillary, cymose, few- to several-flowered; sepals foliaceous, some species with colleters; corolla tubular-campanulate, yellow or violet; stamens inserted just below expanded upper part of the tube; anthers with sagittate, but non-lignified basal appendages; ovary superior; ovules many; style-head spool-shaped with a well-developed wreath at the top and collar at the base. Fruit a septicidal capsule, usually subglobose and spiny, rarely compressed and without spines; seed wing circular. 2n = 18, 36.

14 spp. in tropical America, often at forest margins or river banks, sometimes in cerrado, caatinga or rocky areas, up to 1000 m. Two species, *A. cathartica* L. and *A. schottii* Pohl, widely cultivated and naturalized throughout the tropics.

69. ***Mortoniella*** Woodson

Mortoniella Woodson, Ann. Missouri Bot. Gard. 26: 257 (1939); Morales, Darwiniana, n.s. 43: 138–139 (2005), reg. rev.

Tree up to 30 m. Leaves alternate; petioles at about mid-length with an adaxial cluster of long, tooth-like colleters. Inflorescences terminal, cymose, thyrsiform, several-flowered; bracts small, scarious, caducous, without colleters at the base. Calyx lobes thin, caducous, eglandular at the apex; corolla salverform, white; stamens inserted near the base of the tube; ovary partly inferior; ovules numerous. Follicles long-stipitate, thinly woody. Seeds slightly thickened and angular, naked or with a narrow rudimentary wing. 2n = 32.

One sp., *M. pittieri* Woodson, in Belize, Nicaragua and Costa Rica.

70. ***Plumeria*** L.

Plumeria L., Sp. Pl. 1: 209 (1753); Woodson, Ann. Missouri Bot. Gard. 25: 189–224 (1938), rev.

Small trees or shrubs, stems thick, with corky periderm and conspicuous leaf scars. Leaves alternate, usually clustered near ends of branches. Inflorescences terminal, cymose and thyrsiform, usually many-flowered bracts small scarious, caducous, without colleters at base. Calyx lobes small, ± equal, glandular at apex; corolla salverform, thick, waxy, white, pink or red, often with yellow in throat; stamens inserted near base of tube; ovary partly inferior; ovules numerous. Follicles large, stout. Seeds with thin, papery wing at one end. 2n = 36, 54, 45.

About 7 spp., native to the Caribbean and Central America, but cultivated in tropical and subtropical regions worldwide for their showy flowers.

Despite its wide cultivation, this genus has never been treated in a modern revision.

71. ***Himatanthus*** Willd.

Himatanthus Willd. in Roem. & Schult., Syst. Veg. 5: 221 (1819); Woodson, Ann. Missouri Bot. Gard. 25: 189–224 (1938), rev.; Plumel, Bradea 5, suppl. 1: 1–120 (1991), rev.; Spina et al., Taxon 62: 1304–1307 (2013), synop.

Trees or shrubs; stems thick, with corky periderm and conspicuous leaf scars. Leaves alternate. Inflorescences terminal, cymose to thyrsiform, few- to many-flowered; bracts large and showy, caducous, with many colleters at the base. Calyx of thick, irregular sepal rudiments; corolla salverform, thick, waxy, white, often with a yellow eye; stamens inserted near the base of the tube; ovary partly inferior; ovules numerous. Follicles large, stout, leathery to thinly woody. Seed wing concentric, papery. 2n = 18.

About nine spp. in tropical South America.

This genus has a complex taxonomic history, and the generic limits between it, *Plumeria* and *Mortoniella* need to be tested using modern methods.

72. ***Cameraria*** L.

Cameraria L., Sp. Pl. 1: 210 (1753).

Shrubs. Leaves opposite, usually with numerous parallel veins. Inflorescences terminal, cymose. Corolla relatively small, salverform, white, expanded near orifice, with epistaminal appendages, the lobes large; stamens inserted in upper part of corolla tube; anther connective broad with long filiform apical appendage, included except for apical appendages; dehiscence sublatrorse; ovary superior; ovules 1–2 per carpel; style-head broad, lenticular, with membranous collar and two large apical lobes. Fruit a pair of one-seeded samaroid drupes. Seed margin narrow, wing-like.

Two spp. in the Greater Antilles and Central America.

73. ***Anechites*** Griseb.

Anechites Griseb., Fl. Brit. W. I. 410 (1861); Fallen, Brittonia 35: 222–231 (1983), rev.; Morales, Anales Jard. Bot. Madrid 66: 224 (2009), reg. rev.

Liana. Leaves opposite; blades with recurved hairs with multicellular bases. Inflorescences alternate-axillary, several-flowered dichasia. Calycine colleters present; corolla white, salverform, lobes large, broadly ovate; stamens inserted about midway in the corolla tube, included; anthers with broadened connective and elongate apical appendage; dehiscence sublatrorse; nectary disc annular, adnate, lobed; ovary superior; ovules 2–6 per carpel; style-head lenticular with basal collar and two large lobes at apex. Drupes indehiscent, slender, one-seeded, with recurved hairs and sterile beak at distal end. Seed margin compressed.

One sp., *A. nerium* (Aubl.) Urban, found in the Greater Antilles, Caribbean coast of Central and South America and Pacific coast of N South America, in lowland tropical moist to wet forest usually below 600 m.

74. ***Skytanthus*** Meyen

Skytanthus Meyen, Reise I: 376 (1834).

Shrubs, sometimes prostrate. Leaves alternate and/or opposite, small, coriaceous. Inflorescences terminal and axillary, few-flowered, cymose. Corolla yellow, tubular-campanulate, with small corona lobe above stamens and epistaminal appendages below; stamens inserted at base of expanded part of corolla; anthers with broadened connective and long filiform apical appendage, included except for apical appendages; ovary superior; ovules numerous per carpel; style-head lenticular with collar, and two large apical lobes. Follicles long, slender, terete, woody, corkscrew-like; placentas forming hard articles between the seeds. Seeds with wing at each end.

Three spp. in Brazil and Chile in drier habitats and often higher elevations.

75. *Thevetia* L.

Thevetia L., Opera Var. 212 (1758), nom. cons.; Alvarado-Cárdenas & Ochoterena, Ann. Missouri Bot. Gard. 94: 298–323 (2007), rev.
Cascabela Raf. (1838).

Shrubs or small trees. Leaves alternate. Inflorescences terminal and axillary, few- to several-flowered, cymose. Calycine colleters present; corolla infundibuliform, rarely salverform, yellow to orange, rarely white; tube with infrastaminal appendages and corona lobes behind anthers; stamens inserted near orifice of tube; anthers with broad connective and relatively short apical appendage; dehiscence sublatrorse; nectary disc annular, fleshy; ovary sometimes hemi-syncarpous; ovules 2–4 per carpel. Fruit hemi-syncarpous to syncarpous, red or black, drupaceous, indehiscent; mesocarp fleshy. Seeds broadly ovate, moderately compressed, with a narrow, denticulate margin and apical wing. 2n = 20.

Nine spp. in tropical America, one, *T. peruviana* (Pers.) K. Schum., widely cultivated and naturalized in tropical regions.

Cascabela is sometimes recognized as a distinct genus (Alvarado-Cárdenas et al. 2017) based mainly on fruit structure; but the chromosome number combined with the specialized floral morphology that it shares with *Thevetia* outweigh the differences in fruit structure (Williams and Stutzman 2008).

76. *Cerbera* L. Fig. 35

Cerbera L., Sp. Pl. 1: 208 (1753); Forster, Austrobaileya 3: 569–579 (1992), reg. rev.; Leeuwenberg, Wageningen Agric. Univ. Pap. 98-3: 1–64 (1999), rev.; Middleton, Fl. Males. I, 18: 157–163 (2007), reg. rev.; Middleton, Fl. Cambodia, Laos and Vietnam 33: 72–75 (2014), reg. rev.

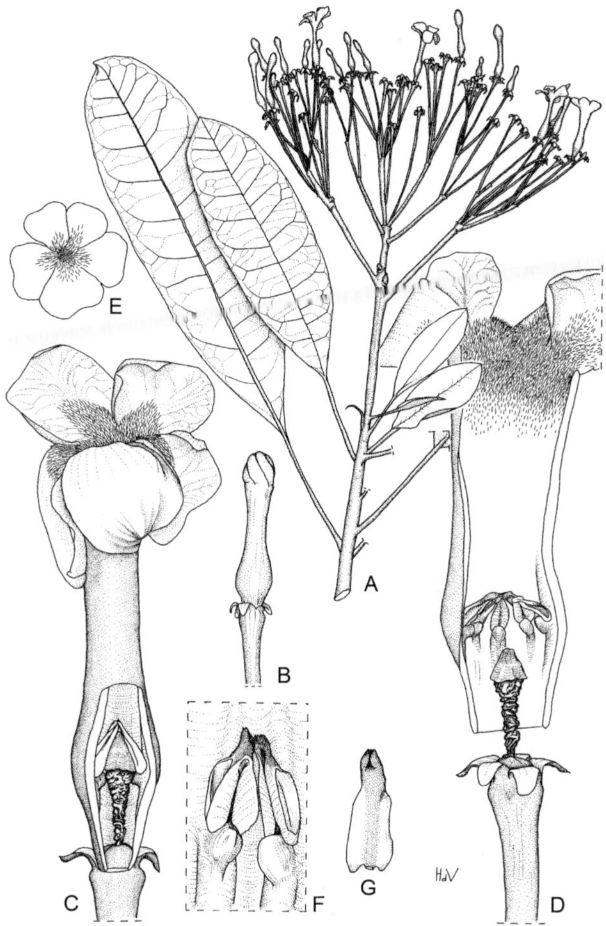

Fig. 35. Apocynaceae-Plumerieae. *Cerbera laeta*. **A** Flowering branch. **B** Bud. **C** Flower, with section removed at base showing position of reproductive organs. **D** Flower with opened, detached corolla and intact gynoecium. **E** Flower from above. **F** Two stamens showing latrorse dehiscence and infrastaminal appendages. **G** Anther dorsal side. (From Leeuwenberg 1999, p. 19, with permission from the Library of Wageningen University, Wageningen; drawn by H. de Vries)

Small trees, sometimes shrubs. Leaves alternate. Inflorescences terminal, several-flowered compound corymbs; bracts and sepals large, foliaceous; calycine colleters absent; corolla salverform, sometimes weakly infundibuliform at top of tube, usually white; tube with infrastaminal appendages and corona lobes behind anthers; stamens inserted near base to near orifice of tube; anthers with broad connective and elongate apical appendage; ovary superior; ovules usually 4 per carpel; style-head with conical upper part and basal collar. Drupes, adapted for water dispersal; mesocarp fibrous. Seeds usually somewhat compressed with narrow flattened rim around margin. 2n = 40, 44.

Six spp. in Madagascar and the Seychelles and from Japan (Ryukus) to Australia (Queensland) and Pitcairn, in rainforest, open forest, woodland, coastal forest or thickets, mostly at low elevations. *Cerbera odollam* Gaertn. is widely cultivated in SE Asia.

77. ***Cerberiopsis*** Vieill. ex Pancher & Sébert

Cerberiopsis Vieill. ex Pancher & Sébert in Sébert, Not. Bois. Nov. Caledonia 187 (1874); Boiteau, Fl. Nouvelle-Cal. 10: 222–232 (1981), rev.

Trees or shrubs; branching monopodial, branches in pseudo-whorls at regular intervals along main stem, often fistulose. Leaves alternate. Inflorescences terminal, few- to many-flowered cymes or panicles. Sepals caducous; corolla tubular-campanulate, white, usually with yellow or reddish markings; stamens inserted about midway in the tube; anthers with broad connective and apical appendage; dehiscence sublatrorse; ovary superior; ovules 2 per carpel. Fruit apocarpous; a pair of one-seeded, flat, dry samaroid drupes, each with two lateral wings. Seeds flat with thin wing-like margin.

Three spp., endemic to New Caledonia.

I.11. **Tribe Carisseae** Dumort. (1829).

Shrubs or small trees, rarely scandent, with or without (often branched) spines in leaf axils; latex white. Leaves opposite (verticillate in some species of *Carissa*). Inflorescences usually axillary, cymose, mostly much contracted, ± umbelliform. Calycine colleters almost always absent; corolla salverform, usually white or cream, sometimes flushed with pink, the lobes not inflexed in bud; corolla lobe aestivation sinistrorse (dextrorse in some species of *Carissa*); corona absent; stamens inserted at the middle of the corolla tube or above; anthers ovate, without lignified guide rails or basal appendages, subsessile, included, or tips exserted; pollen 3-colporate; style-head ovoid to ovoid-cylindrical, body scarcely differentiated, uniformly receptive, without basal collar; nectary disc absent or an indistinct ring adnate to the base of the ovary; ovary congenitally syncarpous, bilocular, usually glabrous or with just a few hairs at the apex; placentation axile; ovules 1–5 per locule, rarely more. Fruit syncarpous, indehiscent, a berry with fleshy, non-fibrous, non-sclerotic pericarp; endocarp not stony, but placentas becoming indurated in fruit forming a pseudostone. Seeds 2–6(–12), rarely more, compressed, without a hilar groove; testa thin, smooth; endosperm smooth, mealy; cotyledons thin, delicate. Cardenolides often present. x = 11. Two genera in the Paleotropics.

78. ***Acokanthera*** G. Don

Acokanthera G. Don, Gen. Hist. 4: 485 (1837); Kupicha, Kew Bull. 37(1): 40–67. 1982, reg. rev.

Trees or shrubs without spines; latex poisonous. Inflorescences dense axillary fascicles. Corolla lobes much shorter than the tube; anthers subsessile, usually with some hairs at the apex, the tips sometimes exserted; ovule one per locule. Fruit a globose, 1–2-seeded, purplish black to black berry. Seeds compressed with a glabrous testa. 2n = 22.

Five spp. in Africa. The toxic latex of *Acokanthera* has long been used by native peoples in Africa as arrow and spear poison.

Pichon (1948a) considered *Acokanthera* to be congeneric with *Carissa*, though they have mostly been treated as distinct genera, especially due to the presence of cardenolide glycosides in *Acokanthera* and their absence in *Carissa*. Phylogenetic analyses place the two genera as sisters, but expanded sampling may show that they are better treated as congeneric.

79. ***Carissa*** L. Fig. 36

Carissa L., Syst. Nat., ed. 12. 2: 135, 189 (1767), nom. cons.; Markgraf, Fl. Madagascar et Comores, Fam. 169: 13–33 (1976), reg. rev.; Forster, Fl. Australia 28: 107–110 (1996), reg. rev.; Leeuwenberg & van Dilst, Wageningen Univ. Pap. 2001-1: 3–109 (2001), rev.; Middleton, Fl. Males. I, 18: 151–154 (2007), reg. rev.; Middleton, Fl. Cambodia, Laos and Vietnam 33: 66–70 (2014), reg. rev.

Shrubs or small trees, sometimes scandent, with paired straight, often branched spines; latex non-poisonous. Leaves sometimes in whorls of 3–4, mostly coriaceous. Inflorescences axillary, sometimes also terminal, few-flowered cymes. Corolla lobe aestivation dextrorse in some species; ovules

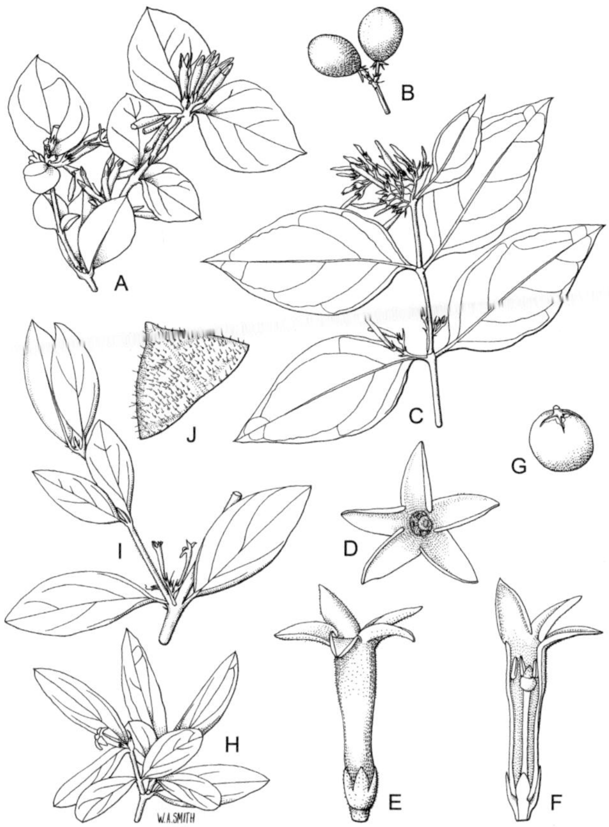

Fig. 36. Apocynaceae-Carisseae, *Carissa*. **A, B** *Carissa ovata*. **A** Flowering branch. **B** Fruit. **C–G** *C. laxiflora*. **C** Flowering branch. **D** Flower from the top. **E** Flower from the side. **F** Longitudinal section through flower showing spatial relationship of anthers and style-head. **G** Fruit. **H** *C. lanceolata*, flowering branch. **I, J** *C. scabra*. **I** Flowering branchlet. **J** Leaf apex showing scabrid indumentum. (From Forster 1992c, p. 584, with permission from the CSIRO Publishing, Clayton; drawn by W.A. Smith)

1–4, rarely more per locule. Fruit globose or oblong, red, purple, bluish black or black. Seeds mostly 1–4, rarely more; seed compressed-discoid, testa rough to velutinous. 2n = 22.

Estimates of species number ranges from seven, in the most recent revision, to about 20 spp. in Asia, Australia, Madagascar, the Mascarenes and Africa. A modern phylogenetic study of the genus is lacking. The fruits of several species are edible, and some are used to make jam. Some species are widely cultivated as ornamentals.

II. Apocynoids

Apocynoideae Burnett (1835).
Echitoideae Luerss. (1882).

Woody lianas, vines, less frequently trees, shrubs or subshrubs, rarely perennial herbs; latex usually white (translucent in many Echiteae and Nerieae, and in some Malouetieae). Leaves almost always opposite, sometimes with cluster of colleters at juncture of petiole and blade adaxially (some Mesechiteae and Baisseeae), less frequently colleters spread along the length of midrib of the leaf blade adaxially (some Mesechiteae), sometimes abaxial domatia in axils of secondary veins (some Apocyneae, Malouetieae, Mesechiteae and Baisseeae). Flowers small, whitish and inconspicuous to large, colorful and showy; calycine colleters normally present; corolla infundibuliform, tubular-campanulate, salverform, urceolate, tubular or rotate; corolla lobe aestivation almost always dextrorsely contorted (sinistrorsely contorted in Wrightieae and in *Urceola* p.p.), rarely valvate (*Parsonsia* p.p., *Urceola* p.p.); corolline corona often present, sometimes annular, often petaloid or dissected (Wrightieae, Nerieae), sometimes lower down in corolla tube behind stamens (Baisseeae, Echiteae), sometimes more than one kind of corona present (mainly Echiteae); rarely corona lobes on dorsal side of anther (some Apocyneae); stamens inserted at various levels; anthers included or exserted, almost always fertile only in upper part, lower part enlarged and sterile, and sides elaborated into lignified guide rails (guide rails poorly developed in some Malouetieae), united with the style-head by a pad of specialized hairs on the enlarged connective via agglutination in most cases (only very weakly so in some Nerieae and Malouetieae), more rarely via postgenital fusion, sometimes also attached at the base of the thecae; pollen porate, most commonly with 3–4 apertures; occasionally grains 2–5-aperturate or polypantoporate; nectaries usually present around base of ovary, sometimes fused into ring (absent in Wrightieae, Nerieae and some Malouetieae); ovary bicarpellate, usually apocarpous, more rarely postgenitally syncarpous (especially in some Echiteae and Wrightieae), normally superior, more rarely partly inferior; ovules mostly numerous per carpel; style mostly slender, sometimes thick; style-head variously shaped, differentiated longitudinally, sometimes with collar at base and crest at the apex, normally with two small free appendages at the apex; epidermis of the body usually radially interrupted by the adnate anthers into 5 secretory and non-secretory zones; stigmatic region on the lower part of the style-head below the adnate anthers or beneath flange or collar, when present. Fruit dehiscent, usually apocarpous, a pair of slender to stout, thinly to thickly woody, ventrally dehiscent follicles with dry pericarp, less frequently the 2 follicles postgenitally fused along their ventral margins, forming a syncarpous "double-follicle", which splits apart at maturity; pericarp dry; endocarp thin, not stony. Seeds usually numerous, compressed, with coma usually on micropylar end, sometimes with a coma at the chalazal end (Wrightieae) or at both micropylar and chalazal ends (some Nerieae and Malouetieae) rarely without a coma (some Malouetieae and Apocyneae); endosperm thin, not ruminate. Steroidal alkaloids or cardenolides often present, pyrrolizidine alkaloids reported from a number of genera of Echiteae, indole alkaloids absent. 77 Genera, most restricted to either the Paleo- or Neotropics.

Key to the Tribes of Apocynoids

1. Corolla lobe aestivation sinistrorse; seeds with a chalazal coma (on the end of the seed directed toward the base of the fruit); nectary absent
   1. **Wrightieae** (p. 283)
- Corolla lobe aestivation dextrorse, very rarely valvate or sinistrorse; seeds with a micropylar coma (on the end of the seed directed toward the apex of the fruit); nectary present (except in some Nerieae and Malouetieae) 2
2. Leaf blades adaxially with colleters clustered at the base of the midrib or scattered along its length, or sometimes colleters scattered along the length of the petiole, (colleters in the axil of the petiole absent or present), abaxial domatia often present in axils of secondary veins 3
- Leaf blades and petioles adaxially without colleters (but colleters in the axil of the petiole often present), abaxial domatia usually absent in axils of secondary veins, rarely present in some taxa from Asia 4
3. Corolla tube mostly with a small knob-like corona lobe above each anther; style-head 5-angled, but without strongly projecting ribs; plants of Africa and Madagascar 9. **Baisseeae** (p. 307)
- Corolla tube lacking knob-like corona lobes above the anthers; style-head with 5 strongly projecting ribs, at least at the base; plants of the New World
   6. **Mesechiteae** (p. 294)
4. Calycine colleters absent; style-head with a large membranous basal collar; seeds rostrate; suberect to scandent perennial herbs or lianas usually growing in mangroves or swampy areas
   4. **Rhabdadenieae** (p. 291)
- Calycine colleters normally present, if absent seeds either not rostrate or plant not growing in mangroves or swamps; style-head with or without basal collar; habit various 5
5. Calycine colleters five, solitary (though sometimes deeply dissected and rarely appearing as a continuous ring), and centered at the base of the sepals (antesepalous position), sometimes absent; nectary disc always present; style-head almost always with a membranous basal collar, the main body typically spool-shaped or clavate; latex often translucent; lianas or vines, rarely erect subshrubs or perennial herbs
   7. **Echiteae** (p. 297)
- Calycine colleters most commonly few (5–10) and quincuncially arranged on the margins at the base of the sepals (alternisepalous position) or several and more or less in a continuous ring, occasionally absent; nectary disc present or absent; style-head variously shaped, with or without a basal collar; latex white or translucent; habit various 6
6. Nectary disc absent; style-head with a basal collar, rarely absent, never 5-lobed at base; latex often translucent; seeds mostly with a deciduous chalazal coma in addition to the micropylar coma 2. **Nerieae** (p. 284)
- Nectary disc present or absent; style-head mostly without a basal collar, more rarely basal collar present, very rarely style-head 5-lobed at the base; latex mostly white, rarely translucent; seeds usually with only a micropylar coma, rarely also with a chalazal coma, rarely without a coma 7
7. Anthers attached to the style-head at one level: the thecae usually not agglutinated to the style-head, the

connective attached at about the middle of the style-head (sometimes only very weakly so) by a circular to ovate pad of hairs; style-head globose with a slender, elongate, conical apex; latex translucent or white; shrubs, trees or perennial herbs, very rarely climbers 3. **Malouetieae** (p. 287)
- Anthers attached to the style-head at two levels: the bases of the thecae mostly agglutinated to the upper style-head, and the connective firmly attached to the style-head at about the middle or near the base; style-head of various shapes; latex white; plants mostly lianas and scrambling shrubs, sometimes perennial herbs 8
8. Style-head globose to ovoid and often with a collar around the base, or spool-shaped or broadly fusiform, the last sometimes weakly pentagonal; plants restricted to the New World 5. **Odontadenieae** (p. 291)
- Style-head broadly ovoid to very broadly fusiform, sometimes pentagonal, almost never with a collar at the base, very rarely present; plants of the Paleotropics and, rarely, temperate Eurasia and North America 8. **Apocyneae** (p. 301)

II.1. **Tribe Wrightieae** G. Don (1837).

Shrubs, trees or woody climbers; latex white. Leaves opposite. Inflorescences mostly terminal, sometimes axillary, few- to many-flowered panicles or aggregate dichasia. Calycine colleters few, alternisepalous; corolla salverform to infundibuliform or rarely tubular-campanulate; corolla lobe aestivation sinistrorse; showy corona of flat petaloid to fimbriate segments at mouth (these fused into a cup in *Stephanostema*), rarely absent or very small; stamens inserted near the top of the corolla tube, mostly exserted (rarely inserted lower in the tube and included in some species of *Wrightia*); filaments slender; anthers with relatively short lignified guide rails, expanded sterile part of connective attached to base of style-head (thecae not attached); pollen 3–4-porate, shed onto upper hair wreath; nectary disc absent, adnate or indistinct; ovary apocarpous (syncarpous in some species of *Wrightia*); style-head usually $\pm$ spool-shaped and with a basal collar and upper wreath of longer hairs; stigma located beneath basal collar. Fruit usually apocarpous and consisting of a pair of follicles, these rather stout or long and slender, rarely the follicles postgenitally fused along their ventral margins (some species of *Wrightia*). Seeds with a chalazal coma. Cardenolides present. $x = 10, 11$. Three genera, two restricted to Africa, one widespread in the Paleotropics.

The generic limits between the genera of Wrightieae need to be tested using modern methods.

80. ***Stephanostema*** K. Schum.

Stephanostema K. Schum., Bot. Jahrb. 34: 325 (1904); Barink, Meded. Landbouwhogeschool Wageningen 83-7: 42–53 (1983), rev.

Small shrub. Inflorescences few-flowered, terminal and axillary panicles. Flowers conspicuously globose in bud, with the appearance of miniature daffodils when open, the tube salverform, white, lobes yellow, about the same length as the tube; corona white, cupular, crenulate; anthers yellow, exposed at the base of the corona; ovules ca. 18 per carpel. Follicles pendent, slender, thinly coriaceous. $2n = 22$.

One sp., *S. stenocarpum* K. Schum., known only from a single locality in Tanzania, in forest and bush.

This genus is very closely related to *Wrightia* and especially to *Pleioceras*. The main difference between *Stephanostema* and *Pleioceras* is that the corona lobes are fused into a cupular structure in the former, and free in the latter. Whether this is worthy of recognition at the generic level needs to be investigated.

81. ***Pleioceras*** Baill. Fig. 37

Pleioceras Baill., Bull. Mens. Soc. Linn. Paris 1: 759 (1888); Barink, Meded. Landbouwhogeschool Wageningen 83-7: 21–53 (1983), rev.

Small trees, shrubs or lianas. Inflorescences terminal, lax, several- to many-flowered panicles. Flowers small; corolla yellow, red or violet, the lobes usually longer than the tube; corona bright yellow, consisting of two types of segments, one type shorter, broader, and irregularly lobed, the other long, filiform and split into 3–4 finger-like lobes; ovules numerous. Follicles pendent, slender, thinly coriaceous. $2n = 22$.

Five spp. in tropical Africa in forest, bush and open places, up to 400 m.

Fig. 37. Apocynaceae-Wrightieae. *Pleioceras zenkeri*. **A** Flowering branch. **B** Flower. **C** Flower opened lengthwise showing exserted gynostegium and fimbriate corona lobes at corolla mouth. **D** Part of corolla showing a stamen and two types of corolline corona lobes. **E** One follicle of the fruit. **F** Base of fruit. **G** Seed showing sessile chalazal coma. (From Barink 1983, p. 40, with permission from the Library of Wageningen University, Wageningen; drawn by Y.F. Tan)

82. *Wrightia* R. Br.

Wrightia R. Br., Asclepiadeae 62 (1810); Brown, Prodr. Fl. Nov. Holland. 467 (1810); Ngan, Ann. Missouri Bot. Gard. 52: 114–175 (1965), rev.; Forster, Fl. Australia 28: 190–194 (1996), reg. rev.; Middleton, Fl. Thailand 7(1): 79–90 (1999), reg. rev.; Middleton, Harvard Pap. Bot. 10: 161–182 (2005), reg. rev.; Middleton, Fl. Males. I, 18: 436–447 (2007), reg. rev.; Middleton, Fl. Cambodia, Laos and Vietnam 33: 234–252 (2014), reg. rev.
Scleranthera Pichon (1951).
Wallida (A. DC.) Pichon (1951).

Shrubs or trees. Inflorescences terminal, few- to many-flowered aggregate dichasia. Corolla salverform (rarely infundibuliform), white, rose, mauve or yellowish; corona of entire or variously dissected petaloid to filiform appendages in the mouth, very rarely absent or very small; anthers usually exserted (rarely inserted lower in the tube and anthers included); ovary apocarpous or syncarpous; ovules numerous. Follicles usually thinly woody, free or fused along the ventral sides, forming a "double-follicle", which splits apart into 2 follicle-like halves at maturity. $2n = 20, 22$.

About 25 spp. from E Africa to the Solomon Islands and India and S China to NE Australia, in evergreen, deciduous and eucalypt forests, thickets and clearings, up to 1650 m. Some species are cultivated for their attractive flowers.

II.2. **Tribe Nerieae** Baill. (1889).

Shrubs, trees or woody climbers; latex typically translucent, sometimes white. Leaves usually opposite (alternate in *Adenium*, ternate in *Nerium*, and exceptionally ternate or quaternate in *Strophanthus*). Inflorescences terminal or in the branch forks (in lianoid taxa sometimes axillary as well), few- to several-flowered dichasial cymes or thyrses. Calycine colleters few, alternisepalous or more numerous, rarely absent (*Adenium*); corolla infundibuliform to tubular-campanulate or rarely salverform (*Isonema*); corolla lobes in *Strophanthus* often with elongate apical appendages; corolline corona almost always present in the petal sinuses of genera with an expanded corolla mouth, sometimes small bifid lobes but often larger, petaloid segments in *Strophanthus*, and these much dissected and fused into a more or less continuous, showy, variously lobed ring in *Nerium*; corolla lobe aestivation dextrorse; stamens inserted at the base of the expanded part of the corolla in species with infundibuliform to tubular-campanulate corollas and from about the middle or above in species with salverform corollas, included to partly or completely exserted; anthers specialized, with short filaments or more commonly more or less sessile on thick ridges, lignified guide rails often relatively short, in *Strophanthus*, *Nerium* and *Adenium* with long, filiform, hairy, interspiraling apical appendages much longer than the anther, connective often only weakly attached to base of style-head at one level (thecae not attached); pollen 3–4-porate; nectary disc absent, adnate or indistinct; ovary apocarpous (in *Strophanthus* congenitally syncarpous at the base up to ca. halfway, apocarpous above and in *Nerium* and one species of *Alafia* a completely syncarpous), usually partly inferior in *Strophanthus*; ovules numerous per carpel; style in most genera gradually thickening toward the apex, thickest just below the style-head, sometimes tuberculate; style-head usually ± spool-shaped and with a basal collar and upper wreath of longer hairs (but these poorly developed to absent in *Alafia* and *Farquharia*); stigma located beneath basal collar; ovary apocarpous or more rarely partially to nearly completely syncarpous. Fruit almost always apocarpous and consisting of a pair of follicles, these often rather stout, sometimes long and slender (follicles postgenitally fused along their ventral margins in *Nerium* forming a "double-follicle"). Seeds usually oblong, elliptic, to narrowly fusiform and somewhat compressed, mostly with a smaller deciduous chalazal coma (this sometimes discernible only in undehisced follicles), and larger persistent micropylar coma, the latter sometimes beaked to long-rostrate (seeds with only a micropylar coma in *Nerium* and *Alafia*). Cardenolides present. x = 10, 11. Six genera, most in tropical Africa, some also in Madagascar, one reaching Asia, two in more arid habitats extending to the Arabian Peninsula or the Mediterranean.

The poisonous properties of plants in this tribe are well known. *Nerium*, *Adenium*, *Strophanthus* and *Alafia* have long been used as arrow poison, or, in smaller doses, as treatment for a diverse array of external and internal ailments by the native peoples where they occur. The responsible compounds are cardiac glycosides. Extracts from the seeds of certain species have been found to have powerful cardiotonic, and more recently, cytotoxic properties that can be utilized in modern medicine. Little is known about the secondary chemistry of *Isonema* or *Farquharia*.

83. ***Nerium*** L.

Nerium L., Sp. Pl. 1: 209 (1753); Pagen, Agric. Univ. Wageningen Pap. 87-2: 1–113 (1987), rev.

Shrub or small tree. Leaves ternate, coriaceous. Inflorescence terminal or in the forks, thyrsoid, lax. Corolla showy, infundibuliform, white, pink, red, yellow or salmon, often with darker markings in the throat, double forms in cultivation; showy petaloid corona present in the mouth; anthers with a conspicuous, ca. 8 mm long hairy apical connective appendage; ovary postgenitally syncarpous, pubescent; ovules numerous; style thick, tuberculate in the upper part. Fruit a woody, glabrous "double-follicle". Seeds covered with short, brown, tomentellous hairs grading into a short micropylar coma. 2n = 22.

One sp., *Nerium oleander* L., native to the Mediterranean region and the Middle East but widely cultivated as well as naturalized in many warmer regions, in exposed sites, usually in wet spots such as along stream banks or in the beds of streams or periodically flooded wadis, up to 2500 m. A number of showy cultivars of *Nerium* make it a popular ornamental.

84. ***Adenium*** Roem. & Schult.

Adenium Roem. & Schult., Syst. Veg. 4: 35, 411 (1819); Plaizier, Meded. Landbouwhogeschool Wageningen 80-12: 1–40 (1980), rev.

Succulent shrubs or trees; latex sometimes recorded as being white. Leaves alternate, congested at the ends of the branches. Inflorescence terminal, thyrsoid, few-flowered. Corolla showy, tubular-campanulate, red, pink or white, sometimes striped, slightly zygomorphic; corona lobes small, bifid; stamens included to exserted; anthers with a long, hairy apical connective appendage; ovary glabrous or pubescent; ovules numerous; style enlarged at apex. Follicles stout, spreading or recurved, pubescent. Testa glabrous, rarely minutely puberulent; micropylar and chalazal comas sessile and about equally developed. 2n = 22.

Five spp. in S Arabia, Africa and Socotra, in savannas or open forests with sandy or rocky soil. A genus with showy flowers, often cultivated in succulent collections, and an important source of arrow poison used by indigenous peoples in Africa.

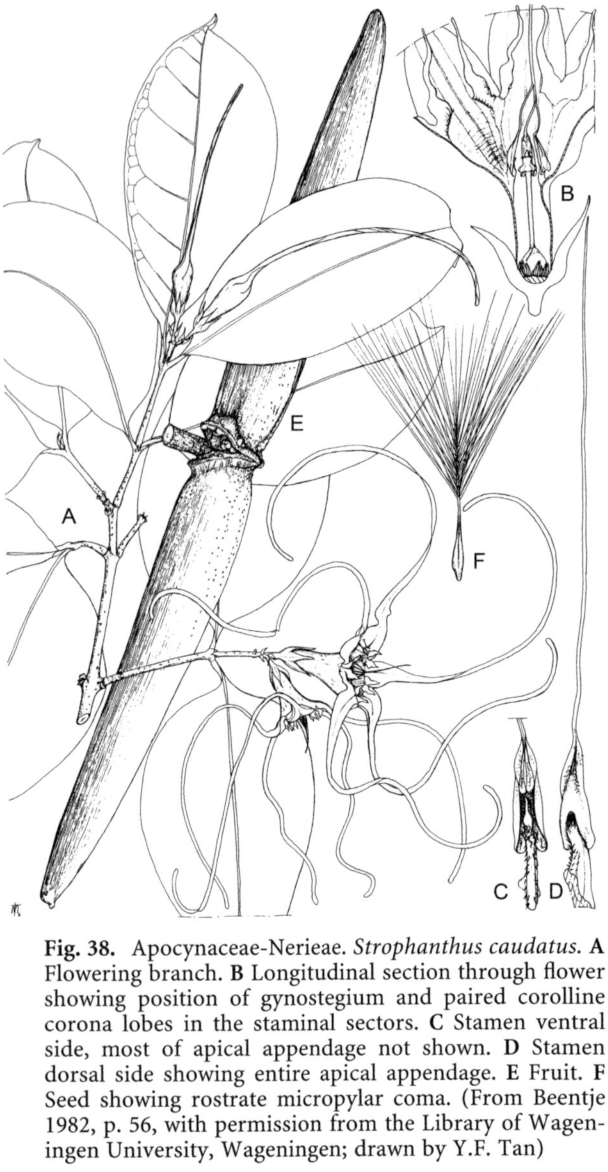

Fig. 38. Apocynaceae-Nerieae. *Strophanthus caudatus*. **A** Flowering branch. **B** Longitudinal section through flower showing position of gynostegium and paired corolline corona lobes in the staminal sectors. **C** Stamen ventral side, most of apical appendage not shown. **D** Stamen dorsal side showing entire apical appendage. **E** Fruit. **F** Seed showing rostrate micropylar coma. (From Beentje 1982, p. 56, with permission from the Library of Wageningen University, Wageningen; drawn by Y.F. Tan)

85. *Strophanthus* DC. Fig. 38

Strophanthus DC., Bull. Sci. Soc. Philom. Paris 3: 122 (1802); Beentje, Meded. Landbouwhogeschool Wageningen 82-4: 1–191 (1982), rev.; Beentje, Fl. Zambesiaca 7: 468–480 (1985), reg. rev.; Middleton, Fl. Males. I, 18: 361–371 (2007), reg. rev.; Middleton, Fl. Cambodia, Laos and Vietnam 33: 176–184 (2014), reg. rev.
Christya Ward & Harv. (1841).
Roupellia Wall. & Hook. (1849).

Shrubs, lianas, or rarely trees. Leaves opposite, occasionally whorled. Inflorescences terminal, less often axillary or in forks of branches. Calycine colleters rarely absent; corolla showy, tubular-campanulate, variously colored, the lobes often undulate and/or with long, slender apical appendages; corona of five well-developed strongly bifid lobes; ovary partly inferior, rarely superior, hemi-syncarpous, glabrous to densely pubescent; ovules numerous; style thick. Follicles thickly to thinly woody, very broadly divergent. Testa mostly pubescent; chalazal coma poorly developed, micropylar coma well-developed, conspicuously long-rostrate. 2n = 18, 20.

38 spp. widely distributed in tropical Africa, Madagascar and Asia, mostly in forest or at forest margins, some species in dry woodlands or more open habitats, at lowland to montane elevations. *Strophanthus* is rich in cardiac glycosides, and is the main ingredient in arrow poison used by several African tribes. It is also the source of the drug ouabain, a cardiac stimulant used to treat heart failure. Several species are cultivated in tropical conditions for their showy flowers with long, dangling petal appendages.

86. *Isonema* R. Br.

Isonema R. Br., Asclepiadeae 52 (1810); Van der Ploeg, Meded. Landbouwhogeschool Wageningen 83-4: 1–20 (1983), rev.

Scrambling shrubs or lianas. Petioles with colleters spread along the adaxial side. Inflorescences terminal, more rarely axillary, few- to many-flowered dichasia. Corolla salverform, variously colored, the lobes with a subapical, crisped appendage; stamens inserted at the orifice of corolla tube; anthers completely exserted; ovary pubescent; ovules numerous; style slender, sometimes tuberculate just below the style-head. Follicles broadly divergent, velvety brown pubescent, narrowly fusiform. Chalazal coma scanty, micropylar coma well-developed, ca. 3–4 cm long. 2n = 22.

Three spp. in W and C Africa, in rainforest, secondary vegetation, often near rivers, or in coastal brush.

87. *Alafia* Thouars

Alafia Thouars, Gen. Nov. Madagasc. 11 (1806); Leeuwenberg, Kew Bull. 52: 769–840 (1997), rev.

Lianas or climbing shrubs; latex sometimes white in some parts and translucent in others. Inflorescences sometimes axillary, usually dense, few- to many-flowered cymes. Corolla salverform, white, pink, mauve, red or yellowish; lobes mostly with long hairs at margins; orifice constricted by a callous ring; stamens inserted in upper part of corolla tube; anthers included or the tips exserted; ovary pubescent; ovules numerous; style gradually enlarged to the apex; style-head poorly developed, without basal collar. Follicles long, slender, cylindrical, glabrous to pubescent. Testa glabrous; coma sessile.

About 23 spp. in tropical Africa and Madagascar, in rainforest, secondary forest, gallery forest, seasonally flooded or riverine forest and bush, up to 1500 m.

88. ***Farquharia*** Stapf

Farquharia Stapf, Bull. Misc. Inform. Kew 1912: 278 (1912); Zwetsloot, Meded. Landbouwhogeschool Wageningen 81-16: 1–46 (1981), rev.

Large woody liana. Petioles with colleters spread along the adaxial side. Inflorescences several- to many-flowered terminal dichasia. Corolla infundibuliform, white to pale yellow; anthers included; ovary rusty brown villous; ovules numerous; style slender, tuberculate for almost the whole length; style-head scarcely developed, without upper wreath or basal collar. Follicles broadly divergent, woody, puberulent rusty brown. Seeds with a deciduous chalazal coma of shorter hairs and a firm, shortly rostrate micropylar coma of longer hairs. $2n = 22$.

One sp., *F. elliptica* Stapf, in tropical Africa.

II.3. **Tribe Malouetieae** Müll. Arg. (1860).

Shrubs, trees, subshrubs, rarely woody lianas or perennial herbs, the latter sometimes with a xylopod; latex usually white (translucent in *Pachypodium* and sometimes so in *Galactophora*). Leaves opposite (rarely verticillate in *Galactophora*) but alternate in *Pachypodium*, often with abaxial domatia in the axils of the secondary veins. Inflorescences terminal or axillary, mostly cymose, few- to several-flowered. Calycine colleters mostly present (absent in *Galactophora* and *Pachypodium* and sometimes absent in *Neobracea*), mostly few, alternisepalous (several and more or less in a ring in *Mascarenhasia* and apparently in *Eucorymbia* and some species of *Carruthersia*); corolla usually infundibuliform to tubular-campanulate, often large, showy and colorful (large and salverform in *Carruthersia* and *Holarrhena*) and smaller, salverform and white or cream in *Allowoodsonia* and *Funtumia*; corolla lobe aestivation dextrorse; corona usually absent, but present as a ring of 5 small bifid protuberances at the orifice of the corolla tube in *Malouetia*; stamens inserted at the base of the expanded part of the corolla in species with infundibuliform to tubular-campanulate corollas in which there is an abrupt division between the expanded upper part and the tubular lower part, and from about the middle or above in species with salverform corollas, included or exserted; anthers sessile or with short filaments, normally clearly morphologically specialized with well-developed lignified guide rails sagittate at the base and firmly attached at about the middle of the style-head (anther base cordate or shortly sagittate and only very weakly attached to the style-head in *Holarrhena*, *Spirolobium*, *Carruthersia* and *Eucorymbia* and additionally non-lignified in *Holarrhena*); pollen quite variable: mostly 3–4-porate, but (1–)2(–4)-porate in *Mascarenhasia* and *Pachypodium*, 2–4-porate in *Malouetia*, and polypantoporate in *Spirolobium*; 5 nectaries usually present around the base of the gynoecium, these often fused into a 5-lobed disc (in *Spirolobium* and *Carruthersia*, 2 separate lobes alternating with the carpels, and nectaries absent in *Holarrhena* and *Malouetiella*); ovary apocarpous; ovules usually numerous per carpel (except in *Malouetiella* and *Allowoodsonia* where only 4–8 per carpel); style-head usually globose with narrowly conical upper part or broadly fusiform, usually without basal collar or upper wreath (in *Eucorymbia* cylindrical and ribbed with a basal collar); stigmatic region on sides of lower cylindrical part below adnation of anthers. Fruit apocarpous, of a pair of follicles mostly long and slender, sometimes stouter, clavate. Seeds flat, linear to fusiform, ellipsoid or ovate, often sometimes elongated at the coma end; seed coma usually micropylar, sessile, of hairs 1–5 cm long (but *Kibatalia* and *Funtumia* with a long-rostrate chalazal coma with retrorse hairs, *Marscarenhasia* with both normally developed micropylar coma and small deciduous chalazal coma (discernible only in undehisced follicles), in *Eucorymbia*, *Allowoodsonia* and *Malouetiella* seeds without a coma, and in *Malouetia* either entire testa glabrous with a few scattered hairs or with an dense indument of long hairs). Steroidal alkaloids. x = 11 (9 in *Pachypodium*). 13 genera, 11 in the Paleotropics, two in the Neotropics.

Most genera have been used for treatment of a broad array of external and internal ailments by the native peoples where they occur; *Pachypodium* and *Malouetia* have long been used as arrow poison.

89. ***Galactophora*** Woodson

Galactophora Woodson, Ann. Missouri Bot. Gard. 19: 49 (1932); Woodson, Ann. Missouri Bot. Gard. 23: 174–178 (1936), rev.; Morales, Sida 21: 2053–2079 (2005), rev.

Subshrubs or perennial herbs, rarely scrambling, with glandular-setose hairs; latex translucent or white. Leaves coriaceous to subcoriaceous, usually revolute at margin. Inflorescences terminal or subterminal, few-flowered. Calyx lobes foliaceous; calycine colleters absent; corolla large, infundibuliform, with 5 longitudinal ridges outside, variously colored; stamens included, weakly attached to style-head; ovary syncarpous at base; style-head conical in upper part, with 5 strongly projecting ribs at base. Follicles slender, terete, united at base into ca. 2 cm stipe. Seed testa glabrous; coma micropylar, tawny, sessile.

Six spp. in NW South America, in sandy savannas and low open forest, from 50–2000 m.

90. ***Neobracea*** Britton

Neobracea Britton in Britton & Millsp., Bahama Fl. 335 (1920); Woodson, Ann. Missouri Bot. Gard. 23: 169–174 (1936), rev.

Shrubs or small, willowy trees. Leaves clustered near the tips of the branches, subcoriaceous, margin revolute. Inflorescences terminal, less frequently subterminal or lateral, few- to several-flowered. Calycine colleters absent or much reduced, rarely few and alternisepalous; corolla infundibuliform, white, pink, rose or purplish; stamens included; nectaries separate or connate at the base; style-head cylindrical with a wreath at the apex. Follicles terete, slender. Seed testa glabrous; coma micropylar, pale yellowish to tawny, sessile.

Four spp. on Cuba, one of them reaching the Bahamas.

91. ***Pachypodium*** Lindl.

Pachypodium Lindl., Edward's Bot. Reg. 16: t. 1321 (1830); Rapanarivo et al., *Pachypodium*, Balkema Publ. (1999), rev.; Burge et al., Peer J. 1: 1–20 (2013), phyl.

Succulent trees, shrubs or perennial herbs with spines at nodes, stems usually swollen at base, often with xylopod; latex translucent. Leaves alternate, crowded at tips of the branches. Inflorescences axillary, cymose, few- to several-flowered. Calycine colleters absent; corolla large, showy, variously colored, tubular-campanulate, infundibuliform or subsalverform; stamens included or exserted; style-head with a basal rim; nectaries free or fused at the base. Follicles stout, glabrous to pubescent. Seeds broad with flattened margin, testa glabrous; coma micropylar white to graywhite, sessile. 2n = 18.

About 21 spp., 12 endemic to Madagascar, the others in Africa. The succulent growth form of *Pachypodium* and its large showy flowers make it a favorite of succulent plant enthusiasts. A number of species are in cultivation, and species are endangered, mainly due to the over-exploitation by hobby gardeners and plant poachers as well as to habitat destruction.

92. ***Spirolobium*** Baill.

Spirolobium Baill., Bull. Mens. Soc. Linn. Paris 1: 773 (1889), nom. cons.; Middleton, Fl. Males. I, 18: 359–361 (2007), reg. rev.; Middleton, Fl. Cambodia, Laos and Vietnam 33: 172–173 (2014), reg. rev.

Small shrub. Leaves narrowly lanceolate, coriaceous. Inflorescences terminal. Flowers large, showy, solitary or in 2–3-flowered cymes; calyx lobes long-lanceolate; calycine colleters few, alternisepalous; corolla infundibuliform, slightly zygomorphic, with long tube curved and gradually widening to the mouth, white with pale yellow markings; stamens inserted in lower part of corolla tube, included; anther bases weakly cordate; nectaries 2, alternate with the carpels; ovary sparsely pubescent on top; style long, slender; style-head cylindrical. Follicles long, erect, slender, glabrous.

Seed testa glabrous to puberulent; coma micropylar, sessile.

One sp., *S. cambodianum* Baill., found in SE Asia, in open grassy areas, scrub, open forests and mangrove thickets, up to 300 m.

93. ***Carruthersia*** Seem.

Carruthersia Seem., Fl. Vit. 155 (1866); Middleton, Blumea 41: 489–498 (1997), rev.; Middleton, Fl. Males. I, 18: 154–157 (2007), reg. rev.

Woody lianas. Leaves long-petiolate, cordate to round at the base. Inflorescences axillary and/or terminal cymes or panicles. Calycine colleters alternisepalous or in a ring; corolla salverform, white to cream, sometimes reddish tinged; stamens inserted near the base of the tube, included, weakly attached to the style-head at the base of the filaments, anther bases short; nectaries 2, alternate with the carpels; ovary glabrous; style slender; style-head fusiform. Follicles terete to fusiform, glabrous. Seed testa glabrous; coma micropylar, sessile.

Four spp. in the Philippines, Solomon Islands, Fiji and Tonga.

94. ***Eucorymbia*** Stapf

Eucorymbia Stapf, Hooker's Icon. Pl. 28: t. 2764 (1903); Pichon, Bull. Mus. Natl. Hist. Nat. II, 21: 270–271 (1949), synop.; Middleton, Fl. Males. I, 18: 196 (2007), reg. rev.; Middleton, Fl. Penins. Mal. II, 2: 144–146 (2011), reg. rev.

Glabrous liana. Inflorescences lax terminal corymbs. Flowers large, showy; sepals subfoliaceous, caducous, with a row of colleters at the base within; corolla tubular, white; lobes broadly obovate; stamens included; filaments short; anthers pubescent dorsally, only weakly attached to the style-head; nectary disc 5-lobed, much shorter than the ovary; ovary glabrous; style filiform; style-head cylindrical with a basal collar. Follicles stout. Seeds elliptic, compressed, testa glabrous; coma absent.

Very little is known about this genus, which contains only one sp., *E. alba* Stapf, known from Sumatra, Peninsular Malaysia and Borneo.

95. ***Mascarenhasia*** A. DC.

Mascarenhasia A. DC., Prodr. 8: 487 (1844); Pichon, Mém. Inst. Sci. Madagascar, Sér. B, Biol. Vég. 2: 68–93 (1949), synop.; Markgraf, Fl. Madagascar, Fam. 169: 248–276 (1976), rev.; Leeuwenberg, Wageningen Agric. Univ. Pap. 97-2: 21–52 (1997), rev.

Trees or shrubs. Inflorescences axillary or terminal, cymose, sometimes reduced to a single flower. Calyx lobes often foliaceous; calycine colleters many and more or less in a ring, or fewer and alternisepalous; corolla infundibuliform to tubular-campanulate, white flushed with rose, tube sometimes yellow, the lobes induplicate in bud, margins often long-ciliate and/or crispate; nectary disc 5-lobed. Follicles long and slender or stouter. Seeds with sessile, micropylar coma and small deciduous chalazal coma; testa glabrous. $2n = 22$.

Eight spp. in Madagascar, one of them reaching tropical Africa, in gallery forest, dry forest, brush or woodland, often on river banks, up to 1250 m.

96. ***Holarrhena*** R. Br.

Holarrhena R. Br., Asclepiadeae 52 (1810); De Kruif, Meded. Landbouwhogeschool Wageningen 81-2: 1–40 (1981), rev.; De Kruif, Fl. Zambesiaca 7: 456–458 (1985), reg. rev.; Middleton, Fl. Males. I, 18: 197–200 (2007), reg. rev.; Middleton, Fl. Cambodia, Laos and Vietnam 33: 96–102 (2014), reg. rev.

Shrubs or trees. Inflorescences terminal or seemingly axillary due to overtopping. Calycine colleters few, alternisepalous; corolla salverform, white to greenish white; stamens inserted in lower third of corolla tube; filaments short; anthers with only rudimentary basal appendages, not lignified, weakly attached to the style-head via hair pads at the base of the filaments; nectary disc absent; ovary superior to partly inferior. Follicles long and slender, thinly coriaceous. Seed testa glabrous; coma micropylar, sessile. $2n = 22$.

Four spp. in S and SE Asia and tropical Africa.

97. ***Malouetia*** A. DC. Fig. 39

Malouetia A. DC., Prodr. 8: 378 (1844); Woodson, Ann. Missouri Bot. Gard. 22: 238–270 (1935), rev. Amer. spp.; Van der Ploeg, Agric. Univ. Pap. 85-2: 70–83 (1985), reg. rev.; Endress, Fl. Venez. Guayana 2: 513–517 (1995), reg. rev.

Shrubs or trees. Leaves mostly with abaxial domatia in the axils of the secondary veins.

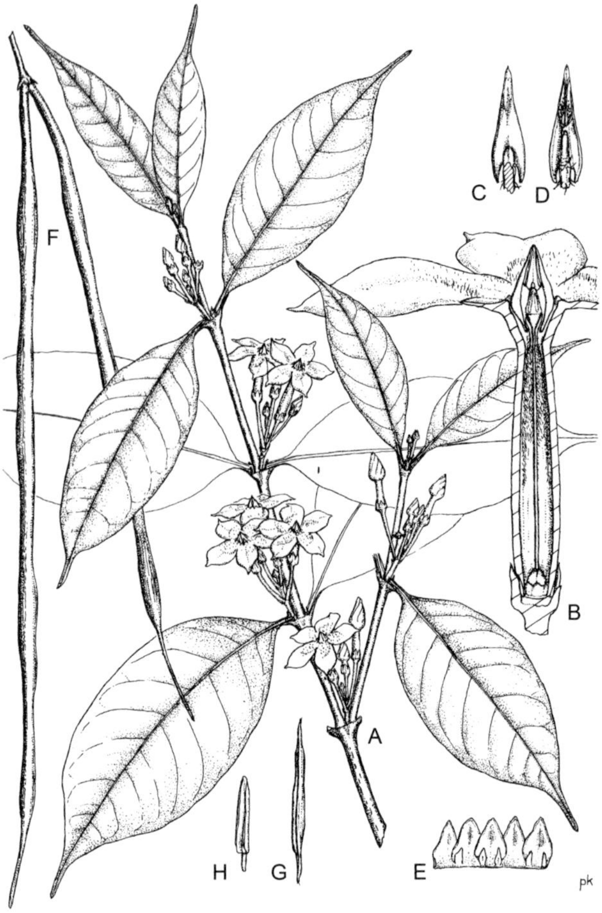

Fig. 39. Apocynaceae-Malouetieae. *Malouetia heudelotii.* **A** Flowering branch. **B** Longitudinal section through flower, showing partially exserted gynostegium. **C** Stamen dorsal side. **D** Stamen ventral side. **E** Opened calyx showing quincuncial arrangement of adaxial colleters. **F** Fruit. **G** Ecomose seed. **H** Embryo. (From Van der Ploeg 1985, p. 78, with permission from the Library of Wageningen University, Wageningen; drawn by P.J. Kostense)

Inflorescences axillary or terminal, fasciculate, few- to many-flowered. Calycine colleters few, alternisepalous; corolla tubular-campanulate, variously colored; corolline corona of five notched lobes behind the anthers often present; anthers exserted to included; nectary disc five-lobed. Follicles thinly to thickly woody, usually slender or stouter, green. Seeds without a coma; testa glabrous or with a few scattered hairs or densely covered with long hairs.

About 27 spp. with a disjunct distribution in the tropics of the New World and tropical W and C Africa, mostly in seasonally inundated forest, riverine forest, swamp forest or mangroves at low altitudes, rarely montane forest. *Malouetia* has long been used as an arrow poison by indigenous peoples in Amazonia.

98. *Kibatalia* G. Don

Kibatalia G. Don, Gen. Hist. 4: 86 (1837); Woodson, Philip. J. Sci. 60: 205–229 (1936), rev.; Rudjiman, Agric. Univ. Wageningen Pap. 86-5: 35–89 (1986), rev.; Middleton, Fl. Males. I, 18: 207–232 (2007), reg. rev.; Middleton, Fl. Cambodia, Laos and Vietnam 33: 112–116 (2014), reg. rev.
Paravallaris Pierre (1898).

Trees or shrubs. Leaves often with abaxial domatia in the axils of the secondary veins. Inflorescences terminal or alternating in the upper leaf axils, fasciculate, few- to many-flowered. Calycine colleters few, alternisepalous, more rarely numerous and spread across the base of the sepals; corolla tubular-campanulate, white or off-white; anthers exserted to included; nectary disc annular, 5-lobed. Follicles thinly to thickly woody, narrowly fusiform to clavate. Seeds with rostrate chalazal coma of long retrorse hairs; testa glabrous.

15 spp. in continental SE Asia and Malesia, particularly in the Philippines, in tropical lowland or montane forest, swamp forest, thickets, savanna, mountain slopes, often on stream banks, road sides, up to 1200 m. Several species from now deforested areas have not been collected for decades and may be extinct.

99. *Allowoodsonia* Markgr.

Allowoodsonia Markgr., Gard. Bull. Singapore 22: 23 (1967).

Tree; latex translucent to white. Inflorescences axillary, few-flowered. Calycine colleters few, alternisepalous; corolla salverform, white; corona absent; stamens inserted ca. midway in the tube; anthers subsessile, just reaching the tube orifice; nectary disc annular, 5-lobed; ovules ca. 8 per carpel. Follicles woody, broadly divergent, long, cylindrical, slightly inflated. Seeds obliquely oblong-lanceolate, without a coma; testa glabrous.

A poorly known genus of one sp., *A. whitmorei* Markgr., from the Solomon Islands, found in primary forest in valley bottoms and in coastal swamp forest.

100. ***Malouetiella*** Pichon

Malouetiella Pichon, Bull. Jard. Bot. État. 22: 131 (1952).

Tree or shrub. Leaves usually with abaxial domatia in at least some of the axils of the secondary veins. Inflorescences axillary or terminal lax dichasia. Calycine colleters few, alternisepalous; corolla tubular-campanulate, white, somewhat thickened in the throat; stamens inserted about midway in the corolla tube, included; nectary disc absent; ovules 4–6. Follicles slender. Seeds obliquely fusiform, longitudinally folded, without a coma; testa very sparsely pilose at ends.

One sp., *M. mildbraedii* Gilg & Stapf, from Nigeria to DR Congo, in montane forest up to 1400 m.

101. ***Funtumia*** Stapf

Funtumia Stapf, Hooker's Icon. Pl. 27: pl. 2694 (1901); Zwetsloot, Meded. Landbouwhogeschool Wageningen 81-16: 1–46 (1981), rev.

Trees or shrubs; latex white. Petioles of leaf pair often connate at the node, forming a short ocrea; abaxial domatia often present in axils of secondary veins. Inflorescences terminal and axillary, congested, cymose, several- to many-flowered. Calycine colleters few to several, alternisepalous; corolla thick, salverform, orifice constricted by a callous ring, white to cream, the tube greenish; stamens inserted at about the middle of corolla tube; anthers subsessile, included; nectary disc 5-lobed; style thick, obconical. Follicles woody, fusiform to clavate, 8–32 cm long. Seed coma chalazal, rostrate, the rostrum also with long retrorse hairs; testa glabrous. $2n = 22$.

Two spp. in tropical Africa.

II.4. **Tribe Rhabdadenieae** Pichon ex M.E. Endress (2014).

Slender woody lianas or perennial herbs with white latex. Leaves opposite. Flowers large and showy; calycine colleters absent; corolla infundibuliform; corolla lobe aestivation dextrorse; anthers with truncate, lignified guide rails fused to filaments, firmly attached to style-head, included; nectaries five separate lobes or fused basally; ovary apocarpous; style filiform; style-head cylindrical, conspicuously pilose apically, with a large membranous basal collar. Fruit a pair of narrowly fusiform follicles. Seeds numerous, narrowly boat-shaped, glabrous; coma micropylar, long-rostrate. One neotropical genus in mangroves or swampy areas.

102. ***Rhabdadenia*** Müll. Arg. Fig. 40

Rhabdadenia Müll. Arg. in Mart., Fl. Bras. 6(1): 173, t. 52 (1860); Woodson, Ann. Missouri Bot. Gard. 23: 205–211 (1936), rev.; Morales, J. Bot. Res. Inst. Texas 3: 541–564 (2009), rev.

Lianas; latex white. Leaves opposite. Inflorescences axillary or subterminal cymes, typically of 2 flowers. Calyx lobes usually foliaceous; corolla large, showy, infundibuliform, white, white with yellow throat, rose, red or purple; stamens sessile to subsessile, inserted at the base of the throat, included; anthers truncate, non-auriculate at the base; ovary apocarpous; ovules many; style-head cylindrical, conspicuously pilose at the apex. Fruit a pair of narrowly fusiform follicles. Seeds narrowly boat-shaped; coma white to cream, up to 4 cm long.

Three spp., two restricted to South America in wet or swampy areas at low elevations; one, *R. biflora* (Jacq.) Müll. Arg., with exceptionally lightweight wood, is frequently found in mangroves, and reaches the Caribbean and Florida.

II.5. **Tribe Odontadenieae** Miers (1878).

Woody climbers or scrambling shrubs; latex white. Leaves opposite, rarely whorled or alternate (some species of *Odontadenia*), the blades lacking colleters adaxially. Inflorescences axillary, sometimes terminal, cymose. Calycine colleters few and alternisepalous or numerous and spread across base of sepals inside (rarely absent); corolla salverform to infundibuliform; corolla lobe aestivation dextrorse; corona absent; stamens inserted in the lower half of the corolla tube; anthers with well-developed lignified guide rails, the bases slender, acuminate; main attachment of the anthers at about the middle of style-head; base of thecae normally agglutinated to upper style-head as well; pollen (2–)3(–4)-porate; nectaries usually 5, these usually fused into a lobed or crenulate ring around the base of the ovary; ovary apocarpous; ovules many; style-head fusiform to ovoid (subcapitate in *Cycladenia*),

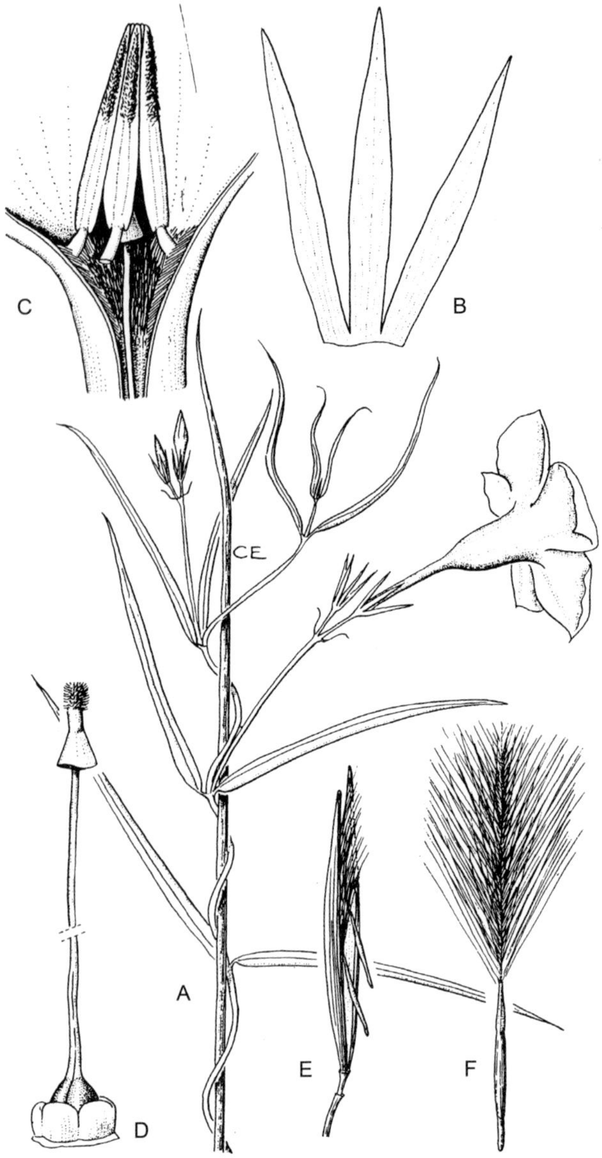

Fig. 40. Apocynaceae-Rhabdadenieae. *Rhabdadenia ragonesei*. **A** Flowering branch. **B** Three sepals showing lack of adaxial colleters. **C** Gynostegium showing nearly truncate basal appendages. **D** Gynoecium and nectary. **E** Dehiscing follicle showing seeds with micropylar coma. **F** Seed showing rostrate coma. (From Ezcurra et al. 1992, p. 102, © Conservatoire et Jardin botaniques de la Ville de Genève and with their permission; drawn by Cecilia Ezcurra)

broadest and often with equatorial flange at about the middle, normally without basal collar or upper wreath; stigmatic region usually on lower cylindrical region below adnation of anthers. Fruit apocarpous, of 2 separate follicles; follicles generally thin-walled. Seeds flat, linear to ellipsoid or ovate, the testa glabrous or hairy, with micropylar (sometimes rostrate) coma. Cardenolides. x = 10, 11 (more rarely 8, 12). Nine genera in the Americas, eight tropical, one restricted to mountain peaks in W and SE USA.

103. *Pinochia* M.E. Endress & B.F. Hansen

Pinochia M.E. Endress & B.F. Hansen, Edinb. J. Bot. 64: 269–274 (2007), nom. cons., synop.

Woody lianas or scrambling shrubs. Leaf blades usually with abaxial domatia in axils of secondary veins. Inflorescences terminal, less often axillary, thyrsiform, many-flowered. Flowers very small; calycine colleters alternisepalous, evenly distributed or rarely absent; corolla tubular to subcampanulate; tube short, usually with hairs in throat; stamens inserted near base of tube, barely included to completely exserted; nectary a 5-lobed disc; style-head ovoid, without ribs. Follicles free. Seeds slender, fusiform, without longitudinal ribs; coma yellowish brown, sessile.

Four spp. in Mexico, Central America and the Greater Antilles, usually in forest, often in coastal thickets or riparian forest up to 1000 m.

104. *Thyrsanthella* (Baill.) Pichon

Thyrsanthella (Baill.) Pichon, Bull. Mus. Nat. Hist. Natl. II, 20: 192 (1948); Livshultz et al., Ann. Missouri Bot. Gard. 94: 324–359 (2007), phyl.
Forsteronia sect. *Thyrsanthella* Baill. (1889).

Woody liana. Leaf blades lacking abaxial domatia. Inflorescences terminal or axillary, many-flowered dichasia. Flowers small; calycine colleters alternisepalous; corolla inconspicuous, whitish, narrowly infundibuliform, with a ring of hairs in the throat; stamens inserted about midway in the corolla tube; filaments short, free; anthers barely included or the tips exserted; nectary of 5 lobes, nearly as high as the ovary; style-head ovoid, without ribs. Follicles free, narrowly terete to subtorulose. Seeds slender, fusiform; coma pale yellowish gray, sessile.

One sp., *T. difforme* (Walt.) Pichon, in the SE USA in a variety of wet to dry habitats, usually at low elevations.

105. *Cycladenia* Benth.

Cycladenia Benth., Pl. Hartw. 322. (1849); Woodson, Ann. Missouri Bot. Gard. 23: 214–217 (1936), rev.; Livshultz et al., Ann. Missouri Bot. Gard. 94: 324–359 (2007), phyl.

Low, subsucculent perennial herb; petioles winged and concrescent at the nodes. Inflorescences lateral, few- to several-flowered cymes. Calycine colleters absent; corolla infundibuliform, with lower cylindrical tube and upper expanded throat, reddish violet to cream; stamens inserted at base of throat, included; filaments short; nectary disc annular; style-head subcapitate. Fruit apocarpous; follicles small, relatively stout, sometimes remaining united at their apices. Seeds somewhat flattened and asymmetrical, brown-papillose with a pale tawny, sessile coma.

One sp., *C. humilis* Benth., with four described varieties, one each restricted to high elevations on disjunct mountain peaks in California, and one to high mesas in Utah and Arizona, respectively.

106. *Stipecoma* Müll. Arg.

Stipecoma Müll. Arg. in Mart., Fl. Bras. 6(1): 175 (1860); Woodson, Ann. Missouri Bot. Gard. 23: 189–191 (1936), rev.; Morales, Candollea 60: 307–311 (2005), synop.

Glabrous liana, woody at base but shoots delicate; latex white. Leaf blades peltate, glaucous, coriaceous, with well-developed colleters in axils of petioles. Inflorescences axillary, alternate, bostrychoid racemose. Calycine colleters few to several in alternisepalous groups; corolla salverform, pink; stamens inserted about midway in corolla tube; anthers with slender basal appendages; filaments short; nectaries connate at very base; style-head fusiform. Fruit apocarpous; follicles slender, terete, broadly divergent. Seeds slender; coma rostrate, cream to tan-cream colored.

One sp., *S. peltigera* (Stadelm.) Müll. Arg., in NE and C Brazil (rare in Bolivia) in open cerrado, campo rupestre and disturbed vegetation from 700–1600 m.

107. *Secondatia* A. DC.

Secondatia A. DC., Prodr. 8: 445 (1844); Woodson, Ann. Missouri Bot. Gard. 22: 224–232 (1935), rev.; Morales, Candollea 58: 305–319 (2003), rev.

Woody lianas; branches usually conspicuously lenticellate. Leaf blades with conspicuously reticulate tertiary venation. Inflorescences terminal or less frequently axillary, few- to many-flowered thyrses. Calycine colleters few, alternisepalous; corolla salverform, usually small, white or greenish yellow; stamens inserted in lower half of corolla tube; anthers with narrowly acuminate sagittate bases; nectary disc thin, 5-lobed; style-head elongate-fusiform without basal collar, sessile or with a very short style. Follicles ovoid-fusiform, stout. Seeds narrowly fusiform; coma cream, sessile to very shortly rostrate.

Four spp., in tropical South America from Colombia and Venezuela to Bolivia and Paraguay, in forest or more open, drier habitats such as savannas, cerrados or granitic outcrops.

108. *Odontadenia* Benth. Fig. 41

Odontadenia Benth., J. Bot. (Hooker) 3: 242 (1841); Woodson, Ann. Missouri Bot. Gard. 22: 270–306 (1935), rev.; Morales, Bull. Jard. Bot. Belg. 67: 381–477 (1999), rev.

Woody lianas; branches often conspicuously lenticellate. Inflorescences axillary or axillary and terminal, thyrsiform to scorpioid, several-flowered. Calycine colleters few and alternisepalous to many and unevenly distributed; corolla infundibuliform or more rarely salverform, cream to yellow, often with red lines, rarely orange to red; stamens inserted at the base of the throat, included; anthers with sagittate bases; nectary disc usually 5-lobed; style-head fusiform. Follicles apocarpous, usually robust. Seeds narrowly fusiform; coma yellowish to pale tawny, sessile. $2n = 24$.

About 20 spp. in Central and South America and the Caribbean.

109. *Salpinctes* Woodson

Salpinctes Woodson, Bull. Torrey Bot. Club 58: 453 (1931); Morales, Novon 8: 429 (1998), synop.; Morales et al., Taxon 66: 623–644 (2017), phyl.

Perennial herb or woody subshrub, all parts glabrous; latex white. Leaves opposite, subsessile, thick-coriaceous. Inflorescences terminal, usually 1-flowered. Calycine colleters numerous, irregularly distributed; corolla salverform, large and showy, bright pink; stamens inserted near the

Fig. 41. Apocynaceae-Odontadenieae. Flowering and/or fruiting branches. **A** *Odontadenia funigera*. **B** *O. macrantha*. (From Zarucchi et al. 1995, p. 541, with permission from the Missouri Botanical Garden Press, St. Louis; drawn by B. Manara)

apex of corolla tube, included; nectaries 2, alternating with the carpels; style-head fusiform with a membranous collar at the base. Follicles terete, relatively stout. Seeds unknown.

One scarcely collected and poorly known sp., *S. kalmiaefolius* Woodson, endemic in savannas in Guyana and Amazonas State, Venezuela.

110. *Angadenia* Miers

Angadenia Miers, Apocyn. S. Amer. 173 (1878), pro parte; Woodson, Ann. Missouri Bot. Gard. 23: 191–198 (1936), rev.; Morales et al., Taxon 66: 623–644 (2017), phyl.

Woody subshrubs; stems erect or sometimes twining; branches alternate; latex white. Inflorescences lateral, infrequently terminal, scorpioid. Calycine colleters few, alternisepalous; corolla infundibuliform, with a lower cylindrical tube and an upper expanded cup, white to pale yellow; stamens inserted somewhat below midway in the corolla tube, included; nectary disc deeply 5-lobed; style-head fusiform with a basal collar. Fruit apocarpous; follicles terete, slender. Seeds narrowly fusiform; coma rostrate, pale yellowish.

Two to three spp. in the Caribbean, usually in sandy areas or in pine or palm barrens near the sea.

This genus has a complex taxonomic history, and has included up to 22 species, all but three of which are now included in genera of diverse tribes of neotropical Apocynoids.

111. *Pentalinon* Voigt

Pentalinon Voigt, Hort. Suburb. Calcutt. 523 (1845); Woodson, Ann. Missouri Bot. Gard. 23: 198–205 (1936), rev. (as *Urechites*); Morales et al., Taxon 66: 623–644 (2017), phyl.
Urechites Müll. Arg. (1860).

Woody lianas or subshrubs with twining stems; latex white. Inflorescences axillary to terminal, simply scorpioid, few- to many-flowered. Calycine colleters few and alternisepalous or more numerous and irregularly distributed; corolla infundibuliform, yellowish or cream, often marked with red; stamens inserted at base of throat, with conspicuous long, spirally coiled, filamentous connective appendages; nectaries separate or connate at base; style-head fusiform with membranous collar at base. Follicles terete, slender. Seeds pubescent with long-rostrate, pale brown coma. $2n = 12$.

Two species, one in Florida and the Caribbean, the other in Mexico and Central America. A popular ornamental in tropical countries.

II.6. **Tribe Mesechiteae** Miers (1878).

Vines, woody lianas or perennial herbs or subshrubs, the latter often with a xylopod; latex usually white. Leaves opposite, rarely whorled in some *Mandevilla* species, mostly with a cluster of sometimes deciduous colleters adaxially at base of the leaf blade (distributed along the length of midrib in some species of *Mandevilla*), sometimes with abaxial domatia in the axils of the secondary veins. Inflorescences axillary and/or terminal. Calycine colleters mostly in alternisepalous groups or in continuous ring, more rarely solitary and centered at the base of the sepals;

corolla usually divided into narrow lower tube and expanded upper throat, mostly infundibuliform or tubular-campanulate, more rarely tubular or salverform; corolla lobe aestivation dextrorse; distinct corona usually absent; stamens mostly inserted at base of the expanded upper part; anthers with long, well-developed lignified guide rails, sometimes ± the entire anther (except for the thecae) strongly lignified and the thecae extending nearly to the base; pollen (3–)5(–6)-porate; nectaries present, usually 5 fleshy lobes (2 lobes in some species of *Mandevilla*), ± separate or fused at the base, forming a ring around the base of the ovary; ovary apocarpous (postgenitally syncarpous in a few species of *Forsteronia*); ovules many; style-head with five arms, which usually project from lower part (arms forming long ridges for most of the length of the style-head in *Mandevilla*) to which the anthers are attached; upper wreath and membranous basal collar absent; stigmatic zone confined to the underside or lower region of the style-head. Fruit apocarpous, of 2 separate follicles (these postgenitally syncarpous in a few species of *Forsteronia* and forming a "double-follicle"), usually slender. Seeds narrowly fusiform to oblong, testa glabrous or minutely puberulent, with a sessile micropylar coma. x = 10, 11. Cardenolide glycosides have been reported from some species of *Mandevilla*. Six neotropical genera.

112. *Elytropus* Müll. Arg.

Elytropus Müll. Arg., Bot. Zeitung (Berlin) 18: 21 (1860); Woodson, Ann. Missouri Bot. Gard. 23: 383–214 (1936), rev.; Ezcurra, Darwiniana 24: 390–391, t. 5 (1981), reg. rev.; Morales, Darwiniana n.s. 1: 40–42, t. 1 (2013), reg. rev.

Woody scrambler with ferruginous pubescence; stems twining to erect; branches alternate above, mostly opposite below. Leaf blades without colleters at base adaxially. Inflorescences opposite lateral cymes of 1 (sometimes 2) pendent flowers; bracts subfoliaceous, pilose. Calycine colleters absent; corolla campanulate, white with rose to lilac markings at base; stamens inserted near base of tube; anthers with pilose apex and slender, acuminate basal appendages; nectaries 5, connate at base; style-head spool-shaped. Follicles relatively stout, ferruginous-hirtellous. Seeds oblong, chalazal end narrowed; testa glabrous.

One sp., *E. chilensis* (A. DC.) Müll. Arg., in Chile and Argentina, shaded rainforest floor on the sides of the coastal mountains near Valdivia up to ca. 2000 m.

Elytropus is an isolated genus and its inclusion in Mesechiteae is tentative (Livshultz et al. 2007).

113. *Allomarkgrafia* Woodson

Allomarkgrafia Woodson, Ann. Missouri Bot. Gard. 19: 45 (1932); Woodson, Ann. Missouri Bot. Gard. 20: 625–628 (1933), rev.; Morales, Brittonia 49: 337–345 (1997), rev.

Glabrous woody lianas. Leaf blades with several colleters clustered at base adaxially. Inflorescences axillary, dichasial cymes. Calycine colleters numerous, uniformly distributed; corolla infundibuliform, with a lower cylindrical tube and an expanded campanulate throat, white to rose-white, cream to greenish white; stamens inserted at base of throat, included; anthers with uniformly fertile thecae; nectaries separate or connate at base; style-head fusiform above, five-ribbed below. Follicles slender, terete. Seeds narrowly oblong, chalazal end narrowed, micropylar end truncate.

Nine spp. in Central America and N South America, in humid forest, montane cloud forests, sometimes in disturbed areas, up to 1500 m.

There is evidence suggesting that *Allomarkgrafia* may be congeneric with *Mesechites* (Simões et al. 2004; Livshultz et al. 2007), but results are inconclusive at present.

114. *Mesechites* Müll. Arg.

Mesechites Müll. Arg. in Mart., Fl. Bras. 6(1): 150 (1860); Woodson, Ann. Missouri Bot. Gard. 20: 625–645 (1933), rev.; Simões et al., Amer. J. Bot. 91: 1409–1418 (2004), phyl.; Morales, Candollea 61: 215–277 (2006), rev.

Woody lianas. Leaf blades with 1–several colleters clustered at base adaxially. Inflorescences axillary, compound cymose. Calycine colleters alternisepalous or numerous and irregularly distributed; corolla salverform to tubular-campanulate or infundibuliform, white to greenish white or rose, often flushed with pink and green; stamens inserted about midway in tube or at base of expanded throat, included; anthers with uniformly fertile thecae; nectaries separate or connate at base; style-head fusiform to pentagonal above, strongly 5-ribbed in lower part. Follicles slender, terete. Seeds slender, fusiform, micropylar end truncate.

Eight spp. in Central and South America, and the Caribbean, in thickets, humid forest, montane cloud forests or cerrado and other more arid vegetation types, sometimes in disturbed areas at low elevations.

115. ***Tintinnabularia*** Woodson

Tintinnabularia Woodson, Ann. Missouri Bot. Gard. 23: 387 (1936); Morales, Novon 6: 392–394 (1996); Williams, Lundellia 2: 136–141 (1999), rev.

Woody lianas. Leaf blades with a few colleters clustered at base of midrib adaxially; abaxial domatia present in axils of secondary veins. Inflorescences axillary, corymbose, several-flowered; bracts and calyx sometimes foliaceous. Calycine colleters in alternisepalous groups; corolla showy, cream, infundibuliform; stamens inserted at base of throat, included except for filiform appendages; anthers with uniformly fertile thecae and obtuse bases; apical connective appendage filiform or absent; nectaries separate or connate at base; style-head fusiform above, strongly five-ribbed at base. Follicles long, slender, torulose. Seeds narrowly fusiform, longitudinally ridged, micropylar end truncate.

Three spp. in Mexico, Guatemala and Honduras; in humid montane forests, usually above 1000 m.

116. ***Forsteronia*** G. Mey.

Forsteronia G. Mey., Prim. Fl. Esseq. 133 (1818), nom. cons.; Woodson, Ann. Missouri Bot. Gard. 22: 153–224 (1935), rev.; Hansen, A monographic revision of *Forsteronia* (Apocynaceae), Ph.D. Thesis, University of South Florida, USA: 1–382 (1985), rev.

Woody lianas or scrambling shrubs. Leaf blades with colleters clustered at base of midrib adaxially; abaxial domatia usually present in secondary vein axils. Inflorescences terminal, less often axillary, thyrsiform, many-flowered. Flowers very small; calycine colleters alternisepalous, evenly distributed or rarely absent; corolla tubular to subcampanulate; stamens inserted near base of tube, barely included to completely exserted; anther bases cordate to truncate, more rarely attenuate; nectary a 5-lobed disc; ovary usually apocarpous, sometimes postgenitally syncarpous; style-head 5-ribbed. Follicles usually free, sometimes postgenitally fused. Seeds narrowly elliptic, longitudinally ridged, micropylar end truncate; testa commonly minutely puberulent.

About 42 spp. in the Neotropics, usually in forest at lower elevations.

117. ***Mandevilla*** Lindl. Fig. 42

Mandevilla Lindl., Edward's Bot. Reg. 26: t. 7 (1840), nom. cons. prop.; Woodson, Ann. Missouri Bot. Gard. 20: 645–790 (1933), rev.; Henrickson, Aliso 14: 179–195 (1996), part. rev. (as *Macrosiphonia* and *Telosiphonia*); Morales, Brittonia 50: 214–232 (1998), synop.; Simões et al., Amer. J. Bot. 91: 1409–1418 (2004), phyl.; Morales, Darwiniana 43: 131–137 (2005), reg. rev.; Morales, Darwiniana 44: 472–474 (2006), reg. rev.
Exothostemon G. Don (1837).
Dipladenia A. DC. (1844).
Macrosiphonia Müll. Arg. (1860).

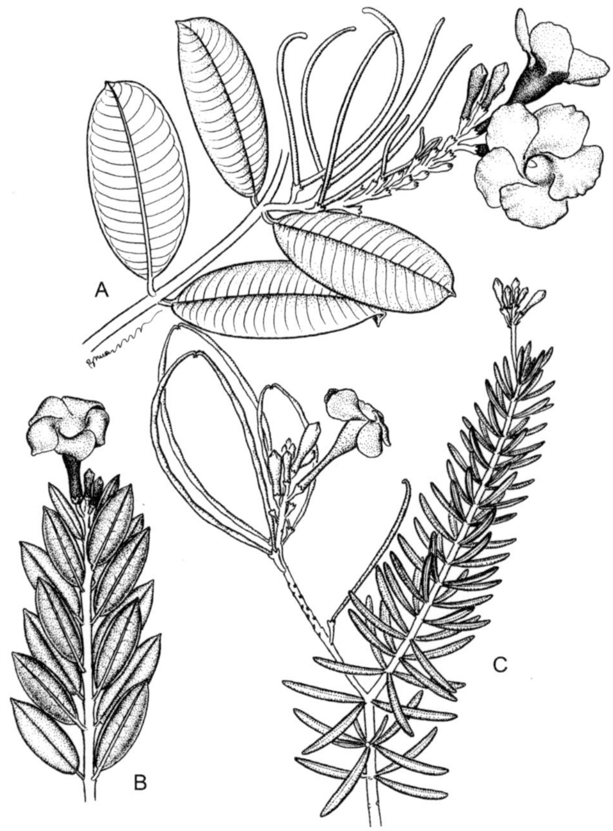

Fig. 42. Apocynaceae-Mesechiteae. Flowering and/or fruiting branches. **A** *Mandevilla subcarnosa*. **B** *M. turgida*. **C** *M. benthamii*. (From Zarucchi et al. 1995, p. 527, with permission from the Missouri Botanical Garden Press, St. Louis; drawn by B. Manara)

Quiotania Zarucchi (1991).
Telosiphonia (Woodson) Henrickson (1996).

Woody lianas, perennial herbs, or rarely subshrubs; stems occasionally winged; sometimes with xylopod. Leaves opposite, more rarely verticillate, blades adaxially with colleters at base of midvein or along length (rarely absent). Inflorescences mostly axillary, occasionally terminal, usually racemose (unbranched). Calycine colleters usually present, variously distributed; corolla infundibuliform, sometimes salverform or tubular-campanulate, corolla often with curved tube, variously colored; stamens usually inserted at base of expanded throat, usually included; anthers strongly fused to style-head ribs; filaments very short; nectaries 2–5, separate or connate at base; style-head 5-ribbed. Follicles slender, terete to somewhat torulose. Seeds narrowly fusiform; testa usually minutely puberulent. $2n = 16, 20$.

About 170 spp. widely distributed from SW USA and northern Mexico, Central America and the Antilles to Paraguay and Uruguay and found in a variety of habitats, from rainforest to dry cerrado, dry rocky areas, desert scrub-grasslands and pine or oak forests, from lowland to montane elevations. A few species are cultivated and in recent years a number of showy cultivars of these are widely available for home gardens.

Although progress has been made, many species are rarely collected and poorly known, and the genus is in need of more in-depth study to better understand species-level relationships.

II.7. **Tribe Echiteae** Bartl. (1830).

Vines or woody lianas, rarely erect shrubs or perennial herbs; latex often translucent, sometimes white. Leaves opposite (verticillate in most *Artia* and some species of *Parsonsia*). Inflorescences mostly axillary, less commonly terminal. Calycine colleters often solitary and antesepalous, in the center of the sepal at the base, or several and spread across base of sepal, rarely absent; corolla salverform, infundibuliform, tubular-campanulate, campanulate (rotate in *Artia* and *Thenardia*, urceolate in *Ecua* and *Thoreauea*); corolla lobe aestivation dextrorse or valvate; stamens mostly inserted in upper part of corolla tube (near base in some *Parsonsia* and *Thenardia*); anthers often partially to almost completely exserted, with large lignified guide rails, attached near base of style-head; pollen 3–5(–8)-porate; nectary present, encircling the base of the ovary, usually of 5 ± free lobes, these sometimes fused at the base, less frequently nectary an (often lobed or crenulate) ring; ovary apocarpous (postgenitally syncarpous in *Parsonsia*, *Artia*, *Ecua*, *Thenardia*, *Thoreauea*, *Temnadenia*, and some species of *Prestonia*); ovules many; style-head cylindrical to narrowly fusiform, broadest and with (usually well-developed, often membranous) collar at base and sometimes upper wreath as well; stigmatic zone located on underside of style-head beneath collar. Follicles typically slender (torulose in *Thenardia*), and postgenitally fused and forming a "double-follicle" in *Parsonsia*, *Artia*, *Thenardia*, *Thoreauea*, *Temnadenia*, probably in *Ecua* (fruit not known) and in some species of *Prestonia*, but splitting apart along suture into 2 follicle-like halves at maturity. Seeds mostly narrowly fusiform but broadly ovate in *Parsonsia* and *Artia* (and probably *Ecua*), the testa glabrous, with a micropylar (often rostrate) coma, the rostrum always glabrous. Pyrrolizidine alkaloids. $x = 6, 7, 8, 9, 11$. 14 genera, 11 in the Neotropics, three mainly in Australasia, of which one reaching continental Asia.

118. ***Laubertia*** A. DC.

Laubertia A. DC., Prodr. 8: 486 (1844); Woodson, Ann. Missouri Bot. Gard. 23: 370–375 (1936), rev.; Morales, Rhodora 104: 170–185 (2002), rev.; Morales et al., Taxon 66: 623–644 (2017), phyl.; Morales, Anales Jardin Bot. Madrid 74(2): e063 (2017), tax.

Woody lianas. Leaves usually opposite, rarely ternate. Inflorescences axillary, rarely terminal, few- to several-flowered scorpioid cymes. Calycine colleters absent; corolla salverform, variously pink, red, purple or white; tube straight or twisted, with thickened annular corona at orifice; stamens inserted midway or above in the corolla tube, usually partly exserted; anthers sessile, narrowly sagittate at the base; nectaries separate or fused at the base; ovary apocarpous; style-head fusiform. Fruit apocarpous, the follicles slender, weakly torulose. Seeds fusiform; coma sessile.

Four spp. with a disjunct distribution: two in Mexico and N. Central America, one in NW South America and one in the Amazon basin, up to 1600 m.

119. *Hylaea* J. F. Morales

Hylaea J. F. Morales, Novon 9: 83–85 (1999); Morales et al., Taxon 66: 623–644 (2017), phyl. Fig. 43

Woody lianas or small trees; latex usually translucent. Leaves glabrous. Inflorescences axillary, many-flowered, often umbelloid. Calycine colleters solitary, antesepalous; corolla salverform to subinfundibuliform, white and pink- to purple-tinged, glabrous, with a corona lobe inserted above each anther; lobes longer to about same length as tube; stamens inserted near orifice of corolla tube, almost completely exserted; filaments short, puberulent; anther bases acute to acuminate; nectaries distinct, as high as or surpassing the ovary; ovary apocarpous, glabrous; style-head fusiform. Fruit unknown.

A rarely collected and poorly known genus of two spp. restricted to Brazilian-Venezuelan Amazonia, often on white sand soils and/or associated with blackwater rivers, below 400 m.

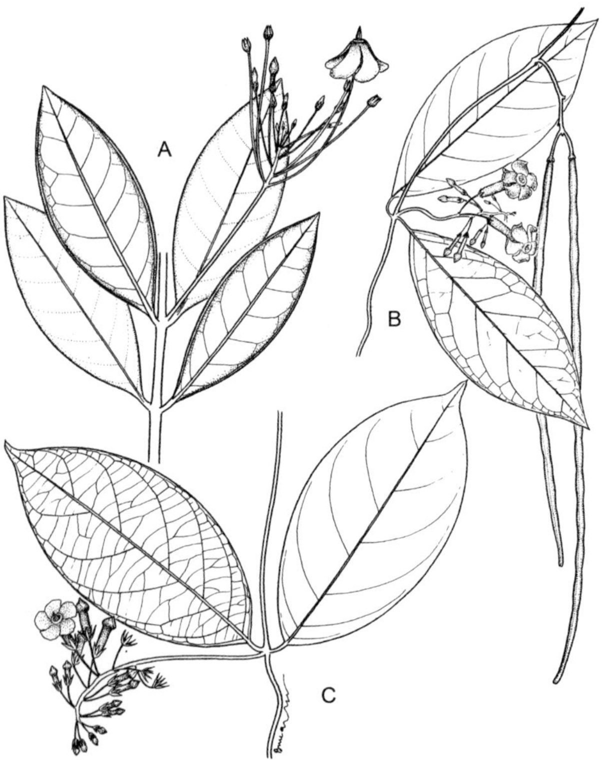

Fig. 43. Apocynaceae-Echiteae. Flowering and/or fruiting branches. **A** *Hylaea arborescens*. **B** *Prestonia acutifolia*. **C** *P. vaupesiana*. (From Zarucchi et al. 1995, p. 551, with permission from the Missouri Botanical Garden Press, St. Louis; drawn by B. Manara)

120. *Temnadenia* Miers

Temnadenia Miers, Apocyn. S. Amer. 207 (1878); Woodson, Ann. Missouri Bot. Gard. 23: 253–260 (1936), rev.; Morales, Candollea 60: 207–231 (2005), rev.; Morales et al., Taxon 66: 623–644 (2017), phyl.

Woody or suffruticose lianas; latex usually translucent. Inflorescences axillary, rarely subterminal, compound, alternate, many-flowered. Calycine colleters solitary, antesepalous, dissected; corolla salverform or infundibuliform, showy rose or purple, rarely greenish or yellowish; stamens with short filaments; anthers with slender basal appendages; nectaries separate or connate at very base; ovary postgenitally syncarpous; style-head fusiform. Fruit syncarpous, of 2 slender, terete follicles postgenitally fused into a "double-follicle", which splits into 2 halves at maturity. Seeds with a tawny rostrate coma. $2n = 18$.

Three spp. in tropical South America, often at forest edges.

121. *Rhodocalyx* Müll. Arg.

Rhodocalyx Müll. Arg. in Martius, Fl. Brasil. 6(1): 172 (1860); Woodson, Ann. Missouri Bot. Gard. 23: 367–370 (1936), rev.; Stranghetti & Kinoshita, Rev. Bras. Bot. São Paulo 19: 133–144 (1996), morph.; Morales, Novon 9: 89–91 (1999), synop.; Morales et al., Taxon 66: 623–644 (2017), phyl.

Small erect subshrubs with xylopod or suffruticose liana; latex translucent. Leaves sessile to petiolate. Inflorescences terminal or lateral, simple racemes of few to several flowers. Calyx petaloid; calycine colleters antesepalous, several or solitary and deeply dissected; corolla salverform, orifice constricted by annular corona; stamens inserted in upper corolla tube; filaments bent inward; anthers with acute basal appendages; nectaries ± separate but fused at base into 5-lobed ring; ovary apocarpous; style-head spool-shaped with membranous basal collar. Fruit apocarpous, follicles falcate, pubescent. Seeds numerous; coma sessile.

Two spp. found from NE Brazil to SE Bolivia and NE Paraguay in savannas, cerrados and campo rupestre vegetation.

122. ***Macropharynx*** Rusby

Macropharynx Rusby, Memo. N.Y. Bot. Gard. 7: 327, t. 6 (1927); Woodson, Ann. Missouri Bot. Gard. 23: 268–271 (1936), rev.; Morales, Rhodora 99: 252–262 (1997), rev.; Morales, Candollea 60: 29–334 (2005), rev.; Morales et al., Taxon 66: 623–644 (2017), phyl.
Peltastes Woodson (1932).

Woody lianas; branches often with rusty brown indument, at least in younger stages; latex translucent, rarely white. Leaves sometimes peltate. Inflorescences axillary, rarely terminal, few- to many-flowered; bracts foliaceous in some species. Calycine colleters solitary and antesepalous or many, irregularly distributed; corolla salverform to infundibuliform, greenish white to pale yellow; tube twisted in some species; stamens inserted in mid-tube or higher, included; anthers with slender basal appendages; filaments short; nectaries 5, ± separate; ovary apocarpous; style-head fusiform. Fruit of two stout, falcate follicles, often with rusty brown indument. Seeds glabrous; coma rostrate.

15 spp. in tropical Central and South America in wet, moist or seasonally dry forest, often at forest margins, some species in cerrados, savannas or disturbed habitats, from low elevations up to 2000+ m.

Macropharynx meyeri (Ezcurra) Xifreda does not fit well with the rest of the genus morphologically and probably would be better placed elsewhere. Until additional material of this species can be obtained for investigation, however, its placement remains speculative.

123. ***Thenardia*** Kunth

Thenardia Kunth in Humboldt et al., Nov. Gen. Sp. (quarto ed.) 3: 209, t. 240 (1818); Woodson, Ann. Missouri Bot. Gard. 23: 271–276 (1936), rev.; Williams, Lundellia 1: 78–94 (1998), rev.; Morales et al., Taxon 66: 623–644 (2017), phyl.

Glabrous woody or suffruticose lianas; latex translucent. Inflorescences axillary, condensed umbelloid cymes. Calycine colleters solitary, antesepalous; corolla rotate or shortly salverform, white to pink; stamens inserted in mouth of corolla, completely exserted, filaments slender, sometimes coiled; anthers narrowly sagittate at base; nectaries separate, as long as or shorter than ovary; ovary postgenitally syncarpous; style-head fusiform. Fruit syncarpous, of two slender, torulose follicles postgenitally fused into a "double-follicle", which splits into halves at maturity. Seeds narrowly fusiform, tapering at coma end; coma white, sessile.

Three spp., endemic to Mexico, in semi-deciduous, montane, and cloud forest, sometimes near water, from 1100–2200 m.

124. ***Thoreauea*** J.K. Williams

Thoreauea J.K. Williams, Lundellia 5: 47–58 (2002); Morales, Brittonia 57: 258–263, synop.; Morales et al., Taxon 66: 623–644 (2017), phyl.

Lianas; latex white. Leaves glabrous. Inflorescences axillary, umbelloid cymes. Calycine colleters solitary, antesepalous; corolla urceolate, whitish to greenish, 5–10-lobed annular corona at mouth; stamens inserted near base of corolla tube, included or tips exserted; anthers with slender auricles; filaments connate in a ring around style; nectaries 5, free or some of them fused at the base; ovary postgenitally syncarpous; style slender; style-head spool-shaped with basal collar. Fruit syncarpous, of two follicles postgenitally fused into a "double-follicle", which splits apart at maturity. Seeds slender, navicular; coma sessile.

Three spp., all from southern Mexico, above 2600 m in montane pine-oak forest.

125. ***Asketanthera*** Woodson

Asketanthera Woodson, Ann. Missouri Bot. Gard. 19: 46 (1932); Woodson, Ann. Missouri Bot. Gard. 23: 263–268 (1936), rev.; Morales et al., Taxon 66: 623–644 (2017), phyl.

Woody or suffruticose lianas. Inflorescences axillary, simply scorpioid, several-flowered; pedicels subtended by single markedly foliaceous bract. Calyx markedly foliaceous; calycine colleters solitary, antesepalous; corolla salverform, greenish white, yellowish, pinkish or cream; stamens inserted near base to midway in corolla tube, anthers included; filaments short; anthers with slender basal appendages; nectaries separate or connate at the base; ovary apocarpous; style-head fusiform. Fruit of two slender, terete follicles. Seeds with a tawny rostrate coma.

Four spp. in the Antilles, often on rocky (limestone) hillsides or cliffs, up to ± 500 m.

126. ***Echites*** P. Browne

Echites P. Browne, Civ. Nat. Hist. Jamaica 182 (1756); Woodson, Ann. Missouri Bot. Gard. 23: 217–252 (1936), rev.; Morales, Brittonia 49: 328–336 (1997), synop.; Williams, Sida 21: 117–131 (2004), part. synop. (as *Allotoonia*); Morales, Darwiniana 44: 467–469 (2006), reg. rev.; Morales, Darwiniana 47: 149–152 (2009), reg. rev.; Morales et al., Taxon 66: 623–644 (2017), phyl.
Fernaldia Woodson (1932).
Allotoonia J.F. Morales & J.K. Williams (2004).

Woody lianas or vines; latex white or translucent. Inflorescences axillary, rarely terminal, alternate, 1- to many-flowered. Calycine colleters solitary, antesepalous, often deeply dissected; corolla salverform to infundibuliform, whitish or orangish; annulus at mouth and corona absent; stamens inserted at base of upper expanded part of corolla tube, included; nectaries separate or connate into a lobed disc; ovary apocarpous; style-head fusiform. Fruit of two slender, terete follicles, which sometimes remain united at apices. Seeds fusiform to broadly ovoid; coma rostrate. 2n = 12.

13 spp. in tropical Mexico, Central America and the Caribbean. At one time several hundred species of Apocynoids were either described or combined in *Echites* from both the Old and New Worlds. Over time these were removed to many and varied other genera leaving a relatively small number of species.

127. ***Bahiella*** J.F. Morales

Bahiella J.F. Morales, Sida 22: 333–353, t. 2, 3 (2006).

Woody lianas; latex white. Leaves coriaceous with revolute margins. Inflorescences many-flowered axillary cymes. Calycine colleters solitary, antesepalous, but usually deeply dissected; corolla large, mostly red to purple, salverform or infundibuliform; stamens inserted about midway in the corolla tube; anthers included, the sagittate bases slender, acuminate; nectaries fused into an irregularly 5-lobed ring; ovary apocarpous; style-head fusiform with basal collar. Fruit of two free follicles. Seeds fusiform; coma rostrate.

Two spp., known at present only from Bahia, Brazil, in humid forest, coastal vegetation, and *restinga*, at low elevations.

Bahiella is only rarely collected and poorly known.

128. ***Ecua*** D.J. Middleton

Ecua D.J. Middleton, Blumea 41: 33–35 (1996); Middleton, Fl. Males. I, 18: 190–191 (2007), reg. rev.

Liana or scrambling shrub. Inflorescences axillary elongated cymes. Calyx lobes small; calycine colleters solitary, antesepalous; corolla urceolate, with inflated tube; lobes ca. a quarter as long as the tube, erect; stamens inserted near orifice of corolla tube, included; filaments curving downward and then up again; anther bases curved strongly outward; nectary lobes cupped, fused at the base; ovary postgenitally syncarpous, pubescent on top; style slender; style-head fusiform with a short collar. Fruit unknown.

One sp., *E. moluccensis* D.J. Middleton, collected only on Morotai and Halmahera in the Moluccas (Indonesia), in thickets below 100 m.

A rarely collected and poorly known genus, which shares most features with *Parsonsia*.

129. ***Parsonsia*** R. Br.

Parsonsia R. Br., Asclepiadeae 53 (1810), nom. cons.; Boiteau, Fl. Nouvelle-Cal. 10: 245–285 (1981); Williams, Fl. Australia 28: 154–189 (1996), reg. rev.; Middleton, Blumea 42: 191–248 (1997), reg. rev.; Middleton, Fl. Thailand 7(1): 135–138 (1999), reg. rev.; Middleton, Fl. Males. I, 18: 310–344 (2007), reg. rev.; Middleton, Fl. Cambodia, Laos and Vietnam 33: 152–156 (2014), reg. rev.
Lyonsia R. Br. (1810).
Delphyodon K. Schum. (1898).
Grisseea Bakh. f. (1948).

Woody lianas or scrambling shrubs; latex often translucent, sometimes white. Leaves rarely verticillate; juvenile leaves often much smaller/narrower. Inflorescences axillary, sometimes terminal, cymose to paniculate. Calycine colleters usually solitary, antesepalous, sometimes absent; corolla usually 5-merous, rarely 4-merous, tubular, variously colored; corolla lobe aestivation dextrorse or valvate; stamen insertion variable; anthers almost always partially to nearly completely exserted; filaments usually long, sometimes interspiraled; nectaries usually free; ovary postgenitally syncarpous, usually glabrous; style filiform; style-head fusiform, with short collar at base. Fruit syncarpous, glabrous, of two follicles postgenitally fused into a "double-follicle", which splits into two follicle-like halves at maturity. Seeds broadly ovate; coma sessile. 2n = 18.

About 82 spp., in S Asia, continental SE Asia, Malesia to the western Pacific Islands, Australia and New Zealand, mostly in forest.

130. ***Artia*** Guillaumin

Artia Guillaumin, Bull. Soc. Bot. France 88: 380 (1941); Boiteau, Flore Nouvelle-Cal. 10: 285–298 (1981), rev.

Lianas; latex translucent to yellowish. Leaves usually in whorls of 3–4, sometimes opposite. Inflorescences axillary or terminal, cymose. Calycine colleters solitary, antesepalous; corolla campanulate to urceolate or subrotate, usually whitish externally, with red-brown, orange, yellow or violet nectar guides at base of lobes internally, corona lobe just behind each anther; corolla lobe aestivation dextrorse; stamens inserted in upper half of corolla tube; filaments relatively long, geniculate; nectary disc annular; ovary postgenitally syncarpous, glabrous; style-head fusiform, with short collar at base. Fruit syncarpous, of two follicles postgenitally fused into a "double-follicle", which splits into two follicle-like halves at maturity. Seeds flat, broadly ovate; coma sessile.

Four spp., endemic to New Caledonia.

This genus is sister to *Parsonsia*.

131. ***Prestonia*** R. Br. Fig. 43

Prestonia R. Br., Asclepiadeae 58 (1810), nom. cons.; Woodson, Ann. Missouri Bot. Gard. 23: 276–367 (1936), rev.; Morales, Novon 7: 59–63 (1997), reg. rev.; Rio & Kinoshita, Hoehnea 32: 233–258 (2005), reg. rev.; Morales & Liede-Schumann, Phytotaxa 265: 204–224 (2016), reg. rev.; Morales et al., Ann. Missouri Bot. Gard. 102: 520–541 (2017), phyl.

Woody lianas; latex translucent to white. Inflorescences axillary, rarely subterminal, simple or compound. Calycine colleters solitary, antesepalous; corolla salverform, rarely infundibuliform; thickened annular corona usually present at orifice of tube, corona of five appendages behind the anthers often present, sometimes reduced to ridges, corolla salverform, rarely infundibuliform, variously colored; stamens included to almost completely exserted; anthers narrowly sagittate at the base; nectaries separate to completely fused; ovary usually apocarpous, more rarely postgenitally syncarpous; style-head fusiform with collar at base. Follicles apocarpous, or of two follicles postgenitally fused into a "double-follicle", which splits into two halves at maturity. Seeds narrowly oblong; coma sessile or scarcely rostrate. 2n = 18.

About 58 spp. ranging from the West Indies through Mexico, Central and South America as far as N Argentina, in a variety of habitats, from rainforest to dry forest, cerrado, open areas, from lowland to montane elevations.

II.8. **Tribe Apocyneae** Rchb. (1831).

Woody climbers or scrambling shrubs (*Apocynum* perennial herbs); latex white. Leaves opposite, rarely whorled (alternate in some species of *Apocynum*), with abaxial domatia often present in the axils of secondary veins in *Urceola* and *Baharuia*. Inflorescences terminal and/or axillary, most commonly cymes, dichasia, thyrses or panicles. Calycine colleters few and alternisepalous or numerous and spread across base of sepals inside (rarely absent); corolla salverform, campanulate or urceolate; corolla lobe aestivation dextrorse (sinistrorse in *Urceola* p.p.), rarely valvate; corona, if present, mostly as small pouches in petal sinuses, sometimes lower down on corolla tube behind stamens (*Apocynum* with alternistaminal appendages near base of corolla tube); stamen insertion variable, commonly near base, included except partly exserted in some spp. of *Trachelospermum* and *Micrechites* and appearing exserted in *Vallaris* and *Beaumontia* but then mostly due to the wide open upper tube rather than because the stamens protrude beyond length of entire tube; anthers with long, well-developed lignified guide rails, bases sagittate; main attachment of the anthers at about the middle of style-head; base of thecae normally agglutinated to upper style-head as well; pollen (2–)3(–4)-porate (polypantoporate in some *Micrechites*, *Trachelospermum* and *Apocynum*); nectaries 5, separate, or these fused into a (usually lobed or crenulate) ring around the base of the ovary; ovary apocarpous (postgenitally syncarpous in *Beaumontia*, *Parepigynum* and *Vallaris*), sometimes partly inferior; ovules almost always numerous; style-head broadly fusiform, broadest and often with equatorial flange at about middle, with or without basal collar (basal collar present in *Streptoechites*,

Sindechites and *Epigynum*); stigmatic region usually on lower cylindrical region below adnation of anthers. Fruit apocarpous, of 2 separate follicles; in *Beaumontia*, *Parepigynum* and *Vallaris* the follicles are postgenitally united along their ventral margins, forming a stout "double-follicle" that splits apart into 2 follicle-like halves at maturity (though sometimes only one carpel matures); follicles generally thin-walled but thick in *Beaumontia*. Seeds flat or not, linear to ellipsoid or ovate, the testa glabrous or hairy with micropylar (sometimes rostrate) coma. Cardenolides. $x = 10$, 11 (more rarely 8, 12). 21 genera, almost all in tropical Asia and/or Oceania, one in temperate regions of the Old and New World.

132. ***Papuechites*** Markgr.

Papuechites Markgr., Nova Guinea 14, 2: 287 (1925); Middleton, Blumea 40: 439–442 (1995), rev.; Middleton, Fl. Males. I, 18: 303–305 (2007), reg. rev.

Liana. Inflorescences terminal and/or axillary thyrses. Colleters few, alternisepalous; corolla small, salverform to tubular-campanulate, pink or red; small corona lobes in mouth in the alternistaminal sectors and with ridges above the anthers, lobes about as long as tube, inflexed in bud; stamens inserted near the middle of corolla tube, included; filaments short; nectaries separate; ovary pubescent on top. Follicles broad at the base, abruptly tapering to a slender apex. Seeds elliptic, flattened, testa glabrous; coma on a long, glabrous rostrum.

One sp., *P. aambe* (Warb.) Markgr., in New Guinea and the Moluccas, in humid forest up to 1300 m.

133. ***Ixodonerium*** Pit.

Ixodonerium Pit. in Lecomte & Humbert, Fl. Indo-Chine 3: 1228 (1933); Lý, Feddes Repert 97: 669–670 (1986), reg. rev.; Middleton, Fl. Cambodia, Laos and Vietnam 33: 110–111 (2014), reg. rev.

Liana. Inflorescences terminal, thyrses with short lateral axes, spike-like, pubescent. Calycine colleters few, alternisepalous; corolla red, tube cylindrical, swollen in the middle, thickened at the orifice, with ridges above the anthers; stamens inserted in lower third of corolla tube, included; anther bases divergent; filaments short; nectary disc fleshy, annular, 5-dentate, higher than ovary; ovary densely pubescent on top; style short; style-head fusiform, pentagonal. Follicles broad at the base, abruptly tapering to a slender apex. Seeds elliptic, flattened, testa glabrous; coma rostrate.

One sp., *I. annamense* Pit., known only from C Vietnam in mountain forest.

134. ***Anodendron*** A. DC.

Anodendron A. DC., Prodr. 8: 443 (1844); Middleton, Blumea 41: 37–68 (1996), rev.; Middleton, Fl. Males. I, 18: 132–145 (2007), reg. rev.; Middleton, Fl. Cambodia, Laos and Vietnam 33: 52–59 (2014), reg. rev.
Formosia Pichon (1948).

Lianas. Inflorescences terminal or axillary cymes, often forming panicles. Calycine colleters few, alternisepalous; corolla salverform, whitish or yellowish; tube with ridges above the anthers; stamens inserted near base of corolla tube, included; filaments short; nectary disc annular; style short, thick, conical; style-head ovoid. Follicles glabrous, stout, ovoid, narrowed at the apex. Seeds flattened, ovoid, with a long rostrum with long hairs both along the rostrum and at its apex.

17 spp., India and China to Japan through SE Asia to Australia and the Solomon Islands, in forest up to 1600 m.

135. ***Amphineurion*** (A. DC.) Pichon

Amphineurion (A. DC.) Pichon, Bull. Soc. Bot. France 95: 215 (1948); Middleton, Fl. Males. I, 18: 129–131 (2007), reg. rev.; Middleton, Fl. Cambodia, Laos and Vietnam 33: 48–51 (2014), reg. rev.
Aganosma (Blume) G. Don sect. *Amphineurion* A. DC. (1844).

Liana or scrambling shrub; stems densely lenticellate. Leaves with strong marginal vein. Inflorescences lax, terminal or axillary panicles. Calycine colleters forming continuous ring; corolla salverform, white, lobes narrowly lanceolate, much longer than tube; stamens inserted about midway in the corolla tube, included; anthers subsessile; nectary disc of 5 lobes or these fused into an ring, shorter than the ovary; ovary minutely puberulent to glabrous; style filiform. Follicles long, terete, rather stout, lenticellate. Seeds narrowly elliptic with sessile micropylar coma.

One sp., *A. marginata* (Roxb.) D.J. Middleton, widely distributed in NE India, SE Asia and

Malesia, in a variety of habitats, especially drier evergreen or deciduous forest or scrubland, up to 850 m.

136. ***Sindechites*** Oliv.

Sindechites Oliv. in Hooker's Icones Pl. 18: t. 1772 (1888); Xu, Agric. Univ. Wag. Pap. 88-6: 25–35 (1988), rev.; Li et al., Flora of China 16: 188 (1995), reg. rev.

Liana. Inflorescences terminal and/or axillary corymbs. Flowers small; calyx lobes small; calycine colleters few, alternisepalous; corolla salverform, white, tube much longer than lobes; lobes broadly ovate; stamens inserted about midway in corolla tube, included; filaments short; anthers with a tuft of hairs at the apex; nectary disc fleshy, entire or 5-lobed, shorter than or as high as the ovary; ovary glabrous or pubescent at the top; style filiform; style-head with basal collar. Follicles slender, cylindrical, glabrous. Seeds linear; coma sessile.

One sp., *S. henryi* Oliv., widespread in C China, in montane forest, brush, roadsides, often near streams, up to 1500 m.

137. ***Streptoechites*** D.J. Middleton & Livsh.

Streptoechites D.J. Middleton & Livsh., Adansonia III, 34: 370 (2012); Middleton, Fl. Cambodia, Laos and Vietnam 33: 173–175 (2014), reg. rev.

Woody climber. Inflorescences terminal or axillary, cymose. Calyx without colleters within; calycine colleters few, alternisepalous; corolla salverform, white, tube much longer than lobes; stamens inserted about midway in the corolla tube, included; anthers with a tuft of hairs at the apex; nectary disc shallowly bilobate, shorter than or as high as the ovary; ovary glabrous; style filiform; style-head with basal collar. Follicles slender, weakly torulose, glabrous; pedicels twisted in fruit. Seeds fusiform, testa glabrous; coma sessile, white.

One sp., *S. chinensis* (Merr.) D.J. Middleton & Livsh., from China and SE Asia.

138. ***Vallaris*** Burm. f. Fig. 44

Vallaris Burm. f., Fl. Indica 51 (1768); Rudjiman, Meded. Landbouwhogeschool Wageningen 81-11: 1–17 (1981), rev.; Li et al., Fl. China 16: 177 (1995), reg. rev.; Middleton, Fl. Males. I, 18: 409–410 (2007), reg. rev.; Middleton, Fl. Cambodia, Laos and Vietnam 33: 229–230 (2014), reg. rev. *Parabeaumontia* Pichon (1948).

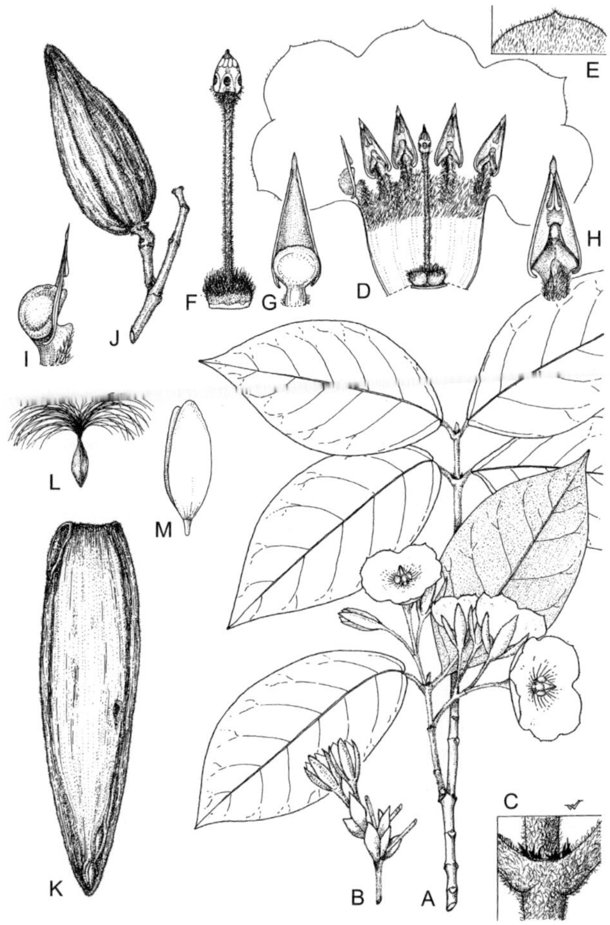

Fig. 44. Apocynaceae-Apocyneae. *Vallaris indecora*. **A** Flowering branch. **B** Inflorescence with flower buds. **C** Node showing colleters at petiole base. **D** Opened flower showing exserted reproductive organs. **E** Corolla lobe apex showing abaxial pubescence. **F** Gynoecium. **G** Anther dorsal side. **H** Anther ventral side. **I** Anther from side showing dorsal swelling. **J** Postgenitally connate immature fruit. **K** One half of dehisced mature fruit. **L** Seed. **M** Embryo. (From Rudjiman 1982, p. 8, with permission from the Library of Wageningen University, Wageningen; drawn by W. Wessel-Brand)

Woody climbers. Inflorescences axillary, rarely terminal, cymose; bracts large, foliaceous. Calycine colleters few, alternisepalous, or absent; corolla of lower cylindrical tube and much-expanded upper cup, white or off-white; stamens inserted at base of upper cup; filaments short; anthers with conspicuous dorsal hump, exserted from lower tube; nectary disc annular, lobed; ovary pubescent; style filiform, pubescent; style-head without upper wreath or basal collar. Fruit of two follicles postgenitally fused into an ovoid "double-follicle", which splits into two halves at maturity (though sometimes only one maturing). Seeds flat, broadly ovate; testa puberulent; coma sessile. 2n = 20, 22.

Three spp. from Kashmir to S China, SE Asia and western Malesia.

139. ***Beaumontia*** Wall.

Beaumontia Wall., Tent. Fl. Napal. 14, pl. 7 (1824); Rudjiman, Agric. Univ. Wageningen Pap. 86-5: 3–35 (1986), rev.; Middleton, Fl. Males. I, 18: 147–151 (2007), reg. rev.; Middleton, Fl. Cambodia, Laos and Vietnam 33: 59–63 (2014), reg. rev.
Muantum Pichon (1948).

Lianas. Inflorescences axillary or terminal, cymose, few-flowered. Calyx lobes usually large, foliaceous; calycine colleters usually many, rarely few and alternisepalous; corolla mostly very large, of cylindrical lower tube and much-expanded upper cup, usually white; stamens inserted at base of upper cup; anthers on long, slender filaments, sometimes projecting beyond corolla tube; nectary disc annular, 5-lobed; carpels postgenitally fused; style slender; style-head fusiform, without upper wreath or basal collar. Fruit of two follicles postgenitally fused into a woody, stoutly fusiform "double-follicle", which splits into two halves at maturity. Seeds broadly ovate; coma sessile. 2n = 24.

Ten spp. in tropical Asia from India to Bali, in thickets and humid lowland, montane, or monsoon forest, often on river banks or coasts, up to 1700 m. One species, *B. grandiflora* Wall., reaching temperate regions in Nepal, Bhutan and Sikkim, and also widely cultivated in Europe, America, and Africa.

140. ***Parepigynum*** Tsiang & P.T. Li

Parepigynum Tsiang & P.T. Li, Acta Phytotax. Sin. 11: 394, t. 54 (1973); Middleton, Fl. Cambodia, Laos and Vietnam 33: 150–152 (2014), reg. rev.

Liana. Inflorescences terminal and axillary, corymbose, long-pedunculate. Calyx pubescent, alternisepalous colleters few; corolla salverform, yellow, the tube densely pubescent, the lobes about as long as the tube; stamens inserted about midway in the corolla tube, included; filaments short; nectary disc fleshy, of 5 ± fused lobes; carpels postgenitally syncarpous, ovary partly inferior to inferior, pubescent; style cylindrical; style-head conical, without upper wreath or basal collar. Fruit of two follicles postgenitally fused into a woody, stoutly fusiform "double-follicle", which splits into two halves at maturity. Seeds narrowly elliptic; coma shortly rostrate.

A rare genus of one sp., *P. funingense* Tsiang & Li, from southern China and northern Vietnam, in dense montane forests, up to 1800 m.

141. ***Cleghornia*** Wight

Cleghornia Wight, Icon. Pl. Ind. Orient. 4(2): 5, t. 1310, 1312 (1848); Xu, Agric. Univ. Wageningen Pap. 88-6: 1–29 (1988), rev.; Middleton, Fl. Males. I, 18: 184–186 (2007), reg. rev.; Middleton, Fl. Cambodia, Laos and Vietnam 33: 88–89 (2014), reg. rev.
Giadotrum Pichon (1948).

Woody climbers. Inflorescences terminal and/or axillary, paniculate-corymbose. Flowers small; calyx lobes small; calycine colleters few, alternisepalous; corolla salverform, variously colored; lobes much shorter to about as long as tube; stamens inserted near base of corolla tube, included; filaments short; anthers pubescent dorsally near apex; nectary a disc or 5 separate fleshy lobes, as long as or shorter than the ovary; ovary glabrous; style short, slender. Follicles slender. Seeds narrowly elliptic, testa glabrous; coma shortly rostrate.

Two spp. in Sri Lanka, China and SE Asia, in montane forest or brush, often along river banks or near streams, up to 1600 m.

142. ***Apocynum*** L.

Apocynum L., Sp. Pl. 1: 213 (1753); Woodson, Ann. Missouri Bot. Gard. 17: 1–213 (1930), rev.; Li et al., Fl. China 16: 181 (1995), reg. rev.
Poacynum Baill. (1888).
Trachomitum Woodson (1930).

Perennial herbs. Leaves opposite, alternate or rarely verticillate. Inflorescences terminal and axillary, few- to many-flowered aggregate dichasia. Calycine colleters absent; corolla greenish to white, sometimes with pink veins, campanulate to urceolate or tubular; lower tube with five appendages opposite the lobes; stamens inserted at base of tube, included; filaments thick, curved inward; pollen in persistent tetrads; nectaries five, fused at the base; ovary partly inferior to superior; style-head thick, fleshy, ovoid-fusiform, sessile. Follicles slender; coma sessile. 2n = 16 or 22.

About nine spp., temperate to subtropical North America, Europe and Asia, in disturbed areas, pastures, cultivated fields, saline regions, dry stream beds, etc.

143. ***Urceola*** Roxb.

Urceola Roxb., Asiat. Res. 5: 169 (1798), nom. cons.; Middleton, Blumea 41: 82–114 (1996), rev.; Middleton, Fl. Males. I, 18: 394–408 (2007), reg. rev.; Middleton, Fl. Cambodia, Laos and Vietnam 33: 212–228 (2014), reg. rev.
Ecdysanthera Hook. & Arn. (1837).
Chavannesia A. DC. (1844).
Parameria Benth. (1876).
Xylinabaria Pierre (1898).
Aganonerion Pierre ex Spire (1905).
Parabarium Pierre ex Spire (1905).
Xylinabariopsis Pit. (1933).
Chunechites Tsiang (1937).
Parameriopsis Pichon (1948).

Woody climbers. Leaves often with abaxial domatia in axils of the secondary veins. Inflorescences axillary and/or terminal, cymose, often forming panicles or spike-like. Flowers small; calycine colleters few to numerous, alternisepalous or continuous; corolla urceolate, salverform or campanulate, variously colored; corolla lobe aestivation sinistrorse, dextrorse or valvate; stamens inserted near base of corolla tube, included; filaments short; nectary disc annular, 5-dentate or 5-lobed; ovary densely puberulent; style short; style-head fusiform. Follicles torulose or not. Seeds narrowly elliptic, testa hirsute; coma sessile, with a crown of shorter hairs at base. 2n = 20.

20 spp. from S Asia, S China, SE Asia and Malesia, in humid evergreen, deciduous or swamp forest (more rarely in open thickets and on dry, sandy soil), up to 1500 m.

144. ***Chonemorpha*** G. Don

Chonemorpha G. Don, Gen. Hist. 4: 76 (1837), nom. cons.; Chatterjee, Kew Bull. 1947: 47–52 (1947), synop.; Li et al., Fl. China 16: 170–171 (1995), reg. rev.; Middleton, Fl. Males. I, 18: 178–184 (2007), reg. rev.; Middleton, Fl. Cambodia, Laos and Vietnam 33: 78–878 (2014), reg. rev.
Rhynchodia Benth (1876).

Woody lianas; plants usually with most parts pubescent. Inflorescences terminal or axillary, few- to several-flowered cymes. Calyx usually cupular; colleters numerous, in continuous ring; corolla usually large, salverform, white or pinkish; stamens inserted in lower half of corolla tube, included; filaments short; nectary annular; style slender; style-head fusiform-pentagonal, with an upturned cup-like structure at the base, to which the anthers are firmly attached. Follicles ovoid, attenuate, rather stout. Seeds compressed ovate; coma rostrate; rostrum glabrous. 2n = 20.

About ten spp. in India, China and SE Asia to western Malesia, mostly in forest at low to mid elevations.

The genus is in need of revision.

145. ***Trachelospermum*** Lem.

Trachelospermum Lem., Jard. Fleur. 1, t. 61 (1851), nom. cons.; Woodson, Sunyatsenia 3: 65–105 (1936), rev.; Li et al., Fl. China 16: 166–168 (1995), reg. rev.; Middleton, Fl. Males. I, 18: 390–394 (2007), reg. rev.; Middleton, Fl. Cambodia, Laos and Vietnam 33: 202–211 (2014), reg. rev.

Woody climbers. Inflorescences terminal or axillary lax or congested cymes. Calycine colleters ± in a continuous ring or few and alternisepalous; corolla salverform, white to yellowish or purplish, the tube 5-angled; stamens inserted in lower or upper half of tube, included or partly exserted; nectary of 5 free lobes or fused in a ring; style short; style-head conical. Follicles linear or fusiform. Seeds linear-oblong; testa glabrous; coma sessile. 2n = 20.

Six to ten spp. from China and Japan through SE Asia and W Malesia, in deciduous, evergreen or gallery forest, often in open areas, climbing in trees or over rocks, sometimes near streams.

This genus is in need of a comprehensive, modern revision.

146. ***Micrechites*** Miq.

Micrechites Miq. Fl. Ned. Ind. 2: 457 (1857); Middleton, Blumea 39: 73–94 (1994), rev.; Middleton, Fl. Males. I, 18: 281–289 (2007), reg. rev.; Middleton, Fl. Cambodia, Laos and Vietnam 33: 139–144 (2014), reg. rev.
Lamechites Markgr. (1926).
Vallariopsis Woodson (1936).

Woody lianas or scrambling shrubs; latex white. Leaves opposite. Inflorescences terminal and/or axillary, often thyrsoid. Calycine colleters in continuous ring or alternisepalous, sometimes absent; corolla salverform, whitish to reddish; lobes strongly inflexed in bud; stamens inserted in lower half to middle of corolla tube, included or slightly exserted; nectary of five lobes or fused in a ring; ovary apocarpous; style slender; style-head with a turbinate lower and cylindrical upper part. Follicles slender and terete to fusiform. Seeds narrowly elliptic; testa glabrous; coma sessile.

Ten spp. in NE India, S China, SE Asia and Malesia, in rainforest, mostly in moist to dry evergreen or deciduous forest up to 1000 m.

147. ***Amalocalyx*** Pierre

Amalocalyx Pierre, Bull. Mens. Soc. Linn. Paris II, 1: 28 (1898); Middleton, Fl. Cambodia, Laos and Vietnam 33: 45–47 (2014), reg, rev.

Woody liana, most parts pubescent. Inflorescences axillary and terminal, dichasial. Flowers showy; calycine colleters in continuous ring; corolla infundibuliform, white to pink, darker inside; lobes much shorter than tube; stamens inserted around middle of corolla tube, at base of expanded part, included; nectary annular, higher than the ovary; ovary postgenitally syncarpous; style slender; style-head cylindrical with a membranous collar at base. Fruit of two follicles weakly postgenitally fused into an ovoid, densely pubescent and unevenly corky "double-follicle", which splits into two halves at maturity. Seeds compressed, broadly ovate; coma sessile.

One sp., *Amalocalyx microlobus* Pierre ex Spire, in S China and SE Asia, in deciduous or dry evergreen forest.

148. ***Baharuia*** D.J. Middleton

Baharuia D.J. Middleton, Blumea 40: 443–447 (1995); Middleton, Fl. Males. I, 18: 145–147 (2007), reg. rev.

Woody climber. Leaves with hair-filled, abaxial domatia in the axils of the secondary veins. Inflorescences terminal and/or axillary cymes. Flowers small; calycine colleters in a continuous ring; corolla urceolate to salverform, yellow or orange, puberulent outside, lobes slightly shorter than tube; stamens inserted near base of corolla tube; filaments short; nectary disc 5-dentate or 5-crenulate; ovary pubescent. Follicles narrow and somewhat torulose. Seeds linear, testa glabrous, coma sessile.

One sp., *B. gracilis* D.J. Middleton, in Borneo and Sumatra, in evergreen forest or forest margins up to 900 m.

149. ***Pottsia*** Hook. & Arn.

Pottsia Hook. & Arn., Bot. Beechey Voy. 198, t. 43 (1837); Middleton, Fl. Males. I, 18: 344–347 (2007), reg. rev.; Middleton, Fl. Cambodia, Laos and Vietnam 33: 156–160 (2014), reg. rev.

Woody climbers; latex white. Inflorescences terminal, large, many-flowered panicles. Calycine colleters in a continuous row; corolla salverform, without corona, pinkish; stamens inserted near mouth of corolla, almost completely exserted; filaments short, slender; nectary of free lobes or in ring; ovary apocarpous, pubescent; style inflated in the middle or near the base; style-head fusiform. Follicles long, slender, terete, glabrous, often somewhat spirally twisted. Seeds narrowly elliptic; testa glabrous; coma sessile.

Three spp. in India, S China, continental SE Asia and W Malesia, in montane forest, open forest, forest borders, thickets, up to 1100 m.

150. ***Aganosma*** (Blume) G. Don

Aganosma (Blume) G. Don, Gen. Hist. 4: 77 (1837); Middleton, Kew Bull. 51: 455–482 (1996), rev.; Middleton, Fl. Thailand 7(1): 104–110 (1999), reg. rev.; Middleton, Fl. Males. I, 18: 30–31 (2007), reg. rev.; Middleton, Fl. Cambodia, Laos and Vietnam 33: 12–18 (2014), reg. rev.
Echites P. Browne sect. *Aganosma* Blume (1826).

Woody climbers. Inflorescences terminal or axillary few- to many-flowered cymes. Calyx lobes mostly large and foliose; colleters alternisepalous; corolla cylindrical or infundibuliform; tube often conspicuously 5-ribbed; stamens inserted in lower part of corolla tube, included; filaments short; nectary tubular, usually higher than the ovary; style slender; style-head fusiform-pentagonal, without a basal collar. Follicles long, cylindrical. Seeds compressed, lanceolate, testa glabrous; coma sessile or scarcely rostrate. $2n = 22$.

Seven spp., from India, S China, continental SE Asia and W Malesia, in forest at low to mid elevations.

151. ***Ichnocarpus*** R. Br.

Ichnocarpus R. Br., Asclepiadeae 50 (1810), nom. cons.; Middleton, Blumea 39: 73–94 (1994), rev.; Middleton, Fl. Males. I, 18: 203–207 (2007), reg. rev.; Middleton, Fl. Cambodia, Laos and Vietnam 33: 105–110 (2014), reg. rev.

Woody climbers. Inflorescences terminal or axillary cymes, thyrsoid, often puberulent to villous. Flowers small; calycine colleters few, alternisepalous (rarely absent); corolla salverform, sometimes with a thickened ring in the throat, lobes much longer than tube, white or yellowish; stamens inserted in lower half of corolla tube, included; filaments short; nectaries five, separate; ovary usually pubescent on top. Follicles long and slender, pubescent or glabrous. Seeds linear, compressed; testa glabrous; coma sessile, fawn, cream or white. 2n = 20.

Three spp. from India, S China, continental SE Asia, Malesia and Australasia, in primary or secondary forest, sometimes more open drier habitats.

152. ***Epigynum*** Wight

Epigynum Wight, Icon. Pl. Ind. Orient. 4(2): 4, t. 1308 (1848); Middleton, Harvard Pap. Bot. 10: 67–81 (2005), rev.; Middleton, Fl. Males. I, 18: 192–195 (2007), reg. rev.; Middleton, Fl. Cambodia, Laos and Vietnam 33: 89–96 (2014), reg. rev.
Nouettea Pierre (1898).
Argyronerium Pit. (1933).

Woody climbers; latex white. Inflorescences congested, flat-topped, terminal and/or axillary corymbs. Calycine colleters in a row at base inside, alternisepalous, or absent; corolla salverform, white to pinkish, lobes obovate, shorter than tube; stamens inserted about middle of corolla tube or below, included; filaments short; nectary 5 separate lobes or crenate ring; ovary apocarpous, pubescent or glabrous; style filiform, style-head long, cylindrical, 5-angled with a (sometimes small) basal collar. Follicles slender. Seeds compressed-ovate; testa glabrous; coma sessile.

Five spp. from S China to W Malesia, in evergreen or deciduous forest below 1000 m.

II.9. **Tribe Baisseeae** (Pichon ex De Kruif) M.E. Endress (2007).

Woody lianas or climbing shrubs (rarely rhizomatous creepers); latex white. Leaves opposite, with colleters adaxially along the petiole and sometimes extending onto the base of the blade, abaxial domatia usually present in the axils of the secondary veins. Inflorescences axillary or terminal, usually many-flowered thyrses. Calycine colleters few, alternisepalous and quincuncially arranged (absent in *Dewevrella*); corolla tube with 5 small knob-like corona lobes in the alternipetalous sectors, one above each anther, and with tufts of stiff hairs alternating with the anthers (in *Dewevrella* corolla tube less than 0.5 mm long, the lobes long and strap-like); corolla lobe aestivation dextrorse; stamens mostly inserted on a thickened ring near the base of the corolla tube; anthers usually sessile, fertile only in the upper part, the enlarged, sterile basal part with well-developed lignified guide rails, the bases usually spreading outward and ending in a conspicuous subglobose thickening (in *Dewevrella*, stamens completely exserted with long, filiform filaments interspiraled around the style); pollen 3(–4)-porate; nectaries fused into a 5-lobed ring around the ovary; ovary apocarpous, usually partly inferior; ovules many; style usually short, thick, but filiform in *Dewevrella*; style-head usually more or less fusiform, greatest in diameter and 5-angled in the middle (where the anthers are attached) with an elongate tapering apex; membranous basal collar and upper wreath absent. Fruit apocarpous, of 2 woody follicles, spreading and joined at the very base; follicles squat and fusiform to narrowly cylindrical. Seeds narrowly ellipsoid, somewhat compressed laterally and navicular, testa glabrous with a sessile micropylar coma. Spermidine alkaloids (but only *Oncinotis* investigated). Four genera in tropical Africa, one also on Madagascar.

153. ***Motandra*** A. DC.

Motandra A. DC., Prodr. 8: 423 (1844); De Kruif, Meded. Landbouwhogeschool Wageningen 83-7: 1–20 (1983), rev.

Climbing shrubs or lianas. Leaves with abaxial domatia in axils of secondary veins. Inflorescences terminal, many-flowered thyrses. Calycine colleters few, alternisepalous; corolla tubular to infundibuliform; corona below petal sinuses, a bulge just behind each anther; stamens included; anthers with tufts of coarse hairs at the apex; nectary disc 5-lobed, adnate to the ovary; ovary partly inferior, syncarpous at base, apocarpous above; style-head thick, sessile or almost so. Follicles thinly to thickly woody, connate at very base, divergent at ca. 180°. 2n = 22.

Three spp. in W and C Africa, in rainforest, secondary deciduous forest, gallery forest, also in open areas, up to 1200 m.

154. ***Baissea*** A. DC.

Baissea A. DC., Prodr. 8: 424 (1844); van Dilst, Bull. Jard. Bot. Nat. Belg. 64: 89–178 (1995), rev.

Climbing shrubs, lianas or, rarely, rhizomatous creepers. Leaves with abaxial domatia usually present in abaxial axils of secondary veins. Inflorescences axillary and terminal, cymose, paniculate sometimes corymbose. Calycine colleters few, alternisepalous; corolla infundibuliform, campanulate or cylindrical, usually with ridges or a distinct corona behind the anthers; stamens inserted near base of corolla tube; filaments stout; anthers exposed; nectary disc crenate to 5-lobed; ovary partly inferior; ovules many; style-head thick, sessile. Follicles narrowly cylindrical. 2n = 22.

18 spp. in tropical Africa, in primary and secondary rainforest, riverine and swamp forest, coastal forests and montane forest, rarely open woodland, up to 2000 m.

155. ***Oncinotis*** Benth. Fig. 45

Oncinotis Benth. in Hook., Niger Fl. 451 (1849); De Kruif, Agric. Univ. Wageningen 85-2: 5–45 (1985), rev.

Climbing shrubs or lianas. Leaves usually with abaxial domatia in axils of secondary veins; petioles mostly with colleters along adaxial side. Inflorescences terminal or axillary, many-flowered thyrses. Calycine colleters few, alternisepalous; corolla urceolate to salverform; corolline corona of five simple lobes in petal sinuses; stamens included; anthers with basal appendages curved backward and globose at base; nectary disc a 5-lobed ring; ovary partly inferior; style-head thick, sessile. Follicles pendulous, woody, connate at very base, broadly divergent. 2n = 22.

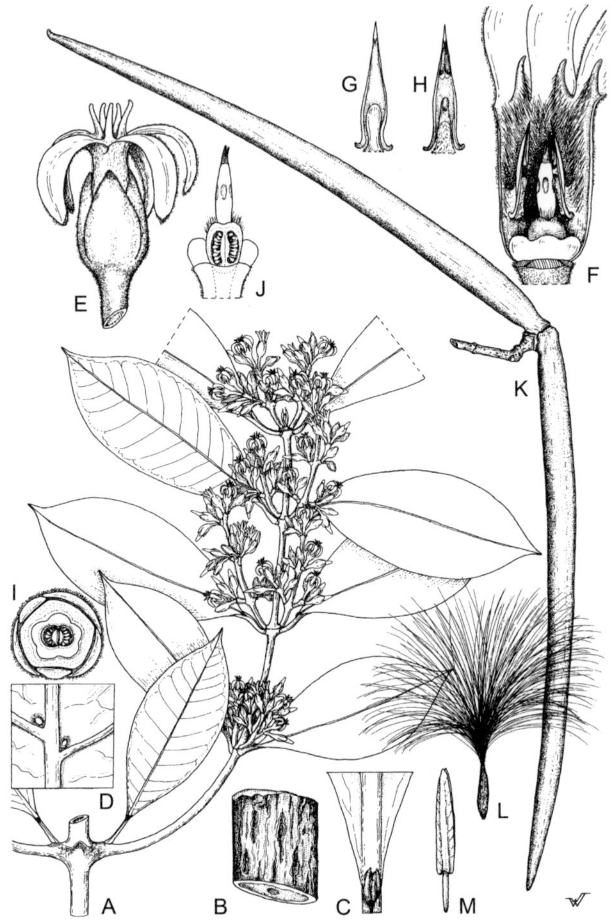

Fig. 45. Apocynaceae-Baisseeae. *Oncinotis nitida*. **A** Flowering branch. **B** Section of branch longitudinally fissured by elongate lenticels. **C** Leaf blade base showing cluster of colleters at juncture of petiole. **D** Close-up of leaf showing abaxial domatia. **E** Flower. **F** Longitudinal section through flower showing gynostegium at base and corolline corona at mouth. **G** Stamen dorsal side. **H** Stamen ventral side. **I** Transverse section through flower at level of ovary. **J** Gynoecium. **K** Fruit. **L** Seed. **M** Embryo. (From De Kruif 1985, p. 26, with permission from the Library of Wageningen University, Wageningen; drawn by W. Wessel-Brand)

Seven spp. in tropical Africa and Madagascar, primary and secondary rainforest, swamp forest and gallery forest, from 0–2200 m.

156. ***Dewevrella*** De Wild.

Dewevrella De Wild., Miss. Em. Laurent 1: 548 (1907); Van der Ploeg, Agric. Univ. Wageningen 85-2: 57–85 (1985), rev.

Liana. Leaves without abaxial domatia in axils of secondary veins. Inflorescences terminal or axillary, many-flowered thyrses. Calyx without colleters within; corolla rotate, white; lobes strap-like; stamens inserted at the base of corolla tube, conspicuously exserted and twisted around the style; anthers white with a yellow apical connective appendage; nectary disc irregularly 5-lobed; ovary partly inferior; ovules 8–15 per carpel; style filiform; style-head fusiform. Follicles slender with a thin wall.

One sp., *D. cochliostema* De Wild., endemic to DR Congo, usually found growing near waterways.

III. Subfam. Periplocoideae Endl. (1838).

Woody perennials; lianas, shrubs to small trees, erect, scrambling or twining herbs or geophytes with underground tuber; latex white (orange in *Cryptolepis sanguinolenta*, translucent in *Raphionacme namibiana*). Leaves opposite, decussate, petiolate, with interpetiolar lines or collars (stipules) and sometimes with cluster of colleters adaxially at juncture of petiole and blade; venation brochidodromous or camptodromous. Inflorescences usually axillary, often paired. Calyx lobes free or basally fused, lanceolate to ovate, acute, usually with few adaxial alternisepalous colleters. Corolla ovoid or conical in bud, rotate to rarely tube longer than lobes; lobes usually spreading, with valvate to dextrorsely imbricate, often contorted aestivation, usually glabrous; corolline corona usually present, usually glabrous (except for *Chlorocyathus* and some species of *Pentopetia*); gynostegial corona absent; gynostegium exposed just above base of corolla; stamens inserted between base of corolla lobes and base of tube on apex of thickened, ridge-like staminal foot running down corolla tube and mostly fusing at base into solid ring around narrow neck in style above carpels, apex of staminal foot also sometimes with corona lobe behind point of insertion of filament, this corona lobe sometimes confluent with corolline corona lobes and more or less in series with them; filaments usually inclined toward style-head, usually more or less cylindrical; anthers usually exposed, four-locular, much broader than filament, usually more or less deltoid, nearly horizontal to ascending with membranous margins not lignified or fashioned into guide rails, often laterally adherent or postgenitally fused into more or less umbrella-like structure over style-head, adaxially fused to style-head usually above its thickest portion (rarely below: *Hemidesmus*); connective appendages usually shorter than anthers; nectaries on sides of staminal foot or in five troughs between staminal feet bearing stamens; pollen usually shed as T-shaped to rhomboidal tetrads, sometimes united into more or less horizontally oriented pollinia (two per locule) not surrounded by waxy outer wall; grains porate, with few to many pores, these scattered or pairwise at the juncture of the four monads; styles glabrous, style-head broad and flat-topped to tapering from broad base into narrow, often bifid apex, with thick and short to long and slender neck joining it to carpels and five stigmatic zones near lower edge between anthers, with five vertically oriented grooves between anthers in which translator is secreted; translator consisting of spoon- to cornet-shaped receptacle above (into which pollen / pollinia is deposited), with small sticky viscidium at lower end projecting vertically over edge of style-head or horizontally beneath it; gynoecium apocarpous, semi-inferior, rarely superior (*Gymnanthera*). Fruit a pair of slender to sometimes very swollen, fusiform, ventrally dehiscent follicles (sometimes one by abortion) with dry, usually thin pericarp. Seeds numerous, compressed, usually narrowly elliptical in outline and without wing-like margin, with micropylar coma (extended around entire margin in *Finlaysonia* and *Raphionacme namibiana*), endosperm thin. 33 genera in the Paleotropics and Australia.

Key to the Genera of Periplocoideae

1. Gynostegium enclosed and concealed within the corolla tube 2
– Gynostegium exposed from short-tubed corolla (corolla usually rotate), or gynostegium in mouth of or exserted from corolla with distinct tube 9
2. Flowers large, 3–6 cm long and 5–9 cm in diameter (Madagascar, widely naturalized elsewhere in the tropics) **165. *Cryptostegia***
– Flowers small, less than 2 cm long and less than 3 cm in diameter 3
3. Corolla subhemispherical to urceolate with five deep vertical cavities (E Asia) **188. *Telectadium***

- Corolla rotate, campanulate or cylindrical, without any cavities 4
4. Corona in mouth of corolla tube or in sinuses of petals 5
- Corona arising lower down in corolla tube, mostly around the middle of the tube 8
5. Erect shrubs with pale green to yellow, 8–12 mm long, salverform corolla (SW Africa) **167. *Ectadium***
- Climbers or lianas; corolla usually shorter, when 8–12 mm long, then not salverform 6
6. Corolla less than 3 mm long, dark red; corolline corona lobes corniculate and bifurcate (West Tropical Africa) **173. *Maclaudia***
- Corolla at least 3 mm long, white, cream or pink to mauve; corolline corona lobes of different shapes, but not corniculate or bifurcate 7
7. Corolline corona lobes subulate, of various colors, but not green; calyx lobes eciliate (Madagascar) **177. *Pentopetia***
- Corolline corona lobes club- or matchstick-like, glossy green; calyx lobes ciliate (southern Africa) **185. *Stomatostemma***
8. Inside of corolla tube and adaxial side of petals papillate; corona lobes adnate to the corolla for most of their length forming vertical ridges with hooded apices, radiating into the corolla tube cavity, without a spongy swelling between corona lobe base and stamen base (tropical Africa) **176. *Parquetina***
- Inside of corolla tube and adaxial side of petals glabrous; corona lobes free from the corolla for most of their length, clavate, filiform or acicular, almost always with a spongy pad between corona lobe base and stamen base (Africa, Arabia and E Asia) **164. *Cryptolepis***
9. Pollen in pollinia 10
- Pollen in tetrads 16
10. Corona lobes fused into a ring outside the stamens (E Africa) **184. *Schlechterella***
- Corona lobes not fused into an ring outside the stamens 11
11. Corolla salverform, tube conspicuous (E Asia to Australia) **170. *Gymnanthera***
- Corolla rotate, tube bowl-shaped to indistinct 12
12. Inflorescence thyrsoid with a total of more than 10 flowers (E Asia) **186. *Streptocaulon***
- Inflorescence cymoid, occasionally with one or two basal dichasial branches, total number of flowers less than 10 13
13. Petals valvate in bud; interstaminal nectary lobes fused into a tube around the style, creating five pockets between the style and stamens (southern Asia) **171. *Hemidesmus***
- Petals contorted in bud; interstaminal nectaries not fused around the style, not forming any pockets between the style and stamens 14
14. Style-head with a flat or concave apex on which five hemispherical depressions contain include the receptacles of the pollen translators (southern Asia) **166. *Decalepis***
- Style-head apex convex, umbonate or rounded, without any hemispherical translator depressions 15
15. Epiphytic climbers; corolla tube with five radial ridges; corona lobes with dilated hairy apex (tropical Africa) **168. *Epistemma***
- Terrestrial climbers; corolla without radial ridges; corona lobes not dilated at apex, glabrous (Asia to Australia) **169. *Finlaysonia***
16. Epiphytic shrubs with aerial tubers (tropical Africa) **183. *Sarcorrhiza***
- Terrestrial climbers, shrubs or herbs; when tuberous, then tubers subterranean or on ground surface 17
17. Corolline corona lobes laterally connate 18
- Corolline corona lobes free 21
18. Corolla tube inside with a ring of horizontally projecting outgrowths (Madagascar) **162. *Camptocarpus***
- Corolla tube inside without such outgrowths 19
19. Gynostegium raised high above corolla base (Asia, Australia) **180. *Phyllanthera***
- Gynostegium sessile or inconspicuously raised above corolla base (Africa, Madagascar) 20
20. Corolline corona 10-partite, fused to an uneven ring-like structure, interpetiolar stipules absent (Madagascar) **158. *Baroniella***
- Corolline corona 5-partite, fused with staminal feet to a collar-like structure; interpetiolar stipules present (tropical Africa) **160. *Batesanthus***
21. Stamens abaxially hairy 22
- Stamens glabrous 23
22. Corolla lobes adaxially basally glandular (Africa, Arabia, Europe and Asia) **178. *Periploca***
- Corolla lobes adaxially not glandular (Madagascar) **177. *Pentopetia***
23. Filaments united with style-head (Asia) **180. *Phyllanthera***
- Anthers, but not filaments, united with style-head 24
24. Subterranean organs tuberous 25
- Subterranean organs not tuberous 32
25. Erect shrubs, more than 50 cm tall 26
- Twiners or lianas, when erect herbs, then less than 50 cm tall 27

26. Corolla more than 8 mm long; leaves in whorls of three (Madagascar) **172. *Ischnolepis***
– Corolla less than 8 mm long; leaves opposite (East Africa) **182. *Sacleuxia***
27. Shoots reddish (South Africa) **179. *Petopentia***
– Shoots green or brown 28
28. Twiners with distinct short shoots (East Africa) **161. *Buckollia***
– Twiners or erect herbs without short shoots 29
29. Strong lianas without tuber from primary root but with tuberous secondary roots, with distinct and dentate interpetiolar collar; leaf bases subcordate **174. *Mondia***
– Woody or herbaceous twiners or erect herbs with one (rarely more) conspicuous root tubers, when interpetiolar stipules present, then not collar-like; leaf bases obtuse or rounded 30
30. Corona inconspicuous, of very small carnose projections underneath stamens; corolla completely reflexed (East Africa) **159. *Baseonema***
– Corona conspicuous, at least as long as the gynostegium; corolla not reflexed 31
31. Plants with numerous tubers, stipules interpetiolar, carnose; corolla semi-succulent **163. *Chlorocyathus***
– Plants with single root tuber, stipules rarely interpetiolar, not carnose; corolla not semi-succulent **181. *Raphionacme***
32. Apical appendages of anther connectives at least as long as the anthers and combined into a long lance-like cone above the style-head (Asia) **157. *Atherandra***
– Apical appendages of anther connectives not longer than the anthers, erect or connivent above style-head 33
33. Interpetiolar collar present, formed of green frills or reddish colleters 34
– Interpetiolar lines present 35
34. Interpetiolar collar green, dentate; flowers large, 15–25 mm in diameter; follicles smooth (tropical and subtropical Africa) **174. *Mondia***
– Interpetiolar collar with reddish colleters; flowers small, up to 10 mm in diameter; follicles winged (E Asia) **175. *Myriopteron***
35. Corona lobes bifid 36
– Corona lobes simple 37
36. Corona lobes apically reflexed (Asia) **189. *Zygostelma***
– Corona lobes erect, often apically curled (Africa) **187. *Tacazzea***
37. Corona only of antesepalous lobes, not longer than the corolla (Madagascar) **177. *Pentopetia***
– Corona with longer antesepalous and shorter antepetalous lobes, when only antesepalous lobes present, then much longer than the corolla (Africa) **187. *Tacazzea***

157. ***Atherandra*** Decne.

Atherandra Decne., Prodr. 8: 497 (1844); Ionta & Judd, Ann. Missouri Bot. Gard. 94: 360–375 (2007), phyl.
Atherostemon Blume (1850).

Liana or herbaceous twiner; glabrous or shortly puberulous. Leaves 4–5 cm long, elliptic to ovate, apically acute to acuminate, glabrous. Inflorescences paired, few-flowered, long-pedunculate. Corolla yellow, lobes lanceolate, carnose, often reflexed, twisted; corolline corona shorter than corolla, equaling to longer than the gynostegium, lobes free, filiform, erect to reflexed; filaments with basal bristly appendage; nectaries lobular, fused to staminal feet; anthers sagittate, adnate to connate to each other; connective appendages forming a lance-like cone above the style-head; spathe spoon-like, apically bifid, viscidium orbicular, stalk present; style-head (elongated-)conical. Follicles 8.5–15 cm long, narrowly oblong. Seeds winged.

One sp., *A. acuminata* Decne., in continental SE Asia and Malesia.

158. ***Baroniella*** Costantin & Gallaud

Baroniella Costantin & Gallaud, Ann. Sci. Nat., Bot., IX, 6: 354 (1907); Klackenberg, Candollea 52: 383–407 (1997), rev.

Suffrutescent twiners to 4 m; entirely glabrous. Leaves occasionally succulent, waxy, 2–6 cm long, linear to ovate. Inflorescences usually paired, 1- to many-flowered, longer than subtending leaves. Corolla small, purplish, lobes ovate; corolline corona very short, 10-partite, laterally connate to an uneven ring-like structure; nectaries lobular, free; anthers free from each other; connective appendages conspicuously long; spathe folded, obovate or rounded, viscidium orbicular, stalk usually absent; style-head umbonate, neck discoid to conical, with five interstaminal lobes protruding between anthers, and occasionally with five additional lobes opposite anthers. Follicles 20–28 cm long, narrowly oblong, moniliform.

Nine spp. in Madagascar, forests along the E coast, ericoid scrub in the central highland, from 400–1700 m.

159. ***Baseonema*** Schltr. & Rendle

Baseonema Schltr. & Rendle, J. Bot. 34: 97 (1896); Venter & Verhoeven, S. Afr. J. Bot. 75: 445–455 (2009), rev.

Suffrutescent twiner with root tubers, glabrous to pubescent. Leaves 4.5–5 cm long, circular; with interpetiolar collar inflated and dentate, or with frills. Inflorescences paired, many-flowered, lax. Corolla ca. 10 mm long, yellow-green, abaxially with trichomes; lobes ovate-oblong, strongly reflexed; corolline corona of very small carnose projections underneath stamens; filaments basally flattened, apically narrow; nectaries lobular, fused to staminal feet; anthers sagittate, connate to each other, fused to upper side of style-head; connective appendages triangular; spathe folded, obovate, viscidium orbicular, stalk short; style-head umbonate. Follicles occasionally solitary, ca. 9 cm long, oblong, obtuse, pubescent. 2n = 22.

One sp., *B. gregorii* Schltr. & Rendle, in Kenya and Tanzania.

160. ***Batesanthus*** N.E. Br.

Batesanthus N.E. Br. in Hooker's Icon. Pl.: ad t. 2500 (1896); Venter & Verhoeven, S. Afr. J. Bot. 75: 445–455 (2009), rev.

Suffrutescent twiners, entirely glabrous; shoots lenticellate. Leaves 7–16 cm long, oblong; interpetiolar collar inflated, dentate. Inflorescences paired, 10–25-flowered, longer than subtending leaves. Corolla to 18 mm long, purple; lobes oblong, reflexed; corolline corona inserted in corolla tube, very short, annular, apically indistinctly five-partite; stamens inserted at base of corolla tube; nectaries lobular, fused to staminal feet; anthers adnate to each other; connective appendages triangular; spathe folded, rhombic and bifid, viscidium cylindrical, stalk present; style-head conical. Follicles basally fused, 11–12 cm long, moniliform. Seeds with coma attached along entire seed margin.

Two spp. in W and C Africa.

161. ***Buckollia*** Venter & R.L. Verh.

Buckollia Venter & R.L. Verh., S. Afr. J. Bot. 60: 97 (1994); Venter, Fl. Ethiopia and Eritrea 4.1: 102–103 (2003), reg. rev.; Venter, Fl. Somalia 3: 135–136 (2006), reg. rev.

Suffrutescent twiners to 4 m, tuberous, with distinct long and short shoots; lenticellate, densely white or rusty puberulous or tomentose. Leaves opposite or fascicled, 2.6–6 cm long, ovate, occasionally undulate. Inflorescences usually solitary, 5–10-flowered, dichasial at base, monochasial in higher order branching. Corolla 5–8 mm long, creamish green, occasionally puberulous; corolline corona ivory to pink, glabrous to puberulous, exceeding gynostegium; lobes free, filiform; filaments filiform; nectaries lobular, fused to staminal feet; anthers ovoid; spathe folded, broadly obovate, viscidium hemispherical or bifid, stalk short; style-head umbonate or bifid, neck ovoid. Follicles 13–19 cm long, tuberculate.

Two spp. in Ethiopia, Kenya, Somalia and Uganda in semi-arid *Combretum–Acacia–Commiphora* scrub, from 1400–1800 m.

162. ***Camptocarpus*** Decne.

Camptocarpus Decne., Prodr. 8: 493 (1844), nom. cons.; Klackenberg, Bot. Jahrb. Syst. 120: 45–85 (1998), rev.
Harpanema Decne. (1844).
Tanulepis Balf. (1877).
Symphytonema Schltr. (1895).

Suffrutescent twiners, glabrous. Leaves occasionally waxy, 1.5–7 cm long, linear to orbicular. Inflorescences solitary or paired, 1- to many-flowered, simple, (sub-)sessile; floral bracts conspicuous, occasionally caducous. Corolla 2–5 mm long, rose or greenish yellow, adaxially with purple center; lobes ovate or triangular, occasionally twisted; corolline corona exceeding gynostegium, five-partite; lobes partially connate to corolla, slender; nectar pockets at base of corolla tube; nectaries lobular, fused to staminal feet; anthers free from each other; spathe folded, (sub-)circular, viscidium (sub-)orbicular; styles distinct; style-head conical. Follicles occasionally solitary, 8–25 cm long, moniliform. 2n = 22.

Nine spp. in two sections (Klackenberg 1998), Madagascar and Mascarene Islands, in forests and forest remnants, some species in xerophytic scrub and dunes; 0–1800 m.

163. ***Chlorocyathus*** Oliv.

Chlorocyathus Oliv. in Hooker's Icon. Pl. 16: t. 1557 (1887); Venter, S. Afr. J. Bot. 74: 288–294 (2008), rev.
Kappia Venter, A.P. Dold & R.L. Verh. (2006).

Lianas to 12 m, with root tubers. Leaves discolorous, 1–6 cm long, lanceolate to ovate with undulate margins, glabrous or sparsely pubescent; interpetiolar collar carnose, dentate. Inflorescences usually solitary, 2–10-flowered, dichasial at base, monochasial in higher order branching. Corolla 6–12 mm long, funnel-shaped, semi-succulent, (light) green, occasionally abaxially pubescent, lobes ovate, occasionally adaxially hirsute; corolline corona yellowish or pinkish, apically minutely sculptured, lobes free, obcordate, occasionally trifid with filiform median segment; gynostegium exserted from corolla; filaments filiform, nectar pockets at base of corolla tube, anthers free from each other; spathe spoon-like, ovate, viscidium (ob)ovate, stalk short; style-head broadly ovoid. Follicles 6.5–11 cm long, ellipsoid.

Two spp., one widespread in S tropical Africa, the other restricted to South African riverine forest.

164. ***Cryptolepis*** R. Br. Fig. 46

Cryptolepis R. Br., Asclepiadeae 58 (1810); Forster, Austrobaileya 4: 67–73 (1993), reg. rev.; Venter & Verhoeven, Ann. Missouri Bot. Gard. 88: 550–568 (2001), phyl.; Ionta & Judd, Ann. Missouri Bot. Gard. 94: 360–375 (2005), phyl.; Venter et al., S. Afr. J. Bot. 72: 139–143 (2006), new taxa; Ionta, Phylogeny reconstruction of Periplocoideae (Apocynaceae) based on morphological characters and a taxonomic revision of *Decalepis*, Ph.D. Thesis, Univ. Florida (2009), phyl.; Joubert et al., Taxon 65: 487–501 (2016), phyl.
Curroria Planch. ex Benth. (1849).
Ectadiopsis Benth. (1876).
Mitolepis Balf. f. (1883).
Socotranthus Kuntze (1903), nom. illegit.
Mangenotia Pichon (1954).

Suffrutescent twiners or lianas, 3–6 m tall, rarely shrubs, glabrous;. Leaves 1.5–18 cm long, oblong, often apiculate, glaucous. Inflorescences occasionally extra-axillary, solitary, 9–15-flowered, thyrsoidal. Corolla 9–20 mm long, campanulate or salverform, creamish yellow; lobes oblong to lanceolate, twisted; corolline corona inserted in corolla tube; lobes free, ovate or clavate, bifid; gynostegium concealed in corolla tube; nectar pockets and the flattened filaments arising in the middle of the corolla tube; nectaries five transversal ridges; anthers sagittate, dorsally occasionally with trichomes; connective appendages occasionally twisted; spathe oblong, acuminate, viscidium cylindrical, stalk absent; style-head umbonate. Follicles (4.5–)7–20 cm long. 2n = 22.

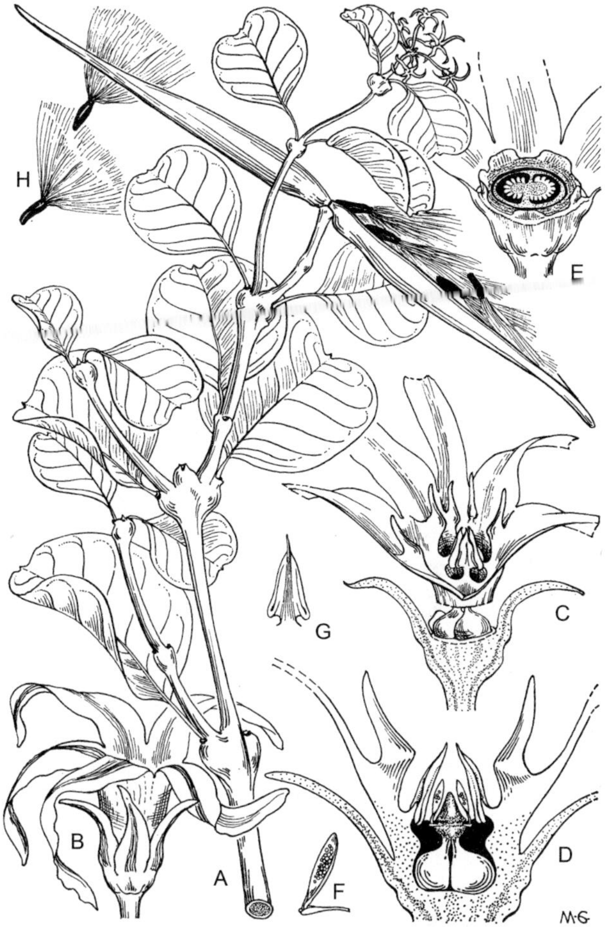

Fig. 46. Apocynaceae-Periplocoideae. *Cryptolepis macrophylla*. **A** Shoot with flowers and fruit. **B** Flower. **C** Flower with corolla cut open and raised to show carpels. **D** Flower, showing gynostegium and carpels. **E** Carpel, with intercalycine scales. **F** Translator. **G** Stamen. **H** Seeds. (From Smith 1971, pl. 3685, with permission from Hooker's Icones Plantarum © Board of Trustees of the Royal Bot. Gard., Kew; drawn by Mary Grierson)

10–15 spp. in Africa, Asia and Australia.

The circumscription of *Cryptolepis* is much disputed; the inclusion of *Phyllanthera* (Forster 1993) is not supported by molecular results (Ionta and Judd 2007; Joubert et al. 2016). The inclusion of *Parquetina*, suggested by Joubert et al. (2016), seems premature and is not followed here.

165. ***Cryptostegia*** R. Br.

Cryptostegia R. Br., Bot. Reg., ad t. 435 (1820); Marohasy & Forster, Aust. Syst. Bot. 4: 571–577 (1991), rev.; Klackenberg, Adansonia III, 23: 205–218 (2001), rev.

Shrubs, suffrutescent twiners or lianas; shoots lenticellate, glabrous or pubescent. Leaves 7–12 cm long, elliptic. Inflorescences terminal, 1–4(–9)-flowered, thyrsoidal; floral bracts conspicuous. Corolla 3–6 cm long, salverform, pale rose or white, occasionally adaxially with trichomes; lobes lanceolate, twisted; corolline corona inserted in corolla tube, reaching the sinus of corolla lobes; lobes free, filiform, deeply bifid; gynostegium concealed in corolla tube; nectar pockets at base of corolla tube; nectaries five transversal ridges, fused to staminal feet; basally connate to each other, connective appendages distinct, lanceolate; spathe flat, stalk absent. Follicles 6–15.5 cm long, obclavate, winged. 2n = 22.

Two spp. in Madagascar, in riverine and seasonally flooded forests, introduced throughout the tropics, noxious weeds in disturbed vegetation and plantations. Plants highly toxic (cardenolides); used as arrow poison in parts of Africa. *C. madagascariensis* Bojer pollinated by Sphingidae.

166. ***Decalepis*** Wight & Arn.

Decalepis Wight & Arn. in Wight, Contr. Bot. India 64 (1834); Venter & Verhoeven, Ann. Missouri Bot. Gard. 88: 550–568 (2001), tax.; Ionta, Phylogeny reconstruction of Periplocoideae (Apocynaceae) based on morphological characters and a taxonomic revision of *Decalepis*, Ph.D. Thesis, Univ. Florida (2009), phyl.; Sharma & Shahzad, J. Pl. Sci. Res. 1: 1–13 (2014), morph., pharm.; Joubert et al., Taxon 65: 487–501 (2016), phyl.; Kambale et al., Taxon 64: 475–477 (2016), tax.
Brachylepis Wight & Arn. (1834), non C.A. Mey. (1829).
Cornacchinia Endl. (1841), non Savi (1837).
Baeolepis Decne. ex Moq. (1849).
Utleria Bedd. ex Benth. (1876).
Janakia J. Joseph & V. Chandras (1978).

Small trees, shrubs or lianas, glabrous; roots often fleshy and fascicled, nodulous. Leaves occasionally alternate to fascicled, 1–15 cm long, linear to ovate, undulate. Inflorescences frequently terminal, 8–10-flowered, thyrsoidal, long-pedunculate. Corolla 2–3 mm long, greenish white; lobes ovate; corolline corona inserted in corolla tube, basally connate to the filaments, fused to lower half of corolla, very short; lobes free, ovate; filaments short; nectaries lobular, fused to staminal feet; anthers fused to underside of style-head; pollen in ovoid pollinia; spathe spoon-like, viscidium orbicular, stalk present; style-head umbonate with hemispherical depressions. Follicles 4.5–5.5 cm long, obclavate. Seeds occasionally winged.

Five spp. in India; the roots of *D. hamiltonii* Wight & Arn. are used medicinally (Sharma and Shahzad 2014).

While the inclusion of *Janakia* and *Utleria* in *Decalepis*, suggested by Venter and Verhoeven (2001) without explanation, was confirmed by molecular analysis (Joubert et al. 2016), the inclusion of *Brachylepis* is not supported by their results.

167. ***Ectadium*** E. Mey.

Ectadium E. Mey., Comm. Pl. Afr. Austr. 188 (1838); Venter et al., S. Afr. J. Bot. 56: 113–124 (1990), rev.

Shrubs, 1–3 m tall, with tap roots, glabrescent or puberulous. Leaves coriaceous to subsucculent, 2–8 cm long, narrowly elliptic, puberulous to tomentose. Inflorescences solitary or paired, 5–20-flowered, simple, subsessile. Corolla salverform, (greenish) yellow; lobes oblong, slightly twisted; corolline corona exceeding gynostegium; lobes free, subulate, basally dilated, fused to corolla lobes; gynostegium concealed in corolla tube; stamens inserted at base of corolla tube; nectaries lobular, fused to staminal feet; anthers free from each other, connective appendages subulate, hirsute, exceeding anthers; spathe flat, oblong, viscidium ellipsoid, stalk short; style-head conical, bifid. Follicles 3.5–9 cm long, obclavate. Seeds small, winged. 2n = 22.

Three spp. in Namibia, very dry and sandy river beds, stony or rocky sandplains.

168. ***Epistemma*** D.V. Field & J.B. Hall

Epistemma D.V. Field & J.B. Hall, Kew Bull. 37: 117 (1982); Huber, Adansonia IV, 11: 447–452 (1989), new taxa; Fischer et al., S. Afr. J. Bot. 77: 680–684 (2011), key, tax.

Twining subsucculent epiphytes; mature shoots with smooth, gray, papery bark. Leaves 8.5–12.5 cm long, broadly elliptic. Inflorescences terminal on lateral branches, few-flowered, simple; peduncles very short and thick. Corolla 5–7 mm long, corolla yellowish green or purple, adaxially with trichomes on the tube and with five pubescent ridges radiating from the center; lobes broadly ovate; corolline corona short; lobes free, clavate; filaments filiform; nectaries lobular, fused to staminal foot; anthers apically adnate to each other; connective appendages triangular, thickened, strongly inflexed; spathe folded, obcordate, viscidium hemispherical, stalk short; styles short; style-head umbonate. Follicles 9–13 cm long. Seeds folded along seta; raphe winged. 2n = 22.

Four spp. in lowland forests of W Africa (Gabon, Ghana, Ivory Coast, Rwanda).

169. ***Finlaysonia*** Wall.

Finlaysonia Wall., Pl. Asiat. Rar. 2: 48, t. 162 (1831); Venter & Verhoeven, Ann. Missouri Bot. Gard. 88: 550–568 (2001); Ionta & Judd, Ann. Missouri Bot. Gard. 94: 360–375 (2007), phyl.; Sidney, M.Sc. Thesis, University of Free State, Bloemfontein (2012), morph.
Atherolepis Hook. f. (1883).
Stelmacrypton Baill. (1889).
Hanghomia Gagnep. & Thérint (1936).
Meladerma Kerr (1938).

Lianas. Leaves 5–10 cm long, linear to ovate. Inflorescences usually solitary, few-flowered, thyrsoidal or dichasial at base, monochasial in higher order branching; flowers malodorous. Calyx urceolate; ciliate; corolla 3–6 mm long, greenish or purplish, adaxially occasionally with trichomes, or five pubescent ridges radiating from the center; lobes ovate; corolline corona inserted in corolla tube, basally connate to the filaments; lobes free; nectaries lobular, fused to staminal feet; anthers ovoid, connective appendages with some trichomes; pollen in pyriform pollinia; spathe hollow, spoon-like, viscidium orbicular, stalk present. Follicles 2.5–9 cm long, occasionally ellipsoid, ribbed. Seeds with coma occasionally along entire seed margin, reflexed.

Eight spp. in India, China, continental SE Asia, Malesia and Australia, often on limestone as well as along tidal rivers and in wetlands at low altitudes.

The inclusion of *Atherolepis*, *Hanghomia*, *Meladerma* and *Stelmacrypton* in *Finlaysonia* was suggested by Venter and Verhoeven (2001), repeated by Sidney (2012), on the basis of morphological characters without established synapomorphic value, and remains questionable until the type species of *Finlaysonia*, *F. obovata* Wall., has been sequenced.

170. ***Gymnanthera*** R. Br.

Gymnanthera R. Br., Asclepiadeae 47 (1810); Forster, Aust. Syst. Bot. 4: 563–569 (1991), rev.; Forster, Austral. Entomol. Mag. 18: 61–64 (1991), ecol.
Dicerolepis Blume (1850).
Cylixylon Llanos (1851).

Shrubs or lianas; shoots often lenticellate. Leaves to 12 cm long, elliptic or ovate; stipules small. Inflorescences solitary, 3–10-flowered, simple. Corolla 9–17 mm long, salverform, yellowish to green; lobes ovate; corolline corona inserted in corolla tube behind the stamens, short; lobes free, often bifid; gynostegium exserted from corolla tube mouth; nectar pockets at base of corolla tube; nectaries five transversal ridges, free; anthers free from each other, fused to underside of style-head; pollen in clavate pollinia; spathe hollow, cup-like, viscidium orbicular, stalk long, thin; style-head (depressed-)conical. Follicles 6–14 cm long.

Two spp. in Indonesia, the Philippines, Australia and Papua New Guinea on banks surrounding areas of permanent or semi-permanent water, mangrove fringes. Food source for larvae of the Common Crow butterfly (*Euploea core corinna*, fide Forster 1991b).

171. ***Hemidesmus*** R. Br.

Hemidesmus R. Br., Asclepiadeae 45 (1810); Ali, Fl. Pakistan 150: 56–57 (1983), reg. rev.; Watson, Fl. Bhutan 2.2: 694 (1999), reg. rev.; Jagtap & Singh, Fasc. Fl. India 24: 301–303 (1999), reg. rev.

Suffrutescent twiner; shoots lenticellate; rootstock woody, occasionally tuberous. Leaves 1.4–17 cm long, linear to (ob-)ovate, ciliate. Inflorescences paired, few-flowered, simple, (sub-)sessile; floral bracts conspicuous, persistent. Corolla 5 mm long, adaxially purplish brown; lobes suborbicular; corolline corona dorsally adnate to anthers, short; lobes free, almost orbicular; gynostegium raised high above corolla; stamens inserted at base of corolla tube, filaments long; nectar pockets on a tube of fused staminal feet; nectaries tubular around style without slits; filaments fused to style-head; pollen in ovoid pollinia; spathe hollow, cup-like, viscidium cylindrical, stalk flattened. Follicles 10–15 cm long.

One sp., *H. indicus* (Willd.) R. Br., in India and Peninsular Malaysia, lowlands. Roots widely used medicinally.

172. ***Ischnolepis*** Jum. & H. Perrier

Ischnolepis Jum. & H. Perrier, Rev. Gén. Bot. 21: 53 (1909); Klackenberg, Candollea 54: 257–339 (1999), rev.; Venter & Verhoeven, Ann. Missouri Bot. Gard. 88: 550–568 (2001), phyl.; Meve & Liede, Ann. Bot. 93: 407–414 (2004), phyl.

Shrub to 2 m; glabrous; shoots red, glossy; bark flaky; roots forming many big tubers. Leaves 3-whorled, 10–25 cm long, linear, often falciform. Inflorescences solitary, 1–7-flowered, simple. Corolla 8–14 mm long, yellowish; lobes slightly twisted; corolline corona inserted in corolla, yellow; lobes free, filiform; filaments filiform; nectaries lobular, fused to stapet; anthers free from each other; spathe folded, elliptic, channeled; viscidium of two broad wings, beaked, stalk indistinct; style-head conical, neck conical with five grooves supporting anthers and five broad knobs at the base. Follicles 12 cm long. 2n = 22.

One sp., *I. graminifolia* (Costantin & Gallaud) Klack., in Madagascar, savanna, grassland or on almost bare granitic, gneissic or sandstone rocks; 0–2000 m.

Venter and Verhoeven (2001) include *Petopentia* in *Ischnolepis*, but a molecular study (Meve and Liede 2004a) has shown that the two genera are not closely allied.

173. ***Maclaudia*** Venter & R.L. Verh.

Maclaudia Venter & R.L. Verh., Bot. J. Linn. Soc. 115: 57–63 (1994).

Suffrutescent twiner or liana; shoots lenticellate, scabrous. Leaves 3–4 cm long, ovate, glabrous, scabrous on the nerves. Inflorescences occasionally paired, many-flowered, thyrsoidal, longer than subtending leaves, condensed; inflorescence bracts two, larger than the persistent subulate floral bracts forming decussate rows along the rachis. Corolla 2.5–3 mm long, campanulate, dark red; lobes broadly elliptic; corolline corona short; lobes corniculate and bifurcate, reflexed; gynostegium concealed in corolla tube; stamens inserted at base of corolla tube; nectaries lobular; anthers ovoid; spathe flat, narrowly ovate and bifid, viscidium hemispherical; style-head ovoid.

One sp., *M. felixii* Venter & R. L. Verh., in Guinea, dry forest; ca. 1000 m.

174. ***Mondia*** Skeels

Mondia Skeels, U.S.D.A. Bur. Pl. Industr. Bull. 223: 45 (1911); Bullock, Fl. W. Trop. Afr. 2: 82 (1963), reg. rev.; Venter et al., S. Afr. J. Bot. 75: 456–465 (2009), rev.; Aremu et al., S. Afr. J. Bot. 77: 960–971 (2011), econ.
Chlorocodon Hook. f. (1871) non (DC.) Fourr. (1869).

Lianas; shoots lenticellate; rootstock woody; secondary roots often tuberous. Leaves 10–20 cm long, (ob)ovate; interpetiolar collar inflated, dentate. Inflorescences usually paired, 3–10-flowered, thyrsoidal, lax, long-pedunculate. Corolla 10–15 mm long, adaxially yellow or purple; lobes valvate, oblong, unilaterally ciliate; corolline corona white, short; lobes free, triangular, occasionally with slender appendage; filaments absent or short, broadened; nectaries lobular, fused to staminal feet; anthers triangular, adnate to each other, fused to upper side of style-head; spathe flat, ovate with revolute margins, stalk absent. Follicles 9–10 cm long, elongate-ellipsoid; pericarp thick, woody. Seeds winged. 2n = 22.

Two spp. in tropical Africa, forest gaps. Roots used medicinally and for ginger beer.

175. ***Myriopteron*** Griff.

Myriopteron Griff., Calcutta J. Nat. Hist. 4: 385 (1844); Jagtap & Singh, Fasc. Fl. India 24: 313 (1999, under *Streptocaulon*), reg. rev.; Li et al., Fl. China 16: 194–195 (1995), reg. rev.; Venter & Verhoeven, Ann. Missouri Bot. Gard. 88: 550–568 (2001), phyl.; Ionta & Judd, Ann. Missouri Bot. Gard. 94: 360–375 (2007), phyl.

Herbaceous twiner, to 10 m, glabrous; shoots lenticellate. Leaves papery, 8–20(–30) cm long, elliptic to ovate; with interpetiolar line, reddish, stipules prominently lobed. Inflorescences paired, 20–30-flowered, thyrsoidal, lax, long-pedunculate. Calyx very short; corolla ca. 3 mm long, lobes oblong, acute; corolline corona short; lobes free, filiform; filaments flattened; nectaries lobular, fused to staminal feet, connate to each other at tips; spathe folded, oblong, viscidium rounded, stalk short; style-head conical. Follicles ca. 6.5 cm long, obclavate, with many membranous wings 3–6 mm wide.

One sp., *M. extensum* (Wight & Arn.) K. Schum., in Asia, thickets, open woods; 600–1600 m. Roots used medicinally in China.

Jagtap and Singh (1999) treat *Myriopteron* as a synonym of *Streptocaulon*, but Li et al. (1995a) and Venter and Verhoeven (2001) consider it a different genus, a view supported by Ionta and Judd (2007).

176. ***Parquetina*** Baill.

Parquetina Baill., Bull. Mens. Soc. Linn. Paris 2: 806 (1889); Venter & Verhoeven, S. Afr. J. Bot. 62: 23–30 (1996), rev.; Ionta & Judd, Ann. Missouri Bot. Gard. 94: 360–375 (2007), phyl.; Venter, S. Afr. J. Bot. 75: 557–559 (2009), tax.; Joubert et al., Taxon 65: 487–501 (2016), phyl.
Omphalogonus Baill. (1890).

Lianas to 20 m; shoots lenticellate, glabrous. Leaves discolorous, 9–11.5 cm long, broadly ovate. Inflorescences solitary, 10–30-flowered, dichasial at base, monochasial in higher order branching, long-pedunculate. Flowers malodorous; corolla 11–16 mm long, cyathiform, carnose, adaxially papillate or velvety, rose to purple-black; lobes elliptic; corolline corona inserted in corolla tube, purplish, basally fused to corolla or fused for half of its length and radiating into the tube's cavity, short; lobes free, rounded; filaments filiform; nectaries five transversal ridges, fused to staminal foot; anthers dorsally with trichomes; spathe folded, oblong, viscidium ellipsoid or triangular, stalk present. Follicles 9–25 cm long, obclavate, black; pericarp woody. 2n = 22.

Two spp. in tropical Africa; at least *P. nigrescens* (Afzel.) Bullock is widely used in traditional medicine. Flowers are visited by flies.

A molecular analysis (Ionta and Judd 2007) has shown that these two Periplocoideae species are nested in *Cryptolepis* unless *C. stefaninii* Chiov. is removed from the latter (Joubert et al. 2016).

177. ***Pentopetia*** Decne.

Pentopetia Decne., Prodr. 8: 500 (1844); Klackenberg, Candollea 54: 257–339 (1999), rev.
Acustelma Baill. (1889).
Gonocrypta Baill. (1899).
Kompitsia Costantin & Gall. (1906).
Pentopetiopsis Costantin & Gall. (1906).

Suffrutescent twiners; shoots lenticellate, occasionally with reddish trichomes. Leaves occasionally fascicled on brachyblasts, 0.5–8 cm long, linear to obovate, occasionally pubescent. Inflorescences solitary, 1–15-flowered, dichasial at base, monochasial in higher order branching. Corolla 3–12 mm long, yellowish or reddish, usually with trichomes; lobes triangular to ovate, occasionally twisted; corolline corona short, rarely absent, occasionally with trichomes; lobes free, subulate; filaments occasionally distinctly arched; nectaries lobular, free; basal part of thecae without pollen; connective appendages absent or short, hirsute; spathe folded; viscidium cylindrical, stalk broad; styles occasionally with trichomes; style-head usually flat. Follicles 5–20 cm long, obclavate or narrowly oblong, occasionally 3–5 ribbed or winged. 2n = 22.

23 spp. in Madagascar, the Comores, Seychelles and Aldabra, dry forest, savanna and scrub vegetation, 0–2000 m.

178. ***Periploca*** L.

Periploca L., Sp. Pl. 211. (1753); Li et al., Fl. China 16: 195–196 (1995), reg. rev.; Venter, S. Afr. J. Bot. 63: 123–128 (1997), rev.; Chaudhari, Fl. Kingd. Saudi Arabia 2.2: 32–33 (2001), reg. rev.; Venter, Fl. Somalia 3: 138–139 (2006), reg. rev.; Heneidak & Naidoo, Turk. J. Bot. 39: 353–363 (2015), flow. morph.
Campelepis Falc. (1842).
Socotora Balf. f. (1883).
Cyprinia Browicz (1966).

Shrubs, suffrutescent twiners or lianas, glabrous. Leaves occasionally caducous or scale-like, 1.5–10 cm long, elliptic or ovate, occasionally with dentate interpetiolar line. Inflorescences occasionally terminal, solitary, (1)3–15(–30)-flowered, thyrsoidal or simple. Corolla 5–12 mm long, adaxially brown or purple, occasionally with dark center, and/or with papillose white spots; lobes oblong or ovate, basally with glandular zone, with trichomes; corolline corona yellow or purplish, as long as corolla; filiform, often segmented; nectaries lobular, fused to staminal feet; anthers apically with trichomes, often bearded; spathe flat, oblong, stalk present; style-head ovoid. Follicles occasionally apically fused, 7–21(–45) cm long, occasionally moniliform. 2n = 22, 44.

13 spp. in two sections, S Europe, Asia, Africa, Arabia and the Canary Islands, humid forests to semi-deserts.

179. ***Petopentia*** Bullock

Petopentia Bullock, Kew Bull. 1954: 362 (1954); Venter et al., African J. Bot. 56: 393–398 (1990), tax.; Venter & Verhoeven, Ann. Missouri Bot. Gard. 88: 550–568 (2001), phyl.; Meve & Liede, Ann. Bot. 93: 407–414 (2004), phyl.; Venter & Venter, S. Afr. J. Bot. 88: 425–431 (2013), rev.

Lianas, 10–15 m tall; shoots reddish with suberose bark, glabrous; root tubers large, occasionally nodulous, semi-subterranean. Leaves 7–11 cm long, broadly oblong. Inflorescences occasionally paired, 5–8-flowered, dichasial at base, monochasial in higher order branching. Corolla 13–15 mm long, greenish yellow; lobes triangular, occasionally twisted; corolline corona exceeding gynostegium; lobes free, filiform, basally broadened; filaments filiform with dilated bases; nectaries lobular, fused to staminal feet; anthers broadly oblong, fused to upper side of style-head; spathe flat, obovate, viscidium orbicular, stalk short and narrow; style-head conical. Follicles (6)9–12 cm long, obclavate. 2n = 22.

Two spp. in South Africa, rocky river bank forests to forested sandstone cliff faces in humid subtropical areas.

180. ***Phyllanthera*** Blume

Phyllanthera Blume, Bijdr. 1048 (1827); Venter & Verhoeven, Ann. Missouri Bot. Gard. 88: 550–568 (2001), phyl.; Ionta & Judd, Ann. Missouri Bot. Gard. 94: 360–375 (2007), phyl.; Takeuchi, Phytotaxa 163: 173–179 (2014), consp., reg. tax.
Pentanura Blume (1850).
Streptomanes K. Schum. ex Schltr. (1905).

Lianas, glabrous. Leaves linear to ovate, discolorous, adaxially occasionally purple-veined. Inflorescences occasionally paired, 1–10-flowered, thyrsoidal, lax. Corolla 8–12 mm long, rotate, basally fused, yellowish brown or purple, occasionally papillate; lobes lanceolate or triangular, twisted; corolline corona inserted in corolla tube, basally connate to filaments, short, annular and/or lobes free, bifid; gynostegium raised high above corolla; stamens inserted at base of corona; nectaries tubular around style with slits, fused to staminal feet; anthers rectangular to trullate, free; filaments fused to style-head; connective appendages exceeding anthers; spathe spoon-like; viscidium rounded; stalk present; style-head conical or semi-globose. Follicles ca. 9 cm long, linear to ovoid-fusiform, woody.

Nine spp., seven of them in New Guinea, the remainder elsewhere in Malesia and Australia, forests and river banks. In recent phylogenies, *Phyllanthera* is retrieved as sister to the remainder of Periplocoideae.

181. ***Raphionacme*** Harv.

Raphionacme Harv., London J. Bot. 1: 22 (1842); Venter, Fl. Ethiopia and Eritrea 4.1: 108–110 (2003), reg. rev.; Venter, Fl. Somalia 3: 139–140 (2006), reg. rev.; Venter, S. Afr. J. Bot. 75: 292–350 (2009), rev.; Pienaar, Phylogeny of the genus *Raphionacme* (Apocynaceae), M.Sc. Thesis, Univ. of Free State, Bloemfontein (2013), phyl.
Apoxyanthera Hochst. (1843).
Zucchellia Decne. (1844).
Zaczatea Baill. (1889).
Mafekingia Baill. (1890).
Pentagonanthus Bullock (1962).

Erect herbs, herbaceous or suffrutescent twiners, puberulous; roots tuberous, single. Leaves (sub-) sessile, linear to (ob-)ovate. Inflorescences often terminal, solitary, few-to many-flowered, simple. Corolla ca. 10 mm long, white, (yellowish) green, purplish or bluish, occasionally papillose, ridged; lobes ovate, occasionally twisted; corolline corona inserted in corolla tube, exceeding gynostegium; lobes free, filiform, bifid or trifid; gynostegium exserted from, stamens inserted in corolla tube; filaments short; nectar pockets at base of corolla tube; nectaries lobular, free; spathe folded, ovate; viscidium hemispherical; stalk present; style-head conical to rostrate. Follicles usually 3.5–8 cm long, ellipsoid to obclavate. Seeds occasionally with copper-colored coma, occasionally attached along entire margin. 2n = 22, 44.

About 35 spp. in Africa, one sp. in Arabia, arid and semi-arid habitats.

182. ***Sacleuxia*** Baill.

Sacleuxia Baill., Hist. Pl. 10: 265 (1890).
Gymnolaema Benth. & Hook. f. (1876) non *Gymnoleima* Decne. (1844).
Macropelma K. Schum. (1895).

Erect shrubs, 0.5–1.5 m tall; rootstock woody or tuberous. Leaves subsessile; ca. 9 cm long, linear to ovate, glabrous; with interpetiolar collar dentate and/or inflated. Inflorescences solitary, many-flowered, simple, long-pedunculate. Corolla 3–8 mm long, campanulate, carnose, yellow or reddish; lobes ovate; corolline corona short; lobes free, oblong, apically bifid or emarginate; filaments absent; nectaries lobular, free; spathe folded, oblong, stalk present; style-head depressed-conical. Follicles 3–6 cm long. Seeds winged. 2n = 22.

Two spp. in Kenya and Tanzania.

183. ***Sarcorrhiza*** Bullock

Sarcorrhiza Bullock in Hooker's Icon. Pl. 36: ad t. 3585 (1962).

Epiphytic shrub, glabrous; root tubers ovoid, reddish brown. Leaves subsessile, to 8 cm long, oblong. Inflorescences solitary, 2–4-flowered, simple, sessile; floral bracts conspicuous, triangular to ovate. Corolla ca. 6 mm long, greenish yellow, with purplish center; lobes ovate; corolline corona white, exceeding gynostegium; lobes free, filiform, basally thickened, inflexed, twisted; filaments filiform; nectaries lobular, fused to staminal feet; anthers adnate to each other, fused to upper side of style-head; spathe folded, oblong, abaxially with trichomes, stalk present; styles short; style-head conical. Follicles ca. 15 cm long.

One sp., *S. epiphytica* Bullock, in DR Congo and Tanzania.

184. ***Schlechterella*** K. Schum.

Schlechterella K. Schum. in Engler & Prantl, Nat. Pflanzenfam. Index 2–4: 462 (1899); Nachtr. 2: 60. (1900); Venter & Verhoeven, S. Afr. J. Bot. 64: 350–355 (1998), rev.; Venter, Fl. Ethiopia and Eritrea 4.1: 110–112 (2003), reg. rev.; Pienaar, Phylogeny of the genus *Raphionacme* (Apocynaceae), M.Sc. Thesis, Univ. of Free State, Bloemfontein (2013), phyl.

Suffrutescent twining geophytes; root tuber cylindrical, napiform-oblongoid. Leaves (sub-)sessile, 3–10 cm long, linear or obovate, occasionally undulate or revolute, discolorous. Inflorescences paired, 9–15-flowered, simple, lax, long-pedunculate; floral bracts conspicuous. Corolla 4–12 mm long, cream or purplish, occasionally with greenish pattern; lobes linear to (ob-)ovate, twisted; corolline corona whitish, exceeding gynostegium; basally connate into a ring; lobes filiform, bi- to pentafid, occasionally connivent; gynostegium much exserted; filaments filiform; nectar pockets at base of corolla tube; nectaries five transversal ridges, free; pollen in small ellipsoid pollinia; spathe folded, oblong, viscidium orbicular; stalk present; style-head umbonate. Follicles ca. 4–7 cm long. 2n = 22.

Two spp. in East Africa, *Acacia-Commiphora* woodland, *Terminalia* grassland with scattered trees, desert scrub.

Schlechterella abyssinica (Schltr.) K. Schum. is nested in *Raphionacme* in the ITS phylogeny of Pienaar (2013), suggesting that *Raphionacme* is paraphyletic without *Schlechterella*.

185. ***Stomatostemma*** N.E. Br.

Stomatostemma N.E. Br. in Thiselton-Dyer, Fl. Trop. Afr. 4(1): 252 (1902); Venter & Verhoeven, S. Afr. J. Bot. 59: 50–56 (1993), rev.

Shrubs or suffrutescent twiners; shoots lenticellate, glabrous; roots tuberous. Leaves 4–11 cm long, linear to (ob-)ovate. Inflorescences paired, 5–25-flowered, thyrsoidal. Flowers fragrant; corolla 9–22 mm long, campanulate; whitish with reddish center, glabrous; lobes ovate, acute, slightly twisted; corolline corona glossy green, exceeding gynostegium; lobes free, basally terete, apically clavate; gynostegium concealed in corolla; stamens inserted at base of corolla tube; filaments short; nectar pockets at base of corolla tube; nectaries five transversal ridges, fused to staminal feet; anthers free; spathe folded, ovate and deeply bifid; viscidium bifid; stalk absent; style-head ovoid. Follicles 6–9 cm long, obclavate. 2n = 22.

Two spp. in SE Africa; dry (Mopane) woodland.

186. ***Streptocaulon*** Wight & Arn.

Streptocaulon Wight & Arn. in Wight, Contr. Bot. India 64 (1834); Li et al., Fl. China 16: 194 (1995), reg. rev.; Jagtap & Singh, Fasc. Fl. India 24: 312–316 (1999), reg. rev.; Sidney, M.Sc. Thesis, University of Free State, Bloemfontein (2012), morph., tax.
Triplolepis Turcz. (1848).
Gongylosperma King & Gamble (1908).

Shrubs or lianas to 8 m, tawny pilose or tomentose; rootstock woody. Leaves 3–20 cm long, ovate or obovate, pubescent. Inflorescences paired, many-flowered, thyrsoidal, lax, shortly pedunculate. Corolla 3–5 mm long, yellowish-greenish to purplish; lobes ovate; corolline corona basally connate to filaments, exceeding gynostegium; filiform, inflexed; filaments (almost) absent; nectaries lobular, fused to staminal feet; pollen in ovoid or ellipsoid pollinia; spathe spoon-like, oblong; stalk present; style-head depressed-conical and bifid. Follicles 7–13 cm long, obclavate, densely pubescent. Seeds occasionally with copper-colored coma.

Widely distributed in India, S China, SE Asia, montane forests and thickets; 0–1000 m.

Seven spp. if the two spp. originally published in *Gongylosperma* are moved to *Streptocaulon* as suggested by Sidney (2012).

187. ***Tacazzea*** Decne.

Tacazzea Decne., Prodr. 8: 492 (1844); Venter et al., S. Afr. J. Bot. 56: 93–112 (1990); Venter, Fl. Ethiopia and Eritrea 4.1: 106–108 (2003), reg. rev.; Venter, Fl. Somalia 3: 140–142 (2006), reg. rev.

Shrubs or lianas to 20 m, hirsute; shoots lenticellate. Leaves occasionally whorled, 6–13 cm long, elliptic to ovate. Inflorescences occasionally paired, many-flowered, thyrsoidal, lax; inflorescence bracts large, ovate. Corolla 5–10 mm long, green or yellowish red; lobes ovate to oblong, occasionally ciliate; corolline corona yellowish, five- or ten-partite; lobes free, antesepalous lobes exceeding gynostegium, filiform or ovate; antepetalous lobes short, subquadrate or bifid, emarginate; nectaries lobular, fused to staminal feet; anthers 2-thecous, ovoid, fused to margin of style-head; spathe flat, apically bifid; viscidium orbicular, ellipsoid or bifid, stalk absent; style-head conical. Follicles 5–14(–20) cm long, obclavate, occasionally keeled; woody. 2n = 22.

Six spp. in Sub-Saharan Africa, lianas in lowland swamp forests along fresh water to mountain forests; shrubby species in more arid areas, but near water.

188. ***Telectadium*** Baill.

Telectadium Baill., Bull. Mens. Soc. Linn. Paris 2: 801 (1889).

Low shrubs, occasionally pubescent. Leaves occasionally whorled; 5–10 cm long, lanceolate. Inflorescences terminal, 4- to many-flowered, dichasial at base, monochasial in higher order branching, long-pedunculate; floral bracts conspicuous. Corolla 14–20 mm long, urceolate, white; lobes oblong, obtuse, twisted; corolline corona inserted in corolla tube, exceeding gynostegium; lobes free, irregularly triangular, erect; gynostegium concealed; stamens inserted at base of corolla tube, filaments flattened; nectar pockets at base of corolla tube; nectaries five transversal ridges, free; anthers sagittate, with two pouches at the base, fused to upper side of style-head; spathe ocreate, oblong when flattened, viscidium orbicular, stalk present; style-head conical. Follicles occasionally ellipsoid, occasionally moniliform. Seeds with winged raphe.

Three spp. in Laos and Vietnam, lowland forests. Flowers of *T. edule* Baill. eaten in SE Asia.

189. ***Zygostelma*** Benth.

Zygostelma Benth. in Benth. & Hook. f., Gen. Pl. 2: 740 (1876).

Suffrutescent twiner; shoots lenticellate, glabrous. Leaves 6–15 cm long, oblong to ovate. Inflorescences solitary, 2–4-flowered, simple, subsessile; floral bracts persistent, conspicuously imbricate, ciliate. Calyx ciliate; corolla 6–8 mm long, glabrous; lobes elliptic, ciliate; corolline corona inserted in corolla tube, short; lobes basally connate, rectangular and deeply bifid, reflexed; stamens inserted at base of corolla tube, filaments very short; nectaries lobular, fused to staminal feet; anthers oblong; connective appendages lanceolate; spathe spoon-like, elliptic, viscidium orbicular, stalk present; style-head umbonate. Follicles 10–11 cm long, obclavate. Seeds slightly winged.

A single species, *Z. benthamii* Baill., in Cambodia, Laos and Thailand.

IV. Subfam. Secamonoideae Endl. (1838).

Large lianas to twining perennial herbs or small erect shrubs; latex white. Leaves paired, usually with colleters at the base. Corolla rotate to campanulate; lobes mostly with dextrorsely, more rarely sinistrorsely contorted or valvate aestivation; corolline corona present only in *Genianthus*, *Goniostemma*, and *Secamone*; gynostegial corona often present, usually of free staminal lobes; gynostegium sessile; stamens inserted on staminal tube around carpels and adaxially fused to sides of style-head (usually in its lower cylindrical portion); anthers 4-locular, sessile on apex of staminal tube, with well-developed, lignified anther wings below fertile part; staminal tube with simple to complex corona attached dorsally (rarely absent) and without clear vascularization; nectaries consisting of 5 vertical alternistaminal troughs behind guide rails on staminal tube; pollen inaperturate, united into minute pollinia, two per locule, not surrounded by waxy outer wall; pollinia more or less erect, formed above guide rails, more or less ellipsoid, without insertion crest, sometimes adjacent pairs adhering to form a single body; style-head very variably shaped, usually with discoid or cylindrical lower part (neck) just above carpels, with the 5 stigmatic zones located on the sides, upper part usually bifid; translator secreted by style-head just above each guide rail, consisting of small, porous, clip-like cream or yellowish corpusculum with flanks and sometimes with floor, sometimes with one or two caudicles (when two, each attaching a pair of pollinia to corpusculum; *Genianthus*, *Secamonopsis*, and some *Secamone* species); gynoecium apocarpous, semi-inferior. Fruit a pair of slender to stout, fusiform follicles (sometimes one by abortion) with dry, thin pericarp. Seeds numerous, compressed, more or less elliptical in outline, usually without wing-like margin, with micropylar coma, endosperm thin. Eight genera in the Paleotropics and Australia.

Key to the Genera of Secamonoideae

1. Staminal corona absent or reduced to five small triangular depressions **197. *Trichosandra***
– Staminal corona more or less well-developed 2
2. Staminal corona lobes fused into a tube for more than half of their length **192. *Goniostemma***
– Staminal corona lobes free or, at the most, basally fused 3
3. Connective appendages completely enclosing style-head 4
– Connective appendages not enclosing style-head 5
4. Connective appendages with a very long filiform extension forming a cone above the style-head or sometimes much broadened and adhering to each other forming a calyptra above the thecae crowned by five distinct clubs on long strings; plants and follicles without conspicuous indumentums **190. *Calyptranthera***
– Connective appendages lacking a very long, filiform prolongation, plants and follicles with a conspicuous, long and soft indumentum **193. *Pervillaea***
5. Abaxial leaf epidermis tuberculate-papillate (observable at ca. × 50) 6
– Abaxial leaf epidermis not tuberculate-papillate 7
6. Pollinia attached in pairs to two long, recurved caudicles **195. *Secamonopsis***
– Pollinia attached directly to the corpusculum, or, when caudicles present, these not long or convexly recurved **194. *Secamone***
7. Staminal corona lobes subulate or falcate; inflorescences frequently axillary, when with an indumentum, then trichomes white or yellowish; upper part of style-head capitate or shortly bifurcate **194. *Secamone***
– Staminal corona lobes dorsiventrally flattened, with or without adaxial appendage, inflorescences normally extra-axillary, when with an indumentum, then trichomes reddish or brown, upper part of style-head cylindrical or ovoid 8
8. Corolla bearded adaxially (except in *G. valvatus* and *G. rectinervis* with five patches of trichomes at the corolla tube mouth), corolla lobe aestivation valvate or only slightly contorted (margins only slightly overlapping), the lobes symmetrical, buds truncate **191. *Genianthus***
– Corolla normally with evenly distributed trichomes adaxially, corolla lobe aestivation distinctly contorted, the lobes asymmetrical, buds acute **196. *Toxocarpus***

190. ***Calyptranthera*** Klack.

Calyptranthera Klack., Novon 6: 27 (1996); Klackenberg, Bull. Mus. Natl. Hist. Nat., B, Adansonia IV, 19: 21–37 (1997), rev.; Klackenberg, Adansonia III, 29: 113–121 (2007), new taxa.

Suffrutescent twiners, 2–5 m tall; younger branches densely tomentose with stiff, occasionally uncinate reddish trichomes. Leaves 6–14 cm long, discolorous, oblong, abaxially pubescent; stipules filiform; petioles often twisted. Inflorescences extra-axillary, solitary, 2–8-flowered, simple. Corolla 5–50 mm long, greenish or purplish, usually with long straight trichomes; lobes sinistrorsely contorted, elliptic to oblong; staminal corona lobes papillose, carnose, filiform or spathulate; filament tube forming cup-like projections underneath the short anther wings; connective appendages triangular, forming a cone or a calyptra above the style-head; pollinia occasionally horizontal, directly attached to the U-folded, soft corpusculum; style-head discoid, slightly depressed. Follicles solitary, ca. 16 cm long.

11 spp. in Madagascar, forests.

191. ***Genianthus*** Hook. f.

Genianthus Hook. f., Fl. Brit. Ind. 4: 15 (1883); Klackenberg, Bot. Jahrb. Syst. 117: 401–467 (1995), rev.; Watson, Fl. Bhutan 2.2: 697 (1999), reg. rev.

Suffrutescent twiners, several meters tall; occasionally densely pubescent. Leaves 7–19 cm long, ovate or oblong. Inflorescences extra-axillary, solitary, many-flowered, indeterminate, thyrsoidal, lax. Corolla 1.5–6 mm long, yellowish, usually with long trichomes; lobes valvate to slightly sinistrorsely contorted, oblong to triangular; gynostegial corona of free staminal lobes; lobes short, narrowly triangular to ovate, basally broadened, with adaxial appendage; connective appendages absent; pollinia usually free, occasionally paired; caudicles cylindrical or flattened; corpusculum soft, ovoid; style-head prominent, discoid, elongate-peltate or ovoid, neck obconical. Follicles occasionally solitary, 9–14 cm long, obclavate to narrowly oblong, occasionally sculptured or with indumentum. Seeds oblong.

16 to 18 spp. in two sections (Klackenberg 1995) distinguished by inflorescence structure; India, Bhutan, China, continental SE Asia, W Malesia, forests.

192. ***Goniostemma*** Wight

Goniostemma Wight, Contr. Bot. India 62 (1834); Li et al., Fl. China 16: 1999 (1995), reg. rev.

Suffrutescent twiners; glabrescent. Leaves discolorous, 6–9 cm long, elliptic to oblong, acute. Inflorescences axillary, paired, many-flowered, thyrsoidal, lax, longer than subtending leaves; peduncles with reddish trichomes. Corolla 5 mm long, yellow, with trichomes; lobes sinistrorsely contorted, oblong, occasionally emarginate; coralline corona of scales; gynostegial corona of connate staminal parts equaling the gynostegium, tubular to campanulate; lobes carnose, lingulate, apically reflexed; connective appendages absent; pollinia directly attached to the minute corpusculum or caudicles very short; style-head fusiform. Follicles occasionally solitary, 10–15 cm long, woody. Seeds pyriform, winged.

Two spp. in China and India, forests.

193. ***Pervillaea*** Decne.

Pervillaea Decne., Prodr. 8: 613. (1844); Klackenberg, Adansonia IV, 19: 21–37 (1997), rev.
Menabea Baill. (1889).

Shrubs or suffrutescent twiners; shoots densely villous with long, curled to wavy trichomes. Leaves linear to broadly ovate, occasionally undulate; stipules filiform. Inflorescences extra-axillary, solitary, many-flowered, simple, condensed. Corolla 5–10 mm long, yellowish, finely pubescent; lobes sinistrorsely contorted, oblong, twisted; staminal corona lobes glabrous, longer than gynostegium, erect, lanceolate to narrowly ovate; connective appendages lanceolate, sometimes long-papillose, forming a cone over the gynostegium; pollinia directly attached to the U-folded, soft corpusculum; style-head discoid. Follicles 7–14 cm long, ellipsoid to obclavate, woolly. Seeds ovate.

Five spp. in S and W Madagascar and Mauritius, dry forests and thickets.

194. *Secamone* R. Br.

Secamone R. Br., Prodr. 464 (1810); Goyder, Kew Bull. 47: 437–474 (1992), reg. rev.; Klackenberg, Kew Bull. 47: 595–612 (1992), reg. rev.; Klackenberg, Opera Botanica 112: 127 pp. (1992), reg. rev.; Klackenberg, Blumea 55: 231–241 (2010), reg. rev.
Rhynchostigma Benth. (1876).

Shrubs, suffrutescent or herbaceous twiners to 6 m, occasionally puberulous. Leaves 0.6–7.5 cm long, narrowly elliptic to ovate; stipules minute. Inflorescences axillary, few- to many-flowered, thyrsoidal or simple. Corolla 3–4.5 mm long, yellowish or greenish, occasionally with trichomes; lobes occasionally sinistrorsely contorted, oblong to ovate to triangular; corolline corona of five paired carnose ridges; staminal corona lobes falcate; connective appendages ovate, occasionally with fimbriate margins; pollinia directly attached to the ovoid, porous corpusculum or caudicles extremely short; style-head capitate or shortly bifurcate. Follicles 4.5–10(–20) cm long. Seeds occasionally winged. 2n = 22.

About 90 spp. (Klackenberg 2001) mainly in Africa and Madagascar, Asia and Australia, dry to humid, lowland to montane forests and bushland.

If the Asian *Toxocarpus* is included in *Secamone*, as suggested by Klackenberg (2010), the species number is ca. 120.

195. *Secamonopsis* Jum.

Secamonopsis Jum., Compt. Rend. Hebd. Séances Acad. Sci. 147: 689 (1908); Civeyrel & Klackenberg, Novon 6: 144 (1996).

Shrubs or herbaceous twiners, with distinct long and short shoots. Leaves occasionally fascicled on opposite brachyblasts, oblong to obovate, occasionally revolute, densely tomentose, abaxial epidermis papillose. Inflorescences extra-axillary on brachyblast, solitary, 1- to few-flowered, simple. Corolla 1.5–2 mm long, urceolate, yellow, with long retrorse trichomes in the tube and bunches of erect trichomes in the sinuses; lobes imbricate, oblong; staminal corona lobes short, carnose, lingulate; anthers basally projecting; thecae apically glandular-papillose; pollinia attached in pairs to two long, recurved caudicles, corpusculum ellipsoid, soft; style-head capitate, slightly bifurcate. Follicles ca. 45 mm long, obclavate. Seeds ovate.

Two spp. in Madagascar, arid scrub.

196. *Toxocarpus* Wight & Arn. Fig. 47

Toxocarpus Wight & Arn. in Wight, Contr. Bot. India 61 (1834); Li et al., Fl. China 16: 198–199 (1995), reg. rev.; Watson, Fl. Bhutan 2.2: 695–696 (1999), reg. rev.; Klackenberg, Blumea 55: 231–241 (2010), reg. rev.

Shrubs, suffrutescent twiners or lianas, 5–10 m tall; shoots rusty tomentose or villous. Leaves 2.5–15 cm long, discolorous, ovate, abaxially occasionally rusty puberulous. Inflorescences axillary or extra-axillary, 3–10–many-flowered, thyrsoidal, lax; floral bracts occasionally persistent. Corolla 4–15 mm long, yellowish, occasionally papillose and with long trichomes; lobes

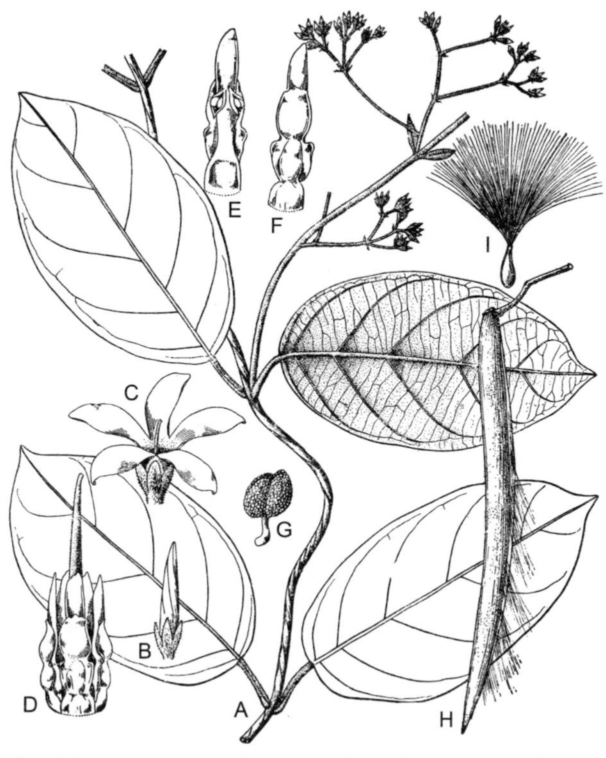

Fig. 47. Apocynaceae-Secamonoideae. *Toxocarpus villosus*. **A** Flowering branch. **B** Flower bud. **C** Flower. **D** Gynostegium with corona. **E** Anther, adaxial view. **F** Anther abaxial view and corona. **G** Pollinarium. **H** Follicle. **I** Seed. (From Flora of China Editorial Committee 1999, Fig. 165, with permission from the Missouri Botanical Garden Press, St. Louis, and Science Press, Beijing; drawn by Chen Guoze)

occasionally sinistrorsely contorted, usually lanceolate, asymmetrical; staminal corona lobes lingulate or lobed, occasionally with adaxial appendage; connective appendages fimbriate or ciliate; pollinia directly attached to the rhomboid soft corpusculum or caudicles very short; style-head ob-infundibuliform, capitate or bifurcate. Follicles 8–15 cm long, obclavate, with dense indumentum; pericarp thick. Seeds ovate.

About 30 spp. in Africa, Madagascar, Asia, Australia and Pacific Islands, forests, scrub, roadsides.

Floral size, style-head shape and flat corona lobes separate *Toxocarpus* well from *Secamone* in Asia and Australia, but not in Madagascar. *Toxocarpus* will, therefore, probably be merged with *Secamone* when more data are available.

197. ***Trichosandra*** Decne.

Trichosandra Decne., Prodr. 8: 625 (1844); Friedmann, Bull. Mus. Natl. Hist. Nat. Paris IV, 12, sect. B, Adansonia 2: 131–138 (1990), rev.

Liana, glabrous. Leaves ovate. Inflorescences extra-axillary, solitary, to 10-flowered, simple. Corolla ca. 5 mm long, subcampanulate, fused for ca. two thirds of their length, with trichomes on the tube and along two V-shaped ridges toward the center of lobes; lobes valvate, ovate, carnose; staminal corona reduced to triangular depressions; connective appendages ovate, fimbriate; pollinia directly attached to the U-folded corpusculum; style-head papillose, clavate or capitate. Follicles 65–85 mm long, obclavate. Seeds ovate, winged.

One sp., *T. borbonica* Decne., endemic to Mauritius.

V. Subfam. Asclepiadoideae Burnett (1835).

Trees to shrubs or herbs, succulents, or small geophytes (rarely annual or even ephemeral), with white, yellow or translucent latex. Leaves opposite (rarely whorled), sometimes reduced to small rudiment or occasionally a spine, often with a cluster of colleters adaxially at base of midrib. Inflorescences usually extra-axillary, solitary, simple. Calyx usually with few to many adaxial, more or less alternisepalous colleters; corolla rotate to tubular; lobes valvate to imbricate; corolline corona rare in sinuses of lobes (except in some Ceropegieae and *Gonolobus* relatives, where present as ring); stamens inserted on staminal tube around carpels and adaxially fused to style-head below its thickest portion near its base; anthers 2-locular, sessile on apex of staminal tube, spreading, erect, or horizontal, rarely descending, deltoid, subquadrate, with or without sterile, membranous, apical appendage, lateral margins sometimes becoming membranous after dehiscence, with well-developed, lignified guide rails alongside or below fertile part; staminal tube with simple to complex corona attached dorsally (rarely absent) and without clear vascularization: outer corona frequently present at or near base, of 5 free lobes to fused into tube; inner corona usually present at or just below bases of anthers at apex of staminal tube, usually of 5 free lobes; nectaries consisting of 5 vertical alternistaminal troughs behind or below guide rails on staminal tube; pollen united into pollinia, one per locule, with waxy outer wall, often with hyaline insertion crest; style-head broad, often with flat or concave apex (rarely conical to slenderly conical), sessile on top of carpels to tapering with narrow neck into them, with 5 stigmatic zones located on sides of lower cylindrical part behind guide rails, secreting translator just above each guide rail; translator consisting of hard, clip-like, brown to black corpusculum, mostly with flanks and floor, with two flexible, translucent caudicles (rarely absent in same Marsdenieae), each of which attaches one pollinium to corpusculum; gynoecium apocarpous, mostly superior. Fruit a pair (often one by abortion) of slender to stout fusiform, more or less obclavate (rarely very stout to more or less spherical), ventrally dehiscent follicles with dry pericarp. Seeds usually numerous, compressed, more or less elliptical in outline, often with wing-like margin, usually with micropylar coma (rarely absent or present around seed margin), endosperm thin. x = 11 (rarely 10, 9). 181 genera worldwide.

Key to the Tribes of Asclepiadoideae

1. Pollinia without well-developed caudicles **1. Fockeeae** (p. 325)
– Pollinia with well-developed caudicles 2
2. Pollinia pendulous in each pollen sac (where the apex of the pollen sac is defined as the upper end of the pollen sac when it is vertical on the stamen, or the outer end when horizontal on the stamen) 3

- Pollinia erect in each pollen sac (where the base of the pollen sac is defined as the lower end of the pollen sac when vertical on the stamen, or the inner end when horizontal) 4
3. Leaves palmately lobed or (at least some) with one or two spreading lateral teeth a little above the base; corona gynostegial, consisting of a ring of very basally fused staminal and deeply tripartite interstaminal parts with an additional inner ring of smaller staminal parts **4. Eustegieae** (p. 353)
- Leaves always entire; corona coralline or gynostegial; when gynostegial, then not consisting of a ring of very basally fused staminal and deeply tripartite interstaminal parts with an additional inner ring of smaller staminal parts **5. Asclepiadeae** (p. 354)
4. Pollinia without pellucid germination zone; or, when present, then on distal margin only; connective appendages present **2. Marsdenieae** (p. 326)
- Pollinia with pellucid germination zone on proximal margin or apex; connective appendages absent (except *Caudanthera*) **3. Ceropegieae** (p. 336)

V.1. **Tribe Fockeeae** H. Kunze, Meve & Liede (1994).

Suffrutescent twiners with large root tubers; latex white. Corolla dextrorsely contorted in bud, rotate to campanulate; coralline corona absent; gynostegial corona of a ring of fused staminal and interstaminal parts, and free inner staminal lobes; gynostegium sessile; anthers with large apical appendages; pollinium consisting of two pollinia without sterile hyaline regions or germination pores, attached dorsally to the corpusculum, caudicles absent; corpusculum with broad outer flanks at least basally isolated and not united by a floor; flanks continuing basally into adhesive pads glued to the anther wings; style-head umbonate. Follicles solitary, obclavate to fusiform. Seeds ovate, winged.

A molecular study (Verhoeven et al. 2003) has shown that Fockeeae are sister to all remaining Asclepiadoideae. Two genera in Africa and the Arabian Peninsula.

Key to the Genera of Fockeeae

1. Corona with free staminal lobes; corolla rotate, adaxially with trichomes; inflorescences many-flowered (> 5 flowers per inflorescence) **198. *Cibirhiza***
- Corona without free staminal lobes; corolla campanulate, adaxially glabrous; inflorescences few-flowered (< 6 flowers per inflorescence) **199. *Fockea***

198. ***Cibirhiza*** Bruyns

Cibirhiza Bruyns, Notes Roy. Bot. Gard. Edinb. 45: 51–54 (1988); Kunze et al., Taxon 43: 367–376 (1994); Verhoeven et al., Grana 42: 70–81 (2003), phyl.; Thulin et al., Kew Bull. 63: 617–624 (2008), new taxon.

Twining geophytes; sparsely pilose; shoots lenticellate. Leaves 4–15 cm long, ovate. Inflorescences 5–25-flowered, sessile. Corolla fused to half of its length, greenish with purple blotches, papillose; lobes triangular-ovate; gynostegial corona ring short, fused for almost half of its length, staminal parts lingulate; inner free staminal lobes brownish, surpassing the gynostegium, subulate; occasionally with lingulate proximal lobe; connective appendages large, membranous, inflexed; pollinia ovoid to rectangular, directly attached to the U-folded corpusculum with a high apical dome and broad flanks; pollen grains single; style-head bifurcate. Follicles 7–10 cm long. 2n = 22.

Three spp.: one in Oman, the other two in E Africa (Ethiopia, Tanzania) and Zambia, respectively, in woodland, bushland or riparian thicket.

199. ***Fockea*** Endl. Fig. 48

Fockea Endl. in Endlicher & Fenzl, Nov. Stirp. Dec. 17 (1839); Court, Asklepios 40: 69–74 (1987), rev.; Bruyns & Klak, Ann. Missouri Bot. Gard. 93: 535–564 (2006), phyl., rev.
Chymocormus Harv. (1842).

Twining geophytes, glabrescent, occasionally succulent. Leaves (sub-)sessile, 1.3–15 cm long, linear-elliptic or ovate. Inflorescences 2–6(–many)-flowered, (sub-)sessile. Corolla 5–15 mm long, creamish yellow, occasionally papillose; lobes linear-lanceolate or ovate, with revolute margins; gynostegial corona ring white, tubular to half of its length, staminal parts trifid; inner staminal lobes white, surpassing the gynostegium, subulate, basally fused to outer ring; connective appendages large and inflated, erect, exceeding the anthers; pollinia directly attached to the ovoid corpusculum with large adhesive pads; pollen in tetrads. Follicles 6–7(–20) cm long, woody.

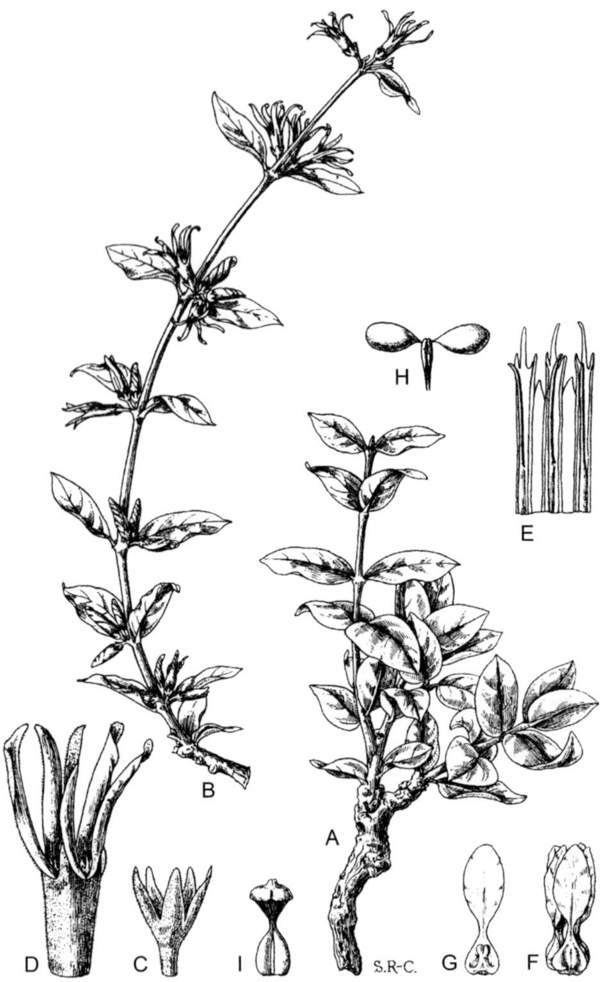

Fig. 48. Apocynaceae-Fockeeae. *Fockea edulis*. **A** Habit, above ground. **B** Flowering branch. **C** Calyx. **D** Corolla. **E** Corona portion in adaxial view. **F** Androecium. **G** Anther adaxial, with pollinarium and terminal anther appendage. **H** Pollinarium. **I** Gynostegium with ovaries and style-head. (From Dyer 1933, pl. 3221, with permission from Hooker's Icones Plantarum © Board of Trustees of the Royal Bot. Gard., Kew; drawn by Stella Ross-Craig)

Seeds with coma occasionally attached along entire margin. 2n = 22.

Six spp. in southern and E Africa, Mopane and *Acacia-Commiphora* woodland, Karoo; *F. edulis* K. Schum. with coastal distribution.

V.2. **Tribe Marsdenieae** Benth. (1868).

Omlor, Generische Revision der Marsdenieae (Asclepiadaceae). Ph.D. Thesis, Univ. Kaiserslautern (1998), rev.

Herbs, vines, lithophytes or epiphytes (these usually leaf succulents), rarely stem succulents, with white, yellow or translucent latex. Leaves herbaceous or coriaceous, with brochidodromous venation, usually apically acute or acuminate. Calyx with one or two colleters in the sinuses, occasionally lacking; corolla rotate to urceolate; lobes with valvate to dextrorsely imbricate aestivation; corolline corona rare, consisting of ridges near sinuses of petals running down corolla tube to near base; gynostegial corona normally of five staminal lobes, attached below or along the back of the anthers; gynostegium usually sessile, exposed just above base of corolla; anthers erect to more or less horizontal, normally with apical appendage, but this never constricted at base, with guide rails below fertile part; pollen in single grains in two pollinia per anther; pollinia erect to horizontal, mostly more or less ellipsoid, occasionally with hyaline insertion crest along inner or lower ledge through which the pollen tubes emerge, pollinia normally basally attached to the caudicles; caudicles (sub-)basally inserted at the corpusculum, horizontal to ascending, usually cylindrical; style-head broad, with thick convex to conical apex, sessile on top of the usually glabrous carpels. Follicles two (often only one by abortion) slender to fusiform or obclavate, rarely nearly spherical, occasionally ornamented with longitudinal ridges on exterior. Seeds ovate to oblong, usually with apically attached coma. x = 11. 26 genera in the Paleotropics and Australia, one in the Neotropics, one extending to southern Europe.

Key to the Genera of Marsdenieae

1. Plants epiphytic 2
– Plants terrestrial 7
2. Adventitious roots conspicuous 3
– Adventitious roots inconspicuous or lacking 5
3. Pollinia without pellucid margin; leaves generally carnose or coriaceous, sometimes concave, a few species additionally with pitcher leaves; corolla urceolate or salver-shaped (rotate in *Dischidia hoyella*), lobes with valvate aestivation; staminal corona lobes anchor-shaped, sagittate, or largely reduced; India to Fiji **207. *Dischidia***
– Pollinia with pellucid margin; plants never with pitcher leaves; corolla of various shapes, staminal corona not anchor-shaped or sagittate 4

4. Inflorescences long-pedunculate, pendent umbels with strongly recurved pedicels; plants with long, narrowly linear, carnose leaves; Borneo **200. *Anatropanthus***
- Inflorescences umbels of various shapes, but pedicels not recurved, leaves, when present (absent in *Hoya spartioides*), much broader **214. *Hoya***

5. Non-twining epiphytic shrub with terminal inflorescences; corolla salverform; tube with prominent folds and dilated between the calyx lobes, adaxially with long trichomes; Java **213. *Hoyella***
- Plants twining or non twining, with extra-axillary inflorescences; corolla of various shapes, but not as above 6

6. Corona with two apical teeth and revolute lateral keels; gynostegium atop a densely villous column; Timor **218. *Oreosparte***
- Corona of various shapes, without two apical teeth; gynostegium mostly sessile, when atop a stipe or column, these not densely villous **214. *Hoya***

7. Pollinia with pellucid germination zone along the outer margin (except *Hoya mitrata* and *H.* sect. *Eriostemma*); corona staminal, lobes generally with a short inner tip that is appressed to the anthers and a prominent outer part that is mostly spreading and has revolute lower margins **214. *Hoya***
- Pollinia without pellucid margin (except *Telosma* with a small curved margin at the base of the outer side); with or without a staminal corona, but corona lobes never with revolute lower margins 8

8. Stems with two opposite longitudinal wings; corolla lobes inside with very long, soft trichomes; follicles always paired, 20–25 cm long, connected at their tips; Laos and Vietnam (apparently very rare) **203. *Campestigma***
- Stems smooth, with lenticels or with a layer of corky ridges, but never with two longitudinal wing-like ridges; follicles not connected at their tips 9

9. Hirsute twiners with thickened or carnose, racemiform inflorescences; flowers small, with a short tube and narrowly elongated corolla lobes extended into a long tip 10
- Plants different; flowers different with corolla lobes not extended into a long tip 11

10. Twiners with yellowish hirsute stems and ovate leaves; inflorescence a long-pedunculate, slightly thickened, unbranched spike, about 9 cm long; flowers small, rotate, corolla lobes reddish purple, ovate; staminal corona lobes bilobed; connective appendages extended into a very long tip; China **208. *Dolichopetalum***
- Hirsute twiners with narrowly obovate leaves (13×5 cm); flowers purplish black, in pairs along a spirally elongated, few-branched, carnose, racemiform inflorescence; staminal corona lobes with an apical tooth; connective appendages not extended into a very long tip; Malacca (apparently very rare or extinct) **219. *Pycnorhachis***

11. Corolla lobes inflexed in bud **216. *Lygisma***
- Corolla lobes not inflexed in bud 12

12. Corolla flattened globose, with a very small opening and very short, triangular lobes; gynostegium enclosed by a cuff-like annular corona; Australia **211. *Gunnessia***
- Corolla rotate, campanulate, urceolate or salverform, gynostegium not enclosed by a cuff-like annular corona 13

13. Staminal corona differentiated into distal and proximal lobule (ligule); distal lobules erect, ovate to oblong or triangular; proximal lobules inflexed, ligules longer than distal lobules, oblong to linear; pollinia with distal-basal sterile hyaline region; SE Asia to Malaya **224. *Telosma***
- Staminal corona simple or lacking; pollinia without hyaline region 14

14. Corolla more than 15 mm long, salverform, rarely urceolate 15
- Corolla less than 15 mm long, rotate or campanulate, rarely salverform or urceolate 18

15. Corolla tube inside, opposite to the stamens, with five corolline ridges, which are covered with two lines of upwardly directed trichomes; staminal corona lobes largely or completely reduced; China to Borneo **215. *Jasminanthes***
- Corolla tube without hairy ridges behind the stamens; staminal corona lobes not completely absent 16

16. Staminal corona lobes elongate-triangular; Madagascar **222. *Stephanotis***
- Staminal corona lobes not elongate-triangular 17

17. Staminal corona lobes hook-shaped and carnose, inserted at the base of the gynostegium; inflorescences long-pedunculate; leaves herbaceous, narrowly ovate, about 10×5 cm; Thailand **210. *Gongronema* (*G. filipes*)**
- Staminal corona of different shapes, attached along the filament tube; inflorescences sessile or shortly pedunculate; leaves coriaceous and elliptic

(*Marsdenia speciosa*, *M. praestans*), or herbaceous and cordate (*M. stenantha*) **217. *Marsdenia***

18. Coralline corona of five scale-like lobes in the sinuses of the corolla lobes or as longitudinal ridges inside the corolla tube 19
– Coralline corona completely absent 22
19. Coralline corona of five longitudinal ridges opposite the stamens, each with two lines of upwardly directed, stiff hairs; in some species these ridges project apically into carnose lobes; style-head massive, cylindrical with blunt apex, always protruding beyond the anthers; staminal corona largely or completely reduced; inflorescences often paired **212. *Gymnema***
– Coralline corona usually in the sinuses of the corolla; when ridges below the sinuses, then style-head flattened, or with well-developed staminal corona 20
20. Corpusculum very narrowly ellipsoid (at least five times longer than wide), caudicles strongly geniculate, ribbon-shaped, at least twice as long as pollinia, style-head flattened, with a papillate hump in the center; guide rails very short; coralline corona glabrous, either as ridges or as bulge-like thickenings in the sinuses of the mostly rotate corolla; mostly mangrove or littoral plants; India to New Caledonia **221. *Sarcolobus***
– Corpusculum less narrow (less than five times longer than wide), caudicles shorter, not ribbon-shaped or strongly geniculate, style-head without a papillate hump in the center, guide rails longer, coralline corona glabrous or with trichomes, between adjacent corolla lobes 21
21. Coralline corona pubescent, domed swellings at the mouth of the corolla tube; Africa **201. *Anisopus*** (*A. mannii*)
– Coralline corona hairy or glabrous small scales between adjacent corolla lobes, carnose obtuse lobes perpendicular to the corolla lobes (*Marsdenia longiloba*), or triangular longitudinal ridges (*M. harmandiella*) **217. *Marsdenia***
22. Gynostegial corona absent **220. *Rhyssolobium***
– Gynostegial corona present 23
23. Style-head rostrate, protruding well beyond the anthers 24
– Style-head short-conical, hemispherical or flattened, not or only slightly protruding beyond the anthers 27
24. Follicles with only 1 or 2 seeds; erect, rigid shrubs; Africa **223. *Stigmatorhynchus***
– Follicles with numerous seeds; lianas or twiners, when erect, then usually apically twining 25
25. Pollinia with warty anticlinal walls in the apical half, corolla lobes narrowly elongate, up to 8 mm long, fused only at the base; SE Mediterranean to Iran **205. *Cionura***
– Pollinia smooth; corolla lobes broader, or when narrowly elongate then with well-developed corolla tube 26
26. Inflorescences axillary and paired, one larger than the other **201. *Anisopus*** (*A. mannii*)
– Inflorescences extra-axillary, alternate (rarely paired) **217. *Marsdenia***
27. Gynostegial corona cupular; staminal lobes obtuse, apically bi- or trifid, laterally fused in lower half; corolla rotate, yellow or pink with a dark red eye; Java **202. *Asterostemma***
– Gynostegial corona not cupular, staminal lobes of various shapes, but not apically bi- or trifid; corolla of various shapes, when rotate, then not yellow or pink with dark red eye 28
28. Corolla rotate or bowl-shaped, lobes at least as long as tube, usually longer 29
– Corolla campanulate or urceolate; lobes generally shorter than tube 37
29. Top of gynostegium truncate, with very short guide rails 30
– Top of gynostegium not truncate, guide rails longer 32
30. Style-head flattened with a small papillate hump in the center; corpusculum very narrowly ellipsoid (at least five times longer than wide); mostly mangrove or littoral plants; India to New Caledonia **221. *Sarcolobus***
– Style-head blunt conical, rounded or flattened, not papillate; corpusculum less narrow 31
31. Corolla and staminal corona lobes carnose, protruding well beyond the short gynostegium, with a very broad basal part and an incurved, finger-like apical part; inflorescences solitary; Sumatra, Philippines **214. *Hoya*** (former *Clemensiella* species)
– Corolla and staminal corona lobes usually not carnose, not protruding beyond the gynostegium and not incurved; inflorescences often paired **217. *Marsdenia*** (*M. secamonoides* group)
32. Inflorescences axillary, paired; C and W Africa **201. *Anisopus*** (*A. efulenis*)
– Inflorescences extra-axillary, solitary; Asia and Australia 33
33. Inflorescences long-pedunculate 34
– Inflorescences short-pedunculate or sessile 36
34. Corolla lobes marginally glabrous, latex translucent **206. *Cosmostigma***

- Corolla lobes marginally ciliate, latex white 35
35. Staminal corona lobes carnose, subglobose, with short inner tips appressed to the anthers; corolla lobes fused only at the base, light-colored; carpels completely or partly hairy; leaves herbaceous; India to China **209. *Dregea***
- Staminal corona lobes laminar, oblong; corolla fused for ca. half of its length, dark purple; carpels glabrous; leaves thick-coriaceous; E Himalayas **225. *Treutlera***
36. Leaves coriaceous, broadly elliptic to obovate; inflorescences sciadioidal, pedicels thread-like, 2–4 cm long; corolla deeply lobed; Java, Philippines **204. *Cathetostemma***
- Leaves herbaceous, elliptic; inflorescences at first sciadioidal, later becoming racemiform with a thickened, contracted rachis; corolla not deeply lobed; New Guinea, Australia **217. *Marsdenia*** (*M. hemiptera* group)
37. Staminal corona of carnose, scale-like or hook-shaped lobes, inserted at the base of the gynostegium; lower surface of corona lobes often bulged **210. *Gongronema***
- Staminal corona of mostly erect lobes with a free apical tip, attached to the filament tube and appressed to the anthers (except *Marsdenia thyrsiflora*, *M. cavalierei* and *M. urceolata* with hook-shaped corona lobes); lower surface of corona lobes not bulged **217. *Marsdenia***

200. *Anatropanthus* Schltr.

Anatropanthus Schltr., Bot. Jahrb. Syst. 40. Beibl. 92: 18 (1908).

Epiphyte with pendent branches and adventitious roots, glabrous. Leaves succulent, linear, mucronate, marginally revolute. Inflorescences 10–20-flowered, sciadioidal, long-pedunculate. Flowers with 180° inwardly bent pedicels; corolla tubular to urceolate; lobes apically abruptly inflexed; free staminal corona lobes differentiated into an erect distal lobule and a slenderly oblong, inflexed, proximal lobule; gynostegium shortly stipitate, concealed in corolla tube; anthers oblong; connective appendages subulate, inflated, erect; pollinia with distal hyaline germination pore; caudicles very short, flattened; corpusculum ellipsoid; style-head elongate-conical.

One sp., *A. borneensis* Schltr., Borneo, forest. Insufficiently known genus with the type specimen destroyed in Berlin (B); possibly related to *Hoya*.

201. *Anisopus* N. E. Br.

Anisopus N. E. Br., Bull. Misc. Inform. 1895: 259 (1895); Goyder, Kew Bull. 49: 737–747 (1994), rev.

Suffrutescent twiners or lianas, glabrous. Leaves 4–15 cm long, obovate(-oblong), basally cordate, truncate or rounded. Inflorescences axillary, paired, 5–20-flowered. Corolla 2.5–8 mm long, rotate, carnose, yellowish or reddish purple, densely covered with trichomes; lobes ovate to oblong; corolline corona, when present, of free, oblongoid lobes, with trichomes; free staminal corona lobes attached along filament tube, almost equaling the gynostegium, obovate, pointed, winged, with a tooth projecting from the top of the inner face; connective appendages broadly ovate; pollinia ovoid; caudicles convexly recurved; style-head conical to semi-globose or rostrate. Follicles 8–16 cm long. Seeds with undulate wing.

Two spp. in W Africa, evergreen tropical rainforest, 0–1800 m.

202. *Asterostemma* Decne.

Asterostemma Decne., Ann. Sci. Nat. Bot. II, 9: 371, t. 10, f. D (1838); Omlor, Gener. Rev. Marsdenieae: 107 (1998), consp.

Suffrutescent twiner; shoots sparsely pubescent. Leaves ovate to oblong, marginally crenulate, hirsute. Inflorescences 4–8-flowered, condensed bostrychoid. Corolla rotate, yellow to rose, center purple; staminal corona lobes connate for about half of their length, slightly carnose, bi- or trifid; pollinia oblongoid to obovoid; caudicles long-geniculate, first horizontal, then erect, articulated; corpusculum ellipsoid-oblongoid, larger than pollinia; style-head umbonate.

One sp., *A. repandum* Decne., in Indonesia (Java).

203. *Campestigma* Pierre ex Costantin

Campestigma Pierre ex Costantin in Lecomte, Fl. Indo-Chine 4: 117 (1912); Omlor, Gener. Rev. Marsdenieae: 103–104 (1998), consp.

Suffrutescent twiner, glabrous; shoots twisted, with two opposite wings. Leaves long-petiolate, broadly ovate, cordate. Inflorescences 5–10-flowered, thyrsoidal, condensed; flowers long-pedicellate. Corolla ca. 6.5 mm long, rotate, lobes densely covered with trichomes to 5 mm long, elongate-ovate; corolline corona annular, with trichomes; free staminal corona lobes dolabriform; pollinia pyriform; caudicles S-shaped; corpusculum elongatedly clavate; style-head rostrate Follicles paired, apically fused when mature, 20–25 cm long, narrowly oblong. Seeds winged.

One sp., *C. purpureum* Pierre ex Costantin, in continental SE Asia.

204. *Cathetostemma* Blume

Cathetostemma Blume, Rumphia 4: 30 (1849); Omlor, Gener. Rev. Marsdenieae 129–130 (1998), consp.

Suffrutescent twiner, sparsely pubescent. Leaves 9–15 cm long, elliptic to obovate. Inflorescences short, sciadioidal, pedicels 2–4 cm long, filiform. Corolla ca. 7.5 mm long, rotate, greenish white; lobes ovate; free staminal corona lobes attached along the filament tube, almost equaling the gynostegium, erect, slightly carnose; gynostegium sessile; anther wings extending beyond the anther; pollinia ovoid; caudicles horizontal; style-head umbonate.

A single species, *C. laurifolium* (Blume) Decne., island of Timor and Philippines.

205. *Cionura* Griseb.

Cionura Griseb., Spicil. Fl. Rumel. 2: 69 (1844); Field, Fl. Iraq 4.1: 561–563 (1980), reg. rev.; Omlor, Gener. Rev. Marsdenieae: 107–108 (1998), consp.

Shrub to 1 m tall, occasionally apically twining, glabrous. Leaves 4–10 cm long, broadly ovate, cordate. Inflorescences 15–20-flowered, lax. Corolla rotate, whitish; lobes oblong; free staminal corona lobes shorter than the gynostegium, erect to reflexed, lanceolate; connective appendages oblong, about twice as long as the anthers; pollinia clavate, with warty anticlinal walls; caudicles S-shaped, cylindrical; corpusculum elongate- bone-shaped; style-head rostrate, apically bifurcate. Follicles solitary, 6.5–8 cm long, obclavate. Seeds winged. 2n = 22.

One sp., *C. erecta* (L.) Griseb., southern Europe and Asia Minor, oak forests, rocky slopes, sandy beaches and dry ravines on limestone.

206. *Cosmostigma* Wight

Cosmostigma Wight, Contr. Bot. India 41 (1834); Omlor, Gener. Rev. Marsdenieae 107–108 (1998), consp.; Jagtap & Singh, Fasc. Fl. India 24: 71–72 (1999), reg. rev.

Suffrutescent twiners or lianas, occasionally yellowish pubescent; latex translucent. Leaves occasionally papery, 5–11 cm long, ovate, cordate. Inflorescences 6–15-flowered; long-pedunculate. Corolla 4–6 mm long, rotate to campanulate, yellowish green, occasionally with trichomes in the throat; lobes valvate or contorted, ovate to oblong; free staminal corona lobes shorter than the gynostegium, erect, ovate or rectangular, deeply bilobed or irregularly dentate; gynostegium sessile, exserted from corolla tube; pollinia clavate; caudicles horizontal or declinate, cylindrical; corpusculum ovoid to oblongoid, larger than pollinia; style-head umbonate. Follicles 12–20 cm long, obclavate; pericarp thick. Seeds broadly winged.

Three spp. in India, Sri Lanka, China and the Philippines, humid forests.

207. *Dischidia* R. Br.

Dischidia R. Br., Prodr. 461 (1810); Rintz, Blumea 26: 81–126 (1980), reg. rev.; Weir & Kiew, Biol. J. Linn. Soc. 27: 113–132 (1986), ecol.; Livshultz et al., Blumea 50: 113–134 (2005), reg. rev.; Peeters & Wiwatwitaya, Asian Myrmecol. 6: 49–61 (2014), ecol.
Dischidiopsis Schltr. (1904).
Oistonema Schltr. (1908).
Dolichostegia Schltr. (1915).
Hoyella Ridl. (1917).

Twining, pendent or creeping herbaceous or succulent epiphytes, usually glabrous, leaves occasionally 3–4-whorled, occasionally reduced, elliptic or ovate; often with modified, hollow leaves invaded by adventitious roots, or shell-shaped leaves appressed to substrate. Inflorescences with persistent, straight, thickened rachis.

Corolla urceolate or tubular, carnose, tube barbate; lobes valvate, ovate, occasionally cucullate; occasionally with corolline corona inserted in tube, annular or 5-lobate, glabrous; free staminal corona lobes, when present, bilobed or sagittate; gynostegium atop a bulge, concealed in corolla; connective appendages conspicuous; pollinia oblongoid; corpusculum very small; style-head conical or ovoid. Follicles 4–8 cm long; pericarp papery. Seeds apically winged. 2n = 22.

About 80 spp. in India, China, continental SE Asia, Malesia and Australia, dry to humid lowland to montane forests, littoral scrub. Many species associated with ants, e.g., housing them under imbricate leaves or in pitchers (Peeters and Wiwatwitaya 2014).

208. ***Dolichopetalum*** Tsiang

Dolichopetalum Tsiang, Acta Bot. Sin. 15: 137 (1973); Li et al., Fl. China 16: 237–238 (1995), reg. rev.; Omlor, Gener. Rev. Marsdenieae: 104 (1998).

Suffrutescent twiner, to 4 m tall; densely yellowish brown villous. Leaves papery, 8–12 cm long, ovate, strongly cordate, basally with 10 colleters. Inflorescences long-pedunculate raceme-like thyrses. Corolla ca. 14 mm long, cyathiform, purple, abaxially villous; lobes valvate, ovate, caudate, ciliate; free staminal corona lobes shorter than gynostegium, erect, rectangular, slightly bifid; gynostegium sessile, concealed in corolla tube; anthers nearly square; connective appendages triangular, extended in a long tip, erect; pollinia oblongoid; caudicles horizontal; corpusculum clavate; style-head conical. Follicles solitary, 6.5–11 cm long, obclavate. Seeds winged.

One sp., *D. kwangsiense* Tsiang, in southern China, mountain forests.

209. ***Dregea*** E. Mey.

Dregea E. Mey. Comment. Pl. Africae Austr. 199 (1838), nom. cons., non *Dregea* Eckl. & Zeyher (1837), Apiaceae, nom. rej.; Omlor, Gener. Rev. Marsdenieae 109–110 (1998), consp.; Jagtap & Singh, Fasc. Fl. India 24: 172–177 (1999), reg. rev.
Wattakaka Hassk. (1857).

Suffrutescent twiners or lianas, 4–8 m tall, pubescent; latex translucent; shoots lenticellate, leaves long-petiolate; 2–20 cm long, broadly ovate, occasionally shallowly cordate. Inflorescences 10–30-flowered, pendulous, long-pedunculate. Flowers long-pedicellate, sweetly fragrant; corolla 5–12 mm long, whitish, yellow or purplish, occasionally veined, occasionally pilose; lobes ovate, ciliate; free staminal corona lobes white or greenish, subglobose and pointed, almost equaling the (sub-)sessile gynostegium; pollinia slenderly ellipsoid; caudicles geniculate, rectangular; corpusculum very elongate-oblongoid; carpels at least partially with trichomes; style-head yellow, conical. Follicles paired, 7–9 cm long, obclavate, with thick pericarp and dense indumentum. Seeds ovate, broadly winged. 2n = 22.

About 12 spp. in China, S Asia, continental SE Asia and Malesia, forests. *Dregea volubilis* (L.f.) Benth. is known for its medicinal properties, much used in Ayurveda and traditional Chinese medicine.

210. ***Gongronema*** (Endl.) Decne.

Gongronema (Endl.) Decne., Prodr. 8: 624 (1844); Li et al., Fl. China 16: 240–241 (1995), reg. rev.; Omlor, Gener. Rev. Marsdenieae 131–132 (1998), consp.; Jagtap & Singh, Fasc. Fl. India 24: 78–82 (1999), reg. rev.; Watson, Fl. Bhutan 2.2: 706–708 (1999), reg. rev.
Gymnema subg. *Gongronema* Endl., Gen. Pl.: 595 (1838).

Suffrutescent twiners or lianas, 5–8 m tall, glabrous. Leaves occasionally papery, 5–14 cm long, elliptic to oblong. Inflorescences condensed, usually long-pedunculate, rachis straight, occasionally thickened (then floral bracts conspicuous), with spirally arranged flowers. Corolla 4–8 mm long, urceolate or campanulate, fused for about half of its length, creamish yellow, occasionally with trichomes, lobes ovate, occasionally ciliate; free staminal corona lobes attached below filament tube, shorter than gynostegium, ovate, lingulate or unguiculate, erect; gynostegium concealed in corolla tube; connective appendages very small; pollinia ovoid; style-head conical. Follicles 4.5–8 cm long. 2n = 22.

At least 12 spp. in Africa, S and SE Asia, China.

211. ***Gunnessia*** P.I. Forst.

Gunnessia P.I. Forst., Austrobaileya 3: 273–289 (1990).

Liana; shoots densely pubescent. Leaves to 15 cm long, ovate, (almost) glabrous, basally with 6 colleters. Inflorescences 1-8-flowered, sciadioidal. Corolla 3–4 mm long, 16 mm diam., broadly urceolate, yellowish green, tips triangular; gynostegial corona yellowish, of a short, completely fused outer ring of connate staminal and interstaminal parts, and of an inner ring of almost completely fused, toothed staminal parts, attached below the filament tube, overtopping the gynostegium; carnose, adaxially papillose; free tips triangular, strongly reflexed; gynostegium sessile, concealed in corolla; pollinia globose; caudicles geniculate, flattened; style-head papillose, semi-globose. Follicles to 8 cm long, ovoid, keeled along suture. Seeds ovate.

One sp., *G. pepo* P.I. Forst., in Australia, endemic to Cape York Peninsula, deciduous vine thickets.

212. ***Gymnema*** R. Br. Fig. 49

Gymnema R. Br., Prodr. 461 (1810); Forster, Austral. Syst. Bot. 8: 703–933 (1995), reg. rev. (under *Marsdenia*); Li et al., Fl. China 16: 238–240 (1995), reg. rev.; Meve & Alejandro, Willdenowia 43: 81–86 (2013), morph., tax.
Bidaria (Endl.) Decne. (1844).

Suffrutescent twiners or lianas, 1–10 m tall; shoots lenticellate. Leaves 1–13 cm long, oblong, occasionally cordate. Inflorescences occasionally paired. Corolla 5–8 mm long, campanulate, fused for about half of its length, yellowish; lobes valvate or slightly contorted; oblong or ovate; corolline corona of five longitudinal ridges below corolla lobe sinuses, sometimes forming carnose appendages, usually with (dense) trichomes along two lines; gynostegial corona absent; gynostegium concealed in corolla; pollinia oblongoid; style-head conical, rostrate or semi-globose. Follicles 3.5–9(–16) cm long, occasionally ellipsoid. Seeds winged. 2n = 22.

About 30 spp., often very similar in flower structure; Africa [only *G. sylvestre* (Retz.) R. Br.] and most Asian-Pacific countries, littoral scrub, bushland, forests.

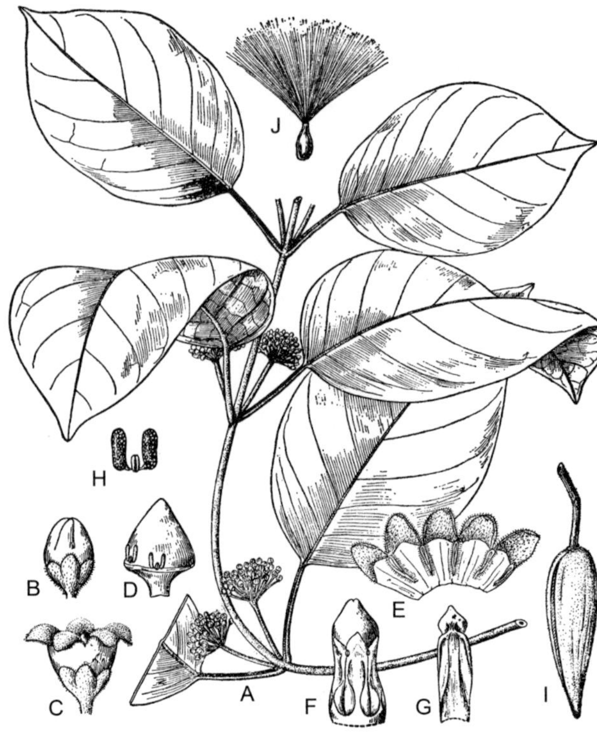

Fig. 49. Apocynaceae-Marsdenieae. *Gymnema latifolium*. **A** Flowering branch. **B** Flower bud. **C** Flower. **D** Style-head with two pollinaria attached. **E** Opened corolla showing corollinic corona. **F** Gynostegium. **G** Anther adaxial view. **H** Pollinarium. **I** Follicle. **J** Seed. (From Flora of China Editorial Committee 1999, Fig. 231, with permission from the Missouri Botanical Garden Press, St. Louis, and Science Press, Beijing; drawn by Chen Guoze)

The inclusion of *Gymnema* in *Marsdenia* (Forster 1995) would make the latter genus too heterogeneous and is therefore not accepted here.

213. ***Heynella*** Backer

Heynella Backer, Blumea 6: 381 (1950); Omlor, Gener. Rev. Marsdenieae 152 (1998), consp.

Epiphytic shrub, 25–30 cm tall. Leaves ovate to oblong. Inflorescences terminal, sciadioidal. Corolla 10–12 mm long, tubular to salverform, fused for about three quarters of length, slightly carnose, cream; tube dilated at base and grooved apically, with long trichomes; lobes valvate, ovate to oblong, erect to patent; free staminal corona lobes laterally compressed, basally widened, with a groove on the back.

One sp., *H. lactea* Backer, in Indonesia (Java). Insufficiently known genus.

214. ***Hoya*** R. Br.

Hoya R. Br., Prodr. 459 (1810); Rintz, Malayan Nat. J. 30: 467–522 (1978), reg. rev.; Wanntorp, Taxon 56: 465–478 (2007), morph.; Wanntorp et al., Mol. Phyl. Evol. 33: 722–733 (2006), phyl.; Wanntorp & Forster, Ann. Missouri Bot. Gard. 94: 36–55 (2007), phyl.; Wanntorp et al., Taxon 60: 4–14 (2011), phyl.; Wanntorp & Meve, Willdenowia 41: 97–99 (2011), tax.; Simonsson Juhonewe & Rodda, Gard. Bull. Singapore 96: 97–147 (2017).
Otostemma Blume (1849).
Absolmsia Kuntze (1891).
Micholitzia N.E. Br., Bull. Misc. Inform. 1909: 358 (1909).
Clemensiella Schltr. (1915), nom. nov. pro *Clemensia* Schltr. (1915), non *Clemensia* Merr. (1908), Meliaceae.
Antiostelma (Tsiang & P.T. Li) P.T. Li (1992); *Hoya* sect. *Antiostelma* Tsiang & P.T. Li (1974).
Madangia P.I. Forst., D.J. Liddle & I.M. Liddle (1997).

(Hemi-)epiphytes, herbaceous or suffrutescent twiners, rarely shrubs; latex occasionally translucent; adventitious roots often conspicuous. Leaves coriaceous, succulent or waxy, occasionally caducous. Inflorescences many-flowered, occasionally long-pedunculate (to 25 cm). Corolla rotate, campanulate, subglobose, rarely tubular, carnose, often pilose or pubescent; lobes valvate, broadly triangular; staminal corona at least equaling the gynostegium, lobes horizontally spreading, solid, often differentiated into a distal lobule with apical tooth and revolute lateral keels forming a deep nectar groove and a dentiform proximal lobule, basally occasionally spurred; gynostegium usually sessile, occasionally concealed; pollinia usually elongate-ellipsoid, mostly with distal germination crest; corpusculum occasionally with basal projections. Follicles usually solitary, 4–10(–30) cm long, occasionally ellipsoid. Seeds (apically) winged, occasionally with coma attached along entire margin. $2n = 22, 44$.

Probably more than 350–450 spp., number constantly increasing due to new discoveries. Tropical Asia and Oceania, littoral areas, various forest types, frequently associated with water courses; on rocks, tree trunks.

215. ***Jasminanthes*** Blume

Jasminanthes Blume, Ann. Mus. Bot. Lugduno-Batavum 1: 148 (1850); Li et al., Fl. China 16: 242–243 (1995), reg. rev.; Omlor, Gener. Rev. Marsdenieae 126–128 (1997), consp.; Tran et al., PhytoKeys 69: 17–22 (2016), key, morph., reg. tax.

Lianas, to 12 m tall; latex occasionally translucent; shoots pubescent. Leaves 7–19 cm long, oblong, long-acuminate, glabrous or glabrescent. Inflorescences to 12-flowered. Flowers sweetly fragrant; corolla 2–10 cm long, urceolate or salverform; tube dilated at base and with five grooves, creamish to green, occasionally with trichomes; lobes oblong to falcate, twisted; corolline corona of two lines with trichomes, fused to corolla tube, occasionally with free tips, opposite the stamens; gynostegial corona absent, gynostegium sessile, concealed in corolla; pollinia ovoid; caudicles horizontal, short; lower part of style-head obconical, upper part conical or capitate. Follicles solitary, 12–21 cm long.

Seven spp. in China, Japan, continental SE Asia and W Malesia, little disturbed tropical forests.

216. ***Lygisma*** Hook. f.

Lygisma Hook. f., Hooker's Icon. Pl. 15: ad t. 1423 (1883); Li et al., Fl. China 16: 262 (1995), reg. rev.; Omlor, Gener. Rev. Marsdenieae 106 (1998), consp.
Costantina Bullock (1965), nom. nov. pro *Pilostigma* Costantin (1912), non *Pilostigma* Tiegh. (1894), Loranthaceae.

Herbaceous or suffrutescent twiners, to 3 m long, pubescent. Leaves papery, 3–6.5 cm long, ovate. Inflorescences occasionally axillary or terminal, occasionally paired, short, to 8-flowered, lax, distinctly pedunculate. Corolla 3–4 mm long, campanulate, white; tube pilose; lobes inflexed in bud, oblong; free staminal corona lobes equaling the gynostegium, erect, slightly carnose, ovate; gynostegium concealed in corolla; anther wings shorter than anthers; pollinia oblongoid; caudicles short, S-shaped; style-head papillose, flat. Follicles solitary, 5–7 cm long, obclavate. Seeds winged.

At least five spp. in China, continental SE Asia and W Malesia, thickets, open woods; 100–300 m.

217. ***Marsdenia*** R. Br.

Marsdenia R. Br., Prodr.: 460 (1810), nom. cons.; Rothe, Bot. Jahrb. Syst. 52: 354–434 (1915); Forster, Austral. Syst. Bot. 8: 703–933 (1995); Li et al., Fl. China 16: 243–249 (1995), reg. rev.; Omlor, Gener. Rev. Marsdenieae (1998), consp.; Jagtap & Singh, Fasc. Fl. India 24: 122–135 (1999), reg. rev.; Watson, Fl. Bhutan 2.2: 708–712 (1999), reg. rev.

Leichardtia R. Br. (1849).
Thozetia F. Muell. ex Benth. (1868).
Stephanotella E. Fourn. (1885).
Verlotia E. Fourn. (1885).
Traunia K. Schum. (1895).
Harmandiella Costantin (1912).
Pseudosarcolobus Costantin (1912).
Dalzielia Turrill (1916).
Dischidanthus Tsiang (1936).
Loniceroides Bullock (1964), nom. nov. pro *Harrisonia* Hook. (1826), non *Harrisonia* R. Br. ex A. Juss. (1825), (Simaroubaceae).
Papuastelma Bullock (1965), nom. nov. pro *Astelma* Schltr. (1913), non *Astelma* R. Br. ex Ker-Gawl. (1821), Asteraceae.
Sinomarsdenia P.T. Li & J.J. Chen (1997).

Lianas, twiners or shrubs, occasionally tuberous; glabrous, or with white, yellow, or rusty indument; latex white, yellow, brown, or translucent. Leaves oblong or linear. Corolla 6–35 mm long, rotate, campanulate, urceolate or salverform, occasionally with trichomes; lobes valvate, imbricate or contorted; corolline corona, when present, of five ridges opposite the stamens, with or without free tips and trichomes; free staminal corona lobes attached along filament tube, scaly or carnose, short; pollinia oblongoid or clavate; caudicles straight; corpusculum smaller than pollinia; carpels occasionally with trichomes; style-head occasionally semi-globose or rostrate. Follicles occasionally winged, occasionally with woody or thick pericarp. Seeds winged; coma rarely absent. $2n = 22$.

About 150 spp. in Africa, Asia, Australia, Central and South America and Oceania. *M. tinctoria* R. Br. yields an indigo dye, and *M. cundurango* Reichenb.f. yields the medicinal drug condurango.

The limits of *Marsdenia* are still unclear; while Forster (1995) advocated a very wide concept for Australia and Papuasia, the concept of Omlor (1998), which is adopted here, is much narrower. Omlor (1998) recognized three sections in Africa; the species of the Americas and Australasia, however, are too little known for a subgeneric concept.

218. ***Oreosparte*** Schltr.

Oreosparte Schltr., Beih. Bot. Centralbl., Abt. 2, 34: 16 (1916); Omlor, Gener. Rev. Marsdenieae: 151–152 (1998), consp.; Rodda & Omlor, Webbia 68: 91–95 (2013), morph.

Large scandent twiner, pubescent. Leaves slightly carnose, 5.5–9 cm long, ovate-oblong, glabrous. Inflorescences long-pedunculate, 1 or 2 flowers open synchronously. Corolla ca. 13 mm long, campanulate, white, glabrous; lobes adaxially thickened opposite the stamens, lanceolate, spreading; free staminal corona lobes surpassing the gynostegium, horizontally spreading, solid, apically bifid, with revolute lateral keels, forming a deep groove containing nectar, basally truncate; gynostegium atop a column, raised high above corolla; pollinia oblongoid, caudicles short; corpusculum thickly rhomboid; style-head depressed-conical, papillate. Follicles ca. 10 cm long. Seeds small, wingless, with long coma.

Only one species, *O. celebica* Schltr., in Indonesia (Sulawesi), mountain forest; 1200 m.

Rodda and Omlor (2013) keep the genus separate from *Hoya*.

219. ***Pycnorhachis*** Benth.

Pycnorhachis Benth. in Benth. & Hook. f., Gen. Pl. 2: 737, 776 (1876); Omlor, Gener. Rev. Marsdenieae: 153 (1998), consp.

Suffrutescent twiner, hirsute. Leaves ca. 13 cm long, ovate, cordate. Inflorescences shortly pedunculate, flowers pairwise along a very long, thickened, occasionally bifurcate rachis. Corolla rotate, blackish purple, lobes valvate, subulate, apically extended into a long tip, with trichomes; tube inflated; free staminal corona lobes obliquely oblongoid with an apical tooth, horizontally spreading; pollinia oblongoid.

One sp., *P. maingayi* Hook.f., peninsular Malaysia only.

220. ***Rhyssolobium*** E. Mey.

Rhyssolobium E. Mey., Comm. Pl. Afr. Austr. 217 (1838); Huber, Prodr. Fl. Südwestafr. 114: 50 (1967), reg. rev.; Omlor, Gener. Rev. Marsdenieae: 85 (1998), consp.

Rigid, strongly branched dwarf shrub, to 40 cm tall; young shoots densely pubescent. Leaves subsessile, leathery-coriaceous, to 1 cm long, elliptic, obtuse, strongly revolute, adaxially glabrescent, abaxially densely pubescent. Inflorescences 1–3-flowered, (sub-)sessile. Corolla 4–5 mm long, campanulate, with long, erect trichomes on lobes and sinuses; lobes oblong; corona absent; anther wings surpassing anther; connective appendages long; pollinia oblongoid to obovoid; caudicles straight, horizontal, short; corpusculum narrowly ellipsoid; style-head conical. Follicles 4–5 cm long, ellipsoid, with thick pericarp. Seeds 1–2 per follicle, almost circular, broadly winged.

One sp., *R. dumosum* E. Mey., in Namibia and South Africa, rocky, arid coastal winter-rainfall areas.

221. ***Sarcolobus*** R. Br.

Sarcolobus R. Br., Asclepiadeae 23 (1810); Rintz, Blumea 26: 65–79 (1980); Forster, Austrobaileya 3: 335–360 (1991), reg. rev.; Omlor, Gener. Rev. Marsdenieae: 132–136 (1998), consp.; McHone et al., Phytotaxa 197: 45–53 (2013), morph., reg. tax.
Dorystephania Warb. (1904).
Quisumbingia Merr. (1936), nom nov. pro *Petalonema* Schltr. (1915), non *Petalonema* Berk. ex Correns (1898), (Cyanophyceae–Scytonemataceae).

Lianas or rheophytic shrubs, 0.5–5 m tall; shoots hollow, glabrescent; bark papery. Leaves 7–13 cm long, oblong, occasionally cordate, acute or retuse. Inflorescences 1–6-flowered, (sub-)sessile. Corolla 4–8 mm long, cream, green, brown or reddish, often patterned; lobes contorted, ovate, occasionally papillose or pilose; corolline corona, when present, of glabrous or hairy patches or ridges fused to corolla; free staminal corona lobes, when present, ovate to rectangular; pollinia ovoid or oblongoid; caudicles usually ribbon-shaped; corpusculum slenderly ellipsoid, equaling pollinia; style-head papillose, umbonate or semi-globose. Follicles solitary, 4–10 cm long, thick, occasionally inflated. Seeds broadly winged; coma short or missing.

About 12 spp. in India, continental SE Asia, Malesia and Australia, littoral and in mangroves.

Omlor (1998) disputed the inclusion of *Papuastelma* Bullock (*Astelma* Schltr.) in *Sarcolobus* as suggested by Forster (1991c) and transferred it to *Marsdenia*. Rintz (1980) recognized four species of *Sarcolobus*, Forster (1991c) found 11 for Australia and Papuasia alone.

222. ***Stephanotis*** Thouars

Stephanotis Thouars, Gen. Nov. Madagasc. 11 (1806); Omlor, Gener. Rev. Marsdenieae: 89–90 (1998), consp.
Isaura Comm. ex Poir. (1913), nom. illegit.

Lianas, glabrous; latex translucent. Leaves carnose, elliptic to ovate, cordate. Inflorescences few-flowered, short-pedunculate. Flowers intensively sweetly fragrant; corolla 3–4 cm long, salverform, carnose; tube rose, with trichomes; lobes white, contorted, ovate, obtuse or emarginate; free staminal corona lobes shorter than gynostegium, elongate-triangular; gynostegium concealed in corolla tube; anther wings following the basal margin of the anther; pollinia ellipsoid to rectangular, caudicles straight, horizontal, shortly trapezoidal; style-head white, conical. Follicles solitary, ellipsoid, with thick pericarp, occasionally inflated. Seeds winged. $2n = 22$.

About five spp. in Madagascar. *Stephanotis floribunda* Brongn. (Madagascar jasmine) is a popular and economically important ornamental.

223. ***Stigmatorhynchus*** Schltr.

Stigmatorhynchus Schltr., Bot. Jahrb. Syst. 51: 141 (1913); Bruyns, Bothalia 25: 170–172 (1995); Omlor, Gener. Rev. Marsdenieae: 72 & 84–85 (1998), consp.

Rigid shrubs, 0.5–2 m tall, glabrescent or shortly pubescent; shoots lenticellate with age. Leaves subsessile; blades 1–3 cm long, ovate, obtuse. Inflorescences few-flowered, subsessile. Corolla ca. 5 mm long, slenderly campanulate, white, bearded in the tube opposite anthers; lobes contorted, oblong, obtuse; staminal corona lobes basally fused, attached along the filament tube, exceeding gynostegium (except for style-head), erect, lanceolate, basally broadened; gynostegium concealed in corolla; pollinia ovoid; caudicles geniculate, broadened at insertion of pollinium; style-head rostrate. Follicles 3.5–5 cm long, obclavate. One seed per follicle, 10–12 mm long, ovate.

Two or three spp. in Namibia, Somalia, Tanzania, savanna and semi-deserts.

224. ***Telosma*** Coville

Telosma Coville, Contr. U.S. Natl. Herb. 9: 384 (1905); Li et al., Fl. China 16: 237–238 (1995), reg. rev.; Omlor, Gener. Rev. Marsdenieae: 136–137 (1998), consp.
Prageluria N.E. Br. (1907), nom. superfl.

Suffrutescent twiners or lianas, to 8 m tall; latex occasionally translucent. Leaves long-petiolate, 4–13 cm long, ovate, cordate. Inflorescences

15–30-flowered, pendulous. Flowers fragrant; corolla 10–20 mm long, salverform with basally urceolate tube, yellowish; throat bearded; lobes twisted, lanceolate or ovate, occasionally ciliate; free staminal corona lobes attached along the filament tube, differentiated into an erect, oblong or triangular distal lobule and a longer, inflexed, oblong to linear proximal lobule; gynostegium sessile, concealed in corolla tube; pollinia obovoid, with distal-basal sterile hyaline region; caudicles shortly rectangular; style-head conical to semi-globose Follicles 4.5–16 cm long, occasionally winged, with thick pericarp. Seeds winged; coma yellowish. 2n = 22.

Six spp. in Africa and Madagascar, widely distributed in India, S China, continental SE Asia and Malesia, open wood- and bushland.

225. ***Treutlera*** Hook. f.

Treutlera Hook. f., Hooker's Icon. Pl. 15: ad t.1425 (1883); Omlor, Gener. Rev. Marsdenieae: 130–131 (1998), consp.; Watson, Fl. Bhutan 2.2: 712–713 (1999), reg. rev.

Suffrutescent twiner, glabrescent. Leaves coriaceous, 7–13 cm long, elliptic, apiculate. Inflorescences 5–7-flowered, pendulous, long-pedunculate. Corolla ca. 4 mm long, purple; lobes triangular-ovate, spreading, ciliate; free staminal corona lobes brownish, slightly shorter than gynostegium, erect, oblong; anthers rectangular; pollinia oblongoid to reniform; caudicles straight, horizontal; corpusculum oblongoid, much smaller than pollinia; style-head conical.

One sp., *T. insignis* Hook.f., in India (Sikkim), Nepal; 2000–3000 m.

V.3. **Tribe Ceropegieae** Orb. (1843).

Herbs, vines, stem succulents or tuberous geophytes; latex translucent, occasionally white-cloudy; roots fibrous, rarely woody; often primary roots (occasionally incl. the hypocotyl) producing tubers or carnose secondary roots, fusiform. Leaves opposite or rarely whorled, herbaceous, but often reduced to minute caducous scales or persistent spines, occasionally absent (*Duvaliandra, Socotrella, White-sloanea*), sometimes with colleters at base of lamina (e.g., *Heterostemma*); stipules membranous or glandular, rarely reduced to few trichomes or absent. Inflorescences usually cymose. Calyx usually basally fused, rotate; flowers with fetid, carrion, dull-fruity or sweetish fragrances; corolla rotate, campanulate, cylindrical (e.g., *Ceropegia*) or urceolate (e.g., *Stapeliopsis*), occasionally with carnose ring around mouth of tube, glabrous or sculptured with emergences and/or papillae, and/or unicellular, mostly verrucose (occasionally vibratile) trichomes; corolla usually fused between a quarter and half of its length; lobes usually valvate, sometimes remaining fused at apices; corolline corona occasionally present in sinuses of corolla lobes (e.g., *Leptadenia*); anthers erect to horizontal, rarely with connective appendage (*Caudanthera, Heterostemma, Sisyranthus*), anthers inversely T-shaped, anther wings of adjacent anthers parallel to each other forming guide rails. Interstaminal ("outer") corona from 5 free, spreading lobes to partially or entirely fused into cup around column; staminal ("inner") corona usually of 5 free, ascending to erect lobes, occasionally divided in distal and proximal lobules, basally often fused with adjacent interstaminal corona, rarely with a third subtending corona series (coronal skirt, *Neoschumannia*); secondary nectaries sometimes on corona lobes; pollinia erect to horizontal, nearly spherical, ellipsoid to D-shaped in outline, dorsiventrally flattened, with hyaline germination crest along proximal (mostly) to apical margin through which pollen tubes emerge, baseo-laterally attached to the caudicles; corpusculum with flanks and floor and (usually) two spreading lateral projections with caudicles attached to the lower side; style-head broad, with flattened to concave apex, white or greenish white. Follicles paired, sometimes solitary by abortion, narrowly oblong to stoutly fusiform. Seeds ovate to pyriform, usually with wing-like margin and apically attached coma. x = 11. Ceropegieae and Marsdenieae are sister groups according to molecular analyses (e.g., Meve and Liede 2004b; Rapini et al. 2007). Characteristic secondary compounds are "Asclepiadaceae bitter principles" (glycosides). 46 genera in the Paleotropics, one extending to Australia, arid temperate and tropical climates.

Key to the Subtribes of Ceropegieae

1. Stems (sub-)succulent (if wiry, then at least peduncles, pedicels and roots fleshy) V.3.d. Stapeliinae
- Stems wiry or suffrutescent to woody (roots fibrous, occasionally secondary roots slightly fleshy in Anisotominae) 2
2. Leafy twiners (Asia only: India to Papua-New Guinea); gynostegial corona of five free or connate staminal corona lobes usually divided into an inner and a spreading outer part; anthers with membranous, reniform sterile apical appendages; two follicles reaching maturity per flower V.3.a. Heterostemminae
- Shrubs, herbs or twiners (Africa, Arabia, Asia); gynostegial corona absent or one-, two or tri-partite; staminal corona solid (rarely bilobed) anther appendage, if present, not reniform; one or two follicles reaching maturity per flower 3
3. Corolline corona lobes often present; gynostegial corona absent or a ring of fused staminal and interstaminal parts; one follicle per flower V.3.b. Leptadeniinae
- Corolline corona lobes absent (corolla sometimes with annulus); gynostegial corona of basally fused or not staminal lobes, with or without distinct interstaminal corona lobes; two follicles reaching maturity per flower (Africa) V.3.c. Anisotominae

Key to the Genera of Ceropegieae

1. Stems woody, suffrutescent or herbaceous, terete, erect to decumbent, scrambling or twining (occasionally stems carnose or (sub-)succulent, but then always with pitfall flowers); leaves usually well-developed 2
- Stems succulent, distinctly (3–)4- to many-angled; leaves reduced to scales or spines (except for the Indian *Boucerosia frerei*), caducous or spiny 11 (Stapeliads)
2. Large liana; gynostegial corona in three series **233. *Neoschumannia***
- Shrubs, herbs or twiners; gynostegial corona in one or two series, or absent 3
3. Ephemeral, erect annual herbs; corolline corona present, subulate; gynostegial corona absent **227. *Conomitra***
- Perennial herbs, twiners or shrubs, corolline corona, when present, usually not subulate, gynostegial corona present or absent 4
4. Erect shrubs or creeping subshrubs, shoots hollow; whole plant frequently densely whitish pubescent **229. *Orthanthera***
- Herbs, shrubs or twiners, shoots not hollow; plants not densely whitish pubescent; often pitfall flowers 5
5. Inflorescences lax, paniculate; rachis filiform, tough **234. *Riocreuxia***
- Inflorescences not lax or paniculate; rachis soft, when wiry then soft and not tough **243. *Ceropegia***
- Plants without tubers or fleshy roots (occasionally with slightly thickened, but rather tenacious secondary roots) 6
6. Staminal corona lobes differentiated into proximal and distal parts **226. *Heterostemma***
- Staminal corona lobes absent or simple 7
7. Corona, when present, formed of free staminal lobes; erect or prostrate herbs 8
- Corona, when present, formed of fused staminal and interstaminal parts; shrubs, twiners or herbs 9
8. Corolla fused only at the base, white; lobes lanceolate **230. *Pentasachme***
- Corolla fused for ca. half of its length, cream; lobes oblong or ovate **235. *Sisyranthus***
9. Gynostegial corona absent or of poorly differentiated fused staminal and interstaminal parts; shrubs or suffrutescent twiners **228. *Leptadenia***
- Gynostegial corona of fused staminal and interstaminal parts with clearly differentiated staminal lobes; delicate herbs or twiners, shoots wiry 10
10. Anthers apically glabrous; pollinia with proximal germination crest **232. *Emplectanthus***
- Anthers apically with trichomes; pollinia with apical germination crest **231. *Anisotoma***
11. Stems with well-developed succulent leaves **240. *Boucerosia* (*B. frerei*)**
- Stems with rudimentary leaves transformed into scales or spines, or leaves absent 12
12. Interstaminal corona with large, free and erect lobes 13
- Interstaminal corona different (when lobes large and erect then fused below to form basal pouches) 14
13. Staminal corona lobes long, subulate, erect; Namibia and South Africa **270. *Tromotriche***
- Staminal corona lobes short, triangular, inflexed; Madagascar **266. *Stapelianthus***
14. Scale-like leaves persistent, tough, becoming spiny 15

- Leaf rudiments inconspicuous, caducous or absent 18
15. Stems cereoid with many parallel angles, with sharp spines 16
- Stems 4- to 5-angled, with short, rigid (occasionally sharp) spines or with inconspicuous scale-like leaves 17
16. Spines 3-partite (a complex of leaf scale and stipules); stems soft, 5–14-angled, 1.5–2.5 cm diam. **268. *Tavaresia***
- Spines simple; stems rather firm, at least 10-angled, 2.5–6.0(–10.0) cm diam. **249. *Hoodia***
17. Plants decumbent to erect; flowers small, less than 2.5 cm in diameter; inflorescences lateral **261. *Quaqua***
- Plants prostrate-creeping; flowers very large, at least 7 cm in diameter; inflorescences basal **248. *Edithcolea***
18. Corolla with small projecting lobes in the sinuses of lobes (occasionally also present in *Tavaresia* and *Stapelianthus*); filament tube with humps in front of guide rail entrance **250. *Huernia***
- Corolla lobe sinuses not tipped; filament tube without humps at entrance of guide rails 19
19. Stems distinctly tessellate (with transverse furrow between the leaf bases) 20
- Stems not tessellate 26
20. Leaf bases irregularly arranged, polygonal, thus stems not angular; staminal corona lobes linear-spathulate 21
- Leaf bases regularly arranged, ovate-rectangular; stems 4- to 6-angled, staminal corona lobes triangular to linear 22
21. Scale-like leaves orbicular; peduncle caducous; top of style-head convex, thickish **236. *Anomalluma***
- Scale-like leaves absent or nearly so; peduncle long-lived; top of style-head flat **260. *Pseudolithos***
22. Scale-like leaves triangular to subulate, stems slender (to 1 cm diam.), prostrate-creeping; E Africa to Arabia **247. *Echidnopsis***
- Scale-like leaves absent or indistinct, stems > 1 cm diam., usually erect or ascending; southern Africa 23
23. Corolla with simple trichomes regularly distributed over surface 24
- Corolla glabrous or papillose or with emergences tipped with a short simple or clavate trichome 25
24. Stems creeping, green, 9–13 mm wide, 6–8-angled; scale-like leaves persistent, minute or absent **254. *Notechidnopsis***
- Stems erect, blue-green, 20–25 mm wide, 8-angled; scale-like leaves forming spines to 4 mm long **263. *Richtersveldia***
25. Stems 10–12-angled; leaf bases flat; scale-like leaves firm, acute **252. *Lavrania***
- Stems 12–20-angled; leaf bases conical; scale-like leaves thickish, sunken in a groove **251. *Larryleachia***
26. Gynostegial corona only in staminal position, interstaminal part reduced to fringe or minute flap 27
- Gynostegial corona biseriate, interstaminal corona well-developed 30
27. Stems 4-angled, without scale-like leaves and stipules 28
- Stems 4–5- (to 6-)angled, with scale-like leaves and stipular glands 29
28. Stems thick (10–15 mm diam.); flowers rotate; gynostegium sessile **246. *Duvaliandra***
- Stems thin (4–7 mm diam.); flowers campanulate; gynostegium and corona atop a column **264. *Socotrella***
29. Stems blue-green, 3–10 mm wide; interstaminal corona conspicuous, spreading, bifid **238. *Australluma***
- Stems green (occasionally brownish green), 8–20 mm wide; interstaminal corona inconspicuous **259. *Piaranthus***
30. Stems rugose and papillose 31
- Stems not rugose or papillose 32
31. Scale-like leaves erect, papillate; inflorescences sunken into dented stem surface **262. *Rhytidocaulon***
- Scale-like leaves spreading to slightly ascending, smooth; inflorescences not sunken **239. *Baynesia***
32. Corolla urceolate, corolla lobes at the most as long as corolla tube 33
- Corolla usually not urceolate, corolla lobes at least as long as corolla tube 34
33. Stems 4-angled; translator not winged **267. *Stapeliopsis***
- Stems 5- to many-angled; translator with wing-like projections **250. *Huernia***
34. Stems 4–9 mm diam., indistinctly 4-angled, creeping 35
- Stems > 8 mm diam. (when less, then ascending to erect), 4- to 6-angled 36
35. Gynostegial corona glabrous, staminal corona abaxially with spreading, pectinate or warty hump **255. *Ophionella***
- Gynostegial corona with trichomes, staminal corona lobes simple, solid, erect **257. *Orbeanthus***

36. Stems 4-, 5- or 6-angled, leaf bases bulging considerably 37
- Stems strictly 4-angled, leaf bases bulging only weakly 42

37. Stems 6-angled; translator without basal projections **258. *Pectinaria***
- Stems 4–5-angled; translator with wing-like basal projections 38

38. Stipules reduced to few multicellular trichomes **269. *Tridentea***
- Stipules glandular or absent 39

39. Stems 4-angled, cylindrical; scale-like leaves deltoid (occasionally absent) **270. *Tromotriche***
- Stems 4–5-angled, cylindrical or clavate; scale-like leaves triangular to subulate 40

40. Filament tube with a little hump in front of entrance of nectarial orifices **250. *Huernia***
- Filament tube without a hump 41

41. Guide rails oblique, embedded in gynostegial tissue **245. *Duvalia***
- Guide rails vertical, not embedded in gynostegial tissue **256. *Orbea***

42. Stems unbranched, leafless, short and clumpy, cuboid to ovoid; inflorescences along (basal) angles of stem **271. *White-sloanea***
- Stems branched, usually with scale-like leaves, less clumpy; inflorescences terminal or on upper half of stem 43

43. Inflorescences pedunculate, mostly arising basally; flowers large, rotate to campanulate; pollinaria large, translator with wing-like basal projections 44
- Inflorescences sessile, arising from apical regions of the stems, lateral or terminal, often many-flowered clusters (sciadioids); flowers rather small, of various shapes; pollinaria small, translator without basal projections 45

44. Stems with pungent scent; scale-like leaves ciliate, with tufts of (stipular) glandular trichomes, or when stems not pungent leaf bases grooved on upper side and without stipular trichomes **253. *Monolluma***
- Stems not with pungent scent; leaf bases not grooved; scale-like leaves ascending-erect, not ciliate but occasionally pubescent, stipules glandular, globose **265. *Stapelia***

45. Inflorescences in terminal sciadioids 46
- Inflorescences not in terminal sciadioids 47

46. Scale-like leaves (broadly) lanceolate, ciliate, stipules absent **240. *Boucerosia***
- Scale-like leaves absent or minute, suborbicular, cordate to lanceolate, rarely ciliate (rarely acute and persistently spiny as in *Desmidorchis foetida* and *D. speciosa*), with scattered (stipular) glandular trichomes **244. *Desmidorchis***

47. Stipular glands present 48
- Stipular glands absent, usually with (tufts of) stipular trichomes instead 49

48. Scale-like leaves 0.5–2 mm long; stems not tapering **238. *Australluma***
- Scale-like leaves 4–12 mm long; stems tapering **242. *Caudanthera***

49. Scale-like leaves subsessile to petiolate, with isolated stipular trichomes only **237. *Apteranthes***
- Scale-like leaves sessile, stipules reduced to tufts of stipular trichomes **241. *Caralluma***

V.3.a. **Subtribe Heterostemminae** Meve & Liede (2004).

Leafy twiners, stems wiry or suffrutescent to woody, latex white or translucent, roots fibrous, gynostegial corona of five free or basally connate staminal corona lobes usually differentiated into a proximal and a spreading to erect distal lobule; anthers with membranous, reniform sterile apical appendages; corpusculum without basal projections; two follicles reaching maturity per flower. Meve and Liede (2004b) and Meve et al. (2017) retrieved Heterostemminae as sister to the remaining Ceropegieae. One genus, India to Australasia.

226. ***Heterostemma*** Wight & Arn.

Heterostemma Wight & Arn. in Wight, Contr. Bot. India 42 (1834); Forster, Austral. Syst. Bot. 5: 71–80 (1992), reg. rev.; Swarupanandan et al., Bot. J. Linn. Soc. 191: 249–259 (1989), rev.; Jagtap & Singh, Fasc. Fl. India 24: 244–258 (1999), reg. rev.; Rodda, Phytotaxa 263: 1–17 (2016), nom.
Symphysocarpus Hassk. (1857).
Oianthus Benth. (1876).
Dittoceras Hook. f. (1883).

Herbaceous to suffrutescent twiners, 1–5 m long; shoots occasionally with trichomes in two lines. Leaves ovate, indistinctly cordate, basally with colleters. Inflorescences pedunculate. Corolla 8–10 mm long, rotate, campanulate or urceolate, yellowish or purple, occasionally maculate, occasionally with carnose ring; lobes triangular; gynostegial corona yellow to green, distal lobules of staminal parts oblongoid, basally often spurred or winged; proximal lobules inflexed, lingulate or dentiform; pollinia ellipsoid, germination crest subapical; caudicles straight or geniculate, cylindrical to spathulate. Follicles rarely fused, 5–15 cm long, fusiform to narrowly oblong. Seeds broadly winged. 2n = 22.

About 30 spp. in Asia and Australia, one in New Caledonia, wet and evergreen forests.

V.3.b. **Subtribe Leptadeniinae** Meve & Liede (2004); Meve et al., Syst. Biodivers. 15: 143–155 (2017).

Shrubs, herbs or twiners, stems wiry or suffrutescent to woody, roots fibrous; corolline corona lobes often present; gynostegial corona absent or a ring of fused staminal and interstaminal parts; pollinia basally or laterally attached to the caudicles; germination crest (sub-)apical. Follicles usually solitary, fusiform or narrowly oblong. Seeds ovate or oblong.

Four genera in Africa, Arabia and Asia.

227. ***Conomitra*** Fenzl

Conomitra Fenzl in Endlicher & Fenzl, Nov. Stirp. Dec. 65 (1839); Field, Kew Bull. 37: 341–347 (1982); Goyder, Fl. Ethiopia and Eritrea 4.1: 156–157 (2003), reg. rev.; Goyder, Fl. Somalia 3: 168 (2006), reg. rev.

Ephemeral, erect herb, 6–35 cm tall, unbranched; shoots papillose. Leaves 1–7 cm long, linear, papillose on midrib and toward margins. Inflorescences 1-flowered, subsessile. Calyx reaching about half of corolla length, campanulate, ciliate; corolla 2–3 mm long, campanulate, green; lobes triangular, extended into a long tip; corolline corona shorter than corolla; lobes antesepalous, basally connate, subulate; gynostegial corona absent; gynostegium sessile; connective appendages minute, subulate, erect; pollinia ovoid; caudicles trapezoidal; style-head elongate-conical. Follicles occasionally paired, 5–6 cm long.

One sp., *C. linearis* Fenzl, widespread across the Sahel (Ethiopia, Kenya, Niger, Somalia, Sudan), but rarely collected, ephemeral in *Acacia-Commiphora* bushland; 200–900 m.

228. ***Leptadenia*** R. Br.

Leptadenia R. Br., Asclepiadeae 23 (1810); Jagtap & Singh, Fasc. Fl. India 24: 258–262 (1999), reg. rev.; Chaudhary, Fl. Kingd. Saudi Arabia 2.2: 41–42 (2001), reg. rev.; Goyder, Fl. Ethiopia and Eritrea 4.1: 155–156 (2003), reg. rev.; Goyder, Fl. Somalia 3: 166–168 (2006), reg. rev.; Masrahi, Saudi J. Biol. Sci. 22: 31–36 (2015), new taxon; Meve et al., Syst. Biodivers. 15: 143–155 (2017), phyl.

Shrubs or suffrutescent twiners, 0.3–2(–5) m tall, glabrous to pubescent; shoots often suberose or finely striate. Leaves occasionally caducous and scale-like, 0.5–10 cm long, elliptic, lanceolate, ovate or hastiform; stipules filiform. Inflorescences 10–20-flowered; flowers sweetly fragrant. Corolla 3–4 mm long, rotate, creamish, with erect, verrucose trichomes, with ridges or five hairy pouches; lobes lanceolate, spreading, ciliate; corolline corona five free, short, creamish, deltoid lobes, dorsally adnate to anthers; gynostegial corona absent or annular, ivory; gynostegium sessile; connective appendages absent; pollinia ovoid; caudicles curved, trapezoidal. Follicles pendulous, 5–11 cm long. 2n = 22.

Six spp. in Africa, Arabia and Asia, (semi-) arid areas, *Acacia-Commiphora* bushland, dry forests.

229. ***Orthanthera*** Wight

Orthanthera Wight, Contr. Bot. India 48 (1834); Huber, Prodr. Fl. Südwestafr. 114: 46–47 (1967), reg. rev.; Jagtap & Singh, Fasc. Fl. India 24: 263–265 (1999), reg. rev.
Barrowia Decne. (1844).

Erect shrubs or creeping subshrubs, 20–150 cm tall, whitish pubescent to tomentose; shoots hollow, greenish. Leaves occasionally caducous, scale-like or absent, 0.1–2 cm long, linear, oblong or ovate, acute; stipules filiform. Inflorescences 3–10-flowered. Corolla 5–15 mm long, tubular, whitish or greenish, glabrous; lobes lanceolate, spreading to erect; corolline corona, when present, short, five-partite, adnate to corolla; lobes free, antepetalous, oblongoid, erect; gynostegial corona a ring of deltoid, erect to spreading staminal and fringe-like interstaminal parts, carnose; gynostegium concealed in corolla; connective appendages, when present, short; pollinia ellipsoid; caudicles horizontal and S-shaped. Follicles 5–12 cm long. 2n = 22.

Five spp. in Africa and Asia. *O. albida* Schinz is moth-pollinated (Nel 1995).

230. ***Pentasachme*** Wall. ex Wight

Pentasachme Wall. ex Wight, Contr. Bot. India 60 (1834); Rahman & Wilcock, Blumea 36: 109–121 (1991), reg. rev.; Jagtap & Singh, Fasc. Fl. India 24: 265–268 (1999), reg. rev.
Spiladocorys Ridl. (1893).
Vietnamia P.T. Li (1994).

Erect or prostrate rheophytic herbs, occasionally pendent, 10–70 cm tall. Leaves 6–13 cm long, linear to ovate, acuminate, occasionally ciliate. Inflorescences 1-5-flowered. Corolla 5–30 mm long, campanulate, white, forming bulges at sinuses; lobes contorted, lanceolate; corolline corona, when present, inserted in the tube; lobes free, short, antesepalous, subulate, corniculate or knobby, erect; gynostegial corona, when present, of short, lanceolate free staminal lobes; gynostegium subsessile or atop a column; anthers occasionally gibbous, erect; connective appendages lanceolate or deltate, strongly inflexed; pollinia ellipsoid, obovoid or pyriform; caudicles S-shaped, trapezoidal. Follicles 5–6 cm long.

Five spp. in S Asia, continental SE Asia and W Malesia. Reminiscent of non-twining *Vincetoxicum* Wolf in habit.

V.3.c. **Subtribe Anisotominae** Meve & Liede (2004); Meve et al., Syst. Biodivers. 15: 143–155 (2017).

Shrubs, herbs or twiners, stems wiry or suffrutescent to woody, roots fibrous or secondary roots slightly carnose; corolline corona lobes absent (corolla sometimes with ring); gynostegial corona of staminal lobes, basally fused or free, with or without distinct interstaminal corona lobes; connective appendages usually absent; pollinia usually ovoid; corpusculum usually without basal projections. Follicles usually paired, fusiform or narrowly oblong. Seeds ovate or oblong. Five genera in Africa.

231. ***Anisotoma*** Fenzl

Anisotoma Fenzl, Linnaea 17: 330 (1844); Meve et al., Syst. Biodivers. 15: 143–155 (2017), phyl., new taxon.
Anisotomaria C. Presl (1844), nom. illegit.
Lophostephus Harv. (1863), nom. illegit.
Aulostephanus Schltr. (1896).

Creeping or erect herbs; with fleshy lateral roots; shoots wiry, pilose in single line. Leaves 1–5 cm long, ovate to circular, cordate, pilose. Inflorescences 2–10-flowered; floral bracts conspicuous. Corolla 3–5 mm long, rotate, yellow, cream with brown center or vice versa, glabrous or pilose; lobes sinistrorsely contorted, triangular; gynostegial corona ivory, campanulate, of inconspicuous ovoid-rectangular, erect staminal parts or of long, erect, filiform adaxial appendages and interstaminal parts basally fused into a ring; gynostegium subsessile, anthers thickish, apically with trichomes; pollinium germination crest apical; caudicles cylindrical. Follicles 8–10 cm long.

Three spp. in South Africa.

232. ***Emplectanthus*** N.E. Br.

Emplectanthus N.E. Br. in Thiselton-Dyer, Fl. Cap. 4(1): 771 (1908); Meve, Bot. Jahrb. Syst. 120: 123–130 (1998), rev.; Styles, Plantlife 39&40: 24–32 (2010), new taxon; Meve et al., Syst. Biodivers. 15: 143–155 (2017), phyl.

Delicate, suffrutescent twiners to 1–2(–4) m tall, sparsely branched, with fleshy lateral roots; shoots wiry, with stiff trichomes in a single line. Leaves horizontally spreading, 5–9 cm long, ovate, cordate; stipules filiform. Inflorescences 3–10-flowered, lax. Corolla 2.5–3.5 mm long, shallowly campanulate, membranous, purple or creamish, spotted pinkish; mouth of tube with ring of trichomes; lobes contorted, triangular, acuminate; gynostegial corona brownish, shallowly cyathiform, of deltate, spathulate or strap-like, erect, long staminal parts and saccate, rarely bilobed, spreading to reflexed interstaminal parts basally fused into a ring; pollinia occasionally D-shaped; caudicles cylindrical. Follicles solitary, 4–7 cm long. 2n = 22.

Three spp. in South Africa, gorges, riverine woodlands.

233. ***Neoschumannia*** Schltr.

Neoschumannia Schltr., Bot. Jahrb. Syst. 38: 38 (1905); Meve, Pl. Syst. Evol. 197: 233–242 (1995), rev.; Meve & Liede, Taxon 53: 61–72 (2004), phyl.; Fischer et al., Phytotaxa 77 (2): 19–26 (2013), new taxon.
Swynnertonia S. Moore (1908).

Lianas, 2–10 m long, roots fibrous. Leaves discolorous, 5–15 cm long, (oblong-)elliptic, acuminate; stipules glandular, ovoid. Inflorescences 3–5-flowered, lax, pedunculate; rachis persistent, thick; flowers nodding from long filiform pedicels, not nectariferous. Corolla 15–20 mm long, apopetalous, membranous, cream with purplish tips or green; tube with clavate, purple trichomes; lobes lanceolate, spreading, occasionally ciliate; gynostegial corona ivory or green, in three series: a basally fused ring of staminal and interstaminal parts extended into a 5-lobed, glabrous skirt; free staminal lobes oblong(-obovate), occasionally bifid, erect, with trichomes; interstaminal lobes unguiculate, spreading-ascending, canaliculate, apically bidentate, with trichomes; gynostegium atop a column; caudicles rectangular. Follicles pendulous, 25–35 cm long. 2n = 22.

Three spp., one each in W, C and E Africa, evergreen forests.

234. ***Riocreuxia*** Decne.

Riocreuxia Decne., Prodr. 8: 640 (1844); Dyer, Fl. Southern Afr. 27.4: 83–88 (1980), reg. rev.; Masinde, Kew Bull. 60: 401–434 (2005), rev.; Meve et al., Syst. Biodivers. 15: 143–155 (2017), phyl.

Erect or twining herbs, 1–3 m tall; rootstock woody or roots slightly fusiform; shoots puberulous. Leaves 1.5–10 cm long, elliptic or ovate, cordate; stipules as pilose interpetiolar line. Inflorescences 10–40-flowered, lax, pedunculate. Corolla 10–30 mm long, elongate-conical (pitfall flower), rarely campanulate, (whitish) green or (orange-)yellow, occasionally streaked white; lobes linear or lanceolate, usually remaining connate at apices, smooth; gynostegial corona of free or basally fused staminal lobes, ivory, erect, ovate, basally with wings fused to interstaminal projections; gynostegium subsessile, concealed in corolla; anther wings basally forming a distinct mouth; pollinia rarely D-shaped. Follicles pendulous, 8–17(–35) cm long, occasionally moniliform. 2n = 22.

Eight spp. in E and SE Africa.

235. ***Sisyranthus*** E. Mey.

Sisyranthus E. Mey., Comm. Pl. Afr. Austr. 197 (1838); Ollerton et al., Ann. Bot. 92: 807–834 (2003), ecol.

Erect, perennial herbs to 25 cm tall, sparsely branched, occasionally densely pubescent; roots thinly fusiform. Leaves linear. Inflorescences terminal or extra-axillary, 2–20-flowered, lax. Corolla 4–10 mm long, campanulate or urceolate, cream, occasionally barbate, occasionally with five adaxial ridges; lobes valvate or imbricate, oblong or ovate, spreading; gynostegial corona of free staminal lobes, adnate to both corolla and gynostegium, carnose, ovoid, occasionally emarginate, basally broadened; gynostegium sessile, concealed in corolla; anthers apically occasionally bearded or with a few trichomes; connective appendages, when present, oblong, erect, occasionally with a few hairs. Follicles horizontal, 7–9 cm long. Seeds occasionally elliptical. 2n = 22.

13 spp. in South Africa. *S. trichostomus* K. Schum. pollinated by Coleoptera (Ollerton et al. 2003).

V.3.d. **Subtribe Stapeliinae** G. Don (1837).

Stems (sub-)succulent (when wiry, then at least peduncles, pedicels and roots carnose), mostly glabrous. Leaves often reduced to sessile, caducous scales; often supported by expanded leaf bases (podaria). Inflorescences usually (sub-)sessile. Corolla usually rotate to campanulate, often carnose, rugose or papillose. Corolla lobes usually spreading to reflexed; gynostegial corona usually carnose, rotate or cyathiform; pollinia ovoid-ellipsoid or D-shaped, corpusculum usually with two basal projections. Follicles usually paired, narrowly oblong. Seeds ovate or pyriform. Fly pollinated. In a molecular analysis (Meve and Liede 2004b), Stapeliinae were sister to Anisotominae. 36 genera in arid areas of Africa, Arabia and Asia.

236. ***Anomalluma*** Plowes

Anomalluma Plowes, Cact. Succ. J. (Los Angeles) 65: 167 (1993); Meve & Liede, Pl. Syst. Evol. 234: 172–209 (2002), rev.

Clump-forming stem succulents to 5 cm tall, often with subterranean shoots, pale green, 10–60 cm long, 4–10 mm wide, sharply 4-angled, irregularly tuberculate or tessellate. Scale-like leaves suborbicular, stipules absent. Inflorescences one to several, apical, 3-flowered, peduncles and pedicels deciduous; flowers with sourish odor. Corolla 3–4 mm long, maroon; lobes triangular; gynostegial corona purplish, of lingulate, inflexed staminal parts and reflexed, bilobed interstaminal parts basally fused into a ring; pollinia D-shaped; caudicles medifixed to the corpusculum. Follicles occasionally solitary, 2–4 cm long, seeds with thick, puffed wing. 2n = 22.

Two spp. in Somalia and Oman.

237. *Apteranthes* J.C. Mikan

Apteranthes J.C. Mikan, Nova Acta Acad. Caes. Leop.-Carol. Nat. 17: 594, t. 41. (1835); Plowes, Haseltonia 3: 63 (1995), rev.; Meve & Liede, Pl. Syst. Evol. 234: 172–209 (2002), phyl.; Meve & Heneidak, Bot. J. Linn. Soc. 149: 419–432 (2005), chem.; Formisano et al., Molecules 14: 4597–4613 (2009), chem. ecol., poll.
Borealluma Plowes (1995).

Clump-forming or creeping stem succulents to 40 cm tall; usually rhizomatous; shoots (blue)-green, 5–70 cm long, 10–25 mm wide, 4-angled. Scale-like leaves reflexed or slightly ascending, basally cordate, apically obtuse; stipules absent. Inflorescences apical, 3–15-flowered; flowers usually fetid. Corolla 6–16 mm long, creamish or purple, often patterned, glabrous, papillose or hirsute; lobes lanceolate or triangular; gynostegial corona yellow or purplish red, of inflexed, subulate or triangular staminal parts and subulate or bilobed interstaminal parts shortly fused into a ring; gynostegium sessile; pollinia D-shaped; caudicles spathulate. Follicles 50–130 mm long. 2n = 22.

Six spp. in Africa, Arabia (incl. Sinai), Europe and Asia.

238. *Australluma* Plowes

Australluma Plowes, Haseltonia 3: 54 (1995); Bruyns, Excelsa 10: 107–111 (1982), morph.; Bruyns, Stapeliads of S. Afr. Madag. 1: 61–65 (2005), rev.

Erect stem succulents to 30 cm tall, rhizomatous; shoots often annual, blue-green, 20–30 cm long, 3–5 mm wide, with 4 rounded angles. Scale-like leaves ascending, ovate; stipules glandular, ovate. Inflorescences several along stems, 1–2-flowered; flowers unscented. Corolla 4–6 mm long, adaxially greenish, purple-maculate, with verrucose trichomes; lobes triangular; gynostegial corona brownish, of trianguloid, spreading, apically reflexed, emarginate staminal parts with inflexed subulate-triangular tooth-like projection and interstaminal parts basally fused into a ring; gynostegium sessile; pollinia broadly ellipsoid, germination crest apical; caudicles rectangular. Follicles obtuse-angled. 2n = 22.

Two spp., one in Namibia (summer rainfall areas), and one in South Africa.

239. *Baynesia* Bruyns

Baynesia Bruyns, Novon 10: 354–358 (2000); Bruyns, Stapeliads of S. Afr. Madag. 1: 66–67 (2005), consp.; Bruyns et al., Mol. Phyl. Evol. 77: 251–263 (2014), phyl.

Creeping stem succulent, 5 cm tall; shoots blue-green, 3–8 cm long, 6–12 mm wide, with 4 rounded angles, papillose. Scale-like leaves persistent, slightly ascending, succulent, ovate; occasionally with stipules. Inflorescences along stems, 1–3(–5)-flowered, corolla 3–4 mm long, reddish(-maroon) with cream center; lobes ovate, acute, keeled; gynostegial corona cream, of ovoid, basally winged, apically inflexed staminal parts longer and thicker than the deltoid-triangular, spreading, purplish interstaminal parts, basally fused into a ring; pollinia ellipsoid; caudicles flattened. Follicles 2.5–3.5 cm long. 2n = 22.

One sp., *B. lophophora* Bruyns, in N Namibia (Baynes Mountains). *Baynesia* was retrieved as sister to the similarly small-stemmed Cape Stapeliad *Ophionella* (Bruyns et al. 2014).

240. *Boucerosia* Wight & Arn.

Boucerosia Wight & Arn. in Wight, Contr. Bot. India 34 (1834); Meve & Liede, Pl. Syst. Evol. 234(3&4): 172–209 (2002), phyl. & tax.
Hutchinia Wight & Arn. (1834).
Frerea Dalz. (1864).

Clump-forming or creeping stem succulents, 2–12 cm tall; shoots green, 3–20 cm long, 3–20 mm wide, 4-angled, smooth. Leaves as succulent scales to 0.3 cm long or with well-developed blades, 2–3 cm long, oblong, ciliate; stipules absent or glandular-ovoid. Inflorescences terminal, (1–)3–60-flowered; flowers fetid, not nectariferous. Corolla 5–15 mm long, whitish to yellowish, often banded brownish, glabrous, papillose or with trichomes; lobes triangular, occasionally ciliate; gynostegial corona purplish red, occasionally campanulate, of triangular or rectangular, inflexed staminal parts and ascending, bilobulate interstaminal parts fused into a ring for more than half of total length; gynostegium sessile; pollinia D-shaped; caudicles spathulate. Follicles 6.5–10(15) cm long. 2n = 22, 44.

Seven spp. in India, Sri Lanka, Nepal and Myanmar.

The inclusion of the sole Stapeliad bearing true leaves, *Boucerosia frerei* (G.D. Rowley) Meve & Liede (= *Frerea indica* Dalzell), is not always accepted.

241. ***Caralluma*** R. Br.

Caralluma R. Br., Asclepiadeae 14 (1810); Gilbert, Bradleya 8: 1–32 (1990), rev.; Plowes, Haseltonia 3: 49–70 (1995); Meve & Liede, Pl. Syst. Evol. 234: 171–209 (2002), phyl.; Lavranos, Fl. Somalia 3: 174–178 (2006), reg. rev.; Bruyns et al., Taxon 59: 2031–1043 (2010), phyl.
Spathulopetalum Chiov. (1912).
Saurolluma Plowes (1995).
Somalluma Plowes (1995).

Erect stem succulents, 10–70 cm tall; shoots (light) green, blue-green or light brown, 10–70 cm long, 5–25 mm wide, 4-angled. Scale-like leaves ovate; stipules reduced to a few trichomes. Inflorescences along tapering stems, 1–4-flowered. Corolla 5–15 mm long, occasionally apopetalous, white, green, brown or purple, occasionally patterned; lobes linear, lanceolate or obovate (often replicate along midrib), rarely apically connate, often ciliate; gynostegial corona membranous, brownish or purplish red, of ovoid or trianguloid, erect or inflexed staminal parts and subulate to deltoid, often bilobed interstaminal parts fused into a ring; gynostegium often atop a column. Follicles 8–15 cm long. 2n = 22.

23 spp. in Africa, Arabia and Asia.

242. ***Caudanthera*** Plowes

Caudanthera Plowes, Haseltonia 3: 58 (1995); Meve & Liede, Pl. Syst. Evol. 234: 171–209 (2002), phyl.; Meve & Heneidak, Bot. J. Linn. Soc. 149: 419–432 (2005), chem.
Cryptolluma Plowes (1995).
Spiralluma Plowes (1995).

Erect stem succulents, 10–60 cm tall; occasionally rhizomatous; shoots blue-green, 5–30 cm long, 5–10 mm wide, with 4 rounded angles. Scale-like leaves slightly ascending, ovate, acute; stipules glandular, ovate. Inflorescences along tapering stems, 1–5-flowered. Corolla 4–8 mm long, creamish, occasionally with red spots and trichomes; lobes lanceolate, acute; gynostegial corona yellowish, of subulate or triangular staminal parts and saccate, deltoid or bilobed interstaminal parts fused into a ring; gynostegium occasionally atop a column; occasionally with lanceolate, long, erect connective appendages; pollinia with proximal to apical germination crest; caudicles ribbon-shaped; corpusculum without basal projections. Follicles 5–6 cm long. 2n = 22.

Three spp. in Africa, Arabia (incl. Sinai) and Asia.

243. ***Ceropegia*** L. Fig. 50

Ceropegia L., Sp. Pl. 211 (1753); Huber, Mem. Soc. Brot. 12: 1–203 (1957); Dyer, Fl. South. Afr. 27.4: 1–82 (1980), reg. rev.; Ansari, Fasc. Fl. India 16 (1984), reg. rev.; Jagtap & Singh, Fasc. Fl. India 24: 178–190 (1999), reg. rev.; Meve, Fl. Kingd. Saudi Arabia 2.2: 42–48 (2001), reg. rev.; Meve & Liede, Pl. Syst. Evol. 228: 89–105 (2001), phyl.; Meve, III. Handb. Succ. Pl.: 20–46, 63–107 (2002), rev.; Gilbert, Fl. Ethiopia and Eritrea 4.1: 158–169 (2003), reg. rev.; Gilbert, Fl. Somalia 3: 168–173 (2006), reg. rev.; de Kock, Checklist *Brachystelma*, *Ceropegia* & Stapeliads: 83–113 (2007), checkl.; Meve & Liede-Schumann, Ann. Missouri Bot. Gard. 94: 392–406 (2007), phyl.; Masinde, Kew Bull. 62: 37–84 (2007), reg. rev., phyl.; Surveswaran et al., Pl. Syst. Evol. 281: 51–63 (2009), phyl.; Bruyns et al., Mol. Phyl. Evol. 90: 49–66 (2015), phyl.
Niota Adans. (1763), nom. illegit.
Apegia Necker (1790), nom. inval.
Microstemma R. Br. (1810), nom. rej. vs. *Brachystelma* Sims (1822).
Brachystelma Sims (1822).
Tenaris E. Mey. (1838).
Eriopetalum Wight (1843).
Macropetalum Burch. ex Decne. (1844).
Dichaelia Harv. (1868).
Micraster Harv. (1868).
Lasiostelma Benth. (1876).
Tapeinostelma Schltr. (1893).
Brachystelmaria Schltr. (1895).
Blepharanthera Schltr. (1913).
Kinepetalum Schltr. (1913).
Siphonostelma Schltr. (1913).

Herbs, lianas or stem succulents, up to 10 m tall; roots often tuberous or fleshy fusiform; shoots 1.5–12 mm wide, terete or 4–(rarely 6)-angled. Leaves occasionally scale-like, linear to ovate, with stipules; inflorescences to 25-flowered, occasionally pedunculate; flowers often with dull-fruity odor or malodorous. Corolla 3–100 mm long, elongate-conical, with basal inflation (ostiolum), tube usually bent and constricted in the middle, highly fused, lobes usually remaining connate at apices, often keeled and with trichomes (lantern-shaped pitfall flower), or radiate

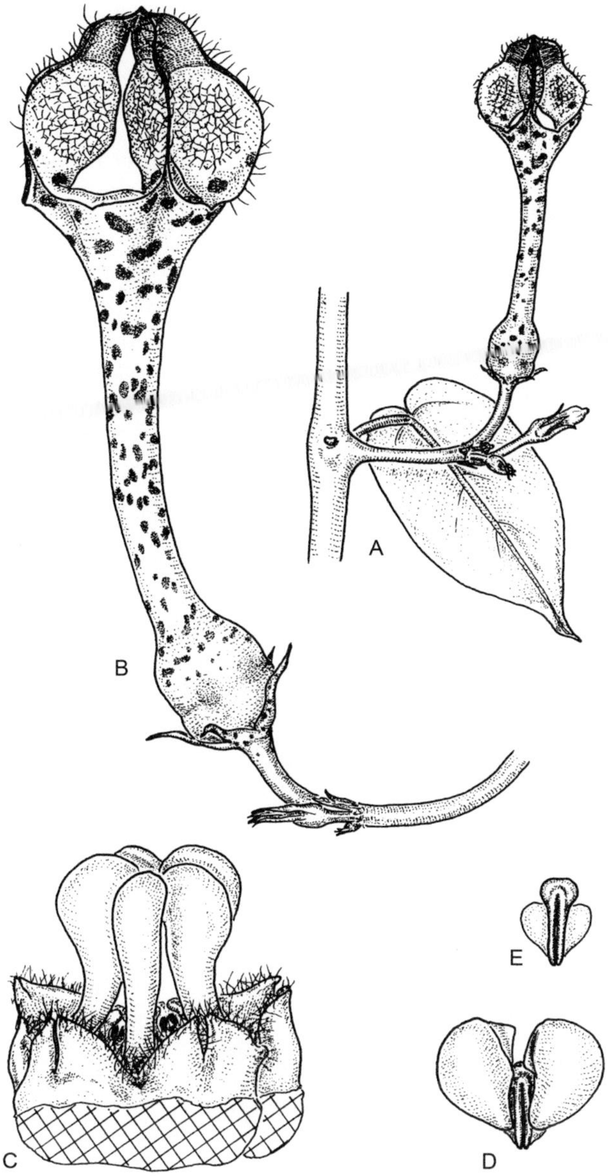

Fig. 50. Apocynaceae-Ceropegieae. *Ceropegia zambesiaca*. **A** Flowering branch (one leaf removed). **B** Flower in lateral view. **C** Corona in lateral view. **D** Pollinarium. **E** Corpuscle with lateral projections. (From Masinde and Meve 2002, p. 207, with permission from Kew Bulletin © Board of Trustees of the Royal Bot. Gard., Kew; drawn by U. Meve)

with lobes reflexed, spreading or connate at apices (*Brachystelma* flower type), occasionally urceolate; variably colored and patterned; gynostegial corona of lingulate, ovoid, deltoid, subulate to filiform staminal parts and occasionally reduced deltoid to subulate, often bilobed, spreading interstaminal parts fused into a ring; gynostegium occasionally atop a column; pollinia subquadrate, globose, ovoid or pyriform. Follicles 3–40 cm long, occasionally obclavate. 2n = 22, 44.

About 330 spp. in Africa, Arabia, Asia and Australia. Horticulturally one of the most popular Asclepiad genera. Tubers occasionally and locally eaten.

Ceropegia has been found to be paraphyletic without *Brachystelma* and without the Stapeliads (Meve and Liede-Schumann 2007; Surveswaran et al. 2009; Bruyns et al. 2015). While the inclusion of *Brachystelma* seems to be justified, considering growth form/habit and the multiple switches between pitfall and open flowers, the suggestion by Bruyns et al. (2017) to include all Stapeliinae in *Ceropegia* is not followed here.

244. ***Desmidorchis*** Ehrenb.

Desmidorchis Ehrenb., Abh. Königl. Akad. Wiss. Berlin 1829: 31, 39 (1832); Lavranos, Fl. Kingd. Saudi Arabia 2.2: 50–53 (2001), reg. rev.; Meve & Liede, Pl. Syst. Evol. 234: 171–209 (2002), phyl.
Sarcocodon N.E. Br. (1878).
Crenulluma Plowes (1995).

Erect or clump-forming stem succulents to 1.2 m tall; shoots light (blue-)green or brown, 20–80 cm long, 10–50 mm wide, sharply 4-angled, smooth. Scale-like leaves spreading, carnose or forming spines, ovate; stipules reduced to a few trichomes. Inflorescences terminal, 10–80(–200)-flowered; flowers with dung-like odor, not nectariferous. Calyx aposepalous; corolla 1.5–3 cm long, yellowish or purplish, often maculate, often sculptured or with trichomes; lobes triangular, occasionally ciliate; gynostegial corona a ring of triangular to subulate, inflexed staminal parts and subulate, bifid, spreading, interstaminal parts, occasionally with trichomes; gynostegium (sub-)sessile, concealed in corolla; caudicles rectangular. Follicles 10–15 cm long. 2n = 22.

15 spp. in N and E Africa and Arabia.

The genus as circumscribed by Meve and Liede (2002a) is not generally accepted, leading Bruyns (2010) to describe a new species under *Caralluma*.

245. ***Duvalia*** Haw.

Duvalia Haw., Syn. Pl. Succ. 44 (1812); Meve, Pl. Syst. Evol. Suppl. 10: 1–133 (1997), rev.; Bruyns, Stapeliads of S.

Afr. Madag. 1: 68–91 (2005), reg. rev.; Lavranos, Fl. Somalia 3: 191–193 (2006), reg. rev.
Ballyanthus Bruyns (2001).

Creeping stem succulents, 2–4 cm tall; shoots (blue-)green, occasionally spotted, conical or club-shaped, 1–10 cm long, 10–25 mm wide, with 4–6 rounded angles. Scale-like leaves succulent, deltoid to subulate; stipules glandular, globose. Inflorescences usually basal, 1–20-flowered, occasionally pedunculate; flowers often fetid, not nectariferous. Corolla 5–25 mm long, cream, yellow or reddish brown, occasionally patterned, occasionally papillose, with central, often trichome-bearing ring; lobes triangular, replicate along midrib; gynostegial corona a flattened ring of staminal and interstaminal parts covering the ring, and free oblongoid, cucullate or subulate staminal lobes with tooth-like projection incumbent on anther; gynostegium atop a column; caudicles spathulate; basal projections of corpusculum wing-like. Follicles 6–18 cm long. 2n = 22, 44 or 66.

17 spp. in two sections; S and NE Africa and SW Arabia.

246. ***Duvaliandra*** M.G. Gilbert

Duvaliandra M.G. Gilbert, Cact. Succ. J. Gr. Brit. 42: 101 (1980); Meve & Liede, Pl. Syst. Evol. 234: 171–209 (2002), phyl.

Clump-forming stem succulent, to 6 cm tall; shoots light greenish brown, 2–10 cm long, 10–15 mm wide, roundly 4-angled. Leaves absent. Inflorescences on flanks of lower half of stems, 1–5-flowered, sessile; flowers with carrion odor. Corolla 9–28 mm long, pale yellow-green or light brown, adaxially rose or reddish, with purple-brown trichomes; lobes ovate, spreading; gynostegial corona of basally fused staminal lobes, rose, very fleshy, lingulate, inflexed; pollinia ovoid, caudicles spathulate; corpusculum nearly the same size as pollinia, with elongated-triangular basal projections; style-head greenish white, cylindrical. Follicles 3–6 cm long. 2n = 22.

One sp., *D. dioscorides* (Lavranos) M.G. Gilbert, endemic to Socotra (Yemen).

247. ***Echidnopsis*** Hook. f. Fig. 51

Echidnopsis Hook. f., Bot. Mag.: ad t. 5930 (1871); Bruyns, Bradleya 6: 1–48 (1988), rev.; Thiv & Meve, Pl. Syst. Evol. 265: 71–88 (2007), phyl., tax.
Virchowia Vatke ex K. Schum. (1893), non *Virchowia* A. Schenk 1852, Scrophulariaceae.
Pseudopectinaria Lavranos (1971).

Creeping stem succulents, to 10 cm tall, rarely rhizomatous; shoots green, 1–10 cm long, 8–25 mm wide, roundly 5–20-angled, papillose or tessellate. Scale-like leaves lanceolate to trianguloid; stipules globose. Inflorescences apical, 1–6-flowered. Corolla 2–10 mm long, occasionally subglobose or tubular, highly fused, creamish to purplish, occasionally maculate, occasionally with trichomes; lobes lanceolate or triangular; gynostegial corona occasionally with trichomes, occasionally subglobose, of triangular to subulate staminal parts and spur-shaped or saccate

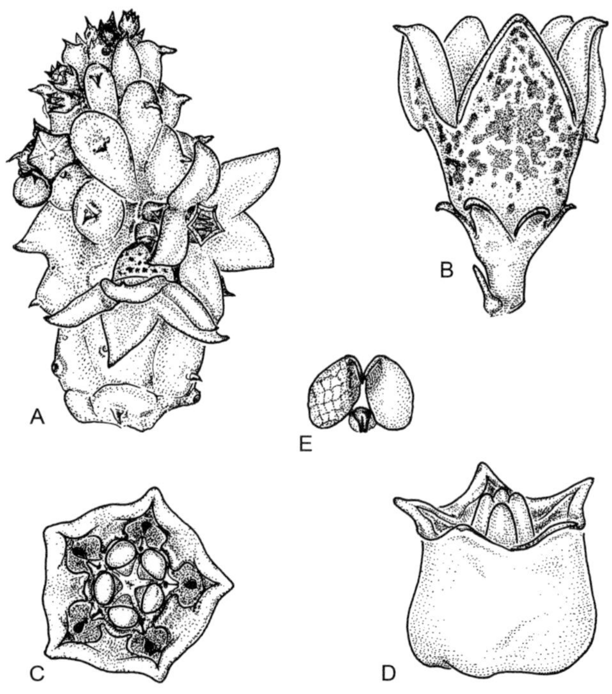

Fig. 51. Apocynaceae-Ceropegieae. *Echidnopsis bentii*. **A** Flowering branch. **B** Flower in lateral view. **C** Corona in top view. **D** Corona in lateral view. **E** Pollinarium. (From Meve and Wolf 2001, p. 117, with permission from Kakteen und andere Sukkulenten; drawn by U. Meve)

interstaminal parts almost completely fused into a ring; gynostegium sessile; caudicles trapezoidal. Follicles 6–10 cm long. 2n = 22, 44.

About 30 spp. in Africa and Arabia.

248. ***Edithcolea*** N.E. Br.

Edithcolea N.E. Br., Bull. Misc. Inform. Kew 1895: 220 (1895); Gilbert, Fl. Ethiopia and Eritrea 4.1: 184–185 (2003), reg. rev.; Lavranos, Fl. Somalia 3: 184–185 (2006), reg. rev.; Bruyns et al., Taxon 59: 2031–1043 (2010), phyl.

Creeping stem succulent, to 15 cm tall; shoots green with darker markings around spines, 2–30 cm long, 10–18 mm wide, roundly 4-angled. Scale-like leaves spiny, spreading; stipules reduced to a few trichomes. Inflorescences usually basal, 1(-2)-flowered. Corolla 4–5 cm long, creamish yellow, maroon-maculate, fused for about half of total length, tube with a small central depression, lobes adaxially with emergences tipped by a brownish trichome; lobes triangular, ciliate; gynostegial corona ivory or brownish, of clavate, connivent-erect, apically warty staminal parts and rectangular interstaminal parts basally fused into a ring; gynostegium sessile. Follicles 9–20 cm long, winged. 2n = 22.

One variable species, *E. grandis* N.E. Br., in E Africa and Arabia.

249. ***Hoodia*** Sweet ex Decne.

Hoodia Sweet ex Decne., Prodr. 8: 664 (1844); Bruyns, Bot. Jahrb. Syst. 115: 145–270 (1993), rev.; Bruyns, Stapeliads of S. Afr. Madag. 1: 92–129 (2005), rev.; Liede-Schumann & Meve, Taxon 67: 446 (2018), nom.
Monothylaceum G. Don (1837), nom. rej. prop.
Scytanthus Hook. (1844), non *Skytanthus* Meyen (1834), (Apocynaceae: Plumerieae).
Trichocaulon N.E. Br. (1878).
Hoodiopsis C.A. Lückh. (1933).

Robust, *Cereus*-like stem succulents, 25–220 cm tall; shoots (light) greenish to brownish, 20–150 cm long, 25–100 mm wide, sharply 11–31-angled. Scale-like leaves spreading acute spines, 0.3–1.2 cm long. Inflorescences 1–5-flowered; flowers occasionally with carrion-odor. Corolla 4–80 mm long, yellowish to reddish, occasionally with emergences tipped by trichomes; lobes broadly triangular; gynostegial corona yellowish or reddish, of ovoid or trianguloid, inflexed staminal parts and rectangular to bilobed interstaminal parts fused into a ring for about half of total length; gynostegium (sub-)sessile; caudicles spathulate. Follicles 4–18(–22) cm. 2n = 22.

13 spp., arid and semi-arid regions of S Africa. *Hoodia* has been listed under CITES (Appendix II) since 2005, the sole genus of Asclepiadoideae. The reputed activity of *Hoodia* ingredients as an appetite suppressant (see Introduction) causes specific threats to wild populations due to uncontrolled collecting.

250. ***Huernia*** R. Br.

Huernia R. Br., Asclepiadeae 11 (1810); Leach, Excelsa Tax. Ser. 4: 1–196 (1988), rev.; Gilbert, Fl. Ethiopia and Eritrea 4.1: 190–193 (2003), reg. rev.; Lavranos, Fl. Somalia 3: 195–196 (2006), reg. rev.; Bruyns, Stapeliads of S. Afr. Madag. 1: 130–211 (2005), rev.
Decodontia Haw. (1812), nom. illegit.

Clump-forming or creeping stem succulents, 1–15 cm tall; shoots (blue) green, 1–40(–1500) cm long, 5–25 mm wide, 4–6(–25)-angled. Scale-like leaves occasionally as spines, 0.1–0.5 cm long; stipules absent. Inflorescences basal, 1–7-flowered, pedunculate; flowers occasionally fetid. Corolla 1–4 cm long, occasionally with central ring, creamish to purplish, often patterned, often sculptured, emergences tipped by trichomes; lobes triangular; gynostegial corona creamish or purplish red, an often emarginate disc of staminal and interstaminal parts connate to free, oblongoid, erect or inflexed staminal lobes; gynostegium sessile, with humps below anther wings; caudicles medifixed, triangular or trapezoidal. Follicles 8–12 cm long, stout. 2n = 22, 44, 66.

About 70 spp. in Africa and Arabia.

251. ***Larryleachia*** Plowes

Larryleachia Plowes, Excelsa 17: 5 (1996); Bruyns, Bot. Jahrb. Syst. 115: 145–270 (1993); Meve & Liede, S. Afr. J. Bot. 67: 161–168 (2001), phyl.; Bruyns, Stapeliads of S. Afr. Madag. 1: 212–228 (2005), rev.
Leachia Plowes (1992), non *Leachia* Cass. (1822), Asteraceae.
Leachiella Plowes (1992), non *Leachiella* Kugrens (1982), Rhodophyceae.

Erect stem succulents, 5–30 cm tall, usually unbranched; shoots (blue-)green 5–30 cm long,

20–60 mm wide, roundly 12–20-angled, tessellate. Scale-like leaves in spirals or whorled, persistent, slightly sunken into the stem-surface, thickish. Inflorescences mostly apical, 1–6-flowered. Flowers occasionally with dung-like odor; corolla 3–8 mm long, creamish to purplish, often patterned, with papillae or emergences tipped by short trichomes; lobes triangular; gynostegial corona occasionally campanulate, ivory or yellow, usually patterned red, of linear (triangular), usually inflexed staminal parts, with long inflexed, adaxial appendage and spreading, bilobed interstaminal parts fused into a ring ; gynostegium sessile; caudicles medifixed, trapezoidal. Follicles 5–10 cm long. 2n = 22.

Five spp. in Namibia and South Africa.

252. ***Lavrania*** Plowes

Lavrania Plowes, Cact. Succ. J. (Los Angeles) 58: 122 (1986); Bruyns, Bot. Jahrb. Syst. 115: 145–270 (1993), rev.; Meve & Liede, S. Afr. J. Bot. 67: 161–168 (2001), phyl.; Bruyns, Stapeliads of S. Afr. Madag. 1: 229–231 (2005), morph.

Clump-forming stem succulent, 10–20 cm tall; shoots (blue-)green; 5–30 cm long, 20–30 mm wide, roundly 10–12-angled, tessellate. Scale-like leaves minute, broadly conical. Inflorescences basal, 5–15-flowered. Flowers with urine odor; corolla 6–8 mm long, very fleshy, yellowish, purple-maculate, warty; lobes triangular, with emergences tipped by papillae; gynostegial corona purplish, of lingulate, inflexed staminal parts and spreading, bilobed interstaminal parts fused into a ring for about half of total length; gynostegium sessile; caudicles spathulate. Follicles ca. 7 cm long. 2n = 22.

One sp., *L. haagnerae* Plowes, in N Namibia only, on cliffs.

253. ***Monolluma*** Plowes

Monolluma Plowes, Haseltonia 3: 64 (1995); Meve & Liede, Pl. Syst. Evol. 234: 172–209 (2002), phyl.
Cylindrilluma Plowes (1995).
Sanguilluma Plowes (1995).
Sulcolluma Plowes (1995).

Erect or clump-forming stem succulents, 3–50 cm tall; shoots occasionally pungent, (light) green, 2–40 cm long, 1–30 mm wide, sharply 4-angled. Scale-like leaves minute, persistent, ciliate; stipules reduced to a few trichomes. Inflorescences terminal, 1–15-flowered. Corolla 5–10 mm long, green, white, yellow or reddish, occasionally dark-maculate, occasionally sculptured; lobes triangular; gynostegial corona yellow or purplish red, occasionally with trichomes, occasionally tubular, of lingulate or triangular, inflexed staminal parts and erect or reflexed, bilobed, occasionally ciliate interstaminal parts fused into a ring; caudicles trapezoidal. Follicles 7–10 cm long, occasionally winged. 2n = 22.

Five spp. in Africa and Arabia.

254. ***Notechidnopsis*** Lavranos & Bleck

Notechidnopsis Lavranos & Bleck, Cact. Succ. J. (Los Angeles) 57: 255 (1985); Bruyns, Kew Bull. 54: 327–345 (1999), rev.; Meve & Liede, S. Afr. J. Bot. 67: 161–168 (2001), phyl.; Bruyns, Stapeliads of S. Afr. Madag. 1: 232–233 (2005), morph.

Creeping stem succulent, 2–6 cm tall, rhizomatous; shoots green, 3–10 cm long, 9–13 mm wide, roundly 6–8-angled, tessellate, glabrous. Scale-like leaves minute or absent. Inflorescences 2–5-flowered, slightly sunken into the stem surface; flowers honey-scented. Corolla 3–4 mm long, yellowish, reddish maculate, with verrucose trichomes; lobes triangular, spreading to slightly reflexed; gynostegial corona yellow, of trianguloid, erect staminal parts and reflexed, bilobed interstaminal parts fused into a ring; gynostegium sessile; caudicles trapezoidal. Follicles 4–6 cm long. 2n = 22.

One sp., *N. tessellata* (Pillans) Lavranos & Bleck, in South Africa.

255. ***Ophionella*** Bruyns

Ophionella Bruyns, Cact. Succ. J. Gr. Brit. 43: 70 (1981); Bruyns, Bot. J. Linn. Soc. 131: 383–398 (1999), rev.; Bruyns, Stapeliads of S. Afr. Madag. 1: 234–235 (2005), rev.

Creeping stem succulents, rhizomatous; shoots brownish green, 4–10 cm long, 3–6 mm wide, with 4 rounded angles. Scale-like leaves persistent, minute, sessile, triangular; stipules absent. Inflorescences 1–2(–5)-flowered. Flowers inodorous; corolla 4–14 mm long, campanulate or subglobose, white or cream, with purplish pattern; lobes lanceolate or triangular, usually remaining connate at apices; gynostegial corona yellow, of trianguloid, inflexed, occasionally denticulate staminal parts abaxially with a pronounced spreading pectinate or warty hump and small, spreading, spur-shaped interstaminal parts basally fused into a ring; gynostegium sessile; pollinia oblongoid; corpusculum without basal projections. Follicles 2.5–4.5 cm long. 2n = 22.

Two spp. in South Africa.

256. ***Orbea*** Haw.

Orbea Haw., Syn. Pl. Succ. 37 (1812); Leach, Excelsa Tax. Ser. 1: 1–75 (1978), rev.; Bruyns, Aloe 37: 72–76 (2000, publ. 2001), tax.; Bruyns, Syst. Bot. Monogr. 63: 1–196 (2002), rev.; Bruyns, Stapeliads of S. Afr. Madag. 1: 240–329 (2005), rev.; Lavranos, Fl. Somalia 3: 193–195 (2006), reg. rev.
Podanthes Haw. (1812), nom. rej., non *Podanthus* Lag. (1816), Asteraceae.
Diplocyatha N.E. Br. (1878).
Stultitia E. Phillips nom. nov. pro *Stapeliopsis* E. Phillips (1932), non *Stapeliopsis* Pillans (1928).
Orbeopsis L.C. Leach (1978).
Pachycymbium L.C. Leach (1978).
Angolluma R. Munster (1990).

Clump-forming stem succulents, 5–15 cm tall; latex rarely yellowish; occasionally rhizomatous; shoots (light) green or blue-green, often maculate, conical or club-shaped, 1–25 cm long, 10–30 mm wide, 4(–5)-angled. Scale-like leaves spreading, triangular; stipules glandular, (sub-)globose. Inflorescences lateral, 1–30-flowered. Flowers fetid; corolla 5–35 mm long, creamish, frequently patterned purplish, usually sculptured, occasionally with verrucose, erect trichomes, occasionally with ring; lobes lanceolate to ovate, often ciliate; gynostegial corona yellow to purplish, of long, occasionally adaxially humped staminal parts and shorter, often bilobed interstaminal parts fused into a ring; gynostegium occasionally atop a column; caudicles trapezoidal to spathulate. Follicles 4–18 cm long. 2n = 22, 44.

About 55 spp., Africa and Arabia.

257. ***Orbeanthus*** L.C. Leach

Orbeanthus L.C. Leach, Excelsa Tax. Ser. 1: 71 (1978); Meve & Liede, Pl. Syst. Evol. 234: 172–209 (2002), phyl.

Creeping stem succulents, 2 cm tall, sparsely branched; shoots blue-green, mottled purple, 5–30 cm long, 8–9 mm wide, with 4 rounded angles. Scale-like leaves triangular, without stipules. Inflorescences usually basal, 1–2-flowered. Flowers slightly fetid, not nectariferous; corolla 15–20 mm long, rotate or urceolate, cream and reddish, unsculptured, with vesicle-like trichomes, rarely with a broad ring; lobes triangular, gynostegial corona creamish, tinged or mottled reddish, of triangular or clavate, apically pilose staminal parts and spreading, bilobed, apically pilose interstaminal parts shortly fused into a ring; gynostegium sessile; pollinia D-shaped; caudicles broadly spathulate. 2n = 22.

Two spp. in South Africa.

Bruyns (2002, 2005) proposed that *Orbeanthus* be included in *Orbea* Haw., a concept rejected by Meve and Liede (2002a).

258. ***Pectinaria*** Haw.

Pectinaria Haw., Suppl. Pl. Succ. 14 (1819), nom. cons., non *Pectinaria* Bernh. (1800), Apiaceae, nom. rej.; Bruyns, Cact. Succ. J. (Gr. Brit.) 43: 61–83 (1981), rev.; Bruyns, Stapeliads of S. Afr. Madag. 2: 331–344 (2005), rev.
Vadulia Plowes (2003).

Clump- to mat-forming stem succulents, 1–5 cm tall; shoots (blue-)green, 1.5–8 cm long, 10–25 mm wide, with 6 rounded angles. Scale-like leaves persistent, triangular; stipules absent. Inflorescences 1–2-flowered. Flowers nodding, inodorous; corolla 3–7 mm long, occasionally subglobose, yellowish to maroon (patterned), occasionally warty by spiculate emergences; lobes lanceolate or ovate, occasionally remaining connate at apices; gynostegial corona yellow, of triangular, occasionally denticulate staminal parts with a pectinate hump and bilobed interstaminal parts highly fused into a ring; gynostegium subsessile or atop a column; caudicles rectangular; corpusculum without basal projections. Follicles 6.5–9 cm long. 2n = 22.

At least three spp. in two sections, South Africa.

259. ***Piaranthus*** R. Br.

Piaranthus R. Br., Asclepiadeae 12 (1810); Meve, Bradleya 12: 57–12 (1994), rev.; Bruyns, Syst. Bot. 24: 379–398

(1999), rev.; Bruyns, Stapeliads of S. Afr. Madag. 2: 345–368 (2005), reg. rev.
Obesia Haw. (1812).
Huerniopsis N.E. Br. (1878).

Clump- to mat-forming stem succulents, 2–10 cm tall; shoots green, usually club-shaped, 1–6 cm long, 8–20 mm wide, with 4–5 rounded angles. Scale-like leaves succulent, deltate; stipules glandular, (sub-)globose. Inflorescences lateral, 1–10-flowered. Flowers with fruity or fetid odor; corolla 8–21 mm long, uniform or patterned, papillate or with trichomes; lobes lanceolate or triangular, occasionally with vibratile trichomes; gynostegial corona ivory or yellow, of subulate to triangular staminal parts abaxially with a pronounced, grooved, rugose, pectinate hump and indistinct interstaminal parts basally fused into a ring; gynostegium sessile or atop a column; caudicles cylindrical. Follicles 6–14 cm long. 2n = 22.

Nine spp., South Africa, Namibia, Botswana.

Inclusion of *Huerniopsis* (sister to *Piaranthus* s.str.) is a matter of taste.

260. *Pseudolithos* P.R.O. Bally

Pseudolithos P.R.O. Bally, Candollea 20: 41 (1965); Bruyns, Bradleya 8: 33–38 (1990), rev.; Lavranos, Fl. Somalia 3: 180–182 (2006), reg. rev.

Clump-forming stem succulents, 2–7 cm tall, unbranched or sparsely branched; shoots blue-green or light brown, cube-like to globose, 1.5–6 cm long, 0.8–6 mm wide, obtusely 4-angled, tessellate. Scale-like leaves (almost) absent; stipules absent. Inflorescences terminal or lateral, often on short shoots, 4–30-flowered. Flowers fetid; corolla 2.5–6 mm long, campanulate or urceolate, creamish to purplish, occasionally maculate, with spiculate papillae or trichomes, adaxially with five ridges; lobes (sub-)triangular; gynostegial corona yellow(-green) or purplish red, occasionally with trichomes, occasionally urceolate, of lingulate or trianguloid, inflexed staminal parts and deltoid or saccate, bilobed interstaminal parts almost completely fused into a ring; caudicles trapezoidal or spathulate. Follicles 5–6 cm long. 2n = 22.

Five to six spp. in Ethiopia and Somalia.

261. *Quaqua* N.E. Br.

Quaqua N.E. Br., Gard. Chron., II, 12: 8 (1879); Bruyns, Bot. Jahrb. Syst. 121: 311–402 (1999), rev.; Bruyns, Stapeliads S. Afr. Madag. 2: 369–415 (2005), rev.
Sarcophagophilus Dinter (1923).

Erect stem succulents, 15–60 cm tall; shoots (blue-)green, 5–50 cm long, 15–40 mm wide, 4-angled (rarely 5–6-angled). Scale-like leaves persistent, forming spines (occasionally nearly absent); stipules glandular, ovate (occasionally absent). Inflorescences apical, 1–30-flowered. Flowers with fetid or sweetish odor; corolla 5–15 mm long, uniform, tinged or patterned, occasionally warty, papillose or with verrucose trichomes; lobes oblong, lanceolate or triangular, occasionally ciliate; gynostegial corona yellow or purplish red, of (small) trianguloid staminal parts occasionally with a spreading dorsal appendage and saccate, subulate, deltoid or bilobed interstaminal parts fused into a ring; gynostegium (sub-)sessile; caudicles trapezoidal and spathulate. Follicles 2–13 cm long. 2n = 22.

19 spp. in Namibia and South Africa.

262. *Rhytidocaulon* P.R.O. Bally

Rhytidocaulon P.R.O. Bally, Candollea 18: 335 (1962); Bruyns, Edinb. J. Bot. 56: 211–228 (1999), rev.; Gilbert, Fl. Ethiopia and Eritrea 4.1: 176–178 (2003), reg. rev.; Lavranos, Fl. Somalia 3: 183–184 (2006), reg. rev.; de Kock, Asklepios 101: 5–15 (2008), morph.

Creeping or erect stem succulents, 5–30 cm tall, sparsely branched from a distinct main stem; shoots dark brownish green, 2–50 cm long, 0.7–2.5 mm wide, with 4 rounded angles, papillose or tessellate. Scale-like leaves strongly ascending; stipules glandular, globose. Inflorescences lateral, 1–3-flowered, slightly sunken into stem surface. Flowers often with dull fruity odor; corolla 2.5–20 mm long, rotate or elongate-conical, cream, green or purple, often patterned, glabrous, papillose or with often vibratile trichomes; lobes linear to ovate, occasionally caudate, or tips remaining connate; gynostegial corona yellow, or purplish red, of lingulate, trianguloid or rectangular staminal parts and subulate, deltoid (to fringe-like) or bilobulate interstaminal parts fused into a ring;

gynostegium sessile; corpusculum without basal projections; follicles ca. 4 cm long; seeds with thick, puffed margin. 2n = 22.

About 14 spp. in E Africa and Arabia.

263. ***Richtersveldia*** Meve & Liede

Richtersveldia Meve & Liede, Pl. Syst. Evol. 234: 204 (2002); Bruyns, Stapeliads of S. Afr. Madag. 2: 416–417 (2005), morph.

Erect stem succulent, 10–20 cm tall; rhizomatous; shoots blue-green, 5–20 cm long, 20–25 mm wide, sharply 8-angled, tessellate. Scale-like leaves persistent, forming spines. Inflorescences apical, 1–5 (–20)-flowered, subsessile. Flowers with dull fruity odor; corolla 5–8 mm long, depressed-rotate, greenish yellow to ochre, reddish maculate, unsculptured, papillose or with trichomes, lobes triangular, spreading; gynostegial corona yellowish, spotted red, of broadly linear, inflexed staminal parts and deltoid-rectangular, occasionally bilobed, spreading interstaminal parts basally fused into a ring; gynostegium sessile; caudicles trapezoidal. Follicles 6–8 cm long. 2n = 22.

One sp., *R. columnaris* (Nel) Meve & Liede, in South Africa, in the Richtersveld only.

264. ***Socotrella*** Bruyns & A.G. Miller

Socotrella Bruyns & A.G. Miller, Novon 12: 330 (2002); Řičánek & Hanáček, Cact. Succ. J. (Los Angeles) 78: 6–9 (2006), ecol.; Bruyns et al., Mol. Phyl. Evol. 77: 251–263 (2014), phyl.

Stem succulent forming scattered clumps, 10 cm tall, rhizomatous; shoots green, 5–15 cm long, 4–7 mm wide, with 4 rounded angles. Leaves and stipules absent. Inflorescences lateral, 1–5-flowered. Flowers long-pedicellate, erect, with vanilla scent; corolla 9–12 mm long, campanulate, yellow, tube streaked reddish to maroon, glabrous; lobes ovate to deltate, spreading, marginally slightly revolute; gynostegial corona yellow, of long, rectangular, basally bulged free staminal lobes and tiny interstaminal flaps; gynostegium atop a column; pollinia ellipsoid; caudicles spathulate; corpusculum rhomboid, the same size as the pollinia.

One sp., *S. dolichocnema* Bruyns, in Yemen (Socotra); sister to the also monotypic Socotran endemic *Duvaliandra*.

265. ***Stapelia*** L.

Stapelia L., Sp. Pl. 217 (1753), nom. cons.; Leach, Excelsa Taxon. Ser. 3: 1–157 (1985), rev.; Bruyns, Stapeliads of S. Afr. Madag. 2: 418–489 (2005), rev.
Stisseria Heister ex Fabr. (1759), nom. illeg.
Gonostemon Haw. (1812).

Clump-forming stem succulents, 10–20 cm tall; occasionally rhizomatous; shoots green, cylindrical, 6–30 cm long, 5–30 mm wide, 4-angled, occasionally pubescent. Scale-like leaves ascending to erect, oblong or lanceolate; stipules glandular, globose. Inflorescences basal, 1–10-flowered, pedunculate. Flowers fetid; corolla 5–200 mm long, globose, rarely campanulate, uniform or banded, mostly rugose, occasionally with flat ring; lobes ovate or triangular, often ciliate; gynostegial corona of subulate staminal parts, occasionally with adaxial appendage, and often apiculate interstaminal parts basally fused into a ring; gynostegium (sub-)sessile or atop a column; pollinia ovoid or reniform; caudicles spathulate; corpusculum with linear to deltoid basal projections. Follicles 9–13 cm long. 2n = 22, 44.

43 spp. in southern Africa.

266. ***Stapelianthus*** Choux ex White & Sloane

Stapelianthus Choux ex White & Sloane, Stapelieae 71 (1933); Bruyns & Klak, Ann. Missouri Bot. Gard. 91: 410–437 (2004), rev.
Stapeliopsis Choux (1931), non *Stapeliopsis* Pillans (1928).

Clump-forming stem succulents, 5–10 cm tall; shoots greenish to brownish, streaked or mottled, 2–30 cm long, 5–12 mm wide, 4–6(–8)-angled, tuberculate or tessellate. Scale-like leaves succulent or spiny. Inflorescences usually basal, 1–5-flowered. Flowers fetid; corolla 6–18 mm long, occasionally cyathiform or urceolate, very carnose, cream or purple, occasionally maculate, occasionally sculptured or with trichome-tipped emergences, occasionally with ring; lobes broadly ovate; gynostegial corona purplish red, cyathiform, of ovoid or trianguloid, inflexed staminal parts and erect, rectangular or bifid interstaminal parts basally fused into a ring; gynostegium occasionally atop a column; caudicles spathulate; corpusculum with small basal projections. Follicles 6–10 cm long. 2n = 22.

Seven spp. in Madagascar.

267. *Stapeliopsis* Pillans

Stapeliopsis Pillans, S. African Gard. 18: 32 (1928); Bruyns, Cact. Succ. J. (Gr. Brit.) 43: 61–83 (1981), rev.; Bruyns et al., Bot. J. Linn. Soc. 148: 125–155 (2005), rev. & phyl.; Bruyns, Stapeliads of S. Afr. Madag. 2: 507–525 (2005), morph.
Hermanschwartzia Plowes (2003).
Neopectinaria Plowes (2003).

Clump-forming stem succulents, 2–8 cm tall; rhizomatous; shoots blue-green, often mottled, 3–15 cm long, 5–20 mm wide, 4-angled, papillose. Scale-like leaves succulent or spiny. Inflorescences basal, 1-3-flowered, pedunculate. Flowers usually fetid; corolla 10–30 mm long, urceolate, subglobose or tubular, white to purple, occasionally sculptured or with trichome-tipped emergences; lobes triangular, occasionally connate at apices; gynostegial corona purplish, of subulate to clavate staminal parts and spreading, deltoid, rectangular or triangular interstaminal parts, occasionally fused into a ring; gynostegium occasionally atop a column, concealed in corolla; caudicles rectangular; corpusculum without basal projections. Follicles 7–11 cm long. 2n = 22.

Eight spp. in Namibia and South Africa.

268. *Tavaresia* Welw.

Tavaresia Welw., Ann. Cons. Ultramarino, Ser. 1: 79 (1854); White & Sloane, Stapelieae 3: 1099–1108 (1937) rev.; Bruyns, Stapeliads of S. Afr. Madag. 2: 526–530 (2005), rev.
Decabelone Decne. (1871).

Clump-forming stem succulents, 8–12 cm tall; green, 2–25 cm long, 15–25 mm wide, sharply 5–14-angled. Scale-like leaves and stipules persistent, forming three spines to 1 cm long, very acute. Inflorescences basal, 1-5-flowered. Flowers with dull pungent or carrion odor; corolla 3–14 cm long, tubular, cream, maculate with (ob-)conical, sometimes trichome-tipped emergences; lobes triangular; gynostegial corona ivory, maculate, cyathiform, forming a ring of sublanceolate staminal parts and saccate, bilobed, subulate, interstaminal parts topped by a red, globose, vibratile trichome; gynostegium atop a short column; pollinia large, obovoid, caudicles spathulate; corpusculum with linear basal projections. Follicles 5–19 cm long. Seeds suborbicular. 2n = 22.

Two spp. in Angola, Botswana, Namibia and South Africa.

269. *Tridentea* Haw.

Tridentea Haw., Syn. Pl. Succ. 34 (1812); Leach, Excelsa Tax. Ser. 2: 1–69 (1980); Bruyns, S. Afr. J. Bot. 61: 180–208 (1995), rev.; Bruyns, Stapeliads of S. Afr. Madag. 2: 531–550 (2005), reg. rev.

Clump-forming stem succulents, 5–10 cm tall; occasionally rhizomatous; shoots blue-green, (2–)5–15(–20) cm long, 10–20(–25) mm wide, 4-angled. Scale-like leaves basally constricted, (sub-)lanceolate; stipules reduced to a few glandular hairs. Inflorescences basal, 1-5-flowered. Flowers fetid; corolla 10–50 mm long, cream or yellow, occasionally patterned purplish, sculptured or with erect conical or spine-like trichome-tipped emergences; lobes triangular; free staminal corona lobes yellow or purplish, subulate(-deltoid), occasionally with basal spreading dorsal spur; interstaminal lobes rectangular, ovate, apiculate, bifid or trifid, spreading; gynostegium (sub-)sessile; caudicles trapezoidal to spathulate. Follicles 11–16 cm long. 2n = 22.

Eight spp. in Namibia and South Africa.

270. *Tromotriche* Haw.

Tromotriche Haw., Syn. Pl. Succ. 36 (1812); Leach, Excelsa Tax. Ser. 2: 1–69 (1980), rev.; Bruyns, S. Afr. J. Bot. 61: 180–208 (1995), rev.; Bruyns, Stapeliads of S. Afr. Madag. 2: 551–574 (2005), morph.
Caruncularia Haw. (1812).

Clump-forming or creeping stem succulents, 3–10 cm tall, rhizomatous; shoots blue-green, 5–50 (–300) cm long, 6–25 mm wide, 4-angled. Scale-like leaves minute, triangular; stipules glandular, globose. Inflorescences 1-5-flowered, pedunculate. Flowers fetid; corolla 7–32 mm long, occasionally cyathiform, cream, yellow, or purplish, occasionally patterned, often rugose and papillate; lobes ovate or triangular, reflexed or spreading, often ciliate; gynostegial corona (black) purplish, cyathiform, of subulate or clavate, staminal lobes with adaxial appendage and rectangular, triangular or bifid interstaminal lobes occasionally basally fused into a ring; gynostegium atop a column with 5-angled foot; caudicles trapezoidal or spathulate. Follicles ca. 8 cm long. 2n = 22.

Nine spp. in Namibia and South Africa.

271. ***White-sloanea*** Chiov.

White-sloanea Chiov., Malpighia 34: 541 (1937); Bruyns, Edinb. J. Bot. 55: 27–37 (1998), morph.; Lavranos, Fl. Somalia 3: 196–197 (2006), reg. rev.
Drakebrockmania White & Sloane (1937), non *Drake-Brockmania* Stapf (1912), Poaceae.

Erect, mostly single-stemmed stem succulent, 3–13 cm tall; shoots light green to brown, 3–13 cm long, 4–5.5 mm wide, 4-angled. Leaves absent. Inflorescences lateral, 5–20-flowered, shortly pedunculate. Flowers fetid; corolla 3 cm long, campanulate, cream, red-maculate, with hemispherical emergences and flexuous trichomes; lobes triangular, with vibratile clavate trichomes near base of lobes; gynostegial corona creamish, striped red, rotate, of oblong-trapezoidal, apically denticulate staminal parts and rectangular, bifid interstaminal parts basally fused into a ring; gynostegium atop a short column; anthers with short, deltoid connective appendages; pollinia rectangular-ellipsoid, caudicles spathulate. Follicles 5–6 cm long, thick fusiform. 2n = 22.

One sp., *W. crassa* (N.E. Br.) Chiov., in Somalia.

V.4. **Tribe Eustegieae** Rchb. ex Meve & Liede (2014).

Leaves palmately lobed or (at least some) with one or two spreading lateral teeth a little above the base. Inflorescences extra-axillary and terminal. Corolla rotate, glabrous; corona gynostegial, consisting of a very basally fused ring of simple staminal and deeply tripartite interstaminal parts with an additional inner ring of smaller staminal lobes; gynostegium (sub-)sessile. Recent molecular study (Surveswaran et al. 2014) has shown that Eustegieae are sister to Asclepiadeae. Two genera in southern Africa.

Key to the Genera of Eustegieae

1. Leaves palmately 5–7-lobed; follicles triangular, indehiscent, seeds single, wingless **272. *Emicocarpus***
– Leaves linear, often with one or two lateral teeth a little above the base; follicles a single, fusiform follicle, seeds several, winged **273. *Eustegia***

272. ***Emicocarpus*** K. Schum. & Schltr. Fig. 52

Emicocarpus K. Schum. & Schltr., Bot. Jahrb. Syst. 29. Beibl. 66: 21 (1900).
Lobostephanus N.E. Br. (1901).

Fig. 52. Apocynaceae-Eustegieae. *Emicocarpus fissifolius*. **A** Flowering branch. **B** Flower. **C** Portion of the outer ring-shaped gynostegial corona of staminal and interstaminal parts. **D** Staminal corona lobes and style-head. **E, F** Pollinarium. **G** Follicle. **H** Follicle in longitudinal section, showing the seed in natural position. (From Brown 1901, pl. 2692, with permission from Hooker's Icones Plantarum © Board of Trustees of the Royal Bot. Gard., Kew; drawn by Matilda Smith)

Prostrate herb, 0.6–1 m tall, almost glabrous. Leaves 1–5 cm long, palmately 5–7-lobed. Inflorescences 5–8-flowered, sciadioidal, long-pedunculate. Corolla 2–3 mm long; lobes oblong to lanceolate; gynostegial corona exceeding gynostegium (except for style-head), forming a ring of connate filamentous to spathulate, erect staminal and shorter tripartite interstaminal parts, with additional inner erect, oblong, free staminal lobes; pollinia oblongoid; caudicles declinate, ribbon-shaped; corpusculum ovoid, minute; style-head rostrate. Follicles 10 mm long, triangular, with woody pericarp, not opening at maturity. Seeds one per fruiting carpel, U-shaped, wingless.

A single species, *E. fissifolius* K. Schum. & Schltr., in Mozambique, sandy areas, sea level; possibly extinct. Closely related to *Eustegia*.

273. ***Eustegia*** R. Br.

Eustegia R. Br., Asclepiadeae 40 (1810); Bruyns, Bot. Jahrb. Syst. 121: 19–44 (1999), rev.; Surveswaran et al., Bot. J. Linn. Soc. 174: 601–619 (2014), phyl.

Erect or prostrate herbs, 20–30 cm tall; sparsely pubescent; latex translucent; rhizomatous, with tap roots; shoots annual. Leaves sessile, 0.5–3 cm long, linear to hastiform. Inflorescences 2–8-flowered, sciadioidal. Corolla 3–4 mm long, grayish green; lobes oblong, revolute; gynostegial corona whitish, equaling gynostegium, forming a ring of basally connate filamentous staminal and deeply tripartite interstaminal parts, with additional inner ring of five small, basally connate staminal lobes; pollinia reniform; caudicles geniculate, articulated, triangular at pollinium end; corpusculum deltoid, frail; style-head cylindrical to rostrate. Follicles 8–10 cm long, fusiform-obclavate. Seeds winged.

Three spp. (only one accepted by Bruyns 1999), in South Africa, winter rainfall area, ephemeral in fynbos and renosterveld; to 2000 m.

V.5. **Tribe Asclepiadeae** Duby (1828).

Herbs, trees (up to 6 m, *Calotropis*), vines, geophytes with subterranean tubers, occasionally succulents, with white (rarely translucent or yellow) latex. Leaves opposite, rarely whorled, herbaceous, often with cluster of colleters adaxially at juncture of petiole and leaf blade. Corolla rotate to occasionally tubular; lobes with valvate to imbricate aestivation; corolline corona rare; anthers generally erect to spreading, with deltoid apical appendage, distinctly constricted at base, with guide rails alongside fertile part (rarely below: some species of *Cynanchum*, *Vincetoxicum*, most *Gonolobus* relatives); gynostegial corona rarely absent (*Astephanus*, *Microloma*, single species in various genera); usually either an outer ring of connate staminal and interstaminal parts, annular (*Gonolobus* relatives) or differentiated in staminal and/or interstaminal position or an inner corona of 5 massive, free, often channeled lobes, often with additional lobules on adaxial surface (rarely partially fused with corolla, *Parapodium*, *Oxypetalum*); occasionally both a ring and an inner corona present (*Cynanchum*, former *Sarcostemma* species); pollinia pendulous or rarely erect (some *Cynanchum* species, former *Karimbolea* species), or more or less horizontal (some *Vincetoxicum* species, *Gonolobus* relatives), small and more or less ellipsoid to long, narrow and dorsiventrally flattened, rarely with hyaline insertion crest along inner edge (*Pergularia*), on upper margin near attachment of caudicles (*Gonolobus* relatives), upper edge (some *Schizoglossum* species) or with caudicles attenuate into small apical spike (*Oncinema*); pollinia usually (sub-)apically, more rarely laterally attached to the straight or S-shaped caudicles; pollen tubes emerging through inner, upper edge or through the concave face of the pollinium (*Gonolobus* relatives); style-head broad, with flat to concave (sometimes conical to slenderly conical) apex, tapering below into narrow neck above carpels; caudicles frequently with projecting hooks and ridges supporting insertion into guide rail; gynoecium apocarpous, mostly superior. Follicles usually one by abortion, occasionally extremely inflated (*Araujia*, *Calotropis*), rarely spherical, more usually fusiform obclavate, occasionally ornamented with spine-like processes or protuberances or longitudinal ridges. Seeds more or less elliptical in outline, with or without wing-like margin, with coma. $x = 11$ (rarely 10, 9); some extra-tropical *Vincetoxicum* species with hypogeal germination. Secondary compounds "Asclepiadaceae bitter principles" (glycosides), more rarely cardenolides; alkaloids only in *Vincetoxicum* and relatives. 105 genera worldwide, mainly tropical and subtropical climates, but several genera extending into the temperate zones.

Key to the Subtribes of Asclepiadeae

1. Plants of Africa, Eurasia, Australia 2
- Plants of the Americas 5
2. Small twiners or rigid shrubs; latex translucent; gynostegial corona usually absent, if present, then leaves linear; plants of southern Africa V.5.a. Astephaninae
- Plants of various habits; latex usually white, if translucent, then plants larger and with gynostegial corona, but without linear leaves 3
3. Gynostegial corona of five staminal lobes, at the most basally fused 4
- Gynostegial corona with a well-developed annular part, rarely with additional free staminal lobes V.5.c. Cynanchinae
4. Plants often erect; inflorescences always sciadioidal; staminal corona lobes three-dimensional, often intricately folded, with or without inner teeth and ornamentations V.5.b. Asclepiadinae
- Plants rarely erect; inflorescences complex or sciadioidal; staminal corona lobes usually laminar, if three-dimensional, then of simple shape and at least partially fused to the anthers V.5.d. Tylophorinae
5. Corolla lobes with vibratile trichomes on the adaxial side; plants of southern South America V.5.l. Topeinae
- Corolla lobes without vibratile trichomes on the adaxial side 6
6. Plants erect; staminal corona lobes three-dimensional, often intricately folded, with or without inner teeth and ornamentations V.5.b. Asclepiadinae
- Plants usually twining, rarely erect; corona structure different 7
7. Anthers usually broader than long, frequently horizontal, with transversally dehiscent membrane; pollinia usually with apical to distal sterile hyaline region V.5.k. Gonolobinae
- Anthers usually longer than broad, never horizontal and without transversally dehiscent membrane; pollinia never with sterile hyaline region 8
8. Inflorescences usually axillary, paired, with one inflorescence shorter, often multi-flowered, thyrsoidal, partial inflorescences often condensed; follicles paired, narrowly oblong, often with a basal thickening; follicles forming an acute angle V.5.i. Tassadiinae
- Inflorescences usually extra-axillary, rarely axillary, never thyrsoidal; follicles solitary or paired, usually fusiform, if narrowly oblong, then without basal thickening; follicles at various angles 9
9. Corolla campanulate, adnate to gynostegium, forming five hairy pouches; gynostegial corona of free staminal lobes with additional peg-shaped parts along filament tube; plants of the northern Andes V.5.e. Pentacyphinae
- Corolla of various shapes, but not adnate to gynostegium and not forming five hairy pouches; gynostegial corona of different construction and without additional peg-shaped parts 10
10. Leaves cordiform, basally deeply cordate V.5.c. Cynanchinae
- Leaves of various shapes, but basally never deeply cordate 11
11. Flowers usually large (> 1 cm diam.), often brightly colored; follicles usually one per flower, with thick pericarp; if two, then plants often erect, and caudicles with conspicuous appendage V.5.j. Oxypetalinae
- Flowers usually smaller (< 1 cm diam.), usually white, cream, yellowish or greenish; follicles one or two per flower, with thin pericarp 12
12. Corolla lobes adaxially usually with trichomes, often barbate; gynostegial corona of five staminal lobes, at the most basally fused; follicles solitary V.5.h. Metastelmatinae
- Corolla lobes adaxially usually glabrous, if with trichomes, then these of equal length; gynostegial corona usually with conspicuous annular part; follicles usually paired, if solitary, then plants of southern South America 13
13. Plants without conspicuous long shoot–short shoot architecture; follicles usually solitary, fusiform; plants of southern South America V.5.f. Diplolepinae
- Plants usually with conspicuous long shoot–short shoot architecture; follicles usually paired, narrowly oblong V.5.g. Orthosiinae

Key to the Genera of Asclepiadeae
(by Sigrid Liede-Schumann and G.N. Morillo)

1. Erect shrubs or small trees, to 3 m tall, branches woody, never twining even apically; corona with a conspicuous basal spur on the prominent staminal lobes; native to tropical Africa and Asia, but widely introduced and naturalized in New World tropics **282. *Calotropis***

- Plants erect or twining, when 3 m or more then not erect or woody only toward the base; corona always lacking a conspicuous coiled basal spur, otherwise variable 2
2. Plants of Africa, Europe, Asia or Australia 3
- Plants of the New World 57
3. Corona of any kind completely absent or at the most forming a small projection on the adaxial side of the corolla 4
- Distinct corona of various composition present 7
4. Leaves cordiform, basally broadly cordate; inflorescences bostrychoid; latex white; plants of Asia **303. *Cynanchum*** (former *Adelostemma*)
- Leaves of various shapes, but basally not broadly cordate; inflorescences sciadioidal; latex translucent; plants of Africa 5
5. Corolla fused only at the base, lobes strongly twisted; style-head rostrate, much longer than connective appendages **306. *Vincetoxicum*** (former *Pleurostelma* species)
- Corolla fused for at least half of its length, lobes usually not strongly twisted; style-head longer or about as long as connective appendages 6
6. Corolla tubular, partly or almost entirely obscuring the gynostegium, corolla occasionally with small projections or trichomes; style-head not longer than connective appendages **275. *Microloma***
- Corolla campanulate, not obscuring the gynostegium; style-head longer than connective appendages **274. *Astephanus***
7. Plants with succulent stems, with or without well-developed leaves 8
- Plants without succulent stems, with well-developed leaves 9
8. Plants with cordate leaves; latex translucent; (outer) corona fused into bowl-shaped tube with 10 lobules around mouth leaving gynostegium well visible from the top; caudicles extended along distal side of pollinia; plants of Africa **304. *Schizostephanus***
- Plants with or without well-developed leaves; latex usually white, white or yellow, when almost translucent, then plants always leafless; corona gynostegial, but not forming a bowl-shaped tube with 10 lobules around mouth; plants of Madagascar, when African, then with scale-like leaves **303. *Cynanchum***
9. Plants erect, not even apically twining 10
- Plants twining, at least apically 45
10. Plants strongly branched, herbaceous or shrubby 11
- Plants unbranched or sparsely branched, erect herbs 20
11. Latex translucent 12
- Latex white 13
12. Small shrublets (< 40 cm); leaves carnose, tiny (to 3 mm long); plants of Somalia **281. *Calciphila*** (*C. gillettii*)
- Plants basitonically branched perennial herbs, leaves herbaceous, ovate to linear, often basally cordate; widespread in Europe and Asia, introduced to N America; occasionally in Africa **306. *Vincetoxicum***
13. Prostrate annual herbs; leaves often with undulate margins, soft, broadly ovoid; follicles thick-walled, with a few soft spines **303. *Cynanchum*** (former *Glossonema* species)
- Plants perennial, leaves normally not with undulate margins and not soft; follicles of various shapes, but not with a few soft spines 14
14. Corona present, but not of free staminal lobes 15
- Corona of free staminal lobes 16
15. Corona of five basally fused lobes adnate to the corolla, forming a cup from which the long-stipitate gynostegium emerges; plants strongly basitonically branched shrubs; leaves slenderly elliptic, strongly ascending; follicles broadly ovoid, thick-walled, smooth; plants of northern Africa and Arabia **297. *Solenostemma***
- Corona gynostegial, not adnate to the corolla, cup-shaped or otherwise, but not with a long-stipitate gynostegium emerging; leaves of various shapes, usually not strongly ascending; follicles thin-walled, narrowly obclavate or fusiform, usually slightly grooved, rarely winged or with slender emergences **303. *Cynanchum***
16. Leaves petiolate and cordate; staminal corona lobes fusiform, small free interstaminal lobules present; plants of northern Africa and Arabia **294. *Pergularia*** (*P. tomentosa*)
- Leaves (sub-)sessile, linear to obovate, at the most indistinctly cordate; staminal corona lobes not fusiform; interstaminal lobules absent 17
17. Leaves broadly obovate; staminal corona lobes with a coiled basal spur **282. *Calotropis***
- Leaves linear to slenderly elliptic; staminal corona lobes without a coiled basal spur 18
18. Inflorescences mostly terminal; rootstock a tuberous caudex **278. *Asclepias***
- Inflorescences extra-axillary; rootstock usually fibrous, occasionally woody and tuberous 19
19. Inflorescences bostrychoid; follicles fusiform to narrowly ovoid, smooth; rheophytes along seasonal water courses **287. *Kanahia***

- Inflorescences sciadioidal; follicles often inflated and covered with soft spines; occurring in various habitats, but not in riverbeds **286. *Gomphocarpus***
20. Gynostegial corona a ring of connate staminal and interstaminal parts 21
- Gynostegial corona of free (or almost free) staminal lobes 22
21. Gynostegial corona adnate to gynostegium only, laminar; interstaminal parts undifferentiated, corolla lobes oblong, ovate, lanceolate or cucullate; plants of Madagascar **303. *Cynanchum***
- Gynostegial corona adnate to both corolla and gynostegium, carnose, interstaminal parts rectangular and bifid; corolla lobes spathulate; plants of southern tropical Africa **290. *Odontostelma***
22. Leaves obovate to oblong, sessile, carnose; corolla lobes to 15 mm long; plants of the Himalayas **282. *Calotropis*** (*C. acia*)
- Leaves of various shapes, but broadest always below the middle, not carnose; corolla lobes mostly smaller 23
23. Staminal corona lobes petaloid, red or pale bluish lilac; as long as to longer than corolla lobes, occasionally apically broadened and emarginate, with a basal tooth projecting into the center of the flower; plants of east tropical Africa **288. *Margaretta***
- Staminal corona lobes not petaloid; not exceeding the corolla lobes; never apically broadened 24
24. Staminal corona lobes laminar or somewhat carnose, flat, never folded (laterally or ventrally) to form a central cavity, lacking conspicuous ventral or apical projections 25
- Staminal corona lobes not laminar, three-dimensional, solid or folded (laterally or ventrally) to form a central cavity, often with apical or ventral teeth or other projections 30
25. Pollinia with a pellucid germination zone on the circumference, but not at the point of attachment to the caudicle **296. *Schizoglossum***
- Pollinia without a pellucid germination zone on the circumference 26
26. Leaves linear; inflorescences long-pedunculate, almost as long as or longer than subtending leaves, inflorescences forming dense, globose heads; marsh and grassland plants 27
- Leaves of various shapes, but not linear; inflorescences sessile or pedunculate, but peduncles distinctly shorter than subtending leaves; inflorescences of various shapes, but not forming dense, globose heads 28
27. Gynostegium cylindrical, connective appendages resting on style-head, caudicles shorter than pollinia, style-head fully exposed, clavate or ovoid **283. *Cordylogyne***
- Gynostegium barrel-shaped, connective appendages connivent over style-head, caudicles at least twice as long as pollinia, style-head fully hidden by connective appendages, flat or umbonate **295. *Periglossum***
28. Corolla lobes adaxially apically densely villous **284. *Fanninia***
- Corolla lobes not densely villous 29
29. Gynostegium sessile; free staminal corona lobes equaling gynostegium, erect, trifid with central lobe pronouncedly keeled; plants of E South Africa **301. *Woodia***
- Gynostegium elevated, free staminal corona lobes overtopping gynostegium, incurved, subrectangular; plants of African woodlands **285. *Glossostelma*** (*G. carsonii*)
30. Style-head distinctly conical, occasionally bifid 31
- Style-head flat, with a central depression or umbonate 35
31. Inflorescences sessile, fasciculate; occasionally with dark brown glandular hairs on the vegetative parts; pollinia often with sterile hyaline regions at the point of caudicle attachment **279. *Aspidoglossum***
- Inflorescences not sessile, or fasciculate; never with dark brown glandular hairs; pollinia always without sterile hyaline regions 32
32. Corolla fused for more than a quarter of its length, broadly campanulate, lobes reflexed; corona adnate to both corolla and gynostegium; follicles with protuberances; plants of South Africa **293. *Parapodium***
- Corolla only basally fused, of various shapes, but lobes normally not reflexed; corona not adnate to corolla; follicles with or without protuberances 33
33. Corolla lobes at least 9 mm long, usually brightly colored, horizontally spreading or reflexed; staminal corona lobes concave-cucullate over at least part of their length, inner apical margins forming a pair of teeth, occasionally with tooth in cavity; pollinaria with conspicuously winged and contorted caudicles; follicles smooth or longitudinally grooved **298. *Stathmostelma***
- Corolla lobes not more than 8 mm long, usually dull colored, not conspicuously spreading nor reflexed; staminal corona lobes not intricately folded; pollinaria without winged and contorted caudicles; follicles smooth or with protuberances 34

34. Staminal corona lobes exceeding gynostegium, usually subulate, S-shaped; plants delicate; leaves always linear, with revolute margins; inflorescences mostly terminal, surrounded by much longer leaves **299. *Stenostelma***
- Staminal corona lobes not exceeding gynostegium, solid, carnose; plants usually stout; leaves ovate, occasionally undulate, or linear; inflorescences often also along the stem **302. *Xysmalobium***

35. Pollinia with sterile hyaline regions 36
- Pollinia without sterile hyaline regions 37

36. Inflorescences pedunculate (rarely sessile), sciadioidal; pollinia dorsiventrally compressed, usually laterally attached to the caudicles, sterile hyaline regions apical, but never at the point of caudicle attachment; plants without glandular hairs on vegetative parts **296. *Schizoglossum***
- Inflorescences sessile, fasciculate; pollinia oblongoid, subapically attached to the caudicles, sterile hyaline regions at the point of caudicle attachment; plants occasionally with dark brown glandular hairs on vegetative parts **279. *Aspidoglossum***

37. Flowers nodding in the sciadioidal, extra-axillary inflorescences, very rarely in subglobose heads along the stem; leaves often non-linear; follicles often with soft spines 38
- Flowers erect, terminal inflorescence usually present; leaves often linear; follicles without soft spines 40

38. Corona lobes at least twice as long as gynostegium **279. *Aspidoglossum***
- Corona at the most slightly exceeding gynostegium 39

39. Staminal corona lobes attached to the back of the anthers, strongly laterally compressed and with a central cavity; rootstock fibrous, occasionally woody, or with a stout woody rhizomatous rootstock but never forming a tuberous caudex; corolla lobes usually ciliate; follicles often inflated **286. *Gomphocarpus***
- Staminal corona lobes always arising from the base of the staminal column and adnate to it over the whole length, highly diverse in shape; rootstock tuberous; corolla lobes usually not ciliate; follicles never inflated **300. *Trachycalymma***

40. Gynostegium elevated 41
- Gynostegium sessile 44

41. Stems usually unbranched; with a stout vertical rhizome from which fusiform lateral tuberous roots arise; leaves semi-succulent, venation obscure; inflorescences terminal and lateral; flowers broadly campanulate; staminal corona lobes carnose, united at the base, when longer than gynostegium then connivent above it **285. *Glossostelma***
- Stems branched or unbranched; without a stout vertical rhizome; leaves linear or herbaceous with apparent venation; inflorescences terminal; flowers not broadly campanulate; staminal corona lobes intricately folded, not united at the base, when longer than gynostegium then laterally expanding 42

42. Staminal corona lobes with conspicuous teeth in the cavities; single-stemmed plants of montane areas in southern Africa **280. *Aspidonepsis***
- Staminal corona lobes without conspicuous teeth in the cavities 43

43. Staminal corona lobes conspicuously laterally extended **277. *Aidomene***
- Staminal corona lobes not laterally extended **278. *Asclepias***

44. Leaves with stiff hairs, scabrid to touch, usually not linear; plants stout; flowers often broadly campanulate; corolla lobes at least 10 mm long; staminal corona lobes usually forming long apical or subapical appendages; follicles often winged **292. *Pachycarpus***
- Leaves glabrous or with soft indumentum, almost always linear; plants delicate; flowers rotate; corolla lobes mostly shorter than 10 mm; staminal corona lobes usually forming long apical or subapical appendages; follicles never winged **289. *Miraglossum***

45. Staminal corona lobes separate, with or without additional annular corolline corona 46
- Staminal corona lobes fused, at least basally, always without additional annular corolline corona 51

46. Staminal corona lobes exceeding gynostegium and connivent above it; corolla ciliate 47
- Staminal corona lobes not connivent above gynostegium, corolla not ciliate 48

47. Leaves narrowly elliptic or ovate; corolla broadly campanulate, almost entirely fused, white with dark red center; follicles ovoid, shortly beaked or inflated, solitary or paired, smooth; plants of riverine vegetation **291. *Oxystelma***
- Leaves broadly ovate, deeply cordate to reniform; corolla salverform, fused to less than half of total length, cream, yellowish or reddish; follicles always paired, narrowly clavate, with soft spines **294. *Pergularia***

48. Free corona lobes alternating with the anthers; latex white; plants of China (Sichuan) **303. *Cynanchum*** (former *Sichuania* species)

- Free corona lobes opposite the anthers; latex white, translucent or yellow 49

49. Flowers large, to 4 cm long, narrowly campanulate, adaxially glabrous; leaves deeply cordate to reniform; latex white; plants of Asia **303. *Cynanchum*** (former *Raphistemma* species)
- Corolla of various shapes, but at the most 2 cm long, adaxially usually with slender trichomes; lobes often conspicuously twisted; leaves basally cuneate, rounded or indistinctly cordate; latex usually translucent, but occasionally white or yellow 50

50. Guide rails conspicuously extended along the carinate corona lobes; corolla lobes almost free, apically often strongly twisted; leaves carnose; latex translucent; plants predominantly of seaside habitats **305. *Pentatropis***
- Guide rails not conspicuously extended along corona lobes; corolla lobes free or fused to various degrees, apically twisted or not, latex usually translucent, but occasionally white or yellow **306. *Vincetoxicum***

51. Inflorescences dichasially branched cymes 52
- Inflorescences simple 53

52. Corolla 2.5–6 mm long, adaxially with trichomes; staminal corona parts connate to the filament for not more than one third of gynostegium length; leaves not ciliate; calyx lobes acute or obtuse **306. *Vincetoxicum***
- Corolla 7–8 mm long, connate to the filament for more than one third of gynostegium length; leaves ciliate; calyx lobes apiculate; insufficiently known plants from Indonesia (Sulawesi) **378. *Mahawoa***

53. Latex translucent, plants without prophylls 54
- Latex white, plants often with prophylls 56

54. Style-head extending well beyond connective appendages, rostrate; plants of South Africa **276. *Oncinema***
- Style-head not extending well beyond connective appendages, flat, umbonate or conical 55

55. Leaves at the most 1.5 cm long, caducous; inflorescences bostrychoid, with conspicuously long rachis; plants of Somalia **281. *Calciphila*** (*C. galgalensis*)
- Leaves usually larger, persistent (but plants often geophytic); inflorescences without conspicuous rachis; plants of Eurasia **306. *Vincetoxicum***

56. Plants often with prophylls; corolla adaxially glabrous, corona usually ring-shaped, with staminal and interstaminal parts of the same thickness (but occasionally staminal parts with ligule) **303. *Cynanchum***
- Plants never with prophylls; corolla adaxially at least with a few trichomes; corona either only of staminal parts, or, when ring-shaped, then staminal parts thicker than interstaminal parts **306. *Vincetoxicum***

57. Plants entirely devoid of a corona 58
- Corona of various origin and composition present 64

58. Leaves linear 59
- Leaves ovate, cordiform or triangular 61

59. Plants strongly branched from a woody rootstock; corolla adaxially glabrous; plants of North America **327. *Funastrum*** (*F. utahense*)
- Plants sparsely branched, grass-like; corolla adaxially papillose or with trichomes; plants of South America 60

60. Leaves whorled or alternate; inflorescences subsessile; follicles obclavate **316. *Hemipogon***
- Leaves opposite; inflorescences long-pedunculate; follicles narrowly fusiform **320. *Morilloa***

61. Leaves less than 5 mm long, strongly recurved, plants erect, sparsely branched; plants of Brazil **319. *Minaria*** (*M. grazielae*)
- Leaves more than 5 mm long, not strongly recurved, when sparsely branched, then not erect 62

62. Plants apically twining subshrubs, leaves coriaceous; seeds almost orbicular, ca. 10 mm wide and to 5 mm thick, with a few short (to 5 mm long) yellowish trichomes; plants of southern Argentina **308. *Diplolepis*** (*D. hieronymi*)
- Plants of various growth forms, leaves herbaceous; seeds ovate, oblong or pyriform, not more than 5 mm wide and 1 mm thick; when coma present, then white, at least 10 mm long 63

63. Plants prostrate, hirsute all over, leaves ovate, basally rounded, sessile; inflorescences few-flowered, sessile; corolla lobes only basally fused; plants of Argentina and southern Brazil **321. *Nautonia***
- Plants erect or twining, but not prostrate, softly puberulous, pubescent or tomentose, leaves pronouncedly cordiform; inflorescences few- or many-flowered, distinctly pedunculate; corolla lobes fused for at least one third of their length, often more **330. *Philibertia***

64. Pollinia without sterile hyaline region 65
- Pollinia with sterile hyaline region 109

---

see Cynanchinae

65. Inflorescences distinctly compound 66
- Inflorescences simple, occasionally a single basal dichasium bearing two monochasia 69
66. Inflorescences forming elongate thyrses with strongly condensed, sessile partial inflorescences; rachis often conspicuously zigzag-shaped; gynostegial corona variable, but not cyathiform or tubular 67
- Inflorescences many-flowered dichasia; gynostegial corona cyathiform or tubular 68
67. Plants hirsute to tomentose; flowers pedicellate; staminal corona lobes distinctly shorter than gynostegium; follicles unknown; endemics of central Brazil **317. *Hypolobus***
- Plants with indumentum, but not distinctly hirsute; flowers sessile; staminal corona lobes equaling gynostegium; follicles usually cylindrical, paired **325. *Tassadia***
68. Plants medium-sized twiners; leaves (narrowly) ovate, often long-acuminate, basally rounded or indistinctly cordate, soft; inflorescences lax; corona equaling or exceeding gynostegium, style-head occasionally rostrate or capitate; follicles usually paired, slender, less than 15 cm long, with thin pericarp **309. *Jobinia***
- Plants large twiners (to 12 m); leaves roundly ovate, basally strongly cordate; inflorescences conspicuously storied, with condensed partial inflorescences in elongate dichasia; corona shorter than gynostegium; style-head flat; follicles solitary, ovoid, large (to 30 cm), with thick pericarp **303. *Cynanchum*** (former *Metalepis* species)
69. Style-head with five to seven pronounced appendages **329. *Oxypetalum*** (former *Schistogyne* species)
- Style-head of various shapes, but not with five to seven pronounced appendages 70
70. Translators with pronounced hyaline teeth (very rarely inconspicuous); style-head rostrate, bifid; corolla lobes often strongly twisted; corona of five free carnose lobes often adnate to corolla and gynostegium **329. *Oxypetalum***
- Translators without pronounced hyaline teeth; style-head variable, but when rostrate then usually not deeply bifid; corolla lobes rarely strongly twisted; corona usually not adnate to both corolla and gynostegium 71
71. Corona of three-dimensional staminal lobes adnate to the back of the anthers, well separated from each other 72
- Corona otherwise, when of staminal origin then laminar and not adnate to the back of the anthers 77
72. Leaves usually glabrous on both sides, occasionally ciliate, linear or narrowly ovate, basally obtuse or cuneate, usually discolorous, lateral veins conspicuously dense and parallel; flowers often nodding; corolla lobes often long-ciliate; staminal corona lobes often bicornous **314. *Blepharodon***
- Leaves with or without indumentum, elliptic, oblong, (ob-)ovate or triangular-deltate, basally often deeply cordate, rarely discolorous, lateral veins not conspicuously dense and parallel; flowers nodding or erect; corolla lobes finely ciliate or glabrous; staminal corona lobes never bicornous 73
73. Flowers nodding; corolla broadly campanulate to salverform 74
- Flowers usually erect; corolla rotate 75
74. Leaves elliptic, oblong or obovate; corolla large (at least 1.5 cm diam.); rare endemics of Andean high altitudes **307. *Pentacyphus***
- Leaves usually cordate, corolla usually smaller (less than 1.5 cm diam.) **330. *Philibertia***
75. Plants erect; staminal corona lobes differentiated into an outer hood and an inner horn **278. *Asclepias***
- Plants twining; staminal corona lobes not differentiated into hood and horn 76
76. Latex with a disagreeable scent; leaves often ovate-elliptical, rarely linear, basally distinctly cordate unless leaves very slender; corolla lobes ciliate, but adaxially glabrous **327. *Funastrum***
- Latex unscented; leaves usually elliptical, never cordate; corolla lobes not ciliate, but adaxially densely pubescent; plants of central South America **324. *Petalostelma***
77. Plants single-stemmed, occasionally very sparsely branched, erect perennials, not more than 1 m tall 78
- Plants usually twiners of variable size when erect then multi-stemmed 81
78. Flowers erect in a terminal bostryx; style-head rostrate, bifid; corolla lobes twisted; follicles solitary, erect, fusiform; swamp plants of Argentina, Bolivia, southern Brazil and Paraguay **329. *Oxypetalum*** (former *Widgrenia*)
- Flowers on long pedicels and peduncles, nodding, arising between the upper, occasionally reduced leaves; corolla lobes straight; follicles pendulous when solitary 79
79. Leaves linear, occasionally subwhorled; staminal corona lobes without a proximal lobule; follicles usually paired **322. *Nephradenia***

– Leaves sessile, broadly to very narrowly cordate; staminal corona lobes often with a proximal lobule; follicles usually single 80

80. Leaves stem-clasping, glabrous, often glaucous, margins straight, with conspicuously crowded and parallel lateral veins; peduncles and pedicels well-developed; plants of central Brazil and Bolivia **313. *Barjonia***

– Leaves not stem-clasping, often with dense indumentum, not glaucous, lateral veins not conspicuously crowded or parallel; flowers almost sessile **319. *Minaria***

81. Leaves distinctly elongate-triangular with truncate base or hastate, often margins slightly undulate; follicles ovoid, to 15 cm long, with thick pericarp, smooth, keeled, or strongly rugose, containing $>$ 100 seeds; large ($>$ 1 m), very lactiferous twiners; plants of southern South America, worldwide weeds, mainly in *Citrus* plantations 82

– Leaves not distinctly elongate-triangular and not pronouncedly truncate or hastate, not undulate; when so, then plants not exceeding 1 m; follicles always smaller, club-shaped or fusiform, containing $<$ 100 seeds 83

82. Style-head usually conspicuously clavate, massive; corolla campanulate to widely salverform, fused for at least one third of its length; gynostegium usually concealed in corolla; corona of free, short, erect, carnose lobes **326. *Araujia***

– Style-head umbonate or rostrate but slender; corolla rotate, only basally fused; gynostegium usually concealed in tubular corona **328. *Morrenia***

83. Style-head distinctly rostrate; mostly plants of southern South America 84

– Style-head of various forms, but not distinctly rostrate 89

84. Plants decumbent; corolla and corona chocolate brown; plants of Argentina and Uruguay **329. *Oxypetalum*** (former *Rhyssostelma*)

– Plants usually twining, more rarely erect; at least corona not chocolate brown 85

85. Leaves linear or narrowly triangular to cordate-sagittate; plants not exceeding 1 m, at the most weakly twining; plants of Argentina and Chile 86

– Leaves ovate, elliptic or cordate; plants usually larger, twining 87

86. Leaves linear; corona gynostegial, annular, much shorter than gynostegium **308. *Diplolepis*** (*D. australis*)

– Leaves mostly narrowly cordate-sagittate, occasionally narrowly triangular-truncate; corona corolline, usually 5-lobed, equaling to exceeding gynostegium (except for style-head) **331. *Tweedia***

87. Leaves pronouncedly cordate, often discolorous, soft; corona of five separate, carnose staminal lobes **330. *Philibertia***

– Leaves ovate or elliptic, but not distinctly cordate, usually concolorous; corona lobes at least basally fused 88

88. Flowers bright yellow, leaves elliptic; follicles single; plants of Chile **308. *Diplolepis*** (*D. menziesii*)

– Flowers of various colors, but not bright yellow; leaves ovate, often long-acuminate; follicles paired; plants of the Andes and Brazil **309. *Jobinia***

89. Inflorescences two per node 90

– Inflorescences one per node 96

90. Inflorescences clearly axillary (both inflorescences of a pair equally developed) 91

– Inflorescences subaxillary to extra-axillary (usually one inflorescence of a pair with a longer peduncle) 94

91. Inflorescences distinctly pedunculate 92

– Inflorescences sessile 93

92. Corona simple **309. *Jobinia***

– Corona double; plants of Brazil **323. *Peplonia***

93. Leaves glabrous, often discolorous; corona either simple, of five free or basally fused lobes, or double, with the outer corona enclosing the gynostegium; plants of Brazil **323. *Peplonia***

– Leaves often softly pilose; corona simple, fused over at least half of its length **311. *Orthosia***

94. Plants large twiners, leaves pronouncedly cordiform; inflorescences long-pedunculate, occasionally a basal dichasium with two bostrychoid partial inflorescences **303. *Cynanchum*** (subgen. *Mellichampia*)

– Plants small twiners, often with distinct long shoot–short shoot architecture, leaves linear, ovate or elliptic, but not pronouncedly cordiform; inflorescences sessile or short-pedunculate 95

95. Leaves not conspicuously distichous; inflorescences often seemingly axillary; corolla lobes adaxially without trichomes, but often papillose; gynostegium usually sessile **311. *Orthosia***

– Leaves conspicuously distichous esp. on short shoots; inflorescences always extra-axillary; corolla lobes adaxially glabrous or with trichomes; gynostegium stipitate or sessile **312. *Scyphostelma***

96. Corolla lobes apically with long vibratile, claviform or capitate trichomes **377. *Topea***

– Corolla lobes glabrous or with an indumentum, but without claviform or capitate trichomes 97

97. Corolla lobes adaxially papillose to barbate; corona of five separate, very rarely basally fused staminal lobes; follicles solitary 98
- Corolla lobes occasionally papillose or pilose, but not barbate; corona lobes at least basally fused to form a ring-, bowl-, or tube-shaped structure; follicles solitary or paired 101
98. Corona lobes without a proximal appendage 99
- Corona lobes with a proximal appendage 100
99. Plants twining, leaves petiolate, herbaceous **318. *Metastelma***
- Plants erect, leaves (sub-)sessile, often coriaceous **319. *Minaria***
100. Plants twining; leaves usually petiolate **315. *Ditassa***
- Plants erect; leaves often sessile; plants of Brazil **319. *Minaria***
101. Prophylls often present at nodes; leaf bases deeply cordate to almost lobate; leaves triangular; free corona parts often intricately folded or extended into long apices; follicles one, clavate **303. *Cynanchum*** (subgen. *Mellichampia*)
- Prophylls always absent; leaf bases indistinctly cordate, rounded or, truncate; corona either without free parts or, when with free parts forming long apices, then follicles always paired 102
102. Leaf blades ovate, often with an extended tip, usually longer than 5 cm 103
- Leaf blades linear to lanceolate, when of other shapes, then distinctly shorter than 5 cm 105
103. Leaves hirsute; corona connivent above gynostegium **310. *Monsanima*** (*M. tinguaensis*)
- Leaves usually glabrous, if with trichomes, then not hirsute; corona not connivent above gynostegium 104
104. Leaves carnose to subsucculent; corona cyathiform, weakly lobed; follicles usually one, fusiform or subclavate with long beak; plants of Argentina and Chile **308. *Diplolepis***
- Leaves soft; free corona parts occasionally extended into long apices; follicles always paired, usually 5 cm or longer, at acute angles, occasionally remaining fused at the tips **309. *Jobinia***
105. Leaves conspicuously whorled; corolla lobes densely pilose, plants of southern South America **324. *Petalostelma*** (*P. robertii*)
- Leaves opposite; corolla lobes glabrous or with trichomes, but not densely pilose 106
106. Corona united into a tube obscuring the gynostegium; tube folded inward apically; anther wings not extending along the entire length of the anther, basally strongly centrifugal; plants of central Brazil **310. *Monsanima*** (*M. morrenioides*)
- Corona either not united into a tube obscuring the gynostegium, or, when so, then the tube not folded inward apically; anther wings usually extending along the entire length of the anther, never strongly centrifugal basally 107
107. Small herbs, prostrate or somewhat twining, or shrublets; leaves usually ovate to orbicular, more rarely lanceolate or ovate-lanceolate; follicles one, fusiform, solitary; plants of Argentina and Chile **308. *Diplolepis***
- Plants profusely branched ramblers with a pronounced long shoot–short shoot architecture; leaves linear to ovate, never orbicular; follicles one or two, fusiform or cylindrical 108
108. Stems usually soon turning brown, leaves on short shoots often smaller and more crowded than on long shoots, when indistinct then long shoots not green; follicles single or paired, at various angles, distinctly fusiform or spindle-shaped and often beaked **312. *Scyphostelma***
- Stems green, leaves on long and short shoots similar in size and spacing, often leafless at flowering time; follicles always in pairs at obtuse angles, more or less cylindrical without distinct beak **311. *Orthosia***
109. Plants of South America 110
- Plants of Central America or the Caribbean 135
110. Mature branches and leaves inconspicuously short-pubescent to almost glabrous 111
- Mature branches and leaves moderate to densely pubescent 117
111. Base of mature leaf blades acute or obtuse to shallowly and broadly subcordate 112
- Base of mature leaf blades cordate 115
112. Abaxial leaf blade all over with dense or sparse short (0.2–0.35 mm long) spreading eglandular trichomes; style-head concave; anthers somewhat bent toward flower axis; pollinia subpendent from the caudicles **333. *Atrostemma***
- Abaxial leaf blade usually glabrous or with few eglandular and sometimes glandular trichomes on midvein or on surface at apex; style-head plane or convex, rarely rostrate; anthers and pollinia horizontal or almost so 113
113. Inflorescences long-pedunculate, peduncles 5–7.5 cm long; corolla campanulate, throat with 5 pubescent interlobular pads; corona gynostegial, staminal corona segments 3-dentate, adnate to the base of stipe; plants of SE Brazil **354. *Malinvaudia***

- Inflorescences short-pedunculate, sometimes subsessile, peduncles up to 1.7 cm long; corolla usually rotate or rotate-campanulate, limb or throat glabrous or pubescent, but then without pubescent pads; corona gynostegial, staminal corona segments of apically ligulate ascending ridges, adnate to the stipe for most of its length, or truncate and adnate to the base of stipe 114

114. Stems and leaves glabrous throughout; leaf colleters 2, in the distal quarter of the petiole; staminal corona of 5 truncate segments -adnate to base of stipe; plants of SE Brazil **339. *Coelostelma***
- Stems and leaves usually short pubescent, with at least some trichomes on nodes and on leaf blades along midvein and/or apex; leaf colleters 2–7 at the adaxial leaf base; staminal corona of apically ligulate ascending ridges adnate to the stipe for most of its length; plants of Central and South America **355. *Matelea***

115. Stem pubescence in two lines, of retrorse eglandular trichomes and spreading glandular trichomes; corolla lobes oblong, tightly contorted in bud; staminal corona segments of 5 fleshy, distinctly bilobed elements, interstaminal segments not differentiated; follicles pentagonal and 5-costate; plants of southern Brazil **352. *Lhotzkiella***
- Stem pubescence generally on entire surface, of spreading or retrorse eglandular trichomes and spreading glandular trichomes; corolla lobes ovate or oblong, moderate or loosely contorted in bud; staminal and interstaminal corona of flat, not distinctly bilobed, laterally connate segments, forming an annular or cup-shaped structure; follicles muricate 116

116. Basal stem internodes thinly suberized, not sulcate; stem nodes without axillary colleters; calyx with 4–5 colleters in each sinus; corolla lobes dull dark purplish, with several long flat and long white trichomes at adaxial apex; pollinia horizontal from the caudicles; follicles muricate, with blunt-tipped non-suberized tubercles; seeds without a coma; plants of riparian wet or rain forests of Amazonia **369. *Riparoampelos***
- Basal internodes with thick sulcate cork layer; stem nodes with several small axillary colleters; calyx usually with 1–2 colleters in each sinus; corolla lobes white, green, brown or purple, adaxially glabrous or pubescent, when pubescent, then trichomes short, on entire surface or on one side; pollinia pendent from the caudicles; follicles muricate, tubercles acute or with a thick irregular corky apex; seeds with a coma; plants of neotropical dry scrub, seasonally dry forests and savannas **349. *Ibatia***

117. Herbs or subshrubs with erect, prostrate, decumbent or short twining stems, usually much less than 4 m long when mature; plants mostly from dry savannas (cerrados), deserts, dry scrub or seasonally dry forests 118
- Subshrubs or shrubs with twining stems, 5–30 m long when mature; plants from seasonally dry, wet or rain tropical forests, or from temperate austral savannas or forests 123

118. Leaves distichous; stems and inflorescences generally prostrate; plants of the savannas of southern Brazil, E Paraguay and NE Argentina 119
- Leaves decussate; stems erect, sprawling or vining; inflorescences spreading or nodding 120

119. Inflorescences one per node, usually prostrate; corolla lobes adaxially verrucose at least at apex; corona gynostegial, of fused staminal and interstaminal segments, cup-shaped, with 10 internal rays; follicles muricate **334. *Austrochthamalia***
- Inflorescences 2–3 per node, erect or spreading; corolla lobes not verrucose; corona of basally connate segments forming a short tube surrounding gynostegium, with 5 erect laminar apical lobes; follicles almost smooth **335. *Brargentina***

120. Leaves early caducous, blades with conspicuously undulate margins; corolla salverform, with 5 fascicles of long white trichomes at the throat; corona gynostegial, annular, adnate to the corolla tube; pollinia horizontal or slightly ascendent; deserts of NW Peru **358. *Peruviasclepias***
- Leaves usually persistent, blades with entire or slightly undulate margins; corolla rotate to broadly campanulate, without fascicles of long white trichomes; corona gynostegial, of 5 free staminal segments, annular and 5-lobed, or cup-shaped, adnate to base of corolla and to the stipe; pollinia pendent 121

121. Stems with dense grayish tomentose pubescence, eglandular trichomes up to 0.45 mm long; corolla lobes erect and apically recurved, glabrous; corona of 5 free, bifid, conspicuously fimbriate segments; style-head truncate or slightly concave, with 5 bluntish scales at top; plants of dry vegetation in Gran Chaco, northern Argentina, SE Bolivia, southern Paraguay, southern Brazil **370. *Rojasia***
- Stems with dense yellowish or white-translucent pubescence, eglandular trichomes 0.25–3 mm long; corolla lobes usually spreading or slightly recurved,

glabrous or pubescent; corona of 5 partly or completely connate staminal and interstaminal segments, entire or lobed, glabrous or puberulous; style-head concave, convex or rostrate, without scales; plants of dry scrub, seasonally dry forests and savannas 122

122. Adaxial petiole base with several small digitate colleters; leaf blades usually ovate or elliptic, subcordate or cordate and with several colleters at base **349. *Ibatia***

- Adaxial petiole base without colleters; leaf blades oblong or oblong-lanceolate, obtuse to cuneate and without colleters at base **347. *Gyrostelma***

123. Corolla campanulate, urceolate, tubular or salverform; pollinia pendent from the caudicles 124

- Corolla rotate; pollinia horizontal or pendent from the caudicles 127

124. Gynostegial corona dark to blackish purple, adnate to the stipe, mostly exserted from the corolla; follicles muricate and hirsute; plants of southern and SE Brazil, Bolivia and northern to NE Argentina **340. *Cristobalia***

- Gynostegial corona usually white, pink or yellow, adnate to the corolla tube and the stipe, included in the corolla or exserted in the upper quarter; follicles winged or with wings and projections, glabrous or pubescent 125

125. Corolla urceolate or tubular, lobes usually not reticulate; corolline corona present; gynostegial corona lobes apically truncate or crenulate; follicles (5–)7-winged, or with 7–9 lines of short conic projections; plants widespread in the Neotropics **353. *Macroscepis***

- Corolla campanulate or salverform, lobes usually conspicuously reticulate; corolline corona absent; gynostegial corona lobes apically bifid-digitate; follicles with 5 wings and several blunt projections 126

126. Nodes usually with 1–2 subaxillary colleters; corolla campanulate; pedicels 2–6.2 times longer than peduncle plus rachis; plants of tropical wet or rain forests of E South America, below 800 m, one species on a sandstone mountain of the Venezuelan Guayana above 1500 m **359. *Phaeostemma***

- Nodes without subaxillary colleters; corolla generally salverform, narrow-campanulate in one species; pedicels 0.5 to 1.5 times longer than peduncle plus rachis; plants mainly of mountain wet forests above 1000 m in western and northern South America, from Peru to northern Venezuela **351. *Lachnostoma***

127. Corolla and calyx lobes strongly reflexed; corona gynostegial, shortly tubular or salverform and as long as or a bit shorter than gynostegium; pollinia pendent from the caudicles; follicles 5-winged, long-attenuate at apex; plants of wet and rain lowland and submontane forests, Central and South America **366. *Pseudolachnostoma***

- Calyx lobes usually ascendent, not reflexed, corolla lobes ascendent, spreading or slightly recurved; corona gynostegial, annular, shallowly cup-shaped, flabellate or of five distinct staminal segments, adnate to the stipe or to the corolla tube, shorter than gynostegium; pollinia horizontal from the caudicles; follicles smooth, with small conical or irregular projections, or winged; when winged, then annular corona present 128

128. Eglandular trichomes on stems, leaves and inflorescences usually translucent or light yellow, sparse to dense, 0.3–2.5(–3.5) mm long; gynostegial and corolline corona present; follicles 3–5-winged, rarely smooth 129

- Eglandular trichomes on stems, leaves and inflorescences usually light brown, yellowish brown or golden-yellow, 1.9–7 mm long; gynostegial corona present; corolline corona absent; follicles smooth or with short inconspicuous projections, unknown for *Orinoquia* 131

129. Lower internodes of mature stems conspicuously lenticellate; anthers with radial laminar dorsal appendages; widespread from northern America to Argentina **345. *Gonolobus***

- Lower internodes of mature stems usually thinly suberized, not lenticellate; anthers without dorsal appendages 130

130. Corolla lobes narrowly ovate to ovate-oblong, adaxial side basally densely covered with erect trichomes; style-head exserted **337. *Chloropetalum***

- Corolla lobes narrowly triangular, adaxial side basally lanose; style-head included in the corolla tube **373. *Tressensia***

131. Inflorescences 7–30-flowered, long-pedunculate; corolla lobes usually apically undulate and crenate, adaxially papillate along a medial line; gynostegium stipitate; anthers convex, sometimes dorsally inflated and vesicular; corona gynostegial, annular or 5-lobed, smooth, striate, sometimes rugose, mainly adnate to the stipe 132

- Inflorescences 2–6-flowered, subsessile or long-pedunculate; corolla lobes usually obtuse or emarginate at apex, not undulate and crenate nor papillate; gynostegium sessile or subsessile; anthers slightly convex or emarginate, not vesicular; gynostegial corona shallowly cup-shaped, flabellate or of five almost distinct staminal segments, not striate or rugose nor verrucose, adnate to base of corolla tube and to gynostegium 133

132. Anthers inflated, thick and vesicular dorsally, hiding most of style-head; staminal corona annular, sometimes slightly 5-lobed, smooth or somewhat striate, without radial projections **344. *Fischeria***

– Anthers convex, not inflated, with a thin translucent apical membrane, not hiding style head; staminal corona of five almost distinct, conspicuously rugose or verrucose lobes, frequently with five laterally flat or conic projections that arise radially from the stipe **368. *Rhytidostemma***

133. Inflorescences subsessile; calyx lobes oblong-elliptic, usually with 1–2 axillary colleters; gynostegial corona purple to black, forming a deeply 5-lobed disc; pollinia oblongoid or narrowly calceolate; follicles somewhat tuberculate, with small conic protuberances; plants of tropical rain forests and lower montane wet or rain forests from Colombia to Bolivia **365. *Pruskortizia***

– Inflorescences long-pedunculate; calyx lobes ovate, usually without colleters; gynostegial corona shallowly cup-shaped or of five almost distinct staminal lobes, yellowish as far as known; pollinia obovoid or obovoid-reniform; follicles smooth or slightly striate when dried (unknown for *Orinoquia*) 134

134. Stems, leaves and inflorescences with sparse to moderate pubescence of dark brown, long (1.9–3 mm) eglandular trichomes, and dense pubescence of dark brown to black glandular trichomes (0.3–0.4 mm long); corolla 17–18 mm diam., with deltoid lobes; pollinia 0.6–0.7 mm long; plants of non-flooded wet forests of French Guiana **346. *Graciemoriana***

– Stems, leaves and inflorescences with moderate pubescence of golden-yellow, eglandular, very long (5–7 mm long) trichomes, and scarce pubescence of translucent glandular trichomes (0.1–0.15 mm long); corolla 50–55 mm diam., with narrowly ovate lobes; pollinia 0.9–1 mm long; plants of wet or rain forest, at headwaters of the Orinoco river, Venezuela **357. *Orinoquia***

135. Plants of Central or North America 136

– Plants of the Caribbean 159

136. Mature branches and leaves inconspicuously short-pubescent or almost glabrous 137

– Mature branches and leaves moderately or densely puberulous to hirsute 141

137. Gynostegial corona and corolline corona (annular or cup-like, entire or discontinuous) present 138

– Only gynostegial corona present 139

138. Lower internodes of mature stems conspicuously lenticellate; anthers with radial laminar dorsal appendages **345. *Gonolobus***

– Lower internodes of mature stems usually thinly suberized, not lenticellate; anthers without dorsal appendages **337. *Chloropetalum***

139. Plants with a thick, erect, conspicuously corky caudex; pubescence of stems, leaves and inflorescence of appressed ferruginous more or less vermiform eglandular trichomes, these trichomes with thin non-sculptured walls, glandular-capitate trichomes few or absent; leaves early caducous, blades suborbicular; follicles narrowly fusiform, with short, obtuse, suberized projections **372. *Suberogerens***

– Plants without a caudex; pubescence of stems, leaves and inflorescence of spreading or retrorse eglandular trichomes and sparse glandular-capitate trichomes, eglandular trichomes with sculptured walls; leaf blades ovate, elliptic or oblong-lanceolate, usually persistent; follicles fusiform or fusiform-attenuate, winged, ridged or smooth, and sometimes with conic or uncinate, non-suberized projections 140

140. Leaf blade abaxial surface frequently with sparse, equally distributed pubescence of short, spreading, eglandular trichomes; style-head concave; anthers somewhat bent toward flower axis; pollinia subpendent; corona of partly connate staminal and interstaminal segments, staminal segments thickly laminar or fleshy, adnate to the stipe only at base, interstaminal segments deeply concave, usually puberulous abaxially; follicles with 5 wings and several conical or conical-uncinate projections **333. *Atrostemma***

– Leaf blade abaxial surface glabrous or with few eglandular trichomes, sometimes short eglandular and glandular trichomes on midvein and/or margins; style-head plane or convex, rarely rostrate; anthers and pollinia horizontal or almost so; corona of partly connate staminal and interstaminal segments, staminal segments fleshy, elevated, ligulate ridges adnate to stipe for most of its length, interstaminal segments plane or slightly concave, glabrous throughout; follicles 5-winged, 5-ridged or almost smooth **355. *Matelea***

141. Stems, leaves, inflorescences and follicles with dense white-silvery woolly indumentum, of intermixed

ribbon-like multicellular eglandular trichomes; corolla lobes with long, plumose, caudate, purple appendages; pollinia pendent, with a slender sterile horn-like basal process connected to the caudicles **374. *Trichosacme***

- Stems, leaves, inflorescences and follicles pubescent to glabrescent, pubescence generally hirsute, hirtellous or puberulous, usually with a mixture of short and long acicular or uncinate eglandular trichomes, or of glandular-capitate and eglandular acicular or uncinate trichomes; corolla lobes without plumose appendages; pollinia horizontal or pendent, without a horn-like basal process 142

142. Stems, leaves and inflorescences with glandular capitate trichomes with white crystalline inclusions; corona of 5 diversely digitately appendaged lobes; follicles long fusiform-cylindrical, smooth, glabrous and mottled **363. *Polystemma***

- Stems, leaves and inflorescences with translucent, glandular, capitate trichomes, yellowish brown or reddish, without white crystalline inclusions, remaining translucent or turning brown to black when dried; corona lobes diverse, not digitately appendaged (except in *Dictyanthus*, but then glandular trichomes translucent and follicles muricate); follicles ovate or fusiform, muricate or winged, or almost smooth and with few sparse conic projections, usually puberulent to hirsute, sometimes glabrous, not mottled 143

143. Corolla lobes undulate and crenate at apex; anthers inflated dorsally, vesicular, frequently pear-shaped; follicles usually obtuse at apex, smooth or with some small conic projections, not muricate or winged **344. *Fischeria***

- Corolla lobes marginally entire or undulate, sometimes marginally revolute but not undulate and crenate; anthers not inflated dorsally; follicles generally acute, acuminate or attenuate at apex, with smooth, muricate or winged surface 144

144. Eglandular long trichomes (5–6 celled) on conic or column-like multicellular bases present on stems, leaves, inflorescences and follicles; style-head long-rostrate, appendage narrowly clavate, apically obtuse; whole plant turning blackish when dried **364. *Prosthecidiscus***

- Eglandular long trichomes usually surrounded by a narrow ring of small modified epidermal cells present on stems, leaves, inflorescence and sometimes on follicles; style-head not rostrate, or short-rostrate and apically rugose or bifid; plants usually not turning blackish when dried 145

145. Corolla urceolate or tubular; follicles 7-winged or with 7–9 lines of conic projections; petioles with several (ca. 3–5) digitate axillary colleters **353. *Macroscepis***

- Corolla rotate to campanulate; follicles muricate or 3–5 winged, rarely smooth; petioles without axillary colleters except for small axillary colleters on petioles of *Ibatia*, but then follicles muricate 146

146. Herbs or subshrubs with erect, prostrate, decumbent or short twining stems, usually less than 3 m long when mature; plants mostly from dry savannas, deserts, dry scrub or seasonally dry forests 147

- Subshrubs or shrubs with twining stems, 5 to 40 m long when mature; plants from seasonally dry, wet or rain tropical forests, or from temperate forests or savannas 153

147. Erect herbs or subshrubs, up to 0.7 m tall; leaf blades attenuate, obtuse, acute, rounded or indistinctly cordate at base; follicles smooth, hirsute, with conspicuous mixed pubescence **360. *Pherotrichis***

- Prostrate, decumbent, sometimes erect herbs or short vining (sub-)shrubs, stems 0.2–3 m long; leaf blades hastate or cordate at base; follicles with obtuse, acute or uncinate projections, glabrous or with homogeneous or mixed pubescence 148

148. Corona digitately 5-lobed, with lobes partly adnate to the corolla tube; pubescence of stems, leaves and inflorescences mixed, consisting of eglandular, uncinate or spreading, long and short trichomes, and spreading glandular, short translucent trichomes; corolla tube internally convoluted with raised parts opposite corona lobes and sacs formed between them; corolla lobes often sharply revolute **342. *Dictyanthus***

- Corona diverse, lobes entire, dentate, fimbriate or ligulate-spathulate, not digitate, adnate to the corolla tube only at the base; pubescence of stems, leaves and inflorescences usually mixed, of eglandular and glandular capitate trichomes, without uncinate trichomes; corolla tube not internally convoluted; corolla lobes plane to somewhat revolute 149

149. Thick erect conspicuously corky caudex present; pubescence of stems, leaves, and inflorescences of appressed ferruginous, more or less vermiform eglandular trichomes, with thin non-sculptured walls; glandular-capitate trichomes few or absent; follicles narrowly fusiform, with short, obtuse, suberized projections **372. *Suberogerens***

- Caudex small or absent; pubescence on stems, leaves and inflorescences of spreading or retrorse translucent eglandular trichomes with sculptured walls, glandular capitate trichomes usually dense; follicles fusiform or ovate-fusiform, with short or long projections, with or without cork 150

150. Corolla tube adaxially with long, linear-spathulate whitish or purplish spreading trichomes 2.5–3.5 mm long; corona gynostegial, conical-truncate, adnate to stipe of gynostegium, marginally differentiated into 10 ligulate-spathulate erect lobes in pairs in interstaminal position, and 5 short, subulate lobes in staminal position; anthers with dorsal-apical dehiscence and without hyaline membranes **348. *Himantostemma***
- Corolla tube adaxially glabrous or pubescent, when pubescent then trichomes acicular or papillate white-translucent, less than 1 mm long; corona gynostegial, of 5 lobes at least basally connate, cyathiform, lobate, crenulate, dentate or fimbriate; anthers with lateral or dorso-lateral dehiscence and with hyaline margins 151

151. Stems usually twining, rarely trailing, not prostrate, usually single, with thick fissured cork on lower internodes; pollinia pendent, with a narrow hyaline sterile margin close to or along the proximal end (close to the caudicles); plants widespread in the Neotropics **349. *Ibatia***
- Stems prostrate or trailing, usually several or many, slightly or not suberized in lower internodes; pollinia horizontal or somewhat bent toward flower axis, with a hyaline furrow or margin along distal half 152

152. Leaf axils with 3–4 small digitate colleters, blades usually with 5 colleters at the base; corolla lobes spreading, adaxially barbate, with a small tooth-like projection at base; gynostegial corona tubular-urceolate, apically with 5 short involute ligulate lobes; plants endemic to sandhills and dry scrub in SE United States **343. *Edisonia***
- Leaf axils without colleters, blades without colleters or these early-caducous; corolla lobes usually ascending to erect, adaxially glabrous or sparsely pubescent, without a basal tooth; gynostegial corona 5-lobed, connate to 3/4 of its length, usually adnate to gynostegium; plants endemic to deserts of Northern Mexico (and probably to SW United States) **338. *Chthamalia***

153. Pubescence of eglandular short and long trichomes present on vegetative structures and inflorescences; corolline corona present 154
- Pubescence of glandular and eglandular trichomes present on vegetative structures and inflorescences; corolline corona present or absent 156

154. Stem and leaf pubescence of long eglandular trichomes 1.5–5.3 mm; corolla rotate, 36–48 mm diam., lobes spreading, ovate-orbicular or broadly deltoid; gynostegium sessile; corolline corona a shallow fleshy cup, roughly pentagonal with prominent radial angles; pollinia horizontal from the caudicles **371. *Rotundanthus***
- Stem and leaf pubescence of short (0.25–0.75 mm) and long (1–2.5 mm) eglandular trichomes; corolla campanulate or rotate-campanulate, 7–28 mm diam., lobes spreading to reflexed, ovate or oblong; gynostegium stipitate; corolline corona cupuliform-urceolate, shortly urceolate or tubular; pollinia pendent from the caudicle 155

155. Corolla lobes ovate, spreading; corolline corona tubular-cupuliform, apically fimbriate; follicles hispid, densely prickled, with long, curved, conspicuous projections; plants from upper mountain cloud forest; plants of northern Guatemala and southern Mexico **376. *Vulcanoa***
- Corolla lobes narrowly ovate to oblong, reflexed; corolline corona tubular or shortly funnel-shaped, apically entire or dentate, not fimbriate; follicles sparsely puberulous, 5-winged or 5-costate; plants from tropical or premontane rain forest; plants of Peru, SW Brazil, extending to Nicaragua **366. *Pseudolachnostomma***

156. Anthers with radial laminar dorsal appendages; gynostegial and corolline corona (annular, entire or discontinuous) present; follicles 3–5-winged, rarely smooth **345. *Gonolobus***
- Anthers without laminar dorsal appendages; only gynostegial corona present; follicles strongly muricate 157

157. Leaf blade adaxial base and interpetiolar line with fascicles of 30–50 colleters; corollas shallowly campanulate, 38–44 mm in diam., with broadly elliptic or ovate-elliptic lobes; pollinia pendent, oblong-spathulate, ca. 1.3 mm long; follicles dark-brown, densely pubescent, 22–27 cm long, with long, strongly curved projections 2.5–3.5 cm long; plants of Central America **336. *Bruceholstia***
- Leaf blade adaxial base with ca. 4 digitate colleters, interpetiolar line with 3–4 colleters; corollas rotate to shallowly campanulate-rotate, 7.5–25 mm in diam., with elliptic, narrowly deltoid to oblong-obovate lobes; pollinia horizontal or almost so, narrowly subquadrate to rounded-rectangular or almost reniform, 0.15–0.2 mm long; follicles green, 6–10 cm long, inconspicuously pubescent or

glabrescent, with antrorse or spreading projections 0.1–0.3 cm long; plants of the United States 158

158. Stout vines; inflorescences umbelliform cymes, 3–6-flowered; corolla rotate, lobes elliptic, reticulate; corona of 5 completely fused staminal and interstaminal elements, forming a marginally undulate low ring adnate to base of corolla, and 5 elevated ridges adnate to the gynostegial stipe; dorsal-external anther margin mostly covered by style-head **341. *Cyclodon***

– Herbaceous vines; inflorescences dichasial or racemiform cymes, usually 8–20-flowered; corollas shallowly campanulate-rotate, lobes usually oblong or spathulate, reticulate or not; corona cup-shaped, arising near the junction of the gynostegial stipe and the corolla, of 5 staminal elements partly united in a ring that reaches or surpasses style-head, each segment usually with two triangular teeth on abaxial side; anthers radially protruding from below style-head **356. *Odontostephana***

159. Stems, leaves and inflorescences sparsely to more or less densely covered with whitish multicellular eglandular trichomes, glandular capitate trichomes absent, or few and inconspicuous 160

– Stems, leaves and/or inflorescences sparsely to more or less densely covered with whitish, yellowish or light brown multicellular eglandular and glandular capitate trichomes 161

160. Leaf blades oblong-lanceolate, basally sagittate; staminal corona segments prominent-convex and cucullate (hooded), not ligulate; plants of Cuba **361. *Poicilla***

– Leaf blades ovate, oblong-ovate to narrowly elliptic, basally cuneate, rounded, obtuse, truncate or indistinctly cordate; staminal corona segments not cucullate, ligulate or not 164

161. Corolla campanulate, lobes glabrous, basally incurved with an eyespot-like concavity and large, white eyespot at apex; staminal corona a costate ridge adnate to stipe along most of its length, apically ligulate; follicles ovoid, 5-ridged; plants of Jamaica **350. *Jacaima***

– Corolla rotate-subcampanulate or urceolate, lobes pubescent at least on the adaxial face, spreading or somewhat curved, without a basal concavity and without ocelli; staminal corona laminate or thick-laminate, adnate to the stipe only at base, not ligulate; follicles fusiform, 10-ridged, smooth or with some isolated protuberances, strongly beaked in some species 162

162. Corolla rotate, lobes reticulate, pubescent on both surfaces; staminal corona lobes oblong, extending ca. 0.5 mm beyond style-head; coralline corona a protuberance at the base of the staminal corona; follicles oblong-fusiform, longitudinally 10-ridged; plants endemic to Hispaniola **332. *Anemotrochus*** (*A. viridivenius*)

– Corolla urceolate, lobes not reticulate, abaxially glabrous; staminal corona lobes well differentiated, or part of a highly complex system of fused staminal-interstaminal coronas, which are folded into the coralline corona, shorter or as long as gynostegium, sometimes apparently absent; coralline corona absent or forming a complicated system of ridges and bulges mostly in the fused part of the corolla; follicles smooth or with some isolated protuberances 163

163. Leaf blades tapering and without colleters at base; corolla lobes twisted, adaxially on left side pubescent with needle-like or short and blunt trichomes; coralline corona absent or a protuberance at base of staminal corona parts; follicles with protuberances; plants of the West Indies **332. *Anemotrochus***

– Leaf blades basally rounded or indistinctly cordate, with 2–4 colleters at base; corolla lobes carnose, not twisted, adaxially without needle-like or short and blunt trichomes; coralline corona part of a complicated system of ridges and bulges mostly in the fused part of the corolla, or apparently absent or reduced; follicles (as far as known) smooth and glabrous; plants of Cuba **375. *Tylodontia***

164. Mature leaf blades 2 × 1 cm to ca. 6 × 3 cm, basally rounded, obtuse or truncate, with glabrous or nearly glabrous surface, pubescent on veins 165

– Mature leaf blades usually from 6 × 4 cm to 20 × 15 cm or larger, basally cordate or subcordate (truncate to cuneate in few *Gonolobus* species, where dorsal laminar anther appendages and coralline coronas occur), usually with whole surface pubescent 166

165. Corolla lobes ovate, adaxially pubescent; staminal corona lobes obovate to suborbicular, rounded to emarginate at apex, with a small internal ligule, as long as the gynostegium; anther wings thick, slightly incurved; style-head flat; follicles subcylindrical-fusiform, not winged; plants of Cuba **362. *Poicillopsis***

– Corolla lobes linear-lanceolate, glabrous; staminal corona lobes swollen at base, subtriangular in front view, ridged, rising vertically and then connecting to the stipe below the anthers, without internal ligule, shorter than gynostegium; anther wings thin, with divergent apices; style-head conical or convex with a slightly raised protuberance; follicles fusiform, with 5 undulating wings (follicles unknown for one species); plants of Hispaniola **367. *Ptycanthera***

166. Mature stems with thick fissured cork on lower internodes; latex with unpleasant smell; pollinia pendent from the caudicles 167
- Mature stems lenticellate or with thin layer of cork on lower internodes; latex with or without an unpleasant smell; pollinia horizontal or slightly bent down from the caudicles 168

167. Inflorescence an umbelliform cyme; corolla urceolate or tubular; gynostegium generally included; staminal corona segments fleshy, basally adnate to the corolla tube over at least half of its length, free at apex and frequently concealing corolla throat; coroline corona present; follicles (5–)7-winged, 2 of the wings incomplete (Neotropics) **353. *Macroscepis***
- Inflorescence a racemiform cyme; corolla subcampanulate or campanulate, gynostegium generally exserted; staminal corona segments usually laminar, adnate to the base of the corolla tube, not concealing corolla throat; corolline corona absent; follicles muricate (Neotropics) **349. *Ibatia***

168. Mature stems 5–10 mm thick or thicker, with light brown pubescence of eglandular, spreading long trichomes (1.5–6 mm), and glandular capitate trichomes (0.15–0.4 mm), these turning black when dried; corolla lobes conspicuously undulate and crenate on one or both margins; gynostegial corona usually forming an annular structure adnate to the stipe, apically 5-lobed in some species; anthers inflated dorsally, vesicular, frequently pear-shaped; corolline corona absent; follicles broadly ovoid to fusiform, smooth or with some small protuberances, not muricate or winged (Neotropics) **344. *Fischeria***
- Mature stems generally 1.5–4 (–5) mm thick, usually with whitish or yellowish pubescence, eglandular trichomes spreading or retrorse (0.4–2 (–2.5) mm), glandular capitate trichomes (0.1–0.4 mm) translucent to light brown when dried; corolla lobes plane, sometimes somewhat concave or marginally undulate, but not undulate and crenate; gynostegial corona of staminal and interstaminal segments; gynostegial corona of staminal lobes adnate to base of corolla tube, usually ridge-like and raised, not ligulate, interstaminal segments flabellate, oblong-sulcate or cup-like; corolline corona present, annular but sometimes interrupted or discontinuous, or of pubescent mounts of tissue subopposite to the anthers; laminar dorsal anther appendages present or absent; follicles 3–5-winged, wings sometimes discontinuous or reduced, absent in one species 169

169. Laminar dorsal anther appendages present; mature stems conspicuously lenticellate (Neotropics and temperate North America) **345. *Gonolobus***
- Laminar dorsal anther appendages absent; mature stems with a thinly suberized layer, not lenticellate (Neotropics) **337. *Chloropetalum***

V.5.a. **Subtribe Astephaninae** Endl. ex Meisn. (1840).

Usually small twiners; latex translucent, leaves small, subcoriaceous. Inflorescences mostly subsessile to shortly pedunculate, sciadioidal; pollinia ovate to clavate; caudicles flattened with conspicuous hyaline margin. Follicles fusiform. In molecular analysis, Astephaninae are sister to the remaining Asclepiadeae (Liede 2001). Three genera in southern Africa.

274. ***Astephanus*** R. Br.

Astephanus R. Br., Asclepiadeae 43 (1810); Bruyns, Kew Bull. 58: 867–887 (2003), rev.
Haemax E. Mey. (1838).

Suffrutescent twiners, glabrous or sparsely puberulous, 0.15–3 m tall, roots fibrous. Leaves 5–20 (–36) cm long, ovate to oblong. Inflorescences 6–8-flowered. Corolla 3–5 mm long, tubular, highly fused; adaxially apically papillose and with a ring of stiff reflexed trichomes in the tube opposite the anthers; lobes oblong, spreading at anthesis; corona absent; gynostegium sessile; anther wings minute; style-head rostrate. Follicles 5–6 cm long. Seeds pyriform, wingless. 2n = 20.

Two spp. in South Africa´s winter rainfall area, Table Mountain sandstone and limestone; at low altitudes.

Hitherto all species of *Astephanus* reported from the New World have been found to belong to other genera, so that the genus, which is closely related to *Microloma*, is restricted to the two African species.

275. ***Microloma*** R. Br.

Microloma R. Br., Asclepiadeae 42 (1810); Wanntorp, Opera Botanica 98: 1–69 (1988), rev.; Bruyns & Linder, Bot. Jahrb. Syst. 112: 453–527 (1991), rev.; Bruyns, Asklepios 52: 71–74 (1991).

Suffrutescent twiners or rigid, spinescent shrubs, to 1 m tall, pubescent or tomentose; rootstock woody. Leaves occasionally caducous, occasionally alternate, subsessile, 0.4–7 cm long, linear to triangular. Inflorescences 1–15-flowered. Calyx lobes occasionally showy; corolla 2–9 mm long, urceolate to tubular, occasionally carnose, green and white, yellow or red, tube with stiff reflexed trichomes, occasionally with antesepalous ridges; lobes orbicular or triangular, twisted, keeled, usually ciliate, usually connivent into mouth of tube, closing it completely; occasionally with corolline corona knobs; gynostegium stipitate, connective appendages lanceolate with intertwined trichomes; style-head (depressed-) conical. Follicles 2–10 cm long. Seeds pyriform, wingless. 2n = 20.

12 spp. in four sections in South Africa and Namibia, winter rainfall area; to 2000 m. Bird-pollinated according to Pauw (1998).

276. ***Oncinema*** Arn.

Oncinema Arn., Edinb. New Philos. J. 17: 261 (1834); Liede, Ann. Missouri Bot. Gard. 88: 657–668 (2001), phyl.
Glossostephanus E. Mey. (1838).

Suffrutescent twiner, 50–75 cm tall, glabrous; rhizomatous. Leaves 4–8 cm long, linear to oblong. Inflorescences 6–12-flowered, pedunculate, bostrychoid. Flowers sweetly fragrant; corolla 6–7 mm long, rotate, greenish yellow, adaxially apically papillose, basally with trichomes; lobes linear, acute, twisted; gynostegial corona equaling gynostegium (except for style-head), of basally fused, subulate, erect staminal lobes; gynostegium sessile, style-head rostrate, neck obconical. Follicles 5.5–7.5 cm long. Seeds ovate, winged.

One sp., *O. lineare* (L. f.) Bullock, in South Africa (Cape Province), winter rainfall area; river banks, valleys; 30–1000 m.

V.5.b. **Subtribe Asclepiadinae** Decne.. ex Miq. (1857).

Usually erect herbs, often with root tubers. Leaves usually sessile, linear or narrowly elliptic. Inflorescences usually terminal and subterminal, extra-axillary, sciadioidal. Corolla usually rotate, basally fused; corona gynostegial, of free staminal lobes; gynostegium usually sessile; style-head flat to conical. Follicles fusiform or obclavate, often paired. Seeds ovate, winged, usually with white apical coma.

Generic circumscriptions in Asclepiadinae are still in a state of flux; following molecular analysis, Goyder (2009) proposed that all genera except for *Calotropis*, *Kanahia* and *Pergularia* be united under a wide concept of *Asclepias*. Here, *Asclepias* is restricted to the monophyletic New World species and the traditional concept for the African species is adopted, even though some of the genera are known to be polyphyletic. 26 genera in Africa, Asia and the Americas.

277. ***Aidomene*** Stopp

Aidomene Stopp, Bot. Jahrb. Syst. 87: 21 (1967).

Usually unbranched herb, ca. 50 cm tall, with root tubers; shoots annual. Leaves slightly ascending; 3–5 cm long, adaxially patchily puberulous, abaxially glabrous. Inflorescences 2–5-flowered, long-pedunculate. Corolla ca. 8 mm long, brown, glabrous; lobes ovate, marginally membranous; staminal corona lobes surpassing gynostegium, horizontally spreading, transversely lingulate, adaxially basally with wings forming a cup-like structure; pollinia clavate; caudicles horizontal, cylindrical. Follicles 4–5 cm long.

A single species, *A. parvula* Stopp, in Angola, moist grassland.

278. ***Asclepias*** L.

Asclepias L., Sp. Pl. 214. (1753); Woodson, Ann. Missouri Bot. Gard. 41: 1–211 (1954); Goyder et al., Ann. Missouri Bot. Gard. 94: 423–434 (2007), phyl.; Fishbein et al., Syst. Bot. 36: 1008–1023 (2011), phyl.
Acerates Elliott (1817).
Anthanotis Raf. (1817).
Podostigma Elliott (1817).
Otaria Kunth (1818).
Podostemma Greene (1897).
Asclepiodella Small (1933).

Herbs or shrubs, usually perennial, to 1 m tall, with fibrous, often carnose, roots. Leaves occasionally alternate or whorled, 2–17 cm long, 0.1–3 cm wide, occasionally triangular-deltate, elliptic or oblong. Inflorescences extra-axillary and terminal, (1–)7–30-flowered. Corolla 5–15 mm long, uniformly white, green, yellow, brownish, reddish or orange-red, glabrous or evenly covered with trichomes, lobes ovate to oblong, usually strongly reflexed; staminal corona lobes light colored, mostly surpassing gynostegium, differentiated into erect, cucullate distal lobules ("hood"), and inflexed, solid, spur-shaped proximal lobules ("horn"); gynostegium highly elevated by column, or, rarely, stipe; pollinia clavate; caudicles S-shaped, cylindrical or ribbon-shaped. Follicles occasionally with soft spines, occasionally inflated. Seeds rarely lacking a coma. 2n = 22.

About 100 spp. in the Americas, with a center of distribution in N America.

The results of Goyder et al. (2007) and Fishbein et al. (2011) suggest that *Asclepias* s. str. is a purely American genus; thus, the inclusion of several African species by Goyder (2009) is not followed here.

279. ***Aspidoglossum*** E. Mey.

Aspidoglossum E. Mey., Comm. Pl. Afr. Austr. 200 (1838); Kupicha, Kew Bull. 38: 599–672 (1984), rev.
Rhinolobium Arn. (1838).

Sparsely branched, apically pendulous, tuberous geophytes, 0.3–1.1 m tall, with dark brown glandular trichomes. Leaves occasionally alternate or whorled, 1–7.5 cm long, occasionally (ob-)ovate. Inflorescences 1–13-flowered, occasionally thyrsoidal, (sub-)sessile; flowers usually nodding. Corolla 2–8 mm long, creamish yellow or green, streaked reddish, occasionally with trichomes; lobes elliptic, occasionally remaining connate at apices, occasionally right-hand margin ciliate; staminal corona lobes usually surpassing gynostegium, erect, basally dorsally compressed, truncate or caudate, with erect or inflexed, triangular or ovate, often caudate proximal lobules; pollinia occasionally with apical germination pore. Follicles 2.5–9 cm long, occasionally with protuberances, with dense indumentum. Seeds pyriform.

35 spp. in four sections across tropical Africa, usually (moist) grassland and open woodland, occasionally in rock outcrops or coastal sandflats; 0–2800 m.

280. ***Aspidonepsis*** Nicholas & Goyder

Aspidonepsis Nicholas & Goyder, Bothalia 22: 24 (1992); Müller et al., Ill. Handb. Succ. Pl.: 18–20 (2002), part. rev.

Herbs, 17–65 cm tall, sparsely branched, (almost) glabrous, root tubers fusiform to globose. Leaves 0.5–13.5 cm long, occasionally lanceolate or falciform. Inflorescences 2–17-flowered. Corolla 4–10.5 mm long, greenish yellow or reddish brown, lobes elliptic or ovate, occasionally reflexed, occasionally revolute; staminal corona mostly surpassing gynostegium, lobes horizontally spreading, cucullate or unguiculate, occasionally basally or apically elongated, occasionally with shorter, erect to inflexed, subulate or liguliform proximal lobules; gynostegium atop a column, pollinia oblongoid or clavate; caudicles articulated; carpels occasionally with trichomes; style-head white or green. Follicles ca. 3.5 cm long, with dense indumentum.

Five spp. in two subgenera, E South Africa, high altitude mountain grasslands; (450–)1000–2100 m.

Closely related to *Aspidoglossum*.

281. ***Calciphila*** Liede & Meve

Calciphila Liede & Meve, Novon 16: 369–370 (2006); Liede-Schumann, Fl. Somalia 3: 138–139 (2006), as "*Calcareophilum*", reg. rev.

Small shrubs or herbaceous twiners, to 40 cm tall, densely pubescent; latex translucent. Leaves caducous, sessile, carnose, 0.15–1.4 cm long, elliptic or obovate. Inflorescences 3–40-flowered. Corolla 1.5–1.7 mm long, creamish, occasionally with red tinge, occasionally with verrucose trichomes; lobes ovate; gynostegial corona ring cyathiform or urceolate, exceeding gynostegium, creamish, of connate, ovate or bifid, sulcate to slightly cucullate staminal and shorter, rectangular or oblong interstaminal parts; pollinia oblongoid, caudicles flattened; style-head white, umbonate or tabular. Follicles 13–30 mm long, fusiform. Seeds oblong, winged, tuberculate.

Two spp. in Somalia, calcareous areas; 800–2000 m.

The genus is weakly associated with Asclepiadinae in molecular analysis (Surveswaran et al. 2014).

282. ***Calotropis*** R. Br.

Calotropis R. Br., Asclepiadeae 28 (1810); Rahman & Wilcock, Nord. J. Bot. 11: 301–308 (1991), rev.
Madorius Kuntze (1891), nom. illegit.

Shrubs to small trees or erect herbs (*C. acia* Buch.-Ham.), to 6 m tall; occasionally densely tomentose when young. Leaves coriaceous or carnose, 9–28 cm long, (ob-)ovate or elliptic, occasionally undulate. Inflorescences 3–17-flowered, occasionally thyrsoidal. Corolla 10–15 mm long, white, green or lilac; lobes valvate, ovate to lanceolate, occasionally undulate; free staminal corona lobes whitish or lilac, maximally equaling gynostegium, erect, solid, laterally compressed, basally with a coiled spur; gynostegium stipitate; pollinia clavate; caudicles declinate, cylindrical; style-head creamish yellow. Follicles 6.5–12.5 cm long, ellipsoid, with thick pericarp, inflated. $2n = 22$.

Three spp., Africa, Asia; naturalized in the arid tropics all over the world, roadsides, savannas, open places at low to medium altitudes.

283. ***Cordylogyne*** E. Mey.

Cordylogyne E. Mey., Comm. Pl. Afr. Austr. 218 (1838); Bester et al., Phytotaxa 321: 114–124 (2017).

Grass-like geophytes, 30–80 cm tall, puberulous; root tubers narrowly cylindrical. Leaves (sub-) sessile, linear, ascending to spreading, 5–12 cm long. Inflorescences 4–15-flowered, long-pedunculate. Corolla 2.5–8 mm long, campanulate, greenish-white to brown; lobes oblong, suberect, revolute; gynostegial corona of basally fused, erect, carnose, oblong staminal corona lobes with a transverse ridge or flap, shorter than gynostegium, with or without interstaminal teeth; gynostegium cylindrical; anther wings short; connective appendages ovate, appressed to style-head; pollinia linear-oblong; caudicles short, ribbon-shaped; corpusculum ovoid, small; style-head ovoid or clavate, protruding. Follicles erect, 9–10 cm long.

Two spp. in Namibia and South Africa, wetlands. Closely related to *Periglossum* (Bester and Nicholas 2016).

284. ***Fanninia*** Harv.

Fanninia Harv., Gen. S. Afr. Pl., ed. 2: 235 (1868); Meve, Ill. Handb. Succ. Pl. Asclepiadaceae: 137 (2002).

Densely villous herb. Leaves 3–5 cm long, lanceolate or elliptic. Inflorescences occasionally paired, 4–8-flowered, long-pedunculate. Corolla 10–15 mm long, white; lobes oblong to elliptic, spreading, with long, soft trichomes on apical parts; staminal corona lobes purplish red, surpassing gynostegium, erect, oblong, emarginate, with conspicuous midrib, with shorter, erect proximal lobules of two basally fused lingulate parts; gynostegium atop a column; connective appendages apically with fimbriate margins. Follicles paired, felty.

One sp., *F. caloglossa* Harv., in South Africa, montane grasslands; 750–2000 m

285. ***Glossostelma*** Schltr.

Glossostelma Schltr., J. Bot. 33: 321 (1895); Goyder, Kew Bull. 50: 527–555 (1995), rev.

Usually unbranched herbs, 0.2–1 m tall, (sub-) glabrous; with stout vertical rhizomes and several fusiform root tubers. Leaves 3–8(–10) cm long, occasionally (ob-)ovate. Inflorescences 2–7-flowered. Corolla 15–20 mm long, campanulate, cream, green or maroon, occasionally maculate, rarely apically papillose; lobes (ob-)ovate; staminal corona lobes occasionally basally fused, white, green, orange, or reddish, mostly surpassing gynostegium, erect, dolabriform or clavate, sometimes dorsally flattened, apically with 1–3 inflexed teeth; rarely with free shorter, rounded interstaminal lobes; gynostegium occasionally atop a column; pollinia occasionally falcate; caudicles occasionally geniculate, or articulated. Follicles paired, 7–13 cm long.

12 spp. in S and E Africa, deciduous woodland (miombo) or montane grasslands; (500–) 1200–2500 m.

286. ***Gomphocarpus*** R. Br.

Gomphocarpus R. Br., Asclepiadeae 26 (1810); Goyder & Nicholas, Kew Bull. 56: 769–836 (2001), rev.

Perennial herbs or shrubs 0.5–1.5(–3) m tall, often with indumentum, with woody rootstock or tap roots. Leaves occasionally whorled, occasionally carnose or coriaceous, 3–12(–18) cm long, linear to ovate. Inflorescences 4–15(–30)-flowered; flowers nodding. Corolla 8–10 mm long, whitish, yellowish, or reddish; lobes ovate, occasionally reflexed, ciliate to barbate on right margin; staminal corona lobes equaling gynostegium, erect, cucullate, occasionally with additional teeth; free interstaminal lobes minute; gynostegium occasionally atop a short column; pollinia obovoid; caudicles occasionally geniculate or articulated. Follicles 2–8 cm long, frequently ellipsoid, often with soft spines, occasionally inflated. 2n = 22.

20 spp. in Africa, naturalized in Madagascar, Australia, Asia, Central and South America, pioneers of open disturbed habitats; (150–)600–2700 m.

Probably a polyphyletic genus.

287. ***Kanahia*** R. Br.

Kanahia R. Br., Asclepiadeae 28 (1810); Field et al., Nord. J. Bot. 6: 787–792 (1986), consp.; Chaudhary, Fl. Kingd. Saudi Arabia 2.2: 35–36 (2001), reg. rev.; Goyder, Fl. Ethiopia and Eritrea 4.1: 121 (2003), reg. rev.

Herbs or shrubs, 1.2–1.5 m tall, strongly basally branched, glabrous; shoots with wide pith. Leaves to 20 cm long, with yellow interpetiolar line and glandular fringes. Inflorescences 20–25-flowered. Flowers occasionally sweetly fragrant; corolla 9–22 mm long, occasionally cyathiform or campanulate, white or greenish; lobes ovate, ciliate to barbate; staminal corona lobes white, maximally equaling gynostegium, erect to inflexed, dolabriform, grooved on upper surface; gynostegium atop a column; pollinia oblongoid or falcate; caudicles convexly recurved or straight, cylindrical. Follicles occasionally paired, 5–6 cm long, occasionally ellipsoid. Seeds pyriform, convex, margin strongly revolute. 2n = 22.

At least two spp. in tropical Africa and Arabia, rheophytic; to 2200 m.

Flowers (corona) very variable.

288. ***Margaretta*** Oliv.

Margaretta Oliv., Trans. Linn. Soc. London 29: 111 (1875); Mwanyambo, Kew Bull. 51: 717–728 (1996), tax.

Sparsely branched herb, 50–70 cm tall, sparsely hirsute; root tuber carrot-shaped. Leaves 2.5–19.5 cm long, lanceolate. Inflorescences 3–10-flowered, pedunculate. Corolla 3.5–10.5 mm long, whitish, orange-red or lilac; lobes lanceolate, reflexed and revolute; free staminal corona lobes orange-red or lilac, obovate, erect, surpassing gynostegium, lower part claw-like, occasionally with two lateral teeth, or with tooth on midrib; upper part petaloid, with entire or lacerate margins; gynostegium concealed between corona lobes; pollinia clavate or oblongoid; caudicles S-shaped, rectangular; style-head occasionally capitate or semi-globose. Follicles 6–9 cm long. Seeds with revolute margin.

One sp., *M. rosea* Oliv., with six subspecies, in tropical Africa.

Closely allied to *Stathmostelma*.

289. ***Miraglossum*** Kupicha

Miraglossum Kupicha, Kew Bull. 38: 625 (1984); Kupicha, Kew Bull. 38: 599–672 (1984), rev.

Unbranched tuberous geophytes, 10–60 cm tall, long-pilose. Leaves occasionally alternate or whorled, 2–5 cm long, occasionally triangular-deltate, revolute. Inflorescences 2–9-flowered, sessile; calyx almost equaling corolla. Corolla 5–15 mm long, (brownish) green, occasionally streaked with purple; lobes elliptic, densely puberulent except for ciliate left-hand margin; free staminal corona lobes surpassing gynostegium, carnose, rectangular, apiculate or cruciform, occasionally with longer, filiform proximal lobules; pollinia oblongoid to falcate; caudicles ribbon-shaped, occasionally with hyaline margin. Follicles 3.5–7 cm long, occasionally with knobby protuberances, with dense indumentum.

Seven spp. in South Africa, grassland, stony slopes; 600–2400 m.

290. ***Odontostelma*** Rendle

Odontostelma Rendle, J. Bot. 32: 161, t. 344 (1894); Meve, Ill. Handb. Succ. Pl. Asclepiadaceae 186 (2002).

Sparsely branched herbs, 15–30 cm tall, glabrous; with carnose fusiform roots or a spindle-shaped root tuber. Leaves slightly ascending, 5–7.5 cm long. Inflorescences 3–6-flowered, pedunculate. Corolla 8–10 mm long, campanulate; lobes spathulate; gynostegial corona carnose, short, adnate to both corolla and gynostegium, of ovate, erect, long staminal parts and short, rectangular, bifid interstaminal parts basally fused into a ring;

pollinia oblongoid; caudicles horizontal, cylindrical. Follicles ca. 9 cm long.

Two spp. in Angola and Zimbabwe.

291. ***Oxystelma*** R. Br.

Oxystelma R. Br., Prodr. 462 (1810); Jagtap & Singh, Fasc. Fl. India 24: 33–35 (1999).

Suffrutescent twiners to 4 m, glabrous; rhizomatous; shoots green, basally often suberose. Leaves papery, 6–11 cm long, linear to ovate. Inflorescences (1–)2–4-flowered. Corolla 8–12.5 mm long, broadly campanulate, glabrous, white, reticulate with purple-red; lobes triangular, densely ciliate; corolline corona inserted near base of corolla, densely covered with trichomes, short, annular; staminal corona lobes ovoid, inflated, extended into a long tip connivent above gynostegium; caudicles declinate, triangular; style-head rostrate. Follicles ca. 5 cm long, narrowly obclavate or broadly ovoid, occasionally inflated. Seeds ovate, winged. 2n = 22.

Two spp. in E and W Africa, tropical Asia, riverine vegetation.

Molecular analyses indicate an association to Asclepiadinae (Liede and Täuber 2000; Surveswaran et al. 2014), and the structure of the staminal corona corroborates this placement.

292. ***Pachycarpus*** E. Mey.

Pachycarpus E. Mey., Comm. Pl. Afr. Austr. 209 (1838); Smith, S. Afr. J. Bot. 54: 399–439 (1988), reg. rev.; Goyder, Kew Bull. 53: 335–374 (1998), rev.

Sparsely branched herbs, 50–70 cm tall, densely pubescent, with slender, vertical root tubers. Leaves spreading, occasionally coriaceous, often discolorous, 7–12(–15) cm long, lanceolate, elliptic (-ovate), scabrous. Inflorescences 5–18-flowered. Corolla 10–20 mm long, occasionally campanulate or cyathiform, green, creamish or purple, often dark-patterned, occasionally papillose; lobes oblong to (ob-)ovate, occasionally with trichomes; free staminal corona lobes equaling gynostegium, spreading, cucullate, consisting of a dorsiventrally flattened plate with 1–2 erect keels; occasionally upper part papillose; pollinia (ob-)ovoid or falcate, occasionally attached to caudicles through a clasping overlap. Follicles 4–12 cm long, ellipsoid or obclavate, occasionally keeled or winged, occasionally inflated. Seeds wingless. 2n = 22.

38 spp. in tropical Africa, montane grassland, open woodland; (100–)700–2500(–2700) m.

Probably polyphyletic (Goyder et al. 2007; Fishbein et al. 2011). Wasp-pollinated (Shuttleworth and Johnson 2009).

293. ***Parapodium*** E. Mey.

Parapodium E. Mey., Comm. Pl. Afr. Austr. 221 (1838).
Rhombonema Schltr. (1895).

Sparsely branched herbs, 20–40 cm tall, sparsely puberulous in two lines. Leaves 4–9 cm long, oblong. Corolla ca. 10 mm long, campanulate; lobes ovate, reflexed; gynostegial corona adnate to both corolla and gynostegium; staminal corona lobes shorter than gynostegium, erect, rhomboid; with minute, free interstaminal lobes; gynostegium concealed in corolla; pollinia oblongoid; caudicles declinate, cylindrical; style-head apically bifid. Follicles paired, ca. 11 cm long, ellipsoid, with soft spines. Seeds with revolute and strongly folded wing.

Three spp. in South Africa, grasslands.

294. ***Pergularia*** L.

Pergularia L., Syst. Nat., ed. 12, 2: 135, 191 & Mant. Pl. 8, 53 (1767); Goyder, Kew Bull. 61: 245–256 (2006).

Suffrutescent twiners or erect herbs, to 4 m tall, pubescent or tomentose. Leaves softly herbaceous, 3–8 cm long, ovate, deeply cordate. Inflorescences 10–25-flowered, peduncles and pedicels long, thin. Flowers nodding, often sweetly fragrant; corolla 10–12 mm long, salverform, whitish to yellowish, occasionally with trichomes in the tube; lobes ovate to oblong, ciliate or barbate; staminal corona lobes surpassing gynostegium, obliquely erect, slenderly fusiform; free interstaminal corona lobes much shorter, bifid; pollinia clavate, with distal germination pore; caudicles horizontal, cylindrical, with pronounced hyaline margin; carpels with trichomes. Follicles paired, 4–9 cm long, with soft spines. 2n = 22.

Two spp., one with three subspecies, Africa, Arabia and Asia, savanna, arid areas. Moth-pollinated (Bhatnagar 1986).

Sister to the remaining Asclepiadinae s. str. (Goyder et al. 2007).

295. ***Periglossum*** Decne.

Periglossum Decne., Prodr. 8: 520 (1844); Bester & Nicholas, Phytotaxa 282: 28–36 (2016).

Grass-like geophytes, 30–80 cm tall, puberulous; root tubers shortly napiform. Leaves (sub-)sessile, linear, ascending to spreading, 2.5–14 cm long. Inflorescences 4–20-flowered, long-pedunculate. Corolla 4–9 mm long, campanulate, white, green or maroon; lobes oblong to lanceolate, erect, apically reflexed, revolute; gynostegial corona of basally fused, erect, fleshy, oblong to ovate staminal lobes, ventrally occasionally with keels or flaps, surpassing gynostegium, with tiny interstaminal teeth; gynostegium barrel-shaped; anther wings short; connective appendages cordiform, connivent over style-head; pollinia falcate, with apical sterile hyaline region; caudicles attached through a clasping overlap, ribbon-shaped; corpusculum ellipsoid, small. Follicles erect, ca. 7 cm long.

Five spp. in Mozambique, South Africa, Swaziland and Zimbabwe; wetlands to grasslands. Closely related to *Cordylogyne* (Bester and Nicholas 2016).

296. ***Schizoglossum*** E. Mey.

Schizoglossum E. Mey., Comm. Pl. Afr. Austr. 218 (1838); Kupicha, Kew Bull. 38: 599–672 (1984), rev.
Lagarinthus E. Mey. (1838).
Mackenia Harv. (1868).

Sparsely branched, tuberous geophytes, 7–130 cm tall. Leaves occasionally whorled or alternate, 2–8.5 cm long, occasionally triangular or oblong, sagittate. Inflorescences 3–15-flowered. Corolla 5–10 mm long, white, greenish brown, or purplish, occasionally streaked darker, occasionally with trichomes; lobes elliptic or cucullate, marginally revolute, occasionally ciliate; free staminal corona lobes whitish, yellow or green, surpassing gynostegium, subulate to subglobose, occasionally deeply bifid, occasionally emarginate; occasionally with deeply bifid proximal lobules exceeding distal lobules; gynostegium often stipitate; pollinia with apical germination pore. Follicles 4.5–8 cm long.

About 15 spp. in E and S Africa, grasslands, often near streams; 10–2700 m.

297. ***Solenostemma*** Hayne

Solenostemma Hayne, Getreue Darstell. Gew. 9: ad t. 38 (1825); Chaudhary, Fl. Kingd. Saudi Arabia 2.2: 39 (2001), reg. rev.
Argelia Decne. (1838), nom. superfl.

Erect herb, 50–75 cm tall, (sub-)glabrous; rootstock woody. Leaves coriaceous, elliptic. Inflorescences many-flowered, thyrsoidal. Corolla ca. 7 mm long, rotate, white, glabrous; lobes lanceolate; corolline corona shorter than both corolla and gynostegium, five-partite; lobes laterally connate, antepetalous, ovate, erect; gynostegium stipitate, raised high above corolla; pollinia oblongoid; caudicles horizontal, cylindrical; style-head flat. Follicles 3–4 cm long, obclavate, with thick pericarp. Seeds marginally with crenulate, revolute margin. 2n = 22.

One sp., *S. oleifolium* (Nect.) Bullock & E.A. Bruce ex Bullock, in N Africa and Arabia, rocky valleys.

298. ***Stathmostelma*** K. Schum.

Stathmostelma K. Schum., Bot. Jahrb. Syst. 17: 129 (1893); Goyder, Kew Bull. 53: 577–616 (1998), rev.

Sparsely branched, (sub-)glabrous geophytes, with stout, vertical tubers,10–150 cm tall. Leaves 10–24 cm long, occasionally ovate. Inflorescences 3–8-flowered. Corolla 9–20 mm long, white, green, yellow or orange-red, (sub-)glabrous; lobes ovate or lanceolate, often reflexed, occasionally revolute; staminal corona lobes yellow, orange or green, glabrous except along midline of cavity, equaling gynostegium, cucullate, inner apical margins forming a pair of teeth, occasionally with tooth in cavity; gynostegium often atop a column; anther wings occasionally forming a contorted basal tail; caudicles articulated with filiform distal portion broadening abruptly into clasping overlap; corpusculum black. Follicles 8–16 cm long. Seeds with inflated wing thicker than body. 2n = 22.

13 spp. in tropical Africa, grasslands, often seasonally waterlogged, *Brachystegia* woodland; 0–2300 m.

Closely allied to *Margaretta*.

299. ***Stenostelma*** Schltr.

Stenostelma Schltr., Bot. Jahrb. Syst. 18, Beibl. 45: 6 (1894); Müller et al., Ill. Handb. Succ. Pl., Asclepiadaceae 263–264 (2001), consp.
Krebsia Harv. (1868), non *Krebsia* Eckl. & Zeyh. (1836), Fabaceae.

Sparsely branched herbs, 10–50 cm tall, pilose; roots a single tuber. Leaves slightly ascending; linear to filiform, revolute. Inflorescences 8–12-flowered. Corolla 3–5 mm long, campanulate to urceolate; lobes lanceolate or obovate, apically reflexed; free staminal corona lobes far exceeding gynostegium, carnose, usually subulate, S-shaped; with free, tiny interstaminal teeth; gynostegium concealed in corolla; pollinia ovoid to falcate; caudicles articulated, twisted; style-head (elongate-) conical, apically bifid. Follicles obclavate. 2n = 22.

About four spp. in southern Africa.

300. ***Trachycalymma*** (K. Schum.) Bullock

Trachycalymma (K. Schum.) Bullock, Kew. Bull. 1953: 348 (1953); Goyder, Kew Bull. 56: 129–161 (2001), rev.; Goyder et al., Ann. Missouri Bot. Garden 94: 423–434 (2007), phyl.; Fishbein et al., Syst. Bot. 36: 1008–1023 (2011), phyl.
Gomphocarpus subsect. *Trachycalymma* K. Schum. (1895).

Sparsely branched tuberous geophytes, 10–60 cm tall, densely spreading-pubescent. Leaves 3–16 cm long, linear-lanceolate or oblong-ovate, ciliate. Inflorescences 5–40-flowered; flowers usually nodding. Corolla 3–10 mm long, white, greenish, or purplish, occasionally dark-veined, occasionally papillose; lobes ovate; free staminal corona lobes equaling gynostegium, erect, cucullate, occasionally with fringed margins, papillose and occasionally with tooth in cavity; gynostegium often atop a column; pollinia obovoid or clavate, caudicles articulated; attached through a clasping overlap. Follicles 5–15 cm long, occasionally obclavate, occasionally with soft spines.

Ten spp. in tropical Africa, montane and plateau grasslands or deciduous woodlands; 900–2600 m.

Goyder et al. (2007) and Fishbein et al. (2011) retrieved *T. pseudofimbriatum* Goyder from Ethiopia as sister to all Asclepiadinae except for *Calotropis*, *Kanahia* and *Pergularia*, while the remaining species seem to form a polyphyletic group.

301. ***Woodia*** Schltr.

Woodia Schltr., Bot. Jahrb. Syst. 18. Beibl. 45: 30 (1894).

Sparsely branched herbs, 15–25 cm tall, sparsely puberulous. Leaves 4–7 cm long, oblong to lanceolate, sharply apiculate, marginally thickened, undulate and minutely scabrous, glabrous. Inflorescences 4–8-flowered; calyx equaling or exceeding corolla, campanulate. Corolla 5–7 mm long, greenish brown; lobes ovate, inflexed; free staminal corona lobes equaling gynostegium, erect, trifid with central lobe pronouncedly keeled, the two lateral ones much shorter; connective appendages transversely lunate, strongly inflexed; pollinia obovoid. Follicles 7.5–12 cm long, with soft spines. Seeds wingless.

Three spp. in South Africa.

302. ***Xysmalobium*** R. Br.

Xysmalobium R. Br., Asclepiadeae 27 (1810); Goyder, Fl. Ethiopia and Eritrea 4.1: 131–133 (2003), reg. rev.

Sparsely branched herbs, 30–80 cm tall, (sub-) glabrous; roots cylindrical, to 1 m long tubers. Leaves 3–15 cm long, frequently undulate. Inflorescences 4–20-flowered. Corolla 4–8 mm long, white, green or brownish purple, occasionally reticulate with brown, occasionally with erect trichomes; lobes triangular, spreading or reflexed, often revolute, rarely barbate; staminal corona lobes basally fused, not exceeding gynostegium, erect, oblongoid, subglobose or dolabriform; pollinia ovoid or clavate. Follicles solitary, 10–20 cm long, ellipsoid, with soft spines and thick pericarp. Seeds with undulate wing. 2n = 22.

46 spp. in southern and E Africa, grasslands.

X. undulatum (L.) W.T. Aiton shows a bimodal pollination system with a chafer beetle and a pompilid wasp acting as pollinators (Shuttleworth and Johnson 2008).

V.5.c. **Subtribe Cynanchinae** K. Schum. (1895).

Usually suffrutescent twiners, often with stipule-like small leaves, constituting prophylls of an extremely reduced short shoot. Leaves ovate, or cordiform, basally usually distinctly cordate. Inflorescences extra-axillary, condensed-bostrychoid or sciadioidal. Corolla rotate to campanulate; lobes ovate, oblong or triangular, glabrous or with equally distributed verrucose trichomes; gynostegial corona usually a ring of fused staminal and interstaminal parts, usually glabrous, white; pollinia ovoid, subapically to laterally attached to the caudicles; caudicles usually declinate, cylindrical. Follicles usually solitary, fusiform to obclavate, usually with thin pericarp. Seeds ovate, often papillose or tuberculate, winged. Two genera in the tropics and subtropics worldwide.

303. ***Cynanchum*** L.

Cynanchum L., Sp. Pl. 212 (1753); Sundell, Evol. Monogr. 5: 1–63 (1981), reg. rev.; Liede, Ann. Missouri Bot. Gard. 83: 283–345 (1996), reg. rev.; Liede & Kunze, Org. Divers. Evol. 2: 239–269 (2002), phyl.; Khanum et al. Taxon 65: 467–486 (2016), phyl., tax.
Holostemma R. Br. (1810).
Metaplexis R. Br. (1810).
Sarcostemma R. Br. (1810).
Ampelamus Raf. (1819).
Raphistemma Wall. (1831).
Decanema Decne. (1838).
Glossonema Decne. (1838).
Odontanthera Wight (1838).
Pentarrhinum E. Mey. (1838).
Pycnoneurum Decne. (1838).
Steinheilia Decne (1838).
Metalepis Griseb. (1866).
Graphistemma (Champ. ex Benth.) Champ. ex Benth. (1876).
Adelostemma Hook. f. (1883).
Telminostelma E. Fourn. (1885).
Mellichampia A. Gray ex S. Watson (1887).
Vohemaria Buchenau (1889).
Flanagania Schltr. (1894).
Platykeleba N.E. Br. (1895).
Decanemopsis Costantin & Gall. (1906).
Folotsia Costantin & Bois (1908).
Mahafalia Jum. & H. Perrier (1911).
Nematostemma Choux (1921).
Karimbolea Desc. (1960).
Seshagiria Ansari & Hemadri (1971).
Rhodostegiella (Pobed.) C.Y. Wu & D.Z. Li (1990).
Sichuania M.G. Gilbert & P.T. Li (1995).

Suffrutescent twiners, lianas, erect or prostrate herbs, stem succulent twiners or erect stem succulents, 0.3–4(–20) m tall, latex white; occasionally rhizomatous; occasionally with root tubers. Leaves often cordate; scale-like in most stem succulents; often with ovate prophylls. Inflorescences (3–)5–15(–50)-flowered. Corolla 3–10(–40) mm long, rarely tubular, elongate-conical, or urceolate, white, cream, green, yellow, brown or rose, occasionally with verrucose trichomes; lobes usually straight; gynostegial corona highly variable, rarely absent, usually with a ring of fused staminal and interstaminal parts, staminal parts occasionally with adaxial appendage; occasionally with additional free staminal lobes connate to ring (former *Sarcostemma* species); gynostegium sessile or stipitate, style-head umbonate to rostrate. Follicles thin-walled, club-shaped, often keeled or winged, or thick-walled, then sometimes with soft spines. 2n = 22.

About 250 spp., cosmopolitan, with centers of distribution in Africa and Asia, very variable, often in slightly disturbed habitats.

The concept of *Cynanchum* has been changed considerably over the last ten years; for circumscription see Liede and Täuber (2002). Recent molecular phylogenetic work (Khanum et al. 2016) has shown that several small genera with atypical morphology have to be included in *Cynanchum*.

304. ***Schizostephanus*** Hochst. ex K.Schum. (1893).

Schizostephanus Hochst. ex K.Schum., Bot. Jahrb. Syst. 17: 139 (1893); Liede, Bot. Jahrb. Syst. 114: 503–550 (1993), rev.; Liede, Fl. Ethiopia and Eritrea 3: 136–137 (2003), reg. rev.; Bruyns & Klak, S. Afr. J. Bot. 75: 532–536 (2009), phyl.

Stem succulent shrubs or twiners, 0.5–4 m tall, glabrous; shoots lenticellate, 5–25 mm diam.; latex translucent. Leaves 3.5–8 cm long, ovate-cordate. Inflorescences 15–30-flowered. Flowers sweetly fragrant; corolla 5–6 mm long, yellow, basally maroon, lobes oblong, occasionally basally with long, slender trichomes; gynostegial corona ring tubular, exceeding gynostegium, staminal and interstaminal parts ovate to triangular, erect, marginally occasionally emarginate; occasionally with additional short inner rectangular staminal lobes; gynostegium stipitate; pollinia clavate, with apical-distal pellucid margin; caudicles declinate, triangular; style-head flat to umbonate, neck conical. Follicles 4.5–5 cm long. Seeds pyriform, winged. 2n = 22.

Two spp. in S and E Africa, steep slopes, rocky sites in open scrub.

Phylogenetic analyses have retrieved *Schizostephanus* as sister to the speciose genus *Cynanchum* (Bruyns and Klak 2009; Khanum et al. 2016).

V.5.d. **Subtribe Tylophorinae** K. Schum. (1895).

Usually suffrutescent twiners; latex usually translucent. Leaves ovate or elliptic, basally occasionally indistinctly cordate. Inflorescences extra-axillary, lax, often dichasial at base, bostrychoid monochasial in higher order branching, often many-flowered. Corolla rotate to campanulate, occasionally with evenly distributed verrucose trichomes, usually with long slender trichomes; gynostegial corona usually of free staminal lobes; gynostegium (sub-)sessile; pollinia ovoid, subapically to laterally attached to cylindrical caudicles; style-head flat to umbonate. Follicles usually fusiform to obclavate, usually with thin pericarp. Seeds ovate, often papillose or tuberculate, winged. Two genera in the Paleotropics and Oceania, also in temperate Eurasia.

305. ***Pentatropis*** R. Br. ex Wight & Arn.

Pentatropis R. Br. ex Wight & Arn. in Wight, Contr. Bot. India 52 (1834); Liede-Schumann et al., Taxon 61: 803–825 (2012), phyl.
Eutropis Falc. (1839), nom. illegit.
Strobopetalum N.E. Br. (1894).
Pseudopentatropis (1912).

Herbaceous or suffrutescent twiners to 4 m. Leaves usually carnose, 1–6.5 cm long, ovate. Inflorescences 1–5-flowered, sciadioidal, (sub-) sessile. Corolla 6–8 mm long, yellowish green or purple, occasionally with yellow apices, with trichomes; lobes contorted, lanceolate, occasionally extended into a long, twisted tip, occasionally revolute; free staminal corona lobes ivory or greenish yellow, at least equaling gynostegium, erect, basally horizontally spreading, oblongoid, carinate; anther wings usually following the basal margin of the anther; pollinia ovoid; caudicles declinate; style-head green. Follicles 5–7 cm long. 2n = 22.

About four spp. in Africa, Madagascar, Arabia, India and Indochina, coastal habitats, sand, rocks, rocky places in dry forests; 0–200 m.

Phylogenetic analyses have retrieved *Pentatropis* as sister to the speciose genus *Vincetoxicum* (Liede-Schumann et al. 2016).

306. ***Vincetoxicum*** Wolf

Vincetoxicum Wolf, Gen. Pl. 130 (1776); Liede, Taxon 45: 193–211 (1996), phyl.; Yamashiro et al., Mol. Phyl. Evol. 31: 689–700 (2004), phylog.; Liede-Schumann et al., Taxon 61: 803–825 (2012), phyl.; Liede-Schumann et al., Mol. Phyl. Evol. 94: 436–446 (2016); Liede-Schumann & Meve, Phytotaxa 369(3): 129–184 (2018), tax.
Tylophora R. Br. (1810).
Blyttia Arn. (1838).
Rhyncharrhena F. Muell. (1859).
Sphaerocodon Benth. (1876).
Pleurostelma Baill. (1890).
Podostelma K. Schum. (1893).
Diplostigma K. Schum. (1895).
Biondia Schltr. (1905).
Ischnostemma King & Gamble (1908).
Merrillanthus Chun & Tsiang (1941).
Pentastelma Tsiang & P.T. Li (1974).
Goydera Liede (1993).

Shrubs, erect herbs or herbaceous twiners, 0.4–4 m tall; latex usually translucent; roots usually wiry, fascicled, rarely forming a woody rootstock. Leaves occasionally carnose, 0.5–10 cm long, linear to ovate. Inflorescences occasionally paired, 1–25-flowered. Corolla 1.5–10 mm long, rarely campanulate, urceolate or elongate-conical, yellowish cream or purplish brown, with long flexuous trichomes on the tube, occasionally additionally with verrucose trichomes; lobes oblong, lanceolate or ovate; gynostegial corona of carnose free staminal lobes or of a ring of connate staminal and laminar interstaminal parts, very rarely (former *Diplostigma canescens*) with both, maximally equaling gynostegium; caudicles erect, horizontal or ascending; style-head rarely rostrate (former *Pleurostelma* species). Follicles 2–12 cm long, rarely ellipsoid. 2n = 22, 44.

At least 150 spp. throughout the tropics, subtropics of Africa, Asia, and Oceania, extending into temperate Eurasia as far north as Sweden, forest margins, steppes, rocky slopes in arid areas. Contains phenanthroindolizidine alkaloids, seco-and disecopregnanes. Some species aggressively invasive in the USA and Canada (e.g., Sheeley and Raynal 1996), capable of selfing (e.g., Lumer and Yost 1995).

Molecular analyses (Liede-Schumann et al. 2012, 2016) have shown that *Tylophora* is inseparable from *Vincetoxicum* and needs to be included in the latter.

V.5.e. **Subtribe Pentacyphinae** Liede & Meve (2014).

Weakly twining subshrubs. Inflorescences sciadioidal, flowers nodding. Corolla campanulate, adnate to gynostegium, forming five hairy pouches; gynostegial corona of free staminal lobes with additional peg-shaped parts along filament tube; pollinia medifixed or (sub-)basally inserted on corpusculum. Follicles solitary, fusiform. One genus in the northern Andes of South America.

307. ***Pentacyphus*** Schltr.

Pentacyphus Schltr., Bot. Jahrb. Syst. 37: 605, f. 3 (1906); Holm, Ann. Missouri Bot. Gard. 37: 477–560 (1950), rev. (under *Sarcostemma*); Liede & Täuber, Pl. Syst. Evol. 225: 133–140 (2000), phyl.; Meve & Liede-Schumann, Pl. Syst. Evol. 301: 997–1004 (2015), rev.
Tetraphysa Schltr. (1906).

Plants glabrous or sparsely finely pilose. Leaves 1.5–9.5 cm long, elliptic, oblong or obovate. Inflorescences 1-several-flowered; flowers nodding. Corolla 18–22 mm long, creamish green to yellow, occasionally with reddish maculation, with trichomes on the entire surface; lobes ovate, obtuse; staminal corona lobes equaling gynostegium, erect, saccate or dolabriform with long apical extension; pollinia oblongoid; caudicles declinate or geniculate, cylindrical; style-head conical and bifid. Follicles ca. 8 cm long. $2n = 22$.

Three spp. in Columbia, Ecuador, Peru and Venezuela, mountain rain forest; 1800–3100 m.

All molecular analyses (e.g., Liede-Schumann et al. 2005; Surveswaran et al. 2014) retrieve *Pentacyphus* as sister to all other New World Asclepiadeae except for *Asclepias* and *Cynanchum*.

V.5.f. **Subtribe Diplolepinae** Liede & Meve (2014).

Suffrutescent twiners, shrublets or shrubs. Inflorescences sciadioidal. Corolla rotate, campanulate or urceolate; corona gynostegial, occasionally vestigial or absent; gynostegium sessile, pollinia ovoid to oblongoid, laterally attached to horizontal, flattened caudicles; style-head often rostrate. Follicles solitary, usually slenderly obclavate. Seeds usually ovate, winged. One genus in southern South America.

308. ***Diplolepis*** R.Br.

Diplolepis R. Br., Asclepiadeae 30 (1810); Liede et al., Syst. Bot. 30: 183–194 (2005), phyl.; Hechem et al., Taxon 60: 638–648 (2011), phyl.
Sonninia Reichenb., nom. illegit. (1828).
Grisebachiella Lorentz (1880).

Plants with coriaceous, marginally revolute leaves, 1–2.5 cm long, ovate to linear. Inflorescences 1–10-flowered. Corolla 6–9 mm long, yellow, pink or whitish, often with purple tinge, usually with trichomes; lobes obovate or oblong; gynostegial corona an urceolate to cyathiform ring of connate oblong staminal parts, often with adaxial appendage, and interstaminal parts; pollinia ovoid to oblongoid, style-head white. Follicles 3–8.5 cm long. Seeds rarely thick and with reduced yellowish coma. $2n = 20, 22$.

14 spp. in Argentina, southern Brazil and Chile.

Following molecular studies (Liede-Schumann et al. 2005; Hechem et al. 2011b), the formerly monotypic *Diplolepis* was recently enlarged to comprise all southern South American former "*Cynanchum*" species, one former *Tweedia* species, and two species without a corona, one previously listed as "*Astephanus*" and one constituting the monotypic *Grisebachiella*.

V.5.g. **Subtribe Orthosiinae** Liede & Rapini (2005).

Suffrutescent or herbaceous twiners. Leaves herbaceous. Inflorescences often (sub-)axillary, occasionally paired. Corolla rotate to cyathiform; corona gynostegial, of free or basally to entirely fused staminal lobes; gynostegium usually sessile;

style-head flat to umbonate. Follicles often paired, fusiform. A molecular analysis retrieved Orthosiinae in an unresolved position with the other New World subtribes. Four genera in South and Central America.

309. ***Jobinia*** E. Fourn.

Jobinia E. Fourn. in Mart., Fl. Bras. 6(4): 327, t. 97 (1885); Araujo Schwarz & Fontella Pereira, Acta Biol. Paran. 24: 49–157 (1995), rev.; Fontella Pereira et al., Fl. Ilustr. Catarinense, Apocináceas-Asclepiadóideas: 165–178 (2004), reg. rev.; Pereira & Capello de Sales, Fl. Fan. Estad. São Paulo 4: 180–190 (2005), reg. rev.; Stevens, Fl. Nicaragua 1: 252–253. (2001), reg. rev.; Liede-Schumann & Meve, Ann. Missouri Bot. Gard. 99: 44–81 (2013), tax.
Cyathostelma E. Fourn. (1885).
Kerbera E. Fourn. (1885).
Dicarpophora Speg. (1926).
Cynanchum L. sect. *Formosum* Liede (1997).

Suffrutescent twiners, glabrous. Leaves 3–10 cm long, ovate to oblong. Inflorescences axillary, occasionally paired, 8–30-flowered, lax. Corolla 1.5–3 mm long, yellowish green; lobes oblong or lanceolate, occasionally extended in a long, twisted tip, with flexuous trichomes; gynostegial corona of basally to highly fused oblanceolate or oblong staminal lobes, tubular or cyathiform, occasionally obscuring the gynostegium; occasionally with short interstaminal lobules or teeth; anthers dorsally rarely with trichomes; pollinia oblongoid, occasionally with warty anticlinal walls, equaling corpusculum in size; style-head umbonate, rostrate or capitate. Follicles 5–15 cm long, seeds ovate, winged.

About 25 spp., one in Central America, the others in Argentina, Bolivia, Brazil, Ecuador and Venezuela, forests, thickets.

310. ***Monsanima*** Liede & Meve

Monsanima Liede & Meve, Ann. Missouri Bot. Gard. 99: 66 (2013); Silva et al., Phytotaxa 173: 11 (2014), new taxon.

Slender twiners, pubescent. Leaves 2–4 cm long, linear. Inflorescences 2–4-flowered. Corolla 5 mm long, brown, glabrous; lobes ovate; gynostegial corona of highly fused staminal lobes folded inward apically over the gynostegium, completely obscuring it; anthers with short, strongly centrifugal anther wings, pollinaria with short, broad, horizontal caudicles; connective appendages strongly inflexed. Follicles solitary, ca. 8 cm long.

Two spp. in Brazil (Bahia), endemic to high altitude cerrado and campo rupestre.

Recent molecular studies consistently retrieved *Monsanima* as sister to the remaining Orthosiinae (Silva et al. 2014; Liede-Schumann and Meve 2015).

311. ***Orthosia*** Decne.

Orthosia Decne., Prodr. 8: 526 (1844); Fontella Pereira et al., Fl. Ilustrada Catarinense, Apocináceas-Asclepiadóideas: 165–178 (2004), reg. rev.; Liede-Schumann et al., Syst. Bot. 30: 184–195 (2005), phyl.; Liede-Schumann & Meve, Novon 18: 202–210 (2008), new taxa.
Amphistelma Griseb. (1862).
Tainionema Schltr. (1899).

Suffrutescent or rarely herbaceous twiners, 0.5–8 m tall, with distinct long and short shoots, often creeping when young, rhizomatous; shoots green, often with suberose bark. Leaves often caducous, 1–9 cm long, linear or oblong. Inflorescences occasionally paired, 1–20-flowered. Corolla 1–2.5 mm long, rarely urceolate, creamish or purple; lobes oblong or lanceolate, usually papillose; gynostegial corona of free or fused staminal lobes, occasionally reduced, white to yellowish, lobes frequently trifid with median tooth prominent; gynostegium rarely stipitate; corpusculum occasionally with hastate basal projections. Follicles 4–7 cm long, narrowly oblong to obclavate. Seeds oblong, wingless. $2n = 20$.

About 25 spp. from SW and SE USA and Mexico through Central America, the Caribbean and South America, forests, often in clearings, thickets, roadsides; to 2000 m.

312. ***Scyphostelma*** Baill.

Scyphostelma Baill., Hist. Pl. 10: 252 (1890), Liede-Schumann & Meve, Ann. Missouri Bot. Gard. 99: 44–81 (2013), tax.
Cynanchum L. sect. *Microphyllum* Liede (1997).
Liedea W.D. Stevens (2005).

Herbaceous or suffrutescent twiners, often densely white, yellow or brown pilose, velvety or villous; often with distinct long and short shoots and adventitious buds forming shoots. Leaves frequently distichous, 0.8–1.5 cm long, linear to circular. Inflorescences occasionally paired, 2–8-flowered. Corolla 1–4 mm long, rarely entirely fused, creamish or reddish; lobes ovate to oblong; gynostegial corona ring tubular or cyathiform, exceeding gynostegium; staminal parts rectangular, ovate, triangular or oblong, erect; gynostegium rarely stipitate. Follicles 35–40 mm long, narrowly oblong to obclavate. Seeds oblong, wingless. 2n = 22.

More than 50 spp. in Costa Rica, Bolivia, Colombia, Ecuador, Peru and Venezuela, mountain forests; 1500–3800 m.

Genus under revision.

V.5.h. **Subtribe Metastelmatinae** Endl. ex Meisn. (1840).

Inflorescences usually extra-axillary, sciadioidal. Corolla rotate to campanulate, small; lobes frequently with trichomes; gynostegial corona of free or fused staminal lobes; gynostegium usually sessile, style-head flat or umbonate. Follicles fusiform, usually solitary. Seeds ovate. Twelve genera in the tropics and subtropics of the New World.

313. ***Barjonia*** Decne.

Barjonia Decne., Prodr. 8: 512 (1844); Marquete, Rodriguésia 31: 7–70 (1979), rev.; Farinaccio, Syst. Bot. 38: 764–768 (2013), key, new taxon.

Subshrubs to erect herbs, 0.4–1 m tall, glabrous, glaucous. Leaves (sub-)sessile, 1.5–5 cm long, oblong triangular-deltate to ovate, with interpetiolar colleters. Inflorescences 3–8-flowered, condensed, occasionally in racemiform, leafless terminal aggregates. Corolla 4–6 mm long, rotate to campanulate, greenish cream or maroon; lobes ovate, with unicellular trichomes and papillae; with multicellular trichomes in sinuses; free staminal corona lobes erect, oblong, rectangular or apiculate, adnate to both corolla and gynostegium, exceeding gynostegium, occasionally with shorter oblongoid or dentiform proximal lobules; pollinia laterally attached to triangular or trapezoidal caudicles. Follicles 4–7.5 cm long. Seeds often with revolute wing.

Seven spp. in C Brazil, Suriname and Bolivia.

314. ***Blepharodon*** Decne.

Blepharodon Decne., Prodr. 8: 603 (1844); Morillo, M.Sc. Thesis, St. Louis University, St. Louis, MO, pp. 1–163 (1976), rev.; Ferreira & Pereira, Fl. Fan. Est. São Paulo 4: 101–104 (2005), reg. rev.
Anomotassa Schltr. (1898).
Vailia Rusby (1898).

Herbaceous or suffrutescent twiners, rarely shrubs or herbs, to 10 m long, usually glabrous. Leaves discolorous, 0.8–21 cm long, ovate to linear, occasionally revolute, occasionally ciliate; with interpetiolar line or colleters. Inflorescences 6–14-flowered. Corolla 4–20 mm long, rotate or campanulate, greenish or yellowish; lobes ovate or triangular, occasionally revolute, occasionally ciliate, occasionally with trichomes; free staminal corona lobes attached along anther backs, exceeding gynostegium, erect, cucullate, bicornate or bucket-shaped; pollinia (sub-)horizontal, laterally attached to cylindrical caudicles with pronounced hyaline margin. Follicles obclavate. Seeds wingless. 2n = 22.

About 15 spp. in Central and South America, forest margins, road sides, dry sandy or rocky savanna; 0–2000 m.

A recent molecular study indicated that the genus in its present circumscription is polyphyletic (Liede-Schumann et al. 2005).

315. ***Ditassa*** R. Br.

Ditassa R. Br., Asclepiadeae 38 (1810); Morillo, Fl. Venez. Guayana 3. 145–152 (1997), reg. rev.; Konno & Pereira, Fl. Fan. Est. São Paulo 4: 107–111 (2005), reg. rev.; Silva et al., Syst. Bot. 37: 795–806 (2012), phyl.
Nematuris Turcz. (1848).
Calathostelma E. Fourn. (1885).
Husnotia E. Fourn. (1885).

Herbaceous or suffrutescent twiners, rarely (sub-)shrubs, occasionally yellow or brown tomentose. Leaves rarely alternate or whorled, occasionally strongly ascending or reflexed, 0.5–5 cm long, linear to obovate or obdeltate, often strongly revolute. Inflorescences 1–10-flowered. Corolla 1–5 mm long, occasionally urceolate, whitish cream, with verrucose or barbate trichomes; lobes oblong, ovate or triangular; staminal corona lobes free, rarely basally fused, erect, filiform, oblong, triangular or apiculate, always with shorter proximal lobes, very rarely absent; gynostegium occasionally stipitate; style-head occasionally capitate. Follicles occasionally paired, occasionally narrowly oblong. Seeds oblong to ovate, usually wingless; coma rarely absent. 2n = 22.

About 140 spp., South America, forest margins, thickets; 100–3000 m.

A recent molecular study indicated that the genus in its present circumscription is polyphyletic (Silva et al. 2012).

316. ***Hemipogon*** Decne.

Hemipogon Decne., Prodr. 8: 509 (1844); Rapini, Bol. Bot. Univ. São Paolo 19: 140–147 (2001), reg. rev.; Pereira & Capello de Sales, Fl. Fan. Estad. São Paulo 4: 116–118 (2005), reg. rev.; Fontella et al., Bonplandia (Corrientes) 23: 25–31 (2014), tax.

Erect herbs or subshrubs, 30–50 cm tall, (sub-)glabrous. Leaves usually alternate or whorled, (sub-)sessile, 15–40 cm long, linear. Inflorescences subsessile, 1–3-flowered. Corolla 3–7 mm long, urceolate, highly fused, creamish or maroon; lobes ovate to triangular, papillose or with erect trichomes, often barbate in sinuses; free staminal corona lobes adnate to both corolla and gynostegium, short, erect, oblong, often absent; gynostegium concealed in corolla tube; caudicles occasionally with pronounced hyaline margin. Follicles obclavate, seeds ovate to pyriform, wingless.

Eight spp. in Bolivia, Brazil and Paraguay, open rocky spaces.

A recent molecular study indicated that the genus in its present circumscription is polyphyletic (Liede-Schumann et al. 2005).

317. ***Hypolobus*** E. Fourn.

Hypolobus E. Fourn. in Mart., Fl. Bras. 6(4): 311 (1885); Fontella Pereira & Konno, Bradea 7: 139–143 (1999), rev.

Suffrutescent twiner, hirsute to tomentose. Leaves 2.5–3 cm long, ovate. Inflorescences occasionally paired, 15–30-flowered, thyrsoidal, condensed, shortly pedunculate. Corolla 1.5–2 mm long, abaxially with trichomes, adaxially glabrous; lobes imbricate, ovate; free staminal corona lobes adnate to both corolla and gynostegium, shorter than gynostegium, erect, laminar, triangular; pollinia broadly ovoid; caudicles geniculate, cylindrical.

One sp., *H. infracta* E. Fourn., in Brazil (Bahia), but possibly extinct.

318. ***Metastelma*** R. Br.

Metastelma R. Br., Asclepiadeae 41 (1810); Liede & Meve, Ann. Missouri Bot. Gard. 91: 31–86 (2004), reg. rev.; Liede-Schumann et al., Syst. Bot. 39: 594–612 (2014), phyl.
Acrocoryne Turcz. (1852).
Stelmation E. Fourn. (1885).
Meresaldia Bullock (1965), nom. nov. pro *Esmeraldia* E. Fourn. (1882), non *Esmeralda* Rchb. (1862), Orchidaceae.

Herbaceous or suffrutescent twiners to 4 m, usually pubescent. Leaves 1.5–6 cm long, linear to ovate. Inflorescences 2–12-flowered, flowers occasionally sweetly fragrant. Corolla 1.5–4.5 mm long, whitish; lobes imbricate, oblong or ovate, with smooth and/or verrucose trichomes, often barbate; free staminal corona lobes attached below the anthers, along the stipe or shifted to corolla base, exceeding gynostegium, white, triangular, oblong or subulate; gynostegium often stipitate or atop a column. Follicles 2–7 cm long. Seeds (almost) wingless. 2n = 22.

About 75 spp., from SW and SE USA and Mexico through Central America, the Caribbean and South America, forest margins, disturbed areas, coastal shrub and riverine vegetation, to 3000 m.

319. ***Minaria*** T.U.P. Konno & Rapini

Minaria T.U.P. Konno & Rapini, Taxon 55: 424 (2006); Ribeiro et al., Mol. Phyl. Evol. 65: 915–925 (2012), phyl.; Ribeiro et al., Taxon 63: 1253–1264 (2014), biogeogr.

(Sub-)shrubs, 10–60(–100) cm tall; usually hirsute or scabrous. Leaves often sessile, often strongly reflexed, coriaceous, 0.5–2 cm long, linear to ovate or triangular-deltate, often revolute to conduplicate. Inflorescences 1–5-flowered, subsessile; flowers usually nodding. Corolla 2–5 mm long, campanulate to urceolate; white to yellowish, rarely rose, usually barbate; lobes ovate; free staminal corona lobes filiform, lanceolate or ovate, usually with shorter proximal lobules, rarely absent; gynostegium stipitate when corona simple or absent; corpusculum smaller than pollinia. Follicles often densely puberulous or velutinous. 2n = 22.

19 spp. in Brazil, extending to Argentina and Bolivia, in open vegetation, on well drained sandy soil; 600–1900 m.

320. ***Morilloa*** Fontella, Goes & S.A. Cáceres

Morilloa Fontella, Goes & S.A. Cáceres, Bonplandia (Corrientes) 23: 28 (2014).

Twiners, glabrous or glabrescent all over. Leaves subsessile, linear. Inflorescences long-pedunculate, 3-9-flowered; flowers white or (greenish) yellow. Corolla urceolate or bottle-shaped, adaxially barbate or whitish puberulent; corona absent or vestigial; gynostegium sessile or shortly stipitate; style-head mamillate. Follicles narrowly linear-lanceolate, seeds verrucose.

Four spp. endemic to C Brazil, campos rupestres, cerrados, forest margins, on stony or sandy soils 1000–1600 m.

321. ***Nautonia*** Decne.

Nautonia Decne., Prodr. 8: 509 (1844); Meyer, Lilloa 7: 379-394 (1942), reg. rev.

Prostrate herb, densely rusty tomentose. Leaves distichous, sessile, ca. 1.5 cm long, ovate. Inflorescences 2–5-flowered, subsessile. Corolla 4–5 mm long, yellowish cream; lobes lanceolate, erect, with trichomes on central parts; corona absent; gynostegium stipitate, but concealed in corolla; pollinia ovoid to oblongoid, laterally attached to the horizontal, rectangular caudicles. Follicles 3.5–4 cm long, obclavate. Seeds winged, tuberculate with long, several-celled fringes.

A single species, *N. nummularia* Decne., in Argentina, Brazil and Paraguay, sandy areas.

Probably a *Ditassa* lacking a corona.

322. ***Nephradenia*** Decne.

Nephradenia Decne., Prodr. 8: 604 (1844); Rapini, Bol. Bot. Univ. São Paolo 19: 149–150 (2001), reg. rev.

Erect herbs, 30–50 cm tall, almost glabrous. Leaves occasionally whorled, (sub-)sessile, 2–4 cm long, linear, with interpetiolar colleters. Inflorescences 2-flowered, thinly pedunculate and pedicellate. Corolla 8–12 mm long, broadly campanulate, reddish to purple; lobes triangular, papillose or with trichomes; free staminal corona lobes adnate to both corolla and filament tube, shorter than gynostegium, erect, subulate or oblongoid; pollinia laterally attached to cylindrical, ascending caudicles with very broad hyaline margin (therefore seemingly horizontal). Follicles occasionally paired, 5–6 cm long. Seeds wingless.

About five spp., centered in C Brazil, extending to Bolivia, Colombia, Venezuela and Suriname; cerrado, open rocky areas; 70–1500 m.

323. ***Peplonia*** Decne.

Peplonia Decne., Prodr. 8: 545 (1844); Rapini et al., Kew Bull. 59: 531–539. (2004), rev.; Silva et al., Syst. Bot. 37: 795–806 (2012), phyl.
Gonioanthela Malme (1927).
Macroditassa Malme (1927).

Suffrutescent twiners, glabrous, with interpetiolar colleters. Leaves 4–10 cm long, elliptic or (ob-) ovate. Inflorescences axillary, paired, 3–10-flowered. Corolla 4–5 mm long, cream or yellowish green, occasionally papillose; lobes ovate to lanceolate, occasionally with erect trichomes or barbate; staminal corona lobes ivory, rectangular or oblong, free or basally fused, very rarely fused in a tube, exceeding gynostegium, occasionally with proximal appendage; style-head occasionally papillose, conical. Follicles 5–12 cm long. Seeds with dentate wing.

Eight spp. in C Brazil, seaside scrub, restinga, riverine vegetation.

Following a molecular study (Liede et al. 2005), the morphologically unique *P. asteria* (Vell.) Fontella & E.A. Schwarz, formerly the only species of *Peplonia*, is sister to *Gonioanthela*. According to Silva et al. (2012), *Macroditassa* is a synonym of *Peplonia*.

324. ***Petalostelma*** E. Fourn. Fig. 53

Petalostelma E. Fourn. in Mart., Fl. Bras. 6(4): 328, t. 98. (1885).

Suffrutescent twiners to 1 m, (sub-)glabrous. Leaves rarely whorled, 3–5 cm long, linear or ovate. Inflorescences 4–8-flowered; peduncles and pedicels filiform. Corolla 2.5–3.5 mm long, green or maroon; lobes ovate, spreading, papillose or with long, soft trichomes; corolline corona, when present, short, five-partite with free antesepalous lobes or annular, occasionally with trichomes; free staminal corona lobes, when present, brownish, erect or spreading, lingulate or cross-shaped, shorter than gynostegium, basally with straight spur,

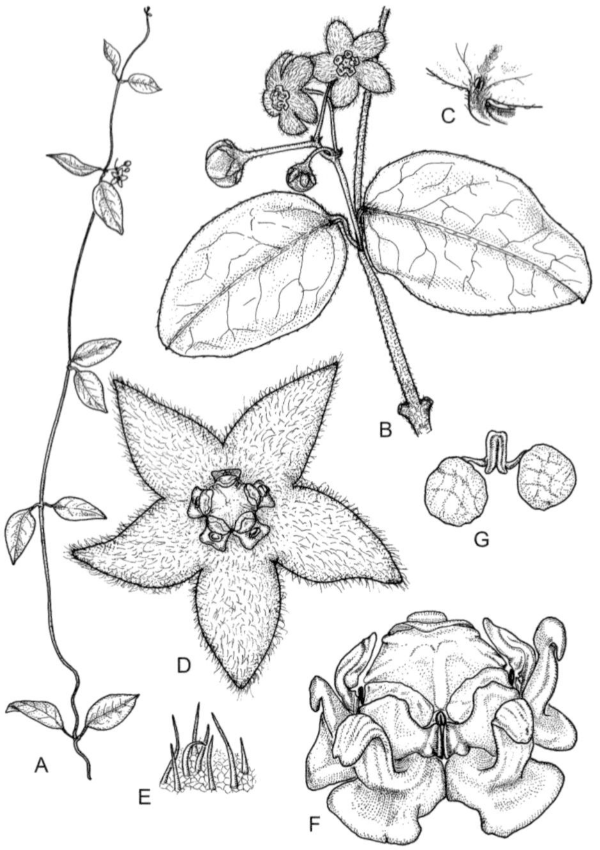

Fig. 53. Apocynaceae-Asclepiadeae. *Vincetoxicum coriaceum*. **A** Flowering plant. **B** Inflorescence. **C** Base of lamina with two colleters on rachis. **D** Flower in top view. **E** Microscopic details of adaxial corolla lobe epidermis with trichomes. **F** Gynostegium with corona. **G** Pollinarium. (From Liede and Meve 1994, p. 753, with permission from Kew Bulletin © Board of Trustees of the Royal Bot. Gard., Kew; drawn by U. Meve)

occasionally with shorter, dentiform, inflexed proximal lobules; gynostegium rarely stipitate; pollinia small, ovoid. Follicles 6–9 cm long, narrowly oblong. Seeds oblong, winged.

Seven spp. in Argentina, Bolivia, Brazil and Paraguay.

V.5.i. **Subtribe Tassadiinae** Liede & Meve (2014).

Suffrutescent twiners. Inflorescences usually axillary, paired, with one inflorescence shorter, often multi-flowered, thyrsoidal, partial inflorescences often condensed. Corolla with trichomes in the tube; lobes frequently with trichomes; gynostegial corona a ring of connate staminal and interstaminal parts; gynostegium usually (sub-)sessile; caudicles cylindrical. Follicles paired, narrowly oblong. Seeds oblong. One genus in South America, extending to the Caribbean and to Central America.

325. ***Tassadia*** Decne.

Tassadia Decne., Prodr. 8: 579 (1844); Fontella Pereira, Arq. Jard. Bot. Rio de Janeiro 21: 235–392 (1977), rev.; Liede-Schumann & Meve, Phytotaxa 202: 35–44 (2015), new taxa.
Stenomeria Turcz. (1852).

Plants often with distinct long and short shoots, occasionally rusty tomentose. Leaves distichous or decussate, 2–9.5 cm long, oblong or ovate. Inflorescences occasionally single, 20–50-flowered; rachis often conspicuously zigzagging. Corolla 3–5 mm long, creamish green; lobes imbricate or contorted, occasionally strongly twisted, ovate to oblong; gynostegial corona ring cyathiform, staminal parts ovate, trifid or apiculate, erect, occasionally with adaxial appendage; pollinia occasionally with apical sterile hyaline region; style-head occasionally papillose, rarely rostrate. Follicles 2–5 cm long, often with basal thickening, rarely inflated. Seeds occasionally without coma.

Probably more than 35 spp. in South America, extending to Central America and the Caribbean, mountain forests, disturbed places, river shores, flood plains; to 1900 m.

V.5.j. **Subtribe Oxypetalinae** E. Fourn. (1885).

Erect herbs or suffrutescent twiners. Gynostegial corona of free or basally fused staminal lobes; gynostegium usually sessile. Follicles usually obclavate or fusiform. Seeds ovate or pyriform, often with distally denticulate wing. Six genera centered in South America, only a few species in the Caribbean and Central America.

326. ***Araujia*** Brot.

Araujia Brot., Trans. Linn. Soc. London 12: 62, t. 4–5 (1817); Malme, Ark. Bot. 8: 1–30, t. 31 (1908), rev.

Suffrutescent twiners, 5–6 m tall, densely hirsute. Leaves discolorous, 5–11 cm long, ovate to hastiform, abaxially felty or tomentose. Inflorescences 1–5-flowered. Corolla 5–20 mm long, urceolate or campanulate, highly fused, white to rose; tube occasionally papillose or with trichomes; lobes triangular, often twisted; staminal corona lobes adnate to corolla and gynostegium, short, erect, carnose, rectangular or triangular, occasionally apically with trichomes; gynostegium concealed in corolla tube, connective appendages rarely fimbriate; corpusculum larger than pollinia, with hyaline margin; style-head rostrate, or capitate and bifid. Follicles 7.5–15 cm long, ovoid, occasionally vaguely tuberculate, with thick woody pericarp. Seeds wingless, denticulate. 2n = 20, 22.

Nine spp. in three sections in Argentina, Bolivia, Brazil, Paraguay and Uruguay, Chaco, dry to moist forest, often in disturbed situations. *A. sericifera* Brot. is invasive in many dry subtropical areas, and a noxious weed particularly in *Citrus* plantations.

327. ***Funastrum*** E. Fourn. Fig. 54

Funastrum E. Fourn., Ann. Sci. Nat. Bot., VI, 14: 388 (1882); Liede & Meve, Nord. J. Bot. 22: 579–591 (2002, publ. 2004), phyl.
Macbridea Raf. (1818), nom. rej., non *Macbridea* Elliott (1818), Lamiaceae, nom. nov. pro *Lyonia* Elliott (1817), non *Lyonia* Raf. (1808), Polygalaceae, nec *Lyonia* Nutt. (1818), Ericaceae.
Seutera Rchb. (1828), nom. illegit. nom. superfl. pro *Macbridea* Raf.
Pattalias S. Watson (1889).
Philibertella Vail (1897).
Ceramanthus (Kunze) Malme (1905), non *Ceramanthus* Hassk. (1844), Phyllanthaceae.

Herbaceous or suffrutescent twiners; frequently rhizomatous; latex with disagreeable garlic scent. Leaves occasionally caducous, often soft, 2–9 cm long, ovate, linear, elliptic, or hastiform. Inflorescences 6–18-flowered. Corolla 6–12 mm long, rotate, whitish, green or reddish; lobes oblong, or ovate(-lanceolate), usually revolute and ciliate; gynostegial corona whitish, usually a completely fused, short ring of connate staminal and interstaminal parts, and, connate to this ring, erect, dolabriform free staminal lobes equaling the gynostegium; pollinia oblongoid to clavate; caudicles horizontal, cylindrical; style-head umbonate or conical. Follicles 5–8.5 cm long. Seeds ovate, winged. 2n = 20, 40, 44.

About 15 spp., from SW and SE USA and Mexico through Central America, the Caribbean and South America, rheophytes, arid and semi-arid areas, plains, pampa, stony slopes, to 1500 m.

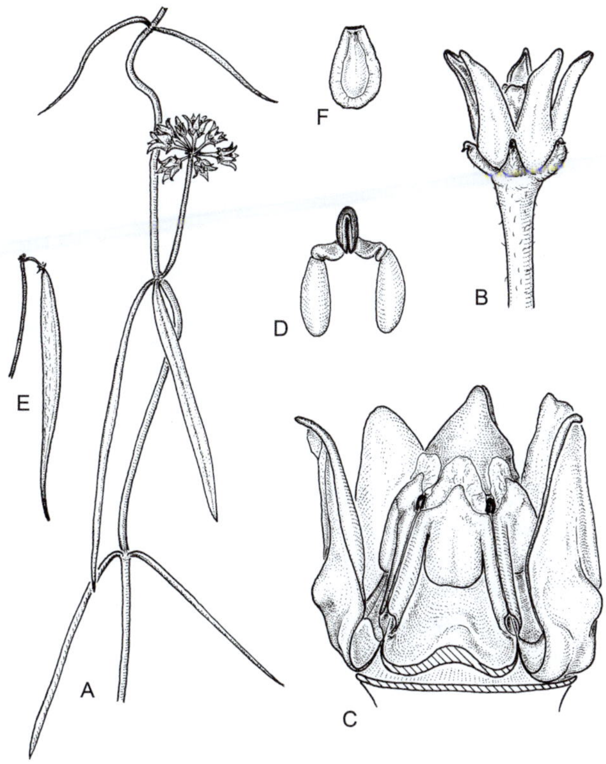

Fig. 54. Apocynaceae-Asclepiadeae. *Funastrum angustifolium*. **A** Flowering branch. **B** Flower. **C** Gynostegium surrounded by staminal corona lobes, front corona lobe removed. **D** Pollinarium. **E** Follicle. **F** Seed. (From Liede and Meve 2002, p. 588, with permission from Nordic Journal of Botany; drawn by U. Meve)

Although the molecular, chemical and karyological results of Liede and Meve (2002, publ. 2004) placed *Seutera* within *Funastrum*, Fishbein and Stevens (2005) continue to recognize it as an independent genus.

328. ***Morrenia*** Lindl.

Morrenia Lindl., Edwards's Bot. Reg. 24, Misc. 71 (1838); Malme, Ark. Bot. 8: 1–30, t. 31 (1908), rev.; Goyder, Kew Bull. 58: 713–721 (2003), rev.; Wiemer et al., Ann. Bot. 109: 77–93 (2012), ecol.
Stuckertia Kuntze (1903), nom. nov. pro *Choristigma* Kurtz ex H. Heger (1897), nom. illegit., non *Choristigma* (Baill.) Baill. (1892), Olacaceae.
Hickenia (1919).

Suffrutescent twiners to 5 m, usually pubescent. Leaves 4–7(–11) cm long, triangular, undulate, abaxially tomentose. Inflorescences 4–10-flowered, subsessile; flowers sweetly fragrant, nectariferous. Corolla 10–15 mm long, rotate, white, brown or greenish yellow; lobes lanceolate, occasionally revolute; gynostegial corona whitish, tubular, extending into ovate, often bifid, staminal parts folding in to obscure gynostegium, occasionally with trichomes; anther wings extending beyond anther, forming an outward curve; pollinia ovoid or oblongoid; caudicles ribbon-shaped; style-head umbonate or rostrate, bifid. Follicles 5–9 cm long, ellipsoid to obclavate, with more than 6 wings, pericarp thick, woody. Seeds pyriform, wingless, denticulate. 2n = 22.

Eight spp. S South America, Chaco, dry to moist forest, often in disturbed situations. *M. odorata* Lindl. is a noxious weed in *Citrus* groves in Florida and Australia; the young fruits are eaten, the latex is used to produce cheese; the plant is said to enhance milk flow in cows and women.

The suggestion of Rapini et al. (2011) to include *Morrenia* in *Araujia* is not followed here.

329. ***Oxypetalum*** R. Br.

Oxypetalum R. Br., Asclepiadeae 30 (1810); Hoehne, Monographia das Asclepiadaceas brasileiras, Fasc. I e II. Relat. Comm. Lin. Telegr., Bot. 38 (1, Suppl.): 1–131, t. 1–59 (1916); Meyer, Lilloa 9: 5–72 (1943); Farinaccio, Fl. Fan. Estad. São Paulo 4: 130–150 (2005), reg. rev.
Schistogyne Hook. & Arn. (1834).
Schizostemma Decne. (1838).
Rhyssostelma Decne. (1844).
Dactylostelma Schltr. (1895).
Widgrenia Malme (1900).
Schistonema Schltr. (1906).
Corollonema Schltr. (1914).
Amblyopetalum (Griseb.) Malme (1927).
Metoxypetalum Morillo (1994).

Often erect herbs, usually with uniform indumentum. Leaves 3–12 cm long, linear, ovate or triangular-deltate. Inflorescences 10–20-flowered. Corolla 5–15 mm long, campanulate or urceolate; lobes linear, lanceolate, or (ob-)ovate, occasionally twisted, with verrucose or barbate trichomes; free staminal corona lobes often adnate to corolla and gynostegium, erect, occasionally carnose, lanceolate, oblong, often bifid, rarely with proximal lobule; rarely with interstaminal rim or teeth; gynostegium occasionally shortly stipitate; caudicles with conspicuous appendage; style-head white, often apically purple, rostrate and bifid or 5-fid (former *Schistogyne* species). Follicles normally two, rarely muricate. 2n = 18, 20, or 22.

About 125 spp. in three subgenera, South America, mostly Argentina and Brazil; one sp. in Mexico and the Caribbean, in forest and open, often slightly disturbed areas.

The genus is in urgent need of revision.

330. ***Philibertia*** Kunth

Philibertia Kunth in Humboldt et al., Nov. Gen. Sp. (4th ed.) 3: 195 & pl. 230 (1819); Holm, Ann. Missouri Bot. Gard. 37: 477–560 (1950), rev. (under *Sarcostemma*); Liede & Täuber, Pl. Syst. Evol. 225: 133–140 (2000), phyl.; Liede et al., Syst. Bot. 30: 183–194 (2005), phyl.; Goyder, Kew Bull. 59: 415–451 (2004), rev.
Melinia Decne. (1844).
Mitostigma Decne. (1844).
Amblystigma Benth. (1876).
Podandra Baill. (1890).
Stigmamblys Kuntze, nom. illegit. (1903).
Steleostemma Schltr. (1906).
Aphanostelma Malme (1933), nom. illegit., non *Aphanostelma* Schltr. (1914).
Fontellaea Morillo (1994).

Suffrutescent twiners, prostrate or erect herbs, frequently white or yellow tomentose; latex with garlic scent, rarely translucent. Leaves usually long-petiolate, blades softly herbaceous, often discolorous, 0.4–16 cm long, ovate or triangular, cordate. Inflorescences 2–10-flowered, often long-pedunculate; flowers nodding. Corolla 5–15 (–45) mm long, (broadly) campanulate or salverform; lobes triangular, or ovate, usually hirsute; often with short corolline corona ring; usually with free, white, ovoid or dolabriform staminal corona lobes equaling gynostegium, occasionally with proximal lobule; gynostegium occasionally stipitate; pollinia oblongoid or clavate; style-head occasionally rostrate and bifid. Follicles 3–10 cm long, rarely ellipsoid, occasionally rugose or spiny, often tomentose, with thick pericarp. 2n = 18, 20, or 22.

43 spp. in South America.

The inclusion of *Amblystigma*, *Melinia*, and *Mitostigma* by Goyder (2004) has rendered the genus morphologically rather heterogeneous, but this clade is supported by DNA sequence data (Liede et al. 2005).

331. ***Tweedia*** Hook. & Arn.

Tweedia Hook. & Arn., J. Bot. (Hooker) 1: 291 (1834); Malme, Ark. Bot. 2: 1–18 (1904), rev.; Rua, Parodiana 5: 375–410 (1989), rev.; Calviño et al., Taxon 63: 1265–1274 (2014), phyl.

Erect herbs or suffrutescent twiners, to 1 m tall. Leaves occasionally whorled, 0.8–7 cm long, linear, obtriangular, or hastiform, revolute. Inflorescences 2–40-flowered. Corolla 3.5–9 mm long, campanulate, white or yellow, veined green; lobes oblong, twisted; corolline corona inserted and fused to tube, occasionally with trichomes in adaxial sinus, short, five-partite; lobes free or laterally connate, rectangular or rounded, bifid, erect; gynostegium (sub-)sessile; pollinia ellipsoid to oblongoid; caudicles trapezoidal; style-head rostrate and bifid. Follicles occasionally paired, 40–95 mm long, fusiform. Seeds pyriform, wingless. $2n = 40$.

Seven spp. in Argentina, Bolivia, Chile and Uruguay, open bushland.

V.5.k. **Subtribe Gonolobinae** Liede (1997).
by S. Liede-Schumann and G.N. Morillo

Usually suffrutescent twiners, densely puberulous to hispid or hirsute, often with short spreading glandular capitate and long or short eglandular sculptured antrorse to retrorse trichomes. Corolla rotate, campanulate, urceolate or tubular; gynostegium sessile or stipitate; anthers usually broader than long, frequently horizontal, with transversally dehiscent membrane; pollinia horizontal or pendent, with apical to distal sterile hyaline region; laterally attached to caudicles; style-head flat or convex, rarely rostrate. Follicles solitary, ovoid or fusiform, usually with thick pericarp. Seeds ovate, winged, mostly with a coma.

Following results of a molecular analysis, Gonolobinae are sister to Oxypetalinae (Rapini et al. 2003). 45 genera in in the tropics and subtropics of the New World.

332. ***Anemotrochus*** Mangelsdorff, Meve & Liede

Anemotrochus Mangelsdorff, Meve & Liede, Willdenowia 46: 452 (2016).

Prostrate to ascending vines with eglandular trichomes. Leaves subsessile, to 1.5 × 0.5 cm, ovate. Inflorescences subsessile, to 6-flowered; flowers to 4.5 mm long. Corolla urceolate or subcampanulate with expanded lobes, uniformly whitish or with greenish reticulation; lobes ovate to linear, often twisted, adaxially pubescent on the left side; corolline corona a protuberance or absent; staminal corona shorter than gynostegium; interstaminal corona inconspicuous; anther wings separated in basal and apical part; pollinaria with pendulous, slightly furrowed pollinia. Follicles one or two per flower, fusiform, occasionally 10-costate, with protuberances.

Three spp., Caribbean, one endemic to Cuba, one to Hispaniola, coastal and in xerophytic thickets, occasionally on ultramafic soils; 0–150 m.

333. ***Atrostemma*** Morillo

Atrostemma Morillo, Pittieria 39: 198 (2015).

Twining shrubs; branches shortly pubescent. Leaves 6–18 cm long, broadly ovate to oblong-elliptic, basally cuneate to broadly cordate, apically apiculate to long-acuminate, both surfaces sparsely puberulous. Inflorescences 3–7-flowered. Corolla rotate, 10–34 mm diam., (yellowish) green and dark-reticulate, lobes oblong-lanceolate to obovate-elliptic; gynostegial corona (annular-)cyathiform, staminal segments incurved, thickly laminar or fleshy, basally adnate to stipe, interstaminal segments concave, usually puberulous abaxially; anthers suberect, subtriangular, pollinia subpendent; style-head concave. Follicles asymmetrically oblong or boat-shaped, obtusely muricate, glabrous.

Ten spp. in Central and South America, forests and thickets; 100–1700 m.

334. ***Austrochthamalia*** Morillo & Fontella

Austrochthamalia Morillo & Fontella, Rev. Biol. Neotrop. 10: 3 (2013).

Subshrubs, with xylopodium ca. 1.5 cm diam.; stems trailing, less than 1.5 m long; hirsute all over, with eglandular trichomes. Leaves distichous, blades to 5 cm long, deltoid or (oblong-)

ovate, basally cordate, apically acute. Inflorescences 2–4-flowered, shortly pedunculate, usually prostrate. Corolla 16–38 mm diam. (sub-)campanulate, green to purple; lobes deltoid, narrowly ovate to broadly oblong, adaxially verrucose and pilose; gynostegial corona a cup of fused staminal and interstaminal segments, with 10 internal rays, purple; gynostegium shortly stipitate; pollinia subhorizontal or pendent; style-head concave. Follicles ovoid, muricate.

Five or six spp. in Brazil, Paraguay and Argentina, savanna, below 600 m.

335. ***Brargentina*** Morillo & H.A. Keller

Brargentina Morillo & H.A. Keller, Bonplandia 25(2): 131 (2016).

Subshrub; stems trailing or vining, 0.15–0.75 m long, corky at base, densely pubescent mainly with eglandular trichomes. Leaves distichous or 3-whorled, blades membranous, up to 3.5 × 2.1 cm, narrow-ovate to oblong, narrowly obtuse to subcordate at base, pubescent throughout. Inflorescences 2–3 per node, 2–4-flowered. Calyx lobes oblong, acute, longer than corolla tube, abaxially pubescent; corolla white, rotate-campanulate, 20–21 mm diam.; lobes strongly contorted, asymmetrically oblong-lanceolate, glabrous except puberulent at adaxial base; corona gynostegial; staminal lobes laminar, oblong-subrectangular, incurved and partly covering the convex style-head; pollinia pendent. Follicles ovoid, 5.5–8 cm long, glabrous and almost smooth. Seeds thick, with a coma.

One sp., *B. bornmuelleri* (Malme) Morillo & H.A. Keller in SE Brazil, NE Argentina and SE Paraguay, cerrados and prairies; 100–500 m.

336. ***Bruceholstia*** Morillo

Bruceholstia Morillo, Pittieria 39: 207 (2015).

Woody vine to 40 m long; with dense (yellowish) brown mixed pubescence all over; with fascicles of 30–50 colleters at the base of petioles and blades; blades to 30 cm long, basally broadly cordate. Inflorescences 4–6-flowered. Corolla ca. 40 mm diam., fleshy, pale orange or yellowish green, reticulate; lobes broadly elliptic; gynostegial corona annular-cyathiform, fleshy, dark brown to purple; staminal segments ligulate, adnate to the base of the anthers; gynostegium sessile; pollinia pendent. Follicles fusiform, with long curved processes and dark-brown pubescence.

A single species, *B. magnifolia* (Pittier) Morillo, in Central America, tropical wet forests; 70–1600 m.

337. ***Chloropetalum*** Morillo

Chloropetalum Morillo, Pittieria 39: 213 (2015).

Suffrutescent twiners or trailers, stems, pedicels and inflorescences hirtellous to hirsute, leaves (oblong-)ovate, basally cordate, apically acuminate, shortly pubescent to almost glabrous. Inflorescences 2–14-flowered. Corolla subcampanulate, 17–40 mm diam., (yellowish) green, reticulate; lobes ovate(-oblong); corolline corona usually present as clusters of short trichomes on corolla tube; gynostegial corona disciform or deeply 5-lobed, adnate to base of corolla and gynostegium; gynostegium (sub-)sessile; pollinia subtriangular or pyriform, horizontal or slightly pendent; style-head flat or slightly concave. Follicles 5-costate or 5-winged.

About seven spp. in Central and South America, and the Antilles, seasonally dry and riparian forests and thickets; 0–1000 m.

338. ***Chthamalia*** Decne.

Chthamalia Decne. Prodr. 8: 605 (1844).
Tetracustelma Baill. (1890).
Amphorella Brandegee (1910).

Prostrate to procumbent suffrutices or herbs, usually with a corky xylopodium or caudex with several short, hirtellous to hirsute stems. Leaves usually distichous; blades to 4 cm long, subsessile, ovate to suborbicular, basally truncate to cordate, pubescent on at least one surface. Inflorescences 2–6-flowered. Corolla 4–13 mm diam., tubular to suburceolate, cream, green, or purple; lobes oblong or lanceolate, veined lengthwise; gynostegial corona 5-lobed, connate to 3/4 of its length, usually adnate to gynostegium; style-head concave to shortly rostrate. Follicles 5–10 cm long, muricate.

Two to ten spp. in Mexico and SW USA, deserts and scrub. Delimitation and number of species presently under study, deserts and scrub.

339. ***Coelostelma*** E. Fourn.

Coelostelma E. Fourn. in Mart., Fl. Bras. 6(4): 320 (1885).

Erect suffrutex, 30–60 cm high, glabrous all over. Leaves shortly petiolate, to 9.5 cm long, oblong-lanceolate, basally rounded to truncate, apically acuminate. Inflorescences shortly pedunculate, 2-flowered. Corolla rotate-campanulate, throat and adaxial basal lobes puberulent, lobes 13–18 mm long, acuminate, conspicuously veined; gynostegial corona of 5 distinct fleshy staminal lobes, these truncate, entire, with inconspicuous external central flap; gynostegium sessile, anthers obtriangular, dorsally without appendages; pollinia horizontal, narrowly obovoid; style-head broadly pentagonal, flat or slightly concave.

One sp., *C. refractum* E. Fourn., endemic to E Brazil.

340. ***Cristobalia*** Morillo, S.A. Cáceres & H.A. Keller

Cristobalia Morillo, S.A. Cáceres & H.A. Keller, Pittieria 40: 132 (2016).

Twining shrubs; stems hirsute, not corky at base. Leaves petiolate, blades to 15 cm long; obovate-elliptic to deltoid-ovate, basally cordate. Inflorescences subsessile, 3–15-flowered. Corolla 14–21 mm diam., rotate to campanulate; lobes ovate, adaxially green or dark pink to purple, reticulate; gynostegial corona cyathiform, 5-lobed, radially sinuate-lobulate, with five broad cavities in interstaminal position, (blackish) purple; gynostegium exserted, shortly stipitate; anthers with prominent, curved, laminar wings; pollinia reniform or asymmetric and shortly calceolate. Follicles fusiform-ellipsoid, muricate, with many conic to unciform projections, hirsute.

Two spp. in Argentina, Bolivia and Brazil, temperate forests; 200–2670 m.

341. ***Cyclodon*** Small

Cyclodon Small, Man. S.E. Fl. 1075 (1933).

Twining shrub; stems with mixed pubescence. Leaves petiolate; blades 6–15 cm long, ovate, basally cordate, apically acuminate, sparingly pilose. Inflorescences 3–6-flowered, pedunculate. Corolla rotate, light-green, reticulate; lobes 6–8 mm long, spreading, narrowly elliptic, abaxially puberulent or sparsely hirtellous; gynostegial corona thinly fleshy, consisting of an outer ring attached to the corolla tube with 5 erect, crest-like apically ligulate staminal lobes adnate to the stipe; gynostegium stipitate; anthers horizontal, with lateral dehiscence; pollinia narrowly obovoid, horizontal; style-head flat. Follicles fusiform, strongly short-muricate and puberulous.

One sp., *C. alabamensis* (Vail.) Small, endemic to SE USA, low elevations.

342. ***Dictyanthus*** Decne.

Dictyanthus Decne., Prodr.8: 604 (1844); Stevens, Ann. Missouri Bot. Gard. 75: 1533–1564 (1988), rev.
Tympananthe Hassk. (1847).
Pachystelma (1920).

Erect, trailing or vining small herbs or subshrubs; usually woody at base, pubescent all over; trichomes eglandular, uncinate or spreading, and glandular, spreading. Leaves petiolate; blades 1.3–13 cm long, ovate, basally cordate, apically acuminate, inflorescences 1- to few-flowered. Corolla campanulate, tubular or urceolate; tube internally convoluted with raised parts opposite corona lobes and sacs formed between them; lobes 2.5–25 mm long, marginally often sharply revolute; gynostegial corona adnate to corolla tube, and partly, by a thin septum, to gynostegium; gynostegium stipitate; pollinia subhorizontal; style-head flat or apiculate. Follicles fusiform, muricate.

About 16 spp. in Mexico to Costa Rica, mainly seasonally dry forests and savanna; 0–1500(–2250) m.

343. ***Edisonia*** Small

Edisonia Small, Man. S.E. Fl. 1078 (1933).

Trailing herb or subshrub; stems to 1.1 m long, hispid with eglandular and glandular translucent trichomes. Leaves petiolate, blades to 5 cm long, broadly ovate to deltoid-reniform, basally cordate, apically acute or acuminate, hispid. Inflorescences (sub-)sessile, 2–6-flowered. Corolla ca. 10 mm diam., campanulate, olive-green or reddish brown; lobes ca. 2 mm long, spreading, deltoid-ovate, with a basal small tooth-like projection, adaxially conspicuously barbate; gynostegial corona tubular-urceolate, apically with 5 short involute ligulate lobes; pollinia horizontal, with longitudinal hyaline sterile lateral margin. Follicles prostrate, ovoid-fusiform, muricate.

One sp., *E. pubiflora* (Decne.) Small, in SE USA, sandhills and dry scrub in coastal plains, 15–100 m.

344. ***Fischeria*** DC.

Fischeria DC., Cat. Horti Pl. Monsp. 112 (1813); Spellman, Ph.D. Thesis, St. Louis Univ. (1975), rev.; Murphy, Syst. Bot. 11: 229–241 (1986), rev.

Suffrutescent twiners, 3–20 m tall, densely puberulous to hispid, trichomes yellow to light brown. Leaves long-petiolate; blades 5–18 cm long, elliptic to ovate, basally cordate. Inflorescences pedunculate, to 25-flowered. Corolla 5–15 mm long; lobes lanceolate, ovate or triangular, papillose, right margin undulate and crenate; gynostegial corona short, carnose, annular, adnate to corolla and gynostegium; abaxially warty or with trichomes, adaxially fimbriate; anthers without transversally dehiscent membrane, dorsally with inflated appendage, free; connective appendages pyriform, inflated; pollinia reniform, horizontal. Follicles 11–15 cm long, obliquely ellipsoid or ovoid, almost smooth.

Eight spp. in Central and South America, forests and thickets; 0–1700 m.

345. ***Gonolobus*** Michx.

Gonolobus Michx., Fl. Bor.-Amer. 1: 119 (1803); Stevens, Fl. Fan. Valle México 2: 236 (1985), reg. rev.; Stevens, Fl. Nicaragua 1: 247–252 (2001), reg. rev.; Stevens, Novon 15: 222–244 (2005); Krings, Harvard Pap. Bot. 13: 209–218 (2007), reg. rev.; Krings et al., Syst. Bot. 33: 403–415 (2008), phyl.
Gonolobium R. Hedw. (1806), nom. illegit.
Fimbristemma Turcz. (1852).
Exolobus E. Fourn. (1885).
Trichostelma Baill. (1890).

Mostly suffrutescent twiners, lenticellate on older internodes, usually pubescent with long, yellow or translucent trichomes, frequently with glandular trichomes. Leaves long-petiolate, to 16 cm long, broadly ovate to linear. Inflorescences 1–15-flowered. Corolla 10–35 mm long, often reticulate, frequently pubescent; lobes linear to ovate-elliptic, tightly twisted; corolline corona inserted in corolla tube, usually annular, ciliate or fimbriate; gynostegial corona annular or cup-shaped, basally adnate to corolla and gynostegium; gynostegium occasionally stipitate, anthers with dorsal appendage; pollinia horizontal, oblong-ovoid to reniform. Follicles 3–5-winged, rarely smooth, basally asymmetric. 2n = 22.

About 120–140 spp. from SW and SE USA and Mexico through Central America, the Caribbean and South America, dry or wet forests and thickets; 0–2500 m.

346. ***Graciemoriana*** Morillo

Graciemoriana Morillo, Pittieria 39: 223–224 (2015).

Twining vine, ca. 10 m long; with mixed dense brownish pubescence of spreading, glandular and eglandular trichomes all over. Leaves petiolate; blades 18–26 cm long, broadly (obovate-)elliptic, basally narrowly cordate, apically obtuse and acuminate. Inflorescences pedunculate, 3–4-flowered. Corolla ca. 18 mm diam., rotate-campanulate, green; lobes broadly deltoid, obtuse, emarginate and reticulate; gynostegial corona of 5 staminal irregularly trapezoidal segments, marginally rugose-caruncolate, fleshy; gynostegium shortly stipitate; pollinia subhorizontal, asymmetrically obovoid-reniform; style-head slightly convex. Follicles broadly ovoid, brown-puberulent.

One sp., *G. graciae* (Morillo) Morillo, French Guiana, non-flooded moist forest, 200–400 m.

347. ***Gyrostelma*** E. Fourn.

Gyrostelma E. Fourn. in Mart., Fl. Bras. 6(4): 302 (1885); Rapini, Bol. Bot. Univ. São Paulo 19: 71–72 (2001), reg. rev.

Erect or creeping subshrub, hirsute all over with stiff trichomes. Leaves ca. 3 cm long, elliptic. Inflorescences ca. 6-flowered, subsessile. Corolla 4–5 mm long, campanulate, glabrous; lobes oblong; gynostegial corona of free, erect, oblong, apically bifid, staminal lobes exceeding gynostegium; anthers rectangular or truncated obtriangular, erect; pollinia ovoid, caudicles cylindrical with pronounced hyaline margin; corpusculum rhomboid, smaller than pollinia; style-head flat to broadly conical.

One sp., *G. oxypetaloides* E. Fourn., Brazil.

348. ***Himantostemma*** A. Gray

Himantostemma A. Gray, Proc. Amer. Acad. Arts 20: 294 (1885).

Spreading herb or suffrutex; branches to 30 cm long, pubescent all over. Leaves to 4.5 cm long, sagittate-cordate to deltate-hastate, apically acute or acuminate. Inflorescences 2-flowered. Corolla 11–13 mm diam., rotate, lobes whitish or green, purple with age, broadly oblong to narrow-lanceolate, reflexed, adaxial tube with long, white, linear-spathulate trichomes; gynostegial corona of basally fused staminal and interstaminal segments; staminal segments dentate, short; interstaminal segments prolonged into 10 erect, strap-shaped lobes; anthers erect, apically dehiscent; pollinia pendent. Follicles fusiform, muricate.

One sp., *H. pringlei* A. Gray, Sonora Desert, NW Mexico and SW USA, 0–250 m.

349. ***Ibatia*** Decne.

Ibatia Decne., Prodr. 8: 599 (1844); Krings & Saville, Syst. Bot. 32: 862–871 (2007), phyl.; Morillo, Pittieria 36: 13–57 (2012).
Callaeolepium H. Karst. (1866).
Omphalophthalma H. Karst. (1866).
Amphidetes E. Fourn. (1885).
Pycnobregma Baill. (1890).
Pseudibatia Malme (1900).

Twining or sometimes erect (sub-)shrubs, usually densely mixed-pubescent all over. Leaves 2.5–15 cm long, ovate to suborbicular, basally usually (sub-)cordate. Inflorescences shortly pedunculate, few to many-flowered. Corolla rotate to campanulate, white or green to purple; lobes 3–6.5 mm long, ovate to oblong; gynostegial corona of connate staminal and interstaminal segments, annular, cup-shaped or 5-lobed; gynostegium stipitate; anthers radially protruding, with dorso-apical dehiscence; pollinia pendent, usually obovate, proximally hyaline; style-head concave to rostrate. Follicles muricate. 2n = 22.

About 25 spp., South and Central America, Caribbean, seasonally dry forests and scrub, 0–2200 m.

Of the genera formerly included in *Matelea*, *Ibatia* is monophyletic with high support by molecular data (Krings and Saville 2007).

350. ***Jacaima*** Rendle

Jacaima Rendle, J. Bot. 74: 340. (1936).

Twining vine; stems with eglandular pubescence, on entire surface or in 2 lines. Leaves 3–9 cm long, ovate, basally rounded to subtruncate, glabrous or sparsely pubescent on midvein. Inflorescences 4–7-flowered, pedunculate. Corolla campanulate, white; lobes 1.5–2.7 mm long, narrowly ovate, basally incurved, with an eyespot-like concavity and large, white eyespot at apex, glabrous; staminal corona a costate ridge adnate to stipe over most of its length, apically ligulate; gynostegium stipitate; pollinia horizontal; style-head umbonate. Follicles ovoid, 5-ridged, glabrous.

A single species, *J. costata* (Rendle) Morillo, endemic to Jamaica, dry rocky thickets and mesophytic forests, 15–200 m.

351. ***Lachnostoma*** Kunth

Lachnostoma Kunth in Humboldt et al., Nov. Gen. Sp. (4. ed.) 3: 198 (1818, publ. 1819); Morillo, Pittieria 36: 24–27 (2012).

Twiners, densely pubescent mainly with eglandular trichomes all over. Leaves to 15 cm long, (oblong-)ovate, basally (sub-)cordate, apically acuminate. Inflorescences 5–15-flowered, pedunculate; pedicels 0.5–1.5 times as long as peduncle plus rachis. Corolla 10–30 mm diam., usually salverform, narrow-campanulate in one species; lobes (yellowish) green or purple, frequently reticulate; gynostegial corona adnate to stipe and to corolla tube over at least half of its length, apically bilobed-digitate; pollinia pendent, oblongoid-pyriform. Follicles narrowly ovoid to fusiform, with 5 obtuse ridges and blunt intercostate projections.

About 12 spp., W and N South America, Peru to Venezuela, montane wet forests, above 1000 m.

352. ***Lhotzkyella*** Rauschert

Lhotzkyella Rauschert, Taxon 31: 557 (1982); nom. nov. pro *Pulvinaria* E. Fourn. (1885), non *Pulvinaria* Bonorden (1851), Fungi–Sphaeriales.

Twiner with short mixed-pubescence in 2 lines; eglandular trichomes retrorse. Leaves 5–7.5 cm long, narrowly ovate or oblong-ovate, basally shortly cordate, sparsely puberulous. Inflorescences short-pedunculate, 4–6-flowered. Corolla ca. 16 mm long, campanulate, dark purple; lobes 12–13 mm long, tightly contorted in bud, oblong, puberulous only at adaxial base; staminal corona segments of 5 distinct fleshy bidentate elements, shorter than gynostegium; gynostegium stipitate, anthers ventricose, with dorso-lateral dehiscence; pollinia subpendent, narrowly obovoid; style-head disciform, slightly convex. Follicles narrowly ovoid, acute, pentagonal and 5-costate.

One sp., *L. lhotzkyana* (E. Fourn.) Rauschert, Mato Grosso, Brazil, savanna (cerrados), below 500 m.

353. ***Macroscepis*** Kunth

Macroscepis Kunth in Humboldt et al., Nov. Gen. Sp. (4. ed.) 3: 201 & ed. fol.: 156 (1819); Stevens, Fl. Nicaragua 1: 254–255 (2001), reg. rev.; Fontella Pereira & Konno, Fl. Fan. Estad. São Paulo 4: 121–122 (2005), reg. rev.
Schubertia Mart. (1824).

Suffrutescent twiners, 5–20 m long, frequently suberized at base; usually hirsute with eglandular trichomes all over. Leaves 6–30 cm long, obovate-oblong to suborbicular, basally narrowly cordate. Inflorescences pedunculate, 2–20-flowered, bracts foliaceous. Corolla 18–25 mm diam., urceolate, tubular or infundibuliform, white, green to brown, fleshy; corolline corona present; gynostegial corona fleshy, partly adnate to gynostegium and corolla, staminal lobes apically truncate or crenulate, occasionally included; pollinia pendent, with thin hyaline margin; style-head convex or concave. Follicles usually (5–)7-winged. 2n = 22.

About 20 spp., the Americas, from Mexico to Argentina, dry or wet forests and thickets, 0–1600 m.

354. ***Malinvaudia*** E. Fourn.

Malinvaudia E. Fourn. in Martius Fl. Bras. 6(4): 312, t. 92 (1885).

Twiner, stems glabrous. Leaves shortly petiolate, blades to 13 × 4.5 cm, oblong-lanceolate, basally subcordate, apically long-acuminate, almost glabrous. Inflorescences 6–9-flowered; peduncles to 7.5 cm long. Corolla campanulate, throat with 5 interlobular pubescent pads; lobes ca. 10 mm long, acuminate, conspicuously veined, adaxially proximally puberulent; gynostegial corona shorter than gynostegium, of 5 distinct, fleshy, 3-dentate staminal segments; gynostegium subsessile, anthers obtriangular, dorsally without appendages; pollinia horizontal, narrowly obovoid; style-head broadly pentagonal, almost flat. Follicles unknown.

One sp., *M. capillacea* E. Fourn., E Brazil and NE Argentina, below 600 m.

355. ***Matelea*** Aubl.

Matelea Aubl., Hist. Pl. Guiane 277, t. 109, f. 1 (1775); Stevens, Flora Fan. Valle México 2: 228–241 (1985), reg. rev.; Stevens, Fl. Nicaragua 1: 236–239 (2001), reg. rev.; Rapini, Bol. Bot. Univ. São Paulo 19: 69–71 (2001), reg. rev.; Konno & Fontella Pereira, Fl. Fan. Estad. São Paulo 4: 118–120 (2005), reg. rev.
Hostea Willd. (1798), nom. illegit.
Peckoltia E. Fourn. (1885).
Acomosperma K. Schum. ex Ule (1908).

Erect or twining (sub-)shrubs; pubescent all over with mixed indumentum of short glandular and eglandular trichomes. Leaves ovate to oblong-lanceolate, basally obtuse to subcordate, apically acute to acuminate. Inflorescences shortly pedunculate, 3–12-flowered. Corolla to 16 mm diam., rotate or shortly campanulate; lobes green, reticulate; gynostegial corona marginally annular; staminal segments apically ligulate ascending ridges, interstaminal segments usually concave; gynostegium stipitate or subsessile; anthers narrowly triangular; pollinia usually horizontal, obovoid or obpyriform; style-head frequently flat or convex. Follicles fusiform to ovoid, 5-costate, 5-winged or almost smooth. 2n = 22.

About 75 spp., Central and South America, wet and riparian transitional forests; 0–2200 m.

356. *Odontostephana* Alexander

Odontostephana Alexander in Small, Man. S.E. Fl. 1076 (1933).

Herbaceous vines; hirtellous or hirsute all over. Leaves ovate to almost orbicular, basally cordate, apically acuminate. Inflorescences dichasial or racemiform cymes, (4–)8–20-flowered. Corolla shallowly campanulate-rotate; lobes 7–18 mm long, oblong or spathulate, occasionally reticulate, adaxially glabrous, abaxially pubescent; gynostegial corona a cup-shaped ring of 5 staminal segments reaching or surpassing style-head, each segment abaxially usually with two triangular teeth; anthers radially protruding from below style-head; pollinia horizontal. Follicles narrowly fusiform or narrowly ovoid, muricate.

About seven spp., USA, mainly in prairies and deciduous temperate forests.

357. *Orinoquia* Morillo

Orinoquia Morillo, Pittieria 39: 229 (2015).

Twining subshrub; with moderate pubescence of golden-yellow, very long (to 7 mm), eglandular trichomes all over, and scarce pubescence of short translucent glandular trichomes. Leaves 10–13 cm long, broadly ovate(-elliptic), basally cordate, apically abruptly acuminate. Inflorescences very long-pedunculate, 3–5-flowered. Corolla 50–55 mm diam., rotate; lobes 18–19 mm long, narrowly ovate; gynostegial corona shallowly cup-shaped, staminal lobes 10, radially disposed, bifid and marginally expanded; interstaminal lobes narrow-ligulate, reflexed; pollinia horizontal, obovoid. Follicles unknown.

One sp., *O. yanomamica* (Morillo) Morillo, endemic to Venezuela, rain forest, headwaters of the Orinoco river, 150–400 m.

358. *Peruviasclepias* Morillo

Peruviasclepias Morillo, Pittieria 39: 232 (2015).

Erect, prostrate or short twining herb or subshrub, 0.4–1.5 m long; stems frequently scarred, densely glauco-villous when young. Leaves 2–4.5 cm long, basally cordate, somewhat fleshy, pubescent on both surfaces. Inflorescences 6–12(–24)-flowered. Corolla 6–7 mm diam., yellowish green, salverform or short-campanulate, with 5 fascicles of white trichomes in throat opposite the anthers; gynostegial corona annular, adnate to corolla tube; gynostegium long-stipitate; anthers protruding from style-head, with white-translucent deltoid apical membrane; pollinia subhorizontal to erect, oblongoid-pyriform. Follicles fusiform, muricate, with short curved projections.

One sp., *P. aliciae* (Morillo) Morillo, NW Peru, deserts, 100–500 m.

359. *Phaeostemma* E. Fourn.

Phaeostemma E. Fourn. in Mart., Fl. Bras. 6(4): 311, t. 91 (1885).

Shrubby vines up to 10 m long; densely pubescent all over with eglandular long and short trichomes, and scarce glandular short trichomes. Leaves 8–20 cm long, broadly (oblong-)ovate or elliptic, basally indistinctly cordate, apically acuminate. Inflorescences shortly pedunculate, 2–12-flowered; pedicels 2–6.2 times longer than peduncle plus rachis. Corolla 23–38 mm diam., broadly campanulate, (yellowish) green; lobes narrowly ovate to deltoid, reticulate; gynostegial corona of 5 apically bifid-digitate staminal segments partly adnate to corolla tube and stipe; pollinia pendent, narrowly pyriform. Follicles with 5 wings and blunt projections.

Six or seven spp., E South America, from Venezuela, Guayana to Argentina, tropical wet, rain or montane forests, 100–700 m; one species on a sandstone mountain of the Venezuelan Guayana above 1500 m.

360. *Pherotrichis* Decne.

Pherotrichis Decne., Ann. Sci. Nat. Bot. 2, 9: 322 (1838); Juárez-Jaimes & Lozada, Fl. Valle Tehuacán-Cuicatlán 37: 53–55 (2003), reg. rev.

Erect herbs, 30–40 cm tall, densely hirsute with erect eglandular and glandular trichomes; with spindle-shaped root tubers. Leaves subsessile, 3.5–8 cm long, elliptic, or obovate. Inflorescences seemingly axillary, subsessile, (3–)5–10(–14)-flowered; bracts linear, conspicuous. Corolla 5–7 mm long, campanulate, creamish green with dense, flexuous or barbate trichomes; lobes ovate, occasionally long-acuminate; gynostegial corona of free, erect, rectangular, emarginate, slightly bifid staminal lobes, adnate to corolla and gynostegium; gynostegium atop a bulge; anthers truncated, obtriangular; pollinia pendent, caudicles with pronounced hyaline margin; style-head elongate-conical to rostrate. Follicles obclavate.

About five spp., Guatemala, Mexico, USA, open grassy slopes, pine forests.

361. *Poicilla* Griseb.

Poicilla Griseb., Cat. Pl. Cub. 176 (1866).

Slender woody vine; stems densely pubescent all over with mixed trichomes; eglandular trichomes retrorse or spreading. Leaves oblong-lanceolate, basally sagittate, apically obtuse to acute. Inflorescences few-flowered, subsessile. Corolla subcampanulate, greenish purple to maroon, lobes 1.9–3.5 mm long, reticulate, ovate to deltoid, obtuse to rounded, abaxially pubescent; staminal corona segments prominent-convex and cucullate, adnate to stipe for most of its length; gynostegium stipitate; pollinia obovoid. Follicles fusiform, 4–5-angled or ridged, (almost) glabrous.

One sp., *P. tamnifolia* Griseb., Cuba; thickets, semi-deciduous forests and woodland pastures.

362. *Poicillopsis* Schltr. ex Rendle

Poicillopsis Schltr. ex Rendle, J. Bot. 74: 343 (1936).

Slender vine; stems pubescent in two lines with mixed retrorse trichomes. Leaves 1–2.7 cm long, ovate to oblong, basally rounded, apically obtuse and apiculate, nearly glabrous. Inflorescences shortly pedunculate, few-flowered. Corolla ca. 7 mm diam., subcampanulate, white to green; lobes 1.8–3.3 mm long, ovate, obtuse to acute, adaxially densely pubescent except along left margin; gynostegial corona of basally fused obovate to suborbicular staminal and interstaminal segments, equaling gynostegium; pollinia lanceolate-oblong to falcate. Follicles subcylindrical-fusiform, pubescent or glabrous.

One sp., *P. ovatifolia* (Griseb.) Rendle, Cuba, Sierra Maestra, 1000 m.

363. *Polystemma* Decne.

Polystemma Decne., Prodr. 8: 602 (1844).
Rothrockia A. Gray (1885).
Urostephanus B.L. Rob. & Greenm. (1895).
Labidostelma Schltr. (1906).
Microdactylon Brandegee (1908).
Heliostemma Woodson (1935).

Shrubby vines, sometimes erect, stems from a thickened taproot; with dense mixed pubescence all over; glandular trichomes with crystalline-white inclusions. Leaves ca. 4–10 cm long, ovate to deltoid, basally cordate, apically acuminate. Inflorescences pedunculate, 2–6-flowered. Corolla broadly to narrowly tubular campanulate; lobes (triangular-)ovate, obtuse, reticulate; gynostegial corona of 5 connate staminal segments, each with filiform or digitate appendages usually in 2 series; gynostegium sessile or stipitate; pollinia subhorizontal or pendent, reniform or obovoid. Follicles fusiform-subcylindrical, long-attenuate, smooth, mottled.

About 20 spp. in northern Mexico to Costa Rica, dry deciduous forests and scrub, 200–1300 m.

364. *Prosthecidiscus* Donn. Sm.

Prosthecidiscus Donn. Sm., Bot. Gaz. 25: 149 (1898).

Woody coarse vine; 8–10 m long, branches densely pubescent with short glandular and eglandular spreading trichomes, and long bristly trichomes. Leaves 8–22 cm long, elliptic, basally cordate, apically abruptly acuminate, adaxially sparsely setulous to glabrous, abaxially moderately to densely puberulent. Inflorescences pedunculate, 3–6-flowered. Corolla green and red or purple; lobes 12–19 mm long, oblong-spathulate, adaxially glabrous, abaxially glandular-puberulous; corolline corona present; gynostegial corona 5-lobed and ligulate, adnate to stipe; style-head long-rostrate; pollinia subpendent, obpyriform. Follicles strongly and densely muricate.

One sp., *P. guatemalensis* Donn. Sm., Central America and Mexico, mainly in dry deciduous forests, 30–1800 m.

365. *Pruskortizia* Morillo

Pruskortizia Morillo, Pittieria 40: 97 (2016); Morillo, Pittieria 39: 236 (2015).

Woody vines; stem pubescence of brown or yellowish long eglandular and short glandular trichomes. Leaves more than 15 cm long, obovate to oblong-elliptic, shortly and narrowly cordate, acuminate. Inflorescences subsessile, 3–6-flowered. Corolla 33–65 mm diam., rotate; lobes spreading to reflexed; gynostegial corona dark purple; staminal corona segments hemidiscoid, with a short central ridge at base of the anthers; gynostegium sessile; pollinia horizontal, oblongoid or narrowly calceolate; anthers with triangular or deltoid apical membrane; style-head flat. Follicles broadly ovoid, thick-walled, pubescent, somewhat tuberculate, with small conic protuberances.

Two spp. in Colombia to Bolivia, wet, rain or montane forests, 100–2550 m.

366. ***Pseudolachnostoma*** Morillo

Pseudolachnostoma Morillo, Pittieria 36: 44 (2012).

Suffrutescent twiners, densely pubescent with yellow trichomes, glandular trichomes absent. Leaves 7–14 cm long, ovate-elliptic, apically acuminate. Inflorescences few-flowered, sciadioidal. Corolla 3.5–7 mm long, yellowish, green or maroon, occasionally veined with green, diminutively hispid; lobes strongly reflexed; gynostegial corona tubular to cyathiform, adnate to corolla and stipe with five laminar lobules forming deep nectaries, subcarnose; gynostegium stipitate; anthers fused to upper side of style-head; connective appendages absent; pollinia pendent, pyriform, larger than corpusculum. Follicles narrowly fusiform, conspicuously 5-costate, attenuate.

About 12 spp., tropical Americas, rain and premontane forests, 200–1350 m.

367. ***Ptycanthera*** Decne.

Ptycanthera Decne., Prodr. 8: 606 (1844).

Slender vines; stem pubescent in two lines. Leaves oblong to ovate, 0.9–5.7 cm long, basally obtuse to truncate, apically obtuse to rounded and apiculate, glabrous. Inflorescences subsessile or pedunculate, 2–3-flowered. Corolla (sub-)rotate; lobes 4.7–7.5 mm long, linear-lanceolate, obtuse; gynostegial corona of basally swollen staminal segments, subtriangular in front view, rising vertically and connecting to the stipe below the anthers; gynostegium stipitate; anther wings with divergent apices; pollinia horizontal, obovate or oblong-obovate; style-head conical or slightly rostrate. Follicles fusiform, with 5 undulating wings.

About five spp. in the Caribbean.

368. ***Rhytidostemma*** Morillo

Rhytidostemma Morillo, Pittieria 37: 127 (2013).

Twiners with a mixed indumentum of short glandular trichomes and long or short non-glandular trichomes all over. Leaves elliptic, basally rounded, apically acute. Inflorescences long-pedunculate, 4–18-flowered. Corolla campanulate; lobes narrowly ovate to oblong, conspicuously undulate and crenate along both margins in the upper third; gynostegial corona of 5 distinct, conspicuously verrucose lobes surrounding base of gynostegium, frequently with 5 laminar laterally flat segments arising radially from upper part of stipe; gynostegium shortly stipitate; carpels densely pubescent. Follicles fusiform, with thick walls, with dense, short, brown glandular pubescence.

About ten spp., Panama to Guianas, Peru and Brazil.

369. ***Riparoampelos*** Morillo

Riparoampelos Morillo, Pittieria 39: 241 (2015).

Suffrutescent twiner; stems shortly pubescent. Leaves 8–13 cm long, lanceolate to obovate-elliptic, basally shortly and narrowly cordate, apically acuminate, abaxially with short erect trichomes. Inflorescences subsessile, 2–3-flowered. Corolla 20–24 mm diam., rotate(-campanulate); lobes dark purplish, spreading, apically with long, flat and white trichomes; gynostegial corona fleshy, dark purple to black, staminal segments incurved, interstaminal segments broadly concave; gynostegium subsessile; pollinia horizontal, asymmetrically pear-shaped. Follicles narrowly ovoid or fusiform, muricate, with irregular blunt-tipped projections.

A single species, *R. amazonica* (Morillo) Morillo, Amazon basin and Guianas, wet or rain forests, mainly in riparian vegetation, 100–300 m.

370. ***Rojasia*** Malme

Rojasia Malme, Ark. Bot. 4(14): 10, t. 2, f. 4 (1905); Goyder, Kew Bull. 61: 31–33 (2006), rev.

Suffrutescent twiner, 3–4 m long; densely grayish tomentose. Leaves 3–6 cm long, ovate to triangular-deltate, basally cordate. Inflorescences 4–6-flowered. Corolla ca. 10 mm long, white; lobes spathulate, glabrous, erect; gynostegial corona of free staminal lobes attached below the filament tube, laminar, erect, branched, fimbriate, exceeding gynostegium; gynostegium stipitate; pollinia ovoid, pendent, laterally attached to the articulated caudicles. Follicles ca. 10 mm long, obclavate, finely pubescent, muricate, projections long and slender.

One sp., *R. gracilis* (Morong) Malme, in Argentina, low Chaco forest, 50–450 m.

371. ***Rotundanthus*** Morillo

Rotundanthus Morillo, Pittieria 39: 245 (2015).

Woody vine to 9 m long, densely pubescent with eglandular, long, yellowish, spreading trichomes. Leaves 8–17 cm long, oblong-lanceolate or narrowly elliptic, basally shortly cordate, apically acuminate. Inflorescences subsessile, 2–5-flowered. Corolla 36–48 mm diam., rotate-campanulate, bright yellow; lobes spreading, ovate-orbicular or broadly deltate; gynostegial corona a shallow fleshy cup, roughly pentagonal with rounded angles, marginally crenulate; staminal lobes shortly ligulate; gynostegium sessile; anthers horizontal, with acicular wings; pollinia horizontal, auriculate; style-head peltate. Follicles fusiform, densely muricate, pubescent.

One sp., *R. fulvidus* (Ballard) Morillo, S Mexico to Costa Rica, wet or rain forests, below 600 m.

372. ***Suberogerens*** Morillo

Suberogerens Morillo, Pittieria 39: 249 (2015).

Suffrutex with woody and corky caudex; stems to 2 m long, appressed-puberulent when young. Leaves 9–17 cm long, broadly ovate to roundly reniform, basally cordate, apically acuminate, adaxially sparsely puberulent or glabrous, abaxially appressed-puberulent. Inflorescences sessile, 2–3-flowered. Calyx as long as corolla; corolla rotate, 18–25 mm diam., green to dark-purple; lobes abaxially appressed-puberulent to glabrous, adaxially white pilose; gynostegial corona purple, disciform, with five small erect ligulate lobes; gynostegium sessile; anthers with white apical membranes; pollinia horizontal; style-head flat. Follicles narrowly fusiform, minutely muricate.

One sp., *S. cyclophylla* (Standl.) Morillo, in Mexico, thorn scrub and dry deciduous forests, 200–1350 m.

373. ***Tressensia*** H.A. Keller

Tressensia H.A. Keller, Lilloa 54: 196 (2017).

Herbaceous twiner, pubescent with eglandular and glandular trichomes. Leaves 2.5–12 cm long, (oblong-)ovate, basally auriculate, apically acute. Inflorescences long-pedunculate, 3–6-flowered; flowers long-pedicellate. Corolla ca. 3 cm diam., shallowly campanulate, green-reticulate; tube and throat lanose; lobes narrowly triangular, reflexed; gynostegial corona annular, yellow; staminal parts fleshy; gynostegium sessile; pollinia ovate, laterally compressed, horizontal; style-head decagonal, umbonate. Follicles obclavate, 5-winged. Seeds verrucose, with dentate wing.

One sp., *T. viridis* H.A. Keller & S.A. Cáceres, in Argentina, riverine forest fragments.

374. ***Trichosacme*** Zucc.

Trichosacme Zucc., Abh. Math.-Phys. Cl. Königl. Bayer. Akad. Wiss. 4(2): 11 (1846).

Twining or erect herb, to 1 m long, with dense white-silvery woolly indumentum all over. Leaves 3.5–9 cm long, (oblong-)ovate, basally subcordate to rounded, apically obtuse or shortly acute. Inflorescences pedunculate, 6–10-flowered; flowers subsessile. Corolla 9–10 mm diam., subrotate, dark purple; lobes ovate, adaxially glabrous or puberulous at base, apically with long, plumose, caudate, purple appendages; gynostegial corona annular, bright orange, adnate to corolla tube, shortly 5-dentate; gynostegium subsessile, pollinia pendent, narrowly clavate, with a slender, sterile, horn-like basal process; style-head convex. Follicles cylindrical-fusiform.

One sp., *T. lanata* Zucc., in Mexico, thorny scrub; 1300–1800 m.

375. ***Tylodontia*** Griseb.

Tylodontia Griseb., Cat. Pl. Cub. 175 (1866); Mangelsdorff et al., Willdenowia 46: 443–474 (2016), rev.

Sparsely branched herbaceous twiners, pubescent with eglandular recurved trichomes. Leaves 1.5–4.5 cm long, lanceolate to ovate, inflorescences few to 70-flowered, thyrsoidal. Corolla 1.8–6 mm long, urceolate, white, carnose, adaxially usually densely pubescent; corolline corona of antepetalous lobules in tube; gynostegial corona of staminal and interstaminal segments, carnose, short; gynostegium occasionally stipitate, concealed in corolla, anther wings separated in a basal, mostly longer, vertically oriented part and an upper, almost horizontal (basally centrifugal) part; connective appendages often strongly inflexed; style-head green, with central depression. Follicles 60–70 mm long, fusiform or narrowly oblong. Seeds pyriform, winged.

Four spp. endemic to Cuba, mountainous regions.

376. ***Vulcanoa*** Morillo

Vulcanoa Morillo, Pittieria 39: 253 (2015).

Woody vine, densely ferruginous-pilose; trichomes eglandular, retrorse or spreading. Leaves narrowly ovate to ovate-elliptic, basally cordate, apically acuminate to cuspidate. Inflorescences shortly pedunculate, few-flowered. Corolla ca. 28 mm diam., campanulate, white, reticulate with green; lobes spreading, narrowly ovate, adaxially hispidulous; corolline corona tubular-cupuliform, apically fimbriate; gynostegial corona adnate to stipe, with 5 horn-like staminal lobes; gynostegium stipitate; anthers suberect, with short membranes; pollinia pendent, narrowly pyriform; style-head convex. Follicles fusiform, densely muricate, with long, curved projections.

One sp., *V. steyermarkii* (Woodson) Morillo, in Guatemala and Mexico, upper montane cloud forests, on volcanos, 1700–3300(–3900) m.

V.5.l. **Subtribe Topeinae** H.A. Keller & Liede (2017).

Twining plants with leaves oblong-lanceolate to lanceolate, basis deeply cordate to auriculate. Inflorescences sciadioidal, pendulous, present all the year. Corolla lobes with vibratile trichomes on the adaxial side. Corona with flattened lobes opposite the anthers. Caudicles sigmoidal, corpusculum with basal hyaline appendix. Follicles solitary, pendulous, narrowly fusiform, with long pointed apex. One genus in South America.

377. ***Topea*** H.A. Keller

Topea H.A. Keller, Bonplandia 26: 134 (2017), nom. nov. pro *Aenigma* H.A. Keller, Lilloa 54: 59 (2017), nom. illeg., non *Enigma* Weber-van Bosse (1932), Rhodophyceae.

Twiners, glabrous or pilose. Leaves 5–12 cm long, oblong-lanceolate, auriculate. Inflorescences long-pedunculate, 3–12-flowered, sciadioidal. Corolla rotate, 3–15 mm diam., lobes lanceolate, patent or reflexed, adaxially with long trichomes, apically with clavate vibratile trichomes; staminal corona lobes lanceolate, basally fused, exceeding the gynostegium; gynostegium sessile, pollinia reniform, caudicles with hyaline projections; style-head flat; follicles solitary, fusiform, to 15 cm long. Seeds obovate, with apically dentate wing.

Two spp. in Argentina and Paraguay, moist forest, lowlands.

Genus of uncertain subtribal placement in Asclepiadeae

378. ***Mahawoa*** Schltr.

Mahawoa Schltr., Beih. Bot. Centralbl. 34(2): 2 (1916).

Suffrutescent twiner; shoots and leaves hirsute with retrorse trichomes. Leaves 4.5–7 cm long, ovate to elliptic. Inflorescences longer than subtending leaves, few-flowered, dichasial at base, bostrychoid monochasial in higher order branching, lax. Corolla 7–8 mm long, rotate, glabrous; lobes ovate; gynostegial corona ring exceeding the gynostegium, fused for about a quarter of its length; staminal parts connate to filaments for more than a third of gynostegium length, lingulate, inflexed; gynostegium sessile, pollinia ovoid, caudicles horizontal, short; style-head flat.

One sp., *M. montana* Schltr., in Indonesia (Sulawesi), forests, at 1200 m.

With an inflorescence reminiscent of Tylophorinae and a corona pointing to Cynanchinae, the position of *Mahawoa* remains uncertain because the type was destroyed in Berlin and no other specimen has ever been encountered.

Selected Bibliography

Abe, F., Yamauchi, T. 1985. Affinosides M and K, cardenolide glycosides from seeds of *Anodendron affine* (*Anodendron* V). Chem. Pharm. Bull. 33: 847–852.

Abe, F., Yamauchi, T. 1989. Pregnane and pregnane glycosides from *Trachelospermum liukiuense*. Chem. Pharm. Bull. 37: 33–35.

Abe, F., Yamauchi, T. 1994. Indole alkaloids from leaves and stems of *Leuconotis eugenifolius*. Phytochemistry 35: 169–171.

Abisch, E., Reichstein, T. 1962. Orientierende chemische Untersuchungen einiger Asclepiadaceen und Periplocaceen. Helv. Chim. Acta 45: 2090–2116.

Abrahamczyk, S., Kessler, M., 2010. Hummingbird diversity, food niche characters, and assemblage composition along a latitudinal precipitation gradient in the Bolivian lowlands. J. Ornithol. 151: 615–625. doi: https://doi.org/10.1007/s10336-010-0496-x

Albers, F., Meve, U. 1997. Taxonomic Groups: Asclepiadaceae. In: Oldfield, S. (ed.) Status Survey and Conservation Action Plan: Cactus and succulent plants. Cambridge: IUCN, pp. 14–17 & 159–163.

Albers, F., Meve, U. 2001. A karyological survey of Asclepiadoideae, Periplocoideae and Secamonoideae, and evolutionary considerations within Apocynaceae s.l. Ann. Missouri Bot. Gard. 88: 624–656.

Albers, P., van der Maesen, L.J.G. 1994. Pollination of Apocynaceae. Wageningen Agric. Univ. Pap. 94: 61–81.

Ali, T., Ali, S.I. 1996. Andromonoecy in *Glossonema varians* (Stocks) Hook.f. (Asclepiadaceae). Pakistan J. Bot. 28(1): 25–29.

Alper, K.R., Lotsof, H.S., Kaplan, C.D. 2008. The ibogaine medical subculture. J. Ethnopharmacol. 115: 9–24.

Alvarado-Cárdenas, L.O., Ochoterena, H. 2007. A phylogenetic analysis of the *Cascabela–Thevetia* species complex (Plumerieae, Apocynaceae) based on morphology 1. Ann. Missouri Bot. Gard. 94: 298–323. doi: https://doi.org/10.3417/0026-6493(2007)94[298:APAOTC]2.0.CO;2

Alvarado-Cárdenas, L.O., Villaseñor, J. L., López-Mata, L., Cadena, J., Ortiz, E. 2017. Systematics, distribution and conservation of *Cascabela* (Apocynaceae: Rauvolfioideae: Plumerieae) in Mexico. Pl. Syst. Evol. 303: 337–369. doi: https://doi.org/10.1007/s00606-016-1375-6

APG IV (Angiosperm Phylogeny Group IV) 2016. An update of the Angiosperm Phylogeny Group classification for the orders and families of flowering plants: APG IV. Bot. J. Linn. Soc. 181: 1–20.

Aremu, A.O., Cheesman, L., Finnie, J.F., Van Staden, J. 2011. *Mondia whitei* (Apocynaceae): A review of its biological activities, conservation strategies and economic potential. S. African J. Bot. 77: 960–971. doi: https://doi.org/10.1016/j.sajb.2011.06.010

Arenas, P. 1999. *Morrenia odorata* (Asclepiadaceae), an edible plant of the Gran Chaco. Econ. Bot. 53: 89–97.

Astaras, C., Waltert, M. 2010. What does seed handling by the drill tell us about the ecological services of terrestrial cercopithecines in African forests? Animal Conserv. 13: 568–578.

Baas, P., Werker, E., Fahn, A. 1983. Some ecological trends in vessel characters. Int. Assoc. Wood Anat. Bull., n.s. 4: 141–159.

Backlund, M., Oxelman, B., Bremer, B. 2000. Phylogenetic relationships within the Gentianales based on *ndhF* and *rbcL* sequences, with particular reference to the Loganiaceae. Amer. J. Bot. 87: 1029–1043.

Bandara, V., Weinstein, S.A., White, J., Eddleston, M. 2010. A review of the natural history, toxicology, diagnosis and clinical management of *Nerium oleander* (common oleander) and *Thevetia peruviana* (yellow oleander) poisoning. Toxicon 56: 273–281.

Baranzelli, M.C., Sérsic, A.N., Cocucci, A.A. 2014. The search for Pleiades in trait constellations: functional integration and phenotypic selection in the complex flowers of *Morrenia brachystephana* (Apocynaceae). J. Evol. Biol. 27: 724–736. doi: https://doi.org/10.1111/jeb.12341

Barink, M.M. 1983. A revision of *Pleioceras*, *Stephanostema* and *Schizozygia*. Series of revisions of Apocynaceae, XII. Meded. Landbouwhogeschool Wageningen 83-7: 21–53.

Barman, C., Singh, V.K., Das, S., Tandon, R. 2018. Floral contrivances and specialized pollination mechanism confer strong influence to elicit mixed-mating in *Wrightia tomentosa* (Apocynaceae). Pl. Biol., 2018 Jan. 13. doi: https://doi.org/10.1111/plb.12690

Beaune, D., Fruth, B., Bollache, L., Hohmann, G., Bretagnolle, F. 2013. Doom of the elephant-dependent trees in a Congo tropical forest. Forest Ecol. Manag. 295: 109–117.

Beentje, H.J. 1982. A monograph on *Strophanthus* DC. (Apocynaceae). Meded. Landbouwhogeschool Wageningen 82-4: 1–191.

Behnke, H.-D. 1981. Sieve-element characters. Nord. J. Bot. 1: 381–400.

Bell, C.D., Soltis, D.E., Soltis, P.S. 2010. The age and diversification of the angiosperms re-revisited. Amer. J. Bot. 97: 1296–1303.

Bentham, G. 1876. Asclepiadaceae. In: Bentham, G., Hooker, J.D. (eds.) Genera Plantarum, Vol. 2(2). London: Williams and Norgate, pp. 739–785.

Bester, S.P., Nicholas, A. 2016. *Periglossum podoptyches* (Apocynaceae-Asclepiadoideae), a new species from KwaZulu-Natal province, South Africa. Phytotaxa 282: 28–36.

Bhatnagar, S. 1986. On insect adaptations for pollination in some asclepiads on Central India. In: Kapil, R.P. (ed.) Pollination Biology – an Analysis. New Delhi: Inter-India Publications, pp. 37–57.

Bierer, D.E., Dubenko, L.G., Zhang, P., Lu, Q., Imbach, P. A., Garofalo, A.W., Phuan, P.-W., Fort, D.M., Litvak, J., Gerber, R.E., Sloan, B., Luo, J., Cooper, R., Reaven, G.M. 1998. Antihyperglycemic activities of Cryptolepine analogues: An ethnobotanical lead structure isolated from *Cryptolepis sanguinolenta*. J. Med. Chem. 41: 2754–2764.

Birkinshaw, C. 2001. Fruit characteristics of species dispersed by the black lemur (*Eulemur macaco*) in the Lokobe Forest, Madagascar. Biotropica 33: 478–486.

Bisset, N.G. 1958. The occurrence of alkaloids in the Apocynaceae. Ann. Bogoriensis 3: 105–236.
Bisset, N.G. 1961. The occurrence of alkaloids in the Apocynaceae. Part II. A review of recent developments. Ann. Bogoriensis 4: 65–144.
Bisset, N.G. 1987. Phytochemistry of *Nerium* L. Agric. Univ. Wageningen Pap. 87-2: 27–38.
Bisset, N.G. 1989. Arrow and dart poisons. J. Ethnopharmacol. 25: 1–41.
Bisset, N.G. 1991. One man's poison, another man's medicine? J. Ethnopharmacol. 32: 71–81.
Bisset, N.G. 1992. Uses, chemistry and pharmacology of *Malouetia* (Apocynaceae, subf. Apocynoideae). J. Ethnopharmacol. 36: 43–50. doi: https://doi.org/10.1016/0378-8741(92)90059-Z
Boiteau, P., Allorge, L. 1978. Morphologie et biologie florales des Apocynacées: I. différences essentielles entre les Plumérioidées et les Tabernaemontanoidées. Adansonia Sér. 2, 17: 305–216.
Bonjean, K., De Pauw-Gillet, M.C., Defresne, M.P., Colson, P., Houssier, C., Dassonneville, L., Bailly, C., Greimers, R., Wright, J., Quétin-Leclerq, J., Tits, M., Angenot, L. 1998. The DNA intercalating alkaloid Cryptolepine interferes with Topoisomerase II and inhibits primarily DNA synthesis in B16 melanoma cells. Biochemistry 37: 5136–5146.
Boppré, M. 1990. Lepidoptera and pyrrolizidine alkaloids: exemplification of complexity in chemical ecology. J. Chem. Ecol. 16: 165–185. https://doi.org/10.1007/BF01021277
Boppré, M. 1995. Pharmakophagie: Drogen, Sex und Schmetterlinge. Biol. Unserer Zeit 25: 8–17. https://doi.org/10.1002/biuz.19950250103
Boppré, M., Schneider, D. 1985. Pyrrolizidine alkaloids quantitatively regulate both scent organ morphogenesis and pheromone bio-synthesis in *Creatonotos* moths (Lep.: Arctiidae). J. Comp. Physiol. 157: 569–577. https://doi.org/10.1007/BF01351351
Brand, E., Leon, C., Nesbitt, M., Guo, P., Huang, R.-Q., Chen, H.D., Liang, L., Zhao, Z. 2017. Economic botany collections: A source of material evidence for exploring historical changes in Chinese medicinal materials. J. Ethnopharmacol. 200: 209–227.
Bremer, B., Jansen, R.K., Oxelman, B., Backlund, M., Lantz, H., Ki-Joong, K. 1999. More characters or more taxa for a robust phylogeny – case study from the coffee family (Rubiaceae). Syst. Biol. 48: 413–435.
Britt, A., Iambana, B.R. 2003. Can captive-bred *Varecia variegata variegata* adapt to a natural diet on release to the wild? Int. J. Primatol. 24: 987–1005.
Brown, R. 1810. On the Asclepiadeae, a natural order of plants separated from the Apocinae of Jussieu. Preprint of: Mem. Wern. Nat. Hist. Soc. 1: 12–78 (1811).
Brown, N.E. 1901. *Lobostephanus palmatus* N.E. Brown. Hook. Icon. Pl. 27: pl. 2692. In: Thiselton-Dyer, W.T. (ed.). London: Dulau & Co.
Bruyns, P.V 1999. The systematic position of *Eustegia* R. Br. (Apocynaceae - Asclepiadoideae). Bot. Jahrb. Syst. 121: 19–44.
Bruyns, P.V. 2002. Monograph of *Orbea* and *Ballyanthus* (Apocynaceae-Asclepiadoideae-Ceropegieae). Syst. Bot. Monogr. 63: 1–196.
Bruyns, P.V. 2005. Stapeliads of Southern Africa and Madagascar, 1st ed. Hatfield, South Africa: Umdaus Press.
Bruyns, P.V. 2010. A new species of *Caralluma* (Apocynaceae-Asclepiadoideae-Ceropegieae) from the Yemen. S. African J. Bot. 76: 249–251.
Bruyns, P.V., Klak, C. 2006. A systematic study of the Old World genus *Fockea* (Apocynaceae-Asclepiadoideae). Ann. Missouri Bot. Gard. 93: 535–564.
Bruyns, P.V., Klak, C. 2009. The rediscovery of *Schizostephanus gossweileri* and its phylogenetic position. S. African J. Bot. 75: 532–536.
Bruyns, P.V., Klak, C., Hanáček, P. 2014. Evolution of the stapeliads (Apocynaceae–Asclepiadoideae) – repeated major radiation across Africa in an Old World group. Mol. Phyl. Evol. 77: 251–263.
Bruyns, P.V., Klak, C., Hanáček, P. 2015. Recent radiation of *Brachystelma* and *Ceropegia* (Apocynaceae) across the Old World against a background of climatic change. Mol. Phyl. Evol. 90: 49–66.
Bruyns, P.V., Klak, C., Hanáček, P. 2017. A revised, phylogenetically-based concept of *Ceropegia* (Apocynaceae). S. African J. Bot. 112 399–2436.
Buhner, S.H. 2012. Herbal Antibiotics (2nd ed.). Massachusetts: Storey Publ.
Burge, D.O., Mugford, K., Hastings, A.P., Agrawal, A.A. 2013. Phylogeny of the plant genus *Pachypodium* (Apocynaceae). PeerJ 1: 1–20. doi: https://doi.org/10.7717/peerj.70
Burkill, H.M. 1985. Apocynaceae, Asclepiadaceae, The useful plants of West tropical Africa, Vol. 1, 2nd ed. Richmond: Royal Bot. Gard. Kew, pp. 135–193, 217–241.
Burrows, G.E., Tyrl, R.J. 2013. Chapter 9, Apocynaceae. Toxic plants of North America, 2nd ed. Ames: Wiley-Blackwell, pp. 81–126.
Burzynski, E.A., Minbiole, K.P.C., Livshultz, T. 2015. New sources of lycopsamine-type pyrrolizidine alkaloids and their distribution in Apocynaceae. Biochem. Syst. Ecol. 59: 331–339. https://doi.org/10.1016/j.bse.2015.02.006
Calviño, C.I., Fernandez, M., Ezcurra, C. 2014. Is the southern South American genus *Tweedia* (Apocynaceae: Asclepiadoideae) monophyletic? Molecular phylogenies, distribution and taxonomy. Taxon 63: 1265–1274.
Candolle, A.P. de 1844. Apocynaceae. In: Candolle, A.P. de (ed.) Prodromus systematis naturalis regni vegetabili, Vol. 8. Paris: Treuttel & Wurtz, pp. 317–489.
Cant, J.G.H. 1979. Dispersal of *Stemmadenia donnell-smithii* by birds and monkeys. Biotropica 11: 122.
Carlquist, S. 1984.Vessel grouping in dicotyledon wood: Significance and relationships to imperforate tracheary elements. Aliso 10: 505–525.
Chang, N., Luo, Z., Li, D., Song, H. 2017. Indigenous uses and pharmacological activity of traditional medicinal plants in Mount Taibai, China. J. Evid. Based Complementary Altern. Med. 2017, 11 pp. doi: https://doi.org/10.1155/2017/8329817
Chaturvedi, S.K. 1988. Abiotic Pollination in *Tylophora hirsuta* Wight (Asclepiadaceae). Asklepios 45: 58–62.
Chen, Z.S., Lee, G.H., Kuo, Y.H. 1993. Disformone and Dischidiol from *Dischidia formosana*. Phytochemistry 34: 783–786.
Chua, L.S.L., Horsten, S.F.A.J. 2001. *Tabernaemontana*. In: Van Valkenburg, J.L.C.H., Bunyapraphatsara, N.

(eds.) Plant Resources of South-East Asia, Vol. 12(2): 530–538. Leiden: Backhuys.

Chuba, D., Goyder, D.J., Chase, J.M., Fishbein, M. 2017. Phylogenetics of the African *Asclepias* complex (Apocynaceae) based on three plastid DNA regions. Syst. Bot. 42: 148–159. doi: https://doi.org/10.1600/036364417X694539

Church, A.H. 1908. Types of floral mechanism, Part 1, Types I-XII. Oxford: Clarendon Press. doi: https://archive.org/details/cu31924000658413

Civeyrel, L. 1994. Variation et évolution des types polliniques du genre *Secamone* (Secamonoideae, Asclepiadaceae). C.R. Acad. Sci. Paris 317: 1159–1165.

Civeyrel, L. 1995. Pollen morphology and ultrastructure of the genus *Secamone* in Africa. 2nd Symposium on African Palynology, Tervuren (Belgium), 1995, Publ. Occas. CIFEG, Orléans, CIFEG: 207–215.

Civeyrel, L., Rowe, N. 2001. Phylogenetic relationships of Secamonoideae based on plastid gene matK, morphology and biomechanics. Ann. Missouri Bot. Gard. 88: 583–602.

Civeyrel, L., Le Thomas, A., Ferguson, K., Chase, M.W. 1998. Critical reexamination of palynological characters used to delimit Asclepiadaceae in comparison to molecular phylogeny obtained from plastid matK sequences. Mol. Phyl. Evol. 9: 517–527.

Cocucci, A.A., Marino, S., Baranzelli, M., Wiemer, A.P., Sérsic, A. 2014. The buck in the milkweed: evidence of male-male interference among pollinaria on pollinators. New Phytol. 203: 280–286. doi: https://doi.org/10.1111/nph.12766

Colegate, S.M., Gardner, D.R., Betz, J.M., Fischer, O.W., Liede-Schumann, S., Boppré, M. 2016. Pro-toxic 1, 2-dehydropyrrolizidine alkaloid esters, including unprecedented 10-membered macrocyclic diesters, in the medicinally-used *Alafia* cf. *caudata* and *Amphineurion marginatum* (Apocynaceae: Apocynoideae: Nerieae and Apocyneae). Phytochem. Anal. 27: 257–276. doi: https://doi.org/10.1002/pca.2624

Collinson, M.E., Manchester, S.R., Wilde, V., Hayes, P. 2010. Fruit and seed floras from exceptionally preserved biotas in the European Paleogene. Bull. Geosci. 85: 155–162.

Collinson, M.E., Manchester, S.R., Wilde, V. 2012. Fossil fruits and seeds of the Middle Eocene Messel biota, Germany. Abhandlungen der Senckenberg Gesellschaft für Naturforschung, Band 570. Stuttgart: Schweizerbart, p. 20.

Coppen, J.J.W., Cobb, A.I. 1983. The occurrence of iridoids in *Plumeria* and *Allamanda*. Phytochemistry 22: 125–128.

Corlett, R.T., Lucas, P.W. 1990. Alternative seed-handling strategies in primates: seed-spitting by long-tailed macaques (*Macaca fascicularis*). Oecologia 82: 166–171.

Corner, E.J.H. 1976. The Seeds of Dicotyledons, Vols.1, 2. Cambridge: Cambridge Univ. Press, 1: pp. 70–73, 2: t. 19–23.

Cronquist, A. 1981. An integrated system of classification of flowering plants. New York: Columbia University Press.

Cullen, J. 1978. A preliminary survey of ptyxis (vernation) in the Angiosperms. Notes Roy. Bot. Gard. Edinb. 37: 161–214.

Darrault, R.O., Schlindwein, C. 2005. Limited fruit production in *Hancornia speciosa* (Apocynaceae) and pollination by nocturnal and diurnal insects. Biotropica 37: 381–388.

Davis, A.R., Gunning, B.E.S. 1992. The modified stomata of the floral nectary of *Vicia faba* L. 1. Development, anatomy and ultrastructure. Protoplasma 166: 134–152.

Decaisne, M.J. 1844. Asclepiadaceae. In: de Candolle, A.P. (ed.) Prodromus Systematis Naturalis Regni Vegetabilis, Vol. 8. Paris: Treuttel & Würtz, pp. 490–684.

Defler, T.R., Defler, S.B. 1996. Diet of a group of *Lagothrix lagothricha lagothricha* in southeastern Colombia. Int. J. Primatol. 17: 161–190.

De Kruif, A.P.M. 1983. Series of revisions of Apocynaceae XI. A revision of *Motandra* A. DC. (Apocynaceae). Meded. Landbouwhogeschool Wageningen 83-7: 1–20.

De Kruif, A.P.M. 1985. A revision of *Oncinotis* Benth. (Apocynaceae). Series of revisions of Apocynaceae XVI. Wageningen Agric. Univ. Pap. 85.2: 1–45.

De Luca, V., Salim, V., Atsumi, S.M., Yu, F. 2012. Mining the biodiversity of plants: A revolution in the making. Science 336: 1658–1661.

Demeter, K. 1922. Vergleichende Asclepiadeenstudien. Flora 115: 130–176.

Denis, M.S., Capuccino, N. 2004. Reproductive biology of *Vincetoxicum rossicum* (Kleo.) Barb. (Asclepiadaceae), an invasive alien in Ontario. J. Torrey Bot. Soc. 131: 8–15.

Domingos-Melo, A., de Lima Nadia, T., Machado, I.C. 2017. Complex flowers and rare pollinators: Does ant pollination in *Ditassa* show a stable system in Asclepiadoideae (Apocynaceae)? Arthropod Pl. Interact. doi: https://doi.org/10.1007/s11829-017-9499-3

Dutt, H.C., Singh, S., Avula, B., Khan, I.A., Bedi, Y.S. 2012. Pharmacological review of *Caralluma* R.Br. with special reference to appetite suppression and anti-obesity. J. Med. Food 15: 108–119. doi: https://doi.org/10.1089/jmf.2010.1555

Dyer, R.A. 1933. *Fockea cylindrica* R.A. Dyer. Hook. Icon. Pl. 34: pl. 3221. In: Hill, A.W. (ed.) London: Dulau & Co.

Edgar, J.A. 1984. Parsonsieae: Ancestral larval food plants of the Danainae and Ithomiinae. In: Vane-Wright, R. I., Ackery, P.R. (eds.) The biology of butterflies. London: Academic Press, pp. 91–93.

Eisikowitch, D. 1986. Morpho-ecological aspects on the pollination of *Calotropis procera* (*Asclepiadaceae*) in Israel. Pl. Syst. Evol. 152: 185–194.

El-Gazzar, A., Hamza, M.K., Badawi, A.A. 1974. Pollen morphology and taxonomy of Asclepiadaceae. Pollen and Spores 16: 227–238.

Endlicher, S. 1838. Asclepiadaceae, Genera Plantarum secundum ordines naturales disposita Vindobonae. Vienna: Beck, pp 586–598.

Endress, M.E. 2001. Apocynaceae and Asclepiadaceae: United they stand. Haseltonia 8: 2–9.

Endress, M.E., Bruyns, P.V. 2000. A revised classification of the Apocynaceae s. l. Bot. Rev. 66: 1–56. doi: https://doi.org/10.1007/BF02857781

Endress, M.E., Hesse, M., Nilsson, S., Guggisberg, A., Zhu, J.-P. 1990. The systematic position of the Holarrheninae (Apocynaceae). Pl. Syst. Evol. 171: 157–185. https://doi.org/10.1007/BF00940603

Endress, M.E., Sennblad, B., Nilsson, S., Civeyrel, L., Chase, M.W., Huysmans, S., Grafström, E., Bremer, B. 1996. A phylogenetic analysis of Apocynaceae s. str. and some related taxa in the Gentianales: a multidisciplinary approach. Opera Bot. Belg. 7: 59–102. https://doi.org/10.3417/0026-6493(2007)94[1:APAOAA]2.0.co;2

Endress, M.E., Lorence, D.H., Endress, P.K. 1997. Structure and development of the gynoecium of *Lepinia marquisensis* and its systematic position in the Apocynaceae. Allertonia 7: 267–272.

Endress, M.E., van der Ham, R.W.J.M., Nilsson, S., Civeyrel, L., Chase, M.W., Sennblad, B., Potgieter, K., Joseph, J., Powell, M., Lorence, D., Zimmerman, Y.-M., Albert, V.A. 2007a. A phylogenetic analysis of Alyxieae (Apocynaceae) based on *rbc*L, *mat*K, *trn*L intron, *trn*L-F spacer sequences, and morphological characters. Ann. Missouri Bot. Gard. 94: 1–35.

Endress, M.E., Liede-Schumann, S., Meve, U. 2007b. Advances in Apocynaceae: The enlightenment, an introduction. Ann. Missouri Bot. Gard. 94: 259–267. doi: https://doi.org/10.3417/0026-6493(2007)94[259:AIATEA]2.0.CO;2

Endress, M.E., Liede-Schumann, S., Meve, U. 2014. An updated classification for Apocynaceae. Phytotaxa 159: 175–194. https://doi.org/10.11646/phytotaxa.159.3.2

Erdtman, G. 1952. Pollen morphology and plant taxonomy. Stockholm: Almqvist and Wiksell.

Everist, S.L. 1981. Asclepiadaceae. In: Poisonous Plants of Australia, 2nd ed. Sydney: Angus and Robertson, pp. 94–109.

Ezcurra, C., Endress, M.E., Leeuwenberg, A.J.M. 1992. Apocynaceae. In: Spichiger, R., Ramella, L. (eds.) Flora del Paraguay 17. Geneva: Editions de Conservatoire et Jardin Botanique de la Ville de Genève.

Fahn, A. 1979. Secretory tissues in plants. New York, NY: Academic Press.

Fallen, M.E. 1983. A systematic revision of *Anechites* (Apocynaceae). Brittonia 35: 222–231. doi: https://doi.org/10.2307/2806018

Fallen, M.E. 1985. The gynoecial development and systematic position of *Allamanda* (Apocynaceae). Amer. J. Bot. 72: 572–579.

Fallen, M.E. 1986. Floral structure in the Apocynaceae: morphological, functional, and evolutionary aspects. Bot. Jahrb. Syst. 106: 245–286.

Farinaccio, M.A., de Mello-Silva, R. 2006. *Oxypetalum gyrophyllum* and *O. oblanceolatum*, new species of Asclepiadoideae (Apocynaceae) from Brazil, and a key for the *O. insigne* group. Novon 16: 235–239.

Farrell, B.D., Mitter, C. 1998. The timing of insect/plant diversification: might *Tetraopes* (Coleoptera: Cerambycidae) and *Asclepias* (Asclepiadaceae) have coevolved? Biol. J. Linn. Soc. 63: 553–577.

Feinsinger, P. 1978. Ecological interactions between plants and hummingbirds in a successional tropical community. Ecol. Monogr. 48: 269–287.

Fishbein, M. 2001. Evolutionary innovation and diversification in the flowers of Asclepiadaceae. Ann. Missouri Bot. Gard. 88: 603–623.

Fishbein, M., Stevens, W.D. 2005. Resurrection of *Seutera* Reichenbach (Apocynaceae – Asclepiadoideae). Novon 15: 531–533

Fishbein, M., Venable, D.L. 1996. Diversity and temporal change in the effective pollinators of *Asclepias tuberosa*. Ecology 77: 1061–1073.

Fishbein, M., Chuba, D., Ellison, C., Mason-Gamer, R.J., Lynch, S.P. 2011. Phylogenetic relationships of *Asclepias* (Apocynaceae) estimated from non-coding cpDNA sequences. Syst. Bot. 36: 1008–1023.

Flora of China Editorial Committee. 1999. Flora of China Illustrations, Volume (16), Gentianaceae through Boraginaceae. Beijing and St. Louis: Science Press, Missouri Botanical Garden Press, pp. 1–383.

Fonseca, L.C.N., Vizentin-Bugoni, J., Rech, A.R., Alves, M. A.A. 2015. Plant-hummingbird interactions and temporal nectar availability in a restinga from Brazil. An. Academica Bras. Ciên. 87: 206–2175.

Forster, P.I. 1991a. A possible identification for "Pollinia attached to adult anopheloine mosquitoes from northern Australia". Entomol. Soc. Queensland News Bull. 18: 113.

Forster, P.I. 1991b. Host records (family Asclepiadaceae) for *Euploea core corinna* (W. S. Macleay) (Lepidoptera: Nymphalidae). Austral. Entomol. Mag. 18: 61–64.

Forster, P.I. 1991c. A taxonomic revision of *Sarcolobus* R. Br. (Asclepiadaceae: Marsdenieae) in Australia and Papuasia. Austrobaileya 3: 335–360.

Forster, P.I. 1992a. Pollination of *Hoya australis* (Asclepiadaceae) by *Ocybadistes walkeri sothis* (Lepidoptera: Hesperidae). Aust. Ent. Mag. 19: 39–44.

Forster, P.I. 1992b. Insects associated with the flowers of *Marsdenia cymulosa* Benth. (Asclepiadaceae) and their possible role in pollination. Aust. Ent. Mag. 19: 45–58.

Forster, P.I. 1992c. A taxonomic revision of *Carissa* (Apocynaceae) in Australia. Aust. Syst. Bot. 5: 581–591.

Forster, P.I. 1993. Conspectus of *Cryptolepis* R.Br. (Asclepiadaceae: Periplocoideae) in Malesia. Austrobaileya 4: 67–73.

Forster, P.I. 1995. New names and combinations in *Marsdenia* (Asclepiadaceae: Marsdenieae) from Asia and Malesia (excluding Papuasia). Austral. Syst. Bot. 8: 691–701.

Frye, T.C. 1901. Development of the pollen in some Asclepiadaceae. Bot. Gaz. 32: 315–331.

Fu, Y.H., He, H.P., Di, Y.T., Li, S.L, Zhang, Y., Hao, X.J. 2012. Mekongenines A and B, two new alkaloids from *Bousigonia mekongensis*. Tetradendron Letters 53: 3642–3646.

Fu, Y.H., Di, Y.T., He, H.P., Li, S.L, Zhang, Y., Hao, X.J. 2014. Angustifonines A and B, cytotoxic bisindole alkaloids from *Bousigonia angustifolia*. J. Nat. Prod. 7: 57–62. doi: https://doi.org/10.1021/np4005823

Gaillard, Y., Krishnamoorthy, A., Bevalot, F. 2004. *Cerbera odollam*: a 'suicide tree' and cause of death in the state of Kerala, India. J. Ethnopharmacol. 95: 123–126.

Galetto, L. 1997. Flower structure and nectar chemical composition in three Argentine Apocynaceae. Flora 192: 197–207.

Galil, J., Zernoni, M. 1965. Nectar system in *Asclepias curassavica*. Bot. Gaz. 126: 144–148.

Gautier-Hion, A., Michaloud, G. 1989. Are figs always keystone resources for tropical frugivorous vertebrates? A test in Gabon. Ecology 70: 1826–1833.

Gautier-Hion, A., Duplantier, J.-M., Quris, R., Feer, F., Sourd, C., Decoux, J.-P., Dubost, G., Emmons, L., Erard, C., Hecketsweiler, P., Moungazi, A., Roussilhon, C., Thiollay, J.-M. 1985. Fruit characters as a basis of fruit choice and seed dispersal in a tropical forest vertebrate community. Oecologia (Berlin) 65: 324–337.

Gentry, A.H., Dodson, C.H. 1987. Diversity and biogeography of Neotropical vascular epiphytes. Ann. Missouri Bot. Gard. 74: 205–233.

Gilani, S.A., Kikuchi, A., Shinwari, Z.K., Khattak, Z.I., Watanabe, K.N. 2007. Phytochemical, pharmacological and ethnobotanical studies of *Rhazya stricta* Decne. Phytotherapy Res. 21: 30–307.

Goh, S.H., Ali, A.R.M., Wong, W.H. 1989. Alkaloids of *Leuconotis griffithii* and *L. eugenifolia* (Apocynaceae). Tetrahedron 45: 7899–7920.

Good, R. 1947. The geography of the flowering plants. London: Longmans, Green & Co.

Good, R. 1952. An atlas of the Asclepiadaceae. New Phytol. 51: 198–209.

Govindchari, T.R. 1967. *Tylophora* Alkaloids. In: Manske, R.H.F. (ed.) The Alkaloids. New York: Academic Press, pp. 518–528.

Goyder, D.J. 2004. An amplified concept of *Philibertia* Kunth (Apocynaceae: Asclepiadoideae), with a synopsis of the genus. Kew Bull. 59: 415–451. doi: https://doi.org/10.2307/4110951

Goyder, D.J. 2006. An overview of Asclepiad biogeography. In: Ghazanfar, S.A., Beentje, H.J. (eds.) Taxonomy and ecology of African plants, their conservation and sustainable use. Kew: Royal Botanic Gardens, pp 205–214.

Goyder, D.J. 2009. A synopsis of *Asclepias* (Apocynaceae: Asclepiadoideae) in tropical Africa. Kew Bull. 64: 369–399.

Goyder, D., Nicholas, A., Liede-Schumann, S. 2007. Phylogenetic relationships in subtribe Asclepiadinae (Apocynaceae: Asclepiadoideae). Ann. Missouri Bot. Gard. 94: 423–434. doi: https://doi.org/10.3417/0026-6493(2007)94[423:prisaa]2.0.co;2

Haber, W.A. 1984. Pollination by deceit in a mass-flowering tropical tree *Plumeria rubra* (Apocynaceae). Biotropica 16: 269–275.

Haber, W.A., Frankie, G.W., Baker, H.G., Baker, I., Koptur, S. 1981. Ants like nectar. Biotropica 13: 211–214.

Hall, W.T.K. 1964. Plant toxicoses of tropical Australia. Austral. Veterin. J. 40: 176–182.

Hechem, V., Acheritobehere, L., Morrone, J.J. 2011a. Patrones de distribución de las especies de *Cynanchum*, *Diplolepis* y *Tweedia* (Apocynaceae: Asclepiadoideae) de América del Sur austral. Revista Geogr. Norte Grande 48: 45–60.

Hechem, V., Calviño, C.I., Ezcurra, C. 2011b. Molecular phylogeny of *Diplolepis* (Apocynaceae-Asclepiadoideae) and allied genera, and taxonomic implications. Taxon 60: 638–648.

Hegnauer, R. 1964. Chemotaxonomie der Pflanzen 3. Basel: Birkhäuser, pp. 124–163, 199–223.

Hegnauer, R. 1970. Cardenolide und Bufadienolide (= Cardadienolide). Verbreitung und systematische Bedeutung. Pl. Med. 19: 137–153.

Hegnauer, R. 1989. Chemotaxonomie der Pflanzen 8. Basel: Birkhäuser, pp. 48–60, 84–95. doi: https://doi.org/10.1007/978-3-0348-9283-4

Heiduk, A., Brake, I., von Tschirnhaus, M., Göhl, M., Jürgens, A., Johnson, A.E., Meve, U., Dötterl, S. 2016. *Ceropegia sandersonii* mimics attacked honeybees to attract kleptoparasitic flies for pollination. Curr. Biol. 26: 2787–2793.

Hendrian. 2001a. *Strophanthus*. In: Van Valkenburg, J.L. C.H., Bunyapraphatsara, N. (eds.) Plant Resources of South-East Asia, Vol. 12(2): 519–523. Leiden: Backhuys.

Hendrian. 2001b. *Voacanga*. In: Van Valkenburg, J.L.C. H., Bunyapraphatsara, N. (eds.) Plant Resources of South-East Asia, Vol. 12(2): 582–585. Leiden: Backhuys.

Hendrian, Middleton, D.J. 1999. Revision of *Rauvolfia* (Apocynaceae) in Malesia. Blumea 44: 449–470.

Herrera, J. 1991. The reproductive biology of a riparian Mediterranean shrub, *Nerium oleander* L. (Apocynaceae). Bot. J. Linn. Soc. 106: 147–172.

Hong, L., Guo, Z.-H., Huang, K.F., Wei, S., Liu, B., Meng, S., Long, C. 2015. Ethnobotanical study on medicinal plants used by Maonan people in China. J. Ethnobiol. Ethnomed. 11: 32. doi: https://doi.org/10.1186/s13002-015-0019-1

Hu, Y.-J., Shen, X.-L., Mu, Q.-Z., Lu, Y., Zheng, Q.-T. 1992. Steroidal constituents from *Amalocalyx yunnanensis*. Phytochemistry 31: 2099–2102.

Hutchings, A.A. 1989. A survey and analysis of traditional medicinal plants as used by the Zulu, Xhosa and Sotho. Bothalia 19: 111–123.

Ionta, G.M., Judd, W.S. 2007. Phylogenetic relationships in Periplocoideae (Apocynaceae s.l.) and insights into the origin of pollinia in the subfamily. Ann. Missouri Bot. Gard. 94: 360–375. doi: https://doi.org/10.3417/0026-6493(2007)94[360:pripas]2.0.co;2

Ivey, C.T., Lipow, S.R., Wyatt, R. 1999. Mating systems and interfertility of swamp milkweed (*Asclepias incarnata* ssp. *incarnata* and ssp. *pulchra*). Heredity 82: 25–35.

Jagtap, A.P., Singh, N.P. 1999. Fascicles of Flora of India: fascicle 24. Asclepiadaceae and Periplocaceae. Calcutta: Botanical Survey of India.

Jensen, S.R. 1992. Systematic implications of the distribution of iridoids and other chemical compounds in the Loganiaceae and other families of the Asteridae. Ann. Missouri Bot. Gard. 79: 284–302.

Johns, S.R., Lamberton, J.A., Price, J.R., Sioumis, A.A. 1968. Identification of coumarins isolated from *Lepiniopsis ternatensis* (Apocynaceae), *Pterocaulon sphacelatum* (Compositae), and *Melicope melanophloia* (Rutaceae). Aust. J. Chem. 21: 3079–3080.

Johri, B.M., Ambegaokar, K.M., Srivastava, P.S. 1992. Comparative Embryology of Angiosperms, Vol. 2. Berlin, Heidelberg: Springer.

Joubert, L., Klak, C., Venter, A.M., Venter, H.J.T., Bruyns, P.V. 2016. A widespread radiation in the Periplocoideae (Apocynaceae): The case of *Cryptolepis*. Taxon 65: 487–501.

Judd, W.S., Sanders, R.W., Donoghue, M.J. 1994. Angiosperm family pairs: Preliminary phylogenetic analysis. Harvard Pap. Bot. 5: 1–51.

Jurgens, A., Dötterl, S., Meve, U. 2006. The chemical nature of fetid floral odours in Stapeliads (Apocynaceae-Asclepiadoideae-Ceropegieae). New Phytol. 172: 452–468.

Jussieu, A.L. de. 1789. Genera Plantarum. Paris: Herissant.

Kahn, A.P., Morse, D.H. 1991. Pollination germination and putative ovule penetration in self- and cross-pollinated common milkweed *Asclepias syriaca*. Amer. Midl. Naturalist 126: 61–67.

Kalimuthu, K., Prabakaran, R. 2013. Preliminary phytochemical screening and GC-MS analysis of methanol extract of *Ceropegia pusilla*. Int. J. Res. Appl. Nat. Soc. Sci. 1: 49–58.

Kephart, S.R. 1981. Breeding systems in *Asclepias incarnata* L., *A. syriaca* L., and *A. verticillata* L. Amer. J. Bot. 68: 226–232.

Khanum, R., Surveswaran, S., Meve, U., Liede-Schumann, S. 2016. *Cynanchum* (Apocynaceae: Asclepiadoideae): A pantropical Asclepiadoid genus revisited. Taxon 65: 467–486. doi: https://doi.org/10.12705/653.3

Kiew, R. 1994. The taxonomy and phytochemistry of the Asclepiadaceae in tropical Asia. Malacca: The Herbarium, Department of Biology, Universiti Pertanian Malaysia, 43400 UPM Serdang, Selangor, Malaysia and BOTANY 2000 ASIA.

Kiew, R. 2001. *Tylophora*. In: Van Valkenburg, J.L.C.H., Bunyapraphatsara, N. (eds.) Plant Resources of South-East Asia, Vol. 12(2): 564–568. Leiden: Backhuys.

Kingston, D.G.I., Reichstein, T. 1974. Cytotoxic cardenolides from *Acokanthera longiflora* Staph and related species. J. Pharm. Sci. 63: 462–464.

Kirchheimer, F. 1957. Die Laubgehölze der Braunkohlezeit. Halle (Saale): Wilhelm Knapp.

Kisakürek, M.V., Leeuwenberg, A.J.M., Hesse, M. 1983. A chemotaxonomic investigation of the plant families of Apocynaceae, Loganiaceae, and Rubiaceae by their indole alkaloid content. In: Pelletier, W.W. (ed.) Alkaloids: chemical and biological perspectives 1. Wiley: New York, pp. 211–376.

Klackenberg, J. 1992. Taxonomy of *Secamone* s.lat. (Asclepiadaceae) in the Madagascar region. Opera Botanica a Societate Botanica Lundensis 112: 1–126.

Klackenberg, J. 1995. Taxonomy and phylogeny of the SE Asian genus *Genianthus* (Asclepiadaceae). Bot. Jahrb. Syst. 117: 401–467.

Klackenberg, J. 1998. Taxonomy and phylogeny of the genus *Camptocarpus* s.l. (Periplocoideae, Asclepiadaceae). Bot. Jahrb. Syst. 120: 45–85.

Klackenberg, J. 1999. Revision of the Malagasy genera *Pentopetia* and *Ischnolepis* (Apocynaceae s.l.). Candollea 54: 257–339.

Klackenberg, J. 2001. Notes on Secamonoideae in Africa. Bull. Mus. Natl. Hist. Nat., B, Adansonia Sér. 3, 23: 317–335.

Klackenberg, J. 2010. New species and combinations of *Secamone* (Apocynaceae, Secamonoideae) from South East Asia. Blumea 55: 231–241.

Kleijn, D., Donkelaar, R. van. 2001. Notes on the taxonomy and ecology of the genus *Hoya* (Asclepiadaceae) in Central Sulawesi. Blumea 46: 157–193.

Koch, I., Bittrich, V., Sumiko Kinoshita, L. 2002. Reproductive biology and functional aspects of the floral morphology of *Rauvolfia sellowii* Müll. Arg. (Apocynaceae; Rauvolfioideae) – a report of dioecy in Apocynaceae. Bot. Jahrb. Syst. 124: 83–104. doi: https://doi.org/10.3417/1055-3177(2007)17[462:TNIRAR]2.0.CO;2

Kress, W.J. 1986. The systematic distribution of vascular epiphytes: an update. Selbyana 9: 2–22.

Krings, A., Saville, A.C. 2007. Two new species and three lectotypifications in the *Ibatia-Matelea* complex (Apocynaceae: Asclepiadoideae) from northern South America. Syst. Bot. 32: 862–871.

Kugler, H. 1973. Zur Bestäubung von *Cynanchum acutum* L. durch Faltenwespen (Vespidae). In: Brantjes, N.B. M., Linsgens, H.F. (eds.) Pollination and Dispersal. Nijmwegen: University Nijmwegen, pp. 61–68.

Kumar, P.S., Suresh, E., Kalavathy, S. 2013. Review on a potential herb *Calotropis gigantea* (L.) R. Br. Sch. Acad. J. Pharm. 2: 135–143.

Kunze, H. 1982. Morphogenese und Synorganisation des Bestäubungsapparates einiger Asclepiadaceen. Beitr. Biol. Pflanzen 56: 133–170.

Kunze, H. 1990. Morphology and evolution of the corona in Asclepiadaceae and related families. Trop. Subtrop. Pflanzenwelt 76: 1–51.

Kunze, H. 1991. Structure and function in asclepiad pollination. Pl. Syst. Evol. 176: 227–253.

Kunze, H. 1993. Evolution of the translator in Periplocaceae and Asclepiadaceae. Pl. Syst. Evol. 185: 99–122.

Kunze, H. 1994. Ontogeny of the translator in Asclepiadaceae *s.str.* Pl. Syst. Evol. 193: 223–242.

Kunze, H. 1995. Floral morphology of some Gonolobeae (Asclepiadeae). Bot. Jahrb. Syst. 117: 211–238.

Kunze, H. 1996. Morphology of the stamen in the Asclepiadaceae and its systematic relevance. Bot. Jahrb. Syst. 118: 547–579.

Kunze, H. 1997. Corona and nectar system in Asclepiadinae (Asclepiadaceae). Flora 192: 175–183.

Kunze, H. 2005. Morphology and evolution of the corolla and corona in the Apocynaceae s.l. Bot. Jahrb. Syst. 126: 347–383.

Kunze, H., Meve, U., Liede, S. 1994. *Cibirhiza albersiana*, a new species of Asclepiadaceae, and establishment of the tribe Fockeeae. Taxon 43: 367–376.

Lahaye, R., Civeyrel, L., Speck, T., Rowe, N.P. 2005. Evolution of shrub-like growth forms in the lianoid subfamily Secamonoideae (Apocynaceae s.l.) of Madagascar: Phylogeny, biomechanics, and development. Amer. J. Bot. 92: 1381–1396.

Lahaye, R., Klackenberg, J., Källersjö, M., Van Campo, E., Civeyrel, L. 2007. Phylogenetic relationships between derived Apocynaceae s.l. and within Secamonoideae based on four chloroplast sequences. Ann. Missouri

Bot. Gard. 94: 376–391. doi: https://doi.org/10.3417/0026-6493(2007)94[376:prbdas]2.0.co;2

Landolt, P.J. 1994. Fruit of *Morrenia odorata* (Asclepiadaceae) as a host for the papaya fruit fly, *Toxotrypana curvicauda* (Diptera: Tephritidae). Florida Entomol. 77(2): 287–288.

Lee, D.U., Kang, S.I., Yoon, S.H., Budesinsky, M., Kasal, A., Mayer, K.K., Wiegrebe, W. 2000. A new steroidal alkaloid from the roots of *Cynanchum caudatum*. Planta Medica 66: 480–482.

Leeuwenberg, A.J.M. 1985. *Voacanga* Thou. Series of revisions of Apocynaceae, XV. Wageningen Agric. Univ. Pap. 85.3: 1–80.

Leeuwenberg, A.J.M. 1991. A revision of *Tabernaemontana*, Vol. 1, The Old World species. Kew: Royal Botanic Gardens Press.

Leeuwenberg, A.J.M. 1994a. Taxa of the Apocynaceae above the genus level. Series of revisions of Apocynaceae, XXXVIII. Wageningen Agric. Univ. Pap. 94 (3): 45–60.

Leeuwenberg, A.J.M. 1994b. A revision of *Tabernaemontana*, Vol. 2, The New World species. Kew: Royal Botanic Gardens Press.

Leeuwenberg, A.J.M. 1999. The genus *Cerbera* L. Series of revisions of Apocynaceae, XLVII. Wageningen Agric. Univ. Pap. 98.3: 1–64.

Leeuwenberg, A.J.M., van Dilst, F.J.H. 2001. *Carissa* L. Series of revisions of Apocynaceae, XLIX. Wageningen Agric. Univ. Pap. 2001.1: 1–64.

Leighton, M. 1993. Modeling dietary selectivity by Bornean orangutans: Evidence for integration of multiple criteria in fruit selection. Int. J. Primatol. 14: 257–313.

Leimu, R. 2004. Variation in the mating system of *Vincetoxicum hirundinaria* (Asclepiadaceae) in peripheral island populations. Ann. Bot. 93: 107–113.

Lens, F., Endress, M.E., Baas, P., Jansen, S., Smets, E. 2008. Wood anatomy of Rauvolfioideae (Apocynaceae): A search for meaningful non-DNA characters at the tribal level. Amer. J. Bot. 95: 1199–1215. doi: https://doi.org/10.3732/ajb.0800159

Lens, F., Endress, M.E., Baas, P., Jansen, S., Smets, E. 2009. Vessel grouping patterns in subfamilies Apocynoideae and Periplocoideae confirm phylogenetic value of wood structure within Apocynaceae. Amer. J. Bot. 96: 2168–2183. doi: https://doi.org/10.3732/ajb.0900116

Li, P.T., Gilbert, M.G., Stevens, W.D. 1995a. Asclepiadaceae. In: Wu, Z.Y., Raven, P.H. (eds.) Flora of China, Vol. 16. Beijing, St. Louis: Science Press & Missouri Botanical Garden, pp. 189–270.

Li, P.T., Leeuwenberg, A.J.M., Middleton, D.J. 1995b. Apocynaceae. In: Wu, Z.Y., Raven, P.H. (eds.) Flora of China, Vol. 16. Beijing, St. Louis: Science Press & Missouri Botanical Garden, pp. 143–188.

Liede, S. 1996a. Anther differentiation in the Asclepiadaceae: Form and Function. In: D'Arcy, W.G., Keating, R.C. (eds.) The Anther: Form, Function and Phylogeny. Cambridge: Cambridge University Press, pp. 221–235.

Liede, S. 1996b. *Cynanchum – Rhodostegiella – Vincetoxicum – Tylophora*: new considerations on an old problem. Taxon 45: 193–211.

Liede, S. 1997. Subtribes and genera of the tribe *Asclepiadeae* (Apocynaceae – Asclepiadoideae) – a synopsis. Taxon 46: 233–247.

Liede, S. 2001. Molecular considerations on the subtribe Astephaninae Endl. ex Meisn. (Apocynaceae – Asclepiadoideae). Ann. Missouri Bot. Gard. 88: 657–668. doi: https://doi.org/10.2307/3298638

Liede, S., Kunze, H. 1993. A descriptive system for corona analysis in Asclepiadaceae and Periplocaceae. Pl. Syst. Evol. 185: 275–284.

Liede, S., Meve, U. 1994. A new species of *Tylophoropsis* (Asclepiadaceae) and notes on the genus. Kew Bulletin 49(4): 749–756.

Liede, S., Meve, U. 2002 (publ. 2004). Dissolution of *Cynanchum* sect. *Macbridea* (Apocynaceae-Asclepiadoideae). Nord. J. Bot. 22: 579–591.

Liede, S., Täuber, A. 2000. *Sarcostemma* R. Br. (Apocynaceae – Asclepiadoideae) – a controversial generic circumscription reconsidered: Evidence from *trn*L-F Spacers. Pl. Syst. Evol. 225: 133–140. doi: https://doi.org/10.1007/bf00985463

Liede, S., Täuber, A. 2002. Circumscription of the genus *Cynanchum* (Apocynaceae – Asclepiadoideae). Syst. Bot. 27: 789–800. doi: https://doi.org/10.2307/2419462

Liede, S., Weberling, F. 1995. On the inflorescence structure of Asclepiadaceae. Pl. Syst. Evol. 197: 99–109.

Liede-Schumann, S., Meve, U. 2015. Synonymy of three South American genera in Apocynaceae, and new combinations in *Oxypetalum* and *Tassadia*. Phytotaxa 202: 35–44. doi: https://doi.org/10.11646/phytotaxa.202.1.4

Liede-Schumann, S., Rapini, A., Goyder, D.J., Chase, M. W. 2005. Phylogenetics of the New World subtribes of Asclepiadeae (Apocynaceae-Asclepiadoideae): Metastelmatinae, Oxypetalinae, and Gonolobinae. Syst. Bot. 30: 184–200. doi: https://doi.org/10.1600/0363644053661832

Liede-Schumann, S., Kong, H.-H., Meve, U., Thiv, M. 2012. *Vincetoxicum* and *Tylophora* (Apocynaceae: Asclepiadoideae: Asclepiadeae)–two sides of the same medal: Independent shifts from tropical to temperate habitats. Taxon 61: 803–825.

Liede-Schumann, S., Nikolaus, M., Soares e Silva, U.C., Rapini, A., Mangelsdorff, R.D., Meve, U. 2014. Phylogenetics and biogeography of the genus *Metastelma* (Apocynaceae-Asclepiadoideae-Asclepiadeae: Metastelmatinae). Syst. Bot. 39: 594–612.

Liede-Schumann, S., Khanum, R., Mumtaz, A.S., Gherghel, I., Pahlevani, A. 2016. Going west – A subtropical lineage (*Vincetoxicum*, Apocynaceae: Asclepiadoideae) expanding into Europe. Mol. Phyl. Evol. 94: 436–446.

Lienau, K., Straka, H., Friedrich, B. 1986. Palynologia Madagassica et Mascarenica, Fam. 167–181. Trop. Subtrop. Pflanzenwelt 55: 1–158.

Lin, S., Bernardello, G. 1999. Flower structure and reproductive biology in *Aspidosperma quebracho-blanco* (Apocynaceae), a tree pollinated by deceit. Int. J. Pl. Sci. 160: 869–878.

Linhart, Y.B., Feinsinger, P. 1980. Plant-hummingbird interactions: Effects of island size and degree of specialization on pollination. J. Ecol. 68: 745–760.

Lipow, S.R., Wyatt, R. 1998. Reproductive biology and breeding system of *Gonolobus suberosus* (Asclepiadaceae). J. Torrey Bot. Soc. 125: 183–193.

Lipow, S.R., Wyatt, R. 1999. Floral morphology and late-acting self-incompatibility in *Apocynum cannabinum* (Apocynaceae). Pl. Syst. Evol. 219: 99–109.

Livshultz, T. 2010. The phylogenetic position of milkweeds (Apocynaceae subfamilies Secamonoideae and Asclepiadoideae): Evidence from the nucleus and chloroplast. Taxon 59: 1016–1030.

Livshultz, T., Middleton, D.J., Endress, M.E., Williams, J. K. 2007. Phylogeny of Apocynoideae and the APSA clade (Apocynaceae). Ann. Missouri Bot. Gard. 94: 324–359. doi: https://doi.org/10.3417/0026-6493(2007)94[324:poaata]2.0.co;2

Lodder, S., Rutten, E.M.J., Van der Ham, R.W.J.M. 2007. Pollen morphology. In: Middleton, D.J. 2007. Apocynaceae (subfamilies Rauvolfioideae and Apocynoideae), Vol. 18. In: Nooteboom, H.P. (ed.) Flora Malesiana, Series I - Seed Plants. Leiden: Foundation Flora Malesiana.

Lopes, A.V., Machado, I.C. 1999. Pollination and reproductive biology of *Rauvolfia grandiflora* (Apocynaceae): Secondary pollen presentation, herkogamy and self-incompatibility. Pl. Biol. 1: 547–553.

Lorence, D.H., Butaud, J.-F. 2011. A reassessment of Marquesan *Ochrosia* and *Rauvolfia* (Apocynaceae) with two new combinations. PhytoKeys 4: 95–107. doi: https://doi.org/10.3897/phytokeys.4.1599

Lorence, D.H., Wagner, W.L. 1997. A revision of *Lepinia* (Apocynaceae), with description of a new species from the Marquesas Islands. Allertonia 7: 254–266.

Ludwig, F. 1880. Über die Bestäubungsvorrichtungen und die Fliegenfalle des Hundskohles, *Apocynum androsaemifolium* L. Kosmos 8: 182–185.

Lumer, C., Yost, S.E. 1995. The reproductive biology of *Vincetoxicum nigrum* (L.) Moench (Asclepiadaceae), a Mediterranean weed in New York State. Bull. Torrey Bot. Club 122: 12–23.

Machado, C.G. 2009. Beija-flores (Aves: Trochilidae) e seus recursos florais em uma área de caatinga da Chapada Diamantina, Bahia, Brasil. Zoologia 26: 255–265.

Maheswari Devi, H. 1964. Embryological studies in Asclepiadaceae. Proc. Indian Acad. Sci., Pl. Sci. 60B: 52–65.

Mahlberg, P. 1980. The latex cells of asclepiads. Asklepios 23: 30–32.

Markgraf, F. 1971. Florae Malesianae Praecursores LI. Apocynaceae I. 1. *Carissa*, 2. *Catharanthus*, 3. *Melodinus*, 4. *Leuconotis*, 5. *Chilocarpus*. Blumea 19: 156–165.

Markgraf, F. 1976. Apocynaceae. In: Leroy, J.-F. (ed.) Flore de Madagascar et des Comores. Fam. 169. Paris: Muséum National d'Histoire Naturelle, Paris.

Martínez-Millán, M. 2010. Fossil Record and Age of the Asteridae. Bot. Rev. 76: 83–135.

Masinde, P.S., Meve, U. 2002. *Ceropegia zambesiaca* (Apocynaceae: Asclepiadoideae-Ceropegieae), a new species from Zambia. Kew Bull. 57: 205–209.

McDiarmid, R.W. 1977. Dispersal of *Stemmadenia donnell-smithii* (Apocynaceae) by birds. Biotropica 9: 9–25.

McFadyen, R.E., Harvey G.J. 1990. Distribution and control of rubber vine, *Cryptostegia grandiflora*, a major weed in northern Queensland. Pl. Protection Quarterly 5: 152–155.

McNeill, J., Barrie, F.R., Buck, W.R., Demoulin, V., Greuter, W., Hawksworth, D.L., Herendeen, P.S., Knapp, S., Marhold, K., Prado, J., Prud'homme van Reine, W.F., Smith, G.F., Wiersema, J.H., Turland, N.J. (eds.) 2012. International Code of Nomenclature for algae, fungi, and plants (Melbourne Code): Adopted by the Eighteenth International Botanical Congress Melbourne, Australia, July 2011. Regnun Vegetabile 154, Königstein: Koeltz Scientific Books. http://www.iapt-taxon.org/nomen/main.php

Metcalfe, C.R., Chalk, L. 1972. Anatomy of the Dicotyledons, Vol. 2 [Apocynaceae, Asclepiadaceae]. Oxford: University Press, pp. 905–925.

Metcalfe, C.R., Chalk, L. 1979. Anatomy of the Dicotyledons, 2nd ed, Vol. 1. Oxford: Clarendon Press.

Metzner, R. 1998. Hallucinogenic drugs and plants in psychotherapy and shamanism. J. Psychoactive Drugs 30: 333–341. doi: https://doi.org/10.1080/02791072.1998.10399709

Meve, U. 1994. The genus *Piaranthus* R. Br. (Asclepiadaceae). Bradleya 12: 57–102.

Meve, U. 1995. A review of phytophagous insects on Stapeliads (Asclepiadaceae). Cimbebasia 14: 103–106.

Meve, U. 1997. The genus *Duvalia* (Stapelieae): stem-succulents between the Cape and Arabia. Pl. Syst. Evol. Suppl. 10. Wien: Springer, 132 pp.

Meve, U., Liede, S. 1994. Floral biology and pollination in Stapeliads - new results and a literature review. Pl. Syst. Evol. 192: 99–116.

Meve, U., Liede, S. 2002a. A molecular phylogeny and generic rearrangement of the stapelioid Ceropegieae (Apocynaceae-Asclepiadoideae). Pl. Syst. Evol. 234: 171–209.

Meve, U., Liede, S. 2002b. Floristic exchange between mainland Africa and Madagascar: A case study of Apocynaceae-Asclepiadoideae. J. Biogeogr. 29: 865–873.

Meve, U., Liede, S. 2004a. Generic delimitations in tuberous Periplocoideae (Apocynaceae) from Africa and Madagascar. Ann. Bot. 93: 407–414. doi: https://doi.org/10.1093/aob/mch057

Meve, U., Liede, S. 2004b. Subtribal division of Ceropegieae (Apocynaceae-Asclepiadoideae). Taxon 53: 61–72. doi: https://doi.org/10.2307/4135489

Meve, U., Liede-Schumann, S. 2007. *Ceropegia* (Apocynaceae, Ceropegieae, Stapeliinae): Paraphyletic but still taxonomically sound. Ann. Missouri Bot. Gard. 94: 392–406.

Meve, U., Liede-Schumann, S. 2015. Taxonomy of the Andean genus *Pentacyphus* (Apocynaceae: Asclepiadeae–Pentacyphinae). Pl. Syst. Evol. 301: 997–1004.

Meve, U., Liede-Schumann, S. 2017. Was ist *Cynanchum* L. (Apocynaceae-Asclepiadoideae)? Schritt für Schritt zu einem erweiterten Gattungskonzept. Avonia 35: 77–85.

Meve, U., Wolf, F. 2001. *Echidnopsis bentii* N.E. Brown (Ceropegieae) auf Sokotra gefunden. Kakt. and. Sukk. 52(5): 113–118.

Meve, U., Jahnke, G., Liede, S., Albers, F. 2004. Isolation mechanisms in the Stapeliads (Apocynaceae-Asclepiadoideae-Ceropegieae). Schumannia 4 / Biodivers. Ecol. 2: 107–126.
Meve, U., Heiduk, A., Liede-Schumann, S. 2017. Origin and early evolution of Ceropegieae (Apocynaceae-Asclepiadoideae). Syst. Biodivers. 15: 143–155.
Meyer, J.-Y. 1996. Espèces et Espaces Menacés de la Sociéte et des Marquises. Contribution à l'Environnement. Délégation à la Recherche, Papeete.
Meyer, J.-Y., Butaud, J.-F. 2009. The impacts of rats on the endangered native flora of French Polynesia (Pacific Islands): Drivers of plant extinction or *coup de grâce* species? Biol. Invasions 11: 1569–1585.
Meyer, J.-Y., Picot, F. 2001. Achatines attack! The impact of giant African land snails on rare endemic plants in La Réunion Island (Mascarene Islands, Indian Ocean). Aliens. (Bull. Invasive Species Spec. Group IUCN Spec. Surv. Comm.) 14: 13–14.
Meyer, B.N., McLaughlin, J.L., Keller, W.J. 1981. Candicine from *Stapelia gigantea*. Pl. Med. 43: 304–306.
Middleton, D.J. 2000. Revision of *Alyxia*, Part 1: Asia and Malesia. Blumea 45: 1–146.
Middleton, D.J. 2002. Revision of *Alyxia* (Apocynaceae). Part 2: Australia and Pacific Islands. Blumea 47: 1–93.
Middleton, D.J. 2007. Apocynaceae (subfamilies Rauvolfioideae and Apocynoideae), Vol. 18. In: Nooteboom, H.P. (ed.) Flora Malesiana, Series I – Seed Plants. Leiden: Foundation Flora Malesiana.
Middleton, D.J. 2010. Three new species of *Wrightia* (Apocynaceae: Apocynoideae) from Thailand. Gard. Bull. Singapore 61: 129–138.
Middleton, D.J. 2014. Apocynaceae, subfamilies Rauvolfioideae and Apocynoideae. Flora of Cambodia, Laos and Vietnam 33. Paris and Edinburgh: Muséum National d'Histoire Naturelle, Royal Botanic Garden Edinburgh, pp 1–276.
Middleton, D.J., Livshultz, T. 2012. *Streptoechites* gen. nov., a new genus of Asian Apocynaceae. Adansonia Sér. 3, 34: 365–375. doi: https://doi.org/10.5252/a2012n2a10
Middleton, D.J., Lindsay, S., Suddee, S. 2006 ('2005'). A new species of *Kamettia* (Apocynaceae: Rauvolfioideae), a genus new to Thailand. Thai Forest Bull. 33: 75–80.
Morales, J.F. 1998. A synopsis of the genus *Mandevilla* (Apocynaceae) in Mexico and Central America. Brittonia 50: 214–232.
Morales, J.F., Zamora, N.A. 2017. A synopsis of *Aspidosperma* (Apocynaceae) in Mexico and Central America with a taxonomic clarification of *Aspidosperma cruentum* and a new cryptic species. Phytoneuron 68: 1–13.
Morales, J.F., Endress M.E., Liede-Schumann, S. 2017a. Sex, drugs and pupusas: Disentangling relationships in Echiteae (Apocynaceae). Taxon 66: 623–644.
Morales, J.F., Endress, M.E., Liede-Schumann, S. 2017b. Systematics of *Prestonia* (Apocynaceae: Apocynoids: Echiteae) eighty years after Woodson. Ann. Missouri Bot. Gard. 102: 520–541.
Moré, M., Sérsic, A.N., Cocucci, A.A. 2007. Restriction of pollinator assemblage through flower length and width in three long-tongued hawkmoth-pollinated species of *Mandevilla* (Apocynaceae, Apocynoideae). Ann. Missouri Bot. Gard. 94: 485–504.
Morillo, G. 2012. Aportes al conocimiento de las Gonolobinae (Apocynaceae- Asclepiadoideae). Pittieria 36: 13–57.
Morillo, G. 2013. Aportes al conocimiento de las Gonolobinae II (Apocynaceae, Asclepiadoideae). Pittieria 37: 101–140.
Morillo, G. 2015. Aportes al conocimiento de las Gonolobinae III (Apocynaceae, Asclepiadoideae). Pittieria 39: 191–258.
Morokawa, R., Mayer, J.L.S., Simões, A.O., Kinoshita, L.S. 2015. Floral development of *Condylocarpon isthmicum* (Apocynaceae). Botany 93: 679–781. https://doi.org/10.1139/cjb-2015-0081
Morse, D.H. 1985. Milkweeds and their visitors. Sci. Amer. 253: 112–119.
Morton, J.F., Alvarez, E., Quinonez, C. 1990. Loroco, *Fernaldia pandurata* (Apocynaceae) – a popular edible flower of Central America. Econ. Bot. 44: 301–310.
Moura, T.N.D., Webber, A.C., Torres, L.N.M. 2011. Floral biology and a pollinator effectiveness test of the diurnal floral visitors of *Tabernaemontana undulata* Vahl. (Apocynaceae) in the understory of Amazon Rainforest, Brazil. Acta Bot. Bras. 25: 380–386.
Mu, Q.Z., Lu, R.J., Zhou, Q.L. 1986. Two new antiepilepsy compounds – otophylloside A and otophylloside B. Sci. Sin. (B) 24: 295–301.
Muller, J. 1968. Palynology of the Pedawan and Plateau Sandstone Formations (Cretaceous-Eocene) in Sarawak, Malaysia. Micropaleontology 14: 1–37.
Muller, J. 1981. Fossil pollen records of extant angiosperms. Bot. Rev. 47: 1–146.
Naumova, T.N. 1992. Apomixis in Angiosperms. Boca Raton: CRC Press.
Nel, M. 1995. Rare and interesting plants of the Namib Desert. Part 2. Three desert plants. Veld Fl. 81: 14–16.
Neuwinger, H.D. 1994a. Asclepiadaceae. In: Afrikanische Arzneipflanzen und Jagdgifte, Vol. 59. Stuttgart: Wissenschaftliche Verlagsgesellschaft, pp. 208–232.
Neuwinger, H.D. 1994b. Fish poisoning plants in Africa. Bot. Acta 107: 263–270.
Nevo, O., Garri, R.O., Hernandez Salazar, L., Schulz, S., Heymann, E.W., Ayasse, M., Laska, M. 2015. Chemical recognition of fruit ripeness in spider monkeys (*Ateles geoffroyi*). Sci. Rep. Oct. 2015. doi: http://dx.doi.org/10.038/srep14895
Nicholas, A., Baijnath, H. 1994. A consensus classification of the order Gentianales with additional details on the suborder Apocynineae. Bot. Rev. 60: 440–482.
Nilsson, S. 1986. The significance of pollen morphology in the Apocynaceae. In: Blackmore, S., Ferguson, I.K. (eds.) Pollen and Spores: Form and Function. J. Linn. Soc. Symp. Ser. 12. London: Academic Press, pp. 359–374.
Nilsson, S. 1990. Taxonomic and evolutionary significance of pollen morphology in the Apocynaceae. Pl. Syst. Evol., Suppl. 5: 91–102.
Nilsson, S., Endress, M.E., Grafström, E. 1993. On the relationship of the Apocynaceae and Periplocaceae. Grana 1993, Suppl. 2: 3–20.
Nishino, E. 1982. Corolla tube formation in six species of Apocynaceae. Bot. Mag. Tokyo 95: 1–17.

Nishino, E. 1983. Corolla tube formation in the Tubiflorae and Gentianales. Bot. Mag. Tokyo 96: 223–243.

Ollerton, J., Liede, S. 1997. Pollination systems in the Asclepiadaceae: a survey and preliminary analysis. Biol. J. Linn. Soc. 62: 593–610.

Ollerton, J., Johnson, S.D., Cranmer, L., Kellie, S. 2003. The pollination ecology of an assemblage of grassland asclepiads in South Africa. Ann. Bot. 92: 807–834.

Omino, E. 1996. A monograph of the subtribe Pleiocarpinae (Apocynaceae-Plumerioideae-Carisseae). Series of revisions of Apocynaceae, XLI. Wageningen Agric. Univ. Pap. 96-1: 81–178.

Omlor, R. 1998. Generische Revision der Marsdenieae (Asclepiadaceae). Kaiserslautern: Shaker Verlag.

Pant, D.D., Nautiyal, D.D., Chaturvedi, S.K. 1982. Pollination ecology of some Indian asclepiads. Phytomorphology 32: 302–313.

Pathania, S., Randhawa, V., Bagler, G. 2013. Prospecting for novel plant-derived molecules of *Rauvolfia serpentina* as inhibitors of aldose reductase, a potent drug target for diabetes and its complications. PloS ONE 8(4): e61327. doi: https://doi.org/10.1371/journal.pone.0061327

Pathania, S., Ramakrishnan, S.M., Randhawa, V., Bagler, G. 2015. SerpentinaDB: A database of plant-derived molecules of *Rauvolfia sepentina*. BMC Complement. Alt. Med. 15: 262. doi: https://doi.org/10.1186/s12906-015-0683-7

Paulo, A., Jimeno, M.L., Gomes, E.T., Houghton, P.J. 2000. Steroidal alkaloids from *Cryptolepis obtusa*. Phytochemistry 53: 417–422.

Pauw, A. 1998. Pollen transfer on birds' tongues. Nature 394: 731–732.

Peeters, C., Wiwatwitaya, D. 2014. *Philidris* ants living inside *Dischidia* epiphytes from Thailand. Asian Myrmecology 6: 49–61.

Pereira, A.S.S., Simões, A.O., Santos, J.U.M. 2016. Taxonomy of *Aspidosperma* Mart. (Apocynaceae, Rauvolfioideae) in the state of Pará, northern Brazil. Biota Neotropica 16(2): e20150080. doi: https://doi.org/10.1590/1676-011öBN-2015-0080

Pereira, A.S.S., Castello, A.C.D., Scudeler, A.L., Simões, A. O., Koch, I. 2017. *Aspidosperma brasiliense* (Apocynaceae), a new and widely distributed species. Phytotaxa 326: 235–244.

Periasamy, K. 1963. Studies on seeds with ruminate endosperm. III. Development of rumination in certain members of the Apocynaceae. Proc. Indian Acad. Sci. 58, sect. B, 1: 325–332, t. 29, 30.

Perry, L.M. 1980. Medicinal Plants of East and Southeast Asia: attributed properties and uses. Cambridge, MA: MIT Press.

Persoon, J., Dilst, F.J.H., Kuijpers, R.P., Leeuwenberg, A.J. M., Vonk, G.J.A. 1992. The African species of *Landolphia*. Series of revisions of Apocynaceae, XXXIV. Wageningen Agric. Univ. Pap. 92.2: 1–232.

Pichon, M. 1948a. Classification des Apocynacées. I. Carissées et Ambelaniées. Mém. Mus. Natl. Hist. Nat., Sér. B, Bot. 24: 111–181.

Pichon, M. 1948b. Classification des Apocynacées. V. Cerbéroïdées. Not. Syst. Paris 13: 212–229.

Pichon, M. 1948c. Classification des Apocynacées. XIX. Le rétinacle de Echitoïdées. Bull. Soc. Bot. France 95: 211–216.

Pichon, M. 1949. Classification des Apocynacées. IX. Rauvolfiées, Alstoniées, Allamandées et Tabernaémontanoidées. Mém. Mus. Natl. Hist. Nat. 27: 153–251.

Pichon, M. 1950a. Classification des Apocynacées XXV, Echitoïdées. Mém. Mus. Natl. Hist. Nat., Sér. B, Bot.1: 1–142.

Pichon, M. 1950b. Classification des Apocynacées: XXVIII, Supplément aux Plumerioïdées, Mém. Mus. Natl. Hist. Nat., Sér. B, Bot. 1: 145–166.

Pienaar, M. 2013. Phylogeny of the genus *Raphionacme* (Apocynaceae). M.Sc. Thesis, Dept. Plant Sciences, Univ. Free State, Bloemfontein, Bloemfontein, South Africa.

Plumel, M.M. 1991. Le genre *Himatanthus* (Apocynaceae). Révision taxonomique. Bradea 5 (suppl.): 1–118.

Poinar, G.O. 2017. Ancient termite pollinator of milkweed flowers in Dominican amber. Amer. Entomol. 63: 52–56.

Potgieter, K., Albert, A.A. 2001. Phylogenetic relationships within Apocynaceae s. l. based on *trn*L intron and *trn*L-F spacer sequences and propagule characters. Ann. Missouri Bot. Gard. 88: 523–549. doi: https://doi.org/10.2307/3298632

Pynee, K., Dubuisson, J.-Y., Hennequin, S. 2013. Flora diversity of Mount Bar Le Duc Volcanic Crater (Ripailles Hill), Nouvelle Découverte, Mauritius. Cahiers Sci. Ocean Ind. Occid. 4: 15–20.

Queiroz, J.A., 2009. Esfingofilia e polinização por engano em *Aspidosperma pyrifolium* Mart., uma Apocynaceae arbórea endêmica de caatinga. Ph.D. Thesis, Universidade Federal de Pernambuco, Recife.

Rahayu, S.S.B. 2001. *Allamanda*. In: Van Valkenburg, J.L. C.H., Bunyapraphatsara, N. (eds.) Plant Resources of South-East Asia, Vol. 12(2): 49–52. Leiden: Backhuys.

Ramakrishna, T.M., Arekal, G.D. 1979. Pollination biology of *Calotropus gigantea* (L.). R. Br. Curr. Sci. 48: 212–213.

Rapini, A., Chase, M.W., Goyder, D.J., Griffiths, J. 2003. Asclepiadeae classification: evaluating the phylogenetic relationships of New World Asclepiadoideae (Apocynaceae). Taxon 52: 33–50. doi: https://doi.org/10.2307/3647300

Rapini, A., van den Berg, C., Liede-Schumann, L. 2007. Diversification of Asclepiadoideae (Apocynaceae) in the New World. Ann. Missouri Bot. Gard. 94: 407–422. https://doi.org/10.3417/0026-6493(2007)94[407:DOAAIT]2.0.CO;2

Rapini, A., Fontella Pereira, J., Goyder, D.J. 2011. Towards a stable generic circumscription in Oxypetalinae (Apocynaceae). Phytotaxa 26: 9–16.

Razafindratsima, O.H., Jones, T.A., Dunham, A.E. 2014. Patterns of movement and seed dispersal by three lemur species. Amer. J. Primatology 76: 84–96.

Reid, E.M., Chandler, M.E.J. 1926. The Bembridge Flora (Apocynaceae, Asclepiadaceae). London: Order of the Trustees.

Ribeiro, P.L., Rapini, A., Damascena, L.S., van den Berg, C. 2014. Plant diversification in the Espinhaço Range: Insights from the biogeography of *Minaria* (Apocynaceae). Taxon 63: 1253–1264.

Rintz, R.E. 1980. A revision of the genus *Sarcolobus* (Asclepiadaceae). Blumea 26: 65–79.

Rodda, M. 2015. Two new species of *Hoya* R.Br. (Apocynaceae, Asclepiadoideae) from Borneo. PhytoKeys 53: 83–93.

Rodda, M., Omlor, R. 2013. The taxonomy of *Oreosparte* (Apocynaceae: Asclepiadoideae). Webbia 68: 91–95.

Rodríguez-Estrella, R., Navarro, J.J.P., Granados, B., Rivera, L. 2010. The distribution of an invasive plant in a fragile ecosystem: The rubber vine (*Cryptostegia grandiflora*) in oases of the Baja California peninsula. Biol. Invas. 12: 3389–3393.

Rosatti, T.J. 1989. The genera of suborder Apocynineae (Apocynaceae and Asclepiadaceae) - Asclepiadaceae. J. Arnold Arb. 70: 443–514.

Rudjiman. 1982. A revision of *Vallaris* Burm. f. (Apocyneae). Series of revisions of Apocynaceae, IX. Meded. Landbouwhogeschool Wageningen 82-11: 1–17.

Rudjiman. 2001. *Kibatalia*. In: Van Valkenburg, J.L.C.H., Bunyapraphatsara, N. (eds.) Plant Resources of South-East Asia, Vol. 12(2): 322–324. Leiden: Backhuys.

Sabir, J.S.M., Jansen, R.K., Arasappan, D., Calderon, V., Noutahi, E., Zheng, C., Park, S., Sabir, M.J., Baeshen, M.N. Hajrah, N.H., Khiyami, M.A., Baeshen, N.A., Obaid, A.Y., Al-Malki, A.L., Sankoff, D., El-Mabrouk, N., Ruhlman, T.A. 2016. The nuclear genome of *Rhazya stricta* and the evolution of alkaloid diversity in a medically relevant clade of Apocynaceae. Sci. Rep. 6, 33782. doi: https://doi.org/10.1038/srep33782

Safwat, F.M. 1962. The floral morphology of *Secamone* and the evolution of the pollinating apparatus in Asclepiadaceae. Ann. Missouri Bot. Gard. 49: 95–129.

Sage, T.L., Williams, E.G. 1993. Self-incompatibility in *Asclepias*. Pl. Cell Incompatibility Newsl. 23: 55–57.

Sage, T.L., Williams, E.G. 1995. Structure, ultrastructure, and histochemistry of the pollen tube pathway in the milkweed *Asclepias exaltata*. L. Sex. Pl. Repro. 8: 257–265.

Sage, T.L., Broyles, S.G., Wyatt, R. 1990. The relationship between the five stigmatic chambers and two ovaries of milkweed flowers: a three-dimensional assessment. Israel J. Bot. 39: 187–196.

Sangat-Roemantyo, H.M., Middleton, D.J. 2001. *Alyxia*. In: Van Valkenburg, J.L.C.H., Bunyapraphatsara, N. (eds.) Plant Resources of South-East Asia, Vol. 12(2): 69–72. Leiden: Backhuys.

Schick, B. 1980. Untersuchungen über die Biotechnik der Apocynaceenblüte.I. Morphlogie und Funktion des Narbenkopfes. Flora 170: 394–432.

Schick, B. 1982a. Untersuchungen über die Biotechnik der Apocynaceenblüte. II. Bau und Funktion des Bestäubungsapparates. Flora 172: 347–371.

Schick, B. 1982b. Zur Morphologie, Entwicklung, Feinstruktur und Funktion des Translators von *Periploca* L. (Asclepiadaceae). Trop. Subtrop. Pflanzenwelt 40: 513–553.

Schill, R., Jäkel, U. 1978. Beitrag zur Kenntnis der Asclepiadaceen-Pollinarien. Trop. Subtrop. Pflanzenwelt 22: 53–170.

Schlechter, F.R.R. 1905. Periplocaceae and Asclepiadaceae. In: Schumann, K., Lauterbach, K. (eds.) Nachträge zur Flora des Deutschen Südseegebiets. Leipzig: Borntraeger, pp. 351–369.

Schlindwein, C., Darrault, R.O., Grisi, T., 2004. Reproductive strategies in two sphingophilous apocynaceous trees attracting pollinators through nectar or deceit. In: Breckle, S.-W., Schweizer, B., Fangmeier, A. (eds.) Proceedings of the 2nd Symposium of the AFW Schimper-Foundation. Stuttgart: Verlag Günter Heimbach, pp. 215–227.

Schnepf, E., Witzig, F., Schill, R. 1979. Über Bildung und Feinstruktur des Translators der Pollinarien von *Asclepias curassavica* und *Gomphocarpus fruticosus* (Asclepiadaceae). Trop. Subtrop. Pflanzenwelt 25: 1–39.

Schroeder, C.A. 1951. Heterostyly and sterility in *Carissa grandiflora*. Proc. Amer. Soc. Hort. Sci. 57: 419–422.

Schultes, R.E. 1979. De plantis toxicariis e mundo novo tropicale commentationes XIX. Biodynamic apocynaceous plants of the northwestern Amazon. J. Ethnopharmacol. 1: 165–192.

Schultes, R.E., Raffauf, R.F. 1990. Historical, ethno-and economic botany series. In: Dudley, T.R. (ed.) The Healing Forest, Vol. 2. Portland: Dioscorides Press, pp. 98–99.

Schumann, K. 1895. Apocynaceae and Asclepiadaceae. In: Engler, A., Prantl, K. (eds.) Die Nat. Pflanzenfam. 4 (2). Leipzig: W. Engelmann, pp. 189–305. doi: https://doi.org/10.5962/bhl.title.4635

Sennblad, B., Bremer, B. 1996. The familial and subfamilial relationships of Apocynaceae and Asclepiadaceae evaluated with *rbc*L data. Pl. Syst. Evol. 202: 153–175. doi: https://doi.org/10.1007/BF00983380

Sennblad, B., Bremer, B. 2000. Is there a justification for differential a priori weighting in coding sequences? A case study from rbcL and Apocynaceae s. l. Syst. Biol. 49: 101–113. doi: https://doi.org/10.1080/10635150050207410

Sennblad, B., Endress, M.E., Bremer, B. 1998. Morphology and molecular data in phylogenetic fraternity: The tribe Wrightieae (Apocynaceae) revisited. Amer. J. Bot. 85: 1143–1158. doi: https://doi.org/10.2307/2446347

Sharaf, M.H.M., Schiff, P.L., Tackie, A.N., Phoebe, C.H., Martin, G.E. 1996. Two new indoloquinoline alkaloids from *Cryptolepis sanguinolenta*: cryptosanguinolentine and cryptotackieine. J. Heterocyclic Chem. 33: 239–243.

Sharma, S., Shahzad, A. 2014. An overview on *Decalepis*: A genus of woody medicinal climbers. J. Plant Sci. Res. 1: 104.

Sheeley, S.E., Raynal, D.J. 1996. The distribution and status of species of *Vincetoxicum* in eastern North America. Bull. Torrey Bot. Club 123: 148–156.

Shuttleworth, A., Johnson, S.D. 2006. Specialized pollination by large spider-hunting wasps and self-incompatibility in the African milkweed *Pachycarpus asperifolius*. Int. J. Pl. Sci. 167: 1177–1186.

Shuttleworth, A., Johnson, S.D. 2008. Bimodal pollination by wasps and beetles in the African milkweed *Xysmalobium undulatum*. Biotropica 40: 568–574.

Shuttleworth, A., Johnson, S.D. 2009. The importance of scent and nectar filters in a specialized wasp-pollination system. Funct. Ecol. 23: 931–940.

Shuttleworth, A., Johnson, S.D. 2012. The *Hemipepsis* wasp-pollination system in South Africa: a comparative analysis of trait convergence in a highly specialized plant guild. Bot. J. Linn. Soc. 168: 278–299.

Sidiyasa, K. 1998. Taxonomy, phylogeny and wood anatomy of *Alstonia* (Apocynaceae). Blumea Suppl. 11: 1–230.

Sidney, N.C. 2012. A taxonomic revision of *Finlaysonia* and *Streptocaulon* (Periplocoideae; Apocynaceae). M.Sc. Thesis, Dept. Plant Sciences, Univ. Free State Bloemfontein, Bloemfontein, South Africa.

Silva, U.C.S., Rapini, A., Liede-Schumann, S., Ribeiro, P. L., Van den Berg, C. 2012. Taxonomic considerations on Metastelmatinae (Apocynaceae) based on plastid and nuclear DNA. Syst. Bot. 37: 795–806.

Silva, U.C.S., Santos, R.G.P., Rapini, A., Fontella Pereira, J., Liede-Schumann, S. 2014. Monsanima tinguaensis (Apocynaceae), an enigmatic new species from Atlantic rainforest. Phytotaxa 173: 11. doi: http://dx.doi.org./10.1600/036364412X648733

Simões, A.O., Endress, M.E., van der Niet, T., Conti, E., Kinoshita, L.S. 2004. Tribal and intergeneric relationships of Mesechiteae (Apocynoideae, Apocynaceae): evidence from three noncoding plastid DNA regions and morphology. Amer. J. Bot. 91: 1409–1418. doi: https://doi.org/10.3732/ajb.91.9.1409

Simões, A.O., Endress, M.E., van der Niet, T., Kinoshita, L. S., Conti, E. 2006. Is *Mandevilla* (Apocynaceae, Mesechiteae) monophyletic? Evidence from five plastid DNA loci and morphology. Ann. Missouri Bot. Gard. 93: 565–591. doi: https://doi.org/10.3417/0026-6493(2006)93[565:IMAMME]2.0.CO;2

Simões, A.O., Livshultz, T., Conti, E., Endress, M.E. 2007. Phylogeny and systematics of the Rauvolfioideae (Apocynaceae) based on molecular and morphological evidence. Ann. Missouri Bot. Gard. 94: 268–297. https://doi.org/10.3417/0026-6493(2007)94[268:PASOTR]2.0.CO;2

Simões, A.O., Endress, M.E., Conti, E. 2010. Systematics and character evolution of Tabernaemontaneae (Apocynaceae, Rauvolfioideae) based on molecular and morphological evidence. Taxon 59: 772–790.

Simões, A.O., Kinoshita, L.S., Koch, I., Silva, M.J., Endress, M.E. 2016. Systematics and character evolution of Vinceae (Apocynaceae). Taxon 65: 99–122. doi: http://dx.doi.org/0000-0003-3256-5922

Smith, A.R. 1971. *Curroria macrophylla* A.R. Smith. Hook. Icon. Pl. 37: pl. 3685. In: Taylor, G. (ed.) London: Bentham-Moxon Trusties.

Solbreck, S. 2000. Ecology and biology of *Euphranta connexa* (Fabr.) (Diptera: Tephritidae) – a seed predator on *Vincetoxicum hirundinaria* Med. (Asclepiadaceae). Entomol. Tidskr. 121: 23–30.

Solereder, H. 1899. Asclepiadaceae. In: Systematische Anatomie der Dicotyledonen. Stuttgart: Enke, pp. 603–609.

Spellman, D.L., Gunn, C.R. 1976. *Morrenia odorata* and *Araujia sericifera* (Asclepiadaceae): weeds in Citrus groves. Castanea 41: 139–148.

Straub, S.C.K., Moore, M.J., Soltis, P.S., Soltis, D.E., Liston, A., Livshultz, T. 2014. Phylogenetic signal detection from an ancient rapid radiation: Effects of noise reduction, long-branch attraction, and model selection in crown clade Apocynaceae. Mol. Phyl. Evol. 80: 169–185. doi: https://doi.org/10.1016/j.ympev.2014.07.020

Struwe, L., Albert, V.A., Bremer, B. 1994. Cladistics and family level classification of the Gentianales. Cladistics 10: 175–206.

Sugiura, S., Yamazaki, K. 2005. Moth pollination of *Metaplexis japonica* (Apocynaceae): pollinaria transfer on the tip of the proboscis. J. Pl. Res. 118: 235–262.

Sukumar, E., Gopal, R.H., Rao, R.B., Viswanathan, S., Thirugnanasbantham, P., Vijayaserkaran, V. 1995. Pharmacological actions of ceropegin, a novel pyridine alkaloid from *Ceropegia juncea*. Fitoterapia 66: 403–406.

Summons, R.E., Ellis, J., Gellert, E. 1972. Steroidal alkaloids of *Marsdenia rostrata*. Phytochemistry 11: 3335–3339.

Surveswaran, S., Sun, M., Grimm, G.W., Liede-Schumann, S. 2014. On the systematic position of some Asian enigmatic genera of Asclepiadoideae (Apocynaceae). Bot. J. Linn. Soc. 174: 601–619.

Suttisri, R., Lee, I.S., Kinghorn, A.D. 1995. Plant derived triterpenoid sweetness inhibitors. J. Ethnopharmacol. 47: 9–26.

Swarupanandan, K., Mangaly, J.K., Sonny, T.K., Kishorekumar, K., Chand Basha, S. 1996. The subfamilial and tribal classification of the family Asclepiadaceae. Bot. J. Linn. Soc. 120: 327–369.

Sylla, T., Albers, F. 1989. Samenentwicklung und Samenmorphologie krautiger und sukkulenter Asclepiadaceae. Bot. Jahrb. Syst. 110: 479–492.

Tanaka, H., Hatano, T., Kaneko, N., Kawachino, S., Kitamura, O., Suzuki, Y., Tada, T., Yaoi, Y. 2006. Andromonoecious sex expression of flowers and pollinia delivery by insects in a Japanese milkweed *Metaplexis japonica* (Asclepiadaceae), with special reference to its floral morphology. Pl. Spec. Biol. 21: 193–199.

Tank, D.C., Eastman, J.M., Pennell, M.W., Soltis, P.S., Soltis, D.E., Hinchliff, C.E., Brown, J.W., Sessa, E.B., Harmon, L.J. 2015. Nested radiations and the pulse of angiosperm diversification: increased diversification rates often follow whole genome duplications. New Phytol. 207: 454–467. doi: https://doi.org/10.1111/nph.13491

Taylor, W.I., Farnsworth, N. (eds.) 1975. The *Catharanthus* Alkaloids: Botany, chemistry, pharmacology, and clinical use. New York: Marcel Dekker, Inc.

Teo, S. 2001. *Alstonia*. In: Van Valkenburg, J.L.C.H., Bunyapraphatsara, N. (eds.) Plant Resources of South-East Asia, Vol. 12(2): 61–68. Leiden: Backhuys.

Thiv, M., Struwe, L., Albert, V.A., Kadereit, J.W. 2000 [1999]. The phylogenetic relationships of *Saccifolium bandeirae* (Gentianaceae) reconsidered. Harvard Pap. Bot. 4: 519–526.

Thomas, V., Dave, Y. 1994. Significance of follicle anatomy of Apocynaceae. Acta Soc. Bot. Pol. 63: 9–20.

Thorne, R.F. 1992. An updated phylogenetic classification of the flowering plants. Aliso 13: 365–389.

Torres, C., Galetto, L. 1998. Patterns and implications of floral nectar secretion, chemical composition, removal effects and standing crop in *Mandevilla pentlandiana* (Apocynaceae). Bot. J. Linn. Soc. 127: 207–223.

Torres, C., Galetto, L. 1999. Factors constraining fruit set in *Mandevilla pentlandiana* (Apocynaceae). Bot. J. Linn. Soc. 129: 239–247.

Tran, C.K. 2001. *Cerbera*. In: Van Valkenburg, J.L.C.H., Bunyapraphatsara, N. (eds.) Plant Resources of South-East Asia, Vol. 12(2): 151–155. Leiden: Backhuys.

Treiber, K. 1891. Anatomischer Bau des Stammes der Asclepiadaceae. Bot. Centralbl. 48: 209–218.

Trigo, J.R., Brown, K.S., Jr. 1990. Variation of pyrrolizidine alkaloids in Ithomiinae: a comparative study between species feeding on Apocynaceae and Solanaceae. Chemoecology 1: 22–29.

Trivedi, B.S., Upadhyay, N. 1984. Cuticular studies of Asclepiadaceae. J. Indian Bot. Soc. 63: 129–147.

Usher, G. 1974. A Dictionary of Plants Used by Man. London: Constable.

Van Beck, T.A., Van Gessel, M.A.J.T. 1988. Alkaloids of *Tabernaemontana* species. In: Pelletier, S.W. (ed.) Alkaloids: Chemical and biological perspectives, Vol. 6. New York: Wiley, pp. 76–226.

Van Beck, T.A., Verpoorte, R., Baerheim-Svendsen, A., Leeuwenberg, A.J.M., Bisset, N.G. 1984. *Tabernaemontana* L. (Apocynaceae): A review of its taxonomy, phytochemistry, ethnobotany and pharmacology. J. Ethnopharmacol. 10: 1–156.

Van de Ven, E.A., Van der Ham, R.W.J.M. 2006. Pollen of *Melodinus* (Apocynaceae): Monads and tetrads. Grana 45: 1–8.

Van der Ham, R., Zimmermann, Y.-M., Nilsson, S., Igersheim, A. 2001. Pollen morphology of the Alyxieae (Apocynaceae). Grana 40: 169–191. doi: https://doi.org/10.1080/001731301317223114

Van der Heijden, R., Jacobs, D.I., Snoeijer, W., Hallard, D., Verpoorte, R. 2004. The *Catharanthus* alkaloids: Pharmacognosy and biotechnology. Curr. Med. Chem. 11: 607–628. doi: https://doi.org/10.2174/0929867043455846

Van der Laan, F.M., Arends, J.C. 1985. Cytotaxonomy of the Apocynaceae. Genetica 68: 3–35.

Van der Ploeg, J. 1985. Revision of genera *Cyclocotyla* Stapf, *Dewevrella* De Wild. and of the African species of the genus *Malouetia* A. DC. (Apocynaceae). Series of revisions of Apocynaceae, XVIII. Wageningen Agric. Univ. Pap. 85.2: 57–83.

Van der Weide, J.C., Van der Ham, R.W.J.M. 2012. Pollen morphology and phylogeny of the tribe Tabernaemontaneae (Apocynaceae, subfamily Rauvolfioideae). Taxon: 61: 131–145.

Van Heerden, F.R. 2008. *Hoodia gordonii*: A natural appetite suppressant. J. Ethnopharmacol. 119: 434–437. doi: https://doi.org/10.1016/j.jep.2008.08.023

Van Roosmalen, M.G.M. 1985. Fruits of the Guianan Flora. Wageningen: Veenman.

Van Valkenburg, J.L.C.H., Hendrian. 2001. *Ochrosia*. In: Van Valkenburg, J.L.C.H., Bunyapraphatsara, N. (eds.) Plant Resources of South-East Asia, Vol. 12 (2): 386–389. Leiden: Backhuys.

Van Valkenburg, J.L.C.H., Horsten, S.F.A.J. 2001. *Thevetia*. In: Van Valkenburg, J.L.C.H., Bunyapraphatsara, N. (eds.) Plant Resources of South-East Asia, Vol. 12 (2): 544–546. Leiden: Backhuys.

Venter, H.J.T. 2009. A taxonomic revision of *Raphionacme* (Apocynaceae: Periplocoideae). S. African J. Bot. 75: 292–350. doi: https://doi.org/10.1016/j.sajb.2009.02.174

Venter, H.J.T., Verhoeven, R.L. 2001. Diversity and relationships within Periplocoideae. Ann. Missouri Bot. Gard. 88: 550–568.

Verhoeven, R.L., Venter, H.J.T. 1998. Pollinium structure in Periplocoideae (Apocynaceae). Grana 37: 1–14. doi: https://doi.org/10.1080/00173139809362633

Verhoeven, R.L., Venter, H.J.T. 2001. Pollen morphology of the Periplocoideae, Secamonoideae and Asclepiadoideae (Apocynaceae). Ann. Missouri Bot. Gard. 88: 569–582. doi: https://doi.org/10.2307/3298634

Verhoeven, R.L., Liede, S., Endress, M.E. 2003. The tribal position of *Fockea* and *Cibirhiza* (Apocynaceae: Asclepiadoideae): evidence from pollinium structure and cpDNA sequence data. Grana 42: 70–81. doi: https://doi.org/10.1080/00173130310012549

Vieira, M.F., Shepherd, G.J. 1999. Pollination of *Oxypetalum* (Asclepiadaceae) in southeastern Brazil. Rev. Brasil. Biol. 59: 693–704.

Waddington, K.D. 1976. Pollination of *Apocynum sibiricum* (Apocynaceae) by Lepidoptera. Southwest. Naturalist 21: 31–36.

Wagenitz, G. 1964. Gentianales. In: Melchior, H. (ed.) Engler's Syllabus der Pflanzenfamilien. Berlin: Borntraeger, pp. 405–425.

Walker, D.B. 1975. Postgenital carpel fusion in *Catharanthus roseus* (Apocynaceae). I. Light and scanning electron microscopic study of gynoecial ontogeny. Amer. J. Bot. 64: 457–467.

Walker, D.B. 1978. Postgenital carpel fusion in *Catharanthus roseus* (Apocynaceae). IV. Significance of the fusion. Amer. J. Bot. 65: 119–121.

Walther, R. 1994. Pollenfracht als Indikator fuer Ressourcennutzung und Einnischung bei madagassischen Schwärmern (Lepidoptera). Ph.D. Thesis, Friedrich-Alexander University, Erlangen-Nürnberg.

Wanntorp, H.-E. 1974. *Calotropis gigantea* (Asclepiadaceae) und *Xylocopa tenuiscapa* (Hymenoptera, Apidae): Studies in flower morphology and pollination biology. Svensk Bot. Tidskr. 68: 25–32.

Wanntorp, H.-E. 1988 [1989]. The genus *Microloma* (Asclepiadaceae). Opera Bot. 98: 1–69.

Wanntorp, L., Forster, P.I. 2007. Phylogenetic relationships between *Hoya* and the monotypic genera *Madangia*, *Absolmsia*, and *Micholitzia* (Apocynaceae, Marsdenieae): Insights from flower morphology. Ann. Missouri Bot. Gard. 94: 36–55. doi: https://doi.org/10.3417/0026-6493(2007)94[36:prbhat]2.0.co;2

Watt, J.M., Breyer-Brandwijk, M.G. 1962. The Medicinal and Poisonous Plants of Southern Africa. Edinburgh: E. and S. Livingstone.

Williams, L.O. 1981. Asclepiadaceae. In: The useful plants of Central America. Ceiba 24: 3–381.

Williams, J.K., Stutzman, J.K. 2008. Chromosome number of *Thevetia ahouai* (Apocynaceae: Rauvolfioideae: Plumerieae) with discussion on the generic boundaries of *Thevetia*. J. Bot. Res. Inst. Texas 2: 489–493.

Winks, C.J., Fowler, S.V. 2000. Prospects for biological control of moth plant *Araujia sericifera* (Asclepiadaceae). Landcare Research Contract Report LC9900/100 (unpubl.), Auckland, New Zealand.

Wong, S.K., Lim, Y.Y., Chan, E.W.C. 2013. Botany, uses, phytochemistry and pharmacology of selected

Apocynaceae species: A review. Pharmacognosy Comm. 3: 2–11.
Woodell, S.R.J. 1979. The role of unspecific pollinators in the reproductive success of Aldabran plants. Philos. Trans., Ser. B 286: 99–108.
Woodson, R.E., Jr. 1936. Studies in the Apocynaceae. IV. The American genera of Echitoideae. Ann. Missouri Bot. Gard. 23: 169–438.
Woodson, R.E., Jr. 1954. The North American species of *Asclepias*. Ann. Missouri Bot. Gard. 41: 1–211.
Wrangham, R.W., Waterman, P.G. 1983. Condensed tannins in fruits eaten by chimpanzees. Biotropica 15: 214–222.
Wyatt, R. 1976. Pollination and fruit-set in *Asclepias*: a reappraisal. Amer. J. Bot. 63: 845–851.
Wyatt, R. 1981. The reproductive biology of *Asclepias tuberosa* II. Factors determining fruit-set. New Phytol. 88: 375–185.
Wyatt, R., Broyles, S.B. 1994. Ecology and evolution of reproduction in milkweeds. Ann. Rev. Ecol. Syst. 25: 423–441.
Wyatt, R., Broyles, S.B. 1997. The weedy tropical milkweeds *Asclepias curassavica* and *A. fruticosa* are self-compatible. Biotropica 29: 232–234.
Wyatt, R., Edwards, A.L., Lipow, S.R., Ivey, C.T. 1998. The rare *Asclepias texana* and its widespread sister species, *A. perennis*, are self-incompatible and interfertile. Syst. Bot. 23: 151–156.
Yaman, B., Tumen, I. 2012. Anatomical notes on *Marsdenia erecta* (Apocynaceae) wood: Is it secondarily woody? Dendrobiology 67: 87–93.
Yamashiro, T., Yamashiro, A., Yokoyama, J., Maki, M. 2008. Morphological aspects and phylogenetic analyses of pollination systems in the *Tylophora* – *Vincetoxicum* complex (Apocynaceae-Asclepiadoideae) in Japan. Biol. J. Linn. Soc. 93: 325–341.
Yamauchi, T., Abe, F. 1990. Cardiac glycosides and pregnanes from *Adenium obesum* (Studies on the constituents of *Adenium* 1). Chem. Pharm. Bull. 38: 669–672.
Yamauchi, T., Abe, F., Santisuk, T. 1990. Cardiac glycosides of *Beaumontia brevituba* and *B. murtonii*. Phytochemistry 29: 1961–1965.
Yang, L.-L., Li, H.-L., Wei, L., Yang, T., Kuang, D.-Y., Li, M.-H., Liao, Y.-Y., Chen, Z.-D., Wu, H., Zhang, S.-Z. 2016. A supermatrix approach provides a comprehensive genus-level phylogeny for Gentianales. J. Syst. Evol. 54: 400–415.
Yoshikawa, M., Murakami, T., Kadoya, M., Yuhao, L.I., Murakami, N., Yamahara, J., Matsuda, H. 1997. Medicinal foodstuffs. IX. The inhibitors of glucose absorption from the leaves of *Gymnema sylvestre* R. Br. (Asclepiadaceae): structures of gymnemosides a and b. Chem. Pharm. Bull. 45: 1671–1676.
Young, J., Weed, A.S. 2014. *Hypena opulenta* (Erebidae): A European species for the biological control of invasive Swallow-worts (*Vincetoxicum* spp.) in North America. J. Lepid. Soc. 68: 162–166.
Zarucchi, J.L. 1987. A revision of the tribe Ambelanieae (Apocynaceae – Plumerioideae). Series of revisions of Apocynaceae, part XXIV. Agric. Univ. Wageningen Pap. 87: 1–106.
Zarucchi, J.L., Morillo, G., Endress, M.E., Hansen, B.F., Leeuwenberg, A.J.M. 1995. Apocynaceae. 2: 471–571. In: Berry, P.E., Holst, B.K., Yatskievych, K. (eds.) Flora of the Venezuelan Guyana. St. Louis, Missouri: Missouri Botanical Garden Press.
Zhu, J.-P., Guggisberg, A., Kalt-Hadamowsky, A., Hesse, M. 1990. Chemotaxonomic study of the genus *Tabernaemontana* (Apocynaceae) based on their indole alkaloid content. Pl. Syst. Evol. 172: 13–34.

Araliaceae

Araliaceae Jussieu, Gen. Pl.: 217 (1789), nom. cons.
Hydrocotylaceae Bercht. & J. Presl (1820), nom. cons.

G.M. Plunkett, J. Wen, P.P. Lowry II, A.D. Mitchell, M.J. Henwood, and P. Fiaschi

Small shrubs to large trees, less commonly lianas or herbs, glabrous or pubescent; mating system hermaphroditic or andromonoecious, rarely dioecious or diphasic; schizogenous secretory canals throughout the plant; plants terrestrial, hemi-epiphytic, or climbing, evergreen or deciduous; stems monocaulous or sparsely to well branched, usually pachycaulous. Leaves alternate (rarely opposite or whorled), frequently heteroblastic; petioles usually present (leaves rarely sessile) and often sheathing at the base, sometimes alate, exstipulate, or with ligulate stipules; blade simple to ternately, palmately or pinnately lobed or compound (or peltate), occasionally bi- or tripinnately compound, with entire, crenate, toothed, or incised margins; venation pinnate or palmate. Inflorescences terminal (rarely also axillary or pseudo-lateral), paniculate, compound-umbellate or simple-umbellate, the ultimate units umbellate, capitulate, racemose, spikate, or flowers rarely solitary; inflorescence axes subtended by foliose to minute bracts (or bracts lacking); flowers subtended by bracteoles (or bracteoles lacking). Flowers perfect, staminate, or pistillate, epigynous (rarely half-epigynous or hypogynous), actinomorphic; sepals and petals typically (3–)5(–12); calyx lobes simple and minute or obscure, but often forming a truncate rim; petals valvate or imbricate, free or rarely united, sometimes calyptrate, the bases broadly inserted; stamens isomerous to several/many times the number of petals (3–250+), in one whorl (and then alternipetalous) to many whorls, anthers dorsifixed, introrse, tetrasporangiate (rarely octosporangiate), dehiscing by longitudinal slits; filaments filiform (to short and stout), inflexed in bud; ovary syncarpous of 2–5(–100+) carpels (or unicarpellate through carpel abortion), each carpel unilocular with apical placentation; stigmas on a distinct style or sessile; stylodia distinct or partially to fully connate, sometimes swollen at the base and confluent with the nectariferous disc of the ovary; ovules anatropous, pendulous, one per locule, unitegmic, crassinucellate or rarely tenuinucellate. Fruits simple or sometimes multiple, fleshy (rarely dry), usually drupaceous (rarely baccate) with a fleshy mesocarp and usually separate pyrenes with variously sclerified endocarps around each locule, or rarely a schizocarp with two mericarps, with or without a free carpophore; one or more secretory canals ("companion canals" or "rib oil ducts") found in association with each vascular strand. Seeds straight; endosperm copious, oily, uniform or variously ruminate; embryo minute but well differentiated.

A widespread family comprising 40 genera and roughly 1,900 species, most diverse in tropical and subtropical Asia, Oceania, South and Central America, and sub-Saharan Africa-Madagascar.

Vegetative Morphology. Typical araliads are terrestrial woody perennials, either shrubs or small trees (more rarely large forest trees), but some species are hemi-epiphytes (usually facultatively, as in many species of *Schefflera* and some of *Astropanax*, *Aralia*, *Oreopanax* and *Raukaua*), lianas (climbing through the use of adventitious roots, as in *Hedera*, or by twining, as in *Cephalaralia* and *Motherwellia* and some *Astropanax*, or through use of grappling prickles, as in some species of *Aralia* and *Eleutherococcus*), perennial herbs with woody underground stems, rhizomes and/or roots (e.g., *Panax* and some *Aralia* species), or fully herbaceous, largely prostrate, and rooting at the nodes (e.g., *Hydrocotyle*) (Philipson 1970; Frodin and Govaerts 2004). The stems of most species are pachycaulous (more rarely

J.W. Kadereit, V. Bittrich (eds.), *Flowering Plants. Eudicots*, The Families and Genera of Vascular Plants 15,
https://doi.org/10.1007/978-3-319-93605-5_4

leptocaulous) and the pith is often well developed and sometimes chambered. The branching system is usually quite simple, typically sympodial and sparsely branched or appearing unbranched, or sometimes truly monopodial (as in *Anakasia simplicifolia* and *Harmsiopanax ingens*); more rarely, the branching may be quite complex (as in some species of *Polyscias* subg. *Arthrophyllum* and *Raukaua*) (Philipson 1978), or with both long and short shoots (*Gamblea*). The leaves of most Araliaceae are alternate (whorled in *Panax*; opposite in *Cheirodendron*) and generally clustered at the apex of the branches. They are usually petiolate (but some species have sessile to subsessile leaves); the petiole bases are frequently sheathing and sometimes alate, with or without laminar or (more typically) ligulate stipules, which may be adnate to the petiole, connate within the petiole, or very rarely free (Hutchinson 1967). The lamina is generally large and ranges from simple to palmately or pinnately lobed or imparipinnately compound (to bi- or tri-pinnately compound), or more rarely peltate (as in some species of *Harmsiopanax* and *Hydrocotyle*) (Philipson 1970). Modifications to compound leaves include reduction to a single leaflet (unifoliolate, as in some species of *Astropanax*, *Aralia*, *Neocussonia*, *Polyscias* and *Schefflera*), webbing among the petiolules (as in some species of *Trevesia* and *Osmoxylon*), or the presence of extra "subsidiary" leaflets (as in some species of *Aralia* and *Polyscias*). The ultimate leaf segments (leaflets or lobes) range from broad and entire to toothed and/or deeply lobed. Domatia are present in *Gamblea*. Venation patterns typically reflect general leaf shape, either pinnate or palmate. Heteroblasty among juvenile vs. mature leaves is common in Araliaceae, and is also sometimes correlated to changes in stem architecture (e.g., in *Hedera helix* and some *Pseudopanax* spp.) (Philipson 1970).

Vegetative Anatomy. In the internodal region of the stem, there is a ring of numerous vascular bundles, each with a prominent, sclerenchymatous outer cap (and sometimes also a smaller inner cap); there may also be a small number of cortical bundles (Mittal 1961). The xylem is characterized by libriform and usually thin-walled fibers (generally with minutely bordered pits with slit-like apertures, but pits with distinct borders occur rarely) that are often septate (sometimes with living content), with scanty paratracheal axial parenchyma, and occasionally vascular tracheids (Oskolski 1996). Vessel elements exhibit considerable diversity, ranging from (370–)600–900(–1370) μm in length, and with exclusively simple to exclusively scalariform perforation plates (1–56 bars), or a mix of simple and scalariform plates (Rodríguez 1971; Oskolski and Lowry 2001). Vessel element end-wall angles are slightly to strongly inclined (Hoar 1915; Rodríguez 1971). The intervessel pits are relatively large, have scalariform to opposite and alternate arrangement, and range from polygonal to rounded in shape and (4–)8–15 μm in size; vessel-ray and vessel-axial parenchyma pits are similar in shape, size and arrangement to intervessel pits (Oskolski 1996). Helical thickenings can be found in the vessels of some Araliaceae (e.g., in *Astrotricha*, *Fatsia*, *Hedera*, *Pseudopanax* and *Schefflera*) (Oskolski 1996). The wood may exhibit growth rings and (semi-) ring porosity, but these characters are limited mostly to species from temperate or subtropical habitats. Rays are both uniseriate and multiseriate, with long to very short uniseriate portions made of upright and square cells (Kribs' heterogeneous I, IIA, or IIB types). Occasionally, the multiseriate rays comprise exclusively procumbent cells (Kribs' homogeneous I types, as in some species of *Aralia*, *Dendropanax* and *Schefflera*), or exclusively square and upright cells (Carlquist's "paedomorphic I" type, as in *Astrotricha* and *Oplopanax*) (Oskolski 1996). The width of multiseriate rays varies from up to 3 rows (as in some species of *Dendropanax*, *Oplopanax* and *Schefflera*) to 25 rows (in *Schefflera digitata*); ray height is typically 1 mm (but may range up to 3 mm in some species of *Cussonia*, *Heteropanax*, *Osmoxylon* and *Schefflera*) (Oskolski 1996). Radial canals are found in some tropical and subtropical taxa (e.g., in some species of *Dendropanax*, *Polyscias* and *Schefflera*), but are absent in most temperate groups (Oskolski 1996; Oskolski and Lowry 2001). The sieve-tube plastids of Araliaceae belong to the S-type (see Wagenitz 1992). Schizogenous secretory canals are found in all parts of the plant (except the periderm), most frequently associated with vascular strands, usually to the outside of the phloem, and in transverse section are lined with an endothelium of 5–9 small, thin-walled, somewhat columnar cells that secrete resinous or gummy substances (Mittal 1961). In the roots, these schizogenous canals are located just interior to the endodermis, opposite both the xylem archs and the phloem groups, leading to a peculiar development of lateral roots, which emerge from the pericycle regions between the resin canals (and thus between the xylem archs and phloem groups), a feature shared with both Apiaceae and Pittosporaceae (van Tieghem 1872; Solereder 1908; Esau 1940).

In the subnodal region, a series of vascular traces diverge through the cortex, typically producing multilacunar nodes (with 7 to 25 traces per leaf), but trilacunar nodes characterize *Hydrocotyle* (Mittal 1961). At the base of the petiole, the vascular bundles are scattered and of various orientations, but in the upper petiole they are usually organized into two rings, with ectophloic outer bundles and smaller, typically endophloic inner or "medullary" bundles (more rarely forming a single ring of ectophloic bundles, as in *Hedera* and *Polyscias*). The petiole has a single epidermal layer covering several (6–10) layers of collenchyma, followed by many layers of parenchyma (Mittal 1961).

The indumentum of Araliaceae often includes simple to multicellular and/or stellate trichomes, which are typically uniseriate (but more rarely multiseriate), and rarely glandular (as in some *Trachymene*) (Mittal 1961; Hutchinson 1967; Wen 2002, 2004, 2011; Hart and Henwood 2006). There are, however, many species that are entirely glabrous, and some taxa exhibit papillae, glaucescence, or epicuticular wax arranged in scales (Metcalfe and Chalk 1983). Many araliads also bear leaf and/or stem prickles, as in *Brassaiopsis*, *Eleutherococcus*, *Harmsiopanax*, *Kalopanax*, *Oplopanax*, *Trevesia*, many species of *Aralia*, and a few species of other genera, such as *Polyscias*. Leaf prickles are located on midribs and veins, and are formed from the epidermis and underlying parenchyma cells. Stem prickles may be derived either from the cortex (e.g., *Oplopanax*, *Polyscias mollis*) or from the periderm (e.g., *Aralia*, *Eleutherococcus*, *Kalopanax*) (Kotina and Oskolski 2010). The stomata are paracytic or anisocytic (rarely anomocytic).

Inflorescence Structure and Floral Morphology. Most Araliaceae have paniculate or umbellate inflorescences, often with several degrees of branching, terminating most typically in umbellules, or sometimes in capitula, racemules, or spicules. The primary inflorescence axis is usually terminal and often erect (sometimes pendulous), but may appear lateral through continued growth of one (or more) adjacent foliar lateral axes; rarely, the inflorescence is truly axillary (e.g., *Anakasia*), or both terminal and axillary (as in *Aralia*; Wen et al. 1998; Wen 2002, 2004, 2011). The primary inflorescence axis and each degree of the lateral axes are often subtended by bracts, typically larger below (sometimes resembling the leaves) and progressively reduced above, and very often caducous. Small bracteoles typically subtend each flower, but may form an involucel if the ultimate inflorescence unit is an umbellule. Pedicels are articulated or unarticulated.

The flowers are almost always epigynous, although *Polyscias gymnocarpa* (= *Tetraplasandra gymnocarpa*) is a notable case of secondary reversion to hypogyny (and several species of the family have half-inferior ovaries) (Eyde and Tseng 1969; Costello and Motley 2004). The calyx is typically a low rim whose margin may be entire to undulate, or crowned by 5 (3–numerous) apiculate to triangular or rounded teeth or lobes. The corolla is sympetalous early in development (Erbar and Leins 2004), but typically appears choripetalous at maturity, formed by 5 (3–numerous) greenish to white, reddish, or purplish to black petals that are usually valvate in bud (or less commonly imbricate, as in *Aralia* and *Panax*). The petals have broad bases and are sometimes calyptrate (e.g., some species of *Schefflera* and *Trevesia*) or even remaining united into a distinct corolla tube (*Osmoxylon*); they are often fleshy, and the apex typically has a pointed adaxial thickening.

The androecium comprises 5 (3–numerous) stamens, usually isomerous and alternate with the petals in a single whorl, or two to many times the number of petals in two or more whorls (as in some species of *Osmoxylon*, *Polyscias* and *Schefflera*). The filaments are short and stout to long and slender, inflexed in bud but ascending to spreading at anthesis. The anthers are dorsifixed and longitudinally dehiscent, with two thecae (tetrasporangiate), or rarely four thecae (octosporangiate, as in some species of *Plerandra*).

The syncarpous gynoecium is formed by 2 to 20 or more carpels (rarely unicarpellate through carpel abortion, as in some species of *Polyscias*, or >100 carpels, as in *Schefflera peuckleri*), with an equal number of locules (or rarely bicarpellate but unilocular due to breakdown of the septum, as in *Seemannaralia*). The apex of the ovary is modified as a nectariferous disc, terminating in free to partially or fully connate styles, or the stigmas may be sessile. The stigmas are papillate and either wet or dry. Each locule encloses a single functional pendulous ovule with apical placentation (plus a second, entirely abortive ovule). Staminate flowers, when present, have abortive carpels.

Embryology. The endothecium of the anther walls develops fibrous thickenings, but the middle one or two layers are ephemeral (Davis 1966). The cells of the tapetum are glandular, becoming multinucleate and finally uninucleate (by fusion). Cell wall formation after meiosis is simultaneous. The microspore tetrads are tetrahedral or isobilateral, and the pollen, when shed, is generally 3-celled (but 2-celled in *Pseudopanax arboreus*) (Davis 1966).

The single functional ovule in each locule is pendulous, epitropous (with a ventral raphe), anatropous, unitegmic, and tenuinucellate or crassinucellate, with a differentiated endothelium, a long micropyle, and commonly a well-developed hypostase (Davis 1966; Mohana Rao 1972; Corner 1976). The funiculus is long and fleshy, and its micropylar side forms an obturator whose surface cells serve as a continuation of the conducting tissue. A second, erect but abortive ovule in each locule also appears to aid in pollen-tube conduction (Philipson 1970). Meiosis and cytokinesis of the megaspore mother cell usually occur together; the resulting tetrads are linear. The embryo sac is typically monosporic (although bisporic embryo sacs have been reported), developing from either the chalazal megaspore (Polygonum-type) or the micropylar megaspore (Oenothera-type) (Davis 1966; Mohana Rao 1972). Alternatively, the homotypic division of meiosis may be suppressed, resulting in a linear triad (three cells, of which only the two micropylar ones are megaspores) (Davis 1966; Bouman 1984). The synergids are pyriform, the polar nuclei fuse just prior to fertilization, and the 3 antipodal nuclei are ephemeral, remaining free or becoming enclosed in separate cells (Davis 1966). Embryogeny is of the onagrad or chenopodiad type (Mohana Rao 1972).

Pollen Morphology. The pollen grains of Araliaceae (Figs. 55, 56) are usually suboblate to subprolate (sometimes prolate) or rhomboid (as in *Hydrocotyle*) in equatorial view (Ting et al. 1964; Wen and Nowicke 1999), and triangular to rounded in polar view, with concave to convex mesocolpia, varying in polar-axis length from 15–40 μm. The exine always comprises an inner (usually thinner) nexine and an outer (usually thicker) sexine, although the nexine is sometimes thicker in some genera (e.g., *Polyscias* and *Schefflera*) and is almost always thicker than the sexine in the area bordering the apertures (Tseng 1971, 1973; Tseng and Shoup 1978; Sosa 1983; Pire 1989), forming costae ectocolpi (Wen and Nowicke 1999). Inner nexinous breaks are sometimes present (Tseng 1971; Tseng and Shoup 1978; Wen and Nowicke 1999), rendering the aperture H-shaped due to polar extensions (Tseng 1971). The sexine is most commonly semitectate or tectate perforate, but intectate (echinate to clavate) in some species of *Polyscias* (Henwood 1991), and sometimes eutectate (smooth or with small granules), especially at the aperture borders (Tseng 1974; Tseng and Shoup 1978; Pire 1989; Wen and Nowicke 1999). Ornamentation is reticulate or striato-reticulate (rugulate) to prominently striate or perforate (Pire 1989; Wen and Nowicke 1999). When reticulate, the lumina can be isometric or larger at the poles (Henwood 1991), and the muri smooth or perforated (Shoup and Tseng 1977; Maguire et al. 1984; Fiaschi 2002; Fiaschi et al. 2010). The relative thickness of the foot layer and columellae varies extensively within and among genera. Most araliad pollen is tricolporate (pororate in some species of *Polyscias* and *Trevesia*), but bicolporate and tetracolporate grains may sometimes be found mixed with typical ones (Ting et al. 1964; Tseng and Shoup 1978; Henwood 1991; Jebb 1998; Fiaschi 2002; Fiaschi et al. 2010). Colpi are usually long, sometimes reaching almost to the poles, or very short and restricted to the medial region (as in *Osmoxylon* and some groups of *Polyscias*), and sometimes have equatorial constrictions (Tseng 1971, 1974; Shoup and Tseng 1977; Tseng and Shoup 1978;

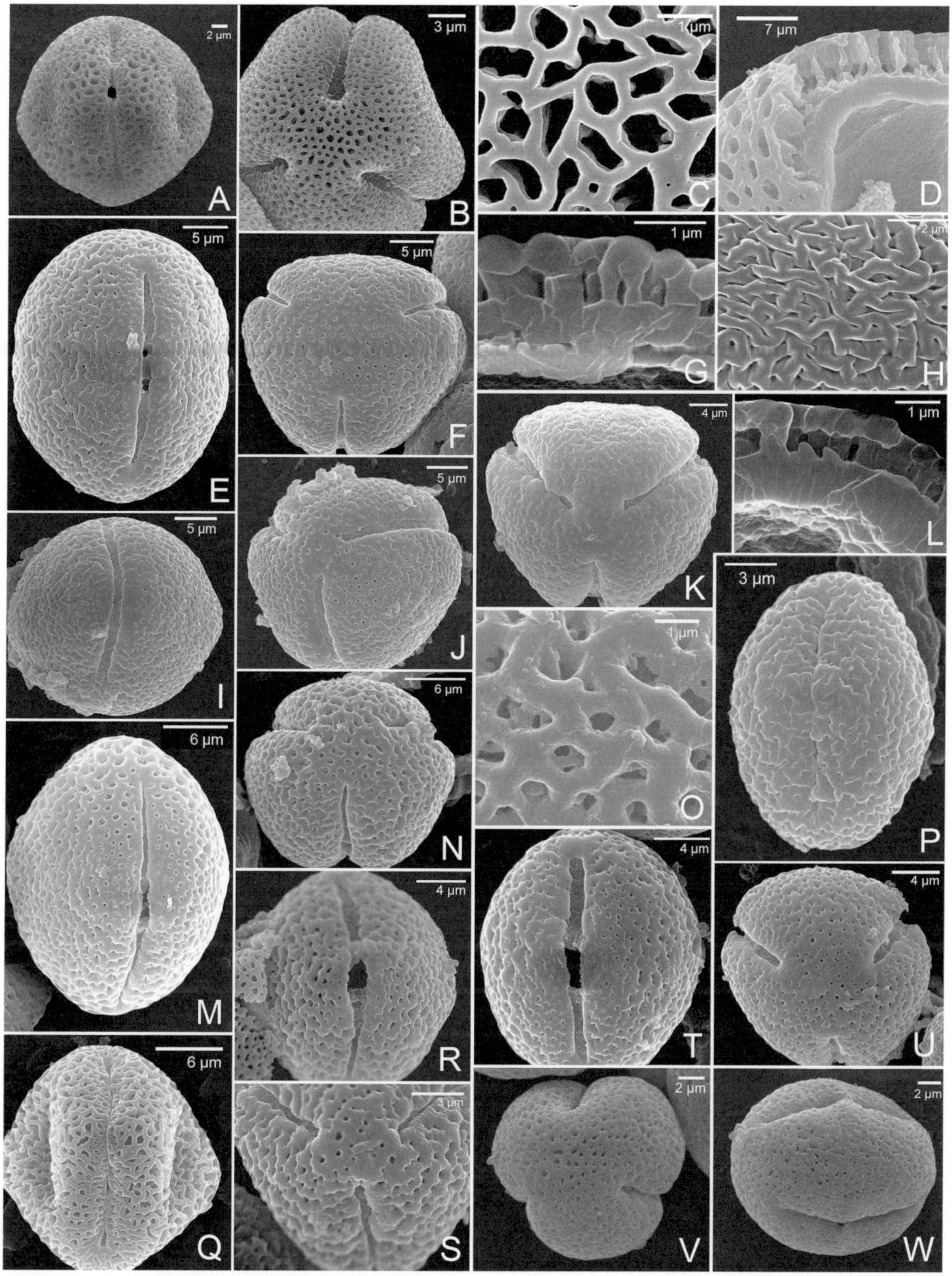

Fig. 55. Araliaceae. Pollen morphology of Neotropical species of *Schefflera*. **A** *S. pimichinensis*, equatorial view. **B** *S. tamatamaensis*, polar view. **C** *S. distractiflora*, detail of apocolpium. **D** *S. decaphylla*, detail of wall. **E–G** *Schefflera* sp. nov. **E** Equatorial view. **F** Polar view. **G** Detail of wall. **H** *S. tamana*, detail of mesocolpium. **I, J** *Schefflera* sp.

Henwood 1991). The ora are highly variable among different species; most are lalongate, but some are lolongate (as in some *Polyscias* spp.) or even compound (Tseng 1974).

Karyology. Chromosome counts have been made for representatives of roughly 60% of araliad genera, but only about 8% of the species. For most Araliaceae, these data have been reviewed by Yi et al. (2004), but *Hydrocotyle* and *Trachymene* (recently transferred from Apiaceae) were also summarized by Pimenov et al. (2003). The vast majority of counts in Araliaceae are $2n = 24$ or 48, clearly suggesting a basic chromosome number of $x = 12$. Numbers deviating from these values are limited primarily to older (and potentially unreliable) reports, and/or counts from commonly cultivated species (e.g., *Hedera* spp., *Aralia racemosa, Eleutherococcus trifoliatus, Polyscias fruticosa* and *P. guilfoylei*). In *Hydrocotyle*, there is a much wider range of values ($2n =$ 18 to $>$200). Yet even among these counts, chromosome numbers based on $x = 12$ ($2n = 24$, 48, 72, 96) are most common (although $2n = 18$ is not uncommon). The basic number for *Trachymene* represents a notable departure from the general trend in Araliaceae, with $x = 11$ as the most common basic number ($2n = 22$, 44, 88, 132; with one count of $2n = 40$). Tetraploidy characterizes most genera from the "Asian palmate" clade (e.g., *Eleutherococcus, Kalopanax, Merrilliopanax, Metapanax* and *Oplopanax*), but a few genera from this group (e.g., *Brassaiopsis, Fatsia, Hedera* and *Macropanax*) have both diploid and tetraploid species, as do several genera from other clades of Araliaceae (e.g., *Schefflera, Polyscias, Aralia, Panax, Hydrocotyle* and *Trachymene*). Aneuploidy is relatively rare, but has been documented in *Hydrocotyle* and *Panax* (Yi et al. 2004).

Pollination and Reproductive Systems. Pollination in Araliaceae appears to be effected primarily by insects, but few species have been studied in great detail. Pollinators are attracted by nectar (often copious, produced by the floral nectary disc) and presumably pollen. A wide range of pollinators have been reported on *Hedera helix* (Knuth 1908; Metcalfe 2005), including various dipterans (e.g., many species of Calliphoridae, Muscidae, and Syrphidae), plus several hymenopterans (Apidae, Ichneumonidae, Vespidae) and lepidopterans (Nymphalidae, Tortricidae). Bees (*Bombus*, Apidae) have been documented as pollinators for *Aralia hispida* (Thomson and Barrett 1981), while syrphid flies (*Toxomerus*) and halictid sweat-bees (*Dialictus* and *Evylaeus*) have been reported for *Panax quinquefolius* (Lewis and Zenger 1983). Schlessman et al. (1990b) documented pollinators of *Polyscias pancheri* (from New Caledonia), on which halictid bees (*Nomia*) and undetermined wasps were found, as well as scarab beetles (*Heteronyx*, Scarabaeidae) and nectar-feeding ants (*Polyrachis*, Formicidae). Pombal and Morellato (1995) report that two species of the fly genus *Morellia* (Muscidae) are the primary pollinators of *Dendropanax cuneatus* (from southeastern Brazil), although additional representatives of Muscidae and 17 other families of flies (plus lepidopterans and hymenopterans) were also recorded as visitors. Davila and Wardle (2008) documented the pollinators of *Trachymene incisa*, which were dominated by hymenopterans, especially colletid and halictid bees (notably *Hylaeus, Homolictus*, and *Lasioglossum*) and ants (Formicidae, particularly *Ochetellus* and *Monomorium*); minor pollinators included various flies (e.g., from Syrphidae and Bombyliidae) and other hymenopterans. Bird pollination has been suggested for some of the Hawaiian species of *Polyscias* with numerous stamens (see Lowry 1990; Costello and Motley 2004) as well as for polyandrous species of *Plerandra* (which may also be bat pollinated). The inflorescence structure of *Osmoxylon* (with sterile pseudo-fruits held below fertile flowers) also appears to be adapted to bird pollination (Philipson 1978).

Fig. 55. (Continued) nov. **I** Equatorial view. **J** Polar view. **K** *S. jahnii*, polar view. **L** *S. harmsii*, detail of wall. **M–O** *S.* cf. *violacea*. **M** Equatorial view. **N** Polar view. **O** Detail of apocolpium. **P** *S. tipuanica*, equatorial view. **Q** *S. tremula*, equatorial view. **R, S** *S. duidae*. **R** Equatorial view. **S** Detail of apocolpium. **T, U** *S. diplodactyla*. **T** Equatorial view. **U** Polar view. **V, W** *S. minutiflora*. **V** Polar view. **W** View of mesocolpium. (Fiaschi et al. 2010; SEM photographs by P. Fiaschi; redrawn with permission from Schweizerbart Science Publishers)

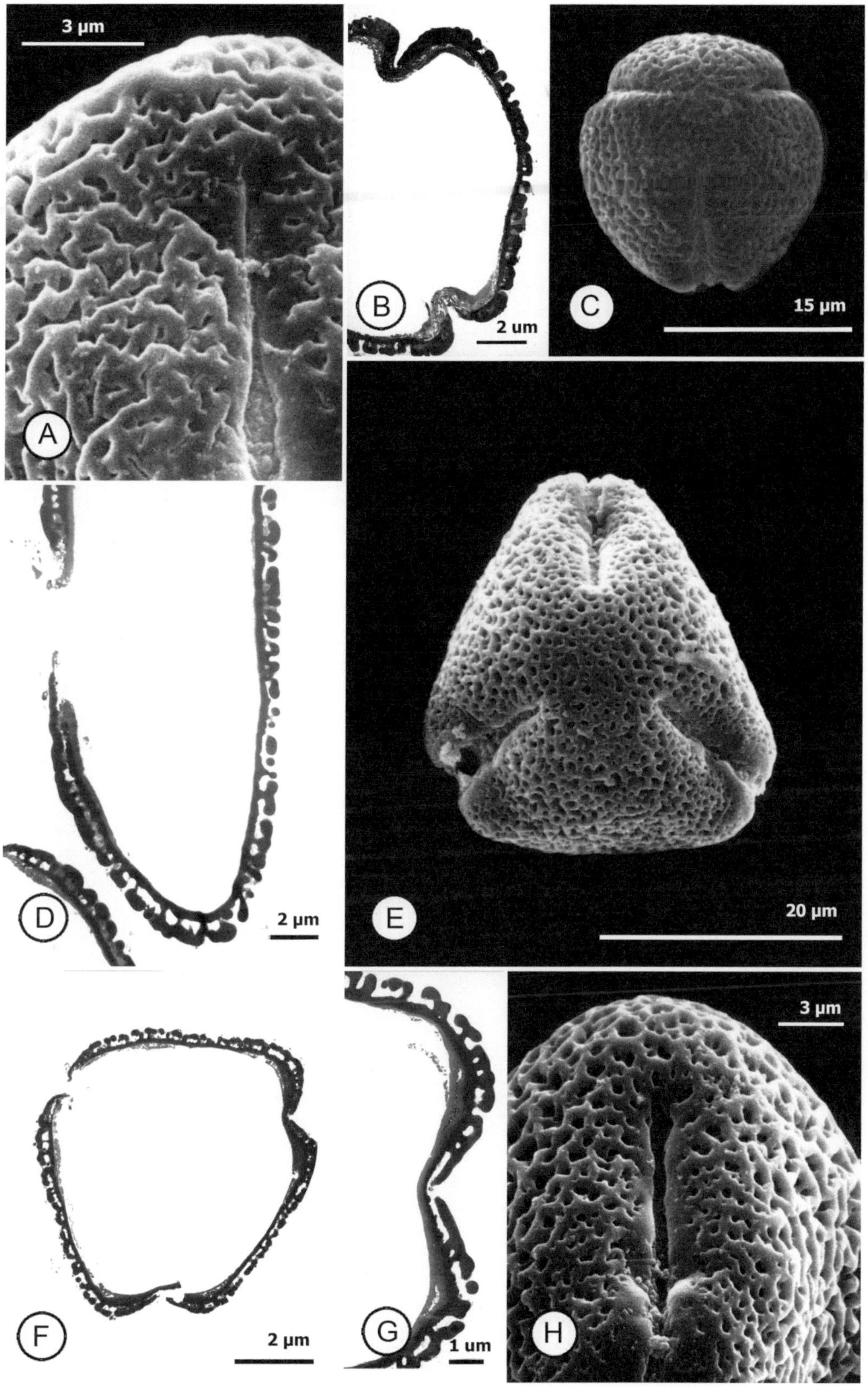

Fig. 56. Araliaceae. Pollen morphology of *Aralia*. **A–D** *A. nudicaulis*. **A** Tectum along colpus. **B** Radial section of grain. **C** Tectum in polar view. **D** Longitudinal section through pole. **E–H** *A. californica*. **E** Tectum in polar

Most araliad species are either hermaphroditic or andromonoecious, and the latter appears to be the ancestral character state in Araliaceae (and the related families Apiaceae and Myodocarpaceae) (see Schlessman et al. 1990a, 2001). Dioecy is less common, but characterizes three genera (*Meryta*, *Pseudopanax*, and most species of *Oreopanax*), as well as some species of *Aralia*, *Panax* and *Polyscias*. Some species of *Trachymene* are gynomonoecious. In diclinous species, imperfect flowers typically retain abortive remnants of the carpels (in functionally staminate flowers) or stamens (in functionally carpellate flowers). In andromonoecious species, staminate flowers or umbellules often subtend perfect flowers or umbellules, or are located at the periphery of mixed umbellules. In hermaphroditic and andromonoecious species, the perfect flowers are strongly protandrous, with stamens (and petals) typically dropping before stigma receptivity. The separation of male and female phases is generally synchronized over the entire plant, leading to one or more cycles of complete synchronous dichogamy, minimizing selfing in these self-compatible species (Schlessman et al. 1990a). The two gender phases of perfect flowers are also typically correlated with two distinct periods of nectar production (Henwood 1986; Gillespie and Henwood 1994; see also Thomson and Barrett 1981). The remarkable *Panax trifolius* is diphasic or "sex changing", exhibiting distinct male stages (with staminate flowers only) when the plants are smaller, and hermaphroditic stages (with perfect flowers) when the plants are larger (Schlessman 1990, 1991). Monocarpy is known only in *Harmsiopanax* (Philipson 1973).

Fruit and Seed. The fruits are usually fleshy, typically drupes with two or more pyrenes. More rarely, they are dry (or scarcely fleshy), as in *Astrotricha*, *Harmsiopanax*, *Hydrocotyle*, *Neosciadium*, and *Trachymene*, all of which have schizocarpic fruits of two mericarps (with or without carpophores). The fruits are typically glabrous and dark purple to blackish at maturity (occasionally pubescent and/or greenish to brownish, red or whitish). They range from transversely terete to strongly compressed laterally (particularly in bicarpellate and some unicarpellate taxa), and often appear distinctly costate, especially when dry. Multiple fruits are known in some genera (e.g., some species of *Meryta* and *Schefflera*), while *Osmoxylon* produces sterile, baccate pseudofruits (in addition to fertile drupes). The dry fruits of *Seemannaralia* are bicarpellate and strongly compressed (superficially resembling the mericarps of dorsally compressed Apiaceae fruits), but they develop into one-seeded, unilocular fruits due to disruption of the septum (Burtt and Dickison 1975; Oskolski et al. 2008). In the fruits of most Araliaceae, each locule is associated with a ventral vascular trace (supplying the ovules) and a dorsal vascular trace (supplying the carpels), in addition to an outer series of peripheral vascular bundles (which supplied the perianth parts and stamens during flowering). The dorsal bundles are often associated with a schizogenous secretory canal, and have thus been referred to as "companion canals" (Philipson 1970). Calcium oxalate crystals may be completely lacking from the fruit wall, but are often present as druses scattered in the mesocarp and/or rhomboidal crystals localized in the narrow zone adjacent to the endocarp (Eyde and Tseng 1971). The endocarps are usually woody, cartilaginous or crustaceous (less commonly membranaceous), and form separately around each locule, resulting in the production of as many individual pyrenes as carpels (Lubbock 1892).

The integument comprises several cell layers (either having or lacking crystals), but these usually begin to collapse after fertilization, ultimately forming a thin, membranaceous testa of crushed and deteriorated cells, typically adnate or closely applied to the endosperm; the outer epidermal layer is composed of cuboidal or tabular cells that are either thin walled or have outer-wall thickenings (Lubbock 1892; Mohana Rao 1972; Corner 1976). The endosperm is copious, nuclear, oily, uniform or ruminate (of the *Spigelia* type, with irregular infoldings caused by meristematic

Fig. 56. (Continued) view. **F** Radial section of whole grain. **G** Radial section across aperture. **H** Tectum along colpus. (Wen 2011; SEM and TEM photographs by J. Wen; reproduced with permission from the Smithsonian Institution)

activity of the integument), with thick-walled cells (Davis 1966; Corner 1976; Boesewinkel and Bouman 1984). The embryo is minute and embedded in the endosperm close to the hilum (Lubbock 1892). Germination is dicotyledonous; the cotyledons are petiolate and round, ovate or oblong (Lubbock 1892, based on *Aralia edulis* Siebold & Zucc., *Hedera helix* L., *Hydrocotyle vulgaris* L., *Trachymene pilosa* Sm., and *Tupidanthus calyptratus* Hook. f. & Thomson [=*Schefflera pueckleri* (K. Koch) Frodin]).

Dispersal. The fleshy drupes that characterize most species of Araliaceae suggest vertebrate dispersal of the pyrenes. In species with small fruits, the disperser likely ingests the entire drupe, allowing some of the pyrenes to pass through undigested, as has been demonstrated in *Hedera helix* for birds (especially thrushes, families Muscicapidae and Turdidae) and possibly mammals (deer and martens) (Metcalfe 2005). Alternatively, when the fruits are quite large, it appears that dispersers may feed on the fleshy mesocarp, discarding the hardened pyrenes. Seeds may also be dispersed under the parental plant simply by gravity, as demonstrated for *Panax trifolius* (Meier et al. 1995).

Phytochemistry. Several species of Araliaceae (especially of *Panax* and *Eleutherococcus*) are well-known adaptogens, and the great interest in their phytochemistry has resulted in a complex literature with descriptions of many novel compounds. Despite this, only Hegnauer (1964, 1971, 1989) has provided any taxonomic order to this information. In general, the family is characterized by triterpenoid saponins, C_{17} polyacetylenes, seed oils containing an abundance of petroselinic acid, and the lack of both iridoids and tannins. In all these respects, Araliaceae are remarkably similar to Apiaceae. These characters, especially the absence of iridoids, were also helpful in confirming the taxonomic distance of Araliaceae from Cornaceae. Triterpenoid saponins are abundant in most Araliaceae, but appear to be lacking in a few species (e.g., *Aralia californica*, *Polyscias tripinnata*). The ginsenosides of *Panax* spp. are the best known saponins, but many other saponins from *Panax* have also been identified. Various other saponins have been identified from genera such as *Aralia*, *Dendropanax*, *Eleutherococcus*, *Fatsia*, *Hedera*, *Kalopanax* and *Polyscias*. Secretions from the schizogenous canals of Araliaceae contain abundant essential oils, balsams, and gum resins. The essential oils are mainly of mono- and sesquiterpenes, and also contain acetylenes, especially C_{17}-polyacetylenes. Falcarinone (a C_{17}-diacetylene) and related C_{17}- and C_{18}-diacetylenes are known from many genera (*Aralia*, *Dendropanax*, *Hedera*, *Panax*, *Polyscias*, and *Schefflera*), and provide another link with Apiaceae. The polyacetylene falcarinol (known from *Hedera helix*, *Schefflera arboricola*, *S. pueckleri*, and the intergeneric hybrid X *Fatshedera*) is an allergen that causes contact dermatitis in humans. Diterpenoids of the pimaran- and kauran-types are also known from the leaves, stems, and/or roots in some species of *Aralia*. The polyphenols of Araliaceae are characterized by chlorogenic acids (sometimes in large quantities). The main flavonoid compounds include simple flavonols (especially kaempferol and quercitin) and flavones (especially apigenin and luteolin), as well as rhoifolin, naringen, and tricine.

The seeds of Araliaceae lack starches but are rich in oils (often 34–45%) and albumin (15–30%). Petroselinic acid is typically the most abundant fatty acid in seed oils (normally 40–83%). Other lipids include myricyl, known from the cuticle of *Polyscias* and *Aralia* species, as well as waxes (alkanes, alkanols, alkanones, fatty-acid esters), phytosterines, and free triterpenes (isolated from *Eleutherococcus* and *Pseudopanax*). Carbohydrates include glucose, galacturonic acid, galactose, arabinose, xylose, rhamnose, maltol, maltoglucoside, innovanoside, raffinose, and scyllo-inositol. Cyanogenic compounds are known from the leaves of *Aralia spinosa*, and acanthopanaxoside has been isolated from *Eleutherococcus sessiliflorus*. Alkaloids have been isolated from several species of *Panax*, including methoxypyrazine, alkylpyrazine, and the neurotoxin dencichine (see also Xie et al. 2007).

Subdivision and Relationships Within the Family. The late nineteenth-century classification systems of Bentham (1867), Seemann (1868), and Harms (1894 and 1897) emphasized the aestivation and insertion of petals (and sometimes also carpel number and endosperm texture), and yielded similar subdivisions of Araliaceae (reviewed in Plunkett et al. 2004b). In Harms' system, which had the greatest influence during the 20th century, three tribes were recognized: Aralieae (petals imbricate and broadly inserted), Schefflereae (petals valvate and broadly inserted), and Mackinlayeae (petals valvate and clawed at the base). To these, a fourth tribe (Myodocarpeae, with imbricate and sometimes clawed petals) is often added, following Calestani (1905) and Viguier (1906). More recent systems (Hutchinson 1967; Tseng and Hoo 1982) broadened the sources of taxonomic characters, but have produced relatively similar classifications. Phylogenetic studies based on molecular characters (e.g., Wen et al. 2001; Plunkett et al. 2004b; Nicolas and Plunkett 2014) have provided a very different picture of relationships in Araliaceae. Taxa traditionally referred to Mackinlayeae (*Mackinlaya* and *Apiopetalum*) have been transferred to Apiaceae and placed in a new subfamily (Mackinlayoideae), together with several historically apiaceous genera (e.g., *Actinotus*, *Centella*, *Xanthosia*). Tribe Myodocarpeae (comprising only *Myodocarpus* and *Delarbrea*) is now recognized as a distinct family, Myodocarpaceae. Other taxonomic transfers include the movement of *Stilbocarpa* from Araliaceae to Apiaceae, and the inclusion of three formerly apiaceous genera (*Hydrocotyle*, *Neosciadium*, and *Trachymene*) in Araliaceae.

An updated infrafamilial system for Araliaceae based on recent phylogenetic progress has not been completed, but several major clades have been recognized (Wen et al. 2001; Mitchell and Wen 2004; Plunkett et al. 2004b). Among these are the *Aralia*-*Panax* group, comprising only those two genera, and the *Polyscias*-*Pseudopanax* group, which also includes *Meryta* and *Plerandra*. Another clade, sometimes called the "Asian Palmate" group, unites genera with leaves having palmate venation, lobing, or divisions, primarily from Asia (e.g., *Brassaiopsis*, *Eleutherococcus*, *Kalopanax* and *Trevesia*) but also from the Neotropics (*Oreopanax*) or both regions (*Dendropanax* and *Schefflera*). Recognition of formal subfamilies and/or tribes awaits the reclassification of *Schefflera*, which is grossly polyphyletic (Plunkett et al. 2004b, 2005; Frodin et al. 2010), and the resolution of a large polytomy comprising several small clades, each of only one or two genera (e.g., *Motherwellia*, *Cephalaralia*, *Osmoxylon*, *Cussonia* + *Seemannaralia*, *Cheirodendron* + *Raukaua*; see also Mitchell et al. 2012, who provided support for a clade uniting *Raukaua*, *Cephalaralia* and *Motherwellia*, *Cheirodendron*, and *Schefflera* s.s.). Some authors have recognized the clade uniting *Hydrocotyle*, *Neosciadium*, and *Trachymene* as "subfamily Hydrocotyloideae" (see Stevens 2001 onwards; Mabberley 2008), but we find this usage problematic due to instability in the phylogenetic placement of some taxa, including *Harmsiopanax*, *Astrotricha* and *Osmoxylon* (see Nicolas and Plunkett 2009).

Affinities. Araliaceae have traditionally been placed in a phyletic progression between the "less specialized" Cornaceae and the "more advanced" Apiaceae (reviewed in Rodríguez 1971; Plunkett 2001) in or near subclass Rosidae (e.g., Cronquist 1981). The relationship to Apiaceae (*quod vide*) is supported by many characters, including flowers with free stamens and petals (but developmentally sympetalous), frequently pentamerous perianths, inferior ovaries, nectariferous floral discs, two ovules per locule (one functional and one abortive), seeds with copious non-starchy endosperm and small embryos, schizogenous secretory canals, and a similar phytochemistry (Hegnauer 1971, 1989; Rodríguez 1971). This relationship has been confirmed by many phylogenetic studies, which place Araliaceae and Apiaceae in an expanded order Apiales (see Plunkett et al. 1996, 2004a; Plunkett 2001; Chandler and Plunkett 2004; Nicolas and Plunkett 2014) in the campanulid or "euasterid II" clade near Asterales and Dipsacales (which all share "early sympetalous" corolla development, unitegmic and mostly tenuinucellate ovules, S-type sieve-tube plastids, and a series of phytochemical characters; reviewed in Plunkett 2001; Erbar and Leins 2004). A close relationship of Araliaceae to Cornaceae has not been supported, but phylogenetic studies have allowed the placement of several genera once considered intermediate between these two families. *Diplopanax*, for example, has been moved to Nyssaceae in Cornales, and *Helwingia* has been recognized as its own family (Helwingiaceae) in Aquifoliales. Four other genera have been moved to Apiales, including *Aralidium*, *Melanophylla* and *Torricellia* (each in its own family, or united as a broadly defined Torricelliaceae), and *Griselinia* (alone in Griseliniaceae). Pittosporaceae have also been included in the expanded Apiales; despite a series of apparent differences (e.g., superior ovaries and entire leaves), this family shares many features with Araliaceae and Apiaceae (e.g., schizogenous secretory canals, ovule structure, phytochemical characters and "early sympetaly" (van Tieghem 1884; Jurica 1922; Jay 1969; Hegnauer 1971, 1989; Erbar and Leins 2004).

Distribution and Habitats. The family is widely distributed, with representatives on every continent except Antarctica, and is found across a broad range of habitats. Its centers of diversity, however, are all tropical or subtropical, particularly in moist and/or high-elevation environments. They are relatively rare in the cooler regions of the northern and southern temperate zones and in arid habitats, and thus there is little geographic and ecological overlap between mainly woody Araliaceae and the closely related mainly herbaceous Apiaceae.

Tropical and subtropical Asia has the greatest generic diversity in the family (25 genera), including many genera endemic to the region (e.g., *Brassaiopsis*, *Chengiopanax*, *Fatsia*, *Gamblea*, *Heteropanax*, *Kalopanax*, *Macropanax*, *Metapanax*, *Sinopanax*, *Tetrapanax* and *Trevesia*). Oceania has 15 genera represented, including Pacific island endemics such as *Cheirodendron* and *Meryta*, and Australian endemics such as *Astrotricha*, *Cephalaralia* and *Motherwellia*, along with many species of the paleotropical genus *Polyscias* (Lowry and Plunkett 2010; Plunkett and Lowry 2010). The Neotropics have only six genera, but many species, particularly in *Dendropanax*, *Schefflera* and the endemic *Oreopanax*. In sub-Saharan Africa and the Indian Ocean basin, there are five genera, four of which are endemic (*Astropanax*, *Cussonia*, *Neocussonia* and *Seemannaralia*). The most depauperate regions are temperate North America, with only four native genera (the widespread *Hydrocotyle*, plus *Aralia*, *Oplopanax* and *Panax*, all shared with eastern Asia), and Europe, with only two genera (*Hedera*, also found in Asia, plus the widespread *Hydrocotyle*). Of the genera spanning more than one region, *Schefflera* (ca. 1000 species) is both the largest and the most widespread, but phylogenetic evidence (Plunkett et al. 2005; Fiaschi and Plunkett 2011; Plunkett and Lowry 2012; Li and Wen 2014) suggests that this genus, as recently circumscribed, comprises five clades corresponding closely to geographic distributions (viz., tropical Asia, the Neotropics, Africa-Madagascar, and two groups in the Pacific). One of the Pacific groups is now recognized as *Plerandra* (Lowry et al. 2013) and the species from Africa and Madagascar are now treated as *Astropanax* and *Neocussonia* (Lowry et al. 2017). *Polyscias* (ca. 210 species) is broadly pantropical, with two large centers of diversity (the Pacific islands and the Indian Ocean Basin, especially Madagascar). *Hydrocotyle* (ca. 130 spp.) is also widespread, and particularly speciose in South America and Australia. Several genera exhibit intercontinental disjunctions, including the well-known eastern Asian–eastern North American disjunctions in *Aralia* and *Panax* (Wen and Zimmer 1996; Wen 1999, 2001a, 2001b, 2001c), and the eastern Asian–western North American disjunction in *Oplopanax* (Wen et al. 2001). *Dendropanax* presents an unusual disjunction between tropical/subtropical Asia and the Neotropics (Li and Wen 2013). This same pattern is also found in the clade that unites the Neotropical *Oreopanax* and the Asian *Sinopanax* (Wen et al. 2001; Plunkett et al. 2004b), and (less perfectly) the Asian and Neotropical clades of *Schefflera*, suggesting a common (perhaps boreotropical) geographical history.

Palaeobotany. Fossils attributed to Araliaceae (and especially *Aralia*) have been reported from the middle Cretaceous (e.g., Berry 1903), but many of these are of doubtful affinity, probably belonging to unrelated families (Dilcher and Dolph 1970). Even *Parafatsia*, a relatively recently described megafossil (Blackburn 1981), has been transferred to Proteaceae (Carpenter et al. 2006). Fossils more reliably assigned to Araliaceae date mostly to the Late Cretaceous to early Tertiary and thereafter, including the microfossil *Tricolporopollenites armatus* (cf. *Aralia*) from the Paleocene of France (Gruas-Cavagnetto and Bui 1976; Muller 1981; Manchester et al. 2015); similar pollen from eastern China (possibly araliaceous) dates to the late Cretaceous (Song et al. 2004). Reliable megafossils from North America include *Aralia* from Banks Island (Northwest Territories, Canada), dated to the early to middle

Miocene (Graham 1999), *Dendropanax* from Tennessee, dated to the middle Eocene (Dilcher and Dolph 1970; but see Manchester et al. 2015), *Oreopanax* from the early Eocene of Wyoming and the late Eocene–early Oligocene of Colorado and Utah (Graham 1999), and *Paleopanax* (with affinities to the Asian genus *Metapanax*) from the middle Eocene of Oregon (Manchester 1994). In Europe, microfossils of *Hedera* have been reported from the upper Miocene of Spain, and the Pliocene of northwestern Germany; pollen comparable to *Panax*, *Eleutherococcus*, *Brassaiopsis*, or *Schefflera* has been reported from the Eocene of France (Muller 1981), and leaf fossils of *Schefflera* have been reported from the upper Oligocene of Hungary and southern France (Rásky 1959). *Hydrocotyle* has also been documented from lower Eocene of France, the upper Eocene in the northeastern English Channel, and the Miocene of New Zealand (Muller 1981; Gruas-Cavagnetto and Cerceau-Larrival 1982). Both megafossils and microfossils comparable to *Polyscias* (but probably also *Pseudopanax* and other Araliaceae) have been reported from the early Oligocene to early Miocene of southeastern Australia (Blackburn and Sluiter 1994). Wood fossils of *Eleutherococcus* have been documented from the Pliocene of the Russian Far East (Bondarenko 2008). Muller (1981) suggests that microfossils of *Nyssapollenites squamosus* from the late Cretaceous of Australia show affinities to recent *Schefflera pickeringii*.

Economic Importance. The ginsengs (*Panax* spp.) are among the most highly prized and best known adaptogens (Wen 2001a, 2001c; Zuo et al. 2011). The complex mixture of triterpenoid saponins known as ginsenosides are derived from the roots, especially from *P. ginseng*, *P. pseudoginseng*, and *P. quinquefolius*. Cultivation of ginsengs has increased in recent decades, but remains difficult, resulting in high prices and overharvesting from wild populations. Siberian ginseng (*Eleutherococcus senticosus*) is often used as a substitute. Several North American species of *Aralia* are used in a similar manner, including *A. nudicaulis* (wild sarsaparilla), *A. racemosa* (American spikenard), and *A. spinosa* (devil's walking stick) (van Wyk and Wink 2004). The leaves and bark of *Hedera* (especially *H. helix*) have also been used for various folk-medicinal purposes (e.g., as an expectorant or spasmolytic), as have many species of *Hydrocotyle* and *Polyscias* (French 1971; van Wyk and Wink 2004). Pith derived from the stems of *Tetrapanax papyrifer* is the traditional source of Chinese rice paper, and this species is often cultivated as an ornamental. Other cultivated ornamentals include *Aralia elata*, *Eleutherococcus* spp., *Fatsia japonica*, *Hedera* spp., *Kalopanax septemlobus*, *Plerandra* spp. (*P. elegantissima* and *P. veitchii*), *Polyscias* spp. (e.g., *P. fruticosa* and *P. scutellaria*) and *Schefflera* spp. (especially *S. actinophylla* and *S. arboricola*).

Key to the Genera of Araliaceae

1. Fruits dry, either schizocarps with 2 mericarps, or unilocular, 1(2)-seeded and flattened 2
– Fruits fleshy drupes 7
2. Herbaceous annuals, biennials or perennials 3
– Trees or shrubs 5
3. Leaves exstipulate **3. *Trachymene***
– Leaves stipulate 4
4. Marginal ribs of fruit enclosing a cavity between the two mericarps. Plants not stoloniferous **2. *Neosciadium***
– Marginal ribs (if present) not enclosing a cavity between the two mericarps. Plants usually stoloniferous **1. *Hydrocotyle***
5. Fruits unilocular, 1(2)-seeded, strongly flattened. Petals imbricate. Endosperm ruminate **12. *Seemannaralia***
– Fruits 2-seeded, schizocarpic. Petals valvate. Endosperm uniform 6
6. Leaves and stems armed with prickles. Leaves palmately lobed **4. *Harmsiopanax***
– Leaves and stems unarmed. Leaves simple and unlobed **5. *Astrotricha***
7. Leaves opposite **9. *Cheirodendron***
– Leaves alternate or whorled 8
8. Leaf bases with 1–several spiral or transverse crests or collars. Inflorescences with both sterile, baccate pseudofruits (in a central umbellule or capitulum) and fertile, drupaceous fruits (in paired, lateral units) **6. *Osmoxylon***
– Leaf bases lacking spiral or transverse crests or collars. Inflorescences with only fertile, drupaceous fruits 9
9. Leaves deeply lobed or compound 10
– Leaves simple and unlobed, or unifoliolate 45

10. Leaves pinnately lobed or compound 11
– Leaves palmately or ternately lobed or compound 14
11. Leaves irregularly pinnately lobed **18. *Meryta***
– Leaves pinnately compound 12
12. Plants deciduous (very rarely evergreen). Petal aestivation imbricate **15. *Aralia***
– Plants evergreen. Petal aestivation valvate 13
13. Leaves 2- to 4-times compound. Endosperm ruminate **25. *Heteropanax***
– Leaves once (rarely 2- to 3-times) compound. Endosperm uniform **20. *Polyscias***
14. Herbaceous perennials. Leaves whorled **16. *Panax***
– Woody plants. Leaves alternate 15
15. Plants armed with prickles 16
– Plants unarmed, prickles lacking 21
16. Fruits red and ovaries glabrous **23. *Oplopanax***
– Mature fruits black, purple or blue, or ovaries tomentose 17
17. Calyx lobes long and prominent. Petals abaxially pubescent and ovaries tomentose **40. *Woodburnia***
– Calyx a low rim or minutely toothed. Petals and ovaries glabrous 18
18. Fruits terete 19
– Fruits laterally compressed 20
19. Carpels 2 (rarely 1–5). Petals and stamens 5 **33. *Brassaiopsis***
– Carpels 6–12 (rarely 5). Petals and stamens (5)6–12 **32. *Trevesia***
20. Leaves palmately lobed **39. *Kalopanax***
– Leaves palmately compound **36. *Eleutherococcus***
21. Plants dioecious 22
– Plants andromonoecious or hermaphroditic 23
22. Inflorescence a panicle of capitula with well-developed bracts. Petals free **27. *Oreopanax***
– Inflorescence a compound umbel (or an umbel of staminate racemules), bracts reduced or lacking. Petals calyptrate **17. *Pseudopanax***
23. Lianas 24
– Trees or shrubs 28
24. Leaves palmately lobed **35. *Hedera***
– Leaves palmately compound 25
25. Stems pachycaulous. Petals valvate 26
– Stems leptocaulous. Petals imbricate 27
26. Ovary 5–9-carpellate. Stamens 5–8; Africa, Madagascar, Comoro Is. **13. *Astropanax***
– Ovary (2–)5–20(–100+)-carpellate. Stamens (4–)5 (–250+); Asia, the Neotropics and SW Pacific Is. **22. *Schefflera***
27. Plants glabrous. Inflorescence a simple umbel **8. *Motherwellia***
– Plants pubescent. Inflorescence a panicle of capitula **7. *Cephalaralia***
28. Leaves palmately compound 29
– Leaves palmately lobed 38
29. Leaves with domatia in axils of secondary veins on lower surface of leaves **29. *Gamblea***
– Leaves lacking domatia 30
30. Leaves deciduous **30. *Chengiopanax***
– Leaves evergreen 31
31. Stems pachycaulous. Pedicels unarticulated 32
– Stems leptocaulous. Pedicels articulated 36
32. Primary inflorescence axes short, secondary axes spicate or racemose (rarely a panicle of spikes) and plants from Africa, Comoro Is., or Arabia **11. *Cussonia***
– Primary inflorescence axes short to elongated, secondary axes umbellules or capitula (or, if spicate or racemose, plants not from Africa, Comoro Is., or Arabia) 33
33. Ultimate inflorescence units umbellules. Plants glabrous; restricted to Melanesia (Fiji, New Caledonia, New Guinea, Solomon Islands, Vanuatu) **19. *Plerandra***
– Ultimate inflorescence units umbellules, capitula, spicate, or racemose. Plants glabrous to pubescent; pantropical 34
34. Ovary (2–)5–20(–100+)-carpellate. Stamens (4–)5 (–250+); Asia, the Neotropics and SW Pacific islands **22. *Schefflera***
– Ovary 2–9-carpellate. Stamens 5–8; Africa, Madagascar, Comoro Is. 35
35. Ovary 5–9-carpellate. Styles 5–9. Apex of leaflets usually acute to acuminate or caudate. Terrestrial, hemi-epiphytic or epiphytic trees **13. *Astropanax***
– Ovary 2–5-carpellate. Styles 2–5. Apex of leaflets usually truncate to emarginate, with an evident retuse notch and a small mucro or boss. Terrestrial trees **14. *Neocussonia***
36. Inflorescence compound umbellate **10. *Raukaua***
– Inflorescence paniculate 37
37. Plants andromonoecious. Styles fully connate (or rarely divided at the apex). Fruits terete **37. *Macropanax***
– Plants hermaphroditic. Styles to 3/4-connate. Fruits compressed laterally **38. *Metapanax***
38. Ultimate inflorescence units racemules or spicules **11. *Cussonia***
– Ultimate inflorescence units umbellules or capitula 39
39. Ultimate inflorescence units capitula 40
– Ultimate inflorescence units umbellules 41

40. Plants glabrous and evergreen **31. *Dendropanax***
– Plants stellate-pubescent and deciduous **28. *Sinopanax***
41. Plants deciduous **24. *Tetrapanax***
– Plants evergreen 42
42. Leaf margins entire to irregularly toothed. Lamina commonly with transparent glandular dots **31. *Dendropanax***
– Leaf margins regularly serrate to serrulate. Lamina never with transparent glandular dots 43
43. Carpels 5 or 10 **26. *Fatsia***
– Carpels 2(–4) 44
44. Leaf bases stipulate. Pedicels unarticulated **34. *Merrilliopanax***
– Leaf bases exstipulate. Pedicels articulated **38. *Metapanax***
45. Leaves unifoliolate, with a distinct articulation and petiolule at the base of the single leaflet 46
– Leaves simple (articulation and petiolule lacking at the base of the blade) 51
46. Lianas. Petal aestivation valvate. Endosperm ruminate **8. *Motherwellia***
– Shrubs or trees (very rarely lianas). Petal aestivation imbricate. Endosperm uniform 47
47. Pedicels unarticulated 48
– Pedicels articulated 50
48. Ovary (2–)5–20(–100+)-carpellate. Stamens (4–)5 (–250+); Asia, the Neotropics and SW Pacific Is. **22. *Schefflera***
– Ovary 2–9-carpellate. Stamens 5–8; Africa, Madagascar, Comoro Is. 49
49. Ovary 5–9-carpellate. Styles 5–9. Apex of leaflets usually acute to acuminate or caudate. Terrestrial, hemi-epiphytic or epiphytic trees **13. *Astropanax***
– Ovary 2–5-carpellate. Styles 2–5. Apex of leaflets usually truncate to emarginate, with an evident retuse notch and a small mucro or boss. Terrestrial trees **14. *Neocussonia***
50. Stems leptocaulous **32. *Raukaua***
– Stems pachycaulous **20. *Polyscias***
51. Plants dioecious 52
– Plants hermaphroditic or andromonoecious 54
52. Ultimate inflorescence units umbellules or racemules **17. *Pseudopanax***
– Ultimate inflorescence units capitula 53
53. Plants glabrous. Endosperm uniform. Plants from the Pacific islands **18. *Meryta***
– Plants usually pubescent. Endosperm ruminate. Plants from the Neotropics **27. *Oreopanax***
54. Ultimate inflorescence units capitula 55
– Ultimate inflorescence units umbellules, racemules, or spicules 56
55. Plants stellate-pubescent **28. *Sinopanax***
– Plants glabrous **31. *Dendropanax***
56. Pedicels articulated 57
– Pedicels unarticulated 58
57. Plants andromonoecious. Inflorescence axillary **21. *Anakasia***
– Plants hermaphroditic. Inflorescence terminal **38. *Metapanax***
58. Lianas **35. *Hedera***
– Trees or shrubs 59
59. Ultimate inflorescence units racemules or spicules **11. *Cussonia***
– Ultimate inflorescence units umbellules 60
60. Leaf margins entire to irregularly toothed. Lamina commonly with transparent glandular dots **31. *Dendropanax***
– Leaf margins serrate to serrulate. Lamina never with transparent glandular dots **33. *Merrilliopanax***

1. *Hydrocotyle* L. Fig. 57

Hydrocotyle L., Sp. Pl. 234 (1753); Mathias, Brittonia 2: 201–237 (1936), reg. rev.; Mathias & Constance, N. Amer. Fl. 28B(1): 51–58 (1944), reg. rev.; Nevling, Fl. Panama 46: 251–254 (1959), reg. rev.; Mathias & Constance, Fl. Peru 13: 62–76 (1962), reg. rev.; Standley & Williams, Fl. Guatemala 8(1): 49–53 (1966), reg. rev.; Mathias & Constance, Fl. Ecuador, 145: 20–34 (1976), reg. rev.; Constance, Fl.

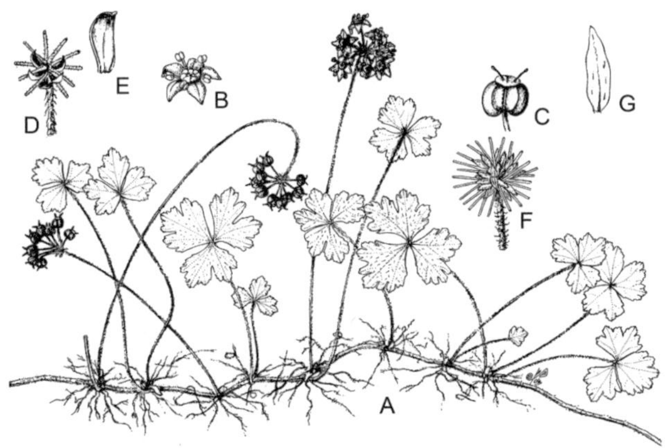

Fig. 57. Araliaceae. *Hydrocotyle*. **A–E** *H. apolobambensis*. **A** Habit. **B** Flower. **C** Fruit. **D** Involucral bracts and pedicels of umbellule, flowers removed. **E** Involucral bract. **F, G** *H. palmata*. **F** Involucral bracts and pedicels of umbellule, flowers removed. **G** Involucral bract. (Mendoza and Fuentes 2010; illustrations by Carlos Maldonado; reproduced with permission from the Missouri Botanical Garden Press)

Patagonica 5: 352–355 (1988), reg. rev.; Eichler, Feddes Rep. 98: 1–2: 1–51; 3–4: 145–196; 5–6: 273–351 (1987), nomenclature; She et al., Fl. China 14: 14–18 (2005), reg. rev.; Constance & Affolter, Fl. Mesoamericana 4(1): 388–390 (2009), reg. rev.

Hermaphroditic (or rarely dioecious), unarmed, glabrous or pubescent, terrestrial or aquatic, low annual or perennial herbs; stems branched, erect to prostrate and often rooting at the nodes, leptocaulous. Leaves alternate, petiolate, the bases stipulate but not sheathing; blades simple (and often peltate and orbicular) to palmately lobed or compound with ovate or obovate lobes or sessile leaflets; membranaceous, the margins entire to crenate or dentate. Inflorescences axillary, erect or lax, often simple umbellate but sometimes compound umbellate or forming continuous to interrupted spikes; bracts present or lacking; pedicels short and stout to long and slender, unarticulated. Calyx minute to lacking; petals 5, valvate, free; stamens 5; carpels 2, ovary inferior, styles free or basally connate; the disc conical to depressed. Fruits dry schizocarps (lacking carpophores), usually 5-ribbed, glabrous to pubescent or spinose, orbicular to ellipsoid, flattened laterally; endocarp sclerified. $2n = 18, 24, 32, 36, 48, 60, 72, 96, 132, 144, 162, 184$.

Circa 180 spp. Cosmopolitan (especially diverse in the Neotropics and Australia), mostly in moist habitats.

2. *Neosciadium* Domin

Neosciadium Domin, Beih. Bot. Centralbl. 23: 291 (1908).

Hermaphroditic, unarmed, nearly glabrous, terrestrial, annual, small herbs; stems branched, leptocaulous. Leaves alternate, petiolate, the base scarcely sheathing, stipulate; blades simple, obovate and slightly succulent, the margins with large dentitions. Inflorescences terminal and axillary, erect, spicate to racemose (often appearing as an elongated capitulum), bracts subtending each flower; pedicels lacking to minute. Calyx a low rim or lacking; petals 5, slightly imbricate, free; stamens 5; carpels 2, ovary inferior, styles free or basally connate; the disc flattened. Fruits dry schizocarps (lacking carpophores), surface with retrorse barbs or occasionally smooth, 5-ribbed, with marginal ribs enclosing a cavity between the two mericarps and the very narrow commissure; orbicular, transversely flattened laterally; endocarp lignified.

One species, *N. glochidiatum* (Benth.) Domin, Australia (S and W), salt pans. Closely related to (and perhaps congeneric with) *Hydrocotyle*.

3. *Trachymene* Rudge

Trachymene Rudge, Trans. Linn. Soc. London 10: 300 (1811); Hart & Henwood, Aust. Syst. Bot. 19: 11–57 (2006), generic rev.
Didiscus DC. ex Hook. (1828).
Uldinia J.M. Black (1922).

Hermaphroditic, andromonoecious, or gynomonoecious, unarmed, typically pubescent (with glandular, multiseriate, stellate, dendritic trichomes), terrestrial, annual, biennial, or perennial herbs; stems branched (rarely unbranched and scapose), leptocaulous. Leaves basal and/or cauline and alternate, petiolate, the bases sheathing and exstipulate; blades simple to ternately lobed or deeply dissected, the lobes linear to ovate or obovate, membranaceous to subcoriaceous, and toothed. Inflorescences terminal and axillary, erect, simple umbellate, often arranged in monochasial or dichasial cymes; bracts whorled at the base of the umbel and basally connate; pedicels long and slender, unarticulated. Calyx lobes 5 and inconspicuous or absent, petals 5, free, imbricate; stamens 5; carpels 2 (or functionally 1 through abortion), ovary inferior, styles connate; the disc flattened or raised. Fruits dry schizocarps with entire or apically bifid (or rarely fully bifurcating) carpophores, glabrous to pubescent, smooth to papillate, rugose, or spiny, orbicular, semicircular, or elliptic, sometimes with dorsal ribs and/or wings, the base often cordate, transversely flattened laterally; endocarp sclerified. $2n = 22, 40, 44, 88$.

56 species, mainly Australia, but also insular SE Asia and Oceania (Philippines, Borneo, New Guinea, New Caledonia, Vanuatu, Fiji), in a wide range of habitats, from deserts to alpine meadows. Several species are known to be poisonous to livestock.

4. *Harmsiopanax* Warb. Fig. 58

Harmsiopanax Warb. in Engl. & Prantl, Nat. Pflanzenfam. ed. 1, Nachtr. 1: 166 (1897); Philipson, Fl. Malesiana 9: 9–14 (1979), rev.

Hermaphroditic or andromonoecious, spiny, pubescent, terrestrial, often monocarpic and proteranthous, weak trees; stems monocaulous to sparsely branched, pachycaulous. Leaves alternate, petiolate and often peltate, bases clasping but not alate, exstipulate; blades simple, palmately lobed, orbicular to ovate, membranaceous to subcoriaceous, the margins sharply, irregularly and/or coarsely incised and toothed. Inflorescences terminal, erect, very large and repeatedly branched panicles, often developing after abscission of the leaves, ultimate units sessile to pedunculate umbellules; bracts often foliaceous below but progressively reduced above; pedicels short, usually with two bracteoles, unarticulated. Calyx a minute, fringed rim, often lacerate; petals 5, valvate, free; stamens 5; carpels 2, ovary inferior, styles free, the disc conical. Fruit a dry schizocarp of 2 three-ribbed, long-ciliate, ellipsoid to ovoid, laterally compressed mericarps, crowned by slightly hooked styles, carpophore fused; endocarp membranous. Endosperm uniform.

Three species, Malesia (Java, Lesser Sundas, Celebes, New Guinea), in mossy and montane forests, and secondary growth of grass-covered hills.

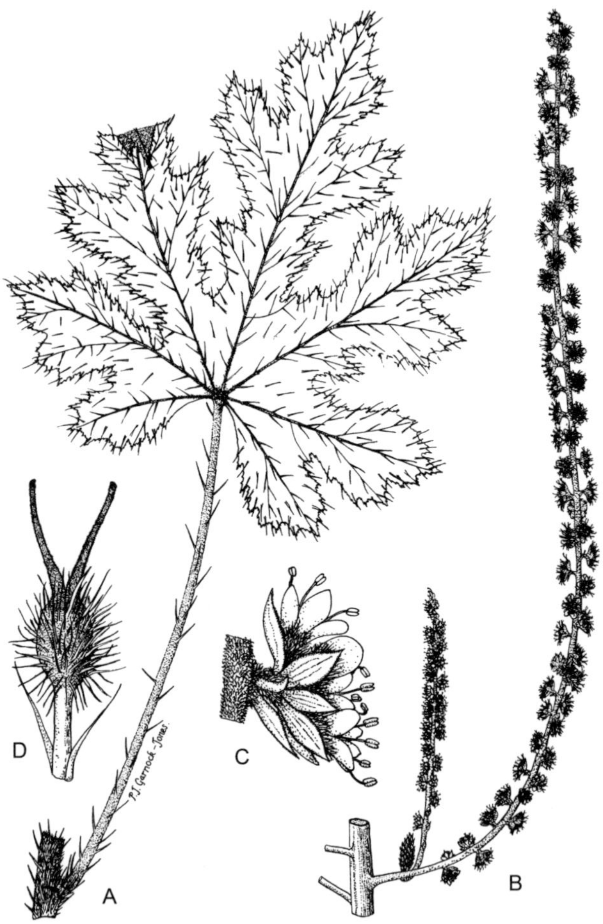

Fig. 58. Araliaceae. *Harmsiopanax ingens*. **A** Leaf. **B** Part of inflorescence. **C** Umbellule of flowers at anthesis. **D** Fruit. (Philipson 1979; illustrations by P.J. Garnock-Jones; reproduced with permission from Flora Malesiana)

5. *Astrotricha* DC.

Astrotricha DC., Coll. Mém. 5: 29 (1829); Bean, Austrobaileya 5: 63–69 (1997), reg. rev.

Andromonoecious, unarmed, stellate-pubescent, terrestrial, evergreen shrubs or small treelets; stems monocaulous to well branched, leptocaulous to pachycaulous. Leaves alternate, petiolate (to subsessile), the bases not clasping or only minutely so, exstipulate; blades simple, ovate, lanceolate, obovate or linear, chartaceous to subcoriaceous, entire or occasionally minutely toothed or basally lobed. Inflorescences terminal (occasionally both terminal and axillary), usually erect, paniculate, the ultimate units umbellules; bracts minute to foliaceous; pedicels usually long and slender, articulated or unarticulated. Calyx minutely toothed, 5-lobed; petals 5, valvate, free; stamens 5; carpels 2, ovary inferior, styles free; the disc flattened. Fruits dry schizocarps (carpophore not free), glabrous to pubescent, elliptic to obovoid, rarely winged, compressed laterally; endocarp membranaceous to fleshy. Endosperm uniform.

20 species, Australia (mostly E, one species NW), dry habitats.

6. *Osmoxylon* Miq. Fig. 59

Osmoxylon Miq., Ann. Mus. Bot. Lugduno-Batavi 1: 5 (1863); Philipson, Fl. Malesiana 9: 31–53 (1979), reg. rev.; Lowry, Bull. Mus. Natl Hist. Nat., sect. B, Adansonia 11: 119–121 (1989), reg. rev.; Conn & Frodin, Handbooks Fl. Papua New Guinea 271–285 (1995), reg. rev.; Costion & Plunkett, PhytoKeys 58: 49–64 (2016), reg. rev.

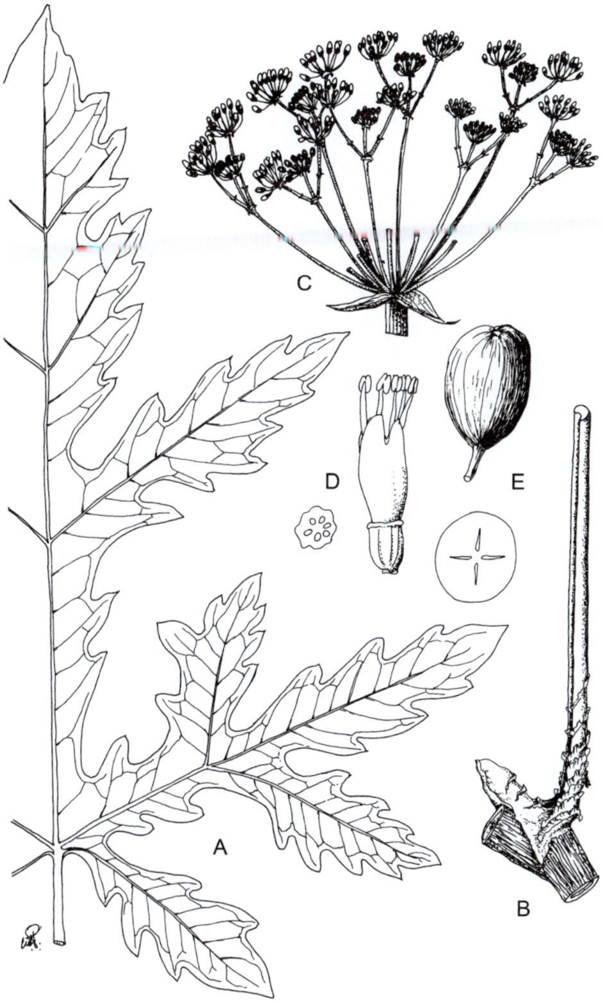

Fig. 59. Araliaceae. *Osmoxylon novoguineense*. **A** Half of leaf. **B** Base of petiole. **C** Inflorescence in bud. **D** Flower and ovary (cross section). **E** Fruit and cross section. (Philipson 1979; illustrations by W.R. Philipson; reproduced with permission from Flora Malesiana)

Boerlagiodendron Harms (1894).

Hermaphroditic (or andromonoecious?), unarmed, glabrous to pubescent, terrestrial, evergreen shrubs or trees (rarely half-woody subshrubs); stems monocaulous to branched, pachycaulous. Leaves alternate, petiolate, the bases clasping, with 1 to several spiral or transverse crests or collars, stipules ligulate; blades simple or palmately lobed to compound or pedate, elliptic to ovate, obovate, lanceolate, oblanceolate, oblong, or linear, membranaceous to subcoriaceous, margins entire or variously to toothed or pinnati-lobed; leaflets (if present) petiolulate. Inflorescence terminal, erect or somewhat pendant, compound umbellate (often on an elongated main axis), each ray terminating in three umbellules or capitula, the central unit formed of sterile flowers (which develop into seedless, baccate pseudo-fruits), and the two lateral units with perfect (or sometimes also staminate?) flowers; bracts present; pedicels slender to stout, or obsolete, unarticulated. Calyx obsolete; petals few to many, valvate, sympetalous with distinct lobes; stamens 4–30; carpels 1–25+, ovary inferior, styles absent (stigmas sessile and pustulate); the disc flattened. Fertile fruits drupes (sterile fruits berries), glabrous, globose to ellipsoid, terete to slightly compressed laterally, the endocarp crustaceous to woody; endosperm uniform or ruminate.

60 species, Malesia (esp., the Philippines), Taiwan, Melanesia (Vanuatu and esp. the Solomon Islands), and Micronesia (Palau and the Mariana Islands), understory trees in primary and secondary rainforests (sometimes rheophytic), low to mid elevations. Two subgroups (*Boerlagiodendron* and *Osmoxylon* s. str.), sometimes recognized as separate genera.

7. *Cephalaralia* Harms

Cephalaralia Harms, Bot. Jahrb. Syst. 23: 22 (1896).

Hermaphroditic, unarmed, pubescent, terrestrial, evergreen lianas; stems well branched and leptocaulous. Leaves alternate, petiolate, the bases clasping but not alate, exstipulate; blades palmately compound (usually ternately so); leaflets ovate, chartaceous, the margins entire (often with stout, ciliate trichomes), petiolulate. Inflorescences terminal on lateral shoots, ascending, paniculate, the ultimate units capitulate; bracts minute, persistent; pedicels obsolete. Calyx of 5 broadly triangular to rounded teeth; petals 5, imbricate, free; stamens 5; carpels 2, ovary inferior, styles free, the disc flattened to depressed. Fruits drupes, glabrous, obovoid, compressed laterally; endocarp crustaceous. Endosperm uniform.

One species, *C. cephalobotrys* (F. Muell.) Harms, Australia (E and NE), humid forests and forest edges.

8. ***Motherwellia*** F. Muell.

Motherwellia F. Muell., Fragm. 7: 107 (1870).

Hermaphroditic, unarmed, glabrous, terrestrial, evergreen lianas; stems well branched, leptocaulous. Leaves alternate, petiolate, the bases slightly clasping and ligulate; blades unifoliolate (to palmately bi- or trifoliolate); leaflets elliptical (to ovate), chartaceous, the margins entire and undulate, petiolulate. Inflorescences terminal on lateral shoots, erect, simple-umbellate; bracts very minute; pedicels long and slender, articulated. Calyx a low, crenate rim; petals 5, imbricate, free but long-connivent; stamens 5; carpels 2, ovaries inferior, styles fully connate, the disc flattened to depressed. Fruits drupes (but the mesocarp scarcely fleshy), glabrous, broadly obovate to broadly elliptical, strongly compressed laterally; endocarp crustaceous. Endosperm ruminate (with infolded margins).

One species, *M. haplosciadea* F. Muell., Australia (NE), closed forests.

9. ***Cheirodendron*** Nutt. ex Seem.

Cheirodendron Nutt. ex Seem., J. Bot. 5: 236 (1867); Lowry, Pacific Sci. 40: 79–87 (1986), reg. rev.; Lowry, Manual Fl. Pl. Hawaii 1: 225–228 (1990), reg. rev.

Andromonoecious, unarmed, glabrous, terrestrial, evergreen trees or shrubs; stems branched, pachycaulous, with a strong carrot-like odor. Leaves opposite, petiolate, bases clasping but not alate, exstipulate; blades ternately or palmately compound; leaflets broadly ovate to elliptic or subreniform, chartaceous to coriaceous, the margins entire to toothed, petiolulate (petiolules often laterally flattened). Inflorescences terminal and erect, oppositely branched panicles, ultimate units umbellules; bracts minute, those subtending the flowers subconnate and forming a pseudocalyx; pedicels slender, articulated. Calyx a short rim with 5 small lobes; petals 5, valvate, free; stamens 5; carpels 2–5, ovary inferior, styles partially or fully connate, the disc conical. Fruits drupes, glabrous, globose to orbicular, oblate, ovoid, or obovoid, terete or laterally compressed; endocarp cartilaginous. Endosperm uniform. $2n = 24$.

Six species, Hawai'i and Marquesas, in mesic to wet forests.

10. ***Raukaua*** Seem.

Raukaua Seem., J. Bot. 4: 352 (1866) ('Raukana'); Mitchell et al., N.Z. J. Bot. 35: 309–315 (1997), rev.

Hermaphroditic or possibly andromonoecious (with strong protandry), unarmed, glabrous or pubescent, terrestrial, evergreen shrubs or small to medium-sized trees; stems well branched, sometimes divaricating, leptocaulous. Leaves alternate, petiolate, bases scarcely clasping, stipules reduced or absent; juvenile leaves palmately compound (leaflets sessile), mature leaves usually unifoliolate by reduction, membranaceous or coriaceous, the margins toothed (or entire). Inflorescence terminal and axillary, erect, twice to thrice compound umbellate; bracts very small or lacking; pedicels slender to stout, articulated. Calyx minutely 5-toothed; petals 5, valvate, free; stamens 5; carpels 2–4, ovary inferior, styles free or basally connate, the disc hemispheric. Fruits drupes, glabrous, obloid to globose, terete or compressed laterally; endocarp crustaceous or hardened, the lateral face flat. Endosperm uniform. $2n = 24$.

Three species, New Zealand, low- to high-elevation forests and forest margins, and lowland scrubs. Frodin and Govaerts (2004) also included the Tasmanian and two Patagonian species of *Pseudopanax* in this genus, a treatment not followed here.

11. ***Cussonia*** Thunb.

Cussonia Thunb., Nova Acta Regiae Soc. Sci. Upsal. 3: 210 (1780); Tennant, Fl. Trop. E. Afr., Araliaceae, 2–11 (1968), reg. rev.; Bamps, Bull. Jard. Nat. Bot. Belg. 44: 102–120 (1974), reg. rev.

Hermaphroditic, unarmed, glabrous to pubescent, terrestrial, evergreen or deciduous shrubs or trees, or caudiciform geophytes; stems monocaulous to well branched, pachycaulous. Leaves alternate, often terminally clustered, petiolate, the bases usually sheathing but not alate, stipulate; blades simple to palmately lobed or compound, lobes or leaflets sometimes more highly dissected, coriaceous (to membranaceous); the margins entire to crenate or serrate, leaflets (when compound) petiolulate or sessile. Inflorescences terminal, erect, simple or compound umbellate with spicate or racemose ultimate units; bracts prominent to minute; pedicels absent to slender, unarticulated. Calyx 5-toothed, undulate, and/or forming a low rim; petals 5, valvate, free; stamens 5; carpels 2, ovary inferior, styles free or connate basally; the disc depressed, flattened, rounded, or conical. Fruits drupes, sometimes densely aggregated, glabrous, subglobose to ellipsoid, terete; endocarp crustaceous or indurated. Endosperm ruminate (rarely uniform).

20 species, Africa, Comoro Is., and SW Arabian peninsula, savannas to woodlands and forests.

12. ***Seemannaralia*** R. Vig.

Seemannaralia R. Vig., Ann. Sci. Nat., Bot., IX, 4: 118 (1906); Bamps, Bull. Jard. Nat. Bot. Belg. 44: 139 (1974), reg. rev.

Hermaphroditic, unarmed, pubescent (esp. in juvenile condition) to glabrous, terrestrial, deciduous trees; stems well branched, pachycaulous. Leaves alternate, petiolate, the bases slightly clasping and/or minutely alate, stipulate or exstipulate; blades simple and palmately lobed, ovate, membranaceous (to subcoriaceous), the margins sharply toothed. Inflorescences terminal, erect, paniculate, the ultimate units umbellules; bracts foliaceous below, reduced above; pedicels long and slender, unarticulated. Calyx lobes 5, short triangular; petals 5, imbricate, free, stamens 5; carpels 2 (but unilocular, with an incomplete septum); ovary inferior, styles 2, free; the disc conical. Fruit dry, glabrous, ellipsoid to obovoid, strongly compressed laterally with 4–5 ribs on each face, 1(2)-seeded, mesocarp thin (1–4 cells) and parenchymatous, endocarp sclerified (fibrous). Endosperm shallowly ruminate.

One species, *S. gerrardii* (Seem.) R. Vig., eastern South Africa, in forests and forest margins or open slopes. The relationship of this genus to other Araliaceae remains unclear, but both morphology/anatomy and molecular data point to an affinity with *Cussonia* (Burtt and Dickison 1975; Plunkett et al. 2004b).

13. ***Astropanax*** Seem.

Astropanax Seem., J. Bot. 3: 174 (1865); Bernardi, Candollea 24: 89–122 (1969); Bamps, Bull. Jard. Bot. Belg. 44: 101–139 (1974), reg. rev. (as *Schefflera*); Gostel et al., Pl. Ecol. Evol. 150: 87–108 (2017), phylog.; Lowry et al., Candollea 72: 265–282 (2017), rev.
Geopanax Hemsl. (1909).

Hermaphroditic, unarmed, mostly glabrous, terrestrial, hemi-epiphytic or epiphytic, evergreen trees or rarely lianas; stems sparsely to well branched or plants monocaulous. Leaves alternate, petiolate (rarely sessile), the bases clasping with connate, ligulate stipules; blades palmately compound (rarely unifoliolate); leaflets lanceolate, ovate, elliptical, oblanceolate, or obovate, coriaceous or subcoriaceous to membranaceous, margins entire, crenate, toothed or incised, apex acute to acuminate or caudate (rarely rounded). Inflorescences terminal, erect; paniculate or compound-umbellate, the ultimate units racemose, spicate, umbellate or capitate; bracts thin (rarely lacking); pedicels slender to stout, unarticulated, or flowers sessile. Calyx forming a low rim or with small lobes; petals 5–8, valvate, free or calyptrate; stamens isomerous with petals; carpels 5–9; ovary inferior, styles free or partially to mostly connate basally to forming a stylopodium or column. Fruits drupes, glabrous, globose, ellipsoid, urceolate, ovoid or obovoid, the endocarps crustaceous to boney. Endosperm uniform.

15 spp. Africa, Madagascar, Comoro Is., Seychelles.

14. ***Neocussonia*** (Harms) Hutch.

Neocussonia (Harms) Hutch., Gen. Fl. Pl., Dicot. 2: 79 (1967); Bernardi, Candollea 24: 89–122 (1969), reg. rev. (as *Schefflera*); Gostel et al., Pl. Ecol. Evol. 150: 87–108 (2017), phylog.; Lowry et al., Candollea 72: 265–282 (2017), rev.

Hermaphroditic, unarmed, glabrous, terrestrial, evergreen trees; stems sparsely to moderately branched or plants monocaulous. Leaves alternate, petiolate, the bases clasping with connate, ligulate stipules; blades palmately compound or unifoliolate; leaflets ovate, lanceolate, elliptical, obovate, oblanceolate, obcordate, obrhombic, obdeltoid or obhastate, coriaceous or subcoriaceous to membranaceous, margins entire, crenate, or serrate-crenate, apex very broadly acute to truncate or emarginate, sometimes rounded or acute to acuminate, generally retuse with an evident notch and a small mucro or minute boss in the sinus. Inflorescences terminal, erect; paniculate or paniculate-umbellate (rarely compound-umbellate), the ultimate units racemose, umbellate, or racemose-umbellate, most flowers in a terminal umbel and 1-several along the peduncle; bracts stiff; pedicels slender to stout, unarticulated, or flowers sessile. Calyx forming a low rim or with small lobes; petals 5, valvate, free; stamens isomerous with petals; carpels 2–5; ovary inferior, styles free, stylopodium short. Fruits drupes, glabrous, globose, ellipsoid, urceolate or ovoid, the endocarps crustaceous to boney. Endosperm uniform.

16 spp. Madagascar and Africa.

15. *Aralia* L. Fig. 60

Aralia L., Sp. Pl. 273 (1753); Wen, Brittonia 45: 47–55 (1993), rev.; Wen, Cathaya 13–14: 1–116 (2002), sect. rev.; Wen et al., Rheedia 12: 1–20 (2002), reg. rev.; Wen, Cathaya 15–16: 1–187 (2004), sect. rev.; Xiang & Lowry, Fl. China 13: 480–489 (2007), reg. rev.; Wen, Contr. U.S. Natl. Herb. 57: 1–172 (2011), sect. rev.
Sciadodendron Griseb. (1858).
Pentapanax Seem. (1864).
Parapentapanax Hutch. (1967).

Andromonoecious or sometimes hermaphroditic (occasionally dioecious), armed or unarmed, terrestrial or sometimes epiphytic, mostly deciduous shrubs, trees, lianas or herbs with rhizomes; stems branched to unbranched, pachycaulous. Leaves alternate, petiolate, the bases mostly stipulate, adnate to clasping; blades 1-4-pinnately compound, the rachis articulated; leaflets usually ovate, oblong, or elliptic, chartaceous to sometimes subcoriaceous or membranaceous, the margins entire to variously toothed, petiolulate or less commonly sessile. Inflorescence terminal or terminal and lateral, usually erect; paniculate, corymbose, or compound umbellate (occasionally simple umbellate), ultimate units umbellules or sometimes capitula or racemules; bracts triangular, lanceolate to semi-rounded; pedicels short to long, or lacking, usually slender, articulated (rarely unarticulated). Calyx lobes 5–10(–12), minutely toothed (rarely forming a low rim); petals 5–10(–12), imbricate (quincuncial), free; stamens 5–10(–12); carpels 5–10(–12), or occasionally aborted to 3, ovary inferior, styles free to basally or entirely connate; the disc flat or projected. Fruits berry-like drupes, glabrous, glo-

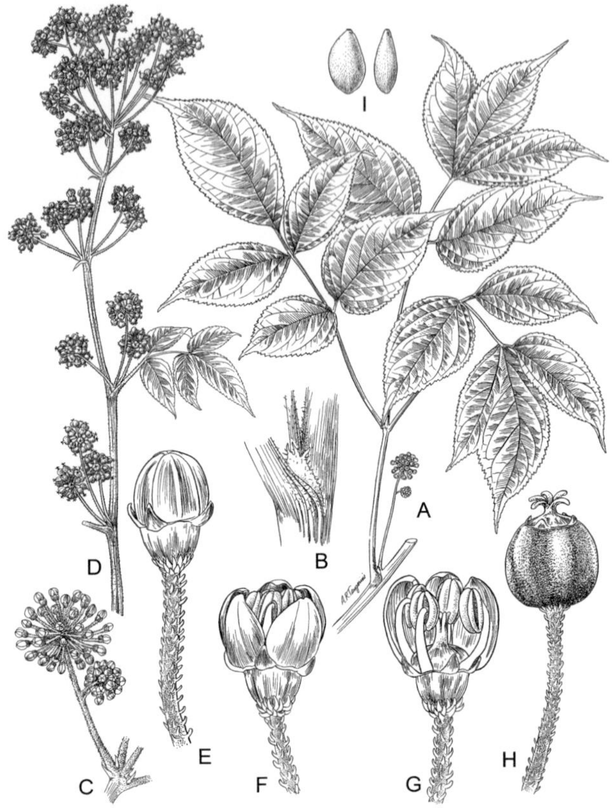

Fig. 60. Araliaceae. *Aralia tibetana*. **A** Leaf with axillary inflorescence. **B** Stipule. **C** Umbel, also showing bracts. **D** Inflorescence. **E** Floral bud. **F** Flower. **G** Flower showing anthers and gynoecium (two petals removed). **H** Fruit. **I** Seed, left showing face view, and right showing side view. (A, D, H & I: *Qinghai-Xizang Team 74-2440*, KUN; B, C & E–G: *Qinghai-Xizang Team 6340*, KUN.) (Wen 2011 (adapted from Fig. 30); illustrations by Alice Tangerini; with permission from the Smithsonian Institution)

bose; endocarp membranous. Endosperm uniform. $2n = 24, 48$.

74 species as defined by Wen (1993, 2011), disjunct in Asia and N and S America (Canada to SE Brazil and N Argentina).

16. *Panax* L.

Panax L., Sp. Pl. 1058 (1753); Smith, N. Amer. Fl. 28B(1): 9–10 (1944), reg. rev.; Wen, in Punja ZK, Utilization Biotechn., Genet. Cult. Approaches North Amer. Asian Ginseng Improvement, 67–88 (2001), rev.; Xiang & Lowry, Fl. China 13: 489–491 (2007), reg. rev.

Hermaphroditic or andromonoecious (rarely diphasic or "sex-changing", *P. trifolius* L.), unarmed, pubescent with simple hairs or glabrous, terrestrial, perennial herbs; stems unbranched, leptocaulous; rhizomes short and upright, or horizontally elongated with thick and short internodes, or horizontally elongated with slender and elongated internodes and globose nodes. Main root carrot-like or slightly fusiform or globose in shape, often decayed in older plants of most species. Leaves 2–5, whorled at the apex of the stem, petiolate, the bases slightly enlarged, exstipulate or occasionally with lanceolate stipules; blades palmately compound with 3–7 leaflets; leaflets ovate, oblong, obovate, to lanceolate, membranaceous to chartaceous, the margins entire, serrate, dentate, or pinnately lobed; petiolulate or occasionally sessile. Inflorescences terminal, erect, simple umbellate or a cluster of a few umbels; bracts triangular to lanceolate; pedicels medium, slender, articulated. Calyx 5-toothed; petals 5, imbricate (quincuncial), free; stamens 5; carpels 2–3(–5), ovary inferior, styles basally connate; the disc flat or convex. Fruits berry-like drupes, glabrous, bright red or red with a black top; endocarp crustaceous. Endosperm uniform. $2n = 24$, 48, occasionally 44.

About 18 species, disjunct in E Asia (about 15 species) and E North America (two species). *Panax ginseng* C.A. Mey., *P. notoginseng* (Burkill) F.H. Chen ex C.H. Chow, and *P. quinquefolius* L. are well-known herbal medicines.

17. ***Pseudopanax*** K. Koch

Pseudopanax K. Koch, Wochenschr. Gärtnerei Pflanzenk. 2: 366 (1859); Allan, Fl. New Zealand 1: 433–440 (1961), reg. rev.
Neopanax Allan (1961).

Dioecious, unarmed, glabrous, terrestrial, evergreen shrubs and trees; stems branched, pachycaulous. Leaves alternate, petiolate, the bases clasping, stipules present or absent; blades simple or palmately compound (with petiolulate to sessile leaflets), coriaceous, the margins toothed (or rarely entire). Inflorescences terminal and axillary, erect or pendant, thrice compound umbellulate (sometimes irregularly so), the ultimate units umbellules (or sometimes racemules of staminate flowers); bracts reduced or absent; pedicels slender to stout, articulated. Calyx a truncate rim or minutely 5-toothed; petals 5, valvate, free; stamens 5; carpels 2–5, ovary inferior, styles basally or entirely connate; the disc hemispherical. Fruits drupes, glabrous, obloid to globose, terete or compressed laterally; endocarp crustaceous or hardened, the lateral faces with deep hollows and grooves along the dorsal margin. Endosperm uniform. $2n = 48$.

12 species, New Zealand and temperate South America (Chile and Argentina). Species having 2 carpels and leaflets with distinct petiolules are sometimes treated in the segregate genus *Neopanax*, and in some treatments, the Tasmanian and two Patagonian species have been transferred to *Raukaua* (see Frodin and Govaerts 2004), but all these species are treated under *Pseudopanax* here.

18. ***Meryta*** J.R. Forst. & G. Forst. Fig. 61

Meryta J.R. Forst. & G. Forst., Char. Gen. Pl. 60 (1775); Cox, J. Arnold Arbor. 66: 133–121 (1985), reg. rev.; Lowry, Ann. Missouri Bot. Gard. 75: 389–391 (1988), reg. rev.; Lowry, Bull. Mus. Natl Hist. Nat., sect. B, Adansonia 11: 121–123 (1989), reg. rev.

Dioecious, unarmed, glabrous, terrestrial, evergreen small to large trees; stems monocaulous to well branched, pachycaulous. Leaves alternate, sessile to subsessile, less commonly distinctly petiolate, the bases clasping with persistent to caducous ligules; blades simple with entire (or pinnatifid in one species) and sometimes undulate margins, linear (often quite long) to obovate, coriaceous to membranaceous; midrib often exhibiting several swollen articulations. Inflorescences terminal, erect, racemose to paniculate, the ultimate units usually capitulate (less commonly spicate, or flowers rarely solitary); bracts conspicuous to small and scale-like, persistent or caducous, entire to dentate; pedicels lacking (flowers sessile) or rarely very short-pedicellate. Calyx obsolete or rarely with several minute triangular teeth; petals 3–13, valvate, free; stamens 3–13; carpels 4–13, ovary inferior, styles present, free and recurved; the disc flattened to subcallose. Fruits drupes, glabrous, ovoid to globose or oblate, terete and sometimes costate when dry, often partially to fully united (forming multiple fruits); endocarp crustaceous or indurated. Endosperm uniform. $2n = 48$.

38 species, Oceania (Melanesia, esp. New Caledonia; Polynesia; Micronesia), wet forests.

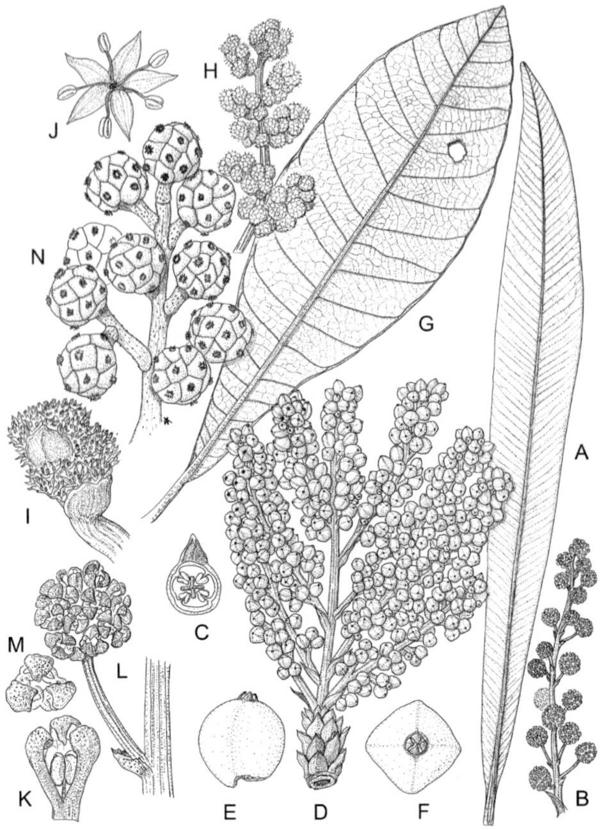

Fig. 61. Araliaceae. *Meryta*. **A–F** *M. coriacea*. **A** Leaf. **B** Branch of male inflorescence. **C** Male flower, cross section. **D** Infructescence. **E** Fruit, lateral view. **F** Fruit, top view. **G–N** *M. denhamii*. **G** Leaf. **H** Branch of male inflorescence. **I** Male umbellules at anthesis. **J** Male flower. **K** Male bud. **L** Female umbellule in late bud. **M** Female flower in late bud. **N** Part of infructescence. (Orig., illustrations by J.K. Myers; © P.P. Lowry II)

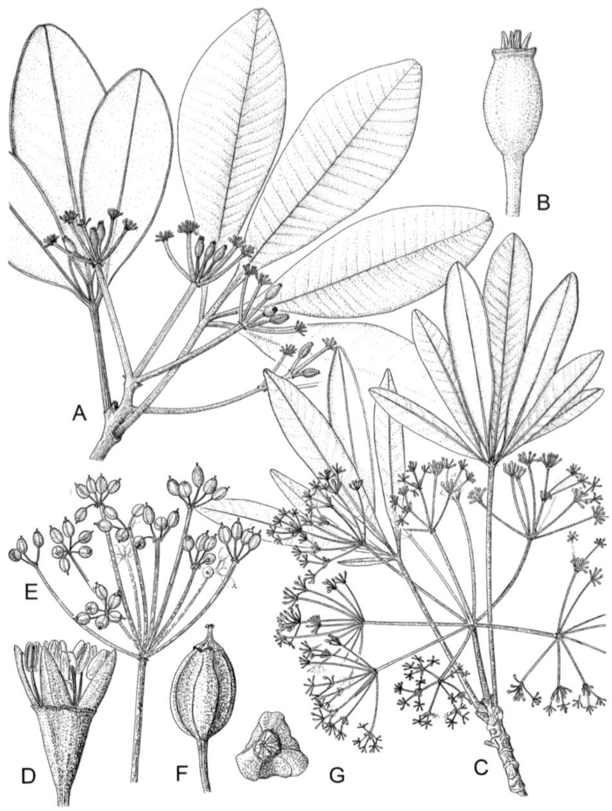

Fig. 62. Araliaceae. *Plerandra*. **A, B** *P. pachyphylla*. **A** Flowering branch with leaves. **B** Young fruit. **C–G** *P. veilloniorum*. **C** Flowering branch. **D** Male flower at anthesis. **E** Branch of infructescence. **F** Fruit, lateral view. **G** Fruit, top view. (Orig., illustrations by J.K. Myers; © P.P. Lowry II)

19. *Plerandra* A. Gray — Fig. 62

Plerandra A. Gray, U.S. Expl. Exped., Phan. 1: 729 (1854); Smith, Fl. Vitiensis Nova 3: 642–648 (1985), reg. rev.; Plunkett & Lowry, Syst. Bot. 37: 279–291 (2012), phylog.; Lowry et al., Brittonia 65: 42–61 (2013), rev.
Dizygotheca N.E. Br. (1892).
Octotheca R.Vig. (1906).

Hermaphroditic or andromonoecious, unarmed, glabrous, terrestrial (rarely epiphytic), evergreen trees; stems monocaulous to well branched, pachycaulous. Leaves alternate, petiolate, the bases clasping with connate, ligulate stipules; blades palmately compound (rarely bundle compound); leaflets ovate to elliptic or oblong, or oblanceolate to obovate or obtriangular, sometimes narrowly so (juvenile foliage often strikingly different); chartaceous to subcoriaceous or coriaceous or somewhat succulent; entire (rarely toothed), sessile or petiolulate. Inflorescences terminal or pseudolateral, erect or pendant, paniculate-racemose or compound-umbellate, the ultimate units umbellules; bracts present (often caducous); pedicels slender to stout, unarticulated. Calyx forming a low rim or with small lobes; petals (4)5(–8), valvate, free; stamens isomerous with petals or up to 250+ in 1–several series, thecae 2 or 4 per anther; carpels (2)3–19; ovary inferior, styles free to fully connate, or stigmas sessile; the disc depressed or flat to conical or nearly hemispherical. Fruits drupes; glabrous; cylindrical or obconical to obloid, globose, ellipsoid, ovoid, obovoid, or oblate-hemispherical; terete to slightly compressed laterally or weakly triangular when fresh (ribbed when dry); the endocarps crustaceous to boney. Endosperm uniform or ruminate.

33 spp. Melanesia, in humid, seasonal, and dry forests and primary scrubland (maquis). Commonly cultivated as ornamentals, especially *P. elegantissima* (Veitch ex Mast.) Lowry, G.M. Plunkett & Frodin and *P. veitchii* (hort. ex Carrière) Lowry, G.M. Plunkett & Frodin.

20. ***Polyscias*** J.R. Forst & G. Forst. Fig. 63

Polyscias J.R. Forst & G. Forst., Char. Gen. Pl.: 32 (1775); Smith & Stone, J. Arnold Arbor. 49: 431–493 (1968), reg. rev.; Bernardi, Candollea 26: 13–89 (1971), reg. rev.; Bamps, Bull. Jard. Nat. Bot. Belg. 44: 120–126 (1974), reg. rev.; Marais, Kew Bull. 39: 809–816 (1984), reg. rev.; Philipson, Fl. Malesiana 9: 53–86 (1979), reg. rev.; Smith, Fl. Vitiensis Nova 3: 634–642 (1985), reg. rev.; Lowry, Manual Fl. Pl. Hawaii 1: 228–237 (1990), reg. rev. (as *Munroidendron, Reynoldsia, Tetraplasandra*); Lowry, Bull. Mus. Natl Hist. Nat., sect. B, Adansonia 11: 137–152 (1989), reg. rev.; Xiang & Lowry, Fl. China 13: 472–473 (2007), reg. rev.; Plunkett & Lowry, Plant Div. Evol. 128: 23–54 (2010), phylog.; Lowry & Plunkett, Plant Div. Evol. 128: 55–84 (2010), rev.
Gastonia Comm. ex Lam. (1788).
Arthrophyllum Blume (1826).
Cuphocarpus Decne. & Planch. (1854).
Reynoldsia A. Gray (1854).
Tetraplasandra A. Gray (1854).
Nothopanax Miq. (1856).
Kissodendron Seem. (1865).
Eremopanax Baill. (1878).
Tieghemopanax R.Vig. (1905).
Munroidendron Sherff (1952).
Gelibia Hutch. (1967).

Fig. 63. Araliaceae. *Polyscias flynnii.* **A** Branch with inflorescence and infructescence. **B** Ultimate inflorescence unit. **C** Anthers. **D** Fruits. (Lowry and Wood 2000; illustrations by J.K. Myers; reproduced with permission of the Missouri Botanical Garden Press)

Hermaphroditic, andromonoecious, or dioecious, unarmed (rarely armed), glabrous or pubescent to stellate-pubescent, terrestrial (occasionally epiphytic), evergreen (rarely deciduous) small to large trees or shrubs (rarely lianas); stems monocaulous to well branched, pachycaulous. Leaves alternate, petiolate, the bases clasping and/or alate or neither clasping nor alate, exstipulate or with rim-like or rudimentary stipules; blades 1(–3)-pinnately compound or unifoliolate; leaflets linear or oblong to (ob-) lanceolate, to (ob-)ovate or elliptical, membranaceous or chartaceous to coriaceous (sometimes somewhat succulent), the margins entire to crenate or variously toothed or lobed; sessile or petiolulate. Inflorescences usually terminal or pseudolateral (sometimes both terminal and truly lateral), erect or pendant, paniculate, simple or compound umbellate, racemose, or verticillate, the ultimate units umbellules, capitula, racemules, spicules, or solitary; bracts absent or present (sometimes opposite), often foliaceous (and compound) below, becoming progressively reduced above; pedicels long and slender to thick and stout, articulated or unarticulated. Calyx rim entire, undulate, lobed, or minutely 4–9-toothed; petals 4–12(–20), valvate, free or less commonly calyptrate or connate in groups of 2–4(5); stamens 4–12(–93), isomerous or 2–several times the number of petals, in 1–several whorls; carpels (1)2–16(–24), ovary inferior (rarely half-inferior or superior), styles present and free or basally connate (with the free distal arms spreading to reflexed) to entirely connate into a (sometimes hollow) stylopodium, or styles absent and stigmas sessile (either capitate or forming a conical stylopodium); the disc depressed to flattened, conical, or nearly hemispherical and sometimes fleshy. Fruits drupes, glabrous to puberulent, densely pubescent, stellate-furfuraceous, or stellate-farinose, obloid to globose, globose-urceolate, cylindrical, ellipsoid, ovoid, or spherical (often asymmetric when unicarpellate), terete to angular or laterally compressed; endocarp chartaceous to cartilaginous, crustaceous or boney. Endosperm uniform to rugose or ruminate. $2n = 24, 48$.

210 species, paleotropical to subtropical, Oceania, SE Asia, Indian Ocean basin (esp. Madagascar), and Africa, in humid forests, seasonal forests, dry to semi-arid habitats, tropical montane forests. Ornamentals and living hedges, folk medicines, and timber trees for light construction.

21. ***Anakasia*** W.R. Philipson

Anakasia W.R. Philipson, Blumea 21: 87 (1973); Philipson, Fl. Malesiana 9: 89 (1979), rev.

Probably andromonoecious, unarmed, glabrous, terrestrial, evergreen shrubs or small trees; stems branched, pachycaulous. Leaves alternate, sessile or short-petiolate, the bases apparently neither clasping nor alate, exstipulate; blades simple, oblanceolate, the margins entire to undulate. Inflorescences axillary, racemose or once forked, the ultimate units umbellules; bracts present; pedicels very short, articulated. Calyx a low rim with 5–6 minute lobes; petals 5–6, valvate, free; stamens 5–6; carpels (4)5–6, ovary inferior, styles free, the disc depressed and fleshy. Fruits drupes, glabrous, broadly obovoid, terete; endocarp crustaceous. Endosperm uniform.

One species, *A. simplicifolia* Philipson, W New Guinea, in primary forests at low elevations.

22. ***Schefflera*** J.R. Forst. & G. Forst. Fig. 64

Schefflera J.R. Forst. & G. Forst., Char. Gen. Pl. 23 (1775), nom. cons.; Smith, N. Amer. Fl. 28B(1): 11–14, 25–29 (1944), reg. rev.; Macbride, Fl. Peru. 13: 26–39 (1959), reg. rev.; Smith & Stone, J. Arnold Arbor. 49: 477–489 (1968), reg. rev.; Bernardi, Candollea 24: 89–122 (1969), reg. rev.; Grushvitzky & Skvortsova, Adansonia II, 9: 369–387 (1969), reg. rev.; Frodin & Royen, Alpine Fl. New Guinea 4: 3040–3083 (1983), reg. rev.; Maguire et al., Mem. New York Bot. Gard. 38: 46–82 (1984), reg. rev.; Smith, Fl. Vitiensis Nova 3: 648–652 (1985), reg. rev.; Frodin, Proc. Acad. Nat. Sci. Philadelphia 138: 403–425 (1986), reg. rev.; Lowry, Bull. Mus. Natl Hist. Nat., sect. B, Adansonia 11: 123–135 (1989), reg. rev.; Frodin, Novon 3: 367–403 (1993), reg. rev.; Frodin, Revised Handbook Fl. Ceylon 10: 1–20 (1996), reg. rev.; Xiang & Lowry, Fl. China 13: 454–463 (2007), reg. rev.; Cannon & Cannon, Fl. Mesoamericana 4(1): 378–385 (2009), reg. rev.
Sciodaphyllum P. Browne (1756), nom. rej.
Heptapleurum Gaertn. (1791).
Brassaia Endl. (1839).
Didymopanax Decne. & Planch. (1854).
Tupidanthus Hook. f. & Thomson (1856).
Crepinella Marchal (1887).

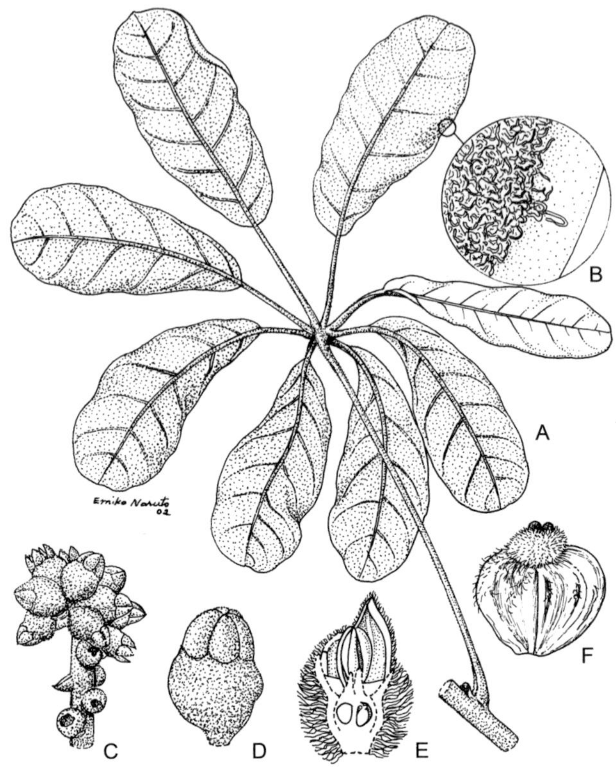

Fig. 64. Araliaceae. *Schefflera villosissima.* **A** Leaf. **B** Detail of indument on the abaxial surface of leaflet. **C** Ultimate inflorescence unit. **D** Floral bud. **E** Longitudinal section of the floral bud. **F** Fruit. (Fiaschi and Pirani 2005; illustrations by Emiko Naruto; reproduced with permission of the Missouri Botanical Garden Press)

Hermaphroditic or andromonoecious, unarmed, glabrous to pubescent, terrestrial, epiphytic or hemi-epiphytic, evergreen trees or shrubs (rarely lianas); stems monocaulous to well branched, pachycaulous. Leaves alternate, petiolate, the bases clasping with connate and often ligulate stipules; blades palmately compound to bundle compound (with leaflets arranged in two or more whorls or spirally from a single point), rarely unifoliolate or twice compound; leaflets linear to lanceolate, ovate, elliptical, obovate, or oblanceolate, membranaceous to coriaceous, entire to variously toothed or lobed, sessile or petiolulate. Inflorescences terminal or pseudolateral, erect or pendant, paniculate, racemose, or simple to compound umbellate, the ultimate units umbellules, capitula, racemules, or spicules; bracts present or absent; pedicels slender to stout; unarticulated. Calyx forming a low rim or with ca. 5 small teeth; petals (4–)5(–7+), valvate, free or occasionally calyptrate; stamens typically isomerous with petals, but up to 250+ in 1–several whorls; carpels (2–)5–20(–100+); ovary inferior to half-inferior, styles free to fully connate; the disc depressed to flat or nearly hemispherical. Fruits drupes, glabrous or pubescent, obloid to globose, ellipsoid, or ovoid, terete to laterally compressed, the endocarp crustaceous to boney. Endosperm uniform, rugose, or ruminate. $2n =$ 24, 48.

582 described species (with perhaps 300+ additional, undescribed species). Pantropical and subtropical, in humid forests, seasonal forests, and dry, semi-arid, tropical-montane, and tropical-alpine habitats. Commonly cultivated as ornamentals (especially *S. actinophylla* (Endl.) Harms and *S. arboricola* (Hayata) Merr.), and in some regions as timber trees (e.g., *S. schweliensis*) and folk medicines (e.g., *S. delavayi* (Franch.) Harms). The largest genus of the family (with many subgeneric and sectional groups recognized informally), but polyphyletic and until recently defined to include species comprising five distinct lineages, one each in the Neotropics, Africa-Madagascar, and SE Asia, plus two in the SW Pacific (see Plunkett et al. 2005). Of these, one of the Pacific clades has been reinstated as a distinct genus, *Plerandra* A. Gray (Lowry et al. 2013), and two subclades of the Africa-Madagascar clade have been recognized as *Astropanax* Seem. and *Neocussonia* (Harms) Hutch. (Lowry et al. 2017).

23. ***Oplopanax*** (Torr. & A. Gray) Miq.

Oplopanax (Torr. & A. Gray) Miq., Ann. Mus. Bot. Lugduno-Batavi 1: 16 (1863); Smith, N. Amer. Fl. 28B(1): 10–11 (1944), reg. rev.; Ohwi, Fl. Japan, 666 (1965); Xiang & Lowry, Fl. China 13: 441 (2007), reg. rev.
Echinopanax Decne. & Planch. ex Harms (1894).

Andromonoecious or hermaphroditic, armed, pubescent and setose, terrestrial, deciduous shrubs; stems usually unbranched, pachycaulous. Leaves alternate, petiolate, the bases sheathing, stipulate; blades simple, 5–9-palmately lobed, orbicular, chartaceous or membranaceous, margins serrate or irregularly toothed. Inflorescences terminal, erect, paniculate, the ultimate units umbellules; bracts present, membranaceous; pedicels medium to short, slender, unarticulated. Calyx nearly entire or with 5 minute teeth; petals 5, valvate, free; stamens 5; carpels 2, ovary inferior, styles basally connate to free; the disc flat to convex. Fruits drupes, glabrous, red, globose and slightly compressed laterally; endocarp crustaceous. Endosperm uniform. $2n = 48$.

Three species, NE Asia, North America (NW and Great Lakes region).

24. ***Tetrapanax*** (K. Koch) K. Koch

Tetrapanax (K. Koch) K. Koch, Wochenschr. Gärtnerei Pflanzenk. 2: 371 (1859); Xiang & Lowry, Fl. China 13: 440 (2007), reg. rev.

Hermaphroditic or andromonoecious, unarmed, densely stellate-pubescent, terrestrial, deciduous large shrubs to small trees; stems branched or unbranched, pachycaulous. Leaves alternate, long-petiolate, the bases clasping and alate, with conspicuous, well-developed elongated stipules; blades simple, palmately 7–11-lobed, subcoriaceous to thick chartaceous, the margins coarsely serrate or irregularly coarsely toothed. Inflorescences terminal, erect, paniculate, the ultimate units umbellules; bracts lanceolate to linear; pedicels medium in length, slender, unarticulated. Calyx nearly entire or obsolete; petals 4(5), valvate, free; stamens 4(5); carpels 2; ovary inferior, stylodia distinct, slender; the disc convex. Fruits drupes, nearly glabrous, globose, somewhat compressed laterally; endocarp crustaceous. Endosperm uniform. $2n = 48$.

One species, *T. papyrifer* (Hook.) K. Koch, Asia (Taiwan and C, S & SW China). The single species is the source plant of rice-paper and is commonly cultivated as an ornamental.

25. *Heteropanax* Seem.

Heteropanax Seem., Fl. Vit.: 114 (1866); Xiang & Lowry, Fl. China 13: 473–475 (2007), reg. rev.

Andromonoecious, unarmed, stellate-pubescent, terrestrial, evergreen shrubs or trees; stems branched, pachycaulous. Leaves alternate, petiolate, the bases stipulate and adnate; blades large, 2–4-pinnately compound; leaflets ovate, elliptic, or oblong, subcoriaceous to chartaceous, the margins entire, subsessile to petiolulate. Inflorescences terminal, erect or spreading, paniculate, the ultimate units umbellules; bracts triangular; pedicels short to medium in length, thick, unarticulated. Calyx of 5 minute teeth; petals 5, valvate, free; stamens 5; carpels 2, ovary inferior, stylodia connate to the middle or distinct and recurved, the disc convex. Fruits drupes, stellate-pubescent, globose, laterally compressed; endocarp crustaceous. Endosperm ruminate.

About eight species, Asia (S and SW China, E Himalaya, and Indochina).

26. *Fatsia* Decne. & Planch.

Fatsia Decne. & Planch., Rev. Hort. IV, 3: 105 (1854); Ohwi, Fl. Japan, 664 (1965); Xiang & Lowry, Fl. China 13: 439 (2007), reg. rev.

Andromonoecious, unarmed, glabrous or pubescent, evergreen small trees or large shrubs; stems branched, pachycaulous. Leaves alternate, petiolate, bases dilated, inconspicuously stipulate; blades simple, palmately 5–9-lobed, coriaceous, the margins serrate to serrulate, sometimes irregularly so. Inflorescences terminal, erect to pendant at the fruiting stage, paniculate, ultimate units umbellules; bracts caducous, membranous; pedicels relatively long, thick, articulated or unarticulated. Calyx truncate to shallowly 5-toothed; petals 5 (rarely 4 or 6), valvate, free; stamens 5; carpels 5 or 10; ovary inferior; stylodia distinct and slender; the disc convex. Fruits drupes, glabrous, globose or ovoid; endocarp crustaceous. Endosperm uniform. $2n = 24, 48$.

Four species, E. Asia (Japan, Korea, Taiwan, and the Ogasawara Islands). *Fatsia japonica* (Thunb.) Decne. & Planch. is widely cultivated as an ornamental.

27. *Oreopanax* Decne. & Planch.

Oreopanax Decne. & Planch., Rev. Hort. IV, 3: 107 (1854); Smith, N. Amer. Fl. 28B(1): 29–41 (1944), reg. rev.; Macbride, Fl. Peru 13: 11–25 (1959), reg. rev.; Nevling, Fl. Panama 46: 238–242 (1959), reg. rev.; Standley & Williams, Fl. Guatemala 8(1): 11–19 (1966), reg. rev.; Borchsenius, Nordic J. Bot. 17: 373–396 (1997), reg. rev.; Frodin, Fl. Venezuelan Guyana 3: 3–4 (1997), reg. rev.; Cannon & Cannon, Fl. Nicaragua 190–191 (2001), reg. rev.; Cannon & Cannon, Fl. Mesoamericana 4(1): 371–378 (2009), reg. rev.

Polygamodioecious, unarmed, pubescent or glabrous, terrestrial or epiphytic, deciduous or evergreen shrubs or trees; stems usually branched, pachycaulous. Leaves alternate, petiolate, the bases enlarged, stipulate or exstipulate; blades simple, palmately lobed or palmately compound, highly variable in shape, membranaceous, chartaceous to coriaceous, the margins entire or variously toothed. Inflorescences terminal, usually erect, paniculate, the ultimate units capitula; bracts well developed, variously shaped; pedicels lacking. Calyx obsolete or forming an annular margin; petals 5–7, valvate, free; stamens 5–7; carpels 2–12; ovary inferior, styles free or connate basally; the disc flat or convex. Fruits drupes, glabrous to stellate, globose or ovoid; endocarp crustaceous. Endosperm ruminate. $2n = 48$ (known only for *O. reticulatus* Decne. & Planch.).

About 85 species, C & S America and West Indies. Several species are cultivated as ornamentals.

28. *Sinopanax* H.L. Li

Sinopanax H.L. Li, J. Arnold Arbor. 30: 231 (1949); Xiang & Lowry, Fl. China 13: 439–440 (2007), reg. rev.

Hermaphroditic, unarmed, stellate-pubescent, terrestrial, deciduous shrubs or small trees; stems branched, pachycaulous. Leaves alternate, petiolate, the bases dilated with deciduous stipules; blades simple and palmately veined, broadly orbicular, subcoriaceous, the margins entire to palmately lobed or coarsely dentate. Inflorescences terminal, erect, paniculate, the ultimate units capitula, bracts three per flower (the central one larger), pedicels very short to obscure, unarticulated. Calyx a rim with 5 teeth;

petals 5, valvate, free; stamens 5; carpels 2, ovary inferior, styles short, free; the disc flattened to very-low conical. Fruits drupes, 1-seeded by abortion, stellate-pubescent, globose, terete. Endosperm ruminate.

One species, *S. formosanus* (Hayata) H.L. Li, Asia (Taiwan), in open areas of montane forests. Occasionally cultivated as an ornamental. Closely related to *Oreopanax* (Wen et al. 2001; Plunkett et al. 2004b).

29. ***Gamblea*** C.B. Clarke

Gamblea C.B. Clarke in J.D. Hooker, Fl. Brit. India 2: 739 (1879); Shang et al., Adansonia III, 22: 44–55 (2000), rev.; Xiang & Lowry, Fl. China 13: 453–454 (2007), reg. rev.
Evodiopanax (Harms) Nakai (1924).

Hermaphroditic, unarmed, glabrous, terrestrial (or occasionally epiphytic), evergreen shrubs to large trees; stems well branched, pachycaulous, with both long and short shoots. Leaves alternate, petiolate, the bases neither clasping nor alate, exstipulate; blades palmately compound (individual leaves rarely unifoliolate); leaflets elliptic to ovate, rhombic or trullate, chartaceous, subentire to serrulate, with domatia in axils of secondary veins on lower surface, margins often with ciliate, hispid teeth, petiolulate. Inflorescence terminal on short shoots, erect, simple or compound umbellate or paniculate, the ultimate units umbellules; bracts minute; pedicels long and slender, unarticulated. Calyx a low rim, the margin subentire or with 4–5 teeth; petals 4–5, valvate, free; stamens 4–5; carpels 2–4(5); ovary inferior; styles free to connate almost to the apex; the disc flattened to conical. Fruit drupes, glabrous, ellipsoid to globose or slightly obloid, terete or sometimes compressed laterally; endocarp crustaceous. Endosperm uniform. $2n = 48$.

Four species, Asia (E Himalaya to S China, Japan, SE Asia, Malesia), humid forests, low to montane elevations; used as timber trees and medicinally (esp. *G. ciliata* C.B. Clarke).

30. ***Chengiopanax*** C.B. Shang & J.Y. Huang

Chengiopanax C.B. Shang & J.Y. Huang, Bull. Bot. Res. (China) 13: 47 (1993); Xiang & Lowry, Fl. China 13: 454 (2007), reg. rev.

Andromonoecious, unarmed, nearly glabrous, terrestrial, deciduous trees; stems well branched, leptocaulous, with both long and short shoots. Leaves alternate, petiolate, the bases stipulate and adnate; blades palmately compound; leaflets (3–)5–7(–9), obovate or oblong to elliptic, chartaceous, the margins serrulate to aristate toothed, petiolulate. Inflorescence terminal, a corymbose panicle of umbels with a short primary axis, ultimate units umbellules; bracts caducous, triangular; pedicels relatively short and slender, unarticulated. Calyx lobes 5, triangularly toothed; petals 5, valvate, free; stamens 5; carpels 2, ovary inferior, styles connate, disc convex. Fruits drupes, glabrous, globose, transversely compressed laterally; endocarp crustaceous. Endosperm uniform. $2n = 48$.

Two species, Asia (Japan and C China).

31. ***Dendropanax*** Decne. & Planch. Fig. 65

Dendropanax Decne. & Planch., Rev. Hort. IV, 3:107 (1854); Smith, N. Amer. Fl. 28B(1): 14–25 (1944), reg. rev.; Macbride, Fl. Peru 13: 39–44 (1959), reg. rev.; Nevling, Fl. Panama 46: 229–234 (1959), reg. rev.; Standley & Williams, Fl. Guatemala 8(1): 4–8 (1966), reg. rev.; Frodin, Fl. Venezuelan Guyana 3: 2 (1997), reg. rev.; Cannon & Cannon, Fl. Nicaragua 189–190 (2001), reg. rev.; Xiang & Lowry, Fl. China 13: 442–446 (2007), reg. rev.; Cannon & Cannon, Fl. Mesoamericana 4(1): 365–371 (2009), reg. rev.

Hermaphroditic or andromonoecious, unarmed, glabrous, terrestrial, evergreen trees or shrubs; stems well branched, leptocaulous. Leaves alternate, petiolate, the bases stipulate and adnate or exstipulate; blades simple and unlobed or palmately lobed, ovate, elliptic, oblong, or lanceolate, coriaceous to chartaceous, commonly with transparent glandular dots, the margins entire to irregularly toothed. Inflorescences terminal, erect, simple (or compound) umbellate or paniculate; ultimate units umbellules or sometimes capitula; bracts often triangular to ovate; pedicels short (or rarely long), unarticulated. Calyx lobes 5–8, minutely toothed, sometimes forming a rim; petals 5–8, valvate, free; stamens 5–8; carpels 2–8, ovary inferior, styles basally to fully connate; the disc convex. Fruits drupes, glabrous, globose to ovoid-globose; endocarp crustaceous. Endosperm uniform.

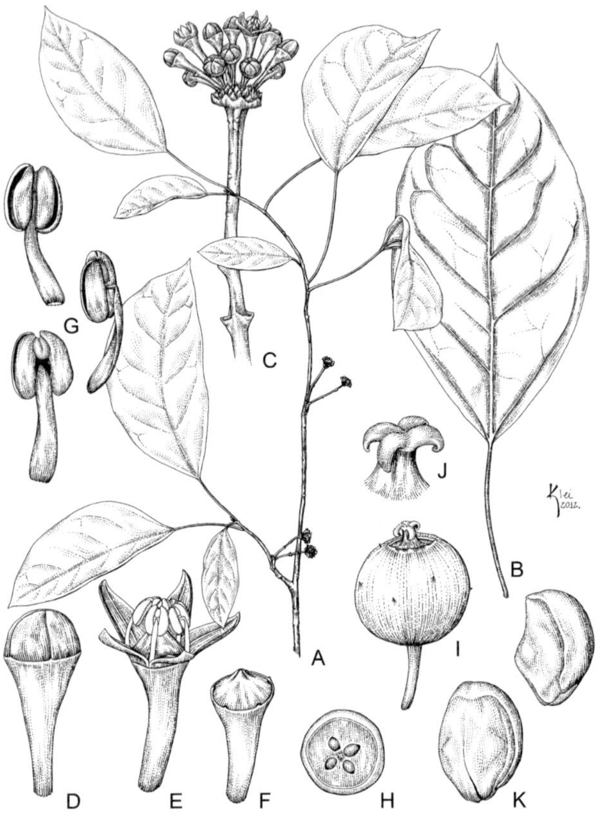

Fig. 65. Araliaceae. *Dendropanax fluminensis*. **A** Flowering branch. **B** Leaf, abaxial view. **C** Detail of a secondary inflorescence axis. **D** Flower bud, lateral view. **E** Flower, lateral view. **F** Detail of the disc. **G** Stamen, adaxial (top), lateral (middle) and abaxial (bottom) views. **H** Ovary, transverse section. **I** Fruit, lateral view. **J** Detail of the styles in fruit. **K** Pyrene, adaxial (left) and lateral (right) views. (A–H: *P. Fiaschi & G.M. Antar 3878*, SPF; I–K: *Braga et al. 2684*, SPF) (Fiaschi 2016; illustrations by Klei Souza; reproduced with permission of the New York Botanical Garden Press)

About 80 species, tropical and subtropical Asia (E, S and SE) and C to S America.

32. *Trevesia* Vis.

Trevesia Vis., Giorn. Tosc. Sci. Med. 1: 72 (1840) and Mem. Reale Accad. Sci. Torino, ser. 2, 4: 262 (1842); Jebb, Glasra 3: 85–113 (1998), rev.; Xiang & Lowry, Fl. China 13: 438–439 (2007), reg. rev.

Hermaphroditic, armed, glabrous or pubescent, terrestrial, evergreen shrubs or trees; stems monocaulous to sparsely branched, pachycaulous. Leaves alternate, petiolate, the bases stipulate and adnate; blades simple, palmately lobed to occasionally palmately compound, leaves of some species webbed-digitate, especially at the juvenile stage, orbicular in outline, subcoriaceous to coriaceous, the margins serrate. Inflorescences terminal, erect or pendant, paniculate, the ultimate units umbellules; bracts triangular to lanceolate; pedicels usually long, thick to stout, unarticulated. Calyx nearly entire or with (5–)6–12 obsolete teeth; petals (5–)6–12, valvate, free or calyptrate; stamens (5–)6–12; carpels (5–) 6–12, ovary inferior to half-inferior, styles connate into a single column; the disc convex. Fruits drupes, glabrous, globose or turbinate; endocarp crustaceous. Endosperm uniform.

About 15 species, Asia (S & SW China, NE India, Indochina to SE Asia).

33. *Brassaiopsis* Decne. & Planch.

Brassaiopsis Decne. & Planch., Rev. Hort. IV, 3: 106 (1854); Xiang & Lowry, Fl. China 13: 447–453 (2007), reg. rev.
Pseudobrassaiopsis R.N. Banerjee (1975).

Hermaphroditic or andromonoecious, armed, glabrous to tomentose, terrestrial, deciduous or evergreen shrubs or trees; stems sparsely branched (less commonly well branched), pachycaulous. Leaves alternate, petiolate, the bases stipulate and adnate; blades simple and palmately lobed, or palmately compound; lobes or leaflets chartaceous to coriaceous (leaflets petiolulate), the margins variously toothed to sometimes entire. Inflorescence terminal or axillary, erect or pendant, paniculate, the ultimate units umbellules of greatly varying sizes among species; bracts sometimes caducous, triangular to lanceolate; pedicels short to long, relatively thick, unarticulated. Calyx lobes 5, minutely toothed; petals 5, valvate, free; stamens 5; carpels mostly 2 (rarely 1 as in *B. simplex* (King) B.C. Stone, 3–5 as in *B. producta*, or 5 as in *B. grushvitzkyi*), ovary inferior to semi-inferior, styles connate into a single column; the disc convex. Fruits drupes, glabrous, subglobose or turbinate, terete; endocarp crustaceous. Endosperm ruminate, nearly ruminate, or uniform. $2n = 48$.

About 40 species, Asia (S and SW China, Himalayas, Indochina to W Malesia).

34. *Merrilliopanax* H.L. Li

Merrilliopanax H.L. Li, Sargentia 2: 62 (1942); Xiang & Lowry, Fl. China 13: 446–447 (2007), reg. rev.

Hermaphrodite to andromonoecious, unarmed, glabrous or stellately pubescent, terrestrial, evergreen shrubs or trees; stems branched, pachycaulous or leptocaulous. Leaves alternate, petiolate, stipules adnate to the petiole; blades simple, unlobed to 2–3-palmately lobed, elliptic, ovate, broadly ovate, to ovately orbicular, subcoriaceous to chartaceous, the margin serrate to serrulate. Inflorescences terminal, erect, paniculate, the ultimate units umbellules; bracts caducous, narrowly triangular to lanceolate; pedicels medium or long, slender, unarticulated. Calyx with 5 minute teeth; petals 5, valvate, free; stamens 5; carpels 2; ovary inferior; styles connate at the base to distinct; the disc convex. Fruits drupes, glabrous, ovoid or globose; endocarp crustaceous. Endosperm uniform. $2n = 48$.

About five species, Asia (Himalaya and W China).

35. *Hedera* L. Fig. 66

Hedera L., Sp. Pl.: 202 (1753); McAllister & Rutherford, Ivy J. 9: 45–54 (1983), reg. rev.; Rutherford. Ivy J. 10: 13–18 (1984), reg. rev.; Rutherford, Ivy J. 15: 7–17 (1989), reg. rev.; Xiang & Lowry, Fl. China 13: 441–442 (2007), reg. rev.

Hermaphroditic or andromonoecious, unarmed, glabrous to stellate-pubescent or scaly, terrestrial, evergreen lianas; stems well branched, pachycaulous. Leaves alternate, petiolate, the bases enlarged, exstipulate; blades simple, lobed or unlobed, coriaceous to subcoriaceous, the margins entire. Inflorescences terminal, erect, paniculate, the ultimate units umbellules; bracts triangular to narrowly so; pedicels medium to long, thick, unarticulated. Calyx obsolete, entire, or shallowly 5-toothed; petals 5, valvate, free; stamens 5; carpels 5, ovary inferior, styles connate, the disc convex. Fruits drupes, pubescent with stellate or scaly hairs, globose or ovoid; endocarp crustaceous. Endosperm ruminate. $2n = 24, 48, 144, 196$.

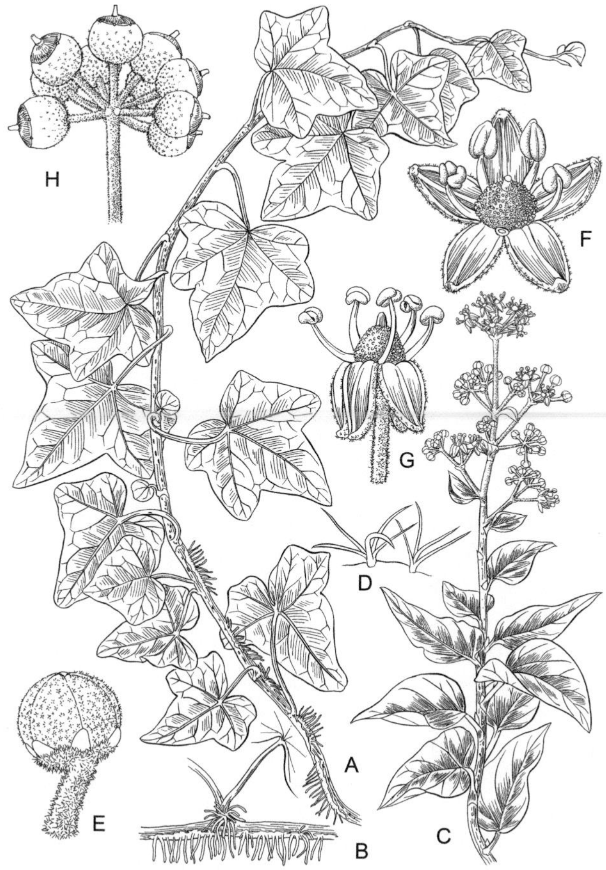

Fig. 66. Araliaceae. *Hedera helix*. **A** Vegetative shoot. **B** Adventitious roots on young stem. **C** Reproductive shoot with inflorescence bearing flowers and floral buds. **D** Stellate trichomes on young leaves. **E** Young floral bud. **F** Young flower. **G** Old flower. **H** Umbel in fruit. (Orig., illustrations by Alice Tangerini; © Smithsonian Institution, reproduced with permission)

About 15 species, Macaronesia, Europe, N. Africa to E and SE Asia. Several species widely cultivated as ornamentals.

36. *Eleutherococcus* Maxim.

Eleutherococcus Maxim., Mém. Acad. Imp. Sci. St.-Pétersbourg Divers Savans 9: 132 (1859); Kim & Sun, Novon 10: 209–214 (2000), rev. reg.; Xiang & Lowry, Fl. China 13: 466–472 (2007), reg. rev.
Acanthopanax (Decne. & Planch.) Miq. (1863).

Hermaphroditic or andromonoecious, armed, pubescent, terrestrial, deciduous shrubs (rarely small trees), sometimes scandent; stems branched, somewhat pachycaulous, often with both long and short shoots. Leaves alternate, petiolate, the bases not clasping, sometimes stipulate and adnate, or exstipulate; blades trifoliolate to palmately compound; leaflets 5–9, elliptic, ovate, lanceolate, oblong, obovate to oblanceolate, chartaceous to membranaceous, the margins serrulate, serrate or doubly so, to variously toothed. Inflorescences terminal, sometimes borne on short shoots, erect or pendant, simple umbellate, or paniculate to compound umbellate, the ultimate units umbellules or occasionally capitula; bracts triangular to lanceolate; pedicels long, slender to thick, or occasionally absent, unarticulated to slightly articulated. Calyx lobes 5, minutely toothed or forming a rim; petals 5 (rarely 4), valvate, free; stamens 5; carpels 2–5; ovary inferior, styles connate to free; the disc convex. Fruits drupes, glabrous, globose or laterally compressed-globose; endocarp crustaceous. Endosperm uniform. $2n = 48$.

About 35 species, Asia (from E Russia to Himalaya, China, Korea, Japan, S to Malaysia and N Philippines).

37. *Macropanax* Miq.

Macropanax Miq., Fl. Ned. Ind. 1: 763 (1856); Xiang & Lowry, Fl. China 13: 464–466 (2007), reg. rev.
Hederopsis C.B. Clarke (1879).

Andromonoecious, unarmed, slightly pubescent or glabrous, terrestrial, evergreen trees or shrubs; stems branched, leptocaulous. Leaves alternate, petiolate, the bases slightly enlarged, inconspicuously stipulate; blades palmately compound with 5–9 leaflets; leaflets elliptic, oblong, ovate to lanceolate, subcoriaceous to chartaceous, the margins serrate to serrulate, petiolulate. Inflorescences terminal, erect, paniculate, the ultimate units umbellules; bracts triangular; pedicels medium to short, thick, articulated. Calyx with 5 small teeth; petals 5, valvate, free; stamens 5; carpels 2(3), ovary inferior, styles connate, rarely divided at the tip; the disc convex. Fruits drupes, glabrous, globose or ovoid, terete; endocarp crustaceous. Endosperm ruminate. $2n = 48$.

About 15 species, Asia (E & SE Asia and Himalaya).

38. *Metapanax* Frodin ex J. Wen & Frodin

Metapanax Frodin ex J. Wen & Frodin, Brittonia 53: 117 (2001); Xiang & Lowry, Fl. China 13: 463–464 (2007), reg. rev.

Hermaphrodite, unarmed, glabrous or nearly so, terrestrial, evergreen shrubs or trees; stems often branched, leptocaulous. Leaves alternate, petiolate, the bases slightly enlarged, exstipulate; blades simple, unlobed, palmately lobed, or palmately compound with 2–5 leaflets; leaves or leaflets ovate, oblong, elliptic to lanceolate, subcoriaceous, the margins serrate to serrulate in upper portion, entire in lower portion. Inflorescences terminal, usually erect, paniculate, the ultimate units umbellules; bracts caducous; pedicels medium, slender, articulated. Calyx of 5 teeth; petals 5, valvate, free; stamens 5; carpels 2 (–4); ovary inferior; styles to 3/4-connate; the disc flat to convex. Fruits drupes, globose, somewhat laterally flattened; endocarp crustaceous. Endosperm uniform. $2n = 48$.

Two species, Asia (C and SW China and northern Vietnam).

39. *Kalopanax* Miq.

Kalopanax Miq., Ann. Mus. Bot. Lugduno-Batavi 1: 16 (1863); Ohwi, Fl. Japan, 666–667 (1965); Xiang & Lowry, Fl. China 13: 441 (2007), reg. rev.

Hermaphroditic to andromonoecious, armed, slightly pubescent, terrestrial, deciduous trees; stems well branched, pachycaulous, with both long and short shoots. Leaves alternate, petiolate, stipules adnate to the petiole; blades simple, palmately lobed with 5–9 triangular-ovate to oblong-ovate lobes, chartaceous, margins serrulate. Inflorescences terminal, a corymbose panicle with umbellately arranged branches on a short main axis, the ultimate units umbellules; bracts minute, triangular; pedicels relatively long and slender, unarticulated. Calyx with 5 minute, triangular teeth; petals 5, valvate, free; stamens 5; carpels 2, ovary inferior; styles connate but divided at the tip; the disc convex. Fruits drupes, glabrous, globose, laterally compressed; endocarp crustaceous. Endosperm uniform. $2n = 48$.

One species, *K. septemlobus* (Thunb.) Koidz., E Asia (China, Japan, Korea, and far-eastern Russia). Cultivated as an ornamental and sometimes used as a timber tree.

40. ***Woodburnia*** Prain

Woodburnia Prain, J. Asiat. Soc. Bengal, Pt. 2, Nat. Hist. 73(2): 23 (1904).

Hermaphroditic, armed, stellate-pubescent, terrestrial small trees; stems unbranched. Leaves alternate, petiolate, the bases broadly alate with entire, cuneate wings; blades palmately compound, the leaflets elliptic, subcoriaceous, serrate, petiolules long and bearing broad, obtriangular, laminar and entire interpetiolular wings. Inflorescences axillary, simple umbellate on a long, stout, tomentose, pendulous peduncle; bracts large, lanceolate; pedicels long and stout, unarticulated. Calyx tomentose, with 7–10 linear, elongated lobes; petals 5, valvate, free, abaxially pubescent; stamens 5; carpels 8–13, ovary inferior, tomentose, styles connate into a cylindrical stylopodium; the disc thick and flat. Fruits (likely fleshy) and seeds unknown.

One species, *W. penduliflora* Prain, Asia (Upper Burma). Known only from the type, its relation to the rest of family remains obscure.

Selected Bibliography

Bentham, G. 1867. Araliaceae. In: Bentham, G., Hooker J. D. (eds.) Genera Plantarum, vol. 1. London: A. Black, W. Pamplin, Lovell Reeve & Co., pp. 931–947.

Berry, E.W. 1903. *Aralia* in American paleobotany. Bot. Gaz. 36: 421–428.

Blackburn, D.T. 1981. Tertiary megafossil flora of Maslin Bay, South Australia: numerical taxonomic study of selected leaves. Alcheringa 5: 9–28.

Blackburn, D.T., Sluiter, I.R.K. 1994. The Oligocene-Miocene coal floras of southeastern Australia. In: Hill, R. S. (ed.) History of the Australian vegetation: Cretaceous to Recent. New York: Cambridge University Press, pp. 328–367.

Boesewinkel, F.D., Bouman, F. 1984. The seed: structure. In: Johri, B.M. (ed.) Embryology of Angiosperms. Berlin: Springer, pp. 567–610.

Bondarenko, O.V. 2008. Fossil wood of *Eleutherococcus* from Southern Primorye (Russian Far East). In: Pimenov, M.G., Tilney, P.M. (eds.) Apiales 2008: The program and proceedings of the 6th International Symposium on Apiales. Moscow: KMK Sci. Press, pp. 24–25.

Bouman, F. 1984. The ovule. In: Johri, B.M. (ed.) Embryology of Angiosperms. Berlin: Springer, pp. 123–157.

Burtt, B.L., Dickison, W.C. 1975. The morphology and relationships of *Seemannaralia* (Araliaceae). Notes Roy. Bot. Gard. Edinb. 33: 449–466.

Calestani, V. 1905. Contributo alla sistematica delle Ombrellifere d'Europa. Webbia 1: 89–280.

Carpenter, R.J., Hill, R.S., Scriven, L.J. 2006. Palmately lobed Proteaceae leaf fossils from the middle Eocene of South Australia. Int. J. Plant Sci. 167: 1049–1060.

Chandler, G.T., Plunkett, G.M. 2004. Evolution in Apiales: nuclear and chloroplast markers together in (almost) perfect harmony. Bot. J. Linn. Soc. 144: 123–147.

Corner, E.J.H. 1976. The seeds of dicotyledons, vol. 1. Cambridge: Cambridge University Press.

Costello, A., Motley, T.J. 2004. The development of the superior ovary in *Tetraplasandra* (Araliaceae). Amer. J. Bot. 91: 644–655.

Costion, C.M., Plunkett, G.M. 2016. A revision of the genus *Osmoxylon* (Araliaceae) in Palau, including two new species. PhytoKeys 58: 49–64.

Cronquist, A. 1981. An integrated system of classification of flowering plants. New York: Columbia University Press.

Davila, Y.C., Wardle, G.M. 2008. Variation in native pollinators in the absence of honeybees: implications for reproductive success of an Australian generalist-pollinated herb *Trachymene incisa* (Apiaceae). Bot. J. Linn. Soc. 156: 479–490.

Davis, G.L. 1966. Systematic embryology of the angiosperms. New York: Wiley.

Dilcher, D.L., Dolph, G.E. 1970. Fossil leaves of *Dendropanax* from Eocene sediments of southeastern North America. Amer. J. Bot. 57: 153–160.

Erbar, C., Leins, P. 2004. Sympetaly in Apiales (Apiaceae, Araliaceae, Pittosporaceae). S. African J. Bot. 70: 458–467.

Esau, K. 1940. Developmental anatomy of the fleshy storage organ of *Daucus carota*. Hilgardia 13: 175–209.

Eyde, R.H., Tseng, C.C. 1969. Flower of *Tetraplasandra gymnocarpa*: hypogyny with epigynous ancestry. Science 166: 506–508.

Eyde, R.H., Tseng, C.C. 1971. What is the primitive floral structure of Araliaceae? J. Arnold Arbor. 52: 205–239.

Fiaschi, P. 2002. Estudo taxonômico do gênero *Schefflera* J.R. Forst. & G. Forst. (Araliaceae) na região Sudeste do Brasil. Dissertação de Mestrado, Universidade de São Paulo.

Fiaschi, P. 2016. A new species of *Dendropanax* (Araliaceae) from the Brazilian Atlantic Forest. Brittonia 68: 103–110.

Fiaschi, P., Pirani, J.R. 2005. Three new species of *Schefflera* J.R. Forst. & G. Forst. (Araliaceae) from Espinhaço Range, Minas Gerais, Brazil. Novon 15: 117–122.

Fiaschi, P., Plunkett, G.M. 2011. Monophyly and phylogenetic relationships of Neotropical *Schefflera* (Araliaceae) based on plastid and nuclear markers. Syst. Bot. 36: 806–817.

Fiaschi, P., Santos, F.A.R., Westbrook, E., Plunkett, G.M. 2010. Taxonomic significance of pollen morphology in Neotropical *Schefflera* (Araliaceae). Plant Div. Evol. 128: 297–323.

French, D.H. 1971. Ethnobotany of the Umbelliferae. In: Heywood, V.H. (ed.) The biology and chemistry of the Umbelliferae. Bot. J. Linn. Soc. 64, Suppl. 1: 385–412.

Frodin, D.G., Govaerts, R. 2004. World Checklist and Bibliography of Araliaceae. Richmond: RBG Kew Publishing.

Frodin, D.G., Lowry II, P.P., Plunkett, G.M. 2010. *Schefflera* (Araliaceae): taxonomic history, overview, and progress. Plant Div. Evol. 128(3–4): 561–595.

Gillespie, L.H., Henwood, M.J. 1994. Temporal changes of floral nectar-sugar composition in *Polyscias sambucifolia* (Sieb. ex DC.) Harms (Araliaceae). Ann. Bot. 74: 227–231.

Gostel, M.R., Plunkett, G.M., Lowry II, P.P. 2017. Straddling the Mozambique Channel: molecular evidence for two major clades of Afro-Malagasy *Schefflera* (Araliaceae) co-occurring in Africa and Madagascar. Pl. Ecol. Evol. 150: 87–108.

Graham, A. 1999. Late Cretaceous and Cenozoic history of North American vegetation. New York: Oxford University Press.

Gruas-Cavagnetto, C., Bui, N.-S. 1976. Présence de pollen d'Araliacées dans le Paléogène Anglais et Français. Rev. Palaeobot. Palynol. 22: 61–72.

Gruas-Cavagnetto, C., Cerceau-Larrival, M.-T. 1982. Présence de pollens d'Ombellifères fossiles dans le Paléogène du bassin Anglo-Parisien: premiers résultats. In: Cauwet, A.M., Carbonnier, J. (eds.) Contributions Pluridisciplinaires à la Systématique; actes du 2ème Symposium International sur les Ombellifères, Centre Universitaire de Perpignan. St. Louis: Missouri Botanical Garden, pp. 255–267.

Harms, H. 1894 and 1897. Araliaceae. In: Engler, A., Prantl, K. (eds.) Die natürlichen Planzenfamilien III, vol. 8. Leipzig: W. Engelmann, pp. 1–62.

Hart, J.M., Henwood, M.J. 2006. A revision of Australian *Trachymene* (Apiaceae: Hydrocotyloideae). Austr. Syst. Bot. 19: 11–55.

Hegnauer, R. 1964. Araliaceae. In: Chemotaxonomie der Pflanzen, vol. III. Basel: Birkhäuser, pp. 173–184.

Hegnauer, R. 1971. Chemical patterns and relationships of Umbelliferae. In: Heywood, V.H. (ed.) The biology and chemistry of the Umbelliferae. Bot. J. Linn. Soc. 64, Suppl. 1. London & New York: Academic Press, pp. 267–277.

Hegnauer, R. 1989. Araliaceae. In: Chemotaxonomie der Pflanzen, vol. VIII. Basel: Birkhäuser, pp. 65–75, 699–700.

Henwood, M.J. 1986. The breeding system of the polymorphic *Polyscias sambucifolia* (Sieb. ex DC.) Harms (Araliaceae). In: Williams, E.G, Knox, R.B., Irvine, D. (eds.) Pollination '86. Melbourne: University of Melbourne School of Botany, pp. 70–78.

Henwood, M.J. 1991. Pollen morphology of *Polyscias* (Araliaceae) – the Malesian and Australian species. Grana 30: 559–576.

Hoar, C.S. 1915. A comparison of the stem anatomy of the cohort Umbelliflorae. Ann. Bot. 29: 55–63.

Hutchinson, J. 1967. The genera of flowering plants, vol. 2. London: Oxford University Press.

Jay, M. 1969. Chemotaxonomic researches on vascular plants. XIX. Flavonoid distribution in the Pittosporaceae. Bot. J. Linn. Soc. 62: 423–429.

Jebb, M.P. 1998. A revision of the genus *Trevesia* (Araliaceae). Glasra 3: 85–113.

Jurica, H.S. 1922. A morphological study of the Umbelliferae. Bot. Gaz. 74: 292–307.

Knuth, P. 1908. Handbook of flower pollination, vol. 2 (translated by Davis, J.R.A.). Oxford: Clarendon Press.

Kotina, E., Oskolski, A.A. 2010. Survey of the bark anatomy of Araliaceae and related taxa. Plant Div. Evol. 128: 455–489.

Lewis, W.H., Zenger, V.E. 1983. Breeding systems and fecundity in the American ginseng, *Panax quinquefolium* (Araliaceae). Amer. J. Bot. 70: 466–468.

Li, R., Wen, J. 2013. Phylogeny and biogeography of *Dendropanax* (Araliaceae), an Amphi-Pacific disjunct genus between tropical/subtropical Asia and the Neotropics. Syst. Bot. 38: 536–551.

Li, R., Wen, J. 2014. Phylogeny and biogeography of Asian *Schefflera* (Araliaceae) based on nuclear and plastid DNA sequences data. J. Syst. Evol. 52: 431–449.

Lowry II, P.P. 1990. Araliaceae, ginseng family. In: Wagner, W.L., Herbst, D.R., Sohmer, S.H. (eds.) Manual of the flowering plants of Hawai'i, vol. 1. Honolulu: University of Hawaii Press and Bishop Museum Press, pp. 224–237.

Lowry II, P.P., Plunkett, G.M. 2010. Recircumscription of *Polyscias* (Araliaceae) to include six related genera, with a new infrageneric classification and a synopsis of species. Plant Div. Evol. 128: 55–84.

Lowry II, P.P., Wood, K.R. 2000. A new species of *Tetraplasandra* (Araliaceae) from Kaua'i, Hawaiian Islands. Novon 10: 40–44.

Lowry II, P.P., Plunkett, G.M., Frodin, D.G. 2013. Revision of *Plerandra* (Araliaceae). I. A synopsis of the genus with an expanded circumscription and a new infrageneric classification. Brittonia 64: 42–61.

Lowry II, P.P., Plunkett, G.M., Gostel, M.R., Frodin, D.G. 2017. A synopsis of the Afro-Malagasy species previously included in *Schefflera* J.R. Forst. & G. Forst. (Araliaceae): resurrection of the genera *Astropanax* Seem. and *Neocussonia* (Harms) Hutch. Candollea 72: 265–282.

Lubbock, J. 1892. A contribution to our knowledge of seedlings. London: Kegan Paul, Trench and Trübner.

Mabberley, D.J. 2008. Mabberley's Plant-book (3rd ed.). Cambridge: Cambridge University Press.

Maguire, B., Steyermark, J.A., Frodin, D.G. 1984. Araliaceae. In: Maguire, B., Cowan, R.S., Wurdack, J.J. et al., The Botany of the Guayana Highland – Part XII. Mem. New York Bot. Gard. 38: 46–82.

Manchester, S.R. 1994. Fruits and seeds of the middle Eocene Nut Beds flora, Clarno Formation, Oregon. Palaeontograph. Amer. 58: 1–205.

Manchester, S.R., Grímsson, F., Zetter, R. 2015. Assessing the fossil record of Asterids in the context of our current phylogenetic framework. Ann. Missouri Bot. Gard. 100: 329–363.

Meier, A.J., Bratton, S.P., Duffy, D.C. 1995. Possible ecological mechanisms for loss of vernal-herb diversity in logged eastern deciduous forests. Ecol. Appl. 5: 935–946.

Mendoza, M., Fuentes, A.F. 2010. *Hydrocotyle apolobambensis* (Apiaceae), una especie neueva andina del Noroeste de Bolivia. Novon 20(3): 303–306.

Metcalfe, D.J. 2005. *Hedera helix* L. (Biological flora of the British Isles). J. Ecol. 93: 632–648.
Metcalfe, C.R., Chalk, L. 1983. Anatomy of the Dicotyledons, 2nd ed. Oxford: Clarendon Press.
Mitchell, A., Wen, J. 2004. Phylogeny of the Asian core Araliaceae clade based on granule-bound starch synthase I (*GBSSI*) sequence data. Taxon 53: 29–41.
Mitchell, A., Li, R., Brown, J.W., Schoenberger, I., Wen, J. 2012. Ancient divergence and biogeography of *Raukaua* (Araliaceae) and close relatives in the southern hemisphere. Austral. Syst. Bot. 25: 432–446.
Mittal, S.P. 1961. Studies in the Umbellales. II. The vegetative anatomy. J. Indian Bot. Soc. 40: 424–443.
Mohana Rao, P.R. 1972. Morphology and embryology of *Tieghemopanax sambucifolius* with comments on the affinities of the family Araliaceae. Phytomorphology 22: 75–87.
Muller, J. 1981. Fossil pollen record of extant angiosperms. Bot. Rev. 47: 1–142.
Nicolas, A.N., Plunkett, G.M. 2009. The demise of subfamily Hydrocotyloideae (Apiaceae) and the re-alignment of its genera across the entire order Apiales. Molec. Phylog. Evol. 53: 134–151.
Nicolas, A.N., Plunkett, G.M. 2014. Diversification times and biogeographic patterns in Apiales. Bot. Rev. 80: 30–58.
Oskolski, A.A. 1996. A survey of the wood anatomy of the Araliaceae. In: Donaldson, L.A., Singh, A.P., Butterfield, B.G., Whitehouse, L.J. (eds.) Recent advances in wood anatomy. Rotorua: New Zealand Forest Institute, pp. 99–119.
Oskolski, A.A., Lowry II, P.P. 2001. Wood anatomy of *Schefflera* and related taxa (Araliaceae): II. Systematic wood anatomy of New Caledonian *Schefflera*. IAWA J. 22: 301–330.
Oskolski, A.A., Sokoloff, D.D., Van Wyk, B.-E. 2008. Floral morphology of *Seemannaralia* (Araliaceae): from bilocular ovary to unilocular fruit. In: Pimenov, M. G., Tilney, P.M. (eds.) Apiales 2008: The program and proceedings of the 6th International Symposium on Apiales. Moscow: KMK Sci. Press, pp. 98–100.
Philipson, W.R. 1970. Constant and variable features of the Araliaceae. In: Robson, N.K.B., Cutler, D.F., Gregory, M. (eds.) New research in plant anatomy. Bot. J. Linn. Soc. 63, Suppl. 1: 87–100.
Philipson, W.R. 1973. A revision of *Harmsiopanax* (Araliaceae). Blumea 21: 81–86.
Philipson, W.R. 1978. Araliaceae: growth forms and shoot morphology. In: Tomlinson, P.B., Zimmermann, M. H. (eds.) Tropical trees as living systems. Cambridge: Cambridge University Press, pp. 269–284.
Philipson, W.R. 1979. Araliaceae. Flora Malesiana ser. 1, 9 (1): 1–105.
Pimenov, M.G., Vasil'eva, M.G., Leonov, M.V., Daushkevich, J.V. 2003. Karyotaxonomical analysis in the Umbelliferae. Enfield, New Hampshire: Science Publishers Inc.
Pire, S.M. 1989. Morfologica polinica de las Araliaceaes de Argentina. Bonplandia 6: 133–150.
Plunkett, G.M. 2001. The relationship of the order Apiales to subclass Asteridae: a re-evaluation of morphological characters based on insights from molecular data. Edinb. J. Bot. 58: 183–200.
Plunkett, G.M., Lowry II, P.P. 2010. Paraphyly and polyphyly in *Polyscias* sensu lato: molecular evidence and the case for recircumscribing the "pinnate genera" of Araliaceae. Plant Div. Evol. 128: 23–54.
Plunkett, G.M., Lowry II, P.P. 2012. Phylogeny and diversification in the Melanesian *Schefflera* clade (Araliaceae) based on evidence from nuclear rDNA spacers. Syst. Bot. 37: 279–291.
Plunkett, G.M., Soltis, D.E., Soltis, P.S. 1996. Higher level relationships of Apiales (Apiaceae and Araliaceae) based on phylogenetic analysis of *rbc*L sequences. Amer. J. Bot. 83: 499–515.
Plunkett, G.M., Chandler, G.T., Lowry II, P.P., Pinney, S. M., Sprenkle, T.S. 2004a. Recent advances in understanding Apiales and a revised classification. S. African J. Bot. 70: 371–381.
Plunkett, G.M., Wen, J., Lowry II, P.P. 2004b. Infrafamilial relationships in Araliaceae: insights from plastid (*trnL-trnF*) and nuclear (ITS) sequence data. Plant Syst. Evol. 245: 1–39.
Plunkett, G.M., Lowry II, P.P., Frodin, D.G., Wen, J. 2005. Phylogeny and geography of *Schefflera*: pervasive polyphyly in the largest genus of Araliaceae. Ann. Missouri Bot. Gard. 92: 202–224.
Pombal, E.C.P., Morellato, L.P.C. 1995. Polinização por moscas em *Dendropanax cuneatum* Decne. & Planch. (Araliaceae) em floresta semidecídua no sudeste do Brasil. Revista Brasil. Bot. 18: 157–162.
Rásky, K. 1959. The fossil flora of Ipolytarnóc (preliminary report). J. Paleontol. 33: 453–461.
Rodríguez, R.L. 1971. The relationships of the Umbellales. In: Heywood, V.H. (ed.) The biology and chemistry of the Umbelliferae. Bot. J. Linn. Soc. 64, Suppl. 1: 63–91.
Schlessman, M.A. 1990. Phenotypic gender in sex changing dwarf ginseng, *Panax trifolium* (Araliaceae). Amer. J. Bot. 77: 1125–1131.
Schlessman, M.A. 1991. Size, gender and sex change in dwarf ginseng, *Panax trifolium* (Araliaceae). Oecologia 87: 588–595.
Schlessman, M.A., Lloyd, D.G., Lowry II, P.P. 1990a. Evolution of sexual systems in New Caledonian Araliaceae. Mem. New York Bot. Gard. 55: 105–117.
Schlessman, M.A., Lowry II, P.P., Lloyd, D.G. 1990b. Functional dioecism in the New Caledonian endemic *Polyscias pancheri* (Araliaceae). Biotropica 22: 133–139.
Schlessman, M.A., Plunkett, G.M., Lowry II, P.P., Lloyd, D.G. 2001. Sexual systems of New Caledonian Araliaceae: a preliminary phylogenetic reappraisal. Edinb. J. Bot. 58: 221–228.
Seemann, B. 1868. Revision of the natural order of Hederaceae. London: L. Reeve & Co.
Shoup, J.R., Tseng, C.C. 1977. A palynological study of *Schefflera paraensis* Huber ex Ducke (Araliaceae). Grana 16: 81–84.
Solereder, H. 1908. Systematic anatomy of the dicotyledons, vol. 1. Oxford: Clarendon Press.
Song, Z.-C., Wang, W.-M., Huang, F. 2004. Fossil pollen records of extant angiosperms in China. Bot. Rev. 70: 425–458.
Sosa, V. 1983. Características palinológicas de las araliáceas de México. Bol. Soc. Bot. México 45: 117–132.

Stevens, P.F. 2001 onwards. Angiosperm Phylogeny Website. http://www.mobot.org/MOBOT/research/APweb/

Thomson, J.D., Barrett, S.C.H. 1981. Temporal variation of gender in *Aralia hispida* Vent. (Araliaceae). Evolution 35: 1094–1107.

Ting, W.S., Tseng, C.C., Mathias, M.E. 1964. A survey of pollen morphology of Hydrocotyloideae (Umbelliferae). Pollen & Spores 6: 479–514.

Tseng, C.C. 1971. Light and scanning electron microscopic studies on pollen of *Tetraplasandra* (Araliaceae) and relatives. Amer. J. Bot. 58: 505–516.

Tseng, C.C. 1973. Systematic Palynology of *Tupidanthus* and *Plerandra* (Araliaceae). Grana 13: 51–56.

Tseng, C.C. 1974. Pollen of *Boerlagiodendron*: a unique type in the Araliaceae. Amer. J. Bot. 61: 717–721.

Tseng, C.-J., Hoo, G. 1982. A new classification scheme for the family Araliaceae. Acta Phytotax. Sinica 20: 125–130.

Tseng, C.C., Shoup, J.R. 1978. Pollen morphology of *Schefflera* (Araliaceae). Amer. J. Bot. 65: 384–394.

Valcárcel, V., Fiz-Palacios, O., Wen, J. 2014. The origin of the early differentiation of ivies (*Hedera* L.) and the radiation of the Asian Palmate group (Araliaceae). Molec. Phylog. Evol. 70: 492–503.

van Tieghem, P. 1872. Sur les canaux oléo-résineaux des Ombellifères et des Araliacées. Bull. Soc. Bot. France 19: 113–129.

van Tieghem, P. 1884. Sur la structure et les affinités des Pittosporées. Bull. Soc. Bot. France 31: 384–385.

van Wyk, B.-E., Wink, M. 2004. Medicinal plants of the world. Pretoria: Briza Publications.

Viguier, R. 1906. Recherches anatomiques sur la classification des Araliacées. Ann. Sci. Nat. Bot. IX, 4: 1–210.

Wagenitz, G. 1992. The Asteridae: evolution of a concept and its present status. Ann. Missouri Bot. Gard. 79: 209–217.

Wen, J. 1993. Generic delimitation of *Aralia* L. (Araliaceae). Brittonia 45: 47–55.

Wen, J. 1999. Evolution of eastern Asian and eastern North American disjunct pattern in flowering plants. Ann. Rev. Ecol. Syst. 30: 421–455.

Wen, J. 2001a. Evolution of the *Aralia-Panax* complex (Araliaceae) as inferred from nuclear ribosomal ITS sequences. Edinb. J. Bot. 58: 183–200.

Wen, J. 2001b. Evolution of eastern Asian and eastern North American biogeographic pattern: a few additional issues. Int. J. Plant Sci 162: S117–S122.

Wen, J. 2001c. Species diversity, nomenclature, phylogeny, biogeography, and classification of the ginseng genus (*Panax* L., Araliaceae). In: Punja, Z.K. (ed.) Utilization of biotechnological, genetic and cultural approaches for North American and Asian ginseng improvement. Vancouver: Simon Fraser University Press, pp. 67–88.

Wen, J. 2002. Revision of *Aralia* sect. *Pentapanax* (Seem.) J. Wen (Araliaceae). Cathaya 11-12: 1–116.

Wen, J. 2004. Systematics and biogeography of *Aralia* L. sect. *Dimorphanthus* (Miq.) Miq. (Araliaceae). Cathaya 15–16: 1–187.

Wen, J. 2011. Systematics and biogeography of *Aralia* L. (Araliaceae): Revision of *Aralia* sects. *Aralia*, *Humiles*, *Nanae*, and *Sciadodendron*. Contr. United States Nat. Herb. 57: 1–172.

Wen, J., Nowicke, J.W. 1999. Pollen ultrastructure of *Panax* (the ginseng genus, Araliaceae), an eastern Asian and eastern North American disjunct genus. Amer. J. Bot. 86: 1624–1636.

Wen, J., Zimmer, E.A. 1996. Phylogeny of *Panax* L. (the ginseng genus, Araliaceae): inference from ITS sequences of nuclear ribosomal DNA. Molec. Phylog. Evol. 5: 167–177.

Wen, J., Shi, S., Jansen, R.K., Zimmer, E.A. 1998. Phylogeny and biogeography of *Aralia* sect. *Aralia* (Araliaceae). Amer. J. Bot. 85: 866–875.

Wen, J., Plunkett, G.M., Mitchell, A.D., Wagstaff, S.J. 2001. The evolution of Araliaceae: a phylogenetic analysis based on ITS sequences of nuclear ribosomal DNA. Syst. Bot. 26: 144–167.

Wen, J., Ickert-Bond, S., Nie, Z.-L., Li, R. 2010. Timing and modes of evolution of eastern Asian - North American biogeographic disjunctions in seed plants. In: Long, M., Gu, H., Zhou, Z. (eds.) Darwin's heritage today: Proceedings of the Darwin 2010 Beijing International Conference. Beijing: Higher Education Press, pp. 252–269.

Xie, G.-X., Qiu, Y.-P., Qiu, M.-F., Gao, X.-F., Liu, Y.-M., Jia, W. 2007. Analysis of dencichine in *Panax notoginseng* by gas chromatography-mass spectrometry with ethyl chloroformate derivatization. J. Pharm. Biomed. Anal. 43: 920–925.

Yi, T., Lowry II, P.P., Plunkett, G.M., Wen, J. 2004. Chromosomal evolution in Araliaceae and close relatives. Taxon 53: 987–1005.

Zuo, Y., Chen, Z., Kondo, K., Funamoto, T., Wen, J., Zhou, S. 2011. DNA barcoding of *Panax* species. Planta Medica 72: 182–187.

Gelsemiaceae

Gelsemiaceae (G. Don) Struwe & V. Albert in Struwe, Albert, and Bremer, Cladistics 10: 206 (1994 [1995]).

L. Struwe

Shrubs or semishrubs, straggling to twining vines, or tall trees, glabrous or pubescent with simple hairs. Leaves opposite (rarely verticillate) or alternate (*Pteleocarpa*), simple, short-petiolate, with interpetiolar lines or small interpetiolar stipules (absent in *Pteleocarpa*), evergreen; blades entire to slightly dentate, with a prominent midvein and pinnate venation. Inflorescence terminal or axillary, few- to many-flowered dichasial or thyrsoid cymes or panicles, or with solitary flowers, bracteate. Flowers perfect, hypogynous, heterostylous (not in *Pteleocarpa*, and not always in *Gelsemium elegans*), 5-merous; sepals free except at very base (*Gelsemium* and *Pteleocarpa*) or fused at the base into a calyx tube (*Mostuea*), campanulate or tubular, calyx lobes often unequal in size, imbricate in bud, often with ciliate margins; corolla sympetalous, funnel-shaped to salver-shaped, sometimes slightly zygomorphic, yellow (most common), white, or orange (less often lilac, sometimes with red spots or patterns), lobes shorter than tube, rounded, imbricate in bud; stamens 5, inserted in middle to lower half of corolla tube or in corolla sinuses, filaments free, anthers free, sagittate, latrorse; ovary bicarpellate, bilocular, syncarpous, without nectary disk; placenta axile, often peltate, ovules several to two per locule, style caducous, dichotomously divided at the apex into 4 long, narrow stigmatic lobes (*Gelsemium* and *Mostuea*) or distinct stylodia (*Pteleocarpa*). Fruit a septicidally or loculicidally dehiscent capsule (2- or 4-valved) or indehiscent, strongly winged samara (*Pteleocarpa*), with 1–8 seeds per fruit. Seeds compressed, winged in *Gelsemium elegans* and *G. sempervirens*. Nectaries present as colleters at the leaf and calyx bases and/or the inside of the sepals (this needs further investigation in *Pteleocarpa*, and it is also uncertain where the nectar production in *Gelsemium* flowers occurs).

A pantropical family of three genera and 13 species, distributed in northern Central America, southern North America, northeastern South America, tropical Africa, Madagascar, and southeastern and eastern Asia.

Vegetative Morphology and Anatomy. Gelsemiaceae have a rather typical vegetative anatomy and morphology for members of the Gentianales, with leaves entire, opposite with interpetiolar stipules/lines (alternate in *Pteleocarpa*), the presence of internal phloem in the stems, vestured vessel pits, and colleters (Struwe et al. 2014 and references therein). Colleters are usually present on the inside base of the bracts, and leaf bases, but are apparently not present in *Gelsemium elegans* and their presence in *Pteleocarpa* is uncertain (Leenhouts 1962; Leeuwenberg and Leenhouts 1980; Struwe et al. 2014). The secondary wood of *Mostuea* and *Gelsemium* is characterized by small intervascular pits, scarcity of parenchyma (absent in *Mostuea*), and fibers with small bordered pits (Mennega 1980). Growth rings are present, vessels are generally solitary, very numerous, perforations are simple (rarely scalariform in *Mostuea*), fibers are mostly non-septate, rays are uni- or multiseriate, and calcium oxalate crystals have been found only in *Gelsemium* wood. All three genera have vestured vessel pits (Jansen et al. 2001).

Vesicular-arbuscular (VA-) mycorrhizae are found in the rhizomes of *Gelsemium rankinii* and *G. elegans* (Tiemann et al. 1994). The hyphae of the VAM-fungi enter the rhizomes at several points and grow from cell to cell with exclusively intracellular vesicles and arbuscles, which is distinctly different from the intercellular hyphae found in most other flowering plants (Tiemann et al. 1994). However, this type of VA-mycorrhiza is also found in the closely related Gentianaceae

J.W. Kadereit, V. Bittrich (eds.), *Flowering Plants. Eudicots*, The Families and Genera of Vascular Plants 15,
https://doi.org/10.1007/978-3-319-93605-5_5

and is referred to as the *Paris*-type (Tiemann et al. 1994).

Inflorescence Structure. The only two species in the family with well-developed, large inflorescences are *Gelsemium elegans* and *Pteleocarpa lamponga* with a terminal thyrse composed of many cymose elements, and *Mostuea surinamensis*, with axillary cymose inflorescences. All other species have few or solitary flowers borne in reduced axillary cymes (sometimes subcapitate) on short branches.

Floral Structure and Anatomy. Gelsemiaceae have flowers typical for the Gentianales, with the exception of being mostly heterostylous in *Gelsemium* and *Mostuea.* Varying length relationships between stamens and style or stylodia are also found in some Rubiaceae, and very few Gentianaceae, Loganiaceae (*Geniostoma* spp.) and Apocynaceae (*Carissa* spp.) in the Gentianales. Placentation is axile, sometimes near-basal, sometimes with peltate placentas. Apparently no floral anatomical or developmental investigations using sectioned material or SEMs have been made of Gelsemiaceae.

Pollen Morphology. *Gelsemium* and *Mostuea* have pollen of the *Gelsemium elegans*-type (Punt 1980). The single pollen grains are tricolporate, prolate spheroidal to subprolate, have pores with lateral extensions, and are striato-reticulate (*Gelsemium*) or finely suprastriate (*Mostuea*: Punt 1980). *Pteleocarpa* also has tricolporate and prolate pollen with small endoapertures that show acuminate lateral sides on the pores (Rueangsawang and Chantaranothai 2014).

Karyology. Reported chromosome numbers are 2n = 20 (*Mostuea*) and 2n = 16 (*Gelsemium*). Reports of 2n = 8–10 in *Gelsemium rankinii* (Duncan and DeJong 1964) are erroneous, but might be explained by unusual mitotic behavior (Wyatt et al. 1993).

Pollination. Pollination in *Gelsemium* occurs primarily by large-bodied bees and bumble bees (Ornduff 1970). The two American species of *Gelsemium* differ in that *G. sempervirens* always has fragrant flowers (except some from the westernmost range in Central America), whereas the sympatric *G. rankinii* usually has odorless flowers (Duncan and DeJong 1964; Ornduff 1970). The yellow corollas of *Gelsemium* also have distinct patterns under ultraviolet light, with the lobes being lighter than the tubes (Eisner et al. 1973).

Reproductive System. The two partially sympatric American species of *Gelsemium* are morphologically similar but differ in sepal size, habitat type, the distribution of bracteoles on the pedicels, stomatal and pollen size, and winged vs. non-winged seeds (Duncan and DeJong 1964). Both species are heterostylous, and the more widespread species from somewhat drier habitats (*G. sempervirens*) has been suggested to be an autotetraploid with 2n = 16 (Duncan and DeJong 1964). Recent data have shown that earlier counts of 2n = 8–10 in *G. rankinii* (Duncan and DeJong 1964), making *G. sempervirens* a tetraploid, were incorrect (Wyatt et al. 1993). Hybridization between the two sympatric species is strongly inhibited and there is probably no introgression between them (Duncan and DeJong 1964; cf. Ornduff 1970, 1979). Artificial crosses between these two species have yielded low seed production, with subsequently lower germination success and less vigorous seedlings (Ornduff 1970).

The breeding system has only been investigated in *Gelsemium*, which has a self-incompatibility system associated with heterostyly. Pin and thrum flowers are self-sterile and occur at an approximate ratio of 1:1 in the populations (Duncan and DeJong 1964; Ornduff 1970). Pin flowers are homozygous recessives and thrum flowers are heterozygous in *Gelsemium sempervirens* (Ornduff 1980). Heterostyly is present in all species in the family except some populations of *Gelsemium elegans.*

Fruit and Seed. The genera of Gelsemiaceae have bilocular fruits, and *Mostuea* differs in having only one to two seeds per locule (Leeuwenberg and Leenhouts 1980). *Mostuea* is characterized by deeply truncate capsules (flattened and heart-shaped), whereas *Gelsemium* have oblong elliptic capsules. The broadly winged samaras of *Pteleocarpa* have given the genus its name (Struwe et al. 2014). Sometimes, no seeds develop in one of the locules in *Mostuea*, resulting in a partly aborted ovary and an asymmetric fruit (Leeuwenberg 1961). The seeds are flattened in all species and winged in *Gelsemium sempervirens* and *G. elegans.*

Dispersal. *Pteleocarpa* has wind-dispersed fruits. For *Gelsemium* and *Mostuea*, no specific reports of seed dispersal have been made but most species are presumably wind-dispersed.

Phytochemistry. Both *Gelsemium* and *Mostuea* contain indole alkaloids of the C-17-type (e.g., gelsemine and sempervirine), which are of a similar type to those found in the Apocynaceae (including Asclepiadaceae) and Rubiaceae (Saxton 1965; Bisset 1980a; Schun and Cordell 1987). Several of these alkaloids have pharmacological effects and are highly toxic (see Economic Importance). Also found in *Gelsemium* are scopoletin, emodin monomethyl ether, higher paraffins, plant sterols, and flavones such as quercetin and kaempferol (Bisset 1980a).

Subdivision and Relationships Within the Family. Only three genera, *Gelsemium*, *Mostuea*, and *Pteleocarpa*, are included in this small family (Struwe et al. 2014). *Pteleocarpa* is monotypic, *Gelsemium* contains only three species, and *Mostuea* has nine recognized species. No tribes are recognized in the family. The enigmatic genus *Pteleocarpa* has variously been classified in Boraginaceae, Sapindaceae, Olacaceae, and as incertae sedis, and its own family (Brummitt 2011), but was recently placed for the first time in Gentianales (and Gelsemiaceae) by the aid of molecular phylogenetics (Struwe et al. 2014). *Pteleocarpa* appears to have been grossly misclassified due to its alternative leaves and winged samaras—indeed, its yellow flowers show clear morphological similarities with other Gelsemiaceae (Struwe et al. 2014).

Affinities of the Family. Until a few decades ago, the Gelsemiaceae were classified as the tribe Gelsemieae in the Loganiaceae (Solereder 1892–95; Leeuwenberg and Leenhouts 1980), but they were raised to family rank by Struwe and Albert in Struwe et al. (1994). The relationships of Gelsemiaceae to other families in the Gentianales are still unclear, but their morphological distinctiveness, monophyly, and placement inside the order are well supported (Jiao and Li 2007; Refulio and Olmstead 2014; Struwe et al. 2014). Very similar indole alkaloids to those present in Gelsemiaceae are also found in Rubiaceae and Apocynaceae, but the family differs from Rubiaceae in having a superior ovary, and from Apocynaceae in lacking latex (Struwe et al. 2014). Suggested synapomorphies for Gelsemiaceae are: flowers with yellow corolla (some *Mostuea* have other colors), latrorse anthers, and syncarpous ovaries with partially or completely separated stylodia that develop into a (usually) flattened or winged fruit with flattened seeds (Struwe et al. 2014).

Distribution and Habitats. *Gelsemium*, *Mostuea*, and *Pteleocarpa* are restricted to tropical and subtropical areas with a few species of *Gelsemium* reaching warm temperate areas in North America. *Mostuea* occurs primarily in tropical Africa and Madagascar but has one additional species present in Surinam and Brazil (*M. surinamensis*; Leeuwenberg 1961). *Gelsemium* has two species in the southern United States, Mexico, and Guatemala (*G. sempervirens* and *G. rankinii*), and one species in Southeast Asia (*G. elegans*; Ornduff 1970). Monotypic *Pteleocarpa* occurs in lowland to mid-elevation forests in Southeast Asia (Struwe et al. 2014).

Gelsemiaceae show interesting disjunct tropical and subtropical distributions in continental areas of Laurasian and Gondwanic origins, respectively (Jiao and Li 2007). *Pteleocarpa* is only known from Southeast Asia, but *Gelsemium* has two species in southeastern North America and one species in Southeast Asia, and *Mostuea* grows mostly in tropical Africa with two species in eastern South America. *Pteleocarpa* has been placed with limited data support as the sister to *Gelsemium*, where the single Asian *Gelsemium* species was placed as sister to the two American species (Wyatt et al. 1993; Struwe et al. 2014). A parallel pattern, but with a more southern disjunction between northeastern South America and tropical Africa plus Madagascar, is seen in *Mostuea* (Jiao and Li 2007). These patterns may be best explained by the boreotropics hypothesis in which a boreotropical continuous distribution was broken up into disjunct and more restricted populations (e.g., Tiffney 1985) by climate cooling around the Eocene-Oligocene boundary (ca. 38 million years ago). Genetic evidence demonstrates that the two lineages comprising the two American *Gelsemium* species split about 3–3.5 million years ago in the late Tertiary at about the same time as the climate became drier and suitable habitats decreased in size (Wyatt et al. 1993).

Economic Importance. *Gelsemium* has extensive ethnomedicinal uses with both positive and negative impact on human lives (see Jin et al. 2014 for review). It has commonly been used in folk remedies but contains highly poisonous alkaloids. *Gelsemium elegans* is renowned for having been used for murder, suicide, and to carry out death-sentences in Asia (Leenhouts 1962). In Mexico and China, *Gelsemium* has been used against fevers and as an antispasmodic, sedative and analgesic (Bisset 1980b). North American *Gelsemium* is deadly to livestock, and honey containing *Gelsemium* nectar is toxic to humans (Gowanloch and Brown 1943; Ornduff 1970). *Gelsemium sempervirens*, known as yellow jasmine/jessamine or Carolina/false jasmine, is the state flower of South Carolina and is commonly cultivated in the southern United States.

Mostuea has been used in Africa as a stimulant, vermifuge, for healing wounds, and also against tooth-aches, colds, and upset stomach (Bisset 1980a). Extracts from *Mostuea* and *Gelsemium* have been shown to have anti-tumor and cytotoxic effects (Schun and Cordell 1987; Dai et al. 1999). There is no information about toxicity for *Pteleocarpa* and there are reports that the seeds are boiled and used as edible flavorings (Heyne 1927; Fern 2018).

Fig. 67. Gelsemiaceae. *Gelsemium elegans*. **A** Flowering branch. **B** Flower. **C** Opened corolla with stamens. **D** Partial stamen. **E** Ovary. **F** Mature fruit. **G** Seed. (From Leenhouts 1962)

Key to Genera

1. Tree with alternate leaves — **3. *Pteleocarpa***
– Shrubs, half-shrubs or lianas and vines — 2
2. Fruits oblong, with rounded to acute apex, longer than wide; more than 2 seeds in each locule; twining lianas (except the shrubby *G. elegans*) — **1. *Gelsemium***
– Fruits bilobed to strongly truncate (rarely ellipsoid), wider than long or about as wide as long; 1–2 seeds per locule; shrubs, sometimes straggling — **2. *Mostuea***

Genera of the Gelsemiaceae

1. *Gelsemium* Juss. Fig. 67

Gelsemium Juss., Gen. Pl. 150 (1789); Leenhouts, Flora Malesiana 6 (2): 343 (1963), reg. rev.; Duncan & DeJong, Sida 1: 346–357 (1964), reg. rev.; Ornduff, J. Arnold Arb. 51: 1–17 (1970), rev.; Wyatt et al., Syst. Bot. 18: 345–355 (1993), reg. rev.; Ping-Tao & Leeuwenberg, Flora of China 15: 329 (1996), reg. rev.

Twining vines to shrubs, glabrous to slightly hairy, rhizomes with adventitious roots present. Leaves opposite. Inflorescence terminal or axillary, dichasial or thyrsoid, large and many-flowered (*G. elegans* (Gardner & Chapm.) Benth.) or flowers few or solitary. Flowers heterostylous; calyx with free, elliptic sepals, caducous or persistent in fruit; corolla broadly salver-shaped to funnel-shaped, yellow to orange (sometimes with red markings). Capsule oblong, not compressed, apex acute. Seeds up to 8, flattened, hairy, warty, or smooth (glabrous), usually winged (not in *G. rankinii* Small).

One species in southeastern United States (*G. rankinii* Small), one in southern United States, Mexico, and Guatemala (*G. sempervirens* (L.) J.

St.-Hil.), and one (*G. elegans* (Gardner & Chapm.) Benth.) in mainland Southeast Asia, Malesia (Sumatra, Borneo), Hainan, and southern and southeast China.

2. ***Mostuea*** Didr.

Mostuea Didr., Vidensk. Medd. Kjoeb. 1853: 86 (1854); Leeuwenberg, Meded. Landbouwhogeschool Wageningen, Nederl. 61: 1–31 (1961), rev.; Sobral & Rossi, Novon 13:325–328 (2003), reg. rev.

Shrubs or lianas, hairy to nearly glabrous. Leaves opposite. Inflorescence axillary or terminal, mostly on short lateral branches, dichasial (sometimes nearly capitate), one- to many-flowered. Flowers heterostylous; calyx with sepals fused for up to half their length, lobes ovate to linear, equal to strongly unequal in length; corolla funnel-shaped, white, cream, or yellow (sometimes only at the base), more rarely orange or red, at least twice as long as the calyx. Capsule oblong to truncate, compressed, not winged. Seeds 1–4, flattened, usually hairy (glabrous in *M. surinamensis* Benth.).

Seven species in tropical Africa and two species in northeastern and eastern South America.

3. ***Pteleocarpa*** Oliv.

Pteleocarpa Oliv, Trans. Linn. Soc. London 28: 515 (1873); Rueangsawang & Chantaranothai, Tropical Natural History 14: 1–6 (2014), rev.

Trees, some vegetative parts often minutely hairy. Leaves alternate; stipules absent. Inflorescences terminal, multi-flowered panicles. Flowers homostylous; calyx fused only at base; corolla with short tube, stamens inserted in corolla lobe sinuses; ovary on thickened stalk, styles two (not fused), persistent in fruit. Fruit a strongly winged and compressed indehiscent samara, truncate at apex, with 1–2 seeds.

One species (*P. lamponga* Bakh. ex K. Heyne), in Southeast Asia (Brunei, Indonesia, Malaysia, and Thailand).

Selected Bibliography

Bisset, N.G. 1980a. Useful plants. In: Leeuwenberg, A.J.M. (ed.) Engler and Prantl's Die natürlichen Pflanzenfamilien, Angiospermae: Ordnung Gentianales, Fam. Loganiaceae, vol. 28b (1): 238–244. Berlin: Duncker and Humblot.

Bisset, N.G. 1980b. Phytochemistry. In: Leeuwenberg, A.J. M. (ed.) Engler and Prantl's Die natürlichen Pflanzenfamilien, Angiospermae: Ordnung Gentianales, Fam. Loganiaceae, vol. 28b (1): 211–233. Berlin: Duncker and Humblot.

Brummitt, R.K. 2011. Valid publication of the family name Pteleocarpaceae. Kew Bull. 66: 1–3.

Dai, J.R., Hallock, Y.F., Cardellina II, J.H., Boyd, M.R. 1999. 20-O-ß-Glucopyranosyl Camptothecin from *Mostuea brunonis*: a potential camptothecin prodrug with improved solubility. J. Nat. Prod. 62: 1427–1429.

Duncan, W.H., DeJong, D.W. 1964. Taxonomy and heterostyly of North American *Gelsemium* (Loganiaceae). Sida 1: 346–357.

Eisner, T., Eisner, M., Hyypio, P.A., Aneshansley, D., Silberglied, R.E. 1973. Plant taxonomy: ultraviolet patterns of flowers visible as fluorescent patterns in pressed herbarium specimens. Science 179: 486–187.

Fern, K. 2018. Tropical Plants Database, tropical.theferns.info. Accessed 2018-07-28.

Gowanloch, J.N., Brown, C.A. 1943. Poisonous snakes, plants and black widow spider of Louisiana. New Orleans: Louisiana Dept. of Conservation.

Heyne, K. 1927. De nuttige planten van Nederlandsch Indië. Batavia: Departement van Landbouw, Nijverheid & Handel.

Jansen, S., Baas, P., Smets, E. 2001. Vestured pits: their occurrence and systematic importance in eudicots. Taxon 50: 135–167.

Jiao, Z., Li, J. 2007. Phylogeny of intercontinental disjunct Gelsemiaceae inferred from chloroplast and nuclear DNA sequences. Syst. Bot. 32: 617–627.

Jin, G.-L., Su, Y.-P., Liu, M., Xu, Y., Yang, J., Liao, K.-J., Yu, C.-X. 2014. Medicinal plants of the genus *Gelsemium* (Gelsemiaceae, Gentianales) — A review of their phytochemistry, pharmacology, toxicology and traditional use. J. Ethnopharmacol. 152: 33–52.

Leenhouts, P.W. 1962. *Gelsemium*. In: van Steenis, C.G.G. J. (ed.) Flora Malesiana vol. 6 (2): 293–387. Groningen: Wolters-Noordhoff.

Leeuwenberg, A.J.M. 1961. The Loganiaceae of Africa II. A revision of *Mostuea*. Meded. Landbouwhogeschool Wageningen, Nederl. 61: 1–31.

Leeuwenberg, A.J.M, Leenhouts, P.W. 1980. Taxonomy. In: Leeuwenberg, A.J.M. (ed.) Engler and Prantl's Die natürlichen Pflanzenfamilien, Angiospermae: Ordnung Gentianales, Fam. Loganiaceae, vol. 28b (1): 8–96. Berlin: Duncker and Humblot.

Mennega, A.M.W. 1980. Anatomy of the secondary xylem. In: Leeuwenberg, A.J.M. (ed.) Engler and Prantl's Die natürlichen Pflanzenfamilien, Angiospermae: Ordnung Gentianales, Fam. Loganiaceae, vol. 28b (1): 112–161. Berlin: Duncker and Humblot.

Ornduff, R. 1970. The systematics and breeding system of *Gelsemium* (Loganiaceae). J Arnold Arb. 51: 1–17.

Ornduff, R. 1979. Features of pollen flow in *Gelsemium sempervirens* (Loganiaceae). J. Arnold. Arb. 60: 377–381.

Ornduff, R. 1980. The probable genetics of distyly in *Gelsemium sempervirens* (Loganiaceae). Canad. J. Genet. Cytol. 22: 303–304.

Punt, W. 1980. Pollen morphology. In: Leeuwenberg, A.J. M. (ed.) Engler and Prantl's Die natürlichen Pflanzenfamilien, Angiospermae: Ordnung Gentianales, Fam. Loganiaceae, vol. 28b(1): 162–191. Berlin: Duncker and Humblot.

Refulio, N.F., Olmstead, R.G. 2014. Phylogeny of Lamiidae. Amer. J. Bot. 101: 287–299.

Rueangsawang, K., Chantaranothai, P. 2014. Studies on Thai Pteleocarpaceae. Trop. Nat. Hist. 14(1): 1–6.

Saxton, J.E. 1965. Alkaloids of *Gelsemium* species. In: Manske, R.H.F. (ed.) The alkaloids, vol. 8: 93–117. New York: Academic Press.

Schun, Y., Cordell, G.A. 1987. Cytotoxic steroids of *Gelsemium sempervirens*. J. Nat. Prod. 50: 195–198.

Solereder, H. 1892–95. Loganiaceae. In: Engler, A., Prantl, K. (eds.) Die natürlichen Pflanzenfamilien, vol. 4(2): 19–50. Leipzig: Wilhelm Engelmann.

Struwe, L., Albert, V.A., Bremer, B. 1994 [1995]. Cladistics and family level classification of the Gentianales. Cladistics 10: 175–206.

Struwe, L., Soza, V.L., Manickam, S., Olmstead, R.G. 2014. Gelsemiaceae (Gentianales) expanded to include the enigmatic Asian genus *Pteleocarpa*. Bot. J. Linn. Soc. 175: 482–496.

Tiemann, C., Demuth, K., Weber, H.C. 1994. Zur VA-Mycorrhiza von *Gelsemium rankinii* und *G. sempervirens* (Loganiaceae). Beitr. Biol. Pflanzen 68: 311–321.

Tiffney, B.H. 1985. Perspectives on the origin of the floristic similarity between eastern Asia and eastern North America. J. Arnold. Arb. 66: 73–94.

Wyatt, R., Broyles, S.B., Hamrick, J.L., Stoneburner, A. 1993. Systematic relationships within *Gelsemium* (Loganiaceae): evidence from isozymes and cladistics. Syst. Bot. 18: 345–355.

Gentianaceae

Gentianaceae Juss., Gen. Pl. 141 (1789), "Gentianae", nom. cons.

L. Struwe and J.S. Pringle

Annual, biennial, or perennial herbs, shrubs, trees, or rarely lianas, nearly always glabrous on vegetative parts, terrestrial or rarely epiphytic, mostly autotrophic, a few genera mycoheterotrophic and then sometimes achlorophyllous. Stems erect, decumbent, or rarely trailing or twining; rhizomes sometimes present. Leaves opposite, rarely whorled or alternate, sometimes in basal rosettes, simple, entire or rarely crenulate-serrulate, sessile to petiolate; stipules generally absent, sometimes present as interpetiolar lines or ocreae in more woody genera or very rarely as petiolar or interpetiolar stipules; colleters (multicellular glands) often present in leaf axils, petiolar glands rarely present. Inflorescences terminal or axillary; flowers solitary or in usually bracteate, dichasial or monochasial cymes, thyrses, or verticillasters, cymes sometimes raceme- or umbel-like, or inflorescences capitate. Flowers perfect and monomorphic, rarely heterostylous or imperfect (plants andromonoecious, gynodioecious, gynomonoecious, trimonoecious, or dioecious), actinomorphic or slightly zygomorphic (especially staminal and stylar parts), generally 4–5-merous, but in a few genera 3- or 6–16-merous; calyx usually green, persistent or rarely caducous, synsepalous (sepals distinct in a few genera), rarely absent; lobes imbricate or valvate, often keeled or winged; colleters or discoid glands often present adaxially, sometimes also with dorsal glandular areas; corolla marcescent or caducous, sympetalous; lobes contorted, rarely imbricate and/or valvate in bud; nectary glands sometimes present adaxially on tube or lobes, diverse in form and appendages; stamens in one whorl, epipetalous, isomerous and alternate with petals, all fertile or rarely with only one or two fertile; filaments separate or less often fused, sometimes only at the base, rarely with appendages or connected by a corona, flattened or terete, straight or declinate, or rarely absent; anthers basifixed or dorsifixed, distinct, rarely connate, dithecal, dehiscent longitudinally or rarely with terminal pores, usually introrse; pistil single, bicarpellate, usually with postgenital fusion of carpels; ovary superior, sometimes stipitate; placentation highly variable, often with strongly intruding placentas creating pseudobilocular (or rarely pseudotetralocular) ovaries from parietal placentation in unilocular ovaries, sometimes unilocular and placentation parietal, or truly bilocular and placentation axile, especially at the base, rarely free-central; nectary disk or separate nectaries at base of ovary common; style single or absent, undivided or shallowly 2-cleft, straight, declinate, or deflexed to one side or lower part of corolla mouth, persistent or caducous; stigma single or two-parted, capitate, funnelform, peltate, slightly to strongly bilobed, rarely decurrent along carpel sutures; ovules few to many, tenuinucellate, unitegmic or in most achlorophyllous species ategmic. Fruit a capsule, usually septicidal and bivalvate, rarely irregularly dehiscent or indehiscent, or occasionally a berry. Seeds non-arillate, usually small, winged or not; testa membranous or crustaceous; embryo small, minute or undifferentiated in mycoheterotrophic genera; endosperm copious, fleshy or oily, nuclear in autotrophic genera, scant and sometimes cellular in achlorophyllous mycoheterotrophic genera.

A family of 102 genera and ca. 1750 species. Worldwide, in arctic, temperate, subtropical, and tropical regions, absent only from Antarctica.

Vegetative Morphology. Most gentians are herbs or subshrubs, with more woody taxa found primarily in tribes Saccifolieae (*Saccifolium*), Helieae (about a dozen genera), and Potalieae (*Anthocleista*, *Cyrtophyllum*, *Fagraea*,

J.W. Kadereit, V. Bittrich (eds.), *Flowering Plants. Eudicots*, The Families and Genera of Vascular Plants 15,
https://doi.org/10.1007/978-3-319-93605-5_6

Limahlania, *Lisianthius*, *Picrophloeus*, *Potalia*, and *Utania*; Struwe et al. 2002; Wong and Sugumaran 2012a, b). Subtribe Potaliinae includes trees to ca. 35 m tall, but outside this subtribe all woody species are shrubs or smaller trees, the largest (*Macrocarpaea* and *Tachia*) to ca. 10 m tall. All are terrestrial, except for one *Lehmanniella* and some *Macrocarpaea* epiphytes (occasionally also found in *Tachia* and *Voyria*, as well as many *Fagraea* that are epiphytic when young, and later rooting in soil). The habit ranges from erect to sprawling; lianas occur rarely in tribes Potalieae (woody lianas in *Anthocleista*, herbaceous twining vines in *Bisgoeppertia*), Chironieae (a trailing sp. in *Geniostemon*), and Gentianeae (*Crawfurdia*, *Pterygocalyx*, and *Tripterospermum*). Mycoheterotrophy arose several times in the family.

The herbaceous species include annuals, winter-annuals, biennials, and continuously flowering perennials, and also monocarps that flower once after several to many years in a rosette stage (some Swertiinae). The growth of perennials is usually sympodial, sometimes monopodial, with both types occurring in *Gentiana*. Perennial species may develop an erect, often branched caudex, which is slightly woody. Some produce new crowns from elongating rhizomes or spreading roots or from the rooting of stems at proximal nodes, producing colonial growth.

Stems range from terete to quadrangular, either unwinged or with two or four decurrent wings, and from unbranched to strongly branched. Phyllotaxy is mostly decussate, but whorled leaves are present in some species. The proximal or all leaves of some *Bartonia*, *Microrphium*, and *Voyria* species, and the distal or all leaves of a few *Swertia* species, are often subopposite or alternate. The leaves of *Saccifolium* also appear to be alternate, although crowding obscures the phyllotaxy. Leaves positioned primarily at the apices of branches occur in several woody genera, such as *Anthocleista* and *Limahlania* (hence the local name "cabbage tree"), *Lisianthius*, *Potalia*, and *Tachia*. The leaves are simple and unlobed; the margins are generally entire, or crenulate to serrulate in a few *Anthocleista*, *Limahlania*, *Metagentiana*, and *Swertia* species. Cartilaginous or revolute leaf margins occur in several genera.

The leaves are either sessile, often with an indistinct pseudopetiole due to an attenuate leaf base, or petiolate. The proximal portion of the cauline leaves (some *Gentianella* and *Halenia*) may be narrowed and erect, clasping the stem (whether or not leaves are basally fused). Distinctly petiolate leaves are frequent among the more woody species. Opposite leaves are sometimes basally united, forming a short sheath around the stem; perfoliate blades occur in species of *Blackstonia*, *Canscora*, *Chelonanthus*, *Deianira*, and *Ixanthus*. Non-connate-sheathing leaf bases are often connected by a raised interpetiolar (stipular) line, which is occasionally ocrea-like, notably in *Macrocarpaea*, *Potalia*, and *Tachia*. In some *Obolaria*, *Gentianella*, and *Swertia* the leaf bases are decurrent. Interpetiolar stipules have been found only in *Macrocarpaea zophoflora*, where they are caducous. Stipule-like lateral expansions of the petiole occur in *Anthocleista*, *Fagraea*, and *Gentianothamnus*. Minute, scale-like leaves are common in the completely mycoheterotrophic species (*Exacum* spp., *Exochaenium oliganthum*, *Voyria*, and *Voyriella*: Figs. 69A, 70B) and partially mycoheterotrophic species (*Bartonia* and *Sebaea*). Colleters (also called squamellae), small multicellular, finger-shaped glands, often occur in the leaf axils and on the inside of calyces, and produce resin or other liquids that potentially provides protection of young developing buds and primordia, but the chemical composition of this is unknown. Large amounts of sticky resin on younger branches are often found in *Potalia* and *Tachia*.

The roots of perennial species are often secondary, relatively stout, and lack root hairs, but persistent tap roots, which may be contractile, also occur. In some *Gentiana* species, the roots can graft to one another, forming a single, massive structure. Endomycorrhizae of the phylum Glomeromycota are probably generally present in Gentianaceae, with a specialized arbuscular mycorrhiza in *Exochaenium oliganthum*, *Voyria*, and *Voyriella* (Imhof 1997; Bidartondo et al. 2002; Merckx et al. 2012). *Exochaenium oliganthum*, *Voyria*, and *Voyriella* and some species of *Exacum* (formerly *Cotylanthera*) are strictly mycoheterotrophic, lacking chlorophyll; their root systems are usually short and stout, including coralloid and "bird's-nest" forms, and are anatomically much reduced (Maas and Ruyters 1986; Imhof 1997, 1999). Reduced root systems are also found in partially mycoheterotrophic species of *Bartonia*, *Exacum*, *Exochaenium*, and *Obolaria* (Cameron and Bolin 2010).

Vegetative Anatomy. The xylem usually forms a continuous cylinder, except in very slender stems, as in some *Voyria* species (Metcalfe and Chalk 1950). Secondary growth is common in some more deeply nested tribes, such as Helieae and Potalieae (Mennega 1980; Carlquist 1984; Carlquist and Grant 2005), and most woody gentians appear to have evolved from a suffrutescent herbaceous ancestor. The primary vascular tissue is bicollateral or less often centrifugal. Rays are usually absent, uniseriate when present or multiseriate in the larger stems of woody species, but regularly present, usually uniseriate, occasionally bi- or multiseriate in subtribe Potaliinae. The vessel segments are round in transection, sometimes solitary, with perforation plates generally simple, rarely pseudoscalariform, scalariform in *Saccifolium*; the lateral wall pits are alternate, oval, with elliptic apertures of about the same length as the pit cavity diameter; helical thickenings do not occur. Vestured pits have been found in several woody genera (Jansen and Smets 1998). The imperforate tracheary elements are fiber-tracheids with vestigial borders or libriform tracheids.

Intraxylary phloem is present in a continuous ring or in isolated bundles at the margin of the pith. Medullary bundles or isolated strands of phloem are often present in the pith. Parenchyma is absent from the secondary xylem. Latex systems are absent except in the fruits of some *Anthocleista*, *Fagraea*, *Cyrtophyllum*, *Picrophloeus*, and *Limahlania*. The nodes are usually unilacunar or trilacunar, occasionally multilacunar, with an arc-shaped bicollateral vascular strand and small lateral bundles entering the petioles. Mucilaginous cells are often present in the stem and leaf epidermis, mesophyll, and roots. The epidermal cells of the stems and leaves are sometimes papillate, especially adaxially.

The leaves are usually dorsiventral, occasionally isobilateral, i.e., without a differentiated palisade layer, as in *Enicostema* and *Obolaria*. Adaxial hypodermis is present or absent. Primary venation (except in spp. with small or narrowly linear leaves) is usually either basally or suprabasally acrodromous, or, especially in woody genera, pinnate and camptodromous. Secondary venation is variable, often ± brochidodromous and sometimes forming an intramarginal vein, or eucamptodromous. The stomata are anomocytic or anisocytic (some paracytic in *Swertia*), on the abaxial or less often on both leaf surfaces, rarely absent (*Voyria* spp.). Kranz anatomy has not been reported.

Most gentians are glabrous on the vegetative parts. Exceptions occur in species of *Chironia*, *Gentiana*, *Gentianella*, *Microrphium*, *Ornichia*, *Orphium*, and *Zeltnera*, which have an indumentum consisting of papillae or simple, often minute hairs, or rarely of longer, uniseriate hairs (Hakki 1984). Glandular hairs are not known on vegetative parts, and sessile glands on the vegetative parts have been reported only in *Ixanthus*. Glaucous leaves occur in *Blackstonia*, *Eustoma*, and rarely in *Chelonanthus*. Spiny stems occur in some *Anthocleista* species and in the monospecific *Limahlania*, and extrafloral nectaries on the petioles, visited by ants, also occur in *Anthocleista*.

Inflorescence Structure. Inflorescences are terminal and/or in leaf axils. The basic and most common type is a dichasial cyme. Lateral dichasia are often produced below the terminal dichasium, forming a thyrse or verticillaster. Branching is sometimes anisocladous, or one lateral branch at each node may be suppressed, resulting in monochasial, corymb-like, raceme-like, or occasionally ± scorpioid cymes. Cyme branches are sometimes reduced, forming terminal or lateral clusters of sessile or subsessile flowers or less often spikes or umbelloid clusters (e.g., in *Faroa*, *Halenia*, and *Swertia*), or flowers may be solitary.

Flower Structure. The flowers are perfect, except as noted below, hypogynous, and actinomorphic or less often slightly zygomorphic (Struwe and Albert 2002). Zygomorphy primarily affects the stamens and ovary and is frequent in tribe Helieae (Fig. 75A, B); bilabiate corollas are found in *Canscora* (Fig. 71B). The parts are borne in four whorls on a small receptacle.

The flowers of most genera are 4- or 5-merous except for the gynoecium, with the sepals, petals, and stamens isomerous. Some variability in merosity in a small percentage of flowers on a plant is not uncommon (for example, (3–)5(6)-merous flowers within two species of *Enicostema*). Exceptions in merosity are 3-merous (*Pycnosphaera*), 4–8-merous (*Gentiana* spp.), 6-merous (*Prepusa* and *Senaea*), 8-merous (*Urogentias*), 6–12-merous (*Blackstonia* and *Gentiana* spp.), and 7–12(–14)-merous (*Sabatia* spp.). Exceptions to perianth isomery are: calyx 2–4-merous, corolla 4-merous in several genera in subtribe Canscorinae; calyx 4-merous, corolla 6-merous in *Chorisepalum*, calyx 4-merous, corolla 8–10-merous in *Potalia* (Fig. 74D); calyx 4-merous, corolla 8–16-merous in *Anthocleista*. Horticultural forms of *Eustoma* and *Gentiana* sect. *Kudoa* derived through breeding and selection can have numerous supernumerary floral parts, usually in combination with irregularities in the stamens and pistil. Floral appendages and nectaries are common and diverse in many tribes (see below).

The calyx is generally persistent in fruit, caducous in *Chorisepalum* and most *Macrocarpaea*. On the inner surface or at the base of the calyx tube there is often a pair or a zone or ring of usually fleshy, multicellular papillae or trichomes (known as colleters or squamellae in Gentianales), flanking each sepal midvein or spread over the inner surface. An annular or cup-like nectary between the calyx and corolla is present in *Exochaenium*, *Orphium*, and *Saccifolium*. Most genera in tribe Helieae have extrafloral, abaxial nectaries on the calyx, and some *Fagraea* have external nectaries at the base of the flowers (Dalvi et al. 2013). In *Crawfurdia* and most sections of *Gentiana*, an invagination of the tube forms an intracalycular membrane, so that the calyx lobes appear to diverge abaxially from the tube slightly below its apex. The calyx is regularly or irregularly lobed (sepals free in *Anthocleista*, *Chorisepalum*, and *Potalia*, nearly so in several other genera) or occasionally spathaceous, usually green and leaf-like in texture, occasionally reduced and hyaline; often coriaceous in Helieae. The calyx lobes are imbricate or valvate in bud, or sometimes reduced or vestigial.

The two bracts subtending each flower of *Obolaria* have often been interpreted as a calyx consisting of two separate sepals, but this needs further study. Since the flowers of many related species are individually subtended by paired bracts, the calyces of some related taxa are much reduced and delicate, and the two leaf-like structures at the base of the *Obolaria* corolla can be either parallel or at right angles to the carpels, they seem better interpreted as bracts. In their venation and decurrent bases and in their shape and texture, they are similar to the leaves and to the unequivocal bracts that subtend the cymules.

The sympetalous corolla is usually marcescent, less often early to tardily caducous. It is generally much larger than the calyx and is the most conspicuous part of the flower, from 5 mm to 50 cm long. The corollas are most often campanulate to funnelform, but range from rotate to cyathiform, cylindric, and salverform. The lobes are usually dextrorsely contorted in bud (sinistrorsely in *Halenia* and some *Swertia*), partially valvate in *Aripuana* and *Jaeschkea*, and imbricate in *Bartonia*, *Canscora*, *Cracosna*, *Obolaria*, *Saccifolium*, and *Schinziella*. A fringe of trichomes or a pair of fimbriate scales near the base of each corolla lobe or at or near the apex of the tube occurs in *Comastoma*, *Djaloniella*, and some *Gentiana* (sect. *Fimbricorona*, now part of sect. *Chondrophyllae* s. l.) and *Gentianella* (Fig. 72). In most *Frasera*, *Lomatogonium*, and *Swertia* fringes of trichomes or tubular or scale-like prolongations of the rim surround the openings of nectary pockets on the adaxial corolla surface (Fig. 73D). Fringed petal edges are present in most *Gentianopsis*. Nectaries on the petals are found in all Swertiinae except *Megacodon*, *Bartonia*, *Latouchea*, and possibly *Obolaria*, and occur inside the petal spurs of *Halenia*. In most genera of Gentianinae (i.e., *Crawfurdia*, most *Gentiana*, *Kuepferia*, *Metagentiana*, *Sinogentiana*, and *Tripterospermum*) each petal has a distally widening appendage on each side (folds or plicae; Fig. 72B), united with that of the adjacent petal, usually extending as free portions between the corolla lobes, which can make a 5-merous corolla appear 10-merous. In mycoheterotrophic species of *Exacum*, the corolla tube and filament bases are fused to the ovary, but the calyx is free.

Many species have corollas that close in cloudy weather and at night, although moth- and bat-pollinated flowers are exceptions; some also close in response to temperature changes. Permanently closed corollas occur in some *Gentiana* and *Gentianella*, and corollas of some

Gentiana close when touched or when the stem is moved.

The stamens are epipetalous, usually inserted on the corolla tube at the same height or in the sinuses between the lobes, and are isomerous and alternating with the petals. In *Canscora*, *Hoppea*, and *Schinziella* some stamens lack fertile anthers or are vestigial (anisomorphy), and the stamens are inserted at different heights in the corolla (Shahina and Nampy 2014; Fig. 71B). Filaments are usually present (absent in some *Voyria*, nearly so in a few other genera), flattened or terete, straight, and equal, less often unequal and/or declinate (mostly in genera with zygomorphic corollas in tribe Helieae, also common in subtribes Gentianinae and Canscorinae). The filaments are sometimes abruptly deflexed near the base, and/or the distal portion is circinately coiled in Helieae (Fig. 75). The filament bases in several genera are connected by a corona (particularly *Symbolanthus*; Fig. 75B) or fused (*Anthocleista* and *Potalia*), or have scales or fringes between the filaments (especially in subtribe Faroinae; Fig. 74E). For example, in *Enicostema* and *Oreonesion* the base of the free portion is expanded into a projection, with the stamen bases dividing the throat into narrow channels. Ridges in the tube below the filaments, forming channels, are also common in some *Gentiana* with tubular corollas. The anthers are basifixed or dorsifixed, usually free, sometimes loosely coherent to firmly connate (*Hockinia* and some *Gentiana* and *Voyria*). Sagittate anther bases are common. In presumably heterostylous species (*Tapeinostemon* and *Hockinia*), the anther connectives are laminar and enlarged between the thecae in some morphs. In several genera of Helieae and Chironiinae, the anthers coil helically or circinately after anthesis. The connectives are often prolonged and sometimes expanded beyond the thecae, with glandular tips or basal glands in many Exaceae genera (Fig. 70D). Caducous anthers occur in *Duplipetala*, *Microrphium*, and *Phyllocyclus*.

The pistil comprises two united carpels, both of which bear functional ovules; fusion is postgenital except in subtribe Potaliinae. The upper part of the developing carpels can be flat or plicate. The ovary is usually oriented in an anteroposterior plane, sessile or less often stipitate, ovoid to ellipsoid or less often subglobose to obovoid, often compressed laterally. Placentation in most genera is parietal, with the ovules usually in two zones adjacent to the sutures, developing spirally in each. In some Swertiinae and *Gentiana*, the placenta covers much of the interior wall. In many genera, the placenta intrudes from the sutures well into the cavity, so that the ovary may appear to be bilocular (hereafter called pseudobilocular), or these projections may be bifid, sometimes forming a pseudotetralocular ovary (Fig. 70E). In some genera, mostly tropical or subtropical, the ovary is basally or completely bilocular, with axile placentation. In the mycoheterotrophic species of *Exacum*, through the loss of the connecting septa, the ovary is unilocular with free-central placentation in the central portion, divided with axile placentation toward the base and apex.

A nectariferous disk, sometimes lobed, at the base of the ovary (or on the stipe if the ovary is stipitate) is common in Helieae, Potalieae, and subtribe Gentianinae (Fig. 74A). Two sessile or stalked glands at or slightly above the base of the ovary are present in some *Voyria*.

The style is simple and single (sometimes absent, especially in tribe Gentianeae), usually undivided or rarely shallowly cleft (more deeply cleft in some *Sabatia*). The style is either sharply demarcated and slender, or short and gradually or poorly differentiated from the ovary, or gradually differentiated from the apexof the ovary but long and slender, straight or declinate. Persistence of the style in the fruiting stage varies, but is usually consistent within genera, and is persistent, caducous, or often only distally caducous.

The terminal stigmas are usually either simple (capitate, peltate, or pointed) or shallowly to deeply bilobed. The receptive surface is wet and papillate or nearly smooth. The stigmatic lobes are rarely decurrent along the sutures of the ovary (some *Lomatogonium*, slightly so in some *Bartonia*), and extra stigmas can be found in *Sebaea*. Long, helically coiled stigmas before and/or after receptivity occur only in *Gentiana lutea* ssp. *lutea* and *Sabatia* sect. *Sabatia*.

Floral Anatomy. Generally, each part of the calyx, corolla, and androecium is supplied by one primary trace (Lindsey 1940; Gopal Krishna and Puri 1962; Wood and Weaver 1982; van Heusden 1986). When each trace has entered the calyx or corolla whorl it divides into three traces, two lateral and one dorsal. Rarely the three traces diverge separately from the stele to each sepal, and in *Bartonia* to each petal as well. Sometimes the lateral traces of each sepal are subsequently fused to those of the adjacent sepal, forming the commissural nerves of the calyx. At the apex, the vascular cylinder usually divides into three traces per carpel, i.e., two ventral traces, which supply the ovules, and one dorsal trace per carpel (six traces total; gentians are always bicarpellate). Less often, each carpel has five traces, with weaker submarginal traces between the ventral and dorsal. Axile placentation is usually associated with five-trace carpels, parietal placentation with three-trace carpels. Fusion between whorls occurs rarely, as when the fused lateral sepal traces are basally fused with the middle trace of the petal directly above, and/or the lateral petal traces are basally fused with the lateral trace of a carpel. In the most extreme cases of fusion, the vascular cylinder is a dictyostele comprising five bicollateral bundles, each containing traces to all whorls. Reduction also occurs in which the lateral sepal traces are absent (*Tripterospermum*).

Floral indumentum is usually absent. Non-glandular simple hairs are found on the calyces in *Chironia* spp., *Microrphium*, and *Orphium*. Trichomes diverse in location, size, and structure, vascularized in some species of *Gentianella* but otherwise generally not vascularized and often minute, are present on the adaxial corolla surfaces of many Swertiinae and a few *Gentiana*. Minute, simple hairs are found on the adaxial or less often the abaxial surfaces of the corollas and/or the filaments in species of *Hockinia*, *Swertia*, and *Voyria*. Calcium oxalate crystals are frequently present in the parenchymatous cells of leaves and stems. They also occur in cells of the receptacle and placentae of tribe Helieae.

Embryology. The anthers are tetrasporangiate and dithecal, with persistent epidermis. The division of the archesporial cells leads to the formation of a primary parietal layer and a sporogenous layer (Bouman et al. 2002). The former, through further division, forms a 3–6-layered anther wall. The endothecium usually develops fibrillar thickenings (absent in mycoheterotrophic *Exacum*). Either the tapetum is derived entirely from the innermost wall layer, or the portion of the tapetum facing the connective is derived from the ground tissue of the connective. The tapetum can also be derived in part or entirely from sterile cells in the sporogenous tissue; in these cases the sterile cells also form trabeculae dividing the locules into chambers. Both glandular and plasmodial (true periplasmodial) tapeta occur in the family. Division of the pollen mother cells is simultaneous; cytokinesis is by furrowing. The initial microspore tetrads are tetrahedral or isobilateral. The mature pollen is usually trinucleate, rarely binucleate. Anther dehiscence is usually longitudinal and introrse (although sometimes functionally extrorse by the inversion of dorsifixed anthers, as in *Gentianopsis* and some *Gentianella*), extrorse in some *Gentiana* and *Gentianella*, or dehiscent only toward the apex, forming 1 or 2 terminal pores per anther, in some *Exacum*.

The ovules are usually anatropous, occasionally amphitropous, orthotropous, hemitropous, or anacampylotropous, sessile, imbedded, or rarely short-funiculate, not vascularized (Bouman and Schier 1979; Hakki 1997; Bouman et al. 2002). They are usually unitegmic with 2 to 20 cell layers in the integument, tenuinucellate, without or sometimes with a differentiated integumentary tapetum (endothelium, e.g., in *Exacum*). Ategmic ovules with a nucellar cap are found in some mycoheterotrophic species of *Exacum* and *Voyria*. The archesporium includes one hypodermal cell (rarely two or three), which functions directly as a megaspore mother cell. At meiosis, the megaspore mother cell forms a linear tetrad of megaspores, of which the chalazal megaspore becomes the embryo sac. The embryo sac is of the Polygonum type. Its orientation is normal (inverse in *Voyria* subg. *Leiphaimos*). Three antipodal cells form, and these are often proliferating. The polar nuclei fuse prior to fertilization.

Fertilization is porogamous, and the pollen tube discharges its contents into one of the synergids through a terminal pore (Shamrov 1996). In the autotrophic taxa, embryo development is of the Physalis I or Physalis II variant of the Solanad type in Johansen's system. The embryo is small, cylindric or conical, and the suspensor is uniseriate or 2–4-seriate. In the mycoheterotrophic species the embryo is minute, undifferentiated or nearly so. The endosperm is copious, fleshy or oily, and nuclear in autotrophic species, scant and cellular in mycoheterotrophic species. Starch and protein crystals are reserve foods in the endosperm. Endosperm haustoria are absent. Germination is phanerocotylar.

Pollen Morphology. Pollen morphological characters show great variation within Gentianaceae and have traditionally been of great taxonomic value (Nilsson 1967, 1970, 2002; Sankara Rao and Chinnappa 1983; Nilsson et al. 2002). The pollen is small to fairly large, 18 to 65 μm in the longest dimension (tending to be smaller in the largely or completely mycoheterotrophic genera), radially symmetrical, and isopolar. It ranges from prolate (the most frequent shape) to spheroid or slightly oblate. Pollen grains are tricolporate, less often tetracolporate, pentacolporate, pericolporate, or triporate; triporate grains prevail in tribe Helieae. Both the exine and intine are well developed, the exine consisting of nexine and sexine, the sexine of endexine and ectexine. In most genera the sexine is striate, striato-reticulate, or finely to coarsely reticulate or occasionally punctuate; rarely it is smooth (*Halenia*). Spines, verrucae, globules, loops, or other elaborate exine projections are often present in tribe Helieae. The tectum is generally supported by columellae, and granules or other subtectal processes are sometimes present. The pores are usually circular, occasionally slightly lalongate or lolongate. In several genera of subtribe Coutoubeinae and tribe Helieae, the pollen is shed as tetrahedral, decussate, or rarely linear tetrads. Tetrahedral and decussate tetrads both often occur within one species. Polyads occur in a few species of tribe Helieae and also in the *Swertia macrosperma* group.

The pollen of most species of *Voyria* is distinctly different from that of the rest of the family (Nilsson and Skvarla 1969; Maas and Ruyters 1986; Nilsson 2002). In this genus, the pollen grains are small, 11–15(–31) μm in the longest dimension, and 1–2(–6)-porate. They vary in shape from spheroid through compressed-ellipsoid, compressed-hemispheroid or hemiellipsoid, reniform, and ± irregular. The exine is usually smooth, occasionally finely scabrous, and is usually not distinctly stratified, but rarely has a thick columellar ectexine and thin endexine.

Karyology. The most common basic numbers are $x = 10$, 11, and 13; $x = 6$, 7, 9 are also relatively frequent. The basic number for the family may be $x = 7$ (Raven 1975), but data are missing for a large majority of the tropical and phylogenetically basal taxa.

Chromosome numbers are constant within some larger genera—for example, $n = 11$ in *Halenia*—or occur in a euploid series only. In some genera, such as *Gentiana*, basic chromosome numbers are usually constant only within groups usually treated as sections, e.g., *Gentiana* sect. *Ciminalis* ($n = 18$) and sect. *Pneumonanthe* ($n = 13$). Euploid and dysploid differences among species are common in other genera, as well as aneuploid differences and variable numbers of B-chromosomes in *Zeltnera* (Chironieae) and tribe Exaceae (Mansion and Zeltner 2004; Kissling 2008; Kissling et al. 2009b).

Pollination and Animal Interactions. Most gentians are insect-pollinated, a large majority by bees. Most flowers are highly productive of pollen and, except in tribe Chironieae, also high in nectar production. Many species have fragrant flowers. Bat-pollinated flowers of some *Fagraea* s.l. are strongly malodorous. Blue, violet, rose-violet, and pink corollas are common, sometimes with contrasting nectar- or pollen-guidelines or marks on the lobes or in the center. These flowers attract bees, especially bumblebees, which can reach nectar inside long corolla tubes or, in the case of bottle gentians' permanently closed corollas, nectar that is inaccessible to smaller insects. Species with abundant nectar and open, rotate corollas, such as *Frasera*, *Gentiana lutea*, and many *Swertia*, are pollinated by a diversity of bees, beetles, flies, and other insects. Many neotropical Helieae have whitish or pale green, long-tubed corollas lacking nectar guides, which attract moths, hawkmoths, or bats (Machado et al. 1998). Pollination by Lepidoptera is also known in *Gentiana* sect. *Calathianae* and *Tachiadenus*. Reddish corollas with long tubes are common among neotropical gentians of several tribes, and these attract hummingbirds, as do some species with yellow or yellowish-green corollas and sometimes with corollas of other colors. Sympatric *Lisianthius* species sometimes contrast markedly in corolla shape and color, and in pollinator adaptions, including adaptations to nectar- and pollen-gathering bees, butterflies, moths, and hummingbirds. Pollen is the only reward provided by the flowers of *Centaurium*, *Orphium*, and related genera in tribe Chironieae and also in *Exacum*, and may be the primary reward in some other genera. In most genera the pollen is moist and sticky, readily adhering to insects or other pollinators and to the moist, papillate stigmas. Dry, non-sticky pollen occurs in most *Exacum*, which have buzz pollination, which is also present in *Chironia* and *Orphium*. Yellow to orange inflorescence branches and calyces frequently contribute to the floral displays of *Potalia* and *Tachia*.

Completely or predominantly autogamous flowers are found in some *Centaurium*, *Sabatia*, and *Zeltnera*, and at least facultatively autogamous species are known in *Cicendia* and arctic/alpine *Comastoma*, *Gentianella*, and *Gentiana* sect. *Chondrophyllae* (Spira and Pollak 1986; Schönswetter et al. 2004). In *Bisgoeppertia*, a delicate projection from the rim of the stigma forms a cylinder around the style, which later splits longitudinally into two parts and adheres to the style (Gilg 1939). Pollen from the introrsely dehiscent anthers adheres to the surface of this cylinder before the flower opens, presumably effecting autogamous pollination. In *Sebaea*, presence of extra stigmas that occurs midway down the style is thought to enable delayed autonomous self-pollination, without limiting opportunities for outcrossing (Kissling and Barrett 2013a).

The extrafloral nectaries of some woody species in Helieae and Potalieae and the colleters in other tropical species attract ants, possibly constituting a defensive adaptation. The hollow stems of *Tachia* are sometimes inhabited by ants.

Fruit and Seed. Most fruits are septicidal, bivalvate capsules, with dry and thin to thick or rarely woody, leathery, or fleshy walls. Dehiscence is usually from the apex downward, but medial dehiscence is common in taxa with persistent styles (some *Bartonia*, *Voyria*, and many Helieae). Indehiscent dry fruits are present in some *Voyria*; subtribe Potaliinae and *Tripterospermum* sect. *Tripterospermum* have berries (Fig. 74C; up to 45 mm in diameter in *Fagraea*).

Seeds of most autotrophic species are small (seed size excluding appendages such as wings usually less than 1.5(–3) mm long), and often numerous (up to several hundred per capsule; Bouman et al. 2002). Ellipsoid or prismatic seeds are common; triquetrous, discoid-lenticular, and irregular forms also occur. The seeds are usually exotestal, i.e., only the outermost of the integumentary layers persists and forms the seed coat, the inner layers becoming compressed and partially or completely resorbed, as is the nucellus at an earlier stage. The seed coat is membranous or crustaceous, often finely reticulate from the collapse of the external walls of the cells, and shows great diversity, which has been taxonomically useful. Partial or circumferential wings or appendages are common in some genera. Most of the mycoheterotrophic species produce minute "dust seeds", which are highly reduced and ategmic in some taxa.

Germination requirements are diverse. The seeds of many temperate, arctic, and alpine perennial species require a cold period, whereas annuals are often capable of germination immediately upon ripening. Some species from dry habitats require a dry period of several weeks (e.g., *Centaurium*), which would be lethal for seeds species that grow in moister habitats. Light is often essential or enhances germination.

Dispersal. Small and light seeds, sometimes winged, that are wind-dispersed occur in a majority of gentians. The marcescent corollas of several genera (e.g., *Gentianopsis*) can become wrapped around the fruit, and thereby obstruct and prolong seed dispersal. The elongation of the ovarian stipe in many species of tribe Gentianeae elevates the capsule above the corolla and renders it more susceptible to breeze action. Capsules forming splash-cups are known in alpine *Gentiana*. Projections on the seed-coats, as in *Gentianopsis* and *Irlbachia*, may function in exozoochorous as well as anemochorous dispersal. Species with berries are dispersed endozoochorically by a variety of animals, from birds to small mammals including bats, and have evolved independently several times within the family as key innovations that appear to have increased speciation rates in those clades (e.g., *Tripterospermum*, Matuszak et al. 2016). Rain-wash and/or melt water from snowfields are probably significant dispersal agents within small areas for alpine *Gentiana*, but long-distance dispersal through muddy bird feet are also likely for *Gentiana* and many other gentians. Mycoheterotrophic species with dehiscent capsules and minute "dust seeds" are anemochorous, but for mycoheterotrophic species with water-repellant seeds, rain-wash and streams are probably more important. In some *Voyria* and perhaps *Curtia*, the entire fruit, although not fleshy, detaches at maturity and functions as a unit of dispersal, with the seeds being released upon the decay of the fruit wall (Maas and Ruyters 1986). Because these seeds are released on the ground, rain-wash is presumably important in their dispersal. Detached vegetative disseminules such as bulblets do not occur in the family.

Reproductive Systems. Bisexual, monomorphic flowers are the standard in Gentianaceae. Self-compatibility is prevalent, often with adaptations restricting self-pollination. Protandry is common, protogyny and adichogamy much less so (Bertin and Newman 1993). Herkogamy, present in some genera, occurs as two types: (1) in horizontally positioned flowers the style is declinate downward with anthers aggregated above the style, and (2) in actinomorphic, upward-facing flowers the style can be deflected to one side and the stamens all curved in the opposite direction. The first, horizontal type is found in some *Exacum*, most Helieae, and *Orphium*. The second type is present in some Chironieae (*Centaurium*, *Gyrandra*, *Sabatia*, and *Zeltnera*). In herkogamous flowers of some *Centaurium*, *Gyrandra*, *Sabatia*, and *Zeltnera*, the style and stamens become erect and closer to each other at a later stage, allowing autogamy if xenogamous or geitonogamous pollination has not occurred previously. Heterostyly is present in *Hockinia*, *Voyriella*, and some *Curtia*, *Exochaenium*, and *Tapeinostemon*. Some South American *Gentianella* are gynodioecious, gynomonoecious, trimonoecious, and perhaps andromonoecious, and some Australian and New Zealand *Gentianella* are gynodioecious. Both species of *Veratrilla* are dioecious. Mycoheterotrophic species of *Exacum* are probably agamospermic, despite their seemingly functional anthers, but there has been no definitive study (Fryxell 1957).

Hybridization. In some large genera, notably *Gentiana* and *Gentianella*, natural hybridization within sections is fairly common, and both diploids and tetraploids are common in *Gentiana* sections *Cruciata* and *Pneumonanthe*. *Gentiana* species, especially in sect. *Kudoa*, have been extensively hybridized in horticulture. Interspecific hybridization in nature is also significant in a few smaller genera, including *Centaurium* and *Zeltnera*, in which some species are of allopolyploid origin. Hybrids have also at least occasionally been reported in *Bartonia*, *Gentianopsis*, *Halenia*, *Macrocarpaea*, and *Sabatia*.

Phytochemistry. As in many Asteridae, iridoid glucosides are produced. These are of the Route I type, including normal and seco-iridoid compounds (Jensen 1992; Jensen and Schripsema 2002). Seco-iridoid glucosides are prominent and universally present, with gentiopicroside (gentiopicrin), swertiamarin, and sweroside being widespread, giving rise to the intense bitter taste of most members of the family. No species are cyanogenic or saponiferous, and caffeoyl esters of phenylethanoid glycosides (CPGs), which are very common in Lamiales, are absent (Jensen 1992). Cardiotonic glycosides or complex indole alkaloids, which are present in related families of the Gentianales, are not produced in Gentianaceae. Xanthones are of general occurrence but are absent in tribe Exaceae; together with Clusiaceae, this is the only family in which xanthones are common (Hostettmann and Wagner 1977). Aluminum is commonly accumulated, and mangiferin and C-glucoflavonoids are common. Tannins, protoanthocyanins, and ellagic acid are absent.

Subdivision, Phylogeny and Biogeography of the Family. The grouping of genera into tribes and subtribes in the present work follows the phylogenetic work and classification presented in Struwe (2014, which builds on Struwe et al. 2002; Fig. 68), with the addition of the acceptance of tribe Voyrieae, and the three new genera *Kuepferia*, *Sinogentiana*, and *Sinoswertia*. This is the most recent and most comprehensive treatment of the whole Gentianaceae, and is an update of Struwe et al. (2002); they replace Gilg's (1895) classification that had been based almost exclusively on pollen characters and was in use until late 20th century. Molecular systematics and the criteria of monophyletic groupings is the base for the new 2002 and 2014 classifications. Informal groupings of genera and phylogenetic work previous to the 2002 classification were based on

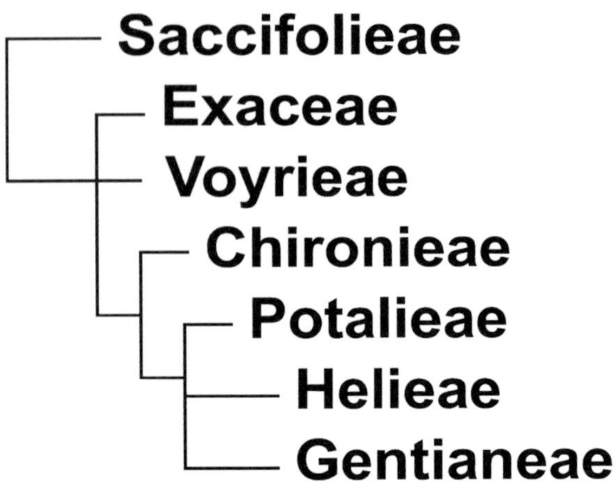

Fig. 68. Gentianaceae. Phylogenetic relationships among Gentianaceae tribes. (Drawing by Lena Struwe)

morphology, xanthone biochemistry, or limited taxonomic sampling (e.g., Wood and Weaver 1982; Mészáros 1994; Ho et al. 1994; Mészáros et al. 1996). Some of the tribes, notably the Gentianeae, still correspond quite closely to groups established by Gilg.

Family-level phylogenetic studies using molecular data have shown the monophyly of the family, and seven well-supported clades within the family, treated as the seven tribes Saccifolieae, Exaceae, Voyrieae, Chironieae, Gentianeae, Helieae, and Potalieae (Thiv et al. 1999; Struwe et al. 2002; Merckx et al. 2013; Yang et al. 2016; Fig. 68). Morphology appears to be of less value in family-level phylogenetic reconstructions because of frequent parallel evolution of similar traits, such as woodiness, anther morphology, and pollen aggregation (Mészáros et al. 2002), but has been useful at lower levels (e.g., von Hagen and Kadereit 2002; Mansion and Struwe 2004; Kissling et al. 2009b; Calió et al. 2017). However, palynological and seed anatomical characters show strong agreement with molecular data (Nilsson 2002; Bouman et al. 2002; Nilsson et al. 2002; Calió et al. 2017), and phytochemical data also include phylogenetically valuable data (Jensen and Schripsema 2002).

The tribe placed as the first branch closest to the ancestral root in Gentianaceae is the neotropical Saccifolieae (Fig. 69), for which detailed tribal and generic level phylogenetic studies are still lacking.

The only strictly paleotropical tribe, Exaceae (Fig. 70), is the sister to the rest of the family above Saccifolieae, and its ancestral area appears to be in Africa with subsequent dispersal to the Arabian Peninsula and South Asia (Yuan et al. 2005; Pirie et al. 2015). The small mycoheterotrophic genus *Cotylanthera* has been found to be an embedded branch within autotrophic *Exacum* and has been transferred to *Exacum* (Klackenberg 2006). Phylogenetic studies have now resolved the large, previously paraphyletic genus *Sebaea* by the acceptance and resurrection of the genera *Lagenias* and *Exochaenium*, and the description of the new genus *Klackenbergia* (Kissling et al. 2009a, b, c; Kissling 2012).

Next up is Chironieae (Fig. 71), which is subdivided into three monophyletic subtribes: Canscorinae, Coutoubeinae, and Chironiinae. Within paleotropical Canscorinae there are evolutionary trends toward reduction of staminal parts, and a complete revision and new classification has been presented (Thiv and Kadereit 2002; Thiv 2003). Subtribe Chironiinae occurs primarily in subtropical and temperate environments, with a historic expansion from the Mediterranean area (Mansion and Struwe 2004). Placed as sister to the rest within this subtribe is a clade formed by *Blackstonia* and *Ixanthus*. The genus *Centaurium* s. lat. was paraphyletic and has recently been reclassified into four monophyletic genera by Mansion (2004), which is reflected here. Very little phylogenetic work has been performed in subtribe Coutoubeinae.

Most phylogenetic work has been done in tribe Gentianeae, home of the well-known blue alpine gentians and mostly confined to temperate or tropical mountain areas. It is clear that there are two major clades in the tribe, identified as subtribe Gentianinae and Swertiinae (Gielly and Taberlet 1996; Thiv et al. 1999; Struwe et al. 2002). In Gentianinae, six genera are now included, the largest gentian genus *Gentiana* with about 370 species, and a group of smaller genera, some of which are recent segregates of *Gentiana*: *Crawfurdia*, *Kuepferia*, *Metagentiana*, *Sinogentiana*, and *Tripterospermum* (Yuan and Küpfer 1995; Chen et al. 2005; Favre et al. 2010, 2014). All these genera most likely originated in the region of the Qinghai-Tibet Plateau, the current centre of diversity for this entire clade (Favre et al. 2016).

Within *Gentiana*, several studies using molecular, morphological, karyological, or palynological data have begun to resolve sectional relationships and circumscription (Gielly et al. 1996; Yuan et al. 1996; Yuan and Kupfer 1997; Favre et al. 2016). Subtribe Swertiinae includes the two large (and, as traditionally circumscribed, paraphyletic) genera *Gentianella* and *Swertia*, and a handful of smaller genera that have been shown to be intermixed within clades of either *Swertia* or *Gentianella* species (von Hagen and Kadereit 2001, 2002, 2003; Chassot et al. 2001; Liu et al. 2001; Chassot and von Hagen 2008). A new classification, preferably dividing *Swertia* into smaller, morphologically identifiably units and transferring a small group of *Gentianella* species to other genera, instead of including genera such as *Halenia* in a large, unmanageable *Swertia*, is needed. *Halenia*, most diverse in the Andes of South America, has evolved from a *Swertia*-like spurless ancestor from Asia, which spread across the Bering Strait into North America (von Hagen and Kadereit 2003). Similarly, long-distance dispersal within *Gentianella* allowed the colonization of New Zealand and Australia and led to a large local radiation (von Hagen and Kadereit 2001). Evolution among European *Gentianella* species has been investigated by Yuan and Kupfer (1997), Jang et al. (2005), and Greimler et al. (2004).

Within the tropical tribes Potalieeae and Helieae, forming a polytomy with tribe Gentianeae, several more detailed evolutionary studies are underway but not yet published. Pantropical tribe Potalieae is divided into three subtribes, Faroinae, Lisianthiinae, and Potaliinae (Struwe et al. 2002; Struwe 2014; Molina and Struwe 2009). DNA sequencing data have placed *Bisgoeppertia*, which was formerly included in the Chironieae, as sister to *Lisianthius* in subtribe Lisianthiinae. In Potaliinae, the group with the largest trees in the family, African *Anthocleista* and American *Potalia* are placed as the sister group of a group of Australasian genera, including *Fagraea*. *Fagraea* was recently split into several new and resurrected genera, viz., *Cyrtophyllum*, *Fagraea*, *Limahlania*, *Picrophloeus*, and *Utania*, as a result of detailed phylogenetic analyses (Wong 2012; Sugumaran and Wong 2012, 2014; Wong and Sugumaran 2012a, 2012b, 2016; Wong et al. 2013).

Neotropical Helieae includes several well-defined clades, as well as many poorly known genera of uncertain position within the tribe (Struwe et al. 2002; Grant et al. 2006; Calió et al. 2017). Most basally placed are *Prepusa* and *Senaea*, endemics of eastern Brazil with pollen tetrads of a unique type. In the core Helieae, three clades are identified, the *Macrocarpaea* clade with *Chorisepalum*, *Macrocarpaea*, *Tachia*, and *Zonanthus*, the monotypic *Irlbachia* clade, and the *Symbolanthus* clade with the remaining genera (Calió et al. 2017). The *Macrocarpaea* group is also identified by pollen in tetrads, whereas *Irlbachia* has polyads, and the *Symbolanthus* group either tetrads or polyads. The biogeographical history of Helieae is very complex, but basal clades all occur on low-nutrient substrates, and Andean species represent more recent speciation events.

Family Circumscription. In the classifications of Dahlgren (1980), Cronquist (1981), Thorne (1992), Takhtajan (1997), and the Angiosperm Phylogeny Group (all four versions; APG 2009), the family Gentianaceae belongs to the order Gentianales, which has consistently appeared monophyletic in cladistic analyses. Family relationships within Gentianales are still being investigated, but most often Rubiaceae is the family placed on the most basal branch, with Gentianaceae on the following branch above (Yang et al. 2016). Based on gross morphology, the Gentianaceae are most similar to Loganiaceae, with both families having superior ovaries, and without the intricate stylar and androecial morphology of many Apocynaceae. The Gentianaceae differ in usually having unilocular ovaries and a markedly different chemistry. Interpetiolar stipules or stipular lines are common in Loganiaceae, but less so in Gentianaceae.

Until the mid-1990s, the genera comprising subtribe Potaliinae were usually included in the Loganiaceae as tribe Potalieae, although their transfer to the Gentianaceae had already been suggested by Perrot (1897) and others. Molecular, morphological, and phytochemical data (presence of specific seco-iridoids) strongly support the placement of this tribe in Gentianaceae within one of its major clades (Jensen 1992; Struwe et al. 1994, 2002; Jensen and Schripsema 2002).

Saccifolium was initially placed in a monotypic family, but was soon thereafter included in the Gentianaceae in Thorne's 1983 classification, which molecular and morphological data support (Thiv et al. 2000; Struwe et al. 2002).

Historically the Menyanthaceae were treated as a subfamily of the Gentianaceae, but are now accorded family status. Morphological, anatomical, embryological, and molecular evidence supports its placement in the Asterales.

Distribution and Habitats. The Gentianaceae are present worldwide except continental Antarctica, and are native to North, Central, and South America, the West Indies, Europe, Asia, Africa, Australia, and the Pacific Islands, with centers of species diversity in montane regions of North and South America, Europe, and southern and eastern Asia (Albert and Struwe 2002). Few genera (but many *Fagraea* and *Gentiana*) are present in the western Pacific Islands, and relatively few species, all of limited distribution, in Australia. On the other continents the family is widely distributed and represented by many genera and species. The largest genera (*Gentiana*, *Gentianella*, *Macrocarpaea*, and *Swertia*) are largely or entirely restricted to temperate regions and montane zones in the tropics. The genera in the low and mid-altitudes of the tropics are smaller or less often medium-sized. Some of this is an artifact of taxonomic bias by botanists in different geographic regions or taxonomic expertise, since some of the large genera have been found to be paraphyletic (and more inclusive) and smaller genera have been favored by some tropical botanists. Species occur in relatively undisturbed habitats as well as in disturbed sites, from dry-mesic to moist sites. Gentians can grow in bogs, marshes, and other wet or periodically inundated habitats, but none is truly aquatic.

Paleobotany. The fossil record of Gentianaceae is very scant and so far is limited to only a few pollen and seed reports. Pollen of middle to late Eocene age in Central America has been identified as that of two species of *Lisianthius*, resembling modern *L. glandulosus* and *L. nigrescens* (Graham 1984). Seeds of Pliocene age have been found from *Gentiana* sect. *Cruciata* (see Pirie et al. 2015, and references therein). Tertiary pollen (*Pistillipollenites macgregorii*; Rouse and Srivastava 1970) resembles that of *Chelonanthus* and related genera, but cannot be placed in a more detailed taxonomic context. Texan floral fossils from the Eocene with *Pistillipollenites macgregorii* pollen have formerly been identified as gentianaceous (Crepet and Daghlian 1981; Stockey and Manchester 1986), but do not fit the general gentian flower bauplan and are not gentians (Struwe et al. 2002).

Economic Importance. Gentians are primarily known for their economic value as medicinal and horticultural plants worldwide. The most commercially important horticultural crops are *Eustoma exaltatum* ssp. *russellianum*, which is used as cut flowers, bedding and pot plants, and the potted plant *Exacum affine*. Recently breeding and genetic transformation in *Eustoma* in Japan and New Zealand has produced many new varieties and color forms. Other species, especially in *Gentiana*, are valued as garden ornamentals and favorites of rock gardeners.

Decoctions of the roots, leaves, or sometimes of whole plants are used medicinally (Jensen and Schripsema 2002). The massive caudex and tap root of *Gentiana lutea* are extensively used in herbal medicine, especially in Europe, and are much used in flavoring of beverages and bitters. *Swertia chirayita* and *Gentiana kurroo* are much used in Asian herbal medicine. Species of *Anthocleista*, *Blackstonia*, *Canscora*, *Centaurium*, *Chelonanthus*, *Chironia*, *Exacum*, *Fagraea*, *Frasera*, *Gentiana*, *Gentianella*, *Gentianopsis*, *Halenia*, *Potalia*, *Sabatia*, *Sebaea*, *Swertia*, *Tachia*, *Zeltnera*, and many other genera have traditional uses in herbal or folk medicine, chiefly for tonics and digestive aids, sometimes as febrifuges, antifungals, antihelminthics, antimalarials, and diuretics, and occasionally for other purposes.

The pharmacological activity of gentians is due to xanthones and secoiridoids (Jensen and Schripsema 2002). The xanthones are strong inhibitors of monoamine oxidase and show antidepressant action and tuberculostatic effects (Pickert et al. 1998; Tomic et al. 2005). Gentianine shows strong antiinflammatory activity. Hypoglycemic and antifungal activity that has inhibitive effects on hepatitis and HIV infections and anticholinergic properties have been found in *Swertia* extracts. Some genera show promise in antimalarial studies, consistent with their use in traditional medicine. Their efficacy against malarial *Plasmodium* is attributed to secoiridoids. Some *Fagraea* s.l. species are important timber trees in Southeast Asia. They are valued for the durability and rot-resistance of their wood, which is used for building construction and pilings.

Key to Tribes and Subtribes of Gentianaceae

1. Shrubs and trees always with fleshy fruits **6c. Potalieae-Potaliinae**
– Herbs with dry fruits (rarely fleshy), if trees and shrubs, then with dry or rarely leathery fruits 2
2. Plants without chlorophyll (and always with calyx tube clearly longer than calyx lobes) **3. Voyrieae**
– Plants with at least some chlorophyll, or if without chlorophyll (*Exacum* spp., *Exochaenium* spp., and *Voyriella*), then with calyx tube shorter or at most as long as calyx lobes 3
3. Calyx with dorsal, thickened, glandular areas on lobes (rarely keeled with or without glands); style flattened and twisted lengthwise after anthesis and in fruit; stigma bilamellate **7. Helieae**
– Calyx without dorsal glandular areas on lobes, but sometimes strongly keeled; style not flattened and not twisted lengthwise after anthesis and in fruit; stigma simple, capitate, peltate, to linear (rarely bilamellate) 4
4. Gynoecium clearly bilocular for at least half of its length (rarely pseudobilocular); testa cells usually star- or puzzle-piece-shaped; anthers often with apical or basal glands/knobs **2. Exaceae**
– Gynoecium unilocular or pseudobilocular (i.e., two locules formed by parietal placentas intruding into center of ovary); testa cells with straight or undulating cell walls; anthers without glands/knobs 5
5. Anthers with broad and often extended connectives; plants usually heterostylous **1. Saccifolieae**
– Anthers without broad or extended connectives; plants never heterostylous 6
6. Corolla with plicae (elongated folds between corolla lobes, rarely absent); with intracalycine membrane between calyx lobes (rarely absent) **5a. Gentianeae-Gentianinae**
– Corolla without plicae; without membrane between calyx lobes (rarely present) 7
7. Ovules usually distributed over entire inner wall of gynoecium (rarely with distinct placentas in *Swertia*); one or two corolline nectaries per petal lobe or nectaries on spurred petals; glands rarely present at the base of the ovary, if no glands at all, then filaments inserted in sinuses of corolla lobes **5b. Gentianeae-Swertiinae**
– Ovules always on distinct placentas or on indistinct placentas along carpel sutures only; corolla never with corolline glands or spurs, but nectariferous disk can be present at base of ovary and then stamens are nearly always inserted in corolla tube 8
8. Pollen released as tetrads; corolla 4-merous **4c. Chironieae-Coutoubeinae**
– Pollen released as monads; corolla usually 5–13-merous (rarely 3–4-merous flowers) 9
9. Corolla with corolline fibers that are easily seen in dried flowers as thin white lines; with nectariferous disk at base of ovary (rarely absent) **6b. Potalieae-Lisianthiinae**
– Corolla without corolline fibers (rarely present); without nectaries at base of ovary 10
10. Flowers sessile or subsessile, solitary or clustered in leaf axils or forming a capitulum; appendages usually present in the corolla tube at the base of each stamen **6a. Potalieae-Faroinae**
– Flowers pedicellate (rarely sessile), in lax inflorescences (rarely not); staminal appendages never present in corolla tubes 11
11. Calyx tubes always longer than calyx lobes; anthers straight after anthesis, never spirally twisted **4b. Chironieae-Canscorinae**
– Calyx tubes shorter than calyx lobes (rarely longer); anthers often spirally twisted or recurved after anthesis **4a. Chironieae-Chironiinae**

Classification of the Gentianaceae

1. Tribe Saccifolieae (Maguire & Pires) Struwe, Thiv, V. A. Albert & Kadereit (2002).
   Genera 1–5
2. Tribe Exaceae Colla (1834).
   Genera 6–13
3. Tribe Voyrieae Gilg (1895).
   Genus 14
4. Tribe Chironieae (G. Don) Endl. (1838).
   a. subtribe Chironiinae G. Don (1838).
      Genera 15–28
   b. subtribe Canscorinae Thiv & Kadereit (2002).
      Genera 29–35

c. subtribe Coutoubeinae G. Don (1838). Genera 36–40

5. Tribe Gentianeae Dumort. (1829).
   a. subtribe Gentianinae G. Don (1838). Genera 41–46
   b. subtribe Swertiinae (Griseb.) Rchb. (1837). Genera 47–61
6. Tribe Potalieae Rchb. (1837).
   a. subtribe Faroinae Struwe & V. A. Albert (2002). Genera 62–70
   b. subtribe Lisianthiinae G. Don (1838). Genera 71–72
   c. subtribe Potaliinae (Mart.) Progel (1865). Genera 73–79
7. Tribe Helieae Gilg (1895). Genera 80–102

Key to Genera

1. Chlorophyll absent (leaves white, yellow, pink, or purple) 2
– Chlorophyll present (leaves green; chlorophyll sometimes scant in *Bartonia*) 5
2. Anthers porate (dehiscent only near apex) **10. *Exacum*** (spp.)
– Anthers longitudinally dehiscent 3
3. Calyx free or nearly so, calyx tube absent or very short and indistinct **5. *Voyriella***
– Calyx lobes united at base into a distinct tube 4
4. Plants with underground cleistogamous flowers; glandular disk present between calyx and corolla; anthers with terminal glands (Africa) **7. *Exochaenium*** (*E. oliganthum*)
– Plants without underground cleistogamous flowers; no glandular disk between calyx and corolla; anthers without terminal glands (Neotropics) **14. *Voyria***
5. Plants woody (with at least primary stems distinctly woody; trees or shrubs) 6
– Plants herbaceous (including perennial suffrutescent herbs that are woody only near the base, non-woody vines, and perennial and monocarpic herbs) 31
6. Inflorescences or solitary flowers axillary; stems and branches do not end in a terminal flower 7
– Inflorescences or solitary flowers terminal (sometimes also with some axillary inflorescence branches) 12
7. Leaves saccate (i.e., with the midrib and the blade curved downward toward apex and margins, the distal part of the leaf thus hood-like) **1. *Saccifolium***
– Leaves flat, not hood-like 8
8. Two bracts fused to calyx **85. *Zonanthus***
– Bracts absent or free, not fused to calyx 9
9. Fruit a berry (tropical Asia) **73. *Cyrtophyllum***
– Fruit a dry capsule 10
10. Stigma distinctly bilobed **84. *Tachia***
– Stigma slightly bilobed to capitate 11
11. Corollas with fiber bundles (seen as white lines in old corollas); leaves sessile or, if petiolate, without stipule-like lateral expansions (Central America and Caribbean) **72. *Lisianthius***
– Corollas without fiber bundles; leaf bases petiolate, with stipule-like lateral expansions (Madagascar) **11. *Gentianothamnus***
12. Fruit a berry 13
– Fruit a dry capsule 18
13. Corollas 8–16-merous (Neotropics, Africa, Madagascar) 14
– Corollas 5-merous (Asia, Australia, Pacific) 15
14. Corollas 8–10-merous (Neotropics) **79. *Potalia***
– Corollas 12–16-merous (Africa, Madagascar) **78. *Anthocleista***
15. Inflorescence a pendulous cyme with all branches condensed along the rachis **74. *Utania***
– Inflorescence an upright cyme, often dichasial 16
16. Style exserted more than 40% from mouth of corolla; stigma capitate **75. *Picrophloeus***
– Style not exserted more than 40% from mouth of corolla; stigma peltate or bilobed 17
17. Young stems with prickles; leaf margins crenulate **76. *Limahlania***
– Stems without prickles; leaf margins entire **77. *Fagraea***
18. Corollas 6-merous 19
– Corollas 4–5-merous 21
19. Calyx 4-merous **82. *Chorisepalum***
– Calyx 6-merous 20
20. Calyx more than 18 mm long and over 10 mm wide **80. *Prepusa***
– Calyx less than 15 mm long and less than 10 mm wide **81. *Senaea***
21. Androecium actinomorphic 22
– Androecium zygomorphic, with stamens curved upward, downward, or laterally in corolla or corolla mouth 24
22. Stamens immediately below sinuses between corolla lobes; pollen in tetrads **89. *Aripuana***
– Stamens inserted in corolla tube, distinctly below sinuses; pollen in monads 23
23. Corollas with fiber bundles (seen as white lines in old corollas); filaments long, distinct; anthers

free (Central America and Caribbean) **72. *Lisianthius***
- Corollas without fiber bundles; filaments short; anthers connate (Madagascar) **13. *Tachiadenus***

24. Corolla tube distinctly inflated below corolla mouth; flowers pendulous 25
- Corolla tube not strongly inflated below corolla mouth; flowers erect to horizontal (rarely nodding) 27

25. Corolla over 9 cm long, with red-and-yellow tube and green lobes; corolla tube abruptly constricted below limb; stamens and style included within corolla tube **92. *Lagenanthus***
- Corolla less than 6 cm long, completely red above apex of calyx; corolla tube not or slightly constricted below limb; stamens and/or style exserted beyond throat of corolla 26

26. Pedicels mostly more than 2 cm long; stigmatic lobes linear, ca. 0.5 mm wide, scarcely wider than style **93. *Lehmanniella***
- Pedicels generally less than 2 cm long; stigmatic lobes ca. 1.0 mm wide, distinctly wider than style and abruptly differentiated **94. *Purdieanthus***

27. Apex of corolla acuminate in bud 28
- Apex of corolla bluntly rounded in bud 29

28. With staminal corona at base of stamens; corollas more than 40 mm long **101. *Symbolanthus***
- Without staminal corona at base of stamens; corollas less than 40 mm long **88. *Neblinantha*** (*N. neblinae*)

29. Both leafy and scale-like bracts in inflorescence **83. *Macrocarpaea***
- Only small, scale-like bracts in inflorescence 30

30. Leaves coriaceous **99. *Rogersonanthus***
- Leaves membranaceous **95. *Adenolisianthus***

31. Plants twining and/or creeping 32
- Plants not twining or creeping (sometimes ± sinuous or thigmonastic in *Bartonia*, with very small, scale-like leaves) 36

32. Corolla 4-merous 33
- Corolla 5-merous 34

33. Calyx tube longer than calyx lobes **50. *Pterygocalyx***
- Calyx lobed nearly to base **21. *Geniostemon*** (*G. artajanus*)

34. Without plicae (folds of extra petal tissue) in sinuses between corolla lobes (Caribbean) **71. *Bisgoeppertia***
- With plicae in sinuses between corolla lobes (Asia) 35

35. Nectaries 5, on gynophores; stamens of equal length, straight **44. *Crawfurdia***
- Nectaries forming a collar around gynophores; stamens of unequal length, recurved **46. *Tripterospermum***

36. With plicae (folds of extra petal tissue, sometimes similar to corolla lobes) in sinuses between corolla lobes (reduced to small folds in *Kuepferia*) 37
- Without plicae in sinuses between corolla lobes (but lobes sometimes fringed) 40

37. Corolla tube much shorter than corolla lobes; plicae small and reduced **42. *Kuepferia***
- Corolla tube longer than corolla lobes; plicae well-developed 38

38. Stamens of equal length, straight **41. *Gentiana***
- Stamens of unequal length, recurved 39

39. Corollas blue, purple, or pink **43. *Metagentiana***
- Corollas yellow or white **45. *Sinogentiana***

40. Androecium zygomorphic, with stamens curved upward, downward, or laterally in corolla 41
- Androecium actinomorphic 58

41. Calyx lobes without thickened, glandular dorsal area; flowers without nectary disk at base of ovary 42
- Calyx lobes dorsally thickened, with glandular area on each lobe (sometimes keeled as well); flowers with nectary disk at base of ovary 44

42. Uppermost leaves not perfoliate **33. *Duplipetala***
- Uppermost leaves perfoliate 43

43. Corolla 5-merous; calyx tube longer than calyx lobes (Asia) **35. *Phyllocyclus***
- Corolla 4-merous; calyx tube shorter than calyx lobes (South America) **40. *Symphyllophyton***

44. Leaves pandurate (fiddle-shaped); flowers oppositely paired in upper leaf axils **102. *Yanomamua***
- Leaves ovate, obovate, lanceolate, elliptic, or linear (never pandurate); flowers in monochasial or dichasial, terminal cymes 45

45. Apex of corolla acuminate in bud 46
- Apex of corolla bluntly rounded in bud 51

46. With staminal corona at base of stamens in corolla tube (corona rarely absent; if so, small staminal pockets are present in corolla tube at insertion site of each filament) **101. *Symbolanthus***
- Without staminal corona at base of stamens, filament simply inserted in corolla tube 47

47. Perennial 48
- Annual 49

48. Flowers solitary **87. *Celiantha***
- Flowers in inflorescences of 2 or more **88. *Neblinantha*** (*N. parviflora*)

49. Corolla more than 40 mm long **96. *Calolisianthus***
- Corolla less than 40 mm long 50

50. Leaves in more than 6 pairs, most of them at or near the base of the stem; corollas (18–)20–36 mm long; pollen in tetrads **91. *Tetrapollinia***
- Leaves in 6 or fewer pairs, not most numerous near the base of the stem; corollas 6–20(–25) mm long; pollen in polyads **86. *Irlbachia*** (*I. pratensis*)

51. Corolla green, yellow, cream, or white 52
- Corolla red, blue, purple, or pink 55

52. Both leaf-like and scale-like bracts present within inflorescence 53
- All bracts in inflorescence small, scale-like 54

53. Corolla funnel-shaped, salver-shaped, or nearly rotate; corolla lobes at least more than 1/3 of the length of the corolla **86. *Irlbachia*** (spp. with greenish-white corollas)
- Corolla urceolate; corolla lobes about 1/6 of the length of the corolla or less **83. *Macrocarpaea*** (*M. rubra*)

54. Inflorescence branches long and monochasial (forming elongated, Y-, V-, or W-shaped inflorescences) **97. *Chelonanthus***
- Inflorescence a compact cyme with short branches **98. *Helia***

55. Leaves thickly coriaceous **100. *Sipapoantha***
- Leaves membranaceous 56

56. Ultimate divisions of inflorescence elongating, racemoid; corolla tube more than twice as long as corolla lobes **86. *Irlbachia*** (*I. tatei*)
- Inflorescence ± dichasial throughout, not racemoid; corolla tube less than twice as long as corolla lobes 57

57. Corolla more than 30 mm long, blue-violet to rose-violet; corolla lobes shorter than corolla tube; persistent style longer than capsule **97. *Chelonanthus*** (*C. purpurascens*)
- Corolla less than 20 mm long, purplish white to white or pink; corolla lobes about as long as corolla tube; persistent style about as long as capsule **86. *Irlbachia*** (*I. nemorosa*)

58. Stamens anisomorphic (of strongly different lengths, some sometimes lacking anthers) 59
- Stamens isomorphic 62

59. Corolla zygomorphic **30. *Canscora***
- Corolla actinomorphic 60

60. Leaves very small, scale-like; stamens inserted in corolla sinuses (Africa) **32. *Schinziella***
- Leaves larger; stamens inserted in corolla tube distinctly below sinuses 61

61. Corolla distinctly longer than calyx (Central America) **72. *Lisianthius***
- Corolla shorter than or about as long as calyx (Asia) **31. *Hoppea***

62. Nectaries on gynophores 63
- Nectaries either epipetalous, a disk at the base of the ovary, or absent 64

63. Flowers pendulous; corolla campanulate, yellowish white with prominent brownish veins **51. *Megacodon***
- Flowers not pendulous; corolla rotate, yellow, with veins not contrastingly colored **41. *Gentiana*** (*G. lutea*)

64. Corolla rotate or nearly so (rarely ± campanulate; if so, with very short corolla tube); corolla lobes more than 5× as long as corolla tube 65
- Corolla shapes various but generally not rotate; corolla lobes less than 5× as long as corolla tube 77

65. Anthers recurving or coiling helically after dehiscence; nectaries absent or indistinct **25. *Sabatia***
- Anthers remaining straight; nectaries prominent, on corolla lobes 66

66. Stigma terminal and decurrent along sutures of ovary **57. *Lomatogonium***
- Stigma only terminal 67

67. Species dioecious, pistillate flowers with sterile stamens, staminate flowers with a sterile pistil (Asia) **53. *Veratrilla***
- Species hermaphroditic, all flowers with functional pistils and stamens 68

68. Nectaries as patches on adaxial surface of corolla lobes, not in pits with raised or appendaged margins (Asia) 69
- Nectaries sharply defined, in pits on adaxial surface of corolla lobes, with ± raised, often fimbriate or tubular margins 70

69. Flowers heteromorphic, two types differing in size, and cleistogamous **61. *Sinoswertia***
- Flowers monomorphic, not cleistogamous **60. *Swertia***

70. Opening of nectary pit with a scale to one side but rim neither fringed nor with a tubular extension (Asia) **60. *Swertia***
- Opening of nectary pit (with or without adjacent part of adaxial surface of corolla lobe) surrounded by a raised, fringed or tubular rim 71

71. Opening of nectary pit distal to nectary itself; area of a different color and texture present on corolla lobe immediately distal to opening of nectary pit or pair

of pits, with raised rim surrounding the pit opening plus this area; styles more than 2 mm long, ± sharply differentiated from ovary; leaves narrowly white-margined (North America) **54. *Frasera***
- Opening of the nectary pit or pits (if more than one per petal lobe) immediately adaxial (above) nectary itself; raised rim surrounding only the opening, not surrounding an area of different color and texture on corolla lobe distal to opening of pit; styles less than 2 mm long, not sharply differentiated from ovary 72

72. Leaves whorled; flowering stems proximally more than 2 cm in diameter (North America) **54. *Frasera***
- Leaves opposite and/or alternate, or, if whorled (in Asian or African spp.), then with flowering stems less than 2 cm in diameter 73

73. Nectaries 2 per corolla lobe **60. *Swertia***
- Nectaries 1 per corolla lobe 74

74. Filaments completely separate, not connected by a staminal corona **60. *Swertia***
- Filaments connected by a staminal corona, the rim of which often bears trichomes or scales extending between the corolla lobes 75

75. Rim of nectary-pit openings smooth, without trichomes, nor prolonged into a tubular extension **60. *Swertia***
- Rim of nectary-pit openings with trichomes or (in 1 North American sp.) a tubular extension 76

76. Leaf margins minutely and irregularly laciniate; corollas brick-red, drying yellowish green (Asia) **60. *Swertia*** (*S. splendens*)
- Leaf margins smooth; corollas white, yellowish-green, or blue **54. *Frasera***

77. Corolla 3-merous (tropical Africa) **68. *Pycnosphaera***
- Corolla 4–12-merous 78

78. Anthers porate (dehiscent only near apex) **10. *Exacum***
- Anthers longitudinally dehiscent 79

79. Calyx coriaceous, deeply divided into 2 parts, each part entire or shallowly 2- or 3-lobed; corolla without adaxial scales or trichomes (Asia) **34. *Microrphium***
- Calyx not deeply divided into 2 parts (or, if ±2-parted in *Gentianella heterophylla*, calyx not coriaceous and corolla with deeply fringed scales or trichomes near base of lobes) 80

80. Corolla 6–12-merous 81
- Corolla 4–5-merous 83

81. Calyx tube longer than calyx lobes; corolla 6-merous, largely concealed by calyx (South America) **80. *Prepusa***
- Calyx deeply lobed, corolla 6–12-merous, showy, much exceeding calyx 82

82. Corolla yellow; corolla lobes much longer than tube, without appendages **15. *Blackstonia***
- Corolla blue; corolla lobes slightly longer than tube, each terminating in two sublobes, between which the midvein is abruptly prolonged into a long-setaceous appendage **70. *Urogentias***

83. Base of filaments (where inserted on corolla tube) swollen or appendaged, or corolla with adaxial scales shortly below bases of filaments 84
- Base of filaments neither swollen nor appendaged (but filaments arise between corolla appendages in *Djaloniella*) 90

84. Simple scale at or shortly below base of filaments (sometimes deeply fimbriate and glandular, but neither spherical nor hooded) 85
- Hooded scale or spherical protuberance at base of filaments 87

85. Scales shortly below base of filaments deeply fringed, glandular; veins not prominently raised on abaxial calyx surface (except sometimes midveins and commissural veins) **64. *Faroa***
- Scales at base of filaments not fringed or glandular; abaxial surface of calyx with prominently raised veins (including secondary veins) 86

86. Calyx tube much longer than calyx lobes **66. *Neurotheca***
- Calyx tube shorter than to about as long as calyx lobes **62. *Congolanthus***

87. Spherical protuberance at base of filaments; calyx tube longer than calyx lobes **67. *Oreonesion***
- Elaborate appendage (hooded scale) present at base of each filament; calyx lobes longer or shorter than calyx tube 88

88. Inflorescences and/or flowers axillary **69. *Enicostema***
- Inflorescences (usually dense spikes) and/or flowers terminal 89

89. Calyx divided nearly to its base; pollen in monads (Africa) **65. *Karina***
- Calyx with distinct tube, sometimes longer than calyx lobes; pollen in tetrads **36. *Coutoubea***

90. Calyx absent (2 bracts subtending each flower alternatively interpreted as a calyx of two separate sepals) **47. *Obolaria***
- Calyx present, of 3 or more, usually ± united sepals 91

91. Nectaries adaxial on corolla tube (nectary disk below ovary never present) 92
- Nectaries absent from corolla (but nectary disk often present below ovary) 98
92. Corolla with spurs or with externally (abaxially) visible protuberances as many as lobes, aligned with petal midveins **55. *Halenia***
- Corolla without spurs or externally visible protuberances 93
93. Adaxial corolla surface with a ring of trichomes or paired, deeply fimbriate scales at bases of lobes or near apexof corolla tube 94
- Adaxial corolla surface glabrous or with small, ± scattered trichomes near stamen insertion 95
94. All sepals separate nearly to base; adaxial corolla surface with 2 non-vascularized, deeply fimbriate scales at base of each lobe; nectaries 2 per petal (Arctic, boreal, and mountainous regions of Northern Hemisphere) **58. *Comastoma***
- All or at least 2 sepals distinctly united; adaxial corolla surface without scales but with a fringe of trichomes (vascularized or non-vascularized) near apexof corolla tube, or rarely with 2 vascularized, deeply fimbriate scales at bases of lobes; nectaries 1 (rarely bilobed) per petal **59. *Gentianella***
95. Corolla lobe margins fringed **49. *Gentianopsis***
- Corolla lobe margins entire or erose 96
96. Flowers tetramerous; corolla greenish; corolla lobes always longer than corolla tube (Neotropics) **55. *Halenia*** (sect. *Swertiella* and extreme forms of *H. brevicornis*)
- Flowers pentamerous (very small individuals rarely tetramerous); corolla with different colors including green; relationship corolla lobe to tube length variable 97
97. Flower buds at least basally valvate; annual; corolla blue to blue-violet; corolla lobes about as long as corolla tube (Asia) **56. *Jaeschkea***
- Flower buds convolute; perennial or annual; corolla color and length relationships variable **59. *Gentianella***
98. Plant a tiny annual, stems filiform **17. *Cicendia***
- Plant more robust, not a filiform annual 99
99. Calyx 3-merous, corolla 5-merous (Asia) **29. *Cracosna***
- Calyx and corolla isomerous, both calyx and corolla 4- or 5-merous 100
100. Corolla adaxially with deeply fringed scales near apex of tube (Africa) **63. *Djaloniella***
- Corolla without scales or trichomes near apexof tube 101
101. Flowers tristylous, some with distinct filaments and subsessile stigmas, others with distinct styles and sessile or subsessile anthers (South America) **3. *Hockinia***
- Flowers monomorphic, with or without distinct styles, but all with well-developed filaments 102
102. Leaves minute, scale-like; corolla lobes imbricate in bud; style indistinct or absent (North America) **52. *Bartonia***
- Leaves generally larger (small in *Curtia*, native from Mexico southward); corolla lobes contorted; style distinct or absent 103
103. Style indistinct or absent; epipetalous nectaries 1 or 2 per petal lobes **59. *Gentianella***
- Style distinct; epipetalous nectaries absent 104
104. Anthers spiralling helically after anthesis; corollas salverform to nearly rotate 105
- Anthers remaining straight or recurving circinately after anthesis; corollas diverse in shape and color 111
105. Plant hairy; corollas violet to blue; anthers glandular-tipped (Madagascar) **12. *Ornichia***
- Plant glabrous or rarely minutely papillose; corollas usually rose-violet, pink, or white, rarely salmon or yellow 106
106. Inflorescences largely spicate, dichasial only at or near base if at all **28. *Schenkia***
- Inflorescences dichasial or partly monochasially divided, divisions sometimes raceme-like or subcapitate 107
107. Stigma 1, capitate or shallowly bilobed with hemispherical lobes 108
- Stigmas 2, or 1 with 2 ± fan-shaped lobes (tardily diverging in a few spp., the stigmas at first appearing capitate) 110
108. Style shorter than ovary at anthesis; fruit an ovoid capsule (Central and North America) **24. *Gyrandra***
- Style as long as or longer than ovary at anthesis; fruit a capsule or berry-like (Africa) 109
109. Calyx not keeled or winged; nectary disk at base of ovary **20. *Orphium***
- Calyx winged or keeled; nectaries absent **19. *Chironia***
110. Stigmas 2, elliptic to ovate; capsules cylindric **18. *Centaurium***
- Stigmas 2, fan-shaped, or 1, with 2 ± fan-shaped lobes; capsule narrowly ellipsoid **26. *Zeltnera***
111. Style abruptly bent to one side or upward **10. *Exacum***
- Style straight or nearly so 112

112. Style has a normal terminal stigmatic area, and an extra non-terminal stigmatic area; anthers with glandular tip **6. *Sebaea***
- Style with normal terminal stigmatic areas; anthers without glands 113
113. Style short, indistinct from ovary; small annuals (Central and South America) 114
- Style long, distinct from ovary; habit various 115
114. Leaves filiform; flowers monomorphic **22. *Zygostigma***
- Leaves ovate or otherwise not filiform; flowers heterostylous **2. *Curtia***
115. Tiny annuals, stem and leaves filiform (Europe) **27. *Exaculum***
- Larger herbs 116
116. Corolla red to orange **90. *Roraimaea***
- Corolla white, blue, purple, pink, or yellow to green 117
117. Corolla tube more than 16 mm long 118
- Corolla tube less than 16 mm long 119
118. Anthers fused or connate **13. *Tachiadenus***
- Anthers free **6. *Sebaea*** (2 African spp.)
119. Glands present on style and/or stamens 120
- Glands absent from style and stamens 123
120. Bracts much longer than flowers **8. *Klackenbergia***
- Bracts shorter than or rarely as long as flowers 121
121. Anthers medifixed, inserted at base of corolla tube **9. *Lagenias***
- Anthers basifixed, inserted at mid-length of corolla tube or higher 122
122. Stigma papillate; stamens always included within corolla tube; style without secondary stigmas **7. *Exochaenium***
- Stigma smooth; stamens often exserted from corolla mouth; style usually with secondary, lower stigmas **6. *Sebaea***
123. Corolla lobes more than 3× longer than corolla tube (Central and North America) **23. *Eustoma***
- Corolla lobes much shorter than to less than 3× as long as corolla tube (Central and South America and Canary Islands) 124
124. Corolla yellow (Canary Islands) **16. *Ixanthus***
- Corolla green, greenish white, white, pink, or pale purple (Central and South America) 125
125. Corolla less than 10 mm long (measured closed), funnelform to salverform, white or pale purple 126
- Corolla more than 10 mm long, urceolate to salverform, green, greenish white, white, pink 128
126. Plants small, generally less than 30 cm tall; corolla tube less than twice as long as corolla lobes **21. *Geniostemon***
- Plants generally more than 30 cm tall; corolla tube at least twice as long as corolla lobes 127
127. Flowers 5-merous **4. *Tapeinostemon***
- Flowers 4-merous **38. *Xestaea***
128. Inflorescences dense or capitate dichasia; corollas pink or white (South America) **39. *Deianira***
- Inflorescences lax dichasia, or flowers few; corollas green or greenish white, white, or pink (Central and South America, Asia) 129
129. Calyx and corolla usually 4-merous (5-merous in 1 sp.) 130
- Calyx and corolla 5-merous 131
130. Corolla pink or white **37. *Schultesia***
- Corolla green **48. *Latouchea***
131. Corolla green or greenish white, salverform **72. *Lisianthius*** (sect. *Omphalostigma*)
- Corolla pink or white, urceolate to funnelform **37. *Schultesia***

Tribes and Genera of the Gentianaceae

1. **Tribe Saccifolieae** (Maguire & Pires) Struwe, Thiv, V. Albert & Kadereit (2002).

Flowers actinomorphic, usually 5-merous; heterostyly frequent. Calyx membranous, not glandular on outside; corolla without appendages or glands; stamens equal, straight after anthesis, inserted in corolla tube, without appendages; connectives often prolonged; pollen in monads; ovary sessile (subsessile in some *Tapeinostemon*); style persistent in fruit (sometimes only at base). Fruit a capsule. Neotropical. Fig. 69.

1. ***Saccifolium*** Maguire & Pires

Saccifolium Maguire & Pires, Mem. New York Bot. Gard. 29: 242 (1978); Struwe et al., Fl. Venez. Guayana 5: 518–519 (1999), rev.; Thiv et al., Harvard Pap. Bot. 4:519–526 (2000), phyl.

Shrubs or subshrubs. Leaves saccate, densely crowded at branch apices, alternate(?), subsucculent, pseudopetiolate, distally curved downward, with each side of the blade also curved downward, the distal portion thus forming a hood. Flowers solitary, axillary, (4)5-merous. Calyx tube much shorter than lobes, sepals nearly free; corolla tubular, white; corolla tube much longer than lobes; stamens inserted in lowest quarter of

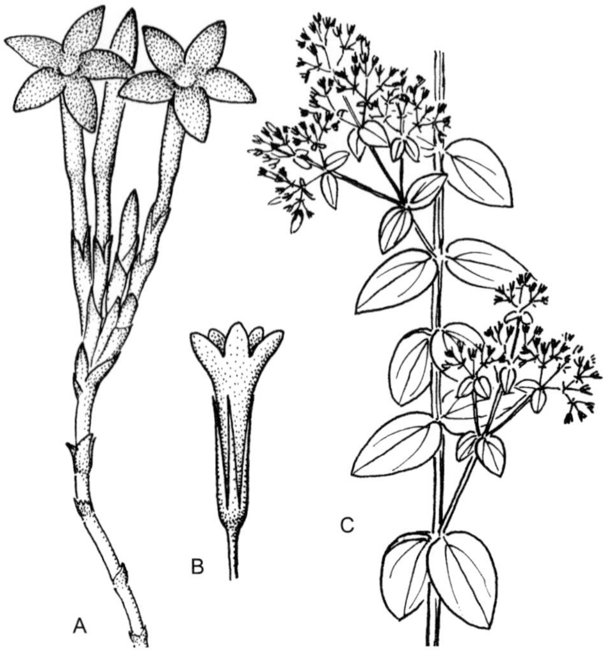

Fig. 69. Gentianaceae. Voyrieae and Saccifolieae. **A** *Voyria caerulea*, habit. **B** *Curtia tenuifolia* (Saccifolieae), flower. **C** *Curtia obtusifolia*, habit. (Drawings by Bruno Manara, reproduced with permission from Missouri Botanical Garden Press, previously published in Struwe et al. 1999)

corolla tube; anthers free, connective mucronate; ovary bilocular; style elongate, probably basally persistent; nectariferous disk at base of ovary; a glandular disk also present between calyx and corolla. Mature fruit unknown; immature fruit thin-walled (capsule?).

One sp., *S. bandeirae* Maguire & Pires, in N South America (Brazil/Venezuela border).

2. *Curtia* Cham. & Schltdl. Fig. 69

Curtia Cham. & Schltdl., Linnaea 1: 209 (1826); Gilg, Notizbl. Bot. Gart. Berlin-Dahlem 14: 66–93 (1938), rev.; Grothe & Maas, Proc. Kon. Ned. Akad. Wetensch., Ser. C, 87 (1): 33–42 (1984), anat.; Struwe et al., Fl. Venez. Guayana 5: 494–496 (1999), reg. rev.; Crespo & Ferreira, Acta Bot. Bras. 20: 273–284 (2006), anat.; Cobb et al., Blumea 52: 5–10 (2007), part. rev.; Crespo et al., Rodriguésia 60: 423–444 (2009), rev.
Schuebleria Mart. (1827) ["1826"] ("*Schübleria*").
Apophragma Griseb. (1838) ["1839"].

Small herbs, annual. Leaves opposite or whorled, often scale-like, membranous, sessile or subsessile. Flowers in dichasial cymes, thyrses, or verticillasters, 5- or 4-merous; sometimes di- or tristylous. Calyx deeply lobed; calyx lobes ± keeled; corolla funnelform to salverform, pink, violet, white, or pale yellow; corolla tube about as long as or longer than lobes, slightly narrowed below throat; stamens inserted near or above middle of corolla tube; anthers free or ± connate, not coiled; connective sometimes slightly prolonged, thecae sometimes with basal appendages; ovary pseudobilocular; style absent or slender but not sharply differentiated, ± persistent; nectaries absent.

Ca. seven spp. in Mexico, Central and tropical regions of South America.

3. *Hockinia* Gardner

Hockinia Gardner, London J. Bot. 2: 12 (1843); Knoblauch, Ber. Deutsch. Bot. Ges. 13: 289–298 (1895), morph.; Gilg, Ber. Deutsch. Bot. Ges. 13: 114–126 (1895), morph.; Grothe & Maas, Proc. Kon. Ned. Akad. Wetensch., Ser. C, 87 (1): 33–42 (1984), anat.; Crespo & Ferreira, Acta Bot. Bras. 20: 273–284 (2006), anat.

Herbs, annual. Leaves membranous, petiolate. Flowers in dichasia, 5-merous, tristylous. Calyx lobed nearly to base; corolla funnelform, blue; corolla tube longer than lobes, proximally pilose adaxially; anthers free or connate; connective proximally pubescent and with prolonged connective, or glabrous with narrow, mucronate connective; ovary pseudobilocular; style persistent.

One poorly known sp., *H. montana* Gardner, in South America (E Brazil).

4. *Tapeinostemon* Benth.

Tapeinostemon Benth., Hooker's J. Bot. Kew Gard. Misc. 6: 194 (1854); Steyermark, Lloydia 14: 58–64 (1951), rev.; Struwe et al., Fl. Venez. Guayana 5: 530–533 (1999), reg. rev.; Pruski & Smith, Brittonia 49: 346 (1997), part. rev.; Grant, J. Bot. Res. Inst. Texas 6: 101–107 (2012), part. rev.

Herbs or subshrubs, perennial or rarely annual. Leaves membranous to coriaceous, petiolate. Flowers usually in a diffuse thyrse, in a head in *T. sessiliflorum*, 5-merous, at least one species heterostylous. Calyx tube as long as to much shorter than lobes; corolla funnelform to nearly salverform, white or greenish white to yellow; corolla tube longer than lobes; stamens inserted near middle of corolla tube; anthers free or initially coherent; connective often widened between thecae and slightly prolonged; ovary sessile or subsessile, pseudobilocular; style distinct, persistent or distally caducous; nectariferous disk at base of ovary.

Ten spp. in tropical South America.

5. ***Voyriella*** (Miq.) Miq.

Voyriella (Miq.) Miq., Stirp. Surinam. Select. 146 (1851); Maas & Ruyters, Fl. Neotropica Monogr. 41 (1986), rev.
Voyria sect. *Voyriella* Miq. (1849).

Herbs, perennial, mycoheterotrophic, lacking chlorophyll. Leaves scale-like. Flowers solitary or in capitate dichasia, 5-merous, heterostylous. Calyx lobed nearly to base; corolla tubular to salverform, white, caducous; corolla tube much longer than lobes, interior papillose-puberulent; anthers coherent or free, connective sometimes prolonged; ovary unilocular; style slender, at least basally persistent; nectaries absent. Fruit an indehiscent capsule.

One sp., *V. parviflora* (Miq.) Miq., in Central and tropical South America.

2. **Tribe Exaceae** Colla (1835) ["1834"].

Herbs and subshrubs. Flowers actinomorphic (sometimes with zygomorphic sexual parts). Corolla without appendages (except *Tachiadenus*); petal epidermis cells rounded and convex; stamens equal (subequal in *Gentianothamnus*); connectives usually prolonged, often with glands; pollen in monads; ovary sessile, usually bilocular at least basally. Fruit a capsule. Seed testa cells often star- or puzzle piece-shaped (Fig. 70F). Paleotropical, mostly in Africa and Madagascar. Fig. 70.

6. ***Sebaea*** Sol. ex R. Br. Fig. 70

Sebaea Sol. ex R. Br., Prodr.: 451 (1810); Schinz, Mitt. Geogr. Ges. Naturhist. Mus. Lübeck, II, 17: 3–55 (1903), rev., Mitt. Bot. Mus. Univ. Zürich 32: 714–746, 801–832 (1906), part. rev.; Hill, Bull. Misc. Inform. 8: 317–341 (1908), part. rev., Ann. Bot. 27: 479–489 (1913); Verdoorn & Marais, Bothalia 7: 458–464 (1961); Nemomissa, Fl. Trop. East Africa, Gentianaceae (2002), reg. rev.; Kissling et al., New Phytol. 184: 303–310 (2009a), rev.; Kissling et al., Molec. Phylogen. Evol. 53: 734–748 (2009b), phyl., rev.
Belmontia E. Mey. (1838).

Herbs, annual or perennial. Leaves cauline, often also basal, membranous or coriaceous, sessile, rarely few and scale-like. Flowers solitary or in

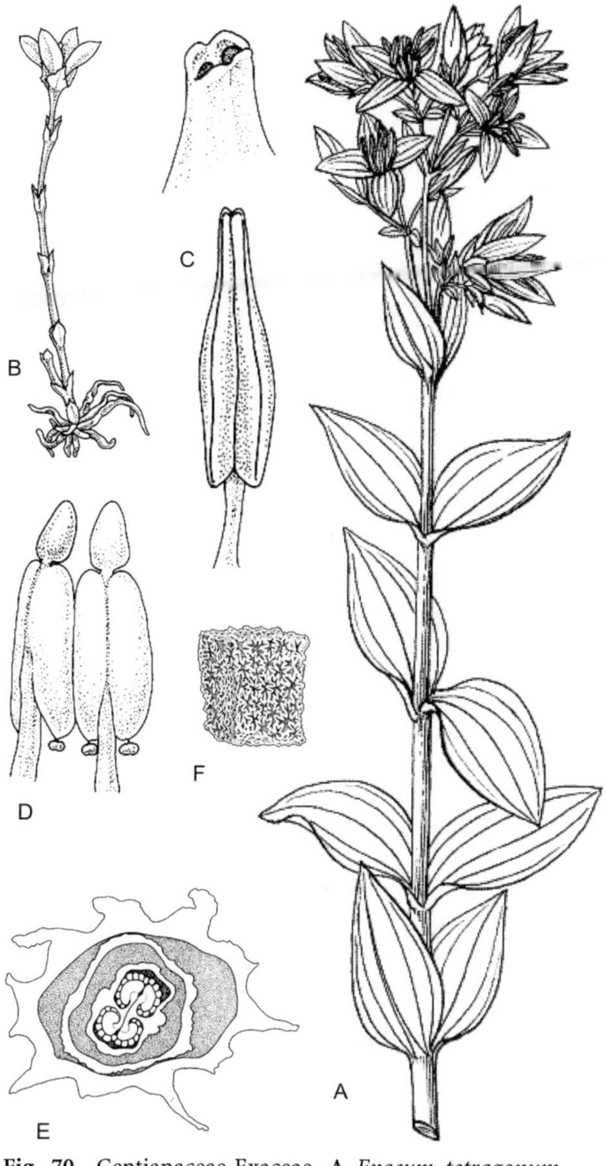

Fig. 70. Gentianaceae-Exaceae. **A** *Exacum tetragonum*, habit. **B** *Exacum tenuis* (formerly *Cotylanthera tenuis*), mycotroph with coralloid roots. **C** Stamen and anther apex of *Exacum axillare*, with porate opening. **D** Anthers of *Sebaea thomasiii* with apical and basal glands. **E** Cross-section through *Tachiadenus longiflorus* flower showing winged calyx tube, corolla tube, and pseudobilocular ovary with intruding placentas and ovules. **F** *Exacum axillare*. Angular seed with puzzle-shaped testa cells. (A. Original drawing by Yan Cuilan, redrawn by Cai Shuquin, reproduced with permission from Missouri Botanical Garden Press and Science Press, previously published in Ho and Pringle 1995); B–F. Drawings by Bobbi Angell, permission for use by Bobbi Angell, previously published in Struwe and Albert 2002)

open or capitate dichasia or thyrses, 4- or 5-merous; a few species heterostylous. Calyx lobes much longer than the calyx tube, usually angled or winged; corolla funnelform, salverform, or rotate, yellow, cream, or white; corolla tube longer or shorter than lobes; stamens in the corolla sinuses or close to sinuses in corolla tube; anthers free, exserted outside of the corolla mouth; connective usually prolonged, often with glandular tip; thecae sometimes with a ± stalked gland at base (Fig. 70D); ovary bilocular; style elongate, straight, caducous, with glandular patches (secondary stigma) near or above middle or confluent with stigma; stigma smooth. Fruit a capsule. n = 14, 21, 28.

Ca. 60 spp. in temperate and tropical regions of Africa, Madagascar, Asia, and Australasia; a few cultivated.

7. ***Exochaenium*** Griseb.

Exochaenium Griseb. (1845); Kissling, Syst. Bot. 37: 238–253 (2012), rev.; Kissling & Barrett, Ann. Bot. 112: 95–102 (2013b), phyl.; Kissling, Phytotaxa 203: 297–300 (2015), rev.

Herbs, annual, rarely achlorophyllous (*E. oliganthum*). Leaves cauline, membranous or coriaceous, sessile, rarely few and scale-like. Flowers in dichasia, 5-merous; a few species heterostylous; flowers rarely underground and cleistogamous (*E. oliganthum*). Calyx with free lobes or only fused at base, angled or winged; corolla tubular, funnelform, or salverform, or rotate, yellow, cream, salmon, or white; corolla tube longer or shorter than lobes; stamens inserted at mid-level in corolla tube; anthers free (fused in *E. grande*), basifixed, included within the corolla mouth; connective usually prolonged, often with glandular tip; thecae with or without stalked glands at base; ovary bilocular; style elongate, straight, caducous, without secondary stigmas; stigma papillate; nectariferous glands or disk often present between calyx and corolla. Fruit a bivalve capsule. Seeds numerous, black, with star-shaped testa cells. n = 21 (n = 14 in *E. oliganthum*).

22 spp. in sub-Saharan Africa.

8. ***Klackenbergia*** Kissling

Klackenbergia Kissling, Taxon 58: 910 (2009); Kissling et al., Taxon 58: 907–912 (2009), rev.

Small herbs. Leaves cauline, linear, acute. Flowers in one-sided cymose fascicles at nodes, with bracts as long as or longer than flowers. Calyx divided nearly to base, keeled to winged; corolla salverform, white; corolla tube longer than lobes; stamens inserted in the middle of corolla tube, not exserted from corolla mouth; anthers arranged in ring; connective prolonged, with gland at apex; ovary bilocular; style elongate, straight, caducous. Fruit a bivalve capsule. Seeds many, with star-shaped testa cells. n = 9 (*K. stricta*).

Two species in Madagascar.

9. ***Lagenias*** E. Mey.

Lagenias E. Mey. (1838); Kissling, Syst. Bot. 37: 238–253 (2012), rev.

Herbs, annual. Leaves cauline, ± succulent, sessile. Flowers in 1–5-flowered terminal cymes, 5-merous. Calyx lobes free; corolla funnelform, yellow to cream; corolla tube longer than lobes; stamens inserted below the middle of the corolla tube; anthers free, medifixed, included in corolla tube; connective prolonged, often with apical gland; thecae with stalked glands; ovary bilocular; style elongate, straight, without secondary stigma; stigma bilobed. Fruit a capsule. Seeds very small, many, with polygonal testa cells. n = 14.

One sp., *L. pusillus* E. Mey., in Cape Province of South Africa.

10. ***Exacum*** L. Fig. 70

Exacum L., Sp. Pl.: 112 (1753); Figdor, Ann. Jard. Bot. Buitenzorg 14: 213–240 (1897); Hara, J. Jap. Bot. 50: 322–328 (1975), rev.; Klackenberg, Opera Bot. 84: 1–144 (1985), rev.; Thulin, Nord. J. Bot. 21: 243–247 (2001), reg. rev.; Nemomissa, Fl. Trop. East Africa, Gentianaceae (2002), reg. rev.; Yuan et al., Molec. Phyl. Evol. 28: 500–517 (2003), phyl., Syst. Biol. 54: 21–34 (2005), phyl.; Klackenberg, Bot. Jahrb. 126: 477–481 (2006), part. rev.
Cotylanthera Blume (1826).

Herbs or subshrubs, annual or perennial, or a few spp. perennial mycoheterotrophic herbs, lacking chlorophyll (Fig. 70A, B). Leaves opposite, rarely subverticillate, membranous to ± succulent, sessile or petiolate (scale-like in mycotrophs). Flowers solitary or in open or capitate dichasia or thyrses, 4–5-merous. Calyx actinomorphic or occasionally zygomorphic, usually angled or winged; calyx tube much shorter than to about as long as lobes; corolla sometimes slightly zygomorphic, rotate or nearly so, blue to violet or white; corolla tube shorter than lobes, usually narrowed toward apex, proximally fused to ovary wall in mycotrophs; stamens inserted in or slightly below corolla sinuses; filaments ± dilated basally, sometimes basally united; anthers free, straight or occasionally curved; thecae dehiscent only toward tip, forming a pore (Fig. 70C), or tardily dehiscent longitudinally; connective sometimes with a dorsal gland near apex; ovary bilocular; style elongate, declinate and curved to one side (straight in mycoheterotrophic species), caducous; nectaries absent or in mycoheterotrophic species present at base of free portion of corolla, in pairs alternating with stamen bases. n = 9, 15, 17, 18, 27, 28, 31, 34.

69 spp. in Africa, Indian Ocean islands, South Asia, and Australasia east to New Guinea; *E. affine* Balf.f. ex Regel is widespread in cultivation. *Cotylanthera* was treated as a separate genus due to its mycoheterotrophic habit, but has been shown to be nested within *Exacum*.

11. ***Gentianothamnus*** Humbert

Gentianothamnus Humbert, Bull. Soc. Bot. France 84: 388 (1937); Klackenberg, Fl. Madagascar et des Comores, Family 168 (1990), rev.

Shrubs. Leaves coriaceous, short-petiolate, petioles with stipule-like lateral expansions. Flowers solitary in axils, 5-merous. Calyx slightly zygomorphic, keeled or narrowly winged; calyx tube about as long as lobes; corolla long-campanulate, yellow to orange with red suffusion at least on lobes; corolla tube much longer than lobes, 3 lower lobes slightly larger than upper 2; stamens inserted near base of corolla tube; ovary unilocular or basally bilocular; style slender, basally persistent; papillate nectariferous disk at base of ovary.

One sp., *G. madagascariensis* Humbert, in Madagascar.

12. ***Ornichia*** Klack.

Ornichia Klack., Bull. Mus. Natl. Hist. Nat., B, Adansonia 8: 198 (1986); Klackenberg, Bull. Mus. Nat. Hist. Nat., Paris, IV, 8, sect. B, Adansonia 2: 195–206 (1986), rev.; Klackenberg, Fl. Madagascar et des Comores, Family 168 (1990), rev.

Subshrubs or herbs, perennial or possibly annual, vegetative parts hairy. Leaves cauline and basal, thin, short-petiolate. Flowers solitary or in dichasial or monochasial cymes, 5-merous. Calyx lobed nearly to base; corolla salverform, violet to blue; corolla tube shorter or longer than lobes; stamens inserted near or above middle of corolla tube; anthers coherent, connective with glandular tip; ovary bilocular; style elongate, caducous. n = 14 (*O. madagascariensis*).

Three spp. in Madagascar.

13. ***Tachiadenus*** Griseb. Fig. 70

Tachiadenus Griseb., Gen. Sp. Gent.: 200 (1838); Klackenberg, Bull. Mus. Natl. Hist. Nat., B, Adansonia 9: 43–80 (1987), rev.; Klackenberg, Fl. Madagascar et des Comores, Family 168 (1990), rev.

Herbs or subshrubs, annual or perennial. Leaves membranous, sessile or short-petiolate. Flowers in dichasial, monochasial, or occasionally umbelloid cymes, 5-merous. Calyx tube shorter or longer than lobes; lobes usually winged and equal, rarely two outer lobes narrower; corolla salverform, white to blue or rose-violet; corolla tube slender, much longer than lobes, expanded near apex, commonly with protuberances alternating with anthers or a callose ring just above anthers, rarely with a ring of papillate trichomes at base of lobes; stamens inserted in expanded throat of corolla tube, below sinuses; anthers coherent, connective ± prolonged, rarely glandular-papillate; ovary pseudobilocular (Fig. 70E); style elongate, basally persistent; nectariferous disk at base of ovary. n = 16.

11 spp. in Madagascar.

3. **Tribe Voyrieae** Gilg (1895).

14. ***Voyria*** Aubl. Fig. 69

Voyria Aubl., Hist. Pl. Guiane 1: 208 (1775); Raynal-Roques, Adansonia II, 7: 64–71 (1967), reg. rev.; Maas & Ruyters, Fl. Neotropica Monogr. 41: 1–93 (1986), rev.; Albert & Struwe, Brittonia 49: 466–479 (1997), phyl.; Imhof, Bot. Acta 110: 298–305 (1997), morph., Mycorrhiza 9: 33–39 (1999), morph.; Cheek, in Ghazanfar & Beentje, Tax. Ecol. African Plants: 693–697 (2006); Hentrich et al., Taxon 59: 298–305 (2010), reg. rev.; Merckx et al., Amer. J Bot. 100: 712–721 (2013), phyl.
Leiphaimos Schltdl. & Cham. (1831).

Herbs, perennial, mycoheterotrophic, without chlorophyll. Leaves scale-like, opposite or lowest alternate. Flowers in dichasia or solitary. Flowers 4–5-merous, actinomorphic. Calyx fused at base, sometimes with discoid scales at adaxial base; corolla funnelform to salverform, white, yellow, orange, salmon, violet, or blue, without appendages; corolla tube longer than lobes, exterior glabrous, sometimes puberulent, interior sometimes papillate in zones; stamens equal, inserted in upper half of corolla tube, without appendages; anthers at least initially coherent, occasionally free, base of thecae sometimes caudate; pollen in monads or very rarely in tetrads, 1–6-porate; ovary sessile or rarely stipitate, unilocular, with 2 sessile or stalked glands near base, or with 2 or 5 glandular scales near base, or eglandular; style usually filiform, occasionally short and thick, caducous or occasionally persistent. Fruit a capsule, completely or only medially dehiscent or indehiscent.

18 spp. in S North America (Florida), Central and tropical South America (including the West Indies), and one species in tropical W Africa.

4. **Tribe Chironieae** (G. Don) Endl. (1838).

Herbaceous or rarely subshrubs, usually annual. Corollas usually ± salverform or rotate, rose-violet, pink, white, or yellow, not blue or blue-violet, often with contrastingly colored centers, but rarely with guidelines, without appendages (except *Coutoubea*), without nectaries (except *Cracosna*); stamens without appendages (except *Chironia* spp.); anthers free; ovary sessile (stipitate in *Symphyllophyton*). Fruit a capsule (berry-like in *Chironia* spp.). Widely distributed in temperate and tropical regions. Fig. 71.

4a. **Subtribe Chironiinae** G. Don (1838).

Herbs or less often shrubs. Leaves often ± succulent and/or glaucous, bases often strongly clasping or perfoliate. Flowers actinomorphic, with styles and stamens often deflexed away from each other at anthesis; stamens equal; anthers often and style branches sometimes coiling helically (Fig. 71D–G); pollen in monads. Widely distributed in temperate and tropical regions.

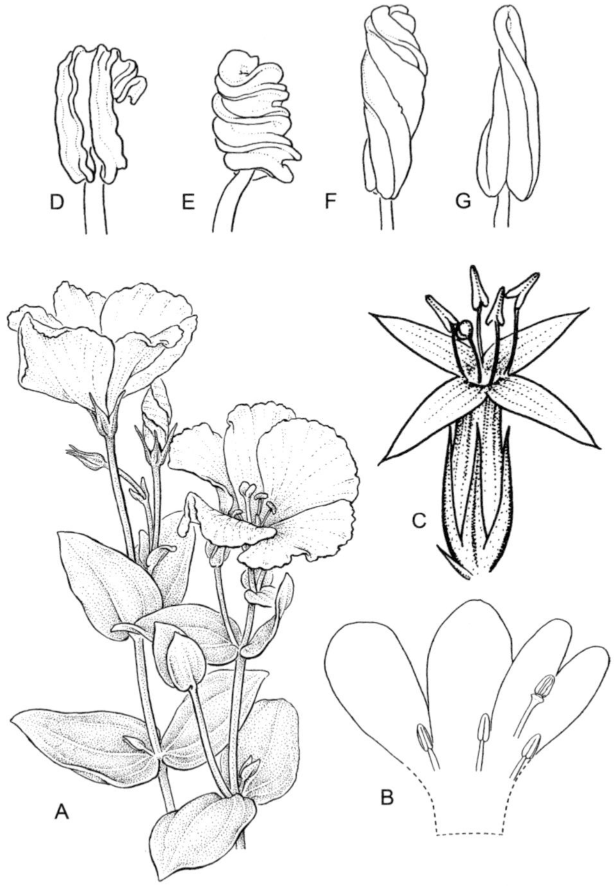

Fig. 71. Gentianaceae-Chironieae. **A** *Eustoma grandiflorum*, habit. **B** Zygomorphic corolla and unequal insertion of anisomorphic stamens in *Canscora roxburghii*. **C** *Coutoubea spicata*, flower with exserted, sagittate anthers. **D** *Sabatia dodecandra*, outwardly twisted anther. **E** *Centaurium beyrichii*, spirally twisted anther. **F** *Chironia laxiflora*, spirally twisted anther. **G** *Orphium frutescens*, spirally twisted anther. (A. Drawing by Bobbi Angell, permission for use by Bobbi Angell; B. Drawing by Bobbi Angell, permission for use by Bobbi Angell, previously published in Struwe and Albert 2002; C. Drawings by Bruno Manara, reproduced with permission from Missouri Botanical Garden Press, previously published in Struwe et al. 1999; D–G. Drawings by Bobbi Angell, permission for use by Bobbi Angell, previously published in Struwe and Albert 2002)

15. ***Blackstonia*** Huds.

Blackstonia Huds., Fl. Angl. 146 (1762); Zeltner, Bull. Soc. Neuchâteloise Sci. Nat. 93: 5–56 (1970), rev.; Tutin, Fl. Europaea 3: 56 (1972); Edmonson, Fl. Turkey 6: 177–178 (1978); van der Sluis, Pl. Syst. Evol. 149: 253–286 (1985); Brys et al., Ann. Bot. doi:10.1093/aob/mct031 (2013), repr. *Chlora* Adans. (1763).

Herbs, annual. Leaves sometimes in basal rosette, ± succulent, lower leaves sessile, upper sessile or perfoliate. Flowers in dichasia, 6–12-merous. Calyx lobed nearly to base; corolla rotate, yellow; corolla tube much shorter than lobes; stamens inserted near apex of corolla tube; anthers not coiling; ovary unilocular; style distally bifid, persistent; nectaries absent. n = 10, 20.

Two spp., in S Europe and W Asia, one of these naturalized in South America, Australia, and New Zealand.

16. ***Ixanthus*** Griseb.

Ixanthus Griseb., Gen. Sp. Gent. 129 (1838); Thiv et al., Pl. Syst. Evol. 218: 299–317 (1999), phyl.; Betzin et al., Ann. Bot. 118: 495–510 (2016), biogeogr.

Suffrutescent herbs or subshrubs, perennial. Vegetative parts and especially the calyx glandular. Leaves membranous, lower leaves sessile with clasping bases, uppermost bracteoles sometimes perfoliate. Flowers in terminal thyrses, 5-merous. Calyx deeply lobed; corolla short-salverform, yellow; corolla tube slightly shorter than lobes; stamens inserted in upper half of corolla tube; anthers not coiling; ovary pseudobilocular; style slender, caducous; nectariferous zone at base of ovary.

One sp., *I. viscosus* (Aiton) Griseb., Canary Islands.

17. ***Cicendia*** Adans.

Cicendia Adans., Fam. Pl. 2: 503 (1763); Çiçek, Phytotaxa 184: 100 –108 (2014), nom.; Lifante & Valdés, Taxon 63: 417–422 (2014), nom.
Microcala Hoffmanns. & Link (ca. 1813) ["1820"].

Small herbs, annual. Leaves basal and cauline, membranous, sessile. Flowers solitary or in small dichasia, 4-merous. Calyx lobes of *C. quadrangularis* in 2 unequal pairs, longer than tube but coherent and appearing united most of their length, with lobes appearing as minute projections; all four calyx lobes of *C. filiformis* equal, shorter than tube; corolla salverform, yellow; corolla tube longer than lobes; stamens inserted near apexof corolla tube; anthers not coiling; ovary unilocular; style distinct, caducous; nectaries absent. n = 13.

Two spp., one in Europe, N Africa, and W Asia, and one in W North America and W South America, both naturalized in Australia.

18. ***Centaurium*** Hill Fig. 71

Centaurium Hill, Brit. Herb. 62 (1756); Gillett, The Gentians of Canada, Alaska and Greenland (1963), reg. rev.; Zeltner, Bull. Soc. Neuchâteloise Sci. Nat., III, 93: 5–56 (1970), reg. rev.; Melderis, Bot. J. Linn. Soc. 65: 224–250 (1972), reg. rev.; Mansion, Taxon 53: 719–740 (2004), rev.; Mansion & Struwe, Molec. Phyl. Evol. 32: 951–957 (2004), phyl.; Mansion et al., Taxon 54: 931–950 (2005), phyl.
Erythraea Borkh. (1796).
Monodiella Maire (1943).

Herbs, annual or rarely perennial. Leaves cauline, sometimes also basal, membranous or ± succulent, sessile. Flowers in dichasial cymes, sometimes in corymbs, heads, panicles or racemes. Flowers 4- or 5-merous. Calyx lobed nearly to base; corolla salverform, pink, white, or rarely yellow; corolla tube about as long as or longer than lobes, often narrowed above apex of ovary; stamens inserted in upper half of corolla tube, sometimes initially curved to one side; anthers usually helically coiled, rarely circinately coiled or straight; ovary unilocular to pseudobilocular, sometimes proximally bilocular; style slender, sometimes initially deflexed away from stamens and later erect, tardily caducous; nectaries absent. n = (9?), 10, 18, 20, 27, 28, 30.

Ca. 20 spp. in Eurasia and North Africa, *C. capensis* C.R. Broome in Mexico; some species naturalized worldwide except sub-Saharan Africa.

19. ***Chironia*** L. Fig. 71

Chironia L., Sp. Pl. 189 (1753); Schoch, Beih. Bot. Centralbl. 14: 177–242 (1903), rev.; Hill, Bull. Misc. Inform. 8: 341–376 (1908), rev.; Nemomissa, Fl. Trop. East Africa, Gentianaceae (2002), reg. rev.

Subshrubs or herbs, annual or perennial. Leaves cauline and basal, membranous or firm and needle-like, sessile or rarely ± petiolate. Flowers solitary or in dichasial or monochasial cymes, 5-merous. Calyx sometimes keeled or winged, divided nearly to base or with tube shorter or longer than lobes; corolla campanulate to salverform or nearly rotate, pink, rose-violet, or white; corolla tube shorter or longer than lobes, ± narrowed above apex of ovary; stamens inserted in upper half of corolla tube, curved to one side, rarely with a gland near base; anthers straight, curved, or helically coiled; ovary unilocular; style elongate, caducous, curved away from stamens or hooked distally; nectaries absent. Fruit a capsule, rarely ± berry-like. n = 34.

Ca. 15 spp. in Africa and Madagascar; occasionally cultivated.

20. ***Orphium*** E. Mey. Fig. 71

Orphium E. Mey., Comm. Pl. Afr. Austr. 1838 ('1837'); Hill, Bull. Misc. Inform. 8: 341–376 (1908), rev.; Hakki, Bot. Jahrb. Syst. 119: 337–383 (1997), morph.

Shrubs or subshrubs, perennial, with bristles on vegetative parts and calyx. Leaves slightly succulent, sessile. Flowers in dichasial cymes, small thyrses, or solitary, 5-merous, with zygomorphic sexual parts. Calyx tube about as long as or shorter than lobes; corolla nearly salverform, pink or rarely white; corolla tube shorter than lobes, slightly sticky; stamens inserted in upper half of corolla tube, curved to one side; anthers helically coiled, poricidal; ovary unilocular; style elongate, deflexed away from stamens, caducous; nectariferous cup-like disk between calyx and corolla. n = 21.

One sp., *O. frutescens* E. Mey., in South Africa; cultivated elsewhere.

21. ***Geniostemon*** Engelm. & A. Gray

Geniostemon Engelm. & A. Gray, Proc. Amer. Acad. Arts 16: 104 (1881); Turner, Phytologia 76: 8–13 (1994), rev.; Rzedowski & Calderón de Rzedowski, Acta Bot. Mex. 32: 1–10 (1995), rev.

Small herbs, annual or perennial. Leaves membranous, sessile. Flowers in a dichasium or solitary, 4-merous. Calyx lobed nearly to base, keeled; corolla salverform, pale purple to white; corolla tube shorter than or about as long as lobes; stamens inserted in corolla sinuses; filaments usually glandular-pubescent; anthers not coiling; ovary unilocular; style elongate, caducous or basally persistent; nectaries absent.

Five spp. in Mexico.

22. ***Zygostigma*** Griseb.

Zygostigma Griseb., Gen. Sp. Gent.: 150 (1838); Williams, J. Bot. 41: 232–234 (1903).

Herbs, perennial or annual(?). Leaves nearly filiform, membranous, sessile. Flowers terminal, solitary or few in dichasia, 5- or rarely 4-merous. Calyx lobed nearly to base; calyx lobes keeled; corolla funnelform, white to pale rose-violet; corolla tube slightly longer than lobes; stamens inserted near apex of corolla tube; ovary unilocular; style slender but not sharply differentiated, ± persistent; nectaries absent.

One poorly known sp., *Z. australe* (Cham. & Schltdl.) Griseb., in South America.

23. ***Eustoma*** Salisb. Fig. 71

Eustoma Salisb., Parad. Lond. 1: pl. 34 (1806); Shinners, Southw. Naturalist 2: 38–43 (1957), rev.

Herbs, annual or short-lived perennial. Leaves basal and cauline, somewhat succulent, sessile. Flowers in partly monochasial cymes, 5-merous. Calyx keeled; calyx tube much shorter than lobes; corolla broadly campanulate, violet-blue with conspicuous yellow throat or rarely rose-violet, white, or yellow; corolla tube shorter than lobes; stamens equal, inserted near middle of corolla tube; anthers not coiling; ovary unilocular; style slender, persistent; nectariferous disk at base of ovary. n = 36.

One variable sp., *E. exaltatum* (L.) Salisb., in S North America, Central America, and the West Indies; widespread in cultivation [subsp. *russellianum* (Hook.) Kartesz], called "Lisianthus", often with supernumerary petals; Fig. 71A.

24. ***Gyrandra*** Griseb.

Gyrandra Griseb. in A. DC., Prodr. 9: 44 (1845); Mansion, Taxon 53: 719–740 (2004), rev.; Mansion & Struwe, Molec. Phylogen. Evol. 32: 951–957 (2004), phyl.
Erythraea sect. *Gyrandra* (Griseb.) A. Gray (1876).

Herbs, annual or perennial. Leaves cauline, membranous. Flowers in a terminal, dichasial cyme, lower branches monochasial, 5-merous. Calyx tube shorter than or about as long as lobes; corolla salverform to nearly rotate, pale blue to pink; corolla tube much shorter than to about as long as lobes; stamens inserted in upper half of corolla tube; anthers helically coiled; nectaries absent. n = 36.

Six spp. in North America (S Texas) and Central America (including Mexico). This genus was recently segregated from *Centaurium* to recircumscribe *Centaurium* as a monophyletic group.

25. *Sabatia* Adans. Fig. 71

Sabatia Adans., Fam. Pl. 2: 503 (1763); Wilbur, Rhodora 57: 1–33, 43–71, 78–104 (1955), rev., 91: 167–171 (1989), part. rev.; Gillett, The Gentians of Canada, Alaska and Greenland (1963), reg. rev.; Perry, Rhodora 73: 309–369 (1971), rev.; Thiv, Fl. Republ. Cuba 6(1) (2002), reg. rev.; Mathews et al., Syst. Bot. 40: 811–825 (2015), phyl.
Lapithea Griseb. (1845).

Herbs, annual or perennial. Leaves cauline, sometimes also basal, membranous or ± succulent, sessile. Flowers in dichasia (sometimes condensed into heads), 5–12(–14)-merous, perfect or sometimes late flowers staminate. Calyx tube shorter or longer than lobes; corolla rotate, pink or white, often with conspicuous yellow center; corolla tube much shorter than lobes; stamens inserted in sinuses between corolla lobes in sect. *Sabatia*, slightly below sinuses in sect. *Pseudochironia*; anthers usually recurved or circinately coiled after dehiscence, rarely straight or only slightly coiled helically; ovary unilocular; style distinct, usually initially deflexed away from stamens, later erect, caducous; stigmas linear, helically twisted before becoming receptive; nectariferous zone at base of ovary, at least in xenogamous spp. n = 13, 14, 16, 17, 18, 19, 20, 32.

Ca. 20 spp. in North America and the West Indies.

26. *Zeltnera* Mansion

Zeltnera Mansion, Taxon 53: 727 (2004); Broome, Brittonia 28: 413–426 (1977), reg. rev. (as *Centaurium*); Holmgren, Intermountain Flora 4: 5–7 (1984), reg. rev. (as *Centaurium*); Turner, Phytologia 75: 259–275 (1993), reg. rev. (as *Centaurium*); Mansion, Taxon 53: 719–740 (2004), rev.; Mansion & Struwe, Molec. Phylogen. Evol. 32: 951–957 (2004), phyl.

Herbs, annual or rarely short-lived perennial. Leaves cauline, often also basal, sessile. Flowers in dichasial or monochasial cymes, 4–5-merous. Calyx tube much shorter than lobes; corolla salverform, with white, yellow, or greenish tube and magenta, pink or white lobes, often with white center; corolla tube shorter to longer than lobes; stamens inserted in upper half of corolla tube; anthers helically coiled; ovary unilocular to pseudobilocular; style slender, sometimes initially deflexed away from stamens and later erect, tardily caducous; nectaries absent. n = 17, 20, 21, 22, 37.

25 spp. in temperate North America to Central and tropical W South America. *Zeltnera* was segregated as a new genus to preserve the monophyly of *Centaurium*.

27. *Exaculum* Caruel

Exaculum Caruel in Parl., Fl. Ital. 6: 743 (1886); Tutin, Fl. Europaea 3: 56–67 (1972), rev.; Lifante & Valdés, Taxon 63: 417–422 (2014), nom.

Small herbs, annual. Leaves membranous, narrow, sessile. Flowers in dichasia, 4-merous. Calyx deeply lobed; corolla salverform, pink or pale yellow; corolla tube longer than lobes, narrowed above apex of ovary; stamens inserted near apex of corolla tube; anthers not coiling; ovary unilocular; style distinct, straight, caducous; nectaries absent. n = 10.

One sp., *E. pusillum* (DC.) Caruel, in S Europe.

28. *Schenkia* Griseb.

Schenkia Griseb., Bonplandia 1: 226 (1853); Mansion, Taxon 53: 719–740 (2004), rev.; Mansion & Struwe, Molec. Phylogen. Evol. 32: 951–957 (2004), phyl.
Erythraea sect. *Spicaria* Griseb. (1839).
Centaurium sect. *Spicaria* (Griseb.) Ronn. (1916).

Herbs, annual. Leaves cauline and basal, membranous or fleshy. Flowers in dense or lax spiciform cymes, rarely in raceme-like monochasia, 5-merous. Calyx lobed nearly to base; corolla salverform, pink, white, or rarely yellow; corolla tube about as long as or longer than lobes; stamens usually inserted in corolla sinuses, sometimes slightly below sinuses in *S. spicata*; anthers usually helically coiled, rarely circinately coiled or straight; ovary unilocular; style slender; nectaries absent. n = 11, 22.

Five spp. in S Europe, N Africa, and W Asia (primarily Mediterranean and Middle East), Australia and the Pacific (New Caledonia and Hawaii). This genus, previously included in *Centaurium*, was recently resurrected to circumscribe *Centaurium* as monophyletic.

4b. **Subtribe Canscorinae** Thiv & Kadereit (2002).

Herbs. Stamens often unequal in level of insertion in corolla tube (anisomorphic; Fig. 71B), length, and/or in number, without appendages; pollen in monads. Fruit a capsule. Paleotropical.

29. ***Cracosna*** Gagnep.

Cracosna Gagnep., Bull. Soc. Bot. France 76: 776 (1929); Thiv, Blumea 48: 1–46 (2003), rev.

Herbs, annual. Leaves cauline and sometimes basal, sessile, Flowers in axillary cymes, sometimes few-flowered, 4-merous (except 2–3-merous calyx in *C. xyridiformis*), actinomorphic. Calyx tube longer than lobes, sometimes winged; corolla salverform to tubular, white to yellow; corolla tube longer than lobes; stamens isomorphic, inserted in upper part of corolla tube; ovary unilocular; style distinct, caducous; nectaries absent.

Three spp. in SE Asia (Cambodia, Laos, and Thailand).

30. ***Canscora*** Lam. Fig. 71

Canscora Lam., Encycl. 1: 601 (1785); Thiv, Blumea 48: 1–46 (2003), rev.; Nemomissa, Fl. Trop. East Africa, Gentianaceae (2002), reg. rev.; Shahina & Nampy, Phytotaxa 164 (4): 201–225 (2014), reg. rev.; Kousalya et al., Bot. Res. Int. 8:59–64 (2015), reg. rev.
Orthostemon R. Br. (1810).
Pladera Sol. ex. R. Br. (1820).
Heterocanscora C.B. Clarke (1875).
Canscorinella Shahina & Nampy (2014).

Herbs, annual. Leaves cauline, rarely basal, sessile or rarely petiolate. Flowers in axillary cymes or rarely spikes, 4-merous; bracts free or perfoliate. Calyx tubular, sometimes strongly winged; calyx tube longer than lobes; corolla zygomorphic with two smaller and two larger lobes (Fig. 71B), salver- or funnelform, white, pink, to purple; corolla tube longer than lobes; stamens inserted in upper part of corolla tube, anisomorphic, with the stamen between the smaller corolla lobes inserted at a higher level and larger than the other stamens (Fig. 71B); ovary unilocular; style distinct, caducous; nectaries absent. n = 18, 19, 38.

Nine spp. in tropical S and SE Asia, Africa, Madagascar, and Australasia (Malesia and N Australia). The genus *Canscorinella* has recently been described as a segregate of *Canscora* (Shahina and Nampy 2014), but since no phylogenetic evidence has yet been presented that shows monophyly of both *Canscora* s. str. and of *Canscorinella*, the latter genus is not accepted in this treatment.

31. ***Hoppea*** Willd.

Hoppea Willd., Ges. Naturf. Freunde Berlin Neue Schriften 3: 434 (1801); Thiv, Blumea 48: 1–46 (2003), rev.

Small herbs, annual. Leaves cauline, membranous, sessile. Flowers in axillary, many-flowered dense or lax cymes, 4-merous, actinomorphic. Calyx tube about as long as lobes; corolla tubular or funnelform, shorter than or about as long as calyx, white; corolla tube longer than lobes; stamens of unequal size, inserted near mouth in corolla tube, 1 fertile, 3 smaller with sterile anthers or filaments only; ovary unilocular; style short but distinct, caducous. n = 19.

Two spp. in S Asia (India, Sri Lanka), introduced in Africa and SE Asia.

32. ***Schinziella*** Gilg

Schinziella Gilg in Engl. & Prantl, Nat. Pflanzenfam. 4(2): 74 (1895); Nemomissa, Fl. Trop. East Africa, Gentianaceae (2002), reg. rev.; Thiv, Blumea 48: 1–46 (2003), rev.

Herbs, perennial. Leaves small, almost scale-like. Flowers in axillary and terminal, many-flowered, dense, head-like cymes, 4-merous, actinomorphic; bracts free. Calyx funnelform, not winged, with distinct veins on lobes; calyx tube longer than lobes; corolla funnelform, white or cream to yellow; corolla tube about as long as lobes; stamens anisomorphic, inserted at unequal levels in corolla tube, 1(2) fertile stamen larger than the sterile stamens; ovary unilocular; style distinct, caducous; nectaries absent.

One sp., *S. tetragona* (Schinz) Gilg, in tropical Africa.

33. ***Duplipetala*** Thiv

Duplipetala Thiv, Blumea 48(1): 25 (2003); Thiv, Blumea 48: 1–46 (2003), rev.

Herbs, annual or perennial. Leaves cauline, petiolate. Flowers in axillary, few-flowered, lax cymes, actinomorphic; bracts free or perfoliate. Calyx 3- or 6-merous, inflated urceolate, winged; calyx tube longer than lobes; corolla (5)6-merous, tubular to salverform, white to cream; corolla tube longer than lobes; stamens inserted in upper part of corolla tube, isomorphic or anisomorphic with filaments of different lengths; ovary unilocular; style distinct, caducous; nectaries absent.

Two spp. in SE Asia (Thailand and W Malaysia).

34. ***Microrphium*** C.B. Clarke

Microrphium C.B. Clarke, J. Asiat. Soc. Bengal, Pt. 2, Nat. Hist. 74: 88 (1906); Regalado & Soejarto, Novon 7: 77–80 (1997), reg. rev.; Thiv, Blumea 48: 1–46 (2003), rev.

Herbs, perennial, suffrutescent, with bristles on vegetative parts. Leaves cauline, sessile, petiolate. Flowers in axillary, many-flowered, lax monochasial cymes, actinomorphic; bracts free. Calyx tubular, shallowly divided into two parts, not winged; corolla 5-merous, salverform, white to cream; corolla tube longer than lobes; stamens isomorphic, inserted near middle of corolla tube; ovary unilocular; style slender, caducous.

Two spp. in SE Asia (W Malaysia, Thailand, and the Philippines).

35. ***Phyllocyclus*** Kurz

Phyllocyclus Kurz, J. Asiat. Soc. Bengal, Pt. 2, Nat. Hist. 42: 235 (1874) ["1873"]; Thiv, Blumea 48: 1–46 (2003), rev.
Euphorbiopsis H. Lév. (1911).

Herbs, annual or perennial. Leaves cauline and basal, cauline leaves perfoliate. Flowers in axillary, few-flowered clusters; bracts perfoliate. Calyx 4–5-merous, tubular or urceolate; corolla tube not winged, longer than lobes; corolla 5-merous, actinomorphic, urn-, salver- or funnelform, white, yellow, to cream; corolla tube longer than lobes; stamens inserted in middle of corolla tube at equal height; filaments of unequal or equal length, broadened at base; ovary unilocular; style distinct, caducous; nectaries absent.

Five spp. in SE and E Asia (Burma to S China).

4c. **Subtribe Coutoubeinae** G. Don (1838).

Herbs. Flowers actinomorphic. Corollas usually 4-merous, ± salverform, usually pink to rose-violet; stamens equal (except *Symphyllophyton*); pollen in tetrads. Fruit a capsule. Nearly all neotropical.

36. ***Coutoubea*** Aubl. Fig. 71

Coutoubea Aubl., Hist. Pl. Guiane: 72 (1775); Guimarães & Klein, Rodriguésia 37(62): 21–45 (1985), rev.; Struwe et al., Fl. Venez. Guayana 5: 491–494 (1999), reg. rev.

Herbs, annual or perhaps sometimes perennial. Leaves opposite or whorled, membranous or ± succulent, sessile or short-petiolate. Flowers paired, sessile or subsessile, usually crowded in spikes, 4(5)-merous. Calyx deeply lobed; calyx lobes keeled; corolla narrowly campanulate to short-salverform, white or pale violet; corolla tube cylindric, about as long as lobes (Fig. 71C), a scale below insertion of each stamen; stamens inserted near apex of corolla tube; ovary unilocular or rarely ± bilocular; style slender, tardily caducous; nectaries absent. n = 15.

Five spp. in tropical Central and N South America (including Mexico and the West Indies).

37. ***Schultesia*** Mart.

Schultesia Mart., Nov. Gen. Sp. Pl. 2: 103 (1826) ["1827"], nom. cons., non Spreng. (1815), nom. rej., nec Schrad. (1821) nec Roth (1827); Struwe et al., Fl. Venez. Guayana 5: 519–522 (1999), reg. rev.; Thiv, Fl. Republ. Cuba 6(1) (2002), reg. rev.; Guimarães & Fontella-Pereira, Bradea 8: 259–263 (2001); Guimarães & Pereira, Bradea 8: 285–287 (2002), reg. rev.; Guimarães et al., Arq. Mus. Nac. Rio de Janeiro 61(3): 151–164 (2003); Guimarães, Rodriguésia 55: 67–72 (2004), reg. rev.; Guimarães et al., Acta Bot. Bras. 21: 309–323 (2007), morph.; Delgado et al., Acta Bot. Bras. 23: 956–967 (2009), morph.

Herbs, annual. Leaves membranous or ± succulent, sessile. Flowers solitary or in terminal dichasia, 4- or 5-merous. Calyx usually 4-angled or winged; calyx tube shorter or longer than lobes; corolla urceolate-salverform to funnelform, pink to rose-violet or white; corolla tube longer than lobes, ± narrowed above apex of ovary; stamens inserted near or below middle of corolla tube; filaments commonly proximally expanded or winged; apex of anthers often short-mucronate; ovary unilocular; style elongate, caducous; nectaries absent or rarely with a nectariferous disk at base of ovary. n = 28.

Ca. 15 spp., Mexico, tropical Central and South America, West Indies, *S. senegalensis* Baker in W Africa (Senegal).

38. ***Xestaea*** Griseb.

Xestaea Griseb., Linnaea 22: 35 (1849); Struwe et al., Gent. Syst. Nat. Hist.: 122 (2002), phyl.; Guimarães, Rodriguésia 55: 67–72 (2004), reg. rev.; Guimarães et al., Acta Bot. Bras. 21: 309–323 (2007), morph.

Herbs, annual. Leaves membranous, sessile. Flowers in dichasia, 4-merous. Calyx not winged; calyx tube much shorter than lobes; corolla funnelform, violet-pink; corolla tube longer than lobes, ± narrowed above apex of ovary; stamens inserted near middle of corolla tube; filaments proximally expanded; apex of anthers prolonged; ovary pseudobilocular or basally bilocular; style elongate, caducous; nectaries absent.

One sp., *X. lisianthoides* Griseb., in Mexico, Central America, and northern South America.

39. ***Deianira*** Cham. & Schltdl.

Deianira Cham. & Schltdl., Linnaea 1: 195 (1826); Guimarães, Arch. Jard. Bot. Rio de Janeiro 21: 45–123 (1977), rev., Arq. Jard. Bot. Rio de Janeiro 26: 215–225 (1983), rev., Daphne 1: 17–22 (1990), rev.; Delgado et al., Acta Bot. Bras. 23: 956–967 (2009), morph.

Herbs or subshrubs, perennial. Leaves cauline, often also basal, membranous or ± succulent, sessile or perfoliate. Flowers in dense or capitate, terminal and axillary dichasia, 4-merous. Calyx lobes shorter or longer than tube, often keeled; corolla salverform, pink or white; corolla tube about as long as or shorter than tube; stamens inserted above middle of corolla tube; ovary pseudotetralocular; style elongate, caducous; nectaries absent. n = 14.

Seven spp. in eastern South America.

40. ***Symphyllophyton*** Gilg

Symphyllophyton Gilg in Engl. & Prantl, Nat. Pflanzenfam., Nachtr. 1: 283 (1897).

Herbs, annual. Leaves membranous, lower leaves sessile, upper perfoliate. Flowers in diffuse dichasia, 4-merous. Calyx lobed nearly to base; corolla salverform, white or cream with a rose-purple and yellow center, or rose-purple throughout; corolla tube about as long as lobes; stamens inserted below middle of corolla tube, two with longer filaments and larger anthers alternating with two smaller, all fertile; connective prolonged; ovary unilocular; style slender, caducous; nectaries at base of ovary.

Two spp. in Brazil.

5. **Tribe Gentianeae** Dumort. (1829).

Herbs and subshrubs. Corollas actinomorphic, diverse in shape but rarely salverform, often variously appendaged, often with guidelines but rarely with contrastingly colored centers; anthers and style branches not coiling; pollen in monads (rarely not); ovary unilocular (rarely proximally bilocular), often stipitate; nectaries well developed, either on petals or as nectary disk. Widely distributed in cold to warm-temperate regions. Figs. 72, 73.

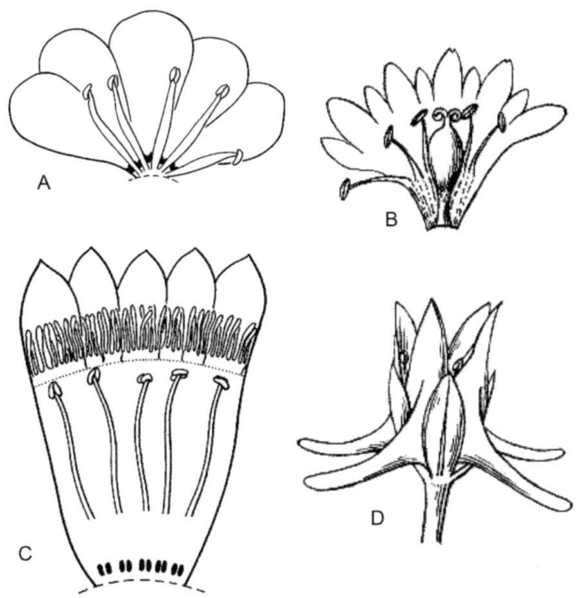

Fig. 72. Gentianaceae. Corollas of Gentianeae. **A** *Gentianella* sp., unfringed and uninectariate at base. **B** *Gentiana vandellioides*, plicae between the 5 corolla lobes. **C** *Comastoma* sp., binectariate at base of corolla tube, fringed in corolla mouth. **D** *Halenia elliptica*, corolla base with spurs. (A. Drawing by Bobbi Angell, permission for use by Bobbi Angell, previously published in Struwe and Albert 2002; B. Original drawing by Wang Ying, redrawn by Cai Shuqin, reproduced with permission from Missouri Botanical Garden Press and Science Press, previously published in Ho and Pringle 1995; C. Drawing by Bobbi Angell, permission for use by Bobbi Angell, previously published in Struwe and Albert 2002; D. Original drawing by Yan Cuilan, redrawn by Cai Shuqin, reproduced with permission from Missouri Botanical Garden Press and Science Press, previously published in Ho and Pringle 1995)

5a. **Subtribe Gentianinae** G. Don (1838).

Corollas usually with lateral petaloid appendages (plicae) or truncate sinuses between lobes (Fig. 72B); ovary stipitate (subsessile in a few *Gentiana* and *Crawfurdia*); styles often short, not sharply differentiated from ovary, persistent; nectaries separate, on ovarian stipe, or in a ring or forming a cup (some *Crawfurdia*). Widely distributed in cold to warm-temperate regions.

41. ***Gentiana*** L.

Gentiana L., Sp. Pl.: 227 (1753); Kusnezow, Trudy S.-Peterburgsk. Obshch. Estestvoisp., Otd. Bot. 24: part 2 (1894), rev.; Gillett, The Gentians of Canada, Alaska and Greenland (1963), reg. rev.; Pringle, Brittonia 19: 1–33 (1967), reg. rev., Sida 7: 174–217 (1977), 8: 14–33 (1979), reg. rev.; Ho, Bull. Bot. Res., Harbin 5(4): 1–22 (1985), reg. rev.; Adams & Williams, Telopea 3: 167–176 (1988), reg. rev.; Ho & Liu, Bull. Brit. Mus. Nat. Hist., Bot. 20: 169–192 (1990), reg. rev.; Ho & Pringle, Fl. China 16: 1–139 (1995), reg. rev.; Halda, The genus *Gentiana* (1996), rev.; Ho et al., Acta Phytotax. Sin. 34: 505–530 (1996); Yuan et al., Amer. J. Bot 83: 641–652 (1996), phyl.; Yuan & Kupfer, Bot. J. Linn. Soc. 123: 25–43 (1997); Hungerer & Kadereit, Persp. Pl. Ecol. Evol. Syst.1: 121–135 (1998); Chen & Wang, Bot. Bull. Acad. Sin. 40: 9–38 (1999), reg. rev.; Chen et al., Ann. Bot. 96: 413–424 (2005), phyl.; Mishiba et al., Breeding Sci. 59: 119–127 (2009), phyl.; Zhang et al., Taxon 58: 862–870 (2009), phyl.; Davitashvili & Karrer, Bot. J. Linn. Soc. 162: 101–115 (2010), morph.; Favre et al., J. Biogeogr. 43: 1967–1978 (2016), phyl.
Ciminalis Adans. (1763).
Dasystephana Adans. (1763).
Tretorhiza Adans. (1763).
Pneumonanthe Gled. (1764).
Calathiana Delarbre (1796).
Ericoila Borkh. (1796).
Varasia Phil. (1860).
Chondrophylla A. Nels. (1904).
Favargera Á. Löve & D. Löve (1972).
Gentianodes Á. Löve & D. Löve (1972).
Kuepferella M. Laínz (1976).
Mehraea Á. Löve & D. Löve (1976).
Qaisera Omer (1989).

Herbs, annual, biennial, or perennial. Leaves opposite or rarely whorled, cauline and often also basal, membranous or ± coriaceous or succulent, sessile or rarely pseudopetiolate. Flowers in cymes or thyrses, sometimes racemoid or capitate, or solitary, 4–5(–9)-merous. Calyx actinomorphic or ± irregularly lobed or occasionally spathaceous; calyx tube longer or shorter than lobes, in most sections with an intracalycular membrane extending above base of lobes; corolla campanulate, funnelform, salverform, tubular (rarely rotate with short tube in *G. lutea*), colors usually blue, purple, or pink, less often white or yellow; corolla tube longer than lobes, with well-developed plicae in corolla lobe sinuses (a truncate gap between lobes in *G. sceptrum*, absent in *G. lutea*); stamens equal, straight, usually inserted below or near middle of corolla tube, without appendages; anthers free or coherent; ovary stipitate or rarely subsessile; style none or indistinct, or short if slender; small nectaries at base of ovarian stipe. Fruit a capsule. n = 6–24, 26, 30, 48.

Ca. 370 spp., worldwide except sub-Saharan Africa, most species in China (ca. 250 spp.); some widely cultivated, some used medicinally.

42. ***Kuepferia*** Adr. Favre

Kuepferia Adr. Favre (2014), Favre et al., Taxon 63: 342–354 (2014); Maity et al., Nord. J. Bot. 34: 416–420 (2016), part. rev.; Dey & Maity, Edinburgh J. Bot. 72: 429–436 (2015), part. rev.
Gentiana sect. *Otophora* Kusn. in Trudy S.-Peterburgsk. Obshch. Estestvoisp., Otd. Bot. 24(2): 102 (1894).

Herbs, perennial. Leaves opposite, cauline and often also basal, membranous. Flowers solitary (or rarely in cymes), 5-merous. Calyx ± irregularly lobed; calyx tube longer or shorter than lobes; corolla campanulate, funnelform, or salverform, blue, less often white or yellow; corolla tube shorter or as long as lobes, with very reduced plicae with entire margins in corolla lobe sinuses; stamens equal, straight, inserted below or near middle of corolla tube, without appendages; anthers free or coherent; ovary stipitate or rarely subsessile; style none or indistinct, or short if slender; small nectaries at base of ovarian stipe. Fruit a capsule. n = 14 (*K. otophoroides*).

13 species in Asia (southeastern part of the Qinghai-Tibet Plateau and in the Himalayas). *Kuepferia* was separated as a new genus from *Gentiana* to preserve the monophyly of *Gentiana*.

43. ***Metagentiana*** T.N. Ho & S.W. Liu

Metagentiana T.N. Ho & S.W. Liu, Bot. Bull. Acad. Sin. 43: 83–91 (2002); Chen et al., Ann. Bot. 96: 413–424 (2005), phyl.; Favre et al., Taxon 63: 342–354 (2014), phyl., rev.

Herbs, annual or perennial; stems erect or decumbent. Leaves membranous, sessile. Flowers solitary at ends of branches, 5-merous. Calyx keeled or winged; calyx tube longer than lobes; corolla funnelform, blue, purple or pink; corolla tube longer than lobes; with plicae, extending into free portions with fringed or erose margins between lobes; stamens unequal, inserted near or below middle of corolla tube, ± declinate; anthers free; style slender, persistent; nectaries small, on ovarian stipe. Fruit a capsule. n = 17, 21, 23.

14 spp. in E Asia. *Metagentiana* was previously recognized as *Gentiana* section *Stenogyne*. *Metagentiana* was recently recircumscribed to remove a clade of species to the new genus *Sinogentiana*, to preserve the monophyly of *Metagentiana*.

44. ***Crawfurdia*** Wall.

Crawfurdia Wall., Tent. Fl. Nepal.: 63 (1826); Smith, Notes Roy. Bot. Gard. Edinburgh 26: 237–258 (1965); Ho & Pringle, Fl. China 16: 1–139 (1995), reg. rev.; Hul, Adansonia 24: 27–41 (2002), reg. rev.; Chen et al., Ann. Bot. 96: 413–424 (2005), phyl.; Dash et al., J. Jap. Bot. 86: 127 (2011), reg. rev.

Herbs, perennial; stems twining or rarely trailing. Leaves membranous or coriaceous, petiolate or sessile. Flowers in small axillary or terminal dichasia or solitary, 5-merous. Calyx not winged; tube longer or rarely shorter than lobes, sometimes with an intracalycular membrane extending above base of lobes; corolla funnelform or campanulate, blue to purple to pink; corolla tube longer than lobes, with well-developed plicae, nearly truncate or slightly prolonged into free portions between lobes; stamens equal, inserted near or below middle of corolla tube, straight or nearly so; filaments asymmetrically winged proximally; anthers free; ovary sessile or stipitate; style short but slender, persistent; nectaries five, basal or on ovarian stipe. Fruit a capsule. n = 21, 23, 24.

Ca. 20 spp. in mountains of eastern Asia.

45. ***Sinogentiana*** Adr. Favre & Y.M. Yuan

Sinogentiana Adr. Favre & Y.M. Yuan; Favre et al., Taxon 63: 342–354 (2014).

Herbs, biennial; stems erect or ascending. Leaves along stem shorter than internodes, with connate petioles, basal leaves withered during flowering. Flowers solitary and sessile at ends of branches, 5-merous. Calyx keeled and winged; calyx tube longer than lobes; corolla funnelform to tubular, yellow or white; corolla tube much longer than lobes, with well-developed asymmetrical plicae, extending into free portions with denticulate or erose margins between lobes; stamens unequal, inserted near or below middle of corolla tube, ± declinate; anthers free; style slender, persistent; nectaries small, on ovarian stipe. Fruit a capsule. Seeds usually winged. n = 23.

Two spp. in south-central China. *Sinogentiana* was recently removed from *Metagentiana* to make *Metagentiana* monophyletic.

46. ***Tripterospermum*** Blume

Tripterospermum Blume, Bijdr.: 849 (1826); Smith, Notes Roy. Bot. Gard. Edinburgh 26: 237–258 (1965); Murata, J. Fac. Sci. Univ. Tokyo, Sect. 3, Bot. 14: 273–329 (1989), rev.; Ho & Pringle, Fl. China 16: 1–139 (1995), reg. rev.; Hul, Adansonia 24: 27–41 (2002), reg. rev.; Chen et al., Ann. Bot. 96: 413–424 (2005), phyl.; Chen et al., Bot. Stud. (Taiwan) 47: 199–205 (2006); Hsu & Chung, Taiwania 57: 183–187 (2012), reg. rev.; Favre et al., Syst. Bot. 38: 224–234 (2013), part. rev.

Herbs, perennial; stems usually twining, less often trailing or rarely suberect. Leaves membranous, ± petiolate. Flowers usually solitary in axils, occasionally in small dichasia. 5-merous. Calyx tube longer or shorter than lobes, sometimes keeled; corolla funnelform or campanulate, with plicae, nearly truncate or ± extending into free portions between lobes, blue to violet or greenish white to pale yellow; corolla tube longer than lobes; stamens ± unequal, inserted near or below middle of corolla tube, declinate; anthers free; style slender, persistent; nectariferous cuplike or lobed disk on ovarian stipe. Fruit a berry or sometimes a capsule. n = 23.

Ca. 34 spp. in East and Southeast Asia.

5b. **Subtribe Swertiinae** Rchb. (1837).

Corollas without plicae between lobes, but often with glands (often in distinct pits, pockets, or spurs), trichomes, or other appendages on petals or in corolla tube; stamens equal, usually without appendages, rarely with fimbriae around filament bases; anthers free or rarely not; nectary disk at base of ovary absent (except *Latouchea*, *Megacodon*). Fruit a capsule

This subtribe suffers from a lack of detailed phylogenetic analyses with broad taxonomic and geographic sampling that can clarify the monophyly or not of many of its genera. It is clear that *Swertia* and *Gentianella* are not monophyletic in their current circumscriptions (von Hagen and Kadereit 2001, 2002; Chassot et al. 2001), but further studies are necessary to create a new, stable generic classification.

47. ***Obolaria*** L.

Obolaria L., Sp. Pl. 632 (1753); Gray, Mem. Amer. Acad. Arts II, 3: 21–31, pl. 3 (1848), J. Proc. Linn. Soc., Bot. 1: 120–130 (1857); Holm, Ann. Bot. (London) I, 11: 369–383, pl. XIX (1897); Gillett, Rhodora 61: 43–62 (1959); Walker, Phytologia 84: 64–68 (1998), reg. rev.; Cameron & Bolin, Amer. J. Bot. 97: 1272–1277 (2010), morph.

Herbs, perennial, partially mycoheterotrophic with coralloid root system, but strongly chlorophyllous. Leaves opposite or lowest occasionally alternate, those below inflorescence much reduced. Flowers in terminal and axillary cymes, solitary, or in cymules of 3. Calyx absent(?); corolla 4-merous, subtended by two foliaceous bracts (often considered to be a reduced calyx), narrowly campanulate, white to pale violet, with glandular trichomes adaxially at or near sinuses and with minute scales on lower part of tube, two per petal; corolla lobes slightly longer than tube, imbricate in bud; stamens inserted slightly below corolla sinuses; ovary sessile; style distinct, persistent; base of ovary glandular but without distinct nectaries. Capsule dehiscing ± irregularly. n = 28.

One sp., *O. virginica* L., in SE United States.

48. ***Latouchea*** Franch.

Latouchea Franch., Bull. Soc. Bot. France 46: 212 (1899); Ho & Pringle, Fl. China 16: 1–139 (1995), rev.

Herbs, perennial. Leaves all basal except for bracts in inflorescence, membranous, ± pseudopetiolate, or stem leaves few, usually one pair, and much reduced. Flowers in a condensed, ± reduced verticillaster of umbelloid clusters, 4-merous. Calyx lobed nearly to base; corolla campanulate, greenish; corolla tube about as long as lobes; stamens inserted in corolla sinuses; ovary sessile, proximally bilocular; style curved, acuminate, not sharply differentiated, persistent; nectaries in a ring on ovary stipe. Capsule curved distally and initially dehiscing along one suture only.

One sp., *L. fokienensis* Franch., in Asia (China).

49. ***Gentianopsis*** Ma

Gentianopsis Ma, Acta Phytotax. Sin. 1: 7 (1951); Gillett, The Gentians of Canada, Alaska and Greenland (1963), reg. rev.; Iltis, Sida 2: 129–154 (1965); Toyokuni, Symb. Asahikawenses 2: 57–72 (1967), 3: 137–146 (1968); Massias et al., Biochem. Syst. Ecol. 10: 319–327 (1982); Garg, Int. Biosci. Monogr. 17 (1987), reg. rev.; Yuan & Kupfer, Caryologia 58: 115–123 (1993); Ho & Pringle, Fl. China 16: 1–139 (1995), reg. rev.; Pringle, Sida 21: 525–530 (2004), reg. rev.; Aitken, Edinburgh J. Bot. 64: 253–260 (2007), reg. rev.
Crossopetalum Roth (1827), non P. Browne (1756), nec *Crossopetalon* Adans. (1763).
Gentiana subg. *Eublephis* Raf. (1828).
Gentiana sect. *Crossopetalum* Froel. ex Griseb. (1838) ["1839"].
Gentiana sect. *Gentianopsis* (Ma) Satake (1957).
Gentianella subg. *Gentianopsis* (Ma) Toyok. (1957).

Herbs, annual, biennial or perennial. Leaves cauline, sometimes also basal, membranous, sessile. Flowers in dichasia or solitary, 4-merous. Calyx tube shorter than lobes; two inner lobes often smaller than outer, with discontinuous intracalycular membrane extending across sinuses; corolla broadly tubular or tubular-campanulate with spreading lobes, blue or rarely rose-violet, yellow, or white; corolla lobes shorter than or about as long as tube, with fringed or dentate margin, minutely puberulent adaxially near stamen bases, often with swollen nectaries on tube near base, one per petal; stamens inserted in lower half of corolla tube; ovary stipitate; style short or indistinct, persistent. n = 13, 22, 26, 39.

Ca. 24 spp. in North America and Eurasia.

50. ***Pterygocalyx*** Maxim.

Pterygocalyx Maxim., Mém. Acad. Imp. Sci. St.-Pétersbourg Divers Savans 9: 198 (1859); Ho & Pringle, Fl. China 16: 1–139 (1995), reg. rev.; Chen et al., Acta Phytotax. Sin. 36: 58–68 (1998).

Herbs, biennial or perennial; stems twining or trailing. Leaves membranous, petiolate. Flowers solitary or in clusters of 2–3 in axils, 4(5)-merous. Calyx winged; calyx tube much longer than lobes, without an intracalycular membrane; corolla narrowly funnelform, blue to violet or white; corolla tube longer than lobes, with nectaries on corolla tube near base, two per petal; stamens inserted near middle of corolla tube, abruptly turned outward near tip; ovary stipitate; style short, slender, persistent. n = (12) 13.

One sp., *P. volubilis* Maxim., in E Asia. *Pterygocalyx* is closely related to *Gentianopsis*.

51. ***Megacodon*** (Hemsl.) Harry Sm.

Megacodon (Hemsl.) Harry Sm. in Hand.-Mazz., Symb. Syn. 7: 950 (1936); Toyokuni, Symb. Asahikawenses 1: 143–146 (1965); Ho & Pringle, Fl. China 16: 1–139 (1995), reg. rev.; Ge et al., Biodivers. Conserv. 14: 849–861 (2005), pop. gen.; Xue & Li, Bot. J. Linn. Soc. 147: 317–331 (2005), anat.
Gentiana sect. *Stylophora* C.B. Clarke (1883).
Gentiana sect. *Megacodon* Hemsl. (1890).

Herbs, perennial. Leaves opposite, cauline and basal, membranous, sessile or lower leaves petiolate. Flowers in thyrses or sometimes in raceme-like monochasia, 5-merous. Calyx tube shorter than lobes; corolla campanulate, white to yellowish green with conspicuous purplish-brown veins; corolla tube shorter than lobes; stamens inserted near or above middle of corolla tube; ovary ± stipitate; style short but slender, persistent; nectaries on ovarian stipe. n = 14.

Two spp. in Asia (Sino-Himalaya region).

52. ***Bartonia*** Muhl. ex Willd.

Bartonia Muhl. ex Willd., Ges. Naturf. Freunde Berlin Neue Schriften 3: 444 (1801); Holm , Ann. Bot. 20: 441–448, pls. 33, 34 (1906), anat.; Fernald & Weatherby, Rhodora 23: 284–300 (1932), rev.; Gillett, Rhodora 61: 43–62 (1959), rev.; Gillett, The Gentians of Canada, Alaska and Greenland (1963), reg. rev.; Mathews et al., Syst. Bot. 34: 162–172 (2009), rev.; Cameron & Bolin, Amer. J. Bot. 97: 1272–1277 (2010), morph.
Centaurella Michx. (1803).
Andrewsia Spreng. (1825).

Herbs, annual, partially mycoheterotrophic, with little chlorophyll. Leaves opposite or sometimes alternate, scale-like. Flowers in small dichasial or racemoid cymes or thyrses, 4-merous. Calyx lobed nearly to base; corolla campanulate, white to yellowish, sometimes purple-tinged, without appendages; corolla lobes longer than tube, imbricate in bud; stamens inserted in corolla sinuses, without appendages; ovary sessile; style none; nectaries absent. Capsule sometimes medially dehiscent. n = 22, 26.

Three spp. in S and E North America.

53. ***Veratrilla*** (Baill.) Franch.

Veratrilla (Baill.) Franch., Bull. Soc. Bot. France 46: 310 (1899); Baillon, Bull. Mens. Soc. Linn. Paris 1: 729–730 (1888), reg. rev.; Smith, Bull. Br. Mus. Nat. Hist. (Bot.) 4: 239–258 (1970), reg. rev.; Ho & Pringle, Fl. China 16: 1–139 (1995), reg. rev.; Xue & Li, Bot. J. Linn. Soc. 147: 317–331 (2005), anat.
Swertia sect. *Veratrilla* Baill. (1888).

Herbs, perennial, dioecious. Leaves opposite and basal, membranous, basal petiolate. Flowers in much-compound thyrses, 4-merous, unisexual (pistillate flowers have sterile stamens; staminate flowers have a sterile pistil). Calyx lobed nearly to base; corolla rotate, deeply lobed, greenish white to greenish yellow with blue or purple veins and nectary pits, sometimes with fringed margins, on each corolla lobe, with one or two nectaries in each pit; stamens inserted in corolla sinuses; ovary sessile; style short, persistent.

Two spp. in Asia (Sino-Himalayan area).

54. ***Frasera*** Walt.

Frasera Walt., Fl. Carol.: 78 (1788); Card, Ann. Missouri Bot. Gard. 18: 245–282 (1931); St. John, Amer. Midl. Naturalist 26(1): 1–29 (1941); Gillett, The Gentians of Canada, Alaska and Greenland (1963), reg. rev.; Pringle, Sida 14: 179–187 (1990); Shah, Sci. Khyber 3: 17–114 (1990), rev.
Tesseranthium Kellogg (1862).
Leucocraspedum Rydb. (1917).
Swertia sect. *Frasera* (Walt.) Zuev (1990).

Perennial herbs, annual or polycarpic. Cauline leaves opposite or whorled, basal rosette present; leaves green or grayish and membranous, flat or conduplicate, often with a distinct white margin. Flowers in axillary panicles, occasionally congested, 4-merous. Calyx deeply lobed with unequal sized pairs; corolla rotate; corolla tube much shorter than lobes, each petal with one (rarely two or partly confluent) nectary pits with a fimbriate, dentate or tubular rim; stamens inserted in sinuses between corolla lobes, connected by a variable intrastaminal corona; ovary sessile; style usually filiform, rarely short and persistent. n = 13, 26, 39.

15 spp. in North America.

55. *Halenia* Borkh. Fig. 72

Halenia Borkh., Arch. Bot. (Leipzig) 1(1): 25 (1796); Gilg, Ber. Deutsch. Bot. Ges. 13: 114–126 (1895), morph.; Allen, Ann. Missouri Bot. Gard. 20: 119–222 (1933), reg. rev.; Gillett, The Gentians of Canada, Alaska and Greenland (1963), reg. rev.; Wilbur, Rhodora 86: 311–337 (1984), reg. rev.; Wilbur, Bull. Torrey Bot. Club 111: 366–374 (1984), reg. rev.; Garg, Int. Biosci. Monogr. 17 (1987), reg. rev.; Ho & Pringle, Fl. China 16: 1–139 (1995), reg. rev.; Pringle, Fl. Ecuador 53: 65–73 (1995), reg. rev.; Chassot et al., Pl. Syst. Evol. 229: 1–21 (2001), phyl.; von Hagen & Kadereit, Syst. Bot. 27: 548–572 (2002), Evolution 57: 2507–2518 (2003), phyl.; Kadereit & von Hagen, Int. J. Pl. Sci. 164: S441–S452 (2003), phyl.; von Hagen, Org. Divers. Evol. 7: 1–11 (2007), phyl., part. rev.

Herbs or subshrubs, annual or perennial. Leaves opposite or whorled, sometimes in basal rosette, membranous, sessile or lower leaves petiolate. Flowers in cymes or thyrses (often umbelloid or racemoid), 4-merous, perfect or lower flowers staminate. Calyx lobed nearly to base; corolla campanulate or scarcely opening, greenish white to greenish yellow or less often purplish, deeply lobed, with nectariferous tubercles on tube, ranging from slight protuberances to spurs, one per petal (Fig. 72D); stamens inserted in corolla sinuses; ovary sessile; style none or indistinct, persistent. n = 11.

39 spp. in W North, Central, and W South America and Eurasia.

56. *Jaeschkea* Kurz

Jaeschkea Kurz, J. Asiat. Soc. Bengal, Pt. 2, Nat. Hist. 39 (2): 230 (1870); Garg, Int. Biosci. Monogr. 17 (1987), reg. rev.; Ho & Pringle, Fl. China 16: 1–139 (1995), reg. rev.; Aitken, Fl. Bhutan 2(2): 612–614 (1999), reg. rev.; Yuan et al., Pl. Syst. Evol. 210: 231–247 (1998), phyl.

Herbs, annual. Leaves basal and cauline, membranous, basal pseudopetiolate, upper sessile. Flowers in thyrses of umbelloid clusters or solitary, (4)5-merous. Calyx lobed nearly to base; corolla narrowly campanulate, blue to blue-violet; corolla lobes proximally valvate in bud, about as long as tube, with nectaries on tube near base, two per petal; stamens inserted near apex of corolla tube; ovary sessile; style short, persistent. n = 8, 9, 10, 11.

Three spp. in Asia (Himalayas region). The monospecific *Kurramiana* Omer & Qaiser has uncertain status and is here treated as a tentative synonym of *Jaeschkea* pending phylogenetic evidence for its placement.

57. *Lomatogonium* A. Braun

Lomatogonium A. Braun, Flora 13: 221 (1830); Fernald, Rhodora 21: 193–198 (1919); Gillett, The Gentians of Canada, Alaska and Greenland (1963), reg. rev.; Garg, Int. Biosci. Monogr. 17 (1987), reg. rev.; Ho & Liu, Acta Phytotax. Sin. 30: 289–319 (1980), rev. (in part as *Lomatogoniopsis*); Liu & Ho, Acta Phytotax. Sin. 30: 289–319 (1992), rev.; Ho & Pringle, Fl. China 16: 1–139 (1995), reg. rev.; Chen & Wang, Bot. Bull. Acad. Sin. 41: 323–326 (2000), reg. rev.
Pleurogyne Eschsch. ex Griseb. (1838) ["1839"].
Swertia subg. *Lomatogonium* (A. Braun) Satake (1944).
Pleurogynella Ikonn. (1970).
Lomatogoniopsis T.-N. Ho & S.W. Liu (1980).

Herbs, annual or perennial. Leaves basal (early withering) and opposite. Flowers in terminal or axillary dichasia (sometimes racemoid), (4)5-merous. Calyx rotate, with short but distinct tube (tube never more than 2.5 mm long, mostly <1 mm); corolla rotate, white or violet-blue; corolla tube much shorter than lobes, with nectaries on corolla tube at or near apex, one or two per petal, in pockets with or without raised, fringed, scale-like, or tubular rims or appendages; stamens inserted in sinuses in the very short corolla tube; ovary sessile; style none, stigma decurrent along sutures. n = 5, 6, 8, 9, 16, 20, 24.

24 spp. in Eurasia and North America.

58. *Comastoma* (Wettst.) Toyokuni Fig. 72

Comastoma (Wettst.) Toyokuni, Bot. Mag. (Tokyo) 74: 198 (1961); Toyokuni, Acta Phytotax. Geobot. 20: 136–138 (1962); Gillett, The Gentians of Canada, Alaska and Greenland (1963), reg. rev.; Garg, Int. Biosci. Monogr. 17 (1987), reg. rev.; Omer, Candollea 47: 539–553 (1992), reg. rev.; Ho & Pringle, Fl. China 16: 1–139 (1995), reg. rev.; Liu & Ho, Acta Phytotax. Sin. 34: 577–585 (1996).
Gentiana sect. *Comastoma* Wettst. (1896).
Lomatogonium sect. *Comastoma* (Wettst.) Á. Löve & D. Löve (1956).
Gentianella subg. *Comastoma* (Wettst.) J.M. Gillett (1957).

Herbs, annual or rarely perennial. Leaves cauline, often also basal, membranous, sessile or lower leaves ± petiolate. Flowers in thyrses or racemes, or solitary, 4–5-merous (calyx 2-lobed in *C. disepalum*). Calyx deeply lobed, without intracalycular membrane, two outer lobes often much wider than inner; corolla tubular, campanulate, or salverform, blue, violet-blue, or occasionally yellow or white; corolla tube longer than lobes, each petal with one or two deeply fimbriate scales and two nectaries at base (Fig. 72C); stamens inserted near middle of corolla tube; ovary sessile; style short, indistinct, persistent. n = 5, 6, 8, 9, 10, 14–16, 18.

Ca. 20 spp. in Eurasia and North America.

59. *Gentianella* Moench — Fig. 72

Gentianella Moench, Methodus: 482 (1794), nom. cons.; Gilg, Bot. Jahrb. Syst. 22: 301–347 (1896), reg. rev.; Gillett, Ann. Missouri Bot. Gard. 44: 195–269 (1957), reg. rev.; Gillett, The Gentians of Canada, Alaska and Greenland (1963), reg. rev.; Garg, Int. Biosci. Monogr. 17 (1987), reg. rev.; Nesom, Phytologia 70: 1–20 (1991), reg. rev.; Aitken & Long, Edinburgh J. Bot. 51: 165–167 (1994); Adams, Austral. Syst. Bot. 8: 935–1011 (1995), reg. rev.; Ho & Pringle, Fl. China 16: 1–139 (1995), reg. rev.; Pringle, Fl. Ecuador 53: 12–65 (1995), reg. rev.; Chassot et al., Pl. Syst. Evol. 229: 1–21 (2001), phyl.; von Hagen & Kadereit, Org. Divers. Evol. 1: 61–79 (2001), phyl., Syst. Bot. 27: 548–572 (2002); Kadereit & von Hagen, Int. J. Pl. Sci. 164: S441–S452 (2003), phyl.; Glenny, New Zealand J. Bot. 42: 361–530 (2004), reg. rev.; Jang et al., Bot. J. Linn. Soc. 148: 175–187 (2005), reg. phyl.
Amarella Gilib. (1782), nom. rej.
Aloitis Raf. (1837).
Eudoxia D. Don ex G. Don (1837).
Arctogentia Á. Löve (1982).
Aliopsis Omer & Qaiser (1991).
Chionogentias L.G. Adams (1995).

Herbs or subshrubs, annual, perennial, or some probably long-lived monocarps. Leaves opposite or whorled, cauline and often also basal, membranous to coriaceous, sessile or lowest pseudopetiolate. Flowers in cymes (dichasial or racemoid, umbelloid, or capitate) or solitary, 4–5-merous, usually perfect, plants sometimes andromonoecious, gynodioecious, gynomonoecious, or perhaps trimonoecious. Calyx actinomorphic or rarely spathaceous; calyx tube longer or shorter than lobes, without intracalycular membrane; corolla campanulate, funnelform, or nearly rotate its colors and depth of lobing diverse, adaxial surface glabrous or puberulent near base or near insertion of stamens, or fimbriate at apex of tube or with paired scales near base of lobes; nectaries on corolla tube near base, one or rarely two per petal (Fig. 72A); stamens inserted at same height in corolla tube (slightly unequal and inserted at slightly different levels in *G. weigendii*); ovary sessile or short-stipitate; style short, indistinct, persistent. n = 9, 18, 19, 27, except *G. auriculata*, n = 24.

Ca. 250 spp. (some estimates at 150 spp.), worldwide except sub-Saharan Africa; some widely cultivated. Chromosome numbers above given for *Gentianella* exclusive of binectariate Asian spp.

60. *Swertia* L. — Fig. 73

Swertia L., Sp. Pl. 226 (1753); Clarke, Gentianaceae in Hooker, Flora of British India, pp. 108–119 (1885), reg. rev.; Fries, Notizbl. Bot. Gart. Berlin-Dahlem 77: 505–534

Fig. 73. Gentianaceae-Gentianeae. **A** *Swertia tibetica*, habit. **B** *Swertia kingii*, leaf with arcuate venation of secondary veins. **C** *Swertia kingii*, side view of flower. **D** *Swertia kingii*, fringed nectary glands at the base of the inside of one petal. (Original drawings by Wang Ying, redrawn by Cai Shuquin, reproduced with permission from Missouri Botanical Garden Press and Science Press, previously published in Ho and Pringle 1995)

(1923); St. John, Amer. Midl. Nat. 26: 1–29 (1941); Gillett, The Gentians of Canada, Alaska and Greenland (1963), reg. rev.; Khoshoo & Tandon, Caryologia (Pisa) 16: 445–477 (1963), reg. rev.; Toyokuni, J. Fac. Sci. Hokkaido Univ. 7(4): 137–259 (1963), reg. rev.; Smith, Bull. Br. Mus. Nat. Hist. (Bot.) 4: 239–258 (1970), reg. rev.; Geesink, Blumea 21: 179–183 (1973), reg. rev.; Garg, Int. Biosci. Monogr. 17 (1987), reg. rev.; Klackenberg, Fl. Madagascar et des Comores, Family 168 (1990), reg. rev.; Pringle, Sida 14: 179–187 (1990); Shah, Sci. Khyber 3: 17–114 (1990), reg. rev.; Shah, Sci. Khyber 5: 117–231 (1992), reg. rev.; Ho & Pringle, Fl. China 16: 1–139 (1995), reg. rev.; Nemomissa, Nordic J. Bot. 17: 145–156 (1997); Sileshi, Kew Bull. 53: 419–436 (1998), reg. rev.; Wang & Lu, Taiwania 43: 273–288, reg. rev. (1998); Chassot et al., Pl. Syst. Evol. 229: 1–21 (2001), phyl.; von Hagen & Kadereit, Org. Divers. Evol. 1: 61–79 (2001), phyl., Syst. Bot. 27: 548–572 (2002); Nemomissa, Fl. Trop, East Africa, Gentianaceae (2002), reg. rev.; Kadereit & von Hagen, Int. J. Pl. Sci. 164: S441–S452 (2003), phyl.
Henricea Lem.-Lis. (1824).
Agathotes D. Don ex G. Don (1837).
Ophelia D. Don ex G. Don (1837).
Anagallidium Griseb. (1838) ["1839"].
Sczukinia Turcz. (1840).
Stellera Turcz. (1840), non L. (1753).
Monobothrium Hochst. (1844).
Rellesta Turcz. (1849).
Blepharaden Dulac. (1867).
Swertopsis Makino (1891).
Kingdon-Wardia C. Marquand (1929).

Herbs, annual, biennial or monocarpic or polycarpic perennial (Fig. 73A). Leaves basal and cauline, rarely some or all alternate or whorled, diverse in texture, basal often ± petiolate, rarely perfoliate. Flowers in dichasia, thyrses, or verticillasters or raceme-like cymes, or solitary, 4–5-merous. Calyx free or connate at the base, rarely (sub-)campanulate; calyx lobes often unequal; corolla rotate, rarely (sub-)campanulate (Fig. 73C), white to greenish or blueish, yellow, purple or violet, rarely brownish red, often purple-spotted and/or suffused; corolla tube much shorter than lobes, with one or two nectaries on each corolla lobe; nectaries protected by scales or fimbriae or both, rarely naked (Fig. 73D); stamens epipetalous, inserted in corolla sinuses; filaments generally free, rarely connate at the base, rarely hairy; ovary sessile or rarely short-stipitate; style absent or short and persistent, less often slender and filiform. n = 6–14, 21, 26.

Ca. 140 spp. in North America, Eurasia, Papuasia, E and W Africa and Madagascar. *Swertia* in this circumscription is highly paraphyletic, and it will eventually be split into many smaller genera.

61. ***Sinoswertia*** T.N. Ho, S.W. Liu & J.Q. Liu

Sinoswertia T.N. Ho, S.W. Liu & J.Q. Liu, Pl. Divers. Resources 35: 393–400 (2013).

Herbs, very small, dimorphic in habit and flowers, cleistogamous. Leaves basal and cauline. Flowers in few flowered cymes, 4–5-merous. Calyx connate at very base; calyx lobes often unequal; corolla subrotate, yellow to greenish, tinged with purple on inside; corolla tube much shorter than lobes, with two nectaries on each corolla lobe; stamens epipetalous, inserted in corolla sinuses; filaments free, very short; ovary sessile; style absent, stigma capitate.

One sp., *S. tetraptera* T. N. Ho, S. W. Liu, & J. Q. Liu, in S China (Qinghai-Tibet Plateau). *Sinoswertia* was segregated from *Swertia* to improve the monophyly of *Swertia*, but *Swertia* is still likely paraphyletic (He et al. 2013).

6. **Tribe Potalieae** Rchb. (1837).

Trees, shrubs, or herbs. Flowers actinomorphic; corolla without nectaries; stamens isomerous with corolla lobes, equal (rarely unequal in *Lisianthius* and subequal in some Faroinae), free, straight after anthesis; pollen in monads (tetrads in *Potalia maguireorum*); ovary sessile (substipitate in some Faroinae, rarely stipitate in *Lisianthius* spp.); stigma often capitate, less often bilobed. Pantropical. Fig. 74.

6a. **Subtribe Faroinae** Struwe & V. Albert (2002).

Herbs. Flowers in usually terminal capitula or spikes, sometimes axillary, often sessile or subsessile. Calyx membranous; corolla often with appendages in form of scales or projections at or near base of filaments, without nectaries (except *Faroa*), without nectariferous disk at base of ovary. Fruit a capsule. Pantropical, but predominantly in Africa and Madagascar.

62. ***Congolanthus*** A. Raynal

Congolanthus A. Raynal, Adansonia II, 8: 56 (1968); Raynal, Adansonia II, 8: 45–68 (1968), rev.

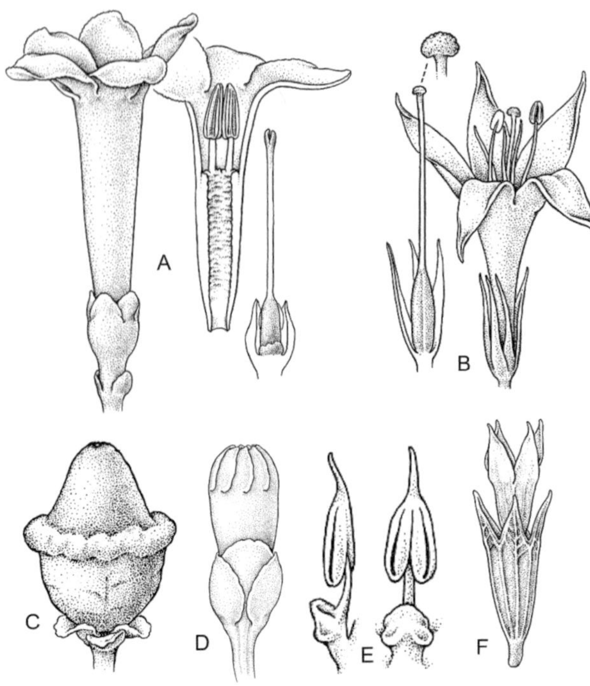

Fig. 74. Gentianaceae-Potalieae. **A** *Fagraea berteroana*, flower (side view, open corolla, and with corolla removed), with fleshy ring below stamens, nectary disk at base of ovary and bilamellate stigma. **B** *Lisianthius laxiflorus*, flower (side view and with corolla removed), with capitate stigma. **C** *Potalia resinifera*, berry with fleshy ring. **D** *Potalia maguireorum*, supermerous flower with eight corolla lobes and four imbricate calyx lobes. **E** *Enicostema verticillatum*, stamens with apical extension and hood-shaped scale at base. **F** *Neurotheca loeselioides*, flower with strong venation on calyx. (A, B. Drawings by Bobbi Angell, permission for use by Bobbi Angell, previously published in Struwe and Albert 2002; C. Drawings by Bobbi Angell, reproduced with permission from Bobbi Angell and Missouri Botanical Garden Press, previously published in Struwe et al. 1999; D. Drawings by Bobbi Angell, reproduced with permission from Harvard University Herbaria, previously published in Struwe et al. 1999; E, F. Drawings by Bobbi Angell, permission for use by Bobbi Angell, previously published in Struwe and Albert 2002)

Herbs, annual. Leaves cauline, membranous, sessile. Flowers solitary in leaf axils, 4-merous, actinomorphic. Calyx membranous, without prominent veins, not reticulate; calyx tube slightly shorter than lobes; corolla funnelform, lilac to mauve; corolla tube much longer than lobes; stamens inserted in middle of corolla tube; ovary sessile, unilocular; style slender, caducous.

One sp., *C. latidens* (N.E. Br.) A. Raynal, in tropical C Africa.

63. *Djaloniella* P. Taylor

Djaloniella P. Taylor, Taxon 12: 294 (1963); A. Raynal, Adansonia II, 8: 45–68 (1968), reg. rev.

Herbs, annual. Leaves membranous, sessile. Flowers in dense terminal dichasia, 4-merous. Calyx tube much shorter than lobes; corolla campanulate, blue; corolla lobes shorter than tube, eight fringed scales in the throat, those of a pair ± confluent; stamens inserted between the pairs of scales in the upper part of the corolla throat; ovary unilocular; style slender but short, persistent.

One sp., *D. ypsilostyla* P. Taylor (or possibly two), in W Africa (Côte d'Ivoire and Guinea).

64. *Faroa* Welw.

Faroa Welw., Trans. Linn. Soc. 27: 45 (1871); Raynal, Adansonia II, 8: 45–68 (1968), reg. rev.; Taylor in Garcia de Orta, Sér. Bot. Lisboa 1: 69–82 (1973); Bamps, Bull. Jard. Bot. Nat. Belg. 52: 486–487 (1982), reg. rev., Bull. Jard. Bot. Nat. Belg. 57: 479–48 (1987), reg. rev.; Nemomissa, Fl. Trop, East Africa, Gentianaceae (2002), reg. rev.

Herbs, annual or rarely perennial. Leaves opposite and basal, membranous, sessile. Flowers in capitula or umbelloid axillary and/or terminal clusters, 4-merous. Calyx lobes about as long as or longer than tube; corolla tubular-campanulate, white or greenish; corolla lobes about as long as tube, with a crescent of glandular hairs below base of each filament; stamens inserted in corolla sinuses; ovary unilocular; style absent or short but slender, persistent.

19 spp. in tropical Africa.

65. *Karina* Boutique

Karina Boutique, Bull. Jard. Bot. Natl. Belg. 41: 261 (1971).

Herbs, annual, perhaps mycoheterotrophic. Leaves small, sessile. Flowers in a dense spike, or solitary in axils, 4-merous. Calyx lobed nearly to base, two unequal pairs of decussate sepals, keeled; corolla tubular, pale blue to white; corolla tube longer than lobes; without appendages; stamens inserted in upper quarter of corolla tube; ovary subsessile, unilocular; style elongate, caducous.

One sp., *K. tayloriana* Boutique, in W Africa (Congo).

66. ***Neurotheca*** Salisb. ex Benth. & Hook.f. Fig. 74

Neurotheca Salisb. ex Benth. & Hook.f., Gen. Pl. 2: 812 (1876); Raynal, Adansonia II, 8: 45–68 (1968), reg. rev.; Klackenberg, Fl. Madagascar et des Comores, Family 168 (1990), reg. rev.; Struwe et al., Fl. Venez. Guayana 5: 509–510 (1999), reg. rev.; Nemomissa, Fl. Trop. East Africa, Gentianaceae (2002), reg. rev.
Octopleura Spruce ex Progel (1865), non Griseb. (1860).

Small herbs, annual. Leaves cauline, sometimes also basal, membranous, sessile. Flowers in terminal, a lax or congested raceme-like cymes (with a terminal flower, sometimes reported as raceme) or rarely 1–3 axillary flowers together along stem, 4-merous, sessile. Calyx coriaceous; calyx tube longer than lobes, with prominent reticulate venation (Fig. 74F); corolla tubular to funnelform, blue (rarely pink or white), without appendages; corolla tube longer than lobes; stamens inserted near or below middle of corolla tube; ovary sessile, unilocular; style slender, caducous. n = 15.

Two spp. in tropical Africa, Madagascar, and NE South America.

67. ***Oreonesion*** A. Raynal

Oreonesion A. Raynal, Adansonia II, 5: 271 (1965); Raynal, Adansonia II, 8: 45–68 (1968), reg. rev.

Herbs, annual. Leaves membranous, sessile. Flowers in terminal and axillary capitate dichasia, 4-merous. Calyx tube longer than lobes; corolla funnelform, blue-violet; corolla tube longer than lobes; stamens inserted in upper part of corolla tube; filaments expanded into a spherical protuberance at base; ovary sessile, unilocular; style slender, caducous. n = 15.

One sp., *O. testui* A. Raynal, in W Africa.

68. ***Pycnosphaera*** Gilg

Pycnosphaera Gilg in Warb., Kunene-Sambesi Exped. 333 (1903); Nemomissa, Fl. Trop. East Africa, Gentianaceae (2002), reg. rev.

Herbs, annual or perennial. Leaves basal and cauline, sessile. Flowers in capitate, terminal cymes, 3-merous. Calyx of three separate sepals, outer one much larger than inner two, inner sepals keeled; corolla tubular, blue fading to white; corolla tube cylindric, narrowed at throat, twice as long as lobes; stamens equal, inserted in upper half of corolla tube; base of filament centrally expanded into a hood-like projection and lobed on each side; ovary sessile, bilocular; style elongate, caducous. n = 15

One sp., *P. buchananii* (Baker) N.E. Br., in Africa.

69. ***Enicostema*** Blume Fig. 74

Enicostema Blume, Bijdr. 848 (1826); Raynal, Adansonia II, 8: 45–68 (1968), reg. rev.; Veldkamp, Blumea 16: 133–136 (1968), rev.; Raynal, Adansonia II, 9: 57–85 (1969), rev.; Klackenberg, Fl. Madagascar et des Comores, Family 168 (1990), reg. rev.; Struwe et al., Fl. Venez. Guayana 5: 496–498 (1999), reg. rev.; Nemomissa, Fl. Trop. East Africa, Gentianaceae (2002), reg. rev.

Herbs, perennial. Leaves membranous or ± succulent, sessile or pseudopetiolate. Flowers in small dense axillary clusters, (3–)5(6)-merous. Calyx deeply lobed; calyx lobes 5, unequal, or in 2 spp. 2 or 3 on inner flowers, 4 on outer; corolla salver- to funnelform, white or cream; corolla tube longer than lobes; stamens inserted near or above middle of corolla tube; filament expanded into a double-hood-like projection at base, alternating with corona scales; connective prolonged (Fig. 74E); ovary unilocular; style short, persistent. n = 19.

Three spp., one in Central America and N South America (including the West Indies), one in Madagascar, and one in tropical Africa, S and SE Asia (India, Malesia region).

70. ***Urogentias*** Gilg & Gilg-Ben.

Urogentias Gilg & Gilg-Ben., Notizbl. Bot. Gart. Berlin-Dahlem 11: 944 (1933); Nemomissa, Fl. Trop. East Africa, Gentianaceae (2002), reg. rev.

Herbs, proximally somewhat woody, annual(?). Leaves sessile or lower leaves short-petiolate. Flowers in large, axillary, often dense dichasia, 8-merous. Calyx deeply lobed; calyx lobes tapering into setaceous tips; corolla salverform, blue; corolla lobes slightly longer than tube, each terminating in two sublobes between which the midvein is abruptly prolonged into a long-setaceous appendage; stamens inserted near middle of corolla tube, with a scale below insertion of each stamen; ovary unilocular; style slender, persistent.

One sp., *U. uluqurensis* Gilg & Gilg-Ben., in E Africa (Tanzania).

6b. **Subtribe Lisianthiinae** G. Don (1838).

Herbs, shrubs, or small trees. Flowers 5-merous. Calyx membranous; corolla without appendages, with nectariferous disk at base of ovary (inconspicuous in *Bisgoeppertia*). Fruit a capsule. Neotropical.

71. ***Bisgoeppertia*** Kuntze

Bisgoeppertia Kuntze, Rev. Gen. Pl. 2: 426 (1891); Thiv, Fl. Republ. Cuba 6(1) (2002), reg. rev.; Greuter & Rankin, Willdenowia 38: 177–185 (2008), rev.

Herbs, annual; stems twining. Leaves membranous, sessile. Flowers in elongate thyrses. Calyx lobed nearly to base; corolla short-salverform, white or pale yellow; corolla tube longer than or about as long as lobes, narrowed above apex of ovary; stamens inserted on very short filaments above middle of corolla tube, without appendages; anthers not coiling; ovary pseudobilocular; style slender, basally persistent; stigma with projection extending downward from rim forming a cylinder around the style, later splitting and adhering to style. n = 18.

Three spp. in the West Indies (Cuba, Hispaniola).

72. ***Lisianthius*** P. Browne Fig. 74

Lisianthius P. Browne, Civ. Nat. Hist. Jamaica: 157 (1756); Weaver, J. Arnold Arbor. 53: 76–100, 234–311 (1972), rev.; Sytsma, Ann. Missouri Bot. Gard. 75: 1587–1602 (1988), reg. rev.; Sytsma & Schaal, Pl. Syst. Evol. 170: 97–106 (1990), pop. gen.; Struwe & Albert, Harvard Pap. Bot. 3: 63–71 (1998); Thiv, Fl. Republ. Cuba 6(1) (2002), reg. rev.
Leianthus Griseb. (1838) ["1839"].

Herbs, shrubs, or small trees, perennial or rarely annual. Leaves membranous to coriaceous, sessile or petiolate. Flowers in terminal or axillary, dichasial cymes or thyrses, sometimes few-flowered, or solitary. Calyx lobed nearly to base; corolla narrowly funnelform to salverform, colors diverse, most frequently yellow, also often red or purple, rarely black; corolla tube longer than lobes, often narrowed above apex of ovary (Fig. 74B); stamens equal or unequal, inserted at various levels in corolla tube, without appendages; anther connectives ± prolonged; ovary unilocular, basally bilocular; style elongate, basally persistent. n = 18.

30 spp. in Central America (including Mexico and the West Indies).

6c. **Subtribe Potaliinae** (Mart.) Progel (1865).

Trees, shrubs, or rarely lianas. Distinct interpetiolar line or low ocrea often present. Flowers in terminal (rarely axillary), dichasial cymes (except some *Fagraea*). Calyx coriaceous, thick; corolla often fleshy, usually 5-merous, less often (6–)8–16-merous, without appendages or nectaries; style caducous in fruit; nectariferous disk at base of ovary. Fruit a berry (Fig. 74C). Pantropical.

73. ***Cyrtophyllum*** Reinw. ex Blume

Cyrtophyllum Reinw. ex Blume (1825), Blume, Bijdr. Fl. Ned. Ind. 1022 (1826); Ridley, Fl. Malay. Pen. 2 421 (1923), rev.; Wong & Sugau, Sandakania 8: 1–93 (1996), reg. rev.; Wong & Sugumaran, Gard. Bull. Singapore 64: 497–510 (2012), rev.
Fagraea sect. *Cyrtophyllum* (Reinw.) Blume (1838).
Fagraea subg. *Cyrtophyllum* (Reinw.) Miq. (1857).

Medium to large-sized trees (to 30 m). Leaves coriaceous and petiolate. Flowers in many-flowered axillary, complete or incomplete dichasia. Calyx tube longer than lobes; corolla funnelform, rarely slightly zygomorphic, creamy white; corolla tube much longer or as long as lobes; stamens inserted in or above middle of corolla tube, filaments not geniculated at base; ovary typically bilocular; style elongate, caducous. n = 6.

Five spp. in S and E Asia and New Guinea.

74. ***Utania*** G. Don

Utania G. Don, Gard. Dict. 4: 663 (1838); Wong & Sugau, Sandakania 8: 1–93 (1996), reg. rev.; Wong & Sugumaran, Gard. Bull Singapore 492 (2012).
Kuhlia Reinwardt (1825), nom. illeg., non Kunth (in Humboldt et al. 1825).
Kentia Steudel (1840), nom. illeg., non Adanson (1763).

Small to medium-sized trees (to 15 m). Leaves coriaceous or fleshy, petiolate. Flowers in terminal, many-flowered, branched pendulous raceme-like cymes. Calyx tube longer than lobes; corolla funnelform, rarely slightly zygomorphic, creamy-white; corolla tube much longer than lobes; stamens inserted in or above middle of corolla tube; filaments not geniculate at base; ovary bilocular; style elongate and caducous.

Ca. 15 spp. in SE Asia and New Guinea.

75. ***Picrophloeus*** Bijdr.

Picrophloeus Bijdr., Fl. Ned. Ind. 1019 (1826); Wong & Sugau, Sandakania 8: 1–93 (1996), reg. rev.; Wong & Sugumaran, Gard. Bull. Singapore 64: 492 (2012).

Medium-sized trees (to 30 m). Leaves coriaceous or fleshy and petiolate. Flowers in terminal, many-flowered branched cymes. Calyx tube longer than lobes; corolla salverform to funnelform, rarely slightly zygomorphic, creamy-white; corolla tube much longer, or as long as lobes; stamens inserted in or above middle of corolla tube; filaments not geniculate at base; ovary bilocular; style elongate and caducous.

4 spp. in SE Asia and New Guinea.

76. ***Limahlania*** K.M. Wong & M. Sugumaran

Limahlania K.M. Wong & M. Sugumaran, Gard. Bull. Singapore 64: 491 (2012).

Medium-sized trees (to 25 m). Leaves coriaceous or fleshy and petiolate. Flowers in terminal, many-flowered branched cymes. Calyx tube longer than lobes; corolla funnelform, rarely slightly zygomorphic, creamy-white; corolla tube much longer than lobes; stamens inserted in middle or above middle of corolla tube; filaments not geniculate at base; ovary bilocular; style elongate and caducous.

One sp., *L. crenulata* (Malingay ex C.B. Clarke) K.M. Wong & Sugumaron, in SE Asia.

77. ***Fagraea*** Thunb. Fig. 74

Fagraea Thunb., Kongl. Vetensk. Acad. Nya Handl. 3: 132 (1782); Leenhouts, Fl. Males. 1(6): 293–387 (1963), reg. rev.; Conn & Brown, Telopea 5: 363–374 (1993), part. rev.; Wong & Sugau, Sandakania 8: 1–93 (1996), reg. rev.; Ping-tao & Leeuwenberg, Fl. China 15: 338 (1996), reg. rev.; Griffin & Parnell, Fl. Thailand 6(3): 197–205 (1997), reg. rev.; Motley, Econ. Bot. 58: 396–409 (2004), ethnobot.; Wong & Sugumaran, Gard. Bull. Singapore 64: 491 (2012), rev., Sandakania 21: 65–130 (2016), reg. rev.

Free-standing trees, erect or scrambling, shrubs, climbers, epiphytes and hemi-epiphytes. Leaves coriaceous or fleshy, petiolate or subsessile. Flowers in terminal, many-flowered branched cymes, or solitary. Calyx tube longer or shorter than lobes; corolla salverform to funnelform, rarely slightly zygomorphic, cream, white, or occasionally pink; corolla tube as long, slightly longer or shorter than lobes; stamens inserted in or above middle of corolla tube, filaments often geniculate at base; ovary bilocular or sometimes distally unilocular; style elongate and caducous. n = 33, 42.

Ca. 50 spp. in S and E Asia, Australasia, and Pacific Islands.

78. ***Anthocleista*** Afzel. ex R. Br.

Anthocleista Afzel. ex R. Br. in Tuckey, Narr. Exped. Zaire, App. 5: 449 (1818); Leeuwenberg, Acta Bot. Neerl. 10: 1–53 (1961), rev.; de Wilde, Blumea 56: 1–3 (2011), reg. rev.

Medium-sized to tall trees (to 35 m) or rarely lianas, perennial; stems sometimes spiny. Leaves coriaceous or membranous, to 1 m long, petiolate, sometimes with extrafloral nectaries. Flowers in terminal, many-flowered, branched cymes. Calyx deeply 4-lobed; corolla salverform to tubular, white, cream, yellow, or rarely blue to violet, 6–16-merous; corolla tube shorter than or about as long as lobes; stamens inserted slightly below corolla sinuses, filaments partially to almost completely united, forming a tube; connective sometimes mucronate; ovary pseudotetralocular; style ± elongate, caducous. n = 30.

14 spp. in Africa, Madagascar, and the Mascarene Islands.

79. ***Potalia*** Aubl. Fig. 74

Potalia Aubl., Hist. Pl. Guiane 1: 394 (1775); Struwe & Albert, Syst. Bot. 29: 670–701 (2004), rev.

Shrubs or small trees, perennial. Leaves membranous to coriaceous, to 1.2 m long, short-petiolate. Flowers in terminal, few to many-flowered branched cymes. Calyx deeply 4-lobed; corolla funnelform to tubular, 8–10-merous, greenish or yellowish white, orange-yellow, or green; corolla tube as long as or longer than lobes (Fig. 74D); stamens inserted near middle of corolla tube, filaments united into a tube; connective mucronate; ovary pseudotetralocular; style much enlarged at base, at least slender distal portion caducous.

Nine spp. in S Central America and tropical South America.

7. **Tribe Helieae** Gilg (1895).

Herbs, shrubs, or rarely smaller trees. Interpetiolar lines or short ocreae common. Colleters in leaf axils and adaxially on calyx often especially well developed. Inflorescences usually cymes with dichasial or monochasial branches (Fig. 75A); bracts either scale-like and minute, or larger and leaf-like. Flowers actinomorphic or slightly zygomorphic, sexual parts often strongly zygomorphic and aggregated in lower part of corolla mouth. Calyx fused at least at base (free in *Chorisepalum*), coriaceous and with thickened and glandular dorsal area (membranous in *Prepusa* and *Senaea*); corolla without appendages (except with corona in *Symbolanthus* and *Chelonanthus hamatus*); corolla tube longer than lobes, rarely as long as lobes or shorter; stamens isomerous with corolla, without appendages; filaments unequal in length (equal in *Aripuana*, *Roraimaea*, some *Prepusa*, and *Senaea*), often strongly bent at apex; anthers often circinately coiled or recurved after anthesis, usually with prolonged connectives resulting in a sterile apical appendage; pollen in monads, tetrads, or polyads, usually triporate; ovaries sessile, often at least basally bilocular or with deeply intruding placentae, pseudobilocular or unilocular in upper part; style when older and dry flattened and longitudinally twisted, often persistent; stigmas bilobed (linear lobes in *Roraimaea* and *Tetrapollinia*); nectariferous disk at base of ovary (possibly absent in *Celiantha* spp., *Neblinantha*, *Irlbachia* spp. and *Tetrapollinia*). Fruit a bivalved capsule (4-valved in *Chorisepalum*), often medially dehiscent (Fig. 75C) or woody (leathery fruitwall in *Symbolanthus*). Neotropical. Fig. 75.

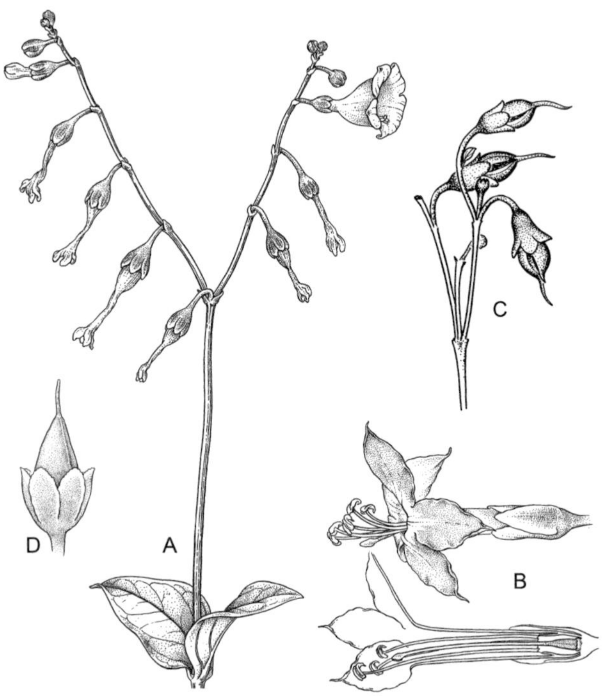

Fig. 75. Helieae. **A** *Chelonanthus alatus*, upper stem with monochasial inflorescence branches and zygomorphic flowers. **B** *Symbolanthus australis*, flower (side view and open), with zygomorphic stamens and style, and corona at the base of the stamens. **C** *Macrocarpaea neblinae*, infructescence with fruits opening medially and persistent styles. **D** *Roraimaea coccineus*, capsular fruit. (A. Drawing by Bobbi Angell, permission for use by Bobbi Angell; B. Drawings by Bobbi Angell, reproduced with permission from Bobbi Angell and Harvard University Herbaria, previously published in Struwe 2003; C. Drawings by Bobbi Angell, reproduced with permission from Bobbi Angell and Missouri Botanical Garden Press, previously published in Struwe et al. 1999; D. Drawings by Bobbi Angell, reproduced with permission from Harvard University Herbaria, previously published in Struwe et al. 2008)

80. ***Prepusa*** Mart.

Prepusa Mart., Nov. Gen. Sp. Pl. 2: 120 (1827) ["1826"]. Calió et al., Kew Bull. 63: 169–191 (2008), rev.; Struwe et al., J. Biogeogr. 38: 1841–1854 (2011), phyl.; De Fraga et al., Phytotaxa 163: 287–294 (2014), part. rev.

Shrubs, small trees, or herbs, perennial. Leaves somewhat succulent, sessile. Flowers in racemoid cymes, or solitary, 6-merous, actinomorphic. Calyx winged or not, large, membranous, concealing much of corolla; calyx tube diameter much greater than that of corolla tube, much longer than lobes; corolla campanulate, white, cream or yellow; stamens equal or unequal, inserted above middle of corolla tube; pollen in tetrads; ovary unilocular; style slender, persistent.

Six spp. in South America (SE and NE Brazil).

81. ***Senaea*** Taub.

Senaea Taub., Bot. Jahrb. Syst. 17: 515 (1893).

Perennial subshrubs. Leaves somewhat succulent, sessile or short-petiolate. Flowers in a dichasial cyme, 6-merous, appearing zygomorphic because two corolla lobes are somewhat reflexed. Calyx not winged; calyx lobes minute; corolla nearly salverform, blue; corolla tube abruptly expanded in upper part; stamens equal, inserted above middle of corolla tube; pollen in tetrads; ovary unilocular; style slender, curved to one side, caducous.

Two spp. in South America (SE Brazil).

82. ***Chorisepalum*** Gleason & Wodehouse

Chorisepalum Gleason & Wodehouse, Bull. Torrey Bot. Club 58: 451 (1931); Ewan, J. Wash. Acad. Sci. 37: 392–396 (1947), rev.; Maguire, Mem. New York Bot. Gard. 32: 330–388 (1981), rev.; Struwe et al., Fl. Venez. Guayana 5: 488–491 (1999), rev.

Perennial small trees or erect to scandent shrubs. Leaves coriaceous, petiolate. Flowers in small, terminal (rarely axillary), sometimes incomplete or umbelloid dichasia, or solitary, actinomorphic. Calyx of four free sepals, outer two larger, rarely winged; corolla tubular to salverform, 6-merous, green, sometimes with purplish suffusion; corolla lobes sometimes caudate-tipped; stamens inserted near base of corolla tube; pollen in monads; ovary bilocular; style elongate, caducous or basally persistent. Fruit splitting into (3)4 parts.

Five spp. in northern South America.

83. ***Macrocarpaea*** (Griseb.) Gilg Fig. 75

Macrocarpaea (Griseb.) Gilg in Engl. & Prantl, Nat. Pflanzenfam. 4(2): 94 (1895); Ewan, Contr. U.S. Natl. Herb. 29: 209–249 (1948), rev.; Weaver, J. Arnold Arbor. 53: 553–557 (1972), 55: 300–302 (1974); Struwe et al., Fl. Venez. Guayana 5: 504–507 (1999), reg. rev.; Grant & Struwe, Harvard Pap. Bot. 5: 521–530 (2001), reg. rev., Harvard Pap. Bot. 8: 61–81 (2003), reg. rev.; Thiv, Fl. Republ. Cuba 6(1) (2002), reg. rev.; Grant & Weaver, Harvard Pap. Bot. 8: 83–109 (2003), reg. rev.; Grant, Harvard Pap. Bot. 7: 423–436 (2003), reg. rev., Harvard Pap. Bot. 9: 11–49 (2004), reg. rev., Harvard Pap. Bot. 9: 305–342 (2005), reg. rev., Harvard Pap. Bot. 11: 129–139 (2007), reg. rev., Harvard Pap. Bot. 13(2): 253–259 (2008), Harvard Pap. Bot. 16: 389–397 (2011), reg. rev., Harvard Pap. Bot. 19: 227–239 (2014), reg. rev., in J. J. Rybczynski et al., The Gentianaceae - vol. 1: Characterization and Ecology, pp. 37–147 (2014), reg. rev.; Struwe et al., Ann. Missouri Bot. Garden 96: 450–469 (2009), phyl.; Grant & Trunz, Harvard Pap. Bot. 16: 399–420 (2011), reg. rev.; Grant & Vieu, Harvard Pap. Bot. 19: 241–246 (2014), reg. rev.
Lisianthius sect. *Macrocarpaea* Griseb. (1838) ["1839"], "*Lisianthus*".
Rusbyanthus Gilg (1895).

Perennial shrubs or small trees, rarely epiphytic, rarely annual and herbaceous. Leaves membranous to coriaceous, sessile or petiolate. Flowers in dichasial cymes, with leaf-like bracts, 5-merous. Calyx tube shorter or longer than lobes; calyx lobes ± equal or outer lobes smaller than inner; corolla campanulate, urceolate, or occasionally nearly salverform, green, yellow, white, or rarely reddish; corolla tube straight or somewhat curved proximally, longer than the lobes; corolla lobes equal or three lower slightly larger than upper two; stamens inserted near or below middle of corolla tube, declinate, inflexed near apex; anthers sometimes tardily coiling; pollen in monads; ovary basally bilocular; style long, slender, basally persistent. $n = 21$.

120 spp. in Central America and tropical South America (including the Greater Antilles).

84. ***Tachia*** Aubl.

Tachia Aubl., Hist. Pl. Guiane 75 (1775); Maguire & Weaver, J. Arnold Arbor. 56: 103–125 (1975), rev.; Cobb & Maas, Brittonia 50: 11–18 (1998), reg. rev.; Struwe et al., Blumea 50: 457–462 (2005), reg. rev.; Struwe & Kinkade, Syst. Bot. 38: 1142–1159 (2013).

Perennial shrubs or small trees. Leaves membranous to coriaceous, petiolate or subsessile. Flowers solitary or paired in leaf axils, sessile or subsessile on a "cushion" (reduced shoot), 5-merous, actinomorphic or slightly zygomorphic. Calyx tube longer or shorter than lobes, sometimes winged or keeled; corolla narrowly funnelform to salverform, green to yellow or white; stamens usually inserted below middle of corolla tube, straight or distally recurved; pollen in monads; ovary unilocular; style long, slender, basally persistent.

13 spp. in Central America and tropical South America.

85. ***Zonanthus*** Griseb.

Zonanthus Griseb., J. Proc. Linn. Soc., Bot. 6: 145 (1862); Thiv, Fl. Republ. Cuba 6(1) (2002), reg. rev.

Perennial shrubs. Leaves coriaceous, petiolate. Flowers solitary in leaf axils, 5-merous, actinomorphic; a pair of bracts proximally fused to calyx tube. Calyx lobed to level of bracts; corolla funnelform, pale green; stamens equal, inserted near middle of tube: pollen in monads; ovary pseudotetralocular; style long, slender, caducous.

One poorly known sp., *Z. cubensis* Griseb., in the West Indies (Cuba).

86. ***Irlbachia*** Mart.

Irlbachia Mart., Nov. Gen. Sp. Pl. 2: 101 (1827) ["1826"]; Struwe & Albert, Harvard Pap. Bot. 3: 63–71 (1998), nom.; Struwe et al., Fl. Venez. Guayana 5: 498–503 (1999), reg. rev.
Pagaea Griseb. (1845).
Lisianthius sect. *Brachycodon* Benth. (1854), "*Lisianthus*".
Brachycodon (Benth.) Progel (1865), non Fed. (1957).

Annual herbs. Leaves cauline, sometimes also in basal rosette, membranous or subcoriaceous, sessile or short-petiolate. Flowers in dichasial or usually distally monochasial cymes, with minute scale-like or occasionally leaf-like bracts, 5-merous, actinomorphic (except sexual parts). Calyx tube shorter than lobes; corolla funnelform, campanulate, or rarely salverform, white, pink, or rose-violet to violet-blue; corolla tube straight, about as long as corolla lobes; stamens inserted in lower half of corolla tube, straight or slightly declinate; pollen in spinose polyads; ovary pseudobilocular; style long, slender, nearly straight, basally persistent; nectariferous disk possibly absent in *I. pratensis*.

Ca. nine spp. in N South America. Analyses of molecular and morphological characters support *I. pratensis* as more closely related to *Chelonanthus purpurascens* and *Calolisianthus amplissimus* than to the rest of *Irlbachia* (Struwe et al. 2009a; Calió et al. 2017).

87. ***Celiantha*** Maguire

Celiantha Maguire, Mem. New York Bot. Gard. 32: 382 (1981); Struwe et al., Fl. Venez. Guayana 5: 481–483 (1999), rev.

Perennial (?) herbs or subshrubs. Leaves often coriaceous, sessile or petiolate. Flowers in cymes or thyrses, rarely solitary, 4- or 5-merous; bracts leaf-like or scale-like. Calyx tube longer to slightly shorter than lobes; corolla narrowly funnelform, pink to purple or yellow; corolla tube straight; stamens inserted in lower half of corolla tube; pollen in polyads; ovary unilocular; style long, slender, persistent.

Three spp. in N South America.

88. ***Neblinantha*** Maguire

Neblinantha Maguire, Phytologia 57: 311 (1985); Struwe et al., Fl. Venez. Guayana 5: 507–509 (1999), reg. rev.

Perennial shrubs or subshrubs. Leaves coriaceous, sessile or subsessile. Flowers in terminal cymes on short or vestigial branches, in the latter case appearing axillary, or solitary, 5-merous. Calyx tube much shorter than lobes; corolla narrowly funnelform or salverform, pink to reddish; corolla tube straight; stamens inserted on lowest quarter of corolla tube, inflexed at apex; anthers straight at anthesis, mucronate; pollen in tetrads; ovary basally bilocular; style long, slender, largely caducous.

Two spp. in N South America (Brazil, Venezuela).

89. ***Aripuana*** Struwe, Maas & V.A. Albert

Aripuana Struwe, Maas & V.A. Albert, Harvard Pap. Bot. 2: 235 (1998).

Perennial shrubs or small trees. Leaves membranous, petiolate. Flowers in dichasial cymes, 5-merous, actinomorphic. Calyx tube longer than lobes; calyx lobes subequal, keeled; corolla salverform, white; corolla tube slender, scarcely expanded distally; corolla lobes proximally valvate, distally contorted in bud, each with a central adaxial ridge; stamens inserted slightly below corolla sinuses; filaments straight; anthers recurving after anthesis; connective mucronate; pollen in tetrads; ovary bilocular; style long, slender, straight, largely persistent.

One sp., *A. cullmaniorum* Struwe, Maas & V. A. Albert, in South America (Brazil).

90. ***Roraimaea*** Struwe, S. Nilsson & V.A. Albert Fig. 75

Roraimaea Struwe, S. Nilsson & V.A. Albert, Harvard Pap. Bot. 13: 36 (2008); Struwe et al., Fl. Venez. Guayana 5: 474–542 (1999), reg. rev. (as *Rogersonanthus*).

Perennial herbs. Leaves cauline, sometimes also basal, sessile or short-petiolate. Flowers in terminal cymes or solitary, 5-merous, actinomorphic; bracts leaf- to scale-like. Calyx campanulate; calyx tube about as long as lobes or shorter; corolla tubular or salverform, orange to red; corolla tube much longer than lobes; stamens equal, inserted on lower third of corolla tube; filaments straight; anthers straight, sometimes with a small sterile tip; pollen in tetrads; ovary pseudobilocular; style slender, caducous in fruit; stigma with two long narrow lobes; nectariferous disk at the base of the ovary (sometimes absent?).

Two spp. in N South America (Brazil, Venezuela). One species was previously included in *Rogersonanthus.*

91. ***Tetrapollinia*** Maguire & B.M. Boom

Tetrapollinia Maguire & B.M. Boom, Mem. New York Bot. Gard. 51: 31 (1989); Struwe et al., Fl. Venez. Guayana 5: 533–534 (1999), reg. rev.

Annual herbs, delicate, perhaps mycoheterotrophic. Leaves membranous, sessile, the distal leaves small and widely spaced. Flowers in small dichasia or larger cymes (distally monochasial), with minute, scale-like bracts, 5-merous. Calyx tube shorter than lobes; corolla narrowly funnelform, blue, violet-blue, rose-violet, or white; stamens inserted on lower half of corolla tube, declinate; pollen in tetrads; ovary pseudobilocular; style long, slender, upturned near apex, at least distally caducous.

One sp., *T. caerulescens* (Aubl.) Maguire & B. M. Boom, in tropical South America.

92. ***Lagenanthus*** Gilg

Lagenanthus Gilg in Engl. & Prantl, Nat. Pflanzenfam. 4 (2): 99 (1895); Ewan, Mutisia 4: 1–5 (1952), rev.

Perennial semi-scandent shrubs, sometimes epiphytic. Leaves subcoriaceous, short-petiolate. Flowers pendulous in few-flowered, terminal corymbs, or solitary, 5-merous. Calyx broadly campanulate; calyx tube about as long as or longer than lobes; corolla 10–15 cm long, tubular, banded or uniform coloring with red, yellow, and green; corolla tube abruptly narrowed at throat, much longer than the short, spreading lobes; stamens inserted near base of corolla tube, straight; pollen in tetrads; style long, slender, basally persistent.

One rare sp., *L. princeps* (Lindl.) Gilg, in NW South America (Colombia and Venezuela).

93. ***Lehmanniella*** Gilg

Lehmanniella Gilg in Engl. & Prantl, Nat. Pflanzenfam. 4 (2): 101 (1895); Ewan, Caldasia 5: 85–98 (1948), rev.; Maas, Ann. Missouri Bot. Gard. 68: 685–686 (1981), part. rev.

Perennial scandent or erect shrubs. Leaves membranous to coriaceous, petiolate. Flowers pendulous in terminal, rarely axillary, dichasial or umbelloid cymes, or solitary, 5-merous. Calyx campanulate; calyx tube longer or about as long as lobes; corolla tubular with small slightly spreading lobes, constricted at throat, red; corolla tube much longer than lobes; stamens inserted in lower quarter of corolla tube, straight; connective prolonged; pollen in tetrads; style long, slender, basally persistent; nectary disk at the base of the ovary.

Two rare spp. in NW South America (Colombia, Peru, Venezuela).

94. ***Purdieanthus*** Gilg

Purdieanthus Gilg in Engl. & Prantl, Nat. Pflanzenfam 4 (2): 99 (1895); Ewan, Caldasia 5: 85–98 (1948), rev.

Perennial shrubs. Leaves membranous, short-petiolate. Flowers pendulous in umbelloid or corymbose cymes, 5-merous. Calyx keeled; calyx tube longer than lobes; corolla tubular, slightly constricted at mouth, red; corolla tube inflated, much longer than lobes; stamens inserted on lower half of corolla tube, straight; connective prolonged; pollen in tetrads; style long, slender.

One sp., *P. pulcher* (Hook.) Gilg, in NW South America (Colombia).

95. ***Adenolisianthus*** (Progel) Gilg

Adenolisianthus (Progel) Gilg in Engl. & Prantl, Nat. Pflanzenfam. 4(2): 98 (1895); Struwe et al., Fl. Venez. Guayana 5: 479–481 (1999), reg. rev.
Lisianthius sect. *Adenolisianthus* Progel in Mart., Fl. Bras. 6(1): 239 (1865), "*Lisianthus*".

Perennial shrubs or subshrubs. Leaves subcoriaceous, sessile. Flowers in cymes, usually basally dichasial, distally monochasial (racemoid), 5-merous; bracts scale-like. Calyx tube shorter or as long as lobes; corolla funnelform, pale green to yellow; corolla tube curved, three lower lobes slightly larger than upper two; stamens inserted on lowest quarter of corolla tube, declinate, deflexed above base, inflexed at apex; anthers recurved, mucronate; pollen in tetrads; ovary pseudobilocular; style long, slender, upturned near apex, basally persistent.

One sp., *A. arboreus* (Spruce ex Progel) Gilg, in N South America. Analysis of molecular and morphological characters supports the inclusion of *Adenolisianthus* and green-flowered species of *Chelonanthus* in *Helia* (Calió et al. 2017).

96. ***Calolisianthus*** Gilg

Calolisianthus Gilg in Engl. & Prantl, Nat. Pflanzenfam. 4 (2): 99 (1895); Harley, Fl. Pico das Almas, Chapada Diamantina–Bahia, Brazil: 321–327 (1995).

Annual or short-lived perennial herbs. Leaves membranous to coriaceous, sessile. Flowers in dichasial cymes or solitary, 5-merous; bracts scale-like. Flowers 5-merous. Calyx actinomorphic; calyx tube shorter than lobes; corolla funnelform, campanulate or salverform, red, scarlet, pink, violet-blue or lilac; corolla tube straight or slightly curved, three lobes slightly larger than two; stamens inserted on lowest quarter of corolla tube, declinate, inflexed at apex; anthers recurving or coiling circinately, mucronate; pollen in tetrads or polyads; ovary sessile, basally bilocular, with basal nectary disk; style slender, basally persistent. n = 40.

Ca. six spp. in tropical to subtropical E and C South America. *Calolisianthus amplissimus* Gilg has recently been shown to be closely related to *Chelonanthus purpurascens* (Aubl.) Struwe, S. Nilsson & V.A. Albert, rendering both *Calolisianthus* and *Chelonanthus* polyphyletic (Calió et al. 2017), and this will result in reclassifications in Helieae in the future.

97. ***Chelonanthus*** (Griseb.) Gilg Fig. 75

Chelonanthus (Griseb.) Gilg in Engl. & Prantl, Nat. Pflanzenfam. 4(2): 98 (1895); Struwe & Albert, Harvard Pap. Bot. 3: 63–71 (1998), nom.; Struwe et al., Fl. Venez. Guayana 5: 483–487 (1999), reg. rev.
Lisianthius sensu Aubl. (1775), "*Lisianthius*" non *Lisianthius* P. Browne (1756).
Lisianthius sect. *Chelonanthus* Griseb. (1838) ["1839"], "*Lisianthus*".

Annual or perennial herbs, sometimes suffrutescent. Leaves membranous to rarely coriaceous, sessile or petiolate. Flowers in cymes, at least basally dichasial, distally generally monochasial (racemoid), rarely ± dichasial throughout, 5-merous, bracts scale-like. Calyx tube shorter than, as long as, or longer than the lobes; corolla funnelform to campanulate, green, white, to yellowish green or blue to purple; corolla tube slightly to strongly curved, longer than corolla lobes, often with a "hump" on one side of the upper part; stamens inserted in lowest half of corolla tube (above middle in *C. grandiflorus*), declinate, deflexed above base, inflexed at apex, corona present in *C. hamatus*; anthers sometimes recurved, mucronate; pollen in tetrads (polyads in *C. purpurascens*); ovary basally bilocular, with nectary disk around base; style long, slender, upturned near apex, basally persistent. n = 20.

Ten spp. in Mexico, Central and tropical South America (including the West Indies). This genus is paraphyletic in its current circumscription as *C. purpurascens* (Aubl.) Struwe, S. Nilsson & V. A. Albert is more closely related to *Calolisianthus amplissimus* Gilg and possibly *Irlbachia pratensis* (Kunth) L. Cobb & Maas, while the green-flowered species of *Chelonanthus* are more closely related to *Adenolisianthus* and *Helia* (Calió et al. 2017). Several species of *Chelonanthus* have previously been included in a broadly defined *Irlbachia* as subspecies of *I. alata* (Aubl.) Maas (Maas 1985; Lepis 2009).

98. ***Helia*** Mart.

Helia Mart., Nov. Gen. Sp. Pl. 2: 122 (1827) ["1826"]; Struwe & Albert, Harvard Pap. Bot. 3: 63–71 (1998), nom.

Annual herbs. Leaves membranous, sessile, mostly closely spaced near base of stem, distal cauline leaves few, widely spaced, and reduced. Flowers in few-flowered (up to 12), compact, proximally dichasial, distally monochasial cymes, 5-merous; bracts scale-like. Calyx actinomorphic; calyx tube shorter than lobes; corolla nearly actinomorphic, salverform, without appendages, pale greenish yellow; corolla tube straight, longer than lobes; stamens subequal, inserted near middle of corolla tube, straight or nearly so, without appendages; pollen in tetrads; ovary sessile, pseudobilocular, with basal nectary disk; style long, slender, nearly straight, basally persistent. Fruit a capsule.

Two spp. in tropical SE South America.

99. ***Rogersonanthus*** Maguire & B.M. Boom

Rogersonanthus Maguire & B.M. Boom, Mem. New York Bot. Gard. 51: 3 (1989); Struwe et al., Fl. Venez. Guayana 5: 515–517 (1999), reg. rev.

Perennial shrubs or small trees. Leaves coriaceous, sessile or short-petiolate. Flowers in 1–3-flowered cymes, 5-merous; bracts scale-like. Calyx tube shorter than lobes; corolla campanulate, green to yellow; stamens inserted on lowest half of corolla tube, declinate, deflexed above base, inflexed at apex; anthers sometimes recurved, mucronate; pollen in tetrads; ovary basally bilocular, with basal nectary disk; style long, slender, basally persistent.

Two spp. in NE South America and the West Indies (Trinidad).

100. ***Sipapoantha*** Maguire & B.M. Boom

Sipapoantha Maguire & B.M. Boom, Mem. New York Bot. Gard. 51: 23 (1989); Struwe et al., Fl. Venez. Guayana 5: 522–523 (1999), rev.; Lepis et al., Blumea 56: 28–32 (2011), part. rev.

Perennial (?) herbs or shrubs. Leaves thickly coriaceous, sessile. Flowers in monochasial or basally dichasial cymes or solitary, 5-merous; bracts scale-like. Calyx tube shorter than lobes; corolla funnel-salverform, violet-blue to purple; corolla tube straight; stamens inserted on lowest quarter of corolla tube, declinate, inflexed at apex; anthers remaining straight, connective prolonged; pollen in tetrads; ovary basally bilocular, with basal nectary disk; style long, slender, largely caducous.

Two spp. in South America (Brazil, Venezuela).

101. ***Symbolanthus*** G. Don Fig. 75

Symbolanthus G. Don, Gen. Hist. 4: 230 (1837); Struwe et al., Fl. Venez. Guayana 5: 524–527, 541–542 (1999), reg. rev.; Struwe, Novon 13: 133–140 (2003), reg. rev., Harvard Pap. Bot. 8: 19–24 (2004), reg. rev.; Gould & Struwe, Ann. Missouri Bot. Garden 91: 438–446 (2004), phyl.; Struwe & Gould, Novon 14: 354–359 (2004), part. rev.; Molina & Struwe, Syst. Biodivers. 6: 477–501 (2008), reg. rev.
Leiothamnus Griseb. 1838 ["1839"].
Wurdackanthus Maguire (1985).

Shrubs, small trees, or perennial herbs. Leaves membranous to coriaceous, petiolate or sessile. Flowers in few-flowered (sometimes umbelloid) cymes or solitary, 5-merous, often large (over 8 cm long); bracts scale-like. Calyx deeply lobed nearly to base, accrescent in fruit, three outer lobes much overlapping 2 inner; corolla salverform to funnelform, pink, red, green, or yellow, often with guidelines in darker or lighter color; corolla tube with corona or scales at the insertion point of stamens (rarely absent, then with "pockets" behind each stamen in the corolla tube); stamens inserted in lowest quarter of corolla tube, declinate, circinately coiled or inflexed near apex; anthers recurved after anthesis, connective prolonged; pollen in tetrads; ovary pseudobilocular, basally bilocular, with basal nectary disk; style long, slender, basally persistent. $n = 40$.

Ca. 37 spp. in S Central America and tropical South America (including S West Indies).

102. ***Yanomamua*** J.R. Grant, Maas & Struwe

Yanomamua J.R. Grant, Maas & Struwe, Harvard. P. Bot. 11: 31–32 (2006).

Herbs with winged stems. Leaves membranous, pandurate, sessile, with acrodromous venation. Bracts subtending flowers leaf-like; bracteoles scale-like. Flowers solitary in upper leaf axils, 5-merous, often large (over 8 cm long). Calyx lobed to middle of calyx, persistent in fruit; corolla, stamens, and gynoecium unknown. Fruit a medially dehiscent capsule with persistent style.

One sp., *Y. araca* J. R. Grant, Maas & Struwe, in northern Brazil (Guayana Shield).

Selected Bibliography

Albert, V.A., Struwe, L. 2002. Gentianaceae in context. In: Struwe, L., Albert, V.A. (eds.) Gentianaceae – Systematics and Natural History. Cambridge: Cambridge University Press, pp. 1–20.

Angiosperm Phylogeny Group. 2009. An update of the Angiosperm Phylogeny Group classification for the orders and families of flowering plants: APG III. Bot. J. Linn. Soc. 161: 105–121.

Bertin, R.I., Newman, C.M. 1993. Dichogamy in angiosperms. Bot. Rev. (Lancaster) 59: 112–152.

Bidartondo, M.I., Redecker, D., Hijri, I., Wiemken, A., Bruns, T.D., Domínguez, L., Sérsic, A., Leake, J.R., Read, D.J. 2002. Epiparasitic plants specialized on arbuscular mycorrhizal fungi. Nature 419: 389–392.

Bouman, F., Schier, S. 1979. Ovule ontogeny and seed coat development in *Gentiana*, with a discussion on the evolutionary origin of the single integument. Acta Bot. Neerl. 28: 467–478.

Bouman, F., Cobb, L., Devente, N., Goethals, V., Maas, P.J. M., Smets, E. 2002. The seeds of Gentianaceae. In: Struwe, L., Albert, V.A. (eds.) Gentianaceae - Systematics and Natural History. Cambridge: Cambridge University Press, pp. 498–572.

Calió, M.F., Pirani, J.R., Struwe, L. 2008. Morphology-based phylogeny and revision of *Prepusa* and *Senaea* (Gentianaceae: Helieae) - rare endemics from eastern Brazil. Kew Bull. 63: 169–191.

Calió, M.F., Lepis, K.B., Pirani, J.R., Struwe, L. 2017 (2016 online). Phylogeny of Helieae (Gentianaceae): resolving taxonomic chaos in a neotropical clade. Molec. Phylogen. Evol. 106: 192–208.

Cameron, D.D., Bolin, J.F. 2010. Isotopic evidence of partial mycoheterotrophy in the Gentianaceae: *Bartonia virginica* and *Obolaria virginica* as case studies. Amer. J. Bot. 97: 1272–1277.

Carlquist, S. 1984. Wood anatomy of some Gentianaceae and ecological conclusions. Aliso 10: 573–582.

Carlquist, S., Grant, J.R. 2005. Wood anatomy of Gentianaceae, tribe Helieae, in relation to ecology, habit, systematics, and sample diameter. Brittonia 57: 276–291.

Chassot, P., von Hagen, K.B. 2008. Pollen morphology of the Swertiinae (Gentianaceae): phylogenetic implications. Bot. J. Linn. Soc. 157: 323–341.

Chassot, P., Nemomissa, S., Yuan, Y.-M., Kupfer, P. 2001. High paraphyly of *Swertia* L. (Gentianaceae) in the *Gentianella*-lineage as revealed by nuclear and chloroplast DNA sequence variation. Pl. Syst. Evol. 229: 1–21.

Chen, S., Xia, T., Wang, Y., Liu, J., Chen, S. 2005. Molecular systematics and biogeography of *Crawfurdia*, *Metagentiana*, and *Tripterospermum* (Gentianaceae) based on nuclear ribosomal and plastid DNA sequences. Ann. Bot. 96: 413–424.

Crepet, W.L., Daghlian, C.P. 1981. Lower Eocene and Paleocene Gentianaceae: floral and palynological evidence. Science 214: 75–77.

Cronquist, A. 1981. An Integrated System of Classification of Flowering Plants. New York: Columbia University Press.

Dahlgren, R. 1980. A revised system of classification of the angiosperms. Bot. J. Linn. Soc. 80: 91–124.

Dalvi, V.C., Meira, R.M.S.A., Azevedo, A.A. 2013. Extrafloral nectaries in neotropical Gentianaceae: occurrence, distribution patterns, and anatomical characterization. Amer. J. Bot. 100: 1779–1789.

Delgado, M.N., Azevedo, A.A., Valente, G.E., Kasuya, M.C. M. 2009. Morfo-anatomia comparada de espécies da subtribo Coutoubeinae (Chironieae - Gentianaceae). Acta Bot. Bras. 23: 956–967.

Delgado, M.N., Da Silva, L.C., Báo, S.N., Morais, H.C., Azevedo, A.A. 2011. Distribution, structural and ecological aspects of the unusual leaf nectaries of *Calolisianthus* species (Gentianaceae). Flora 206: 676–683.

Favre, A., Yuan, Y.M., Küpfer, P., Alvarez, N. 2010. Phylogeny of subtribe Gentianinae (Gentianaceae): biogeographic inferences despite limitations in temporal calibration points. Taxon 59: 1701–1711.

Favre, A., Matuszak, S., Muellner-Riehl, A.N. 2013. Two new species of the Asian genus *Tripterospermum* (Gentianaceae). Syst. Bot. 38: 224–234.

Favre, A., Matuszak, S., Sun, H., Liu, E., Yuan, Y.M., Muellner-Riehl, A.N. 2014. Two new genera of Gentianinae (Gentianaceae): *Sinogentiana* and *Kuepferia* supported by molecular phylogenetic evidence. Taxon 63: 342–354.

Favre, A., Michalak, I., Chen, C.H., Wang, J.C., Pringle, J. S., Matuszak, S., Sun, H., Yuan, Y.M., Struwe, L., Muellner-Riehl, A.N. 2016. Out-of-Tibet: the spatio-temporal evolution of *Gentiana* (Gentianaceae). J. Biogeogr. 43: 1967–1978.

Frasier, C., Albert, V.A., Struwe, L. 2008. Amazonian lowland, white sand areas as ancestral regions for South American biodiversity: biogeographic and phylogenetic patterns in *Potalia* (Angiospermae: Gentianaceae). Org. Divers. Evol. 8: 44–57

Freitas, L., Sazima, M. 2009. Floral biology and mechanisms of spontaneous self-pollination in five neotropical species of Gentianaceae. Bot. J. Linn. Soc. 160: 357–368.

Fryxell, P.A. 1957. Mode of reproduction of higher plants. Bot. Rev. (Lancaster) 23: 135–233.

Gielly, L., Taberlet, P. 1996. A phylogeny of the European gentians inferred from chloroplast *trn*L (UAA) intron sequences. Bot. J. Linn. Soc. 120: 57–75.

Gielly, L., Yuan, Y.M., Kupfer, P., Taberlet, P. 1996. Phylogenetic use of noncoding regions in the genus *Gentiana* L.: Chloroplast *trn*l (UAA) intron versus nuclear ribosomal internal transcribed spacer sequences. Molec. Phylogen. Evol. 5: 460–466.

Gilg, E. 1895. Gentianaceae. In: Engler, A., Prantl, K. (eds.) Die natürlichen Pflanzenfamilien 4(2). Leipzig: W. Engelmann, pp. 50–108.

Gilg, C. 1939. Beiträge zur Morphologie und Systematik der Gentianoideae-Gentianeae-Erythraeinae. Notizbl. Bot. Gart. Berlin-Dahlem 14: 417–430.

Gopal Krishna, G., Puri, V. 1962. Morphology of the flower of some Gentianaceae, with special reference to placentation. Bot. Gaz. (Crawfordsville) 124: 42–57.

Graham, A. 1984. *Lisianthius* pollen from the Eocene of Panama. Ann. Missouri Bot. Gard. 71: 987–993.

Grant, J.R., Maas, P.J.M., Struwe, L. 2006. *Yanomamua araca* (Gentianaceae: Helieae), a new genus and species from outliers of the Guayana Shield on Serra Aracá in Amazonian Brazil. Harvard Pap. Bot. 11: 33–41.

Greimler, J., Hermanowski, B., Jang, C.-G. 2004. A re-evaluation of morphological characters in European *Gentianella* section *Gentianella* (Gentianaceae). Pl. Syst. Evol. 248: 143–169.

Hakki, M.I. 1984. Die Entwicklung der Trichome bei *Orphium frutescens* E. Meyer (Gentianaceae). Bot. Jahrb. Syst. 104: 361–368.

Hakki, M.I. 1997. Embryological and morphological studies on plants from South Africa 2. On floral morphology and embryology of *Orphium frutescens* (L.) E. Meyer (Gentianaceae). Bot. Jahrb. Syst. 119: 337–383.

He, T.N., Liu, S.W., Liu, J.Q. 2013. A new Qinghai-Tibet plateau endemic genus *Sinoswertia* and its pollination mode. Pl. Divers. Resources 35: 393–400.

Heusden, E.C.H., van. 1986. Floral anatomy of some neotropical Gentianaceae. Proc. Kon. Ned. Akad. Wetensch. C, 89 (1): 45–59.

Ho, T.-N., Pringle, J.S. 1995. Gentianaceae. In: Wu, Z.-Y., Raven, P.H. (eds.) Flora of China vol. 16, St. Louis. Missouri Botanical Garden Press, and Beijing: Science Press, pp. 1–139.

Ho, T.-N., Liu, S.W., Xi, Y.Z., Ning, J.C. 1994. Pollen morphology and phylogenetic analysis of *Gentiana*. Cathaya 6: 93–114.

Hostettmann, K., Wagner, H. 1977. Xanthone glycosides. Phytochemistry 16: 821–829.

Imhof, S. 1997. Root anatomy and mycotrophy of the achlorophyllous *Voyria tenella* Hook. (Gentianaceae). Bot. Acta 110: 298–305.

Imhof, S. 1999. Root morphology, anatomy and mycotrophy of the achlorophyllous *Voyria aphylla* (Jacq.) Pers. (Gentianaceae). Mycorrhiza 9: 33–39.

Jang, C.-G., Muellner, A.N., Greimler, J. 2005. Conflicting patterns of genetic and morphological variation in European *Gentianella* section *Gentianella*. Bot. J. Linn. Soc. 148: 175–187.

Jansen, S., Smets, E. 1998. Vestured pits in some woody Gentianaceae. IAWA J. 19: 35–42.

Jensen, S.R. 1992. Systematic implications of the distribution of iridoids and other chemical compounds in the Loganiaceae and other families of the Asteridae. Ann. Missouri Bot. Gard. 79: 284–302.

Jensen, S.R., Schripsema, J. 2002. Chemotaxonomy and pharmacology of Gentianaceae. In: Struwe, L., Albert, V.A. (eds.) Gentianaceae - Systematics and Natural History. Cambridge: Cambridge University Press, pp. 573–632.

Kadereit, J.W., von Hagen, K.B. 2003. The evolution of flower morphology in Gentianaceae-Swertiinae and the roles of key innovations and niche width for the diversification of *Gentianella* and *Halenia* in South America. Int. J. Pl. Sci. 164: S441–S452.

Kissling, J. 2008. Cytogeography of Gentianaceae–Exaceae in Africa, with a special focus on *Sebaea*: the possible role of dysploidy and polyploidy in the evolution of the tribe. Bot. J. Linn. Soc. 158: 556–566.

Kissling, J. 2012. Taxonomy of *Exochaenium* and *Lagenias*: Two resurrected genera of tribe Exaceae (Gentianaceae). Syst. Bot. 37: 238–253.

Kissling, J., Endress, P.K., Bernasconi, G. 2009a. Ancestral and monophyletic presence of diplostigmaty in *Sebaea* (Gentianaceae) and its potential role as a morphological mixed mating strategy. New Phytol. 184: 303–310.

Kissling, J., Yuan, Y.M., Küpfer, P., Mansion, G., Huber, H. 2009b. The polyphyletic genus *Sebaea* (Gentianaceae): a step forward in understanding the morphological and karyological evolution of the Exaceae. Molec. Phylogen. Evol. 53: 734–748.

Kissling, J., Buerki, S., Mansion, G. 2009c. *Klackenbergia* (Gentianaceae - Exaceae), a new endemic genus from Madagascar. Taxon 58: 907–912.

Kissling, J., Barrett, S.C.H. 2013a. Diplostigmaty in plants: a novel mechanism that provides reproductive assurance. Biol. Lett. 9: 20130495.

Kissling, J., Barrett, S.C.H. 2013b. Variation and evolution of herkogamy in *Exochaenium* (Gentianaceae): implications for the evolution of distyly. Ann. Bot. 112: 95–102.

Klackenberg, J. 2006. *Cotylanthera* transferred to *Exacum* (Gentianaceae). Bot. Jahrb. Syst. 126: 477–481.

Lepis, K.B. 2009. Evolution and systematics of *Chelonanthus* (Gentianaceae). Ph.D. dissertation, Rutgers University, New Brunswick, NJ, USA.

Lindsey, A.A. 1940. Floral anatomy in the Gentianaceae. Amer. J. Bot. 27: 640–652.

Liu, J., Chen, Z., Lu, A. 2001. A preliminary analysis of the phylogeny of the Swertiinae (Gentianaceae) based on ITS data. Israel J. Pl. Sci. 49: 301–308.

Maas, P.J.M., 1985. Nomenclatural notes on neotropical Lisyantheae (Gentianaceae). Proc. Kon. Ned. Akad. Wetensch., Ser. C, 88: 405–412.

Maas, P.J.M., Ruyters, P. 1986. *Voyria* and *Voyriella* (saprophytic Gentianaceae). Fl. Neotropica Monogr. 41.

Machado, I.C.S., Sazima, I., Sazima, M., 1998. Bat pollination of the terrestrial herb *Irlbachia alata* (Gentianaceae) in northeastern Brazil. Pl. Syst. Evol. 209: 231–237.

Mansion, G. 2004. A new classification of the polyphyletic genus *Centaurium* Hill (Chironiinae, Gentianaceae): description of the New World endemic *Zeltnera*, and reinstatement of *Gyrandra* Griseb. and *Schenkia* Griseb. Taxon 53: 719–740.

Mansion, G., Struwe, L. 2004. Generic delimitation and phylogenetic relationships within the subtribe Chironiinae (Chironieae: Gentianaceae), with special reference to *Centaurium*: evidence from nrDNA and cpDNA sequences. Molec. Phylogen. Evol. 32: 951–977.

Mansion, G., Zeltner, L. 2004. Phylogenetic relationships within the New World endemic *Zeltnera* (Gentianaceae-Chironiinae) inferred from molecular and karyological data. Amer. J. Bot. 91: 2069–2086.

Mansion, G., Zeltner, L., Bretagnolle, F. 2005. Phylogenetic patterns and polyploid evolution within the Mediterranean genus *Centaurium* (Gentianaceae-Chironieae). Taxon 54: 931–950.

Matuszak, S., Muellner-Riehl, A.N., Sun, H., Favre, A. 2015. Dispersal routes between biodiversity hotspots in Asia: the case of the mountain genus *Tripterospermum* (Gentianinae, Gentianaceae) and its close relatives. J. Biogeogr. 43: 580–590.

Matuszak, S., Favre, A., Schnitzler, J., Muellner-Riehl, A. N. 2016. Key innovations and climatic niche divergence as drivers of diversification in subtropical Gentianinae (Gentianaceae) in the region of the Qinghai-Tibetan Plateau. Amer. J. Bot. 103: 899–911.

Mennega, A.M.W. 1980. Anatomy of the secondary xylem. In: Leeuwenberg, A.J.M. (ed.) Engler and Prantl's Die natürlichen Pflanzenfamilien, Angiospermae: Ordnung Gentianales, Fam. Loganiaceae, vol. 28b(1). Berlin: Duncker and Humblot, pp. 112–161.

Merckx, V., Stöckel, M., Fleischmann, A., Bruns, T.D., Gebauer, G. 2010. ^{15}N and ^{13}C natural abundance of two mycoheterotrophic and a putative partially mycoheterotrophic species associated with arbuscular mycorrhizal fungi. New Phytol. 188: 590–596.

Merckx, V.S.F.T., Janssens, S.B., Hynson, N.A., Specht, C. D., Bruns, T.D., Smets, E.F. 2012. Mycoheterotrophic interactions are not limited to a narrow phylogenetic range of arbuscular mycorrhizal fungi. Molec. Ecol. 21: 1524–1532.

Merckx, V.S., Kissling, J., Hentrich, H., Janssens, S.B., Mennes, C.B., Specht, C.D., Smets, E.F. 2013. Phylogenetic relationships of the mycoheterotrophic genus *Voyria* and the implications for the biogeographic history of Gentianaceae. Amer. J. Bot. 100: 712–721.

Mészáros, S. 1994. Evolutionary significance of xanthones in Gentianaceae: a reappraisal. Biochem. Syst. Ecol. 22: 85–94.

Mészáros, S., de Laet, J., Smets, E. 1996. Phylogeny of temperate Gentianaceae: a morphological approach. Syst. Bot. 21: 153–168.

Mészáros, S., De Laet, J., Goethals, V., Smets, E., Nilsson, S. 2002. Cladistics of the Gentianaceae: a morphological approach. In: Struwe, L., Albert, V.A. (eds.) Gentianaceae – Systematics and Natural History. Cambridge: Cambridge University Press, pp. 310–376.

Metcalfe, C.R., Chalk, L. 1950. Anatomy of the Dicotyledons. London: Oxford University Press.

Molina, J., Struwe, L. 2009. Utility of secondary structure in phylogenetic reconstructions using nrDNA ITS sequences—an example from Potalieae (Gentianaceae: Asteridae). Syst. Bot. 34: 414–428

Nilsson, S. 1967. Pollen morphological studies in the Gentianaceae-Gentianinae. Grana Palynol., Ser. 2, 7: 46–145.

Nilsson, S. 1970. Pollen morphological contributions to the taxonomy of *Lisianthus* s. lat. (Gentianaceae). Svensk Bot. Tidskr. 64: 1–43.

Nilsson, S. 2002. Gentianaceae – a review of palynology. In: Struwe, L., Albert, V.A. (eds.) Gentianaceae – Systematics and Natural History. Cambridge: Cambridge University Press, pp. 377–497.

Nilsson, S., Skvarla, J.J. 1969. Pollen morphology of saprophytic taxa in the Gentianaceae. Ann. Missouri Bot. Gard. 56: 420–438.

Nilsson, S., Hellbom, M., Smolenski, W. 2002. A reappraisal of the significance of pollen in classifications of the Gentianaceae. Grana 41: 90–106.

Perrot, M.E. 1899 ["1897"]. Anatomie comparée des Gentianacées. Ann. Nat. Sci. Bot., Ser. 8, 7: 105–292, pl. 1–9.

Pickert, M., Schaper, K.J., Frahm, A.W. 1998. Substituted xanthones as antimycobacterial agents. Part 2. Antimycobacterial activity. Arch. Pharm. (Weinheim) 331: 193–197.

Pirie, M.D., Litsios, G., Bellstedt, D.U., Salamin, N., Kissling, J. 2015. Back to Gondwanaland: can ancient vicariance explain (some) Indian Ocean disjunct plant distributions? Biol. Lett. 11: 20150086.

Raven, P.H. 1975. The basis of angiosperm phylogeny: cytology. Ann. Missouri Bot. Gard. 62: 724–764.

Rouse, G.E. Srivastava, S.K. 1970. Detailed morphology, taxonomy, and distribution of *Pistillipollenites macgregorii*. Canad. J. Bot. 48: 287–292.

Sankara Rao, K., Chinnappa, C.C. 1983. Studies in Gentianaceae. Microsporangium and pollen. Canad. J. Bot. 61: 324–336.

Schönswetter, P., Tribsch, A., Niklfeld, H. 2004. Amplified fragment length polymorphism (AFLP) suggests old and recent immigration into the Alps by the arctic-alpine annual *Comastoma tenellum* (Gentianaceae). J. Biogeogr. 31: 1673–1681.

Shahina, P.M., Nampy, S. 2014. A taxonomic revision of the genus *Canscora* in South India, and the erection of the new genus *Canscorinella* (Canscorinae, Gentianaceae) with two new combinations. Phytotaxa 164: 201–225.

Shamrov, I.I. 1996. Ovule development and significance of its features for Gentianaceae systematics. Opera Bot. Belg. 7: 113–118.

Spira, T.P., Pollak, O.D. 1986. Comparative reproductive biology of alpine biennial and perennial gentians (*Gentiana*: Gentianaceae) in California. Amer. J. Bot. 73: 39–47.

Stockey, R.A., Manchester, S.R. 1986. A fossil flower with in situ *Pistillipollenites* from the Eocene of British Columbia. Canad. J. Bot. 66: 313–318.

Struwe, L. 2014. Classification and evolution of the Gentianaceae. In: Rybczyński J.J., Davey, M.R., Mikula, A. (eds.) The Gentianaceae: Characterization and Ecology, vol. 1. Heidelberg & New York: Springer Verlag, pp. 13–35.

Struwe, L., Albert, V.A. (eds.) 2002. Gentianaceae – Systematics and Natural History. Cambridge: Cambridge University Press.

Struwe, L., Kinkade, M. 2013. Revision of *Tachia* (Gentianaceae: Helieae). Syst. Bot. 38: 1142–1159.

Struwe, L., Albert, V.A., Bremer, B. 1995 ["1994"]. Cladistics and family level classification of the Gentianales. Cladistics 10: 175–206.

Struwe, L., Maas, P.J.M., Pihlar, O., Albert, V.A. 1999. Gentianaceae. In: Berry, P.E., Yatskievych, K., Holst, B.K. (eds.) Flora of the Venezuelan Guayana, vol. 5, St. Louis: Missouri Botanical Garden Press, pp. 474–542.

Struwe, L., Kadereit, J.W., Klackenberg, J., Nilsson, S., Thiv, M., von Hagen, K.B., Albert, V.A. 2002. Systematics, character evolution, and biogeography of Gentianaceae, including a new tribal and subtribal classification. In: Struwe, L., Albert, V.A. (eds.) Gentianaceae – Systematics and Natural History. Cambridge: Cambridge University Press, pp. 21–309.

Struwe, L., Nilsson, S., Albert, V.A. 2008. *Roraimaea* (Gentianaceae: Helieae) – a new gentian genus from white sands and tepuis of Brazil and Venezuela. Harvard Pap. Bot. 13: 35–45.

Struwe, L., Albert, V.A., Calió, M.F., Frasier, C., Lepis, K. B., Mathews, K.G., Grant, J.R. 2009a. Evolutionary patterns in neotropical Helieae (Gentianaceae): evidence from morphology, chloroplast and nuclear DNA sequences. Taxon 58: 479–499.

Struwe, L., Haag, S., Heiberg, E., Grant, J.R. 2009b. Andean speciation and vicariance in neotropical *Macrocarpaea* (Gentianaceae–Helieae). Ann. Missouri Bot. Gard. 96: 450–469.

Struwe, L., Smouse, P.E., Heiberg, E., Haag, S., Lathrop, R.G. 2011. Spatial evolutionary and ecological vicariance analysis (SEEVA), a novel approach to biogeography and speciation research, with an example from Brazilian Gentianaceae. J. Biogeogr. 38: 1841–1854.

Sugumaran, M. 2010. A systematic study of *Fagraea* sensu lato and a revision of the Peninsular Malaysian taxa. Ph.D. Thesis, University of Malaya, Kuala Lumpur, Malaysia.

Sugumaran, M., Wong, K.M. 2012. Studies in Malesian Gentianaceae, I. *Fagraea* sensu lato: complex genus or several genera? A molecular phylogenetic study. Gard. Bull. Singapore 64: 301–332.

Sugumaran, M., Wong, K.M. 2014. Studies in Malesian Gentianaceae VI. A revision of *Utania* in the Malay Peninsula with two new species. Pl. Ecol. Evol. 147: 213–223.

Takhtajan, A.L. 1997. Diversity and Classification of Flowering Plants. New York: Columbia University Press.

Thiv, M. 2003. A taxonomic revision of *Canscora*, *Cracosna*, *Duplipetala*, *Hoppea*, *Microrphium*, *Phyllocyclus* and *Schinziella* (Gentianaceae-Canscorinae). Blumea 48: 1–46.

Thiv, M., Kadereit, J.W. 2002. A morphological cladistic analysis of Gentianaceae-Canscorinae and the evolution of anisomorphic androecia in the subtribe. Syst. Bot. 27: 780–788.

Thiv, M., Struwe, L., Albert, V.A., Kadereit, J.W. 2000 ["1999"]. The phylogenetic relationships of *Saccifolium bandeirae* (Gentianaceae) reconsidered. Harvard Pap. Bot. 4: 519–526.

Thorne, R.T. 1992. An updated classification of the flowering plants. Aliso 13: 365–389.

Tomic, M., Tovilovic, G., Butorovic, B., Krstic, D., Jankovic, T., Aljancic, I., Menkovic, N. 2005. Neuropharmacological evaluation of diethylether extract and xanthones of *Gentiana kochiana*. Pharm. Biochem. Behav. 81: 535–542.

von Hagen, K.B., Kadereit, J.W. 2001. The phylogeny of *Gentianella* (Gentianaceae) and its colonization of the southern hemisphere as revealed by nuclear and chloroplast DNA sequence variation. Org. Divers. Evol. 1: 61–79.

von Hagen, K.B., Kadereit, J.W. 2002. Phylogeny and flower evolution of the Swertiinae (Gentianaceae-Gentianeae): Homoplasy and the principle of variable proportions. Syst. Bot. 27: 548–572.

von Hagen, K.B., Kadereit, J.W. 2003. The diversification of *Halenia* (Gentianaceae): Ecological opportunity versus key innovation. Evolution 57: 2507–2518.

Wong, K.M. 2012. Studies in Malesian Gentianaceae II: a revision of *Picrophloeus*. Gard. Bull. Singapore 64: 511–522.

Wong, K.M., Sugau, J.B. 1996. A revision of *Fagraea* (Loganiaceae) in Borneo, with notes on related Malesian species and 21 new species. Sandakania 8: 1–93.

Wong, K.M., Sugumaran, M. 2012a. Studies in Malesian Gentianaceae II: a taxonomic framework for the *Fagraea* complex, including the new genus *Limahlania*. Gard. Bull. Singapore 64: 481–495.

Wong, K.M., Sugumaran, M. 2012b. Studies in Malesian Gentianaceae III. *Cyrtophyllum* reapplied to the *Fagraea fragrans* alliance. Gard. Bull. Singapore 64: 497–510.

Wong, K.M., Sugumaran, M. 2016. Studies in Malesian Gentianaceae VII: A revision of *Fagraea* in the Malay Peninsula with five new species. Sandakania 21: 65–130.

Wong, K.M., Sugumaran, M., Sugau, J. 2013. Studies in Malesian Gentianaceae V: The *Fagraea* complex in Borneo: New generic assignments and recombinations. Gard. Bull. Singapore 65: 235–239.

Wood, C.E. Jr., Weaver, R.E. Jr. 1982. The genera of Gentianaceae in the southeastern United States. J. Arnold Arbor. 63: 441–487.

Yang, L.L., Li, H.L., Wei, L., Yang, T., Kuang, D.Y., Li, M. H., Liao, Y.Y., Chen, Z.D., Wu, H., Zhang, S.Z. 2016. A supermatrix approach provides a comprehensive genus-level phylogeny for Gentianales. J. Syst. Evol. 54: 400–415.

Yuan, Y.M., Küpfer, P. 1995. Molecular phylogenetics of the subtribe Gentianinae (Gentianaceae) inferred from the sequences of internal transcribed spacers (ITS) of nuclear ribosomal DNA. Pl. Syst. Evol. 196: 207–226.

Yuan, Y.-M., Kupfer, P. 1997. The monophyly and rapid evolution of *Gentiana* sect. *Chondrophyllae* Bunge s.l. (Gentianaceae): Evidence from the nucleotide sequences of the internal transcribed spacers of nuclear ribosomal DNA. Bot. J. Linn. Soc. 123: 25–43.

Yuan, Y.M., Küpfer, P., Doyle, J.A. 1996. Infrageneric phylogeny of the genus *Gentiana* (Gentianaceae) inferred from nucleotide sequences of the internal transcribed spacers (ITS) of nuclear ribosomal DNA. Amer. J. Bot. 83: 641–652.

Yuan, Y.M., Kupfer, P., Zeltner, L. 1998. Chromosomal evolution of *Gentiana* and *Jaeschkea* (Gentianaceae), with further documentation of chromosome data for 35 species from western China. Pl. Syst. Evol. 210: 231–247.

Yuan, Y.-M., Wohlhauser, S., Moller, M., Chassot, P., Mansion, G., Grant, J., Kupfer, P., Klackenberg, J. 2003. Monophyly and relationships of the tribe Exaceae (Gentianaceae) inferred from nuclear ribosomal and chloroplast DNA sequences. Molec. Phylog. Evol. 28: 500–517.

Yuan, Y.-M., Wohlhauser, S., Moeller, M., Klackenberg, J., Callmander, M.W., Kuepfer, P. 2005. Phylogeny and biogeography of *Exacum* (Gentianaceae): A disjunctive distribution in the Indian Ocean Basin resulted from long distance dispersal and extensive radiation. Syst. Biol. 54: 21–34.

Zhang, X.L., Wang, Y.J., Ge, X.J., Yuan, Y.M.,Yang, H.L., Liu, J.Q. 2009. Molecular phylogeny and biogeography of *Gentiana* sect. *Cruciata* (Gentianaceae) based on four chloroplast DNA datasets. Taxon 58: 862–870.

Griseliniaceae

Griseliniaceae Takh., Sist. Magnol. 209 (1987).

M.O. Dillon

Evergreen, dioecious trees, erect shrubs or trailing or vining, New Zealand species are epiphytes or hemi-epiphytes, aerial roots enlarging upon contact with the ground; stems erect to scandent or climbing, glabrous to hispidulose, rarely winged. Leaves simple, evergreen, entire to dentate or dentate-spinose, alternate and distichous or subopposite; subsessile or petiolate; petioles rugulose, glabrous or hispidulose, exstipulate, the bases dilated and subsheathing; blades coriaceous, basally oblique to equal, apically rounded to acuminate, abaxial surface penninerved or prominently 3–5-nerved from near the base, reticulate, minutely pellucid-punctate, adaxial surface smooth, drying rugulose, nitid, margins revolute or plane, strictly entire to dentate or spinose, occasionally subentire with only the apex minutely 2–3-fid. Inflorescences racemose to paniculate, terminal or axillary, subglabrous to hispidulose or glandular, drying nigrescent; bracts and bracteoles caducous. Flowers unisexual, actinomorphic, 5-merous; sepals free, valvate, squamellate, margins ciliolate to glabrous, occasionally caducous; petals 5, free, occasionally absent, subimbricate, oblong to ovate, reflexed to patent at anthesis, caducous, margins ciliolate to glabrous; stamens 5, antisepalous, filaments erect, attached outside of lobulate epigynous disc, ovary remnants obscure; anthers dithecate, tetrasporangiate, dorsifixed, dehiscence longitudinal; staminodia sometimes present in female flowers; gynoecium 3-carpellate, syncarpous, ovary inferior, trilocular in upper part, unilocular in lower part; styles 2 to 3 or rarely 4, simple, terminal, free to connate basally, persistent, subulate, divergent or recurved, placentation apical, one pendulous ovule, anatropous. Fruit 1-locular, fleshy, baccate or coriaceous, drupaceous, oval to oblong, glabrous to hispidulose-puberulent; 1-seeded. Seeds with copious endosperm; embryo minute, straight; $n = 18$.

One genus with seven spp., trans-Pacific disjunct (Argentina, Brazil, Chile, New Zealand).

Vegetative Morphology and Anatomy. All members of the family are woody, ranging from small trees or large shrubs reaching a maximum height of 10–20 m to shrubs seldom over 2 m. The growth habit is typically erect and self-supporting, but epiphytic, hemi-epiphytic, and climbing habits have been recorded (Dawson 1966). Leaves are evergreen, leathery and often shining. Stomata are generally confined to the abaxial surface of the leaves and encyclocytic. Mucilage is generated as epistomatal plugs in New Zealand species, but are absent in the one South American species investigated (Westhoff et al. 2009). An adaxial hypodermis is present, the mesophyll contains sclerenchymatous idioblasts, and minor leaf veins are without phloem transfer cells. Secretory cavities are absent, a cork cambium is present, nodes are tri-lacunar or multi-lacunar. Secondary thickening develops from a conventional cambial ring, the xylem contains vessels with horizontal, scalariform end-walls, fibers with bordered pits, and the wood parenchyma is apotracheal (Li and Chao 1954; Noshiro and Baas 1998). The effect of vesicular-arbuscular mycorrhizas on the growth of *Griselinia littoralis* has been investigated and shown to increase branching in the root system (Baylis 1958).

Inflorescences. The flowers are arranged in open, racemose to paniculate inflorescences which are borne either terminally or axillary on branchlets. The peduncles are typically subglabrous to hispidulose or glandular, nigrescent; bracts and bracteoles are caducous.

J.W. Kadereit, V. Bittrich (eds.), *Flowering Plants. Eudicots*, The Families and Genera of Vascular Plants 15,
https://doi.org/10.1007/978-3-319-93605-5_7

Floral Morphology and Anatomy. Flowers are dimorphic, minute, rarely more than 5 mm in diameter. The staminate flowers have a campanulate hypanthium and are always petaliferous; ovary remnants obscure; epigynous disc nitid and producing nectar. The petals are yellow to greenish-yellow, reflexed at anthesis, and quickly caducous. Anthers are opposite the sepals and alternating with the petals. The pistillate flowers also have a hypanthium and in a few species do not have petals (*G. lucida*). Rudiments of a single whorl of stamens (staminodia) are found in the female flowers of New Zealand species (*G. littoralis* and *G. lucida*). The epigynous disc is not unlike that found in Araliaceae and trans-sepal bundles are present in the ovary (Eyde 1964). Nectar production has not been recorded. Ovary of three carpels (Kubitzki 1963), trilocular in upper part, unilocular in lower part, with 1 pendulous, anatropous ovule.

Embryology. Ovule morphology has only been investigated in one species and the ovules were found to be unitegmic and crassinucellate (Philipson 1977).

Pollen Morphology. Pollen shed as single grains. Pollen grains are 3-aperturate; colporate, ellipsoidal, and with a complete tectum with striate surface (Ferguson and Hideux 1978; Tseng 1980). Images of *Griselinia* pollen are available at Australian Pollen and Spore Atlas (APSA 2007).

Karyology. Only one reported chromosome count of $n = 18$ exists for the family (Federov 1969).

Reproductive Systems and Pollination. All members of the family are dioecious. The gynoecium of staminate flowers is obscure, but rarely rudimentary stamens are found in the pistillate flowers. *Griselinia littoralis* is reported to be wind-pollinated; however, insect pollination probably is the most common mode of pollination when nectar is present (Percival 1961).

Fruit, Seed, Dispersal. The presence of small (<10 mm long, 5 mm in diameter), fleshy, red to purple fruits may suggest some type of animal mediated dispersal (Takhtajan 2000). Testa probably many-layered, outer two (and esp. third) layers with thickened walls.

Phytochemistry. Jensen and Nielsen (1980) investigated five *Griselinia* species and isolated the iridoid glucoside, griselinoside, from *G. littoralis*, *G. lucida*, *G. ruscifolia*, and confirmed that iridoids are elaborated in both New Zealand and South American species. The occurrence of griselinoside in *Aralidium*, *Torricellia* and *Griselinia* suggests affinities; however, the same compound was also isolated from *Verbena* (Damtoft et al. 1983). Reports indicate that the family is not cyanogenic and lacks proanthocyanidins and ellagic acid (Watson and Dallwitz 1992). Petroselinic acid has been isolated and is present as a major component in many species of Apiaceae and Araliaceae (Breuer et al. 1987).

Systematics and Affinities. The family is monogeneric with the seven species of *Griselinia*. The history of familial placement has been complex. Early accounts suggested placement in families such as Euphorbiaceae and Juglandaceae (see Taubert 1892). Endlicher (1850) placed *Griselinia* in the Araliaceae and Hooker (1867) was the first to position it as a member of the Cornaceae. Taubert (1892) accepted the placement in the Cornaceae and established two sections, *Decostea* Taub. and *Griselinia* ("*Eugriselinia*" Taub.). Wangerin (1910) in his treatment of the Cornaceae placed *Griselinia* in tribe Griselinieae Wangerin where it was grouped with *Melanophylla*. Philipson (1967) was the first worker to question the placement of *Griselinia* in the Cornaceae, contending that it shared many features with the Araliaceae and Escalloniaceae, and differed from the Cornaceae by the alternate arrangement of the leaves, nodal anatomy, the presence of a hypodermis below the upper epidermis of leaves, encyclocytic stomata, the presence of spiral thickenings in both vessels and fibers, imbricate petals, free or only shortly connate stylodia, a usually 1-locular ovary, and baccate fruits. Takhtajan (1980) treated *Griselinia* as a monogeneric family and allied it with *Garrya* (Garryaceae). Takhtajan (1987) proposed the valid familial name, since Cunningham's (1791–1839) published name for the father and son team of Johann Reinhold Forster (1729–1798) and Johann Georg Adam Forster (1754–1794) in the Annuals of Natural History was a typographic error (Earp 2013). Earp (2014) has submitted a proposal to conserve the generic name. Takhtajan (1997) established the order Griseliniales for the family, while

Thorne (1968), Dahlgren (1980), and Cronquist (1981) considered it an anomalous member of the Cornaceae. Thorne (1992, 2001) placed it as a monogeneric family with relationships outside of the Cornales. Eyde (1987) also recognized that *Griselinia* was misplaced in the Cornaceae and suggested familial rank.

Xiang et al. (1993) and Xiang and Soltis (1998) examined *rbc*L sequence data and suggested *Griselinia* be removed from the Cornaceae and grouped it with *Aralidium* and *Torricellia* (both Torricelliaceae) and more distantly with *Aralia* (Araliaceae), *Pittosporum* (Pittosporaceae), and *Sanicula* (Apiaceae). Plunkett et al. (1996) also examined *rbc*L sequences and confirmed that the Apiales form a monophyletic group with Pittosporaceae along with *Griselinia*, *Melanophylla* (Malagasy endemic), *Torricellia* (E. Himalayas to W. China) and *Aralidium* (W. Malaysia). Plunkett (2001) and Lundberg (2001) both suggested that Torricelliaceae and Griseliniaceae are successive clades near the base of Apiales (Soltis et al. 2011).

Clokie (2001) investigated infrageneric relationships with an array of molecular markers (*rpo*A, *trn*L-F, *trn*H-K and ITS) and confirmed the monophyly of the genus. Her phylogenetic results suggested that the present-day distribution of *Griselinia* represents a dispersal event from New Zealand to South America, a view shared by Dillon and Muñoz-Schick (1993) based on comparative morphology. Clokie (2001) placed the timing of a putative dispersal at ca. 52 million years ago, after the estimated 95–80 million years ago for the physical separation of South America from Australasia. Nicolas and Plunkett (2014) have dated the crown-group in the Griseliniaceae at around 12.1 m.y.o.

Palaeobotany. The first appearance of pollen in the fossil record is from New Zealand in the Miocene (Mildenhall 1980).

Distribution and Habitats. The family has a trans-Pacific disjunct distribution between austral South America and New Zealand. All the South American species are confined to Chile and adjacent Argentina (Prov. Chubut) with an additional disjunction to south and southeastern Brazil, including the states of Minas Gerais, Paraná, Rio Grande do Sul, Rio de Janeiro, São Paulo, and Santa Catarina. The local environments vary greatly, from mesic, evergreen rain forest environments of austral Chile (sea level–1300 m) to dry coastal quebradas of the Atacama Desert in northern Chile (400 m). Southern Brazilian populations are found in *Araucaria* forests and range north to remnants of the Atlantic coastal forest (400–2700 m). The New Zealand species are distributed from near-ocean sites to montane forests of North and South Islands. One species, *G. littoralis*, is introduced in Australia as an ornamental and naturalized in Tasmania (Baker 2009).

Only one genus:

1. ***Griselinia*** G. Forst. Fig. 76

Griselinia G. Forst., Fl. Ins. Austr.: 75 (1786), nom. cons., non Scop. (1777).
Decostea Ruiz & Pav., Fl. Peruv. Prodr.: 130 (1794).
Pukateria Raoul, Ann. Sci. Nat., Bot. III, 2: 120 (1844).

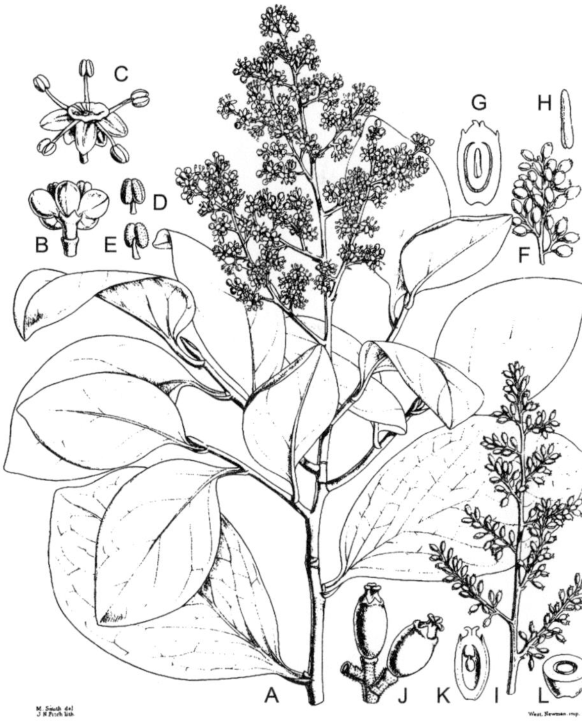

Fig. 76. Griseliniaceae. *Griselinia lucida.* **A** Flowering male branch. **B** Male flower-bud. **C** Fully expanded male flower. **D** Anther ventral view. **E** Anther dorsal view. **F** Fruiting female branchlet. **G** Longitudinal section of fruit. **H** Embryo. **I** Flowering female inflorescence. **J** Female flowers. **K** Section of ovary showing the pendulous ovule. **L** Transverse section of ovary. (Cheeseman 1914; drawn by Matilda Smith)

Description as for family.

Seven species with a trans-Pacific disjunct distribution between New Zealand (two spp.) and Chile (five spp.); two species also in Argentina (Prov. Chubut); one variety of a Chilean species disjunct in southern and southeastern Brazil. Dillon and Muñoz-Schick (1993) monographed *Griselinia*, considering Griseliniaceae a monogeneric family and described a new northern Chilean desert species.

Selected Bibliography

APG IV. 2016. An update of the Angiosperm Phylogeny Group classification for the orders and families of flowering plants: APG IV. Bot. J. Linn. Soc. 181: 1–20.

APSA. 2007. The Australasian Pollen and Spore Atlas V1.0. Canberra: Australian National University. http://apsa.anu.edu.au/

Baker, M.L. 2009. 136 Griseliniaceae, ver. 2017: 1. In: Duretto, M.F. (ed.) Flora of Tasmania Online. 2 pp. http://demo1.tmag.tas.gov.au/treatments/families/Griseliniaceae/Griseliniaceae.html

Baylis, G.T.S. 1958. Effect of vesicular-arbuscular mycorrhizas on growth of *Griselinia littoralis* (Cornaceae). New Phytol. 58: 274–279.

Breuer, B., Stuhlfauth, T., Fock, H., Huber, H. 1987. Fatty acids of some Cornaceae, Hydrangeaceae, Aquifoliaceae, Hamamelidaceae and Styracaceae. Phytochemistry 26: 1441–1445.

Cheeseman, T.F. 1914. Illustrations of the New Zealand Flora. Vol. 1. Wellington: John Mackay, Govt. Printer.

Clokie, M.R.J. 2001. Molecular studies of southern hemisphere disjunction in three plant genera: *Eucryphia*, *Griselinia* and *Coriaria* (Doctoral dissertation). Retrieved from University of Leicester Research Archive. http://hdl.handle.net/2381/29823

Cronquist, A. 1981. An integrated system of classification of flowering plants. New York: Columbia University Press.

Dahlgren, R.M.T. 1980. A revised system of classification of the angiosperms. Bot. J. Linn. Soc. 80: 91–124.

Damtoft, S., Jensen, S.R., Nielsen, B.J. 1983. Biosynthesis of the iridoid glucosides cornin, hastatoside, and griselinoside in *Verbena* species. J. Chem. Soc. Perkin Trans. 1: 1943–1948.

Dawson, J.W. 1966. Vegetative features of *Griselinia lucida* – a New Zealand shrub epiphyte. Tuatara 14: 121–129.

Dillon, M.O., Muñoz-Schick, M. 1993. A revision of the dioecious genus *Griselinia* (Griseliniaceae) including a new species from the coastal Atacama Desert of northern Chile. Brittonia 45: 261–274.

Earp, C. 2013. The date of publication of the Forsters' *Characteres Generum Plantarum* revisited. New Zeal. J. Bot. 51: 252–263.

Earp, C. 2014. (2282) Proposal to conserve the name *Griselinia* G. Forst. (Griseliniaceae) against *Griselinia* Scop. (Fabaceae). Taxon 63: 438.

Endlicher, S.L. 1850. Genera plantarum. Suppl. 5. (*Griselinia*, p. 16, n. 4562). Wien: Fr. Beck.

Eyde, R.H. 1964. Inferior ovary and generic affinities of *Garrya*. Amer. J. Bot. 51: 1083–1092.

Eyde, R.H. 1987. The case for keeping *Cornus* in the broad Linnaean sense. Syst. Bot. 12: 505–518.

Federov, A. 1969. Chromosome numbers of flowering plants. Leningrad: Academy of Sciences of the USSR.

Ferguson, I.K., Hideux, M.J. 1978. Some aspects of the pollen morphology and its taxonomic significance in Cornaceae *sens. lat.* Proc. IV Int. Palynol. Conf., Lucknow (1976–77) 1: 240–249.

Hooker, J.D. 1867. Cornaceae. In: Bentham, G., Hooker, J. D. (eds.) Genera Plantarum 1(3): 947–952. London: Reeve & Co.

Jensen, S.R., Nielsen, B.J. 1980. Iridoid glucosides in *Griselinia*, *Aralidium* and *Toricellia*. Phytochemistry 19: 2685–2688.

Kubitzki, K. 1963. Zur Kenntnis des unilokulären Cornaceen-Gynözeums (Cornaceen-Studien I). Ber. Dtsch. Bot. Ges. 76: 33–39.

Li, H., Chao, C. 1954. Comparative anatomy of the woods of the Cornaceae and allies. Quart. J. Taiwan Mus. 7: 119–136.

Lundberg, J. 2001. Polyosmaceae. Chapter III. In: Lundberg, J., Phylogenetic Studies in the Euasterids II with Particular Reference to Asterales and Escalloniaceae. Uppsala: Acta Universitatis Upsaliensis.

Mildenhall, D.C., 1980. New Zealand Late Cretaceous and Cenozoic plant biogeography: a contribution. Palaeogeogr. Palaeoclimatol. Palaeoecol. 31: 197–233.

Nicolas, A.N., Plunkett, G.M. 2014. Diversification times and biogeographic patterns in Apiales. Bot. Review 80: 30–58.

Noshiro, S., Baas, P. 1998. Systematic wood anatomy of Cornaceae and allies. IAWA J. 19: 43–97.

Percival, M. 1961. Types of nectar in angiosperms. New Phytol. 60: 235–281.

Philipson, W.R. 1967. *Griselinia* Forst. fil. – Anomaly or link. New Zeal. J. Bot. 5: 134–165.

Philipson, W.R. 1977. Ovular morphology and the classification of the dicotyledons. Plant Syst. Evol., Suppl. 1: 123–140.

Philipson, W.R., Stone, B.C. 1980. The systematic position of *Aralidium* Miq. – A multidisciplinary study. 1. Introduction and floral and general anatomy. Taxon 29: 391–403.

Plunkett, G.M. 2001. Relationship of the order Apiales to subclass Asteridae: a re-evaluation of morphological characters based on insights from molecular data. Edinb. J. Bot. 8: 183–200.

Plunkett, G.M., Soltis, D.E., Soltis, P.S. 1996. Higher level relationships involving Apiales (Apiaceae and Araliaceae) based on phylogenetic analysis of *rbc*L sequences. Amer. J. Bot. 83: 499–515.

Plunkett, G.M., Chandler, G.T., Lowry II, P.P., Pinney, S. M., Sprenkle, T.S. 2004. Recent advances in understanding Apiales and a revised classification. S. Afr. J. Bot. 70: 371–381.

Puntieri, J.G., Brion, C.A.M. 2000. *Griselinia racemosa* (Cornaceae), a new record for the Argentinian flora. Hickenia 3(19): 65–68.

Soltis, D.E., et al. 2011. Angiosperm phylogeny: 17 genes, 640 taxa. Amer. J. Bot. 98: 704–730.

Takhtajan, A. 1980. Outline of the classification of flowering plants (Magnoliophyta). Bot. Rev. (Lancaster) 46: 225–359.

Takhtajan, A. 1987. Systema Magnoliophytorum. Leningrad: NAUKA. [In Russian]

Takhtajan, A. 1997. Diversity and Classification of Flowering Plants. New York: Columbia University Press.

Takhtajan, A. (ed.). 2000. Anatomia seminum comparativa. Tomus 6. Dicotyledones. Rosidae II. Leningrad: NAUKA. [In Russian]

Taubert, P. 1892. Revision der Gattung *Griselinia*. Bot. Jahrb. Syst. 16: 386–392.

Thorne, R.F. 1968. Synopsis of a putatively phylogenetic classification of the flowering plants. Aliso 6(4): 57–66.

Thorne, R.F. 1992. Classification and Geography of the Flowering Plants. Bot. Rev. (Lancaster) 58: 225–348.

Thorne, R.F. 2000 (2001). The classification and geography of the flowering plants: dicotyledons of the class Angiospermae (subclasses Magnoliidae, Ranunculidae, Caryophyllidae, Dilleniidae, Rosidae, Asteridae, and Lamiidae). Bot. Rev. (Lancaster) 66: 444–647.

Tseng, C.C. 1980. The systematic position of *Aralidium* Miq. – A multidisciplinary study. 3. Pollen morphology. Taxon 29: 407–409.

Wangerin, W. 1910. Cornaceae. Series IV, family 229 (Heft 41). In: Engler, A. (ed.) Das Pflanzenreich. Leipzig: W. Engelmann.

Watson, L., Dallwitz, M.J. 1992 onwards. The families of flowering plants: descriptions, illustrations, identification, and information retrieval. Version: 19th October 2016.

Westhoff, M., Zimmermann, D., Zimmermann, G., Gessner, P., Wegner, L.H., Bentrup, F.-W., Zimmermann, U. 2009. Distribution and function of epistomatal mucilage plugs. Protoplasm 235: 101–105.

Xiang, Q.Y. Soltis, D.E. 1998. *Rbc*L sequence data define a cornaceous clade and clarify relationships of Cornaceae sensu lato. In: Boufford, D.E., Ohba, H. (eds.) Sino-Japanese flora, its characteristics and diversification. Tokyo: The University of Tokyo Press, pp. 123–139.

Xiang, Q.Y., Soltis, D.E., Morgan, D.R., Soltis, P.S. 1993. Phylogenetic relationships of *Cornus* L. sensu lato and putative relatives inferred from *rbc*L sequence data. Ann. Missouri Bot. Gard. 80: 723–734.

Zuloaga, F.O., Morrone, O., Belgrano, M.J., Marticorena, C., Marchesi, E. (eds.) 2008. Catálogo de las Plantas Vasculares del Cono Sur (Argentina, Sur de Brasil, Chile, Paraguay y Uruguay). Monogr. Syst. Bot. Missouri Bot. Gard. 107(1): i–xcvi, 1–983; 107(2): i–xx, 985–2286; 107(3): i–xxi, 2287–3348.

Loganiaceae

Loganiaceae R. Br. ex Mart., Nov. Gen. Sp. Pl. 2: 133 (1827) ("Loganieae"), nom. cons.
Strychnaceae DC. ex Perleb, Vers. Arzneikr. Pfl.: 244 (1818) ("Strychneae").
Spigeliaceae Bercht. & J. Presl, Přir. Rostlin 1: 40 (1823).
Gardneriaceae Wall. ex Perleb, Clav. Class.: 23 (1838) ("Gardnereae").
Antoniaceae Hutch., Fam. Fl. Pl., ed. 2: 375 (1959).
Geniostomaceae Struwe & V. Albert, Cladistics 10: 206 (1995).

L. Struwe, K.L. Gibbons, B.J. Conn, and T.J. Motley

Terrestrial herbs, shrubs, lianas, or trees, sometimes epiphytic, glabrous or hairy; plants bisexual (or less often unisexual and dioecious or gynodioecious). Leaves decussate, rarely subverticillate or rosulate, petiolate, with interpetiolar lines, stipules, or sheaths (sometimes inconspicuous, rarely absent); blade entire, with either a prominent midvein and pinnate venation, or acrodromous venation with 1 or 2 pairs of distinct, basally divergent secondary veins curved similarly to Melastomataceae (some *Mitrasacme* and *Spigelia*, and most *Strychnos*). Colleters (small, multicellular, and finger-shaped glands) often present on the adaxial side of leaves and/or at the adaxial base of sepals. Inflorescence terminal and/or axillary (ramiflorous in some *Geniostoma*), cymose, in dichasial or thyrse-like cymes with a terminal flower, sometimes with cincinnous branches (*Mitreola* and *Spigelia*), or pleiochasia (some *Mitrasacme*). Flowers bisexual (sometimes unisexual), hypogynous, actinomorphic (zygomorphic calyx and androecium in *Usteria*), 5-merous (4-merous in *Gardneria*, some *Logania*, most *Mitrasacme*, *Phyllangium*, *Schizacme*, some *Strychnos* spp., *Usteria*); sepals fused at base (in *Antonia* free and subtended by 3 or 4 series of scale-like bracts [epicalyx]), persistent in fruit, campanulate, funnelform, rotate, or tubular, imbricate, open, or valvate in bud; corolla sympetalous, campanulate, rotate, tubular, funnelform, or salverform, often with hairs on inner surface of corolla mouth, corolla lobes imbricate or valvate in bud. Stamens inserted in corolla tube or in sinuses between corolla lobes, isomerous and alternating with corolla lobes (one stamen in *Usteria*); filaments free (partially fused to corolla tube in some *Mitrasacme*); anthers free (connate in *Gardneria* spp.), basifixed or subdorsifixed, often sagittate and versatile, longitudinally dehiscent, introrse (extrorse or latrorse in some *Mitrasacme*, latrorse in *Norrisia*), tetrasporangiate. Ovary bicarpellate, bilocular (rarely unilocular in *Strychnos* spp.), superior (semi-inferior in *Adelphacme*, *Mitreola* and *Phyllangium*), syncarpous (secondarily apically apocarpous in *Adelphacme*, species of *Mitreola*, *Mitrasacme*, *Phyllangium* and *Schizacme*; Fig. 77B); without distinct nectary disk but glandular cells sometimes present on ovary wall; placenta axile (basal in *Strychnos* spp.), often peltate; ovules many. Style(s) terminal, fused (free in *Schizacme* and *Mitrasacme secedens*, sometimes free in *Mitreola* and *Phyllangium*, partially separating during fruit development in *Spigelia*). Stigma simple, peltate, capitate, or slightly bilobed. Fruit a septicidal capsule (septicidal, loculicidal and circumscissile near base in *Spigelia*), berry (*Gardneria* and *Strychnos*; Fig. 78C), or drupe with fibrous mesocarp (*Neuburgia*); placenta dry or fleshy (*Gardneria*, *Geniostoma*, and *Strychnos*), with few to many seeds. Seeds

Dr. Motley was the J. Robert Stiffler Distinguished Professor of Botany at Old Dominion University and Director of Science at the Norfolk Botanical Garden in Norfolk, Virginia, USA, when he suddenly passed away on March 23, 2013. LS had been working with Tim on this treatment for several years and it was in its final stages. Since then, new research findings from 2012–2015 have been added from Loganieae research by new co-authors BC and KG. We miss you!

J.W. Kadereit, V. Bittrich (eds.), *Flowering Plants. Eudicots*, The Families and Genera of Vascular Plants 15,
https://doi.org/10.1007/978-3-319-93605-5_8

Fig. 77. Loganiaceae. Examples of Australian Loganiaceae. *Mitrasacme exserta*. **A** Habit. **B** Fruiting capsule. *Geniostoma huttonii*. **C** Habit. **D** Fruiting capsule. *Logania minor*. **E** Habit. **F** Fruiting capsule. (Drawings by Catherine Wardrop (A, B), used with permission; David Mackay (C, D), previously published in Conn (1993), reproduced with permission; David Mackay (E, F, previously published in Conn (1995), all images copyright Royal Botanic Gardens and Domain Trust)

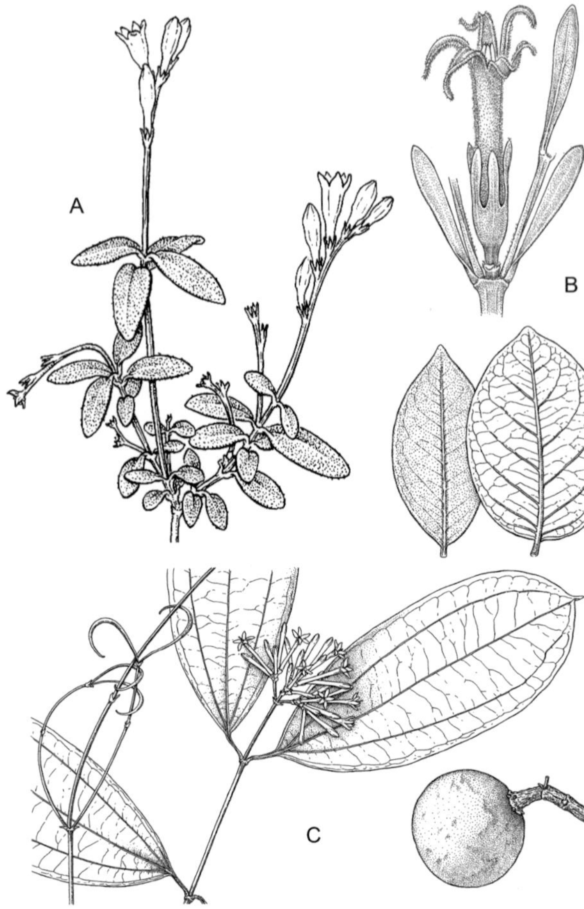

Fig. 78. Loganiaceae. Examples of South American Loganiaceae. **A** *Spigelia genuflexa*, habit. **B** *Bonyunia superba*, flower (above), leaves (below). **C** *Strychnos melinoniana*, branch with flowers and fruit. (A: Popovkin et al. 2011; B: Grant 2009; C: Mori et al. 2002; drawings by Bobbi Angell, used with permission)

rounded to angular (*Adelphacme*, *Geniostoma*, *Logania*, *Mitrasacme*, *Mitreola*, *Orianthera*, *Phyllangium*, *Schizacme* and *Spigelia*), concave-convex (*Gardneria*, *Neuburgia*, and *Strychnos*), or strongly flattened and winged (*Antonia*, *Bonyunia*, and *Usteria*), with fleshy (*Geniostoma*, *Mitrasacme*, *Mitreola*, and *Norrisia*), horny (*Gardneria* and *Strychnos*), or starchy (*Antonia*, *Bonyunia*, *Logania* and *Orianthera*) endosperm; embryo straight (curved in *Gardneria*).

A pantropical family of 16 genera and ca. 460 species, with the highest species diversity in tropical South America, tropical Africa, Southeast Asia, and Australia.

Vegetative Morphology and Anatomy. The habit varies from ephemeral, annual herbs (aquatic in *Mitrasacme secedens* and sometimes in *Phyllangium*) to perennial herbs, shrubs, lianas, to large trees. The branches vary from terete to quadrangular and winged branches are common. The leaves are simple and entire, decussate (reduced to scales in *Orianthera* spp. and, rarely, in *Mitrasacme*), subverticillate in some *Spigelia* and *Strychnos* spp., and rosulate in some *Mitrasacme* spp. The venation is pinnate, and most species of *Strychnos* (and some *Mitrasacme* and *Spigelia*) are easily distinguished by their Melastomataceae-like acrodromous venation with one to two pairs of basally divergent and curved secondary veins. Characteristic lignified and curved tendrils, derived from leaves, are present in climbing *Strychnos* spp. Distinct interpetiolar stipules, lines, or sheaths are common in all genera, but are never very large. Intrapetiolar buds have been found in *Strychnos nux-vomica* and several *Strychnos* species have straight or curved axillary spines (Leeuwenberg and Leenhouts 1980).

Colleters (multicellular, finger-like glands) are common at the adaxial base of leaves, bracts, and calyx in most genera. In some taxa the colleters are present as multicellular cilia on the interpetiolar sheath or line. Many species lack an indumentum on vegetative parts, but simple, unbranched hairs occur in *Bonyunia* and species of *Mitrasacme*, *Mitreola*, *Logania*, *Orianthera*, *Phyllangium*, *Schizacme*, *Spigelia* and *Strychnos*. Stellate hairs are present in three *Spigelia* spp. (Gould 1997), and domatia have been found in *Neuburgia novocaledonica* (Molina and Struwe 2004).

Only unilacunar nodes have been reported from the Loganiaceae, but the number of investigated taxa is limited (Sinnott 1914; Bendre 1975). An extensive survey of the wood anatomy was presented by Mennega (1980). All genera of Loganiaceae have simple vessel perforations or occasionally perforation plates with scalariform perforations (e.g., *Geniostoma* and *Spigelia*), alternate intervascular pits, and vasicentric tracheids. Included phloem is present in some *Strychnos* species and members of tribe Antonieae (except the liana *Usteria*, which also has wide rays composed of procumbent cells containing styloid crystals). Additionally, genera in Antonieae have uniseriate, homocellular rays. The remaining genera lack included phloem and contain both uni- and multiseriate rays (with the exception of *Neuburgia* with only uniseriate rays, which also contain abundant crystal sand). Internal phloem is present in investigated species of all tribes (Mennega 1980). Genera in tribe Loganieae have multiple radial vessels. *Logania* is unique in having spiral thickenings in the vessels in a flame-like pattern, and helical thickenings have been found in *Usteria*. *Geniostoma* has heterocellular rays, long vessel elements, partly septate fibers, and unique flat fibers with sharp-pointed tips. *Spigelia*, *Neuburgia*, and most *Strychnos* species have coenocytic fibers. Anatomical characteristics are very diverse in the family and there are few synapomorphies that support each tribe except Antonieae. For example, *Geniostoma* is anatomically more similar to *Spigelia* and *Neuburgia* than to *Logania*, but molecular phylogenetic studies (Backlund et al. 2000; Gibbons et al. 2012; Foster et al. 2014) provide strong support for the inclusion of *Geniostoma* in Loganieae. Anatomical characters do not provide clear support for tribal relationships and circumscriptions, but they have lent support for the exclusion of the Buddlejaceae, *Desfontainia* (Columelliaceae), and tribe Potalieae (Gentianaceae) from the Loganiaceae (Bremer et al. 1994; Struwe et al. 1994).

Inflorescences, Floral Structure, and Floral Anatomy. The inflorescences of the Loganiaceae are terminal, axillary, ramiflorous, or cauliflorous. They are cymose, although in *Schizacme*, and occasionally in *Mitrasacme* and *Phyllangium*, they are reduced to a single flower (Conn et al. 1996). Inflorescences of *Adelphacme*, *Gardneria* and *Strychnos* spp. are nearly strictly dichasial, but the inflorescences of *Mitreola* and *Spigelia* have cincinnous branches with dichasial branching at the base (Figs. 77, 78).

Loganiaceae have a typical Asteridae flower morphology with 4- or 5-merosity, with sepals basally fused to form a calyx (rarely with free sepals), and a sympetalous corolla with isomerous stamens (except for the mono- (to di-)merous androecium in *Usteria*) inserted in the corolla tube or in sinuses between corolla lobes. Slightly to largely unequal calyx lobes are common and are most distinct as zygomorphic calycophylls in *Usteria* (Hakki 1998). Corolla aestivation is valvate in tribes Antonieae, Spigelieae, and Strychneae, but is variable in Loganieae, sometimes within genera. Genera of Strychneae and Spigelieae often have corolla lobes that are thickened towards the apex. The bicarpellate and bilocular ovaries are superior (or semi-inferior in *Adelphacme*, *Mitreola* and *Phyllangium*) with axile placentation. The ovules are usually numerous and scattered on a peltate placenta. Secondary pollen presentation has been found in *Geniostoma* and *Spigelia*. *Spigelia* has partly fused and early caducous styles, with the pollen deposited onto stylar hairs immediately below the apical stigmatic surfaces. It has been hypothesized that the separation of the style in *Spigelia* affects the seed set by limiting the growth of pollen tubes

(Erbar and Leins 1999). In *Geniostoma* the pollen is deposited on stylar head papillae (Endress et al. 1998).

Only a limited number of species have been investigated for floral anatomy (Tiagi and Kshetrapal 1972; Endress et al. 1983, 1996). The ovaries are bicarpellate and congenitally syncarpous (sometimes only at base), but cases of 3-carpellate ovaries have been found in Hawaiian *Geniostoma* (formerly *Labordia*). Postgenital (and pre-anthesis) stylar fusion and secondary apocarpy has been demonstrated in *Mitrasacme alsinoides* by Endress et al. (1983), and *Mitrasacme*, *Mitreola* and *Phyllangium* show varying degrees of carpel and stylar fusion. The stigmas of *Strychnos* and *Geniostoma* have globose distinct papillae, and a viscid stigmatic surface is found in *Geniostoma* (Endress et al. 1996). The latter genus also has postgenital stylar fusion. A prominent nectary disk at the base of the ovary is lacking in Loganiaceae, but glandular cells are sometimes present on the ovary wall. In species of *Strychnos*, *Spigelia*, and *Geniostoma* the outer ovary wall has raised stomata presumably involved in nectar production (Endress et al. 1996; Erbar and Leins 1999).

Many species have a ring of (often penicillate) hairs in the corolla tube or mouth (e.g., *Geniostoma*, *Mitrasacme*, *Mitreola*, *Neuburgia*, and *Strychnos*), and hairs can also be present on the outer surface of the corolla, on the anthers, and on the calyx. Colleters (specialized glands, see above) are often present on the inner surface of the calyx lobes.

Embryology. Few studies exist on the embryology of the family (e.g., Hakki 1980, 1998; Reddy et al. 1999). The anthers are tetrasporangiate, the walls are comprised of 4 or 5 layers, and have a glandular tapetum. The tapetum is of dual origin and the cells are both uni- and binucleate at maturity. Orbicules (or Ubish bodies) are granules of sporopollenin on the walls of the tapetum cells. They are known to occur in *Antonia*, *Gardneria*, *Geniostoma*, *Logania*, *Orianthera*, *Mitrasacme*, *Mitreola*, *Neuburgia*, *Norrisia*, *Spigelia*, *Strychnos*, and *Usteria* (Vinckier and Smets 2002). Orbicules also occur in Apocynaceae, Gelsemiaceae, Gentianaceae, Rubiaceae, and Solanaceae.

The ovules are hemianatropous (*Mitrasacme* and *Spigelia*), atropous (*Mitreola*) or anatropous (*Strychnos*), and unitegmic and tenuinucellate (Hakki 1980, 1998; Reddy et al. 1999). The endosperm development is nuclear in *Strychnos* and *Usteria*, and *Mitrasacme* has been reported to have endosperm intermediate between cellular and nuclear (Yamazaki 1963). Yamazaki (1963) reported Solanad-type embryo development in *Mitrasacme*. The seed coat varies in the family from 1 or 2 cell layers in *Spigelia* (Dahlgren 1922), 3 or 4 cell layers in *Mitrasacme indica* (Yamazaki 1963), to 4–6 layers in *Geniostoma randianum* (Corner 1976).

Palynology. Palynological characters are diverse in the Loganiaceae. Investigated species have monads of spheroidal shape (sometimes angular to elliptic) and with a reticulate, psilate, tectate, and/or perforated exine (Punt 1980). The most common pollen type is 3-colporate, but 3- or 4-porate grains are found in *Geniostoma*. Pollen characters also support the current family circumscription, and the exclusion of Buddlejaceae, *Retzia*, and *Plocosperma* from the Loganiaceae (Punt 1980). Most *Geniostoma* have pollen grains with a smooth exine and prominent pores. However, some Hawaiian *Geniostoma* (formerly *Labordia* sect. *Labordia*) have pollen grains with an irregular, coarse exine and shallow pores (Selling 1947).

Karyology. The chromosomes are small (0.7–1.8 μm). Polyploidy and dysploidy is common in *Spigelia* and *Strychnos* and the base number varies from 8–12 (13 in *Spigelia splendens*; Gadella 1980). Four *Spigelia* and thirty-four *Strychnos* species show interspecific chromosome number variation; in *Spigelia* as $2n = 32, 48, 64$, or 26 (Moore 1947; Gadella 1962, 1978; Mangenot and Mangenot 1962; Rogers 1986) and in *Strychnos* $2n = 24, 36, 44, 88$, or 110 (Mohrbutter 1936; Mangenot and Mangenot 1957, 1958, 1962; Raghavan 1959; Miège 1960; Gadella 1962, 1963, 1967, 1969, 1970, 1972, 1977, 1980; Bedi et al. 1980). *Spigelia* has base numbers of 8, 10, and 13 and *Strychnos* has base numbers of 11 and 12. Counts have been published for *Usteria* ($2n = 22$; Miège 1960) and *Mitreola* ($2n = 20$; Lewis et al. 1962). Among the genera of the tribe Loganieae, a ploidy difference is noted between Hawaiian *Geniostoma* and species from other areas. Studies of nine Hawaiian species of *Geniostoma* yielded counts of $2n$ = ca. 80 (Carr 1978; Motley and Carr 1998), whereas four non-Hawaiian specimens had counts of $2n$ = ca. 38 (*G. antherotrichum* var. *archboldianum*, Papua New Guinea; Borgmann 1964) and $2n = 40$ (*G. ligustrifolium*, New Zealand; Gadella 1963; Hair 1963; *G. rupestre*, Guam and Tonga; Motley and Carr 1998), each with a base number of 10. *Logania* has a base number of 8, based on a count of a single species, *L. flaviflora* ($2n = 32$; Keighery in Löve 1975).

Pollination. Pollination biology is grossly understudied in Loganiaceae. At least some *Spigelia* species are pollinated by hummingbirds (Machado et al. 2007). The two sections of Hawaiian *Geniostoma* (sects. *Darbolia* and *Labordia*) have different floral morphologies. Based on this divergence, Motley and Carr (1998) speculated that birds pollinate species in sect. *Labordia* and insect-pollination occurs in sect. *Darbolia*. No native pollinators had been seen visiting flowers during field investigations, and many native Hawaiian birds are now extinct or absent from forests below 1200 m. Instead, two non-indigenous insects, European honey bees (*Apis mellifera*) and flower flies (*Allograpta exotica*) visited flowers of *G. tinifolia* (sect. *Darbolia*). Additionally, flowering individuals of the species known as *Labordia hedyosmifolia* (now in *Geniostoma*, but the name has not been combined) documented by Motley and Carr (1998) were being visited by the native Hawaiian honeycreeper Apapane in the bird sanctuary at Honolulu Zoo's Apapane (*Himatione sanguinea*) enclosure (T. Motley, pers. obs.). Three species of honeyeaters (tui, *Prosthemadera novaseelandiae*; bellbird, *Anthornis melanura*; and hihi, *Notiomystis cincta*) have been observed visiting flowers of *Geniostoma ligustrifolium* on Kapiti Island, New Zealand, over several years (Castro and Robertson 1997). An unidentified ant species has also been consistently observed on the cauliflorous flowers of the Fijian endemic *G. macrophyllum* (Castro and Robertson 1997), but it is unclear if the ants pollinate the flowers.

Reproductive Systems. Bisexual flowers are the most common condition in the family, but unisexual flowers are found in tribe Loganieae—in *Geniostoma*, *Logania* and, occasionally, in *Orianthera* and *Mitrasacme*. Motley and Carr (1998) found that all individuals of Hawaiian *Geniostoma* were either functionally entirely female (flowers with ovules and no pollen grains in the anthers, the latter small and necrotic) or functionally male (pollen grains present, no ovules in the ovary locules), and this confirmed the occurrence of strict functional dioecism in all investigated species. Although the presence of monomorphic hermaphroditic flowers is the most common condition in non-Hawaiian *Geniostoma* (Conn 1980; Motley and Carr 1998), Valeton (1902) and Smith and Stone (1962) reported heterostyly in a species from Java, and Rattenbury (1980), Endress et al. (1998) reported gynodioecious populations in New Zealand, and heterostyly and gynodioecy are common throughout Papua New Guinea (Conn, pers. obs.). However, in a detailed investigation, Smith and Stone (1962) could find no definite example of strict dioecy within *Geniostoma* in the Pacific, with the possible exception of one variety of *G. rupestre* from Fiji. Gynodioecy also occasionally occurs in a few species of *Mitrasacme* (Conn et al. 1996). The closely related genus *Logania* has unisexual flowers with dioecious distribution, whereas most species of *Orianthera* have bisexual flowers (Conn et al. 1996).

Fruit and Seed. Dry, apically dehiscent, capsular fruits are present in Antonieae and Loganieae. *Adelphacme* and most species of *Mitreola*, *Mitrasacme*, *Phyllangium* and *Schizacme* have mitre-shaped capsules, often with persistent styles. Distinct woody capsules, with abscising valves exposing the seeds, which are embedded in pulpy, orange placentas are found only in *Geniostoma*, and are a synapomorphy for this genus (Struwe et al. 1994). *Spigelia* has unique bilobed to rounded and basally circumscissile (and septicidally and loculicidally dehiscent as well) capsules with the style inserted in the middle of the truncated apex. Tribe Strychneae has fleshy-walled fruits with fibrous 1- to few-seeded drupes in *Neuburgia* and fleshy or leathery berries in *Gardneria* and *Strychnos*.

The seeds are flattened or angular and winged in Antonieae (albeit *Norrisia* seeds are described as spindle-shaped and appear slightly flattened).

Solereder (1892–95) reported the presence of chlorophyll in *Spigelia* seeds. Seeds of *Adelphacme, Logania* and *Orianthera* are more or less ellipsoid (Conn 1995; Foster et al. 2014), whereas seeds of *Geniostoma* and *Schizacme* are subglobose to ellipsoid, seeds of *Phyllangium* are angular and seeds of *Mitrasacme* are subglobose to ellipsoid or angular (Conn 1980). Ruminate seeds (i.e., testa deeply folded) are only known from *Spigelia* (Dahlgren 1922; Hakki 1980), and in *Mitrasacme*, *Mitreola*, and *Phyllangium* the seed testa is variously textured, or smooth. Netolitzky (1926) stated that the outer epidermal cells of the seed coat of some *Strychnos* species have hair-like projections (with those of *S. icaja* being long enough to become appressed to the seed surface).

Dispersal. *Antonia*, a tall lowland rainforest tree in the Neotropics, has winged seeds in dry capsules that facilitate wind dispersal (Mori and Brown 1994). Similarly, *Bonyunia* and *Usteria* are presumed to be wind-dispersed. *Strychnos camptoneura* seeds have a wing that is narrow and irregular, but the seeds are enclosed in a hard, sometimes deciduous fruit that is still considered a berry. In *Geniostoma* the seeds are positioned axillary in a colorful, fleshy placenta that is exposed when the capsule valves open and break off, and these seeds are recorded as bird-dispersed (Conn 1980). Other genera of Loganiaceae have either animal-dispersed fleshy fruits or small dry seeds in capsules on erect branches, which can be dispersed over short distances by wind or rain.

Phytochemistry. Seco-iridoids are present in Loganieae and Strychneae but absent from Antonieae and Spigelieae (Jensen 1992). Seco-iridoids are characteristic of the Gentianales, and are a precursor to complex indole alkaloids. In Loganiaceae, complex indole alkaloids (among them the toxic compounds strychnine and brucine) are largely restricted to Strychneae, but similar alkaloids are also found in Apocynaceae, Gelsemiaceae, and Rubiaceae. Aluminum accumulation has been reported for *Antonia* (Bisset 1980a). For further information on phytochemistry, see Bisset (1980a).

Classification and Relationships Within the Family. The Loganiaceae contain four tribes, one of which is monogeneric (Spigelieae). Strong support for monophyly exists for Loganieae and Antonieae, but support is weaker for Strychneae. DNA sequence data from the nuclear and chloroplast genomes (as well as morphological data) have been analyzed and published for at least one representative of all genera (except *Norrisia*) and form the basis for the tribal classification presented here. Studies of tribe Antonieae show *Usteria* as sister to *Antonia* and *Bonyunia*, and support by morphological and molecular data for this tribe is strong (Struwe and Albert 1997; Backlund et al. 2000). The most species-rich tribe is Strychneae with three genera and an estimated 220 spp. (*Gardneria*, *Neuburgia*, and *Strychnos*), and *Strychnos* appears to be monophyletic (Frasier 2008; Adebowale et al. 2016). The position of Spigelieae (*Spigelia*) relative to other tribes is still uncertain. Chloroplast and mitochondrial data support inclusion of *Spigelia* in the Strychneae (Backlund et al. 2000; Yang et al. 2016), but morphological data do not unambiguously support this position (Struwe et al. 1994; Struwe and Albert 1997). Studies using nuclear data have placed *Spigelia* within Loganieae, sister to *Mitrasacme* (previously placed in Spigelieae), but without support (Oxelman and Bremer 2000; Popovkin et al. 2011). Considering the contradicting evidence for *Spigelia*, we here treat Spigelieae conservatively as a separate tribe until more data have been gathered and analyzed, and we do not propose any major classificatory changes for these four tribes. Further studies with additional data likely will show that this classification needs to be updated. Tribe Loganieae (in part, Leeuwenberg and Leenhouts 1980) includes *Adelphacme*, *Geniostoma*, *Logania*, *Orianthera*, *Mitrasacme*, *Mitreola*, *Phyllangium*, and *Schizacme*. The latter four genera were previously placed in Spigelieae (see Conn et al. 1996; Struwe 2004). Molecular phylogenetic studies support a close relationship among *Adelphacme*, *Mitrasacme*, *Phyllangium* and *Schizacme*, but relationships among this group and the remaining genera of Loganieae are still unclear (Gibbons et al. 2012; Foster et al. 2014). The inclusion of *Labordia* within *Geniostoma* also has strong phylogenetic support and is followed here.

Circumscription. The circumscription of the Loganiaceae has varied dramatically during its history, with treatments including from one genus to as many as 29 genera. The Loganiaceae in the broadest interpretation sensu Leeuwenberg and Leenhouts (1980) is highly para- and polyphyletic as was shown by many analyses of morphological, anatomical, phytochemical, and DNA data (Mennega 1980; Jensen 1992; Olmstead et al. 1993; Bremer et al. 1994, 2002; Struwe et al. 1994; Struwe and Albert 1997; Oxelman et al. 1999; Backlund et al. 2000). In general, Loganiaceae was thought of as a stipule-less, opposite-leaved group with largely superior ovaries and actinomorphic flowers, and it served as a 'grab bag' for many genera that did not really fit anywhere else. Based on these studies, tribe Buddlejeae (*Buddleja*, *Nuxia*, now often in Scrophulariaceae s.str. or Buddlejaceae-Lamiales), *Desfontainia* (Columelliaceae-Bruniales), tribe Gelsemieae (*Gelsemium* and *Mostuea*; Gelsemiaceae-Gentianales), *Polypremum* (Tetrachondraceae-Lamiales), *Plocosperma* (Plocospermataceae-Lamiales), tribe Potalieae (*Anthocleista*, *Fagraea*, and *Potalia*; Gentianaceae-Gentianales), *Retzia* (Stilbaceae-Lamiales), and *Peltanthera* and *Sanango* (Gesneriaceae-Lamiales) have been excluded from the Loganiaceae (Struwe et al. 1994; Backlund et al. 2000; Olmstead et al. 2001; Struwe and Albert 2002; APG 2003).

Distribution and Habitats. The genera of Loganiaceae show primarily tropical distribution patterns with most genera confined to a major landmass/continent. An exception is *Strychnos*, which is pantropical, and *Mitreola*, with one widespread species (*M. petiolata*) in the tropics and several narrow endemics in southern North America, Madagascar, and Southeast Asia (Leeuwenberg and Leenhouts 1980). Three genera are exclusively neotropical (*Antonia*, *Bonyunia*, and *Spigelia*; the latter has a few temperate species as well). In contrast, *Gardneria*, *Neuburgia*, and *Norrisia* are distributed in the Old World regions of Southeast Asia, the Pacific Basin, and Malesia. The Australian endemic genera *Adelphacme*, *Logania*, *Orianthera* and *Phyllangium* are unusual in Loganiaceae by occurring predominantly at temperate latitudes, as does *Schizacme* (Australia and New Zealand). Although a few species of *Mitrasacme* also occur in southeastern Australia, *Mitrasacme* is most species-rich in the monsoon tropics of northern Australia, with a few disjunct and geographically isolated species in Malesia/continental Asia. *Geniostoma* is restricted to islands in Malesia, the Pacific and easternmost Indian Ocean, with one species recorded from mainland Australia. Interestingly, Africa and Madagascar contain very few native genera of Loganiaceae (*Strychnos*, *Usteria*) and a few naturalized species (e.g., *Spigelia anthelmia*, Asia and Africa). Africa has only one endemic genus, *Usteria*, which occurs in western tropical Africa.

Most genera occur in lowland tropical forests (e.g., *Antonia*, *Bonyunia*, most species of *Geniostoma*, *Strychnos*, *Usteria*). Genera with species growing at higher altitudes are *Bonyunia* (tepuis of Guayana Highlands), *Mitreola* (Himalayas and Mt. Kinabalu), *Geniostoma* (New Guinean and Hawaiian mountains), *Schizacme* (alpine and montane southeastern Australia and New Zealand) and *Spigelia* (Brazil). The herbaceous genera *Adelphacme*, *Logania* (some species), *Mitrasacme*, *Mitreola*, *Orianthera*, *Phyllangium* and *Spigelia* are common in open habitats such as grasslands, savannas, thickets, and disturbed areas.

Economic Importance. The Loganiaceae contain many ethnobotanically important plants. The most well-known is *Strychnos*, which is also the economically most important genus (Bisset 1974). *Strychnos* is the source of several poisonous complex indole alkaloids, such as strychnine and brucine. *Strychnos* is one of the sources of curare (a mixture of plant extracts used as arrow poison), is used for poisoning of fish, and also has many pharmacological uses (Bisset 1974, 1980b; Smith 1991). The pulp of the fruits of *Strychnos* is often edible and eaten, and has been used by several different indigenous groups in the Northern Territory, Australia, to treat skin conditions (Smith 1991) and to treat diabetes using the roots (Wightman et al. 1992). The poisonous genus *Spigelia* has been used against parasitic intestinal worms (thereby rendering the common name "worm grass"), but might often have killed the patient as well (Rogers 1986). Just touching a plant can be dangerous in some species, and *Spigelia* has been used to kill criminals, domestic animals, fish and insects (e.g., Quattrocchi 2016: 3537–3538; Schmelzer and Ameenah Gurib-Fakim 2008: 534–536). *Geniostoma* species are reported to be a cure for stomach disorders in Tonga, Samoa, and Fiji (Zepernick 1972). The fragrant flowers of *Geniostoma rupestre* from the Cook Islands are used to scent coconut oil used for body adornment (Whistler 1990). Bark infusions of *Geniostoma* are used in New Zealand (Booker and Cooper 1961) and New Caledonia (Rageau 1973) to treat skin irritations and boils. In the Solomon Islands *Neuburgia* is used for similar purposes (Leenhouts 1962). In Senegal, children with fevers are bathed in an infusion of *Usteria* bark, and the bark was also made into rope used especially in beehive construction (Kerharo and Adams 1974: 507–513). *Antonia* has been utilized as a fish poison in Guyana and Brazil (Schultes 1977), and the wood of *Norrisia* and *Neuburgia* is used as timber (Leenhouts 1962). *Mitrasacme pygmaea* has been used in traditional Chinese medicine (Li and Leeuwenberg 1996).

Conservation. Since most members of the family have little economic importance, most major conservation efforts have focused on endangered species occurring on islands. In Hawaii, one species of *Geniostoma* (currently still referred to as *Labordia lydgatei*) is federally listed as an endangered species and six others are categorized as "species of concern" or threatened (USFWS 2002). In French Polynesia seven taxa of *Geniostoma* are protected. These are *G. clavatum*, endemic to Raiatea (Meyer 1996), *G. astylum* from Moorea and Tahiti, *G. hallei* (varieties *hivaoense*, *fatuivense*, and *hallei*), and *G. gagneae* from the Marquesas (Hilton-Taylor 2000). Another protected species is *Geniostoma rapense*, which is endemic to Rapa Iti in the Austral Archipelago (Hallé and Florence 1986). Furthermore, three species of *Geniostoma* are rare in the Fijian Islands (*G. stipulare*, *G. macrophyllum*, and *G. clavigerum*) and two additional rare Melanesian taxa occur in New Guinea (*G. rupestre* var. *rouffaeranum*) and on Guadalcanal, Solomon Islands (*G. umbellatum*; Hilton-Taylor 2000). *Neuburgia tubiflora* is a rare species occurring in New Guinea, while the following three rare species (*N. alata*, *N. macrocarpa*, and *N. macroloba*) occur in Fiji (Hilton-Taylor 2000). In Sri Lanka, *Strychnos tetragona* and *S. benthamii* are protected. *Strychnos millepunctata* and *S. chromatoxylon* from the Ivory Coast and the widespread *S. mellodora* (Kenya, Mozambique, Tanzania, and Zimbabwe) are known only from very restricted and threatened habitats on the African continent (Hilton-Taylor 2000). In New Zealand, *Logania depressa* (a species of uncertain status, possibly attributable to *Geniostoma*) is listed as extinct and *Schizacme helmsii* is listed as nationally endangered (de Lange et al. 2013). In Australia, *Logania diffusa* and *L. insularis* are nationally listed as vulnerable, and in Australia's western and northern areas a number of narrow endemic species are noted to be data deficient. Lastly, *Spigelia gentianoides*, native to Alabama and Florida of the southeastern United States, is an endangered and protected species in the United States (USFWS 2002).

Key to Tribes

1. Fruit a berry or drupe; indole alkaloids present **IV. Strychneae**
– Fruit a capsule (with a fleshy placenta in *Geniostoma*); indole alkaloids absent 2
2. Corolla aestivation imbricate (exduplicate-valvate in *Adelphacme*, *Mitrasacme*, *Phyllangium*, *Schizacme*, and, rarely, *Mitreola*, but then with mitre-shaped fruits); seco-iridoids present **II. Loganieae**
– Corolla aestivation valvate; not with mitre-shaped fruits, seco-iridoids absent 3
3. Woody plants; fruits elliptic, septicidal capsules with two persistent valves; style persistent; seeds flattened, winged (not winged in *Norrisia*) **I. Antonieae**
– Herbaceous plants (rarely subshrubs); fruits bilobed, circumscissile, septicidal or loculicidal capsules with four caducous valves; style partly caducous; seeds angular to rounded **III. Spigelieae**

Key to Genera

1. Herbs or undershrubs 2
– Shrubs, trees or woody climbers 9
2. Corolla aestivation valvate; fruit a circumscissile, septicidally and loculicidally dehiscent capsule with four caducous valves; corolla without hairs on the inside; style partly caducous **13. *Spigelia***
– Corolla aestivation imbricate or exduplicate-valvate; fruit a septicidally dehiscent capsule with two persistent valves; corolla with hairs in the corolla mouth (rarely not); style persistent 3

3. Capsule without horns, not bilobed 4
– Capsule two-horned (horns occasionally adherent along almost their entire length, appearing continuous with base of styles) or bilobed 5
4. Stamens inserted in middle of corolla tube, anthers included in corolla tube; apex of calyx lobes obtuse; flowers unisexual **7. *Logania***
– Stamens inserted in mouth of corolla tube, anthers exserted from corolla tube; apex of calyx lobes acute; flowers bisexual **10. *Orianthera***
5. Calyx, corolla and androecium 4-merous, or calyx absent (flowers of some *Mitrasacme* rarely 5-merous, then inflorescence cymose to umbel-like, never incompletely dichasial nor cincinnate) 6
– Calyx, corolla and androecium 5-merous, inflorescence incompletely dichasial, sometimes with cincinnate branches 8
6. Calyx absent; corolla and capsule enclosed in a two-lobed foliaceous involucre; ovary semi-inferior; southern Australia (including Tasmania) **11. *Phyllangium***
– Calyx present; involucre absent; ovary superior 7
7. Calyx tube indistinct, up to 1 mm long; calyx lobes generally unequal; capsule laterally compressed, appearing broadly cuneiform in lateral view; placenta elongate, seeds few; Australia (Tasmania and alpine Victoria) and New Zealand **12. *Schizacme***
– Calyx tube distinct; calyx lobes equal; capsule generally globular, ovoid or ellipsoid; placenta hemispherical, seeds many; Southeast and East Asia, New Guinea, northern and eastern Australia (including Tasmania), New Caledonia **8. *Mitrasacme***
8. Leaves $<$ 4 mm long; stipules arranged as a persistent membranous interfoliar sheath; corolla mouth glabrous or papillose; styles retained in fruit, connate at their apices (sometimes separating after maturity); southwestern Australia **5. *Adelphacme***
– Leaves at least 10 mm long (except *M. sessilifolia* $\geq$ 6 mm long and *M. petiolatoides* $\geq$ 5 mm long); stipules well-developed (mostly triangular); corolla mouth with penicillate ring of hairs; styles not persisting in fruit or stigmas free, subsessile; Americas, Africa, Madagascar, Southeast and East Asia, New Guinea, northern Australia **9. *Mitreola***
9. Fruit a dry capsule, sometimes with a fleshy placenta 10
– Fruit a fleshy, fibrous, or leathery berry or drupe 16
10. Seeds embedded in a pulpy placenta; corolla aestivation imbricate **6. *Geniostoma***
– Seeds attached to a dry placenta; corolla aestivation valvate (imbricate in *Logania*) 11
11. Corolla aestivation imbricate; placenta not peltate **7. *Logania***
– Corolla aestivation valvate; placenta peltate 12
12. Subshrub; leaves membranous; capsule bilobed, circumscissile, septicidally and loculicidally dehiscent with four caducous valves; style partly caducous; seeds angular to rounded, not winged **13. *Spigelia***
– Tree, shrub, or liana; leaves (sub)coriaceous; capsule elliptic, septicidally dehiscent with two persistent valves; style persistent or completely caducous; seeds flattened, winged (not winged in *Norrisia*) 13
13. Liana or climbing shrub; flower 4-merous; one calyx lobe distinctly enlarged; corolla lobes unequal (two larger); flowers with only one stamen; capsule valve without torn apex **4. *Usteria***
– Erect shrub or tree; flower 5-merous; calyx lobes of slightly unequal to equal size; corolla lobes equal; stamens as many as corolla lobes; capsule valve with torn apex 14
14. Epicalyx present and 2–3 pairs of bracteoles below flower; calyx with free sepals; gynoecium glabrous **1. *Antonia***
– Epicalyx absent and 0–1 (2 in *Norrisia*) pairs of bracteoles below flower; calyx lobes fused at base; gynoecium hairy 15
15. Corolla lobes about as long as corolla tube; corolla purplish at anthesis; stamens included in corolla tube or corolla mouth, with very short filaments and introrse anthers; seeds few, flattened, winged **2. *Bonyunia***
– Corolla lobes longer than the corolla tube; corolla yellow to cream-colored at anthesis; stamens exserted from the corolla, with long filaments and latrorse anthers; seeds many, slenderly spindle-shaped, not winged **3. *Norrisia***
16. Leaves with 1 or 2 (or 3) pairs of secondary main veins (Melastomataceae-like), rarely pinnately multi-veined; tendrils often present **16. *Strychnos***
– Leaves with more than 2 pairs of pinnate secondary veins; tendrils absent 17
17. Fruit a globose berry; inflorescence axillary; flower pedicellate; interpetiolar lines present, sheaths or stipules absent **14. *Gardneria***
– Fruit a spindle-shaped drupe; inflorescence terminal; flower sessile; prominent interpetiolar sheaths or triangular stipules present **15. *Neuburgia***

Genera of the Loganiaceae

I. Tribe Antonieae

Tribe Antonieae Endl., Gen. Pl.: 575 (1838).
Tribe Usterieae Endl. (1838).
Subtribe Antoniinae Benth. & Hook.f. (1876) ("Antonieae").

1. ***Antonia*** Pohl

Antonia Pohl, Pl. Bras. Icon. Descr. 2: 13–14, t. 109 (1831).

Shrubs or trees up to 30 m, usually hairy. Inflorescence terminal, dichasial or thyrsoid, many-flowered. Flowers 5-merous; calyx with free sepals, surrounded by 3–4 series of scale-like bracts, corolla salver-shaped, white, cream, or yellow; corolla lobes about as long as corolla tube; stigma capitate. Capsule oblong with one to few flattened, winged seeds. Colleters present in leaf and bract axils and on the inner base of calyx.

One sp., *A. ovata* Pohl, in Guianas and Brazil. Closely related to *Bonyunia.*

2. ***Bonyunia*** M.R. Schomb. ex Progel Fig. 78B

Bonyunia M.R. Schomb. ex Progel in C.F.P. Martius, Fl. Bras. 6(1): 267, t. 72. 1868; Leeuwenberg, Acta Bot. Neerl. 18: 152–158 (1969), rev.; Grant, Ann. Missouri Bot. Gard. 96: 541–563 (2009), rev.

Shrubs or small trees up to 40 m high, hairy to glabrous. Inflorescence terminal, dichasium of cymes, 3–7-flowered per cyme. Flowers 5-merous; calyx fused at the base; calyx lobes slightly unequal; corolla salver-shaped, white at anthesis, turning pink to purple with age; corolla lobes about as long or slightly shorter than corolla tube; stigma bilobed. Capsule ellipsoid to obovoid with 2–40 flattened winged seeds. Colleters present in leaf axils only.

Ten spp. in tropical South America (Guianas, Bolivia, Brazil, Colombia, Peru, and Venezuela). Closely related to *Antonia.*

3. ***Norrisia*** Gardn.

Norrisia Gardn., Hooker's J. Bot. Kew Gard. Misc. 1: 326 (1849); Leenhouts, Fl. Mal. 6: 341–343 (1962), rev.

Trees up to 45 m, rusty-pubescent (at least young branches, calyx, and outside of corolla). Inflorescence terminal, dichasial or thyrsoid, many-flowered. Flowers 5-merous; calyx campanulate with sepals fused at base; corolla salver-shaped, cream or yellow; corolla lobes shorter than corolla tube; stigma entire, small. Capsule obovate with few to many flattened elongated seeds. Colleters present in leaf and bract axils, absent from calyx.

Two species in Malaysia, the Philippines, and Indonesia (Kalimantan and Sumatera). Probably closely related to *Antonia* and *Bonyunia.*

4. ***Usteria*** Willd.

Usteria Willd., Disp. Veg. Meth. 1. 179: 151 (1790); Leeuwenberg, Acta Bot. Neerl. 12: 112–118 (1963), rev.

Climbing and twining shrubs or lianas up to 12 m high, glabrous (flowers partly hairy). Inflorescence terminal, dichasial, many-flowered. Flowers 4-merous; calyx fused at the base; calyx lobes heteromorphic, 3 small and triangular, 1 prominently enlarged and spathulate; corolla salver-shaped, white to purplish; corolla lobes shorter than corolla tube; stigma small, entire. Capsule oblong with several flattened and prominently winged seeds. Colleters present in leaf and bract axils, absent from calyx.

One sp., *U. guineensis* Willd, in western tropical Africa (from Senegal to Angola). Closely related to *Antonia* and *Bonyunia.*

II. Tribe Loganieae

Tribe Loganieae Endl., Gen. Pl.: 576 (1838).
Tribe Labordieae Endl. (1838).
Tribe Mitrasacmeae Meisn. (1840).
Tribe Geniostomateae Baill. (1889) ("Geniostomeae").

5. ***Adelphacme*** K.L. Gibbons, B.J. Conn & M.J. Henwood

Adelphacme K.L. Gibbons, B.J. Conn & M.J. Henwood, Telopea 15: 38 (2013).

Annual herb to 5 cm high, minutely scabrous. Inflorescences terminal, incompletely dichasial. Flowers 5-merous; calyx deeply divided; corolla slightly urceolate, white; corolla lobes ± obovate; styles 2, united apically at anthesis and in fruit; stigma capitate. Capsule 2-horned (mitre-shaped), dehiscing along ventral suture (septicidal). Seeds many, ellipsoid.

The single species *A. minima* (B.J. Conn) K.L. Gibbons, B.J. Conn & M.J. Henwood is only known from southwestern Australia and is closely related to *Mitrasacme*, *Phyllangium* and *Schizacme*.

6. ***Geniostoma*** J.R. Forst. & G. Forst. Fig. 77C, D

Geniostoma J.R. Forst. & G. Forst., Char. Gen. Pl. 23, t. 12 (1776); Smith & Stone, Contr. U. S. Natl. Herb. 37: 1–42 (1962), reg. rev.; Leeuwenberg, Meded. Landbouwhogeschool 77 (8): 1–14 (1977), rev.; Conn (1980), Blumea 26: 245–364, reg. rev.
Labordia Gaudich. (1829).
Nautophylla Guillaumin (1953).

Shrubs or small trees up to 12 m high, glabrous or hairy. Inflorescences axillary (terminal in *G. rapense* and Hawaiian species), ramiflorous or cauliflorous, dichasial, few- to many-flowered. Flowers 5-merous, occasionally 6–9-merous, perfect or plant dioecious and flowers functionally unisexual; calyx fused at base; corolla campanulate, salver-shaped, funnel-shaped, urceolate to rotate, orange-yellow, white to pale yellow, or yellowish-green; corolla lobes shorter to longer than the tube; stigma capitate, ellipsoid, clavate, or globose. Capsule globose to elliptic, dehiscing septicidally to expose numerous black, subglobose to ellipsoid, unwinged seeds embedded in an orange, fleshy placenta. Colleters often present inside the calyx.

Approximately 40 species, distributed through the Pacific from Polynesia (excluding Rapa Nui), Micronesia, to Melanesia, Solomon Islands and eastern Australia, extending to the Mascarene Islands in the Indian Ocean (Mauritius and Réunion). *Labordia* is included within the circumscription of *Geniostoma* even though the nomenclatural changes may not be available yet for all of the Hawaiian taxa. Infrageneric classification and species limits in *Geniostoma* require revision.

7. ***Logania*** R. Br. Fig. 77E, F

Logania R. Br., Prodr. Fl. Nov. Holl. 454–456 (1810), nom. cons.; Conn, Austr. Syst. Bot. 8 (1995); Conn et al., Fl. Austr. 28 (1996), reg. rev.

Herbs, semi-woody perennials, or shrubs up to 2 (–4) m high, dioecious, glabrous or hairy. Inflorescences terminal, often appearing axillary, cymose, one- to many-flowered. Flowers unisexual, usually 5-merous; calyx fused at base; corolla rotate, campanulate, or with a cylindrical tube, white or yellow; corolla lobes shorter to longer than tube; stigma subglobose. Capsule subglobose to ellipsoid, septicidal. Seeds many, ellipsoid, brown, held loosely in two valves. Colleters present on inside of calyx.

23 species, one of these (extinct) in New Zealand is of uncertain status, possibly attributable to *Geniostoma*, the remainder confined to Australia.

8. ***Mitrasacme*** Labill. Fig. 77A, B

Mitrasacme Labill., Nov. Holl. Pl. 1: 39 (1804); Leenhouts, Fl. Mal. 6: 378–387 (1962), reg. rev.; Dunlop, in Conn et al., Fl. Austr. 28: 29–62 (1996), reg. rev.

Annual or perennial herbs or less often subshrubs up to 1 m high, glabrous or hairy. Inflorescence terminal, cymose, one-, few- or many-flowered, often a pleiochasium (umbel-like), occasionally frondose botryoidal. Flowers 4-merous (very rarely 5-merous); calyx fused at the base; corolla tubular, campanulate to urceolate or salver-shaped, white or yellow; corolla lobes shorter than corolla tube; styles 2, united apically at anthesis and in fruit (rarely free); stigma(s) small, truncate to slightly bilobed. Capsule 2-horned (mitre-shaped) apically, dehiscing along ventral suture (septicidal). Seeds numerous, angular to ellipsoid, unwinged. Colleters present in axils of leaves and bracts.

About 55 species primarily in Australia, also in New Caledonia, Malesia, Southeast Asia, China, Japan, and India. Closely related to *Phyllangium* and *Schizacme*.

9. ***Mitreola*** L.

Mitreola L., Opera Varia: 214 (1758); Leeuwenberg, Meded. Landbouwhogeschool 74(23): 1–28 (1974); reg. rev.
Cynoctonum J.F. Gmel. (1791).

Annual or perennial herbs, erect or creeping, up to 1 m high, glabrous (except floral parts). Inflorescence terminal and/or axillary, dichasial, with long cincinnate branches, many-flowered. Flowers 5-merous, (sub-)sessile; calyx fused at the base only; corolla urceolate, white to pinkish; corolla tube often constrained at mouth; corolla lobes somewhat shorter than corolla tube; stigma small, subcapitate to slightly bilobed. Capsule bilobed or mitre-shaped with 2 horns apically. Seeds many, ellipsoid to globose, unwinged. Colleters present in leaf and bract axils, absent from calyx.

Six species, with one widespread tropical species (Neotropics, west Africa, tropical Asia, to Australia) and five species endemic to northern Vietnam, Mt. Kinabalu (Borneo), northern Madagascar, central Asia, and southern United States, respectively.

10. ***Orianthera*** C.S.P. Foster & B.J. Conn

Orianthera C.S.P. Foster & B.J. Conn, Telopea 16: 149–158 (2014); Conn, Telopea 5: 657–692 (1994) (as *Logania* sect. *Stomandra*).

Herbs or small shrubs. Leaves sometimes appearing bract- to scale-like. Inflorescences terminal, often appearing axillary, one- or 2–15-flowered variously reduced dichasial or modified botryoidal clusters. Flowers usually bisexual, hypogynous, 5-merous; calyx deeply divided; corolla campanulate, white or yellow; corolla lobes longer than tube; stigma subglobose (capitate) or obloid. Capsule subglobose or ellipsoid, septicidal with two carpels almost completely separating. Seeds many, ± ellipsoid, brown.

13 species, all endemic to Australia, closely related to *Mitreola* (Foster et al. 2014).

11. ***Phyllangium*** Dunlop

Phyllangium Dunlop in B.J. Conn, E.A. Brown & C. Dunlop (eds.) Loganiaceae. Fl. Austr. 28: 315 (1996).

Annual terrestrial or aquatic herbs to 25 cm high, glabrous or hairy. Inflorescences terminal, incompletely pleiochasial, few- to many-flowered. Flowers 4-merous, enclosed in a 2-lobed foliaceous involucre; calyx absent; corolla white; corolla lobes shorter than tube; styles 2, separate, sometimes cohering in bud or connate distally; stigma(s) capitate or truncate. Capsules enclosed in persistent involucre, 2-horned (mitre-shaped) apically, horns (style bases) short, dehiscing along ventral suture (septicidal). Seeds many, angular.

Five species endemic to temperate Australia, closely related to *Schizacme*.

12. ***Schizacme*** Dunlop

Schizacme Dunlop in B.J. Conn, E.A. Brown & C. Dunlop (eds.) Loganiaceae, Fl. Austr. 28: 314 (1996); Gibbons, Conn & Henwood, Telopea 17: 363–381 (2014), rev.

Perennial mat- or cushion-forming subshrubs to 8 cm high, glabrous or hairy. Flowers solitary, terminal, subtended by paired, sepal-like or leaf-like bracts, 4-merous; calyx deeply divided; calyx lobes heteromorphic or ± equal in size and shape; corolla tubular or campanulate, white or translucent; styles 2, erect, connivent; stigmas capitate. Capsule mitre-shaped, distal (free) portion of carpels somewhat or very broadly divaricate, distally winged or with wings forming a sheath between the locules, dehiscing along ventral suture (septicidal). Seeds few, ± elliptic.

Five species of predominantly alpine and montane habitats in southeastern Australia and New Zealand; closely related to *Phyllangium* (Gibbons et al. 2014).

III. Tribe Spigelieae

Tribe Spigelieae Dum., Anal. Fam. Pl.: 26 (1829).
Tribe Coelostyleae Endl. (1840).
Subtribe Spigeliinae Griseb. (1861) ("Spigelieae").

13. ***Spigelia*** L. Fig. 78A

Spigelia L., Sp. Pl. 1: 149 (1753); Leeuwenberg, Acta Bot. Neerl. 10: 460–465 (1961), reg. rev.; Smith, Wrightia 2(2): 90–102 (1960), reg. rev.; Guimarães & Fontella-Pereira, Loefgrenia 34: 1–18 (1969), reg. rev.; Henrickson, Sida 17: 89–103 (1996); Gould & Jansen, Lundellia 2: 1–13 (1999), reg. rev.; Popovkin et al., Phytokeys 6: 47–65 (2011), reg. rev.
Coelostylis Torr. & A. Gray ex Endl. (1839).
Pseudospigelia Klett (1923).

Annual or perennial herbs (rarely subshrubs or shrubs) up to 1 m high, glabrous or sparsely hairy (3 spp. densely stellate-pubescent). Inflorescence terminal, monochasial (dichasial at the base), one-, few- or many-flowered. Flowers 5-merous; calyx fused at the base or with free sepals; corolla tubular to funnel-shaped, often constricted at mouth, white, yellow, purplish, to red; corolla lobes distinctly shorter than corolla tube; stigma small, entire to slightly bilobed. Capsule bilobed apically, septicidally, loculicidally, and circumscissily dehiscent, leaving a hardened, persistent disk on rachis, and slightly flattened with many compressed, angular to globose unwinged seeds. Colleters present in leaf and bract axils, absent from calyx.

The number of species in this genus is uncertain (60–85) due to the large number of recently described species from South America and no recent global revision. *Spigelia* is distributed from the southern United States of America to Argentina, including the Caribbean, with high diversity in eastern Brazil; *Spigelia anthelmia* L. has spread to tropical Africa, Malaysia, and Indonesia. The phylogenetic position of *Spigelia* is uncertain (see discussion above).

IV. Tribe Strychneae

Tribe Strychneae Dumort., Anal. Fam. Pl.: 26 (1829).
Tribe Gardnerieae Endl. (1841).
Subtribe Strychninae Progel (1865) ("Strychneae").

14. *Gardneria* Wall.

Gardneria Wall., Fl. Ind. 1: 400 (1820).

Shrubs, mostly creeping or climbing, glabrous. Inflorescence axillary, dichasial, few-flowered (rarely many-flowered or flowers solitary). Flowers 4- or 5-merous; calyx fused at the base; corolla rotate, creamy to yellow; corolla lobes much longer than corolla tube; stigma capitate to bilobed. Berry globose, red, few-seeded. Seeds elliptic to orbicular, concave on the hilar side, convex on the other, not winged. Colleters present inside calyx.

About five species in Southeast and East Asia, from India to Japan, Malaysia, and Indonesia (Java).

15. *Neuburgia* Blume

Neuburgia Blume, Mus. Bot. 1: 156 (1850); Leenhouts, Fl. Mal. 6: 363–369 (1962), reg. rev.; Conn, Handbk Fl. Papua New Guinea 3: 174–182; Molina & Struwe, Austr. Syst. Bot. 17: 399–406 (2004), reg. rev.
Couthovia A. Gray (1859).

Trees or shrubs, up to 40 m high, glabrous or with simple hairs. Stipules interpetiolar, deltoid to obdeltoid. Leaves subsessile or petiolate, decussate. Inflorescence terminal, cymose, usually compact, distal flowers forming a cincinnus, many-flowered; bracts semi-amplexicaulous, shortly fimbriate. Flowers more or less sessile, calyx, corolla and androecium 5-merous (calyx rarely 4-merous); sepals fused at the base; corolla rotate to salver-shaped, white; corolla lobes shorter to longer than tube; stamens included; ovary mostly glabrous, 2-locular; ovules many; style often caducous; stigma mostly ellipsoid to club-shaped, faintly grooved, hollow (in *N. sarcantha* (Gilg & Gilg-Ben.) Leenh. broadly truncate). Drupe spindle-shaped, orange to red or white, with one, spindle-shaped, winged seed per locule (usually only one seed per fruit). Colleters present inside the calyx.

About 12 species, in Malesia (from the Philippines and Indonesia to New Guinea) and western Pacific (Caroline Islands, Solomon Islands, Vanuatu, New Caledonia, and Fiji).

16. *Strychnos* L. Fig. 78C

Strychnos L., Sp. Pl. 1: 189 (1753); Leenhouts, Fl. Mal. 6: 343–361 (1962), reg. rev.; Leeuwenberg, Meded. Landbouwhogeschool 69: 1–316 (1969); Krukoff, Lloydia 35(3): 193–271 (1972), reg. rev.; Adebowale et al. (2016), reg. phyl.
Scyphostrychnos S. Moore (1913).

Lianas, shrubs, or trees up to at least 20 m, glabrous or with simple hairs, unarmed or armed with axillary or terminal spines, sometimes with tendrils formed by leaves, venation mostly acrodromous. Inflorescence axillary or terminal, dichasial or thyrsoid, 1- to many-flowered, lax or congested. Flowers 4- or 5-merous, actinomorphic or with only sepals unequal; calyx fused at base, equal to unequal; corolla rotate to salver-shaped, white to yellowish, greenish, pale green, or rarely orange or yellowish-brown; corolla lobes shorter or longer than corolla tube; stigma capitate, rarely slightly bilobed or conical. Berry globose or nearly so, variable in size, yellow to red, green, brown or bluish-black, sometimes woody. Seeds one to many, disk-shaped to subglobose, rarely winged. Colleters sometimes present in leaf and bract axils and on the inner side of the calyx.

About 200 species, circumtropical, genus divided into 12 sections, many of which do not appear to be monophyletic (Frasier 2008; Adebowale et al. 2016).

Selected Bibliography

Adebowale, A., Lamb, J., Nicholas, A., Naidoo, Y. 2016. Molecular systematics of southern African monkey orange *Strychnos* L. (Loganiaceae). Kew Bull. 71: 1–16.

APG [Angiosperm Phylogeny Group] 2003. An update of the Angiosperm Phylogeny Group classification for the orders and families of flowering plants: APG II. Bot. J. Linnean Soc. 141: 399–436.

Backlund, M., Oxelman, B., Bremer, B. 2000. Phylogenetic relationships within the Gentianales based on *ndh*F and *rbc*L sequences, with particular reference to Loganiaceae. Amer. J. Bot. 87: 1029–1043.

Baillon, H.E. 1880. Sur la tribu des Labordiées. Bull. Mensuel, Soc. Linnéenne de Paris 1: 238–240.

Bedi, Y.S., Bir, S.S., Gill, B.S. 1980. In: Löve, A. (ed.) IOPB Chromosome reports LXVII. Taxon 29: 353–355.

Bendre, A.M. 1975. Patterns of nodal anatomy in Loganiaceae. Curr. Sci. 44(3): 99.

Bisset, N.G. 1974. The Asian species of *Strychnos*. Part III. The Ethnobotany. Lloydia 37: 63–107.

Bisset, N.G. 1980a. Loganiaceae. Phytochemistry. In: Natürl. Pflanzenfam. ed. 2, 28b(1). Berlin: Duncker & Humblot, pp. 211–237.

Bisset, N.G. 1980b. Loganiaceae. Useful plants. In: Natürl. Pflanzenfam., ed. 2, 28b (1). Berlin: Duncker & Humblot, pp. 238–244.

Booker, S.G., Cooper, R.C. 1961. New Zealand Medicinal Plants. Auckland: Unity Press.

Borgmann, E. 1964. Anteil der Polyploiden in der Flora des Bismarckgebirges von Ostneuguinea. Z. Bot. 52: 118–177.

Bremer, B., Olmstead, R.G., Struwe, L, Sweere, J.A. 1994. *rbc*L sequences support exclusion of *Retzia*, *Desfontainia*, and *Nicodemia* from the Gentianales. Pl. Syst. Evol. 190: 213–230.

Bremer, B., Bremer, K., Heidari, N., Erixon, P., Olmstead, R.G., Anderberg, A.A., Källersjö, M., Barkhordarian, E. 2002. Phylogenetics of asterids based on 3 coding and 3 non-coding chloroplast DNA markers and the utility of non-coding DNA at higher taxonomic levels. Mol. Phyl. Evol. 24: 274–301.

Carr, G.D. 1978. Chromosome numbers of Hawaiian flowering plants and the significance of cytology in selected taxa. Amer. J. Bot. 65: 236–242.

Castro, I., Robertson, A.W. 1997. Honeyeaters and the New Zealand forest flora: The utilization and profitability of small flowers. New Zeal. J. Ecol. 21: 169–179.

Conn, B.J. 1980. A taxonomic revision of *Geniostoma* subg. *Geniostoma* (Loganiaceae). Blumea 26: 245–364.

Conn, B.J. 1993. A new species of *Geniostoma* (Loganiaceae) from Lord Howe Island. Telopea 5: 301–304.

Conn, B.J. 1995. Taxonomic revision of *Logania* section *Logania* (Loganiaceae). Austr. Syst. Bot. 8: 585–665.

Conn, B.J., Brown, E.A, Dunlop, C.R. 1996. Loganiaceae. In: Wilson, A. Flora of Australia, vol. 28: Gentianales. Melbourne: CSIRO, pp. 1–72.

Corner, E.J.H. 1976. The seeds of dicotyledons. 2 vols. London: Cambridge University Press.

Dahlgren, K.V.O. 1922. Die Embryologie der Loganiaceengattung *Spigelia*. Sv. Bot. Tidskr. 16: 77–87.

de Lange, P.J., Rolfe, J.R., Champion, P.D., Courtney, S.P., Heenan, P.B., Barkla, J.W., Cameron, E.K., Norton, D.A., Hitchmough, R.A. 2013. Conservation status of New Zealand indigenous vascular plants, 2012. New Zealand Threat Classification Series 3. Wellington: Department of Conservation.

Endress, P.K., Jenny, M., Fallen, M.E. 1983. Convergent elaboration of apocarpous gynoecia in higher advanced dicotyledons (Sapindales, Malvales, Gentianales). Nord. J. Bot. 3: 293–300.

Endress, M.E., Sennblad, B., Nilsson, S., Civeyrel, L., Chase, M.W., Huysmans, S., Grafström, E. Bremer, B. 1996. A phylogenetic analysis of Apocynaceae s. str. and some related taxa in Gentianales: a multidisciplinary approach. In: Robbrecht, E., Puff, C., Smets, E. (eds.) Second Int. Rubiaceae Conference Proceedings, Opera Bot. Belg. 7: 59–102.

Endress, M.E., Igersheim, A., Garnock-Jones, P. 1998. Sexual dimorphism and secondary pollen presentation in *Geniostoma ligustrifolium* (Geniostomaceae). Amer. J. Bot. 85, suppl.: 126.

Erbar, C., Leins, P. 1999. Secondary pollen presentation and a curious rupture of the style in *Spigelia* (Spigeliaceae, Gentianales). Pl. Biol. (Stuttgart) 1: 389–402.

Foster, C., Conn, B.J., Henwood, M.J., Ho, S. 2014. Molecular data support *Orianthera*: a new genus of Australian Loganiaceae. Telopea 16: 149–158.

Frasier, C.L. 2008. Evolution and systematics of the angiosperm order Gentianales with an in-depth focus on Loganiaceae and its species-rich and toxic genus *Strychnos*. Ph.D. Thesis, Rutgers University, New Brunswick, NJ, USA.

Gadella, T.W.J. 1962. Some cytological studies in the Loganiaceae. Acta Bot. Neerl. 11: 51–55.

Gadella, T.W.J. 1963. Some cytological studies in the Loganiaceae II. Verh. Proc. Kon. Ned. Akad. Wetensch., Ser. C, 66: 265–269.

Gadella, T.W.J. 1967. Some cytological studies in the Loganiaceae IV. Verh. Proc. Kon. Ned. Akad. Wetensch., Ser. C, 70: 302–304.

Gadella, T.W.J. 1969. Chromosome numbers of some Angiospermae collected in Cameroun and the Ivory Coast. Proc. Kon. Ned. Akad. Wetensch., Ser. C, 72: 306–310.

Gadella, T.W.J. 1970. Chromosome numbers of some angiosperms collected in Cameroun and the Ivory Coast II. Acta Bot. Neerl. 19: 431–435.

Gadella, T.W.J. 1972. Cytological studies on some flowering plants collected in Africa. Bull. Jard. Bot. État 42: 393–402.

Gadella, T.W.J. 1977. In: Löve, A. (ed.) IOPB Chromosome number reports LVI. Taxon 26: 257–274.

Gadella, T.W.J. 1978. In: Löve, A. (ed.) IOPB Chromosome number reports LXI. Taxon 27: 392.

Gadella, T.W.J. 1980. Loganiaceae. Cytology. In: Natürl. Pflanzenfam., ed. 2, 28b(1). Berlin: Duncker & Humblot, pp. 202–210.

Gibbons, K.L., Conn, B.J., Henwood, M.J. 2012. Phylogenetic relationships in Loganieae (Loganiaceae) inferred from nuclear ribosomal and chloroplast DNA sequence data. Austr. Syst. Bot. 25: 331–340.

Gibbons, K.L., Conn, B.J., Henwood, M.J. 2014. The Australasian genus *Schizacme* (Loganiaceae): new

combinations and new species in the New Zealand flora. Telopea 17: 363–381.
Gould, K.R. 1997. Systematic studies in *Spigelia*. Ph.D. dissertation, The University of Texas at Austin, Austin, TX.
Grant, J.R. 2009. A revision of neotropical *Bonyunia* (Loganiaceae: Antonieae). Ann. Missouri Bot. Gard. 96: 541–563.
Hair, J.B. 1963. Contributions to a chromosome atlas of the New Zealand flora. 6. New Zeal. J. Bot. 1: 243–257.
Hakki, M.I. 1980. Loganiaceae. Embryology. In: Natürl. Pflanzenfam., ed. 2, 28b(1). Berlin: Duncker & Humblot, pp. 192–201.
Hakki, M.I. 1998. On floral morphology and embryology of *Usteria guineensis* Willd. (Loganiaceae). Bot. Jahrb. Syst. 120: 275–293.
Hallé, N., Florence, J. 1986. Description de 10 espèces rares de plantes à fleures de l'ile de Rapa. In: Rapa: Direction des centres d'expérimentations nucléaires, pp. 129–149. Service Mixte de Contrôle Biologique, ORSTOM, France.
Hilton-Taylor, C. 2000. 2000 IUCN Red List of Threatened Species. International Union for Conservation of Nature and Natural Resources, Gland, Switzerland and Cambridge, UK.
Jensen, S.R. 1992. Systematic implications of the distribution of iridoids and other chemical compounds in the Loganiaceae and other families of the Asteridae. Ann. Missouri Bot. Gard. 79: 284–302.
Keighery, G.J. 1975. In: Löve, A. (ed.) IOPB Chromosome number reports XLIX. Taxon 24: 501–516.
Kerharo, J., Adams, J.G. 1974. La pharmacopée sénégalaise traditionnelle. Plantes médicinales et toxiques. Paris: Vigot.
Leenhouts, P.W. 1962 [1963]. Loganiaceae. In: van Steenis, C.G.G.J. (ed.) Flora Malesiana, Ser. 1, vol. 6(2). Groningen: Wolters-Noordhoff, pp. 293–387.
Leeuwenberg, A.J.M., Leenhouts, P.W. 1980. Loganiaceae. Taxonomy. In: Natürl. Pflanzenfam., ed. 2, 28b(1). Berlin: Duncker & Humblot, pp. 22–96.
Lewis, W.H., Stripling, H.L., Ross, R.G. 1962. Chromosome numbers for some angiosperms of the southern United States and Mexico. Rhodora 64: 147–161.
Li, P.-T., Leeuwenberg, A.J.M. 1996. Loganiaceae. In: Raven, P.H., Wu, Z. (eds.) Flora of China 14. Beijing: Science Press, pp. 320–338.
Machado, C.G., Coelho, A.G., Santana, C.S., Rodrigues, M. 2007. Hummingbirds and their flowers in the 'campos rupestres' of Chapada Diamantina, Bahia, northeastern Brazil. Rev. Brasil. Ornith. 15: 267–279.
Mangenot, S., Mangenot, G. 1957. Deuxième liste nombres chromosomiques nouveaux chez diverses dicotylédones et monocotylédones d'Afrique occidentale. Bull. Jard. Bot. État 27: 639–654.
Mangenot, S., Mangenot, G. 1958. Deuxième liste nombres chromosomiques nouveaux chez diverses dicotylédones et monocotylédones d'Afrique occidentale. Bull. Jard. Bot. État 28: 315–329.
Mangenot, S., Mangenot, G. 1962. Enquête sur les nombres chromosomiques nouveaux dans une collection d'espèces tropicales. Rev. Cytol. Biol. Vég. 25: 411–447.
Mann, H. 1866. Enumeration of Hawaiian plants. Proc. Amer. Acad. Arts Sci. 7: 196–197.
Mennega, A.M.W. 1980. Loganiaceae. Anatomy of the secondary xylem. In: Natürl. Pflanzenfam., ed. 2, 28b(1). Berlin: Duncker & Humblot, pp. 112–161.
Meyer, J.-Y. 1996. Espèces et espaces menacés de la Société et des Marquises. Contribution à la biodiversité de Polynésie française N°1–5. Délégation à l'Environnement/Délégation à la Recherche, Papeete, Tahiti.
Miège, J. 1960. Nombres chromosomiques de plantes d'Afrique occidentale. Rev. Cytol. Biol. Vég. 21: 373–384.
Mohrbutter, C. 1936. Embryologische studien an Loganiaceen. Planta 26: 64–80.
Molina, J., Struwe, L. 2004. *Neuburgia novocaledonica*, comb. nov. and the first record of domatia in the family Loganiaceae. Austr. Syst. Bot. 17: 399–406.
Moore, R.J. 1947. Cytotaxonomic studies of the Loganiaceae I. Chromosome numbers and phylogeny in the Loganiaceae. Amer. J. Bot. 34: 527–538.
Mori, S.A., Brown, J.L. 1994. Report on wind dispersal in a lowland moist forest in central French Guiana. Brittonia 46: 105–125.
Mori, S.A., Cremers, G., Gracie, C., de Granville, J.-J., Hoff, M., Mitchell, J.D. 2002. Guide to the Vascular Plants of Central French Guiana. Part 2. Dicotyledons. Mem. N. Y. Bot. Garden 76(2).
Motley, T.J., Carr, G.D. 1998. Artificial hybridization in the Hawaiian endemic genus *Labordia* (Loganiaceae). Amer. J. Bot. 85: 654–660.
Motley, T.J., Cross, H.B. 1999. Phylogeny, biogeography, and breeding system evolution among Hawaiian *Labordia* and Pacific (New Zealand) *Geniostoma* species (Loganiaceae). XVI International Botanical Congress, St. Louis, MO: abstract 66.
Netolitzky, F. 1926. Anatomie der Angiospermensamen. Handbuch der Pflanzenanatomie, II. Bd. 10 (Berlin): 266–267.
Olmstead, R.G., Bremer, B., Scott, K.M., & Palmer, J.D. 1993. A parsimony analysis of the Asteridae sensu lato based on rbcL sequences. Ann. Missouri Bot. Gard. 80: 700–722.
Olmstead, R.G, dePamphilis, C.W., Wolfe, A.D., Young, N. D., Elisons, W.J., Reeves, P.A. 2001. Disintegration of the Scrophulariaceae. Amer. J. Bot. 88: 348–361.
Oxelman, B., Bremer, B. 2000. Discovery of paralogous nuclear gene sequences coding for the second-largest subunit of RNA polymerase II (RPB2) and their phylogenetic utility in Gentianales of the asterids. Mol. Biol. Evol. 17: 1131–1145.
Oxelman, B., Backlund, B., Bremer, B. 1999. Relationships of Buddlejaceae s.l. investigated using parsimony jackknife and branch support analysis of chloroplast *ndh*F and *rbc*L sequence data. Syst. Bot. 24: 164–182.
Popovkin, A.V., Mathews, K.G., Mendes Santos, J.C., Carmen Molina, M. Struwe, L. 2011. *Spigelia genuflexa* (Loganiaceae), a new geocarpic species from the Atlantic forest of northeastern Bahia, Brazil. PhytoKeys 6: 47–65.
Punt, W. 1980. Pollen morphology. In: Natürl. Pflanzenfam., ed. 2, 28b(1). Berlin: Duncker and Humblot, pp. 162–191.
Quattrocchi, U. 2016. CRC World Dictionary of Medicinal and Poisonous Plants: Common Names, Scientific

Names, Eponyms, Synonyms, and Etymology. CRC Press: Boca Raton.
Rageau, J. 1973. Les plantes médicinales de la Nouvelle-Calédonie. Trav. Doc. ORSTOM 23: 91.
Raghavan, R.S. 1959. Chromosome numbers in Indian plants III. Proc. Indian Acad. Sci. B 49: 239–244.
Rattenbury, J.A. 1980. Dioecy in *Geniostoma ligustrifolium*. Auckland Bot. Soc. Newsletter January 1980: 8–10.
Reddy, A.S., Rao, Y.B.N., Khanam, A., Durga, I.S. 1999. A contribution to the embryology of *Mitreola oldenlandioides* Wall. Phytomorphology 49: 283–287.
Rock, J.F. 1913. The indigenous trees of the Hawaiian Islands. Published privately, Honolulu. Pp. 401–406, 536. [Republished, with introduction by S. Carlquist and addendum by D.R. Herbst, 1974, by Charles E. Tuttle Co., Rutland, VT]
Rogers, G.K. 1986. The genera of Loganiaceae in the southeastern United States. J. Arnold Arbor. 67: 143–185.
Scherff, E.E. 1939. Genus *Labordia*. Field Mus. Nat. Hist., Bot. Ser. XVII (6): 449–546.
Schmelzer, G.H., Ameenah Gurib-Fakim (eds.) 2008. Medicinal Plants, vol. 1. Wageningen, Netherlands: PROTA.
Schultes, R.E. 1977. De plantis toxicariis e mundo novo tropicale commentationes XV. *Desfontainia*: a new Andean hallucinogen. Harvard Bot. Mus. Leafl. 25: 99–104.
Selling, O.H. 1947. Studies in Hawaiian pollen statistics, Part II. B. P. Bishop Museum Special Publ. 38: 287–290.
Sinnott, E.W. 1914. Investigations on the phylogeny of the angiosperms. I. The anatomy of the node as an aid in the classification of angiosperms. Amer. J. Bot. 1(7): 303–322, pl. 30–35.
Skottsberg, C. 1936. Vascular plants from the Hawaiian Islands. II. Acta Horti Gotoburgensis 10: 97–193.
Smith, N.M. 1991. Ethnobotanical field notes from the Northern Territory, Australia. J. Adelaide Bot. Gard. 14: 1–44.
Smith, A.C., Stone, B.C. 1962. Studies of Pacific Island plants. XVII. The genus *Geniostoma* (Loganiaceae) in the New Hebrides, Fiji, Samoa, and Tonga. Contr. U. S. Natl. Herb. 37: 1–42.
Solereder, H. 1892–1895. Loganiaceae. In: Natürl. Pflanzenfam. IV, 2: 19–48 (1892), 49–50 (1895). Leipzig: W. Engelmann.
Struwe, L. 2004. Loganiaceae (Logania or Strychnine family). In: Smith, N., Mori, S.A., Henderson, N., Stevenson, D.W., Heald, S.V. (eds.) Flowering plants of the Neotropics. Princeton, New Jersey: The New York Botanical Garden Press and Princeton University Press, pp. 219–221.
Struwe, L., Albert, V.A. 1997. Floristics, cladistics, and classification: three case studies in Gentianales. In: Dransfield, J., Coode, M.J.E., Simpson, D.A. (eds.) Plant diversity in Malesia III. Kew: Royal Botanic Gardens, pp.321–352.
Struwe, L., Albert, V.A. (eds.) 2002. Gentianaceae: Systematics and natural history. Cambridge: Cambridge University Press.
Struwe, L., Albert, V.A., Bremer, B. 1994 [1995]. Cladistics and family level classification of the Gentianales. Cladistics 10: 175–206.
Tiagi, Y.D., Kshetrapal, S. 1972. Studies on the floral anatomy, evolution of the gynoecium and the relationships of the family Loganiaceae. Adv. Pl. Morph. 1972: 408–416.
USFWS (United States Fish and Wildlife Services) 2002. Species Information Threatened and Endangered Animals and Plants homepage, http://endangered.fws.gov/ [accessed June 12, 2002]
Valeton, T. 1902. Einige Notizen über neue und schon bekannte Arten der Gattung *Geniostoma*. Bull. Jard. Bot. Buitenzorg 12: 1–29.
Vinckier, S., Smets, E. 2002. Morphology, ultrastructure and typology of orbicules in family Loganiaceae s. l. and related genera, in relation to systematics. Rev. Palaeobot. Palyn. 119: 161–189.
Wagner, W.L., Herbst, D.R., Sohmer, S.H. 1990. Loganiaceae. In: Wagner, W.L., Herbst, D.R., Sohmer, S.H. Manual of the flowering plants of Hawaii. Honolulu, HI: University of Hawaii Press and Bishop Museum Press, pp. 850–864.
Whistler, W.A. 1990. Ethnobotany of the Cook Islands: The plants, their Maori names, and their uses. Allertonia 5: 393–394.
Wightman, G., Dilkbarri, D., Williams, L., Dalywaters, I. 1992. Mudburra ethnobotany Aboriginal plant use from Kulumindini (Elliot), Northern Australia. NT Bot. Bull. 14: 1–44.
Yamazaki, T. 1963. Embryology of *Mitrasacme alsinoides* var. *indica*. Sci. Rep. Tohoku Imp. Univ., Ser. 4, Boil. 29: 201–205.
Yang, L-L., Li, H-L., Wei, L., Yang, T., Kuang, D-Y., Li, M-H., Liao, Y-Y., Chen, Z-D., Wu, H., Zhang, S-Z. 2016. A supermatrix approach provides a comprehensive genus-level phylogeny for Gentianales. J. Syst. Evol. 54: 400–415.
Zepernick, B. 1972. Arzneipflanzen der Polynesier. Baessler Arch. Beih. 8: 69.

Myodocarpaceae

Myodocarpaceae Doweld, Tent. Syst. Pl. Vasc. lii (2001).

P.P. Lowry II and G.M. Plunkett

Shrubs to small trees, glabrous and unarmed; schizogenous secretory canals throughout the plant; plants terrestrial, evergreen; stems monocaulous or sparsely to well branched, pachycaulous; mating system andromonoecious (rarely hermaphroditic). Leaves alternate, frequently heteroblastic; petioles sheathing at the base, exstipulate, sometimes alate; blade simple or pinnately compound, with entire or toothed margins; venation pinnate. Inflorescences terminal, paniculate, the ultimate units umbellate; inflorescence axes subtended by small to foliose bracts; flowers subtended by an involucel of bracteoles; pedicels articulate. Flowers either perfect or staminate, epigynous, actinomorphic; perianth 5-merous; calyx forming a short low tube with evident lobes, valvate or imbricate; petals imbricate (rarely valvate), sometimes calyptrate, the bases clawed or broadly inserted; stamens 5, in a single whorl, alternipetalous, anthers dorsifixed, introrse, tetrasporangiate, each pollen sac dehiscing by a longitudinal slit; filaments filiform to stout, inflexed in bud; ovary syncarpous, of 2 carpels (vestigial in staminate flowers), each carpel unilocular with apical placentation; stigmas on a distinct style; styles free, geniculate or not, sometimes swollen at the base and confluent with the nectariferous disc (reduced in staminate flowers); ovules anatropous, pendulous, one per locule, unitegmic. Fruits simple, either drupaceous with a fleshy mesocarp and a separate, variously sclerified endocarp (pyrene) around each locule, or dry schizocarpic with two mericarps borne on a carpophore; numerous secretory vesicles present in the exocarp and canals in the mesocarp and sometimes the endocarp. Seeds straight; endosperm copious, oily, uniform; embryo minute but well differentiated.

A family comprising two genera and 17 species, all endemic to New Caledonia, except *Delarbrea michieana* (Queensland, Australia) and *D. paradoxa* subsp. *paradoxa* (widespread in Melanesia and eastern Malesia).

Vegetative Morphology and Anatomy. The two genera of the family are terrestrial woody perennials, either shrubs or small trees (Lowry 1986a, 1986b; Lowry et al. 2004). The stems are pachycaulous, lacking discernible pith. The growth pattern is sympodial, in which each shoot terminates in an inflorescence, leaving subsequent growth to develop from the axillary meristems of the upper leaves. In monocaulous individuals, sympodial growth involves production of a single axillary branch (Chamberlain's model; Hallé et al. 1978), whereas a branched habit results from growth involving two or more axillary meristems (Leeuwenberg's model). The leaves are alternate and often clustered at the apex of the branches. They are petiolate, and the bases are sheathing and sometimes alate. The lamina is generally large and may be either simple (*Myodocarpus*) or pinnately compound (both genera). The leaf or leaflet margins are entire to serrate, dentate, or incised, with pinnate venation. Leaf heteroblasty is found in all species of both genera (Lowry 1986a, 1986b; Lowry et al. 2004), with juvenile foliage often deeply lacerate.

The plant contains schizogenous secretory canals. In the leaves of *Myodocarpus* (Viguier 1906), the petiole and laminar midrib bear an epidermis with a strong cuticle surrounding a region of very thick collenchyma cells. To the inside of these layers is found an irregular ring of discrete vascular bundles separated by large bands of parenchyma. This ring surrounds a series of additional, scattered vascular bundles. Secretory canals are scattered throughout the outer collenchyma (where the canals are 30–50 µm in diam.) and the inner parenchyma (where

J.W. Kadereit, V. Bittrich (eds.), *Flowering Plants. Eudicots*, The Families and Genera of Vascular Plants 15,
https://doi.org/10.1007/978-3-319-93605-5_9

they are 100–200 μm in diam.). In the lamina, the epidermis has a raised cuticle and a subepidermal hypodermis of collenchymatous cells, which typically ranges from one to several layers (in especially coriaceous taxa), but is sometimes absent. The leaf anatomy of *Delarbrea* is similar, but in species such as *D. harmsii*, the petiole has a subepidermal meristem that produces a thick corky and lenticellate tissue more typical of stems, and the vascular bundles form a continuous ring and develop abundant secondary tissues (similar to stems). Both the bundle caps and the outer pith tissue to the inside of the vascular bundles are strongly lignified. Secretory canals are sparse in the pith and bundle caps, and are very small (20–30 μm diam.). In the stem of *Myodocarpus* (Viguier 1906), there is a poorly differentiated collenchyma layer whose cells bear numerous druse crystals. Large secretory canals (100–150 μm) are found in the inner bark, the vascular bundle caps (which are differentiated into thin, fibrous arcs), and the pith (which is very early lignified toward the periphery), but are lacking in the phloem. In the wood (Oskolski et al. 1997), growth rings are lacking or indistinctly marked. Vessels are rounded (rarely angular), small, solitary or solitary mixed, with variable proportions of radial and sometimes also tangential multiples of 2–12; vessel walls 2–6 μm thick; tyloses lacking; vessel elements ca. 300–1500 μm; perforation plates simple or scalariform with many bars; intervessel pits and vessel ray pits alternate or opposite, rarely scalariform; helical thickenings almost always absent. Vasicentric and vascular tracheids are almost always lacking. Fibers libriform, having thick to very thick walls, septate or aseptate; bordered pits simple to minutely bordered. Axial parenchyma is apotracheal, diffuse and diffuse-in-aggregates. Rays are 2–5 per mm, uniseriate or multiseriate (2–6 cells wide), mostly of procumbent cells, occasionally with 1 or 2 marginal rows of square cells; pits on tangential cells rounded, oval, or elongate. Radial canals are present, bordered by thin-walled epithelial cells. Crystals are lacking (Oskolski et al. 1997).

Inflorescence Structure and Floral Morphology. Myodocarpaceae have paniculate inflorescences with several degrees of branching, terminating in umbellules. The primary axis is terminal and erect or pendulous. Axes of all degrees are subtended by bracts, typically larger proximally (sometimes foliaceous and resembling small leaves) and progressively reduced distally, and sometimes caducous. Small to moderately sized bracteoles form an involucre subtending each umbellule. The pedicels are articulated below the flower.

The flowers are always epigynous. The calyx forms a short tube with five lobes whose margins are often scarious. The corolla is formed by five white to greenish-white (rarely yellowish, yellowish-green, or rose-pink) petals that are keeled within, imbricate (but valvate in *Delarbrea balansae*), with broad or clawed bases, and sometimes calyptrate (*Myodocarpus*); this calyptra is formed through postgenital fusion of petal surfaces by interlocking of epidermis papillae and cuticular ledges (Baumann 1946), and completely encloses the inner parts of the flower; the apex is plane, lacking the pointed adaxial thickening common in related families. The androecium comprises five alternipetalous stamens in a single whorl. The filaments are long and slender (*Myodocarpus*) to short and stout (*Delarbrea*), inflexed in bud but ascending to spreading at anthesis. The anthers are dorsifixed and longitudinally dehiscent, with two thecae (tetrasporangiate). The syncarpous gynoecium is formed of two carpels with an equal number of locules. The apex of the ovary is modified as a nectariferous disc (Erbar and Leins 2010), terminating in two free styles, either geniculate (*Myodocarpus*) or not (*Delarbrea*; Lowry 1986a). In *D. balansae*, an evident stipe is present at the base of the ovary and fruit (Lowry et al. 2004). Each locule encloses a single functional pendulous ovule with apical placentation. The ovary is vestigial in staminate flowers.

Fruit and Seed. The fruits of Myodocarpaceae differ markedly between the two genera (Lowry 1986a; Liu et al. 2010). In *Delarbrea*, they are drupes, terete or subterete in transverse section, and maturing to a dark purple or blackish color (an iridescent blue in *D. michieana*; Lee et al. 2000) and containing two dorsally compressed pyrenes. In *Myodocarpus*, the fruits are dry and schizocarpic; each of the two mericarps is strongly compressed laterally, and has a basal median wing. In both genera, the mesocarp is parenchymatous, but in *Delarbrea*, sclereids also occur, and in some species, the inner layer (closest to the endocarp) is lignified with thin-walled cells; sclereids and lignified parenchyma are lacking in the mesocarp of *Myodocarpus* (Liu et al.

2010). Each carpel is fed by a single ventral vascular bundle located at the commissure between the two carpels (except *D. balansae*, which has four ventral bundles, and *D. michieana*, which has two) and by a series of anastomosing peripheral vascular bundles (typically 7–15 per carpel, but indistinct in *M. fraxinifolius*). In *Myodocarpus*, the ventral bundles form a simple, uncleft carpophore (Liu et al. 2012), to which the mericarps are attached basally (unlike Apiaceae, in which the mericarps are attached apically). Companion canals (equivalent to the rib oil ducts of Apiaceae) are located to the outside of some of the peripheral vascular bundles. There are two additional series of secretory structures, including anastomosing canals (equivalent to anastomosing vittae in some Apiaceae) dispersed in the mesocarp, and large secretory vesicles, found in the innermost layer of the mesocarp, adjacent to the endocarp, deep within the fleshy fruits of *Delarbrea*, but just below the surface in the dry fruits of *Myodocarpus*. All three sets of secretory structures are lined by a single layer of secretory cells. Calcium oxalate crystals have been detected as druses scattered in the mesocarp in all species studied (Liu et al. 2010). The endocarps of both genera are woody, formed by 2 or 3 layers of fibers that are arranged longitudinally and transversely (sometimes also plagiotropically; Liu et al. 2010).

Each seed has copious, uniform endosperm, although the expansion of the secretory vesicles against the seed cavity may produce an appearance of "false rumination" (see Konstantinova and Yembaturova 2010). The testa is thick, well developed, and two-layered in *Delarbrea*, but compressed with no discernible layers in *Myodocarpus* (Konstantinova and Yembaturova 2010). The embryo is apical, small, and dicotyledonous.

Karyology. Chromosome numbers in both genera have been reported as $n = 12$ (Lowry 1986a; Yi et al. 2004).

Reproductive Systems. All species of Myodocarpaceae are andromonoecious, with the exception of *Delarbrea harmsii*, which is hermaphroditic (Lowry 1986a, 1986b), and andromonoecy appears to be ancestral in the family (see Schlessman et al. 1990, 2001; Schlessman 2010). In andromonoecious species, umbellules of perfect flowers terminate the lower order inflorescence axes and are subtended by umbellules of staminate flowers (rarely a few staminate flowers may also be present in otherwise perfect umbellules, and vice versa). The perfect flowers are strongly protandrous, with stamens (and petals) typically dropping before stigma receptivity. The separation of male and female phases is generally synchronized over the entire plant, leading to duodichogamy (i.e., 1.5 cycles of complete synchronous dichogamy), minimizing selfing in these apparently self-compatible species (Schlessman et al. 1990).

Dispersal. The dry, winged, schizocarpic fruits of *Myodocarpus* are wind dispersed, whereas the fleshy drupes characterizing *Delarbrea* appear to be adapted to zoochory.

phytochemistry. Very few data are available for the family. Lebouvier et al. (2014) analyzed the leaf, wood, and bark oils of three *Myodocarpus* spp. and found monoterpene hydrocarbons, various long chain alcohols, sesquiterpenic oils, and terpene alcohols, among other compounds.

Affinities. *Myodocarpus* and *Delarbrea* were described in separate families (Apiaceae and Araliaceae, respectively), but were later placed together as the sole members of a tribe of Araliaceae (Viguier 1906). Reflecting this early history, the two genera have often been viewed as intermediates that bridge the morphological gap between Araliaceae and Apiaceae (Baumann 1946; Thorne 1973). Vegetatively, they more closely resemble Araliaceae, with a woody habit and large, mostly subcoriaceous to coriaceous leaves (in contrast to most Apiaceae, which are largely herbaceous, often with membranaceous leaves). Reproductively, however, these genera show a mixture of features typical of each of these two families. Like most Araliaceae, the flowers of *Myodocarpus* have broadly inserted petals, but the fruits are dry schizocarps with free carpophores, similar (at least superficially) to many Apiaceae. Conversely, the flowers of *Delarbrea* have clawed petals (as in most Apiaceae), but fleshy, drupaceous fruits (typical of most Araliaceae). In some ways, the wood anatomy of Myodocarpaceae shows greater similarity to Cornaceae (Rodríguez 1957, 1971; Oskolski et al. 1997), and Rodríguez (1957) preferred to view *Myodocarpus* as a "primitive" group in Apiales rather than as a bridge between the other families.

Early phylogenetic studies based on molecular data (e.g., Plunkett et al. 1997) suggested that *Delarbrea* represented a distinct lineage in the order, placing it outside the core clades of both Araliaceae and Apiaceae. Plunkett and Lowry (2001), using much broader sampling from both genera, confirmed the monophyly of the group and suggested it be recognized as a new family. These and subsequent phylogenetic studies (Chandler and Plunkett 2004; Nicolas and Plunkett 2009) have placed Myodocarpaceae in a clade that also includes Apiaceae, Araliaceae, and Pittosporacae, most commonly as sister to Apiaceae. The four families comprise the core group of Apiales, defined by Plunkett et al. (2004) as suborder Apiineae. Despite the mixture of features found in Myodocarpaceae, the family can be defined by several synapomorphies, including wood anatomy (diffuse or diffuse-in-aggregates apotracheal axial parenchyma; Oskolski et al. 1997) and fruit structure (secretory vesicles located immediately adjacent to the endocarp; Konstantinova and Yembaturova 2010; Liu et al. 2010).

Distribution and Habitats. All but two species of the family are endemic to the Pacific island of New Caledonia. The only exceptions include *Delarbrea michieana*, which is endemic to Queensland, Australia, and *D. paradoxa* subsp. *paradoxa*, which ranges from Norfolk Island in the south, to New Caledonia and Vanuatu and northwest to the Solomon Islands and the Bismarck archipelago, through New Guinea to the Lesser Sunda Islands (Timor and Wetar) and the Moluccas (Aru and Halmahera). Most members of *Delarbrea* occur in low- to mid-elevation humid forests, although *D. paradoxa* subsp. *depauperata* is restricted to low-elevation maquis vegetation in NE New Caledonia. The species of *Myodocarpus* occur primarily in humid forests or maquis vegetation, from sea level to some of the highest summits of the island (over 1500 m elevation); several species are somewhat fire tolerant and are frequently found in secondary vegetation and early successional communities. On New Caledonia, all species are restricted to either ultramafic or non-ultramafic substrates (except *D. paradoxa* subsp. *paradoxa*, which occurs on both substrates in New Caledonia, and on non-ultramafic substrates elsewhere in its range).

Key to the Genera of Myodocarpaceae

1. Fruits fleshy, drupaceous, unwinged; the seeds dorsally compressed; petals free, abscising individually; styles not geniculate, leaves pinnately compound **1. *Delarbrea***
– Fruits dry, schizocarpic, each mericarp bearing a basal median wing; seeds laterally compressed; petals calyptrate, abscising as a single unit; styles geniculate; leaves simple or pinnately compound **2. *Myodocarpus***

1. *Delarbrea* Vieill. Fig. 79

Delarbrea Vieill., Bull. Soc. Linn. Normandie 9: 342 (1865); Lowry, Allertonia 4: 169–201 (1986), generic rev.
Porospermum F. Muell. (1870).
Pseudosciadium Baill. (1878).

Andromonoecious or hermaphroditic. Leaves pinnately compound. Inflorescence erect or

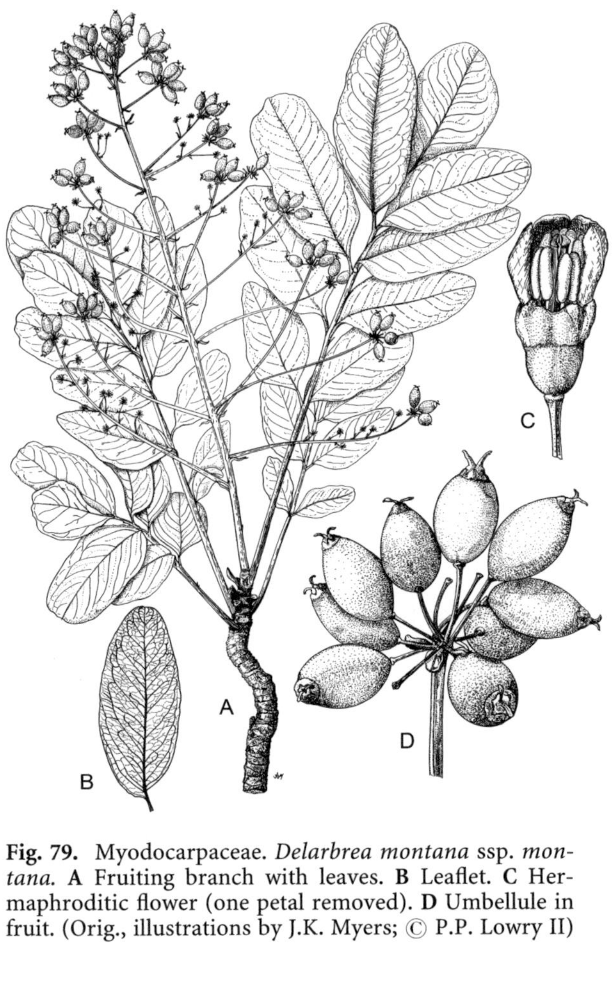

Fig. 79. Myodocarpaceae. *Delarbrea montana* ssp. *montana*. **A** Fruiting branch with leaves. **B** Leaflet. **C** Hermaphroditic flower (one petal removed). **D** Umbellule in fruit. (Orig., illustrations by J.K. Myers; © P.P. Lowry II)

pendent, the pedicels free or basally united into groups of 2–4. Petals free, distinctly clawed toward the base. Styles not geniculate. Fruits fleshy, drupaceous, terete, lacking wings, turning dark purple to black or iridescent blue at maturity, the seeds compressed dorsally. $n = 12$.

Seven species. Australasia (Queensland, Malesia, and Melanesia, especially New Caledonia).

2. ***Myodocarpus*** Brongn. & Gris Fig. 80

Myodocarpus Brongn. & Gris, Bull. Soc. Bot. France 8: 123 (1861).

Andromonoecious. Leaves simple or pinnately compound. Inflorescence erect, the pedicels free. Petals with a broad base, forming a calyptra that falls off at anthesis. Styles geniculate. Fruits dry, schizocarpic with an evident carpophore, strongly compressed laterally, bearing a basal median wing; turning brownish at maturity, the seeds compressed laterally. $n = 12$.

Ten species. Endemic to New Caledonia.

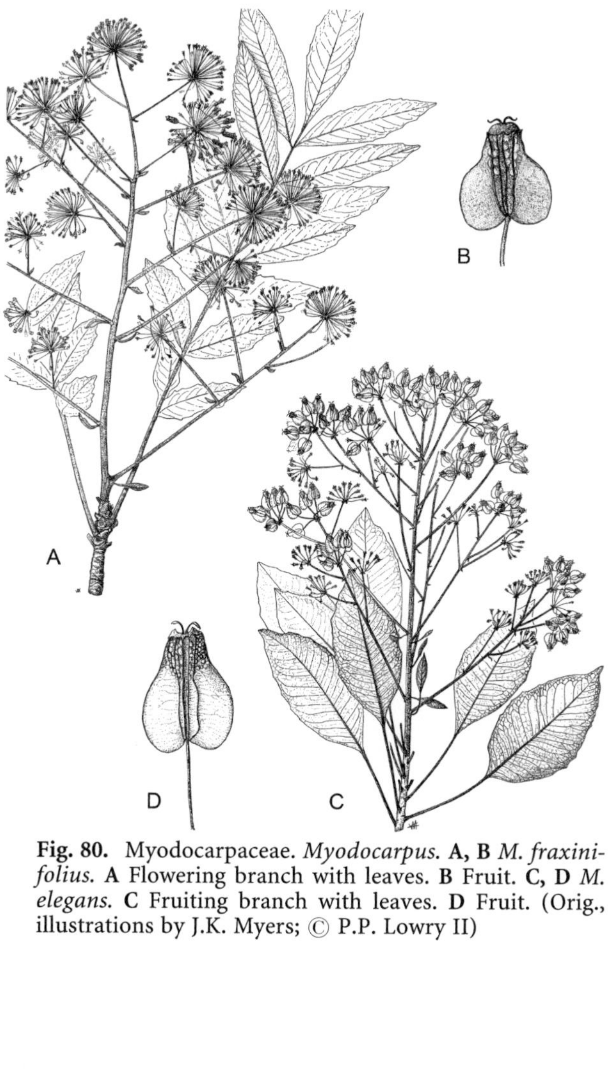

Fig. 80. Myodocarpaceae. *Myodocarpus.* **A, B** *M. fraxinifolius.* **A** Flowering branch with leaves. **B** Fruit. **C, D** *M. elegans.* **C** Fruiting branch with leaves. **D** Fruit. (Orig., illustrations by J.K. Myers; © P.P. Lowry II)

Selected Bibliography

Baumann, M. 1946. *Myodocarpus* und die Phylogenie der Umbelliferenfrucht. Ber. Schweiz. Bot. Ges. 56: 13–112.

Chandler, G.T., Plunkett, G.M. 2004. Evolution in Apiales: nuclear and chloroplast markers together in (almost) perfect harmony. Bot. J. Linn. Soc. 144: 123–147.

Erbar, C., Leins, P. 2010. Nectaries in Apiales and related groups. Pl. Div. Evol. 128: 269–295.

Hallé, F., Oldeman, R.A.A., Tomlinson, P.B. 1978. Tropical trees and forests. An architectural analysis. Berlin, Heidelberg, New York: Springer-Verlag.

Konstantinova, A.I., Yembaturova, E.Y. 2010. The family Myodocarpaceae: looking at the system from the standpoint of comparative carpology. Pl. Div. Evol. 128: 347–367.

Lebouvier, N., Lawes, D., Hnawia, E., Page, M., Brophy, J., Nour, M. 2014. The leaf, wood and bark oils of three species of *Myodocarpus* (Myodocarpaceae) endemic to New Caledonia. Nat. Prod. Comm. 9: 1223–1227.

Lee, D.W., Taylor, G.T., Irvine, A.K. 2000. Structural fruit coloration in *Delarbrea michieana* (Araliaceae). Int. J. Pl. Sci. 161: 297–300.

Liu, M., Plunkett, G.M., Lowry, P.P., II. 2010. Fruit anatomy provides structural synapomorphies to help define Myodocarpaceae (Apiales). Syst. Bot. 35: 675–681.

Liu, M., Plunkett, G.M., Van Wyk, B.-E., Tilney, P.M., Lowry, P.P., II. 2012. The phylogenetic significance of the carpophore in Apiaceae. Ann. Bot. 110: 1531–1543.

Lowry, P.P., II. 1986a. "A Systematic Study of Three Genera of Araliaceae Endemic to or Centered on New Caledonia: *Delarbrea, Myodocarpus*, and *Pseudosciadium*". Ph.D. dissertation, Washington Univ., St. Louis.

Lowry, P.P., II. 1986b. A systematic study of *Delarbrea* Vieill. (Araliaceae). Allertonia 4: 169–201.

Lowry, P.P., II, Plunkett, G.M., Raquet, V., Sprenkle, T.S., Jérémie, J. 2004. Inclusion of the endemic New Caledonian genus *Pseudosciadium* in *Delarbrea* (Apiales, Myodocarpaceae). Adansonia III, 26: 251–256.

Nicolas, A.N., Plunkett, G.M. 2009. The demise of subfamily Hydrocotyloideae (Apiaceae) and the re-alignment of its genera across the entire order Apiales. Molec. Phylogen. Evol. 53: 134–151.

Oskolski, A.A., Lowry, P.P., II, Richter, H.G. 1997. Systematic wood anatomy of *Myodocarpus, Delarbrea*, and *Pseudosciadium* (Araliaceae). Adansonia III, 19: 61–75.

Plunkett, G.M., Lowry, P.P., II. 2001. Relationships among "Ancient Araliads" and their significance for the systematics of Apiales. Molec. Phylogen. Evol. 19: 259–276.
Plunkett, G.M., Soltis, D.E., Soltis, P.S. 1997. Clarification of the relationship between Apiaceae and Araliaceae based on *mat*K and *rbc*L sequence data. Amer. J. Bot. 84: 565–580.
Plunkett, G.M., Chandler, G.T., Lowry, P.P., II, Pinney, S. M., Sprenkle, T.S. 2004. Recent advances in understanding Apiales and a revised classification. S. African J. Bot. 70: 371–381.
Rodríguez, R.L. 1957. Systematic anatomical studies on *Myrrhidendron* and other woody Umbellales. Univ. Calif. Publ. Bot. 29: 145–318.
Rodríguez, R.L. 1971. The relationships of the Umbellales. In: Heywood, V.H. (ed.) The biology and chemistry of the Umbelliferae, Bot. J. Linn. Soc. 64, Suppl. 1: 63–91.
Schlessman, M.A. 2010. Major events in the evolution of sexual systems in Apiales: ancestral andromonoecy abandoned. Pl. Div. Evol. 128: 233–245.
Schlessman, M.A., Lloyd, D.G., Lowry, P.P., II. 1990. Evolution of sexual systems in New Caledonian Araliaceae. Mem. New York Bot. Gard. 55: 105–117.
Schlessman, M.A., Plunkett, G.M., Lowry, P.P., II, Lloyd, D.G. 2001. Sexual systems of New Caledonian Araliaceae: A preliminary phylogenetic reappraisal. Edinb. J. Bot. 58: 221–228.
Thorne, R.F. 1973. Inclusion of the Apiaceae (Umbelliferae) in the Araliaceae. Notes Roy. Bot. Gard. Edinb. 32: 161–165.
Viguier, R. 1906. Recherches anatomiques sur la classification des Araliacées. Ann. Sci. Nat. Bot. IX, 4: 1–210.
Yi, T., Lowry, P.P., II, Plunkett, G.M., Wen, J. 2004. Chromosomal evolution in Araliaceae and close relatives. Taxon 53: 987–1005.

Pennantiaceae

Pennantiaceae J. Agardh, Theoria Syst. Pl. 301 (1858).

M.J. Potgieter

Small to medium-sized trees or shrubs; vegetative parts and inflorescence axes puberulent with unicellular, straight to curved, uncinate or non-uncinate hairs and also clavate hairs with a short, several-celled stalk and a globular (glandular?), usually several-celled head; new shoots usually conspicuously lenticellate, distinct resting buds not formed but a few cataphylls sometimes present at base of a flush in at least *P. baylisiana* and *P. corymbosa*. Leaves alternate in 2/5 spiral phyllotaxy, ptyxis curved to almost conduplicate; blades chartaceous to subcoriaceous, brown to dark brown when dry; margins of adult leaves mostly entire or obscurely sinuous but sometimes crenately toothed (at least in seedlings, on reversion shoots, and in *P. corymbosa* adult foliage); domatia (pocket or pit types) variously present, fringed by non-uncinate hairs (domatia obscure and glabrous in *P. endlicheri*); lateral nerves 4–6 pairs, diverging at ca. 45° from midrib, further venation only slightly prominent above or not at all (*P. corymbosa*). Inflorescences terminal on leafy shoots (but mostly ramiflorous to cauliflorous in *P. baylisiana*), paniculate; branches spirally arranged or ± disjunct-opposite; bracts bracteose, narrow, up to ca. 4 mm long; pedicels articulate to calyx base. Flowers probably functionally unisexual, plants ± dioecious; calyx a collar with 5 minute triangular lobes (*P. cunninghamii*) or some or all lobes lacking; petals 5, free, valvate, white to greenish, adaxially with a fleshy-papillate apical projection and a low central longitudinal ridge; stamens 5, epipetalous, filaments free, glabrous; in mature bud filaments inflexed in their distal part or bent out laterally; anthers versatile, dehiscence latero-introrse; ovary 1-locular, apparently 3-carpellate; ovule solitary, pendulous, anatropous; stigmas 3, each on a short apical style (*P. cunninghamii*) or sessile and latero-apical in a ring at the truncate ovary top, free (*P. corymbosa*, *P. endlicheri*) or not (*P. baylisiana*). Fruit a drupe, dark purple; stone ± smooth, either bony and somewhat trigonous in cross-section, or (*P. cunninghamii*) chartaceous and ± terete. Embryo non-chlorophyllous, minute, endosperm copious, oily.

Monogeneric with four species, distributed in eastern Australia, Norfolk Island, Three Kings Islands (Great Island only) and New Zealand.

Vegetative Morphology and Anatomy. *Pennantia cunninghamii*, a forest tree of rather open sites, can attain ca. 30 m in height, albeit with a trunk which may be leaning and fluted (Guymer 1984; Floyd 1989). The three other species are all small- to medium-sized trees but are rather different from one another in architecture. *Pennantia corymbosa* and *P. endlicheri* are generally single-trunked but *P. baylisiana* is almost shrubby. Young plants soon produce large-diameter, orthotropic new shoots near the base of existing stems. In older trees, these new shoots grow up to supplement or replace the main trunk (Gardner and de Lange 2002). White (1922) and Francis (1970) recorded that *P. cunninghamii* has a suckering habit.

The alternately arranged leaf blades of all *Pennantia* species have a slightly crenate margin in the seedling and juvenile phases and on reversion shoots. In *P. baylisiana*, the first formed leaves of vigorous, basally produced orthotropic shoots have up to 15 minute teeth per side; leaves in the crown of adult trees are entire (Baylis 1977). In *P. endlicheri*, toothed leaves are maintained from the seedling stage; the leaves of canopy twigs, at first denticulate, mature to have sinuous margins. The leaves of *P. corymbosa* (Fig. 81A) are rather strongly crenate, and those of *P. cunninghamii* entire. The apices of the teeth are not differentiated into water- or mucilage-secreting

J.W. Kadereit, V. Bittrich (eds.), *Flowering Plants. Eudicots*, The Families and Genera of Vascular Plants 15,
https://doi.org/10.1007/978-3-319-93605-5_10

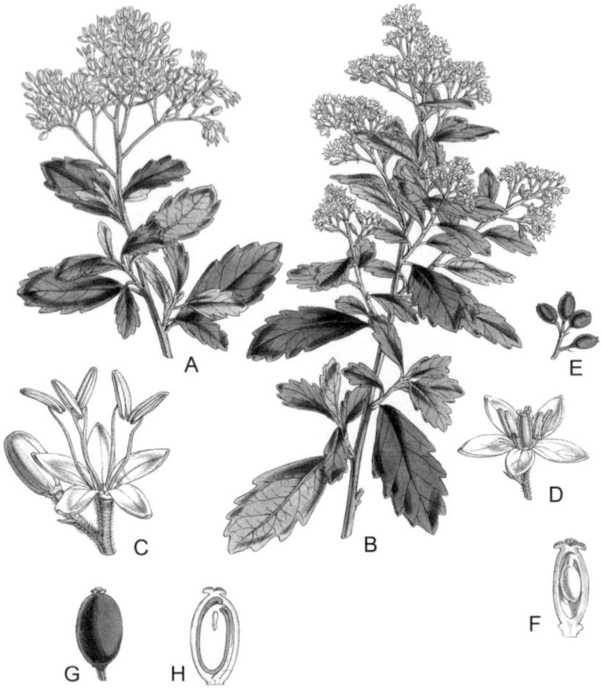

Fig. 81. Pennantiaceae. *Pennantia corymbosa*. **A** Branch with male inflorescence. **B** Branch with female inflorescence. **C** Male flower. **D** Female flower. **E** Partial infructescence. **F** Ovary in longitudinal section. **G** Drupe. **H** Drupe and seed in longitudinal section. (Modified from Hooker 1853, t. 12; drawn by W.H. Fitch; reproduced with kind permission of the Peter H. Raven Library/Missouri Botanical Garden)

hydathodes. The vein which terminates in the tooth is characteristically fan-shaped and does not reach the epidermis of the tooth. Of the four species, *P. baylisiana* has the largest and most coriaceous leaves. The blades of *P. baylisiana* are obovate and, when exposed to the sun, develop strongly recurved sides and thus appear saddle-shaped or almost cylindrical. In shaded conditions, however, the leaves of *P. baylisiana* are almost plane (Baylis 1977). Floyd (1989) noted that the blades of *P. cunninghamii* have wax on their lower surfaces, a feature lacking in other species.

The domatia of *P. cunninghamii* are pit domatia, with short, non-uncinate hairs on the floor of the domatium and around its rim. In *P. corymbosa*, they are of the pocket type, with more or less patent, slender, non-uncinate hairs. The domatia of *P. baylisiana* are axillary pocket domatia with rather long and slender non-uncinate hairs. The domatia of *P. endlicheri* can be considered to be present only because of a slight undercutting of the axils at the largest lateral veins; hair aggregations are never seen.

Baas (in Sleumer 1970), confirmed by Gardner and de Lange (2002), found a striking anatomical feature in the petiole of *P. baylisiana*. Near the junction with the blade the vascular system of the petiole consists of a central bundle enclosed by a cylindrical one, external to which are two small cylindrical "wing bundles". Adult leaves of the four species all have the large cylindrical bundle and two wing bundles, these deriving each from the three traces which leave the axis.

The mature leaves of *P. baylisiana* and *P. endlicheri* have an adaxial hypodermis of one cell layer; the cells are larger than those of the epidermis and are thickened on their outer and radial walls (Gardner and de Lange 2002). Such tissue has been recorded for *P. cunninghamii* by Hamilton (1897) and for *P. corymbosa* by Wylie (1954).

Stomata are of the paracytic type (van Staveren and Baas 1973). The guard cells have conspicuous outer stomatal ledges with rather distinct polar T-pieces. Stomata size ranges from 21–28 μm long to 16–20 μm wide.

Heintzelman and Howard (1948) found *Pennantia* to have "normal hairs" (termed non-uncinate by Gardner and de Lange 2002) which are straight to curved, rather thick-walled and acute, and "uncinate hairs" which are similar but hook-tipped. Uncinate hairs are abundant in *P. corymbosa*, with non-uncinate hairs present in the other three species. Uncinate hairs are absent in *P. baylisiana* and *P. cunninghamii* but present, albeit sparsely, on the inflorescence axes of *P. endlicheri*.

Gardner and de Lange (2002) also found that all species have a kind of "gland hair". This extends the record of "multicellular hairs with a rounded apex" in *P. corymbosa* (van Staveren and Baas 1973).

Rhombic crystals in the cortex have been reported by Heintzelman and Howard (1948).

In a study of the wood anatomy of *Pennantia*, Lens et al. (2008) found the growth ring boundaries to vary from indistinct to distinct. Vessels were (48–)70(–30)/mm^2, predominantly solitary with an angular outline. The perforation plates are exclusively scalariform (Bailey and Howard

1941) with (18–)34(–56) bars, sometimes with reticulate portions in *P. cunninghamii*. Intervessel pits are opposite, with the pits 3–4 μm in horizontal diameter, and non-vestured. Vessel ray pitting is comparable to intervessel pitting in shape and size. Helical thickenings are present throughout vessel elements and in fibres in *P. corymbosa*, and sometimes in tails of vessel elements in *P. cunninghamii*. Tyloses are absent. The tangential diameter of vessels ranges from 30 to 95 μm, with the vessel elements 700–2250 μm long. Tracheids when occasionally present are between 1400 and 2700 μm long. Non-septate fibres have distinctly bordered pits concentrated in tangential and radial walls. The axial parenchyma is diffuse or diffuse-in-aggregates and scanty paratracheal with 5–10 cells per parenchyma strand. Uniseriate rays are common, 5–9 rays/mm, consisting of square to upright cells of between 200 and 1100 μm length. Multiseriate rays are often 4–8-seriate, consisting of procumbent body ray cells with 1–4 (> 20) rows of upright to square marginal ray cells. Indistinct sheath cells are sometimes present in *P. cunninghamii*.

Inflorescence Structure. All species have a paniculate inflorescence, the branches being spiral to opposite-disjunct and terminated by a flower. In *P. corymbosa*, *P. cunninghamii* and *P. endlicheri* these inflorescences, which are hand-sized or less, are borne terminally on leafy shoots and are produced without pause between the shoot's vegetative and reproductive phases. In *P. cunninghamii* there are sometimes smaller inflorescences in the axils of the 1–3 uppermost (little reduced) foliage leaves below the terminal inflorescence. The inflorescences of *P. baylisiana* are usually ramiflorous or cauliflorous and found at all levels of the branches and main trunks. Ramiflorous inflorescences are very occasionally seen in *P. endlicheri*.

Flower Structure. Cheeseman (1906, 1925) noted that the inflorescences and flowers are larger in male than in female individuals. There is some controversy about the structure and sexual functioning of *Pennantia* flowers, particularly in clones of *P. baylisiana* (Murray and de Lange 1995; Webb 1996). Although these clones have pollen-bearing anthers which open, the grains have been reported to be abnormal by having very thick walls and poorly developed apertures (Cranwell 1962; Müller in Sleumer 1970; Moar 1993). However, fruits are produced by clones, including the original Three Kings tree (Wright 1989).

Pennantia corymbosa is generally described as a dioecious species. The male flower has a rudimentary ovary which lacks a stigma (Fig. 81B). The female flowers have anthers shorter than those of the male, as in *P. baylisiana* (Fig. 81C). The sexes of *P. corymbosa* can be distinguished in bud. Buds of the male flowers are substellate in cross-section, while those of the female flowers are perfectly circular. The same, although less conspicuously, can be observed in *P. endlicheri*. In *P. endlicheri*, *P. cunninghamii* and *P. corymbosa*, the male flowers usually have a very small, conical, rudimentary ovary. In the female flowers of *P. endlicheri*, the anthers are even more reduced than in *P. baylisiana*, being shorter and obtuse rather than broadly apiculate, and the filaments are slightly shorter, and sometimes briefly inflexed but never laterally bent.

The calyx in *Pennantia* is articulate to the pedicel and glabrous (whereas the pedicel may have a sparse pubescence). In all species the calyx is reduced to a collar less than 1 mm high. In *P. cunninghamii* this collar may be slightly expanded around its upper circumference, and the five or sometimes fewer small triangular calyx teeth are borne on this wing; sometimes the lobes alone are present. In the other species the lobes may occur as minute outgrowths but usually they and the wing are lacking. The five petals are free and valvate. They have a longitudinal, low, fleshy ridge, and the apex forms a papillate and inward-directed projection. The most notable feature of the stamens of the male flowers are the distally inflexed (*P. corymbosa*, *P. cunninghamii*, *P. endlicheri*) or laterally bent (sometimes *P. corymbosa*) filaments. The filaments straighten as the flowers open or just after, and in the male flowers slightly exceed the petals (Gardner and de Lange 2002). The anthers are versatile and open introrsely but are often presented extrorsely (Dawson 1988). Species differ in the anther surface. In *P. cunninghamii* the epidermis is relatively thick, stiff and dark brown, and has scattered prominent papillae especially towards the apices of the thecae; these are hardly developed in the other species.

The family is regarded as 3-carpellate and *P. cunninghamii* has three major longitudinal furrows on its oblong-conical ovary; each of the three lobes of the ovary narrows into a very short style which is more or less punctate. The ovule is solitary, apical and anatropous, with the micropyle directed towards the arch of the funicle where this originates from the ovary wall (Miers 1852). The ovule is supplied by a vascular bundle which is independent from and considerably more robust than the other bundles in the ovary wall. In *P. baylisiana*, *P. corymbosa* and *P. endlicheri*, the ovary in the female flower is barrel-shaped and marked only by weak and irregular grooves arising from pressure from the stamens (Gardner and de Lange 2002). A top view of the truncate apex of the ovary shows a low-crested, papillate, sessile stigmatic ring and a smaller central non-papillate region. In these species there are only three bundles in the ovary wall in addition to the ovule-supplying bundle. In *P. corymbosa* and *P. endlicheri* the ovary has three distinct (i.e. not laterally confluent) subapical stigmas, each shaped like a horseshoe with an inwardly directed sinus (Dawson 1988). In *P. baylisiana*, however, the three sinuses are present but obscure, and the ring appears more or less continuous and made up of three triangular-shaped units each with a somewhat crested central line (Gardner and de Lange 2002).

Embryology. According to Mauritzon (1936) and Fagerlind (1945), embryo sac development is of the Polygonum-type, polar nuclei fuse prior to fertilization, 3 antipodal cells are formed which do not proliferate and are ephemeral or persistent. Synergids are hooked (with filiform apparatus) and endosperm formation is nuclear. The above information, provided for Icacinaceae (a former placement of *Pennantia*), must be verified for Pennantiaceae.

Pollen Morphology. Pollen grains are ellipsoidal to oblately spheroidal. Pollen varies in size from 22.5–(25.5 × 20.9)–25.3 μm (Dahl 1952). Polar/equatorial quota is > 0.9 (Kårehed 2001). Grains are tricolporate with the furrows elongated and extending almost to the poles. Kodela (2006) reported tetracolporate pollen in *P. cunninghamii*. The colpi are intruding. The furrow margin is somewhat irregular. Pores are small (< 1 μm in diameter), circular to somewhat rectangular with ragged, irregular margins, and often appear costate (Lobreau-Callen 1972, 1973). Frequently, fragments of foveolate exine occur on the poral membrane. The exine is foveolate (very finely pitted), 0.5–0.8 μm thick, though the general appearance is smooth (Dahl 1952). The sexine is thicker than the nexine. Baculation is vestigial (Erdtman 1952).

Pennantia baylisiana forms a fair quantity of pollen grains (Gardner and de Lange 2002) which, according to Baylis (1977), appear "perfect" but have (relatively) thickened walls and occluded apertures. Moar (1993) suggested that, although 95% of the pollen grains of *P. baylisiana* have a well-stained cytoplasm, germination is inhibited because of poorly developed, non-functional apertures as found in female flowers of *P. corymbosa*.

Karyology. A chromosome number of 2n = 50 has been reported for *P. baylisiana*, *P. corymbosa* and *P. endlicheri* (Hair and Beuzenberg 1959; Murray and de Lange 1995).

Pollination and Reproductive Systems. Gardner and de Lange (2002) did not find nectar in the unisexual flowers of *P. baylisiana*, *P. corymbosa* and *P. endlicheri*. Flowers of these species have a pleasant, light scent. Thomson (1927) described the flowers of *P. corymbosa* as being "very fragrant" and "evidently dependent on insects for pollination". No data on nectar or scent are available for *P. cunninghamii*. In *P. endlicheri* flies and butterflies have been seen visiting the male flowers (de Lange, pers. obs.). In *P. baylisiana* and *P. corymbosa* the anthers of female flowers contain some pollen grains of unknown fertility. Gardner and de Lange (2002) speculate that pollen may be the reward for insect visitors.

Fruit and Seed. The fruits of all species are drupes with a fleshy but rather thin mesocarp and a thickened and bony or chartaceous endocarp. The seed-coat is papery. Fruit size varies from ca. 8 mm (*P. corymbosa*) to ca. 15 mm (*P. cunninghamii*). The fruit of *P. cunninghamii* is beaked by the slightly enlarged styles, whereas the fruit apex of *P. corymbosa* is crowned by a stigmatic circle of ca. 1.5 mm diameter in which the three stigmas have neither enlarged nor moved apart laterally. The non-confluent stigmatic ring of *P. endlicheri* is retained in the fruit, as is the confluence of the stigmatic ring in *P. baylisiana* (Gardner and de Lange 2002).

The surface of the dried fruit of *P. cunninghamii* tends to be longitudinally wrinkled. Its wall is thin (ca. 0.05 mm) and two-layered, the outer layer consisting of shortly branched, pitted, thick-walled stone cells, and the inner layer of thick-walled fibres which run circumferentially. In the other *Pennantia* species the stone is unequally trigonous. Its shortest and least curved side contains the ovule's vascular supply (Gardner and de Lange 2002). In *P. corymbosa* the endocarp is considerably thicker and harder than in *P. cunninghamii* (Miers 1852). It has an outer layer of branched stone cells and an inner layer, about half as deep, of thick-walled fibres.

Dispersal. Fruits of *P. cunninghamii* are fleshy and bird-dispersed, amongst others by Brown Cuckoo-dove and *Pteropus poliocephalus*.

Phytochemistry. Bate-Smith and Lerner (1954) reported leucoanthocyanins in the leaves of *P. cunninghamii*.

Affinities. On the basis of analyses of DNA sequences and morphological data, Kårehed (2001) showed that the Icacinaceae is not monophyletic but polyphyletic, and should be split into at least four not closely related families. These are Icacinaceae s.s., Stemonuraceae and Cardiopteridaceae (both belonging in Aquifoliales), and Pennantiaceae. Pennantiaceae belong in Apiales. The data presented by Kårehed (2001) provided no support for the inclusion of *Pennantia* in any other family, which led him to suggest that Pennantiaceae must be recognized as a family on its own which is sister to the remainder of Apiales (Kårehed 2003; Soltis et al. 2011; Nicolas and Plunkett 2014). Nicolas and Plunkett (2014) estimated that the divergence of the Pennantiaceae lineage from the rest of the order occurred in the mid-Cretaceous (ca. 117 Ma).

Distribution and Habitats. Gardner and de Lange (2002) stated that in the wild *P. baylisiana* is known only as a single tree on Great Island of the Three Kings Islands. In 1971, however, two juveniles grown from cuttings were planted near it, and one of these has survived. *Pennantia corymbosa* grows in relatively low-stature, open, windswept coastal scrub and forest, and is found almost throughout New Zealand but with some local gaps. *Pennantia cunninghamii* grows in Australia, especially around Atherton Tableland, Queensland, and south to near Milton, New South Wales. According to Harden (1992), the species is widespread in subtropical and warm-temperate rainforest, especially in rock gullies and along watercourses. *Pennantia endlicheri* is endemic to Norfolk Island in sheltered forest sites. The analyses of Nicolas and Plunkett (2014) suggest that Australia + Zealandia is the likely origin of the Pennantiaceae lineage. The current presence of *Pennantia* in New Zealand, however, is probably due to long-distance dispersal from Australia, as most or all of New Zealand suffered submergence periods until the early Miocene.

Economic Importance. No species of economic importance exists. However, in New Zealand the species is cultivated in numerous private gardens as an ornamental (Gardner and de Lange 2002). According to Kirk (1889) and Best (1941), a short length of the hard wood of *P. corymbosa* was commonly used by the pre-European Maori in the "plough method" of fire-making.

Only one genus:

1. ***Pennantia*** J.R. Forst. & G. Forst. Fig. 81

Pennantia J.R. Forst. & G. Forst., Charact. Gen. 133, t. 67 (1776); Gardner & de Lange, J. Roy. Soc. N.Z. 32: 669–695 (2002), rev.
Plectomirtha W.B.R. Oliv. (1948).

Description as for family.

Four species, eastern Australia, Norfolk Island, Three Kings Islands (Great Island only), New Zealand.

Divided into sect. *Pennantia* (*P. baylisiana* (W.R.B. Oliv.) G.T.S. Baylis, *P. corymbosa* J.R. Forst. & G. Forst., *P. endlicheri* Reiss.) and sect. *Dermatocarpus* (*P. cunninghamii* Miers) by Gardner and de Lange (2002).

Selected Bibliography

Allan, H.H. 1961. Flora of New Zealand. Vol. I. Wellington: Government Printer.

Bailey, I.W., Howard, R.A. 1941. The comparative morphology of the lcacinaceae. II. Vessels. J. Arnold Arbor. 22: 171–187.

Bate-Smith, E.C., Lerner, N.H. 1954. Leuco-Anthocyanins. 2. Systematic distribution of Leuco-Anthocyanins in leaves. Biochem. J. 58: 126–132.

Baylis, G.T.S. 1977. *Pennantia baylisiana* (Oliver) Baylis comb. nov. N.Z. J. Bot. 15: 511–512.

Best, E. 1941. The Maori. Vol. 5. Wellington: Polynesian Society Memoir.

Cheeseman, T.F. 1906. Manual of the New Zealand flora. Wellington: Government Printer.

Cheeseman, T.F. 1925. Manual of the New Zealand flora. 2nd edn. Wellington: Government Printer.

Cranwell, L.M. 1962. Endemism and isolation in the Three Kings Islands, New Zealand—with notes on pollen and spore types of the endemics. Records of the Auckland Institute and Museum 5: 215–232.

Dahl, A.O. 1952. The comparative morphology of the lcacinaceae. 4. The pollen. J. Arnold Arbor. 33: 252–291.

Dawson, J. 1988. Forest vines to snow tussock. Wellington: Victoria University Press.

Erdtman, G. 1952. Pollen morphology and plant taxonomy: An introduction to the study of Angiosperm pollen. Stockholm: Almqvist and Wiksell.

Fagerlind, F. 1945. Bau des Gynöceums, der Samenanlage und des Embryosackes bei einigen Repräsentanten der Familie Icacinaceae. Svensk Bot. Tidskr. 39: 346–364.

Floyd, A.G. 1989. Rainforest trees of mainland south-east Australia. Melbourne: Inkata Press.

Francis, W.D. 1970. Australian rain-forest trees. 2nd edn. Canberra: Australian Government Publishing Service.

Gardner, R.O., de Lange, P.J. 2002. Revision of *Pennantia* (Icacinaceae), a small isolated genus of Southern Hemisphere trees. J. Roy. Soc. N.Z. 32: 669–695.

Guymer, G.P. 1984. Icacinaceae. In: George, A.S. (ed.) Flora of Australia. Vol. 22. Canberra: Australian Government Publishing, pp. 204–211.

Hair, J.B., Beuzenberg, E.J. 1959. Contributions to a chromosome atlas of the New Zealand flora. 2. Miscellaneous families. N.Z. J. Bot. 2: 148–156.

Hamilton, A.G. 1897. On domatia in certain Australian and other plants. Proc. Linn. Soc. N.S. Wales, ser. 2, 9: 758–792.

Harden, G.J. 1992. Icacinaceae. In: Harden, G.J. (ed.) Flora of New South Wales. Vol. 3. Sydney: University of Sydney Press, pp. 437–438.

Heintzelman, C.E., Howard, R.A. 1948. The comparative morphology of the Icacinaceae. V. The pubescence and the crystals. Amer. J. Bot. 35: 42–52.

Hooker, J.D. 1853. The botany of the Antarctic voyage of H.M. discovery ships Erebus and Terror in the Years 1839–1843, under the command of Captain Sir James Clark Ross, vol. 2(1): t. 12 (1853).

Kårehed, J. 2001. Multiple origin of the tropical forest tree family Icacinaceae. Amer. J. Bot. 88: 2259–2274.

Kårehed, J. 2003. The family Pennantiaceae and its relationship to Apiales. Bot. J. Linn. Soc. 141: 1–24.

Kirk, T. 1889. The Forest Flora of New Zealand. Wellington: Government Printer.

Kodela, P.G. 2006. Pollen morphology of some rainforest taxa occurring in the Illawarra region of New South Wales, Australia. Telopea 11: 346–389.

Lens, F., Kårehed, J., Baas, P., Jansen, S., Rabaey, D., Huysmans, S., Hamann, T., Smets, E. 2008. The wood anatomy of the polyphyletic Icacinaceae s.l., and their relationships within asterids. Taxon 57: 525–552.

Lobreau-Callen, D. 1972. Pollen des Icacinaceae. 1. Atlas (1). Pollen et Spores 14: 345–388.

Lobreau-Callen, D. 1973. Le pollen de Icacinaceae: 2. Observations en Microscopie électronique, Corrélations, Conclusions. Pollen et Spores 15: 47–89.

Mauritzon, J. 1936. Embryologische Angaben über Stackhousiaceae, Hippocrateaceae und lcacinaceae. Svensk Bot. Tidskr. 30: 541–550.

Miers, J. 1852. On some genera of the Icacinaceae. Ann. Mag. Nat. His., ser. 2, 9: 485–492.

Moar, N.T. 1993. Pollen grains of New Zealand dicotyledonous plants. Lincoln: Manaaki Whenua Press.

Murray, B.G., de Lange, P.J. 1995. Chromosome numbers in the rare endemic *Pennantia baylisiana* (W.R.B. Oliv.) G.T.S. Baylis (Icacinaceae) and related species. N.Z. J. Bot. 33: 563–564.

Nicolas, A.N., Plunkett, G.M. 2014. Diversification times and biogeographic patterns in Apiales. Bot. Review 80: 30–58.

Sleumer, H. 1970. The identity of *Plectomirtha* Oliv. with *Pennantia* J.R. & G. Forster. Blumea 18: 217–218.

Soltis, D.E., Smith, S.A., Cellinese, N., Wurdack, K.J., Tank, D.C., Brockington, S.F., Refulio-Rodriguez, N. F., Walker, J.B., Moore, M.J., Carlsward, B.S., Bell, C. D., Latvis, M., Crawley, S., Black, C., Diouf, D., Xi, Z., Rushworth, C.A., Gitzendanner, M.A., Sytsma, K.J., Qiu, Y.-L., Hilu, K.W., Davis, C.C., Sanderson, M.J., Beaman, R.S., Olmstead, R.G., Judd, W.S., Donoghue, M.J., Soltis, P.S. 2011. Angiosperm phylogeny: 17 genes, 640 taxa. Amer. J. Bot. 98: 704–730.

Thomson, G.M. 1927. The pollination of New Zealand flowers by birds and insects. Trans. N.Z. Inst. 57: 106–125.

van Staveren, M.G.C., Baas, P. 1973. Epidermal leaf characters of the Malesian Icacinaceae. Acta Bot. Neerl. 22: 329–359.

Webb, C.J. 1996. The breeding system of *Pennantia baylisiana* (Icacinaceae). N.Z. J. Bot. 34: 421–422.

White, C.T. 1922. An elementary text-book of Australian forest botany. Vol. 1. Melbourne: John Spence.

Wright, A. 1989. *Pennantia baylisiana* fruiting on Great Island, Three Kings. N.Z. Bot. Soc. Newslett. 15: 16.

Wylie, R.B. 1954. Leaf organization of some woody dicotyledons from New Zealand. Amer. J. Bot. 41: 186–191.

Pittosporaceae

Pittosporaceae R. Br. in Flinders, Voy. Terra Austr. 2: 542 (1814), nom. cons. ("Pittosporeae").

R.C. Carolin and V. Bittrich

Evergreen trees, shrubs, rhizomatous or not, sometimes procumbent or climbing or twining, occasionally epiphytic, often heteroblastic, glabrous or pilose, trichomes uniseriate, often with elongated or T-shaped terminal cell, with resin canals particularly in the bark, sometimes ± aromatic; sometimes branches ending in thorns. Leaves simple, usually coriaceous and entire, rarely dentate (especially when juvenile), spirally arranged, sometimes pseudoverticillate, without stipules. Inflorescences cymose or racemose, often panicles forming more or less dense corymbs, or single flowers. Pedicels usually bibracteolate; flowers actinomorphic, rarely zygomorphic, bisexual, sometimes unisexual (at least functionally), then plants monoecious, (sub)dioecious, gynodioecious or rarely polygamous; flowers 5-merous (except pistil); sepals imbricate or valvate, free or fused at the base, generally caducous; petals often clawed, completely free or fused at the very base, spreading from the base or coherent into a tube, imbricate, rarely valvate; stamens hypogynous, alternating with the petals; filaments hairy or glabrous, sometimes coherent to the petals when these are coherent, anthers sagittate or not, free or rarely apically coherent, dehiscing introrsely by longitudinal slits or (in *Cheiranthera* and a few *Billardiera* spp.) through apical pores or short slits, tetrasporangiate; ovary superior, sometimes stipitate, glabrous or pilose, 2(3–5)-carpellate, 1 or 2(–5)-locular, often incompletely so, with parietal, parietal-basal, or more rarely axile placentae; style simple, stigma inconspicuous, capitate or slightly lobed, nectary a sometimes strongly lobed ring at the base of the ovary, rarely vestigial or absent; ovules 1–many in each locule, anatropous, tenuinucellate, unitegmic. Fruit a loculicidal and sometimes also ± septicidal capsule, or a fleshy or fibrous berry, rarely woody indehiscent. Seeds 1–several in each locule, sometimes brightly coloured or blackish and coated with sticky material, rarely winged or arillate; embryo tiny, with 2, rarely 3, 4 or 5 cotyledons; endosperm well-developed, oily and/or proteinaceous. $x = 12$.

Nine genera and ca. 250 species. In temperate to tropical regions of Pacific Islands, New Zealand, Asia, Africa and particularly Australia.

Vegetative Morphology. Most species are small trees or shrubs, sometimes scandent or twining, but some *Pittosporum* species are up to 20 m high and a few species start as epiphytes. *Bentleya* and, to a lesser extent, *Pittosporum angustifolium* produce suckers from underground rhizomes. Cataphylls are usually present at the base of new shoots. Plants show considerable phenotypic plasticity, mostly due to different ecological conditions, but juvenile and adult leaves can also be rather different. Such heteroblasty has caused some taxonomic confusion, and different developmental stages were sometimes even considered different species (Chandler et al. 2007). Seedlings and intermediate leaves are reported to have sometimes a different form and a more or less deeply lobed, crenate or toothed margin, rarely (e.g. *Auranticarpa ilicifolia*) also with bristles, while the margin of mature leaves is mostly entire, very rarely undulate, toothed or crenate. In *Pittosporum* the branches (i) may be vegetative in the first year and produce a terminal inflorescence in the second year and have lateral vegetative branches, (ii) produce a pseudoterminal inflorescence from one of the terminal cluster of buds in the second year or (iii) are strictly sympodial (Gowda 1951).

Vegetative Anatomy. A review of most anatomical data can be found in Solereder (1898). Wood anatomy was studied in detail by Carlquist (1981), leaf anatomy by Wilkinson (1992). Four

J.W. Kadereit, V. Bittrich (eds.), *Flowering Plants. Eudicots*, The Families and Genera of Vascular Plants 15,
https://doi.org/10.1007/978-3-319-93605-5_11

types of trichome can be distinguished: (i) uniseriate with 1–3(–8) basal cells and an elongated, sometimes curling terminal cell, (ii) 2–6-celled basal stalk and 2-armed terminal cell, which might be curling or not, (iii) uniseriate without extended terminal cell, and (iv) uniseriate with 1–6 short or elongated terminal glandular cells. Crystals of calcium oxalate are common in the parenchyma, mostly as druses; in the phloem of *Pittosporum* and *Bentleya* small styloids and rhombohedra were found. The vessels of the wood are small compared to other families, perforations are generally simple, oblique. Paratracheal wood parenchyma is scanty, fibres usually are septate. *Pittosporum paniense* from New Caledonia has a primitive wood structure (Carlquist 1981). Schizogenous secretory canals are found in several parts of the plant, most notably in the bark and in the root, where they show a characteristic distribution similar to Apiaceae and Araliaceae, as already reported by van Tieghem (1884). He also noted the correlated typical duplication of the number of lateral roots localized on either side of the xylem poles, as the place at the apex of each xylem pole is occupied by a resin canal. Canals are found neither in the primary cortex nor in the pith. In the leaves they form a network and are located adjacent but not inside the phloem (Wilkinson 1992). Sticky resin with ethereal oils exudates when the canals are broken, and crushed leaves produce a characteristic resinous odour. Periderm origin was found to vary within the family, directly subepidermal (*Pittosporum*), in the third subepidermal layer (*Hymenosporum*), and on the border of primary to secondary cortex (*Billardiera*). Nodes are unilacunar or trilacunar with one trace per gap. Leaf venation is camptodromous or more rarely craspedodromous, veinlets are simple to 2- or 3-branched, stomata are paracytic. Epicuticular wax is mostly present as platelets of varying density and form, more rarely rodlets, the cuticle varies from smooth to striate. Epidermal anticlinal walls are usually straight, rarely undulate. A double leaf epidermis occurs in *Pittosporum* spp. Mesophyll is bifacial to less commonly isobilateral, veins imbedded or the larger ones frequently transcurrent in *Pittosporum*. Palisade cells are typically short and in several layers.

Flower Structure and Floral Biology. Many species of *Pittosporum* have functionally unisexual flowers (see, for example, Gowda 1951; Cooper 1956; Haas 1977), with monoecy, dioecy (sometimes leaky), gynodioecy, rarely polygamy reported for various species of *Pittosporum* (e.g. Sakai et al. 1995; Cayzer et al. 2000; Clarkson et al. 2012; Schlessman et al. 2014). Anthesis of the species of other genera seems to be in general protandrous. Zhou et al. (2005) made a detailed comparison of functionally male and female flowers of *Pittosporum tobira*, which may occur in the same inflorescence. All floral organs are initiated in a strictly acropetal succession and the petal primordia join laterally at the time of initiation (early sympetaly; Erbar and Leins 1995, 2004). The flowers are usually actinomorphic, but more or less zygomorphic in *Marianthus*, and in *Cheiranthera* the anthers at anthesis are placed in a line on one side of the ovary by twisting of the filaments. The petals of flowers at anthesis are free or only very shortly connate at the base and are spreading from the base or coherent into a tube often longer than the ovary. Corolla aestivation is variable. Narayana and Radhakrishnaiah (1982) studied five species of *Pittosporum* and found contorted, quincuncial and cochlear ascending aestivation. Anthers are rarely apically coherent; their length seems to be more or less correlated with the length of the style, sometimes they cluster at three different heights (1 + 2 + 2 in *Hymenosporum* and *Marianthus*; Cayzer and Crisp 2004). The ovary can be distinctly stipitate; it can be 1-locular, or partly or completely septate, placentation is axile or parietal to parietal-basal. The sometimes deeply intruding parietal placentae are giving the appearance of axile placentation. The stigma is wet, capitate or lobed, rarely inconspicuous as in *Marianthus* (Cayzer et al. 2004). A more or less prominent nectary is present in most species, usually dark-green in fresh flowers. It does not develop as an outgrowth of the receptacle but at the dorsal base of the carpels, the nectar is exuded by nectary slits (Narayana and Radhakrishnaiah 1982; Erbar and Leins 1995). The nectary usually forms a more or less conspicuous ring, sometimes with conspicuous lobes alternating with the stamens, rarely restricted to the interstaminal areas (some

Billardiera spp.) or completely lacking in a few *Billardiera* spp. and in *Cheiranthera* (Cayzer et al. 1999, 2004). Many species have strongly scented flowers and a nectary indicating insect pollination, and some have brilliantly white flowers which may indicate moth pollination. For *Pittosporum dasyphyllum* with bisexual but self-incompatible flowers, bees and butterflies were reported as pollinators (Gopalakrishnan and Thomas 2014). In *Cheiranthera* the apical pores or short slits of the basifixed anthers and the absence of the nectary indicate "buzz-pollination", and *Stenotritis* (Stenotritinae) bees were observed agitating flowers of a *Cheiranthera* species in Western Australia (Thorp 2000).

Pollen. The pollen grains of *Pittosporum* are tricolporate, 2–3-nucleate, exine more or less reticulate with little sculpturing (Gowda 1951; Haas 1977; Judd 1996).

Karyology. Several species of *Pittosporum* have been reported as having 2n = 24, a number also found in *Billardiera* and *Auranticarpa* (cf. Cooper 1956; IPCN 2010).

Embryology. The ovule throughout the family is anatropous, unitegmic, tenuinucellar. The embryo-sac is of the Polygonum-type and the endosperm is nuclear at first (Mauritzon 1939; Narayana and Sundari 1981).

Fruit, Seed and Dispersal. The capsules open loculicidally, sometimes also septicidally, lacking a persistent central column. Indehiscent fruits are mostly fibrous or fleshy berries, the epicarp thin to leathery. The contrasting colours of the seeds coated with a sticky substance and surrounding capsule endocarp of *Pittosporum* species attract birds, often in large numbers (R.C. Carolin, personal observation; Judd 1996). The sticky, edible nature of the pulp is responsible for the birds dispersing the seeds internally or externally. The substance is produced by multicellular epidermal hairs in the septal region (Erbar and Leins 1995). It is often called resinous, while Erbar and Leins (1995) use the term 'latex', but the chemical composition seems to be unknown. The fruits of *Billardiera* are berries or berry-like with varying amounts of mucilaginous pulp which attracts birds and presumably assists in dispersal. *Rhytidosporum alpinum* has flattened dry and indehiscent fruits. The exotestal seeds are more or less smooth, the tangentially elongated epidermis cells have thick outer walls, the parenchymatous endotesta contains tannins (Netolitzky 1926), the raphe is conspicuous. The seeds of *Hymenosporum* are winged and those of *Bursaria* spp. are flattened and are dispersed over short distances by wind. Two species of *Cheiranthera* have been reported to be arillate (Cayzer et al. 2007). The oily and starchless endosperm is usually hard, the embryo minute and located near the hilum. The seeds of *Pittosporum* occasionally have up to five cotyledons. The oily seeds of *Pittosporum hosmeri* are eaten (and dispersed?) by the Hawaiian Crow (Haas 1977).

Phytochemistry. Hegnauer (1969, 1990) gave a valuable overview. Many Pittosporaceae contain triterpenoid saponins which may be responsible for the medicinal and poisonous properties reported by some authors (Cooper 1956). The essential oils from roots and leaves contain monoterpenes, sesquiterpenes, diterpenes, alkanes and polyacetylenes, C_{15}-acetylene is common, falcarinone and falcarinol acetylene also present; flower and fruit oils may contain benzyl acetate, coumarins and furanocoumarins, monoterpenes and sesquiterpenes as major constituents. No iridoid compounds occur in the family, nor ellagic or gallic acids, proanthocyanidins seem to be absent or restricted to the seed coat (Jay 1969). Sesquiterpene lactones, acetate-derived anthraquinones, chlorogenic and quinic acids were found. Quercetin and kaempferol are flavonols observed in the family, in the seed oils C_{18}, C_{20}, and C_{22} fatty acids are common, but petroselinic acid is lacking (Stuhlfauth et al. 1985).

Relationships Within the Family. Pritzel (1930) divided the family into tribe Pittosporeae with a capsular fruit and tribe Billardiereae with a fleshy fruit. As Bennett (1972, 1978) pointed out, however, there is a morphological series from fleshy to dry fruits in *Billardiera*. Pritzel's tribal distinction, therefore, is not satisfactory.

Cayzer et al. (2004), based on morphological characters, and Chandler et al. (2007), using DNA sequence data, recently studied infrafamiliar relationships and the monophyly of hitherto recognized genera. Based on these studies, *Citriobatus* had to be included in *Pittosporum* and *Sollya* and *Pronaya* (morphological data only) in *Billardiera*. The monophyly of *Marianthus* was weakly supported by the morphological analysis, but not by the DNA sequence data (albeit based on insufficient sampling). The DNA sequence data suggest a sister group relationship of *Pittosporum* with a clade formed by *Auranticarpa*, *Rhytidosporum* and *Bursaria*. This clade forms a trichotomy with a clade composed of *Billardiera*, *Cheiranthera*, *Marianthus*, and *Bentleya* and the monotypic *Hymenosporum*.

Affinities. The relationships of Pittosporaceae for some time were discussed controversially. Carlquist (1981) provided a useful summary of this. A rosid affinity has been suggested, and Carlquist (1981) made the point that the trichomes support such a relationship. However, this would make Pittosporaceae the only family in that group with secretory canals and unitegmic ovules, and the absence of ellagic and gallic acids in Pittosporaceae also argues against this hypothesis. The aralioid group of families had also been suggested as possible relatives [see Takhtajan (1997) for a summary]. Van Tieghem (1884) already early proposed a close relationship to Apiaceae and Araliaceae after studying root anatomy and noticing the schizogenic secretory canals and their distribution and the similar development of lateral roots (one on either xylem pole) in the three families. The unitegmic ovules also supported such a relationship. Floral structure, especially the superior ovary, on the other hand, appeared rather different from other Apiales and prevented most taxonomists to follow van Tieghem's proposal (e.g. Pax 1891; Cronquist 1981). Erbar and Leins (1995, 2004) found early sympetaly in Pittosporaceae similar as in Araliaceae (*Hedera*, *Hydrocotyle*), however, and the gynoecial development as being only gradually different. Also, the position of the gynoecial nectary basically corresponds to that of other Apiales; it is always formed at the dorsal base of the carpels (Erbar and Leins 2004, 2010). The secretory canals are absent from the rays of Pittosporaceae whilst they are present in the aralioid group of families. Chemical characters, however, especially the lack of iridoids and the presence of falcarinone polyacetylenes, clearly support a relationship with Apiaceae and Araliaceae (cf. Judd 1996), although petroselinic acid, common in these families, was not found in the fatty acids of seed oils of Pittosporaceae. The sequence analysis of the chloroplast gene *rbc*L by Chase et al. (1993) and more recent DNA sequence data (e.g. Nicolas and Plunkett 2014) basically confirmed van Tieghem's (1884) old hypothesis. The data show that Pittosporaceae belong to a clade together with Apiaceae, Araliaceae and Myodocarpaceae, being sister to a clade formed by the other three families. Stevens (2001 onwards) lists several non-macromolecular and possibly synapomorphic characters supporting this clade, besides the typical development of lateral roots—e.g. absence of iridoids, presence of ethereal oils and schizogenous secretory canals.

Distribution and Ecology. The family is centred in the Australian region, where all genera occur, most of them are endemic there. However, most species occur outside that area, belonging to *Pittosporum*. As *Pittosporum* is not basal within the family, an Australian origin of the family is most plausible (Chandler et al. 2007). According to Nicolas and Plunkett (2014), Pittosporaceae diverged from the other Apiineae (suborder also including Araliaceae, Myodocarpaceae, Apiaceae) about 98 Ma ago. The minimum age of the Pittosporaceae crown group was estimated at about 19 Ma, clearly post-Gondwanan. *Pittosporum* is by far the most widely distributed genus. It occurs throughout the Indo-Pacific region, with species endemic to individual islands or groups of islands, into Africa and even to Madeira and Tenerife. This distribution may be the result of seed dispersal by birds. The Hawaiian species (11 spp.) form a monophyletic group, suggesting a single colonization event from South Pacific islands, while those of New Zealand (26 spp.) and New Caledonia (50 spp.) and many areas in the Pacific islands seem to belong to different unrelated lineages (Gemmill et al. 2002; Chandler et al. 2007). *Pittosporum* species of Hawaii show high levels of interspecific gene flow indicating lack of reproductive isolation (Bacon et al. 2011). *Hymenosporum* occurs also in New Guinea. Species of *Pittosporum* and *Hymenosporum* grow mainly in mesic habitats, although *Pittosporum* intrudes into moist sclerophyllous habitats of southern Australia when the nutrient status is raised. The other genera are usually found in sclerophyllous communities. A few species of *Pittosporum* are considered weedy.

Economic Importance. Species of various genera are used in floriculture and landscape gardening in tropical and Mediterranean climate regions. Some *Pittosporum* spp. in the Pacific region (Java, Fiji) are used as a fish poison. The fruits of some *Billardiera* species are edible ("apple berries"). *Pittosporum undulatum*, native to Australia, is sometimes considered a noxious weed elsewhere.

Key to the Genera

1. Anthers longer or as long as the filament or nearly so — 2
– Anthers clearly shorter than the filaments — 3
2. Androecium zygomorphic; ovary stipitate; fruit a capsule — **8. *Cheiranthera***
– Androecium actinomorphic; ovary not stipitate; fruit indehiscent — **6. *Billardiera***
3. Procumbent or prostrate shrublets; young stems with cobweb-like indumentum of uniseriate hairs with long curling end cell — **3. *Rhytidosporum***
– Trees or shrubs, not procumbent; indument absent or of different hair type — 4
4. Plants spinescent — 5
– Plants not spinescent — 7
5. Flowers appearing at base of the plant below the foliage; petals hairy outside; style twice the length of the ovary — **9. *Bentleya***
– Flower position different; petals mainly glabrous outside; style generally shorter than ovary — 6
6. Petals spreading from the base, apically not strongly recurving; placentation axile; seeds dark red-brown, not immersed in glutinous material — **2. *Bursaria***
– Petals basally mostly cohering, apically often strongly recurved, placentation parietal with placentae sometimes protruding; seeds mostly orange to red, immersed in glutinous material — **1. *Pittosporum***
7. Flowers zygomorphic; plants usually twining — **7. *Marianthus***
– Flowers actinomorphic; trees or shrubs, rarely twining — 8
8. Petals 3–4 cm long; filaments hairy; seeds winged — **5. *Hymenosporum***
– Petals mostly shorter; filaments usually glabrous; seeds not winged — 9
9. Hairs all or partly T-shaped; flowers functionally unisexual, plants monoecious or dioecious; seeds immersed in glutinous material — **1. *Pittosporum***
– Hairs all uniseriate; flowers bisexual; seeds not immersed in glutinous material — 10
10. Small shrublets (5–20 cm high); flowers at base of plants; base of pedicels with several whorled bracts, bracteoles absent — **9. *Bentleya***
– Plants generally larger; flower position different; pedicel without bracts at the base, bracteoles present or absent — 11
11. Hairs on vegetative parts without extended terminal cell; filaments widened at base or middle, not angular; ovary stipitate; capsule with seeds in 1 row per locule, seeds black and glossy — **4. *Auranticarpa***
– Hairs on vegetative parts with extended terminal cell; filaments sometimes more widened at base, angular; ovary not stipitate; fruit indehiscent, seeds in 2 rows per locule, red-brown — **6. *Billardiera***

Genera of Pittosporaceae

1. ***Pittosporum*** Banks ex Gaertn. Fig. 82

Pittosporum Banks ex Gaertn., Fruct. Sem. Pl. 1: 286, t. 59, f. 7 (1788), nom. cons.; Cufodontis, Feddes Repert. 55: 27–113 (1952), rev. Afr. spp.; Zhang & Turland, Fl. China 9: 1–17 (2003), reg. rev.; Cayzer et al., Austr. Syst. Bot. 13: 845–903 (2000), reg. rev.
Citriobatus A. Cunn. ex Putt. (1839).

Shrubs or small trees, glabrous or with uniseriate or T-shaped hairs, sometimes spinescent. Leaves sometimes appearing opposite or whorled at the ends of branchlets, margin entire or subserrate, often undulate. Flowers in terminal or axillary panicles, rarely solitary, often functionally unisexual, the plants monoecious or dioecious. Sepals free or connate at base; petals free or shortly fused at base, connivent or coherent to about the middle or higher, glabrous or rarely hairy; anthers dorsifixed, dehiscing by longitudinal slits, shorter than the glabrous filaments, more or less strongly reduced in female flowers; ovary often stipitate, glabrous or pubescent, unilocular or imperfectly 2–5-locular, placentation parietal or parietal-basal; style short, stigma capitate or lobed, more or less reduced in male flowers; nectary at ovary base present, sometimes lobed. Fruit a thick-walled capsule or berry-like

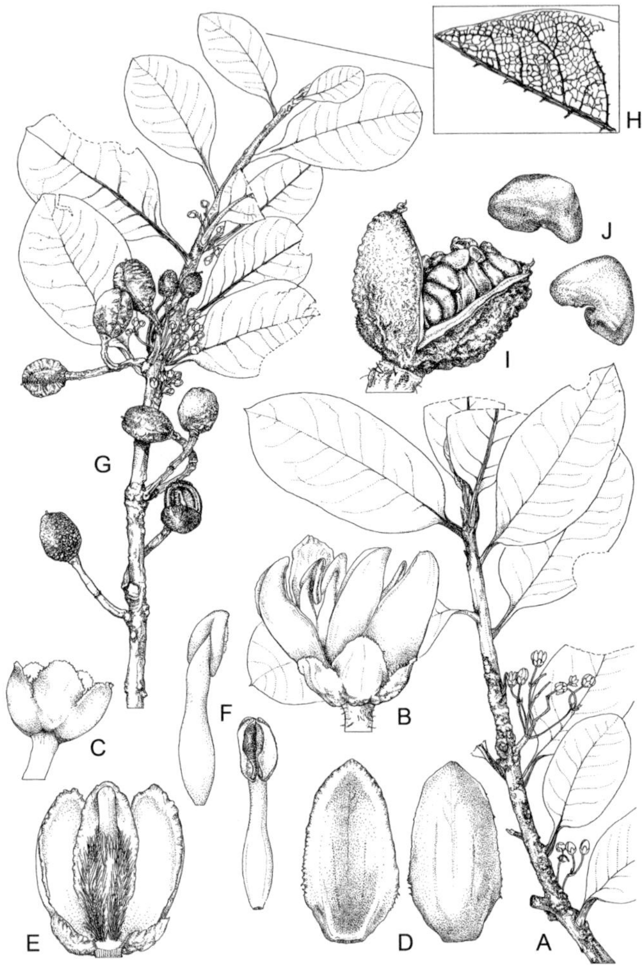

Fig. 82. Pittosporaceae. *Pittosporum spissescens*. **A** Flowering branch. **B** Flower. **C** Calyx. **D** Petals, adaxial and abaxial view. **E** Flower with sepals, petals and stamens removed. **F** Stamens, adaxial and abaxial view. **G** Fruiting branch. **H** Abaxial surface of leaf, showing venation. **I** Fruit. **J** Seeds. (Utteridge 2000; drawn by Kathleen McKeehen; reproduced with permission from Kew Bulletin)

(pericarp hard or thin). Seeds mostly red, orange or blackish, usually surrounded by a sticky fluid, embryo sometimes with 3, 4 or 5 cotyledons. $2n = 24$.

Up to 200 species, mostly southern parts of Africa and the Indo-Pacific region, northwards to Japan. Mostly tropical and subtropical.

There have been several attempts to divide *Pittosporum* into subgeneric groups (see especially Schodde 1972) but they either appear to be quite unnatural or they only deal with parts of the overall distribution of the genus. Cayzer et al. (2000) found several well-supported infrageneric clades in Australian *Pittosporum*.

2. ***Bursaria*** Cav.

Bursaria Cav., Icones 4: 30 (1797); Cayzer et al., Austr. Syst. Bot. 12: 117–143 (1999), rev.

Much-branched shrubs or small trees often showing heteroblasty, lateral branches spinose at least when young, epidermis of young shoots with uniseriate hairs with elongated terminal cell or sometimes T-shaped hairs. Leaves alternate, sometimes clustered, entire or rarely toothed. Flowers very aromatic, bisexual or rarely unisexual, small, glabrous, usually in axillary or terminal racemes or panicles. Sepals free; petals free, spreading from the base, white or cream; stamens free, anthers versatile, dehiscing by longitudinal slits, shorter than the filament; ovary 2(3 or more)-celled, glabrous or hairy, compressed, distinctly stipitate; style short, stigma slightly cleft, minute; nectary present, lobed. Fruit a thin-walled, strongly compressed loculicidal and partly septicidal capsule. Seeds often winged.

Eight species, eastern and south-western Australia, mostly temperate. *Bursaria spinosa* Cav. is a source of the medicinal drug aesculin and planted as a street tree.

3. ***Rhytidosporum*** F. Muell.

Rhytidosporum F. Muell., Plants Indigenous to the Colony of Victoria 1: 75 (1862); Cayzer et al., Austr. Syst. Bot. 12: 689–708 (1999), rev.

Erect to mostly prostrate undershrubs to 0.5 m high, often with underground shoots, young parts densely or sparsely covered with cobweb-like indumentum of uniseriate hairs with long curling terminal cells, rarely also with T-shaped hairs. Leaves entire or toothed at the apex. Flowers bisexual, solitary or several together in leaf axils or sometimes terminal. Sepals free; petals free, spreading from the base, glabrous, white, tinged pink; stamens free; anthers versatile, dehiscing by longitudinal slits, shorter than filaments; ovary glabrous, sometimes stipitate, 2–3-locular; style short, stigma usually minute; nectary at base of the ovary present. Fruit a loculicidal chartaceous capsule with fibrous compartments, or strongly flattened and indehiscent (*R. alpinum*). Seeds not winged, small.

Five species, temperate eastern and south-eastern Australia incl. Tasmania.

4. ***Auranticarpa*** L.W. Cayzer, Crisp & I. Telford

Auranticarpa L.W. Cayzer, Crisp & I. Telford, Austr. Syst. Bot. 13: 904 (2000); Cayzer et al., Austr. Syst. Bot. 13: 903–917 (2000), rev.

Small to medium-sized trees, indumentum exclusively of uniseriate hairs, terminal cell not elongated. Leaves alternate, sometimes appearing verticillate, margins entire and undulate, or slightly crenulate to toothed. Inflorescences multi-flowered, corymbose. Flowers bisexual, very fragrant. Sepals usually spreading from the base, very rarely 6; petals free, spreading from the base, mostly glabrous, cream or white; ovary strongly stipitate, placentation axile, style stout, stigma minute; nectary small, at base of the stipe. Fresh capsule usually with red-orange pericarp, 2 (3)-locular. Seeds few, black, shiny, not immersed in sticky pulp. $2n = 24$.

Six species, mostly monsoonal northern Australia, one species southwards to New South Wales.

5. ***Hymenosporum*** F. Muell.

Hymenosporum F. Muell., Fragm. 2: 77 (1860).
Chelidospermum Zipp. ex Blume, Mus. bot. Lugd. batav. 1: 162 (1850), nom. inval.
Quinsonia Montrouzier (1860).

Shrub or small tree, hairs uniseriate without elongated terminal cell. Leaves alternate, often in pseudo-whorls towards the ends of the branches, entire, glabrous or slightly hairy on the lower surface. Flowers bisexual, fragrant, in loose terminal panicles. Sepals 5, free, hairy; petals 5, cohering into a tube to above the middle, hairy on the outside, colour changing during anthesis from white to yellowish; stamens free, filaments hairy; anthers clustering at three different heights at anthesis, dehiscing by longitudinal slits, shorter than the filaments. Ovary hairy, incompletely 2-locular, narrow; style long. Fruit a compressed, thick-walled loculicidal capsule. Seeds flat, winged.

One species, *H. flavum* (Hook.) F. Muell., eastern Australia and New Guinea, temperate to tropical, ornamental.

6. ***Billardiera*** Sm.

Billardiera Sm., Spec. Bot. New Holland 1, t. 1 (1793); Cayzer et al., Austr. Syst. Bot. 17: 83–125 (2004), phylog., rev.
Sollya Lindl. (1832) ("1831").
Pronaya Hügel (1837).

Scrambling or climbing rhizomatous shrubs; hairs uniseriate with 1–3 basal cells and an elongated terminal cell. Leaves sometimes clustering, entire. Flowers bisexual, solitary or in terminal cymes. Sepals free, usually unequal; petals clawed, briefly coherent into a tube to above the middle, or elliptic and spreading from the base, rarely hairy outside; stamens free; anthers versatile, dehiscing by longitudinal or rarely apical slits, much shorter than the filaments or ± equal to longer, sometimes recurving, rarely cohering apically; ovary sessile, 1- or 2-locular, hairy or not; style short; nectary conspicuous or vestigial to absent. Fruit a succulent or somewhat fibrous berry. Seeds not winged, embedded in a mucilaginous pulp or in fibrous tissue.

23 species, temperate eastern and southern Australia, some ornamental. The inclusion of *Sollya* as proposed by Cayzer et al. (2004) needs further study. The three species of *Sollya* are characterized by anthers which are apically coherent and equal to longer than the filaments.

7. ***Marianthus*** Hügel ex Endl.

Marianthus Hügel ex Endl. in Endl. et al., Enum. Pl. Hügel 8 (1837); Cayzer & Crisp, Austr. Syst. Bot. 17: 127–144 (2004), rev.

Twining woody perennials, hairs uniseriate with elongated terminal cell. Leaves alternate, shortly petiolate or sessile, stem-clasping and sometimes clustered into a funnel-shaped arrangement. Flowers solitary or in dense corymbose or umbellate axillary cymes, zygomorphic, bisexual. Sepals free, coloured similarly to petals; petals free to the base or coherent to above the ovary and becoming free with age, white, red, blue, cream-yellow, mostly glabrous; stamens free, at anthesis anthers clustering at 3 different heights (1 + 2 + 2), dehiscing by longitudinal slits, shorter than the flat slender filaments; ovary stipitate, glabrous or hairy, 2-locular; style slender. Fruit a loculicidal capsule with or without fibrous compartments. Seeds discoidal, reniform, rarely globose.

14 species, south-western Australia, one species into south-eastern Australia.

8. ***Cheiranthera*** A. Cunn. ex Lindl.

Cheiranthera A. Cunn. ex Lindl., Edwards' Bot. Reg. 20, sub. t. 1719 (1834); Cayzer et al., Austr. Syst. Bot. 20: 340–354 (2007), rev.

Shrubs with twining or flexuose stems, with erect, uniseriate trichomes without elongated terminal cell or glabrous. Leaves sometimes clustering, often narrow. Flowers bisexual, solitary or in lax corymbs. Sepals free, purple-green or blue; petals spreading from the base, mostly deep blue; androecium zygomorphic at anthesis; anthers dehiscing by 2 apical pores or when cohering apically by slits, longer than filaments; ovary glabrous, 2-locular, stipitate, placentation axile; style subulate, curved; nectary absent. Fruit a capsule opening septicidally by 2 valves, sometimes additionally loculicidally. Seeds numerous, small, sometimes arillate.

10 species, temperate Australia, in eucalypt communities. As regards authorship of the name of the genus, see Cayzer et al. (2007).

9. ***Bentleya*** E.M. Bennett

Bentleya E.M. Bennett, Nuytsia 5: 401 (1986); Crisp & Taylor, Bot. J. Linn. Soc. 103: 109–115 (1990), rev.

Rhizomatous small shrubs 5–20 cm high, sometimes producing suckers, spinose or not, hairs uniseriate with elongated terminal cell. Leaves small, sometimes clustered or rosulate. Flowers bisexual, solitary, at the base of the plants below the foliage, with a whorl of bracts at the base of the pedicel. Sepals free, imbricate; petals hairy on both sides, greenish cream, cohering into a tube when young but separating at the base and remaining coherent in the middle or becoming completely free; stamens free, anthers dehiscing by longitudinal slits, shorter than filaments; nectary prominent, 5-lobed; ovary silky hairy, 2–3-locular, with numerous ovules on axile or parietal placentae; style hairy. Fruit a septicidal-loculicidal or loculicidal capsule. Seeds compressed.

Two species, south-western Australia.

Selected Bibliography

Bacon, C.D., Allan, G.J., Zimmer, E.A., Wagner, W.L. 2011. Genome scans reveal high levels of gene flow in Hawaiian *Pittosporum*. Taxon 60: 733–741.

Bennett, E.M. 1972. New taxa and new combinations in Western Australian Pittosporaceae. Nuytsia 2: 266–269.

Bennett, E.M. 1978. New taxa and new combinations in Australian Pittosporaceae. Nuytsia 2: 184–199.

Carlquist, S. 1981. Wood anatomy of Pittosporaceae. Allertonia 2: 355–392.

Cayzer, L.W., Crisp, M.D. 2004. Reinstatement and revision of the genus *Marianthus* (Pittosporaceae). Austr. Syst. Bot. 17: 127–144.

Cayzer, L.W., Crisp, M.D., Telford, I.R.H. 1999. *Bursaria* (Pittosporaceae): a morphometric analysis and revision. Austr. Syst. Bot. 12: 117–143.

Cayzer, L.W., Crisp, M.D., Telford, I.R.H. 2000. Revision of *Pittosporum* (Pittosporaceae) in Australia. Austr. Syst. Bot. 13: 845–902.

Cayzer, L.W., Crisp, M.D., Telford, I.R.H. 2004. Cladistic analysis and revision of *Billardiera* (Pittosporaceae). Austr. Syst. Bot. 17: 83–125.

Cayzer, L.W., Crisp, M.D., Donaldson, S. 2007. *Cheiranthera* (Pittosporaceae). Austr. Syst. Bot. 20: 340–354.

Chandler, G.T., Plunkett, G.M., Pinney, S.M., Cayzer, L.W., Gemmill, C.E.C. 2007. Molecular and morphological agreement in Pittosporaceae: phylogenetic analysis with nuclear ITS and plastid *trn*L-*trn*F sequence data. Austr. Syst. Bot. 20: 390–401.

Chase, M.W. et al. 1993. Phylogenetics of seed plants: an analysis of nucleotide sequences from the plastid gene *rbc*L. Ann. Missouri Bot. Gard. 80: 528–580.

Clarkson, F.M., Clarkson, B.D., Gemmill, C.E.C. 2012. Biological flora of New Zealand 13. *Pittosporum cornifolium*, tāwhiri karo, cornel-leaved *Pittosporum*. New Zeal. J. Bot. 50(2): 185–201.

Cooper, R.C. 1956. The Australian and New Zealand species of *Pittosporum*. Ann. Missouri Bot. Gard. 43: 87–188.

Cronquist, A. 1981. An Integrated System of Classification of Flowering Plants. New York: Columbia Press.

Erbar, C., Leins, P. 1995. An analysis of the early floral development of *Pittosporum tobira* (Thunb.) Aiton and some remarks on the systematic position of the family Pittosporaceae. Feddes Repert. 106: 463–473.

Erbar, C., Leins, P. 2004. Sympetaly in Apiales (Apiaceae, Araliaceae, Pittosporaceae). S. Afr. J. Bot. 70: 458–467.

Erbar, C., Leins, P. 2010. Nectaries in Apiales and related groups. Plant Divers. Evol. 128: 269–295.

Gemmill, C.E.C., Allan, G.J., Wagner, W.L., Zimmer, E.A. 2002. Evolution of Insular Pacific *Pittosporum* (Pittosporaceae): origin of the Hawaiian Radiation. Mol. Phyl. Evol. 22(1): 31–42.

Gopalakrishnan, K.K., Thomas, T.D. 2014. Reproductive biology of *Pittosporum dasycaulon*, a rare medicinal tree endemic to Western Ghats. Bot. Studies 55: 15.

Gowda, M. 1951. The genus *Pittosporum* in the Sino-indian region. J. Arnold Arb. 32: 263–343.

Haas, J.E. 1977. The Pacific species of *Pittosporum* Banks ex Gaertn. (Pittosporaceae). Allertonia 1: 73–167.

Hegnauer, R. 1969. Chemotaxomie der Pflanzen. Pittosporaceae. Band 5: 325–330. Basel: Birkhäuser.

Hegnauer, R. 1990. Chemotaxomie der Pflanzen. Pittosporaceae. Band 9: 240–244. Basel: Birkhäuser.

Index to Plant Chromosome Numbers (IPCN). 1979–. Goldblatt, P., Johnson, D.E. (eds.) Missouri Botanical Garden, St. Louis. http://www.tropicos.org/Project/IPCN (accessed May 2010).

Jay, M. 1969. Chemosystematic researches on vascular plants. XIX. Flavonoid distribution in the Pittosporaceae. Bot. J. Linn. Soc. 62: 423–429.

Judd, W.S. 1996. The Pittosporaceae in the southeastern United States. Harvard Pap. Bot. 8: 15–26.

Mauritzon, J. 1939. Contribution to the embryology of the orders Rosales and Myrtales. Lunds Univ. Arssk. 35: 1–121.

Narayana, L.L., Radhakrishnaiah, M. 1982. Floral anatomy of Pittosporaceae: 5 species of *Pittosporum*. Can. J. Bot. 60(10): 1859–1867.

Narayana, L.L., Sundari, K.T. 1981. Embryology of Pittosporaceae. Ind. J. Bot. 4: 123–131.

Netolitzky, F. 1926. Anatomie der Angiospermen-Samen. Berlin: Gebrüder Bornträger.

Nicolas, A.N., Plunkett, G.M. 2014. Diversification times and biogeographic patterns in Apiales. Bot. Review 80: 30–58.

Pax, F. 1891. Pittosporaceae. In: Engler, A., Prantl, K. (eds.) Die natürlichen Pflanzenfamilien ed. 1, vol. 3 (2a). Leipzig: W. Engelmann, pp. 106–114.

Pritzel, E. 1930. Pittosporaceae. In: Engler, A., Prantl, K. (eds.) Die natürlichen Pflanzenfamilien, ed. 2, 18a. Leipzig: W. Engelmann, pp. 265–285.

Sakai, A.K., Wagner, W.L., Ferguson, D.M., Herbst, D.R. 1995. Origins of dioecy in the Hawaiian flora. Ecology 76(8): 2517–2529.

Schlessman, M.A., Vary, L.B., Munzinger, J., Lowry II, P.P. 2014. Incidence and evolution of dioecy in the island flora of New Caledonia. Int. J. Plant Sci. 175: 271–286.

Schodde, R. 1972. Papuasian species of *Pittosporum*. Austr. J. Bot. Suppl. no. 3: 3–60.

Solereder, H. 1898. Systematische Anatomie der Dicotyledonen. Stuttgart: Ferdinand Enke.

Stevens, P.F. 2001 onwards. Angiosperm Phylogeny Website. Version 12, July 2012 [and more or less continuously updated since]. http://www.mobot.org/MOBOT/research/APweb/. Accessed September 2016.

Stuhlfauth, T., Fock, H., Huber, H., Klug, K. 1985. The distribution of fatty acids including petroselinic and tariric acids in the fruit and seed oils of the Pittosporaceae, Araliaceae, Umbelliferae, Simarubaceae and Rutaceae. Biochem. Syst. Ecol. 13: 447–453.

Takhtajan, A. 1997. Diversity and classification of flowering plants. New York: Columbia University Press.

Thorp, R.W. 2000. The collection of pollen by bees. Plant Syst. Evol. 222: 211–223.

Utteridge, T.M.A. 2000. The subalpine members of *Pittosporum* (Pittosporaceae) from Mt Jaya, New Guinea. Contributions to the Flora of Mt Jaya, II. Kew Bull. 55 (3): 699–710. [Fig. 2: *Pittosporum spissescens*].

Van Tieghem, P. 1884. Sur la structure et les affinités des Pittosporées. Bull. Soc. Bot. France 31: 383–385.

Wilkinson, H.P. 1992. Leaf anatomy of the Pittosporaceae. Bot. J. Linn. Soc. 110: 1–59.

Zhou, Q., Fu, D., Jin, X. 2005. Floral morphology and anatomy of *Pittosporum tobira* (Pittosporaceae). Nord. J. Bot. 23: 345–352.

Torricelliaceae

Torricelliaceae Hu, Bull. Fan Mem. Inst. Biol. Bot. 5: 313 (1934).

G.M. Plunkett, Q.-Y. Xiang, P.P. Lowry II, and G.E. Schatz

Small trees or shrubs, evergreen or deciduous; plants dioecious or hermaphroditic, lacking secretory canals. Leaves spiral and simple; margins often dentate, serrate, crenate to pinnatifid, more rarely entire; leaf venation pinnate or palmate; petioles exstipulate and distinctly sheathing; blades chartaceous to subcoriaceous or somewhat succulent, glabrous or trichomes unicellular or multicellular, unbranched, glandular and sometimes also nonglandular. Inflorescence terminal (occasionally in the upper leaf axils) and paniculate with racemules or cymules as the ultimate inflorescence unit (or bifid of two racemes), erect or pendulous. Flowers minute to medium-sized, perfect or imperfect, actinomorphic, epigynous, pedicel articulated or not, bracteolate; calyx a low rim or tube with (3–)5 small teeth or subequal lobes; corolla imbricate or induplicate-valvate; petals free 5 (sometimes lacking), broad at base, plane or inflexed at apex, ovate to long-elliptical; stamens 5 (lacking or present as staminodia in carpellate flowers), alternipetalous; anthers tetrasporangiate, basi- or dorsifixed, with longitudinal dehiscence; pollen tricolporate; carpels 3, connate; ovary inferior; epigynous disc absent to flat or forming a gibbous stylopodium; styles 3, free, erect to recurving (sometimes bifid); locules 3, sometimes all but one aborting; a single functional ovule per ovary (others abortive); ovules pendant and anatropous, placentation apical or axile. Fruits drupaceous, single-seeded crowned by the persistent calyx and styles; exocarp smooth, mesocarp fleshy, endocarp sclerified or lignified. Seeds with copious endosperm (ruminate or uniform) and a small embryo.

A family comprising three highly heterogeneous genera and 10 species distributed in Madagascar (*Melanophylla*), the Malay Peninsula and western Malesia (*Aralidium*), and the central and southern Himalayas to SW China (*Torricellia*).

Vegetative Morphology and Anatomy. The three genera of the family are deciduous or evergreen trees or sometimes shrubs, 2.5–17 m tall, and are monocaulous to sparsely or much branched, and have thick or slender stems with sympodial branching. The mature leaves are small to large and spirally arranged. The petioles are exstipulate, but the base is well developed and sheathing. The lamina may be thin or coriaceous to subsucculent, with pinnate or palmate venation, and is glabrous or has unicellular or multicellular, unbranched, glandular and sometimes also nonglandular trichomes. The lamina margins may be revolute in some species and are entire or dentate, serrate or crenate (often along the distal third or half), or pinnatifid with deep sinuses, or palmately lobed and serrate. The stomata are anomocytic (or anisocytic in *Aralidium*). The nodes are multilacunar in *Aralidium* (Philipson and Stone 1980) and *Torricellia*, but unknown in *Melanophylla*.

Torricelliaceae lack secretory canals. The wood (Adams 1949; Li and Chao 1954; Philipson and Butterfield 1980; Noshiro and Baas 1998) is diffuse-porous or (semi-)ring porous and lacks growth rings except in *Torricellia*, where the rings are distinct. The vessels are angular or rounded, solitary or in multiples or clusters of 2–9; tyloses are present and conspicuous in *Torricellia*. The vessel elements are 1070–1240 μm long (or 480–570 in *Torricellia*), 39–127 μm in diameter. The perforation plates are simple or scalariform with few to many bars. Intervessel pits are opposite or alternate, often scalariform, crowded, and lacking vestures. The vessel ray pits have reduced borders and are round to horizontal or opposite. Helical thickenings are absent (or present and thick in *Torricellia*). The fibers have thin to thick walls and minutely bordered pits in the radial walls; they are mostly or entirely septate, and 1820–2090

J.W. Kadereit, V. Bittrich (eds.), *Flowering Plants. Eudicots*, The Families and Genera of Vascular Plants 15,
https://doi.org/10.1007/978-3-319-93605-5_12

µm long. The axial parenchyma is scanty paratracheal. The rays are heterocellular, uniseriate or multiseriate (2–12 cells wide), up to 5.5 mm tall, and comprise mostly procumbent cells, or procumbent, square, and upright cells (sometimes with 1 or 2 marginal rows of upright cells). Crystals are lacking in the wood, but present as crystal sands in the rays, pith, and phloem parenchyma, as well as in the subepidermal mesophyll and sometimes in the epidermal cells of the leaves of *Aralidium* (Philipson and Butterfield 1980; Philipson and Stone 1980).

Inflorescence and Floral Morphology. *Melanophylla* is hermaphroditic, whereas both *Aralidium* and *Torricellia* are dioecious. All three genera have paniculate inflorescences whose ultimate units are either cymules or racemules (but reduced to a bifid inflorescence of two racemes in one species of *Melanophylla*; Schatz et al. 1998). The pedicels are bracteolate and either articulated (*Aralidium*) or unarticulated.

The flowers are very small to medium-sized, perfect or imperfect, and always epigynous. The calyx is (3–)5-merous, with small persistent teeth or subequal lobes. The corolla is 5-merous (but lacking in carpellate flowers of *Torricellia*), induplicate-valvate (*Torricellia*) or imbricate, the petals free and broadly inserted. The androecium has 5 stamens (or 5 staminodes in the carpellate flowers of *Aralidium*, and lacking in those of *Torricellia*), alternipetalous, basi- or dorsifixed, tetrasporangiate with longitudinal dehiscence, and the pollen is tricolporate. The gynoecium has 3 carpels, only 1 of which is fertile, and an equal number of locules, except in *Aralidium*, where only one locule develops. The styles are three, free, erect to recurved (but vestigial in the staminate flowers of *Torricellia*), and the epigynous disc is obsolete to flat, or forming a cushion or gibbous stylopodium in the carpellate flowers of *Aralidium*.

Floral Anatomy and Embryology. In *Aralidium* (Philipson and Stone 1980), the floral vasculature originates as a stele of several (usually 3) vascular bundles in the pedicel, which are then rearranged into an inner ring (of 5 bundles) and an outer ring (of 8–10 bundles) at the base of the ovary. The outer ring continues upward (dividing to form 10 bundles if there were fewer below), and supplying the sepals, petals, and stamens/staminodia. The inner ring divides to form an anastomosing network with 12 principal bundles. Three sets of 3 bundles supply the dorsal (stylar) regions of the gynoecium, and the remaining three (ventral) bundles supply the single ovule. The inner ring of bundles is lacking in staminate flowers. In *Torricellia*, there is no vascular tissue in the central axis of the ovary; instead, the ventral bundles run through the ovary wall to supply the pendulous ovule (trans-septal vasculature; Eyde 1967). Nothing is known of the floral anatomy of *Melanophylla*.

In *Melanophylla* (Sokoloff et al. 2018), the corolla aestivation is contort, a feature previously unknown among the campanulids, and corolla handedness is labile within a single inflorescence, a situation that is rare within asterids as a whole.

In all three genera, the ovary has a single anatropous, pendulous ovule located in the single functional locule, with apical or axile placentation. In *Aralidium*, the ovule is unitegmic, with a massive integument and funiculus, a dorsal raphe, a *Polygonum*-type embryo sac (developing from a linear tetrad), and a tenuinucellate nucellus (with a single cell-layer laterally, although there may be as many as four layers apically; Philipson and Stone 1980). The nucellus in *Melanophylla* and *Torricellia* is also tenuinucellate, but less is known about their ovules other than the presence of a dorsal raphe in *Melanophylla* and a thickened funiculus in *Torricellia*.

Fruit and Seed. The fruits of Torricelliaceae are ovoid to ellipsoid drupes crowned by the persistent styles and calyx, with a smooth exocarp and fleshy mesocarp. The endocarps from each carpel are united into a single unit (not forming distinct pyrenes, as in Araliaceae). In each genus the typically tricarpellate fruits exhibit distinct locular conditions. In *Melanophylla*, the two ventral carpels form large, empty locules, whereas the third, dorsal locule is smaller and filled by the single seed. *Torricellia* is similar, but the locules are subequal in size, but in some individuals one locule may form a U-shaped chamber, giving transverse sections of the ovary the appearance of being four-locular (see also Yembaturova and Konstantinova 2013). In *Aralidium*, the fruits have a single, large locule, within which the vestiges of the two undeveloped carpels may be seen towards the apex in mature fruit. In all three genera, the endocarp is hardened. In *Aralidium*, it has four or more distinct ridges visible from the inside of the locule (Philipson and Stone 1980; Takhtajan 1997).

The seed is large and ridged in *Aralidium*, but smaller in *Melanophylla* and *Torricellia*. In all three genera, the endosperm is copious (ruminate in *Aralidium*, uniform in *Melanophylla*, and unknown in *Torricellia*). The embryos are small and dicotyledonous. In *Melanophylla*, the testa lacks lignin, but is differentiated into two layers, an exotesta (with numerous stomata and several cell types) and a destroyed endotesta (Trifonova 1998).

Pollen Morphology. Pollen in Torricelliaceae is tricolporate. In *Aralidium* (Tseng 1980), it has narrow, shallow, and inconspicuous colpi, with pores (ora) that are somewhat longitudinally elongate. The grains are suboblate or sometimes spheroidal, with a 2-layered exine (sexine and nexine) that is thickened at the mesocolpium. The sexine is smooth but tectate with minute pores. In *Melanophylla* (Ferguson and Hideux 1978), the pollen is spheroidal to prolate and the exine is thickened at the mesocolpium and more narrow at the apertures and poles. The tectum is partially reticulate or perforate. Pollen grains of *Torricellia* (Ferguson and Hideux 1978) are spheroidal; the ectoapertures are simple, narrow, and less than half as long as the polar axis. The tectum is either granulose or superficially rugulate with processes or short head-like projections that are denticulate in appearance in section.

Karyology. *Aralidium pinnatifidum* has been reported to have a chromosome number of $2n = 40 \pm 2$ (Hellmayer et al. 1994), whereas that of both species of *Torricellia* is $n = 12$ and $2n = 24$ (Tang et al. 1984; Oginuma et al. 1994a, b). The chromosome number of *Melanophylla* is unknown.

Phytochemistry. In *Aralidium*, two iridoid glycosides (griselinoside and aralidioside) have been detected. Flavonoids are rare, but include caffeic acid (and a derivative), but not myricetin, quercetin, ferulic acid, proanthocyanidins, gallo- or ellagitannins (Bate-Smith 1980). *Araliadium pinnatifidum* uses the C_3 photosynthetic pathway. In *Torricellia*, the iridoids griselinoside and torricellate have been isolated (Jensen and Nielsen 1980; Liang et al. 2009). Nothing is known of the phytochemistry of *Melanophylla* except for the lack of gallo- and ellagitannins and procyanidins (Bate-Smith et al. 1975).

Affinities. The three genera of Torricelliaceae share many features with Araliaceae and/or Cornaceae, but their taxonomic placements have historically been problematical. *Aralidium* was traditionally assigned to Araliaceae (Miquel 1856; Bentham 1867; Harms 1897), with which it shares a general floral plan and fruit type, as well as a single ovule per locule, a massive integument, a disintegrating nucellus, and wood with diffuse vessels and heterogeneous rays (Philipson and Stone 1980). The gynoecium of *Aralidium*, however, is somewhat more like that of Cornaceae, with no floral vasculature in the central axis of the ovary and a dorsal raphe on the ovule (in contrast to a ventral raphe in Araliaceae). Moreover, the distinctive secretory canals found in the wood of Araliaceae and its close relatives (Apiaceae, Myodocarpaceae, and Pittosporaceae) are completely lacking in *Aralidium*, whereas iridoids (characterizing both *Aralidium* and Cornaceae) are altogether absent in Apiales. The difficulty in placing *Aralidium* led Philipson and Stone (1980) to erect a monotypic family.

Both *Melanophylla* and *Torricellia* have long been treated as members of Cornaceae (Baker 1884; Harms 1897; Wangerin 1910), although their relationship to that family has often been thought of as tenuous. For example, *Melanophylla* and *Torricellia* both differ from Cornaceae in having alternate leaves, sheathing petioles, glandular trichomes, and free styles (characters more typical of Araliaceae). Eyde (1967) had originally suggested that the trans-septal vasculature in the ovaries of both *Torricellia* and Cornaceae provided an important link between the two groups, but in his later studies, he observed that this feature was more broadly distributed among the dicots (see Eyde 1988). Although Cronquist (1988) maintained a cornaceous placement of *Torricellia*, many of its features suggested segregation (e.g., vessels with simple perforations, the presence of typical collenchyma, a thickened funiculus, $x = 12$ [not 11], and the lack of cornaceous nectaries; see Eyde 1988; Takhtajan 1997). For these reasons, both *Torricellia* and *Melanophylla* were eventually recognized as monogeneric families, most typically placed in or near Apiales (Dahlgren 1980; Eyde 1988; Takhtajan 1997; Thorne 2000).

Phylogenetic studies based on molecular data now suggest that these three genera form a single clade most closely related to Araliaceae, Apiaceae, Myodocarpaceae, and Pittosporaceae (Plunkett et al. 1996, 1997; Plunkett 2001; Chandler and Plunkett 2004; Nicolas and Plunkett 2009). To reflect these relationships, Plunkett et al. (2004) recircumscribed Torricelliaceae to include *Aralidium* and *Melanophylla* as well as *Torricellia*, and placed this family in an expanded order Apiales, together with the monogeneric Griseliniaceae. This expansion of Torricelliaceae (and its relationship to Griseliniaceae) is supported by many characters, including the distinctive iridoid griselinoside (except in *Melanophylla*, for which this character is unknown), alternate phyllotaxy, sheathing petiole bases, mostly dioecious mating systems (except hermaphroditic in *Melanophylla*), mostly pentamerous perianths and androecia, tricarpellate gynoecia, a single functional, pendulous ovule with apical or axile placentation, single-seeded drupaceous fruits, and seeds with copious endosperm and a small embryo.

Palaeobotany. Fossil fruits of *Torricellia* have been reported from the Paleocene of North Dakota, the Eocene of Washington, Oregon, and Germany, and the Miocene of Austria (Manchester 1999; Meller 2006; Manchester et al. 2009).

Economic Uses. The leaves and young branches of *Torricellia* are used as a fertilizer by native farmers. The roots and bark of this genus are also often used medicinally by local people (e.g., as a tincture to cure physical injuries). Macerated root and stem bark is used to treat fractured bones, tonsillitis, and asthma, as well as injured nerves of livestock (Hu 1990; Liang et al. 2009). Philipson (1979) indicates that the leaves of *Aralidium pinnatifidum* are used as "ghost medicine" in Brunei, but no other use is known. No information is available on whether *Melanophylla* is economically useful.

Distribution and Habitats. The family is found across a wide portion of the Paleotropics, but each genus is more narrowly distributed and their ranges are mutually exclusive. The single species of *Aralidium* has an Australasian distribution, from the Malay Peninsula (peninsular Thailand, peninsular Malaysia, Singapore) to western Malesia (Sumatra, Anambas Islands, Borneo). It is primarily a lowland species, where it can be found along shady river banks and in bamboo thickets or secondary forests, but it can reach elevations of 1250 m and has been reported up to 1500–1800 m in Borneo (Philipson 1979; Philipson and Stone 1980). The seven species of *Melanophylla* are all endemic to Madagascar, where three are critically endangered and one is vulnerable (Schatz et al. 1998; Catalogue of the Vascular Plants of Madagascar 2018). Most grow in humid forest, ranging from near sea level to 1800 m elevation. *Torricellia* has two species, distributed in the eastern Himalayas (Tibet and western Nepal through NE India to Bhutan) and E Asia (SE and SC China, N Burma, and Vietnam), which grow in broadleaf forests and forest margins, stream sides and open slopes, at 900–2600 m elevation (Xiang and Boufford 2005).

Key to the Genera of Torricelliaceae

1. Leaves palmately veined and often palmately lobed, deciduous; petals induplicate-valvate (Himalayas to southern China, Burma, and Vietnam) **3. *Torricellia***
– Leaves pinnately veined and pinnately lobed or unlobed, evergreen (to semi-deciduous in *Aralidium*); petals imbricate 2
2. Plants dioecious; pedicels articulated; fruits 1-locular; leaves pinnatifid (sometimes unlobed when juvenile) (Malay Peninsula to western Malesia) **1. *Aralidium***
– Plants hermaphroditic; pedicels unarticulated; fruits 3-locular; leaves unlobed (Madagascar) **2. *Melanophylla***

1. ***Aralidium*** Miq. Fig. 83

Aralidium Miq., Fl. Ned. Ind. 1: 762 (1856); Philipson, Fl. Malesiana I, 9: 1–105 (1979), rev.; Philipson & Stone, Taxon 29: 391–404 (1980), anat.

Evergreen (to semi-deciduous) small trees or shrubs, dioecious, pubescent (trichomes glandular and nonglandular). Leaves chartaceous,

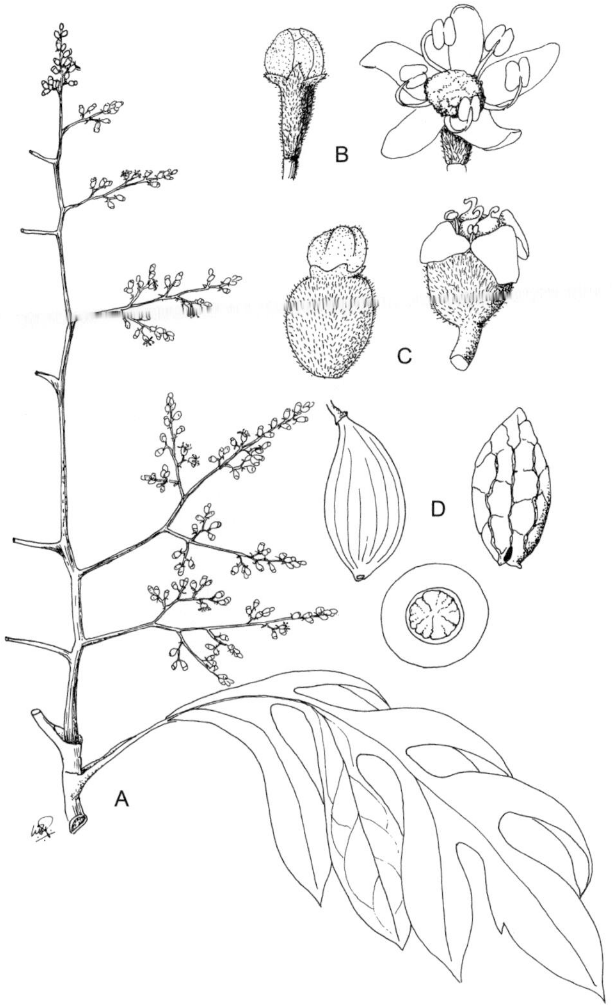

Fig. 83. Torricelliaceae. *Aralidium pinnatifidum.* **A** Habit. **B** Male bud and flower. **C** Female bud and flower. **D** Fruit, seed and cross-section of fruit. (Philipson 1979; illustrations by W.R. Philipson; reproduced with permission from Flora Malesiana)

usually pinnatifid, margins deeply lobed or less often entire or strongly dentate; venation pinnate; stomata anisocytic. Inflorescence terminal (occasionally also in the upper leaf axils), paniculate, with cymules as the ultimate units. Flowers minute, pubescent; pedicels articulated, bracteolate; staminate flowers with a calyx of 5 very small teeth; petals imbricate, cream-white to red-tinged, triangular to ovate, broadly inserted; stamens 5; epigynous disc intrastaminal, forming a cushion; vestigial gynoecium lacking; carpellate flowers with similar calyx and corolla; staminodia 5; epigynous disc intrastaminodial, forming 3 gibbous stylopodia; ovary 3-carpellate, all but one aborting (leaving a single developed locule); styles 3, short; ovule with a massive funiculus. Fruit a single-seeded, obliquely ellipsoid drupe, white when immature, blackish at maturity; exocarp smooth, mesocarp fleshy, endocarp hard with 4–8 ridges; vestiges of the 2 undeveloped carpels may be found at the apex of the single developed locule, inside the ovary and mature fruit. Endosperm ruminate. $2n = 40 \pm 2$.

One species, *A. pinnatifidum* (Jungh. & de Vriese) Miq., Malay Peninsula and western Malesia.

2. *Melanophylla* Baker — Fig. 84

Melanophylla Baker, J. Linn. Soc., Bot. 21: 352 (1884); Schatz et al., Adansonia III, 20: 233–242 (1998), rev.; McPherson & Rabenantoandro, Adansonia III, 24: 263–265 (2002), taxon.

Evergreen small trees or shrubs, hermaphroditic, glabrous or pubescent (trichomes glandular). Leaf chartaceous to subcoriaceous or slightly succulent, turning black upon drying, margins dentate, serrate, or crenate, often just in the distal portion, or entire; venation pinnate; stomata anomocytic. Inflorescences terminal (sometimes appearing lateral), paniculate, with racemes as the ultimate units (or bifid and reduced to two racemes). Flowers medium-sized, with an unarticulated pedicel bearing 2 subequal bracteoles; calyx saucer-shaped with 5 minute teeth; petals imbricate, white, yellow or pink, long-elliptical to strap-like, recurved to reflexed at maturity; stamens 5; epigynous disc absent; ovary 3-carpellate and 3-locular (but only one bearing a functional ovule). Fruit an obliquely ellipsoid drupe with two large, sterile ventral locules and a third fertile dorsal locule bearing the single seed; green when immature, purple to black at maturity. Endocarp lignified (but only weakly so around ventral locules). Endosperm uniform.

Seven species, endemic to Madagascar.

3. *Torricellia* DC. ("*Toricellia*") — Fig. 85

Torricellia DC., Prodr. 4: 257 (1830), nom. cons.; Hu, in Fang & Hu, Fl. Reipub. Pop. Sin. 56: 35–38 (1990), reg. rev.; Xiang & Boufford, Fl. China 14: 233–234 (2005), reg. rev.

Fig. 84. Torricelliaceae. *Melanophylla modestei.* **A** Fruiting branch with leaves. **B** Branch of inflorescence. **C** Flower at anthesis and gynoecium (perianth and androecium removed). (Schatz et al. 1998; illustrations by L. Nivoarintsoa; reproduced with permission from Adansonia)

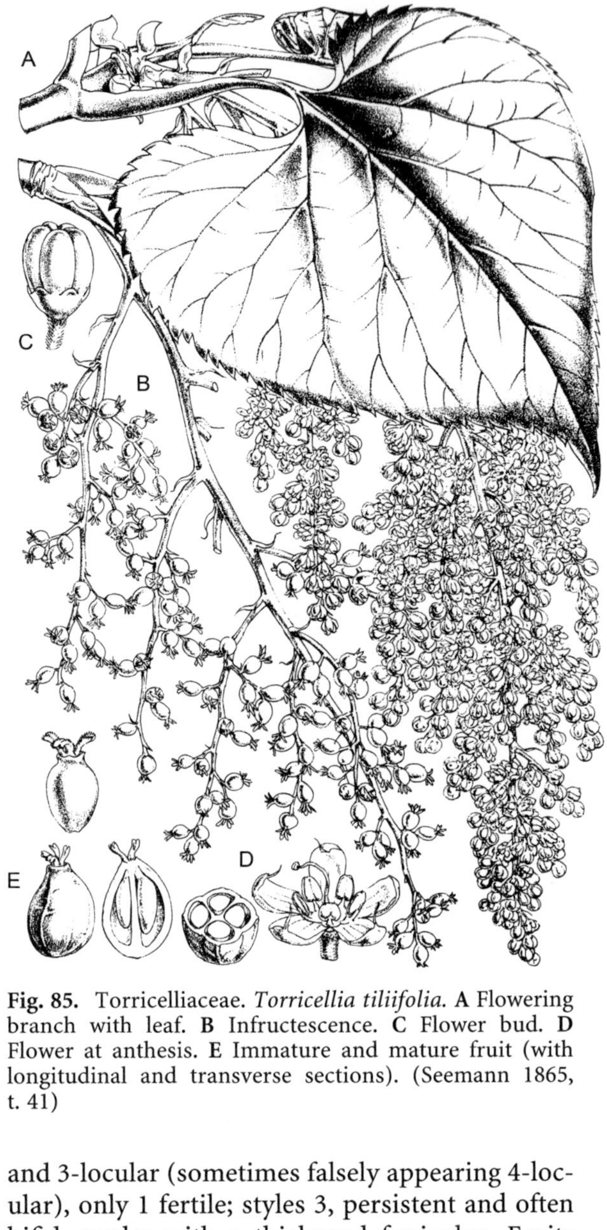

Fig. 85. Torricelliaceae. *Torricellia tiliifolia.* **A** Flowering branch with leaf. **B** Infructescence. **C** Flower bud. **D** Flower at anthesis. **E** Immature and mature fruit (with longitudinal and transverse sections). (Seemann 1865, t. 41)

Deciduous trees, dioecious, pubescent (trichomes glandular). Leaves chartaceous, margins serrate to dentate, or slightly 5–7-lobed (with entire or serrate margins); venation palmate; stomata anomocytic. Inflorescences large, terminal, pendulous panicles. Flowers small, imperfect, epigynous, pedicels short, unarticulated, with minute, linear, alternate bracteoles; staminate flowers with a persistent calyx forming a shallow tube with 5 small, ± subequal lobes; petals induplicate-valvate, white, long elliptical, inflexed at apex; stamens 5; epigynous disc ± flat, with vestigial styles lacking or forming 1–3 subulate processes; carpellate flowers with a calyx of 3–5 small, irregular teeth; petals and stamens lacking; epigynous disc inconspicuous; ovary 3-carpellate and 3-locular (sometimes falsely appearing 4-locular), only 1 fertile; styles 3, persistent and often bifid; ovules with a thickened funiculus. Fruits single-seeded, ovoid or obliquely ovoid drupes, green when immature, reddish-purple or black at maturity; mesocarp fleshy, endocarp hardened and bearing a triangular germination valve. $n = 12$, $2n = 24$.

Two species, central and southern Himalayas to SW and SC China, in mid- to high-elevation forests and forest edges, or near streams. Plants of *T. angulata* Oliv. with dentate margins were sometimes recognized as a third species (*T. intermedia* Harms in Diels).

Selected Bibliography

Adams, J.E. 1949. Studies in the comparative anatomy of the Cornaceae. J. Elisha Mitchell Sci. Soc. 65: 218–244.

Baker, J.G. 1884. Further contributions to the flora of central Madagascar. J. Linn. Soc., Bot. 21: 317–353.

Bate-Smith, E.C. 1980. The systematic position of *Aralidium* Miq. – a multidisciplinary study. 5. A note on the phenolic constituents. Taxon 29: 412.

Bate-Smith, E.C., Ferguson, I.K., Hutson, K., Nielsen, B.J., Swain, T. 1975. Phytochemical interrelationships in the Cornaceae. Biochem. Syst. Ecol. 3: 79–89.

Bentham, G. 1867. Araliaceae. In: Bentham, G., Hooker, J. D. (eds.) Genera Plantarum, 1. London: Reeve & Co, pp. 931–947.

Catalogue of the Vascular Plants of Madagascar. 2018. http://www.tropicos.org/Name/40016214 [accessed 31 Aug. 2018].

Chandler, G.T., Plunkett, G.M. 2004. Evolution in Apiales: nuclear and chloroplast markers together in (almost) perfect harmony. Bot. J. Linn. Soc. 144: 123–147.

Cronquist, A. 1981. An integrated system of classification of flowering plants. New York: Columbia University Press.

Cronquist, A. 1988. The evolution and classification of flowering plants. 2nd ed., New York: New York Botanical Garden.

Dahlgren, R.M.T. 1980. A revised system of classification of the angiosperms. Bot. J. Linn. Soc. 80: 91–124.

Eyde, R.H. 1967. The peculiar gynoecial vasculature of Cornaceae and its systematic significance. Phytomorphology 17: 172–182.

Eyde, R.H. 1988. Comprehending *Cornus*: puzzles and progress in the systematics of the dogwoods. Bot. Rev. 54: 233–351.

Ferguson, I.K., Hideux, M.J. 1978. Some aspects of the pollen morphology and its taxonomic significance in Cornaceae sens. lat. Proc. IV Int. Palynol. Conf., Lucknow (1976–1977) 1: 240–249.

Harms, H. 1897. Araliaceae. In: Natürl. Pflanzenfam. III, 8. Leipzig: Engelmann, pp. 1–62.

Harms, H. 1898. Cornaceae. In: Natürl. Pflanzenfam. III, 8. Leipzig: Engelmann, pp. 250–270.

Hellmayer, E.M., Kiehn, M., Weber, A. 1994. Chromosome numbers of Malayan rain-forest Angiosperms. Beitr. Biol. Pflanzen 68: 51–71.

Hu, W. 1990. *Toricellia*. In: Fang, W., Hu, W. (eds.) Fl. Reipubl. Popularis Sin. 56: 35–38.

Hutchinson, J. 1967. The genera of flowering plants (Angiospermae). Dicotyledones, Vol. 2. Oxford: Clarendon Press, p. 46.

Jensen, S.R., Nielsen, B.J. 1980. Iridoid glucosides in *Griselinia*, *Aralidium* and *Toricellia*. Phytochemistry 19: 2685–2688.

Li, H.-L., Chao, C.-Y. 1954. Comparative anatomy of the woods of the Cornaceae and allies. Quart. J. Taiwan Mus. 7: 119–136.

Liang, G., Xu, B., Pan, W., Cao, P., Zhang, Y., Lu, Y., Wu, Y., Hao, X. 2009. A novel iridoid from *Torricellia angulata* var. *intermedia*. Nat. Prod. Res. 23: 1–4.

Manchester, S.R. 1999. Biogeographical relationships of North American Tertiary floras. Ann. Missouri Bot. Gard. 86: 472–522.

Manchester, S.R., Chen, Z.-D., Lu, A.-M., Uemura, K. 2009. Eastern Asian endemic seed plant genera and their paleogeographic history throughout the Northern Hemisphere. J. Syst. Evol. 47: 1–42.

Meller, B. 2006. Comparative investigation of modern and fossil *Toricellia* fruits – a disjunctive element in the Miocene and Eocene of Central Europe and the USA. Beitr. Paläont. 30: 315–327.

Miquel, F.A.W. 1856. Araliacearum indicarum genera et species aliquot novae. Bonplandia 4: 137–139.

Nicolas, A.N., Plunkett, G.M. 2009. The demise of subfamily Hydrocotyloideae (Apiaceae) and the re-alignment of its genera across the entire order Apiales. Molec. Phylogen. Evol. 53: 134–151.

Noshiro, S., Baas, P. 1998. Systematic wood anatomy of Cornaceae and allies. IAWA J. 19: 43–97.

Oginuma, K, Gu, Z., Yue, Z., Kondo, K. 1994a. Chromosomes of some woody plants native to Yunnan, China. Kromosomo 73: 2491–2497.

Oginuma, K., Tobe, H., Ohba, H. 1994b. Chromosomes of some woody plants from Nepal. Acta Phytotax. Geobot. 45: 15–22.

Philipson, W.R. 1979. Araliaceae. In: van Steenis, C.G.G.J. (ed.) Fl. Malesiana I, 9. The Hague: W. Junk Publ., pp. 1–105.

Philipson, W.R., Butterfield, B.G. 1980. The systematic position of *Aralidium* Miq. – a multidisciplinary study. 2. Wood anatomy. Taxon 29: 404–406.

Philipson, W.R., Stone, B.C. 1980. The systematic position of *Aralidium* Miq. – a multidisciplinary study. 1. Introduction and floral and general anatomy. Taxon 29: 391–404.

Plunkett, G.M. 2001. Relationship of the order Apiales to subclass Asteridae: a reevaluation of morphological characters based on insights from molecular data. Edinb. J. Bot. 58: 183–200.

Plunkett, G.M., Soltis, D.E., Soltis, P.S. 1996. Higher level relationships of Apiales (Apiaceae and Araliaceae) based on phylogenetic analysis of *rbc*L sequences. Amer. J. Bot. 83: 499–515.

Plunkett, G.M., Soltis, D.E., Soltis, P.S. 1997. Clarification of the relationship between Apiaceae and Araliaceae based on *mat*K and *rbc*L sequence data. Amer. J. Bot. 84: 565–580.

Plunkett, G.M., Chandler, G.T., Lowry, P.P. II, Pinney, S. M., Sprenkle, T.S. 2004. Recent advances in understanding Apiales and a revised classification. S. African J. Bot. 70: 371–381.

Schatz, G.E., Lowry, P.P. II, Wolf, A.-E. 1998. Endemic families of Madagascar. I. A synoptic revision of *Melanophylla* Baker (Melanophyllaceae). Adansonia III, 20: 233–242.

Seemann, B.C. 1865. Revision of the natural order Hederaceae. IX. On the genera *Toricellia*, *Decostea*, and *Adoxa*. J. Bot. 3: 361–363, t. 41.

Sokoloff, D.D., Karpunina, P.V., Nuraliev, M.S., Oskolski, A.A. 2018. Flower structure and development in *Melanophylla* (Torricelliaceae: Apiales): lability in direction of corolla contortion and orientation of pseudomonomerous gynoecium in a campanulid eudicot. Bot. J. Linn. Soc. (early online publ.).

Takhtajan, A. 1997. Diversity and classification of flowering plants. New York: Columbia University Press.

Tang, Y.-C., Xiang, Q.-Y., Cao, Y.-L. 1984. Cytological studies on some plants of Sichuan and neighboring regions - I. Acta Phytotax. Sin. 22: 343–350.
Thorne, R.F. 2000. The classification and geography of the flowering plants: Dicotyledons of the class Angiospermae. Bot. Rev. 66: 441–647.
Trifonova, V.I. 1998. Fruit and seed anatomy of the genus *Melanophylla* (Melanophyllaceae) in relation to its taxonomical position. Bot. Zhurn. 83: 97–103.
Tseng, C.C. 1980. The systematic position of *Aralidium* Miq. – a multidisciplinary study. 3. Pollen morphology. Taxon 29: 407–409.
Wangerin, W. 1910. Cornaceae. In: Pflanzenreich IV, 229. 101 pp. Leipzig: Engelmann.
Xiang, Q., Boufford, D.E. 2005. Toricelliaceae. Fl. China 14: 233–234.
Xiang, Q.-Y., Soltis, D.E. 1998. *Rbc*L sequence data define a cornaceous clade and clarify relationships of Cornaceae sensu lato. In: Boufford, D.E., Ohba, H. (eds.) Sino-Japanese flora – its characteristics and diversification. Tokyo: University of Tokyo Press.
Yembaturova, E.Y., Konstantinova, A.I. 2013. Fruit structure of the genus *Toricellia* DC. (Toricelliaceae) and its taxonomic position in the order Apiales. Izv. Timiryazev Agric. Acad., special issue, pp. 197–205.

Index

References to accepted names in **bold**-faced print, to synonyms in upright print, to illustrations in *italics*.

J.W. Kadereit, V. Bittrich (eds.), *Flowering Plants. Eudicots*, The Families and Genera of Vascular Plants 15,
https://doi.org/10.1007/978-3-319-93605-5

Volumes published in this series

Volume I Pteridophytes and Gymnosperms
Edited by K.U. Kramer and P.S. Green (1990)
Date of publication: 28.9.1990

Volume II Flowering Plants. Dicotyledons. Magnoliid, Hamamelid and Caryophyllid Families
Edited by K. Kubitzki, J.G. Rohwer, and V. Bittrich (1993)
Date of publication: 28.7.1993

Volume III Flowering Plants. Monocotyledons: Lilianae (except Orchidaceae)
Edited by K. Kubitzki (1998)
Date of publication: 27.8.1998

Volume IV Flowering Plants. Monocotyledons: Alismatanae and Commelinanae (except Gramineae)
Edited by K. Kubitzki (1998)
Date of publication: 27.8.1998

Volume V Flowering Plants. Dicotyledons: Malvales, Capparales and Non-betalain Caryophyllales
Edited by K. Kubitzki and C. Bayer (2003)
Date of publication: 12.9.2002

Volume VI Flowering Plants. Dicotyledons: Celastrales, Oxalidales, Rosales, Cornales, Ericales
Edited by K. Kubitzki (2004)
Date of publication: 21.1.2004

Volume VII Flowering Plants. Dicotyledons: Lamiales (except Acanthaceae including Avicenniaceae)
Edited by J.W. Kadereit (2004)
Date of publication: 13.4.2004

Volume VIII Flowering Plants. Eudicots: Asterales
Edited by J.W. Kadereit and C. Jeffrey (2007)
Date of publication: 6.12.2006

Volume IX Flowering Plants. Eudicots: Berberidopsidales, Buxales, Crossosomatales, Fabales p.p., Geraniales, Gunnerales, Myrtales p.p., Proteales, Saxifragales, Vitales, Zygophyllales, Clusiaceae Alliance, Passifloraceae Alliance, Dilleniaceae, Huaceae, Picramniaceae, Sabiaceae
Edited by K. Kubitzki (2007)
Date of publication: 6.12.2006

Volume X Flowering Plants. Eudicots: Sapindales, Cucurbitales, Myrtaceae
Edited by K. Kubitzki (2011)
Dates of publication: 10.12.2010 (online); 10.1.2011 (print)

Volume XI Flowering Plants. Eudicots: Malpighiales
Edited by K. Kubitzki (2014)
Dates of publication: 27.11.2013 (online); 17.12.2013 (print)

Volume XII Flowering Plants. Eudicots: Santalales, Balanophorales
By J. Kuijt and B. Hansen (2015)
Dates of publication: 21.10.2014 (online); 18.11.2014 (print)

Volume XIII Flowering Plants. Monocots: Poaceae
By E.A. Kellogg (2015)
Dates of publication: 18.5.2015 (online); 12.6.2015 (print)

Volume XIV Flowering Plants. Eudicots: Aquifoliales, Boraginales, Bruniales, Dipsacales, Escalloniales, Garryales, Paracryphiales, Solanales (except Convolvulaceae), Icacinaceae, Metteniusaceae, Vahliaceae
Edited by J.W. Kadereit and V. Bittrich (2016)
Date of publication: 10.4.2016 (online); 10.5.2016 (print)

Volume XV Flowering Plants. Eudicots: Apiales, Gentianales (except Rubiaceae)
Edited by J.W. Kadereit and V. Bittrich (2018)

MIX
Papier aus verantwortungsvollen Quellen
Paper from responsible sources
FSC® C105338

If you have any concerns about our products,
you can contact us on
ProductSafety@springernature.com

In case Publisher is established outside the EU,
the EU authorized representative is:
Springer Nature Customer Service Center GmbH
Europaplatz 3, 69115 Heidelberg, Germany

Printed by Libri Plureos GmbH
in Hamburg, Germany